NUCLEAR PHYSICS

Experimental and Theoretical

NUCLEAR PHYSICS

Experimental and Theoretical

Revised Second Edition

H S Hans
Professor Emeritus
Panjab University
Chandigarh

Publishing Globally

NEW AGE INTERNATIONAL (P) LIMITED, PUBLISHERS
LONDON • NEW DELHI • NAIROBI
Bangalore • Chennai • Cochin • Guwahati • Hyderabad • Kolkata • Lucknow • Mumbai
Visit us at **www.newagepublishers.com**

Published by New Age International (P) Ltd., Publishers
First Edition: 2001
Revised Second Edition: 2013
Reprint: 2019

GLOBAL OFFICES

- **New Delhi** **NEW AGE INTERNATIONAL (P) LIMITED, PUBLISHERS**
  7/30 A, Daryaganj, New Delhi-110002, (INDIA)
  Tel.: (011) 23253771, 23253472, **Telefax:** 23267437, 43551305
  E-mail: contactus@newagepublishers.com • Visit us at www.newagepublishers.com
- **London** **NEW AGE INTERNATIONAL (UK) LTD.**
  27 Old Gloucester Street, London, WC1N 3AX, UK
  E-mail: info@newacademicscience.co.uk • Visit us at www.newacademicscience.co.uk
- **Nairobi** **NEW AGE GOLDEN (EAST AFRICA) LTD.**
  Ground Floor, Westlands Arcade, Chiromo Road (Next to Naivas Supermarket), Westlands, Nairobi, KENYA
  Tel.: 00-254-713848772, 00-254-725700286
  E-mail: kenya@newagepublishers.com

BRANCHES

- **Bangalore** 37/10, 8th Cross (Near Hanuman Temple), Azad Nagar, Chamarajpet, Bangalore- 560 018
  Tel.: (080) 26756823, **Telefax:** 26756820, **E-mail: bangalore@newagepublishers.com**
- **Chennai** 26, Damodaran Street, T. Nagar, Chennai-600 017, **Tel.:** (044) 24353401, **Telefax:** 24351463
  E-mail: chennai@newagepublishers.com
- **Cochin** CC-39/1016, Carrier Station Road, Ernakulam South, Cochin-682 016, **Tel.:** (0484) 2377303, **Telefax:** 4051304
  E-mail: cochin@newagepublishers.com
- **Guwahati** Hemsen Complex, Mohd. Shah Road, Paltan Bazar, Near Starline Hotel, Guwahati-781 008,
  Tel.: (0361) 2513881, **Telefax:** 2543669, **E-mail: guwahati@newagepublishers.com**
- **Hyderabad** 105, 1st Floor, Madhiray Kaveri Tower, 3-2-19, Azam Jahi Road, Near Kumar Theater, Nimboliadda Kachiguda, Hyderabad-500 027, **Tel.:** (040) 24652456, **Telefax:** 24652457
  E-mail: hyderabad@newagepublishers.com
- **Kolkata** RDB Chambers (Formerly Lotus Cinema) 106A, 1st Floor, S N Banerjee Road, Kolkata-700 014
  Tel.: (033) 22273773, **Telefax:** 22275247, **E-mail: kolkata@newagepublishers.com**
- **Lucknow** 16-A, Jopling Road, Lucknow-226 001, **Tel.:** (0522) 2209578, 4045297, **Telefax:** 2204098
  E-mail: lucknow@newagepublishers.com
- **Mumbai** 142C, Victor House, Ground Floor, N.M. Joshi Marg, Lower Parel, Mumbai-400 013
  Tel.: (022) 24927869, **Telefax:** 24915415, **E-mail: mumbai@newagepublishers.com**
- **New Delhi** 22, Golden House, Daryaganj, New Delhi-110 002, **Tel.:** (011) 23262368, 23262370, **Telefax:** 43551305
  E-mail: sales@newagepublishers.com

ISBN: 978-81-224-3141-4
C-18-09-11590

Printed in India at Ajit Printing Press, Delhi.
Typeset at Goswami Associates, Delhi.

NEW AGE INTERNATIONAL (P) LIMITED, PUBLISHERS
7/30 A, Daryaganj, New Delhi-110002
Visit us at **www.newagepublishers.com**
(CIN: U74899DL1966PTC004618)

Preface to the Revised Second Edition

The response to the first edition of *Nuclear Physics—Experimental and Theoretical* has been quite encouraging. Though it was confined only to the readership in India because of the conditionality of sponsorship of National Book Trust, India, the personal interactions and reviews showed a heartening welcome from different parts of India.

It seems, the book was used as a first text for the special paper on Nuclear Physics, and second text for the general course in Nuclear Physics. Of course, it became a reference book for the teachers of Nuclear Physics.

The second edition, which is meant for readership, for both the students in India and abroad, has been upgraded by adding to each chapter a page, or two, the latest developments in the subject.

It was interesting to note that because of the availability of a large number of the most modern accelerators in the world, a new culture of collaborative research work has come out, involving in a single paper a large number of authors from different countries of Europe, and of course, USA, Russia, China and India. The number of authors in a single paper ranged from a few to more than fifty. This has allowed, the quality and the complexity of the measurements and calculations, to grow phenomenally. We have mentioned the work of many papers involving authors from India, because of author's familiarity.

A chapter, 'Theory of Nuclear Matter and Finite Nucleus' (Appendix-A) has been added, at the end, to complete the contents of the book in accord with syllabi of many universities.

The flavour of the reading matter in the book, of course, remains the same. The theoretical part is 'basic' to meet the needs of the students, but the experimental part has been brought up-to-date. The subject of nuclear physics—at intermediate and low energies—somewhat follows traditional lines, though recent studies—both theoretical and experimental have added new dimensions. This is especially true of problems in nuclear structure. The interest in studies involving Pre-Equilibrium has continued; while there has been a great increase in nuclear reactions involving heavy ions.

Because of need of nuclear technology, in our general advancement of modern life, the subject of nuclear physics remains quite popular among students. We hope this book will fulfill this need of society.

We thank the publishers for pointing out the reactions of readership to bring out this revised edition in an up-to-date manner.

—H.S. Hans

Preface to the First Edition

The subject of nuclear physics, embodying the discussion of nuclear forces, nuclear structure and nuclear reaction-mechanism has always been a part of undergraduate and postgraduate studies in physics. Along with the subjects of atomic and molecular spectroscopy, solid state physics and particle physics, the subject of nuclear physics, allows the students to apply the basic laws of classical and quantum physics to the physical problems encountered in these subjects.

In nuclear physics, we deal with the smallest of the constituents of matter—the nuclei with size of $10^{-12} - 10^{-13}$ cm, containing nucleons (protons and neutrons), which interact with each other through strong forces.

Though the various phenomenon in nuclear physics can only be understood through quantum mechanics, its direct application through various perturbation techniques is not so easy. Hence various models—both in nuclear structure and nuclear reactions—have been developed; though the fundamental approaches, through more comprehensive mathematical techniques applied to many-body system, interacting through strong forces have also been tried with some limited success.

On the experimental side, there have been developed a large number of techniques of detection and energy discrimination along with acceleration to arrive at a large amount of data about nuclear properties. In the last few decades, nuclear physics has entered a new phase. More and more precise measurements are being made of the unknown quantities and very new methods have been developed for extending the range of measurements, especially of the properties of excited states of nuclei. The recent development of modern accelerator laboratories for heavy ions, *e.g.,* pelletrons and super-conducting linear accelerators and cyclotrons, throughout the world, have added a new thrust to nuclear studies.

The present book—basically meant as a textbook for senior level undergraduate (say B.Sc. Hons.) and postgraduate—endeavours to capture this new spirit of the subject for the students, by including the topics like pre-equilibrium reaction mechanism and heavy ion reactions, etc. The attempt has been made to give the subject a balanced treatment of experiment and theory: to bring out, on one side, the physical insight into the various theoretical concepts in nuclear structure and nuclear reactions and on the other side, the basic principles of mathematical treatment. In every topic, subject has been brought upto date, starting from fundamentals. For advanced topics, the results have been written, explaining the physical concepts, but sometimes not giving mathematical derivation for which references are provided. This book is expected to be the second exposure of students in their carrier, to the subject of Nuclear Physics. Hopefully the first exposure of Introductory Nuclear Physics has been carried out at undergraduate level earlier. In the present book, we have devoted chapter one to develop an over all perspective—both theoretical and experimental. This gives the student an upto-date bird's eye-view of the whole subject before he goes in details in the various chapters. In Chapter 2, we have described the various physical phenomena in nuclear physics, including experimental methods of measurements and the basic concepts at a somewhat introductory level, along with systematics of some physical properties.

The next four chapters deal with nuclear forces as encountered in the two-body problems. Chapter three on deuteron problem, Chapter 4 on low energy nucleon-nucleon scattering, Chapter 5 on higher energy nucleon-nucleon scattering, including polarisation and Chapter 6 on the expressions for nucleon-nucleon forces, based on the concepts of earlier three chapters. A section on quarks is included, to connect nuclear forces to quarks and gluons.

The next three chapters—7, 8 and 9—deal with the problems of radioactive decay. Chapter 7, deals with gamma transitions. Here we have given the classical as well as quantum mechanical treatment of electromagnetic transitions. Internal conversion and angular correlation and distributions have also been dealt with. Chapter 8 deals with beta decay, both in theory and experiments. Chapter 9 deals with particle decay including alpha decay, cluster decay and spontaneous fission. The theory of the last two topics forms a very interesting extension of alpha decay. For the theoretical understanding of the unstable nuclei, especially those away from beta-stability line, we have added a topic on delayed proton decay, which is partly an extension of cluster decay, though, in practice, it takes place from the highly excited nuclei produced by beta decay. In Chapters 10–12, we have dealt with the nuclear structure models, *i.e.*, single particle shell model (Chapter 10), the collective model (Chapter 11) and Nilsson model of shell structure for deformed nuclei (Chapter 12), involving both collective and particle motion.

Chapters 13–17, deal with nuclear reactions and various models of reaction mechanism, *e.g.*, (*i*) compound nucleus formation (Chapter 13); (*ii*) direct reaction model (Chapter 14); (*iii*) optical model (Chapter 15); (*iv*) pre-equilibrium model (Chapter 16) and (*v*) heavy ions induced reactions (Chapter 17). The last two chapters are intended to introduce the student to these modern topics, as they form a serious part of the research studies these days.

This book can be used as a two semester course for general students at M.Sc. level or a one semester course by selecting the topics carefully. We hope, the book meets the needs of the students, in understanding the various aspects of nuclear physics, at graduate level.

—H.S. Hans

Acknowledgement

The completion of this book over the years would not have been possible, without the incessant encouragement of my family, especially my wife—Tripta Hans.

—H.S. Hans

Contents

CHAPTER 1

The Perspective

1.1 EARLY HISTORY

The subject of Nuclear Physics has a very large range—both in concepts and techniques. Dealing with the smallest of the physical entities—nucleons and nuclei—interacting through the strongest forces available in nature; the subject requires the application of quantum mechanics in a particular manner—which is somewhat different from that required in atomic or solid state physics. Also the experimental techniques developed are special, because of the higher energies involved, by a factor of 10^3 to 10^6 compared to the problems of atomic or molecular physics. We will try to convey the perspective of both the concepts and techniques to the reader, in this chapter, before we plunge into the actual subject.

Nuclear Physics, like many other branches of physics, had a very humble beginning. In 1896, Becquerel[1], at the suggestion of Poincáre, was investigating the uranium salts to find out a relationship between the property of optical fluorescence and the newly discovered X-rays. While a search for this relationship proved illusory, he found, accidently, that the uranium salts emitted some penetrating radiation; which could fog the photographic plates even when they were covered with a good amount of wrapping material. Since then, a large number of experimental and theoretical developments, have brought the subject of nuclear physics to its present status where one is able to understand to a large extent, the various properties of nuclei in their ground as well as excited states.

In understanding the nature of the structure of the nucleus, first breakthrough came when Rutherford along with Geiger and Marsden[2] in 1913, performed the famous experiments on α-scattering from thin gold and platinum foils. They discovered that the number of α-particles scattered from a thin platinum foil at backward angles were one in 8000 compared to 1 in 10^{14} expected on the basis of the J.J. Thomson's[3] melon-seed model of the atom. Rutherford[4], successively explained these experiments on the basis of a model in which the positively charged heavy nucleus sits at the centre of the atom, surrounded by electrons. This model was supplemented by N. Bohr[5] in the same year, by the assumption of stationary orbits of electrons, thereby giving birth to the presently accepted, Bohr-Rutherford model of the atom or the more commonly called Bohr's Atomic Model.

The experiments on positive rays by J.J. Thomson[6], in 1912, and afterwards by Aston[7] in 1919 showed that hydrogen nuclei were protons and that nuclear masses were nearly the integral number of the proton mass. Aston also discovered that many nuclei consisted of more than one isotope, which had, the same amount of charge but different masses. Each isotope, however, consisted of the integral number of the proton masses approximately. The proper interpretation of the isotopes had, of course, to wait for some years. In the beginning, it was surmised that the nucleus consists of protons and electrons, with the number of protons about twice the number of electrons. This could account for the charge and the mass of both the nucleus and an atom. It seemed to be further supported by the observation that some radioactive nuclei emitted electrons. However, this hypothesis was rejected on the basis of statistics. Also the de Broglie wavelength of an electron inside a nucleus is expected to be of the same size as the nucleus. This requires the energy of electrons, emitted from nuclei (called β^--rays) to be of the order of more than 10 MeV; while the experimentally measured values of energies of beta rays are only of the order of a couple of MeV.

The above mystery was resolved in 1932, by the discovery of neutrons by Chadwick[8]. This discovery proved to be a landmark in the development of nuclear physics. Neutron was, at once, recognised as the 'other' particle, besides proton, which constituted the nucleus. Various isotopes were, then, understood to be as nuclei, with the same number of protons, but different number of neutrons. This neutron-proton model was later on confirmed by many observations on nuclear reactions. It also resolved the difficulties of the electron-proton model in a natural manner.

In the meantime, a lot of developments took place on the theoretical side. Especially quantum mechanics was developed from 1900 to 1928; by stalwarts like Planck, de Broglie, Schrödinger, Heisenberg and Dirac[9]. A theoretical framework for the understanding of the many-body microscopic structures like atoms and molecules was thus created. Application of these concepts to nuclei was a logical consequence. This meant development of the theoretical framework to explain the observed properties of nuclei in terms of the interaction between neutrons and protons (called nucleons) inside the nucleus, assuming that the same theoretical framework which explains atomic phenomena quite satisfactorily is also applicable to a nucleus which is a much smaller system.

The phenomenon of radioactivity, was quite well established by this time. It was known, for example, that there existed among heavier nuclei, a large number of naturally occuring radioactive isotopes which emitted nuclei of He^4 (called alpha particle), electrons (called β^--rays), and electromagnetic radiations of very short wavelengths (called gamma (γ) rays). Also, whereas the observed spectra of alpha and γ-rays were discrete, corresponding to the discrete excited states of the residual nuclei, the spectrum of β-rays was continuous. In the beginning, this gave rise to many speculations; one of them even envisaged a breakdown of the law of conservation of energy[10]. This paradox was, however, solved in 1933 by Pauli's hypothesis[11] of the existence of neutrino—a massless and a chargeless particle with an angular momentum (or spin) of $1/2\hbar$. Fermi[12] in 1934, gave his theory of β-decay, assuming the simultaneous emission of an electron and anti-neutrino (e– $\bar{\nu}$) in a negative beta (β^-) decay and the emission of positron and a neutrino (e^+–ν) in a positive beta decay (β^+). The two particles share their energies and give rise to a continuous spectrum of electrons or positrons (see Chapters 2 and 8).

The next relevant question which arose was: 'What is the nature of the nuclear forces, which bind the nucleons inside the nucleus?' For this purpose, it was necessary to study not only the properties of ground states of nuclei, but also those of the excited states; and the phenomenon of break up of nuclei. It had been earlier (1919) demonstrated by allowing α-particles from radioactive nuclei to fall on stable nuclei, that nuclei can be broken; emitting protons and neutrons. A frantic search was, therefore, started to find out artificial methods to excite or to break the nucleus, by artificial means.

1.2 ACCELERATORS

This resulted in designing and constructing different types of particle accelerators.

First accelerator was designed and fabricated in 1931 by R.J. Vande Graff at MIT, Cambridge[13] (U.S.A.). It was an electrostatic accelerator known after his name (Vande Graff Accelerator). Another accelerator around the same time was designed by J.D. Cockcroft and E.T.S. Walton at Cavandish Lab, Oxford (U.K.)[14], based on the principle of multiplying the voltage by charging the condensers in parallel and discharging them in series. As a matter of fact, the first nuclear reaction, by using any accelerator was conducted with the Cockcroft-Walton accelerator[15] in 1932, by accelerating protons to 300 KeV and allowing them to fall in Li^7. In the same year of 1932, E.O. Lawyence[16] built and tested the first cyclic machine called the Cyclotron at Berkley (U.S.A.) for protons. Later in 1945–46, Veksler and McMillan[17] modified the Cyclotron principle to include phase-stability to develop the so-called synchrocyclotron for higher energies, which can now go up to an energy of some 700 MeV for protons. These electrostatic and cyclic machines, with their variations, are now quite commonly[18] used for accelerating protons, deuterons, tritons, He^3, alpha particles and the heavier ions from Li^7 right upto uranium to several MeV per particle (say up to more than 50 MeV/A).

A combination of the principle of phase-stability and the application of alternating gradient was developed by Christophilos[19] (1950) and independently, by Courrant Livingston and Snyder[20] (1952) to apply to a doughnut type of cyclic machine called synchrotron. This machine has no apparant energy limits. Already energies of more than 10^6 MeV for protons have been achieved[25]. The synchrotrons are now being used also for heavy ions, as well as for electrons.

The acceleration of electrons, however, presented some special problems, because, their motion becomes relativistic even around one MeV of energy. Therefore, the cyclotron concept could not be easily adapted for them. D.W. Kerst[23], in 1941, used the principle of electromagnetic induction to develop the betatron. With this machine, it has been possible to accelerate electrons up to several hundred MeV's. Another important development in accelerator technology has been the evolution of the concept of linear accelerator. These are based on the principle of multiple acceleration on an approximate straight trajectory[21]. The first linear accelerator for protons was developed by D.H. Sloan and E.O. Lawrence[22] in 1931. Energies of 20 GeV for electrons; and 10 GeV for protons and greater than 10 MeV/A for heavier ions has been achieved using the principle of linear accelerators, especially using the technology of superconducting linear accelerators.[21,26]

Some very recent interesting developments in the cyclotron[24] (including synchrotrons) technology have opened up new ranges of intensities and energies. The concept of pulsed storage rings has increased the intensity of protons at the highest energy enormously[25]. The protons from synchrotron are injected into a ring, to which more and more protons are added at regular intervals from the main synchrotron. They are, then, ejected from the storage ring by applying a pulsed electric field at suitable times. Also new ion-sources (*e.g.* E.C.R. type) and the use of superconducting[24] magnets has made it possible to design very high energy cyclotrons for heavy ions. These developments have now made it possible to have electrons up to 10–20 GeV; protons for more than 1000 GeV; and heavy ions for more than 500 MeV/A[25,26]. Many accelerators are under development in the world, in these energy ranges.[25,26]

Recently[27] heavy ions, as projectiles have become very much popular for experiments on nuclear reactions or nuclear structure. Energies up to 10 MeV per nucleon have already been achieved for uranium and up to about 50 MeV/A for lighter nuclei. Still higher energies are expected from the new accelerators under development. The heavy ion induced nuclear reactions may be the major activity in nuclear physics in the near future.

The latest entry[28 (a)] into this field in accelerators (linear accelerators and (or plus) cyclotrons), accelerating exotic beams, *e.g.* radioactive ions or cluster-molecules, resulting in very new research fields in nuclear physics and material sciences.

The nuclear studies, with which we are concerned in this book; are generally carried out with accelerators up to say a few hundred MeV per particle. Still higher energy accelerators are, in general, employed for the production of fundamental particles like mesons, etc.

1.3 REACTORS

Another phenomenon, in nuclear physics, which has gained importance since 1939, was 'fission'. The phenomenon of fission of nuclei induced by thermal neutrons was discovered, experimentally by O. Hahn and F. Strassman[28] in 1939 and is one of the great discoveries in nuclear physics (Chapter 9). Afterwards[28], in fifties and sixties, the phenomenon of spontaneous fission was discovered for very heavy nuclei, beyond uranium. This principle of nuclear fission was used by Enrico Fermi[29] in 1942, for designing the first nuclear reactor. Later this principle resulted in the first nuclear explosion in 1945. Apart from their use as a source of power; the reactors form a major category of machines used extensively for research in nuclear physics. They are a copious source of thermal or fast neutrons and are used not only for producing new species of nuclei through neutron-capture but also for studying neutron reactions at these low energies and the structure of materials through neutron diffraction. At present[30], there are many research reactors in the world with neutron flux of the order of 10^8 to 10^{14} neutrons/cm^2/sec. Apart from these research reactors, there are several power reactors for providing electric power. The power reactors may range from a few megawatts to many hundred megawatts.[30]

1.4 COMPLEX NUCLEI

Various properties of complex nuclei have been studied using these instruments and machines, *e.g.*, (1) The masses and the binding energies of various nuclei in their ground state (2) Nuclear radii (3) Energies of the excited states (4) Angular momenta (spins), parities, magnetic moments and quadrupole moments of the ground and excited states of nuclei (5) Transition probabilities between the various excited states; and (6) The various cross-sections involving elastic and inelastic scattering and reactions leading to states of different nuclei.

The experimental techniques involved, for the measurement of the various properties of both the ground states and excited states of nuclei are:

1.4.1 The Ground State

(*i*) Mass spectrometry for measuring the masses of various nuclear species and hence their binding energies[31] (*ii*) Various atomic-beam methods as developed by Rabii and Coworkers[32], for measuring angular momenta and magnetic moments (*iii*) Nuclear magnetic resonance (NMR) and nuclear quadrupole resonance (NQR) for the measurement of the magnetic and quadrupole moments[33], respectively (*iv*) Some techniques based on atomic spectroscopy[34] for angular momenta and other properties of nuclei, and (*v*) The various scattering techniques for the measurements of nuclear radii[35] [Chapter 2].

1.4.2 The Excited States

(*i*) Nuclear spectrometry[36] using magnetic spectrometers, scintillation crystals, and solid state detectors like Ge-Li, Si-Li and surface-barrier detectors; along with sophisticated electronics like multi-channel and multi-parameter analysers or online computing systems for measuring energies of particles or gamma rays. (*ii*) Various techniques of angular distribution, angular correlation or polarisation for measuring the spins, parities, magnetic moments and quadrupole moments of the excited states[37] [Chapter 2].

These techniques combined with the techniques of accelerators and reactors have made experimental nuclear physics as one of the most challenging and exciting subjects of physics. A large amount of data has now been collected and published from time to time[38], in Nuclear Data Sheets, etc.

1.5 NUCLEAR FORCES

Any theoretical attempt to correlate these experimental facts and to understand them in terms of the motion of nucleons in the nuclei, requires the knowledge of nuclear forces operating between proton-neutron, proton-proton and neutron-neutron. The study of two-nucleon system of deuteron, and *n-p* or *p-p* scattering up to about 100 MeV provides information on free nucleon-nucleon interaction. Information about nuclear forces between *n-n* has been obtained either from *n-d* scattering or from comparison of the mirror nuclei. Analysis of the three-body systems like He^3, H^3, or *n-d* or *p-d* scattering has further contributed a great deal to the detailed knowledge about the nuclear forces. As a result, it has been possible to draw the following conclusions about the nuclear forces. (1) They have a short-range, of the order of 2 Fermis (2) They are predominantly central, but with a small, though significant, tensor term (3) The nuclear potential has a hard repulsive core of the range of $\approx$ 0.5 fermis, and an attractive part of the range of about a couple of fermis (4) Central forces are spin-dependent (5) The nuclear forces have an exchange character, which gives rise to the property of saturation of nuclear forces, which explains in a natural manner, the property of binding energy per nucleon being independent of number of nucleons in the nucleus; and the constant nuclear density. (6) They are charge-independent, *i.e.* they are intrinsically the same for *n-p, p-p* or *n-n* interaction. This property has given rise to a new concept of isotopic spin. (7) They depend on the spin-orbit coupling of a nucleon. (8) A detailed study of the complex nuclei exhibits, many-body character of the nuclear forces. (9) They may also depend on relative angular momenta and hence on relative velocities. (see Chapters 3–6).

Though the broad features of the nuclear forces, as mentioned above, are established, the exact quantitative expression for nuclear potential which should be applicable to the free nucleon-nucleon scattering as well as to nuclear structure problems is still, less than settled. Based on two-body interactions, certain effective nucleon-nucleon interactions have been proposed[39] whose application to real complex nuclei have shown limited successes (Chapter 10). Three-body forces have also been considered[40]. Discovery of mesons and latter on quarks, and their relationship with nuclear forces, has brought the subject to the present status.[40]

1.6 NUCLEAR DECAY

The nuclear structure problems based on our understanding of nuclear forces, can be studied either through nuclear decay using radioactive nuclei, or through nuclear reactions.

Nuclear decay, in radioactive nuclei involves three modes: (*i*) β-decay, (*ii*) γ-decay and (*iii*) α-decay. Out of these three modes, β-decay corresponds to 'Weak' interaction; γ-decay to electromagnetic interaction; and α-decay to 'nuclear' and coulomb-interaction. The strength of the β-decay is governed by weak interaction constant – $\left[(g_\beta c^2/\hbar^2)\, m_\pi^2)/\hbar c\right]^2 \approx 10^{-13}$, that of γ-decay, intrinsically by electromagnetic interaction constant $(e^2/\hbar c) = 1/137$, which is called the fine structure constant; and that of α-decay by nuclear interaction constant $(g_N^2/\hbar c) \approx 1$, and the Coulomb interaction.

The interest in α-decay arises, both because of its relationship with nuclear structure; as well as for the mechanism of decay. The characteristics of α-decay were explained by Gamow's theory[41] which proved to be one of the earliest successes of the quantum mechanics. This explained the phenomenon of tunnelling of α-particles through coulomb barrier on the basis of W.K.B. approximation, as discussed in Chapter 9.

The electromagnetic transitions, also include internal conversion, apart from γ-decay. In these cases, the form of the interaction is very well understood. The interest in this decay process basically arises because of the information that one gets about the properties of the various nuclear levels and the transitions, *e.g.* the spins and parities of levels. The mixing ratios of the transitions help in understanding the detailed wave functions of the nuclear states (Chapter 7). The beta decay includes electron (β^-) and positron (β^+) emission; as well as electron capture (EC). In this case, the interest is not only in the problem of nuclear structure but also in the basic interaction itself, because it represents one of the less understood fundamental interactions. The theory of beta decay was earlier developed by Fermi and Gamow-Teller.[12] The discovery of non-conservation of parity in β-decay has created a lot of interest among physicists because of its effect on weak interactions in particle physics. Detailed theory of beta decay; and its implications for nuclear structure are developed in Chapter 8.

Apart from the studies of decay of radioactive nuclei, a lot of information about nuclear structure has come from the excitation and de-excitation of nuclei involved in various reactions.

Because of the still existing ambiguities in the knowledge of nuclear forces, the solution to the problem of a nuclear structure-involving the complex nuclei, has not been an easy one. It gets further complicated by the fact that for obtaining the theoretical solution of these problems, one has actually to solve a many-body problem with strong internucleon forces, which do not lend easily to the various perturbation techniques, used, say in atomic physics.

1.7 NUCLEAR STRUCTURE MODELS

Historically, the problem was circumvented by postulating various models of nuclear structure. The first nuclear structure model extensively developed was the liquid drop model. This model was inspired by Neil Bohr's ideas of the compound nucleus, according to which, once a nucleon enters a nucleus, it loses the properties of its individual motion; because of the extremely strong nucleon-nucleon interaction inside the nucleus. Because of this reasoning, it was assumed in the liquid drop model, that the motion of individual nucleons in a nucleus are not important. Rather the whole nuclear matter in the nucleus, behaves like charged liquid drop, and one should consider the general motion of the liquid for calculating the various properties of the nucleus. The nuclear drop model was developed by Weiszãcker[42] for obtaining the nuclear masses and the binding energies in terms of macroscopic parameters like volume energy, surface energy, coulomb energy and pairing energy, etc. of the nucleons, considering the nucleus as a liquid drop. This model was later come handy in explaining the phenomenon of nuclear fission. Its latest version, *i.e.* collective model as developed by Rainwater[43] and by A. Bohr and B. Mottleson[44] has helped us in understanding the vibrational and rotational motions in nuclei. This is described in Chapter 11.

The collective and the liquid drop models, however, could not explain the properties of nuclei which exhibited the extra stability, for nuclei having neutrons or protons equal to the magic numbers of 2, 8, 20, 50, 82 and 126. This was successfully explained by the shell model, as developed by Mayor and Jensen[45] and later on modified by many other workers. This model requires that the nucleons in a given nucleus arrange themselves in groups of energy states—shells—so that the magic number nuclei correspond to the closed shells in the same manner as the closed shells in atoms correspond to noble gases. For creating such a shell structure, each nucleon is supposed to move in the common potential, $[-V(r)]$ created by other nucleons to which is added a spin-orbit coupling term $\mathbf{l.S}\ V_{ls}$. In this model, the magic numbers are explained in a natural manner, as well as the spins of the ground state of almost all nuclei. This simple shell model, also called, extreme single particle model, is, however, incapable of explaining, the magnetic moments, quadrupole moments, and the binding energies of nuclei in the ground state and also many properties of the excited states of nuclei. For this purpose, many extended versions of the simple shell model, have been used and developed in Chapter 10.

Basically the various modifications take into account the nucleon-nucleon interactions of 'loose' nucleons, outside the closed shells. The shape of the common potential itself has been modified in the case of the deformed nuclei, so that the loose nucleons move in an ellipsoidal common potential, rather than in a spherical potential, of the simple shell model. The introduction of ellipsoidal potential corresponds to the recognition of the collective effect of the 'loose' nucleus. This 'marriage' between the shell model and collective model, as developed by Nilsson and others[46], especially Davidov and Filipov, has helped in explaining the properties of excited states in deformed nuclei and has been in use for quite some time in nuclear structure calculations (Chapter 12). Core excitation and core polarisation imposed in the above mentioned conditions, are other sophistications, which have proved very useful in explaining the many anamalous moments and transition rates. Attempts have also been made to develop microscopic theories to take into account the fact that a nucleus is a many-body system, with a large number-(ranging from a few to many hundred) of nucleons which interact with each other strongly. Two approaches have been made in this direction.

1.8 MICROSCOPIC THEORIES

(*i*) *Theory of nuclear matter:* In this approach, developed by Brüeckner et al. and Bethe[47], the properties of nuclear matter (the infinite nucleus) were investigated assuming that the nuclear wave functions could be taken as plane-waves for infinite matter. One attempts to derive, in a self-consistent manner, the common potential in which each nucleon moves in an infinite matter, using the two-body interaction in accord with the scattering experiments as an input data, in the form of the reaction matrix (K matrix). Application to finite nuclei was developed by Brüeckner, Gammel and Weitzner[48] and others, where the effect of the boundary conditions of a finite nucleus was taken into account by appropriately modifying the K-matrix, so that it corresponds to the local density (which is uniform and independent of space-coordinates in infinite nuclear matter) which will vary as a function of radius in a finite nucleus. Calculations of binding energies, etc. for some nuclei on the basis of this theory have yielded results of the right order but the success is limited.

(*ii*) *The Hartree-Fock self-consistent theory:* Essentially this method reduces the problem of many interacting particles to one of non-interacting particles in a field, which is obtained in a self-consistent Hartree-Fock procedure[49], using the two-body nuclear potential as the input parameter. For light nuclei, with a few nucleons of say, up to $A \approx 20$, the method can be used to treat the whole nucleus in this manner. For medium and heavy nuclei, however, one uses this method to take into account only the interactions of 'loose' nucleons outside a 'Core' which may be assumed to be unperturbed in the excitation under consideration. This method, though very useful in simplifying the problems is, however, an approximation and neglects a large part of the long range internucleon-forces, called the 'residual interaction'. Various attempts of the recent calculations are essentially directed towards the inclusion of these residual interactions to better approximation. Notable among these attempts is the quasi-particle or BCS theory[50], which directly takes the short range part of the nuclear forces (the pairing energy) into account and the long range part is treated by perturbation methods. This theory has been borrowed from the theory of superconductivity as developed by Bardeen-Cooper-Schrieffer (BCS).[50] TamDancoff Approximation (T-D)[51] is a fancy name for about the simplest realistic microscopic treatment of nuclear excitations, based on H-F approximation. A variation of the Hartree-Fock (H-F) theory is the time dependent HF theory, and is called the Time Dependent Hartree-Fock-Approximation Theory (TDHF)[51]. This is used for calculating the time dependent phenomenon, involving excited states, and is designed to take into account the long range part of the residual interactions. The Random-phase Approximation theory (RPA)[52] is an alternative formulation of the time dependent Hartree-Fock theory and is borrowed from the theory of plasma oscillation as developed by Bohm and Pines.[53] This theory gives a lower order solution (and hence is a better approximation), than TDHF[51] theory for time dependent phenomena. Another model, applied to the description of quadrupole collective properties of low lying states in nuclei, is termed as Interaction Boson Model (IBM)[53] where bosons are assumed to be made up of correlated pairs of valence nucleons, carrying even angular momenta $l = 0, 2, 4$, etc. These and still more generalised theories have been developed and applied to deformed nuclei in a limited manner.[53] We have, however, not dealt these topics in this book.

Interacting Boson Approximation (IBA) Model has been a major tool for calculating energies, the transition probabilities and quadrupole moments of even-even nuclei and has been used extensively. Recently an extensive work was reported for even-even $Cd^{110, 112, 114}$, $Pd^{100-116}$ and Pu^{94-114} chains[54], using IBA model, where values of energies, B (M1), mixing ratios and magnetic dipole moments was calculated and compared with experimental data, with reasonable agreement. An extension of this model was used, where IBA model plus broken pair description was used for high spin dipole in ten bands[55], especially applying to $_{60}Nd^{136}$ nucleus, up to $l > 20$. For odd nuclei, or odd-odd nuclei, an Interacting Boson-Fermion model, with or without broken pair has been[56] used say for $_{39}Y^{97}$, or $_{51}Sb^{117}$ or $_{29}Cu^{62, 64, 66}$ nuclei.

Similarly extensive use of Hartree-Fock calculation have been combined with large basis shell model. A recent[57] calculation is that of $_{20}Ca^{47-60}$ where detailed comparison in made for $_{20}Ca^{48}$ to give the parameters. A detailed review of this method is given in Annual Review nuclear and particle Sciences[58].

A modification of this method, *i.e.* Cranked Hartree-Fock Bogoliubov model has been applied for a large number (more than forty) of even-even nuclei from Xe to Ba recently.[59]

1.9 NUCLEAR REACTION MODELS

Theoretically, the problems in nuclear reactions contain two basic components (*i*) The reaction mechanism (*ii*) The nuclear structure associated with the properties of nuclear states involved in the reaction. The problem of nuclear structure will, in principle, be the same as discussed earlier, except that in cases where highly excited states are involved, the states are so close to each other, that one may use the statistical model[60], rather than the individual properties of the levels. The statistical model deals with the nuclear level-density on the basis of statistical considerations. For the low excited states, however, the detailed properties of individual states can be taken into account, and dealt with in the manner discussed earlier. On the other hand, the reaction mechanism requires specific models or theories. One adopts either the compound nucleus model[61] or the direct reaction model[62] or some intermediate mechanism[63], depending on the type of the projectile, the target nucleus, and the incident energy. These are macroscopic models. As for example, in the compound nucleus models, one assumes that in entering the nucleus, the incident particle shares its energy with other nucleons and forms an intermediate state called the compound nucleus. In this case, the decay of the nucleus depends on the properties of the compound nuclear state rather than on the mode of the production of that state. These assumptions are based on the existence of very strong nuclear forces between nucleons. The decay time involved in this case is of the order of 10^{-14} to 10^{-16} seconds and corresponds to the time of many traversals of the incident particle in compound state of the nucleus. The compound and the statistical models have been dealt with in Chapter 13.

The direct reaction corresponds to the condition where the incident nucleon interacts with the nucleons in such a manner that the emitted particle comes out as a result of a single direct encounter of the projectile with a nucleon in the nucleus. In this case the time of interaction is shorter, of the order of 10^{-22} seconds, which is approximately the time taken by the incident particle to travel the incident nucleus, once. Typical examples of direct reaction are (d, p), (d, n), (H^3, d), (He^3, d), etc. In practice, the reaction may go through both the direct and the compound (and also through some intermediate processes) where the energy of the incident particle is shared by two or three particles in contrast to one particle in direct reaction or many particles in the compound nucleus. This topic has also been discussed in Chapter 14.

Another model is the optical model[64] in which one replaces the nucleus by a potential which is complex. The real part of the potential gives rise to elastic scattering and the imaginary part produces the absorption through reaction and inelastic scattering. This model has been extensively used for elastic and non-elastic scattering of particles, in cases where only the average properties of the scattering nucleus are involved and not the properties of any individual levels: Chapter 15.

Recently[63, 65], there have been carried out many studies of nuclear reactions experimentally and theoretically—where reaction mechanism corresponds to an intermediate status; between compound nucleus formation and single-hit direct reaction. This reaction mechanism, called the pre-compound, or pre-equilibrium model of nuclear reactions, assumes that the incident projectile, interacts with nucleons inside the nucleus successively in such a manner; that either the ejectile is emitted comparatively with high energy, leaving behind low energy projectile which shares its energy with other nucleons forming a compound nucleus, which then decays through statistical process. Or if the first ejectile is of low energy; the projectile then proceeds to either come out directly; or hits another nucleon; and again starts a reaction with low energy and high energy particles sharing energy. In this manner, after a few such encounters, say about 4 to 6; the energy shared between the two particles is low enough, that only compound nucleus is formed. The ejected particles then have an energy spectrum or angular distribution, which is a combination of these successive steps. A lot of experiments with both light and heavy incident particles have been carried out, at somewhat higher energies—10 to 100 MeV/nucleon—which can be understood on the basis of pre-compound or pre-equilibrium reaction model[65]; and have been analysed with theoretical models, developed in the last two decades. These are described in Chapter 16.

Analysis of the nuclear reactions based on these macroscopic models, yields broadly, the nuclear parameters like (*i*) the nuclear level densities (*ii*) nuclear radii (*iii*) the real and imaginary parts of the potential in the optical model (*iv*) the level widths or decay rates in the compound nucleus (*v*) the orbital angular momenta of the various levels in direct reaction and (*vi*) the spectroscopic factors which basically determine the strength of the direct reaction involving the particular level. These parameters and models are only indirectly related to the basic two-body nuclear forces and one requires a more detailed analysis to connect them directly to the nuclear forces. Attempts have been made to develop the generalised theories of nuclear reactions. As for example, a general theory, based on the collision or scattering matrix (called *S*-matrix) theory has been developed.[66] The reaction or scattering cross-sections in this general theory are expressed in terms of *S*-Matrix which gives the asymptotic forms of the wave functions of the system. To determine *S*-matrix, however, one requires to know the properties of the system, in the interaction region, where the two particles collide, so that the asymptotic wave functions of *S*-matrix can be connected with this region through the use of the continuity properties of the wave function. The wave function in the interaction region are expressed in the formalism of *R*-Matrix which, essentially involves various quantities evaluated in or just outside the region, within which the particles may interact, and outside which there is no further reaction. Kapur and Pierls[66] gave the basic form of the rigorous theory of nuclear reaction on the basis of *S*-Matrix and *R*-Matrix. Wigner[67] has given another representation which is very generally used. The inputs to the theory are the properties of the various states of the interaction system in the interaction region which provide *R*-Matrix. For understanding the basic features of these theories, the reader should refer to the above mentioned references; and 'Nuclear Theory V.I, II and III, J.M. Eisenberg and W. Greiner, North Holland Publishing[68] Co. Again, these generalised theories of nuclear reactions have not been dealt in this book.

1.10 HEAVY-ION REACTIONS

Reactions with heavy projectiles like ^{12}C, ^{16}O, ^{40}Ca, ^{84}Kr, ^{132}Xe, and many other heavy ions, up to even ^{238}U have opened a new vista in nuclear physics[68]. Many new phenomena are observed and many theoretical suggestions are advanced. As for the reaction mechanism involved in heavy ion collision, many authors have looked at heavy-ion physics as a playground for nuclear physicists to distinguish amongst, distant collisions, grazing collisions, hit-and-run, formation of two-body system, formation of a composite system and finally the formation of a compound system. Different degrees of contact can be classified by studying the density overlaps and interaction times between the two colliding nuclei. For energies well below coulomb energies, only Coulomb scattering takes place, and Rutherford scattering model holds good. As the energy is somewhat increased, the diffraction phenomena occur from the edges. So Fresnel and Fraunhauffer diffraction are observed. Inelastic scattering at higher energies, through Coulomb excitation, yields a lot of information about collective modes of excitement. Because of the transfer of high angular momenta by heavy ions, one observes very high spin states. Next comes one or two-nucleon transfer reactions, and one observes an interplay of nuclear structure and reaction mechanism like direct reaction or deep inelastic scattering. The latter is especially significant, for cluster-transfer by heavy ions-induced reactions. Fusion-fission and compound nucleus reaction mechanisms are observed at higher energies. At very high energies, the shock-waves or nuclear matter density isomers may be observed.

Heavy-ion nuclear physics, therefore, tends to become a very interesting subject (Chapter 17). As a matter of fact, a large amount of work is being conducted these days—both experimentally and theoretically—on the various aspects of heavy-ion reactions. One of the important fields, which has opened up because of these activities is, the availability of high spin states in nuclei at higher energies, because of the collective excitations. States of angular momenta up to $I \geq 50\ \hbar$ have been observed. The excitation of these states offers interesting insight into the excitations of many deformed and super deformed nuclei.[69]

With the availability of Tandem accelerators, superconducting linear accelerators and cyclotrons; heavy-ion projectiles are increasingly being made available in many laboratories. This has led to the experimental and theoretical[70] studies of heavy-ion reactions in the energy range from 2–3 MeV/A to 35 or 40 MeV/A, depending on the heavy nucleus in the projectile and using synchrotrons, sometimes going to greater than 10 GeV/A.

In a typical experiment, using $_{62}Sm^{152}$ $(_3Li^7\ 4n)_{65}$ Tb,[155] at 45 MeV and $_{50}Sn^{124}$ $(_{15}P^{35}, 4n)_{65}$ Tb^{155} at 165 MeV; using a Tandem accelerator[71], nuclear states up to angular momenta $I = 95/2\hbar$ were excited. This is an example of fusion-evaporation, at a medium incident energy.

At very high energies of 2 GeV protons and 3 GeV. He^3-induced reactions on Ag, Bi and U, one observed the phenomenon of nuclear cascade process of classical step by step evaporation and fission. At still higher energies, *i.e.* 11–6 GeV/A, for central, Au + Au reaction, the proton rapidity distribution, showed the possibility of formation of state of matter with baryon density substantially greater than normal nuclear matter.[72]

At lower energies, one observes the phenomenon of fission fragments as was investigated[73] by a group at Bombay using 14 UD tandem, using C^{12}, O^{16} and F^{19} projectiles on Th^{232} target.

The subject of nuclear physics, thus, provides a large field of interplay of theory with physical phenomenon, in nuclear interactions, using many fascinating experimental and theoretical techniques. It gives an insight into systems (Nuclei) of a limited number of constituents (Nucleons), governed by strong short-range forces.

REFERENCES

1. H. Becquerel: Comp. Rend. 122, 422, 501 (1896).
2. H. Geiger and E. Marsden: Phil. Mag. 25, 604 (1913); Proc. Roy. Soc. (London); A. 82, 495 (1909).
3. J.J. Thomson: Phil. Mag. 44, 293 (1897).
4. E. Rutherford: Phil. Mag. 11, 1661 (1906); 12, 134 (1900); 21, 669 (1911); 25, 10 (1913).
5. N. Bohr: Phil. Mag. 26 (1913); Phil. Mag. 26, 476 (1913).
6. J.J. Thomson: Phil. Mag. 13, 561 (1907); 20, 752 (1910); 21, 225 (1911), 24, 204 (1912).
7. F. W. Aston: Nature, 123, 313 (1919).
8. J. Chadwick: Proc. Roy Soc. (London); A 136, 692 (1932).
9. (*i*) M. Planck: Ann. Physik, 4, 553 (1901).
   (*ii*) L. de Broglie: Suggestions (1924); L. de Broglie and L. Brillouin: Selected Papers on Wave-mechanics, London, Blackie (1929).
   (*iii*) E. Schrödinger: Ann. d. Physik, 79, 361, 484 (1926); 81, 109 (1926).
   (*iv*) W. Heisenberg: Zeits of Physik, 43, 172 (1927); Physical Principles of Quantum Theory; Univ. of Chicago Press, New York (1929), Dover Publications.
   (*v*) P. A.M. Dirac: Proc. Royal Society; A 112, 661 (1926), Section 5, 114, 243 (1927); Proc. Comb. Phil. Soc. 30, 150 (1934).
10. C.D. Elliot and N.F. Mott: Proc. Roy. Soc. (London); A 141, 502 (1933).
11. W. Pauli: Rapports di Septieme Council de Physique Solvay; Brussels, (1933); Ganthier-Villars and Cie, Paris (1934).
12. E. Fermi: Z. Physik, 88, 161 (1934); G. Gamow and E. Teller, Phy. Rev. 49, 895 (1936).
13. R.J. Vande Graff: Phy. Rev. 38, 1919 (1931).
14. J.D. Cockcroft and E. Walton: Proc. Roy. Soc. (London); A. 136, 619 (1932).
15. J.D. Cockcroft and E. Walton: Proc. Roy. Soc. (London); A. 137, 229 (1932).
16. E.O. Lawrence and N.E. Egelfereri: Science 72, 376 (1930); E.O. Lawrence, H.S. Livingston and M.G. White: Phy. Rev. 42, 1501 (1932).
17. E.M. McMillan: Phy. Rev. 68, 143 (1945). Veksler, V.J. Phy. (U.S.S.R.), 9, 153 (1953).
18. R.K. Bhandari, K.P. Nair and A.S. Divatia: Survey of Medium and High Energy Accelerators and Storage Rings, VEC Project, Calcutta (1979).
19. W. Christophilos and H.S. Livingston: Ann. Rev. Nuclear Science, 1, 169 (1952).
20. E.D. Courrant, H.S. Livingston and H.S. Snyder: Phy. Rev. 88, 1190 (1952).

Linear Accelerators

21. L.W. Alverz., H. Bradner, J.V. Frank, H. Gordon, J.D. Gow, L.C. Marshall, F. Oppencheimer, W.K.H. Panofsky, C. Richman and J.R. Woodyard: Rev. Sc. Inst. 26, 111 (1955).

    P.M. Lapostolle and A.L. Septter: Linear Accelerators, North Holland Publishing Co., Amsterdam (1970).

22. D.H. Sloan and E.O. Lawrence: Phy. Rev. 38, 2021 (1931).

23. D.W. Kerst: Phy. Rev. 60, 47 (1941); Amer. J. of Physics, 10, 224 (1942).

Superconducting Cyclotrons

24. G.S. Brown and I. Vindan: N.I. and M. A. 246 (1986).

Pulsed Storage Rings and Synchrotron Radiation

25. John. H. Omrod: Nuclear Instruments and Methods in Physics Research; A. 244, 236–245 (1986).

Superconducting Linear Accelerators

26. L.M. Bollinger: Nuclear Instruments Methods; A. 244, 246–258 (1986).

High Energy, Heavy-Ion Linear Accelerators

27. Jose R. Alfonso: N.I. and M.A. 244, 262–272 (1986).

28. O. Hahn and F. Strassman: Naturwiss 27, 11 (1939).

A. Chrisko, G.H. Higgins, A.E. Larsh, G.T. Seaborg and S.G. Thompson: Phy. Rev. 87, 163 L (1952); G.T. Seaborg. Phy. Rev. 85, 157 L (1952).

J.R. Huizenga, C.L.Rao and D.W. Egel Kerneir: Phy. Rev. 107, 319 (1957).

Exotic Beams

28 (*a*) B. Waast, et al.: Nuclear Instruments and Methods; A. 382, 348 (1996); Ch. Fomaschkko, Ch. Shoppman, D. Frandt and M. Viot: Instr. and Methods, B 88, p. 6 (1994).

Reactor—Ist

29. Enrico Fermi: Science, 105, 27 (1947); W.H. Zinn: Rev-Modern Physics, 27, 263 (1955); Brugger, Evans, Joki and R.S. Shakland, Phy. Rev. 104, 1054 (1956).

High Flux Reactors

30. A.P. Oslon: N.I.M.A., 249, 77–90 (1986); W.H. Zinn: Nucleonics, 15, 100 (1957).

Mass Spectrometers

31. H.E. Duckworth, S.D. Barker and V.S. Venkatsubramaniam: Mass Spectroscopy, Cambridge University Press, Cambridge (1990).

32. I.I. Rabic, J.M.B. Kellog and J.R. Zachiaras: Phy. Rev. 40, 157 (1934); J.M.B. Kellog, I.I. Rabi and J.R. Zachiaras: Phy. Rev. 50, 2 (1936); I.I. Rabi, J.R. Zachiaras, S. Millman and P. Rash: Phy. Rev. 53, 315 L (1938).

Magnetic Quadrupole Moments

33. 1. Norman F. Ramesey: Nuclear Moments, (1953), John Wiley & Sons, Inc. New York.

2. H. Koperfermann: Nuclear Moments, (1958), Academic Press Incorporated, New York.

3. H.A. Enge: Nuclear Physics, (1966), Addison Wesley, Publishing Co.

Atomic Spectroscopy

34. H.E. White: Introduction to Atomic Spectra, McGraw-Hill Book Company, Inc. New York (1934); Robert S. Shakland, Atomic and Nuclear Physics: McMillan Company, New York (1960).

Nuclear Radius Techniques

35. R.C. Barret and D.F. Jackson: Nuclear Sizes and Structure; Clarenden Press, Oxford (1979).

Nuclear Spectroscopy—Magnetic Spectrometers

36. K. Siegbahn: Beta-ray Spectrometer; Theory and Design; Beta and Gamma Rays Spectroscopy—North Holland Publishing Co., Amsterdam (1955).

Magnetic Moments and Quadrupole Moments of Excited States

37. Norman F. Ramsey: Nuclear Moments, John Wiley & Sons (1953).
    Handbuch der Physik, V. 38 and 39. Springer Verlag (1958).
38. Nuclear Data Sheets: Academic Press Inc. New York, V. 1 to 85 (1966–1998).
39. B.R. Berret: Phy. Rev. 154, 955 (1967).
40. Three Body Forces: T. Veda Progress Theory, Phy. 29, 829 (1963); D.W.E. Brorts, E.B. H.J. McKeller Phy. Rev. C II, 614 (1975); T.Veda et al.: Few Body Dynamics by A.N. Mitra, North Holland, Phy. Rev. D7, (1973), R.D. Amado: Phy. Rev. Lettes 339, New York (1975); A.S. Kronfeld and P.B. Markenzii: Annual Rev. Nuclear and Particle Science, 43, 793 (1993), Concepts of Particle Physics, V.1, Kurt Godttfried and V.F. Weisskopf; Amsterdam Press, Oxford (1984).
41. G. Gamow: Z. Physik, 51, 204 (1928); P.W. Gurney, E.U. Condon: Nature, 122–439 (1928) Phy. Rev. 33, 127 (1929).
42. C.F. Von and Weizsäcker: Z. Physik, 96, 431 (1935); Naturwiss 24, 813 (1936).
43. J. Rainwater: Phy. Rev. 79, 432 (1950).
44. A. Bohr. Dan Mart. Fys. Medd: 26, No. 14. (1952); A. Bohr and B.R. Mottleson; Den. Mat. Fys, Medd, 27, No. 16 (1953).
45. M.G. Mayer: Phy. Rev. 75, 1969 (1949); Phy. Rev. 75, 1766 (1949); Z. Phy. 178, 295 (1952). Haxel O., J.H.D. Jenson and H.E. Suess: Phy. Rev. 75, 1766 (1949).
46. S.G. Nilsson: K.D. Denske Vidensk Selesk, Matfys. Medd. 29, No. 16 (1955). A.S. Davydov and G.F. Fillipov: Nucl. Physics 87, 237 (1958). A.S. Davydov and V.S. Rostovsky: Nuclear Phy., 12, 58 (1959). T.D. Newton: Can. J. of Phy. 38, 700 (1960).
47. K.A. Brüeckner and J.L. Gammel: Phy. Rev. 109, 1023 (1958).
    H.A. Bethe: Phy. Rev. 103, 1353 (1956).
    H.A. Bethe and J. Goldstone: Proc. Roy. Soc. (London); A. 238, 551 (1956).
48. K.A. Brüeckner, J.L. Gammel and H. Weitzner: Phy. Rev. 110, 431 (1958).
    K.A. Brüeckner et al.: Phy. Rev. 118, 438, (1960): Phy. Rev. 118, 1442 (1960).
49. M.K. Pal and A.P. Stamp: Phy. Rev. 158, 924 (1967).
    K.A. Brüeckner et al.: Phy. Rev. 121, 255 (1961).
    R.E. Pierls and J. Yoccoz, Proc. Phy. Royal Soc. (London); A. 70, 381, (1957), 50.
50. J. Bardeen, L.N. Cooper and R. Schrieffer: Phy. Rev. 108, 1175 (1957).

51. D.J. Rowe: Nuclear Collective Models, Metheune & Co. London (1970).
J.P. Elliot and B.H. Flower: Proc. Roy. Soc. 242A, 57 (1957).
G.E. Brown, J.A. Evami and D.J. Thomas; Nuclear Physics V. 24, 1 (1961).

52. R.A. Fessel: Phy. Rev. 107, 450 (1957).

53. D. Bohm and D. Pines: Phy. Rev. 92, 609, 626 (1953), F. Ichello and A. Arima, The Interacting Boson Model, Cambridge Univ. Press, Cambridge, England (1987).

54. A. Nannini, Giannatiempo., P. Sona et al.: Phy. Rev. C. 44, 1508 (1991), Phy. Rev. C 44, 1841 (1991). Phy. Rev. C 52, 2969, (1995), Phy. Rev. C 58, 3316 (1998), C 58, 3335 (1998).

55. F. Ichello and D. Vretener: Phy. Rev. C 43, R 945, (1991), D. Vretener et al., Phy. Rev. C 57, 675 (1998).

56. G. Lherronneau, S. Brandt, V. Paar and D. Vretener: Phy. Rev. C 57, P. 681, (1998), Yu N. Lobach and D. Bucurescu: Phy. Rev. C 58, 1515 (1998), A.K. Singh and G. Gangopadhyay: Phy. Rev. C 55, P 726 (1997).

57. B.A Brown and W.A Richter: Phy. Rev. C 58, P. 2099 (1998).

58. B.A. Brown and B.H. Wildenthal: Annual Rev. Nuclear Particle Science 38, 29 (1998) and B.H. Wildenthal, Progress in Particle and Nuclear Physics II Edited by D.H. Wilkinson; Pergamon Oxford (1984), P. 5.

59. M.S. Sarkar and S. Sen: Phy. Rev. C 56, 3140 (1997).

60. N. Bohr, Nature 137, 344, (1936), H.A. Bethe, Rev. of Modern Physics 9, 69 (1937).

61. C.F. Porter and R. Thomas: Phy. Rev. 104, 483 (1956).
H. Feshbach: N-Spectroscopy ed: F. Ajzenberg-Selov. Chapter VA, New York, Acad. Press (1960).
B.L. Cohen: Phy. Rev. 120, 925 (1960).

62. N. Austern, S.T. Butler and H. McManus: Phy. Rev. 92, 353 (1953).
M.K. Bannerjee: Nuclear Spectroscopy ed. F. Ajzenberg, Selov. New York , Acad. Press (1960).

63. M. Blann: Annual Rev. Nuclear Science, V. 25, P. 123 (1975).

64. H. Feshbach, C. Roster and V.F. Weisskopf: Phy. Rev. 96, 448 (1954). F. Bjork Lund and S. Feshbach: Phy. Rev. 109, 1295 (1958).

65. H. Feshbach, A. Kerman and S. Koonin: Ann. Phy. (N), 125, 429 (1980).
C. Kalbach and F.M. Mann: Phy. Rev. C 23, 112 (1981), C25, 3197 (1982).

66. G. Breit: Encyclopaedia of Physics ed. S. Flügge V. 41/1, Berlin, Springer Verlag (1959).
P.L. Kapur and R.E. Pierls: Proc. Roy. Soc. (London) A-166, 277 (1938).

67. E.P. Wigner and L. Eisenbud Phy. Rev. 72, 29 (1947).

68. J.M. Eisenberg and W. Greiner, Nuclear Theory: V.I.II and III, North Holland Publishing Co., Amsterdam (1972).

69. P.E. Hodgson: Heavy-Ion Interaction, Clarenden Press, Oxford (1978).

70. D.M. Milazze et al. (26 Authors): Phy. Rev. C 58, 953 (1998).

71. D.J. Hartley, T.B. Brown, F.G. Kondev, J. Pfohl, and M.A. Riley, S.M. Fischer, R.V.F. Jenssens, D.T. Nissivs, P. Fallon, W.C. Ma, and J. Simpson Phy. Rev. C 58, P. 2720 (1998).

72. X. Ledoux et al. (22 authors): Phy. Rev. C 57, P. 2375 (1998).

73. D.V. Shetty, R.K. Chowdhary, B.K. Nayak, D.M. Nadkarni and S.S. Kapoor: Phy. Rev. C.V. 58, P.R. 616 (1998), V.S. Ramamurthy et al.: Phy. Rev. C 41, 2702 (1990).

SUGGESTED READING

1. Eder Gernet: Nuclear Forces, M.I.T. Press, Cambridge, Mass (U.S.A.) (1968).
2. Rose David: Nuclear Fusion, Plasmas and Controlled Fusion, MIT (1961).
3. Nuclear Interactions: De Benendeite Sargre, John Wiley & Sons, New York (1968).
4. Nuclear Heavy-Ion Reactions:
   Hodgsons P.E. Oxford, Clarendon (1978).
5. Nuclear Magnetic Resonance: E.R. Andre (Cambridge Univ. Press) (1995).
6. Nuclear Matter and Nuclear Reactions: Kikchuchi Ken Amsterdam, N. Holland (1968).
7. Nuclear Physics. R.K. Bhaduri (Addison Wesley) (1988).
8. W.E. Burcham and M. Jobes: Nuclear and Particle Physics, Longman, Scientific Technical Burnt Hill (UK) (1995).
9. Frankfelder Hans: Nuclear and Particle Physics, (Advanced Book, Programme W.A. Bougamin, Inc.) (1975).
10. R.J. Gupta: Heavy-Ion and Related New Phenomenon, (W.Sc. Singapore) (1999).
11. Philosophical Problems of Nuclear Sciences: Heisenberg Wesner Lowon: Faber (1969).
12. Heyde K. Basic Ideas and Concepts in Nuclear Physics. (Overseas) (2005).
13. International Conference on Three Body Problems in Nuclear and Particle Physics (Lowon, N. Holland) (1970).
14. New Trends in Theoretical and Experimental Nuclear Physics W.Sc. (1992).
15. E.B. Paul: Nuclear and Particle Physics: North Holland Publishing Co., Amsterdam (1969).
16. M.A. Preston and R.K. Bhaduri: Physics of the Nucleus, Addison Wesley Publishing Company, Inc. Reading (1962) Mass. (U.S.A.).
17. V.G. Soloviev: Theory of Complex Nuclei, (Oxford: Pergamon) (1976).
18. Hyperfine Structure and Nuclear Radiation, E. Mathias (1968) Amsterdam, N. Holland.
19. Direct Nuclear Reaction Theories: Austern Morman N.Y., John Wiley (1970).
20. Bass Reiner: Nuclear Reactions with Heavy-Ions, (N.Y., Springer) (1980).
21. D.H. Wilkinson: Isospins in Nuclear Physics, Amsterdam, N. Holland (1969).
22. Theory of Nuclear Structure: M.K. Paul, Affiliated East West Press Pvt. Ltd. New Delhi/Chennai (1982).
23. J.M. Edenberg and W. Greiner: Nuclear Theory, V-I, II and III, North Holland Publishing Co. (1972).
24. A Survey of Medium and High Energy Accelerators and Storage Rings: R.K. Bhandari, K.P. Nair and A. S. Divatia BARC 1-562, Mumbai (India) (1979).
25. Rudolph Kotah: Particle Accelerators, Sir Issac Pitman & Sons Ltd. (London) (1967).
26. Accelerator Particle Physics:
   Klapder Kleingrothans H.V. I.O.P. (1995).
27. Casten Richard F.: Nuclear Structure, Algebraic Approaches to Nuclear Structure, N.Y. Howard Academic (1993).
28. Relativistic Dynamics and Quark – Nuclear Physics (1986).
29. D.A. Bromley: Treatise on Heavy-Ion Science, Premium Press, New York (1984).
30. Emilus Segre, W.A. Benjamin: Nuclei and Particles – An Introduction to Nuclear and Subnuclear Physics, Reading Mass (U.S.A.) (1977).
31. Nuclear Physics, Longman London, (1963): Burcham W.E.

2

CHAPTER

Static and Dynamical Properties of Nuclei

The various physical phenomena in nuclear physics studied for the last fifty years or so, have yielded a large amount of information, to enable us to obtain systematics; and to develop theories. In this chapter, we will define and explain quantitatively the various physical quantities encountered in nuclear physics; and describe, in principle, methods of their measurement and present the data (or its source) and the systematics for each quantity.

We may divide the physical phenomena in nuclear physics in two parts: (*i*) Static properties of nuclei and (*ii*) Dynamical phenomena and nuclear properties based on such phenomena. The static nuclear properties consist of: (*i*) Nuclear masses, (*ii*) Nuclear sizes, (*iii*) Angular momenta (I) and parity (π) (*iv*) Dipole magnetic moments (μ) and (*v*) Quadrupole electric moments (Q), of the ground states of the stable nuclei. Very precise experimental methods have been developed, to measure each of these quantities, and data is available for nearly all the ground states of stable nuclei for these quantities. Also, enough data is available for the ground states of some unstable nuclei.

The dynamical phenomena, on the other hand, cover a larger range of quantities, concerned with the excited states of nuclei, as well as the ground states, obtained through dynamic processes like the study of radioactive decay or the nuclear reactions. Apart from the energetics, which are provided by the spectra of the emitted radiation in radioactive decay; we can also measure the lifetimes (τ) of the decaying states. This provides the information about the transition probability ($1/\tau$) of the decay from one state to the other; connecting the wave-functions of the two states, through a Hamiltonian, for a given interaction. It is, also possible, by the detailed studies of angular distribution, or angular correlation of the emitted radiation to get information about the quantum numbers (I,1, π) of the two states. This information helps in understanding the implication of different nuclear structure models. More involved angular correlation techniques like perturbed angular correlations (PAC) yield the values of μ and Q of the excited states, if the information of the perturbing fields is available. The study in radioactive decay can also provide information about the properties of the Hamiltonian connecting the two nuclear states. This has particularly yielded interesting results in beta decay, which involves weak interaction.

Nuclear reactions involve a larger number of parameters, *e.g.* the incident energy, the type of incident particle, the type and energy of the outgoing particle and the angular distributions and correlations. The detailed studies of nuclear reactions can yield, not only the energetics of the connected states; but also the wave-functional properties of these states, including their quantum numbers. The nuclear reactions have therefore become very important tools for nuclear spectroscopy, *i.e.*, for measuring the quantities like I, π, μ and Q of different states. But it is possible to obtain this information about nuclear spectroscopy, only if one knows the reaction mechanism, through which the reaction proceeds. The measurement of absolute cross-sections and differential cross-sections as a function of incident energies and various outgoing particles provides this information, through a proper theoretical analysis.

The study of these cross-sections for nucleon-nucleon, [*i.e.*, $(n-p)$ and $(p-p)$] scattering, has provided the detailed information about nuclear forces.

2.1 STATIC PROPERTIES OF NUCLEI

2.1.1 Nuclear Masses

The nuclei form the core of atoms, with electrons revolving around them. Except for very light nuclei, it is difficult to ionise the atoms completely, so as to 'bare' the nuclei. One, generally, ionises the atoms by removing only one or two electrons, in various experimental methods used for measuring atomic nuclear masses. It is therefore, more convenient to measure directly the atomic masses including the masses of revolving atomic electrons, rather than nuclear masses. One obtains the nuclear mass, by subtracting from the atomic mass, the mass of the electrons and making correction for the binding energy of electrons, *i.e.*,

$$M_N = M_A - [Zm_e - (B.E.)_e] \qquad ...(2.1)$$

where M_N is the mass of the nucleus, M_A is the mass of the neutral atom, m_e is the mass of one electron, Z is the atomic number of atoms, and $(B.E.)_e$ is the total binding energy of all electrons. The binding energy of all the electrons $(B.E.)_e$ for a given atom can be obtained from literature from X-ray spectra or ionisation studies[1]. The atomic masses, M_A may be obtained from various methods, expressed in such a way that the atomic mass of carbon-12 atom is defined as 12 atomic mass units (12 A.M.U.) or mass number of carbon-12 is 12. This is the scale used now for atomic or nuclear masses. Before 1961, atomic mass unit was based on O^{16} which is larger by about 0.03% than the atomic mass unit based on C^{12} scale. According to the present scale, the atomic mass unit (A.M.U.) has a value of:

$$\begin{aligned} 1 \text{ A.M.U.} &= 931.748 \text{ MeV/c} \\ &= 1.6602 \times 10^{-24} \text{ gm} \end{aligned} \qquad ...(2.2)$$

The mass, referred hereafter, for atomic masses will be based on C^{12} unless especially mentioned, otherwise. There are three derived quantities, connected with nuclear masses, which requires special mention.

(*i*) Mass Defect (Δ): defined as:

$$\Delta \equiv M(A, Z) - A \qquad ...(2.3)$$

where A is the mass number, and $M(A, Z)$ is the atomic mass of the neutral atom with mass number A, and charge Z. A related quantity, *i.e.* mass defect per unit mass number is called the packing fraction (P) and is given by:

$$P = \frac{\Delta}{A} = \frac{M(A, Z) - A}{A} \qquad ...(2.4)$$

(*ii*) Binding Energy (B.E.): This is the energy required to break an atom completely into hydrogen atoms and neutrons. This can be expressed as:

$$\text{B.E.}\ (A, Z) = ZM_H + (A - Z)\ M_n - M\ (A, Z) \qquad ...(2.5)$$

where M_H is the mass of neutral hydrogen atom and M_n is the mass of neutron. It may be noted that the binding energy, as defined above, is the energy required to break an atom (A, Z) into Z hydrogen atoms; and $(A - Z)$ neutrons. This is a bit different from the binding energy of a nucleus (A, Z) breaking into Z protons and $(A - Z)$ neutrons; by the difference of electronic binding energies. This difference may be neglected for most of the purposes, but may be taken into account for most precise measurements. This argument applies subsequently also.

(*iii*) Separation Energy (S.E.): This is the energy required to remove a specific particle from a nucleus, *e.g.* the last neutron or the last proton or alpha particle, etc. The separation energy is defined as:

For a neutron (S_n):

$$\begin{aligned} S_n &= B\ (A, Z) - B\ (A - 1, Z) \\ &= M\ (A - 1, Z) - M\ (A, Z) + M_n \end{aligned} \qquad ...(2.6)$$

For a proton (S_p):

$$\begin{aligned} S_p &= B\ (A, Z) - B\ (A - 1, Z - 1) \\ &= M\ (A - 1, Z - 1) - M\ (A, Z) + M_H \end{aligned} \qquad ...(2.7)$$

And for an α-particle (S_α):

$$\begin{aligned} S_\alpha &= B\ (A, Z) - B\ (A - 4, Z - 2) \\ &= M\ (A - 4, Z - 2) - M\ (A, Z) + M_{\text{He}^4} \end{aligned} \qquad ...(2.8)$$

and similarly for any particle. The most upto date mass tables are available in literature[2].

2.1.1.1 Measurement of Nuclear Masses

The measurements of nuclear masses is, essentially, based on three methods (*i*) Mass spectrometry (*ii*) Nuclear reactions and (*iii*) Radioactive decay.

We will discuss the method of nuclear reactions and radioactive decay in the subsequent sections. Here we confine ourselves to the methods of mass spectrometry. The final compilation as available[2] today, is a result of an elaborate procedure, involving least square fitting to the weighted experimental values, from different methods.

The mass spectrometry has been developed since 1922; when Aston[3] developed his mass spectrograph, which used electric field as an energy filter and magnetic field as momentum filter. One applies an electric field perpendicular to the emerging ions from an ion-source, so that ions are bent with a radius of curvature of R_E, obeying the relation (Fig. 2.1):

$$ne\varepsilon = \frac{Mv^2}{R_E} \text{ or } R_E = \frac{Mv^2}{ne\varepsilon} \qquad ...(2.9)$$

where ε is the electric field in electrostatic energy filter, ν is the velocity of the ion entering the filter, *ne* is the charge of ion; and *M* is the mass: and R_E is the radius of the filter.

The ions emerging from the electrostatic energy filter, are then made to pass through a magnetic field, which further bends the ions in the appropriate direction; continuous to the one due to electric field (Fig. 2.1). The radius of curvature of the ions, after passing through the magnetic field (momentum filter) is given by:

$$Bne\nu = \frac{M\nu^2}{R_M} \text{ or } R_M = \frac{M\nu}{neB} \quad \text{...(2.10)}$$

where magnetic field *B* is measured in gausses.

Combining Eqs. 2.9 and 2.10, we can write:

$$\frac{ne}{M} = \frac{R_E \varepsilon}{R_M^2 B^2} \quad \text{...(2.11)}$$

Hence for a given ε, R_E and *B*; R_M is proportional to $(M/ne)^{1/2}$ and hence different masses will have different values of R_M. A photographic plate put in the focussing plane of the magnetic field will register different masses at different points in the focal plane (for mass spectrograph). The resolution of such an instrument[3] is only one in 10^5.

The more modern version is the mass spectrometer based on the principles developed by Dempster and later on by Bainbridge and Nier, involving single and double magnetic focussing[4]. Here one allows the ions produced in the ion-source, to pass through an electric field applied along the direction of the beam, so as to accelerate the ions. The accelerated ions, then obey the relation (Fig. 2.2):

$$\frac{1}{2} M\nu^2 = neV \text{ or } \nu = \left(\frac{2\,neV}{M}\right)^{\frac{1}{2}} \quad \text{...(2.12)}$$

where *V* is the voltage between the two slits right after the ion-source.

The ions then pass through a magnetic field which bends them to an orbit with a fixed and definite radius of curvature R_M related to the velocity and mass of the particle, through the relation:

$$BR_M = \frac{M\nu}{ne} = \sqrt{2MV/ne} \quad \text{...(2.13)}$$

or

$$\frac{M}{ne} = \frac{B^2 R_M^2}{2V} \quad \text{...(2.14)}$$

In this manner, for a fixed *B* and *V*; the ions with a given M/ne will be able to have the fixed R_M. To allow different masses to have the same R_M, one can change *B* or *V* or both. For avoiding the hysteresis effect in magnetic field, one generally keeps *B* constant; and only changes *V*. The single or double focussing modes are provided by creating proper field gradients in the magnetic field.[4]

By using matched doublet method, it is possible to obtain an accuracy of 1 in 10^6 by mass spectrometer[5], as was developed by Bhanot, Johnson and Nier in 1960.

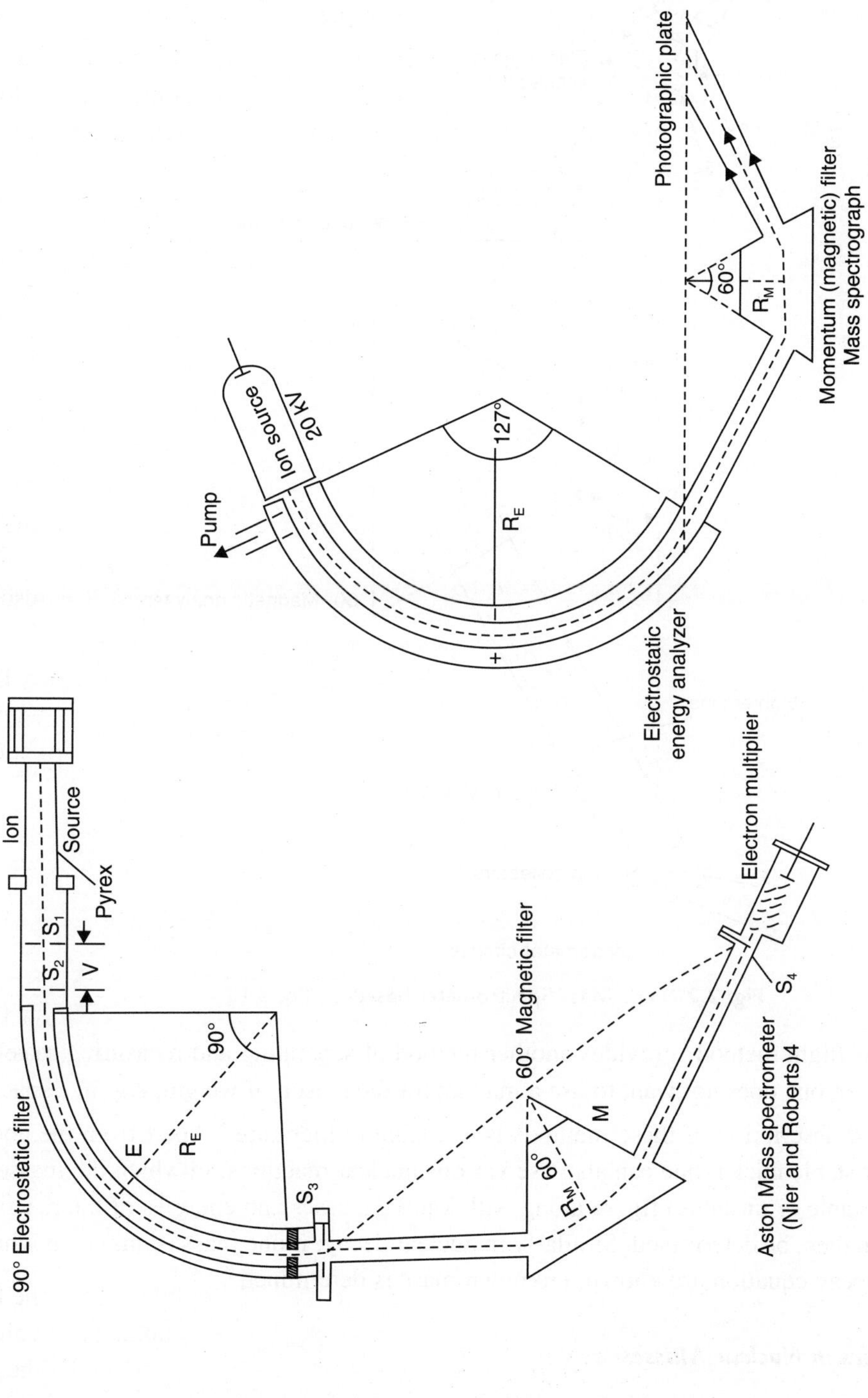

Fig. 2.1 Mass Spectrograph and Mass Spectrometer based on Eqs. 2.10 and 2.11.

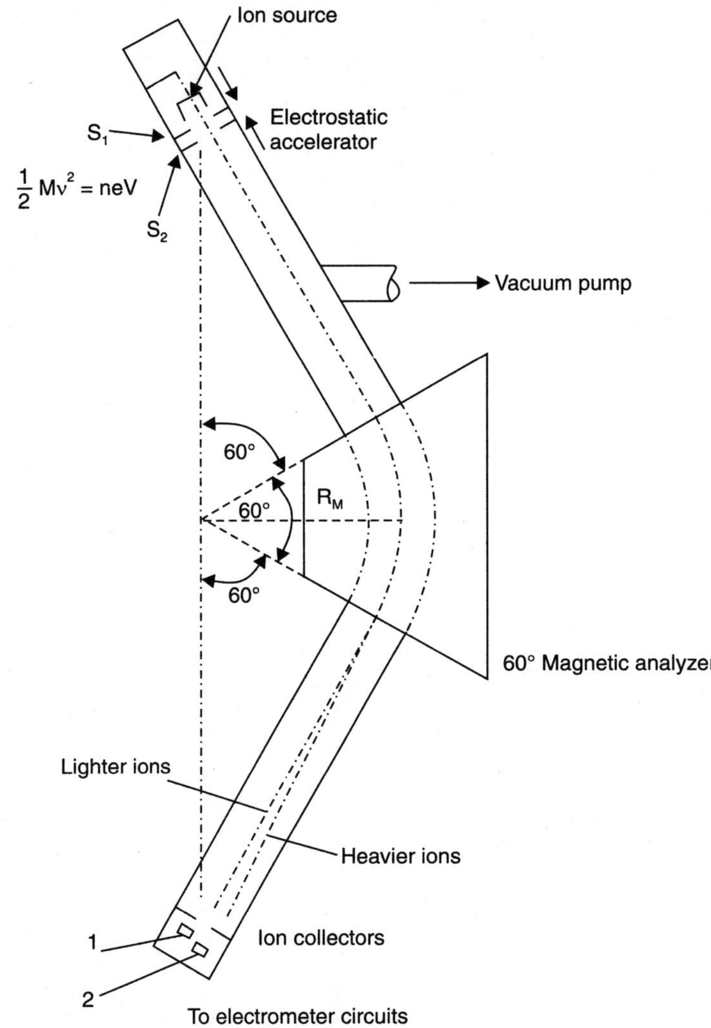

Fig. 2.2 Nier's Mass Spectrometer based on Eq. 2.13.

The time-of-flight method[5] provides another method of separating and measuring masses. This is utilized, whenever, one does not want to use a magnet for the reason of weight, *e.g.* in a space flight.

The detailed description of these methods is available in literature.[6] Apart from the above two direct methods for stable nuclei; one can also use various nuclear reactions; in which the masses of the unknown nuclei (stable or unstable) figure along with known masses and energies of the reaction. The unknown mass can then, be determined. Similarly in nuclear decay; if one mass is unknown while other quantities in the decay equation are known, unknown mass is determined.[6]

2.1.1.2 Systematics in Nuclear Masses

A large amount of data exists on the nuclear masses of both stable and unstable nuclei.[7] It has been possible to obtain, from this data the derived quantities, *i.e.,* binding energies, separation energies and the packing fraction. A few facts of great significance emerge from these systematics.

1. *Nuclear Stability*: The detailed studies of mass-measurements, have provided the information about the number of protons and neutrons in different nuclear isotopes-both stable and unstable. In Fig. 2.3, where the number of neutrons in a given nucleus is plotted against the number of protons, for all known nuclei, are indicated certain features about the ratio of neutrons/protons required for stability in a nucleus.[8] As for example, for low values of N and Z, the stable isotopes have $N/Z = 1$. On the other hand for heavy nuclei, *e.g.* for $A = 238$, $N/Z = 1.6$ and for intermediate values of A, $1.6 \geq N/Z \geq 1$. This indicates that when coulomb effect due to electric charge is not large, protons and neutrons behave similarly, in nuclear attractive interactions. For heavier nuclei, the coulomb energy due to protons, becomes significant, and more neutrons have to be added to create extra attractive interaction, to counter-balance the repulsive coulomb energy.

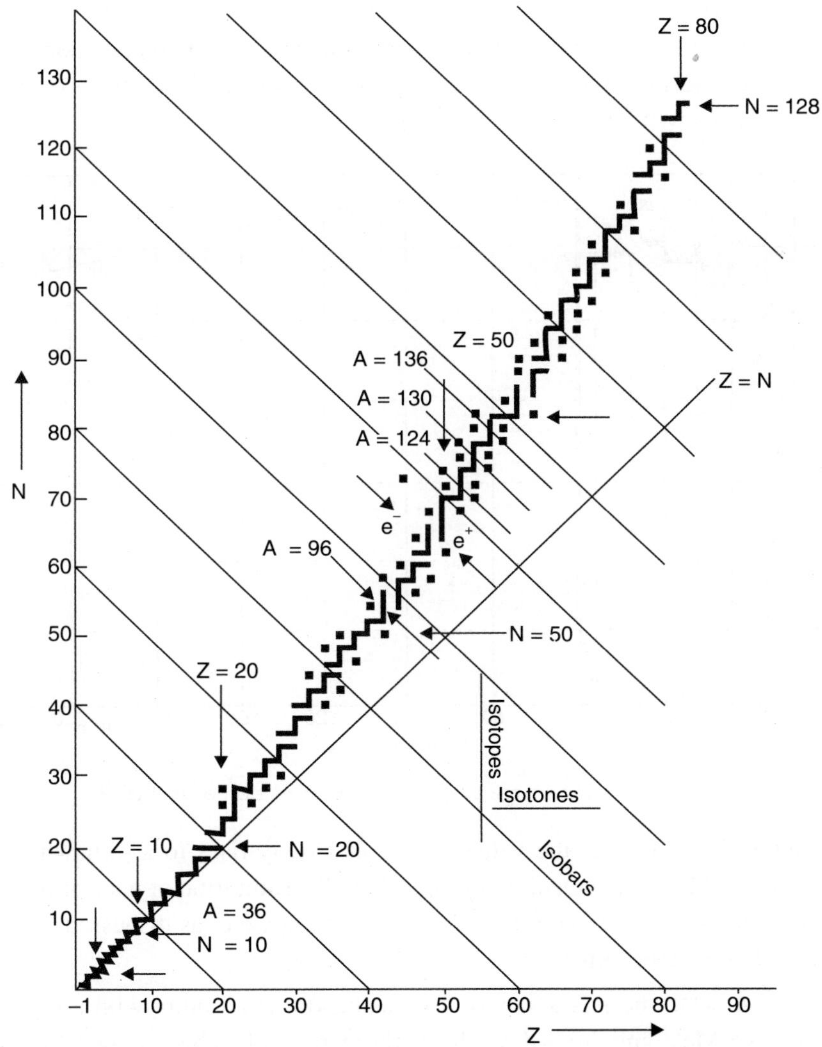

Fig. 2.3 Segre-Chart, showing relationship of neutron number and charge number of stable and unstable nuclei (Ref. 8).

Isotopes on both sides of the stability curve are radioactive, those on the right are neutron-deficient and hence are b⁺ emitters; and those on the left are proton deficient and hence β^- emitters. The radioactive decay brings the nuclei to the stability line. For very heavy nuclei, α-decay becomes predominant.

2. *Binding Energies*: The curve for binding energy per nucleon, as plotted in Fig. 2.4, gives a great insight into some of the properties of nuclei and nuclear forces. As for example, one can see from the curve that:

(*i*) For very low atomic weight, *i.e.,* for $A \leq 20$, the binding energy per nucleon (B.E./A) rises very fast. This is the region; where only a limited number of nucleons interact with each other, and saturation of nuclear forces has yet not set in and hence each new bond increases the total energy.

But above $Z \approx 10$, the binding energy per nucleon is approximately constant, because the number of nucleons is large; and hence saturation of nuclear forces takes over. The saturation of nuclear forces will be discussed in details in Chapter 6.

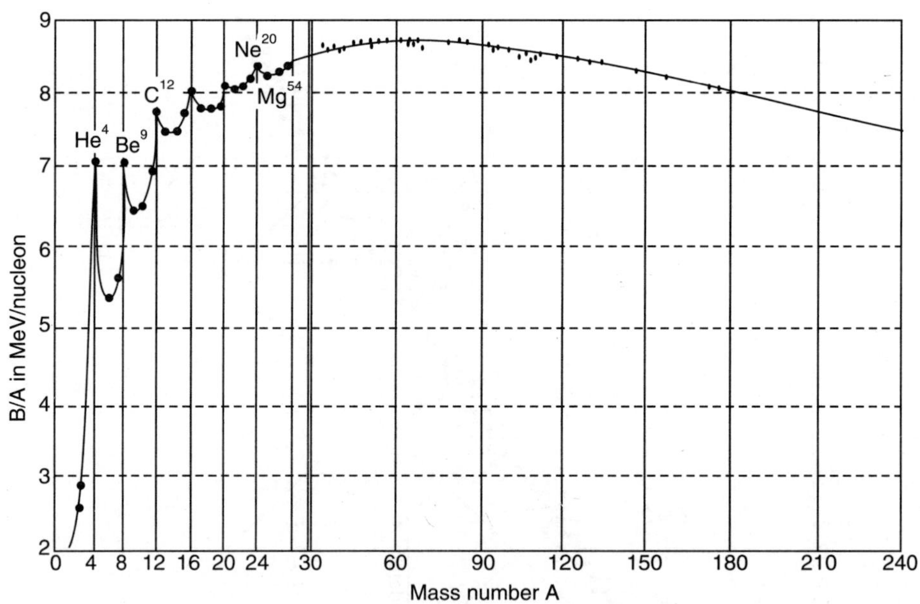

Fig. 2.4 The binding energy per nucleon as a function of A (Ref. 9).

(*ii*) In the region $20 \leq A \leq 60$, the value of B.E./A slowly rises to a somewhat constant value of ≈ 8.5 MeV. This is the transition region, where the saturation of nuclear bonds is setting in, and other effects, *e.g.* surface effects, coulomb energy, etc. as discussed in the next section are slowly having their effect.

(*iii*) For around $A \approx 60$, there is a slowly decreasing flat maximum, with a binding energy per nucleon as 8.5 MeV, and drops down for higher values of A, becoming 8.4 MeV, for $A = 140$.

(*iv*) Above $A = 140$, the binding energy per nucleon keeps on decreasing smoothly and reaches a value of 7.6 MeV at $A = 238$ for U^{238}.

The smooth decrease in the binding energy per nucleon is because of coulomb energy of protons in the nucleus.

These general features of the binding energy curve; are explainable by considering the nucleus like a liquid drop. As explained above, the first portion of the curve is explained by the fact that the saturation of nuclear forces has yet not started playing its role in this region of nuclear bonds. The general constancy, above $A \approx 60$ is explainable by the behaviour of the volume and surface energy coming from intrinsic nuclear forces. To these main features of a liquid drop; one adds some corrective terms, coming from symmetry energy and pairing energy.

Weizsäcker Semi-empirical Mass Formula[10]: Weizsäcker has given a mass-formula, written semi-empirically from the above considerations which is expressed by the relation:

$$M = ZM_p + NM_n - a_1A + a_2A^{2/3}$$
$$+ a_3 \frac{Z^2}{A^{1/3}} + a_4 \frac{(Z-N)^2}{A} + \delta(A) \quad \text{...(2.15)}$$

The first two terms describe the total mass of protons and neutrons respectively as if they are free and other five terms give various corrections due to the binding energy in a nucleus.

The term $a_1 A$ denotes the main nuclear binding energy term. As will be seen subsequently, A is proportional to R^3 and hence to the volume of the nucleus. This term is, therefore, the volume energy term; and represents the empirical fact that the B.E./A is, broadly speaking, constant for most of the range of the atomic weight. This term takes into account most of the binding energy due to nuclear forces.

The term $a_2 A^{2/3}$ is proportional to surface area of the nucleus; and its positive value corresponds to the fact; that it reduces the binding energy. Physically this term represents the fact, that the nucleons on the surface of the nucleus have less binding energy, because they have nucleons only on one side.

The term $a_3 Z^2/A^{1/3}$ represents coulomb energy and is again positive, because of the repulsive nature of coulomb forces between protons. The value of a_3 is given by $a_3 = (3/5\, e^2/r_0)$. The value of the term for coulomb energy can be easily calculated, by the following reasoning:

Assuming the uniform charge density (ρ) of the nucleus, ρ can be evidently written as:

$$\rho = \frac{Ze}{\frac{4\pi}{3} R^3} \quad \text{...(2.16)}$$

The total electrostatic energy, due to this charge-density, in a sphere can then, be written as:

$$E_C = \int_0^{R0} \frac{(4\pi r^3 \rho/3)(4\pi r^2 \rho)\, dr}{r} \quad \text{...(2.17a)}$$

which on integration yields:

$$E_C = \frac{16}{15}\pi^2\rho^2 R_0^5 = \frac{3}{5}\frac{Z^2e^2}{R_0} = \frac{3}{5}\frac{e^2}{r_0}\left(\frac{Z^2}{A^{1/3}}\right) \quad \text{...(2.17b)}$$

The next term, *i.e.,* $a_4\,(A/2 - Z)^2$ depends on the relative number of protons and neutrons. It is a term, which will tend to be zero for $A = 2Z$. The Fermi gas model explains this term partially, taking into account the Pauli's Exclusion principle. This term, basically, is based on an empirical observation that stable nuclei prefer, as far as possible, to have equal number of protons and neutrons, as is somewhat evident from the stability curve in Fig. 2.3. This term is, therefore, called the symmetry energy term.

The last term *i.e.,* $\delta\,(A)$ is called the pairing energy term; and arises out of the physical fact that neutrons as well as protons prefer to pair with the particles of their own type with opposite spins. This gives more stability to the even-even nuclei, compared to even-odd or odd-odd nuclei. This effect is called even-odd effect. This term is negative for even-even nuclei, zero for odd A, *i.e.,* for even-odd or odd-even nuclei, and positive for odd-odd nuclei.

The values of the various constants in the semi-empirical formula, *i.e.,* a_1, a_2, a_3, a_4 and $\delta\,(A)$ are obtained empirically by making the best fit of the Weizsäcker's formula to the experimental curve of mass versus Z or A.

It is expected from Eq. 2.15 that:

(*i*) A plot of $M\,(A, Z)$ versus Z for a fixed value of A; *i.e.,* for the same number of nucleons, but different Z (isobars) should give a parabola; (Fig. 2.5), and the minimum in the parabola should correspond to the most stable nucleus. This means:

$$\frac{\partial M}{\partial Z} = 0$$

$$= (M_p - M_n) + 2a_3\,\frac{Z}{A^{1/3}} - 2a_4\,\frac{\frac{A}{2} - Z}{A} \qquad ...(2.18)$$

We get from Eq. 2.18

$$Z = \frac{M_n - M_p + a_4}{2\,[a_3 A^{-1/3} + a_4 A^{-1}]} \qquad ...(2.19)$$

One can solve Eq. 2.19 for different set of isobars, and obtain the best fitted values of a_3/a_4, which turns out to be

$$\frac{2a_3}{a_4} = 0.014989 \qquad ...(2.20)$$

If one plots the masses of odd A nuclei versus Z, one may neglect $\delta\,(A)$, (Fig. 2.5), then from the least-square fit of Eq. 2.15, the values of a_1, a_2, a_3 and a_4 can be obtained by simultaneously solving the mass formula for at least four points. The result, obtained semi-empirically, by this method are:

$$a_1 = 0.0169123 \text{ A.M.U.}$$

$$a_2 = 0.19114 \text{ A.M.U.}$$

$$a_3 = 0.0007626 \text{ A.M.U.}$$

$$a_4 = 0.10175 \text{ A.M.U.} \qquad ...(2.21)$$

From the mass data of odd-odd or odd-even and even-even nuclei, one can obtain the value of $\delta\,(A)$ in A.M.U. which turns out to be

$$\delta(A, Z) = \begin{cases} -0.36A^{-3/4} & \text{for even-even nuclei} \\ 0 & \text{for odd-even nuclei} \\ +0.36A^{-3/4} & \text{for odd-odd nuclei} \end{cases} \quad ...(2.22)$$

There have been many semi-empirical attempts, to obtain more precise formulations of the mass formula, by making better fits over the whole range of periodic table[4].

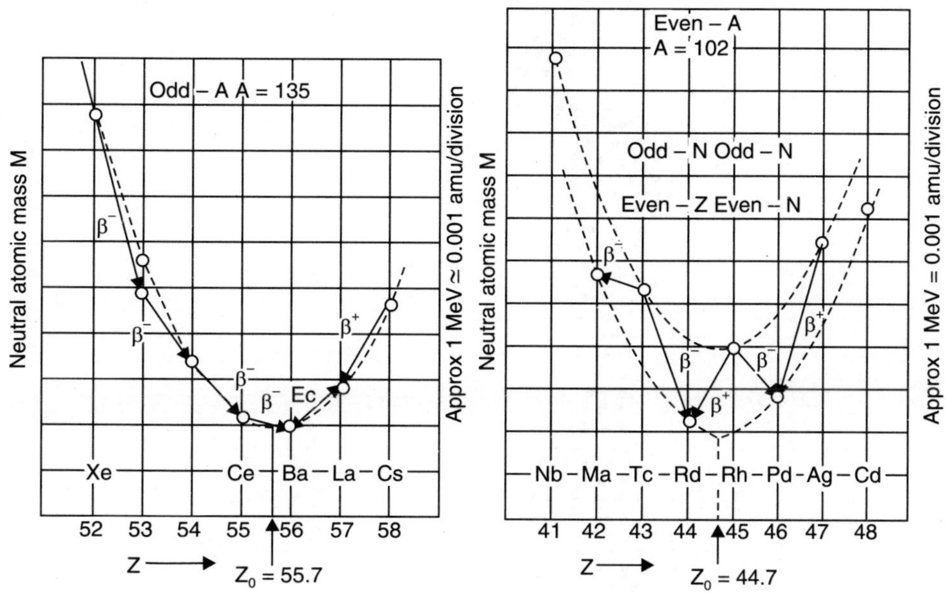

Fig. 2.5 Masses of odd and even A nuclei versus Z, to obtain the values of a_1, a_2, a_3 and a_4.

A formulation by Wapstra[12] (handbuch der physik, XXXVIII/I) has yielded the binding energy $B(Z, A)$ semi-empirically as:

$$B(Z, A) = 15.835\, A - 18.33\, A^{2/3} - \frac{0.1785\,(A-1)^2}{A^{1/3}} + \frac{-23.2 I^2}{A} \pm \delta\ (A - \text{even})\ \text{MeV}$$

where $I \equiv N - Z$, $A = N + Z$ and $\delta = 1.2A^{-1/2}$ ...(2.23)

2.1.1.3 Nuclear Separation Energies

The nuclear separation energies as defined in Eqs. 2.6 to 2.8 are like the ionisation energies in an atom; and should exhibit the characteristics of the nuclear structure in the nuclear energies. As a matter of fact very striking regularities are found in the separation energies, if one plots them for the same neutron number N or the same proton number Z against Z or N. Figure 2.6 reproduces the separation energies for odd-N, and even-Z nuclei, and Fig. 2.7 for odd-Z, and even-N nuclei. It is interesting to note that:

(*i*) 83rd of 84th nucleon (proton or neutron) is considerably less strongly bound than 81st or 82nd neutron. This corresponds to the tightly bound structure for the magic number of 82. Similar effects are seen for other magic numbers of 2, 8, 20, 28, 50, 82 and 126.

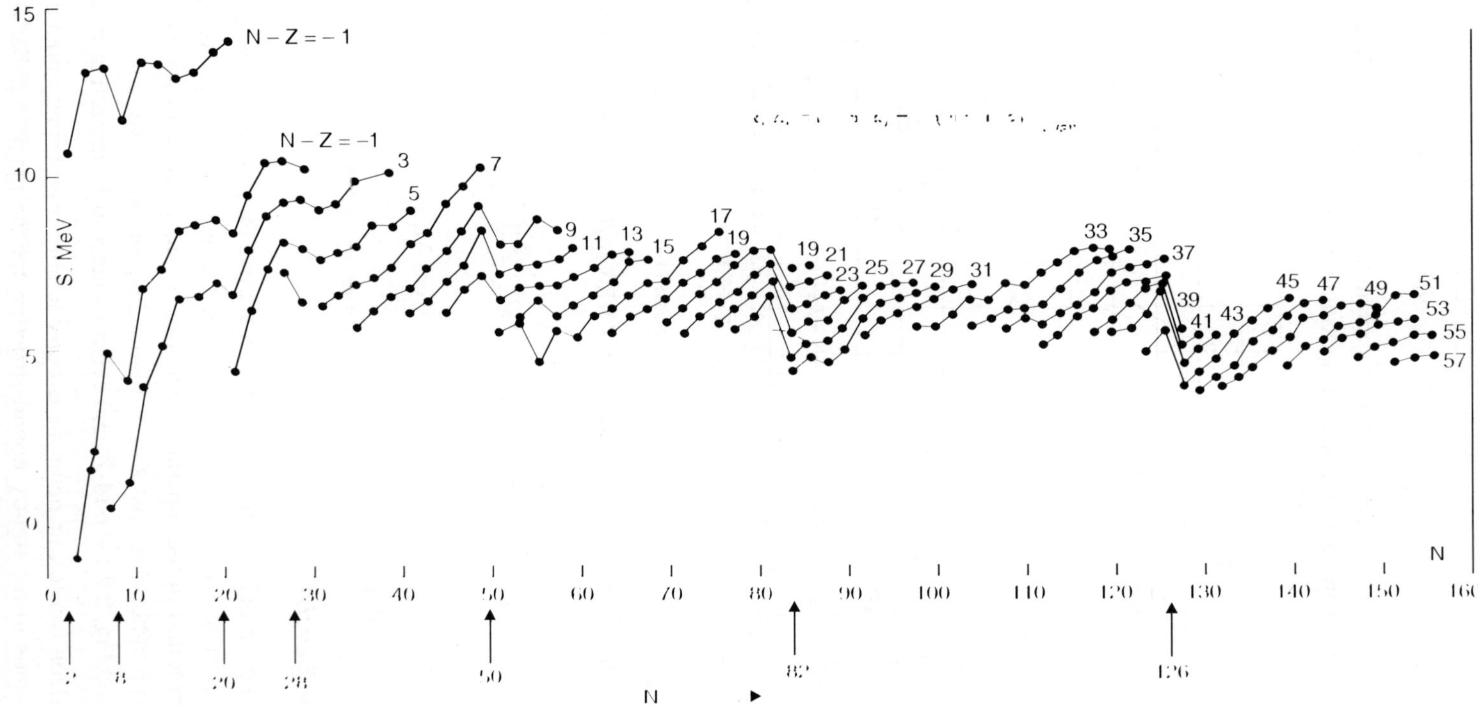

Fig. 2.6 Separation energies for odd N and even Z nuclei against N (Ref. 13)

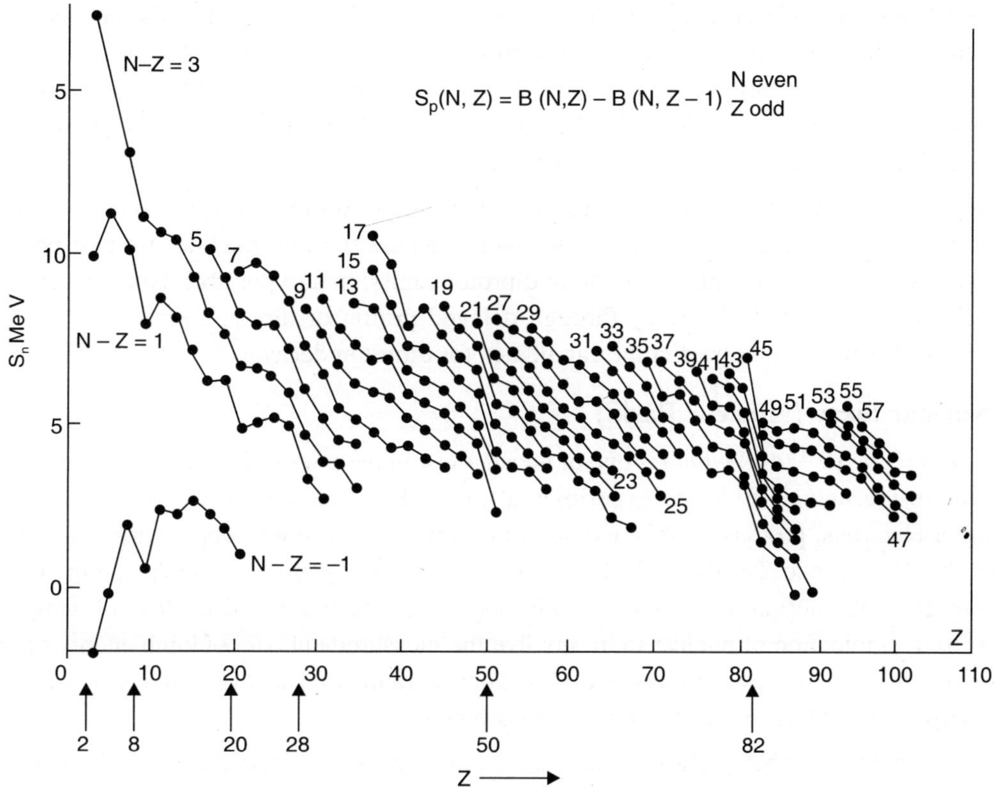

Fig. 2.7 Separation energies for odd Z and even N nuclei against Z (Ref. 13).

(*ii*) There is even-odd structure, so that there are regular linking lines only if one connects odd N (or Z) or even N (or Z) nuclei separately. This is due to the pairing term; represented by δ (A) in the Weizsäcker formula, and corresponds to the fact that neutrons (or protons) like to pair, with each other with their spins opposite to each other. This is brought out more strongly, if we plot S_{2n} or S_{2p} against A, given by:

$$S_{2n} = B\,(Z, N) - B\,[Z - (N - 2)]$$

$$S_{2p} = B\,(Z, N) - B\,(Z - 2; N) \qquad \text{...(2.24)}$$

The quantity S_{2n} or S_{2p} gives the difference of energy of two nuclei, with a neutron or proton difference of 2, and therefore, corresponds to neutron (or proton) pair separation energies. It brings out the shell effects very clearly and many other regularities. (For detailed discussion, there is a large amount of recent data available in literature.[13])

We may summarise the systematics in the nuclear data on masses as follows:

1. In general, the nuclear masses may be understood on the basis of liquid drop model, which indicates the existence of the strong inter-nucleon forces in the nuclei.

2. The initial increase in the value of B.E./A, show the effect of saturation of nuclear forces, which sets in at $Z \geq 10$.

3. The separation energies [Figs. 2.6 and 2.7] show clearly the effect of closed shells. As a matter of fact, the B.E./A curve also shows the shell effects, as shown in Fig. 2.4.

4. Except the coulomb-energy term, all other terms in Weizsäcker Mass formula correspond to the property of nuclear forces. The value of term containing a_4 is independent of the fact, whether protons are in excess or neutrons are in excess. Similarly a_2 is the same for odd A, whether Z is odd and N is even or vice versa. These facts show that except Coulomb effects, the protons and neutrons have the same nuclear interaction strength and properties.

Recently, a systematics has been developed for proton separation energies and diproton separation energies. The latter corresponds to the cases which separate the two nuclides by two protons[14]. Its necessity arose, because of the phenomenon of diproton emission as a possible exotic decay mode[15]. It is known now that A^{22} and A^{31} decay through delayed proton or diproton emission[16]. Using these systematics[14], it is possible to predict the candidates for diproton decay.

2.1.2 Nuclear Sizes (Nuclear Radii)

Nuclear size (10^{-13}–10^{-12} cm) which is one of the smallest in the microscopic domain, can be discussed only quantum-mechanically. Some experimental methods for determining nuclear[17] radii involve scattering of electrons, protons, alphas and neutrons which are all microscopic particles, with radii of the order of 10^{-13} cm or so. The theoretical analysis of the scattering is carried out quantum mechanically which also gives the definition of nuclear radii appropriate to the actual method of analysis. Other methods for determination of nuclear radii, involve the measurements of coulomb energy of nuclei, or the X-ray energies of μ-mesic X-rays, etc. The nuclear radii used in the analysis of such data, are suitably defined according to the method of measurements.

These methods either yield the radii corresponding to the electric charge distribution in the nuclei, or to the nuclear density distribution. As for example, the scattering by electrons, and the measurements of Coulomb energy and X-rays from μ-mesic atoms, yield radii appropriate to the Coulomb charge distribution; while scattering of neutrons, protons and alpha-particles yield radii corresponding to the nuclear density, after the Coulomb effects have been accounted for, in the case of protons and alphas. Evidently the electric charge distribution, corresponds to the distribution of protons; and the distribution of nuclear density, corresponds to the distribution of both protons and neutrons. There is some evidence that neutron density distribution requires somewhat larger radii than the proton-density distribution, as will be discussed subsequently.

Theoretical treatment of most of these methods require us to define, the mean square radius of the nucleus, as:

$$\langle r^2 \rangle \equiv \frac{\int_0^\infty r^2 (4\pi r^2)\, \rho(r)\, dr}{\int_0^\infty 4\pi r^2\, \rho(r)\, dr} \qquad ...(2.25)$$

where $\rho(r)$ is the charge or nuclear density and is defined so that

$$\int_0^\infty 4\pi r^2\, \rho(r)\, dr = Z \qquad ...(2.26)$$

i.e., total nuclear charge is assumed to be Z. Then,

$$\langle r^2 \rangle = \left(\frac{4\pi}{Z}\right) \int_0^\infty \rho(r)\, r^4\, dr \quad ...(2.27)$$

Many times, the quantity $\langle r^2 \rangle^{1/2}$ is taken as the radius R of an equivalent uniform spherical distribution (S.D.) assuming constant density distribution of the charge or nuclear density. It can, then, be seen that R^2 and $\langle r^2 \rangle_{\text{S.D.}}$ are related by:

$$\langle r^2 \rangle_{\text{S.D.}} = \left(\frac{4\pi}{Z}\right) \int_0^R \frac{3Z}{4\pi R^3}\, r^4\, dr \quad ...(2.28)$$

If total charge is Z, then,

$$\langle r^2 \rangle_{\text{S.D.}} = \frac{3}{5} R^2 \quad ...(2.29)$$

Empirically, both charge and nuclear density distribution seem to fit the Fermi distribution, given[17] by:

$$\rho(r) \approx \frac{\rho_0}{1 + \exp{(r - R_0)/a}} \quad ...(2.30)$$

where ρ, R_0 and a are constants. Figure 2.8 shows this function $\rho(r)$ for the realistic case when R_0 corresponds to gold nucleus.

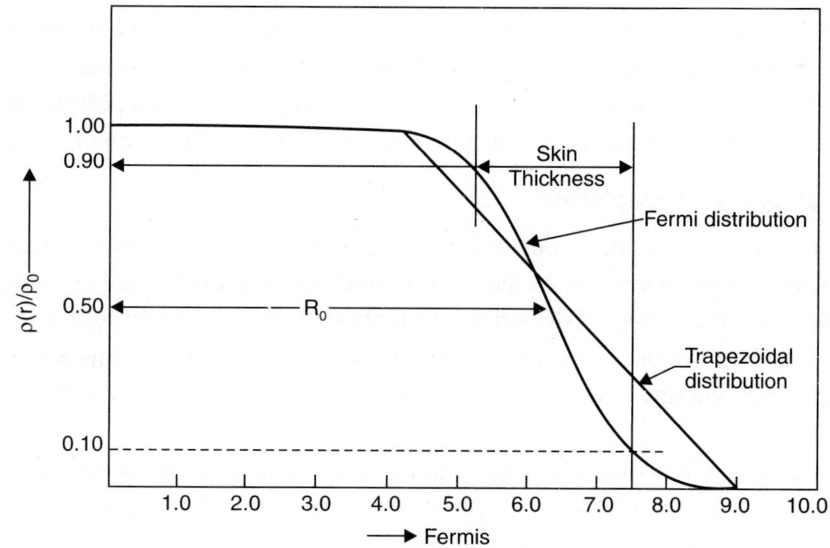

Fig. 2.8 The function $\rho(r) = \dfrac{\rho_0}{1 + \exp.\,(r - R_0)/a}$ as a function of r, called Fermi distribution, for a gold nucleus. Also shown in the trapezoidal shape (Ref. 17).

This figure shows that nuclei have nearly a constant density upto a certain distance from the centre, then the density falls according to the Fermi distribution given in Eq. 2.30, so that when, $r = R_0$;

$\rho(r) = \rho_0/2$. Sometimes R_0 is also referred to as the radius of the nucleus. One, sometimes, defines a quantity $S = 4a \ln 3 \approx 4.4a$; which physically corresponds to the skin-thickness of the nucleus. It can be seen from Eq. 2.30, that S is equal to the distance over which the density $\rho(r)/\rho_0$ falls from 0.9 to 0.1. If we assume the normalisation condition, *i.e.*,

$4\pi \int \rho(r)\, r^2\, dr = Z$; the value ρ_0 can be written as:

$$\rho_0 = \frac{3Z}{4\pi R_0^2}\left(1 + \frac{\pi^2 a^2}{R_0^2}\right)^{-1} \qquad \text{...(2.31)}$$

The quantities $\langle r^2 \rangle^{1/2}$ and R_0 are, however, inter-related. It is possible to find[18], by actual integration, that they are related by (for charge distribution Eq. 2.30):

$$\langle r^2 \rangle = \frac{3}{5} R_0^2 \left[1 + \frac{7\pi^2 a^2}{3R_0^2}\right] \qquad \text{...(2.32)}$$

In this manner, $\langle r^2 \rangle^{1/2}$, R_0 and R are all inter-related, and the measurement of one can lead to the other quantity. The quantity referred to mostly as nuclear radius is either $\langle r^2 \rangle^{1/2}$ or R.

2.1.2.1 Methods of Measurement of Nuclear Radii

As stated earlier, the methods of measurements of nuclear radii are based on two principles (*i*) Scattering; and (*ii*) Binding Energy and X-rays energy determination, requiring the measurements of energies and energy differences. Further, we determine either the charge radius or the nuclear potential radius depending on whether the interaction responsible for the phenomenon is purely Coulomb or nuclear.

A. Elastic Scattering of Electrons

Out of the various experiments carried out on scattering, most extensive and most precise have been the ones using electrons at high energies[15]. As the electron-nucleus interaction is only Coulomb, the electron scattering probes only the charge distribution. At high energies, the de Broglie wavelength (λ_e) of electrons is shorter than the nuclear radii. (λ_e at 200 MeV of electron energy is $\approx 10^{-13}$ cms). The electrons at these high energies may be looked upon as point charges and can probe in details, the charge distribution.

Classically the electron-scattering can be understood, through invoking the concept of the impact parameter. The expression for cross-sections for elastic scattering of the point charge is, then, given by (See Classical Mechanics by E. Goldstein, Page 84)[67]:

$$\sigma(\theta) = \left(\frac{Ze^2}{4E}\right)^2 \operatorname{cosec}^4\left(\frac{\theta}{2}\right) \qquad \text{...(2.33)}$$

On the other hand, the quantum mechanical expression for point-scatterer is based on Born-approximation, according to which, the elastic scattering cross-section is given by (See Quantum Mechanics by L.I. Schiff, Page 193)[43]:

$$\sigma = \frac{1}{v}\frac{2\pi}{\hbar} |M|^2 \rho \qquad ...(2.34)$$

where $1/v$ is the flux factor appropriate to the problem; v being the relative velocity of electron with respect to the target and ρ is the density of states in the final state.

The matrix-element $|M|$ is given by:

$$|M| = \left| \iint e^{-i\mathbf{K}'\cdot\mathbf{r}} \left[\left(\frac{Ze^2}{r}\right)\right] e^{-i\mathbf{K}\cdot\mathbf{r}} \frac{d\mathbf{r}^3\, d\mathbf{r}_t^3}{V} \right| \qquad ...(2.35)$$

where V is the volume of box of integration, $\mathbf{K}$ and $\mathbf{K}'$ are electron propagation vectors before and after interaction, $\mathbf{r}_t$ is the radial vector within target, and $\mathbf{r}$ is the radial vector outside target. Writing $\frac{\mathbf{q}}{h} = \mathbf{K}' - \mathbf{K} = \frac{\mathbf{p}' - \mathbf{p}}{\hbar}$, we can write the matrix element $|M|$ as:

$$|M| = \left| Ze^2 \int_0^\infty \frac{e^{i\mathbf{q}\cdot\mathbf{r}/\hbar}}{r} dr^3 \right| \qquad ...(2.36)$$

where the interaction Coulomb energy is taken to be Ze^2/r. Putting

$$q^2 \equiv 2p^2 (1 - \cos\theta) = 4p^2 \sin^2 \theta/2 \qquad ...(2.37)$$

and $dr^3 = r^2 \sin(\pi - \chi)\, d\chi\, dr\, d\phi = r^2\, dr\, d(\cos\chi)\, d\phi \qquad ...(2.38)$

where $\mathbf{q}\cdot\mathbf{r} = qr\cos\chi$

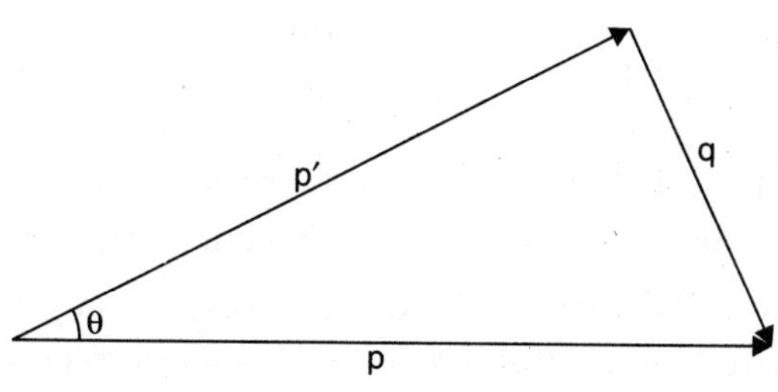

Fig. 2.9 Relationship of q, p and θ, as given in Eq. 2.37.

We can, then, write:

$$|M| = \left| e^{iqr\cos\chi/\hbar} \left(\frac{Ze^2}{r}\right) 2\pi r^2\, dr\, d(\cos\chi) \right|$$

$$= \left| 2\pi Ze^2 \int_{-1}^{+1}\int_0^\infty e^{iqr\cos\chi/\hbar}\, r\, d(\cos\chi)\, dr \right|$$

$$= \left| 2\pi Ze^2 \int_0^\infty \left[\frac{\hbar}{iq} e^{iqr\cos\chi/\hbar}\right]_{-1}^{+1} dr \right|$$

$$= \left| 2\pi\, Ze^2 \left(\frac{\hbar}{q}\right)^2 \int_0^\infty \sin y\, dy \right| ; \left[y \equiv \frac{qr}{\hbar} \right]$$

$$= 4\pi\, Ze^2 \left(\frac{\hbar}{q}\right)^2 \qquad \text{...(2.39)}$$

Now, for a two-body final state, the density of states in phase-space is given by:

$$d\rho = \frac{p^2\, dp\, d\Omega}{(2\pi\hbar)^3\, dE} \text{ and } \frac{dp}{dE} = \frac{1}{v} \qquad \text{...(2.40)}$$

Then one can write:

$$d\sigma(\theta) = \frac{1}{v}\frac{2\pi}{\hbar} \left| 4\pi\, Ze^2 \left(\frac{\hbar}{q}\right)^2 \right|^2 \frac{p^2\, d\Omega}{(2\pi\hbar)^3\, v}$$

$$= \frac{4p^2}{v^2} \frac{(Ze^2)^2}{q^4}\, d\Omega \qquad \text{...(2.41)}$$

From which we obtain:

$$\sigma(\theta)_{\text{point}} = \frac{d\sigma(\theta)}{d\Omega} = \frac{1}{4}\left(\frac{Ze^2}{mv^2}\right)^2 \operatorname{cosec}^4\left(\frac{\theta}{2}\right)$$

$$= \left(\frac{Ze^2}{4E}\right)^2 \operatorname{cosec}^4\left(\frac{\theta}{2}\right) \qquad \text{...(2.42)}$$

This expression is exactly the same, as the one derived classically. This is so, because we have used only the non-relativistic quantum mechanical method of Born-approximation.

At higher energies, say at 200 MeV above which energy, as many as ten phase-shifts are involved, one uses relativistic procedure.[19] Then one cannot get a closed expression for even a point-charge scattering. However, analytic forms for relativistic expression for the first terms of a series in $Z/137$ has been obtained[19, 20] as:

$$\sigma(\theta)_{\text{point}} = \left(\frac{Ze^2}{2E}\right)^2 \frac{\cos^2 \theta/2}{\sin^4 \theta/2} \left[1 + \frac{\pi Z}{137} \frac{\sin\theta/2\,(1-\sin\theta/2)}{\cos^2(\theta/2)} \right] \qquad \text{...(2.43)}$$

For realistic cases, however, the nuclei have, a definite radius, say R; and we cannot write $V(r) = Ze^2/r$ up to $r \to 0$. In that case, the Coulomb potential, which the electron experiences, is given by assuming the nucleus to be a sphere of uniform charge-density[21], and write:

$$V(r) = \frac{Ze^2}{r}\left[\frac{3}{2} - \frac{1}{2}\left(\frac{r}{R}\right)^2\right] \qquad \text{for } r \le R$$

$$V(r) = \frac{Ze^2}{r} \text{ for } r \geq R \qquad \text{...(2.44)}$$

where one has used the general expression of $V(r)$ as:

$$V(r) = Ze^2 \int [e\,(r') / 1\mathbf{r} - \mathbf{r}'\,]\, d^3\mathbf{r}$$

The scattering Matrix, in general, can then, be written as:

$$|\,M\,| = \left| \iint \frac{e^{i\mathbf{q}\cdot\mathbf{r}}}{\hbar} (Ze^2) \left| \frac{\rho\,(r')}{\mathbf{r} - \mathbf{r}'} \right| d^3\mathbf{r}'\, d^3\mathbf{r} \right| \qquad \text{...(2.45)}$$

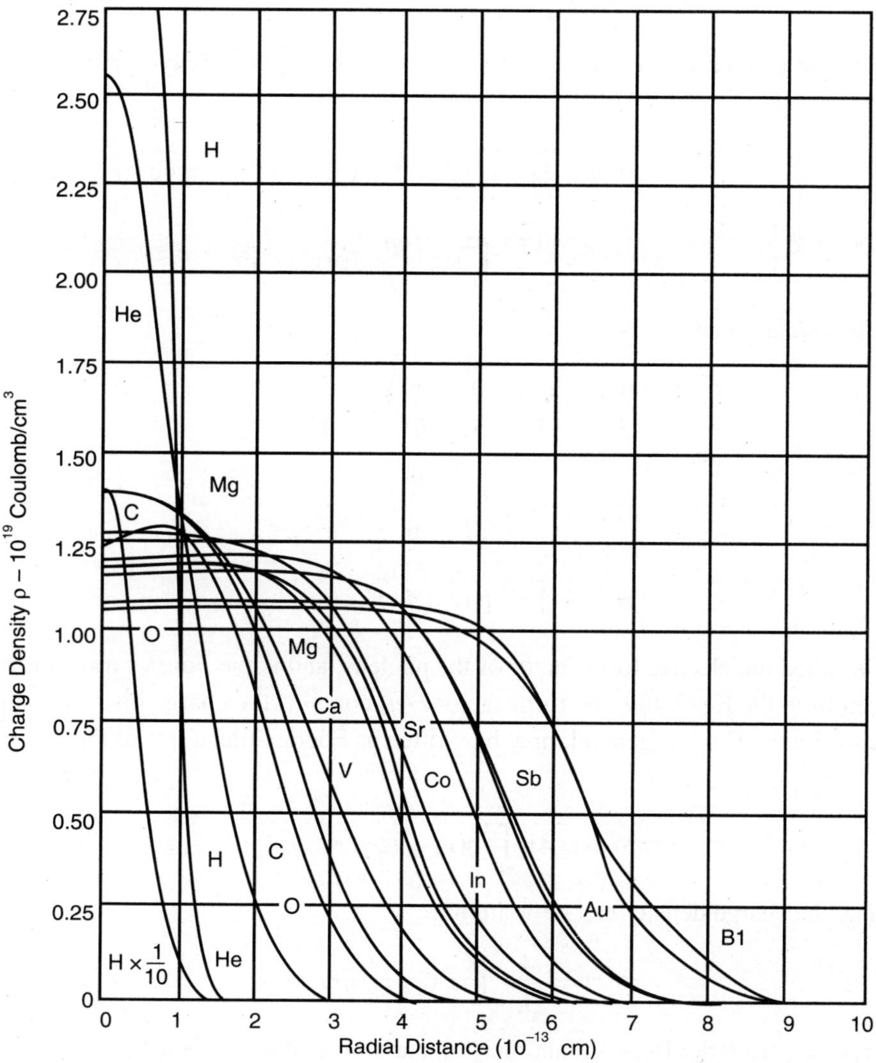

Fig. 2.10 The value of $\rho(r)$ as a function of r, for many nuclei from electron scattering (22).

Putting $\mathbf{r} - \mathbf{r}' \equiv \mathbf{R}'$ we can write:

$$M = \iint \frac{e^{i\mathbf{q}\cdot\mathbf{r}'}}{\hbar} e^{i\mathbf{q}\cdot\mathbf{R}'} (Ze^2) \frac{\rho(r')}{R'} d^3\mathbf{r}'\, d^3\mathbf{R}'$$

$$= 4\pi Ze^2 \left(\frac{\hbar}{q}\right)^2 \int \frac{e^{i\mathbf{q}\cdot\mathbf{r}'}}{\hbar} \rho(r')\, d^3r'$$

$$= 4\pi Ze^2 \left(\frac{\hbar}{q}\right)^2 F(q) \qquad ...(2.46)$$

where $$F(q) \equiv \int \frac{e^{i\mathbf{q}\cdot\mathbf{r}'}}{\hbar} \rho(r')\, d^3r' \qquad ...(2.47)$$

$F(q)$ can be evaluated somewhat easily for the case of spherically symmetric charge density $\rho(r')$. Then

$$F(q) = \iint \frac{e^{i\mathbf{q}\cdot\mathbf{r}'} \cos\chi'}{\hbar} \rho(r')\, 2\pi r'^2\, dr'\, d\cos(r')$$

$$= \frac{4\pi\hbar}{q} \int_0^\infty \sin\left(\frac{qr'}{\hbar}\right) \rho(r')\, r'\, dr' \qquad ...(2.48a)$$

If $qr'/\hbar$ is small, then

$$\frac{\sin q'}{\hbar} \approx \frac{qr'}{\hbar} - \frac{1}{3!}\left(\frac{qr'}{\hbar}\right)^3 + \ldots$$

so that $$F(q) = 1 - \left(\frac{q}{\hbar}\right)^2 \frac{\langle R^2 \rangle}{6} \qquad ...(2.49a)$$

where $$\langle R^2 \rangle = \int r'^2 \rho(r')\, d^3 r' \qquad ...(2.49b)$$

$F(q)$ is called the electric form-factor of the nucleus, and is the Fourier transform of the source. As can be seen from the Eq. 2.48*a*, the form-factor $F(q)$ contains the charge distribution $\rho(r)$ (assumed to be spherical). Hence $F(q)$, in general, may be written as Fourier-transform of the charge distribution, *i.e.*,

$$F(q) = 4\pi \int_0^\infty \rho(r)\, j_0(qr)\, r^2\, dr \qquad ...(2.48b)$$

and, in reverse, the charge density can be written as:

$$\rho(r) = \frac{1}{2\pi^2} \int_0^\infty F(q)\, j_0(qr)\, dq \qquad ...(2.48c)$$

where $j_0(qr)$ is the zero order Bessel function. Comparison of Matrix-elements M in Eqs. 2.39 and 2.46 and using Eq. 2.34, one can express the form-factor as:

$$|F(q)|^2 = \frac{\sigma(\theta)}{\sigma(\theta)_{\text{point}}} \quad \text{...(2.50)}$$

One can use Eq. 2.50 to derive $F(q)$ from the experimentally observed values of $\sigma(\theta)$. One can, then, determine the charge density distribution from Eq. 2.48 or 2.49. It is pertinent to remark that $\sigma(\theta)_{\text{point}}$ is also referred to as Mott Scattering and can be calculated theoretically from Eq. 2.33, or 2.42.

A large amount of date especially by R. Hofstader[22] exists, for scattering of electrons from various nuclei, right from hydrogen to bismuth. Figure 2.10 gives the results of such an analysis, giving the value of $p(r)$, as a function of r. It is seen that the shapes of the various curves, in general, confirm to Eq. 2.30 for Fermi distribution. An example of the actual shapes of the differential scattering cross-section as function of q is shown in Fig. 2.11 along with the theoretical fit. The detailed fit of the experimental data shows the accuracy of the data and parameterisation of the theoretical calculations. The values of radii from electron scattering is the most precise and acceptable, compared to ones obtained from other methods.

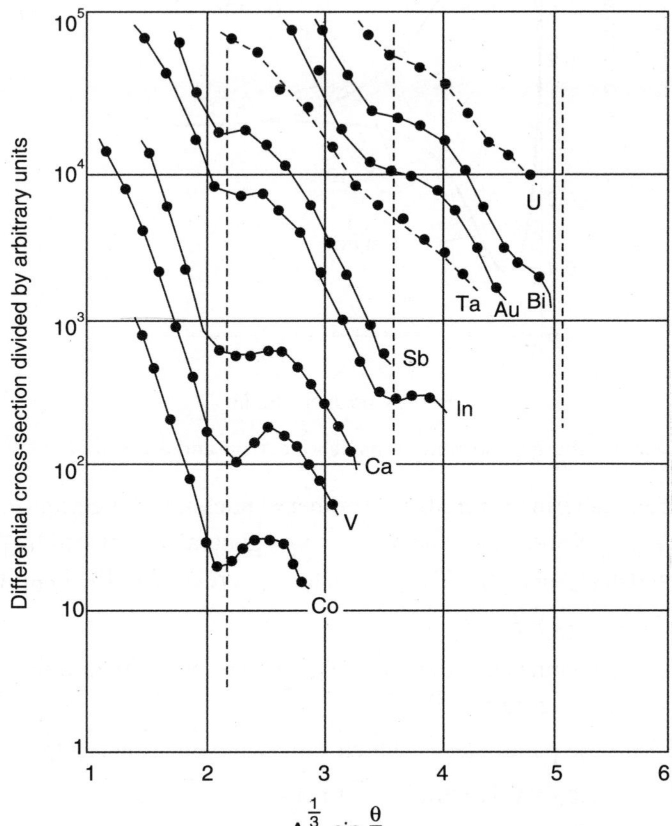

Fig. 2.11 Differential scattering cross-section as a function of $A^{1/3} \sin \theta/2$ for electrons, for many targets, along with theoretical fits. (Ref. 23).

An interesting case is that of the charge distribution of proton and neutron, as obtained from the electron scattering experiments on hydrogen and deuterium by Hofstader and co-workers[22].

Figure 2.12 shows the results for charge distribution inside a neutron and a proton. While the positive charge distribution on proton in understandable; but the negative and positive charge distribution of neutron is interesting. While it explains the zero total charge on neutron and a negative magnetic moment for it; as we shall discuss later, it also brings out the difference in the internal charge structures of neutron and proton. In Fig. 2.12, the charge distribution contains three components; a core of positive charge within a radius unexplored by electrons of energy used, for both proton and neutron and a negative plus positive charge distribution for neutron (so that total charge is zero) and a positive charge distribution for proton.

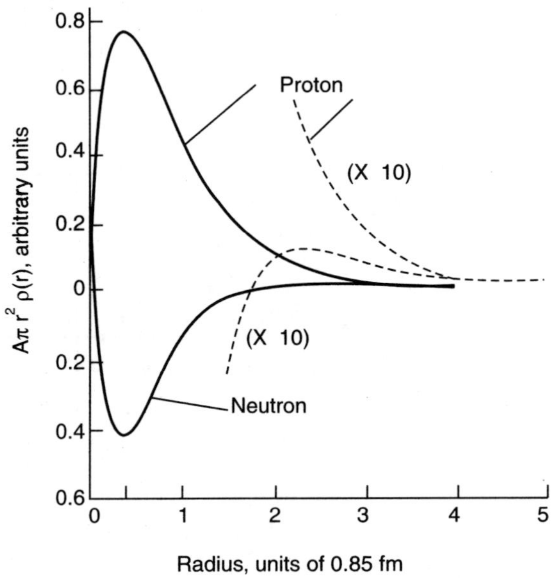

Fig. 2.12 The charge distribution inside a neutron and a proton (Ref. 22).

The data on electron scattering from a large number of nuclei, yields certain consistent information about the parameters ρ_0 R_0, a_0, S and r_0. The parameter ρ_0 is generally, replaced by $A\ \rho_0$ and corresponds to the maximum nucleon density per fm^3. The parameter r_0 is related to the nuclear radius R_0 by

$$r_0 \equiv R_0\, A^{-1/3} \qquad \text{...(2.51)}$$

assuming the proportionality of nuclear radius to $A^{1/3}$. This corresponds to uniform nuclear density. S represents, the thickness of nuclear surface[24].

The latest values of these parameters[24] which satisfy the electron scattering are:

$$A\rho_0 \approx 0.168 \text{ fm}^{-3} \qquad \text{for } A \geq 12$$

$$S \approx 2.5 \pm 0.1 \qquad \text{for } A \geq 16$$

$$\approx 1.9 \pm 0.1 \qquad \text{for } A \leq 16$$

$$r_0 \approx 1.2 \pm 0.1 \qquad \text{for } A \geq 12$$

and

$$R_0 A^{-1/3} \approx 1.0 \pm 0.1 \qquad \text{for } A \geq 12 \qquad \text{...(2.52)}$$

The near constancy of $A\rho_0$ and r_0 corresponds to the fact that the inner part of nuclei behave like nuclear matter, whose density is independent of any surface effects. The constancy of *S*, corresponds to the fact that the thickness of the nuclear surface is nearly the same for all nuclei. Such a behaviour is expected, if one assumes short range nature of nuclear forces. Because of this, a nucleon in the inner region of the nucleus is unaffected by surface effects, and the surface effects are independent of the size of inner nuclear core. These consideration, however, break down for light nuclei.

The aspherical nuclei like Hf, TaW, Th and U do not fall smoothly on these systematics. These nuclei have large quadrupole moments and hence are elliptically deformed. When averaged over all directions, this ellipsoidal distortion yields a spherically symmetrical distribution, with a much increased effective surface thickness. Such nuclei, therefore, yield much larger value of effective radii and also skin thickness.

B. Other Methods

Scattering of protons, neutrons and alphas: In this case, there is a short range nuclear interaction, over and above the long range Coulomb interaction. The nuclear interaction is, in general, written in terms of the parameters of the Optical Model, in which a nucleon or any incident particle expresses a potential *V*, given by

$$V = V(r) + i\,W(r) \qquad ...(2.53)$$

where $V(r)$ is a real part of the potential and corresponds to pure elastic scattering, while the imaginary part $W(r)$ corresponds to actual nuclear reaction, including inelastic scattering. The shapes and parameters of these potentials have been discussed in chapter on Optical Model; and are given in references (25), (26) and (27) for protons, alphas and neutrons respectively.

Other methods, which are used for measurements of nuclear radius are: (*i*) X-ray energies from meson capture and isotopic measurements from atomic spectra[30] (*ii*) Isomer shifts from Mössbaur measurements[31] (*iii*) Coulomb energy differences for mirror nuclei[32] like $3Li_4^7$, $4Be_3^7$; $7N_8^{15}$, $8O_7^{15}$; and $8O_9^{17}$, $9F_8^{17}$.

It has also been possible to measure the radius of trapezoidal nuclei taking into account the deviation form spherical shape by this method [see ref. (28) and (29)]. If ΔB is the binding energy difference, from say an even-even nucleus, then experimentally it can be seen that for $8O_9^{17}$ and $9F_8^{17}$, ΔB is positive as both of them contain an extra proton ($9F_8^{17}$) or a neutron ($8O_9^{17}$); while for $8O_7^{15}$ and $7N_8^{15}$; ΔB is negative, because they contain one proton less ($7N_8^{15}$), or one neutron less ($6C_7^{15}$), than the reference nucleus $8O_8^{16}$.

The radii, obtained, through scattering by alphas, proton and neutrons, take into account, the radii of projectiles, while comparing with those, obtained from electron scattering[33].

In a recent work[34], precise measurements of Ka X-ray energies has been used to determine the difference in nuclear charge radii of U^{235} and U^{238}. Basically it was possible to measure:

$$\delta\,(r^2) = (r_{rms}^2)^A - (r_{rms}^2)^{238}$$

It has been further possible to use, the muonic atoms for such a purpose earlier[35]. In these cases, the atoms are embedded in a flux of muons, which get captured and provide muonic X-rays. Such data gives similar but more detailed results. Laser spectroscopy, through the measurement of hyperfine structure can also provide not only magnetic dipole moments and quadrupole moments, but also charge radii, (especially the difference of radii of neighboring nuclei). As for example, recently mean square nuclear charge radii for Hf^{175} – Hf^{174} have been measure by this method[36]. This method was quite extensively used earlier also[37].

An interesting experiment was conducted by which one measures reaction cross-section σ_R at energies from 50–70 MeV/A for Ar, K and Sc isotopes as projectiles which are neutron-rich, and hence are obtained from radioactive beams. The nuclear radii are obtained from the relationship:

$$\sigma_R(E) = \pi r_0^2 f(E)$$

where $f(E)$ is a calculable function of A's of targets and projectiles[38].

2.1.3 Angular Momenta, Parity and Statistics

I. Angular Momenta

Nuclei possess total angular moments due to the motion of nucleons in the nuclei. An individual nucleon possesses an orbital angular momentum $\mathbf{l}_i$ which is related to its linear momentum $\mathbf{p}$ by:

$$\mathbf{l}_i = \mathbf{r}_i \times \mathbf{p}_i \qquad \text{...(2.54)}$$

The intrinsic angular momentum of each nucleon, on the other hand, is expressed as $\mathbf{s}_i$. The total angular momentum of the nucleus $\mathbf{I}$ can, then be expressed by one of the alternative coupling schemes.

L-S Coupling: In this case, the orbital angular momenta of individual nucleons are coupled together to get the total orbital angular momentum $\mathbf{L} = \sum_i \mathbf{l}_i$ and the spin angular momenta $\mathbf{s}_i$ of nucleus are, again coupled together, to obtain the total spin angular momentum $\mathbf{S} = \sum_i \mathbf{s}_i$ and the total angular momentum of the nucleus $\mathbf{I}$ is given by:

$$\begin{aligned} \mathbf{I} &= \mathbf{L} + \mathbf{S} \\ &= \sum_i \mathbf{l}_i + \sum_i \mathbf{s}_i \end{aligned} \qquad \text{...(2.55)}$$

This scheme of coupling is also referred to a Russel-Saunders coupling and is resorted to when the forces dependent on orbital angular momentum are stronger than the spin-orbit force. This is the case in electron-electron interaction in atomic spectroscopy. It is seldom used for nuclei except for very light nuclei. In this case, L^2, S^2, M_L and M_S are good quantum numbers, *i.e.,* they are the constant of motion. Their operational properties can be written as:

$$L^2\psi = L(L+1)\hbar^2\psi$$

$$S^2\psi = S(S+1)\hbar^2\psi$$

and

$$L_z\psi = M_L\hbar\,\psi$$

$$S_z\psi = M_S\hbar\,\psi \qquad \text{...(2.56)}$$

where ψ is the total wave function $\psi_{I,L,S,M}$. Of course, for any isolated quantum system, $I^2 \equiv J^2$; and $M_I \equiv M_J$ are good quantum numbers, *i.e.,*

$$I^2\psi \equiv J^2\psi = J(J+1)\hbar\psi$$

and

$$M_I\,\psi \equiv M_J\psi = M_J\hbar\,\psi \qquad \text{...(2.57)}$$

j-j Coupling: Alternatively, if the spin-orbit couplings is much stronger, then $\mathbf{l}_i$ and $\mathbf{p}_i$ of each nucleon may couple to $\mathbf{j}_i$, *i.e.*,

$$\mathbf{j}_i = \mathbf{l}_i + \mathbf{s}_i \qquad ...(2.58)$$

so that the total angular momentum of the nucleus is given by:

$$\mathbf{I} \equiv \mathbf{J} = \sum_i \mathbf{j}_i \qquad ...(2.59)$$

$$\mathbf{M}_j = \sum_i m_i \qquad ...(2.60)$$

where m_i is the Z-component of $\mathbf{j}_i$ along a fixed direction, and similarly $\mathbf{M}_i$ is the Z-component of the total angular momentum, $\mathbf{J}$. This scheme of coupling is called *j-j* coupling and becomes more valid when the orbital angular momentum is large, as for medium and heavy atomic weight nuclei, and spin-orbit coupling is very strong. In *j-j* coupling; apart from the fact that J^2 and M_J are constants of motion; j_i and m_j of each nucleon are also constant of motion, *i.e.*, they are good quantum numbers. This requires that:

$$j_i^2 \, \psi_j = j\,(j+1)\,\hbar\,\psi_j$$

$$m_j\,\psi_j = m\hbar\,\psi_j \qquad ...(2.61)$$

Intermediate Couplings: When the forces dependent on orbital angular momentum are of the same order, as the spin-orbit forces, as may be the case for light nuclei, we may have a situation in between the *L-S* and *j-j* coupling called the intermediate coupling. In atomic spectroscopy of atoms with only two valence electrons, outside the closed shell, *e.g.* 6 C, 14 Si, 32 Ge, 50 Sn and 82 Pb, it has been found that the energy separation of the fine structure levels lies intermediate between that expected from *L-S* and *j-j* coupling.

It may be pointed out here that, different types of couplings give rise to different energy separations between separate fine structure levels; while the number of levels remains the same.

In the above discussion, only internal interactions are involved *i.e.*, the inter-nucleon interactions and **l.s** couping. In these cases, if **l.s** coupling is much stronger than inter-nucleon interaction, we have *j-j* coupling. Then orbital (**l**) and spin (**s**) angular momenta vectors precess around the direction of **j**'s; the vector **j** precesses around **J**. In the case of *L-S* coupling; the individual **l**'s precess around **L** and individual **s** precesses around **S** and **L** and **S** precess around **J**. The direction of the quantisation in each case is the direction of total angular momentum around which other angular momenta precess.

If, however, external fields are turned on, *e.g.* the magnetic fields, then *Z*-direction is physically defined along the direction of applied field, and the total angular momentum **J** will precess around this direction. If the interaction of individual angular momenta **j**'s with the external field is stronger, than inter-nucleon interaction between them, these *j*'s will also precess around the external field. If, on the other hand, the inter-nucleon forces are stronger, the individual **j**'s will precess around the direction of **J**, which in turn will precess around the external field.

It may, further, be realised that if the field is central, *e.g.* a Coulomb field, the orbital angular moments of individual nucleons are good quantum numbers. On the other hand, if the external field is non-central, *e.g.* a tensor field, the different values of 1, get mixed so that the orbital angular momentum may not be a good quantum number, but its Z-component will be. (for its proof see Quantum Mechanics L.I. Schiff, Page 140).[43] The total angular momentum **J** and its **Z**-component M_J , along an arbitrary Z-direction are, of course, always constants of motion.

The Wave Functions: The composite wave function of a total angular momentum **J**, or **L** or **S** involving the coupling of any two angular momenta, ($\mathbf{j}_1$, $\mathbf{j}_2$ in j-j coupling, or $\mathbf{l}_1$, $\mathbf{l}_2$ for getting **L**, or $\mathbf{s}_1$, $\mathbf{s}_2$ for getting **S**; or **L** and **S** for obtaining **J** in **L-S** coupling), can be written; say for j-j coupling:

$$\psi_{j_1 j_2}, JM = \sum_{m_1 = j_1}^{-j_1} \sum_{m_2 = j_2}^{-j_2} C\,(j_1 j_2, m_1 m_2, MJ) \times \psi_{j_1 m_1}\, \psi_{j_2 m_2} \qquad ...(2.62)$$

where $C\,(j_1\, j_2,\, m_1 m_2,\, MJ)$ are called Clebsch-Gordon coefficients. For more than two particles, the coupling is more complicated.

Experimental Methods: Total angular momenta of the ground states of all the stable nuclei have been measured by a combination of various techniques of atomic spectroscopy, involving direct and indirect methods. Some of the widely used methods are:

The Number of Hyperfine Structure Components in Atomic Spectra[39]: The total angular momentum **I** of the nucleus; and total angular momentum **J** of the electrons in the atom, may couple to produce a total angular momentum **F** of the atom. The vector **F** can take any value between [**I** – **J**] and [**I** + **J**]. The total number of possible values of **F** gives the multiplicity or the number of hyperfine states and is given by:

$$(2I + 1) \text{ when } I \le J$$

$$\text{and} \qquad (2J + 1) \text{ when } I > J \qquad ...(2.63)$$

If **J** = 0, as it will be the case for closed shell atoms, *e.g.* for Ne, A, etc. the number of hyperfine levels is always single.

The hyperfine structure (hfs) of many atomic spectra has been analysed, using high resolution Fabry-Perot interferomenter, and the values of **I** have been determined for many nuclei, for which the valence electrons are not in closed shells, by counting the number of hyperfine components.

Breit-Rabi Method of Atomic Beams[40]: The principle of the method is related to Zeeman effect-splitting in atomic spectrum, when the excited atom is located in a magnetic field. In cases of a single valence electron, *e.g.* in alkali atoms like Li, N, Rb and Cs; the magnetic field produced by the atomic electrons is of the order of 10^5 to 10^7, guass at the position of the nucleus. The energy of the nuclear dipole in this magnetic field is given by:

$$W = \mu\, H_J \cos\,(I, J) = -\,\mu\,.\,\mathbf{H}_J \qquad ...(2.64)$$

Where W is the interaction energy, due to the interaction of nuclear magnetic moment μ and the magnetic field H-created by the external electrons at the nucleus, which is parallel to **J**. If an external field H_0 is also applied, then one takes into account both the internal and external fields.

Breit-Rabi's magnetic resonance—molecular beams method of measuring **I**, is based on the interaction of atomic, (*i.e.*, nuclear and electronic) magnetic moments with external magnetic field in a certain specific arrangement. The term 'molecular beams' here is a generic term and it denotes the neutral atoms as well as neutral molecules. Figure 2.13, shows a typical molecular beam apparatus. The

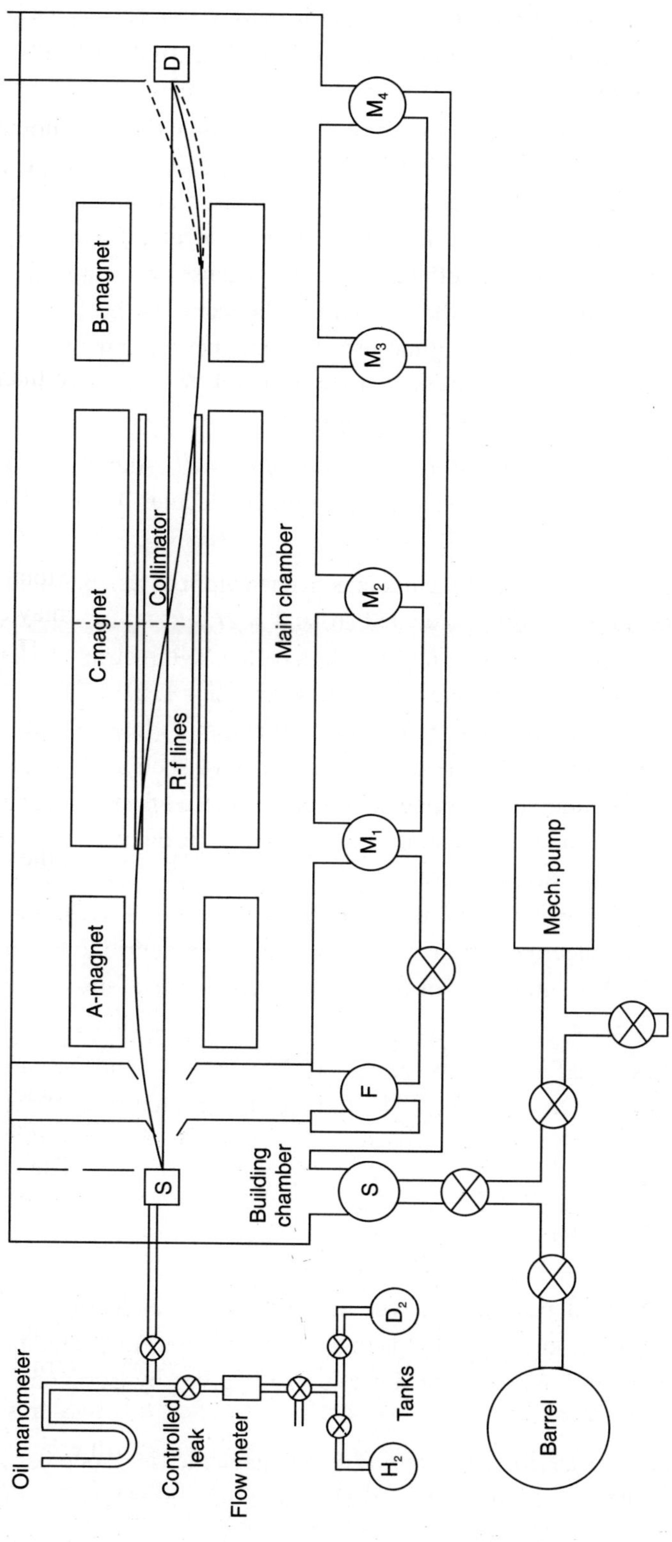

Fig. 2.13 The molecular beam method by Rabi and co-workers (Ref. 40).

neutral beam of the atoms created in the oven is made to pass through three magnets, before being finally detected by an electron multiplier detection system. The first and third magnets produce inhomogeneous magnetic fields in opposite directions; and hence deflect the atoms in opposite directions; the exact amount of deflection, depending on the strengths of the fields and atomic magnetic moments. The fields of the two inhomogeneous fields are adjusted, so that with no magnetic field applied in the second magnetic, the beam is detected by the detector. The second magnetic field **H** is a homogeneous magnetic field, in the field of which a loop is introduced, producing an *r-f* field normal to **H**. The field **H** produces a Zeeman effect; which will deflect the beam because of change in effective magnetic moment of the atom. If the frequency f of the *r-f* field is the same as the Larmor frequency due to Zeeman effect, there is a transition to the original effective magnetic moment, and the beam is again detected. By sweeping the *r-f*, all the transitions can be scanned, for a hyperfine splitting of a level, whose number is $2I + 1$ for $I \leq J$, and hence I is determined.

Alternating Intensities in Diatomic Molecular Band Spectra[41]: Another method of determining I, is by observing the band spectrum of homonuclear diatomic molecules like H^1H^1, $C^{12}C^{12}$, $N^{14}N^{14}$ and $O^{16}O^{16}$, etc.

If I is the total angular momentum of the nucleus in each atom of the diatomic molecule, then the total nuclear angular momentum T of the molecule can be $T = 2I, 2I - 1, 2I - 2, ... 0$. The rotational states with these angular moments either belong to a class called 'symmetric', or 'anti-symmetric'. As for example $2I, 2I - 2, 2I - 4, ...$ may belong to one class and $2I - 1, 2I - 3, 2I - 5, ...$ may belong to another class. Each state with a total nuclear angular momentum T, has $2T + 1$ hyperfine structure components in the presence of a magnetic field. In the absence of such a field, they contribute to the intensity of the line in the molecular spectrum because each of these *hfs* states has equal probability of occurring. But rotational states of one class, can make transition to its own class only if accompanied by an electronic transition. The transition of a rotational state from one class to another class is almost completely forbidden. So only lines observed in the molecular spectra of such diatomic molecules are those, which correspond to the electronic transition either from symmetric to symmetric or from anti-symmetric to anti-symmetric. The intensities of these two types of transitions will be $\Sigma\, 2T + 1$ for each case.

For one class of lines, $\Sigma_a 2T + 1 = 2I + (2I - 2) + (2I - 4) + \cdots\cdots$ and for another class it will be $\Sigma_b\, 2T + 1 = (2I - 1) + (2I - 3) + (2I - 5) + \cdots\cdots$. The ratio of these intensities of these lines will be

$$\frac{\sum_a 2T + 1}{\sum_b 2T + 1} = \frac{I + 1}{I} \text{ or } \frac{I}{I + 1} \qquad ...(2.65)$$

The class of a and b will depend on the type of electronic states involved. Hence regardless of which type of class the state of the nucleus belongs in the homogeneous diatomic molecule; the average ratio of the intensity of the more intense to less intense line is given by Eq. 2.65 form which I can be measured.

Other Radio Frequency Spectroscopy Methods: In a magnetic field; one has Zeeman splitting of atomic levels. A radio-frequency is impinged on such a sample, corresponding to Larmor frequency given by:

$$f = gH \frac{e}{4\pi mc} \quad ...(2.66a)$$

where g is the gyromagnetic ratio of the state, given by:

$$g = \frac{\mu}{I} \quad ...(2.66b)$$

where μ is magnetic moment in nuclear magnetons.

There is a transition between two hyperfine structure levels. If one can measure this frequency, one knows g. If μ is known, we can obtain I. For details of such measurements, see Ref. (70). For measuring g and μ in Eq. 2.66, one uses any of the following methods: (*i*) Nuclear Resonance Induction Method, developed by Bloch and Co-workers. (*ii*) Nuclear paramagnetic Absorption Method and (*iii*) Molecular Beam Resonance Method.

We have already discussed the method (*iii*) form which I is obtained. The methods (*i*) and (*ii*) give μ and g, so that I is obtained from Eq. 2.66*b*. For knowing the details of these methods, see Nuclear Moments by Norman F. Ramsay; John Wiley & Sons[70]. Some of these methods have been discussed in the section on nuclear magnetic moments [See Section (2.1.4.1)].

II. Parity

The parity, is a quantum mechanical concept, applicable to the wave function of a system. A wave function may have either a positive or a negative parity. For a positive parity, the wave function does not change sign on reflection through the centre of the coordinate system, while for a negative parity, the wave function changes sign. This depends on the initial space-properties of the wave function, and as we shall see later on, the orbital angular momentum plays an important part in this.

The parity of a wave formation is, conserved for electromagnetic and strong interactions [See reference (42)].

This means that for a given state, where only strong and electromagnetic interaction play their role, the parity of the state stays constant with time.

For a given eigen state in a nucleus, the parity of the state is given by $(-1)^l$, where l is the orbital angular quantum number. (see Quantum Mechanics by L.I. Schiff; page 73)[43]. Hence for even values of l of a state, the parity is positive and for odd-l-values, it is negative.

In nuclear strong interactions, the parity is always conserved, but the orbital angular momentum may not be conserved. Then, the mixing of different 1's will take place, in a manner that parity is always conserved. This results in the mixing of either even 1's among themselves or only odd 1's but not odd with even. This mixing affects the angular dependence of nuclear reactions. For a given projectile and target in a given reaction, the definite values of spins and parities of ground states are involved. Then depending on the 1-value of the incident particle, the final states produced in the reactions can either have positive or negative parity. This can be seen from the relation:

$$\Pi(1)\,\Pi(2)\,\Pi_l = \Pi_f \quad ...(2.67)$$

where Π (1) and Π (2) are the intrinsic parities of the projectile and targets, which are fixed and $\Pi_l = (-1)^l$, give the contribution to the parity due to the orbital angular momentum, of the incident particle. The parity of the final state Π_f is, then, determined by Eq. 2.67. If more than one 1 value is involved (which will be the case for higher energies) leading to a given definite final state, then the 1's involved in the creation of the final state will be either even or odd depending on the parity of the initial state.

III. Statistics

In a many-body quantum system of the identical particles, the total wave function (which is a solution of two or more than two identical particles), in the Schrödinger wave equation is either symmetric, (*i.e.*, does not change sign) or anti-symmetric, (*i.e.*, changes sign), on the exchange of the coordinates of any two particles. In other words, if there is a system with identical particles, which has a wave function given by:

$$\psi\,(x_1, x_2 \ldots\ldots x_i \ldots\ldots x_j \ldots\ldots x_n) \qquad \ldots(2.68)$$

where x_1, ...x_i ......; denote the position of the particles; then on an exchange of say x_i and x_j, the resultant wave function will either be positive or negative of the initial wave function, *i.e.*,

$$\psi\,(x_1, x_2 \ldots\ldots x_j \ldots\ldots x_i \ldots\ldots x_n)$$
$$= \pm\,\psi\,(x_1, x_2 \ldots\ldots x_i \ldots\ldots x_j \ldots\ldots x_n) \qquad \ldots(2.69)$$

This is in the nature of the solution of the many-body Schrödinger equation. (See Quantum Mechanics by L.S. Schiff, Page 221)[43]. A given system will, however, have only one of these two types of symmetries. It will either have symmetric wave function or anti-symmetric wave function with respect to the exchange of coordinates of the two particles. The transition between these two types of wave functions is forbidden. Hence, the system will continue having its class of wave functions-symmetric or anti-symmetric—whichever it had possessed intrinsically. The intrinsic nature of the system depends on the number and the nature of quantum numbers of elementary particles, which constitute the system. It has been shown[43], (see Quantum Mechanics, L.I. Schiff, Page 216, 267) in statistical quantum mechanics, that the elementary particles with half-integer spins of I = 1/2, 3/2, ... like electrons, positrons, neutrons, protons and μ-mesons, etc. result in anti-symmetric wave function on exchange of quantum numbers, while the particles with integral spins of l = 0, 1, 2, ... like photons, Π-mesons, etc. result in symmetric wave functions.

These intrinsic properties of the wave functions result in specific physical properties of the system. The particles which lead to anti-symmetric wave functions, on exchange of coordinates of any two particles, occupy one quantum state per particle, and the occupancy of each phase space cell is 0 or 1. In other words, they obey Pauli Exclusion Principle, according to which two particles cannot have all the same quantum numbers. This will lead to a certain type of distribution of particles in different states, and gives rise to Fermi-Dirac (FD) statistics. Evidently electrons and nucleons and all other elementary particles having half-integral spins obey Fermi-Dirac statistics.

On the other hand, the particles which lead to symmetric wave function, on exchange of coordinates of any two identical particles, have the property that many particles can occupy one quantum state. These particles obey the Bose-Einstein (B.E.) statistics. Photons, Π-mesons and all particle having integral spins obey Bose-Einstein statistics.

If there are two systems; each system constituted of Fermi-Dirac particles, then each system can be overall treated either like a Fermi-Dirac or Bose-Einstein system, for the purpose of statistical behaviour, depending on whether the number of Fermi-Dirac particles in each system is odd or even.

This can be understood, if one exchanges the coordinates of all the particle-pairs in the system, one by one and changes the sigh of the system on each exchange.

Some interesting physical systems of identical particles are the diatomic homonuclear molecules—each atom having at the centre a nucleus with odd or even nucleons of one type, *i.e.,* neutrons or protons. As for example, for odd protons, we have ${}_1H^1$, ${}_3Li^7$, ${}_9F^{19}$, ${}_{11}Na^{23}$, ${}_{15}P^{31}$ and ${}_{17}Cl^{35}$, and for even protons we have, ${}_2He^4$, ${}_6C^{12}$, ${}_8O^{16}$ and ${}_{16}S^{32}$. The nuclei containing odd protons, of course, will obey Fermi-Dirac statistics, and those containing even protons will obey Bose-Einstein statistics. On both these cases, nuclei have even number of neutrons, which obey Bose-Einstein statistics on exchange of coordinates among themselves. If these nuclei are contained in diatomic molecules, interesting physical properties result in these two cases, in the intensities of line emission, due to statistics. We have already seen in the previous section that the ratio of the intensities of the two alternate lines in both these cases is given by $\frac{I+1}{I}$. It should be, of course, pointed out that if in the case of diatomic homonuclear molecules with nuclei having even relevant nucleons, this ratio for odd to even lines is $\frac{I+1}{I}$, then for molecules with nuclei with odd relevant nucleons, this ratio is $\frac{I}{I+1}$.

Experimentally, such a ratio has been measured in the study of band spectra in homonuclear molecules containing nuclei like H^1, H^2, He^4, Li^7, C^{12}, C^{13}, N^{14}, N^{15}, O^{16}, F^{19}, Na^{23}, P^{31}, S^{32} and Cl^{35}. The experimental results agree with the conclusions discussed above and also yield the total angular momentum I of the nucleus.

A review of the values of I and statistic being obeyed by different homomolecules is given in reference 44.

2.1.4 Nuclear Magnetic Moments

The protons and neutrons, which constitute the nuclei, both possess intrinsic magnetic moments, given by[45]:

$$\mu_p = 2.798278 \pm 0.000017 \text{ n.m.}$$

$$\mu_n = -\ 1.91315 \pm 0.00013 \text{ n.m.} \qquad \text{...(2.70)}$$

One may express these intrinsic values of magnetic moments as:

$$\mu_{p,\,n} = g_{p,\,n} \frac{e\hbar}{2M_p c} S_{p,\,n} \qquad \text{...(2.71)}$$

where $g_{p,\,n}$ is called the gyromagnetic ratio, and $S_{p,\,n}$ is the intrinsic angular momenta of protons or neutrons and $e\hbar/2M_pc$ is called the nuclear magneton (n, m) given by[46]:

$$\frac{e\hbar}{2M_p c} = 3.1525 \times 10^{-18} \text{ MeV/Gauss} \qquad \text{...(2.72)}$$

As the intrinsic angular spins of both proton and neutron, $S_{p,\,n}$ is given by $1/2\hbar$, the gyromagnetic ratios of proton and neutron $g_{p,\,n}$, as obtained from Eq. 2.71, are expressed as twice the values of magnetic moments expressed in nuclear magnetons (*n.m*). The gyromagnetic ratio basically represents the ratio of magnetic moment in nuclear magnetons (*n.m*) and the angular momenta quantum numbers; and for proton and neutron is given by[47]:

$$g_p = 5.595564 \pm 0.000034$$

and $$g_n = -\,3.82630 \pm 0.00013 \qquad \text{...(2.73)}$$

According to Dirac theory of electrons[47] based on quantum electrodynamics, the magnetic moment of electron can be written as:

$$\mu_e = g_e \frac{e\hbar}{2m_e c} S \qquad \text{...(2.74)}$$

where $g_e = 2$. The quantity $e\hbar/2m_e c$ is called Bohr magneton (B.M.), where m_e is the mass of the electron. Nuclear magneton is, on the other hand, obtained by replacing electron mass by a nuclear mass.

Evidently, the Bohr magneton is much larger than the nuclear magneton by a factor of $m_p/m_e \approx 1836{:}1$. The fact that the gyromagnetic ratio of proton is not 2 but much larger; shows that proton cannot be looked upon as a point-charge interaction with virtual particles as was assumed in quantum electrodynamics, for electron. As a matter of fact, we know it from electron scattering from hydrogen that proton has a finite size, and there is a specific charge distribution inside the proton. As a matter of fact, the nucleons interact with the short range virtual mesonic field responsible for nuclear forces, giving rise to the anomalous magnetic moments of protons and neutrons. The fact that neutrons have negative magnetic moment, and have negative charge-distribution, for $r \leq 10^{-13}$ cm. (Fig. 2.12) shows that intrinsic magnetic moment of neutrons also cannot be treated (on the basis of quantum electromagnetics) as of a point charge.

On the other hand, the magnetic moments of protons and neutrons due to their orbital angular momentum, can be treated in the same way as for electrons. The neutrons are electrically neutral, and therefore cannot give rise to any electric currents and hence no magnetic moment due to orbital angular momentum of neutrons. On the other hand, the orbital angular momentum of protons does give rise to the magnetic moment of nuclei. This can be seen, by considering the proton as a positive charge orbiting around centre of mass of the nucleus. This will give rise to the vector potential **A(r)** given by:

$$\mathbf{A(r)} = \int \frac{\mathbf{I}(r')\, dr'^3}{|\mathbf{r} - \mathbf{r}'|} \qquad \text{...(2.75)}$$

where $\mathbf{I}(r')$ is the current density due to the orbiting proton. Since $|\mathbf{r}| >> |\mathbf{r}'|$, we carry out the expansion of $|\mathbf{r} - \mathbf{r}'|^{-1}$, in powers of r'/r. The first term in the expression vanishes. If we only take the next term after that, we get[45]:

$$\mathbf{A}\,(\mathbf{r}) = \oint \frac{\mathbf{I}\,(\mathbf{r}')}{2} [\mathbf{r}' \times d\mathbf{r}'] \times \frac{\mathbf{r}}{r^3} \qquad \text{...(2.76)}$$

From Eq. 2.76, we can write:

$$\mathbf{A}\,(\mathbf{r}) = \boldsymbol{\mu}_1 \times \frac{\mathbf{r}}{r^3} \qquad \text{...(2.77)}$$

where μ_1 the magnetic moment, is given by:

$$\mu_l = \oint \frac{\mathbf{I}(\mathbf{r}')}{2} [\mathbf{r}' \times d\mathbf{r}'] \qquad ...(2.78)$$

Here $\mathbf{I}(\mathbf{r}')$ is, current density, given by:

$$\mathbf{I}(\mathbf{r}')\, d\mathbf{r}' = \left(\frac{\rho}{c}\right) \mathbf{v}' dV' \qquad ...(2.79)$$

so that μ_l is given by:

$$\mu_l = \int \frac{\rho}{2c} [\mathbf{r}' \times d\mathbf{v}']\, d^3 r'$$

$$= \int \frac{\rho}{2c} \left[\frac{\mathbf{L}}{M_p}\right] d^3 r' \qquad ...(2.80)$$

where $\mathbf{L}$ is the angular momentum of the system and ρ is the charge density. Of course, M_p is the mass of the proton and c is the velocity of light.

For a single charged particle, we can write the quantum-mechanical expression for μ_l as:

$$\mu_l = \frac{e\hbar}{2M_p c} \int \psi^* \mathbf{L} \psi\, d^3 \mathbf{r}' \qquad ...(2.81)$$

Experimentally, it is the z-component of the magnetic moment, aligned parallel to the applied magnetic field, which is measured. One may therefore, express from (Eq. 2.81):

$$\langle \mu_l \rangle = \frac{e\hbar}{2M_p c} \int \psi^* (r') \mathbf{L}_z\, \psi (r')\, d^3 \mathbf{r}'$$

$$= \frac{e\hbar}{2M_p c} \langle \mathbf{L}_z \rangle = \frac{e}{2M_p c} l_z \qquad ...(2.82)$$

where l_z is the z-component of the orbital quantum number $\mathbf{l}$. For the magnetic moment being parallel to the magnetic field, $l_z = \mathbf{l}$. Hence one can write for a proton:

$$\mu_p^l = \frac{e\hbar}{2M_p c} \mathbf{l} \qquad ...(2.83)$$

Comparing Eq. 2.71 with Eq. 2.83, we can see that gyromagnetic ratio g_l for the angular momenta for protons, is 1. Evidently, for neutrons, it is zero; we may, therefore, write:

$$g_l\,(p) = 1$$
$$g_l\,(n) = 0 \qquad ...(2.84)$$

As magnetic moments are parallel to the angular momenta, the total magnetic moment μ_I of the nucleus consisting of A nucleons may be written as:

$$\mu_I = \frac{e\hbar}{2Mc} \sum_{k=1}^{A} [g_k\,(l)\, \mathbf{l}_k + g_k\,(s)\, \mathbf{s}_k] \qquad ...(2.85)$$

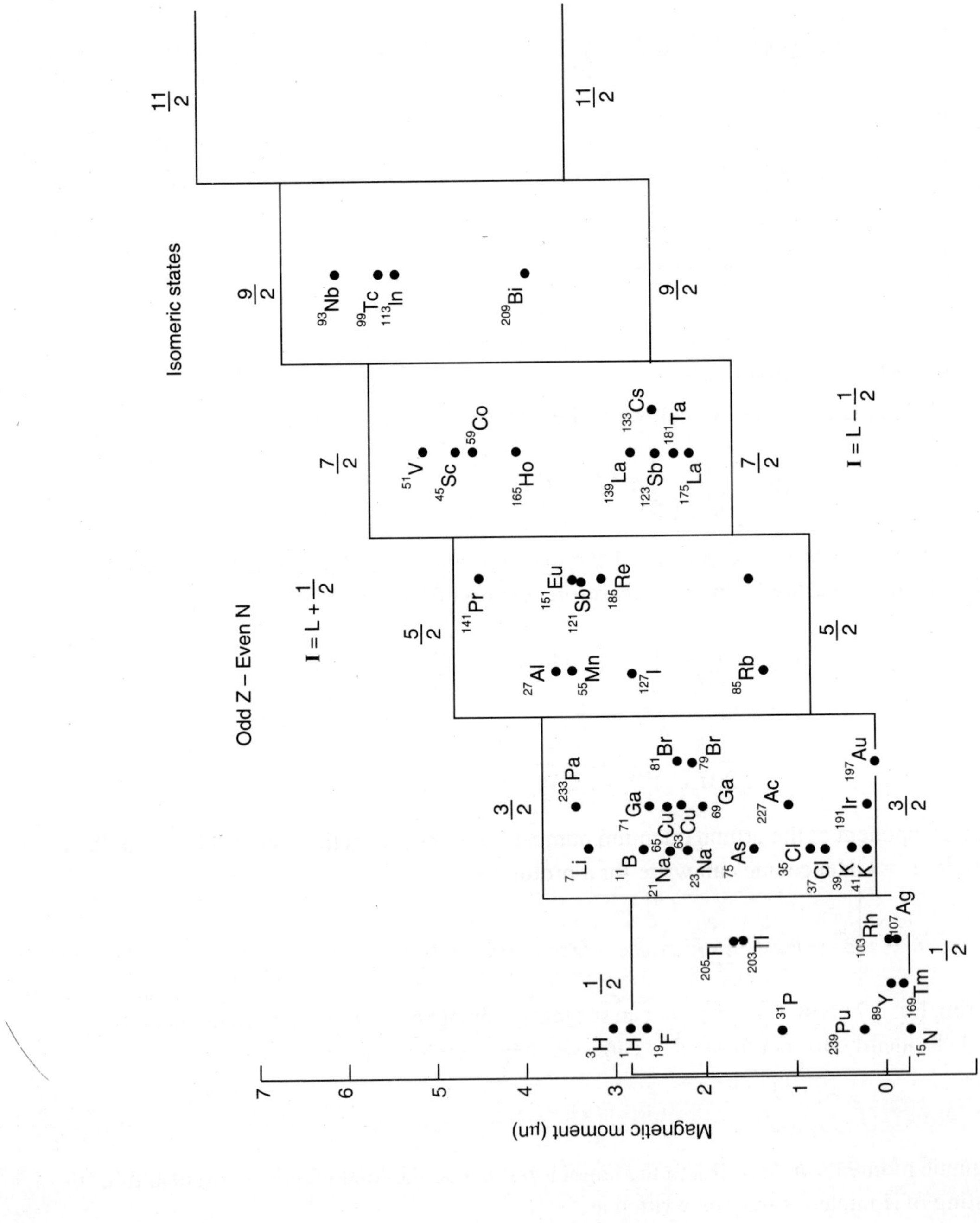

Fig. 2.14 Experimental magnetic moments and Schmidt lines for odd-Z and even-N nuclei.

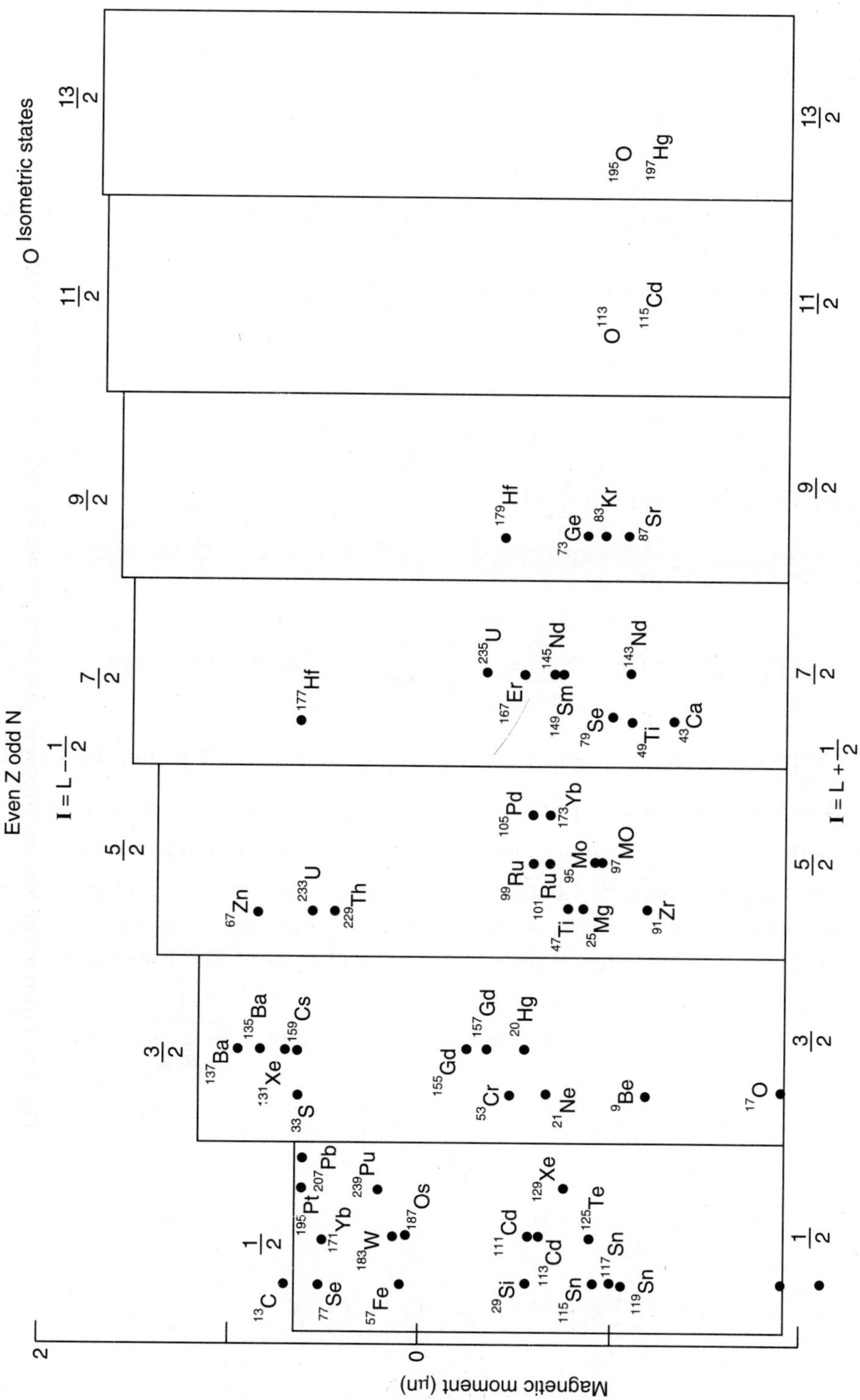

Fig. 2.15 Experimental magnetic moments and Schmidt lines for even Z-odd N (Ref. 47).

where summation variable k covers both protons and neutrons, and M is the mass of a free nucleon. We should use proper values of gyromagnetic ratio, as given in Eqs. 2.73 and 2.84. From Eq. 2.85, we can also express the effective magnetic moment as:

$$\mu_{eff} \equiv \frac{1}{\mathbf{I}^2} \langle \mu_I \,.\, \mathbf{I} \rangle \, \mathbf{I}$$

$$= \frac{e\hbar}{2Mc} \left[g(l) \frac{\langle \mathbf{l} \,.\, \mathbf{I} \rangle}{\mathbf{I}^2} + g(s) \frac{\langle \mathbf{s} \,.\, \mathbf{I} \rangle}{\mathbf{I}^2} \right] \mathbf{I} \qquad \text{...(2.86)}$$

where
$$\frac{\langle \mathbf{l} \,.\, \mathbf{I} \rangle}{\mathbf{I}^2} = \frac{I(I+1) + l(l+1) - s(s+1)}{2I(2I+1)}$$

and
$$\frac{\langle \mathbf{s} \,.\, \mathbf{I} \rangle}{\mathbf{I}^2} = \frac{I(I+1) - l(l+1) + s(s+1)}{2I(2I+1)} \qquad \text{...(2.87)}$$

If we put proper values of g_k and use $\sum_k (\mathbf{l}_k + \mathbf{s}_k) = \mathbf{I}$

then we can express Eq. 2.85 as:

$$\mu_I = \frac{1}{2}\left[1.38\,\mathbf{I} + 0.38 \sum_{k=1}^{A} \mathbf{s}_k \right] + \left[\frac{1}{2} \sum_k \tau_{k_3} \, (0.38 l_k + 9.41 \mathbf{s}_k \right] n.m. \qquad \text{...(2.88)}$$

where τ_{k_3} is the 3-component of iso-spin quantum number of nucleon (anticipating the section 6.2 about isospin). Sufficient to say here that for proton $\tau_{k_3} = +\,1/2$ and for neutron $\tau_{k_3} = -\,1/2$. In Eq. 2.88, we have used the numerical values of g_1's and g_s for protons and neutrons in proper units.

Figure 2.15 gives the experimental[47] values of the magnetic moments of nuclei for 'odd protons and odd neutrons'. The interesting point to note is, that nearly all (except a few cases like N^{15}, Np^{237} for odd protons and He^3 and C^{13} for odd neutrons) these values lie within the Schmidt lines, drawn on the basis of extreme single particle model, given by Eqs. 2.86 to 2.88.

The expected values for Schmidt lines, can be calculated as follows, from Eq. 2.86.

$$\mu_{eff} = \frac{e\hbar}{2Mc} \left[\frac{\langle \mathbf{l} \,.\, \mathbf{I} \rangle}{\mathbf{I}^2} g_l + \frac{\langle \mathbf{s} \,.\, \mathbf{I} \rangle}{\mathbf{I}^2} g_s^2 \right] \mathbf{I}$$

$$= \frac{e\hbar}{2Mc} g\mathbf{I} = \mu_n \, g\mathbf{I} \qquad \text{...(2.89)}$$

where
$$g = \frac{1}{I}\left[\frac{1}{2} g_s + (I-1)\, g_1 \right] \text{ for } l = I - \frac{1}{2}$$

or
$$g = \frac{1}{I+1}\left[-\frac{1}{2} g_s + \left(I + \frac{3}{2} \right) g_l \right] \text{ for } l = I + \frac{1}{2} \qquad \text{...(2.90)}$$

(*i*) For odd proton, we can write:

$$(a)\ \mu = \frac{1}{I}\left[\frac{1}{2}g_p + \left(I - \frac{1}{2}\right)\right]\mu_N \mathbf{I} \qquad \text{for } l = I - \frac{1}{2}$$

$$(b)\ \mu = \frac{1}{I+1}\left[-\frac{1}{2}g_p + \left(I + \frac{3}{2}\right)\right]\mu_N \mathbf{I} \qquad \text{for } l = I + \frac{1}{2} \qquad ...(2.91)$$

(*ii*) For odd neutron:

$$(c)\ \mu = \frac{1}{I}\left(\frac{1}{2}g_n\right)\mu_N \mathbf{I} \qquad \text{for } l = I - \frac{1}{2}$$

$$(d)\ \mu = \frac{1}{I+1}\left(-\frac{1}{2}g_n\right)\mu_N \mathbf{I} \qquad \text{for } l = I + \frac{1}{2} \qquad ...(2.92)$$

Schmidt lines, corresponding to these two cases are drawn in Figs. 2.14 and 2.15.

The reason, why experimental points lie not on Schmidt lines, but between the Schmidt lines, can be understood qualitatively by assuming that nuclear states cannot be described by extreme single particle model. Collective effects exist, which give rise to deviations. One has also to take into account the unfilled states alongwith the occupied single particle state.

Further, the experimental points lie closer to either $I = l + 1/2$ or $I = l - 1/2$; therefore, one can predict the value of the ground state configuration of the nucleus under consideration. Predictions of almost all the cases for 1, have come true. This means that the single particle model is applicable here to a large extent.

2.1.4.1 Measurement of Nuclear Magnetic Moments of the Ground States

There are intrinsically two principles, which are used to measure the nuclear magnetic moments of the ground states of stable nuclei, or of unstable nuclei, with long life-times: (*i*) The use of Eq. 2.64; in a manner so that by applying a suitable magnetic field, one can measure the interaction energy *W*, through the techniques of Nuclear Magnetic Resonance (NMR), Mössbauer effect or perturbed angular correlation (PAC). (*ii*) In a given atom, there is an interaction between the magnetic field, created by electronic angular momentum and the magnetic moment of the nucleus. This gives rise to the hyperfine structure of emission spectra in atomic spectroscopy. The measurement of the hyperfine structure yields the magnetic moment.

(*i*) Nuclear magnetic resonance method (NMR)[48] is based on the principle of the precession of a magnetic moment dipole in a magnetic field. If a radiofrequency is, then impinged on this magnetic dipole, with the same frequency as the precession frequency; enhanced resonance absorption takes place.

The precession of a nuclear magnetic moment in the presence of a magnetic field is shown in Fig. 2.16. Classically, if a torque of $\mu \times \mathbf{H}$ is applied on the magnetic moments; the rotational movements of **I** around the magnetic field **H** and the torque $\mu \times \mathbf{H}$, get coupled and a precession occurs of **I** around **H** with a constant angle of precession β (see Classical Mechanics by Goldstein, page 161–182). Quantum mechanically, angle β is given by:

$$\cos \beta = \frac{m}{I} \qquad \text{...(2.93)}$$

where m is the quantum number corresponding to the z-component of the angular momentum I.

The angular frequency of precession ω is given by:

$$\omega = \frac{\text{Torque}}{\text{Angular momentum}}$$

$$= \frac{|\mu| H \sin \beta}{|\mathbf{I}| \hbar \sin \beta} = \left(\frac{|\mu|}{|\mathbf{I}|}\right)\frac{H}{\hbar} = \frac{g\,\mu_N}{\hbar} H \qquad \text{...(2.94)}$$

The Larmour frequency $\nu = \omega/2\pi$ is, then given by:

$$\nu = \frac{g\,\mu_N}{2\pi\hbar} H \qquad \text{...(2.95)}$$

The energy of interaction, W, between the magnetic field $\mathbf{H}$ and the magnetic moment μ which gives rise to the precession above, is given by:

$$W = -\mu \,.\, \mathbf{H} = -|\mu| H \cos \beta$$

$$= -\frac{m|\mu|}{|\mathbf{I}|} H = -m\hbar\omega = -m\hbar\nu \qquad \text{...(2.96)}$$

The neighbouring m-states are separated by $\Delta m = \pm 1$. Hence, the energy separation between neighbouring states is given by:

$$\Delta W = \hbar\nu = \frac{|\mu|}{I} H = g\mu_N H \qquad \text{...(2.97)}$$

Experimentally (in NMR technique), the sample containing the nuclei, for which we want to measure the magnetic moment is placed in the radio-frequency field applied in x-y plane, while H is along the Z-axis (see Fig. 2.16). This creates a resonance between the applied radio-frequency and the precession frequency and will change the value of β, corresponding to $\Delta m = \pm 1$, if the frequency of applied field is the same as the precession frequency. In practice a varying field H at the mains-frequency of 50 or 60 cycles is imposed over the main magnetic field, so that an exact resonance is created for a fixed radio-frequency. The magnetic moment $|\mu|$ is, then, obtained from Eq. 2.97, if $|\mathbf{H}|$ and $|\mathbf{I}|$ are known. As has been seen earlier, the value of $|\mathbf{I}|$ may be obtained from Rabi's atomic beam method, or from optical spectroscopic methods.

(*ii*) Nuclear magnetic induction[49] is a variation of the above technique of NMR. Here one applies at right angles to the main magnetic field H_z, an alternating magnetic field say:

$$H_x = H \cos \omega t. \text{ (see Fig. 2.17)} \qquad \text{...(2.98)}$$

This creates a pick-up current in a pick-up coil wound over the sample at right angles to both H_z and H_x. When resonance condition is satisfied, *i.e.*, $H_z = \hbar\omega/g\mu_N$, there is increase in signal in pick-up coil because of induction.

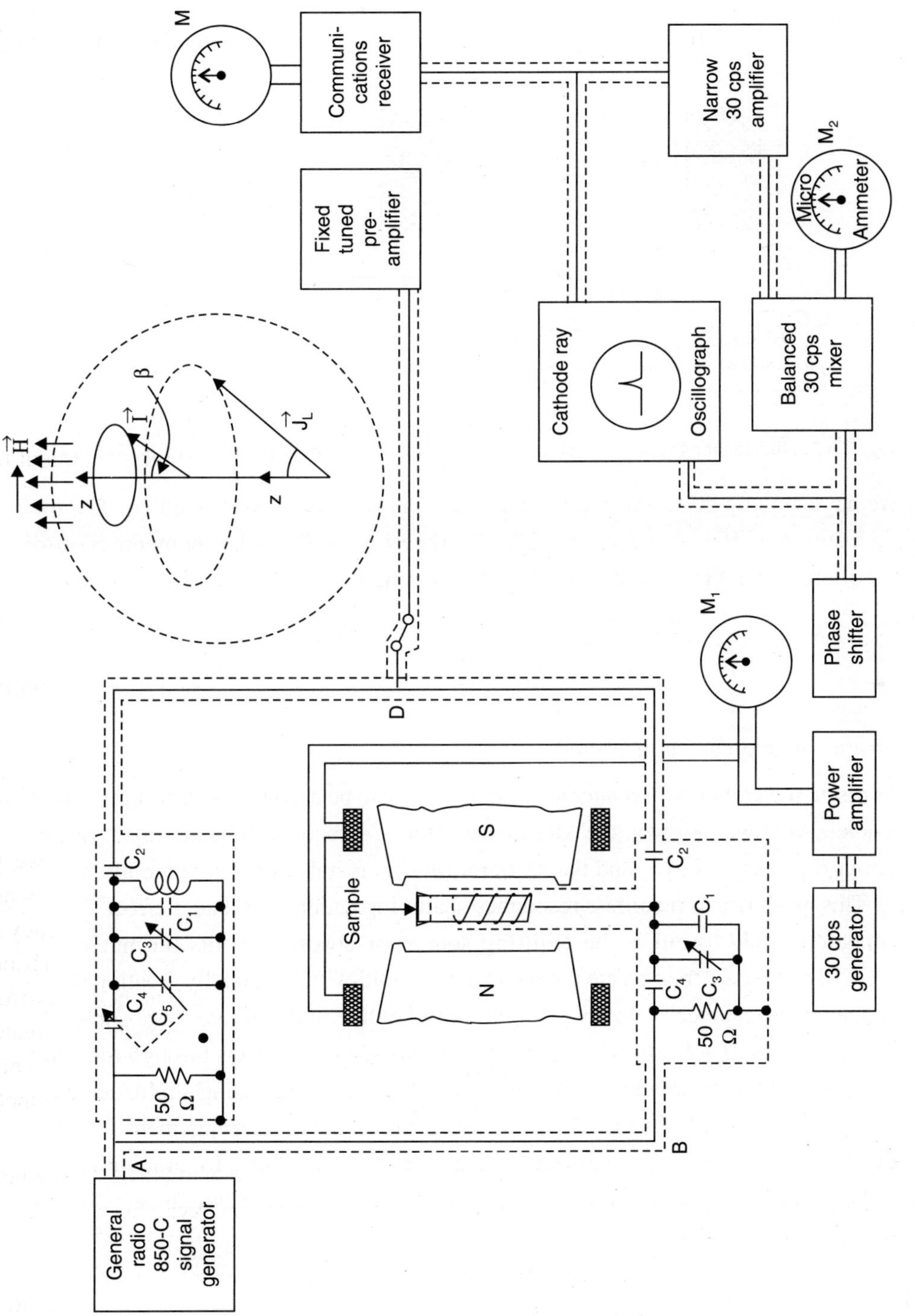

Fig. 2.16 The technique of NMR in determining magnetic moments of nuclei in their ground state (Ref. 48).

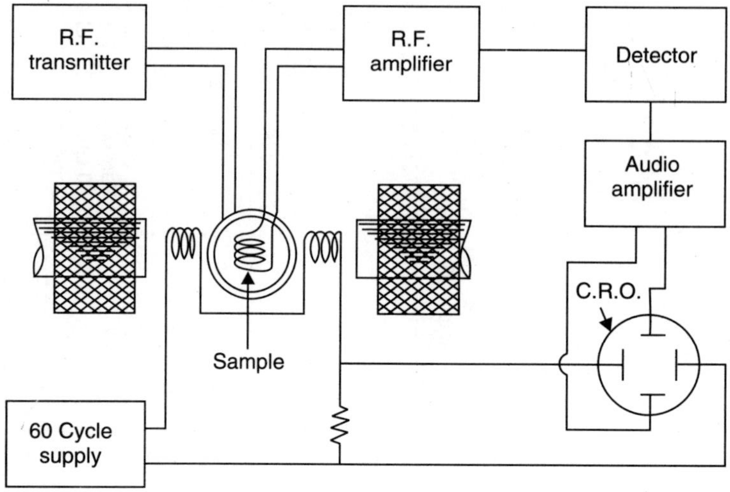

Fig. 2.17. Nuclear magnetic induction technique for determining the value of μ (Ref. 49).

(*iii*) We have already discussed the Rabi's molecular or atomic beam method. The total number of minima, in a sweep of H or ν is given by $2I + I$, where I is total angular momentum of the nucleus. The magnetic moment of the nucleus can then be obtained from the resonance frequency and the relationship,

$$h\nu = \frac{\mu H}{I} \qquad ...(2.99)$$

holds good, from which μ can be measured.

(*iv*) As discussed earlier, Mössbauer technique[50] can also be used for the measurement of magnetic moment of nuclei. As is well known, the Mössbauer effect *i.e.* recoil-less scattering is observed, when scattered gamma-ray energy is low; and the scatterer (*e.g.* a crystal) as a whole recoils, with negligible recoil energy. This gives rise to recoilless resonance scattering called Mössbauer effect. The application of external magnetic field to either the emitting source or absorber, splits the nuclear levels into *m*-states, (Fig. 2.18). One can, then create the resonance condition by artificially creating the momentum or energy changes by mechanical velocity, being imposed either on the emitter or absorber. For a given value of I, one can obtain $2I + I$ values of resonance conditions, if one scans through the various values of velocities. These velocities are created by electromechanical means using sophisticated electronics for stability and reproducibility. In Fig. 2.18, is shown the schematic arrangement for observing the Mössbauer effect in time-mode; and the hyperfine structure of Fe^{57}, in FeF_2, for predominantly magnetic interaction of the applied magnetic field. The energy difference between any two fingers of the Mössbauer absorption spectrum, can be written as:

$$\Delta E = \Delta m \mu H \qquad ...(2.100)$$

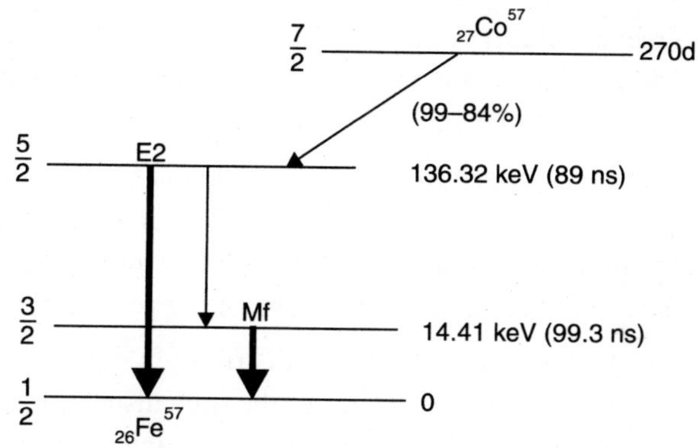

The Y decay Scheme of 57 Co Showing the 14.41 keV and 136.32 keV Mössbauer transitions

(a)

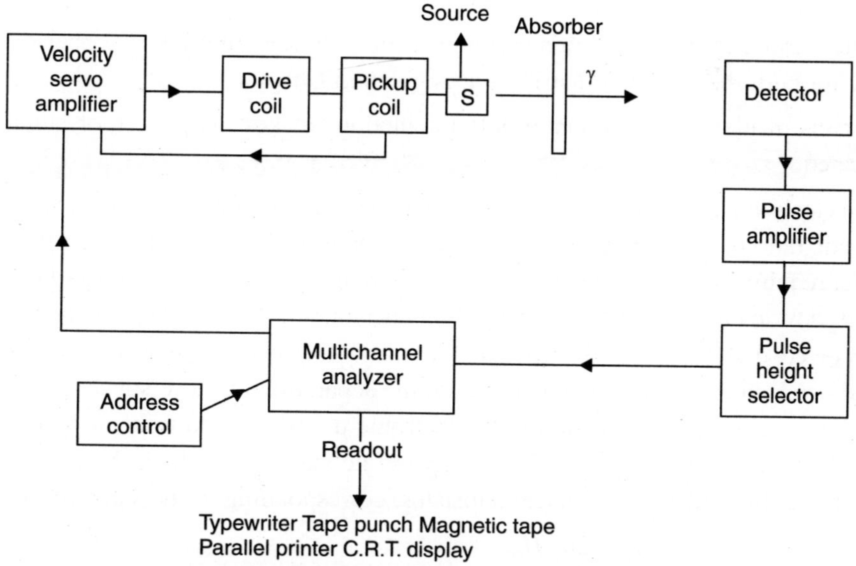

Schematic arrangement for a time-mode spectrometer

(b)

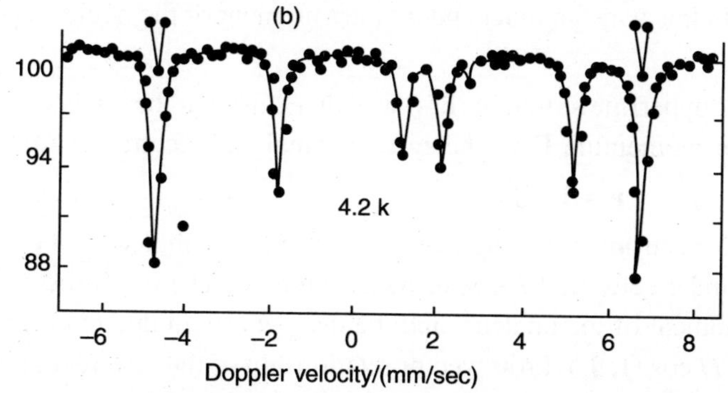

Hyperfine structure of Fe^{57}, in FeF_2 where magnetic coupling is stronger than the quadrupole coupling.

(c)

Fig. 2.18 Mössbauer effect method for measuring l or μ (Ref. 50).

where H is the applied magnetic field, μ is the magnetic moment of the excited state of 14.41 keV for Fe^{57}, and $\Delta m = \pm 1$. Any Mössbauer source can be used in this manner; making it possible to measure magnetic moments of excited states of different nuclei.

(v) As discussed in the previous section, perturbed angular correlation (PAC) technique[51] can be used for measuring magnetic moment of the excited states. The angular correlation pattern for integral angular correlation is shifted by an angle proportional to μH as shown in Fig. 2.19, when a field H is applied to the emitting nuclear perpendicular to the plane of the two detectors.

In differential perturbed angular correlation or distribution mode (DPAD) one can measure the differential attenuation coefficient directly, which gives the time distribution of gama-rays, with respect to zero time of emission. The frequency of this pattern is then given by:

$$\omega_L = 2\pi\nu_L = \frac{-\mu H}{I\hbar} = -\frac{g\mu_N H}{\hbar} \qquad ...(2.101)$$

where ω_L is the Larmor frequency of precession measured directly in a DPAD experiment (Fig. 2.19). Simple theory for perturbed angular distribution is given in Chapter 7.

The effects of magnetic and quadrupole interactions makes the pattern of angular correlation, more complicated.

(vi) The electron paramagnetic resonance[52] (EPR) is another technique, where the larger value of the gyromagnetic ratio of g of electron is made use of. The resonance conditions for the same magnetic field, therefore, requires much larger frequencies in the range of thousands of mega-hertz (MHz) rather than a few megacycles as in NMR. The paramagnetic ions in a solid often behave as though their magnetic effect were due to a single electron and the state is characterised by spin $S = 1/2\hbar$ and substates, $m_s = + I/2$ and $- I/2$. In an external magnetic field, the separation of these substates can be measured, through resonance technique corresponding to electronic μ_e. But in addition, each substate is split up due to the nuclear spin interaction into $2I + 1$, hyperfine states, for which the resonance conditions correspond to hyperfine splitting of nuclear substates, corresponding to the same resonance condition,

$$h\nu_e = 2\mu_e H_0 \qquad ...(2.102)$$

If we change H_0 there will be resonance condition for every transition between hyperfine settings whose number is $2I + I$. In this way, an independent measurement of the nuclear spin can be made. This is explained in Fig. 2.20.

(vii) The method of hyperfine structure (hfs) was first[53] used in the analysis of atomic spectra. For an atom, the total angular momentum **F**, of the system containing electrons and nucleus is given by:

$$\mathbf{F} = \mathbf{I} + \mathbf{J}_e \qquad ...(2.103)$$

where $\mathbf{J}_e = \mathbf{L}_e + \mathbf{S}_e$, is the electronic total angular momentum, contributed by the electronic total orbital angular momentum $\mathbf{L}_e$ and electronic total spin momentum $\mathbf{S}_e$. There will be a hyperfine interaction between the magnetic moment of the nucleus, and the magnetic field due to electrons. This interaction is proportional to $\mathbf{I}.\mathbf{J}_e$ or $IJ \cos(\mathbf{I}, \mathbf{J}_e)$. From vector relationship of the different angular moments, it can be seen that

$$\langle 2\mathbf{I}\cdot\mathbf{J}_e\rangle = |\mathbf{F}|^2 - |\mathbf{I}|^2 - |\mathbf{J}_e|^2$$

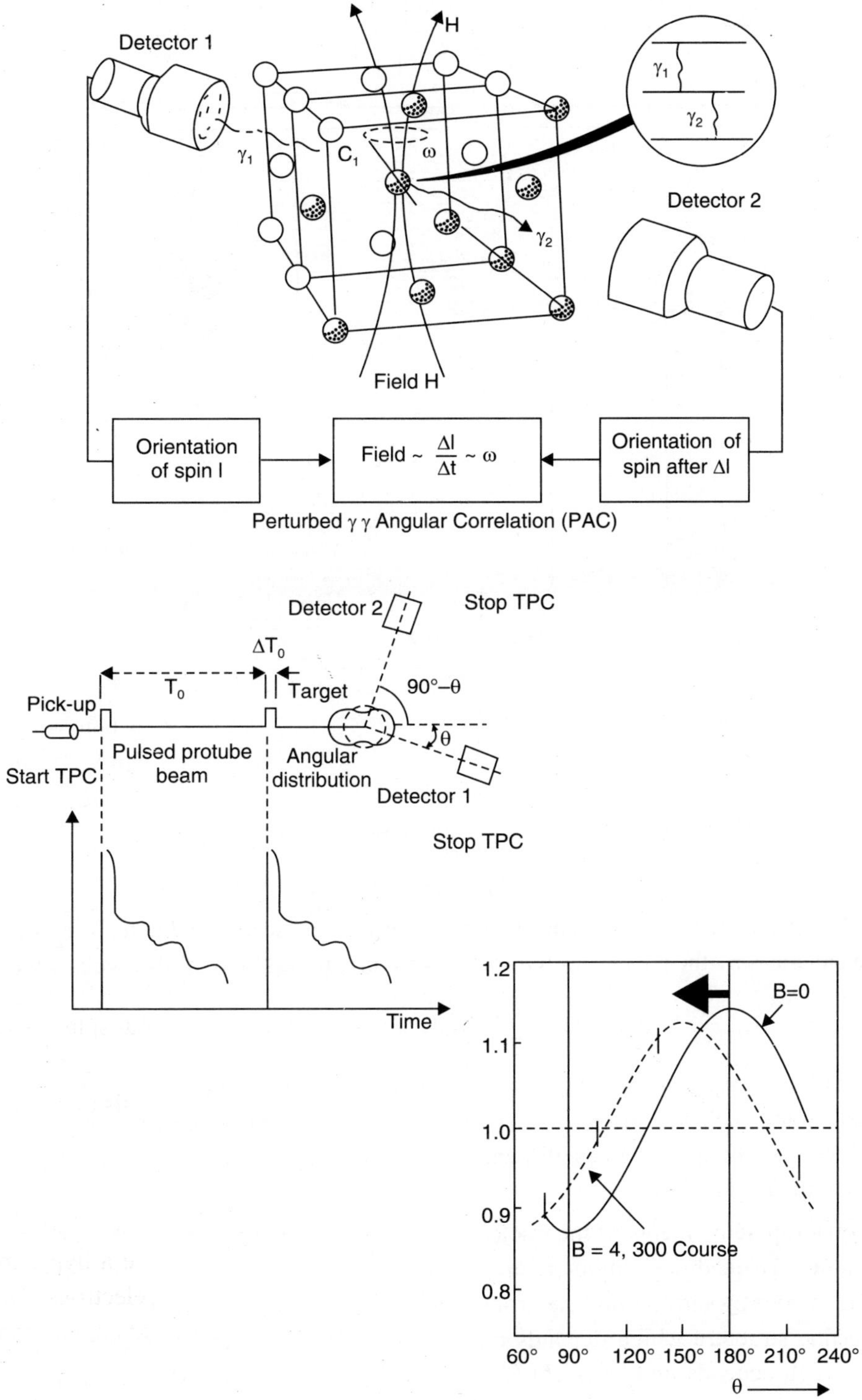

Fig. 2.19 Perturbed Angular Correlation (PAC) technique for measuring μ.

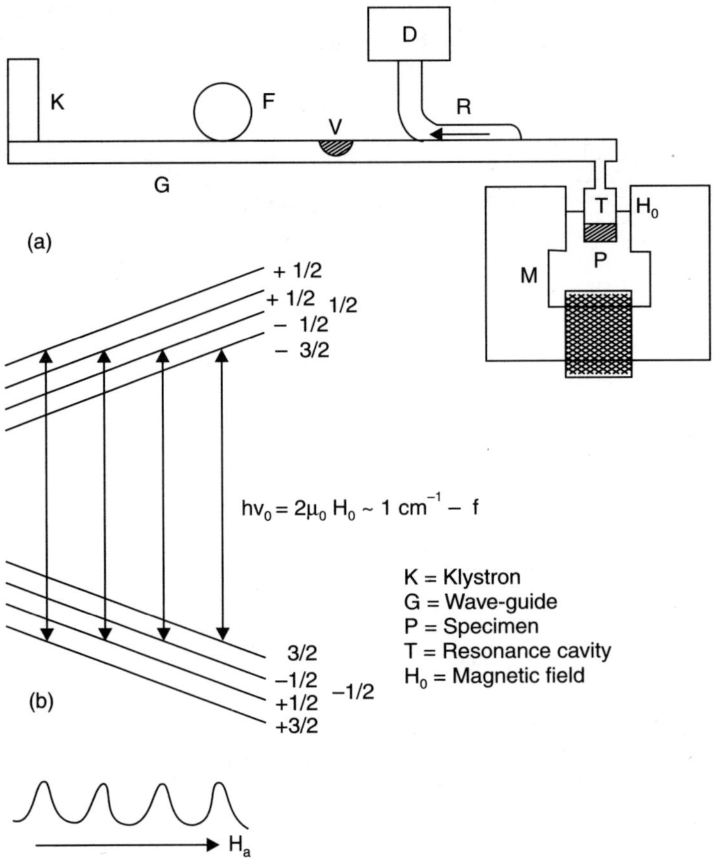

Fig. 2.20 Paramagnetic resonance absorption apparatus (Ref. 52).

$$= F(F+1) - I(I+1) - J_e(J_e+1) \qquad ...(2.104)$$

This hyperfine interaction is superimposed over the fine structure due to *l-s* splitting. In atomic spectra, one can measure the energy for hyperfine structure, for a group of hfs levels, which is given by:

$$E_{hf} = E_0 + \frac{1}{2} h\,\Delta\nu_0[F(F+1) - I(I+1) - J_e(J_e+1)] \qquad ...(2.105)$$

The value of $h\Delta\nu_0$, of course, is proportional to μ, the nuclear magnetic moment, and the electronic magnetic field at the nucleus. For $I < J_e$, we expect $2I + I$ lines due to hyperfine structure. If these lines are resolved, we can obtain the values of I, and from the separation $h\Delta\nu_0$, the value of nuclear magnetic moment is obtained.

The quadrupole moments of the nuclear ground states are obtained by the modifications of the above technique. The quadrupole moment of the nucleus through its interaction with electric quadrupole field-gradient created by surrounding electrons will give rise to an additional term in the energy splitting due to hyperfine structure which depends on $\cos^2(I, J)$ over and above the term due to magnetic interaction, which depends on $\cos(I, J)$ (for details[54] see: Experimental Nuclear Physics Volume I edited by E. Segre).

(*viii*) *Coulomb Excitation*: One can measure magnetic moments of excited states, by coulomb exciting a nucleus in the beam when it passes by a target. In a recent experiment[55], Se^{74-78} beams of 230 MeV and 262 MeV, were made to fall on a target composed of several materials say a layer of 10–95 mg/cm^2 of natural Si evaporated on 4.4 mg/cm^2 of gadolinium which itself was evaporated on 1 mg/cm^2 of tantalum foil backed by 1.35 mg/cm^2 of aluminium when a 7.5 mg/cm^2 of copper was placed behind the target. The target was subjected to an external magnetic field, which resulted in target magnetisation.

Finally the precession was measured from which magnetic moment could be determined by using differential perturbation angular correlation technique. Such experiments have been carried out[56] since 1983.

2.1.5 Nuclear Electric Moments

The nuclei, contain, electrically charged protons and are, thus, expected to have not only electric charge, but because of a certain charge-distribution, also higher electric moments. If nuclei were only spherical, the only interaction energy between an applied constant electric potential, say ϕ_0 and the nucleus will be $q\phi_0$ where q is the electric charge of the whole nucleus. In other words, nucleus could have behaved like a point charge. But as it is well known, nuclei in many cases are deformed. We should, therefore, consider a general case of interaction of a charge distribution $\rho\ (x, y, z)$ with an electric potential $\phi\ (x, y, z)$. Then the interaction energy W can be written as:

$$W = \int \rho\ (x, y, z)\ \phi\ (x, y, z)\ d\tau \qquad \text{...(2.106)}$$

In general, we can expand $\phi\ (x, y, z)$ in Taylor series *i.e.*,

$$\phi\ (x, y, z) = \phi_0 + \sum_j \left(\frac{\partial \phi}{\partial x_j}\right) x_j + \frac{1}{2} \sum_{k,j} \left(\frac{\partial^2 \phi}{\partial x_k \partial x_j}\right) x_k\, x_j \qquad \text{...(2.107)}$$

where x_j and x_k ($j, k = 1, 2, 3$) stand for x, y and z.

One can, then, write Eq. 2.106 as:

$$W = \phi_0 \int \rho\ d\tau + \sum_j \left(\frac{\partial \phi}{\partial x_j}\right) \int \rho\ \mathbf{x}_j \,.\, d\tau$$

$$+ \frac{1}{2} \sum_{j,k} \left(\frac{\partial^2 \phi}{\partial x_k \partial x_j}\right) \int \rho x_j \,.\, \mathbf{x}_k\ d\tau \qquad \text{...(2.108)}$$

Considering, $q \equiv \int \rho\ d\tau$, the total charge as a scalar quantity; and defining, $\int \rho \mathbf{x}_j \,.\, d\tau \equiv \mathbf{P}$; the electric dipole moment, which is a radial vector and $\int \rho \mathbf{x}_j \,.\, x_k\ d\tau \equiv Q_{j,k}$ the nuclear quadrupole moment which is a tensor; one can, then rewrite Eq. 2.108 as:

$$W = \phi_0\, q - \mathbf{P} \,.\, \mathbf{E} + \frac{1}{2} Q_{jk} \left(\sum_{j,k} \frac{\partial^2 \phi}{\partial x_j \partial x_k}\right) \qquad \text{...(2.109)}$$

where
$$\mathbf{E} = -\left(\sum \frac{\partial \phi}{\partial x_j}\right)_0.$$

A nucleus possesses a definite charge q and, therefore, the first term in Eq. 2.109 is non-zero. The second term, however, is zero for the nucleus; as the electric dipole moment of the nucleus, in any state, is zero. This is so because the charge density ρ has a symmetry of reflection. But the radial vector x_j changes sign on reflection, thus making the expectation value of **P** *i.e.*, $\int \psi^* \mid P \mid \psi \, d\tau$ as zero. This, of course, assumes that the wave function has a definite parity (either positive or negative) so that $\psi^* \psi$ has always positive parity.

It may be pointed out that, many times molecules of even number of atoms possess a permanent electric dipole moment, because the charge density can be made asymmetrical. The third term, *i.e.*,

$$\frac{1}{2}\left(\sum_{j,k} \frac{\partial^2 \phi}{\partial x_j \, \partial x_k}\right) Q_{jk}$$

gives rise to the quadrupole interaction. This term exists, if both quadrupole moment Q_{jk} and electric field gradient (EFG)

$$\frac{\partial^2 \phi}{\partial x_j \partial x_k}$$

due to applied electric field are finite. In practice, large values of EFG's are available in the crystalline or molecular surroundings of the host material in which the nucleus is generally embedded. These EGF's can also be created by the motion of a charged particle near a nucleus, which will happen, when say a proton or alpha or a heavy ion, is used as a projectile to excite a nucleus. Experimental details about EFG's will be discussed in the section.

Writing,

$$\frac{\partial^2 \phi}{\partial x^2} \equiv \phi_{xx}; \frac{\partial^2 \phi}{\partial y^2} \equiv \phi_{yy} \text{ and } \frac{\partial^2 \phi}{\partial z^2} \equiv \phi_{zz};$$

and assuming that the applied electric field possesses the cylindrical symmetry, we get the condition that $\phi_{xx} = \phi_{yy}$ and ϕ_{zz} are the only two components of the field which exist. Further let us assume that the intrinsic shape of the nucleus is also cylindrically symmetrical. When considering the body axes, we can assume that $Q_{x'x'} = Q_{y'y'}$ and $Q_{z'z'}$ are the two components of the quadrupole tensor. Because the components of the tensor are simply related through geometry only, we can assume that in lab system also, we have only $Q_{xx} = Q_{yy}$ and Q_{zz} components. We can, then, write the energy of a quadrupole moment and field gradient as:

$$W = \frac{1}{2} [Q_{xx} \phi_{xx} + Q_{yy} \phi_{yy} + Q_{zz} \phi_{zz}] \qquad \text{...(2.110)}$$

Further we assume that the Poisson equation for fields holds good, *i.e.*, $\nabla^2\phi = 0$ and therefore, we can write:

$$\phi_{zz} = -2\phi_{xx} = -2\phi_{yy}$$

Then expressing:

$$Q_{rr} \equiv Q_{xx} + Q_{yy} + Q_{zz}$$

where
$$Q_{rr} = \int \rho r^2 \, d\tau; \; Q_{zz} = \int \rho \, z^2 \, d\tau \qquad \text{...(2.111)}$$

One can now write:

$$W = \frac{1}{4}\phi_{zz}\{3Q_{zz} - Q_{rr}\}$$

$$= \frac{1}{4}\phi_{zz}\,Q \qquad \text{...(2.112)}$$

where
$$Q \equiv 3Q_{zz} - Q_{rr}$$

$$= \int \left(3z^2 - r^2\right) \rho \, d\tau$$

$$= \int P_2 \, (\cos\theta) \, \rho \, r^2 \, \mathrm{d}\,\tau \qquad \text{...(2.113)}$$

where θ is the angle which r makes with z-axis and where we have used.

$$r^2 = x^2 + y^2 + z^2$$

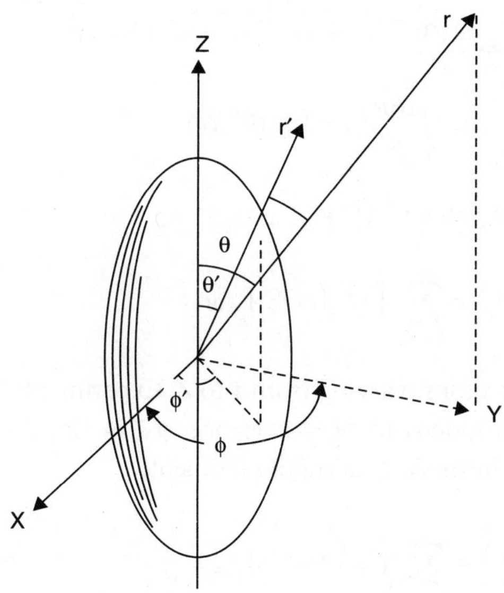

Fig. 2.21 Relationship of r, θ, φ with r′, θ′ and φ′.

Q is called the quadrupole moment of the nucleus and ρ is the charge density defined by

$$\rho = \frac{\int \psi^* \, (Ze) \, \psi \, d\tau}{\int \psi^* \, \psi \, d\tau} \qquad \text{...(2.114)}$$

where Z is the number of protons.

We may similarly define intrinsic quadrupole moment Q' as:

$$Q' \equiv \int \left(3z'^2 - r'^2\right) \rho' \, d\tau' = \int P_2 \ (\cos\theta') \, \rho' \, r'^2 \, d\tau'$$

$$= \int r'^2 \ (3\cos^2\theta' - 1) \, \rho' \, d\tau' \qquad ...(2.115)$$

where r', θ' and ρ' and τ' correspond to the body-axes. Referring to Fig. 2.21, we can develop a relationship between Q' and Q. It can be seen that

$$\rho d\tau = \rho' \, d\tau' \qquad ...(2.116)$$

and $\rho' = \rho' \, (r', \theta')$ is independent of ϕ'. It is, then, possible to write, (Fig. 2.21),

$$Q = \frac{1}{2} \, (3\cos^2\Theta - 1) \, Q' \qquad ...(2.117)$$

and $$\cos^2\Theta = [\sin\theta' \cos\phi' \sin\theta \cos\phi + \sin\theta' \sin\phi' \sin\theta \sin\phi + \cos\theta' \cos\theta] \qquad ...(2.118)$$

where Θ is the angle between $\mathbf{r}'$ and $\mathbf{r}$, and Q and Q' are the quadrupole moments in the lab and body coordinates respectively. They should not be confused with the total charge of the system.

Till now, we have dealt with the problem of quadrupole moment only classically. To get the expression for the intrinsic quadrupole moment quantum mechanically, we write operator Q'_{op} as:

$$Q'_{op} = (3z'^2 - r'^2) = r'^2 \, (3\cos^2\theta' - 1)$$

$$= \sqrt{\frac{16\pi}{5}} \, r'^2 \, Y_{20} \, (\theta', \phi') \qquad ...(2.119)$$

and write the expectation value of Q' as:

$$\langle Q' \rangle = \sum_{k=1}^{Z} \int \psi^* \left(eQ'_{op}\right)_k \psi \, d\tau \qquad ...(2.120)$$

where k denotes the number of protons, varying from 1 to Z. Experimentally, one defines the quadrupole moment, Q, as that which corresponds to $M = I$. Hence ψ's in Eq. 2.120 should correspond to this condition, and we write for the intrinsic quadrupole moment:

$$\langle Q' \rangle = \sum_{k=1}^{Z} \int \psi_{\text{II}}^* \left(eQ'_{op}\right)_k \psi_{\text{II}} \, d\tau$$

$$\langle Q' \rangle = \sqrt{\frac{16\pi}{5}} \sum_{k=1}^{Z} \int \psi_{\text{II}}^* \, e[\, r'^2 \, Y_{20} \, (\theta', \phi')]_k \, \psi_{\text{II}} \, d\tau \qquad ...(2.121)$$

Using Eqs. 2.119 and 2.121, we can write expected value of the quadrupole moment as:

$$q' \equiv \langle Q \rangle \, eZ \left[\left\langle 3z'^2 \right\rangle - \left\langle r'^2 \right\rangle\right]$$

$$= eZ\left[2\langle z'^2\rangle - \langle x'^2 + y'^2\rangle\right] \qquad ...(2.122)$$

Assuming an ellipsoid of rotation as the shape for the nucleus, with uniform charge-density, and choosing z'-axis as the axis of symmetry, we see that

$$\langle x'^2\rangle = \langle y'^2\rangle$$

and Eq. 2.122 becomes:

$$q' = 2\,eZ\left[\langle z'^2\rangle - \langle x'^2\rangle\right] \qquad ...(2.123)$$

Carrying out the actual integration in Eq. 2.123 and defining

$$\Delta R \equiv \left|\left[\langle z'^2\rangle - \langle x'^2\rangle\right]^{\frac{1}{2}}\right|$$

and

$$R \equiv \left|\left[\frac{\langle z'^2\rangle - \langle x'^2\rangle}{2}\right]^{\frac{1}{2}}\right|$$

Equation 2.123 reduces to

$$q' = \frac{4}{5}\,eZ\,R\,\Delta\,R$$

$$= \frac{4}{5}\,eZ\,R^2\,\Delta R/R \qquad ...(2.124)$$

One generally defines a deformation parameter β as

$$\beta \equiv \frac{4}{3}\sqrt{\frac{\pi}{5}}\,\frac{\Delta\,R}{R} \approx 1.6\,\frac{\Delta\,R}{R} \qquad ...(2.125)$$

Then q' can be written as:

$$q' = \frac{3}{\sqrt{5\pi}}\,eZR^2\beta \qquad ...(2.126)$$

For prolate charge distribution which are elongated along z'-axis; $\langle z'^2\rangle >> \langle x'^2\rangle$, q' is positive; and for oblate shape charge distribution, which is squeezed along z'-axis, *i.e.*, $\langle z'^2\rangle << \langle x'^2\rangle$, q' is negative. We have assumed in Eq. 2.123, that q', which depends on M; has $M = I$, here. As a matter of fact, the angle χ determines the value of M and is related to it as:

$$\cos^2\chi \equiv \frac{M^2}{I(I+1)} \qquad ...(2.127)$$

The measurement of q', in Eq. 2.126 is always done for the smallest value of χ, *i.e.*, when $\cos\chi \approx 1$. Classically, this corresponds to the situation where the direction of the applied field is along the axis of symmetry of the charge distribution, of the nucleus. Then $\cos\chi = 1$. Quantum mechanically, exact alignment of the type is not possible, because of uncertainty principle, and the smallest angle between **I** and z' axis is given by:

$$(\cos\chi_{\text{smallest}}) = \frac{I}{\sqrt{I(I+1)}} \qquad ...(2.128)$$

Under these conditions, one can get from Eq. 2.117, which on $\theta \to 0$, and $\theta' \to 0$; gives:

$$Q \approx Q'\left[\frac{1}{2}(3\cos^2\chi - 1)\right] \qquad ...(2.129)$$

(because $\chi = \Theta$); from which we obtain, the relation between $\langle Q \rangle$ and $\langle Q' \rangle$, for $M = I$ as:

$$\langle Q \rangle = \langle Q' \rangle \frac{[3M^2 - I(I+1)]_{M=I}}{2I(I+1)} \qquad ...(2.130)$$

or

$$\langle Q \rangle = \langle Q' \rangle \frac{2I-1}{2I+2} \qquad ...(2.131)$$

It is easy, to see from Eq. 2.131 that for $I = 1/2$, $\langle Q \rangle = 0$. This does not mean that $\langle Q' \rangle$ is necessarily zero. It basically means, that for $I = 1/2$, the extent, to which the vector I can be aligned with the Z-axis is so slight, that, $\langle Q \rangle = 0$.

Again for $I = 0$; $\int \psi_{\text{II}}^* (P_2 \cos\theta)\, \psi_{\text{II}}\, d\tau = 0$ ...(2.132)

This means from Eq. 2.113, that the value of $\langle Q \rangle = 0$ for $I = 0$, also. Again this does not imply that $\langle Q' \rangle = 0$; but it only shows that the quadrupole moment parameter $\langle Q \rangle$ does not provide a measure of its deviation from spherical charge distribution for $I = 1/2$ or 0. Then, we have a general rule, that $\langle Q \rangle = 0$ unless $I \geq 1$.

These two results can also be derived from the vector relations. Realising that in the calculation of $\langle Q \rangle$, one couples a wave function of angular momentum I; with another wave function of angular momentum I, through $P_2(\cos\theta)$. [See Eq. 2.132], which carries an angular momentum $I = 2$, we expect for a complete over lap of two wave function in Eq. 2.132; that

$$\mathbf{I} = \mathbf{1} + \mathbf{1} \qquad ...(2.133)$$

This relation is possible, only if $|\mathbf{I}| \geq 1$. This shows that nuclei with $|\mathbf{I}| = 1/2$ or $|\mathbf{I}| = 0$ will yield zero value for $\langle Q \rangle$.

It is instructive to calculate the theoretically expected values of $\langle Q \rangle$, using the single particle model, which assumes that only the last odd proton, contributes to the quadrupole moment. Neglecting the dashed notation, then one can use from Eq. 2.119:

$$(Q)_{op} = r^2(3\cos^2\theta - 1) = 2r^2 P_2(\cos\theta) \qquad ...(2.134a)$$

and for ψ's, the single particle wave functions which, say for $j = 1 \pm 1/2$, and $j_z = m$, can be written as [see Eq. 2.62]:

$$\psi\left(l, s = \frac{1}{2}, j = l \pm \frac{1}{2}, m\right)$$

$$= \sum_{m_1 m_s} C(l, s, m_1, m_s; j, m)\, Yl_{ml}\, \chi_{s,ms} \qquad \text{...(2.134}b\text{)}$$

Using the theorem by Bethe:

$$\langle Y_l^{\text{ml}} \mid P_2 \cos\theta \mid Y_l^{\text{ml}} \rangle = \frac{[l(l+1) - 3m_l^2]}{(2l+3)(2I-1)} \qquad \text{...(2.135)}$$

we can write $\langle Q \rangle_{s-p}$ from Eq. 2.121,

$$\langle Q \rangle_{s-p} = \frac{\langle r^2 \rangle}{2j(j+1)} \{j(j+1) - 3m^2\} \qquad \text{...(2.136)}$$

where $$\langle r^2 \rangle \equiv \int r^2 u^2 (r)\, r^2 dr \qquad \text{...(2.137)}$$

This yield the result:

$$\langle Q \rangle_{s-p} = \langle r^2 \rangle \frac{-(2I-1)}{2(I+1)} \qquad \text{...(2.138)}$$

For odd neutron nucleus, there is a quadrupole moment due to the recoil motion of the rest of the nucleus, which may be represented by a charge Z at a distance r_n/A from the centre of mass. Hence effective r_n now is $1/A$ of the effective r for protons, but now charge is multiplied by Z, as the whole nucleus recoils. Hence,

$$\langle Q \rangle_{s-n} = \frac{Z}{A^2} \langle Q \rangle_{s.p} \qquad \text{...(2.139)}$$

Equation 2.138 gives the single proton value of the quadrupole moment; and Eq. 2.139 for a single neutron.

In Fig. 2.22, we show the experimental values of ($\langle Q \rangle/R^2$) versus odd N or Z; showing that single particle model does not explain the results. As a matter of fact, experimental results are many times larger that $\langle Q \rangle_{s-p}/R^2$ for which; the expected values are – 0.4 to – 0.5. From Eq. 2.126; $\langle Q \rangle/R^2$ represents the deviation from sphericity, hence we have plotted the curves in Fig. 2.22 in this particular manner. The large discrepancy shows that we have to evolve some other model, involving the collective motion of many protons to explain the experimental results. We will deal with it in Chapter 12.

It may be seen from Fig. 2.22 that for Z or N equal to magic numbers; the experimental values of $\langle Q \rangle$ are nearly zero, showing that these nuclei are nearly spherically symmetrical. The addition of a proton or a neutron, gives it a negative values of $\langle Q \rangle$ as expected from Eq. 2.138, and also gives the same value as expected from $\langle Q \rangle_{s-p}$. The nuclei with one less proton or neutron than magic number, is

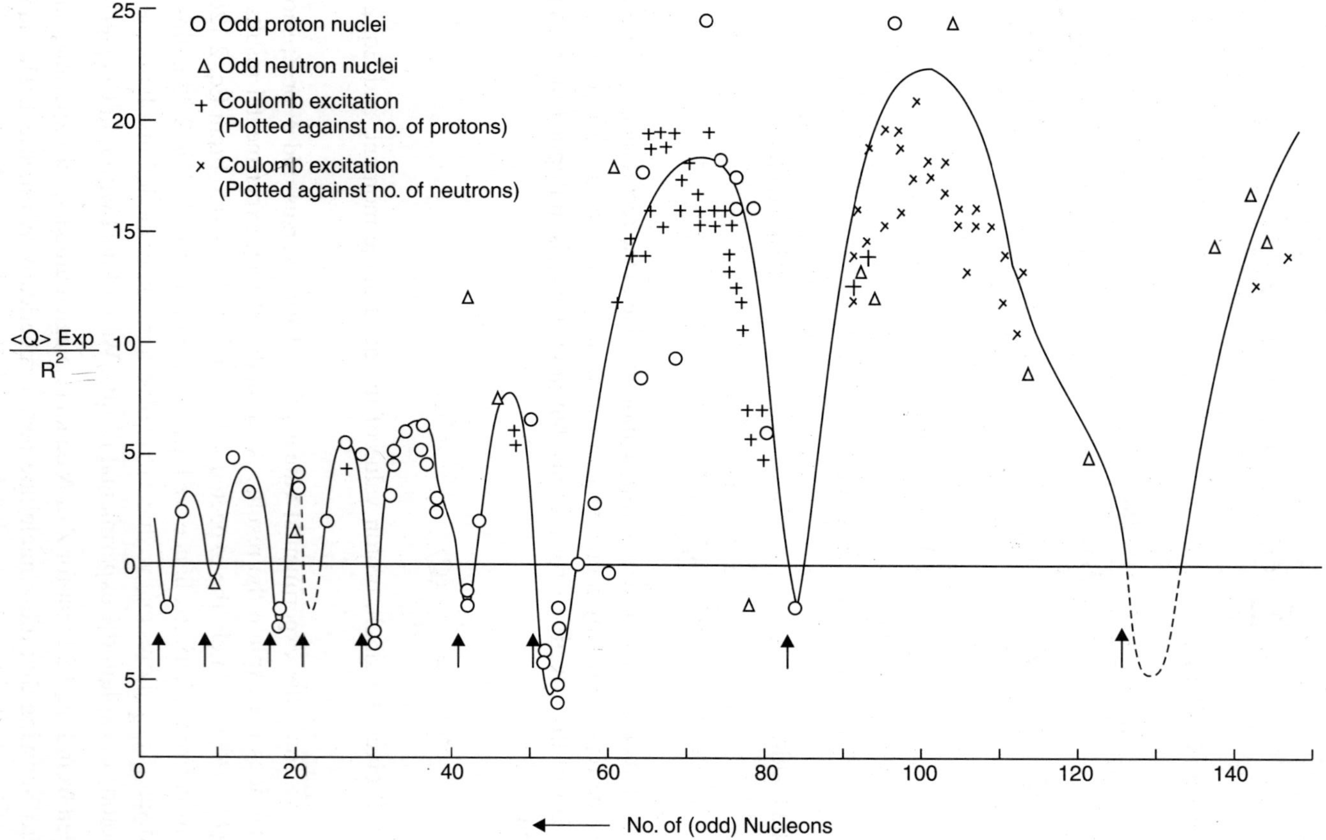

Fig. 2.22 The values of $\langle Q\rangle_{Exp}/R^2$ versus odd N or Z; showing that single particle model does not explain the experimental results (Ref. 71).

expected to give quadrupole moment of opposite sign, as this corresponds to the case of a 'hole'. The absolute values are, however, expected to be the same, as seen to be the case, experimentally. But between the magic numbers, the measured values are much larger than the 'single particle' values.

The largest measured[71] value of Q for stable nuclei is that of Lu^{175} [$\langle Q \rangle = 15.68 \times 10^{-24}$ cm^2] and the smallest value is that of deuteron [$\langle Q \rangle = 0.003 \times 10^{-24}$ cm^2]. Other cases of very small values of $\langle Q \rangle$ are that of O^{17} and Cs^{133} [$\langle Q \rangle O^{17} = -0.0205 \times 10^{-24}$ cm^2] ; and [$\langle Q \rangle Cs^{133} = -0.003 \times 10^{-24}$ cm^2]. Both of these cases are of unstable nuclei. A case of largest $\langle Q \rangle$ among unstable nuclei is that of Hg^{203} [$\langle Q \rangle Hg^{203} = \pm 13.00 \times 10^{-24}$ cm^2].

[The units of experimental values of $\langle Q \rangle$ in the above measurements correspond to the definition of

$$\langle Q \rangle = \frac{\rho}{e} \int (3z^2 - r^2)\, d\tau ,$$

in contrast to the definitions in Eq. 2.113 or 2.122. Hence the dimensions are in cm^2, with no charge connected to it.]

2.1.5.1 The Methods of Measurement of Electric Quadrupole Moments

Basically, the experimental techniques used to measure $\langle Q \rangle$ are based on the relationship of interaction energy E_2 of the quadrupole moment with the electric field gradient, from Eq. 2.112:

$$E_2 = (W)_Q = \frac{e}{4} \langle Q \rangle_{\text{exp}} \frac{\partial^2 \phi}{\partial z^2} \qquad ...(2.140)$$

where we are assuming the electric field gradient to be cylindrically symmetrical.

There are basically only two ways to create large enough and symmetrical E.F.G's to measure E_2. One is based on the electric charge provided by electrons around the nucleus. If they provide completely a spherically symmetrical charge distribution, then $\partial^2\phi/\partial z^2$ is zero and quadrupole interaction given in Eq. 2.140 is zero. This will be the case, say in free atoms of noble gases which have closed shells. In other atoms, the finite E.F.G. may exist at the nucleus. But as free atoms are oriented in all directions; the average EFG experienced at any given point in the gas may still be zero. On the other hand, if these atoms are embedded in crystalline fields, then the surrounding fields may polarise electric charge density around the atoms to create a symmetrical E.F.G. at the nucleus of the atom, the axis of symmetry being provided by the crystalline axis of symmetry. If the crystal structure is exactly known, and the electronic configuration of the atomic electrons around the nucleus is known then one can calculate $\partial^2 \phi/\partial z^2$ for a perfect crystal. This is, however, not easy because of the difficulty of calculations and physical difficulties in growing crystals with different atoms embedded in them. For a few cases, however, reliable values of E.F.G. are available[57].

For measuring the nuclear quadrupole moments of excited states, one allows an excited nucleus in an in-beam experiment to be injected into a crystal, by recoil techniques and studies the perturbation in angular correlation of suitable gamma-rays due to the interaction of EFG at a given site in the crystal, where the nucleus gets embedded. For details, see reference (58).

The other method of measuring E.F.G. is, by allowing a charged projectile, say an alpha or even a heavier nuclear projectile, to pass by a nuclear target say in Coulomb excitation. This creates an electric field gradient, near the nucleus for a small time, which may be used to create an interaction with the quadrupole moments of the nucleus.

There are two experimental methods employed, for measuring the quadrupole moments of the excited states of nuclei, using this principle.

Re-orientation by Coulomb Interaction: Here one measures Coulomb excitation cross-section or transition probability due to Coulomb excitation as a function of energy, for heavy ions, and compares it with the theoretical values expected on the basis of re-orientation. If the quadrupole moment of the ground state is known, and the excited state belongs to the same rotational band, the quadrupole moment of the excited state can be determined.[59,60]

The other method is called Re-Orientation Precession technique (REPREC): Here one determines the change in the angular distribution of gamma rays; due to the passing of heavy ions. In practice, one plots $W(22.5°)/W(157.5)$ as a function of the incident energy, and compares it with the theoretically expected values for the quadrupole moment of the ground state is known.[59]

Other experimental method of measuring $(W)_Q$ of the ground states are based on the following techniques: (*i*) The modification of nuclear magnetic resonance (N.M.R.). (*ii*) The nuclear quadrupole resonance (N.Q.R.) (*iii*) Mössbauer or perturbed angular correlation techniques.

We will describe below, only in principle, the methods associated with these techniques.

The nuclear magnetic resonance as well as nuclear quadrupole resonance methods are based on the principle of resonance absorption. The principle of N.M.R. has been described in the previous section on nuclear magnetic moments. We know here that N.M.R. measures the interaction energy ΔE of the magnetic moment μ of the nucleus and the external magnetic field, **H**, which is classically given by:

$$\Delta E = \boldsymbol{\mu} \cdot \mathbf{H} \qquad \text{...(2.141)}$$

This interaction energy will depend on the orientation of μ which points along I and the direction of the magnetic field **H**, which can be expressed quantum mechanically in terms of magnetic quantum number m, *i.e.*,

$$\Delta E = \left[\left(\frac{e\hbar}{2\,M_p\,c}\right) gH\right] m$$

If the interaction had only the magnetic interaction as given by Eq. 2.141, then splitting between different m-states would be the same. However, if there is quadrupole interaction also present, then there will be the interaction energy term:

$$\Delta E = \frac{e}{4}\,\langle Q_{\text{exp}}\rangle_o\,\frac{\partial^2\phi}{\partial z^2} \qquad \text{...(2.142)}$$

From Eq. 2.130 one can, however, see that the measured value of $\langle Q \rangle$ depends on M^2. Therefore, the splitting is not the same for different values of M. One may first, subtract, the magnetic effect, and then obtain the residual splitting for the largest value of $M_I = l$ corresponding to the quadrupole interaction given by Eq. 2.142. One can determine the value of $\langle Q \rangle$ from this; and then extract the intrinsic value $\langle Q \rangle$ from Eq. 2.131. For details see, Reference (61).

The Mössbauer Technique[50,62] again measures splitting of an interaction energy into different m-states, for pure magnetic interaction. If, however, a quadrupole interaction is present, (which means the presence of both E.F.G. of the host material and Q of the nuclear level) we have an extra splitting, in the same way as in NMR case, and one can obtain the quadrupole interaction.

A large amount of date on deformed nuclei, like that of Ir^{77}, Pt^{78}, Au^{79} and Hg^{108} has been obtained with hyperfine studies, using laser spectroscopy[63]. Another method which is used most recently is based on quadrupole-interaction resolved, nuclear magnetic resonance on oriented nuclei (QI-NMR-ON). Recent measurements[64] using this method, have been reported on Pt^{185}, Pt^{189} and Pt^{194}. The atoms containing these nuclei were embedded in hpe-co which provided magnetic hyperfine field along with one axial symmetric electric field gradient (EFG). By measuring the sub-resonance, corresponding to *rf* transition between state $m > m + 1 >$, one can get magnetic and quadrupole interaction frequencies and from them, the values of $\langle Q \rangle$ as given in Eq. 2.142, and μ as given in Eq. 2.141. Similar measurements have been earlier carried out on some platinum nuclei[65].

2.2 DYNAMICAL PROPERTIES OF NUCLEI

2.2.1 Introduction

We have discussed till now the nuclear properties of only stable nuclei. There are, however, a whole class of nuclei, which are unstable to β-decay or alpha-decay or even fission. One can obtain, not only the ground state properties of such unstable nuclei through various special methods developed for unstable nuclei, but also the properties of excited states.

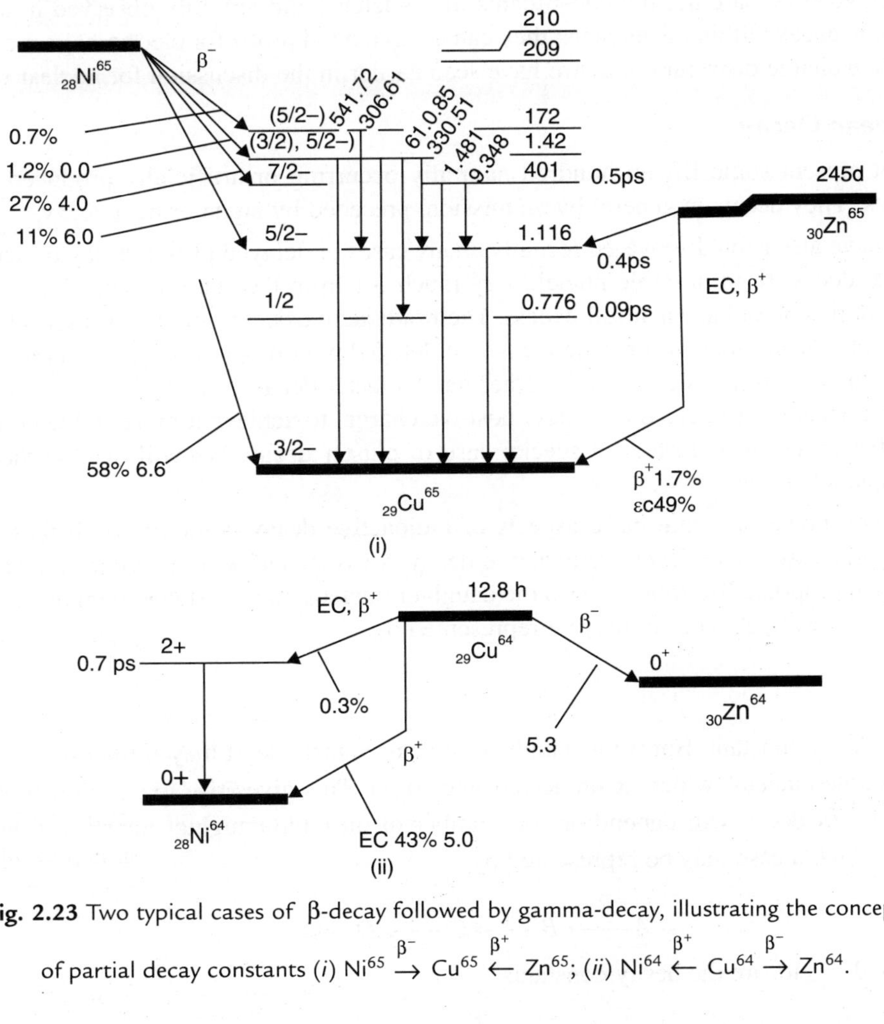

Fig. 2.23 Two typical cases of β-decay followed by gamma-decay, illustrating the concept of partial decay constants (*i*) $Ni^{65} \xrightarrow{\beta^-} Cu^{65} \xleftarrow{\beta^+} Zn^{65}$; (*ii*) $Ni^{64} \xleftarrow{\beta^+} Cu^{64} \xrightarrow{\beta^-} Zn^{64}$.

All such methods are based either on the study of radiation (α, β or γ-rays) of radioactive nuclei; or the study of particles or radiation in a nuclear reaction. We will discuss the dynamical properties of these particles or radiations and nuclear properties of excited and ground states derived from such studies, in this section.

In the case of radioactive nuclei, existing naturally or produced artificially through nuclear reactions, (Fig. 2.23), one can study the energetics and spatial and temporal distribution and correlations of the emitted radiations, which in turn yield, energies, spins, parity and magnetic and quadrupole moments of the excited and ground states. One of the earliest studies in this category was carried by J. Varma and C.E. Mandeville (Ref. 71).

In nuclear reactions, one generally studies, the total or partial cross-sections for a particular incident particle; or for a specific emitted particle, the angular distributions, or correlations of the emitted particles. These studies yield information not only about the properties of excited and ground states involved; but also about the reaction mechanism. The emitted particles may involve not only the light particles like protons, neutrons and alphas, etc; but also the heavy clusters like Li^7 and C^{12} or even heavier clusters. Similarly the incident particles also may involve not only light particles; but also the heavy ions. Gamma-rays are also involved many times, as incident projectiles or emitted radiation.

Though electrons are not the constituents of a nucleus, and are only observed in a beta decay through a proton decay within the nucleus; they can act as a good probe for electromagnetic structure of the nucleus, like charge distribution, as we have seen earlier in the discussion for nuclear radii.

2.2.2 Nuclear Decay

There exist at present some fifteen hundred naturally occurring or artificially produced radioactive nuclear species. They decay, in general by γ-emission, preceded by alpha or beta decay.

There have also been discovered recently, many cases of delayed proton decay or neutron decay or even cluster decay, from unstable nuclei; very much off from the stability line (Fig. 2.3), and are either neutron deficient; or have neutron-excess. Their radioactive decay, infact, represents $\beta^{\pm}$ emissions followed by proton/neutron emission. These cases of delayed proton-neutron decay, therefore, basically represent special cases of $\beta^{\pm}$ decay. On the other hand, cluster-decay is a special case of alpha-decay, arising out of a necessity to get rid of excess positive charge, to achieve a more stable condition. The spontaneous fission-decay is further, a special case of cluster decay. We will discuss these cases in details in Chapters 8 and 9.

There are, however, some basic aspects of radioactive decay common to all these categories which may be discussed here. Every radioactive decay is associated with a certain radioactive decay constant λ and appropriate life-time, τ. Also the daughter nucleus may be stable, then one requires only the simple decay law. Such a decay may be represented by:

$$A \xrightarrow{\lambda} B \qquad \text{...(2.143)}$$

where λ is the decay constant. But if the daughter nucleus is unstable, it may further decay, giving rise to another unstable nucleus, which again decays and so on. This gives rise to a radioactive series, for which the stages of decay will depend on the number of unstable daughter-nuclei, till we reach the stable nucleus. Such a case may be represented by:

$$A \xrightarrow{\lambda_1} B \xrightarrow{\lambda_2} C \xrightarrow{\lambda_3} D, \text{ etc.} \qquad \text{...(2.144)}$$

where λ_1, λ_2 and λ_3 etc. are the decay constants.

It is, of course, apparent, that a nucleus will decay to another nucleus only if the parent nucleus has a higher mass than the sum of the masses of daughter nucleus and the decay products. Such energy relationship will be discussed in the following sections.

It has been experimentally found that radioactive parent nuclei decaying to stable daughter nuclei, obey a simple decay law, given by:

$$\frac{dN}{dt} = -\lambda N \qquad ...(2.145)$$

where N is the number of radioactive parent nuclei at a given time, so that dN/dt is the rate of decay or the number of radioactive nuclei decaying per unit time. Equation 2.145, therefore, defines the radioactive decay-constant as the probability that any particular nucleus will disintegrate in unit time. Rewriting Eq. 2.145 and integrating, we get:

$$\int_{N_0}^{N} \frac{dN}{N} = -\lambda t \qquad ...(2.146)$$

From which, we get:

$$N = N_0\, e^{-\lambda t} \qquad ...(2.147)$$

where N_0 is the number of radioactive nuclei at $t = 0$.

Equations 2.145 – 2.147 are basically based on laws of probability, irrespective of the exact mechanism responsible for decay. The basic assumption for any laws of probability is that the decay constant is constant for a given species and is also independent of the age of the parent nucleus. The definition of λ gives the probability of decay of a given nucleus in unit time. Then if $\Delta t << 1/\lambda$; $\lambda \Delta t$ gives the probability that a particular nucleus will decay in time Δt. Chance of its survival is, then, $(1 - \lambda \Delta t)$. Applying this logic to successive intervals, survival probability, after a time $t = n \Delta t$, will be $(1 - \lambda \Delta t)^n = (1 - \lambda \Delta t)^{t/\Delta t}$.

It can be seen that, for $\frac{\Delta t}{t} \to 0$;

$$\underset{\frac{\Delta t}{t} \to 0}{\text{Lt}} \; (1 - \lambda \Delta t)^{t/\Delta t} \to e^{-\lambda t} \qquad ...(2.148)$$

as derived in Eq. 2.147. This shows that decay of a nucleus is governed by laws of probability. Out of N nuclei, one nucleus may decay immediately, the other may decay after many times the value of I/λ. As we shall see subsequently, the laws of quantum mechanics explain this in a natural manner.

It may be realised that $1/\lambda$ has the dimensions of time. As a matter of fact, from Eq. 2.148, if $t = 1/\lambda$; $N = N_0 e^{-1}$. In other words, in time $1/\lambda$, the nuclei which survive are $1/e$ of the original number of nuclei. One, therefore, defines:

$$\tau = \frac{1}{\lambda} \qquad ...(2.149)$$

as the life-time (or mean life-time) of the state. A related concept is the half life-time $T_{1/2}$, which corresponds to the time in which $N = N_0/2$. From Eq. 2.148, it is apparent that

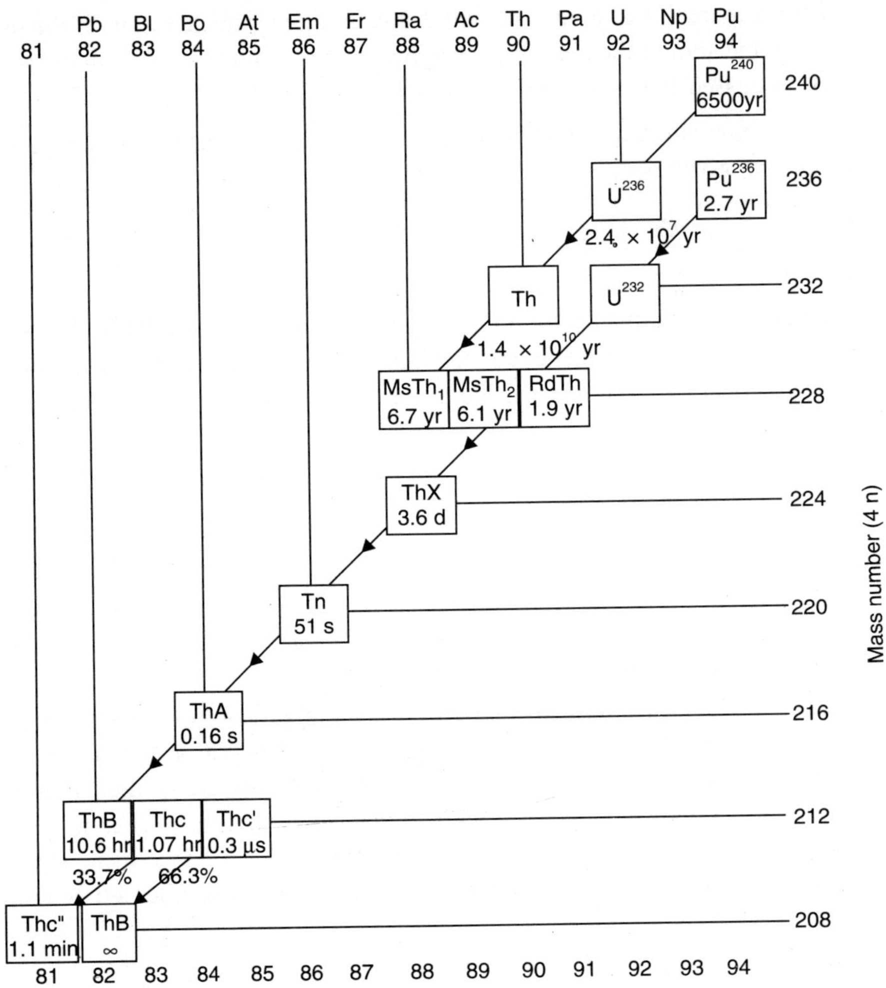

Fig. 2.24 Radioactive series decay connecting many nuclei.

$$T_{1/2} = \frac{\ln 2}{\lambda} = \frac{0.693}{\lambda} \qquad \text{...(2.150)}$$

The decay constants of known radioactive nuclides extend between

$$\lambda = 3 \times 10^6 \text{ sec}^{-1} (P_0{}^{212}) \text{ to } \lambda = 1.58 \times 10^{-15} \text{ sec}^{-1} (Th^{232}). \qquad \text{...(2.151)}$$

In terms of mean life-times, this corresponds to $\tau = 0.3 \times 10^{-6}$ secs; to $\tau = 6 \times 10^{17}$ secs $= 2 \times 10^{10}$ years, a range of 10^{24}, for the decay constant λ, or $1/\tau$. The mean life-time of a particular nucleus or nuclear state, is its intrinsic property, and remains constant under all external environments of chemical and physical nature.

As discussed earlier, the decay constant λ gives the probability of decay of a nucleus. There are, however, instances, when there may be many modes of decay, *e.g.* the decay of a certain nuclear state through transitions to different states, through the emission of either gamma-rays or beta-rays of different end energies, [Fig. 2.23]. One can, then talk of partial decay constants, corresponding to each mode of decay. As the total probability of decay is equal to the sum of partial probabilities, one can write,

$$\lambda = \lambda_1 + \lambda_2 + \lambda_3 + \ldots \qquad \text{...(2.152)}$$

In the case of radioactive series decay, representing say two decays, *i.e.* for

$$A \xrightarrow{\lambda_1} B \xrightarrow{\lambda_2} C$$

One can write the transitions as follows:

$$\frac{dA}{dt} = -\lambda_1 A$$

$$\frac{dB}{dt} = \lambda_1 A - B\lambda_2 \qquad \text{...(2.153)}$$

It can be, then, seen that

$$A = A_0 e^{-\lambda_1 t}$$

and

$$B = A_0 \frac{\lambda_1}{\lambda_2 - \lambda_1} (e^{-\lambda_1 t} - e^{-\lambda_2 t}) \qquad \text{...(2.154)}$$

The activity of B, *i.e.*, $B\lambda_2$ is given by

$$B\lambda_2 = A_0\lambda_1 \frac{\lambda_2}{\lambda_2 - \lambda_1} (e^{-\lambda_1 t} - e^{-\lambda_2 t}) \qquad \text{...(2.155)}$$

As $A\lambda_1 = A_0\lambda_1 e^{-\lambda_1 t}$; Eq. 2.155 can be written as:

$$B\lambda_2 = \frac{(A\lambda_1)\lambda_2}{\lambda_2 - \lambda_1} \times (1 - e^{-(\lambda_2 - \lambda_1)t}) \qquad \text{...(2.156)}$$

For a general case, (Fig. 2.24) when at time $t = 0$, A_0 radioactive nuclei with decay constant λ_A are present, but the subsequent decay products are represented by $B, C, D, \ldots M, N$, with decay constant $\lambda_B, \lambda_C, \lambda_D, \ldots \lambda_M, \lambda_N$ respectively, then at any time t, the number of nuclei N, in the last daughter present are given by:

$$dN/dt = M\lambda_M - N\lambda_N \qquad \text{...(2.157)}$$

where M, itself is evaluated from a series of equations similar to Eq. 2.153 for the amounts of preceding products. One can obtain the values of N, by integrating the series of such equations[66].

An interesting case of nuclear decay is the production of radio-active nuclei in a nuclear reaction, *e.g.*

$${}_{11}\text{Na}^{23} + {}_1\text{H}^2 \rightarrow {}_1\text{H}^1 + {}_{11}\text{Na}^{24} \left(\xrightarrow[\beta_1]{\lambda_1} {}_{11}\text{Mg}^{24} \right) \qquad \text{...(2.158)}$$

The differential equation for this process is given by:

$$dN/dt = Q - \lambda_1 N \qquad \text{...(2.159)}$$

where Q is the number of nuclei formed per unit time in the nuclear reaction . If at $t = 0$, $N_0 = 0$, then one obtains:

$$N\lambda_1 = \frac{Q}{\lambda_1}\,(1 - e^{-\lambda_1 t}) \qquad ...(2.160)$$

With a flux I of incident beam, falling on a target nuclei, producing the radioactive nuclei with a cross-section σ, one gets:

$$Q = I\,\sigma\,N_s \qquad ...(2.161)$$

where N_s is the number of target nuclei ($_{11}Na^{23}$), exposed to the beam. In this manner, one can obtain the decay constants of radioactive decay under various conditions.

For various applications, with different situations of decay constants, and a number of decay products, see H. Bateman Phil. Mag. 20, 704 (1910); H.P. Knauss, Science 107, 324, (1948)[66].

2.2.3 Nuclear Reactions

A nuclear reaction, in general, corresponds to a case of a projectile 'a' striking a target 'A' resulting in an interaction between the two, following the emission of other light projectiles, b_1, b_2, b_3 ... leaving behind a residual nucleus B^*. This may be expressed as:

$$a + A \rightarrow B^* + \Sigma\, bi^* \qquad ...(2.162)$$

The (*) on B and bi denote the fact that the residual nucleus B and emitted particles 'b', may be left in an excited state. If bi's are very light particles, like protons, neutrons, alphas, etc., they will not be excited. On the other hand, if they are somewhat heavier, they may be excited.

A typical example can be alpha-particles at say $_{50}$MeV, falling on $_{28}Ni^{60}$, resulting in the emission of a proton and a neutron, leaving behind the nucleus $_{29}Cu^{69}$, which may be expressed as:

$$_2He^4 + {}_{28}Ni^{60} \rightarrow {}_{29}Cu^{*62} + {}_1p^1 + {}_0n^1 \qquad ...(2.163)$$

The excited nucleus $_{29}Cu^{*62}$, may decay through the emission of gamma-rays. In Eqs. 2.162 and 2.163, we have only shown the relationship of the particles, before and after the reaction, where the charge $\sum_i Z_i$ and mass number $\sum_i A_i$ on the two sides have been balanced.

2.2.3.1 Invariances in Nuclear Reactions

The physical quantities (*i*) charge and (*ii*) mass number are the two well known physical quantities among many others, which remain invariant in a nuclear reaction. Other invariants are (*iii*) mass-energy (*iv*) total linear momentum (*v*) total angular momentum (*vi*) parity and (*vii*) isotopic spin. We will discuss one by one the effect of these invariances on the nuclear reactions.

(*i*) *Charge*: The total charge, on the two sides of the reaction equation must be equal under all circumstances. There has been found no exception to it.

(*ii*) *Mass Number*: The number of nucleons before and after the reaction remain invariant, up to a very high energy say up to $E \approx 1000$ MeV of the incident energy. For higher energies, the number of nucleons may be different on the two sides, but number of baryons which includes nucleons, but also encompasses the particles with higher mass than nucleons, say,

$$\Lambda^\circ \ (m_{\Lambda^\circ} = 1115 \text{ MeV}),$$

$$\Sigma^\circ \ (m_{\Sigma^\circ} = 1193 \text{ MeV})$$

or $$\Sigma^+ \ (m_{\Sigma^+} = 1189 \text{ MeV})$$

remains invariant. In other words, the invariance of the number of baryons above $E \approx 1000$ MeV, is a general case of invariance, and the invariance of number of nucleons at somewhat lower energies, is its special case.

(*iii*) *Mass-Energy*: In any nuclear reaction at any incident energy, the total energy (including the kinetic energy and the rest mass-energy given by $E = mc^2$), remains invariant, both in the lab system and the centre of mass system. While considering the energetics of the nuclear reactions, we will use the lab-system.

To include the conservation of mass-energy, Eq. 2.163 should be written as:

$$_2He^4 + {}_{28}Ne^{60} + E_i \longrightarrow {}_{29}Cu^{62} + {}_1p^1 + {}_0n^1 + E_t \qquad \text{...(2.164)}$$

where $$E_f = E_{recoil}\,({}_{29}Cu^{62}) + E_p + E_n + E_\gamma$$

and $$E_\gamma = {}_{29}Cu^{*62} - {}_{29}Cu^{62}$$

Q-value: One defines the *Q*-value of the reactions as:

$$Q \equiv M_i - M_f = E_f - E_i \qquad \text{...(2.165)}$$

where $$M_i \equiv M({}_2He^4) + M({}_{28}Ne^{60})$$

and $$M_f \equiv M\,({}_{29}Cu^{62}) + M({}_1p^1) + M({}_0n^1)$$

E_f is total recoil energy plus energy of gamma-rays and E_i is incident energy.

The *Q*-value of a nuclear reaction represents the nuclear disintegration energy. The reaction is said to be "exoergic" if $E_f > E_i$ and it is 'endoergic', if $E_f < E_i$. In the case of exoergic reaction, some of the mass of the particles on the left side of the reaction equation is converted into kinetic energy of the reaction-products, while in the case of 'endoergic' reaction, some of the kinetic energy E_i is converted into the mass of the reaction products.

In the centre of mass-system, *Q*-value is defined similar to Eq. 2.165 and is expressed as:

$$Q = T_0\,(\text{out}) - T_0\,(\text{in}) \qquad \text{...(2.166)}$$

where T_0 (out) is the total kinetic energy of exist channel of all reaction products and T_0 (in) the total kinetic energy of incident-channel in c.m. system.

It is self-evident, that the value of Q is the same, in both c.m. and lab-systems.

Threshold Energy: In the endoergic reactions the minimum energy of the bombarding particle in the lab system, which is just sufficient to produce a given reaction is called the threshold energy. It is evident from Eq. 2.166, that this energy corresponds to $T(\text{out}) = 0$. Hence in the centre of mass system, this corresponds to $T_0\,(\text{in})_{\min} \equiv -\,Q$. Under these conditions, the ratio of the incident kinetic energies in lab (T_1), and c.m. system $(T_0)_{\text{inc}}$ is given by (See Mechanics by H.S. Hans and S.P. Puri; Page 180)[67]:

$$\frac{T_1}{(T_0)_{\text{inc}}} = \frac{(m_1 + m_2)\,m_1}{m_1 m_2} = \frac{m_1}{\mu} \qquad \text{...(2.167)}$$

where $\mu = m_1 m_2/(m_1 + m_2)$ is the reduced mass. For the threshold energy, in the lab system, T_1 corresponds to T_0 (in)$_{min}$ = – Q. Hence T_1 in Eq. 2.167 can be expressed, for threshold energy as:

$$T_1 = T\,(\text{Threshold}) = -\frac{m_1}{\mu}\,Q \qquad ...(2.168)$$

It may be realised, that though the energy T_0 (out) here, for threshold energy, is zero in c.m. system, the reaction products are moving with the finite velocities in the lab system.

(iv) Conservation of Linear Momentum and Q-values: In any nuclear reaction the linear momentum is conserved, both in lab and centre of system. This is a universal law and depends on the properties of free space and has no exceptions. In the lab system, the linear momentum of the incident particle is vectorially distributed between the linear momenta of the reaction products. In the centre of mass system, the linear momentum in the incident channels is zero, and so it is after the reaction.

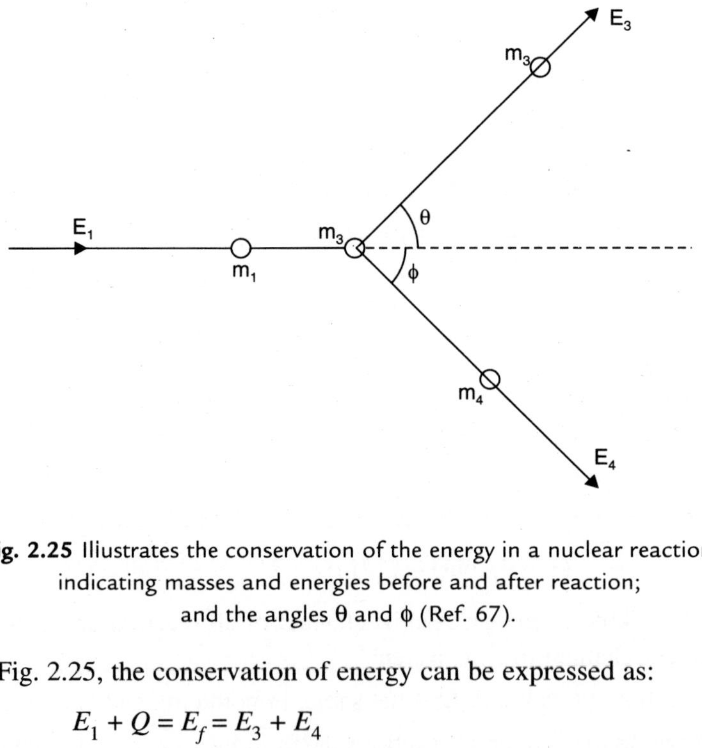

Fig. 2.25 Illustrates the conservation of the energy in a nuclear reaction; indicating masses and energies before and after reaction; and the angles θ and ϕ (Ref. 67).

Referring to Fig. 2.25, the conservation of energy can be expressed as:

$$E_1 + Q = E_f = E_3 + E_4 \qquad ...(2.169)$$

and conservation of linear momentum in the direction of the incident particles is expressed as:

$$p_1 = p_3 \cos\theta + p_4 \cos\phi$$

where $p_1 = \sqrt{2m_1 E_1}\,,\ p_3 = \sqrt{2m_3 E_3}$

and $p_4 = \sqrt{2m_4 E_4} \qquad ...(2.170)$

For the direction perpendicular to the incident particles, the linear momentum is zero, and one can write:

$$0 = \sqrt{2m_3 E_3} \sin\theta - \sqrt{2m_4 E_4} \sin\phi \qquad \text{...(2.171)}$$

Combining Eqs. 2.169 to 2.171 one can write:

$$Q = E_3\left(1 + \frac{m_3}{m_4}\right) - E_1\left(1 - \frac{m_1}{m_4}\right) - \frac{2\sqrt{m_1 E_1 m_3 E_3}}{m_4}\cos\theta \qquad \text{...(2.172)}$$

Equation 2.172 is called the Q-equation, where all the kinetic energies have been expressed in the lab system.

A few interesting features of Eq. 2.172 may be listed below:

(*i*) For exoergic reaction, *i.e.*, when $Q > 0$, then even when $E_1 \to 0$,

$$E_3 = Q\frac{m_4}{m_3 + m_4} \quad (\text{for } Q > 0) \qquad \text{...(2.173)}$$

and is independent of angle θ.

(*ii*) For certain values of E_1, E_3 may have two values for the same angle.

(*iii*) In the endoergic reaction ($Q < 0$), the particles first appear, at the threshold energy of the reaction in $\theta = 0$ direction, with the kinetic energy:

$$E_3 = (E_1)_{\text{thr}}\left[\frac{m_1 m_3}{(m_3 + m_4)^2}\right] \qquad \text{...(2.174)}$$

where $(E_1)_{\text{thr}} = -\,Q(m_3 + m_4)/(m_3 + m_4 - m_1)$, which is the minimum possible value of E_1 at which reaction can take place.

In the endoergic reaction, as the incident energy E_1 is raised above threshold, the emitted particles with mass m_3 can appear in the forward direction, with two discrete values of kinetic energy E_3. This can be seen from the Q-equation, Eq. 2.172.

In endoergic reactions, E_3 becomes single-valued for all values of Q when E_1 exceeds $-Qm_1/(m_4 - m)$.

(*v*) *Conservation of Total Angular Momentum and Parity*: In any collision between two particles, total angular momentum, before and after the collision, has to be the same as required from the general properties of symmetry of free space. In a nuclear reaction, the total angular momentum $\mathbf{J}_i$ in the incident channel is composed of the total angular momentum of the projectile $\mathbf{j}$ (inc) the total angular momentum of the target (j_t) and the relative angular momentum of the two ($\mathbf{L}_i$), arising out of the linear momentum of the projectile with respect to the target. Similarly, in the exit channels, the total angular momentum ($\mathbf{J}_f$) consists of $\mathbf{j}$(out) for the emitted light particles, $\mathbf{j}_R$ for the residual nucleus and ($\mathbf{L}_f$) the relative angular momentum of the emitted particle and the residual nucleus. We can express the relationship of these quantities as:

$$\mathbf{j}\text{ (inc)} + \mathbf{L}_i + \mathbf{j}_T = \mathbf{J}_i \qquad ...(2.175)$$

$$\mathbf{j}\text{ (out)} + \mathbf{L}_f + \mathbf{j}_R = \mathbf{J}_f \qquad ...(2.176)$$

Then
$$|\mathbf{J}_i| = |J_f| \text{ and } J_{iz} = J_{fz} \qquad ...(2.177)$$

In other words, the total angular moments of the interacting particles and their components along a given direction are both conserved in a nuclear reaction.

Taking the example of the reaction given in Eq. 2.163, we see that the ground state spin of both He^4 and Ni^{60} are zero. Hence the total angular momentum of the incident channel is equal to the relative orbital momentum l_i. At higher energies, one has, besides $l = 0$, higher orbital angular momenta also, say $I = 1, 2, ...$ Any reaction will, however, go through an intermediate composite state which, in most of the cases, will correspond to the compound nucleus formation, but in some cases there may be a state leading to direct reaction, or any other mode of reaction. In a given reaction, this intermediate state will have specific total angular momentum J_o and specific parity.

In the above case, using Eq. 2.175 the allowed value of I_i's are determined by the property of this intermediate state. Only $J_o - l_i$ are allowed in a reaction like in Eq. 2.163 and if the parity of this intermediate state is positive, only even l_i's are allowed and if it is negative only odd l_i's are allowed. Though may l_i's may be possible energetically, only those will be responsible for reaction, which obey the above conditions. The vector sum of the angular momentum in the final channel, should be equal to $\mathbf{J}_o$, *i.e.*,

$$\mathbf{J_o} = \mathbf{J_f} = \mathbf{j}\,({}_{29}Cu^{*62}) + \mathbf{j_p} + \mathbf{j}_n + \mathbf{l}_f \qquad ...(2.178)$$

where l_f is the sum of the relative orbital angular momenta of the reaction products, with respect to the centre of mass of the three particles in the exist channel. Again those l_i's are allowed, which obey the rule for conservation of parity, *i.e.*,

$$\pi_0 = \pi({}_{29}Cu^{*62})\ \pi_p \pi_n\ (-1)^{l_f} \qquad ...(2.179)$$

where π_0 is the parity of intermediate state. Intrinsic parities of protons and neutrons are even. Hence one can write:

$$\pi_0 = \pi({}_{29}Cu^{*62})\ (-1)^{l_f} \qquad ...(2.180)$$

If ${}_{29}Cu^{62}$ is left in the ground state, its parity is positive, hence

$$\pi_0 = (-1)^{l_f} \qquad ...(2.181)$$

It is apparent from the above discussion that if π_0 is positive, then only even values of l_i and l_f are allowed. If π_0 is negative, then only odd values of l_i and l_f are allowed, provided Cu^{62} is left in the ground state.

(*vi*) *Isotopic Spin*: We will discuss, somewhat more in detail, the concept of Isotopic Spin in Chapter 6 on Nuclear Forces. Enough to say here, that it is now accepted that nuclear forces are charge independent. This concept is expressed as:

$$(n - n) = (p - p) = (n - p) \qquad ...(2.182)$$

It should, therefore, be possible to associate a quantum number which is the same for both the nucleons, *i.e.* neutrons and protons, and write the nuclear forces in terms of such a quantum number.

Such a quantum number is called isotopic spin quantum number T. The neutron and the proton are said to be the two states of the nucleon, in this system of the iso-spin quantum numbers.

This can formally be expressed by assigning to a nucleon, a value of $T = 1/2$, and assigning to proton the z-component T_Z equal to $+ 1/2$ and to the neutron $T_Z = - 1/2$ so that when a proton changes to a neutron, as it happens inside the nucleus in β^+-decay, or when a neutron changes to a proton, as in the case of the decay of the free neutron, then $\Delta T_Z = \pm 1$; but $\Delta T = 0$.

This formalism for expressing the nucleons is similar to the formalism for expressing the intrinsic spins of nucleons, except that instead of having a real space in which intrinsic spin is expressed, the isotopic spin is expressed in an imaginary isotopic space.

Then for a two-nucleon system, T can be either 0 or 1. We can have, for $T = 1$, three possible two-nucleon systems, *i.e.* diproton, dineutron and proton-neutron with $T_Z = 1, - 1$ and 0, respectively and for $T = 0$ only proton-neutron system. This fact is expressed in detail in Chapter 6.

For a complex nucleus, it is easy to see that:

$$T_Z = \frac{Z - N}{2} \qquad ...(2.183)$$

so that a nucleus, with say 2 protons and 4 neutrons, *e.g.* ($2He_4^6$) has $T_Z = - 1$, $3Li_3^6$ has $T_Z = 0$ and $4Be_2^6$, has $T_Z = + 1$. The value of $T_Z = + 3$ corresponds to six protons and no neutrons, and $T_Z = - 3$ corresponds to six neutrons and no protons; and as we know, such nuclei do not exist. Again, $T_Z = - 2$, for six nucleons will mean $1Li_5^6$; and $T_Z = + 2$ will mean $5Be_1^6$, which again physically do not exist. So only allowed values of T_Z are $T_Z = - 1$, 0 and + 1 corresponding to He^6, Li^6 and Be^6, for six nucleons. If we treat T_Z as the z-component in isotopic-spin space, then we can define a quantum number called 'isotopic spin' or 'iso-spin' T in the above case, with a value of $T = 1$.

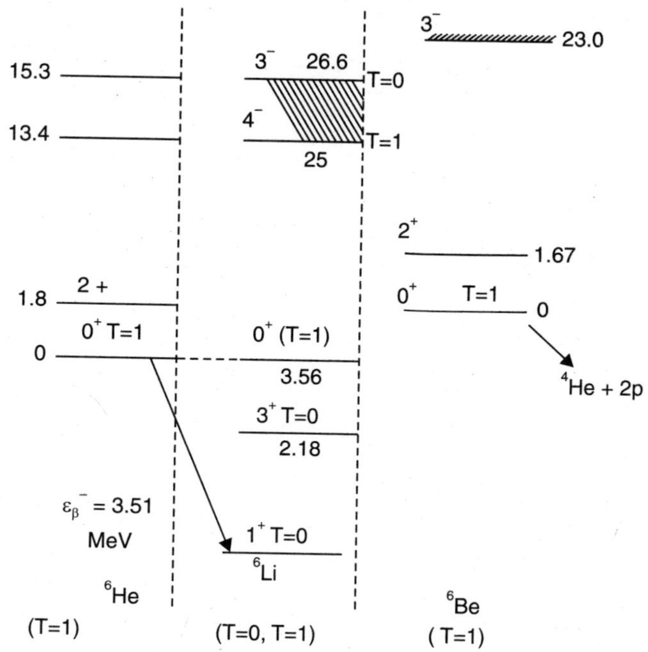

Fig. 2.26 The energy levels of $2He_4^6$, $3Li_3^6$ and $4Be_2^6$ illustrate the concept of iso-spin T (Ref. 68).

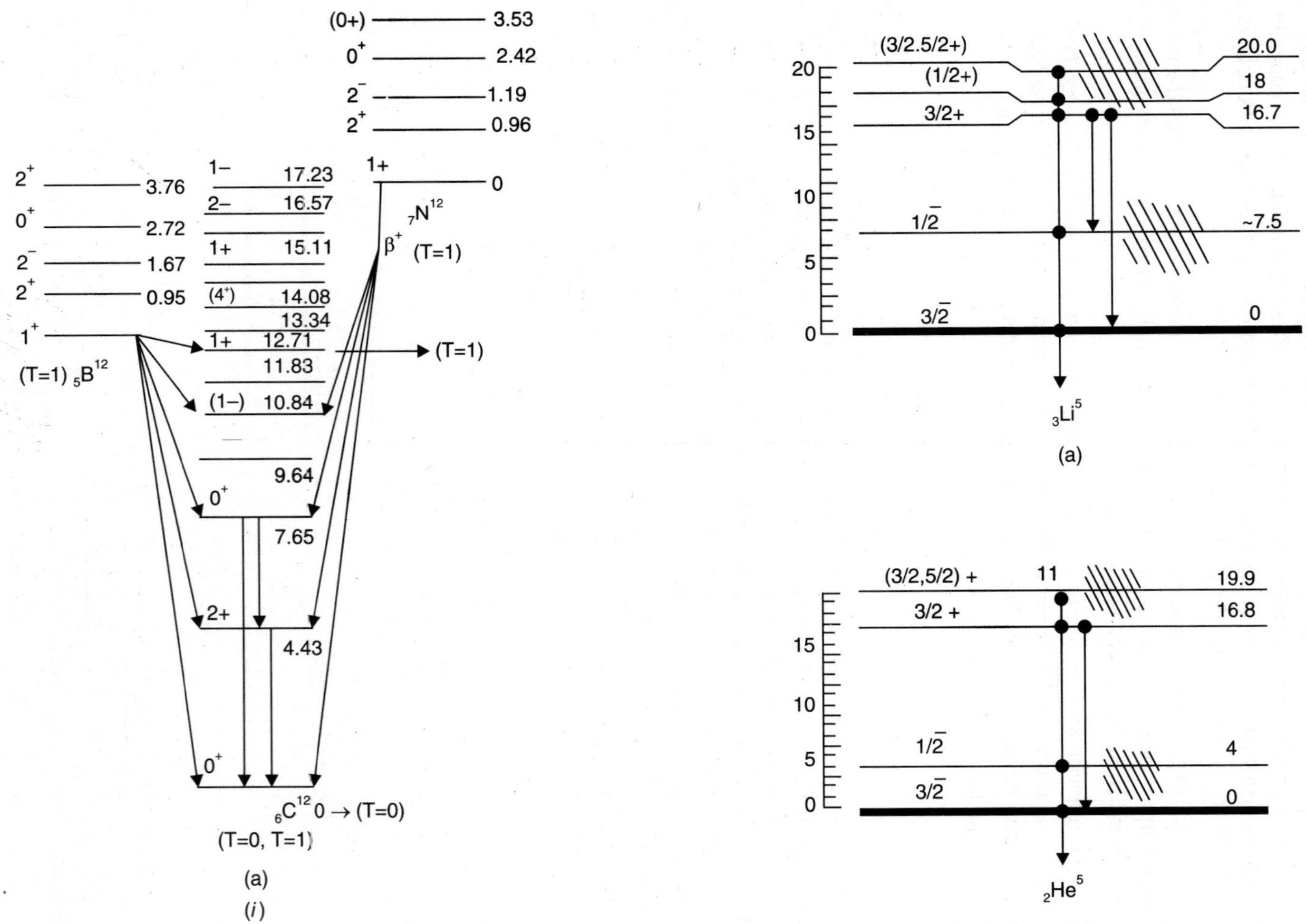

Fig. 2.27 (*i*) The decay of N^{12} and B^{12} to C^{12} with $\Delta T = 0, +1$ (*ii*) Similarity of excited states of two iso-spin doublets $_3Li^5$ and $_2He^5$ (Ref. 69).

The extreme possible value of $T = 3$, for a system of nucleons is not possible, because as described above, the system with $T_Z = +3$ does not exist physically. Similarly, $T = 2$ is not possible. So for ground states of He^6, Li^6 and Be^6, $T = 1$, is a possible quantum number. For Li^6, $T = 0$ is also possible because $T_Z = 0$. In practice, the detailed analysis shows[55] (as shown in Fig. 2.26) that the ground states of He^6 and Be^6, belong to $T = 1$; and while the first two states of Li^6 belong to $T = 0$, the third state belongs to $T = 1$.

There are many examples where iso-spin plays an important role. Such examples are: $3Li_5^8$, $4Be_4^8$ and $5B_3^8$; $5B_7^{12}$, $6C_6^{12}$ and $7N_5^{12}$; and $6C_8^{14}$, $7N_7^{14}$, and $8O_6^{14}$ and many others, where the two extreme nuclei decay to the middle nucleus by $\beta^{\pm}$ decay, obeying the selection rules of angular momentum and iso-spin. Similarly, mirror nuclei $3Li_4^7$ and $4Be_3^7$ and $2He_3^5$ and $3Li_2^5$ show iso-spin characteristcs. As for example, for $A = 12$, the three states, *i.e.* ground state of $5B^{12}$, the excited state of $6C^{16}$ (at 12.7 MeV excitation) and the ground state of $7N_5^{12}$, have all angular momenta $I = 1^+$ and similar properties of wave-functions. They correspond to $T = 1$, for which three *z*-components are $T_Z = -1[5B_7^{12}]$, $T_Z = 0$ $[6C_6^{12}]$; and $T_Z = +1$ $[7N_5^{12}]$. The ground state of $6C_6^{12}$ with $I = 0$ corresponds to $T = 0$, or $T = 1$ and $T_Z = 0$. The β-decay corresponds to $\Delta T = 0, \pm 1$.

For $A = 5$, the two mirror nuclei ($3Li_2^5$ and $2He_3^5$) represent the cases of iso-spin doublet with $T = 1/2$, $T_Z = +1/2$ and $T = 1/2$, $T_Z = -1/2$.

Figure 2.27 (*a*) shows the decay characteristics of $A = 12$ nuclei and Fig. 2.27 (*b*) shows the iso-spin doublets $3Li_2^5$ and $2He_3^5$.

In nuclear physics, therefore, the concept of iso-spin is useful in classifying the nuclear states in multiplets of iso-spin. The charge independence of nuclear forces is, then expressed as conservation of iso-spin so that the wave-functions of various members of these multiplets are identical, apart-from Coulomb effects.

2.2.3.2 Basic Concepts of Cross-sections

One of the most important quantities which one defines in a nuclear reaction is the probability of the occurrence of a reaction or the cross-section, which is closely related to the probability. The probability (P_r) is defined as the total number of reactions per unit time/total number of incident particles per unit time. If the target area on which particles are incident is A, and if I is the number of incident particles per unit area per unit time (flux) and N_r is the number of reactions per unit time, then

$$P_r = \frac{N_r}{1A} \qquad \text{...(2.184)}$$

Alternatively, one can express the value of P_r in terms of effective cross-section or area, which the target particle offers to the incoming beam. Suppose each target particle in the target, offers an effective cross-section σ_r to the incoming beam, then the total effective area which the particles N_s in the target will offer to the incident beam will be $N_s \sigma_r$. But then total area of the target is A. Hence one can represent P_r as:

$$P_r = \frac{N_s \sigma_r}{A} \qquad \text{...(2.185)}$$

From Eqs. 2.184 and 2.185 we can write:

$$\frac{N_s \sigma_r}{A} = \frac{N_r}{IA} \quad \text{or} \quad \sigma_r = \frac{N_r}{IN_s} \qquad ...(2.186)$$

In the above discussion, the term nuclear reaction has been used in a general manner, including in it, phenomenon like elastic scattering, inelastic scattering, or nuclear reaction proper, in which the emitted reaction products are very different from the ones in the incident channel. We discuss below some specific features of these cases.

(*i*) *Elastic Scattering*: In this case, particles in the exit channels are exactly the same as in the incident channels, with no changes in any quantum number. As for example:

$$_2He^4 + {}_{28}Ni^{60} + E_0 = {}_{28}Ni^{60} + {}_2He^4 + E_0 \qquad ...(2.187)$$

This means that the internal states of He^4 and Ni^{60} are not changed after scattering. The direction of He^4 may, however, be different[28] after scattering from the incident direction and this may result in sharing of linear momenta and energy after the scattering between Ni^{60} and He^4. In practice, of course, only a small amount of linear momentum and energy are shared by Ni^{60}, giving rise to the recoil and most of the initial energy and linear moments are carried by alpha-particle (see Eq. 2.172 for $\theta = 0$). On the other hand, if the incident particle and target nucleus have similar masses, these quantities will be shared more equally between them after scattering. It is evident that elastic scattering is characterised by no change in the internal quantum numbers and configuration of the participating particles.

(*ii*) *Inelastic Scattering*: In inelastic scattering, one or both of the particles taking part in the interaction can change their internal state, but keeping the constituents of the two particles the same. As for example:

$$\begin{array}{ccc} _2He^4 + {}_{28}Ni^{60} & \longrightarrow & {}_{28}Ni^{*60} + {}_2H'e^4 \\ & & \downarrow \\ & & {}_{28}Ni^{60} + \gamma \end{array} \qquad ...(2.188)$$

In this example, after the interaction, the nucleus $_{28}Ni^{*60}$ is in an excited state, and gets de-excited by the emission of a gamma-ray. The constituents, (*i.e.*, number of protons and neutrons) in the two particles, *i.e.*, $_2He^4$ and $_{28}No^{60}$ are, however, the same after the reaction, as before it. The interaction has, however, changed the internal quantum numbers of Ni^{60}, including its internal energy; and the outgoing alpha particles denoted by $_2H'e^4$ has lesser energy than the energy of incident $_2He^4$.

(*iii*) *Nuclear Reactions*: In a nucleus interaction, the outgoing particles may be quite unlike the incident projectiles, *e.g.*,

$$_2He^4 + {}_{28}Ni^{60} \longrightarrow \begin{cases} {}_{30}Zn^{*63} + n \\ {}_{30}Zn^{62} + 2n \\ {}_{29}Cu^{62} + {}_1H^1 + {}_0n^1 \end{cases} \qquad ...(2.189)$$

In the above reaction, outgoing channels are completely different from the incident channel. Such types of reactions are referred to as nuclear reactions proper. Though elastic scattering and inelastic scattering are also a part of nuclear reactions, they have been given special names, because the outgoing channel are the same or very similar to the incident channel.

The conservation law of charge, mass number, angular momentum, parity, linear momentum, mass energy and isotopic spin hold good for all nuclear reactions.

It is easy to see from Eqs. 2.185–2.189 that, in general, when a particle say He^4 falls on a target say $_{28}Ni^{60}$, a large number of outgoing channels are opened. One describes the cross-section for different channels as follows:

1. *Total Cross-section*: This is the cross-section, corresponding to all the channels, summed over all directions. It is measured, basically, by counting the incoming projectiles first directly, and then with the target as absorber. Let n be the number of projectiles at distance x from the surface along the direction of projectile and Δn the number absorbed by the absorber. Assuming that:

$$\Delta n \propto n$$

$$\propto dx \quad \text{(where } dx \text{ is the thickness through which the absorption is measured).}$$

One can, then, write:

$$\Delta n = -\mu_1 \, n \, dx \qquad \text{...(2.190)}$$

where μ_1 is called the linear absorption coefficient. The minus sign corresponds to the fact, that n is decreasing, while passing through the thickness dx. Equation 2.190 assumes that the absorption is based on the laws of probability.

We can, alternatively see, that $\Delta n/n$ is the probability $P(r)$, and from Eqs. 2.185 and 2.186 can be expressed as:

$$\frac{\Delta n}{n} = \frac{dn}{n} = -\sigma_t \frac{Ns}{A} = -\sigma_t \frac{\rho A \, dx}{A} = -\sigma_t \rho \, dx \qquad \text{...(2.191)}$$

Ns = number of scatters in thickness dx, for the area A.

A = Area of the absorber over which the projectiles are incident

ρ = the density of nuclei in the absorber.

σ_t = total cross-section

Then, from Eqs. 2.190 and 2.191, we can see that

$$\mu_l = \sigma_t \rho \qquad \text{...(2.192)}$$

Another equivalent quantity, $\mu_m = \mu_l/\rho$ is sometimes, defined as mass absorption coefficient, so that we can write:

$$\frac{dn}{n} = -\mu_l \, dx = -\mu_m \, dm \qquad \text{...(2.193)}$$

where dm is measured in gms/cm^2. Evidently, $dm = \rho dx$; Hence $\mu_l = \mu_m \rho$.

Integrating Eq. 2.193, one can write:

$$n = n_0 \exp - \mu_l x$$

or

$$n = n_0 \exp - \mu_m \, m \qquad \text{...(2.194)}$$

where x is the total linear thickness of the absorber expression in cm, and m is the mass thickness of the absorber expressed in gm/cm^2, after travelling which; the number n is measured.

2. *Partial Cross-sections*: As discussed earlier, the total cross-section may be contributed by various partial cross-sections, say by elastic scattering, inelastic scattering, and many channels in nuclear reactions proper. The cross-sections, being related to probabilities, are additive. Hence one can write:

$$\sigma = \sigma_1 + \sigma_2 + \sigma_3 + \ldots = \sum_i \sigma_i \qquad \text{...(2.195)}$$

For measuring the various partial cross-sections, one directly measures the particular particles, corresponding to the particular partial cross-section. As for example, for elastic scattering, one measures the alpha-particles of the same energy, as that of incident particles: for the reaction given in Eq. 2.187. For inelastic scattering, one detects the alphas of lower energies, [Eq. 2.188] corresponding to residual energies of the alphas for the various excited states of Ni^{60}. For other reactions, one measures the particular particle of different energies, as expected in that reaction, Eq. 2.189. Evidently, one uses the suitable detectors and energy discriminating devices for this purpose. One uses Eq. 2.186 for obtaining the partial cross-section corresponding to a particular channel.

3. *Differential Cross-sections*: For each particular channel, there is expected to be an angular distribution which may be different for different channels. One may, therefore, express the cross-section for a given channel as the sum of differential cross-sections, summed over all the zenith (θ) and a azimuthal (φ) angles.

One always measures the outgoing particles emitted over a certain solid angle. The differential cross-section is, then, characterised by a function $\sigma(\theta, \varphi)$, so that the detected particles N_r, over a solid angle $d\Omega$ are related to $\sigma(\theta, \varphi)$ as:

$$N_r(\theta, \varphi) = IN_s\sigma(\theta, \varphi)\, d\Omega$$

where $$d\Omega = \sin\theta\, d\theta\, d\varphi \qquad \text{...(2.196)}$$

The differential cross-section $\sigma(\theta, \varphi)$, is defined as the cross-section per unit solid angle, for particles going along θ and φ direction. The angular spreads $d\theta$ and $d\varphi$ are angles subtended by the detector over the scatterer or target.

In general, if one is not including the polarisation studies in the measurements, the distribution is uniform over the azimuthal angle φ, for the same zenith angle. One is, then, interested in $\sigma(\theta)$ only, which one can obtain by measuring the integrated counts N_r over the spread of 2π, of φ. One can, then, write:

$$N_r(\theta) = IN_s\sigma(\theta)\, d\Omega \qquad \text{...(2.197)}$$

Hence $$\sigma(\theta) = \frac{N_r(\theta)}{IN_s} \qquad \text{...(2.198)}$$

Sometimes, one wants to measure the energy dependence of a cross-section, as well as the angular dependence. Then one writes:

$$N_r(\theta, \varphi, E) = IN_s\sigma(\theta, \varphi, E)\, d\Omega\, dE \qquad \text{...(2.199)}$$

Because of the double dependence of the cross-section sometimes one expresses equation 2.199 as:

$$\frac{d^2N(\theta, \varphi, E)}{d\Omega\, dE} = \frac{I\, N_s\, d^2\sigma(\theta, \varphi, E)}{d\Omega\, dE} \qquad \text{...(2.200)}$$

It may be realised that the measurement of different cross-sections as a function of angles and energies gives information about the reaction mechanism involved in a particular reaction.

The cross-sections in nuclear reactions may be expressed quantum mechanically in terms of the transitions between the initial and final states. For details see the chapters on nuclear reactions.

2. Static and Dynamical Properties of Nuclei
2000–2008

Nuclear Masses

Using $Q_{E.C}$ values from β^+ spectrum and end points, mass values have been determined for Se^{70}, Br^{71}, Br^{72} and B^{73}, using β-γ coincidence spectroscopy, by a group of 19 authors, from U.S. and Israel [Phy. Rev. (63, 034314 (2001)].

From charge radius predictions of two modern mass formulas by Nakjakev, et al. [Nuclear Data sheets, 56, 133 (1994)] it has been found that finite range droplet model of mass formula leads to rms deviation of 0-045 fm, and the corresponding quantity for Skyrme-Hartree-Fock mass formula is 0.024fm [Phy. Rev. 0.65, 0.67303 (2002)].

In another interesting paper, atomic masses of radioactive Zr isotopes (Zr^{96}, Zr^{96}, Zr^{98}, Zr^{99}, Zr^{107}), were measured using isotope separator-on-line system with an accuracy to $\leq \times 10^{-7}$. [Phy. Rev. 70.011301(R) 2004)].

This is a continuation of a local formula for binding energy for heavy and superheavy nuclei, published in [Phy. Rev. C.73 064331 (2005)] by authors from China. In this new formula, neutron-proton correction has been considered for higher precision. With this new formula α-decay energies Q_α, and half lives τ_α of nuclei with $Z = 102$-118 are reproduced quite well [Phy. Rev. 77, 064310 (2008)].

Magnetic Moments

A group of eleven authors in U.S.A. has measured $\mu = 1.114$ HN, as magnetic moment of ground state of Cr^{32}, using β-NMR spectroscopy. Polarized Cl^{32} nuclei were produced by fragmentation of 100 MeV beam of Ar^{36} on Nb^{93} target. Isovector and Iso-scalar moments for $A = 32$, $T = 1$, isopin were extracted and compared with shell model calculations [Phy. Rev. C.62, 0.44312 (2001)].

In another paper, published in 2005 [Phy. Rev. C.72, 064316 (2005)], 16 authors from Czech Republic, Germany, Switzerland and Isolede (CERN), have measured nuclear moment of As^{69}, for ground state of $5/2^-$ using on-line β-NMR technique on oriented nuclei, with NICOLE, He^3 – He^4 dilution refrigeration set-up. The result for g differs from theoretical expectation for single particle value for $\pi f/5$; for a free proton, but is in reasonable agreement, with the value obtained from effective g factors, and with value from a core polarization.

Anti Protonic Atoms

Anti-protons were produced at CERN in the range of (2.4–7.3) × 10^8 in the beam when beam of anti-protons was made to fall on a target. After being slowed down; in the target, the anti-proton is captured by a target atom, to form an anti-proton atom. This creates a nucleus (N, Z) and anti-protons circuiting around the nucleus. This will change the atoms configuration giving different anti-protonic X-ray energies and shapes, only by the electromagnetic interaction but also by the strong interaction between the nucleons in the nucleus and anti-protons. In a paper [Phy. Rev. C.65, 014306 (2002)], a group of 17 authors, used

Ca^{40}, Ca^{42} and Ca^{48} as targets. In 2004, [Phy. Rev. 69. , 014311 (2004)] in another collaborative effort, Te^{122}, Te^{124}, Te^{126}, Te^{129} and Te^{130} were studied, using anti-protons of 106 MeV/C. The measured lower and upper level widths, corrected for $E2$ effects were used to determine the properties of nuclear periphery, by using Hartree-Fock Bouglibouv calculations. In another paper of collaborative effort, 17 authors from Germany and Poland [Phy. Rev. C.67, 0144308, (2003)], used Cd^{106}, Cd^{116}, Sn^{112}, Sn^{116}, Sn^{120} and Sn^{124} as targets for anti-protons beam.

REFERENCES

1. L.L. Foldy: Phy. Rev. 83, 397 (1951); C.E. Moore: Atomic Energy Levels: Circular 467, V.1, p.XL. National Bureau of Standards, Washington D.C. (1949).
2. H.H.E. Mattauch, M. Thiele and A.H. Wapstra: Nuclear Physics 67, 1, 32, 76 (1965).
3. F.W. Aston: Nature, 123, 313 (1929); Mass Spectra and Isotopes: Edward Arnold & Co., London (1933); Longman Green and Co. In. New York (1942). K.T. Bainbridge and E.B. Jordon: Phy. Rev. 50, 282 (1936).
4. K.T. Bainbridge and A.O. Nier: National Research Council (U.S.). Preliminary Report 9, (1950). K.T. Bainbridge and E.B. Jordon, Phy. Rev. 50, 282, (1936). A.J. Dempster: Phy. Rev. 53:64 (1938). A.O. Nier and T.R. Roberts. Phy. Rev. 81, 507 (1951).
5. T.L. Collins, A.O. Nier and W.H. Johnson: Jr. Phy. Rev. 86, 408 (1952): 84, 717 (1951), 94, 398 (1954), V.B. Bhanot, W.H. Johnson Jr. and A.O. Nier: Phy. Rev. 120, 235, (1960).

   Times of flight method of mass spectrometry:

   F.A. White: Mass Spectrometry in Science and Technology; John Wiley & Sons Inc. (N.Y.) (1968).
6. S.M. Rothstein: Advances in Mass Spectrometry, Vol. 78, ed. Daly N.R. P. 913, London, Heydon & Son (1978). Mass Spectroscopy: H.E. Duckworth, Barker S.D. Venkatsubramaniam U.S., Cambridge University Press, Cambridge (1990).
7. Table of Isotopes: edit. Virginia S. Shirley, Lawrence Berkly Lab., John Wiley & Sons, Inc. New York. (1986).
8. T.P. Kohman: Phy. Rev. 73; 16(1948) M.G. Bowler, Nuclear Physics: Pergamon Press. W.H. Sullivan; Trilinear Chart of Nuclear Species, John Wiley & Sons, Inc, New York (1949).
9. G. Eder: Nuclear Forces, M.I.T. Press (1968). J. Mattauch and S. Flügge ~ ~ "Nuclear Physics Tables"; Interscience Publishers K.T. Baingridge in E. Segre Experimental Nuclear Physics, John Wiley & Sons Inc., New York (1953).
10. C.F. Von Weizsäcker: Z Physik, 96, 431 (1935): Natur Coiss 24, 813: (1936).
11. W.D. Myers and W.O. Swaitecki (1966): Nuclear Physics 81.1 (1966).
12. Handbuch der Physik: Wapstra A.H. V. XXXVIII/I (Springer-Verlag): (1958).
13. H.B. Levy: Phy. Rev. 106, 1265 (1957). Saeger P.A. Nuclear Physics 25, 1 (1961).
14. B.J. Cole: Phy. Rev. C. 56, 1866 (1997).
15. V.I. Goldansky: Nuclear Phy. 19, 482 (1960).
16. V.Borrel et al.: Nuclear Physics A 531, 353 (1991), D. Bazen et al.: Phy. Rev C 45, 69 (1992), B. Blank et al: Nuclear Physics A 615, 52 (1997).
17. R. Hofstader: Reviews of Mod. Physics 30, 412 (1958).
18. R. Hofstader: Nuclear and nucleon scattering of high-energy electrons: Annual Rev. Nuclear Science 7, 231 (1957).

19. W. A. Mckinley and H. Feshbach: Phy. Rev. 74, 1759 (1948).
20. D.R. Yennie, D.G. Ravenhall and R.R Wilson: Phy. Rev. 121, 283 (1961).
21. R.A. Uher and R.A. Sorensen (1966): Nuclear Physics 86, 1. Glassgold A.E.: Rev. of Mod. Physics 30, 419 (1963).
22. R. Hofstader: Electron Scattering and Nuclear and Nucleon Structure; Stanford University; Benjamin Inc. New York (1963).
23. Robert Hofstader: Electron Scattering and Nuclear Structure: Rev. Mod. Physics, 28, 214 (1956).
24. P. Brix and H. Kopfermann: Rev. of Mod. Physics 30, 517 (1958): 507, (1958). H. Kopfermann: Nuclear moments, Academic Press, New York (1958). P. Herman and R. Hofstader: High Energy Electron Scattering Tables, Stanford University Press; Stanford, California (1960).
25. S. Fernbach: Rev. of Mod. Physics 30, 414 (1958).
26. R.M. Eisberg and C.E. Porter: Rev. Mod. Physics 33, 190, (1961), G. Farewell and G. Wagner: Phy. Rev. 95, 1212 (1954), D.O. Kerlee et al.: Phy. Rev. 107, 1343 (1957); G. Igo and R.M. Thaler: Phy. Rev. 106, 126 (1957).
27. Kofoed-Hansen: O. Rev. Mod. Phys. 30, 448 (1958).
28. Pappademos, J.N.: Nuclear Physics 42, 122 (1963).
29. D.F. Jackson: Ann. Physics, 105, 151 (1977).
30. V.L. Fitch and J. Rainwater: Phy. Rev. 92, 789 (1953) H.L. Anderson. C.S. Johnson and E.P. Hincks: Phy. Rev. 130, 2468 (1963) G. Backenstoss et al.,: Nuclear Physics, 62, 449 (1965).
31. K.W. Ford and D.L. Hill: Annual Rev. of Nuclear Science 5, 25 (1955), L.N. Cooper and E.M. Henley: Phy. Rev. 92, 801 (1953).
32. A.L. Schwalow and C.H. Townes: Phy. Rev. 100, 1273 (1955).
33. Barret R.C. and Jackson D.F.: Nuclear Sizes and Structures: Clarenden Press, Oxford (1977).
34. S.R. Elliot, P. Beiresdorfer, M.H. Chen, V. Decaux and D.A. Knapp: Phy. Rev. C 57, P. 583 (1998).
35. J.D. Zimbro, E.B. Shera, Y. Tanaka, C.E. Bemis Jr., R.A. Naumann, M.V. Hoen, W. Reuter and R.M. Steffen: Phy. Rev. letters, 53, 1888 (1984).
36. W.G. Jenet et al. (12 authors): Phy. Rev. C 55, P. 1545 (1997).
37. N. Boosse et al. (21 authors): Phy. Rev. letters 72, 2689 (1994).
38. I. Licot et al. (19 authors): Phy. Rev. C V. 56, P. 250 (1997).
39. H.E. White: Introduction to Atomic Spectra, McGraw-Hill Book Company, Inc., New York (1934).
40. I.I. Rabii, J.M.B. Kellog and J.R. Zachiaras; and S. Millman and P. Kush: Phy. Rev. 46, 157 (1934). 53, 318L (1938).

    G. Breit and I.I. Rabii: Phy. Rev. 38, 2082L (1931).
41. S.C. Brown and L.G. Elliot: American J. of Physics 11, 311 (1943). D.P. Jones, P.G. Murphy and P.L.O' Neill; Prac. Phy. Society (London), 72, 429 (1958); T.D. Lee and C.N. Yang: Phy. Rev. 104, 254 (1956).
42. D.H. Wilkinson: Phy. Rev. 109, 1603, 1610, 1614, (1958): D.E. Alburger, R.E. Rexley, D.H. Wilkinson, and P. Donovan: Phil. Mag. 6, 171 (1961).
43. L.I. Schiff: Quantum Mechanics, McGraw-Hill Book Company. Inc. (1949).
44. S.C. Brown and L.G. Elliot: American J. of Physics 11, 311 (1943).
45. J.H. Sander: Nuovo Cemento Supp. 6, No. 1. 242, (1957), (For μ_0) Cohen V, N.R. Corngold and N.F. Ramsey: Phy. Rev. 104, 283 (1956).
46. R.D. Evans: The Atomic Nucleus, P. 149, McGraw-Hill Book Company, Inc. (1955).

47. (*a*) J.H. Gardener and E.M. Purcelli: Phy. Rev. 76, 12672 (1949) F. Bloch, D. Nichodomus, and H.H. Staub: Phy. Rev. 74, 1025 (1948), P.A.M. Dirac: Quantum Mechanics, Oxford (1947).
(*b*) Data by G.H. Fulley, and V.W. Cohen: Nuclear Moments; Nuclear Data Sheets (1960).
48. E.M. Purcell, H.C. Torrey and R.V. Pound: Phy. Rev. 69, 371, (1946); F. Bloch, O. Hansen and M. Packard: Phy. Rev. 69, 127 (1946); 70, 474 (1946).
49. F. Bloch: Phy. Rev. 70, 460 (1946).
50. R.L. Mössbauer: Z. Physik. (1958) 124, Mössbauer Spectroscopy: L.F. Greenwood and T.C. Gibb, Chapman and Hall Ltd. London (1971).
51. E. Karlson, E. Matthias, K. Siegbahn: Perturbed Angular Correlations; North Holland Publishing Company, Amsterdam (1964).
52. E.J. Zavoisky: J. Phy U.S.S.R. 10, 197 (1946) Cummerow, O. Halliday and C.E. Moore: Phy. Rev. 75, 148 (1949).
53. G. Breit and I.I. Rabii: Phy. Rev. 38, 20822 (1931). H. Taub and P. Kush: Phy. Rev. 75, 148 (1949).
54. E. Segre: Experimental Nuclear Physics V.1; John Wiley & Sons, Inc., New York (1959).
55. H.H. Speidel et al. (12 authors): Phy. Rev. C.V. 57, 2181 (1998).
56. A.E. Kavka et al.: (17 authors): Nuclear Physics A 593, 1711 (1995).
57. R. Inglass: Phy. Rev. (1964), 133, 1787 (1964), C.E. Johnson, Proc. Phy. Society, (1966), 88; 943, A.S. Nozek and M. Kaplan: Phy. Rev. (1967), 159, 273: Mössbauer Spectroscopy, L.E. Greenwood and T.C. Gibb; Chapman and Hall; Ltd. (London (1971).
58. Goldansky, and R.H. Herber: Application of Mössbauer Effect in Nuclear Physics; Academic Press, New York (1968).
59. L. Hasselgren; C. Fahlander: F. Falk ; L.O. Edvardson, J.E. Thum and B.S. Ghuman: Nuclear Physics. A 264, 391 (1976).
60. G. Breit et al.: Phy. Rev. 100, 942 (1959).
61. K.A. Alder, E. Matthias, W. Schneider and R.M. Steffan: Phy. Rev. 129, 1199 (1963); J.R. Gabriel and S.L. Ruby: Nuclear Inst., Methods (1965) 36, 23; Ibid 70, 209, (1969).
62. J.L. Gabriel and S.L. Ruby: Nuclear Instruments Methods (1965), 36, 23: (1969), 79, 209.
63. E.W. Otten: Nuclear Physics A 354, 471 C, (1998).
64. Hinfurtner, H. Ratai, G. Seewald, E. Hagn and E. Zech and R. Eder: Phy. Rev. C. V. 57, 2165 (1998).
65. S. Ohya, K. Nisihimura, N. Okabe, and N. Mutsuso: Hyperfine Interactions, 22, 585 (1985). R. Eder, E. Hagn and E. Zech: Phy. Rev. Lett. B. 158, 371 (1985).
66. H. Bateman: Phil. Mag. 20, 704 (1910): H. Knauss: Science 107.
67. H.S. Hans and S.P. Puri: Mechanics; Tata McGraw-Hill, (1994); H. Goldstein: Classical Mechanics, Addison-Wesley Publishing Co., London (1959).
68. See Ref. (7): Also Table of Isotopes. M. Lederer, J.M. Hollander, and I. Perlman, John Wiley & Sons (N.Y.) (1967).
69. W.A. Schier and C.P. Browne: Phy. Rev. 138, B857 (1965).
70. F. Norman Ramsay: Nuclear Moments; John Wiley & Sons (1953).
71. C.H. Townes: Handbuch der Physik XXXVIII/1 P1 (Springer-Verlag) (1958); Nuclear Data Sheets and Table of Isotopes, John Wiley & Sons, New York (1986). J. Varma and C.E. Mandeville: Phy. Rev. 97, P. 977 (1955).

PROBLEMS

1. When Ni^{60} is bombarded with He^4; Ni^{60} (*a, pn*) Cu^{62} reaction takes place, forming Cu^{62} which decays through β^+ with a half-period of 10 minutes. The nucleus emits a β^+ spectrum with E_{max} = 2.91 MeV followed by gamma-rays. The neutral atomic mass excesses; $\Delta = M - A$ are (1). Ni^{60}, $\Delta = -64.471$ (2) Cu^{62} $\Delta = -62.81$ (3) Cu^{63}, $\Delta = -65.583$.
   (*i*) Find the *Q*-value of Ni^{60} (α, *pn*) Cu^{62}.
   (*ii*) Find the separation energy of last neutron in $_{30}Zn^{64}$, and compare it with the binding energy per nucleon.
   (*iii*) Determine the increase in the total nuclear binding energy, when one neutron is added to Cu^{62}.
   (*iv*) If the pairing energy (*A, Z*) is given by Eq. 2.22: what is expected to be the mass of Zn^{64}.
2. Using the nuclear density expression $\rho(r) = \rho/1 + \exp{(r - R_0)/a}$, obtain F(*q*), from Eqs. 2.48 and 2.50; and hence , the experimental cross-section $\sigma(\theta)$, using the expression of Eq. 2.43 for $\sigma(\theta)$ (point).
3. From the Weizäcker mass formula, explain, the difference of energy relationship of mirror nuclei, ($7N_8^{15}$, and $8O_7^{15}$) ; and ($8O_9^{17}$ and $9F_8^{17}$).
4. Write down, the single-particle wave function for lf 7/2, *i.e.* (*J* = 7/2, *L* = 3, *S* = 1/2) and lf 5/2, *i.e.* (*J* = 5/2, *L* = 3, *S* = 1/2). Show by evaluation of $\int \psi_1{}^* \psi_2 \, d\tau$, that these two wave functions are orthogonal to each other.
5. Compare and contrast the Mössbauer and Perturbed angular correlation (PAC) method, to measure the magnetic moments of the excited states of nuclei. Which method is applicable for higher excited states and why?
6. Why don't the nuclei have permanent electric dipole moments; but have magnetic dipole moments, while the molecules can have both electric and magnetic dipole moments? Give the arguments in details.
7. What is the effect of quadrupole moment of the excited state, (*i*) on the Mössbauer spectrum and (*ii*) on the Differential Perturbed Angular Correlation (see ' Perturbed Angular Correlations' E. Karleson; E. Mathias, and Kai Siegbahn; North Holland (1964)).[51]
8. The half life of alpha-decaying nucleus of Pu^{239} is determined by increasing the rate of evaporation of the liquid. The evaporation rate corresponded to a power supply of 0.231 Watts. Calculate, to the nearest hundred years, the half life of Pu^{239}, given that the energy of its decay alpha-particles is 5.144 MeV. (Take into account the recoil energy of the product nucleus).
9. Calculate the kinetic energies and linear momenta of the product nuclei in Ni^{60} (*d* , *n*) Cu^{61} reaction when neutrons are emitted at 45° to the direction of the incoming beam of 100 MeV deuterons.
10. Construct a momentum diagram and find the internal momentum of the proton in the deuteron, at the moment of stripping. What is the probability of finding the proton in the deuteron with this momentum? (Atomic masses are: $_1H^1$: 1.008145; H^2: 2.014640; $_8O^{17}$: 17.004534).
11. (*i*) Show that the quadrupole moment *Q* of a uniformly chargely ellipsoid is,
$$2/5\, Z\,(b^2 - a^2).$$
   (*ii*) Show that the elliptical parameter η is related to the ratio of the semi-axes (*b*/*a*) of the ellipsoid; as:
$$\eta = 2\,\frac{(b/a) - 1}{(b/a) + 1}$$

3

CHAPTER

Bound State Problem: The Deuteron

The theoretical understanding of the various physical properties of nuclei, requires the detailed knowledge of the nuclear forces; operating between nucleons of a nucleus at energies of, say up to a couple of hundred MeV. Beyond this energy, the excited states lose their discrete character; and also meson production sets in; which takes us into a new field, away from the nuclear structure aspects of nuclear physics. Though there are indications of many-body forces in nucleon-nucleon interactions in the nucleus, they are believed to be small. The two-body forces in nucleon-nucleon interactions play the dominant role.

This requires us to understand, in depth, the two-body nucleon-nucleon forces both for free nucleon-nucleon system as well as for complex nuclei. Because of the effect of other nucleons, arising out of the various restrictions of Pauli Exclusion Principle and many-body forces, it is expected that one may require somewhat different effective nucleon-nucleon interaction in the complex nuclei, compared to the case in free space. Attempts have, however, been made to explain the effective nucleon-nucleon interaction, with the help of self-consistent procedures of solving the many-body problem of the nucleus using the basic free nucleon-nucleon interaction. It is, therefore, important to understand the free two-body nucleon-nucleon forces.

The two-nucleon systems available are (*i*) The deuteron consisting of a proton and a neutron in an abound state and (*ii*) *n-p* and *p-p* scattering systems at different incident energies. Except the deuteron, there is no other nucleon-nucleon bound system in nature. On the other hand, the data on *n-p* and *p-p* scattering, both at low and medium energies, provides information about the unbound positive states of the nucleon-nucleon system.

The deuteron is the nucleus of deuterium or heavy hydrogen, first discovered in 1932 by Urey[1]. Other nucleon-nucleon systems i.e., the proton-proton (di-proton) or neutron-neutron (di-neutron), are not found in the bound state in nature and hence their properties can only be studied through scattering experiments, as we shall see in the next two Chapters. On the other hand, the properties of deuteron as a stable bound system, have been studied very precisely and in details, and one of the most accurate measured set of quantities in nuclear physics.

3.1 PHYSICAL PROPERTIES OF DEUTERON

1. *Mass (M)*: According to the latest estimate[2], the mass of the deuterium atom has been found to be 2.0147 atomic mass units (AMU) on physical scale. Taking into account the electron mass m_e and the binding energy of electron, in the deuterium atom: we get the mass of deuteron as: $M_d = 2.0136$ A.M.U.

2. *Binding Energy (B.E.)* of the deuteron is given by[3]:

$$B.E. = (M_p + M_n - M_d)\, c^2 = 2.226 \text{ MeV} \qquad ...(3.1)$$

where M_p, M_n and M_d are the masses of proton, neutron and deuteron respectively. The most recently adopted masses of proton[4] and neutron[5] are:

$$M_p = 1.0072766 \text{ A.M.U.} = 938.2796 \text{ MeV}$$

$$M_n = 1.0086654 \text{ A.M.U.} = 939.573 \text{ MeV}$$

Most recent attempts for measuring the binding energy use the process of capture of neutron by proton, to form a deuteron. One allows the thermal neutrons (energy of the order of 0.025 MeV), to be incident on protons, and measures the energy $h\nu$ of the capture gamma rays, obeying the energy equation:

$$_1H^1 + _0n^1 \longrightarrow _1D^2 + h\nu \qquad ...(3.2)$$

Obviously the value of $h\nu$ gives the binding energy, if the very small kinetic energy of the neutron or deuteron-recoil is neglected. The energy of the gamma rays has been measured very accurately by high resolution gamma rays spectrometers[6].

3. *Spin or Total Angular Momentum (I)*: It is now well established[7], that the deuteron has a total angular momentum of $1\hbar$ in the ground state. It was measured by orthohydrogen to parahydrogen conversion method.

4. *Parity*: The parity of the deuteron ground state wave function is inferred to be positive from the fact that only $l = 0$ and $l = 2$ are effective, for which parity is positive[8].

5. *Magnetic Moment* (μ_d): The presently accepted value of magnetic moment[9] measured by the molecular beam method, is:

$$\mu_d = 0.8574376 \pm 0.000009 \text{ n m} \qquad ...(3.3)$$

6. *Electric Quadrupole Moment* (Q_d): The fact that the deuteron has an electric quadrupole moment, was first revealed by the experiments[10] of Auffray J.P. These and other measurements have shown that the quadrupole moment of deuteron is positive and the most accepted value based on the measurements of Glenderning N.K. and Krammer[11] is given by:

$$Q_d = (2.875 \pm 0.00002) \times 10^{-27} \text{cm}^2 \qquad ...(3.4)$$

The value of the angular momentum of the deuteron and positive parity, suggests that the proton and neutron are oriented with their intrinsic spins predominantly in the same direction and that their relative orbital angular momentum $1\hbar$ is equal to zero in the ground state of deuteron. The magnetic moment, on the basis of this assumption will simply be the algebraic sum of the intrinsic magnetic moments of protons and neutrons *i.e.*,

$$\mu_d = \mu_p + \mu_n \qquad ...(3.5a)$$

where $$\mu_p = 2.7928 \pm 0.000001 \text{ nm}^{12} \qquad ...(3.5b)$$

and $$\mu_n = -1.913042 \pm 0.000001 \text{ nm}^{13} \qquad ...(3.5c)$$

From which one gets $\mu_d = 0.8798$ nm which is approximately the same as the measured value of 0.8574376 nm. The difference between the experimental and theoretical values is, however, much larger than the errors involved. This, therefore, require the modification of the above theoretical assumption that the ground state of deuteron contains only $l = 0$ state. Further, the fact that in the ground state; the deuteron has a definite quardrupole moment, also suggests that besides $l = 0$ there should exist higher angular momenta in the ground state. The value of $l = 0$ alone will give a wave function without any angular-dependence; leading to a spherical shape of the deuteron and hence zero quadrupole moment. The detailed calcuIations (as shown subsequently) show that though $l = 0$ wave function is very predominant (about 96%) in the ground state deuteron, admixture of about 4% of $l = 2$ state exists. Odd 1 values are not allowed since the parity of the ground state is measured to be positive.

3.2 THE GROUND STATE OF DEUTERON (SOLUTION FOR $l = 0$)

Since, the measured quadrupole moment of deuteron is small, we assume to the first approximation, that the deuteron is spherical with $l = 0$. The assumption of pure $l = 0$ state for the ground state requires that the potential $V(r)$ between the neutron and the proton is central *i.e.*, it depends only on the distance between them and not on angle θ or ϕ. This follows from the fact that for $l = 0$, the angular part of the wave function. $Y_0^0 = 1/\sqrt{4\pi}$ is independent of angles θ and ϕ. The two-body problem may be reduced to a problem of single particle of reduced mass μ moving in a central potential $V(r)$ where,

$$\frac{1}{\mu} = \frac{1}{M_p} + \frac{1}{M_n} \approx \frac{2}{M} \qquad ...(3.6)$$

[Reduced mass μ of a single body problem should not be confused with μ_d; the magnetic moment of deuteron, discussed in Eq. 3.5].

Here $M \approx M_p \approx M_n$ is the mass of the nucleon. The Schrödinger equation for the centre of mass may, then, be written as:

$$-\frac{\hbar^2}{2\mu}\nabla^2 \psi + V(r)\,\psi = E\psi \qquad ...(3.7)$$

where r is the distance between the centres of the two nucleons and E is the total energy of the system, which for the ground state of the deuteron will be negative of the binding energy of the deuteron. The potential $V(r)$ may have any one of the forms; given below:

(*i*) Square Well: $V(r) = -V_0$ for $r < r_0$ ...(3.8*a*)

$= 0$ for $r > r_0$

(*ii*) Gaussian: $V(r) = -V_0 \exp\left(-\frac{r^2}{\alpha^2}\right)$ ...(3.8*b*) (*iii*) Exponential: $V(r) = -V_0 \exp\left(-\frac{r}{\alpha}\right)$ ...(3.8*c*)

(*iv*) Yukawa: $V(r) = -V_0 \exp\frac{(-r/\alpha)}{r/\alpha}$ ...(3.8*d*)

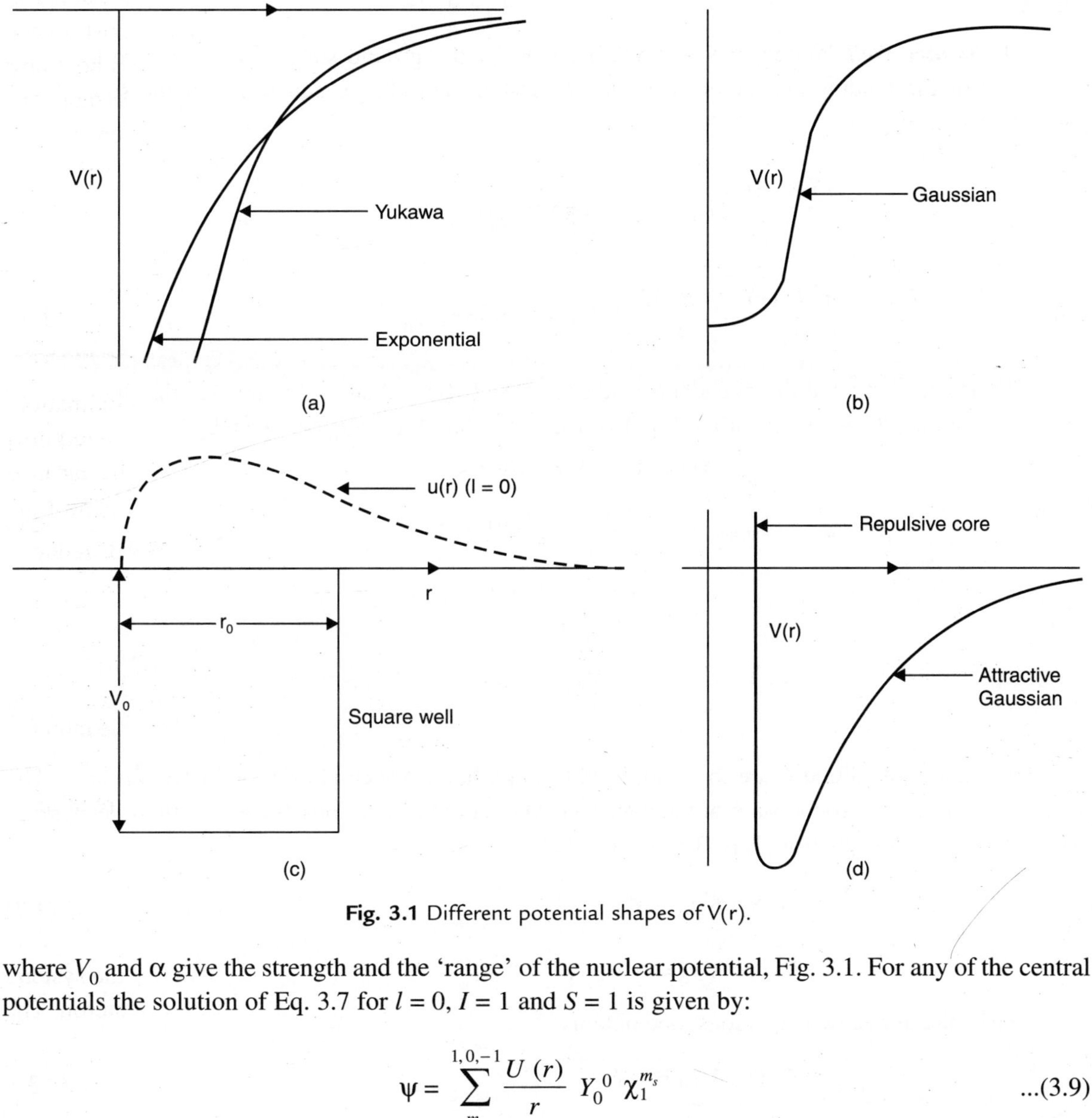

Fig. 3.1 Different potential shapes of V(r).

where V_0 and α give the strength and the 'range' of the nuclear potential, Fig. 3.1. For any of the central potentials the solution of Eq. 3.7 for $l = 0$, $I = 1$ and $S = 1$ is given by:

$$\psi = \sum_{m_s}^{1,0,-1} \frac{U(r)}{r} Y_0^0 \chi_1^{m_s} \qquad ...(3.9)$$

where $\chi_1^{m_s}$ is the spin part of the wave function. The function $U(r)$ obeys the radial part of the Schrödinger equation:

$$\frac{d^2 U(r)}{dr^2} + \frac{M}{\hbar^2}[E - V(r)]\, U(r) = 0 \qquad ...(3.10)$$

The solution of Eq. 3.10 for square well and exponential shapes of $V(r)$, can be obtained analytically, while for other potentials, only numerical solutions are possible. Here we will deal only with square well and Yukawa potentials. (For a complete discussion see Handbuch[14] der Physik V.XXXIX (1958).

A. Square Well Potential: Remembering that for the ground state $E = -|B|$ where $|B|$ is the magnitude of the binding energy of deuteron, the radial Schrödinger equations for the square well potential become:

$$\frac{d^2 U(r)}{dr^2} + \frac{M}{\hbar^2}(V_0 - |B|)\, U(r) = 0 \quad \text{for } r < r_0 \qquad ...(3.11a)$$

and

$$\frac{d^2 U(r)}{dr^2} + \frac{|B|M}{\hbar^2}\, U(r) = 0 \quad \text{for } r > r_0 \qquad ...(3.11b)$$

Taking into account the boundary conditions that $U(r) = 0$ at $r = 0$; and also that it is zero as $r \to \infty$ (Fig. 3.1*d*); the solutions of the Eq. 3.10 for $r < r_0$; and for $r > r_0$ are given by:

$$U_{\text{inside}}(r) = A \sin K_g r \quad \text{for } r < r_0 \qquad ...(3.12a)$$

$$U_{\text{outside}}(r) = Ce^{-\gamma_g r} \quad \text{for } r > r_0 \qquad ...(3.12b)$$

where

$$K_g = \frac{\sqrt{M(V_0 - |B|)}}{\hbar} \qquad ...(3.13a)$$

and

$$\gamma_g \equiv \frac{\sqrt{M|B|}}{\hbar} \qquad ...(3.13b)$$

The constants A and C are to be adjusted through boundary conditions and normalisation. The suffix 'g' denotes the ground state of the deuteron. Matching the solutions $U(r)$ and their derivatives dU/dr, at $r = r_0$, one gets from Eqs. 3.12*a* and 3.12*b*.

$$A \sin K_g r_0 = Ce^{-\gamma_g r_0} \qquad ...(3.14a)$$

$$A K_g \cos K_g r_0 = -C\gamma_g e^{-\gamma_g r_0} \qquad ...(3.14b)$$

From the above two equations, one obtains,

$$K_g \cot K_g r_0 = -\gamma_g \qquad ...(3.14c)$$

$$\frac{-K_g}{\gamma_g} = \tan K_g r_0$$

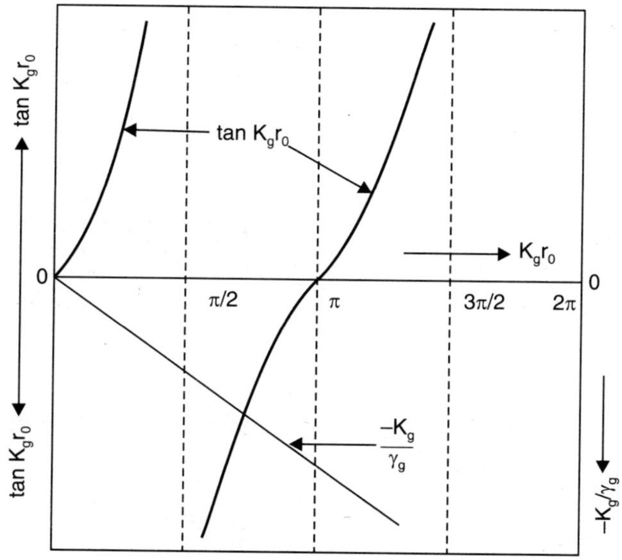

Fig. 3.2 Intersections of a graph of tan K_gr_0 versus K_gr_0, with the graph of $-K_g/\gamma_g$ versus K_gr_0. Intersection point corresponds to $\frac{\pi}{2} \leq K_gr_0 \leq \pi$.

Using Eqs. 3.13*a* and 3.13*b*, Eq. 3.14*c* may be alternatively written as:

$$K_gr_0 = \left[\text{Arc}\tan\left(-\frac{K_g}{\gamma_g}\right)\right] \qquad ...(3.15a)$$

Further, we define

$$K_0^2 \equiv \frac{M V_0}{\hbar^2} \qquad ...(3.15b)$$

So that,

$$K_0^2 \equiv K_g^2 + \gamma_g^2 \qquad ...(3.15c)$$

Equation 3.14 together with Eq. 3.15 gives a relationship between r_0 and V_0. The separate values of r_0 and V_0, of course, cannot be found from these relationships.

If we plot tan K_gr_0 versus K_gr_0 and also plot on the same graph $-K_g/\gamma_g$ versus K_gr_0, the first intersection will give the ground state while the second intersection will give the Ist excited state and so on. We will see from this curve that for the ground state of the deuteron, [Fig. 3.2]:

$$\frac{\pi}{2} \leq K_gr_0 \leq \pi \qquad ...(3.16)$$

We can get an order of magnitude of the values of r_0 and V_0 from the following approximate considerations. In Fig. 3.3*a* where the matching of the wave function for $l = 0$, is shown at $r = r_0$ for a square well potential, it is evident that the insider wave function should have just crossed the peak to match with the exponentially decaying outside wave functions. We might realise here that this is true only for $l = 0$ solution; contribution from $l = 2$, forces the matching of the wave function at a value of $r < r_0$. This requires that:

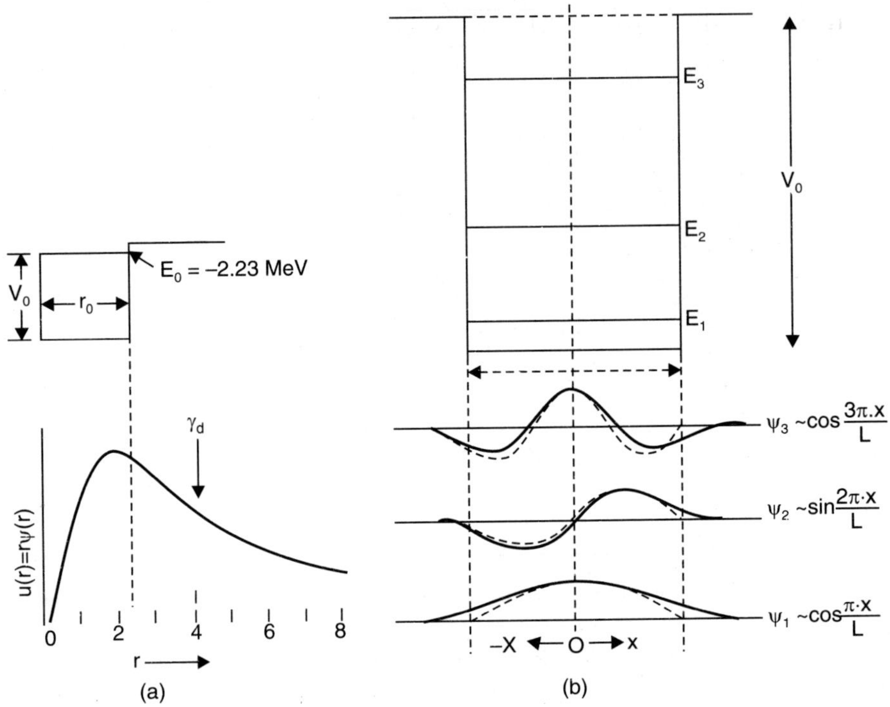

Fig. 3.3 Eigen-energies and eigen functions of deuteron in square well potential.

$$r_0 \approx \frac{\lambda}{4} = \frac{1}{4}\frac{2\pi}{K_g} = \frac{2\pi\hbar}{4}\,[M\,(V_0 - |B|)]^{-\frac{1}{2}} \qquad \text{...(3.17)}$$

If we assume that $|B| << V_0$, one may write:

$$r_0^2\,\overline{V_0} = \frac{\pi^2\,\hbar^2}{4\,M} = 1.44 \times 10^{-24}\ \text{MeV-cm}^2 \qquad \text{...(3.18)}$$

where $\overline{V_0}$ denotes the nuclear potential between neutron and proton, in the bound state assuming the binding energy to be zero.

In Figure 3.3*b*, we have shown, for a general case, (dashed lines) the values of ψ for various values of *n*, related to *L* and *k* approximate conditions of $kL = n\pi$, and $\psi = 0$ at $x = \pm L/2$. In Fig. 3.3*b*, for the first state *i.e.*, for $n = 1$, we have shown in dashed lines $U(r) = r\,\psi$, for $V_0 \approx 36$ MeV. The solid lines for ψ are for a more correct solution in Fig. 3.3*b*.

An estimate of r_0 and V_0 separately may be made by combining Eq. 3.14 or 3.18, with the following set of equations:

(*i*) The normalisation of the total wave function for deuteron requires that:

$$\int_0^\infty U^2(r)\,dr = 1$$

or

$$\int_0^{r_0} (A \sin K_g\,r)^2\,dr + \int_{r_0}^\infty \left[C e^{-\gamma_g r}\right]^2 dr = 1 \qquad \text{...(3.19)}$$

Also using Eqs. 3.14*a*, 3.14*c* and 3.15*c* one gets:

$$A \sin K_g r_0 = A\,(1 + \cot^2 K_g r_0)^{-\frac{1}{2}}$$

$$= \frac{AK_g}{(K_g^2 + \gamma_g^2)^{\frac{1}{2}}} = Ce^{-\gamma_g r_0}$$

or

$$\frac{AK_g}{K_0} = Ce^{-\gamma_g r_0} \qquad ...(3.20)$$

Using Eqs. 3.19 and 3.20 one can get:

$$A = \left[\frac{2\gamma_g}{1+\gamma_g r_0}\right]^{\frac{1}{2}} \qquad ...(3.21a)$$

$$C = \frac{K_g}{K_0}\left[\frac{2\gamma_g}{1+\gamma_g r_0}\right] e^{\gamma_g r_0} \qquad ...(3.21b)$$

(*ii*) The square of the radius of the deuteron, R^2, may be expressed as:

$$R^2 \equiv \langle r_d^2 \rangle = \int \psi^* \, r^2 \, \psi \, dr$$

$$= \int_0^{r_0} (A^2 r^2 \sin^2 K_g r)\, dr + \int_{r_0}^{\infty} C^2 r^2 e^{-2\gamma_g r}\, dr \qquad ...(3.22)$$

An estimate of R may be made by defining it as the value of r at which the radial wave function outside the nuclear potential reduces to $1/e$ of its maximum value C, see Fig. 3.3*a* and Eq. 3.14*a*. This is obviously given by ($R \equiv r_d$):

$$R = \frac{1}{\gamma_g} = \frac{\hbar}{\sqrt{M\,|\,B\,|}} = 4.31 \times 10^{-13}\ \text{cm} \qquad ...(3.23)$$

The root mean square (r.m.s.) distance between neutron and proton, has been obtained from scattering experiments of neutron from protons[15], with incident energy of the order of $|B|$, which corresponds to the relative kinetic energy of neutron and proton in a deuteron. This value may, therefore, be taken as equal to the root mean square radius of the deuteron. It comes out to be 4.2×10^{-13} cm, which is nearly the same as given in Eq. 3.23.

Using this value of R in Eq. 3.22 and Eq. 3.14 it is possible to obtain sets of r_0 and V_0. Typically, if $r_0 = 2 \times 10^{-13}$ cm; then $V_0 = 36$ MeV, for a square well potential.

B. Yukawa and Other Potentials: Though the square well potential Eq. 3.8*a* is easier to handle analytically, this potential is not expected to be physical for nucleon-nucleon interaction. In nature, even for short ranges, the potentials are not expected to be so sharply cut-off. They are expected to have the smooth variation as exists in exponential, Gaussian or Yukawa potentials, as written in Eq. 3.8*d*. Among these, Yukawa potential is physically more plausible, because of its origin in the well-accepted meson exchange theory of Yukawa. But, then deuteron problem for such a potential cannot be solved analytically.

Numerical and some approximate equivalent analytical methods have been developed to solve the deuteron problem with Yukawa and other potentials, with diffused edges. In one of the attempts, Húlthen[16] has used a potential of the type:

$$V(r) = -V_0\left[\exp.-\left(\frac{r}{R}\right) - \exp.\frac{-r}{\alpha}\right] \qquad ...(3.24)$$

It resembles Yukawa potential, Eq. 3.8*d*. Analytical solution of such a potential has been obtained[16]. For details see Feshbach H and V. F, Weisskopf, Phy. Rev. 76,1550, (1949), also Gartenhaus. Phy.Rev. 100,903, (1955).[16]

By numerical analysis, it is possible to obtain a relationship between $K_g r_0$ and $(K_0 r_0)^2$ for the above potential. It is interesting to notice that for the same binding energy, one requires a shallower potential in this case, compared to square well. Similar will be the case with other potentials, having diffused edges.

3.3 EXCITED STATES OF DEUTERON

Experimentally, we know that there are no excited bound states of deuteron. As we shall see in the next chapter, there is at ≈ 60 KeV, a quasi-bound state observed in *n-p* scattering, corresponding to $l = 0$ but with neutron and proton having their spin antiparallel, and hence having $I = 0$.

But we can also see, from theoretical considerations, that for the same potential well V_0, as for explaining the ground state, it is not possible to have a bound excited state. We have already seen in Eq. 3.14 that for the ground state:

$$K_g \cot K_g r_0 = -\gamma_g$$

or

$$-\tan K_g r_0 = \frac{K_g}{\gamma_g}$$

and also from Eq. 3.15*c*:

$$\frac{K_g}{\gamma_g} = \left[\frac{K_0^2}{\gamma_g^2} - 1\right]^{\frac{1}{2}} \approx \left[\frac{V_0}{|B|} - 1\right] \qquad ...(3.25)$$

Therefore $-\tan K_g r_0 >> 1$.

We see from this, that the lowest solution, corresponding to the ground state demands that :

$$\frac{\pi}{2} \le K_g r_0 \le \pi$$

Which is the same as expected and given in Eq. 3.16.

Now if an excited state of the deuteron exists, it should be possible to obtain for the same V_0, a non-contradictory condition for the excited state. Such a state can be for (*i*) $I = 0$, $l = 0$, $S = 0$, (*ii*) $I = 1$, $l = 1$, $S = 0$ (*iii*) $I = 1$, $l = 0$, $S = 1$, etc. But for all these states V_0 should be the same . In general, it can be shown[17] that for any value of $l > 0$, the condition:

$$j_{l-1}(K_0 r_0) \approx 0 \qquad ...(3.26)$$

is satisfied. So for $l = 1$,

$$j_0 (K_0 r_0) = \frac{\sin K_0 r_0}{K_0 r_0} \approx 0 \qquad ...(3.27)$$

and hence $K_0 r_0$ can be found out.

Then,
$$K_0 r_0 = \pm \pi, \pm 3\pi, ... \qquad ...(3.28)$$

and hence
$$K_0^2 r_0^2 = \pi^2, 9\pi^2, \text{etc.} \qquad ...(3.29)$$

so that
$$V_0 = \frac{\pi^2 \hbar^2}{M r_0^2}, \text{ etc.}$$

when
$$|V_0| >> E \qquad ...(3.30)$$

This gives $V_0 = 144$ MeV, for $r_0 \approx 2 \times 10^{-13}$ cm. This is in contradiction of $V_0 = 36$ MeV required for $l = 0$, with the same value of r_0. As a matter of fact, for higher l, the value of V_0 is still higher.

It can also be seen from Eq. 3.14c, that the first excited state corresponding to K_ε for $l = 0$ should correspond to the next root above K_g. This corresponds to the condition:

$$K_\varepsilon^2 \cot K_\varepsilon r_0 = -\gamma_\varepsilon \qquad ...(3.31a)$$

where $K_\varepsilon^2 = K_0^2 - \gamma_\varepsilon$ using $K_0^2 = MV_0 / \hbar^2$ in Eq. 3.15b. Then realising that $K_\varepsilon / \gamma_\varepsilon$ will now lie in the fourth quadrant; compared to K_g / γ_g of Eq. 3.14c lying in the second quadrant, we can write:

$$\frac{3\pi}{2} \leq K_\varepsilon r_0 \leq 2\pi \qquad ...(3.31b)$$

On the other hand, $K_g r_0 < \pi$, from Eq. 3.16 for the ground state. From Eqs. 3.14, 3.15 and 3.16, it is easy to see then that:

$$K_0^2 r_0^2 - \pi^2 \leq \gamma_g^2 r_0^2 \qquad ...(3.32a)$$

If both Eqs. 3.31b and 3.32a are true, then

$$K_\varepsilon^2 r_0^2 \geq \frac{9\pi^2}{4}$$

or
$$K_0^2 r_0^2 - \gamma_\varepsilon^2 r_0^2 \geq \frac{9\pi^2}{4}$$

or
$$K_0^2 r_0^2 - \pi^2 \geq \frac{5\pi^2}{4} + \gamma_\varepsilon^2 r_0^2$$

i.e.,
$$K_0^2 r_0^2 \geq \frac{5\pi^2}{4}$$

Then from Eq. 3.15c, $\gamma_g^2 r_0^2 \geq \dfrac{5\pi^2}{4}$ ...(3.32b)

or from Eq. 3.23; $r_0 > R$. This is physically not possible.

We therefore, conclude that it is, theoretically, not possible to have a bound excited state of deuteron, both for $I = 0$ and $l \neq 0$ for the same nuclear potential.

3.4 THE QUADRUPOLE MOMENT OF THE DEUTERON AND TENSOR FORCES

From the experimental value of the quadrupole moment of the deuteron *i.e.*, $Q = 2.875 \times 10^{-27}\ \text{cm}^2$, it is possible to estimate the shape of the deuteron. Classically, the quadrupole moment of a uniformly

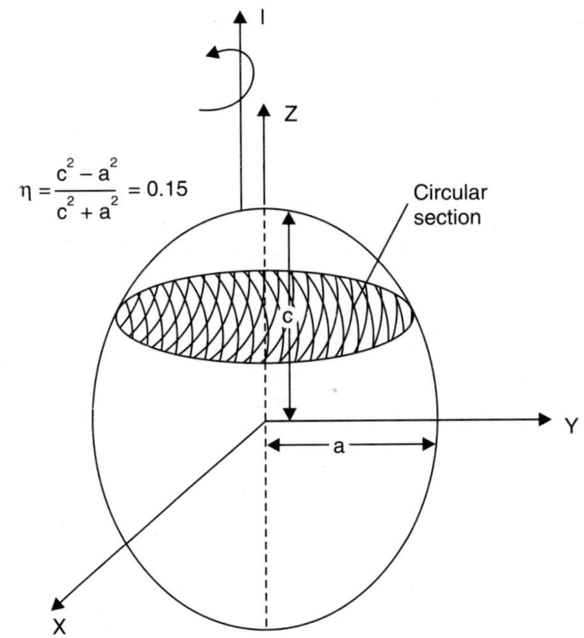

Fig. 3.4 Shape of the deuteron, which is prolate (Cigar shape), with major axis c along 1.

charged body with charge *Ze* may be expressed as:

$$Q = \frac{2}{5}\ Ze\,(c^2 - a^2) = \frac{4}{5}\ Ze\,\eta\,R^2 \qquad ...(3.36)$$

where eccentricity $\eta \equiv (c^2 - a^2)/\,(c^2 + a^2)$ and equivalent radius, $R = (c^2 + a^2)/\,2$; c and a being the two semi-axes of the ellipsoid and *Ze* is the total charge on the body, Fig. 3.4. Using $R = 4.2 \times 10^{-13}$ cm, one gets $\eta = 0.15$, for the experimentally measured value of Q, for deuteron.

This shows, that the deviation from the spherical shape is small. As the measured value of Q, represents the measurement of Q along the direction of the total angular momentum; the positive value of Q implies that the major axis with lenght $2c$ points along I-axis and is larger than $2a$. The deuteron is, therefore, prolate or cigar-shaped—its longer axis pointing along I.

As briefly mentioned in Section 3.1, the existence of the quadrupole moment for deuteron requires, besides $l = 0$; higher values of l for the ground state of deuteron. A non-central potential, between the neutron and proton in deuteron, in which l is not conserved, meets this requirement.

How do we write such a potential? What are the constraints for such a potential? From general considerations of invariance properties for any potential, we should demand that;

(*i*) The potential should be scalar; which means that it should be invariant under the operation of reflection and rotation of the coordinate system used to describe the relative motion of the particles. This is required, since potential is energy, which is a scalar quantity. Such a potential conserves parity, which condition is assumed for nuclear forces. The other possibility is that the potential is pseudo-scalar *i.e.*, it changes sign on reflection. But a pseudo-scalar potential leads to non-conservation of parity which is not expected of the nuclear forces.

(*ii*) The nucleon-nucleon potential should be invariant under time reversal. This is believed to be a universal law in classical and quantum mechanics and leads to the conservation of total energy.

(*iii*) We assume for this discussion, that the nuclear forces are velocity-independent. As a first approximation, this assumption may be taken as correct.

(*iv*) Such non-central potential should be built out of the separation vector **r** and the spin vectors σ_1 and σ_2 for the two nucleons, in such a manner, that the above laws are obeyed. Spin of nucleons is given by $\mathbf{S}_{1,2} = -\hbar\sigma_{1,2}$ where $\sigma = (\sigma_x, \sigma_y, \sigma_z)$ are the usual triple Pauli spin matrices. Because of the condition of 'no velocity dependence; the potential should not depend on the relative linear momentum of the particles. One may now examine the various possibilities:

(*a*) While **r** is a vector, σ_1 and σ_2 are axial vectors *i.e.*, though they have a direction, their signs do not change on reflection. This means that expressions like $\sigma_1 \cdot \mathbf{r}$ and $\sigma_2 \cdot \mathbf{r}$ are pseudo-scalar *i.e.*, they do not have any direction associated with them but change sign on reflection. Same is the case with $\sigma \,.\, \mathbf{p}$. Such terms, therefore, do not give the required form of the potential.

(*b*) Similarly $(\sigma_1 \times \sigma_2)$ is a pseudo-vector.

(*c*) $(\sigma_1 \cdot \mathbf{r})^2$ or $(\sigma_2 \cdot \mathbf{r})^2$ or r^2, are, on the other hand, scalar but lead to the central potential, where,

$$\mathbf{S} = \mathbf{S}_1 + \mathbf{S}_2 = \frac{1}{2}\hbar\,(\sigma_1 + \sigma_2)$$

is the total spin vector. Higher powers *e.g.*, $(\mathbf{S} \,.\, \mathbf{r})^n$ where for $n > 2$ may, on the other hand, be reduced to $(\mathbf{S} \cdot \mathbf{r})$ or $(\mathbf{S} \cdot \mathbf{r})^2$ and hence either lead to the same difficulties as mentioned in 'a', or lead to a central potential.

(*d*) Terms like $(\mathbf{r} \cdot \mathbf{p})$, though scalar are excluded; because on time reversal **r** does not change, but the sign of **p** changes—hence the product is not invariant on time reversal. Also as mentioned earlier, any term containing **p** directly, is excluded; because this leads to the velocity dependent nuclear forces. Similarly $\mathbf{s} \cdot \mathbf{r}$ and $\mathbf{I} \cdot \mathbf{r}$ are also not invariant with time reversal and hence are unacceptable. The dependence on I may also be excluded, because that leads to velocity-dependent forces.

(*e*) The term like $(\sigma_1 \cdot \mathbf{r})\,(\sigma_2 \cdot \mathbf{r})$ though scalar, are non-central because it contains the angle between **r** and σ_1 and σ_2. Similarly,

$$(\sigma_1 \times \mathbf{r}) \cdot (\sigma_2 \times \mathbf{r}) = \sigma_1 \cdot \sigma_2 - (\sigma_1 \cdot \mathbf{r})\,(\sigma_2 \cdot \mathbf{r}) \qquad ...(3.37a)$$

is partly scalar and central (the first part) and partly scalar and non-central (the second part). One may, therefore, build a non-central scalar potential or the tensor potential from the expression $(\sigma_1 \cdot \mathbf{r})\,(\sigma_2 \cdot \mathbf{r})$.

(*f*) Other possible terms are:

(*i*) **r** ·**r** ; This leads to only central potential.

(*ii*) (**S** ·**S**); This also leads to central potential.

(*iii*) (**S** ·**S**) (**S** ·**r**)^2, leads to tensor forces similar to one described in Eq. 3.37a.

We conclude from the above that $(\sigma_1 \cdot \mathbf{r})(\sigma_2 \cdot \mathbf{r})$ should necessarily be contained in the expression for a non-central tensor potential.

It should be further demanded that the tensor potential should be such that its average over the directions of **r** vanishes. This condition meets the requirement of the isotropy of space.

It may be easily seen that the space average of $(\sigma_1 \cdot \mathbf{r})(\sigma_2 \cdot \mathbf{r})$ is given by:

$$\frac{\int (\sigma_1 \cdot \mathbf{e})(\sigma_2 \cdot \mathbf{e})\, d\Omega}{\int d\Omega} = \frac{1}{3}\, \sigma_1 \cdot \sigma_2 \qquad ...(3.37b)$$

where $\mathbf{r} = |\, r \,|\, \mathbf{e}$

In summary, one, therefore, introduces a tensor potential V_T such that like:

$$V_T = V_T(r)\, S_{12} \qquad ...(3.38a)$$

where

$$S_{12} \equiv \frac{3(\sigma_1 \cdot \mathbf{r})(\sigma_2 \cdot \mathbf{r})}{r^2} - \sigma_1 \cdot \sigma_2 \qquad ...(3.38b)$$

The potential in Eq. 3.38, meets the requirements mentioned above *i.e.*, (*i*) it is scalar, (*ii*) it is non-central *i.e.,* depends on the relative orientation. (*iii*) It does not depend on the velocities or momenta (**p**) of nucleons (*iv*) It is invariant under time reversal and (*v*) Its average over direction of **r** vanishes.

We may, therefore, write the nucleon-nucleon potential, as applicable to deuteron as:

$$V = V_c(r) + V_T(r)\, S_{12} \qquad ...(3.39)$$

where $V_c(r)$ is the central potential and $V_T = V_T(r)\, S_{12}$ is the tensor potential. We will see, later, that such a potential is a special case of the general expression for nucleon-nucleon potential.

3.4.1 Constants of Motion with Non-central Tensor Forces

The potential V is written on the assumption that the space is isotropic. This leads to the conservation of total angular momentum. Hence **I** = **I** + **S**, the total angular momentum of the deuteron; is a constant of motion.

Like any quantum-mechanical system in a stationary state involving electromagnetic and nuclear forces, the ground state of deuteron has a definite parity, which is positive. The parity will, therefore, also be of a constant motion, under tensor forces, as the parity (π) commutes with S_{12} *i.e.*,

$$(\pi, S_{12}) = (S_{12}, \pi) \text{ or } [\pi, S_{12}] = 0 \qquad ...(3.40)$$

The potential $V_T(r)$ in Eq. 3.38 is not invariant on the rotation of space coordinates and spin coordinates separately; but is invariant on the rotation of both the coordinate systems simultaneously. This leads to the non-conservation of orbital and spin angular moment separately. Hence **I** and **S**, separately are not constant of motion.

Though $\mathbf{S}$ is not a constant of motion in the two-body system with tensor forces, S^2 is still a constant of motion. This may be seen from the fact that S^2 commutes with S_{12} writing:

$$\mathbf{S} = \frac{1}{2}(\sigma_1 + \sigma_2);\ S^2 = \frac{1}{2}(3 + \sigma_1 \cdot \sigma_2) \qquad ...(3.41)$$

It may be useful to write the following relation, on the basis of which Eq. 3.41 is derived:

$$\mathbf{S} = \mathbf{S}_n + \mathbf{S}_p = \frac{1}{2}\sigma_1 + \frac{1}{2}\sigma_2$$

$$|\mathbf{S}|^2 = |\mathbf{S}_n|^2 + |\mathbf{S}_p|^2 + 2\mathbf{S}_n \cdot \mathbf{S}_p$$

$$|\mathbf{S}_n|^2 = |\mathbf{S}_p|^2 = \frac{1}{2}\left(1 + \frac{1}{2}\right) = \frac{3}{4}$$

and

$$2\mathbf{S}_n \cdot \mathbf{S}_p = |\mathbf{S}|^2 - \frac{3}{2}$$

It is easy to see that

$$S^2 S_{12} = S_{12} S^2 \quad \text{or} \quad [S^2, S_{12}] = 0 \qquad ...(3.42)$$

It may be seen, that for a singlet state,

$$\mathbf{S} = \frac{1}{2}(\sigma_1 + \sigma_2) = 0$$

and

$$S^2 = 0 = \frac{1}{2}(3 + \sigma_1 \cdot \sigma_2)$$

and hence

$$\sigma_1 = -\sigma_2$$

Then

$$S_{12} = \frac{3[(\sigma_1 \cdot \mathbf{r})(\sigma_2 \cdot \mathbf{r})]}{r^2} - \sigma_1 \cdot \sigma_2$$

$$= -3 + 3 = 0 \qquad ...(3.43)$$

This also may be seen qualitatively, by noting that in the singlet spin state, there is no preferred direction, hence we expect S_{12} in the singlet state to be zero. This means, that in singlet state, the nuclear forces are only central; and tensor force is zero.

For triplet state *i.e.*, for $S = 1$, there are, three possible values of $\mathbf{I}$ *i.e.*, $\mathbf{I} = \mathbf{I} - \mathbf{S} = 0$, 1 and 2; hence three states are possible *i.e.*, $3S_1$, $3P_1$ and $3D_1$. If parity is conserved, then one can either have only $3P_1$ or $3S_1 + 3D_1$. For positive parity only $3S_1 + 3D_1$ are possible because the parity of a state with an angular momentum $1\hbar$ is given by $(-1)^1$. This will lead to a mixture of $I = 0$ and $l = 2$ state. One can, therefore, write the total wave function ψ as:

$$\psi = \psi_{3S_1} + \psi_{3D_1}$$

$$= \psi_S + \psi_D \qquad ...(3.44)$$

where we have neglected the subscripts for simplicity.

It may be further seen from Eq. 3.38 that if σ_1, σ_2 and $\mathbf{r}$ are in the same direction, then $S_{12} = 2$. If σ_1 and σ_2 are in the same direction, but if both are perpendicular to $\mathbf{r}$, then $S_{12} = -1$ (see Fig. 3.5). It is

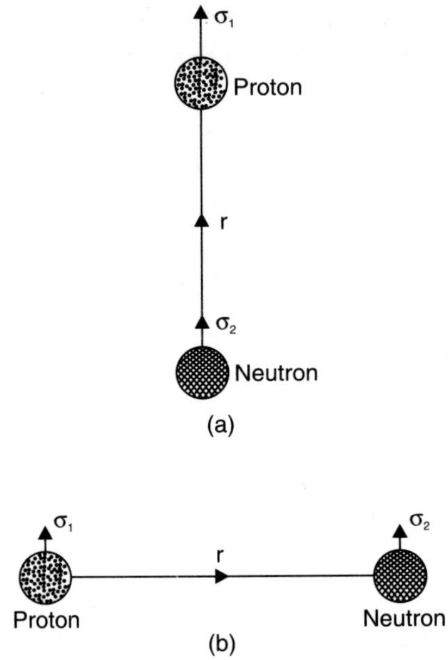

Fig. 3.5 (*a*) When spins of protons and neutrons are parallel *to* **r**; S_{12} = 2 (From Eq. 3.38) and the tensor force is attractive. (*b*) when spins of protons and neutrons are perpendicular to **r**; S_{12} = – 1 from Eq. 3.38 and the tensor force is repulsive.

evident from the figure that $S_{12} = 2$ corresponds to the configuration of deuteron, stretched along the total angular momentum which gives it a positive value of Q_d resulting in prolate shape, and $S_{12} = -1$ corresponds to an oblate shape. We know experimentally that Q_d; the quadrupole moment of the deuteron is positive; and hence the shape of the deuteron is prolate. $S_{12} = 2$; therefore, meets this requirement. From Eq. 3.39, it may, therefore, be concluded that $V_T(r)$ should be negative, to get an overall negative V, a necessary condition for an overall attractive potential.

3.5 SOLUTION OF DEUTERON WITH CENTRAL AND TENSOR POTENTIAL

One writes the Schrödinger equation for the deuteron system as:

$$-\frac{\hbar^2}{M}\nabla^2 \psi_{IMS} + [E - \{V_c(r) + V_T(r)\, S_{12}\}]\, \psi_{IMS} = 0 \qquad ...(3.45)$$

The spin part of ψ_{IMS} is now important, because of the spin dependence of the potential. We will assume ψ_{IMS} as a mixture of $l = 0$ and $l = 2$, only.

Under these conditions, we can write the wave function for the ground state of deuteron as:

$$\psi_{IMS} = \psi_{111} = \sum_{1} \frac{U_1(r)}{r}\, \phi_{IMS}$$

$$= \frac{U_0(r)}{r}\phi_{1101} + \frac{U_2(r)}{r}\phi_{1121} \qquad ...(3.46)$$

The angular part of the wave function *i.e.*, ϕ_{IMIS} is composed of the orbital angular part and spin part in such a manner, as to give the total angular momentum $I = 1$, and $M = 1$. The wave function ϕ_{IMIS} is given by:

$$\phi_{IMIS} = \sum_{M_1, M_S} \langle Is\ M_1\ M_S \mid IM\rangle Y_1^{M_1} \chi_S^{M_S} \qquad ...(3.47)$$

$$M = M_1 + M_S$$

where $Y_1^{M_1}$ and $\chi_S^{M_S}$ are spherical harmonics and spin wave function respectively. The quantity $\langle Is\ M_1\ M_S \mid IM\rangle$ are the Glebsch-Gorden coefficients. For detailed expression of Glebsch-Gorden coefficients and Φ'_s see Blatt and Weisskopf[18] (Page 780) Eqs. 3.45 and 3.46 relate $U_0(r)$ and $U_2(r)$; and the central and tensor potentials. The following operational relationship are useful in writing the coupled equation which is as follows:

$$(i)\ S_{12}\,\chi_0^0 = 0 \qquad ...(3.48a)$$

$$(ii)\ (\sigma_1 \cdot \sigma_2)\,\chi_1^M = \chi_1^M \qquad ...(3.48b)$$

$$(iii)\ S_{12}\,\Phi_{IM01} = \sqrt{8}\,\Phi_{IM21} \qquad ...(3.48c)$$

$$(iv)\ S_{12}\,\Phi_{IM21} = \sqrt{8}\,\Phi_{IM01} - 2\,\Phi_{IM21} \qquad ...(3.48d)$$

These relation may be readily proved; (Appendix). It may also be mentioned that the following normalisation conditions hold good, for the wave functions:

$$(a)\ \int \psi^* \psi\, d\tau = 1 \qquad ...(3.49)$$

Because the Φ'_s are normalised; this leads to

$$(b)\ \int \left[U_0^2(r) + U_2^2(r)\right] dr = 1 \qquad ...(3.50)$$

Equation 3.45 may, now, be written as:

$$-\frac{\hbar^2}{M}\nabla^2\left[\frac{U_0(r)}{r}\Phi_{1101} + \frac{U_2(r)}{r}\Phi_{1121}\right] - \left[E - \{V_0(r) + V_T(r)\,S_{12}\}\right]$$

$$\times\left[\frac{U_0(r)}{r}\Phi_{1101} + \frac{U_2(r)}{r}\Phi_{1121}\right] = 0 \qquad ...(3.51)$$

Operator ∇^2 can be expressed as:

$$\nabla^2 \equiv \frac{1}{r^2}\frac{\partial}{\partial r}\left(\frac{r^2\,\partial}{\partial r}\right) + \frac{1}{r^2 \sin\theta}\frac{\partial}{\partial\theta}\frac{\sin\theta\,\partial}{\partial\theta} + \frac{\partial^2}{r^2 \sin^2\theta\,\partial\phi^2}$$

Using the above relations and the ones in Eq. 3.48; one writes Eq. 3.51 as:

$$\left[\frac{-\hbar^2}{M}\frac{d^2 U_0(r)}{dr^2} + V_c(r)U_0(r)\right]\Phi_{IM01} + \sqrt{8}\, V_T(r)\, U_0(r)\,\Phi_{IM21}$$

$$+\left[\frac{-\hbar^2}{M}\left(\frac{d^2 U_2(r)}{dr^2} + \frac{6\,U_2(r)}{r^2}\right) + (V_c(r) - 2\,V_T(r))\right]\Phi_{IM21}$$

$$+\sqrt{8}\, V_T(r)\, U_2(r)\,\Phi_{IM01} = E\left[U_0(r)\,\Phi_{IM01} + U_2(r)\,\Phi_{IM21}\right] \qquad ...(3.52)$$

Keeping in mind, the relationship in Eq. 3.48, and equating coefficients in Φ_{IM01} and Φ_{IM21} on the two sides of Eq. 3.52; one obtains the following coupled equations:

$$-\frac{\hbar^2}{M}\left(\frac{d^2U_0(r)}{dr^2}\right) + [V_c(r) - E]\,U_0(r) = -\sqrt{8}\;V_T(r)\,U_2(r) \qquad ...(3.53)$$

and

$$-\frac{\hbar^2}{M}\left[\frac{d^2U_2(r)}{dr^2}\right] + \left[\frac{6}{r^2} + V_c(r) - 2\,V_T(r) - E\right]$$

$$= -\sqrt{8}\;V_T(r)\,U_0(r) \qquad ...(3.54)$$

For Eqs. 3.53 and 3.54, no exact analytical solution exists even for square well shapes. They have been solved only numerically.

Attempts have, however, been made to obtain some approximate analytical solutions by variational methods. One can draw some qualitative and some quantitative conclusions about the wave functions and potentials; on the basis of their analysis. Equations 3.53 and 3.54 may be rewritten as:

$$-\frac{\hbar^2}{M}\frac{d^2 U_0(r)}{dr^2} - [V_1(r) - E]\,U_0(r) = 0 \qquad ...(3.55)$$

and

$$-\frac{\hbar^2}{M}\frac{d^2 U_0(r)}{dr^2} + \left[\frac{6}{r^2} + V_2(r) - E\right]U_2(r) = 0 \qquad ...(3.56)$$

where

$$V_1(r) \equiv V_c(r) + \sqrt{8}\;V_T(r)\,\frac{U_2(r)}{U_0(r)} \qquad ...(3.57a)$$

and

$$V_2(r) \equiv V_c(r) - 2\,V_T(r) + \sqrt{8}\;V_T(r)\,\frac{U_0(r)}{U_2(r)} \qquad ...(3.57b)$$

It is clear from Eqs. 3.55, 3.56 and 3.57 that to make $V_1(r)$ and $V_2(r)$ negative; so that proper binding energies are obtained; one can have $V_T(r)$ either positive or negative keeping, of course, $V_c(r)$ always negative. We have, however, seen earlier, that for a positive value of quadrupole moment, $V_T(r)$ should be negative.

Anticipating the expression of Q_d in terms of $U_0(r)$ and $U_2(r)$ as given in Eq. 3.72; we see that a positive Q_d requires, that $U_0(r)$ and $U_2(r)$ should have the same sign. The ratio is about 0.03 from Eqs. 3.72 and 3.75*a*.

As we have seen earlier, the value of the central potential required for $l = 1$, is four times that of the value for 1 = 0. The value for 1 = 2, is expected to be still larger. This means $V_2(r)$ is expected to be at least four times the value of $V_1(r)$. If we put $U_2(r)/U_0(r) = 0.02$, then $V_T(r)$, comes out to be at least as large as $V_c(r)$, from Eqs. 3.55, 3.56 and 3.57. Though $V_c(r)$ could be even repulsive, if one takes $V_T(r)$ as very large and negative, on the basis of these equations alone; the wave functions obtained in this manner do not reproduce the deuteron moments properly.

Hence we conclude that $V_c(r)$ and $V_T(r)$ are both negative; and $U_0(r)$ and $U_2(r)$ have the same signs, with a ratio of $U_2(r)/U_0(r) = 0.03$.

The asymptotic behaviour of $U_0(r)$ and $U_2(r)$ can be analytically calculated for different types of potentials. It may be mentioned here, that the asymptotic wave function is used, [see Eq. 3.73], to estimate the quadrupole moment. Hence the importance of having the exact asymptotic solution.

(*i*) *Square Well*: it is seen that for $r >> r_0$; $V_c(r) = V_T(r) \to 0$ and Eqs. 3.55 and 3.56 are simplified to:

$$-\frac{\hbar^2}{M}\frac{d^2U_0(r)}{dr^2} + EU_0(r) = 0 \qquad \text{...(3.58a)}$$

and

$$-\frac{\hbar^2}{M}\frac{d^2U_2(r)}{dr^2} - \frac{6U_2(r)}{r^2} + EU_2(r) = 0 \qquad \text{...(3.58b)}$$

The solution of these equations, then, acquire the forms:

$$U_0(r) = N_S e^{-r/R} \quad \text{for } r >> r_0 \qquad \text{...(3.59a)}$$

and

$$U_2(r) = N_D e^{-r/R}\left[1 + \frac{3R}{r} + \frac{3R^2}{r^2}\right] \qquad \text{...(3.59b)}$$

where,

$$R = \frac{1}{\gamma_g} = \frac{\hbar}{\sqrt{M|B|}} \qquad \text{...(3.59c)}$$

and N_S and N_D are determined from the conditions of normalisations. As $U_2(r)$ is expected to be much smaller than $U_0(r)$, we may get from Eq. 3.21*b*:

$$N_S = C \approx \frac{K_g}{K_0}\left[\frac{2\gamma_g}{1+\gamma_g r_0}\right]^{\frac{1}{2}} e^{\gamma_g r_0}$$

$$\approx (2\gamma_g)^{\frac{1}{2}} \qquad \text{...(3.60)}$$

where we have assumed:

$$\gamma_g r_0 = \frac{r_0}{R} << 1;\ K_g \approx K_0$$

and

$$e^{\gamma_g r_0} \approx 1 \qquad \text{...(3.61)}$$

The coefficient N_D, on the other hand, may be determined from the expression of the quadrupole moment Q in Eq. 2.21 in Chapter 2. One should, however, realise that the function $U_2(r)$ as given in

Eq. 3.59*b* cannot be used as a substitute for the whole wave functions, even as an approximation, because it is singular at $l = 0$. In general, one substitutes it by:

$$U_2(r) \approx \sqrt{2\gamma_g}\, ne^{-nr} \qquad ...(3.62)$$

by neglecting singular terms, and determining 'n' empirically from the value of the quadrupole moment. The value of $n = 0.22$ gives $P_D = 0.048$, which is not very different from the value obtained from other methods; *see* Section 3.8.

3.6 THEORETICAL DETERMINATION OF QUADRUPOLE MOMENT 'Q'$_d$ OF THE DEUTERON

The quadrupole moment 'Q'$_d$ of a charge system like deuteron is defined as:

$$Q_d = \langle Q \rangle_{IM}\, Ze \int \psi^* (3z_p^2 - r_p^2)\, d\tau \qquad ...(3.63a)$$

where Ze is the electric charge of the system and z_p and r_p are coordinates of a charge point (proton in these case), in the cylindrical coordinate system centred at the centre of mass of the system. It may be pointed out that the quadrupole moment of the deuteron exists only due to the charge distribution of the proton in the deuteron. In the case of the deuteron, the r used in the Schrödinger equation is the distance between neutron and proton which is twice the value of r_p *i.e.*, $r_p = r/2$. If we, therefore, replace r_p of Eq. 3.63*a* by $r/2$ for deuteron, the expression for quadrupole moment Q_d for deuteron for which $Z = 1$ is given by:

$$Q_d = Ze \int \psi^* \left(\frac{3z^2}{4} - \frac{r^2}{4} \right) \psi\, d\tau \qquad ...(3.63b)$$

which can be, further, expressed as:

$$Q_d = Q_1 + Q_2 + Q_3 \qquad ...(3.64a)$$

where

$$Q_1 = \int \frac{e}{4} r^2 P_2^0 (\cos\theta) |\psi_s|^2\, d\tau \qquad ...(3.64b)$$

$$Q_2 = \int \frac{e}{4} r^2 P_2^0 (\cos\theta) |\psi_0|^2\, d\tau$$

and

$$Q_3 = \int \frac{e^2}{4} r^2 P_2^0 (\cos\theta) (\psi_S^* \psi_D + \psi_D^* \psi_S)\, d\tau \qquad ...(3.64c)$$

where we have used, the relationship

$$(3z_p^2 - r_p^2) = \frac{1}{4}(3z^2 - r^2) = \frac{r^2}{4}(3\cos^2\theta - 1) = \frac{r^2}{4} P_2^0 (\cos\theta)$$

According to a general theorem by Bethe[19], average value of $(3\cos^2\theta - 1)$ in a state of angular momentum 1 and z-component M_1 is given by:

$$[(3\cos^2\theta - 1)]_{AV} = \frac{\int (3\cos^2\theta - 1)\,|Y_l^{M_l}|^2\,d\Omega}{\int |Y_l^{M_l}|^2\,d\Omega}$$

$$= \frac{1(l+1) - 3M_l^2}{(2l+3)(2l-1)} \qquad \text{...(3.65)}$$

where $Y_l^{M_l}$ denotes the normalised angular part of the wave functions in ψ's.

From Eq. 3.65, we see that for $l = 0$, $M_l = 0$, $(3\cos^2\theta - 1)_{AV} = 0$; and hence, for $l = 0$; we can write:

$$Q_1 = \int \frac{e}{4} r^2 (3\cos^2\theta - 1)_{AV} \frac{U_0^2(r)}{r^2} \left| Y_0^0\, \chi_1^1 \right|^2 d\tau$$

$$= \frac{1}{\sqrt{4\pi}} \int (3\cos^2\theta - 1)_{AV} \left| \chi_1^1 \right|^2 \times \left[\frac{e}{4} U_0^2(r)\, d\tau \right] = 0 \qquad \text{...(3.66)}$$

$$\left(\text{Here we have used} \quad \psi_S \equiv Y_0^0\, \chi_1^1 \frac{U_0(r)}{r} \right)$$

$Q_1 = 0$ is also expected qualitatively since ψ_S corresponds to $l = 0$, which refers to the spherical shape.

From Eq. 3.64*b* we can write: for Q_2,

$$Q_2 = \int \frac{e}{4} r^2 (3\cos^2\theta - 1)\,|\psi_D|^2\,d\tau$$

$$= \int \frac{e}{4} r^2 (3\cos^2\theta - 1) \frac{U_2^2(r)}{r^2} \Big[\langle 2, 1, 2, -1 | 1, 1 \rangle Y_2^2\, \chi_1^{-1}$$

$$+ \langle 2, 1, 1, 0 | 1, 1 \rangle Y_2^1\, \chi_1^0 + \langle 2, 1, 0, 1 | 1, 1 \rangle Y_2^0\, \chi_1^1 \Big] \times r^2\, dr\, d\Omega$$

$$= \frac{e}{4} \int r^4\, U_2^2(r)\, dr \Big[\left(\langle 2, 1, 2, -1 | 1, 1 \rangle \right)^2 \int (3\cos^2\theta - 1) \left| Y_2^2\, \chi_1^{-1} \right|^2 d\Omega$$

$$+ \left(\langle 2, 1, 1, 0 | 1, 1 \rangle \right)^2 \int \left| Y_2^1\, \chi_1^0 \right|^2 (3\cos^2\theta - 1)\, d\Omega$$

$$+ \left(\langle 2, 1, 0, 1 | 1, 1 \rangle \right)^2 \int \left| Y_2^0\, \chi_1^1 \right|^2 (3\cos^2\theta - 1)\, d\Omega \Big] \qquad \text{...(3.67a)}$$

where we have used:

$$\psi_D = \frac{U_2(r)}{r}\, \Phi_{1121} \qquad \text{...(3.67b)}$$

and $$\Phi_{1121} = \langle 2,1,2,-1|\,1,1\rangle Y_2^2\,\chi_1^{-1} + \langle 2,1,1,0,|\,1,1\rangle Y_2^1\,\chi_1^0 + \langle 2,1,0,1|\,1,1\rangle Y_2^0\,\chi_1^1 \quad ...(3.67c)$$

Using Eq. 3.64 and the orthogonality property of the functions and putting the values of the Glebsch-Gorden coefficients, Eq. 3.67*a* may be written as:

$$Q_2 = \frac{e}{4}\int r^2\,U_2^2\,(r)\,dr\left[\left(\frac{6}{10}\right)\times\left(\frac{-12}{21}\right)+\left(\frac{3}{10}\right)\times\left(\frac{6}{21}\right)+\left(\frac{1}{10}\right)\times\left(\frac{12}{21}\right)\right]$$

$$= \frac{e}{4}\int r^2\,U_2^2\,(r)\,dr\left[\frac{-1}{5}\right]$$

$$= -\frac{1}{20}\int_0^\infty e r^2\,U_2^2\,(r)\,dr \quad ...(3.68)$$

Similarly, we can write the expression for Q_3 as:

$$Q_3 = \int \frac{e}{4} r^2\,(3\cos^2\theta - 1)\,(\psi_S^*\,\psi_D + \psi_D^*\,\psi_S)\,d\tau$$

$$= \int e r^2\,\frac{U_0\,(r)\,U_2\,(r)\,r^2\,dr\,(3\cos^2\theta - 1)}{r^2}$$

$$\times\int\left[\langle 0,1,0,1|\,1,1\rangle Y_2^0\,\chi_1^1 + (\langle 2,1,2,-1|\,1,1\rangle)\,Y_2^0\,\chi_1^{-1} + (\langle 2,1,1,0|\,1,1\rangle)\,Y_2^1\,\chi_1^0 + (\langle 2,1,0,1,|\,1,1\rangle)\,Y_2^0\,\chi_1^{-1}\right]d\Omega \quad ...(3.69)$$

Remembering that

$$\int \chi_1^1\,\chi_1^{-1} \text{ and } \int \chi_1^1\,\chi_1^0$$

are zero, because of the property of orthogonality; and putting the values of Glebsch-Gorden coefficients, we find that:

$$Q_3 = \frac{e}{2}\int r^2\,U_0\,(r)\,U_2\,(r)\,dr$$

$$\times\int (3\cos^2\theta - 1)\left|\frac{1}{\sqrt{4\pi}}\times\left(\frac{1}{10}\right)^{\frac{1}{2}} Y_2^0\right| d\Omega \quad ...(3.70a)$$

Putting $\frac{1}{4}(3\cos^2\theta - 1) = \left(\frac{\pi}{5}\right)^{\frac{1}{2}} Y_2^0$, we obtain:

$$Q_3 = 2e\int r^2\,U_0\,(r)\,U_2\,(r)\,dr\int\left(\frac{\pi}{50}\right)^{\frac{1}{2}}\times\frac{1}{\sqrt{4\pi}}\,|\,Y_2^0\,Y_2^0\,|\,d\Omega$$

$$= \frac{e}{\sqrt{50}}\int r^2\,U_0\,(r)\,U_2\,(r)\,dr$$

Since $\int \left| Y_2^0 \right|^2 d\Omega = 1$ ...(3.70*b*)

Summing the various terms, we get:

$$Q_d = \frac{e}{\sqrt{50}} \int r^2\, U_0(r)\, U_2(r)\, dr - \frac{e}{20} \int r^2\, U_2^2(r)\, dr \qquad ...(3.71)$$

We notice that the quadrupole moment arises due to $l = 2$ state and its interaction with $l = 0$. However, $U_2(r)$ is expected to be much smaller that $U_0(r)$ as the total wave function is predominantly S-wave function. One can, therefore neglect the second term compared to the first term; and write:

$$Q_d \approx \frac{e}{\sqrt{50}} \int_0^\infty r^2\, U_0(r)\, U_2(r)\, dr \qquad ...(3.72)$$

It is clear from Eq. 3.72 that large values of r^2 contribute more to Q. As a matter of fact, we may only use the asymptotic wave function; assuming it to continue up to $r \to \infty$. As the contribution to Q from the wave function inside the nuclear potential is very small, it does not matter very much, even if we use, somewhat incorrect wave function for $r < r_0$. Hence, as a first approximation, we may write, from Eqs. 3.59*a* and 3.59*b*:

$$Q_d \approx \frac{e}{\sqrt{50}} \int_0^\infty N_S\, N_D r^2 \left[\exp\left(\frac{-r}{R}\right)\right]^2 \left[3\left(\frac{R}{r}\right)^2 + \frac{3R}{r} + 1\right] dr$$

$$\approx \frac{e}{\sqrt{8}} N_S\, N_D\, R^3 \qquad ...(3.73)$$

Using $N_S \approx (2\,\gamma_g)^{\frac{1}{2}} = \frac{2}{\sqrt{R}}$ from Eq. 3.60, we write:

$$Q_d \approx \frac{e}{\sqrt{8}} \left(\frac{2}{R}\right)^{\frac{1}{2}} N_D\, R^3 \qquad ...(3.74a)$$

So that

$$N_D \approx \frac{2\,Q_d}{e\,R^{5/2}} \approx \frac{2\,Q_d}{e\,R^2} \qquad ...(3.74b)$$

We now define

$$P_D \equiv \frac{\int_0^\infty \psi_D^*\, \psi_D\, d\tau}{\int_0^\infty \psi^*\, \psi\, d\tau} = \frac{\int U_2^2(r)\, dr}{\int \left[U_0^2(r) + U_2^2(r)\right] dr} \qquad ...(3.75a)$$

So that,

$$P_D \equiv \int_0^\infty U_2^2(r)\, dr$$

Physically P_D, gives the probability for the deuteron to exist in the D-state. P_D, can be, now, written as:

$$P_D = \int_0^{b_T} U_2^2(r)\,dr + \int_{b_T}^{\infty} U_2^2(r)\,dr \qquad \text{...(3.76)}$$

where b_T is the nuclear range of the tensor potential. As a rough approximation, if the total contribution of the wave function inside the potential is of the same order as outside, we may write:

$$\int_0^{b_T} U_2^2\,(r)\,dr \approx \int_{b_T}^{\infty} U_2^2\,(r)\,dr \qquad \text{...(3.77)}$$

This is justified approximately because $U_2\,(r)$ has a very sharp maximum at $r \approx b_T$. Substituting the value of the asymptotic solution of $U_2\,(r)$ in the second part of the right side of Eq. 3.76 and putting $e^{-r/R} \approx 1$, and neglecting smaller terms compared to $(R/b_T)^4$ we obtain, from Eqs. 3.59*b* and 3.77,

$$P_D \approx 2 \int_{b_T}^{\infty} U_2^2\,(r)\,dr$$

$$\approx 2 \int_{b_T}^{\infty} 9\,N_D^2 \left(\frac{R}{r}\right)^4 dr \approx 6\,(N_D)^2 \left(\frac{R}{b_T}\right)^3 \qquad \text{...(3.78a)}$$

Also from Eq. 3.74, we can write:

$$6\left[N_D \left(\frac{R}{b_T}\right)^3\right] \approx 24\left(\frac{Q}{eR^2}\right)^2 \left(\frac{R}{b_T}\right)^3 \qquad \text{...(3.78b)}$$

Then

$$P_D \approx C\left(\frac{Q}{eR^2}\right)^2 \left(\frac{R}{b_T}\right)^3 \qquad \text{...(3.78c)}$$

From Eq. 3.78*c*, it may be seen that, b_T cannot have an orbitrarily small range; otherwise p_D will become too large to be physically realistic.

3.7 MAGNETIC MOMENT OF DEUTERON

The operator μ_d for the magnetic moment, when we neglect the meson exchange currents is, given by:

$$\mu_d\,(op) = \frac{e\hbar}{2\,MC}\left[\sum_{p=1}^{Z} (g_1^p\,\mathbf{l}_p + g_s^p\,\mathbf{S}_p) + \sum_{n=1}^{A-Z} (g_L^n\,\mathbf{l}_n + g_s^n\,\mathbf{S}_n)\right] \qquad \text{...(3.79)}$$

where $g_1^p = 1$ and $g_1^n = 0$, are the gyromagnetic ratios for proton and neutron respectively due to orbital angular momenta; and $g_s^p = 5.59$ and $g_s^n = -3.83$ are the gyromagnetic ratios for proton and neutron respectively due to spin. Further, $\mathbf{l}_p = \frac{1}{2}\mathbf{l}$, is the orbital angular momentum of the proton, $\mathbf{S}_p$ and $\mathbf{S}_n$ are the spin angular momenta of proton and neutron respectively. In the case of deuteron $Z = 1$, $A = 2$.

We have used the relationship $\mathbf{l}_p = \frac{1}{2}\mathbf{l}$; because the angular momentum of the proton $\mathbf{l}_p$, which contributes to the orbital part of the magnetic moment, is half that of the total angular momentum $\mathbf{l}$; the other half being contributed by the neutron. We note that,

$$g_s^p \mathbf{S}_p + g_s^n \mathbf{S}_n = \frac{1}{2} g_s^p \sigma_p + \frac{1}{2} g_s^n \sigma_n \qquad ...(3.80)$$

$$\left(\because \ \mathbf{S}_{p,n} = \frac{1}{2}\sigma_{p,n}\right)$$

Therefore for deuteron, we can write the relation:

$$\frac{e\hbar}{2\,MC}\left[g_s^n \mathbf{S}_n + g_s^p \mathbf{S}_p\right]$$

$$= [(\mu_p + \mu_n)(\mathbf{S}_p + \mathbf{S}_n) + (\mu_p - \mu_n)(\mathbf{S}_p - \mathbf{S}_n)] \qquad ...(3.81)$$

So that,

$$\mu_d(op) = \left[(\mu_p + \mu_n)(\mathbf{S}_p + \mathbf{S}_n) + (\mu_p - \mu_n)(\mathbf{S}_p - \mathbf{S}_n) + \frac{1}{2}\mathbf{l}\right] \qquad ...(3.82)$$

The expectation value $\langle\mu\rangle_{IM}$ for $I = M = 1$ for deuteron is, therefore, given by:

$$\langle\mu\rangle_{IM} = \int \psi_{IM}^* (\mu_{op})_z \,\psi_{IM}\, d\tau \qquad ...(3.83)$$

where $(\mu_{op})_z$ has been used, because the experimentally quoted value of magnetic moment corresponds to the case, when $I = M$ *i.e.*, when $\mathbf{I}$ point along z-axis. It can be proved that:

$$\int \psi_{IM}^* (\mu_{op})_z \,\psi_{IM}\, d\tau = \frac{\int \psi_{IM}^* [\mathbf{I}\cdot\mu_d(op)]\,\psi_{IM}\, d\tau}{I+1}$$

Therefore,

$$\langle\mu\rangle_{IM} = \frac{1}{I+1}\int \psi_{IM}^* (\mathbf{I}\cdot[(\mu_p + \mu_n)(\mathbf{S}_p + \mathbf{S}_n) + (\mu_p - \mu_n)(\mathbf{S}_p - \mathbf{S}_n) + \frac{1}{2}\mathbf{I}])\,\psi_{IM}\, d\tau$$

$$= \frac{1}{I+1}\int \psi_{IM}^* (\mu_p + \mu_n)[\mathbf{I}\cdot(\mathbf{S}_p + \mathbf{S}_n)]\,\psi_{IM}\, d\tau + \int \psi_{IM}^* [(\mu_p - \mu_n)\,\mathbf{I}\cdot(\mathbf{S}_p - \mathbf{S}_n)]\,\psi_{IM}\, d\tau + \int \psi_{IM}^* \left[\frac{1}{2}\mathbf{I}.1\right]\psi_{IM}\, d\tau \qquad ...(3.84)$$

The second term on the right side of Eq. 3.84 vanishes because $\mathbf{S}_p - \mathbf{S}_n$ will have an average value of zero over the whole volume, with respect to a fixed direction of **I**.

Using the relations $\mathbf{I} = \mathbf{S} + \mathbf{l}$ and $\mathbf{S}_p + \mathbf{S}_n = \mathbf{S}$ we can, therefore, write:

$$\langle\mu\rangle_{IM} = \frac{1}{I+1}\int \psi^* \left\{ \mathbf{S}+\mathbf{l}\cdot\left[\frac{1}{2}\left(\mu_p+\mu_n+\frac{1}{2}\right)(\mathbf{S}+\mathbf{l}) + \frac{1}{2}\left(\mu_p+\mu_n-\frac{1}{2}\right)(\mathbf{S}-\mathbf{l})\right]\right\}\psi_{IM}\, d\tau \qquad ...(3.85)$$

or for $I = M = 1$, we have:

$$\langle\mu\rangle_{11} = \frac{1}{2}\left[\left(\mu_p+\mu_n+\frac{1}{2}\right)+\frac{1}{2}\left(\mu_p+\mu_n-\frac{1}{2}\right)\left\langle S^2-1^2\right\rangle\right] \qquad ...(3.86)$$

We are now, in a position to calculate the value of the magnetic moment of deuteron for different possible configurations. We have already seen that the quadrupole moment can be explained on the basis of a mixture of $l = 0$ and $l = 2$ wave function. Therefore, the magnetic moment of deuteron, with this mixture can be calculated. This is a case of a mixture of $3S_1$ and $3D_1$ configurations, with $I = 1, l = 0$ and $S = 1$ in the first case and $I = 1, l = 2, S = 1$, in the second case. Hence, we can write, from Eq. 3.85 and Eq. 3.86:

$$\langle\mu_d\rangle = \frac{1}{2}\left\{\left(\mu_p+\mu_n+\frac{1}{2}\right)+\left[\frac{1}{2}\left(\mu_p+\mu_n-\frac{1}{2}\right)\right] \times\left[(2-6)\int\psi_D^*\psi_D\, d\tau+(2-0)\int\psi_S^*\psi_S\, d\tau\right]\right\} \qquad ...(3.87)$$

Now keeping, in mind, that we define:

$$P_D \equiv \int (\psi_D^*\psi_D)\, d\tau = \int U_2^2(r)\, dr \qquad ...(3.88a)$$

and

$$\int\psi^*\psi\, d\tau = \int(\psi_S^*\psi_S+\psi_D^*\psi_D)\, d\tau = 1 \qquad ...(3.88b)$$

We can write Eq. 3.87, as:

$$\langle\mu_d\rangle = \frac{1}{2}\left\{\left(\mu_p+\mu_n+\frac{1}{2}\right)+\frac{1}{2}\left(\mu_p+\mu_n-\frac{1}{2}\right)[-4P_D+2(1-P_D)]\right\}$$

$$= 0.879 - \frac{3}{2}(0.879-0.500)\,P_D$$

$$= [0.879 - 0.569\, P_D]\ \text{nm} \qquad ...(3.88c)$$

It is easy to see that one can adjust P_D to obtain the experimental value of $\langle\mu_d\rangle$. The value $P_D = 0.04$ gives $\langle\mu_d\rangle = 0.857$ nm which is the observed value of the magnetic moment of deuteron.

It will be interesting to calculate the values of $\langle\mu_d\rangle$ for other possible configurations of deuteron. As for example:

(*i*) If only $I = 1$, $l = 1$ and $S = 0$ (lP_1) configuration were present,

the $\langle\mu_d\rangle$ comes out to be 0.5 nm ...(3.89*a*)

(*ii*) For $I = 1$, $l = 0$, $S = 1$, ($3S_1$), configuration,

$$\langle\mu_d\rangle = \mu_p + \mu_n = 0.879 \text{ nm} \quad \text{...(3.89}b\text{)}$$

This is close to the experimental value of $\langle\mu_d\rangle$ but still lies very much outside the experimental error. This shows that the deuteron may exist predominantly in $3S_1$ state; but not completely. Similarly the configurations $3P_1$, gives $\langle\mu_d\rangle = 0.689$ nm which differs very much from experimental value, and the configuration $1P_1 + 3P_1$ with any mixing ratio gives:

$$\langle\mu_d\rangle = [0.500 + 0.189\, P_{3P}] \text{ nm} \quad \text{...(3.89}c\text{)}$$

This cannot give us the experimental value of 0.857 nm even if $P_{3P} = 1$. Also the configurations, $1P_1$, $3P_1$ and $1P_1 + 3P_1$ correspond to negative parity, which does not agree with the experimentally measured value of the parity of the deuteron.

We conclude from this discussion, that no configuration except $3D_1 + 3S_1$ is capable of explaining the value of experimental magnetic moment. This supports our previous arguments of not accepting $I = 1$, $l = 1$ or $S = 0$ states for the ground state of deuteron. The pure $3S_1$ or pure $3D_1$ states also do not lead to the correct magnetic moment.

As mentioned earlier, we have neglected the effects due to relativistic meson-exchange currents which have been found to be[20] 2–3% if detailed calculations are carried out. Though this does not change the above discussion essentially, the exact value of P_D for the $3D_1 + 3S_1$ mixture becomes uncertain. The presently accepted value of P_D is given by:

$$P_D = (3 \pm 1)\% \quad \text{...(3.90)}$$

This is an important quantity, obtained from the analysis of the deuteron data, which, not only supports the tensor nature of nuclear forces, but also sets limit to the range of tensor forces, as was seen earlier; Eq. 3.78.

Appendix

A. ***The Spin Wave Functions:*** One can write the singlet (Spins antiparallel) and triplet (Spins parallel) state functions of two particles as:

$$\chi_0^0 = \frac{1}{\sqrt{2}} [\alpha(1)\beta(2) - \beta(1)\alpha(2)] \text{ Singlet State} \qquad \text{...(A-1)}$$

and

$$\left.\begin{aligned} \chi_1^1 &= \alpha(1)\alpha(2) \\ \chi_1^0 &= \frac{1}{\sqrt{2}}[\alpha(1)\beta(2) + \beta(1)\alpha(2)] \\ \chi_1^{-1} &= \beta(1)\beta(2) \end{aligned}\right\} \text{ Triplet State} \qquad \text{...(A-2)}$$

where

$$\alpha = \begin{pmatrix}1\\0\end{pmatrix} \text{ and } \beta = \begin{pmatrix}0\\1\end{pmatrix}$$

and 1 and 2 refer to the individual particles, α and β are to be interpreted as the spin wave functions for the individual particle with spin up [$S_2 = 1/2$] and down [$S_2 = -1/2$], respectively. The operators σ's and α, β wave functions are related by the relations:

$$\begin{aligned} \sigma_x \alpha &= \beta & \sigma_x \beta &= \alpha \\ \sigma_y \alpha &= i\beta & \sigma_y \beta &= -i\alpha \\ \sigma_z \alpha &= \alpha & \sigma_z \beta &= -\beta \end{aligned} \qquad \text{...(A-3)}$$

where $\sigma_x = \begin{pmatrix}0 & 1\\1 & 0\end{pmatrix}$, $\sigma_y = \begin{pmatrix}0 & -i\\i & 0\end{pmatrix}$ and $\sigma_z = \begin{pmatrix}1 & 0\\0 & -1\end{pmatrix}$

are Pauli matrices.

Remembering that

$$S_{12} = 3(\sigma_1 \cdot \hat{\mathbf{r}})(\sigma_2 \cdot \hat{\mathbf{r}}) - \sigma_1 \cdot \sigma_2 \qquad \text{...(A-4)}$$

By taking σ_1 along z-axis, σ_2 in the x-z plane making an angle α with z-axis, and the vector **r** with polar angles θ and ϕ ; we represent:

$$\sigma_1 = \sigma_1 \hat{z}$$

$$\sigma_2 = \sigma_2 (\sin\alpha\, \hat{x} + \cos\alpha\, \hat{z})$$

and $\hat{\mathbf{r}} = \sin\theta\cos\phi\, \hat{x} + \sin\theta\sin\phi\, \hat{y} + \cos\theta\, \hat{z}$ ...(A-5)

It is, then easy to show that,

$$S_{12}\, \chi_0^0 = 0$$

and $$(\sigma_1 \cdot \sigma_2)\, \chi_1^{M_S} = \chi_1^{M_S} \qquad \text{...(A-6)}$$

B. Proof of

$$S_{12}\, \Phi_{IM01} = \sqrt{8}\, \Phi_{IM21} \qquad \text{...(A-7a)}$$

$$S_{12}\, \Phi_{IM21} = \sqrt{8}\, \Phi_{IM01} - 2\, \Phi_{IM21} \qquad \text{...(A-7b)}$$

Starting from Eq. 3.47, one can write for $I = M = S = 1$, and $l = 0, 2$ as follows:

$$\Phi_{1101} = (4\pi)^{-\frac{1}{2}}\, \chi_1^1 = Y_{00}\, \chi_1^1$$

$$\Phi_{1121} = \left(\frac{6}{10}\right)^{\frac{1}{2}} Y_{22}\, \chi_1^{-1} - \left(\frac{3}{10}\right)^{\frac{1}{2}} Y_{2,1}\, \chi_1^0 + \left(\frac{1}{10}\right)^{\frac{1}{2}} Y_{20}\, \chi_1^1 \qquad \text{...(A-8)}$$

A straight forward, but tedious method of proving Eq. A-7 is to use the definitions of S_{12} and Φ_{IMLS} as given in Eqs. A-4 and A-8; and using the properties of spherical harmonics Y_{22}, $Y_{2,1}$ and $Y_{2,0}$. We get the answer, when we further use the equations:

$$(\sigma \cdot \hat{\mathbf{r}}) = \sigma_x \sin\theta \cos\phi + \sigma_y \sin\theta \sin\phi + \sigma_z \cos\theta \qquad \text{...(A-9)}$$

A less tedious method requires using some physical insight. For $I = 1$, and even parity, we can have two possible triplet states $3S_1$ and $3D_1$. Because of the operation of S_{12} on $3S_1$ wave function, this leads to:

$$S_{12}\, \Phi_{IM01} = a\, \Phi_{IM01} + b\, \Phi_{IM21} \qquad \text{...(A-10)}$$

Now it is assumed that operator S_{12} vanishes when averaged over the angles of orientation of $\hat{\mathbf{r}}$ because of the isotropic properties of space. In Eq. A-10, the first term on the right-hand side corresponds to $l = 0$ i.e., belongs to a spherically symmetrical state. Hence it should be zero, because if it is not zero, then this will be the only state to give average of zero. Hence $a = 0$.

For the second term, let us select $M = 1$, and pick the direction of vector $\hat{\mathbf{r}}$ as our z-direction. Then left side in Eq. A-10 after direct evaluation gives:

$$[3\sigma_{1x}\sigma_{2x} - \sigma_1 \cdot \sigma_2]\, (4\pi)^{-\frac{1}{2}}\, \alpha(1)\, \alpha(2)$$

$$= 2\, (4\pi)^{-\frac{1}{2}}\, \alpha_1(1)\, \alpha_2(2) \qquad \text{...(A-11)}$$

On the other hand, the right-hand side of Eq. A-10 with $a = 0$, assumes a simple form in this case, because $Y_{22} = Y_{2,1} = 0$, at the pole of the sphere $\theta = 0$. Then the right-hand side, according to Eq. A-8, becomes:

$$b\,\Phi_{1121} = \sqrt{\frac{1}{10}}\,Y_{2,0}\,(\theta = 0)\,\chi_1^1 = b\,(8\pi)^{-\frac{1}{2}}\,\alpha\,(1)\,\alpha\,(2) \qquad ...(A\text{-}12)$$

Comparison of Eqs. A-11 and A-12 gives $b = \sqrt{8}$ and Eq. A-7a is proved.

To prove the second equation of Eq. A-7, we write:

$$S_{12}\,\Phi_{IM21} = b\,\Phi_{IM01} + c\,\Phi_{IM21} \qquad ...(A\text{-}13)$$

where b must be the same as in Eq. A-10, because of the Hermitian nature of tensor operator S_{12}. We again evaluate both sides for the special case when $\hat{r}$ points in the z-direction *i.e.,* $\theta = 0$; and solve for the unknown. The result is $c = -2$ proving the second equations of Eq. A-7b.

3. The Deuteron

2000–2008

Though, there is no literature, about the bound system of Deuteron, in these nine years, there are some papers, concerned with proton-deuteron interaction, to investigate if there are three-body forces, in the nucleon-nucleon interaction, where deuteron is involved.

In a paper, on proton-deuteron scattering at energies $E_{c.m.} = 431.3$ KeV and for energies at 431.3 KeV $\leq E_{c.m.} \leq 2000$ KeV; the values of Ay, T_{20}, T_{21}, T_{22} and iT_{11} have been measured and compared to nucleon-nucleon interaction with and without three-body forces. $A\gamma$ and iT_{11} data are found, to be unpredicted by the calculations [Phy. Rev. C. 63, 044013 (2001)].

In another paper on $p\,\bar{d}$ elastic scattering, performed at Osaka University of Japan, by 24 authors, the angular distribution of cross section; the proton analysing power and transfer coefficient of $p\,\bar{d}$ elastic scattering were measured at 250 MeV incident energies. The results call for a better understanding of the spin structure of three nuclear force. [Phy. Rev C. 66, 04002 (2002)].

In two more papers on proton-deuteron scattering *n-d* and deuteron-proton break up [*H*(*d*, *pp*)*n*] by a group of 20 authors from Germany and Poland, Netherlands and U.S.A.; using Netherland facility of superconducting cyclotron AGOR; the role of three-body nuclear forces was investigated. In the first experiment on proton-deuteron scattering differential scattering cross section were measured at proton energies of 108, 120, 135, 150, 170 and 190 MeV at c m angles between 30° and 170° . The comparison with Fedeev equations showed the effect of the model with three nucleon forces; and the shortcoming of calculations, using two-body forces was exhibited [Phy. Rev. 68, 05100 (R) (2003)]. In an other paper, by the same group, the break up of deuteron in H^1 (α, *pp*) reaction at 130 MeV energy; the breakup cross sections were measured at nine kinematic configurations and compared with Fedeev model of three nucleon force (3 *NF*). The results show a significant influence of 3 NF; on the break-up cross section of the measurements. [Phy. Rev. C. 68, 021004 (2003)].

Thus the existence of three nuclear force seems to be proved in p-d interaction, giving a new dimension to nuclear forces as available from the properties of deuteron.

In another paper published in [Phy. Rev. C. 72061002 (R) (2005)] evidence of three-body force in neutron-deuteron scattering at somewhat lower energy of 95 MeV, has been found, by comparing the full angular distribution, with the theoretical description including three nuclear forces.

In a theoretical paper [Phy. Rev. (77) 021001 (R) (2008)] general formulas describing deuteron spin dynamics in storage ring with allowance for tensor electric and magnetic polarisation are derived. These calculations show that tensor magnetic polarisation of the deuteron causes the spin rotation with two frequencies and experiences beating for polarised deuteron beams in storage ring. This is a case of application of detailed forces in a deuteron beam in storage ring.

REFERENCES

1. H.C. Urey, F.G. Brickwedde and G.N. Murphy: Phy. Rev. 39, 1642 (1932); 40, 1 (1932).
2. C.W. Li: Phy. Rev. 88 (1952), erratum 90: 1131E (1953).
3. C.W. Li, W. Whaling, W.A. Fowler and C.C. Lauretsen: Phy. Rev. 82: 512 (1951).
4. A.O. Nier et al.: Phy. Rev. 102, 107 (1956); 105, 1014 (1957); 407, 1664 (1957); T.L. Collins et al.: Phy. Rev. 84, 717 (1957).
5. J.Chadwick: Proc. Royal Society (London), A 136: 692, (1932) and Ref. (2) and (3).
6. R. E. Bell and R. G. Elliot: Phy. Rev. 79, 282 (1950).
7. S.C. Brown and L.G. Elliot: American J. of Physics 11: 311 (1943).
8. Determination of $I = 1$ and $l = 0$ and $l = 2$, from experiments given in the Chapter 2, for determination of μ and Q and I finally determine parity by the logic of quantum mechanics.
9. F. Ramsey: Nuclear Moments, John Wiley & Sons, New York (1953).
10. J.M.B. Kelloy, I.I. Rabii, N.F. Ramsey Jr. and J.R. Zachiaras: Phy. Rev. 55, 318L (1939); 57, 677, (1940). J.P. Auffray: Phy. Rev. Letters 6, 120 (1961).
11. N.K Glenderning and G. Krammer: Phy. Rev. 126, 2159 (1962).
12. J.H. Gardener and E.M. Purcell: Phy. Rev. 126, 2159, (1962). H. Somoner, H.A. Thomas and J.A. Hipple: Phy. Rev. 80, 487L, (1950). Table of Nuclear Spins and Moments (*I*), Alpha, Beta and Gamma Ray Spectroscopy Appendix 4, edited by K. Siegbahn, North-Holland Publishing Company, Amsterdam (1964). J.H. Sanders Nuovo Cemeto Suppl. 6, No. I, 242m (1957); 57, 677, (1940).
13. F. Bloch, D. Nichodemus and H.H. Staub, Phy. Rev. 74, 1025, (1948). Cohen V.W., N.R. Corngold and N.F. Ramsey, Phy. Rev. 104, 283 (1956).
14. Handbuch der Physik V.XXXIX (Springer-Verlag), (1958).
15. Blatt and V.F., Weisskopf: Theoretical Nuclear Physics: Chapter 2. H.A. Bethe; Phy. Rev. 76:38 (1949). M. Camac and H.A. Bethe: Phy. Rev. 73, 191 (1948).
16. H. Feshbacch and V.F. Weisskopf: Phy. Rev. 76, 1550 (1949). S. Gartenhaus: Phy. Rev. 100, 900, 903 (1955). L. Húlthen and M. Sugawara; Encyclopedia of Physics, ed. S. Flügge. V. 39, Berlin, Springer (1957).
17. R.R. Roy and B.P. Nigam: Nuclear Physics, Page 53, New Age International (P) Limited, New Delhi (1987).
18. J.M. Blatt and V.F. Weisskopf: Theoretical Nuclear Physics (Appendix A, Page 789) John Wiley & Sons, New York (1952).
19. H.A. Bethe: Handbuch der Physik V. 24, 1, Chapter 3, (Springer-Verlag), (1933).
20. H. Margenau: Phy. Rev. 57, 383 (1940) G. Breit and I. Bloch, Phy. Rev. 72, 135, (1947) R.G. Sachs: Phy. Rev. 72, 91 (1947), H. Primakoff, Phy. Rev. 72, 118, (1947). Young H.D. and R.E Kutkosky; Phy. Rev. 117, 595 (1960) M. Sugawara: Phy. Rev. 117, 614 (1960).

PROBLEMS

1. A neutron-proton system (deuteron) forms a bound state; but a di-neutron (neutron-neutron) and a di-proton (proton-proton) does not, though *n-p*; *n-n* and *p-p* nuclear forces are charge independent. Why?
2. Calculate the magnetic moment of deuteron, for $I = 1$, $S = 1$ and $l = 2$, from general expressions of μ_d in Eqs. 3.79 and 3.86.
3. What will be the value of the quadrupole moment of deuteron for the quantum numbers of Eq. 3.89*a*?
4. If the binding energy of deuteron was five times, the measured binding energy of 2.2247 MeV; what would be the approximate depth of the potential $V(r)$ for a square well potential in Fig. 3.1; and what would be r_0 using the logic of section 3.2?
5. Calculate the energy due to interaction of the magnetic moments of proton and neutron, for $S = 1$, when spins of proton and neutron are perpendicular to **r** or when they are parallel to **r**. Compare these with tensor force of Eq. 3.39.
6. Consider a nucleon-nucleon potential of the form

$$V = -V_0 (a + b\, \sigma_1 \cdot \sigma_2) f(r)$$

where r is the relative distance of the two nucleons. Find the strength of potentials for singlet and triplet states.

7. Suppose, the quadrupole moment of deuteron was increased by a factor 2; what will be its effect on the value of $V_T(r)$ and $V_T(r)/V_c(r)$?
8. Prove that; if,

$$S = \frac{1}{2}(\sigma_1 + \sigma_2), \text{ then } S^2 = \frac{1}{2}(3 + \sigma_1 \cdot \sigma_2)$$

9. Prove Eq. 3.65 *i.e.*,

$$(3\cos^2\theta - 1)_{av} = \frac{1(l+1) - 3M_1^2}{(2l+3)(2l-1)}$$

10. In order that 66 keV singlet bound state may exit in (n, p) systems, what should be the depth of the potential V? [see Ma S.T., Med. Phy. 25, 853 (1953)].

4

CHAPTER

Nucleon-Nucleon Scattering at Low Energies

4.1 INTRODUCTION

The deuteron problem, as discussed in Chapter 3, has given us information about nuclear forces between neutron and proton, when their spins are parallel; *i.e.,* for $S = 1$, $l = 0$, and $l = 2$. Information about other spin and orbital angular momentum states can come from scattering of neutrons or protons from protons at various incident-energies yielding details about neutron-protons, and proton-proton interactions. Scattering of neutrons and protons from deuterons, further, has yielded information about the neutron-neutron interaction. The study of complex nuclei, has also been useful in the understanding of some of the properties of nucleon-nucleon interaction.

We divide the problem into (*i*) scattering at low energies and (*ii*) scattering at high energies. By low energies, we essentially mean energies where only $l = 0$ is effective. Higher energies correspond to higher values of l. This division corresponds to both experimental and theoretical aspects. A lot of detailed experimental data is available at low energies, with very small intervals of energies and with very high resolution. The analysis, can, therefore, be carried out in details. At higher energies, new phenomena like polarization effects appear, which further help in understanding the behaviour of nuclear forces at higher l-values. Incident energies below 10 MeV or so, are generally, referred to as low energies. From the relationship:

$$|\mathbf{L}| = |\mathbf{r} \times \mathbf{p}| = l\hbar \qquad ...(4.1)$$

it may be seen that for low energies, the effective l is small for a given r, say nuclear range r_0. Many experiments have shown that below 10 MeV, predominantly $l = 0$ is effective for nuclear part of the scattering.

Monoenergetic protons with these energies may be made available directly from accelerators like Cockcroft-Walton, Van de Graff or Cyclotrons. For the study of proton-proton scattering, one generally uses energies above 100 keV because below this energy, the Coulomb effect is predominant and the protons may not come near enough to be within nuclear range.

For the study of neutron-proton scattering, it is not so easy to produce monoenergetic neutrons. For thermal energies ($E = 3/2\ kT$ where $T = 300$ K) of $E \approx 0.025$ eV, one generally uses the reactors, where arrangements are made to make the neutrons attain the thermal temperature of $T = 300$ K before they are brought out for experimentation. These neutrons are, however, not mono energetic, but have a Maxwellian distribution around $E = 0.025$ eV. For obtaining mono energetic neutrons around these energies, one, either, uses the crystal spectrometers for obtaining 'cold' neutrons, *i.e.,* below $E = 0.002$ eV; or choppers as well as crystal spectrometers for above-thermal (epithermal) energies, *i.e.,* greater than 0.3 eV–0.4 eV. For both purposes, one generally uses the thermal Column in a reactor as the source of neutrons. Sometimes the pulsed beam of a cyclotron along with time of-flight method is used to produce neutrons in this energy range from an endoergic neutron-producing reactions say:

$$_1p^1 + {}_1\text{H}^3 \rightarrow {}_2\text{He}^3 + {}_0n^1 \qquad (Q = -\,0.764 \text{ MeV})$$

or

$$_3\text{Li}^7 + {}_1p^1 \rightarrow {}_4\text{Be}^7 + {}_0n^1 \qquad (Q = -\,2.225 \text{ MeV})$$

It may be noted that one can conveniently study *n-p* scattering at these low energies because of the absence of Coulomb forces, in contrast to *p-p* scattering. For high energies one has to take recourse to the various exoergic reactions like:

$$_1\text{D}^2 + {}_1\text{D}^2 \rightarrow {}_2\text{He}^3 + {}_0n^1 \qquad (Q = 3.265 \text{ MeV})$$

or

$$_1\text{D}^2 + {}_1\text{H}^3 \rightarrow {}_2\text{He}^4 + {}_0n^1 \qquad (Q = 17.6 \text{ MeV})$$

One, however, faces a difficulty, as one goes to higher incident energies in these reactions. At one given angle between the incident and outgoing particles, one gets two energies for neutrons in the lab-system because of the kinematics. One generally uses time-of-flight techniques to select the desired energies.

The detection and energy-determination of neutrons in *n-p* scattering, is more difficult than protons in *p-p* scattering. In the latter case, the protons may be detected and measured directly by various detectors based on ionisation, *e.g.* proportional counters, scintillation detectors or recently the surface barrier detectors. But for the detection of neutrons, one has to go through a reaction, producing at first a charged particle, whose energy can be then measured by conventional methods based on ionisation as mentioned above. For low energies say from thermal to a few MeV, one normally uses counters containing Boron^{10}(B^{10}). A reaction like $\text{B}^{10} + n \rightarrow \text{He}^4 + \text{Li}^7$ is used for producing the alpha particle for detection. For higher energies say above 2–3 MeV, the counters containing hydrogenous materials, like plastic scintillators are used, where one uses the principle of the recoil of protons from *n-p* scattering for detection and energy measurements.

At low energies, in *p-p* interaction, only elastic scattering is possible; while in *n-p* interaction, apart from elastic scattering, part of the reaction may go through neutron capture forming deuteron. One normally measures different cross-sections at these low energies: *e.g.*, total cross-section; (σ_{total}) which will be the sum of elastic scattering and capture-cross sections in the case of *n-p* interaction; and only elastic scattering for *p-p* scattering, *i.e.*,

$$\sigma_{\text{total}} = \sigma_{sc} + \sigma_{\text{cap}} \qquad \text{(for } n\text{-}p \text{ interaction)} \qquad \text{...(4.2)}$$

$$= \sigma_{sc} \qquad \text{(for } p\text{-}p \text{ interaction)} \qquad \text{...(4.3)}$$

Also one generally measures the total scattering cross-section σ_{sc} which is given by:

$$\sigma_{sc} = \int \frac{d\,\sigma_{sc}(\theta)}{d\Omega}\, d\Omega$$

which requires measuring at different angles, the values of differential scattering cross-section, *i.e.,*

$$\frac{d\sigma_{sc}(\theta)}{d\Omega}$$

Under certain conditions, the scattering may lead to polarisation (as will be discussed later) in which case one may measure $P(\theta)$—the amount of polarisation at an angle.

It is not easy to study *n-n* scattering directly. Till now the knowledge about *n-n* scattering has been obtained only indirectly from the study of (*n-d*) reaction using,

$$\sigma(n, d) \approx \sigma(n, n) + \sigma(n, p) + I(\theta) \qquad \text{...(4.4)}$$

where $I(\theta)$ represents an interference term.

4.2 NEUTRON-PROTON SCATTERING

4.2.1 Experimental Facts

Total Cross-sections: One may measure σ_{total} under conditions of incoherent scattering where incident neutrons have their spins oriented randomly and the distance between the various protons in the scattering target is larger than de Broglie wavelength of the incident neutrons. If the distance between the two protons in the target is less than the wavelength of the incident neutrons, then the scattering may be coherent. Because of this condition, the coherent scattering is possible only for thermal or below thermal neutrons (0–0.025 eV).

4.2.1.1 Incoherent Scattering

Experimentally these cross-sections have been measured with hydrogenous materials like water (H_2O) and paraffin as targets. For thermal and below thermal energies, chemical binding energy effects, as well as, the coherent scattering effects are important, for which corrections should be applied. Figure 4.1 gives the observed values[1] of the total cross-sections from 0.001 eV to 1000 eV. It has been found experimentally that the contribution of coherent scattering and chemical binding are large at smaller (below thermal) neutron energies.

At 10 eV, which is very much above the chemical binding energies in water, there are no chemical effects, showing that for these neutron energies, protons get knocked out of the molecule and scattering takes place as if from free protons.

The accepted value of the total *n-p* cross-section for free protons, *i.e.,* without any effects of chemical binding and for incoherent scattering is 20.692 barns at thermal energy of 0.025 eV.

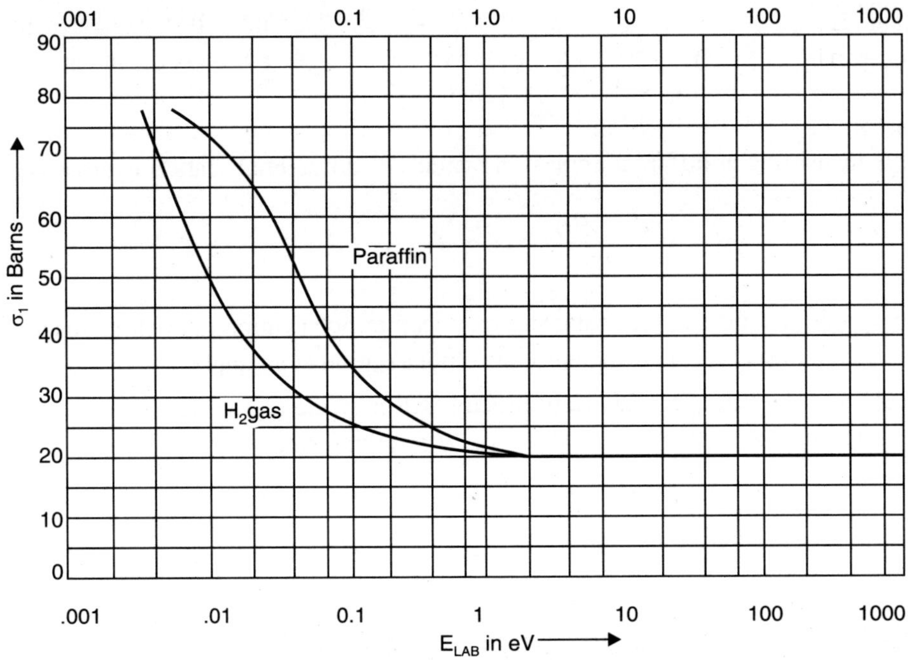

Fig. 4.1 Experimental values of total neutron-proton scattering, using (*i*) H_2 gas and (*ii*) Paraffin, showing chemical effects (Ref. 1).

4.2.1.2 Coherent n-p Scattering (Experimental)

It is possible, to perform the scattering experiments, so that the scattering amplitudes have definite phase differences, making the scattering coherent. In molecular form of hydrogen; the distance between the two protons is 0.78×10^{-8} cm, so that the de Broglie wavelength of the slow neutrons (8–10 Å) is much larger than interproton distance in H_2 molecule. Under these conditions, it is expected that the scattering amplitudes from the two protons of the H_2 molecule will add coherently.

It is experimentally known that molecular hydrogen H_2 can exist in two forms: (1) Parahydrogen, in which the spins of the two protons are anti parallel and (2) Orthohydrogen, in which the spins are parallel. Conventionally, the prefix 'ortho', is applied to the more abundant species and 'para' to the other species and hence the confusion of applying 'para' to the case where spins are anti-parallel. In molecular hydrogen gas; at normal temperatures, the ratio of ortho to para hydrogen is 3 : 1; given by the ratio of the statistical weight $2S + 1$ in the two cases where S is the total spin of the two protons in each case. The value of $\mathbf{S} = \mathbf{S}_1^p + \mathbf{S}_2^p$ is 1 for ortho-hydrogen and zero, for para-hydrogen; where $\mathbf{S}_1^p$ and $\mathbf{S}_2^p$ are the spin values of the two protons in the hydrogen molecule.

The space-part of the wave functions corresponding to rotation is symmetric for para-hydrogen (since it is anti-symmetric in nuclear spin space); and anti-symmetric for ortho-hydrogen (because of its being symmetric in nuclear spin space). This requires that the permitted rotational angular momenta for para-hydrogen should be 0, 2, 4, etc. while for ortho-hydrogen they should be 1, 3, 5, etc. At liquid hydrogen temperatures, the ratio of ortho to para-hydrogen is again 3:1, unless one applies a mechanism to change the spin directions. At around 20 K, the lowest rotational state of rotational angular momentum 'zero' is obtained for para; and 'one' for ortho-hydrogen. With these values for rotational angular moments, the para-hydrogen has lower energy than the ortho-hydrogen, and therefore, one should expect a transition from ortho to para. In pure hydrogen, however, this conversion, is very slow, unless one provides artificially some magnetic field gradients. This is done by introducing some paramagnetic atoms when the whole liquid hydrogen can be changed to para hydrogen state. Without the introduction of such paramagnetic atoms; the liquid hydrogen may be assumed to contain both ortho-and para-hydrogen in the ratio 3:1.

Through experiments on scattering of neutrons from liquid para hydrogen, it has been possible to measure σ_{sc} (para–H_2), for cold neutrons, and its value is found to be:

$$\sigma_{sc}\ (\text{para–}H_2) = 4 \times 10^{-24}\ \text{cm}^2 \qquad (\text{for slow neutrons}) \qquad ...(4.5)$$

Similarly, by scattering from pure liquid hydrogen, (liquid mirror) where the ratio of ortho-to para-hydrogen is 3:1 one can obtain the cross-section of (ortho–H_2) from the equation:

$$\sigma\ (\text{Total–}H_2) = \frac{3}{4}\ \sigma\ (\text{ortho–}H_2) + \frac{1}{4}\ \sigma\ (\text{para–}H_2) \qquad ...(4.6)$$

from which we obtain

$$\sigma\ (\text{ortho–}H_2) = 125 \text{ barns} \qquad ...(4.7)$$

It may be realised that in writing Eq. 4.6, we have assumed that while scattering from the atoms of a given molecule is coherent, because of short distance between protons of a molecule; it is incoherent from different molecules of the gas; because they will be, in general, at a large distance from each other.

Another quantity of interest in *n-p* coherent scattering is 'f' the coherent scattering length. It is defined as:

$$f \equiv 2\left(\frac{3}{4}\,a_t + \frac{1}{4}\,a_s\right) \qquad ...(4.8)$$

where a_t and a_s are the triplet and singlet scattering lengths defined as:

$$\sigma_t = 4\pi a_t^2 \text{ and } \sigma_s = 4\pi a_s^2 \qquad ...(4.9\ a)$$

So that a_t and a_s behave like scattering amplitudes as will be explained in section (4.2.2.2) σ_t is the cross-section when the incoming neutron and target proton have their spins parallel (triplet state), and σ_s is the cross-sections when, incident neutron and target proton have their spin anti-parallel (singlet state). The factor 2 as given in Eq. 4.8 takes care of the two protons in H_2 molecule. One is assuming above, that the energies are low enough so that only $l = 0$ is effective. We notice that the value of f can be obtained indirectly from the measurements of σ_t and σ_s.

It is also evident that the total cross-section for *n-p* scattering (incoherent) from free protons, may be written as:

$$(\sigma_{sc})_{\text{incoh}} = \frac{3}{4}\,\sigma_t + \frac{1}{4}\,\sigma_s$$

$$= 3\pi a_t^2 + \pi a_s^2 \qquad ...(4.9\ b)$$

The value of f has also been measured accurately by studying the total reflection of thermal neutrons from a pure liquid hydrocarbon triethyl benzene ($C_{12}H_{18}$). The relationship between the refractive index n (for neutron waves) of a scattering medium and the coherent scattering length of amplitude f is given by:[2]

$$n^2 = 1 - \frac{4\pi\, Nf(0)}{k^2} \qquad ...(4.10)$$

where $f(0)$ is the forward scattering amplitude and $k = 2\pi/\lambda$ for neutrons, so that:

$$n \approx 1 - \frac{4\pi\, Nf(0)}{k^2}\,;\text{ if } \frac{2\pi\, Nf(0)}{k^2} << 1 \qquad ...(4.11\ a)$$

Here N is the number of lattice points/cm^3 in the crystalline scattering or reflecting surface. The quantity $f(0)$ denotes the fact that f is measured at approximately zero angle for θ_c, the critical angle.

It may be recognised, that $f(0)$ as used above corresponds to the total scattering amplitude. For total reflection; we have $\cos\theta_c/n \approx 1$, where θ_c is the critical angle, so that one can write n as:

$$n = \sqrt{1-\sin^2\theta_c} \approx \sqrt{1-\theta_c^2} \text{ for small angles} \qquad ...(4.11\ b)$$

This leads to (from Eq. 4.11b):

$$n - 1 = \frac{1}{2}\theta_c^2 = -\frac{2\pi\, Nf(0)}{k^2}$$

or

$$\theta_c = \sqrt{\frac{4\pi\, Nf(0)}{k^2}} = \lambda\sqrt{\frac{Nf(0)}{\pi}} \qquad ...(4.12)$$

where λ is the wavelength of the neutrons, related to k as: $k^2 = 1/\lambda^2$. In practice $n - 1 \approx 10^{-6}$, and the critical angle θ_c is around 10 minutes. It can be seen from Eq. 4.12 that total reflection can only be observed if $f(0)$ is positive.

It is clear from the definition of $f(0)$ in Eq. 4.8 that if a_s is negative but large and a_t is positive but small (as will be seen subsequently for n-p scattering); a pure hydrogenous material cannot give positive $f(0)$. It is, therefore, necessary, to introduce some other material say carbon along with the hydrogen. The value of $f_c(0)$ for carbon can be separately determined and is known to have a positive value of $6.63 \pm 0.03 \times 10^{-13}$ cm. A carbo-hydrogenous material is, therefore, selected for observing the total reflection of neutrons. In the experiment by Hughes, Burgy and Ringo[3], neutrons were reflected from triethyl benzene ($C_{12}H_{18}$) for which one can write:

$$[Nf(0)]_{\text{eff}} = N_m\,[12\,f_C(0) + 18\,f_H(0)] \qquad ...(4.13)$$

where N_m, is the number of triethyl benzene molecules/cm^3 so that:

$$\theta_c = \lambda\sqrt{\frac{N_m[12\,f_C(0) + 18 f_H(0)]}{\pi}} \qquad ...(4.14)$$

The experiments were conducted with λ ranging from 8 Å to 15 Å: from which experimental value of $f_C(0)/f_H(0)$ comes:

$$\frac{f_C(0)}{f_H(0)} = -1.753 \pm 0.005 \qquad \text{...(4.15a)}$$

$$\text{giving } f_H(0) = -3.78 \pm 0.02 \times 10^{-13} \text{ cm} \qquad \text{...(4.15 b)}$$

4.2.1.3 Angular Distribution (n-p scattering)

A large number of experiments have been performed on angular distribution of neutrons after they are scattered from protons. Figure 4.2 gives[4] the cross-section σ(θ) for different incident energies. An interesting feature of these curves is their isotropic symmetry around 90° for low energies. As we shall see later, this isotropy is explained on the assumption that only $I = 0$ part of the wave function is effective at such low energies.

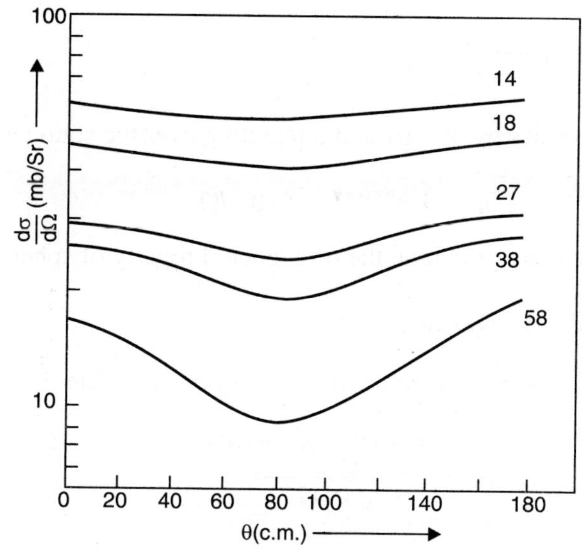

Fig. 4.2 The differential cross-sections of n-p scattering at various incident energies shown as numbers (in MeV) on the curves (Ref. 4).

4.2.2 Theoretical Treatment

One generally treats the low energy nucleon-nucleon scattering problems either in terms of (*i*) phase-shift analysis or (*ii*) in terms of effective range theory. The first approach is somewhat more general and is also applied at high energies. The second approach has the advantage that no assumption for a particular form of the potential function is required.

4.2.2.1 Phase Shift Analysis or Partial Wave Analysis

The scattering experiments are performed with parallel incident beams. A parallel beam travelling in z-direction may be represented by a wave-function e^{ikz}. We neglect the spin of the incident and target nucleon in this treatment. The function e^{ikz} is independent of x and y, which corresponds to the physical condition that the intensity of the beam is the same for any value of x and y for a given target nucleon.

Such a plane wave is the solution of the Schrödinger equation:

$$(\nabla^2 + k^2)\,\psi_{\text{inc}}\,(r) = 0 \qquad ...(4.16)$$

where
$$k^2 = \frac{ME}{\hbar^2};$$

E being the energy of the incident beam in the centre of mass system and M is the mass of a nucleon. In this equation the potential on the incident particle is taken to be zero, so that it represents a plane wave when the scattering centre is not present. The solution of this equation may be represented by:

$$\psi_{\text{inc}}(r) = e^{ikz} = e^{ikr\cos\theta} \qquad ...(4.17\ a)$$

The function is independent of ϕ which represents the physical condition of the symmetry of the beam around Z-axis, *i.e.,* independent of x and y as mentioned above. It may, therefore, be expanded in terms of the spherical harmonics $Y_1^0\,(\theta)$ for $m = 0$, so that:

$$e^{ikr\cos\theta} = \sum_1 A_1 Y_1^0(\theta) \qquad ...(4.18)$$

Multiplying the two sides by $Y_1^{*0}(\theta)$ and integrating over the solid angle, it may be seen that:

$$A_1 = \int e^{ikr\cos\theta}\,Y_1^{*0}\,(\theta)\,d\Omega \qquad ...(4.19\ a)$$

We have used in the above equation the orthogonal property of spherical harmonics, *i.e.,*

$$\int Y_1^0(\theta)\,Y_1^{*0}\,(\theta)\,d\Omega = \delta_{11} \qquad ...(4.20)$$

Remembering that the spherical Bessel function $J_1(kr)$ are related to spherical harmonics, by:

$$J_1\,(kr) = \frac{\int e^{ikr\cos\theta}\,Y_1^0\,(\theta)\,d\Omega}{i^1\,\sqrt{4\pi\,(2l+1)}} \qquad ...(4.21\ a)$$

one obtains,

$$A_1 = i^1\,\sqrt{4\pi\,(2l+1)}\,J_1\,(kr) \qquad ...(4.19\ b)$$

Writing Eq. 4.17 *a* in terms of Legendres' polynomials, one gets:

$$e^{ikz} = \sum i^1\,(2l+1)\,J_1\,(kr)\,P_1\,(\cos\theta) \qquad ...(4.17\ b)$$

$$\left[\because\ Y_1^0(\theta) = \sqrt{\frac{(2l+1)}{4\pi}}\,P_1\,(\cos\theta)\right]$$

Remembering that,

$$J_1\,(kr) \underset{r\to\infty}{=} \frac{1}{kr}\sin\left(kr - \frac{l\pi}{2}\right)$$

$$= \frac{1}{2ikr}\left[e^{-i\left(kr - \frac{l\pi}{2}\right)} - e^{+i\left(kr - \frac{l\pi}{2}\right)}\right] \qquad ...(4.21\ b)$$

One obtains for the field-free scattering, the plane-wave given by:

$$e^{ikz} = \sum_{1} \frac{i^1 (2l+1)}{2kr} \left[e^{-i\left(kr - \frac{l\pi}{2}\right)} - e^{+i\left(kr - \frac{l\pi}{2}\right)} \right] P_1 (\cos\theta) \quad ...(4.22)$$

On introducing the scattering centre, [*i.e.*, scattering from a potential $V(r)$]; one has to obtain a solution of the Schrödinger equation:

$$[\nabla^2 + k^2 - V'(r)]\ \psi(r) = 0 \quad ...(4.23)$$

where $$V'(r) = \frac{M}{\hbar^2}\ V(r);$$

$V(r)$ being the nucleon-nucleon potential. If $V(r)$ is spherically symmetrical, physically it is expected that the asymptotic solution of Eq. 4.23, should consist of a plane wave over which is superimposed a spherically outgoing wave; so that:

$$\psi \underset{r\to\infty}{(r)} = \psi_{\text{inc}} + \psi_{\text{scatt}}$$

$$= e^{ikz} + f(\theta, \phi)\frac{e^{ikr}}{r} \quad ...(4.24)$$

If the scattering potential is independent of ϕ, so will be the function $f(\theta, \phi)$, so that we can replace it by $f(\theta)$.

The differential cross-section $\sigma_{sc}(\theta)$ at a given angle (θ) is given by:

$$\sigma_{sc}(\theta)\ d\Omega = \frac{N_{sc}\ (\theta)\ d\Omega}{\text{Incident flux}} \quad ...(4.25)$$

where $N_{sc}(\theta)$ is the number of scattered particles per unit solid angle per unit time from a sphere of radius R, in the average direction θ, and the incident flux defined as the number of incident particles per unit area per unit time. Quantum-mechanically one can write:

$$N_{sc}(\theta)\ d\Omega = \frac{\hbar}{2iM}\left(\frac{\partial \psi_{sc}}{\partial r}\psi^*_{sc} - \frac{\partial \psi^*_{sc}}{\partial r}\psi_{sc}\right) r^2\ d\Omega \quad ...(4.26\ a)$$

(see Quantum Mechanics by Schiff, page 23)[12]

where, $$\psi_{sc} \equiv f(\theta)\ \frac{e^{ikr}}{r} \text{ for } r \to \infty$$

From this we get,

$$N_{sc}\ (\theta)\ d\Omega = \frac{2\hbar k}{M} |\ f(\theta)|^2\ d\Omega$$

$$= \left(\frac{2E}{M}\right)^{1/2} |\ f(\theta)|^2\ d\Omega \quad ...(4.26\ b)$$

The flux is given by:

$$v = \left(\frac{2E}{M}\right)^{1/2}$$

so that $$\sigma_{sc}(\theta) = |f(\theta)|^2 \, d\Omega \quad ...(4.27\ a)$$

and $$\sigma_{sc} = 2\pi \int |f(\theta)|^2 \sin\theta \, d\theta \quad ...(4.27\ b)$$

Defining

$$f(\theta) \equiv \frac{1}{2ik} \sum_{l=0}^{l=\infty} (2l+1) \, f_1 \, P_1 \, (\cos\theta) \quad ...(4.28)$$

One can rewrite Eq. 4.24 (using Eqs. 4.28 and 4.17 *b*) as:

$$\underset{r\to\infty}{\psi\,(r,\theta)} = \frac{1}{2ikr} \sum_{l=0}^{l=\infty} i^l (2l+1) \left[(1-f_1) e^{i\left(kr - \frac{l\pi}{2}\right)} - e^{-i\left(kr - \frac{l\pi}{2}\right)} \right] P_1 \, (\cos\theta)$$

$$= \frac{1}{2ikr} \sum_{l=0}^{l=\infty} [(1-f_1)\, e^{ikr} - (-1)^1 \, e^{-ikr}] \, P_1 \, (\cos\theta) \quad ...(4.29)$$

We further know that Eq. 4.29 is a solution of Eq. 4.23 and therefore, may be written as:

$$\underset{r\to\infty}{\psi\,(r,\theta)} = \sum_{l=0}^{\infty} i^l \, (2l+1) \, \psi_1(r) \, P_1(\cos\theta) \quad ...(4.30)$$

where $\psi_1(r)$ is the radial part of the solution and is given by:

$$\psi_1(r) = \frac{U_1(r)}{r} \quad ...(4.31)$$

where $U_1(r)$ is a solution of the equation:

$$\left[\frac{d^2}{dr^2} + k^2 - V(r) - \frac{l(l+1)}{r^2}\right] U_1(r) = 0 \quad ...(4.32)$$

Asymptotically ($r \to \infty$) the solution of $U_1(r)$ is given by:

$$\underset{r\to\infty}{U_1(r)} = C_1 \sin\left(kr - \frac{l\pi}{2} + \delta_1\right) \quad ...(4.33)$$

where δ_1 is the phase shift due to scattering from the potential. Comparing Eq. 4.29 with 4.30, 4.31 and 4.33 it is easy to see that [by substituting $U_1(r)$ from Eq. 4.33 into Eqs. 4.30 and 4.29]:

$$C_1 = \frac{1}{k} e^{i\delta_1} \quad ...(4.34\ a)$$

and $$f_1 = [1 - e^{2i\delta_1}] \quad ...(4.34\ b)$$

So that

$$f(\theta) = \frac{1}{2ik}\sum_{l=0}^{\infty}(2l+1)[e^{2i\delta_1} - 1]\, P_1(\cos\theta) \quad ...(4.35)$$

and $$\sigma_{sc}(\theta) = |f(\theta)|^2$$

$$= \frac{1}{4k^2}\left|\sum_{l=0}^{\infty}(2l+1)[e^{2i\delta_1} - 1]\, P_1(\cos\theta)\right|^2 \quad ...(4.36)$$

$$= \frac{4\pi}{k^2}\sum_{l=0}^{\infty}(2l+1)\sin^2\delta_1 \quad ...(4.37)$$

For low energies, where only $l = 0$ is effective, the value of $\sigma_{sc}(\theta)$ may be given by:

$$\sigma_{sc}(\theta) = \frac{1}{4k^2}|e^{2i\delta_0} - 1|^2$$

$$= \left|\frac{e^{i\delta_0}\sin\delta_0}{k}\right|^2 = \frac{1}{k^2}\sin^2\delta_0 \quad ...(4.38\ a)$$

and $$\sigma_{sc}(l = 0) = \int \sigma_{sc}(\theta)\, d\Omega = 4\pi\sigma_{sc}(\theta) = \left(\frac{4\pi}{k^2}\right)\sin^2\delta_0 \quad ...(4.38\ b)$$

The angular distribution for $l = 0$ is evidently isotropic. The concept of phase shift is illustrated in Fig. 4.3 and is evidently related to the parameters of the potential $V(r)$. It is a quantity which can be experimentally measured.

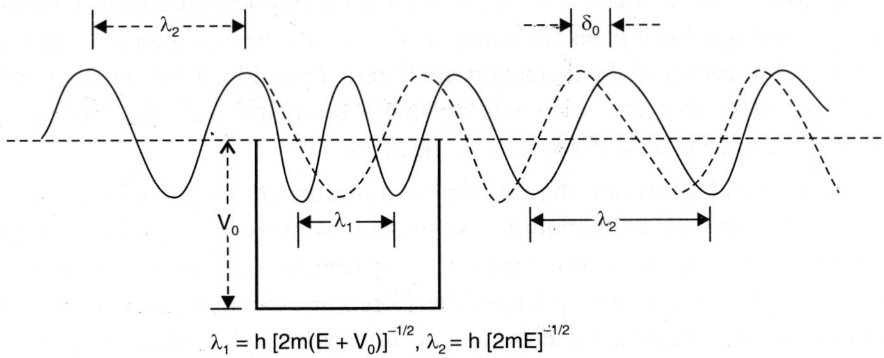

Fig. 4.3 Illustration of the concept of the phase shift δ_0.

4.2.2.2 Scattering Length

For incident energies approaching zero, *e.g.*, for thermal or cold neutrons, one normally uses a quantity called 'Scattering length' 'a' which is related to δ_0 as follows:

$$k \cot \delta_0 \equiv -\frac{1}{a} \qquad \text{...(4.39)}$$

for $$k \to 0;\ \delta_0 \to 0$$

The phase shift is also expected to go to zero, as $k \to 0$; otherwise the value of σ_{sc} ($l = 0$) will become infinite [see Eq. 4.38], which is not indicated by experiments.

The physical significance of 'a' may be seen from the behaviour of the scattered wave-function asymptotically outside the range of nuclear forces. Using relation,

$$\psi(r) = \frac{U_1(r)}{r}$$

and Eqs. 4.31 and 4.33, we can write:

$$\underset{\substack{l=0 \\ k=0}}{\psi(r)} = \frac{e^{i\delta_0} \sin(kr + \delta_0)}{kr}$$

$$= \frac{kr + \delta_0}{kr} e^{i\delta_0} = \frac{U_0(r)}{r} \qquad \text{...(4.40 } a\text{)}$$

So that,

$$U_0(r) = r \underset{k \to 0}{\psi_{sc}} = \left(r + \frac{\delta_0}{k}\right) e^{i\delta_0}$$

$$\underset{\substack{k \to 0 \\ \delta_0 \to 0}}{\to} \left(r + \frac{\delta_0}{k}\right) \underset{\substack{k \to 0 \\ \delta_0 \to 0}}{\to} (r - a) \qquad \text{...(4.40 } b\text{)}$$

It is seen from Eq. 4.40 that at $r = a$; $U_0 = 0$. Thus the physical significance of the scattering length 'a' is given by the fact that it gives the value of 'r' at which the asymptotic solution $U_0(r)$ of the scattered wave at very low energy of the incident beam is zero. From Fig. 4.4 a and 4.4 b it may be seen that if the internal wave function at $r = 0$ tends to bend towards the axis representing r, then 'a' is positive while if it tends to bend away from it, 'a' is negative.

A wave-function bending towards the axis represents the bound state, in which case it is expected that outside the potential, the wave-function decays to zero, as in the case of deuteron. On the other hand the wave-function bending away corresponds to scattering, where one expects the wave-function at $\gamma \to \infty$ to remain large. Positive scattering length, therefore, corresponds to attractive enough potential to give a bound state, while a negative scattering length corresponds to less attractive potential so that it gives rise to unbound scattering state at low incident energies.

We further see, from the definition of $f(\theta)$ and $\sigma(\theta)$ [Eqs. 4.8, 4.24 and 4.27] that:

$$\sigma_{sc}(\theta) = |f(\theta)|^2 = a^2 \qquad \text{...(4.41 } a\text{)}$$

and $$\sigma_{sc} = \int \sigma_{sc}(\theta)\, d\Omega = 4\pi a^2 \qquad \text{...(4.41 } b\text{)}$$

The value of 'a' is thus expected to be of the order of the radius of the bound system (in the case when it is positive), because beyond this, the wave-function is supposed to vanish. Classically, one knows that the maximum value of the cross-section is πa^2. The enhancement by a factor of 4 in Eq. 4.41b is a result of the quantum mechanical considerations. Classically, the particle grazing at the boundaries go straight, casting a geometrical shadow of the target, while quantum-mechanically there will be diffraction, at the boundaries giving rise to the large amount of small angle scattering. It is this small angle scattering which makes the quantum mechanical cross-section four times the classical value.

From detailed phase-shift analysis, it is possible to extract the values of phase shifts δ's and the scattering lengths as under various conditions from the experimental data; and then to correlate these values through the various theoretical equations as mentioned above with the properties of $V(r)$. This will be accomplished in the next section.

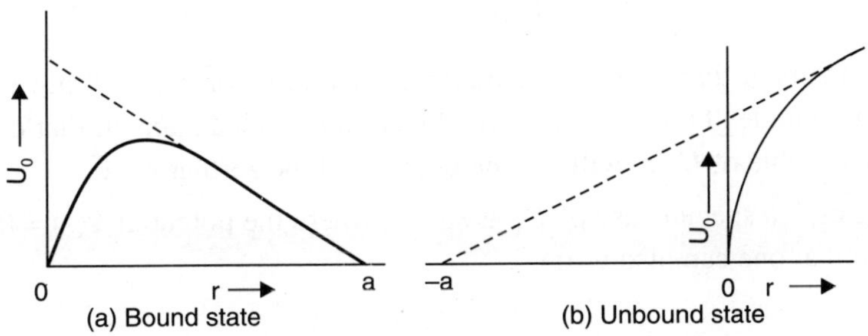

Fig. 4.4 Radial; wave-function U_0 under conditions of (a) Bound state and (b) Unbound state, demonstrating the concept of scattering length.

4.2.2.3 Effective Range Theory

Here, we reformulate the theory[4] of scattering in a manner, that the range over which the potential is especially effective, comes explicitly in the expression of the scattering cross-section along with the phase shift and scattering length. The theory is applicable only for low energies.

Let us consider the radial equations for low energy neutron-proton scattering at kinetic energies E_1 and E_2. Then, the Schrödinger equation at these two energies will be given by:

$$\frac{d^2U_1(r)}{dr^2} + \left[k_1^2 - \frac{M}{\hbar^2}V(r)\right]U_1(r) = 0 \qquad \text{...(4.42 }a\text{)}$$

and

$$\frac{d^2U_2(r)}{dr^2} + \left[k_2^2 - \frac{M}{\hbar^2}V(r)\right]U_2(r) = 0 \qquad \text{...(4.42 }b\text{)}$$

where subscripts 1 and 2 correspond to E_1 and E_2 respectively. Multiplying Eq. 4.42 a by $U_2(r)$ and Eq. 4.42 b by U_1 (r) and subtracting the latter from the former, one gets:

$$U_2U_1'' - U_1U_2'' = U_1U_2(k_2^2 - k_1^2) \qquad \text{...(4.43)}$$

or

$$\frac{d}{dr}(U_2U_1' - U_1U_2') = U_1U_2(k_2^2 - k_1^2) \qquad \text{...(4.44)}$$

where U_1' and U_2' denote the first order differentiation w.r.t. time and U_1'' and U_2'' denote the second order differentiation. Integrating the two sides, one obtains:

$$| U_2U_1' - U_1U_2' |_0^\infty = (k_2^2 - k_1^2) \int_0^\infty U_1U_2 \, dr \quad ...(4.45)$$

We now introduce an auxiliary function v_1 which represents the asymptotic behaviour of U_1 at large distance, given by:

$$v_1 \equiv C_1 \sin (k_1 r + \delta_0) \quad ...(4.46)$$

(compare Eq. 4.46 with Eq. 4.33 for $l = 0$)

We choose C_1 so that $v_1 = 1$ at $r = 0$, which gives:

$$v_1 = \frac{\sin (k_1 r + \delta_{10})}{\sin \delta_{10}} \quad ...(4.47)$$

One should remember that actually the internal wave function U_1, will never be 1 at $r = 0$. We have only fixed the amplitude of the external wave-function in such a manner, that if extended back, its value is 1 at $r = 0$. The value of $U_1(r)$ will, on the other hand, be zero at $r = 0$.

Now v_1 being only a special case of U_1 at $r \to \infty$ where the potential $V(r) = 0$, Eq. 4.45 may be applied to v's also so that one can also write:

$$\left| v_2 v_1' - v_1 v_2' \right|_0^\infty = (k_2^2 - k_1^2) \int_0^\infty v_1 v_2 \, dr \quad ...(4.48)$$

We now subtract Eq. 4.45 from Eq. 4.48. Remembering that:

$$(U_1)_{r\to\infty} = (v_1)_{r\to\infty} \,; (U_2)_{r\to\infty} = (v_2)_{r\to\infty}$$

and

$$(U_1)_{r=0} = (U_2)_{r=0} = 0; (v_1)_{r=0} = (v_2)_{r=0} = 1 \quad ...(4.49)$$

We obtain,

$$(v_1 v_2' - v_2 v_1') |_{r=0} = (k_2^2 - k_1^2) \int_0^\infty (v_1 v_2 - U_1 U_2) \, dr \quad ...(4.50\ a)$$

or

$$(v_2' - v_1') |_{r=0} = (k_2^2 - k_1^2) \int_0^\infty (v_1 v_2 - U_1 U_2) \, dr \quad ...(4.50\ b)$$

The left hand side of Eq. 4.50 *b* may be further written in terms of the phase shifts δ_{10} and δ_{20} by using Eq. 4.47 so that one gets:

$$k_2 \cot \delta_{20} - k_1 \cot \delta_{10} = (k_2^2 - k_1^2) \int_0^\infty (v_1 v_2 - U_1 U_2) \, dr$$

$$= (k_2^2 - k_1^2) \left[\frac{1}{2} \rho(E_1, E_2) \right] \quad ...(4.51)$$

where

$$\rho(E_1, E_2) \equiv \int_0^\infty (v_1 v_2 - U_1 U_2) \, dr \quad ...(4.52)$$

$\rho(E_1, E_2)$ is called 'effective range' for energies E_1 and E_2. The quantity $\rho(E_1, E_2)$ has the dimension of length, and is of the order of the effective range of nuclear potential, because outside the nuclear range, $v_1, v_2 = U_1, U_2$ and the integral in Eq. 4.52 vanishes so that the integral is only effective inside the nuclear range. The integral is a difference quantity and quite sensitive to the properties of nuclear potential. One should remember that v's also depend on the potential. As $k \to 0$, $\sin \delta_{10} \to \delta_{10}$ and $\cos \delta_{10} \to 1$, so that, [Eq. 4.40 *b*]:

$$\underset{k_1 \to 0,\, \delta_{10} \to 0}{k_1 \cot \delta_{10}} \to \frac{k_1}{\delta_{10}} = -\frac{1}{a} \qquad \text{...(4.53)}$$

Replacing δ_{20} by δ, for $k_2 = k$ and also v_2 by v and U_2 by U, and δ_{10} by δ, for $k_i \to 0$, and also v_1 by v_0 and U_1 by U_0, we rewrite Eq. 4.51 as:

$$k \cot \delta = -\frac{1}{a} + k_2^2 \int_0^\infty (v_0\, v - U_0 U)\, dr$$

$$= -\frac{1}{a} + \frac{1}{2} k^2 \rho\,(0, E) \qquad \text{...(4.54)}$$

Also

$$\underset{k \to 0}{v} = \frac{\sin (kr + \delta)}{\sin \delta} = 1 + \underset{k \to 0}{kr} \cot \delta \to 1 - \frac{r}{a} \qquad \text{...(4.55)}$$

So that, for small values of k, v is independent of k, and may be equated to v_0. Again inside the nuclear range, the potential $(M/\hbar^2)\, V(r)$ is much larger than k^2, so that k^2 may be neglected in comparison to $(M/\hbar^2)\, V(r)$ and hence $V(r)$ may also be, in the first approximation, independent of k when $k \to 0$.

One may also write $U = U_0$ and $v = v_0$ for $k \to 0$.

Therefore for $k \to 0$, one may write:

$$k^2 \int_0^\infty (v_0\, v - U_0 U)\, dr = k^2 \int_0^\infty (v_0^2 - U_0^2)\, dr \equiv \frac{1}{2} k^2 r_0 \qquad \text{...(4.56)}$$

where

$$r_0 \equiv 2 \int_0^\infty (v_0^2 - U_0^2)\, dr = \rho(0, 0) \qquad \text{...(4.57)}$$

The quantity r_0 gives an alternative definition of the effective range. Eq. 4.54 may now be written as:

$$\underset{k \to 0}{k \cot \delta} = -\frac{1}{a} + \frac{1}{2} k^2 r_0 \qquad \text{...(4.58)}$$

We, thus, obtain a relation between the phase-shift, scattering length and the effective range. One should realise that effective range contains the properties of the potential $V(r)$; so also δ is related to $V(r)$. For higher approximations, one expands U and v around $k^2 = 0$ so that:

$$U(r) = U_0(r) + k^2 \left(\frac{dU\,(r)}{dk} \right) + \ldots$$

$$v(r) = v_0(r) + k^2 \left(\frac{dv(r)}{dk}\right) + \ldots \qquad \text{...(4.59)}$$

Remembering that $r = 0$,

$$U_0(0) = 0, \; \frac{dU}{dk} = 0; \; v(0) = 1 \text{ and } \frac{dv}{dk} = 0 \qquad \text{...(4.60)}$$

We can write:

$$\underset{k \to 0}{v(r)} = \frac{\sin (kr + \delta)}{\sin \delta} = 1 + kr \cot \delta$$

$$= 1 + \left[-\frac{1}{a} + \frac{1}{2} k^2 \rho (0, E)\right] \qquad \text{...(4.61)}$$

where

$$\rho(0, E) = 2 \int_0^\infty [v_0(r)\, v(r) - U_0(r)\, U(r)]\, dr$$

$$= \int_0^\infty [v_0^2(r) - U_0^2(r)]\, dr$$

$$+ 2k^2 \int_0^\infty \left[v_0(r)\left(\frac{dv}{dk}\right)_{k=0} - U_0\left(\frac{dU}{dk}\right)_{k=0}\right] dr \qquad \text{...(4.62)}$$

Defining P as:

$$P \equiv -\frac{1}{r_0^3} \int_0^\infty [v_0(r)\, v_0'(r) - U_0(r)\, U_0'(r)]\, dr$$

$$\left[v_0'(r) = \left(\frac{dv}{dk}\right)_{k=0}; U_0'(r) = \left(\frac{dU}{dk}\right)_{k=0}\right] \qquad \text{...(4.63)}$$

We obtain,

$$\rho(0, E) = r_0 - 2Pk^2 r_0^3 \qquad \text{...(4.64 a)}$$

and

$$k \cot \delta = -\frac{1}{a} + \frac{1}{2} k^2 r_0 - Pk^4 r_0^3 \qquad \text{...(4.64 b)}$$

It may be added that Eq. 4.58, which does not involve P is less sensitive to the exact shape of the potential chosen, than Eq. 4.46 *b*. Hence Eq. 4.58 is called shape-independent equation.

4.2.2.4 Effective Range Theory and Deuteron Problem

It is possible to relate the above equations with the parameters of deuteron problem. Like Eq. 4.42, we write the radial equation of the ground state of the deuteron as:

$$U_g'' + \left[\frac{M|E_B|}{\hbar^2} - \frac{M}{\hbar^2} V(r)\right] U_g(r) = 0 \qquad \text{...(4.65)}$$

As before, $U_g(0) = 0$ and $v_g\,(0) = 1$. Remembering that $v_g(r) = C_1 e^{-\gamma_g r}$

where
$$\gamma_g = \frac{\sqrt{M|E_B|}}{\hbar}$$

For $v_g(0) = 1$, we see that $C_1 = 1$. We, now, define:

$$\rho(E_0, E) \equiv 2\int_0^\infty [v_g(r)\,v(r) - U_g(r)\,U(r)]\,dr \qquad \text{...(4.66 } a)$$

and
$$\rho(0, -E_B) \equiv \rho(-E_B, 0)$$

$$= \int_0^\infty [v_g(r)\,v_0(r) - U_g(r)\,U_0(r)]\,dr \qquad \text{...(4.66 } b)$$

Now it can be seen that if one starts with Eqs. 4.65 and 4.66 *a* and 4.66 *b* instead of 4.42 and 4.52 and proceeds in the same manner as before, one gets a relationship,

$$k\cot\delta = -\gamma_g + \frac{1}{2}(k^2 + \gamma_g^2)\int_0^\infty [v_g(r)\,v(r) - U_g(r)\,U(r)]\,dr \qquad \text{...(4.67 } a)$$

or
$$k\cot\delta = -\gamma_g + \frac{1}{2}(k^2 + \gamma_g^2)\,\rho(-E_B, E) \qquad \text{...(4.67 } b)$$

If, $k \to 0$, the above equation is reduced to:

$$k\cot\delta = -\gamma_g + \frac{1}{2}\gamma_g^2\,\rho(-E_B, 0)$$

or
$$\frac{1}{a} = \gamma_g - \frac{1}{2}\gamma_g^2\,\rho(-E_B, 0) \qquad \text{...(4.68 } a)$$

If $|E_B| << |V|$ one may replace $\rho(-E_B, 0)$ by $\rho(0, 0) \equiv r_0$. Then (4.68 *a*) reduces to:

$$\frac{1}{a} = \gamma_g - \frac{1}{2}\gamma_g^2 r_0 \qquad \text{...(4.68 } b)$$

Similarly treating the problem to a higher approximation, it may be seen that:

$$\frac{1}{a} = \gamma_g - \frac{1}{2}\gamma_g^2 r_0 - P\gamma_g^4 r_0^3 \qquad \text{...(4.69 } a)$$

where
$$P \equiv \frac{1}{\rho(0, 0)}\int_0^\infty (v_0\,v_0' - U_0 U_0')\,dr$$

In general, we may replace γ_g by γ to take into account other situations of bound state, and write:

$$\frac{1}{a} = \gamma - \frac{1}{2}\gamma^2\,r_0 - P\gamma^4 r_0^3 \qquad \text{...(4.69 } b)$$

Equation 4.69 *b* is the counterpart of Eq. 4.64 *b* for the deuteron case, where the energy of the *n-p* system, is taken as $-E_B$.

4.2.2.5 Neutron-Proton Scattering Cross-sections in Effective Range Theory

In general, when unpolarised neutrons are scattered from protons in a target of atomic hydrogen, and we have incoherent scattering, both triplet and singlet states of *n-p* system are formed, with a weightage of 3 : 1. The total cross-section, is, then given by:

$$\sigma_T = \frac{3}{4}\sigma_t + \frac{1}{4}\sigma_s \quad \text{...(4.70 a)}$$

$$= \frac{3}{4}\frac{4\pi}{k^2}\sin^2\delta_t + \frac{1}{4}\frac{4\pi}{k^2}\sin^2\sigma_s$$

$$= \frac{3\pi}{k^2 + k^2\cot^2\delta_t} + \frac{\pi}{k^2 + k^2\cot^2\delta_s}$$

$$= \frac{3\pi}{k^2 + \left[\frac{1}{a_t} + \rho_t\,(0, E)\frac{k^2}{2}\right]^2} + \frac{\pi}{k^2 + \left[\frac{1}{a_s} + \rho_s\,(0, E)\frac{k^2}{2}\right]^2} \quad \text{...(4.70 b)}$$

In the approximation of very low energies, one may write:

$$\rho_{t,\,s}\,(0, E) = \rho_{t,\,s}(0, 0) \equiv r_{t,\,s} \quad \text{...(4.70 c)}$$

However, if we use the deuteron parameters, and use Eq. 4.68 *b*, we get the cross-section corresponding to *n-p* scattering in the triplet state at very low energies, as the deuteron state corresponds to the triplet state of *n-p*. We then have from Eq. 4.68 *b*:

$$\sigma_t = 4\pi a_t^2 = \frac{4\pi}{\left(\gamma_g - \frac{1}{2}\gamma_g^2 r_t\right)^2} \quad \text{...(4.71)}$$

It may be emphasised that the quantity γ_g does not represent the energy of the neutrons which is given by $k \to 0$. Equation 4.71, only used the deuteron parameters, because they represent the relationship for triplet state for which we want to obtain the scattering cross-section.

Assuming that deuteron represents a triplet state of *n-p* system; the nuclear range ($r_0 \approx 2.0$ fm) of its potential, may be taken approximately as for the triplet range, r_t. We will, however, see later on, that the exact value of r_t comes out to be 1.7×10^{-13} cm; which we can use in Eq. 4.71.

***Determination of Scattering Lengths*:** From the deuteron problem, as discussed in Chapter 3, we obtain the values;

$$\gamma_g = \frac{\sqrt{M\,|\,E_B\,|}}{\hbar} \text{ and } \frac{1}{\gamma_g} = R = 4.3 \times 10^{-13}\text{ cm} \quad \text{...(4.72 a)}$$

Using the above parameters in Eq. 4.71 one obtains

$$\sigma_t = 4\pi a_t^2 \approx 3.6 \times 10^{-24}\text{ cm}^2 \quad \text{...(4.72 b)}$$

giving

$$a_t = \pm\, 5.3 \times 10^{-13}\text{ cm} \quad \text{...(4.72 c)}$$

Further using Eq. 4.70, one may write for the case of incoherent scattering:

$$\sigma_{\text{total}} = 20.69 \times 10^{-24} \text{ cm}^2$$

$$= \frac{3}{4}(3.6 \times 10^{-24}) + \frac{1}{4}\sigma_s \qquad ...(4.73\ a)$$

From which we get:

$$\sigma_s \approx 70.5 \times 10^{-24} \text{ cm}^2 \qquad ...(4.73\ b)$$

giving

$$a_s \approx \pm\, 24.0 \times 10^{-13} \text{ cm} \qquad ...(4.73\ c)$$

The signs of scattering lengths can be obtained only from coherent scattering.

The experimental data of incoherent scattering of neutrons from free protons could be analysed in terms of either Eqs. 4.58 or 4.64 *b*. In practice, one can determine the values of r_{ot} and P_t for *n-p* scattering from the analysis of Eqs. 4.68 *b* and 4.69*a* based on deuteron properties; as the binding energy of deuteron have been measured more accurately than the scattering cross-section of neutron from protons. The value of r_{ot} may be first determined from Eq. 4.68 *b* for shape-independent case and P_t is determined from definition in Eq. 4.63. Then by reiterative process one can determine the self-consistent value of r_{ot} and P_t from Eq. 4.69 *a*. The value of a_t as determined earlier with the positive sign (as will be seen later in the analysis of coherent scattering) is used in this analysis.

Table 4.1, gives the values of r_{ot} and P_t for different potentials obtained in this manner[5].

Table 4.1 *The effective range r_{ot} for shape independent case, and for different potentials along with P_t– for triplet neutron-proton system, determined from zero energy scattering data for three potentials*[5].

Well Shape	r_{ot} *(10^{-13} cm)*	P_t
1. Square well	1.726 ± 0.028	– 0.04
2. Experimental	1.689 ± 0.027	0.029
3. Yukawa	1.639 ± 0.026	0.137
4. Shape Independent	1.704 ± 0.028	0

From Eq. 4.68 *b*, one may neglect the much smaller second term of $-1/2\,\gamma_g^2\, r_0$ on the right side, so that

$$a_{s,\,t} = \left|\frac{1}{\gamma_{s,t}}\right| \qquad ...(4.74)$$

The value of E_{BS} the energy of the *n-p* system in the singlet state may be obtained from:

$$\gamma_s = \frac{\sqrt{ME_{BS}}}{\hbar}$$

using the value of a_s in Eq. 4.73. A value of E_{BS} of 66 keV satisfies this equation. The experiments as described later show that the sign of a_S is negative which means that this corresponds to the unbound state, of *n-p* system. The energy of 66 keV may be taken to be the energy above the zero-energy for the virtual singlet state of the *n-p* system. Experimentally[5], a resonance in *n-p* scattering has been observed at 66 keV, corresponding to this virtual singlet state.

The alternative method of determining r_s from coherent scattering is given in the next section.

4.2.2.6 Coherent Scattering (Theoretical)

We shall now see, how the coherent scattering data yields the values and signs of the two scattering lengths. The coherent scattering, as mentioned before, is observed under conditions when distance between the two protons is less or of the order of the wavelength of the incident neutrons. The scattering cross-section for hydrogen molecule, under these conditions, is given by:

$$\sigma_{coh}\text{ (molecular)} = 4\pi\left(\sum_i a_i\right)^2 \qquad \text{...(4.75)}$$

where a_1 and a_2 are the scattering lengths from proton 1 and 2 for scattering from a hydrogen molecule. We now define the two projection operators, P_s and P_t as follows:

$$P_s \equiv \frac{1}{4} - \mathbf{S}_n + \mathbf{S}_p$$

and

$$P_t \equiv \frac{3}{4} + \mathbf{S}_n + \mathbf{S}_p \qquad \text{...(4.76)}$$

Remembering that:

$$|\,\mathbf{S}\,|^2 = |\,\mathbf{S}_n\,|^2 + |\,\mathbf{S}_p\,|^2 + 2\mathbf{S}_n\,.\,\mathbf{S}_p$$

with

$$|\,\mathbf{S}_n\,|^2 = |\,\mathbf{S}_p\,|^2 = \frac{1}{2}\left(\frac{1}{2}+1\right) = \frac{3}{4}$$

It can be seen that:

$$|\,\mathbf{S}\,|^2 = 1(1 + 1) = 2 \qquad \text{for triplet state} \qquad \text{...(4.77)}$$

$$= 0(1 + 0) = 0 \qquad \text{for singlet state} \qquad \text{...(4.78)}$$

It is, then easy to see that:

$$P_s = 1 \qquad \text{for singlet state}$$

$$= 0 \qquad \text{for triplet state} \qquad \text{...(4.79 a)}$$

and

$$P_t = 0 \qquad \text{for singlet state}$$

$$= 1 \qquad \text{for triplet state} \qquad \text{...(4.79 b)}$$

we now define, $a_{1,\,2}$ as:

$$a_{1,\,2} = a_s\,P_{s_{1,2}} + a_t\,P_{t_{1,2}} \qquad \text{...(4.80 a)}$$

Then Eq. 4.80 *a* coupled with Eq. 4.79 fixes the physical meaning of the projection-operators P_s and P_t. They switch on or off the a_s or a_t for a given scattering, depending on whether the scattering is from a singlet state or triplet state. We can, therefore, write:

$$a_1 = a_s\left(\frac{1}{4} - \mathbf{S}_n \cdot \mathbf{S}_{p_1}\right) + a_t\left(\frac{3}{4} + \mathbf{S}_n \cdot \mathbf{S}_{p_1}\right) \qquad \text{...(4.80 } b\text{)}$$

and

$$a_2 = a_s\left(\frac{1}{4} - \mathbf{S}_n \cdot \mathbf{S}_{p_2}\right) + a_t\left(\frac{3}{4} + \mathbf{S}_n \cdot \mathbf{S}_{p_2}\right) \qquad \text{...(4.80 } c\text{)}$$

so,

$$a = a_1 + a_2 = \frac{a_s + 3a_t}{2} + (a_t - a_s)\,\mathbf{S}_n \cdot (\mathbf{S}_{p_1} + \mathbf{S}_{p_2}) \qquad \text{...(4.81)}$$

Now for parahydrogen, $\mathbf{S}_H = \mathbf{S}_{p_1} + \mathbf{S}_{p_2} = 0$

so that, from Eqs. 4.75, 4.81 and 4.82 we get:

$$\sigma\,(\text{para-molecule}) = 4\pi\left(\frac{a_s + 3a_t}{2}\right)^2 \qquad \text{...(4.82)}$$

For orthohydrogen $\mathbf{S}_H = \mathbf{S}_{p_1} + \mathbf{S}_{p_2}$

$$\text{giving } \sigma\,(\text{ortho-molecule}) = 4\pi\left|\frac{(a_s + 3a_t)}{2} + (a_t - a_s)\,\mathbf{S}_n \cdot \mathbf{S}_H\right|^2 \qquad \text{...(4.83)}$$

We note that $|\mathbf{S}_n \cdot \mathbf{S}_H|_{AV=0}$

because the neutron spin is randomly oriented and is as many times positive as negative, for any given direction-say that of the $\mathbf{S}_H$. Further,

$$\begin{aligned}|\mathbf{S}_n \cdot \mathbf{S}_H|^2_{AV} &= (S^2_{nx}\,S^2_{Hx} + S^2_{ny}S^2_{Hy} + S^2_{nz}\,S^2_{Hz})_{AV} \\ &+ (S_{nx}S_{ny}S_{Hx}S_{Hy} + S_{nx}S_{nz}S_{Hx}S_{Hz})_{AV} \\ &+ (S_{ny}S_{nx}S_{Hy}S_{Hx} + S_{ny}S_{nz}S_{Hy}S_{Hz})_{AV} \\ &+ (S_{nz}S_{nx}S_{Hz}S_{Hx} + S_{nz}S_{ny}S_{Hz}S_{Hy})_{AV}\end{aligned} \qquad \text{...(4.84)}$$

Average of each of the six terms, in the last three brackets is zero, because each component of these terms will be as many times positive as negative, so that,

$$|\mathbf{S}_n \cdot \mathbf{S}_H|^2_{AV} = (S^2_{nx}\,S^2_{Hx} + S^2_{ny}S^2_{Hy} + S^2_{nz}\,S^2_{Hz})_{AV} \qquad \text{...(4.85)}$$

Now

$$S^2_{nx} = S^2_{ny} = S^2_{nz} = \frac{1}{4}$$

and

$$(S^2_{Hx} + S^2_{Hy} + S^2_{Hz})_{AV} = (S^2_H)_{AV} = 1(1 + 1) = 2$$

Therefore,

$$\sigma\,(\text{ortho-molecule}) = 4\pi\left(\left(\frac{a_s + 3a_t}{2}\right)^2 + (a_t - a_s)^2\,|\mathbf{S}_n \cdot \mathbf{S}_H|^2_{AV}\right)$$

$$= \pi (a_s + 3a_t)^2 + 2\pi (a_t - a_s)^2 \qquad ...(4.86)$$

Combining Eqs. 4.82 and 4.86, we obtain:

$$\sigma_{ortho} = \sigma_{para} + 2\pi (a_t - a_s)^2 \qquad ...(4.87\ a)$$

and

$$\frac{\sigma_{ortho}}{\sigma_{para}} = 1 + \frac{2(a_t - a_s)^2}{(3a_t + a_s)^2} \qquad ...(4.87\ b)$$

Note that here we are considering an unbound *n-p* system.

One should, however, take into account the effect of chemical binding energy of the hydrogen atoms in the molecule at these low energies of incident neutrons. It may be seen that in the case of neutron scattering from H_2 molecules:

$$\frac{a_{free}}{a_{bound}} = \frac{2\mu_{free}}{2\mu_{bound}}$$

where

$$\frac{1}{\mu_{free}} = \frac{1}{M_p} + \frac{1}{M_n} \approx \frac{2}{M} \text{ and } \frac{1}{\mu_{bound}} = \frac{1}{M_p} + \frac{1}{2M_n} \approx \frac{3}{2M}$$

Hence

$$\frac{\sigma_{bound}}{\sigma_{free}} = \frac{a_{bound}^2}{a_{free}^2} \approx \frac{16}{9} \qquad ...(4.88)$$

The values of σ_{ortho} and σ_{para} which correspond to the bound states are obtained theoretically by multiplying by 16/9 the free values calculated in Eqs. 4.82 and 4.87. Using the experimental values of σ para-molecular[5] ≈ 4 barns, and σ (ortho-molecular) ≈ 125 barns; we obtain the accepted values of $a_s/a_t \approx -4.55$. The other solution, for a_s/a_t gives too large value of a_t and does not agree with the magnitude in Eq. 4.72*c*. Knowing that a_t is positive (from the fact that the triplet state of *n-p*, *i.e.*, deuteron is a bound state), we conclude, that a_s is negative.

The presently accepted values of a_s and a_t from the best masurements on incoherent and coherent scattering are[5],

$$a_s = -23.73 \pm 0.07 \times 10^{-13} \text{ cm.}$$

$$a_t = 5.37 \pm 0.004 \times 10^{-13} \text{ cm.} \qquad ...(4.89)$$

The negative sign of the scattering length a_s, for singlet state, shows that we cannot have a bound state of *n-p* with antiparallel spins; while there exists, in nature, the bound state of *n-p* with parallel spins (deuteron), for which a_t is positive. The experimental measure value of $f_H = -3.78 \times 10^{-13}$ cm, confirms the negative value of a_s and the positive value of a_t, as well as their absolute magnitudes. The accepted zero energy *n-p* cross-sections for triplet and singlet state scattering are:

$$\sigma_{ot} = 4\pi a_t^2 = 3.63 \pm 0.029 \text{ barns}$$

$$\sigma_{os} = 4\pi a_s^2 = 70.52 \pm 0.35 \text{ barns} \qquad ...(4.90)$$

One may combine the results of coherent and incoherent scattering to obtain the singlet effective range r_{os}. One substitutes in Eq. 4.70 *b*, the values of a_t and a_s; as obtained from coherent scattering; and the value of r_{ot}, as obtained in Table 4.1, and obtain the values of r_{os} for shape independent case. The values of *P* may, then, be obtained by using the reiterative procedure in Eq. 4.64 *a*:

$$\rho_s(0, E) = r_{os}(1 - 2P_s r_{os}^2 k^2) \qquad ...(4.91)$$

and Eq. 4.70 *b* for different potentials using the data on incoherent scattering for different values of *k*, for very low values of *k*.

The results of such calculation is given[6] in Table 4.2.

Table 4.2 *The effective, and intrinsic ranges, and shape dependent parameters, in the singlet neutron-proton system, determined from low energy data for three potential types:*

Well Shape	r_{os} $(10^{-13}$ *cm*)	P_s
Shape independent	2.40 ± 0.23	0
Square well	2.47 ± 0.23	– 0.03
Experimental	2.37 ± 0.23	+ 0.01
Yukawa	2.21 ± 0.23	+ 0.058

4.3 PROTON-PROTON SCATTERING

4.3.1 Experimental Facts

4.3.1.1 Total Cross-sections

The proton-proton scattering data is obtained more precisely than the neutron-proton sacttering data, because of the availability of precise resolved mono-energetic incident beams of protons; as well as because of the detection sensitivity of nearly 100 per cent compared to 30–40 per cent the case of neutral neutrons. As mentioned earlier, the data is, however, available only above 100 keV due to the Coulomb barrier effects, which precludes the possibility of making measurements about coherent scattering. Low energy data up to 20–40 MeV is obtained from measurements using Van de-Graff, and is comparatively more precise. Data for higher energies is obtained from cyclotron and has poorer resolutions.

There is only singlet scattering at low energies in *p-p* scattering because due to the Pauli-exclusion principle, the two protons cannot have their spins in the same direction for $l = 0$. The data on total cross-section at low energies, therefore, directly gives information about the parameters in the singlet state. Figure 5.2, in the next chapter shows the behaviour of $\sigma(p, p)$, both at low and high energies. The data for low energies has been collected by Hess[7]; and others.

4.3.1.2 Angular Distributions (p-p scattering)

Figure 4.5 gives the angular distribution of protons scattered from proton targets upto 2.42 MeV of incident energies[7]. It also gives, angular distribution curves at 9.68 MeV and 25.63 MeV energies. The following feature of these curves are note-worthy.

The angular distribution has a very strong forward peaking, for very small angles up to 10°. As will be seen later, this is due to Coulomb scattering. There is a minimum at around 30° in many curves: this is shown clearly in Fig. 4.5. As the subsequent analysis will show, this minimum arises due to the interference between the nuclear and Coulomb scattering. The angular distribution is found to be isotropic from 30° to 90° and beyond. As will be explained later, this isotropy arises due to fortuitous cancellations of angle dependent interference terms in the expression of differential cross-section.

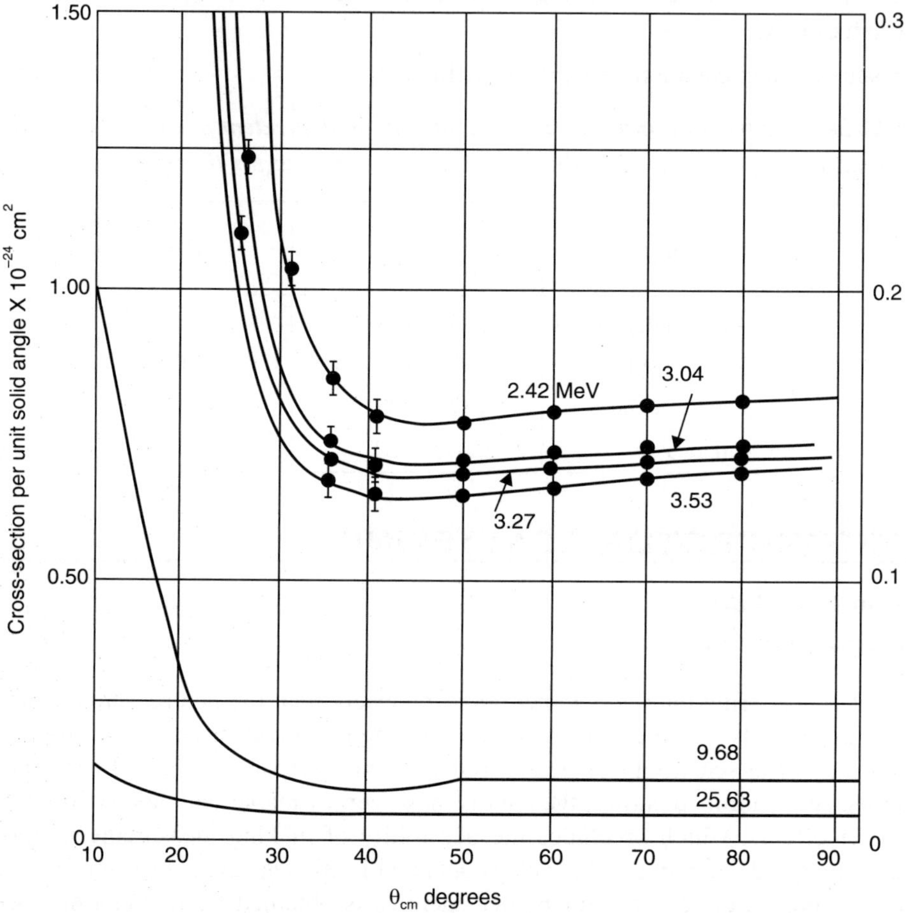

Fig. 4.5 Angular distribution for p-p scattering for energies 2.42, 3.04, 3.27 and 3.53 MeV (low energies) and for 9.68 and 25.63 MeV (higher energies).The lower curves (for higher energies) are plotted to the vertical scale on the left and higher curves (for low energies) to that on right (Ref. 7).

The angular distribution is symmetric around 90°. This is similar to *n-p* scattering and arises due to exchange part in the nuclear forces, and also due to indistinguishability between the two protons in *p-p* scattering.

4.3.2 Phase Shift Analysis of *p-p* Scattering

Broadly speaking, the problem of phase-shift analysis of *p-p* scattering is expected to be similar, in principle, to that of *n-p* scattering. But there are important differences of details. In proton-proton scattering, one should take into account the coulomb forces, apart from the nuclear forces. For low energies for which $l = 0$, the *p-p* scattering can only be in the singlet state. For higher values of l, it can either be a singlet state or a triplet state; depending on whether l is even or odd respectively. In the centre of mass system the scattered protons at say angle θ from the incident beam and the recoil protons from the target at an angle $\pi - \theta$ and vice versa cannot be distinguished. This position is explained and may be understood in Fig. 4.6. Because of this reasoning, the resultant scattering amplitude is given by:

$$f(\theta) + f(\pi - \theta) \text{ or } f(\theta) - f(\pi - \theta)$$

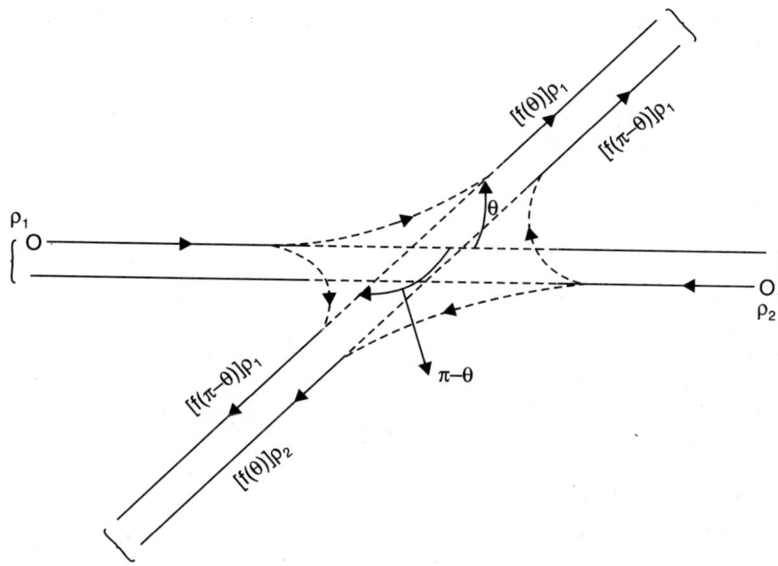

Fig. 4.6 Illustration of the scattering of p-p for angles of θ, and π - θ.

The first combination is symmetric with exchange of space coordinates, (*i.e.,* for $\theta \to \pi - \theta$); and hence is expected to be anti-symmetric with the exchange of spin-coordinates, *i.e.,* it should correspond to the singlet state of *p-p* system. On the other hand the second combination is anti-symmetric with the exchange of space coordinates and, hence, should be symmetric with the exchange of spins; and, therefore, corresponds to the triplet state of *p-p* system. This is the effect of Pauli Exclusion principle.

Keeping these points in mind, the phase shift analysis of *p-p* scattering is discussed below:

The Schrödinger equation for a proton-proton system may be written as:

$$\left[-\frac{\hbar^2}{2M}\nabla^2 + \frac{e^2}{r} + V(r)\right]\psi(r,\theta) = E\psi(r,\theta) \qquad \text{...(4.92 } a\text{)}$$

It differs from Eq. 4.23 of *n-p* interaction, essentially in the term, e^2/r which is due to the repulsive coulomb interaction. The term $V(r)$ represents the nuclear potential between the two protons. We assume it to be central for simplicity of our analysis, here. The solution of the above equation may, then be expressed as:

$$\psi(r,\theta) = \sum_{l} \frac{U_l(r)}{r} P_l(\cos\theta) \qquad \text{...(4.92 } b\text{)}$$

where $U(r)$ is given by the usual radial equation:

$$\left[\frac{d^2}{dr^2}+k^2-\frac{M}{\hbar^2}V(r)-\frac{M}{\hbar^2}\frac{e^2}{r}-\frac{l(l+1)}{r^2}\right]U_1(r)=0 \quad ...(4.93\ a)$$

with
$$k^2=\frac{ME}{\hbar^2}$$

The boundary condition $U_1(0) = 0$ should be satisfied here too. Asymptotically

$$V(r) \to 0, \text{ as } r \to \infty$$

and the solution at large distances should correspond to a pure coulomb potential. Without going into the mathematical details of solving the differential equations, with these conditions, (for which see (Ref. 8)) and defining:

$$I(r) \equiv \left[1+\frac{\eta}{kr}+\ln 2kr\right]e^{i(kr-\eta\ln 2kr)}\cos(\theta) \quad ...(4.94\ a)$$

$$S(r) \equiv \frac{e^{i(kr-\eta\ln 2kr)}}{r} \quad ...(4.94\ b)$$

and
$$f(\theta) \equiv \frac{1}{2ik}\sum_{l=0}^{\infty}(2l+1)(e^{2i(\sigma_1+\delta_1)}-1)\,P_1\cos(\theta) \quad ...(4.94\ c)$$

where
$$\eta \equiv \frac{Me^2}{2R^2K} \text{ and}$$

σ_1 is the phase shift due to coulomb scattering and δ_1 is the phase shift due to nuclear scattering. Then, we can write:

$$\psi_{r\to\infty} = I(r) + f(\theta)\,S(r) \quad ...(4.95)$$

Physically speaking, $I(r)$ may be regarded as the incident plane wave modified by the Coulomb forces, while $S(r)$ is the scattered wave modified by the Coulomb force (compare with Eq. 4.24). The function $f(\theta)$ is, of course, the amplitude of the scattered wave. The cross-section for *p-p* scattering is, therefore, given by:

$$\sigma(\theta) = |f(\theta)|^2 \text{ and } \sigma = \int \sigma(\theta)\,d\Omega \quad ...(4.96)$$

We have, of course, to take into account the fact of indistinguishability of the two protons as mentioned earlier. As the target material as well as incident protons have the protons with randomly oriented spins, one writes the cross-section for *p-p* scattering, after taking into account the proper weights of singlet and triplet scattering cases, so that

$$\sigma(\theta) = \frac{1}{4}|f_s(\theta)+f_s(\pi-\theta)|^2 + \frac{3}{4}|f_t(\theta)-f_t(\pi-\theta)|^2 \quad ...(4.97)$$

where
$$|f_s(\theta)+f_s(\pi-\theta)|^2 = |f_s(\theta)|^2 + |f_s(\pi-\theta)|^2 + f_s^*(\theta)f_s(\pi-\theta) + f_s(\theta)\,f_s^*(\pi-\theta) \quad ...(4.98\ a)$$

and $$| f_t(\theta) - f_t(\pi - \theta) |^2 = | f_t(\theta) |^2 + | f_t(\pi - \theta) |^2 - f_t^*(\theta) f_t(\pi - \theta) - f_t(\theta) f_t^*(\pi - \theta) \quad ...(4.98\ b)$$

where s, and t stand for singlet and triplet states. One should remember while summing over 1, that for odd 1 for which parity is negative; only triplet state is allowed, while for even 1, for which parity is positive, only the singlet state is allowed. This means that for the singlet state, one only sums over even l's while for triplet state, one sums over odd l's. For low energies only a few nuclear phase shifts will be effective. One, however, should take all the Coulomb phase shifts into account, because the Coulomb forces are effective up to large distances. One may divide $f(\theta)$ into two parts, so that

$$f(\theta) = f_c(\theta) + f_N(\theta) \quad ...(4.99\ a)$$

where $$f_c(\theta) \equiv \frac{-1}{2ik} \sum_{l=0}^{\infty} (2l+1)\,[e^{2i\sigma_l} - 1]\, P_l(\cos\theta) \quad ...(4.99\ b)$$

and $$f_N(\theta) \equiv \frac{-1}{2ik} \sum_{l=0}^{l=\infty} (2l+1)\, e^{2i\sigma_l}\, (e^{2i\delta_l} - 1)\, P_l(\cos\theta) \quad ...(4.99\ c)$$

It may be seen that $f_c(\theta)$ can be regarded as the coulomb scattering amplitude and $f_N(\theta)$ as the nuclear scattering amplitude. The phase shifts δ_1 are due to nuclear scattering; while σ_1 correspond to the Coulomb scattering.

For low energies say up to 5 MeV, the Coulomb barrier reduces the effect of nuclear interaction, so that the values of δ_1 for higher l's are quite small. Also for higher l's, the centrifugal barrier further reduces the nuclear interaction. For low energies one may, therefore, take nuclear phase shift only for $l = 0$ but Coulomb phase shift σ_1 for all l's. One then obtains,[8, 9] the following expression for the combined nuclear and Coulomb scattering for low incident energies ($l = 0$ for nuclear part; but all l's for Coulomb part), which we write without proving as:

$$\sigma(\theta) = \left(\frac{e^2}{2E}\right)^2 \left\{\left[\operatorname{cosec}^4\frac{\theta}{2} + \sec^4\frac{\theta}{2} - \operatorname{cosec}^2\frac{\theta}{2}\sec^2\frac{\theta}{2}\cos\left(2\eta\ln\tan\frac{\theta}{2}\right)\right]\right.$$

$$- \frac{2}{\eta}\sin\delta_\theta \left\{\operatorname{cosec}^2\frac{\theta}{2}\cos\left(\delta_0 + 2\eta\ln\sin\frac{\theta}{2}\right)\right.$$

$$\left.\left. + \sec^2\frac{\theta}{2}\cos\left(\delta_0 + 2\eta\ln\cos\frac{\theta}{2}\right)\right] + \frac{4}{\eta^2}\sin^2\delta_0\right\} \quad ...(4.100\ a)$$

One may obtain the pure Coulomb scattering (commonly called Mott scattering) by putting $\delta_0 = 0$ in the above expression so that:

$$\sigma_M(\theta) = \left(\frac{e^2}{2E}\right)^2 \left[\operatorname{cosec}^4 \frac{\theta}{2} + \sec^4 \frac{\theta}{2} - \operatorname{cosec}^2 \frac{\theta}{2} \sec^2 \frac{\theta}{2} \cos\left(2\eta \ln \tan \frac{\theta}{2}\right)\right] \quad ...(4.100\ b)$$

$\sigma(\theta) - \sigma_M(\theta)$ will, therefore, give the expression for *S*-wave differential cross-section involving nuclear interaction only, *i.e.*,

$$\sigma_N(\theta) = \sigma(\theta) - \sigma_M(\theta) = \left(\frac{e^2}{2E}\right)^2 \left\{-\frac{2}{\eta}\sin\delta_0 \left[\operatorname{cosec}^2 \frac{\theta}{2} \cos\left(\delta_0 + 2\eta \ln \sin\frac{\theta}{2}\right) + \sec^2\frac{\theta}{2}\cos\left(\delta_0 + 2\eta\ln\cos\frac{\theta}{2}\right)\right] + \left(\frac{4}{\eta^2}\right)\sin^2\delta_0\right\} \quad ...(4.100\ c)$$

Equation 4.100 *c* may be regarded as consisting of two parts: (1) the second term consisting of the pure nuclear part of $(4/\eta^2)\sin^2\delta_0$, and (2) the first term involving the interference between the nuclear and Coulomb scattering. It is this interference term that gives rise to the minimum in the angular distribution, around 20 to 30 degrees. The knowledge of the cross-section at this minimum yields the sign of δ_0. The magnitude of δ_0 may be obtained by comparing the experimental value of *p-p* scattering with Eq. 4.100*c*. The scattering due to pure Coulomb interaction is given by Eq. 4.100 *b*. Theoretically one expects infinite scattering for $\theta = 0$ or $\theta = \pi$. In practice, however, one does not achieve infinite cross-section for these angles, because the Coulomb potential produced at great distance and for small forward angles scattering is shielded by the electrons in the target atom.

4.3.3 Effective Range Theory of *p-p* Scattering[10]

Proceeding in a similar manner as for *n-p* scattering; but introducing e^2/r for Coulomb potential along with the nuclear potential $V(r)$; it has been possible to obtain the relationship, which we write down without proving, as:

$$K = R\left[-\frac{1}{a_p} + \frac{1}{2} r_{op}\, k^2 + P r_0^3\, k^4 + \ldots\right] \quad ...(4.101\ a)$$

where

$$P \equiv -\frac{1}{r_{op}^2}\int_0^\infty [\phi_0(r)\,\phi_0'(r) - U_0(r)\,U_0'(r)]\,dr \quad ...(4.101\ b)$$

Though the value of *P* is small, as shown in Table (4.3), it does have a definite value for different potentials[10]. It may be mentioned here that $\phi_0(r)$ is the auxiliary function like $v(r)$ in *n-p* scattering, representing the asymptotic behaviour of $U(r)$. Of course $\phi_0(r)$ is a wave function for $l = 0$, and $\phi_0^1(r)$ is the first-order differential with respect to time.

Experimentally, *p-p* scattering is dominated by pure Coulomb scattering. However, by using extrapolation procedure, one can subtract the Coulomb part from the total; and obtain the values of r_{op}, using the value of a_s, from *n-p* experiments. This yields $r_p = 2.65 \times 10^{-13}$ cm, which compares, quite favourably with the values of r_{os} given in Table (4.2) for r_{os} for *n-p* scattering. One can, therefore, infer that nuclear force is charge-independent.

Table 4.3 *The scattering parameters i.e., r_{os}, P, and a_s for singlet p-p scattering system[10]:*

Well Shape	r_{os} (10^{-13} cm)	P	a_s (for p-p) (10^{-13} cm)
Square well	2.55 ± 0.17	0.003313	– 7.68 ± 0.012
Exponential	2.673 ± 0.017	0.00907	– 7.698 ± 0.012
Yukawa	2.774 ± 0.017	0.05540	– 7.723 ± 0.012
Shape Independent	2.656 ± 0.017	0	– 7.694 ± 0.012

It is interesting to note, that while r_{os} is the same for *p-p* scattering as for *n-p* scattering; a_s is quite different, while the value of *P* is very small in both cases. For a_s this great difference is expected, since it depends explicitly on the *S*-wave nuclear phase shift for zero energy, which is strongly affected by the Coulomb force. Separating out this Coulomb effect is difficult and hence explicit calculation of the scattering parameters for a given potentials has to be carried out, if we want to compare the proton-proton and neutron-proton forces in IS-states.

It is possible to make a direct comparison of the singlet scattering length a_s of the *n-p* system; and the apparent scattering length a_p of the *p-p* scattering, where a_p corresponds to the scattering length that would be obtained in the absence of Coulomb interaction[11].

Without going into details (for which, the reader should refer to Ref. (11)), it may be stated that by writing relationship between a_p and *k*, and assuming the proton-proton nuclear potential to be the same as for neutron-proton singlet potential; it turns out[12] that,

$$a_p \approx -16 \times 10^{-13} \text{ cm.} \qquad \text{...(4.102)}$$

This is much larger, than $\approx -7 \times 10^{-13}$ cm. for a_s for *p-p* scattering as given in Table (4.3), but compares favourably with $a_s \approx -23 \times 10^{-13}$ cm for *n-p* scattering. Even the small difference with a_s for *n-p* scattering is explained, on the basis of the choice of the potential-depth or the difference in the magnetic interaction between two protons and a neutron and a proton. The values of a_s in Table (4.3) are strongly affected by Coulomb interaction, while the value in Eq. 4.102 is nuclear wave-function-dependent. Hence this difference.

We may therefore, presume for all practical purposes, that the hypothesis of charge independence for nuclear forces is supported by the values of r_{os} and a_s.

We may conclude that low energy *n-p* scattering as well as *p-p* scattering, intrinsically obey the same law of nuclear forces.

Also the discussion of this chapter shows, that nuclear forces are spin-dependent; and one can write the central forces as:

$$V_c(r) = V_0(r) + (\sigma_1 \cdot \sigma_2)\, V_1(r) \qquad \text{...(4.103)}$$

Of course, one should add the tensor forces, obtained from deuteron problem to obtain the complete expression for the nucleon-nucleon potential.

4.4 POLARISATION AT LOW ENERGIES

Recently there have become available beams of polarised protons, deuterons, and polarised light nuclei like say Li^6, etc. While the polarised protons and deuterons are obtained by using the atomic-beam polarised ion source[13], the $\vec{Li}^6$ (polarised Li^6) beam is obtained from a colliding beam polarised ion-source, as developed by G.S. Masson et al[14]. It has been possible to measure: (*i*) The analyzing power *Ay,* (*ii*) Analyzing powers T_{11}, T_{20}, T_{21}, and T_{22}, for say $d(\vec{p}, \gamma)$. He^3 (Polarised proton capture by deuteron at energies[15] for $40 < E < 110$ keV), or for $\vec{d} + d \rightarrow d + p + n$ break-up at 12 MeV[16], or Li^7 $(\vec{p}, \gamma)$ Be^8 at Ep = 80 – 0 keV (integrated from 0 to 80 keV)[17]. Similarly these analyzing powers have been measured for $\vec{L}i^6$ – He^4 scattering at an incident $\vec{L}i^6$ energy[18] of about 5.5 and 19.6 MeV. The meaning and physical significance of *Ay* and T_{11}, T_{20}, T_{21}, and T_{22}, have been given in the Appendix in Chapter *V*, and Ref. (19).

The motivation behind these measurements was two-fold (*i*) To compare the experimental values of analysing power *Ay* and T_{11}, T_{20}, T_{21}, and T_{22}, with theoretical models, *e.g.*, the role of meson exchange currents[20] (MEC), to have a check on an exact three body model, using realistic *N-N* potentials. (*ii*) For astrophysical interests, like *D*/*H* ratio in high red shift hydrogen clouds, in the theory of big bang[21], or to measure *S,* the astrophysical *S* factor. For details see reference[22].

The data, on H^2 $(\vec{p}, \gamma)$ and $(\vec{d}, \gamma)$ reactions; shows that meson-exchange calculations (MEC) are very important, while comparing the results of *S*, *A*2*y* and T_{20} with detailed theory involving the three particle wave function. Also the detailed comparison of *S*, with theories of cross-sections using MEC, helps determine stellar deuterium to hydrogen ratio (*D*/*H*), which when compared with astrophysical[23] theories, leads us to the model of the formation and expansion of protostellar core. These new developments, show how the low energy data is being used for detailed understanding of nuclear forces.

4. Nucleon-Nucleon Scattering at Low Energies 2000–2008

Measurements of *n-p* elastic scattering angular distribution at $E = 10$ MeV has been carried out and compared with charge dependent Brown and Nijmeegen Potential [Phy. Rev. C. 65, 014004, (2002)]. Also angular measurements of *p-d* scattering at $E_{c\text{-}m} = 667$ keV; (*i*) $\sigma(\theta)$ have been measured along with (*ii*) *Ay* (*iii*) iT_{11} (*iv*) T_{20}, (*v*) T_{21} (*vi*) T_{22}. Result are compared with nucleon-nucleon potential and also with the three nucleon (3N) potential [Phy. Rev. C. 65, 034002 (2002)].

In a theoretical analysis, of proton-deuteron scattering at a low energy of $E_p = 3$ MeV; using three nucleon, forces (3 *NF*), employing Fedeev equations, with central and tensor forces, it was found that there is a discrepancy with tensor part of 3NF. [Phy. Rev. C. 67, 06100 (R) (2003)].

In still another theoretical paper, the thermal behaviour of isoscalar ($\tau = 0$) and iso-vector ($\tau = 1$) proton-neutron pairing energies, at thermal energies, has been reported with shell model calculations, by Japanese authors (Kanako and Haregawa), in [Phy. Rev. C.72, 031302 (2005)]. It was interestingly found that at a finite temperature, the delicate balance between iso-scalar and iso-vector, *p-n* pairing energies on zero temperature disappears, and when the temperature rises, the iso-vector *p-n* pairing energy decreases and iso-scalar *p-n* energy increases.

In a paper authorized by 19 authors; from Europe, Canada and USA, this article reviews the new data taken at KVl, Netherlands and compares it with old data. Differential cross-section and analyzing power for $H^2(\vec{p}\,d)\,p$ and $\mathrm{H}^1(\vec{p}\,d)\,p$, reaction at 135 MeV/N and 65 MeV/N are measured. The differential data differs from previous measurement and consistently follows the energy dependence as expected from an interpolation of one taken over a long range at intermediate energies [Phy. Rev. C.78, 014006 (2008)].

REFERENCES

1. L.J. Rainwater, W. Heavens, C.C. Wu and J. Dunning: Phy. Rev. 73, 733 (1948), E. Melkonian: Phy. Rev. 76, 1750 (1949).
2. E. Fermi and H.W. Zinn: Phy. Rev. 70, 103 (1946), E. Fermi and L. Marshall: Phy. Rev. 71, 666 (1947).
3. D.J., Huges, M.T. Burgy, G.R. Ringo: Phy. Rev. 77, 291 (1950), Ibid 84, 1160 (1951).
4. J.S. Schwinger: Phy Rev. 72, 742 (1947), H.A. Bethe: Phy. Rev. 76, 38 (1949).
5. J.M. Blatt: Phy. Rev. 74, 92 (1948); J.M. Blatt and J. Jackson: Phy. Rev. 76, 18 (1949); Chew G. and M. Goldberger: Phy. Rev. 75, 16 (1949). G. Breit: Rev. Mod. Physics 23, 238 (1951). Handbuch der Physik. V. 39 (Springer-Verlag).
6. Shull, C.G., E.O. Wollan, G.A. Morton and W.L. Davidson: Phy. Rev. 73, 842 (1948).
7. J.M. Blair, G. Frier, E.E. Lamp, W. Sleator and J. Williams: Phy. Rev. 74, 553 (1948), M.J. Moravcsik: The Two Nucleon Interaction, Clarendon Press, Oxford, (1963), W.N. Hess: Rev. Mod. Physics, 30, 368 (1958).
8. M.K. Pal: Theory of Nuclear Structure, (Page 53-113) Affiliated East-West Press Pvt. Ltd., (1982). Encyclopaedia of Physics, Reference (4). Jackson J. and Blatt J.M.: Phy. Rev. 77, 122 (1950); M.C. Yovites; R.L. Smith, J. Bengston and G. Breit: Phy. Rev. 85, 540 (1952).
9. M.H. Hull, Jr., F.L. Yost, J.A. Wheeler and G. Breit: Phy. Rev. 49, 174, (1937), Mott N.F. and S.W. Massy: The Theory of Atomic Collisions: 2nd Ed. Oxford Press, (1949), Page 45 ff.
10. H.H. Hall, J.L. Powell, Phy. Rev. 90, 912, (1953), H.R. Worthington, J.N. McGruer and D.E. Findley: Phy. Rev. 90, 899 (1953).
11. G. Sachs Roberts: Nuclear Theory, (1953), Addison Wesley Publishing Company Inc., Cambridge 42, Mass: J.S. Schwinger: Phy. Rev. 78, 135 (1950).
12. V.P. Dzalepov, Yu M. Kazarinov, B.M. Golovin, V.B. Flyagin and V.I. Starov, Nuovo Cimento III, supplement No. 1, 61 (1956).
13. T.B. Clegg, H.J. Karawowsky, S.K. Lemieux, R.W. Sayer, E.R. Crosson, W.M. Hooke, C.R. Howell, H.W. Lewis, A.W. Lovette, H.J. Pfutzner, K.A. Sweton and W.S. Wilkburn: Nuclear Instruments. Methods A 357, 200 (1995).

14. G.S. Massen, T. Wise, P.A. Quin, and W. Haeberli: Nuclear Instruments Methods, A 242, 196 (1986).

15. L. Ma, H.J. Karawowski, C.R. Brune, Z. Ayer, T.C. Blackmon, and E.J. Ludwing, M. Viviani, A. Kievsky, and R. Schiavilla: Phy. Rev. C.V. 55, Page 588, (1997), G.T. Schmidt et al.: Phy. Rev. C-56, P-2565, (1997).

16. P.D. Flesher, C.R. Howell, W. Tonznov, M.L. Roberts, J.M. Hanly, G.J. Weisel, M.L. Oheli, and R.I. Walter, I. Slaus, and J.M. Lambert, P.A. Treado, G. Mertens, A.C. Fonseca, A. Soldi, B. Vlahovic: Phy. Rev. C. 56, p. 38 (1997).

17. M.A. Godwin, C.M. Laymen, R.L. Prior, D.R. Trilley and H.R. Weller: Phy. Rev. 56, p. 1605 (1997).

18. E.A. George, D.D. Pomcasavant, and L.D. Knutson: Phy. Rev. C. 56, p. 270 (1997).

19. K. Stephensen and W. Haeberli: Nuclear Instruments and Methods V. 169, p. 483 (1980).

20. A.C. Phillips: Nuclear Physics, A 184 (1972).

21. D. Tytter, X–M, Fan and S. Burles, Nature (London), 381, 207 (1996).

22. F.E. Cecil, D. Ferg, H. Liu, J.R. Scorby, J.A. Nc-Neil and P.D. Kunz: Nuclear Physics A. 536, 96 (1992).

23. S.W. Stahler: Astrophysics, J 322, 804 (1988).

PROBLEMS

1. Starting with about 10 MeV neutrons (as available in the case of a reactor); how much thickness of water, they have to pass approximately before they become thermalised, *i.e.,* $E = -\ 0.025$ eV. [For *n-p* cross-sections at various energies; consult, Fig. 3.1].

2. Using $V(r) = -\ V_0 = -\ 40$ MeV, calculate $\delta_1\ (r)$ for $r \rightarrow \infty$; and from it Co and δ_0 for $l = 0$; [Eqs. 4.33 and 4.34]; and hence determine the cross-section of scattering for thermal neutrons. [Eq. 4.38].

3. For incident neutrons of say 10 MeV; what is the maximum values of l, which will be effective for targets of iron and uranium ? Calculate the values of δ_1 and hence f_1 for these maximum values of l.

4. Calculate $\rho(0, E)$ for $E = 1$ MeV and 5 MeV [Eq. 4.54]; for square well potential V_0 of 30 MeV and 40 MeV. Recalculate $\rho(0, E)$ for these energies from Eq. 4.64, for Yukawa and Guassian potentials, with $V_0 = 40$ MeV and other appropriate potential parameters.

5. Using Eq. 4.70, calculate σ_T for $a_s = -\ 23.71\ fm$ and $a_t = 5.38\ fm$; $\rho(0, E) = 1.71\ fm$; and various values of $\rho(0, E)$ for various energies and compare them with σ_T of Fig. 4.1.

6. Analytically solve the *s*-state radial Schrödinger equation for an unbound state of energy + ε. Using $V(r) = -\ V_0 \exp^{-r/\alpha}$ (Exponential potential); and $V(r) = -\ V_0 e^{-r/\alpha}/(1 - e^{-r/\alpha})$ (Hurlthen potential) and obtain δ_1 from it.

7. Assuming from Born approximation that the scattering amplitude is given by

$$f_B = \frac{1}{4\pi} \int d^3r \exp(-i\,\mathbf{k}_f \,.\, \mathbf{r})\, v(r) \exp(i\,\mathbf{k}_i \,.\, \mathbf{r})$$

where $$v(r) = -\frac{2\mu}{\hbar^2}\, V(r)$$

Calculate f_B for a central, tensor and spin-orbit-potential $V(r)$.

8. Using Eqs. 4.31 and 4.32 calculate the values of, $U_1(r)$ for values of V_0 from – 10 MeV to – 40 MeV and $r_0 = 2.02\ fm$ and show from the behaviour of $U_1(r)$ the bound and unbound.

9. For a central potential, we can express the scattering amplitude $f(\theta)$ as given in Eq. 4.28; prove that $\sigma_{total} = 4/k\ lm\ f(0)$, where $f(0)$ is the scattering amplitude for $\theta = 0$.

10. Show that the cross-section for s-wave (n, p) scattering for spin flip and without spins flip are given by

$$\sigma_{flip} = 4\pi \times \frac{1}{2}\left\{|f_0|^2 + \frac{1}{4}|f_{00}|^2\right\}$$

and
$$\sigma_{no\text{-}flip} = 4\pi \times \frac{1}{2} \times \frac{1}{4}|f_{01} - f_{00}|^2$$

where f_{01} corresponds to spins parallel and f_{00} corresponds to spins anti-parallel.

5

CHAPTER

Nucleon-Nucleon Scattering at High Energies

5.1 INTRODUCTION

The low energy data for the two body problem of nucleon-nucleon interaction, can provide us with only a limited amount of information about the nucleon-nucleon potential. The analysis of the deuteron problem and the low energy nucleon-nucleon scattering, has indicated that nucleon-nucleon potential has both central and tensor components and that the potential is stronger in the triplet compared to that in the singlet state. We, however, cannot find from that data the exchange character or the velocity dependence of the nuclear forces. Also many other features of nuclear forces like spin-orbit coupling or the presence of short range repulsive core, cannot be ascertained from the low energy data. This is understandable if we realise that at low energies, the incident nucleons have large wavelengths, and therefore, can only 'see' the gross features of the nucleon-nucleon potential. It is useful, therefore, to extend these studies to high energy scattering say between 10 MeV and 300 MeV for which the phase shifts for $l > 0$ are effective. It is useful to know that the average relative energy of the two nucleons in a nucleus is about 100 MeV or so. The scattering at these energies, therefore, is expected to give those features of the nuclear forces, which are still applicable to the problems of nuclear structure. Above 290 MeV or so, the π-meson production starts and the normal phase-shift analysis and the relationship of the phase-shifts with the nucleon-nucleon potential loses its significance.

5.2 EXPERIMENTAL DATA

The experimental quantities relevant at high energy nucleon-nucleon scattering are: (1) Total cross-section $\sigma_t\,(E)$ as a function of incident energy (2) Angular distribution or differential scattering cross-section $d\sigma/d\Omega$ or $\sigma\,(\theta)$ at various energies for both the unpolarised incident beam and target. The differential scattering cross-section is sometimes called single scattering parameter as it involves single scattering (3) Polarisation parameters like $P\,(\theta)$, $D\,(\theta)$, $R\,(\theta)$, $R'\,(\theta)$, $A\,(\theta)$ and $A'\,(\theta)$, etc. and the related spin correlations coefficients C_{nn}, C_{kp}, C_{pp} and C_{kk} . The polarisation $P\,(\theta)$ is also called the double scattering parameter as it involves scattering twice. The depolarisation parameter $D\,(\theta)$, the rotation parameters $R(\theta)$ and $R'(\theta)$, and the parameters $A(\theta)$ and $A'(\theta)$—are called the triple scattering parameters, as they involve scattering thrice. The various polarisation parameters essentially are concerned with the excess of scattered nucleons with spin orientation in a given direction, over the average number of nucleons with spin-orientation in all directions. The exact definitions and the physical significance of these polarisation parameters will be given below.

Such studies have been carried out for both *n-p* and *p-p* scattering. Experimentally, while the protons from a few MeV to several hundred MeV can be either obtained from Van de Graffs; (up to say 30 MeV) and from Cyclotrons or Synchro-cyclotrons (say up to 300 MeV); the monoenergetic neutrons of high energy are generally obtained, from many (p, n) or (d, n) reactions; or even from (γ, n) reactions, as described earlier in Chapter 4, section 4.1.

For obtaining very high energy neutrons, heavier nuclei are also sometimes used as targets. In those cases, one gets a continuous spectrum from evaporation over which is super-imposed a narrow peak of a well defined energy from strippling reaction. The energy selection is done through proper detecting techniques.

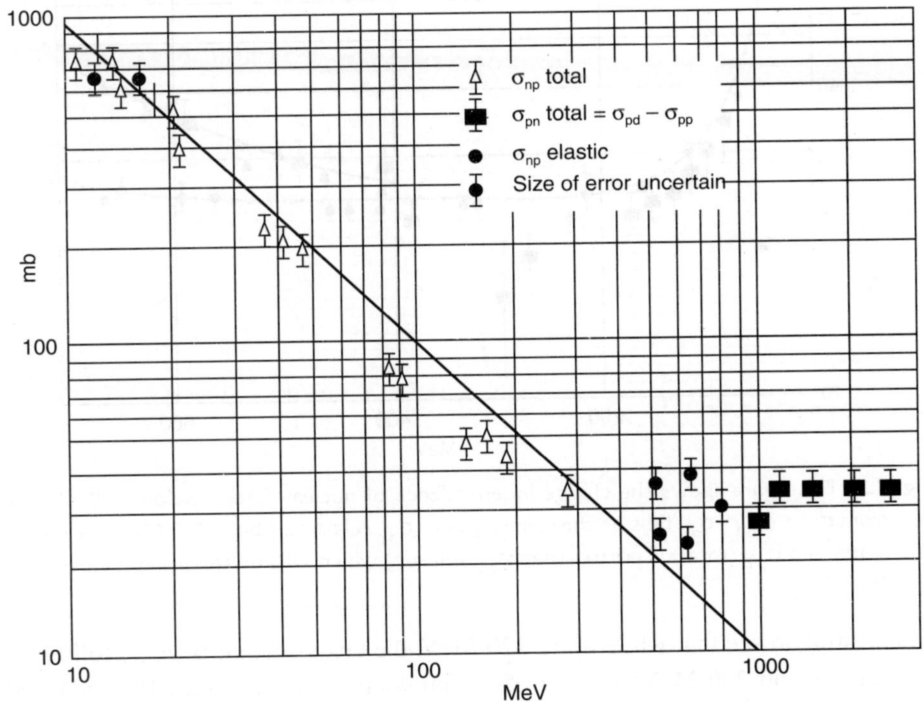

Fig. 5.1 The experimental data on (n, p) up to 300 MeV incident energy and beyond (Ref. 1).

5.2.1 Total Cross-Sections

Figures 5.1 and 5.2 give the experimental data[1] for total cross-sections for *n-p* and *p-p* interactions up to 300 MeV and beyond, including the elastic scattering data. In Fig. 5.2 are also included the points for neutron-neutron scattering. The cross-section $\sigma_{n,n}$ is obtained from:

$$\sigma_{n,n} \approx \sigma_{n,d} - \sigma_{n,p} \qquad ...(5.1)$$

where $I(\theta)$ of Eq. 4.4 has been neglected, because at high energy its contribution is less than 0.6 mbs/st out of about 25 mbs/st at 100 MeV and higher. Also in Fig. 5.1, there are points plotted corresponding to:

$$\sigma_{p,\,n} \approx \sigma_{p,\,d} - \sigma_{p,p} \qquad ...(5.2)$$

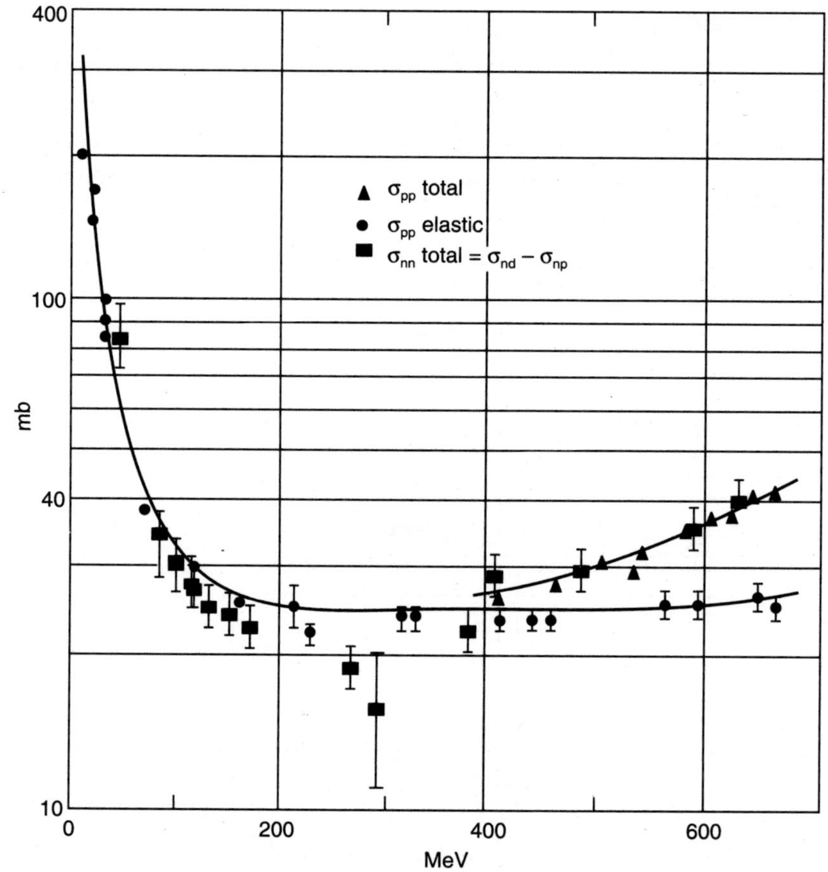

Fig. 5.2 This figure shows the charge independence of nuclear forces. Below 400 MeV σ_{nn}, obtained from $\sigma_{n\,n} = \sigma_{n\,,d} - \sigma_{n\,p}$ falls on the same curve as $\sigma_{p\,p}$ (elastic). Above 400 MeV, the meson production starts and $\sigma_{p\,p}$ (total) is greater than $\sigma_{p\,p}$ (elastic) which falls on the general curve for $\sigma_{n\,n}$ and $\sigma_{p\,p}$ (Ref. 1).

It is interesting to see that while up to 220 MeV, the cross-section $\sigma\,(n,\,p)$ follows *I/E* law, the curve flattens out beyond 290 MeV because of the production of π-mesons. Between 290 MeV and 400 MeV, the curve is flat for $\sigma_{(p,\,p)}$ and then at still higher energies, it slowly slopes upwards, because of still higher modes of meson production at these energies. The *I/E* law corresponds to elastic scattering. We also see in Fig. 5.2 that for *p-p* interaction, up to around 300 MeV, $\sigma_{p\,p}$ (elastic) $\approx \sigma_{p\,p}$ (total) $\approx \sigma_{n\,n}$ (total), while above this energy; $\sigma_{p,\,p}$ (elastic) $\leq \sigma_{p\,p}$ (total) $\approx \sigma_{n,\,n}$ (total) because, above 300 MeV or so, as in the case of (*n*, *p*); π-mesons get created both in *p-p* and *n-n* interaction. It also shows that *p-p* and *n-n* interaction forces are of the same order, thus indicating the charge symmetry and charge independence of nuclear forces. The cross-sections for *n-p* scattering, for a given energy is, in general, higher than for *p-p* scattering. As we will see later, this happens because all the states which are possible for *n-p* scattering are not allowed in *p-p* scattering due to Pauli exclusion principle. The charge independence hypothesis, however, governs the *n-p* scattering as it does the *p-p* and *n-p* scattering.

Some values of σ (*n, p*), σ (*p, p*) and σ (*n, n*) are shown in Table (5.1) which gives an idea of the order of the magnitude of cross sections[2].

Table 5.1

No.	*Incident Energy*	σ_{np} (*mb*)	σ_{pp} (*mb*)	σ_{nn} (*mb*)
1.	40 MeV	170	65	85
2.	156 MeV	50	25	24
3.	440 MeV	35	24 ± 2	28
4.	1000 MeV	28	19 ± 3	–

5.2.2 Angular Distributions or Differential Cross-Sections

The studies on the angular distribution of elastically scattering neutrons from protons and from deuterons and of protons from protons, have been carried out by many workers.

In Figure 5.3 we have shown, the angular distribution of (*n, p*) scattering[3], at 90 MeV and 300 MeV. Below the threshold energy of 290 MeV, for π-production; only elastic scattering is possible; while above it; differential cross-sections also contain the contribution from π-meson production.

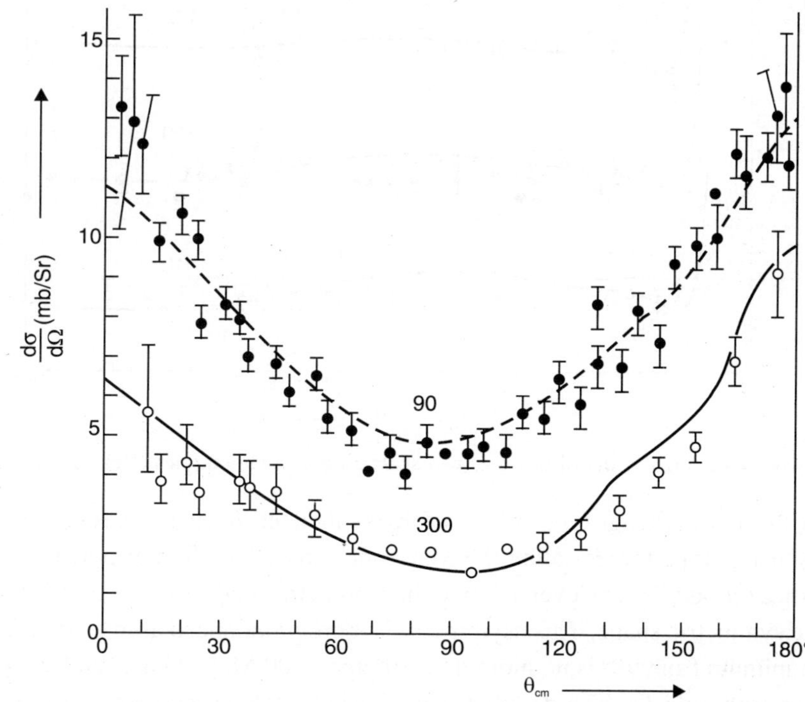

Fig. 5.3 Angular distribution of (n, p) scattering, at 90 and 300 MeV (Ref. 3).

It is interesting to note that in these curves shown in Fig. 5.3, the angular distribution at both 90 MeV and 300 MeV neutron energies, is anisotropic, though somewhat symmetric around 90° in the centre of mass system. The very low energy isotropic distribution in Fig. 4.2 may be interpreted as due to $l = 0$ while the symmetric but anisotropic distribution at higher energies in Fig. 5.3, is due to higher values of l. The near symmetric nature of the curve around 90° is understood on the basis of the exchange character of nuclear forces, *i.e.* the neutron and proton exchange their position in the process of their interaction. This symmetry is specifically interpreted to be due to Serber type of exchange forces containing only even l's. The nuclear potential $V(r)$ for Serber forces may be represented as:

$$V(r) = -V_o\left[1 + \frac{1}{4}(1 + \sigma_1 \cdot \sigma_2)(1 + \tau_1 \cdot \tau_2)\right] \qquad ...(5.3)$$

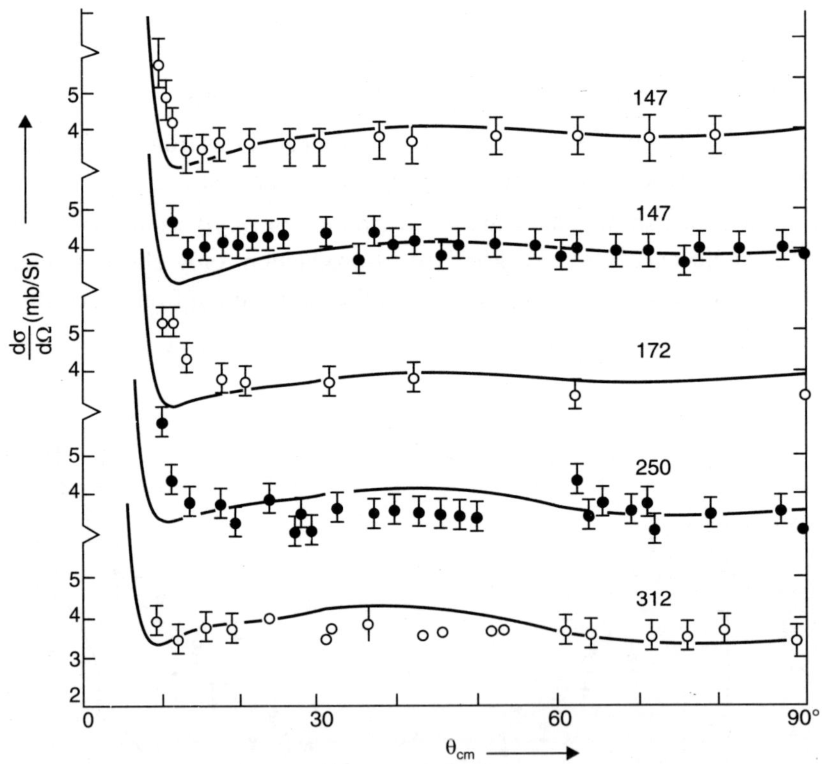

Fig. 5.4a Angular distribution of p-p scattering at high energies (Ref. 1, 3).

In this case the non-exchange type ordinary forces called the Wigner forces have the same strength as the position exchange type forces called the Majorona forces. The Serber potential is attractive for even l's and vanishes for odd l's. The even l's give the symmetric shape. Above 30 MeV there is a slight deviation form the symmetric shape. This asymmetry is more pronounced above 100 MeV or so but the deviation of the minimum from 90° is no more than 10° up to 400 MeV. This is understood to be caused by the contribution from odd l's. The detailed behaviour of exchange forces is discussed in the next chapter.

The angular distribution of *p-p* scattering at low energies has already been discussed in the previous chapter. Curves for higher energies[4] are presented in Fig. 5.4*a*.

Apart from a symmetry around 90° in the angular distribution which arises due to exchange forces and due to the indistinguishability of protons; one sees forward angle peaking for high energies due to Coulomb interaction. The flat portion of the curve for all energies is believed to be due to the hard repulsive core of the nuclear potential.

This feature of flatness in the angular distribution of *p-p* scattering has provided a very sensitive condition to be satisfied for the nucleon-nucleon potential. The minimum around 10° arises basically due to interference between Coulomb and nuclear potentials.

At still higher energies, say, above 400 MeV, the flatness character in the angular distribution is destroyed, because of the effect of meson production.

In Fig. 5.4*b*, are shown the dependence of the variation of phase shifts δ's for $l = 0$, singlet (IS_0) and $l = 2$ singlet (ID_2), on the incident energy. The detailed analysis[5], shows that the behaviour of these curves is seen to conform to the repulsive-core hypothesis.

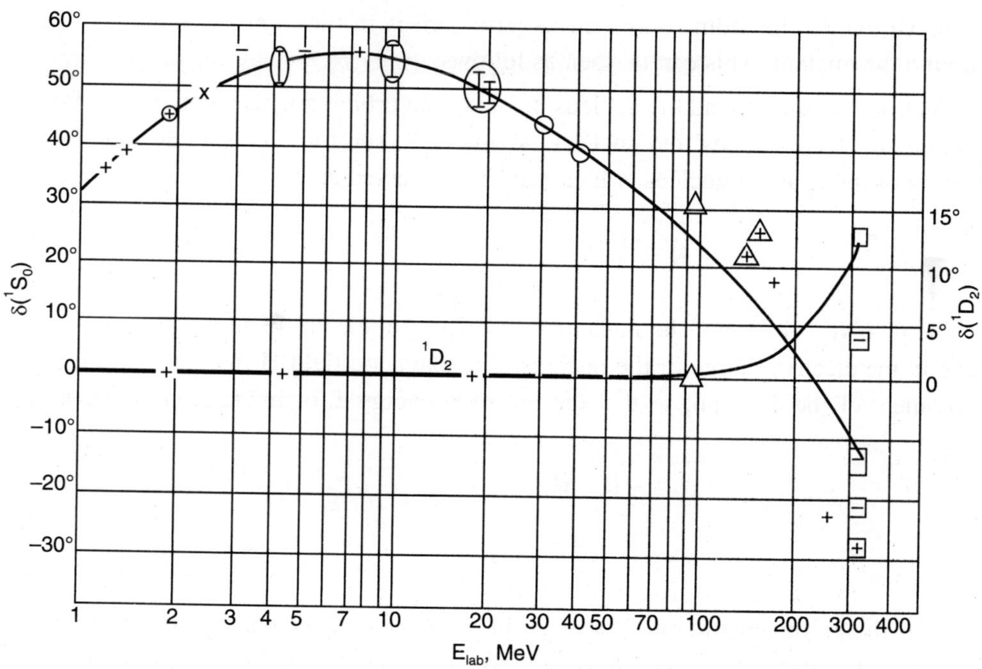

Fig. 5.4b Singlet phase-shifts δ (IS_0) and δ (ID_2) as a function of incident energy. The experimental points are from experimental data of scattering—both pp and np—and the continuous curves are theoretical curves, based on a potential with repulsive core (Ref. 4.5).

5.2.3 Polarisation

The angular distribution and the total cross-sections of unpolarised nucleons do not provide all the parameters, required for the complete and detailed phase-shift analysis of nucleon-nucleon scattering at high energies. On the other hand, if one studies polarisation, introduced by the scatterer in an unpolarised incident beam of nucleons, one obtains extra information about phase-shifts, which help in understanding the detailed, nucleon-nucleon potential. The polarisation caused in the scattered beam of neutrons or protons, due to scattering from protons is understood to be mainly due to a spin-orbit interaction arising from tensor part or **l.s** coupling, in nucleon-nucleon potential. The behaviour of $P(\theta)$, the polarisation parameter obtained experimentally from double scattering experiments, can be explained, on this basis. The triple scattering polarisation parameters $D(\theta)$, $R(\theta)$, $R'(\theta)$, $A(\theta)$ and $A'(\theta)$ and spin correlation coefficients C_{nn}, C_{kp}, C_{pp} and C_{kk}, give detailed information about the spin-space and spin-spin dependence of the nucleon-nucleon potential respectively. The significance and the measurements of these quantities have been discussed in the subsequent pages. Suffice to say, that all these quantities are related to that part of the interaction-potential between two nucleons which is responsible for the preferential spin orientation and hence contains terms in which the spin-spin and space interaction of the interacting nucleons is essentially involved.

Experimental data is available for polarisation parameters $P(\theta)$, $D(\theta)$, $A(\theta)$, and $R(\theta)$, and the spin-correlation coefficients C_{nn} and C_{kp} for *p-p* and *n-p* scattering, for some of the energies.[5, 6, 8]

5.2.3.1 The Spin-orbit Force in Nucleon-Nucleon Interaction and Polarisation

It is well-known that in the atomic case, an **l.s** terms arises in the interaction energy of an electron orbiting around the nucleus. This can be seen as follows:

An electron moving around the nucleus possesses an orbital angular momentum. Because of the relative motion, an electron 'sees' the nucleus moving and, therefore, feels a magnetic field H, due to relative motion of the charged nucleus, which may be expressed as:

$$\mathbf{H} = -\frac{1}{c}\,\mathbf{v} \times \boldsymbol{\varepsilon} \qquad \text{...(5.5}a\text{)}$$

where **v** is the relative velocity of the nucleus with respect to electron and ε is the electric field at the electron due to the electric charge of the nucleus. The magnetic field **H** interacts with $\boldsymbol{\mu}_s$, the intrinsic magnetic moment of the electron; so that the potential energy E of interaction between **H** and $\boldsymbol{\mu}_s$ is given by:

$$E = -\boldsymbol{\mu}_s \cdot \mathbf{H}$$

$$= -\frac{g_s \mu_B}{\hbar}\,\mathbf{S} \cdot \mathbf{H} \qquad \text{...(5.5}b\text{)}$$

where $g_s = 2$, is the gyromagnetic ratio for electron and $\mu_B = e\hbar/2m_o c$ is the Bohr-magnetron and **S** is the spin-vector for electron. The energy E, is, however, calculated in the frame of reference in which the electron is at rest. This should be transferred back to the frame of reference in which the nucleus is at rest. This introduces a factor 1/2 because of the effect of the relative motion; so that in the frame of reference of nucleus at rest:

$$E = \frac{1}{2}\frac{g_s \mu_B}{c\hbar}\,\mathbf{S} \cdot (\mathbf{v} \times \boldsymbol{\varepsilon}) \qquad \text{...(5.6)}$$

writing $\boldsymbol{\varepsilon} = -\frac{dv}{dr}\frac{\mathbf{r}}{r}$, and $m\mathbf{v} \times \mathbf{r} = -\hbar L$

We finally obtain,

$$E = \frac{1}{2}\frac{g_s \mu_B}{\hbar c}\mathbf{S}\cdot\left(\frac{\hbar}{m}\frac{1}{r}\frac{dV}{dr}\mathbf{L}\right)$$

$$= \frac{e\hbar}{2m^2c^2}\left(\frac{1}{r}\frac{dV}{dr}\right)\mathbf{L}\cdot\mathbf{S} \qquad ...(5.7)$$

Equation 5.7 shows that in the interaction of the two particles, one possessing an electric charge and the other possessing a magnetic moment moving relative to each other; there is a spin-orbit coupling. Qualitatively one could extend this argument to the interaction of a neutron and a proton possessing a relative orbital angular momentum. The motion of proton may provide the magnetic field, which may interact with the magnetic moment of the neutron. Such a purely electromagnetic interaction, however, turns out to be much smaller than the experimentally observed spin-orbit interaction in nucleon-nucleon potential by many orders of magnitude. Also, while an electromagnetic interaction may provide **l.s** coupling in *n-p* and *p-p* nuclear potential; it cannot give any **l.s** coupling term in *n-n* interaction. On the other hand, the charge-independence of nuclear forces requires that *n-n* potential should also have, in general, similar features as for *p-p* potential.

The meson theory, as discussed in the next chapter also does not provide a satisfactory explanation of **l.s** coupling in a free nucleon-nucleon interaction. In the complex nuclei, however, the genesis of **l.s** term in nucleon-nucleon potential is expected to arise from changes in nuclear potential at the surface. When a nucleon is passing through a nucleus, in the body of the nucleus, it is acted upon by nucleons from all sides; and hence there is no net force on the projectile. If it is passing in the region near the surface, there will be a net force pointing inwards because the nuclear-matter density is now more on one side. It is expected qualitatively, that for this region, the interaction potential, which is a scalar, can be built out of **p**, the linear momentum of the particle, the spin **S** of the projectile, and the density gradient ∇_ρ of the nuclear matter at the surface. Such a scalar potential is, therefore, proportional to:

$$V = \text{constant } \mathbf{S}\cdot(\nabla\rho\times\mathbf{p})$$

where

$$\nabla\rho = \frac{d\rho}{dr}\hat{r} = \frac{1}{2}\frac{d\rho}{dr}\mathbf{r}$$

Hence, one may write

$$V = \text{constant } \mathbf{S}\cdot\frac{1}{r}\frac{d\rho}{dr}(\mathbf{r}\times\mathbf{p})$$

$$= \text{constant } \frac{1}{r}\frac{d\rho}{dr}(\mathbf{S}\cdot\mathbf{L}) \qquad ...(5.8)$$

The presently accepted expression of the spin-orbit coupling term in nucleon-nucleon potential (*see* Chapter 6, Section 6.4); is given by:

$$V_{L\cdot S} = \frac{1}{\hbar}\frac{W(r)\,\mathbf{L}\cdot\mathbf{S}}{2} \qquad ...(5.9)$$

where $$W(r) = -\left(\frac{\hbar}{\mu c}\right)\frac{1}{r}\frac{dY(r)}{dr} \quad ...(5.10)$$

$Y(r)$ which is related to density of nuclear matter, has to be determined by comparison with experiments and μ is the induced mass of π-meson. An early experimental[7] evidence for spin-orbit coupling in nuclear potential was provided by the scattering of neutrons and protons from He^4. The nucleus He^4 has spin zero. Therefore, for $l = 1$; the system $n + He^4 \rightarrow He^5$ can be formed either in $P_{3/2}$ or in $P_{1/2}$ state. In the first case, the **l** and **s** are parallel and in the second case they are antiparallel. Now,

$$\mathbf{l}\cdot\mathbf{s} = \frac{1}{2}\,[j(j+1) - l(l+1) - s(s+1)]$$

$$= \frac{1}{2}\,l \text{ for } j = l + \frac{1}{2}$$

$$= \frac{-l(l+1)}{2} \text{ for } j = l - \frac{1}{2} \quad ...(5.11)$$

Therefore if $W(r)$ in Eq. 5.10 is negative, $\mathbf{l}\cdot\mathbf{s}$ is larger for $j = l - 1/2$ and smaller for $j = l + 1/2$. One should, therefore, expect a $P_{1/2}$ level in He^5 to be higher than $P_{3/2}$ level. Experimentally it was found that the resonance in $n + He^4$ corresponding to $j = 3/2$ occurred at the incident neutron energy of 0.95 MeV, with a half-width of 1/2 MeV and the analysis showed, that this maximum is the result of the neutrons in a $P_{3/2}$ state.

The contribution of $P_{1/2}$ state is also resolved in the same analysis. So the experiment proves a ground state of He^5 with $P_{3/2}$ configuration and an excited state at 2.6 MeV excitation, with $P_{1/2}$ configuration. Very similar results are obtained for the scattering of protons from He^4, for which the resonance, corresponding to $P_{1/2}$ occurs at 2.5 MeV excitation in Li^5 and the one corresponding to $P_{3/2}$ occurs for the ground state. These experiments, therefore, show that: there exists a spin-orbit coupling in nuclear-interaction and the interaction potential denoted by $W(r)$ in Eq. 5.10 is negative, giving rise to inverted doublets, where the higher l-state is lower in energy in contrast to the atomic case.

The $\mathbf{l}\cdot\mathbf{s}$ coupling in the above two examples is basically a result of such coupling, existing in nucleon-nucleon interaction. We will show in subsequent analysis that such coupling produces polarisation which also has been experimentally measured directly in nucleon-nucleon scattering at higher energies.

It has been found that in $p + He^4$ system, there is a resonance at about 1.8 MeV energy of protons corresponding to $P_{3/2}$ of the ground state for which the intrinsic spin vector and orbital spin vectors are parallel. This means that at this energy, the scattered protons will have their intrinsic spins preferentially oriented along the direction of their orbital angular momentum. Referring to Fig. 5.5, it may be seen, that those scattered to the right are expected to have their spins predominantly oriented into the paper, and those scattered to the left with their spins oriented out of the paper, with spins parallel to the angular momentum. On the other hand for the resonance at 2.5 MeV excitation of Li^5 corresponding to $P_{1/2}$, the spin vector and orbited angular momentum vector should be antiparallel so that the spin-directions of the scattered protons are expected to be, predominantly, opposite to the previous case. The spins of the protons in the above two cases, have, therefore, been oriented predominantly in certain definite directions after the first scattering, *i.e.* the scattered protons have been polarised as a consequence of the existence of the spin-orbit coupling term in the nuclear forces. To detect the degree of polarisation, one can perform the second scattering from the same type of target, as the first one. At energies corresponding to $P_{3/2}$ resonance, the protons scattered to the right from the first scatterer, have their spins oriented into the paper. If these protons are now scattered at the second scatterer with the resonance characteristics of $P_{3/2}$ again, the protons will get predominantly scattered to the right. This happens because the polarisation after the first scatterer makes the proton with spins oriented into the paper preponderant in the scattered beam. Their spins are parallel to the orbital angular momentum. To maintain the $P_{3/2}$ characteristics, (because, of their energy corresponding to this resonance), they will be predominantly scattered again towards the right after the second scattering. In this manner, a left-right asymmetry develops after the second scatterer if the first scatterer polarises the protons. The first scatterer is, therefore, called a 'polariser' and the second scattterer is called an 'analyser'. One can see by similar arguments, that for the second scatterer, on the left side, left-right asymmetry will be of the opposite sign, compared to the previous case. This means that for the left hand second scatterer, there will be more protons on the left side compared to the right side, in contrast to the previous case. Such an experiment was carried[7] out, and the expected polarisation was detected for protons scattering from helium (Fig. 5.5).

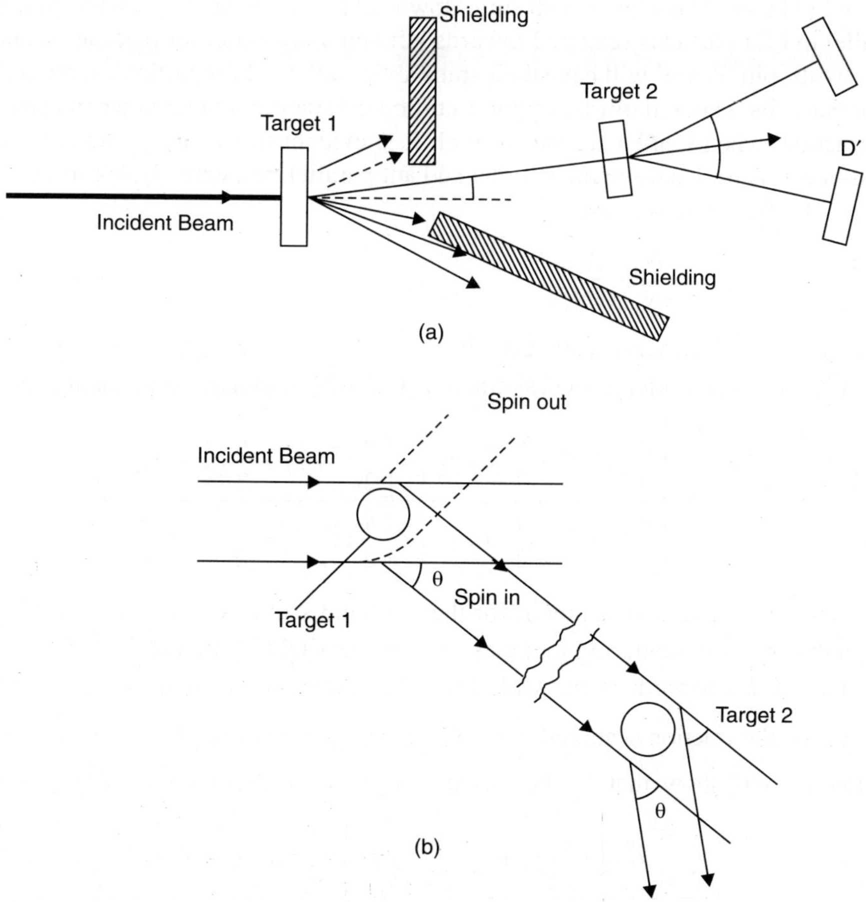

Fig. 5.5 (*a*) Double scattering experiment to measure P (θ); (*b*) shows the spin directions, in the first scattering, for the incident beam energy near $P_{3/2}$ resonance.

We have, thus shown, that $\mathbf{l}\cdot\mathbf{s}$ coupling leads to polarisation; and a double scattering experiments can lead to the detection of polarisation from the left-right asymmetry.

5.2.3.2 Scattering Parameters

As mentioned earlier, apart from the differential scattering cross-section $\sigma(\theta)^1$, called the single scattering parameter and the polarisation $p(\theta)$, called the double scattering parameter, we can also measure the triple scattering parameters, $D(\theta)$, $R(\theta)$, $R'(\theta)$, $A(\theta)$ and $A'(\theta)$, etc. and the spin correlation coefficients $C_{n\,n'}$, $C_{k\,p'}$, $C_{p\,p'}$, $C_{k\,k'}$ etc. The exact definitions of these parameters and coefficients and their physical significance are given below:

***(i) The polarisation P* (θ):** We have qualitatively explained in the previous article the physical significance of polarisation $P(\theta)$. In a completely polarised beam, the spins of the polarised nucleons are either parallel or antiparallel to the orbital angular momentum and are perpendicular to the plane containing the incident and scatterer beam. As we have seen before, the left-right asymmetry after the second scattering indicates the polarisation after the first scattering.

If we have an unpolarised beam of, say 'n' nucleons, it will have on the average $n/2$ nucleons with spin 'up' $n/2$ (↑) and $n/2$ nucleons with spin 'down' $n/2$ (↓). The nucleons with spin 'up' will have their spin parallel to $\mathbf{l}$ for particles scattered towards left and antiparallel for particles scattered towards right. Nucleons with spin 'down' will have their spins antiparallel to $\mathbf{l}$ for particles scattered towards left and parallel for particles scattered towards right. Let f_1 be the fraction scattered when $\mathbf{l}$ and $\mathbf{s}$ are parallel and f_2 be the fraction, when $\mathbf{l}$ and $\mathbf{s}$ are antiparallel. It is evident that f_1 and f_2 are connected with the scattering cross-sections for $\mathbf{l}$ and $\mathbf{s}$ being parallel and antiparallel, respectively. We now, define the left-right asymmetry 'A' after scattering as:

$$A = \frac{N_L - N_R}{N_L + N_R} \qquad \text{...(5.12}a\text{)}$$

where N_L is the number of particles scattered towards left and N_R is the number of particles scattered towards right. After the first scatterer, the asymmetry A_1 for the unpolarised incident beam is given by:

$$A_1 = \frac{\left[\frac{n}{2}(\uparrow) f_1 + \frac{n}{2}(\downarrow) f_2\right] - \left[\frac{n}{2}(\uparrow) f_2 + \frac{n}{2}(\downarrow) f_1\right]}{\left[\frac{n}{2}(\uparrow) f_1 + \frac{n}{2}(\downarrow) f_2\right] + \left[\frac{n}{2}(\uparrow) f_2 + \frac{n}{2}(\downarrow) f_1\right]} \qquad \text{...(5.12}b\text{)}$$

We now introduce a second scatterer, for the left hand scattered neutrons capable of producing polarisation by preferentially scattering particles of spins parallel to $\mathbf{l}$. Figure 5.5b, however shows the first and second scattering towards right, while Eq. 5.12 onwards correspond to the scattering towards left. Again let f_1' be the fraction scattered when $\mathbf{l}$ and $\mathbf{s}$ are parallel and f_2' when they are antiparallel then a little reflection will show, that A_2, the left-right asymmetry after the second scatterer is given by:

$$A_2 = \frac{\left[f_1'\frac{n}{2}(\uparrow) f_1 + f_2'\frac{n}{2}(\downarrow) f_2\right] - \left[f_2'\frac{n}{2}(\uparrow) f_1 + f_1'\frac{n}{2}(\downarrow) f_2\right]}{\left[f_1'\frac{n}{2}(\uparrow) f_1 + f_2'\frac{n}{2}(\downarrow) f_2\right] + \left[f_2'\frac{n}{2}(\uparrow) f_1 + f_1'\frac{n}{2}(\downarrow) f_2\right]} \qquad \text{...(5.12}c\text{)}$$

$$= \frac{(f_1' - f_2')(f_1 - f_2)}{(f_1' - f_2')(f_1 - f_2)} \text{ since } \frac{n}{2}(\uparrow) = \frac{n}{2}(\downarrow)$$

for the unpolarised beam. We now define, the polarisation P, of a beam of nucleons as:

$$P \equiv \frac{N_+ - N_-}{N_+ + N_-} \equiv \frac{N(\uparrow) - N(\downarrow)}{N(\uparrow) + N(\downarrow)} \quad ...(5.13a)$$

where N_+ or $N(\uparrow)$ represents the number of nucleons with spin up; and N_- or $N(\downarrow)$ represents the number of nucleons with spin down. After the first scattering, the polarisation P_1 of the left scattered beam is given by:

$$P_1 \text{ (left side)} = \frac{\frac{n}{2}(\uparrow) f_1 - \frac{n}{2}(\downarrow) f_2}{\frac{n}{2}(\uparrow) f_1 + \frac{n}{2}(\downarrow) f_2} = \frac{f_1 - f_2}{f_1 + f_2} \quad ...(5.13b)$$

Similarly, if the second scatterer was placed in the place of first scatterer, then polarisation introduced by it will be given by:

$$P_2 \text{ (left side)} = \frac{f_1' - f_2'}{f_1' + f_2'} \quad ...(5.13c)$$

So that the total asymmetry A_2, as given in Eq. 5.12*c* is given by:

$$A_2 = P_1 P_2 \quad ...(5.13d)$$

The total polarisation, of an unpolarised beam after being first scattered by the polariser (scatterer *I*) and then by analyser (scatterer *II*), is given by (from Eq. 5.13*a*):

$$P_{\text{total}}^{\text{left}} = \frac{f_1 f_1' - f_2 f_2'}{f_1 f_1' + f_2 f_2'} = \frac{P_1 + P_2}{1 + P_1 P_2} \quad ...(5.14)$$

If the first and second scatterer are the same and the particles are scattered through the same angles then,

$$P_1 = P_2 = P; \text{ and } A_2 = P^2 \text{ or } P = \pm\sqrt{A_2} \quad ...(5.15)$$

Hence the measure of asymmetry gives a determination of the polarisation parameter $P(\theta)$.

It is easy to see from the above discussion that though that first scatterer causes the polarisation, it is only through the left-right asymmetry after the second scattered that we can detect the polarisation. If the incident beam is already polarised, say with polarisation P the left-right asymmetry at the very first scatterer will be given by P.

If P_1, the polarisation caused by the first scatterer is known from previous independent experiments; the polarisation of the beam can be found, from one scattering.

The above arguments are strictly valid only for targets with spins zero. For targets with spins different from zero, one has to take into account, not only the spin-orbit interaction; but also the spin-spin interaction. The definition given in Eqs. 5.12, 5.13*a*, 5.13*b* and 5.14*a*, are however, in general valid.

Figure 5.6 represents some of the typical data about the polarisation parameters.[7] Some of the salient features about experimental curves may be summarised as follows:

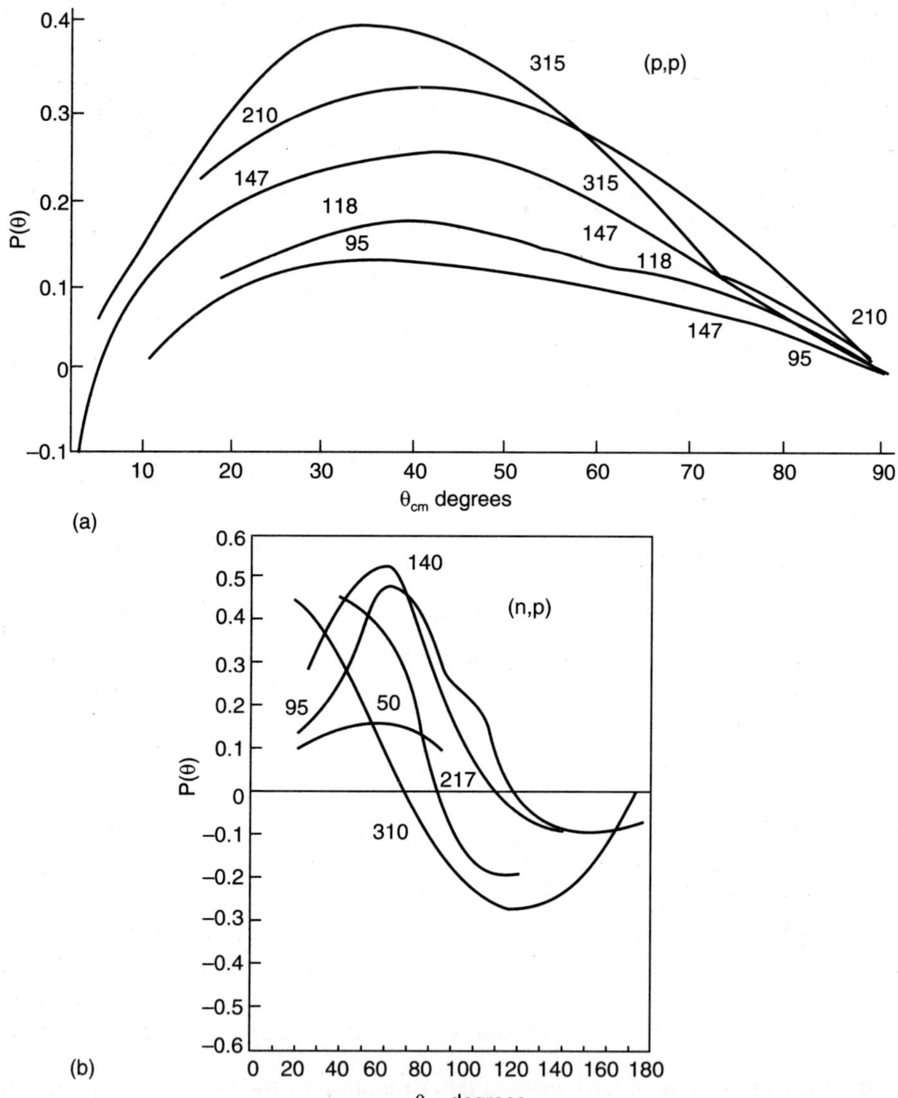

Fig. 5.6 The polarisation parameters p (θ); for (a) p-p and (b) n-p scattering at different energies in MeV (marked as numbers on the curves), as a function of scattering angle (Ref. 1,7).

The magnitude of these parameters are less than one. The polarisation $P(\theta)$ is always positive for *p-p* from 0° to 90° for all energies, including quite low energies. The values of $P(\theta)$ for (*n-p*) is also generally positive for these angles, but for high energies, *i.e.* at 310 MeV; it becomes negative near 90°. The values of $P(\theta)$ for 90° to 180° is anti-symmetry to those from 0° to 90° in *p-p* scattering. For *n-p* scattering, the values undergo a change of sign between 45° and 135° and do not have as good a symmetrical behaviour as in *p-p* case. The value of $P(\theta)$ for 90° is zero for *p-p* scattering for all energies. For *n-p* case it is positive up to say 100 MeV or so; and slowly becomes negative as the energies are raised up to 300 MeV. The value of $P(\theta)$ for a given angle are, in general; larger for higher energies, especially in *p-p* scattering. In *n-p* scattering, this is only true for small angles.

***(ii) Depolarisation parameter D* (θ):** This is a triple scattering parameter, giving essentially the depolarisation caused by the second scatterer. To understand the definition of $D(\theta)$, let us assume an unpolarised beam falling on the first scatterer T_1, which introduces the polarisation P_1. We now introduce second scatterer T_2, in the scattered beam, which not only causes a further polarisation P_2, but also causes depolarisation D. Let the third scatterer T_3 similarly introduce a polarisation P_3. Then left-right asymmetry A_{23} after the third scatterer is defined by:

$$A_{23} = \frac{P_2 \pm DP_1}{1 \pm P_1 P_2} P_3 \qquad ...(5.16)$$

where positive sign refers to the left side and negative sign to the right side in all the scattering. Depolarisation D has thus been defined in such a manner that for the scattering towards left in all the three scatterers, if the first scatterer completely polarises, *i.e.* $P_1 = 1$ and if P_2 is also 1, which means that both T_1 and T_2 are scatterers of the same type; then $D = 1$ will imply that $A_{23} = P_3$ and there is no depolarising in the second scatterer. Figure 5.7 illustrates the phenomenon of depolarisation.

In fact, the parameter D, determines the amount of polarisation perpendicular to the scattering plane in the beam after the first scattering; which is converted into the same direction of polarisation after the second scattering. Generally D is less or equal to one. A value of $D < 1$ implies that the second scatterer has converted part of the polarised beam with polarisation perpendicular to the scattering plane after the first scattering; in some other plane $D = 1$, implies the polarisation of the beam with polarisation perpendicular to scattering plane after the first scatterer, has been completely converted by the second scatterer, into its original direction. If the initial beam is completely polarised, then, $P_1 = 1$, and Eq. 5.16 reduces to:

$$A_{23} = \frac{P_2 \pm D}{1 \pm P_2} P_3 \qquad ...(5.17a)$$

For left-hand scattering, in all scatterers the above expression reduces to:

$$A_{23} = \frac{P_2 + D}{1 + P_2} P_3 \qquad ...(5.17b)$$

If $D < -P_2$, then A_{23} is negative, which corresponds to the polarisation being reversed. We can rewrite Eq. 5.16 as:

$$D = \frac{A_{23}}{A_3}(A_2 \pm 1) \pm \frac{A_2}{A_1} \qquad ...(5.18a)$$

where $A_3 = P_1 P_3$, $A_2 = P_1 P_2$ and $A_1 = P_1^2$ ...(5.18*b*)

The positive sign in Eq. 5.18*a* refers to the scattering to the left and negative sign to the right in all scattering. Equation 5.18, provides a practical method for the measurement of D. One has to measure polarisation parameters P_1, P_2 and P_3 and left-right asymmetry A_{23} after the third scatterer, performing all scattering experiments in the scattering plane.

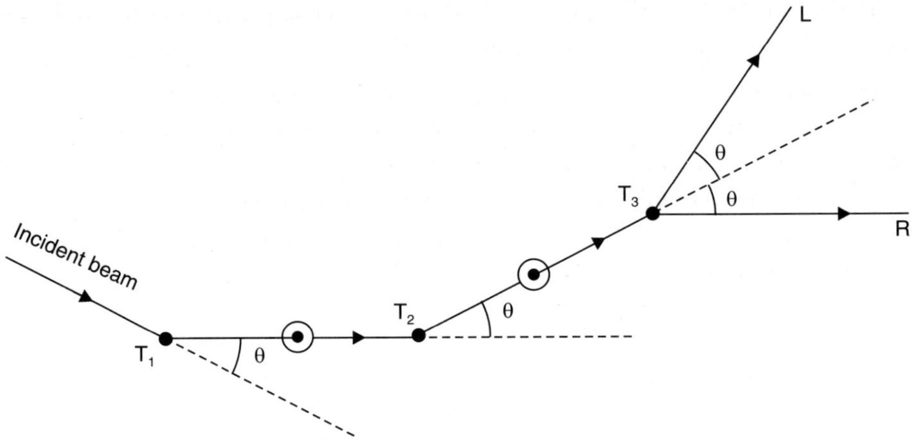

Fig. 5.7 The figure illustrates the phenomenon of depolarisation and the measurement (After Ref. 9).

***(iii) The Rotation parameters R* (θ), *R′* (θ), *A* (θ) *and A′* (θ):** These parameters determine the rotation of the spin-orientation by the second scatterer from one direction to another direction; but keeping them in the scattering plane. Their physical meaning is shown in Figs. 5.8 and 5.9. The parameters R (θ) and R' (θ) describe the process in which the direction of polarisation vector is converted from a direction perpendicular to the beam (but in the scattering plane) after the first scatterer; to perpendicular of parallel to the beam, respectively, after the second scatterer but polarisation remaining in the scattering plane. Similarly the parameters A (θ) and A' (θ) describe the process in which the direction of polarisation vector is converted from a direction parallel to the beam after the first scatterer to perpendicular or parallel to the beam, respectively, after the second scatterer, the polarisation remaining in the scattering plane.

Measurements of these parameters is carried out as follows:

(*a*) Measurement of R (θ) and R' (θ):

The value of R (θ) is given by

$$A_{3s} = P_3 \, R\,(\theta) \, P_1 \qquad ...(5.19a)$$

where A_{3s} is the left-right asymmetry measured after the third scatterer when P_1 and P_3 are polarised by the first and the third scatterer, the successive scattering planes being perpendicular to each other and A_{3s} is measured in the same plane as P_3. Then R (θ) gives the rotation due to the second scatterer (Fig. 5.8).

In a very similar manner one can measure R' (θ) using the equation,

$$A'_{3s} = P_3 \, R'\,(\theta) \, P_1 \qquad ...(5.19b)$$

where A'_{3s} is the left-right asymmetry after the third scatterer, which is measured after orienting the spin of the particles in beam, scattered from the second scatterer to left of the beam. This is done by applying an appropriate magnetic field perpendicular to the spin and the beam direction between the second and the third scatterer which bends the beam through and angle say θ_s and also turns the spin through an angle say θ_b. In general $\theta_s > \theta_b$, because of the large magnetic field interacting with the magnetic moment of the nucleus. One should, of course, realise that the third scattering plane should be perpendicular to the direction of the plane containing new spin direction and the beam, incident on the third scatterer.

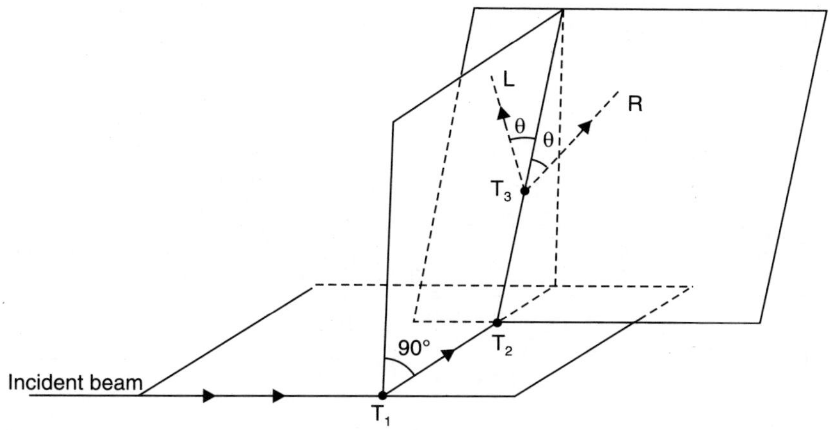

Fig. 5.8 The figure shows the triple scattering set-up for measuring R (θ) (After Ref. 9).

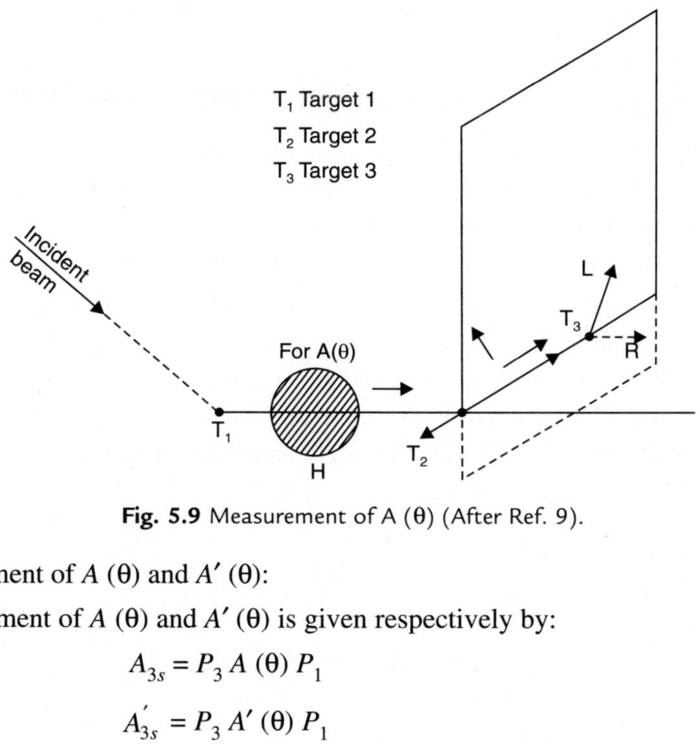

Fig. 5.9 Measurement of A (θ) (After Ref. 9).

(*b*) Measurement of A (θ) and A' (θ):

The measurement of A (θ) and A' (θ) is given respectively by:

$$A_{3s} = P_3\, A\,(\theta)\, P_1 \qquad ...(5.20a)$$

$$A'_{3s} = P_3\, A'\,(\theta)\, P_1 \qquad ...(5.20b)$$

Here the magnetic field perpendicular to the spin and beam directions is applied between the first and the second scatterer for the measurement of A (θ) (Fig. 5.9). Similarly for the measurement of A' (θ) we apply the magnetic field between the second and third scatterer, in a manner that the magnetic field direction are perpendicular to the spin and the beam directions. The scattering planes after the second and third scatterer should again be chosen, such that, they are perpendicular to the spin direction and the incident beam directions in the two cases.

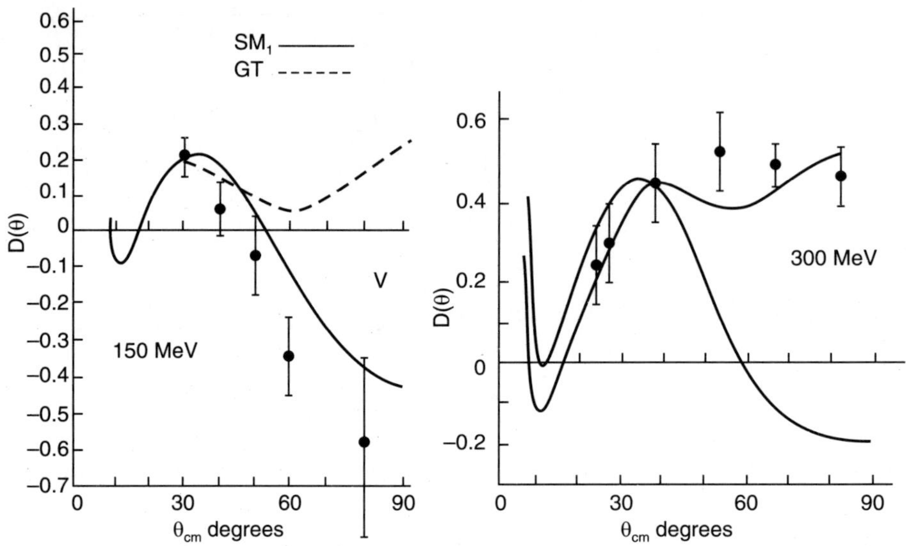

Fig. 5.10 Depolarisation parameter D, for 150 MeV and 300 MeV energies for p-p scattering, as a function of the angle θ cm, and the comparison with various theoretical potentials (Ref. 11).

The double scattering parameter P (θ) is understood on the basis of somewhat long range spin-orbit term; arising on the surface of nuclear matter. On the other hand, the triplet scattering rotation parameters; and spin correlation coefficients are understood theoretically, on the basis of a short range spin-orbit potential near the core; introduced by Signal and Marshak[8] (SM).

The triplet scattering parameters D (θ), A (θ) and R (θ) shown in Figures 5.10, 5.11 and 5.12 increase for higher values of energies for a given angle and go from negative to positive, as the energy is increased. These parameters, have been, in general, measured only for p-p scattering.[8] There are some measurements, however, available for p-n scattering (by scattering protons from deuterons; and subtracting it from p-p scattering) at 140 MeV for D (θ) and R (θ) (Ref. 11).

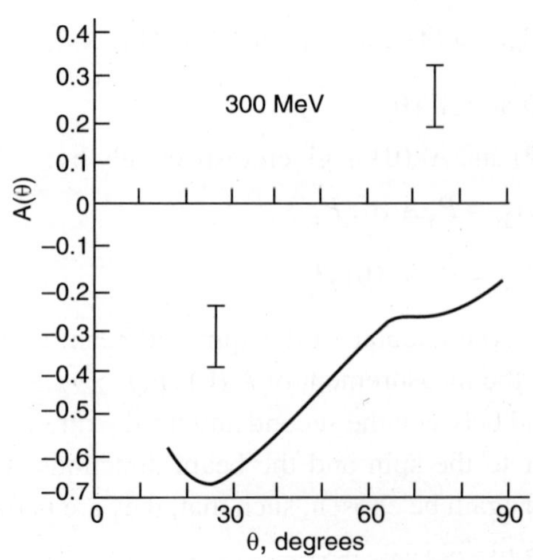

Fig. 5.11 The behaviours of A (θ) at energies of 300 MeV (Ref. 11).

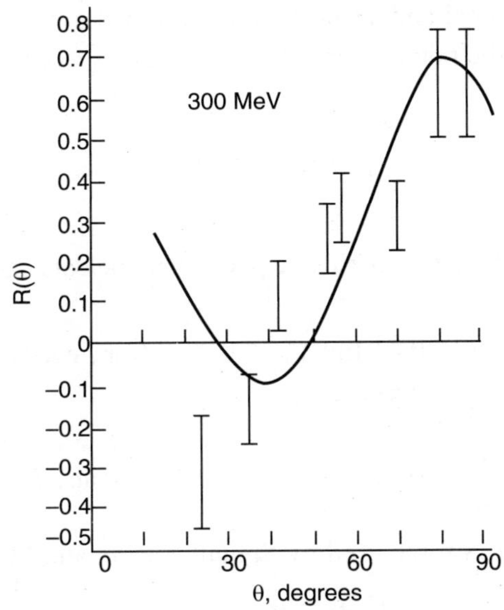

Fig. 5.12 The behaviour of R (θ) at energies of 300 MeV (Ref. 11).

***(iv) The spin correlation coefficients C_{nn}, C_{kp}, C_{pp}, C_{kk}*:** These coefficients, basically correlate the spin direction of the scattered and recoil nucleons. For example, C_{nn} gives the expectation value that the scattered projectile is polarised in direction $\hat{n}$ and recoil particle is also polarised in direction $\hat{n}$. This is obtained by measuring the coincidences between the scattered particles (polarised along $\hat{n}$ direction) and recoil particles (also polarised along $\hat{n}$-direction). Evidently this involves many fast coincidences so that the polarisation of the two particles and the correlation are measured simultaneously for which fast coincidence techniques are involved. The meaning of $\hat{p}, \hat{n}$ and $\hat{k}$ vectors is given in Eq. 5.21. Their exact definition, determines the mode of measuring the coincidences. For methods of measurements see references (6) and (9).

The unit vectors $\hat{p}$, $\hat{k}$ and $\hat{n}$ correspond to the first scattering from target N. These vector directions are given by:

$$\hat{p} = \frac{\mathbf{k}_i + \mathbf{k}_f}{|\mathbf{k}_i + \mathbf{k}_f|}, \quad \hat{k} = \frac{\mathbf{k}_f - \mathbf{k}_i}{|\mathbf{k}_f + \mathbf{k}_i|}$$

and

$$\hat{n} = \frac{\mathbf{k}_i \times \mathbf{k}_f}{|\mathbf{k}_i \times \mathbf{k}_f|} \qquad \text{...(5.21)}$$

***Spin-orbit force and polarisation experiments*:** It seems that at energies below 40 MeV; there is very little, evidence for an **l.s** force. This implies, that **l.s** force must be of short range compared with central and tensor forces. Detailed calculations, using nuclear forces, with different force parameters, and comparison with phase shift and polarisation data, show **l.s** force should be restricted to about 0.7 fm.

As a matter of fact Gammal and Thaler[10] have assumed a mixture of central, tensor and spin-orbit forces; with the shape of the potential in all the three cases to be of the form:

$$V(r) = -\frac{Ve^{-\mu_L r}}{\mu_L r} \qquad r > r_c$$

$$= \infty \qquad r < r_c \qquad ...(5.22)$$

where r_c is the radial distance of the first repulsive part. Then the data can be fitted with spin-orbit force parameters as:

$$r_c = 0.41 \text{ fm}; V_{LS} = 0 \qquad (\text{for } l \text{ even})$$

$$V_{LS} = 7318 \text{ MeV} \qquad (\text{for } L \text{ odd})$$

$$\frac{1}{\mu_L} = 0.27 \qquad (\text{for any } L) \qquad ...(5.23)$$

Another expression for **L.S** force used by Signall,[12] Zinn and Marshak is:

$$V_{\mathbf{L.S}} = C\frac{1}{x}\frac{d}{dx}\frac{e^{-2x}}{x} \qquad r > r_c$$

$$= V_{LS}(r_c) \qquad r < r_c$$

where $C = 21$ MeV ...(5.24)

This expression is similar to Eq. 5.9 expected from arguments in section 5.13. The spin-orbit forces introduced above give reasonable fit to the main features of the nuclear interactions up to 300 MeV but do not provide a detailed agreement. But it is clear that they do exist and they are of very short range and are immediately adjacent to the edge of the repulsive core.

5.2.4 Recent Developments

The measurements of angular distributions, $\sigma(\theta)$, the analysing power Ay, and spin correlation coefficients *Axx, Axy,* and *Ayz* have been carried out at higher energies of 197.4 MeV, 200 MeV and 250–450 MeV for *p-p* interaction by many workers.[13] These measurements provided a very precise data, due to the carefully evolved experimental methods. The results were compared with *p-p* potential models and *p-p* wave analysis. These comparisons are progressively able to distinguish between the various potential models[14]. Similar measurements of *Ay* carried out for break-up reactions H^2 *(p, ppn)* at 65 MeV, also seem to agree with these potentials. An interesting work on precision measurement on *n-p* elastic scattering has been carried by J. Zhao et al.[15], at 347 MeV. The neutron beams and protons target were alternatively polarised for measurement of analysing power for polarised protons from unpolarised protons and vice versa. The data was analysed, and compared with the theoretical potentials of Iqbal and Naskanen called Argonn V 18 (AV 18), C.D. Bonn, and Nijgemen.[14, 16]

The difference between the two analysing powers, as discussed above, shows a clear evidence for charge symmetry-breaking and agrees well with the theory, which predicts such a charge symmetry-breaking. On the other hand, when rigorous calculations were carried out for *Ay* (θ) for the incident energies from 1 to 3 MeV, no significant charge dependence was observed. It seems charge dependence comes into play at higher energies. Such studies of nucleon-nucleon elastic scattering have been extended up to 2.5 GeV[17].

Appendix

Analysing Power and Related Quantities

We have used the terms *Ay* and iT_{11} (θ), T_{20} (θ), T_{21} (θ) and T_{22} (θ) which are collectively called analysing powers and it_{11}, t_{21} and t_{22} (the beam polarisation moments) and spin correlations A_{xx}, A_{xy} and A_{xz} in the main text of Chapters 4 and 5.

Earlier, the double scattering quantities like *P* (θ) and the triple scattering quantities like *D* (θ), *R* (θ), *R*′ (θ), *A* (θ) and *A*′ (θ) have been discussed in Chapter 5, which are also related to polarisation properties, which are used when the initial beam is unpolarised. Recently,[18] the polarised beams of protons, deuterons and Li^6 have become available and in the recent literature the measurement of analysing powers and spin correlations for the target have been carried out.

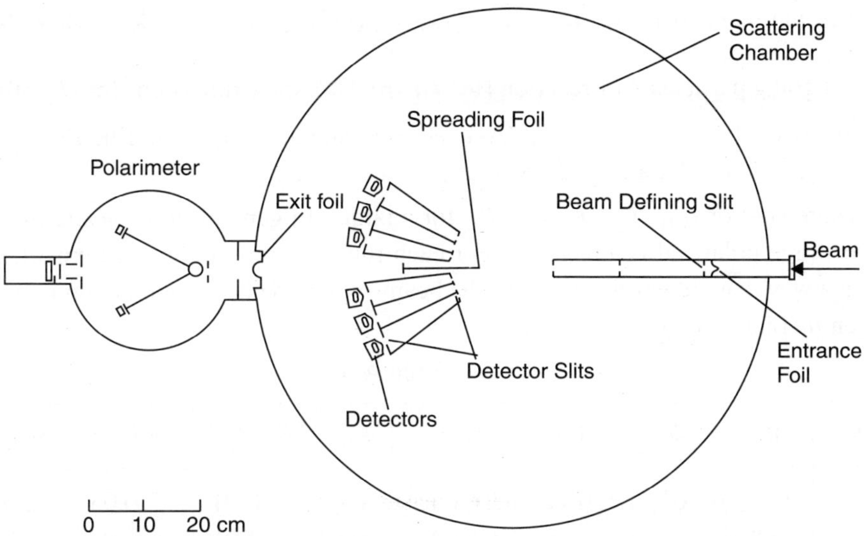

Fig. A-1 Schematic diagram of the experimental arrangement.

To understand the physical significance of these quantities, we reproduce in Fig. A-1 the experimental arrangements, typical of such measurements[19]. Using this arrangement, it has been possible to measure T_{11} (θ), T_{20} (θ), T_{21} (θ) and T_{22} (θ) for the reaction H^1 $(\vec{d}, d)$ H^1, *i.e.* elastic scattering of polarised deuterons from the target of protons. One can also adapt it for measuring *Ay* and A_{xx}, A_{xy} and A_{xz}. In this arrangement, the polarised beam of deuterons, produced by crossed beam of polarised ion-source enters a scattering chamber, filled with hydrogen (the target) and gets scattered by protons in hydrogen. The six detectors, in the scattering chamber, detect the scattered deuterons and recoil protons, at different angles. The polarised deuteron-beam (which is not scattered by hydrogen) enters the polarimeter, which makes use of He^3 $(\vec{d}, p)$ He^4 reaction and provides direct measurements of four relevant moments, *i.e.* $Re\,(it_{11})$, t_{20}, $Re\,(t_{21})$ and $Re(t_{22})$ as described in references (20).

We give below the principle of these measurements. We start by writing the expression for differential cross-section $\sigma(\theta)$ for a reaction by a polarised deuteron beam[21] as:

$$\sigma(\theta) = \sigma_0(\theta)\,[1 + 2iT_{11}(\theta)\,Re\,(it_{11}) + T_{20}(\theta)\,t_{20} + 2\,T_{21}(\theta)\,Re\,(t_{21}) + 2T_{22}(\theta)\,Re\,(t_{22})] \quad \text{...(A-1)}$$

where $\sigma_0(\theta)$ is the cross-section for unpolarised beam. The quantities t_{qk} describe the spin-state of the incident beam and are referred to as beam polarisation moments. In general t_{11}, t_{21} and t_{22} are complex quantities, because of parity conservation. Only the real part of t_{21} and t_{22} and only the imaginary part of t_{11} enter the cross-section. The beam moments of t_{qk} at the target are related to the moments τ_{qk} at the ion-source by the transformation equation:

$$t_{qk} = \sum_{q'} \tau_{q'k}\, D^k_{q'q}\,(\Phi, \Theta, \Psi) \quad \text{...(A-2)}$$

where Φ, Θ and Ψ are Euler's angles and $D^k_{q'q}$ are the well-known rotation matrices. According to the Madison[20] convention the beam moments are specified in a right-handed co-ordinate system, whose z-axis is parallel to the incident beam, $\mathbf{K}_{in}$ and whose y-axis is along $\mathbf{K}_{in} \times \mathbf{K}_{out}$ where $\mathbf{K}_{out}$ gives the momentum of outgoing particles. For reaction $He^3\,(\vec{d}, p)\,He^4$, the expressions for $iT_{11}(\theta)$, $T_{21}(\theta)$, and $T_{22}(\theta)$ are known (Ref. 19). Then using Eq. (A-1) at different angles, one can find $Re\,(it_{11})$, t_{20}, $Re\,(t_{21})$ and $Re\,(t_{22})$. For details see Godard et al. (Ref. 20).

For measurement of iT_{11}, T_{20}, T_{21} and T_{22} for a target, in general in the scattering chamber one takes a series of three polarised beam-runs.[19] For each run, the spin alignments axis of the beam was oriented in such a way that either one or two of the beams would be large and remaining moments close to zero, as given in Table (A-1).

Table A-1

Run	β	ϕ	τ_{10}	t_{20}	t_{20}	$Re\,[t_{21}]$	$Re\,[t_{22}]$	$Re\,[it_{11}]$
a	90°	0°	± 0.72	± 0.04	∓ 0.02	≤ 0.01	∓ 0.03	± 0.51
b	0°		0	± 0.62	± 0.62	≤ 0.01	≤ 0.01	≤ 0.01
c	90°	270°	0	± 0.62	± 0.08	∓ 0.37	± 0.22	≤ 0.01

The quantities β and θ in Table (A-1) are angles, which specify the normal orientation of the spin alignment axis, while τ_{qk} give the magnitude of the vector and tensor polarisation as given in Eq. A-2.

The quantities $T_{11}(\theta)$, $T_{20}(\theta)$, $T_{21}(\theta)$ and $T_{22}(\theta)$ at a given angle θ are measured by the yield for a detector on that left or right side of the detectors in the scattering chamber. The yield on the left (Y_1) or on the right side (Y_r) is related to the analysing powers as follows:

$$Y_{1,r} = Y^0_{1,r}\,[1 \pm 2\,Re\,(it_{11})\,iT_{11} + t_{20}\,T_{20} + 2\,Re\,(t_{21})\,T_{21} + 2\,Re\,(t_{22})\,T_{22}] \quad \text{...(A-3)}$$

where the upper and lower signs are for Y_1 and Y_r respectively. The quantities Y^0_1 and Y^0_r are yields for an un-polarised beam. The values of T_{11}, T_{20}, T_{21} and T_{22} are then related to the yield and $< t_{qk} >$ as follows:

$$iT_{11} = \frac{1}{4}\frac{[\varepsilon_1(a) - \varepsilon_r(a)]}{\langle it_{11}\rangle}$$

$$T_{20} = \frac{1}{2}\frac{[\varepsilon_1(b) + \varepsilon_r(b)]}{\langle t_{20}\rangle}$$

$$T_{21} = \frac{1}{4}\frac{[\varepsilon_1(c) - \varepsilon_r(c)]}{\langle t_{21}\rangle}$$

$$T_{22} = \frac{1}{4}\frac{[\varepsilon_1(c) - \varepsilon_r(c)]}{\langle t_{22}\rangle} \qquad \text{...(A-4)}$$

where $$\varepsilon_1(a) \equiv \frac{Y_1^+(a) - Y_1^-(a)}{Y_1^+(a) + Y_1^-(a)} \qquad \text{...(A-5}a\text{)}$$

and $$\varepsilon_r(a) \equiv \frac{[Y_r^+(a) - Y_r^-(a)]}{[Y_r^+(a) + Y_r^-(a)]} \qquad \text{...(A-5}b\text{)}$$

and where $$\langle it_{11}\rangle = \frac{1}{2}\left\{Re\,[it_{11}^+(a)] - Re\,[it_{11}^-(a)]\right\}$$

$$\langle t_{20}\rangle = \frac{1}{2}\left\{t_{20}^+(b) - t_{20}^-(b)\right\}$$

$$\langle t_{22}\rangle = \frac{1}{2}\left\{Re\,[t_{21}^+(c)] - Re\,[t_{21}^-(c)]\right\}$$

$$\langle t_{22}\rangle = \frac{1}{2}\left\{Re\,[t_{22}^+(c)] - Re\,[t_{22}^-(c)]\right\}$$

For the un-polarised beam, one uses the relationship:

$$Y_{1,r}^0(a) = \frac{1}{2}[Y_{1,r}^+(a) + Y_{1,r}^-(a)] \qquad \text{...(A-6)}$$

where + and – sign refer to two spin states of the incident beam.

It is evident that the analysing powers are related to the polarisation power of the target for a given projectile and can be compared with the expected theoretical values calculated using the given nucleon-nucleon potentials.

Similarly the analysing power $Ay(\theta)$ were calculated from spin up (+) and spin down (–) yields, according to the relation[21].

$$Ay(\theta) = \left(\frac{1}{P}\right)^* \frac{Y^+(\theta) - Y^-(\theta)}{Y^+(\theta) + Y^-(\theta)} \qquad \text{...(A-7)}$$

where P is the beam polarisation, as defined in Fig. 5.5. Physically the quantity $Ay(\theta)$ represents the relative power of the target to polarise the scattered beam relative to the initial polarisation of the incident beam.

For the definition of A_{xx}, A_{xy} and A_{xz}, see Ref. (13). They are somewhat involved and are related to the components of polarisation in a frame where z points along the beam, y points upwards and **x = yxz**. The analysing power Ay and spin correlation coefficients A_{mn} are function of scattering angle, θ.

5. Nucleon-Nucleon Scattering at High Energies 2000–2008

In an interesting experimental paper, involving polarisation at high energies, the angular distribution of longitudinal $\vec{p}\ \vec{p}$ spin-correlation parameter A_2 has been measured at 1974 MeV beam energy, of polarised protons, at Indiana University Cyclotron Facility (IUCF). The results are compared to recent partial wave analysis and *n-n* potential models. [Phy. Rev. C. 61, 054002 (2000)]

In another experiment, nucleon-nucleon scattering has been measured, up to 3 GeV. [Phy. Rev. 62, 0.34005; (2000)]

In a theoretical paper, by Ramachandran, Deepak and Vidya; pion production at high energies of 280–300 MeV, in *n-n* collisions has been calculated [Phy. Rev C. 62, 0.11001 (R) (2000)].

In another interesting theoretical paper several nucleon-nucleon potentials, *e.g.*, Paris, Vijimeagan, Argonne, and others are extended from 1300 MeV to 3 GeV by using latest phase shift analysis of Ardnt given in [Phy. Rev. D. 97 (1983)] [Phy Rev. C. 64, 054003 (2002)].

Proton-Proton Parity violating asymmetry in logitudinal asymmetry is calculated in lab. energy range of 0–300 MeV, using latest generation strong interaction potentials.

- Argonne V18; Bonn-200; and Vijimeagan.
- In combination with week interaction potential consisting of ρ and ω meson exchange.
- In a model known as *DDH*; the theoretical results are compared with measured, asymmetrics, at 13.6 MeV; 45 MeV, and 22 MeV.

The values of ρ and ω mesons weak-coupling h_ρ^{pp} and h_ω^{pp} are determined.

In another paper from Poland, Germany and Japan, (4 authors); the authors have solved the (3*N*) three nucleon-equation of Feddev; including relativistic features at incident energies of neutron at $E_n^{\text{lab}} \equiv 28$, 65, 135 and 250 MeV. [Phy. Rev. 71, 054001, (2005)].

In a theoretical paper, [Phy. Rev. C 77, 014001 (2008)] the authors generate an energy dependence Lorentz Convariant parameterisation of on-shell nucleon-nucleon scattering amplitude, in terms of Yukawa type meson exchanges in first, order Bom approximation. This parameterisation provides a good description on *n-n* scattering in the energy range of interest, and can be successfully extrapolated to energies between 420 and 300 MeV.

REFERENCES

1. W. N. Hess: Rev. Mod. Phy. 30, 368 (1958).
2. Same as reference (1), Lock, W.O.: High Energy Nuclear Physics, p. 146, Metheun and Co., London, (1960); Signall, P. et al., Phy. Rev. 1358, 1128 (1964).

3. T. Hamada, and Johnson, J. T.: Nuclear Physics 34, 382 (1962).
4. R.J.N. Phillips: Prec. Phy. Soc. (London) A70, 721 (1950).
5. E. Clamental, C. Villi, and L. Jess: Nuovo Cimente 5, 907 (1957).
6. Oxley, C. L., W. F. Caruright, and J. Rouvina, Phy. Rev. 93, 806, (1954). Wouters, L. F., Phy Rev. 84, 1069, (1951). Wolfenstein, L., Annual Rev. Nuclear Sciences, 6, 43, (1956). R.K. Adair, Phy. Rev. 86, 160, (1952), D. C. Doddar and J. Gammel, Phy. Rev. V. 88, 520, (1952).
7. Reference (1), J.N. Palmieic A. M. Commack, N. F. Ramsey, and R. Wilson, Annual Physics (New York), 5, 299, (1958). M. J. Moravcsik, The Two Nucleon Interaction Clarendon Press, Oxford, (1963). O. Chamberlain, E. Segre, R. D. Trip, C. Wiegand, and T. Ypsilantis, Phy. Rev. 105, Rev. 105, 288, (1957). R. Wilson: The Nucleon-Nucleon Interaction Interscience, New York, (1962).
8. Signall, P. and R. Marshak, Phys. Rev. 106, 832, (1957), ibid, 109, 1229, (1958) M. H. Hull Jr. K. D. Pyatt, Jr., C. R. Fischer, and G. Breithy, Rev. Letters 2,264, (1959).
9. Theory of Nuclear Structure: M. K. Pal, Afflicated East-West Press Pvt. Ltd. New Delhi/Chennai, (1972).
10. Gammel J. and R. Thaler, Phy. Rev. 107, 291, (1957).
11. J. N. Palmieic, A. M. Commack, N. F. Ramsey, and R. Wilson, Ann. Phy. (New York), 5, 299, (1958).
12. P. S. Signall, R. Zinn, and R. E. Marshak, Phy. Rev. Letters, 1, 416, (1958).
13. F. Rathman, D. Yon Prezenworski, W. A Dezarn, J. Deskow et al., Phy. Rev. C. V. 58, P. 658, (1998), W. Haeberli, B. Lorentz et al., Phy. Rev. C 55, P. 597, (1997), B. V. Przewoske, F. Rathman et al. (14 authors), Phy. Rev. C 58, 1897, (1998).
14. R. B. Wiringa, V. G. T. Stoks, R. Schiavilla, Phy. Rev. C 51, 38, (1995) (for AV 18) M. Lacombe, B. Loisean, J. M. Richard, R. Vinh Mau, J. Cote, P. Piers, and R. de Toureil, Phy. Rev. C. 21, 860, (1980), V. G. J. Stoks, R. A. M-Klemp, C. P. F. Terheggen and J. J. De Swart, Phy. Rev. C. 49, 2950, (1994). (for Nijgemen 93, 1, 2 and 3). R. B. Wiringa, R. A. Smith and J. C. Ainsworth, Phy. Rev. C 29, 1207, (1984). (AV – 14 potential), R. Machlaidt, F. Sammumeuruca, and Y. Song, Phy. Rev. C 53, 1483, (1996) (for C. D > Bonn), L. Jaide and H. V. Von Geramb, Phy. Rev. 57, 496, (1998).
15. J. Zhao, B. Abegg et al. (26 authors), Phy. Rev. C 57, P. 2126, (1998).
16. M. J. Iqbal and J. A. Niskanen, Phy. Rev. C 38, 2259, (1998).
17. Richardt A. Arndt, Change Heon Oh, Igor I. Starkovsky, Ron L. Workman and Frank Dobramann, Phy. Rev. C 56, P. 3005, (1997).
18. W. Haeberli et al., Nuclear Instruments and Methods 196, 319, (1982).
19. J. Sowinski, D. D. Pun Casavani and L. D. Knustson, Nuclear Physics A, 464, p. 223, (1987).
20. R. R. Cadmus and W. Haeberli, Nuclear Instruments and Methods, V. 129, p. 403, (1975). I. G. Godara, N. Rolering and L. D. Knustson and W. Haebarli, Nuclear Instruments, Methods V. 137, p. 451, (1976). K. Stephens and W. Haeberli, Nuclear Instruments, Methods V. 169, p. 483, (1980).
21. R. M. Chastelar, H. R. Waller, D. R. Tilley and R. L. Prior, Phy. Rev. Letter, V. 72, p. 3949, (1994).

PROBLEMS

1. Show that the Serber potential, $V(r)$ given by:

$$V(r) = -g^2 \frac{e^{-kr}}{r} \times \frac{1}{2}\left[1 + \frac{1}{2}(1 + \sigma_1 \cdot \sigma_2)(1 + \vec{\tau}_1 \cdot \vec{\tau}_2)\right]$$

with $$\frac{g^2}{c\hbar} = 0.405, \quad \frac{1}{k} = 1.2 \text{ fm}.$$

gives no force in the odd angular moment **l**.

2. If Serber forces hold good for *p-p* scattering, what will be the shape of angular distributions of *p-p* scattering from 90° to 180° (Fig. 5.4*a*).

3. Show qualitatively, that

$$V = \text{Const}\ \sigma \cdot (\nabla\rho \cdot \mathbf{p})$$

in Eq. 5.8 represents the physical conditions on the surface of a nucleus, for a nuclear potential.

4. Show, qualitatively, that a positive polarisation $P(\theta)$ for *p-p* at high energies means, that spin-orbit-coupling potential V **l.s** confined to triplet spin states; and the nucleons with spin up scatter preferentially to the left. What will be the properties of V **l.s** for this ?

5. Calculate the phase shift of an impenetrable sphere of radius γ_c . What does this result suggest about the value of the phase-shift of a repulsive core surrounded by an attractive potential?

6. Show that, applied to any singlet state, the operator S_{12} gives zero.

7. Consider, for scattering amplitude of a central potential, the partial wave-expression:

$$f(\theta) = \frac{1}{2ik} \sum_{l=0}^{\infty} (2l+1)\,[\exp(2i\delta_1) - 1]\, P_1(\cos\theta)$$

Compute the corresponding total cross-section and show that

$$\sigma = \frac{4\pi}{k} I_m\, f(\theta).$$

8. When a 10 MeV proton interacts with an electron, which is initially at rest; what is the maximum angle by which the proton can be deflected? Treat the problems non-relativistically.

9. If we express the polarisation of the scattered beam in a general equation as:

$$\frac{\sigma(\theta)}{\sigma_0(\theta)} P_1^{\text{scatt}} = (P + P_1^{\text{inc}} \cdot \hat{n})\, \hat{n} + (A\, P_1^{\text{inc}} \cdot \hat{e} + RP_1^{\text{inc}} \cdot \hat{s})\, K + (A' P_1^{\text{inc}} \cdot \hat{e} + R'\, P_1^{\text{inc}} \cdot \hat{s})\, P$$

Then show Eqs. 5.19 and 5.20 can be derived from it.

10. Assuming $\langle \sigma_f \rangle$ depends linearly on $\langle \sigma_i \rangle$ and that σ is pseudoscular (provided parity is conserved), one can write the general expression, of the final intensity for all cases as:

$$I_f \langle \sigma_f \rangle = I_i\, [P + D\, \langle \sigma_i \rangle \cdot (\mathbf{k}_l \times \mathbf{k}_f)]\, (\hat{\mathbf{k}}_i \times \hat{\mathbf{k}}_f) +$$
$$\left[A \langle \sigma_i \rangle . \hat{\mathbf{k}}_i + R \langle \sigma_i \rangle . \hat{\mathbf{k}}_i \times (\hat{\mathbf{k}}_i \times \hat{\mathbf{k}}_f)\right] \hat{\mathbf{k}}_f \times (\hat{\mathbf{k}}_i \times \hat{\mathbf{k}}_f) +$$
$$\left[A' \langle \sigma_i \rangle . \hat{\mathbf{k}}_i + R' \langle \sigma_i \rangle . \hat{\mathbf{k}}_i \times (\hat{\mathbf{k}}_i \times \hat{\mathbf{k}}_f)\right] \hat{\mathbf{k}}_f \times (\hat{\mathbf{k}}_i \times \hat{\mathbf{k}}_f)$$

where $\langle \sigma_f \rangle$ has been expressed in terms of components of $\langle \sigma_i \rangle$ along the three mutually perpendicular directly $\mathbf{k}_i$, $\mathbf{k} \times \mathbf{k}_f$ and $\mathbf{k}_i \times (\mathbf{k}_i \times \mathbf{k}_f)$.

Show that the above equation represents a general case of polarisation (P); depolarisation (D) and the rotation parameter (R, R', A and A').

6

CHAPTER

Nuclear Forces

6.1 INTRODUCTION

We have discussed in the last three chapters[1], the two body problem involving nucleon-nucleon interaction, both in the bound state of deuteron as well as in the unbound states of *n-n* and *p-p* scattering at low and high energies. While the low energy data from deuteron or nucleon–nucleon scattering yield the basic features of nuclear forces, it is the high energy scattering say above 20 MeV or so, which gives detailed structure of nuclear forces.

We have already[1] discussed the special cases of exchange type nucleon-nucleon potential, *e.g.*

$$V(r) = V_c(r) + V_T(r)\, S_{12} \qquad \text{...(6.1}a\text{)}$$

where $$V_c(r) = V_0(r) + \frac{1}{2}(1 + \sigma_1 \cdot \sigma_2)\, V_\sigma(r) \qquad \text{...(6.1}b\text{)}$$

is the central potential; $V_0(r)$ being independent of spin and $1/2\,(1 + \sigma_1 . \sigma_2)\, V_\sigma(r)$ being spin dependent. This is a case of central potential involving spin-exchange (Eq. 4.103). On the other hand, the tensor potential (Eqs. 3.39 and 3.43), $V_T(r)\, S_{12}$, is dependent on the direction relation of spin and space coordinates. We have also, seen a case, (Eq. 5.3), showing that there can be Serber type potential, where both spin and iso-spin are exchanged, given by:

$$V(r) = -\,V_0(r)\left(1 + \frac{1}{4}(1 + \sigma_1 \cdot \sigma_2)(1 + \tau_1 \cdot \tau_2)\right) \qquad \text{...(6.2)}$$

We will now discuss, the general case, where such exchange forces form an integral part the formalism.

6.2 ISO-SPIN FORMALISM

We have already shown, while discussing the *n-p* and *p-p* scattering, that nuclear forces are charge-independent. We will discuss some more details about this property of nuclear forces in the succeeding sections. Formally this concept can be put into practice by looking upon protons and neutrons as two almost degenerate states of a single entity called a nucleon. The two states are ascribed to a two-valued internal degree of freedom of a nucleon, and therefore, one treats iso-spin formalism in exactly the same way as we treat two z-states of angular momentism 1/2 $\hbar$. We have discussed the concept of isotopic spin T earlier in Section 2.2.3.

In analogy to the formalism for spin matrices, in physical space, we express the iso-spin states of proton and neutron respectively as:

$$\eta\,(p) \equiv \gamma = \begin{pmatrix} 1 \\ 0 \end{pmatrix} \qquad \text{...(6.3)}$$

and

$$\eta\,(n) \equiv \delta = \begin{pmatrix} 0 \\ 1 \end{pmatrix} \qquad \text{...(6.4)}$$

In this iso-spin space, γ-matrix behaves like α or 'up' spin state of the physical space and δ-matrix behaves like β or 'down' spin-state of the physical space.

Similarly the iso-spin operators, corresponding to the spin operators are:

$$\tau_1 = \begin{pmatrix} 0 & 1 \\ 1 & 0 \end{pmatrix} \qquad \text{...(6.5)}$$

τ_1, thus behaves like σ_x. Further,

$$\tau_2 = \begin{pmatrix} 0 & -i \\ i & 0 \end{pmatrix} \qquad \text{...(6.6)}$$

τ_2, then, behaves like σ_y, and

$$\tau_3 = \begin{pmatrix} 1 & 0 \\ 0 & 1 \end{pmatrix} \qquad \text{...(6.7)}$$

τ_3 behaves like σ_z.

We thus, see that τ_1, τ_2 and τ_3 iso-spin matrices behave like Pauli matrices σ_x, σ_y and σ_z of the angular momentum spin. Again in analogy to the case of angular momentum spin, only the eigen-values of τ_3 (or $t_3 = \tau_3/2$) are measurable, so that:

$$\frac{1}{2}\tau_3\gamma \equiv t_3\gamma = \frac{1}{2}\gamma \qquad \text{...(6.8)}$$

and

$$\frac{1}{2}\tau_3\delta \equiv t_3\,\delta = -\frac{1}{2}\gamma \qquad \text{...(6.9)}$$

Hence the eigen-values of the operator τ_3 are +1 and –1 and for t_3, they are + 1/2 and – 1/2 for proton and neutron respectively. It is, these eigen-values, for $t_3 \equiv \tau_3/2$ which have been denoted as T_3 or T_z, earlier. They are z-components of the iso-spin quantum number T, which is eigen-value of a nuclear eigen state in iso-spin space.

One also defines, the creation and annihilation operators τ_+ and τ_- respectively so that:

$$2t_+ \equiv \tau_+ \equiv (\tau_1 + i\tau_2) = 2\begin{pmatrix} 0 & 1 \\ 0 & 0 \end{pmatrix} \quad ...(6.10)$$

and $$2t_1 \equiv \tau_- \equiv (\tau_1 - i\tau_2) = 2\begin{pmatrix} 0 & 0 \\ 1 & 0 \end{pmatrix} \quad ...(6.11)$$

It can be easily seen that:

$$t_+ \gamma \equiv \frac{1}{2}\tau_+ \gamma = 0 \quad ...(6.12a)$$

$$t_+ \delta \equiv \frac{1}{2}\tau_+ \delta = \gamma \quad ...(6.12b)$$

$$t_- \gamma \equiv \frac{1}{2}\tau_- \gamma = \delta \quad ...(6.12c)$$

$$t_- \delta \equiv \frac{1}{2}\tau_- \delta = 0 \quad ...(6.12d)$$

In this manner operator t_+ acting on proton annihilates a proton state, but converts a neutron into a proton state. On the other hand, the operator t_- annihilates a neutron state, but converts a proton state to a neutron state.

The isotopic spin quantum number T for any nuclear state is an eigen-value of the state in the iso-spin space; which can be seen from the expression of the Hamiltonian H_{NN} in terms of T, for a two-nucleon system,[2] (for pure nuclear part without Coulomb interaction).

$$H_{NN} = H_o + H_2 T^2 \quad ...(6.13)$$

for condition of charge independence.

Thus the interaction H_{NN} can only depend on the iso-spin variable through T^2; as the iso-spin states are eigen-states of T^2. This means that the energy of nuclear interaction is independent of T_z as it does not occur explicitly in Eq. 6.13. If we start a system in a specified state of iso-spin and then apply to it a strong nuclear interaction; the system will continue in the same iso-spin state, T.

The behaviour of T will be similar to S, the spin quantum number proton or a neutron $T = 1/2$; so that:

$$T^2\psi_p = \frac{1}{2}\left(\frac{1}{2}+1\right)\psi_p = \frac{3}{4}\psi_p;\ T_z\,\psi_p = \frac{1}{2}\psi_p \quad ...(6.14a)$$

and $$T^2\psi_n = \frac{1}{2}\left(\frac{1}{2}+1\right)\psi_n = \frac{3}{4}\psi_n;\ T_z\,\psi_n = -\frac{1}{2}\psi_n \quad ...(6.14b)$$

We will discuss the properties of the iso-spin dependence of the wave functions in detail in Section 6.2.1. Enough to say here, that in any nuclear reaction involving any specific eigen-states in nuclei, only those states will be involved, where T^2 is conserved, within generalised Pauli Exclusion Principle in such a way that, if ψ in:

$$\psi = \psi_{\text{space}}\, \psi_{\text{spin}}\, \psi_{\text{isospin}} \qquad ...(6.15)$$

is anti-symmetric on exchange of all quantum numbers. Consider a system of two nucleons in *p*-state. In this case, space wave-function is anti-symmetric on exchange of position; so that symmetric spin wave-function must be accompanied by symmetric isospin wave-function, which means either *pp* or *nn* state. In *pn* state, both spin wave-functions, singlet and triplet, are possible; so a triplet *p-n* state will have the same energy as triplet *pp* or *nn* state. Thus isospin plays an important part in the behaviour of nuclear reactions.

6.2.1 Generalised Pauli Exclusion Principle and Wave-Functions

It is evident from the above discussion, that in iso-spin formalism, we treat neutron and proton as the two states of the same particle. This facilitates the treatment of the many body problems quantum-mechanically in nuclear physics, as we shall see subsequently. To illustrate the power of the iso-spin formalism, we take the case of the two-nucleon system. Generally, we consider the proton and neutron as two distinct particles; write the wave-function for proton-neutron system, without introducing the antisymmetrisation; while for the proton-proton and neutron-neutron systems, we can write anti-symmetric wave-functions. By introducing the iso-spin formalism, however, we can treat protons and neutrons as the same particles with different quantum numbers. The nucleons are then said to obey Generalised Pauli Exclusion Principle; according to which, the wave-function containing both space, angular momentum, as well as iso-spin quantum numbers becomes anti-symmetric; on the exchange of all the quantum numbers. It may be emphasised that Generalised Pauli Exclusion Principle does not introduce any new principle. It only combines the Pauli Exclusion Principle for neutron-neutron; proton-proton system and the intrinsic sameness of the particles in neutron-proton system, with different quantum numbers. This can be seen below:

We denote, the position, spin and iso-spin variables by **r**, m_s and T_3. The wave-function of a nucleon may, then, be written as:

$$\psi = \phi(\mathbf{r})\, \chi_s^{ms}\, \eta_{T_3} \qquad ...(6.16)$$

where $\phi(\mathbf{r})$ represents the space part of the wave-function which may be written as:

$$\phi(\mathbf{r}) = \sum_{m_1} \sum_{1} \frac{U_1(r)}{r} Y_1^{m_1}(\theta, \phi) \qquad ...(6.17)$$

Similarly χ_s^{ms} is the physical spin part of the wave-function, $\phi(\mathbf{r})$ and χ^{ms} being both in the real space, may couple to give J. So that:

$$\phi(\mathbf{r})\, \chi_s^{ms} = \frac{U(r)}{r} Y_J^{m_J} \qquad ...(6.18a)$$

where

$$Y_J^{m_J} \equiv \sum_{m_1 m_s} \sum_{1} C(J, m_J \,|\, l, s, m_1, m_s)\, Y_1^{m_1}\, \chi_s^{m_s} \qquad ...(6.18b)$$

The wave-function η_{T_3} of course, represents either a proton or a neutron, depending, on the value of T_3; so that:

$$\eta_{T_3=+\frac{1}{2}} \equiv \eta(p) \qquad ...(6.19a)$$

is the wave-function for a proton and

$$\eta_{T_3=-\frac{1}{2}} \equiv \eta(n) \qquad ...(6.19b)$$

is the wave-function for a neutron.

For the two nucleons, we write below the anti-symmetrical wave-function obeying generalised Pauli Exclusion Principle. The quantum numbers now are related to:

$$\mathbf{r}_1,\ m_{s_1},\ T_3\ (1);\ \text{and}\ \mathbf{r}_2,\ m_{s_2},\ T_3\ (2) \qquad ...(6.20)$$

so that the wave-function is a function of these six variables. To write the wave-function, obeying the Generalised Pauli Exclusion Principle, we require that (*i*) for a proton-proton or a neutron-neutron pair, the space part of the wave-function should be anti-symmetric on the exchange of the space quantum numbers; while the iso-spin part may remain symmetric on the exchange of iso-spin quantum numbers and (*ii*) for a proton-neutrons, the space part may be symmetric; while the iso-spin part is anti-symmetric on the exchange of the relevant quantum numbers; so that the overall wave-function is anti-symmetrised on the exchange of all the quantum numbers. This can be achieved, if we can write the wave-function ψ_{12} for two nucleons, as:

$$\psi_{12} = \frac{1}{\sqrt{2}}(\psi_I - \psi_{II})$$

$$= \frac{1}{2}\left[\phi(r_1, m_{s_1}; r_2, m_{s_2}) + \phi(r_2, m_{s_2}; r_1, m_{s_1})\right] X_0$$

$$+ \frac{1}{2}\left[\phi(r_1, m_{s_1}; r_2, m_{s_2}) - \phi(r_2, m_{s_2}; r_1, m_{s_1})\right] X_{10} \qquad ...(6.21a)$$

where the iso-spin function X are defined as:

$$X_0 = \frac{1}{\sqrt{2}}[\eta(n_1)\,\eta(p_2) - \eta(n_2)\,\eta(p_1)] \qquad ...(6.21b)$$

and

$$X_{10} = \frac{1}{\sqrt{2}}[\eta(n_1)\,\eta(p_2) + \eta(n_2)\,\eta(p_1)] \qquad ...(6.21c)$$

The expression for ψ_{12} as written in Eq. 6.21*a* brings out the basic features of the generalised Pauli Exclusion Principle. The first term in Eq. 6.21*a* consists of two parts; the first part corresponding to physical space and the second to iso-spin space. The wave-function corresponding to physical space is evidently symmetric on the exchange of r and m_s; while X_0 is evidently anti-symmetric on the exchange iso-spin variables for particle (1) and (2). On the other hand the second term in Eq. 6.21*a* consists anti-symmetric part of the wave function corresponding to exchange of r and m_s; while it is symmetric for X_{10} exchange of the iso-spin variables of the two particles. In this manner, it is seen that the wave-function is anti-symmetric, with the exchange of two nucleons, using iso-spin formalism and we can treat proton and neutron in the same manner.

By a similar process, we can write the anti-symmetric wave-function for three, or four A nucleons. It is evident, from Eq. 6.21 that the wave-function has to be written in such a manner that if space-spin parts are symmetric, iso-spin parts will be anti-symmetric and vice versa. We will, only show below, how to write the anti-symmetric part. Symmetric part can be written by replacing the minus in antisymmetric part by plus.

It may be seen, that for two nucleons, (1) and (2) with quantum numbers, expressed as a and b; one can write the anti-symmetric wave-functions as:

$$\phi_{ab} = \frac{1}{\sqrt{2}}[\phi_a(1)\,\phi_b(2) - \phi_a(2)\,\phi_b(1)]$$

$$= \frac{1}{\sqrt{2}}\begin{vmatrix} \phi_a(1) & \phi_b(1) \\ \phi_a(2) & \phi_b(2) \end{vmatrix} \qquad \text{...(6.22}a\text{)}$$

$$\equiv \frac{1}{\sqrt{2}}[\pi - P(1,2)][\phi_a(1)\,\phi_b(2)] \qquad \text{...(6.22}b\text{)}$$

where operator π leaves the wave-functions unexchanged, while P (1, 2) exchanges nucleon (1) to (2) and vice-versa. We can define an anti-symmetrising operator A, so that:

$$A_{(1)} = \frac{1}{\sqrt{2}}[\pi - P(1,2)] \text{ (for two particles, } i.e. \text{ one pair)}$$

and $$A_{(p)} = \frac{1}{\sqrt{A!}}\sum_P \delta(P)\,P \text{ (for } A \text{ particles, having } P \text{ pairs of exchanges)} \qquad \text{...(6.22}c\text{)}$$

δ (P) = +1, when the permutation operator P involves an even number of particle exchanges and δ (P) = – 1, when it involves odd number of particle exchanges. Then we can write:

$$\phi_{ab} = A\,[\phi_a(1)\,\phi_b(2)] \qquad \text{...(6.22}d\text{)}$$

For three particles, the anti-symmetric wave function can be written as:

$$\phi_{abc} = \frac{1}{\sqrt{3!}}\begin{vmatrix} \phi_a(1) & \phi_b(1) & \phi_c(1) \\ \phi_a(2) & \phi_b(2) & \phi_c(2) \\ \phi_a(3) & \phi_b(3) & \phi_c(3) \end{vmatrix}$$

$$= \frac{1}{\sqrt{3!}}\sum_P \delta(P)\,P[\phi_a(1)\,\phi_b(2)\,\phi_c(3)] \qquad \text{...(6.23}a\text{)}$$

For A nucleons, we can similarly write:

$$\phi_{abc\ldots\xi} = \frac{1}{\sqrt{A!}}\begin{vmatrix} \phi_a(1) & \phi_b(1)\ldots\phi_\xi(1) \\ \phi_a(2) & \phi_b(2)\ldots\phi_\xi(2) \\ \vdots & \vdots \\ \phi_a(A) & \phi_b(A)\ldots\phi_\xi(A) \end{vmatrix}$$

$$= \frac{1}{\sqrt{A!}} \sum_P \delta(P) P[\phi_a(1)\phi_b(2)\ldots\phi_\xi(A)] \qquad \ldots(6.23b)$$

One can use Eq. 6.23*b*, when one is required to write the anti-symmetric wave-function of A fermions, on which exchange operator representing exchange forces, operate.

6.2.2 Exchange Operators and Exchange Forces

While we will discuss in the next section, the exact nature and physical implications of exchange forces; we discuss below the properties of exchange operators, which, operate on a wave-function, of say a two-nucleon system and bring about the exchange of various quantum numbers. They are required, for the discussion of exchange forces. Their discussion will also bring about the implication of generalised Pauli Exclusion Principle.

We define three operators, P^B, P^H and P^M called Bartlett, Heisenberg and Majorona exchange operators, respectively as follows:

$$P^B \equiv P^\sigma \equiv \frac{1}{2}(1 + \sigma_1 \cdot \sigma_2)$$

$$P^H \equiv -P^\tau \equiv -\frac{1}{2}(1 + \tau_1 \cdot \tau_2)$$

and

$$P^M \equiv -\frac{1}{4}(1 + \sigma_1 \cdot \sigma_2)(1 + \tau_1 \cdot \tau_2) \qquad \ldots(6.24)$$

It can be easily seen, that P^σ exchanges spin quantum numbers; P^τ exchanges iso-spin quantum numbers and P^M exchanges both spin and iso-spin, which corresponds to change of space.

This can be illustrated by considering the role of P^σ in details. Two fermions, having spins σ (1) and σ (2) can have the following possible situations:

(1) ***For singlet S = 0:***

$$|\mathbf{S}| = \left|\frac{1}{2}(\sigma_1 + \sigma_2)\right| = 0;\ S^2 = S(S+1) = \frac{1}{2}(3 + \sigma_1 \cdot \sigma_2) = 0$$

Hence $\sigma_1 \cdot \sigma_2 = -3$; and $P^\sigma = \frac{1}{2}(1 + \sigma_1 \cdot \sigma_2) = -1$ ...(6.25*a*)

(2) ***For triplet S = 1:***

$$|\mathbf{S}| = \left|\frac{1}{2}(\sigma_1 + \sigma_2)\right| = 1;\ S^2 = S(S+1) = 2$$

Hence $\sigma_1 \cdot \sigma_2 = 1$

and

$$P^\sigma = \frac{1}{2}(1 + \sigma_1 \cdot \sigma_2) = 1 \qquad \ldots(6.25b)$$

Therefore $P^{\sigma}\chi_0^0 = -\chi_0^0$ (for singlet state)

and $P^{\sigma}\chi_1^{1,0,-1} = \chi_1^{1,0,-1}$ (for triplet state)

where χ's are the spin wave-functions of Eq. 6.16.

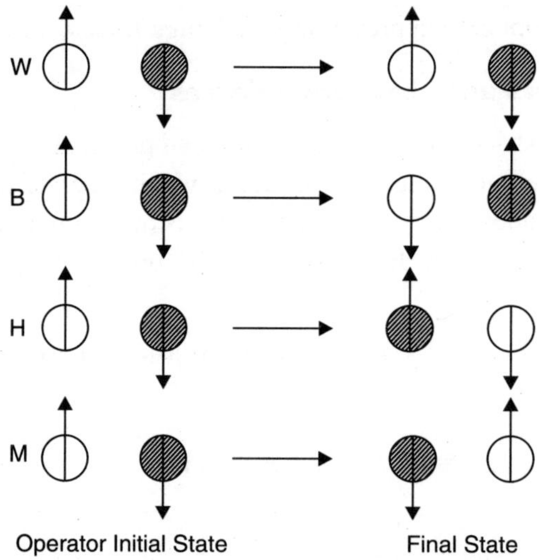

Fig. 6.1 Effect of the Wigner (W) Bartlett (B) Heisenberg (H) and Majorona (M) operators on two-nucleon state.

So the operator P^{σ} reverses the sign of the wave-function of anti-parallel spin particles; but we know from Pauli Exclusion Principle, that this will happen, if we exchange the spin of the two particles, with anti-parallel spins. On the other hand, P^{σ} does not reverse the sign of the wave-function, when spins of the two particles have their spins parallel; as the Pauli Exclusion Principle also demands. So the effect of P^{σ} is the same, as the exchange of spins, on the wave-function. Operator P^{σ} is, therefore, said to exchange the spins.

In a similar manner, P^{τ} exchanges the iso-spin of the two nucleons. We can now, show that $P^H = -P^{\tau}$ corresponds to the exchange of both mechanical spin and space coordinates. Take the case of two protons, with anti-parallel spins. Their iso-spins are parallel, hence P^{τ} will give positive sign on exchange of isospin of two protons. But according to generalised Pauli Exclusion Principle, the exchange of all the three quantum numbers changes the sign of the wave-function. Hence the negative of operator P^{τ}, will correspond to the simultaneous exchange of space and spin quantum numbers. This will, similarly be true for other cases also.

By a similar logic it can be easily seen, that P^M can be interpreted either as exchange of iso-spin and mechanical spin simultaneously or only as exchange of space coordinates. This is shown in Fig. 6.1.

We have thus seen, that the simultaneous operation of the three operators P^{σ}, P^{τ} and P^M reverses the sign of a wave-function, according to generalised Pauli Exclusion Principle, *i.e.*,

$$P^{\sigma} P^{\tau} P^M \psi = -\psi \quad \text{...(6.26)}$$

Also, from discussion above, it may be seen that:

$$P^{\tau} P^H \psi = -\psi$$

$$P^M P^\tau \psi = -\psi \qquad \text{...(6.27a)}$$

and
$$P^H = P^M P^\sigma = P^M P^B$$

$$P^M = -P^\sigma P^\tau = P^B P^H \qquad \text{...(6.27b)}$$

Further
$$(P^B)^2 = (P^H)^2 = (P^\tau)^2 = (P^M)^2 = 1 \qquad \text{...(6.28)}$$

Equation 6.28 can be easily understood, by releasing, that any of these exchange operators, where they convert the original wave-function to positive or negative, will always bring it to the original state, when repeated.

If we define a Majorona exchange potential so that:

$$P^M V_M(r)\,\phi(\mathbf{r}_1, \mathbf{r}_2) \equiv V_M(r)\,\phi(\mathbf{r}_2, \mathbf{r}_1) \qquad \text{...(6.29)}$$

then, it is evident that for negative $V_M(r)$,

if
$$\phi(\mathbf{r}_1, \mathbf{r}_2) = \phi(\mathbf{r}_2, \mathbf{r}_1) \qquad \text{...(6.30a)}$$

i.e. it is a symmetric wave-function; then the exchange potential $P^\sigma V_M(r)$ creates an attractive force. On the other hand,

if
$$\phi(\mathbf{r}_1, \mathbf{r}_2) = -\phi(\mathbf{r}_2, \mathbf{r}_1) \qquad \text{...(6.30b)}$$

i.e. it is an anti-symmetric wave-function, then the exchange potential creates a repulsive force. This means, that Majorona exchange potential for nuclear forces (negative V_M) creates an attractive force for symmetric pairs with respect to the exchange of coordinates and repulsive force for anti-symmetric pairs.

Similarly if we define Bartlett exchange potential $P^\sigma V_\sigma(r)$, so that:

$$P^\sigma V_\sigma(r)\,\phi(\mathbf{r}_1, m_{s_1}\,;\mathbf{r}_2, m_{s_2})$$

$$= V_\sigma(r)\,\phi(\mathbf{r}_2, m_{s_2}\,;\mathbf{r}_1, m_{s_1}) \qquad \text{...(6.31)}$$

then again for negative $V_\sigma(r)$, the Bartlett exchange potential creates an attractive force for symmetric pairs and repulsive nuclear force for anti-symmetric pairs.

Some reasoning is applicable to Heisenberg exchange potential. The behaviour of these operators which can be deduced from the above logic under different conditions is given, in Table (6.1).

Table 6.1 *Values of spin-operators for different conditions*

Operator	*Even Parity Triplet*	*(Even l) Singlet*	*Odd Parity Triplet*	*(Odd l) Singlet*
$P^B = P^\sigma \equiv \frac{1}{2}(1+\sigma_1\cdot\sigma_2)$	1	– 1	1	– 1
$-P^H = P^T \equiv \frac{1}{2}(1+\tau_1\cdot\tau_2)$	– 1	1	1	– 1
$P^M \equiv -P^\sigma P^T$	1	1	– 1	– 1
$P^W \equiv 1$	1	1	1	1

It may be seen from Table 6.1, that generalised Pauli Exclusion Principle, [Eq. 6.26] holds good for every case. $P^W = 1$ is called Wigner force and corresponds to non-exchange part of the nuclear force. It does not exchange the particles at all.

All the potentials, corresponding to different exchange or non-exchange parts, may not contain the tensor part S_{12}. As for example, operator S_{12} commutes with the Bartlett exchange operator P^B; so that:

$$P^B S_{12} = S_{12} P^B = S_{12} \qquad ...(6.32)$$

Operator P^B gives +1 in triplet state; but S_{12} transforms triplet state to triplet state; so the net result is no effect of P^B. Again for singlet state, $P^B = -1$; but S_{12} gives zero. Hence for a tensor potential, the Bartlett exchange forces do not operate. As Heisenberg forces contain Bartlett and Majorona exchange forces therefore, only Majorona exchange forces and Wigner parts will be effective for tensor potentials. Keeping these arguments in mind; one writes, the most general potential of exchange[3] type as:

$$V = V_W(r) + V_M(r) P^M + V_B(r) P^B + V_H(r) P^H + V_{TW}(r) S_{12} + V_{TM}(r) S_{12} \qquad ...(6.33)$$

The various potential terms $V_i(r)$ may not be of the same shape or strength and some of them may even vanish. We will discuss this in subsequent sections.

6.3 EFFECT OF EXCHANGE FORCES

We have already seen, while discussing *p-p* scattering that its angular distribution can be understood if we assume the exchange of protons in interaction. In neutron-proton interaction, on the other hand, we have seen effect of exchange forces in the angular distribution of neutrons in the *n-p* scattering. The forward-backward symmetry is explained on the basis of equal amount of Wigner and Majorona forces, Eq. 5.3.

Other experimental results which require the presence of exchange of forces are: (*i*) Saturation of binding energy per nucleon. (*ii*) The constant density of nuclear matter in nuclei. (*iii*) Of course, we will see later that many features of phase shift analysis and polarisation at high energies also require the exchange forces in nuclear interaction.

6.3.1 Exchange Forces and Saturation Properties of Nuclear Forces

We, now show that exchange forces, give rise to the constancy of density of nuclear matter in the nuclei, and the saturation of binding energy per nucleon; as is experimentally well known (*see* Chapter 2, Section 2.2).

First of all, we realise that exchange forces give rise to either a symmetric pair, or anti-symmetric pair, depending on the state of the two nucleons, as shown in Table 6.1. We have already seen [Eq. 6.26] that symmetric pairs give rise to an attractive potential and anti-symmetric pairs, lead to an repulsive potential; if we use the Majorona operator on a wave-function. Other operators also behave the way as indicated under various operations, shown in Table 6.1.

If we now, define P_+ as the probability of finding a symmetric pair within the range of nuclear forces, and p_- as the probability of finding an anti-symmetric pair within the same distance; n_+ as the number of symmetric pairs and n_- as the number of anti-symmetric pairs, then potential energy $\overline{V}$ may be expressed as:

$$\overline{V} \approx -(n_+ p_+ - n_- p_-) V_0 \qquad ...(6.34)$$

where V_0 is the well-depth of the interaction potential, for both the symmetric and anti-symmetric pairs.

On the other hand, if there were no exchange forces, and we assume that all nucleons interact with an attractive potential energy, when they come within their range of interaction b, then the potential energy $\overline{V}$, may be written as:

$$\overline{V} \approx -n\,p\,V_0 \qquad ...(6.35)$$

where n is the number of total number of pairs in the nucleus given by:

$$n = \frac{1}{2} A(A-1) = n_+ + n_- \qquad ...(6.36)$$

and p is the probability of finding a pair within a distance b.

We first calculate the expression for potential energy, if all the nucleon pairs have only attractive force. This requires us to calculate p for the whole nucleus. Let us assume, in this case, that the nucleons move indepently of each other, within a sphere of radius R. Then, the average probability of finding a pair of nucleons closer than the range b of nuclear force, is given by:

$$p = \frac{1}{\left(\frac{4\pi}{3} R^3\right)^2} \int_0^R \int_0^R e\,(b - r_{12})\, dV_1\, dV_2 \qquad ...(6.37)$$

Here r_{12} is the distance between any two nucleons1 and 2 and the function $e(x)$ is defined by:

$$e(x) = 1, \text{ if } x > 0$$

$$e(x) = 0, \text{ if } x < 0 \qquad ...(6.38)$$

This means physically that the nucleon-pairs within the nuclear range b attract each other, and outside b they do not interact at all.

The integral in Eq. 6.37 has been calculated[3, 5] to be:

$$p = \left(\frac{b}{R}\right)^3 \left[1 - \frac{9}{16}\left(\frac{b}{R}\right) + \frac{1}{32}\left(\frac{b}{R}\right)^3\right] \quad \text{for } R > \frac{b}{2}$$

$$= 1 \quad \text{for } R < \frac{b}{2} \qquad ...(6.39)$$

Using these values of p in Eq. 6.35, it is evident that for $R < b/2$; $V = 1/2\, A\,(A-1)\, V_0$ while for $a > b/2$, it goes on decreasing (in magnitude); approximately by $1/R^3$ plus higher terms.

We further estimate the kinetic energy $\langle T \rangle$ of all the nucleons. It can be seen that when we squeeze the nuclear gas, into a sphere of radius R, then kinetic energy of each nucleon is governed by the uncertainty principle, so that:

$$pR \approx \hbar \text{ or } p = \frac{\hbar}{R}$$

Hence $$\langle T \rangle = \frac{p^2}{2m} \approx \frac{\hbar^2}{2mR^2} \qquad \text{...(6.40}a\text{)}$$

So that, we can approximately express the kinetic energy of all nucleons as:

$$T = \frac{A\hbar^2}{2mR^2} \qquad \text{...(6.40}b\text{)}$$

Addition of Eqs. 6.35 and 6.40, will yield the total binding energy of A nucleons. It is evident that such an expression will have minimum at $R = b/2$; hence all the nucleons will tend to collapse to this size for all values of A, Fig. 6.2*a*. This will give the total binding energy proportional to A^2, contrary to the experimental facts.

On the other hand, if we use Eq. 6.34, for the potential energy, with exchange forces, the results come to be more in agreement with experimental facts as shown below:

Let us assume a mixture of Wigner and Majorona forces between any two nucleus i and j. Then

$$V_{ij}(\mathbf{r}_{ij}) = V_{ij}^{W}(\mathbf{r}_{ij}) + V_{ij}^{M}(\mathbf{r}_{ij})\, P_{ij}^{M}$$

$$= V_0^W + V_0^M\, P_{ij}^M \qquad \text{...(6.41)}$$

where P_{ij}^M is the Majorona operator. Then, it can be seen, that:

$$\bar{V} \approx -npV_0^W - (n_+\,p - n_-\,p)\,V_0^M \left\langle P_{ij}^M \right\rangle \approx \bar{V}_W - \bar{V}_M \qquad \text{...(6.42}a\text{)}$$

where $\bar{V}_W$ and $\bar{V}_M$ are the average potential depths of Wigner and Majorona forces. In Eq. 6.42*a*, we have, written the average Wigner and Majorona potentials as:

$$\bar{V}_W = -\,npV_0^W$$

and expressing $\left\langle P_{ij} \right\rangle^M$ as:

$$\left\langle P_{ij}^M \right\rangle = \int \psi^*\, P_{ij}^M\, \psi\, d^3\tau_3 = 1 \qquad \text{...(6.42}b\text{)}$$

for i and j in the same level. And $\left\langle P_{ij}^M \right\rangle = -\,1$, for i and j in different levels, with particles i and j being of the same kind and same spin state.

$$= 0, \text{ otherwise.} \qquad \text{...(6.42}c\text{)}$$

We can, then, calculate $\bar{V}_M$; as:

$$\bar{V}_M \equiv +\left(n_+\, V_0^M \left\langle P_{ij}^M \right\rangle - n_- \left\langle P_{ij}^M \right\rangle V_0^M \right) p \qquad \text{...(6.43)}$$

if we know n_+ and n_- in this particular case. Assuming that $N > Z$; levels near the ground state contain both neutrons and protons; while the higher levels contain only neutron pairs. Each level near the ground state will have four nucleons—two protons and two neutrons, with opposite spins for each pair. There will be $(Z/2)$ such levels and six bonds in each level; which correspond to symmetric mode with respect to the exchange of space-coordinates. So that there are 6 $(Z/2)$ such bonds. For the rest of $(N - Z)/2$ levels, there is one symmetric bond for each level. So the total number of symmetric pairs n_+ may be written as:

$$n_+ = 6\left(\frac{Z}{2}\right) + \frac{(N-Z)}{2} \qquad \text{...(6.44)}$$

Similarly space anti-symmetric bonds are created between protons and neutrons of different levels with parallel spins, exchanging space-coordinates only. Such pairs n_-, can, therefore be written[5] as:

$$n_- = 2\left[\frac{1}{2}\left(\frac{Z}{2}\right)\left(\frac{Z}{2}-1\right)+\frac{1}{2}\left(\frac{N}{2}\right)\left(\frac{N}{2}-1\right)\right] \qquad \text{...(6.45)}$$

where factor 2 appears for two types of protons and neutrons, with spins 'up' and 'down'. Hence, Eq. 6.42*a* may, then be:

$$\bar{V} \approx \frac{1}{2}A(A-1)V_0^W\, p + V_0^W\left\{\left[6\left(\frac{Z}{2}\right)+\frac{1}{2}(N-Z)\right]-2\left[\frac{Z}{2}\left(\frac{Z}{2}-1\right)+\frac{N}{2}\left(\frac{N}{2}-1\right)\right]\right\}p$$

$$= \frac{1}{2}A\left[(A-1)V_0^W - \left(\frac{A}{4}-4\right)V_0^M\right]p - \left[\frac{T_Z^2}{2}+2T_Z\right]V_0^M\, p \qquad \text{...(6.46)}$$

where we have used, $A = N + Z$, and $T_z \equiv 1/2\,(N - Z)$; since $A << A^2$ and $T_Z << T_Z^2$, we can write Eq. 6.46, as:

$$\bar{V} = \left[\frac{A}{2}\left(V_W - \frac{V_M}{4}\right) - \frac{T_Z^2}{2A}V_M\right]\left[\frac{b}{r_0}\right]^3\left[1 - \frac{9}{16}\frac{b}{r_0 A^{1/3}} + \frac{1}{32}\frac{b^2}{r_0^3 A}\right] \qquad \text{...(6.47a)}$$

where we have replaced V_0^W and V_0^M by V^W and W^M for the sake of convenience.

Neglecting smaller terms, we can write $\bar{V}$ as:

$$\bar{V} = -a_v A + a_s A^{2/3} + C_1 \frac{T_Z^2}{A} \qquad \text{...(6.47b)}$$

where

$$a_v = -\left(V_W - \frac{V_M}{4}\right)\left[\frac{1}{2}\left(\frac{b}{r_0}\right)^3\right]$$

$$a_s = -\left(V_W - \frac{V_M}{4}\right)\left[\frac{1}{2}\left(\frac{b}{r_0}\right)^3\left(\frac{9b}{16r_0}\right)\right]$$

and
$$C_1 = -V_M\left[\frac{1}{2}\left(\frac{b}{r_0}\right)^3\right] \quad ...(6.48)$$

It is evident that the first terms in Eq. 6.47*b* is the strongest. It shows that if

$$|V_M| \leq |4V_W| \text{ or } 4|V_W| - |V_M| \geq 0 \quad ...(6.49)$$

then the potential energy V is negative, because V_W and V_M are negative.

The detailed expression of kinetic energy can be obtained from Fermi gas model consideration of N neutrons and Z protons obeying Pauli Exclusion Principle. The kinetic energy of the system can be calculated on the simple assumption, that nucleons in the nucleus, can be treated on the Fermi gas model, in which the particles move without interaction in a sphere of radius R and volume V, like the particles of an ideal gas, but obeying Pauli exclusion momentum principle. The number of states of say protons (or neutrons) with momenta between p and $p + dp$ in a volume V is given by:

$$\frac{2V \times 4\pi p^2\, dp}{(2\pi\hbar)^3}$$

where the factor 2 is the spin weight factor from the two spin states of the protons (and neutron). We determine the kinetic energy of these two Fermi gases, by calculating:

$$(K.E)_p = \int_0^{P_{FP}} \left(\frac{p^2}{2M}\right)\left(\frac{2V \times 4\pi p^2 dp}{(2\pi\hbar)^3}\right) = \frac{8\pi V}{2M\hbar^3}\frac{(P_{Fp})^5}{5}$$

$$= \frac{4\pi}{5M}\left(\frac{3}{8\pi}\right)^{5/3}\frac{h^2}{V^{2/3}}Z^{5/3} \quad ...(6.50)$$

where P_{Fp} is the Fermi momentum given by:

$$P_{Fp} = \hbar K_{Fp} = h\left(\frac{3}{8\pi}\frac{Z}{V}\right)^{1/3}$$

The ground state corresponds to the momentum spread of P_{Fp}, and the number of protons Z in this ground state is given by:

$$Z = \int_0^{P_{Fp}} \frac{2V \times 4\pi p^2 dp}{(2\pi\hbar)^3} = \frac{8\pi V}{\hbar^3}\frac{(P_{Fp})^3}{3} \quad ...(6.51)$$

Summing up, the two kinetic energies, it is possible to write:

$$(K.E.)_T = (K.E.)_p + (K.E.)_n$$

$$= \frac{4\pi}{5M}\left(\frac{3}{8\pi}\right)^{5/3} \frac{h^2}{V^{2/3}}(Z^{5/3} + N^{5/3}) \qquad ...(6.52)$$

where we have used for $(K.E.)_n$ a similar expression as for $(K.E.)_p$, except, we replace $Z^{5/3}$ by $N^{5/3}$.

Putting

$$V^{2/3} = \left(\frac{4\pi r_0^3}{3}\right)^{2/3} \alpha \left(\frac{4\pi}{3} A\right)^{2/3} \qquad ...(6.53)$$

So that, we can write:

$$(K.E.)_T \ \alpha \ bA + C_2 \frac{T_Z^2}{A} \qquad ...(6.54)$$

where $$A = Z + N; \text{ and } T_z = \frac{N - Z}{2} \qquad ...(6.55a)$$

and $$b = \frac{4\pi}{5M}\left(\frac{3}{8\pi}\right)^{5/3} \frac{h^2}{(4\pi r_0^3)^{2/3}} \frac{2}{2^{5/3}}$$

$$C_2 = \frac{4\pi}{5M}\left(\frac{3}{8\pi}\right)^{5/3} \frac{h^2}{(4\pi r_0^3)^{2/3}} \frac{2}{2^{5/3}} \frac{20}{9} \qquad ...(6.55b)$$

Comparing Eq. 6.54, with the approximate expression 6.40, we see that, the approximate expression contains the first term of Eq. 6.54.

Adding Eqs. 6.47 and 6.54, we obtain, the expression for total binding energy as:

$$(B.E.) = -a_v A + bA + C\frac{T_Z^2}{A} + a_s A^{2/3} \qquad ...(6.56)$$

where $$C = C_1 + C_2$$

It is found that b and C_2 are independent of V_M or V_W and C_1 depends only on V_M. Also in Eq. 6.56, the first term is the strongest [Fig. 6.2*b*].

Condition (6.49), therefore, represents the relationship between V_M and V_W; if we want to have a negative binding energy and Eq. 6.56 shows that, if this condition is true, the *B.E./A* is approximately independent of A if we negelect third and fourth terms, which are smaller than the first two terms. Equation 6.49, therefore, represents one saturation condition, if we assume only Wigner and Majorona forces.

By a similar logic, if we assume the addition of Heisenberg and Bartlett forces in the above discussion, the saturation condition turns out[5] to be:

$$4|V_M| + 2|V_B| - 2|V_H| - |V_M| \geq 0 \qquad ...(6.57)$$

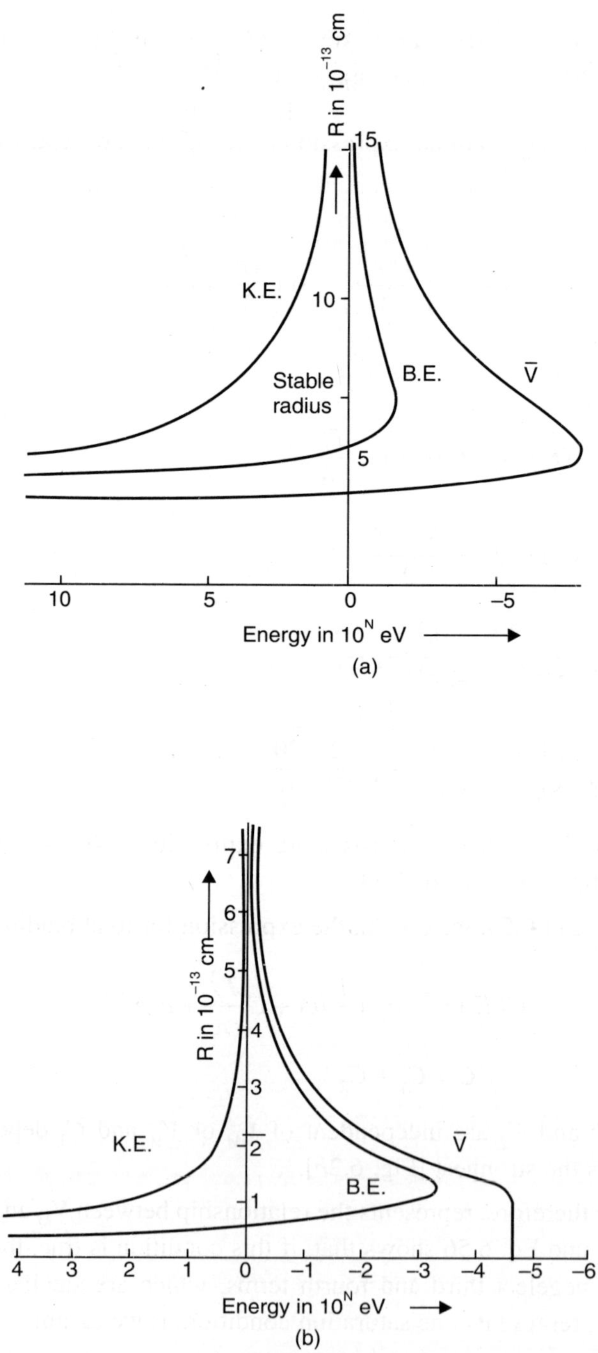

Fig. 6.2 (*a*) A schematic diagram of the potential energy V, kinetic energy T and total energy E = T + V, as a function of nuclear radius R, based on Eqs. 6.35, 6.36 and 6.40; (b) Same as (a), but based on Eqs. 6.34, 6.47a, 6.54 and 6.56.

Till now, we have not considered the tensor part of the potential. First of all, it should be realised that average of the tensor operator is zero over a sphere; hence there is always a shape for the collapsed state, where the tensor forces, on the average are always attractive, no matter whether operator S_{12} is positive or negative. Thus tensor forces, by themselves never lead to saturation. In order to get saturation, the central forces must satisfy the saturation condition and the tensor forces must be less strong than the central forces. Tensor forces, can only make saturation worse, not better.

The exchange forces also lead to saturation of nuclear densities. As the collapsed state is not the stable state now, because of the opposite nature of forces form symmetric and anti-symmetric pairs; a stable state is reached, at a distance R, such that the average distance between nucleons is of the order of b, the range of nucleon-nucleon forces. For much shorter distances, the repulsive forces due to anti-symmetric pairs will dominate because though the probability of finding symmetric and anti-symmetric pairs, within a distance $d << b$ is the same, the number of anti-symmetric pairs is larger, for a given A, than the symmetric pairs. However, at larger distances, *i.e.* $d >> b$; the probability of anti-symmetric pair decreases fast with distance. So at an intermediate distance $d \approx b$; we get a stable system. This corresponds to $R \approx R_0$ at which radius, the nucleon pack themselves side by side.

This evidently leads to the normal density of nuclei, which is independent of any particular nucleons.

$$\rho = \frac{A}{\frac{4\pi}{3} R_0^3} = \frac{A}{\frac{4\pi}{3} r_0^3 A} = \frac{3}{4\pi} r_0^{-3} \text{ a.m.u.}$$

where $r_0 = 1.25 \times 10^{-13}$ cms.

6.4 MESON THEORY OF EXCHANGE FORCES—INTRODUCTORY DISCUSSION

The real impetus to the concept of exchange forces was given by Yukawa's meson theory[4] of nuclear forces, which has basic characteristics of exchange forces.[6] According to this theory, if we treat the problem of nuclear forces in the frame-work of relativistic quantum mechanics, it follows in a natural manner, that the range of nuclear forces is short; *i.e.* of the order of $\hbar/m_\pi c$ where m_π is the mass of the π-mesons. This theory was supported by the experimental facts, which came later. We know now, that π-meson or pion (a mass of $m_\pi = 266\ m_e$; lifetime $\tau_\pm^\pi \approx 2.5 \times 10^{-8}$ secs and $\tau_0^\pi \approx 2.0 \times 10^{-16}$ secs and spin $S = 0$) is responsible for the interaction of the nucleons, giving rise to the various properties of the nuclear forces in a natural manner; as we shall see subsequently.

6.4.1 Scalar Klein-Gordon Equation[7]

The basis of the Yukawa's meson theory is the relativistic Schrödinger equation called Klein-Gordon equation. We know that the relativistic equation for energy of a free particle may be written as:

$$E^2 = c^2p^2 + m_\pi^2 c^4 \qquad ...(6.58)$$

Quantum Mechanically, we can write E and $\mathbf{p}$ operationally as:

$$E \to i\hbar \frac{\partial}{\partial t}; \mathbf{p} \to - i\hbar \, \mathbf{grad.} \qquad ...(6.59)$$

Hence using the operators of Eq. 6.59 operating on wave-function ϕ, one can write Eq. 6.58 as:

$$-\hbar^2 \frac{\partial^2 \phi}{\partial t^2} = -\hbar^2 c^2 \nabla^2 \phi + m_\pi^2 c^4 \phi \qquad ...(6.60)$$

or using x_1, x_2, x_3 for three space-coordinates and ict for x_4; one can write Eq. 6.60 as:

$$\left[-\hbar^2 c^2 \left[\frac{\partial^2 \phi}{\partial x_1^2} + \frac{\partial^2 \phi}{\partial x_2^2} + \frac{\partial^2 \phi}{\partial x_3^2} \phi + \frac{\partial^2 \phi}{\partial x_4^2}\right] - m_\pi^2 c^4 \right] \phi = 0 \qquad ...(6.61)$$

or

$$\left(\Box^2 + \frac{m_\pi^2 c^2}{\hbar^2}\right) \phi = 0 \qquad ...(6.62a)$$

or

$$(\Box^2 + \mu^2)\, \phi = 0 \qquad ...(6.62b)$$

where

$$\Box^2 \phi \equiv -\left(\frac{\partial^2 \phi}{\partial x_1^2} + \frac{\partial^2 \phi}{\partial x_2^2} + \frac{\partial^2 \phi}{\partial x_3^2} + \frac{\partial^2 \phi}{\partial x_4^2}\right) \qquad ...(6.63a)$$

and

$$\mu \equiv \frac{m_\pi c}{\hbar} \qquad ...(6.63b)$$

Equation 6.62 is called the Klein-Gordon equation.

For static case; $\frac{\partial}{\partial x_4}$

does not exist, so that for such a case,

$$\Box^2 = \nabla^2 = -\left(\frac{\partial^2}{\partial x_1^2} + \frac{\partial^2}{\partial x_2^2} + \frac{\partial^2}{\partial x_3^2}\right) \qquad ...(6.64)$$

and Eq. 6.62 becomes:

$$(\nabla^2 + \mu^2)\, \phi = 0 \qquad ...(6.65)$$

This has a spherically symmetrical solution of the form:

$$\phi = \frac{e^{-\mu r}}{r} \qquad ...(6.66)$$

The solution has a range of $1/\mu$ or $\hbar/m_\pi c$ which is the Compton wavelength of a particle with mass m_p. If we equate $\hbar/m_\pi c$ with the range of nuclear force, say 10^{-13} cms., we get $m_\pi \approx 200\, m_e$. This is comparable with the experimental value of the mass of π-meson which is 266 m_e. It is recognised now that π-mesons are the particles of quantised field for nuclear forces. In this manner it is seen that a very simple application of Klein-Gordon equation leads to the explanation for the existence or π-mesons.

Further we realise that experimentally there are known to be three types of π-mesons, *i.e.* π^+, π^- and π_0 as mentioned earlier. While the neutral meson π^0 can act as quantum of mesic-field between neutron-proton; proton-proton and proton-neutron; the π^+-meson can act only between proton-proton and proton-neutron and π^--meson only between neutron-proton. These mesons can be assumed to be acting as exchange particles between nucleons. *A* π^+-meson field acting between proton-neutron, can be emitted by a proton making it a neutron, and absorbed by the neutron, making it a proton and the process can be repeated back and forth. Similarly π^--meson can also be emitted by neutron and absorbed by the proton. In a proton-proton case, both of the protons may emit π^+ which can be absorbed by the other nucleon. In this case it is not possible for π^--meson to be emitted, because of the conditions of electric charge. This meson-exchange language is only illustrative, but it should be understood that the emitted or absorbed mesons are only virtual. Their masses (converted into energy) and the life time of the exchange must obey the uncertainty principle so that:

$$n\,(m_\pi c^2)\,\Delta t \approx \hbar \qquad \text{...(6.67)}$$

where n is number of mesons exchanged.

Hence
$$c\,\Delta t \approx \frac{\hbar}{n\,m_\pi\,c} \qquad \text{...(6.68)}$$

gives the range of the nuclear forces, which comes out to be the same as the Compton wavelength for $n = 1$, *i.e.* for one meson-exchange. For $n > 1$, it is less, which means that for exchange of more than one particle, the range is smaller. Also heavier the mass, less is the range. We know, that *K*-mesons with $m_K \approx 500\ m_e$ can also be emitted by nucleons. Because of their heavy mass, however, their range is much smaller than the measured nuclear range, which is contributed mainly by π-mesons.

The Klein-Gordon Equation 6.62 represents a one component field φ and corresponds to scalar mesons.

6.4.2 Vector Mesons and Nuclear Forces

The Yukawa's theory of nuclear forces in terms of exchange of mesons between the nucleons requires that the spin of the meson be integral and the mesons could be scalar or vector. A scalar meson corresponds to one component, whose behaviour is fully described by klein-Gordon equation, as shown in the previous section.

On the other hand, the equation for a vector meson can be obtained from the extension of Dirac formalism to Maxwell equations (*see* Quantum Mechanics by L.I. Schiff, McGraw-Hill, p. 242). It is known that for an electromagnetic field in vacuum, the four vector potential A_λ satisfies the wave-equation:

$$\nabla^2 A_\lambda - \frac{1}{c^2}\,\partial^2 \frac{A_\lambda}{\partial t^2} = 0 \qquad \text{...(6.69}a\text{)}$$

or
$$\Box^2 A_\lambda = 0 \qquad \text{...(6.69}b\text{)}$$

Also the Lorentz's conditions require (i = 1, 2, 3 corresponding to three space components and $\nu = t$, the time):

$$\partial_\lambda^2 A_\lambda = 0; (\partial_\lambda = -\nabla_i), \partial_\nu = \left(\frac{\partial}{\partial t}\right) \qquad ...(6.70)$$

Following the formalism of contra-variant components (*see* Quantum Theory of Radiation by W. Heitler, Oxford, Page 10)[5] the double contra-variant components of the electromagnetic field tensor are:

$$F_{\lambda\nu} = \partial_\lambda A_\nu - \partial_\nu A_\lambda \qquad ...(6.71)$$

and satisfy the following relation obtained from Eqs. 6.70 and 6.71 and 6.69*b*.

$$\partial_\nu F_{\lambda\nu} = \partial_\nu \partial_\lambda A_\nu - \Box^2 A_\nu = 0 \qquad ...(6.72)$$

In analogy to the above relations, *i.e.* (Eq. 6.72); one writes for the components of the vector mesons, (which have integral spin, like photons for which Maxwell equations hold good).

$$(\Box^2 + m^2)\phi_\lambda = 0 \qquad ...(6.73a)$$

Also

$$\partial_\lambda \phi_\lambda = 0 \quad \text{(like Lorentz's condition)} \qquad ...(6.73b)$$

and m^2 is obtained from Eqs. 6.72 and 6.73*a*.

This leads to a similar r-dependence of ϕ, as given in Eq. 6.65. However, if we introduce a source of vector mesons, with spin (integral), it has been proved [Ref. (4)], that the wave-function of vector mesons can be written (without giving proof):

$$(\Box^2 + \mu^2)\,\phi = \frac{\sqrt{4\pi}}{\mu} f\sigma\cdot\nabla\phi \qquad ...(6.74)$$

where f is a coupling constant and $\mu \equiv \dfrac{m_\pi c}{\hbar}$

and m_π is the mass of the vector meson (say π-meson), and ρ is the field distribution. Now without going into mathematical details of derivation, we describe qualitatively, some of the interesting result connecting the properties of ϕ in Eq. 6.74 and the nuclear forces, based on the so-called pseudo-scalar theory.

If we take into account only the spin properties of the two nucleons, the expression for the energy of meson field due to the second nucleon in the field of the first nucleon, *i.e.* the interaction potential U, or Hamiltonian for interaction H_{int}, can be written as (Ref. 8):

$$U = H_{\text{int}} = \frac{\sqrt{4\pi}}{\mu} f\sigma\cdot\nabla\,\phi\,(r)$$

$$= f^2\,(\sigma_2\cdot\nabla)(\sigma_1\cdot\nabla)\frac{e^{-\mu_\pi r}}{r} \qquad ...(6.75)$$

which, after the detailed calculation of the gradients yields:

$$U = f^2\left[\left(\frac{3}{\mu r^2}+\frac{3}{\mu r}+1\right)S_{12}+\sigma_1 . \sigma_2\right]\frac{e^{-\mu r}}{3\mu r}$$

$$-\frac{4\pi}{3}f^2(\sigma_1 . \sigma_2)\,\delta(r) \qquad ...(6.76)$$

where S_{12} is the well-known tensor force operator [*see* Eq. 3.38]. The last term (containing δ-function) gives the attractive or repulsive core if we assume the finite size of the nucleons. It is interesting to see that in Eq. 6.76 representing the interaction using pseudo-scalar theory; the interaction terms contain $\sigma_1 \cdot \sigma_2$ and the tensor force operator S_{12} occurs in a natural manner. Also a pseudo-scalar theory leads to an interaction similar to an e.m. interaction between two loops of electric currents, as if the two nucleons behave like loops of mesic currents. The Compton-wavelength of pion, *i.e.* $\hbar/m_\pi c$ comes out to be somewhat larger, but of the same order of magnitude as the size of the nucleon. The term containing $\sigma_1 \cdot \sigma_2$ in the expression of U is the spin exchanging term in spin exchange or of Bartlett character. Hence the phenomenon of spin-exchange and tensor forces is explained in a natural manner in the meson theory of nuclear forces. One can further introduce the isotopic spin in the above formalism, and write the expression for nucleon-nucleon potential—which automatically leads to a charge-symmetric coupling; so that the nucleon-nucleon interaction based on exchange of pions is charge-independent. For detailed theory, involving isotopic spin *see* Ref. (8).

One can also consider the two pion or three pion exchange, between two nucleons. This will give rise to fourth (for two pion exchange) or higher order terms, in addition to the 2nd order terms as given in Eq. 6.76 . Also Eq. 6.76 was derived on the basis of static field. One can introduce the dynamical condition by considering the recoil of the meson-emitting nucleon. This can give rise to the velocity dependent terms like **L.S** terms in the nucleon-nucleon interaction.

Till, now we have considered only the exchange of one π-meson between the two nucleons. The potential obtained in this manner as given in Eq. 6.76, is called one pion-exchange potential (OPEP). The Compton wavelength of π-mesons, *i.e.* $1/\mu = \hbar/m_\pi c$ is about 1.4 fm. The potential described by Eq. 6.76 therefore, corresponds to this range. It is well known, that there are heavier mesons in nature which also can play a role in nuclear forces. These are all bosons like π-mesons. Their masses and Compton-wavelength are given by:

Particles	*Masses*	*Compton wavelength*
η-meson	$M_\eta c^2 = 549$ MeV	0.35 fm
ρ-meson	$M_\rho c^2 = 769$ MeV	0.25 fm
ω-meson	$M_\omega c^2 = 783$ MeV	0.25 fm
π- meson	$M_\pi c^2 = (266\ m_e)\ c^2$	1.2 fm

One can write expressions similar to Eq. 6.76 to express the nucleon-nucleon potential due to these heavy bosons. As a matter of fact, such an approach has been made using a one-boson-exchange potential (OBEP) in which single exchange of each of the mesons listed above, as well as π-mesons is considered. In practice, however, an OBEP approach has not proved to be so successful. But it is, understood that the repulsive core in the nucleon-nucleon potential is due to the heavy mesons listed above.

Instead of considering only one meson exchange, one can consider the simultaneous emission and absorption of many mesons. This can give rise to many-body forces. As for example, if there are three nucleons, and one of them emits two mesons, they can be absorbed by the other two nucleons, giving rise to three body nuclear forces. Similarly, one can consider four body or five body forces, etc. Collectively called many-body-forces, these forces have, however, short ranges. Since many mesons have to be created simultaneously, in many-body-forces, the range of the forces decreased as the number of bodies increase as only at very short distances is there some probability of many nucleons being within the nuclear range. This range, for n-body force is crudely given by $1/1 - n$ times the range of a two body force. If the range of the two body force is taken to be 1.4 fm, then the range for 3, 4, 5 body force can be taken to be 0.7, 0.47, 0.35 fm, etc. respectively. In practice, however only three-body (*i.e.* $n = 3$) forces, out of these, are important. They are estimated to contribute about 15–20 per cent of the two-body forces, to the total nuclear force.

Equation 6.76 contains the tensor operator with $1/r^3$ term, which diverges too strongly at the origin for an acceptable solution. If we include $1/r^3$ term, it results in an infinitely small deuteron with infinitely large binding energy—because very small values of r will then dominate, with very large value for U. This difficulty is caused by the fact that Eq. 6.76 is a second order non-relativistic formula, and is not expected to hold at short distances, where high momentum transfers and many meson exchanges become important.

This difficulty has been removed, by resorting to a cutoff device. This corresponds to the recognition of the physical fact that source of mesons is expected to spread over a finite volume. This corresponds to replacing the δ-function repulsive force over a finite distance giving rise to a repulsive core, that cannot be ignored. In practice this means, introduction of two parameters, to fit the experimental data: (*i*) A coupling constant f suitable for an extended source and, (*ii*) A cut-off parameter such as a minimum radius r_{min} or maximum momentum K_{max}.

In conclusion, it is interesting to note that the meson theory of nuclear forces gives the expression for nuclear potential, which is quite in agreement with the expression expected on the basis of experiments. For further detailed understanding of meson theory of nuclear forces, *see* Ref. (8).

6.5 NUCLEON-NUCLEON POTENTIAL

We can now discuss in an integrated manner the form of the nucleon-nucleon potential as deduced from the deuteron problem, the low and high energy-scattering data and the meson theory, where spin-dependence of nuclear forces has been discussed, involving the concepts of iso-spin and exchange forces. The meson theory discussed above also has been used in the analysis of the high energy scattering data.

6.5.1 Phase-Shift Analysis and Nucleon-Nucleon Potential

We have already discussed the nucleon-nucleon scattering at low energies (*i.e.* for $l = 0$) in terms of the nucleon-nucleon potential both for n-p and p-p scattering. For higher values of l, higher energies are involved. We have already discussed the phase-shifts obtained at these energies in the last chapter. How do we connect these phase shifts to the nucleon-nucleon potential? For this purpose, we basically, use Eqs. 4.16 to 4.37 for *n-p* scattering and from Eqs. 4.92 to 4.100 for *p-p* scattering for high energy scattering which relate the values of δ's with the potential, through the asymptotic behaviour of the wave function.

To obtain the form of the nucleon-nucleon potential from phase-shift analysis, one can use a phenomenological potential based on the above empirical considerations. One can also base the form of the potential on the detailed meson-theory as discussed earlier or combine, the two approaches. Alternatively, one can write a general form of the nucleon-nucleon potential, on the basis of general principles of conservation and symmetry laws as shown below.

There are four vectors associated with two nucleons, *i.e.* $\mathbf{r}_1, \mathbf{r}_2, \boldsymbol{\sigma}_1, \boldsymbol{\sigma}_2$. The Hamiltonian expressing the interaction between the two nucleons, involves these vectors and should be a scalar, so that the space-conservation laws are obeyed. Two particles of spin 1/2 $\hbar$ may be found in four different states. Any expression involving their spins is a 4 × 4 matrix in spin-space. There are 16 linearly independent 4 × 4 matrices involving spins, *i.e.*

1. One scalar matrix: 1
2. One scalar matrix: $1/2\ (1 + \sigma_1 \cdot \sigma_2)$
3. Three components of pseudo-vector matrix: $\sigma_1 + \sigma_2$
4. Three components of pseudo-vector matrix: $\sigma_1 - \sigma_2$
5. Three vectors of the pseudo-vector matrix: $\sigma_1 \times \sigma_2$
6. Five components of symmetric tensor matrix with zero trace:

$$T_{ij} = \frac{1}{2}\left[\sigma_i\,(1)\,\sigma_j\,(2) + \sigma_j\,(1)\,\sigma_i\,(2) - \frac{1}{3}\delta_{ij}\,\sigma_{(1)}\cdot\sigma_{(2)}\right] \qquad ...(6.77)$$

It may be realised that in the last expression for the tensor, some of the tensor components have already been taken into account, *e.g.* one in (2) and three in (5). This leaves five independent components in the tensor, out of nine possible components.

Again considering the displacement components; we should realise that the vector **r** (1) and **r** (2) must occur only in the combination **r** (1) – **r** (2) = **r**, to conserve linear momentum. We can form the following expressions out of **r**.

(*i*) **r** = **r** (1) – **r** (2); One vector function

(*ii*) $f(|\mathbf{r}|) = f(r)$; Scalar function

(*iii*) $r_i\, r_j$; One tensor function ...(6.78)

One should form a scalar for the Hamiltonian by composing the products of expressions 6.77 and 6.78. One can obtain only three types of scalar expressions from these:

(*i*) Scalar function $f(r)$ multiplied by 1.

(*ii*) Scalar function $f(r)$ multiplied by $1/2\ (1 + \sigma_1 \cdot \sigma_2)$

(*iii*) Tensor T_{ij} dotted with $r_i\, r_j$ and multiplied by the scalar function.

Combining these scalar parts of the Hamiltonian we may write the interaction Hamiltonian V, as:

$$V = V_W(r)\,1 + V_B(r)\,\frac{1}{2}\,[1 + \sigma_1 \,.\, \sigma_2] + V_T(r) \sum_{ij} T_{ij} \,.\, r_i\, r_j \qquad ...(6.79)$$

Equation 6.79 can be readily seen to be reduced to:

$$V = V_W + V_B(r)\, P^\sigma + V_T(r)\, S_{12} \qquad ...(6.80a)$$

where
$$P^\sigma \equiv \frac{1}{2}\,(1 + \sigma_1 \,.\, \sigma_2) \qquad ...(6.80b)$$

and
$$S_{12} \equiv 3\,(\sigma_1 \,.\, \hat{r})\,(\sigma_2 \,.\, \hat{r}) - \sigma_1 \,.\, \sigma_2 \qquad ...(6.80c)$$

and V_W, V_B and V_T represent the Wigner, Bartlett and Tensor potentials.

If we take into account, the possibility of velocity dependent forces we may include momenta **p** (1), and **p** (2) and **p** = **p** (1) – **p** (2), and then from three scalars from **p** and **r**, *i.e.* p^2, r^2 and **p . r**. Then function $f(r)$ in Eq. 6.78 may be replaced by $f(p^2, r^2, \mathbf{p.r})$ with the condition that f is an even function of **p.r**, so that one has time-reversal invariance. We can then form a scalar and time invariant quantity, by writing a scalar product of the pseudo vector **r** × **p** with any of the pseudo vectors of Eq. 6.78. One of these scalar products will be:

$$V_s(r)\,\frac{1}{2}[\sigma(1) + \sigma_2] \,.\, [\mathbf{r} \times \mathbf{p}] = V_s(r)\,\mathbf{l} \cdot \mathbf{s} \qquad ...(6.81)$$

where **l** = **r** × **p** is the orbital angular momentum. Equation 6.81 represents the spin-orbit coupling term. One can combine the expression 6.80 and 6.81, thus taking into account both the potentials.

A very general expression of the potential has also obtained by Okubo and Marshak[8] under the restriction of (1) Translational Invariance; (2) Rotational Invariance; (3) Gallinan Invariance; (4) Space Reflection Invariance; (5) Time-Reversal Invariance (The potential obtained by combining Eqs. 6.79 and 6.81 corresponds to these first five restriction). The extra restrictions put by Okubo and Marshak[8] were: (6) Charge Independence; (7) Permutation Symmetry and (8) Hermicity. They arrived at the following expression for nucleon-nucleon potential.

$$V = V_0 + (\sigma_1 \,.\, \sigma_2)\, V_1 + S_{12}\, V_2 + (\mathbf{L \cdot S})\, V_3 +$$

$$\frac{1}{2}\,[(\sigma_1 \cdot \mathbf{L})\,(\sigma_2 \cdot \mathbf{L}) + (\sigma_2 \cdot \mathbf{L})\,(\sigma_1 \cdot \mathbf{L})]\, V_4 + (\sigma_1 \cdot \mathbf{p})\,(\sigma_2 \cdot \mathbf{p})\, V_5 + h{\cdot}c \qquad ...(6.82)$$

For *n-p* interaction, we sum up $T = 0$ and $T = 1$ potentials, so that:

$$V_i = V_i^0\,(r, p, L) + V_i^\tau\,(r, p, L)\,(\tau_1 \,.\, \tau_2)\; i = 1, 2, 3, 4, 5 \qquad ...(6.83)$$

where V_i^0, and V_i^τ are the real scalar function of the magnitudes of the three vectors **r, p** and **L**.

For elastic nucleon-nucleon scattering, *i.e.* where there is no meson production, it can be shown that[8] V_5 in Eq. 6.82 can be dropped out.

This means that there are no *p*-dependent forces till about 300 MeV or so. Further, as we shall see subsequently, even the term containing V_4 is, in general, not required for the analysis of the data up to 300 MeV or so, and for all practical purposes, we can write the nucleon-nucleon potential in terms of the first four terms of Eq. 6.82 with $V_i = V_i^0\,(r) + V_i^\tau\,(\tau_1 \cdot \tau_2)$. It is easy to see that one can, then write Eq. 6.82 in terms of various exchange terms, with first five terms as:

$$V = V_d\,(r) + V_\sigma\,(r)\,(\sigma_1 \cdot \sigma_2) + V_\tau\,(r)\,(\tau_1 \cdot \tau_2) + V_{\sigma\tau}\,(r)\,(\sigma_1 \cdot \sigma_2)\,(\tau_1 \cdot \tau_2) + V_{Td}\,(r)\,S_{12} +$$
$$V_{\tau T}\,(r)\,(\tau_1 \cdot \tau_2)\,S_{12} + V_{LS}\,(\mathbf{L \cdot S}) + V_{LS,\,\tau}\,(r)\,(\mathbf{L \cdot S})\,(\tau_1 \cdot \tau_2) \qquad ...(6.84)$$

In Eq. 6.84 different subscripts of V stand for various types of contributions, *e.g.* V_d stands for direct (no exchange), V_σ stands for spin-exchange terms; V_τ stands for iso-spin exchange direct term, $V_{\sigma\tau}$ for position exchange direct-terms; V_{Td} stands for tensor direct term with no exchange; $V_{\tau T}$ for tensor term with iso-spin exchange, V_{LS} for spin orbit-coupling no-exchange term and $V_{LS,\,\tau}$ for spin-orbit iso-spin exchange term.

Alternatively, one can write nuclear potential in terms of Wigner, Majorona, Bartlett and Heisenberg exchange operators and potentials, so that:

$$V = V_W\,(r) + V_M\,(r)\,P^M + V_B\,(r)\,P^B + V_H\,(r)\,P^H + V_{TW}\,(r)\,S_{12}$$
$$+ V_{TM}\,(r)\,S_{12}\,P^M + V_{LS}\,(\mathbf{L \cdot S}) \qquad ...(6.85)$$

Equation 6.85 is the same as Eq. 6.33, except the addition of last term in Eq. 6.85. It may be realised that Eqs. 6.78 to 6.85 are equivalent; if one takes terms in Eq. 6.82 upto V_3. Many attempts have been made to fit the phase shifts, to the phenomenological potential shape as given in Eqs. 6.79, 6.82 and 6.84. The first attempt of such an analysis was made by Gammel, Christians and Thaler[9] to fit the nucleon-nucleon scattering data at 170 MeV and 310 MeV by using only central and tensor potentials of the Yukawa type outside the hard core; the radius of the hard core being taken as independent of parity. The results fitted the *n-p* data well, but not for *p-p* data. In other words, triplet ($S = 1$) odd parity ($1 =$ odd) available in *p-p* scattering are not capable of being described, by only central and tensor forces; while triplet, even parity ($1 =$ even) in *n-p* scattering can be described by such forces.

Next, Gammel and Thaler[10] analysed Srapp's phase-shift data[15] of *p-p* scattering at 310 MeV with the addition of strong short range spin-orbit force, along with central and a tensor force, which has long range and is attractive in the 3*P* state and has a hard core, using general form of the Gammel-Thaler potential:

$$V = \infty; \qquad \text{for } r < r_0$$
$$= V_c\,(r) + S_{12}\,V_T\,(r) + (\mathbf{L \cdot S})\,V_{LS}\,(r) \qquad \text{for } r > r_0 \qquad ...(6.86)$$

where $$V_T\,(r) = V_i\,\frac{e^{-\mu r}}{\mu r}$$

has the Yukawa shape. The calculations fit quite well, with the data of *p-p* scattering at 310 MeV if one assumes charge independence of nuclear forces and a spin-orbit term in the triplet-even-parity potential of the same short-range, with lesser depth, as the triplet-odd-parity spin-orbit term. The following parameters[8, 9] fit the data quite well [Fig. 6.3]:

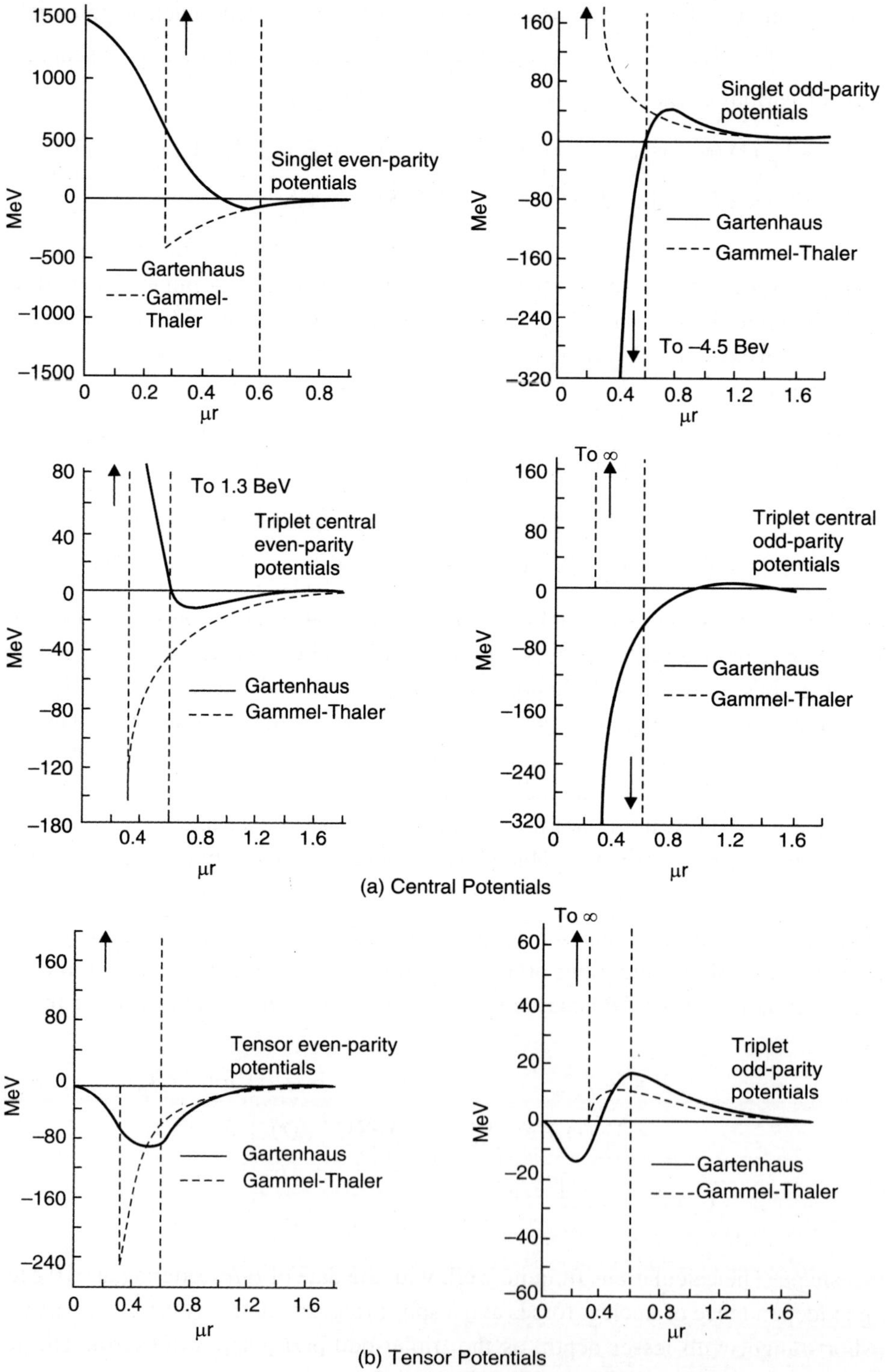

Fig. 6.3 Gammel-Thaler phenomenological nuclear potentials and Gartenhaus theoretical (*a*) Central (*b*) Tensor potentials (Ref. 10).

Triplet odd parity

$r_0 = 0.4125 \times 10^{-13}$ cm

$V_c = 0$

$V_T = -22$ MeV

$V_{\mathbf{L.S}} = 7317.5$ MeV

$\mu_T = 0.8 \times 10^{13}$ cm^{-1}

$\mu_{LS} = 3.7 \times 10^{13}$ cm^{-1}

Singlet even parity

$r_0 = 0.4 \times 10^{-13}$ cm

$V_c = 425.5$ MeV

$\mu_c = 1.45 \times 10^{13}$ cm^{-1}

...(6.87)

One of the most extensive studies of the two nucleon phenomenological phase shifts with energy has been carried out by Breit[11] and co-workers at Yale; with the analysis of the data right from low energies up to 345 MeV, starting with one of the phase shift solution—either the Signall-Marshak[13] type or Gammel-Thaler[10] types. Suitable corrections were introduced into the preliminary phase-parameter so that the mean weighted sum of the squares of deviations from experimental values was minimised. The so-called Yale potential,[11] which gave a good fit to the experimental data is expressed as:

$$V = V_c(r) + V_T(r)\, S_{12} + V_{LS}(r)\,(\mathbf{L{\cdot}S}) + V\,[(\mathbf{L{\cdot}S})^2 + (\mathbf{L{\cdot}S}) - L^2] \quad ...(6.88)$$

The term $V_c(r)$ is different for even or odd states and also depends on S. An infinite repulsive hard core is also used for:

$$r \leq \frac{0.35\hbar}{m_\pi c}$$

A very well-known potential, including terms corresponding to ranges of one, two and three-pion Compton wavelengths, was developed by Hamada[12], which was found to reproduce *n-p* experimental data below 300 MeV. This included both $T = 0$ and $T = 1$ terms in the potential. Finally Hamada-Johnson[12] wrote the following potential, which is now more commonly used:

$$V(r) = +\infty \text{ for } X \leq 0.343$$
$$= V_c(r) + V_T(r)\, S_{12} + V_{\mathbf{L{\cdot}S}} + V_{LL}(r)\, L_{12}$$

for $X > 0.343$ ...(6.89)

In Table 6.2 are given the various parameter used in Hamada-Johnson potential.

Table 6.2 *The parameters of the Hamada-Johnson potential*[12]

	Singlet-even	*Triplet-odd*	*Triplet-even*	*Singlet-odd*
a_c	8.7	– 9.07	6.0	– 8.0
b_c	10.6	3.48	– 1.0	12.0
a_T	–	– 1.29	– 0.5	–
b_T	–	0.55	0.2	–
G_{LS}	–	0.1961	0.0743	–
b_{LS}	–	– 7.12	– 0.1	–
G_{LL}	– 0.000891	– 0.000891	+ 0.00267	– 0.00267
a_{LL}	0.2	– 7.26	1.8	2.0
b_{LL}	– 0.2	6.92	– 0.4	6.0

where

$$L_{12} \equiv (\sigma_1 \cdot \sigma_2)\, L^2 - \frac{1}{2} \{(\sigma_1 \cdot \mathbf{L})(\sigma_2 \cdot \mathbf{L}) + (\sigma_2 \cdot \mathbf{L})(\sigma_1 \cdot \mathbf{L})\}$$

$$V_C \equiv 0.08 \left(\frac{\mu_\pi}{3}\right) (\tau_1 \cdot \tau_2)(\sigma_1 \cdot \sigma_2)\, Y(X)\,[1 + a_c\, Y(X) + b_c\, Y^2(X)]$$

$$V_T \equiv 0.08 \left(\frac{\mu_\pi}{3}\right) (\tau_1 \cdot \tau_2)\, Z(X)\,[1 + a_T\, Y(X) + b_T\, Y^2(X)]$$

$$V_{LS} \equiv \mu_\pi\, G_{LS}\, Y^2(X)\,[1 + b_{LS}\, Y(X)]$$

$$V_{LL} \equiv \mu_\pi\, G_{LL}\, X^2\, Z(X)\,[1 + a_{LL}\, Y(X) + b_{LL}\, Y^2(X)]$$

$$Z(X) \equiv \left(1 + \frac{3}{X} + \frac{3}{X^2}\right);\; Y(X) = \frac{e^{-X}}{X};\; X = r\frac{\mu c}{\hbar} \qquad \text{...(6.90)}$$

where μ = pion mass.

Table 6.3 indicates the strengths of the dominant forces at 1.4 fm and 1.8 fm, which correspond to the average distance in the nuclear matter.

Table 6.3 *Values of various potentials in Hamada-Johnson potential*[12] *at separation distance of 1.4 fm and 1.8 fm*

		V at 1.4 fm (MeV)	*V at 1.8 fm (MeV)*
Singlet-even (S = 0, 1 = even)	V_c	– 45	– 18
Triplet-odd (S = 1, 1 = odd)	V_T	+ 11	+ 8
	V_c	– 4	– 2
	V_{LS}	– 12	– 2
Triplet-even	V_T	– 52	– 24
	V_c	– 25	– 11
Singlet-odd		Small	Small

Figure 6.4 illustrates the shape of Hamada-Johnson potential obtained from *p-p* scattering data at various energies.

6.5.2 Realistic and Effective Potentials

Local and Non-Local Realistic Potentials

We have discussed in the previous section the nucleon-nucleon potentials, derived or used by different authors, based on phase shift analysis of nucleon-nucleon scattering data. These are called the realistic, local potentials. A local potential $V(r)$ has a single valued dependence on r; the separation distance between the two nucleons and is connected[14] to the phase shift δ_1 by:

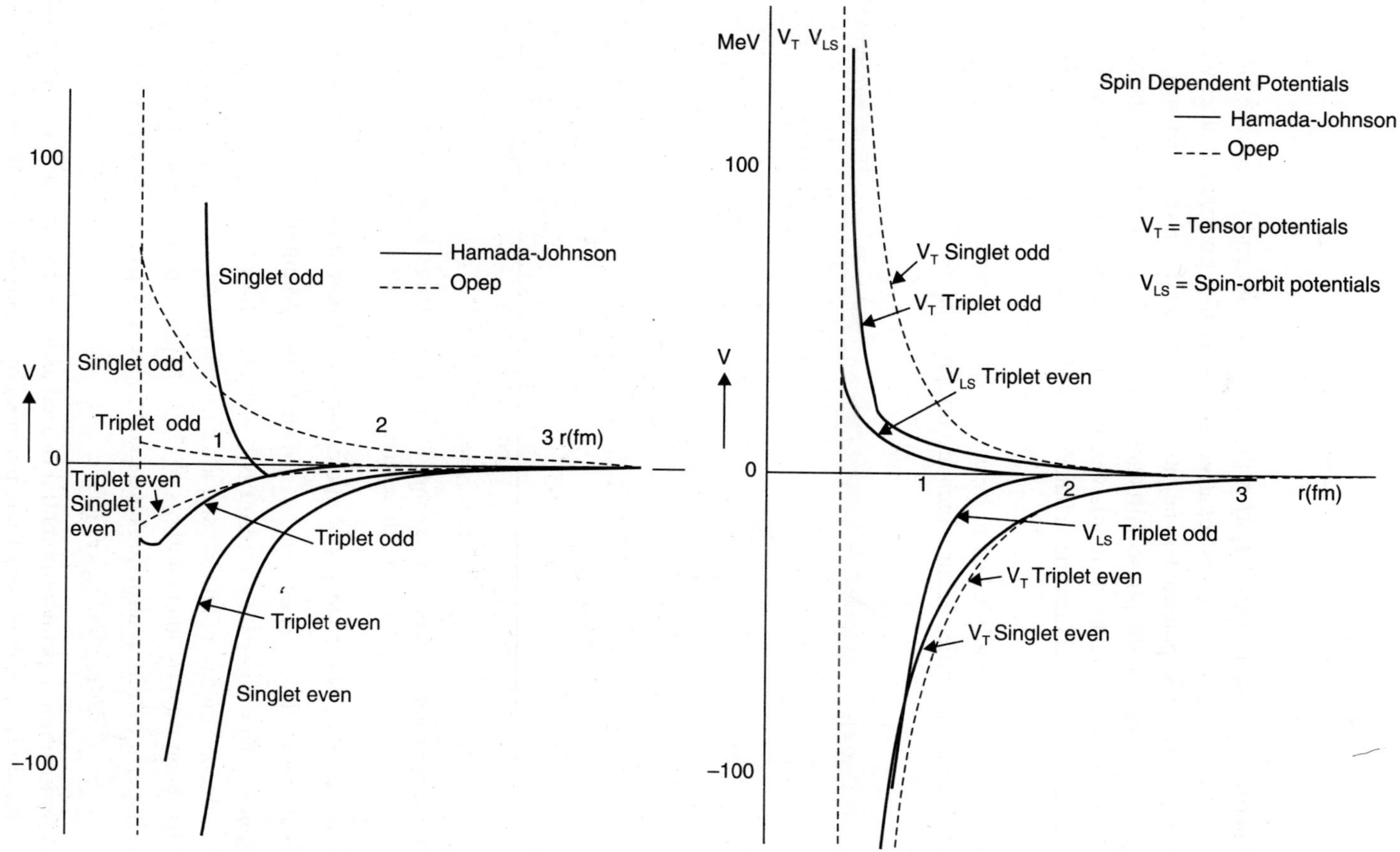

Fig. 6.4 Hamada-Johnson phenomonological[12] potentials (central, tensor and spin-orbit) are compared with one pion exchange potential (OPEP), based on meson theory (Ref. 12).

$$\tan \delta_1 = -\frac{Mk}{\hbar^2} \int V_1(r)\, j_1^2(kr)\, r^2\, dr \qquad ...(6.91)$$

where $$E_{lab} = \frac{k^2 \hbar^2}{2M}$$

is the laboratory energy of incident-particles, M is the nucleon mass and j_1's are the Bessal functions. Equation 6.91, shows that if the phase shift is negative, the potential is positive, in other words repulsive. If the phase shift is positive, the potential is negative, *i.e.* attractive. Experimentally,[15] as shown in Fig. 6.5, for low incident energies, the phase shifts are positive, while for energies above 250 MeV, the phase shift is negative corresponding to the repulsive core of the nucleon-nucleon potential. The realistic potentials of Hamada-Johnson[12] and Gammel-Thaler[9, 10] as shown in Figures 6.3 and 6.4, contain these features.

The repulsive core is generally taken to be a soft-core potential, so that perturbation treatment can be employed, otherwise infinities occur which are not physical.

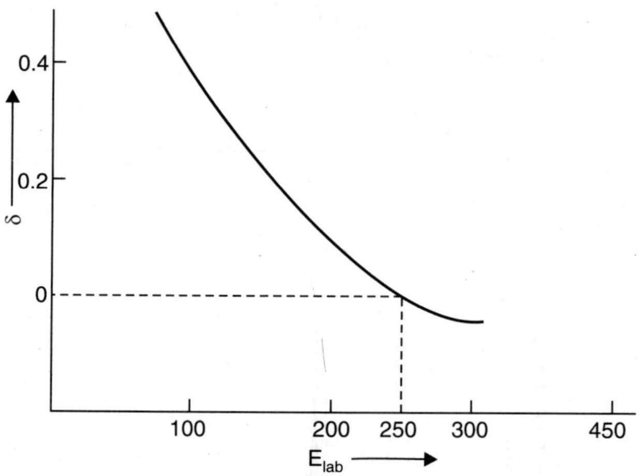

Fig. 6.5 Experimental values of phase shifs for IS_o as a function of laboratory energy. At $E_{lab} \approx 250$ MeV, the phase shifts change sign (Ref. 15, 16).

We will now introduce two concepts of nucleon-nucleon potential only qualitatively, *i.e.*

(*i*) Non-local potential; (*ii*) Effective potential.

A. *Non-local Potential*: This takes into account two different co-ordinates for particles 1 and 2, *i.e.* before potential is switched on say $\mathbf{r}_1$ and $\mathbf{r}_2$; and after it is switched off say $\mathbf{r}_1'$ and $\mathbf{r}_2'$. Such a potential $V(\mathbf{r}_1 \cdot \mathbf{r}_2 ; \mathbf{r}_1', \mathbf{r}_2')$ has been used by many authors,[17, 18] *e.g.* Mitra, Tabakins and others. Mathematically it avoids infinities in perturbation treatment. A typical non-local, separable potential can be written as:

$$V(\mathbf{r}, \mathbf{r}') = \lambda f(\mathbf{r})\, g(\mathbf{r}') \qquad ...(6.92)$$

Sometimes momentum dependent potential is used as given by Green[19]. Further Moszkowski[20] has shown that it is possible to simulate a hard core potential by means of momentum dependent potential. Also Moszkowski-Scott[21] have developed a separation method for dealing with core problem, for which *see* Reference (21).

***B. Realistic and Effective Interactions*:** The potential V between the two free nucleons is different from the one experienced by the two nucleons inside a nucleus, basically because of the Pauli Exclusion Principle and the interaction with other nucleons inside the nucleus. The potential between two nucleons inside a nucleus, is called an 'Effective' potential, compared to the 'realistic' potential used for free nucleon-nucleon interaction.

Many-body theory for effective interaction has been developed by Brueckner.[22] We will, however, not go in details about it. It is basically based on the development of K-matrix or reaction matrix equation which connects K—the effective potential matrix, with V, the two body potential through Q-the Pauli operator, *i.e.*,

$$K = V + V\frac{Q}{E}K \qquad ...(6.93)$$

The Pauli-operator Q, disallows the intermediate virtual excitation for particles to occupied states and E is the energy denominator appropriate for the intermediate states. For details *see* Ref. 22 and 23.

6.5.3 Conclusive Observations

It is evident from the above discussion, that there exists strong repulsive core of radius from 0.4 to 0.5 fm. This occurs in all states of Gammel-Thaler[10] potential and in the potentials used in work by Japanese group[25] but is absent from some of the states for Signall-Marshak[13] potential. Such cores were also suggested by earlier meson-theoretic calculation. The meson theories, however, do not give quantitatively correct results for these cores. It is believed that, these repulsive cores are contributed by heavy meson, *i.e.* ρ-mesons with $T = 1$, and $m_\rho c^2 = 769$ MeV and ω-mesons ($T = 0$), $M_\omega c^2 = 783$ MeV, and η-mesons ($T = 0$) and $m_n c^2 = 549$ MeV, along with two pion exchange. The contribution from heavy mesons is believed to be dominant, in the repulsive core.

For $r > 0.6\ \hbar/m_\pi c$, all the potentials have similar shapes. In this range, the meson-theoretic calculations based on one-pion-exchange are reliable. Here the meson theory predicts a tensor force, along with the central part, as required by experiment data. **L·S** term, as obtained from meson-theory, however, is much smaller than used in phenomenological spin-orbit forces, which have been suggested on the basis of experimental data—by a factor of 10, or more. As a matter of fact, there is a controversy about **L·S** term in nucleon-nucleon potential. According to Japanese school, this term is not required and the data can be explained on the basis of central and tensor forces only; if we consider three different regions and takes different potentials for these regions. The de-polarisation data on D (θ), however, seems to requires an **L·S** term.

We will conclude that though **L·S** term is not understood quantitatively on the basis of the meson theory, its necessity is very much felt empirically. Such a term is also not in contradiction to the general form of the potential given in Eq. 6.84.

The **L·S** term required to explain the high energy data is required to have a short range, though for low energies, it is required even at larger ranges, *i.e.* up to 2 or 3 fermis.

The exchange terms of spin-exchange, iso-spin exchange and position exchange type, as well as terms, with no exchange properties are built into the meson-theory and are also required phenomenologically.

The tensor force is attractive in 3 S_1 and 3 P_0 state and is an important contribution to the total nuclear force.

The even parity (1 = 0.2.4) potentials are attractive outside the repulsive core. The odd parity potentials on the other hand, are considerably weaker than the even parity potential especially for the central potentials.

In complex nuclei, the relative momentum of two nucleons rarely exceeds the equivalent of the 200 MeV in lab system in a scattering experiment.

We may, therefore, consider for calculations of nuclear structure, a nucleon-nucleon potential, of the type as given say in Eq. 6.82 with only terms up to V_3; with exchange terms of different types included. For many cases, for which greater precision is not required, still simpler forces may, sometimes, be used; say Serber exchange potential. As has been stated earlier, the three-body forces contribute up to 15 to 20 per cent of the overall nucleon-nucleon potentials. Therefore, the parameters used for free nucleon-nucleon scattering are not expected to reproduce the nuclear structure properties exactly. One uses an effective potential in these cases, with general characteristics of the free nucleon-nucleon potential, but with parameters to be adjusted to fit the experimental data.

The one-pion exchange potential (OPEP) based on the second order static extended source mesic theory does reproduce the central, tensor and exchange characteristics of the phenomenological nucleon-nucleon potential for $r > 0.6\hbar/m_\pi c$, *i.e.* beyond about 1 fm. The short-range, charge independence and charge symmetry—all these can be explained by the meson theory. Higher order approximations, and one-boson-exchange potential (OBEP) are yet not as well understood as OPEP.

However, it is well established that a repulsive core exists for $r \leq 0.5$ fm for many or most states. An **L·S** force of a short range seems to be required definitely for higher energies. At low energies, the indications are that an **L·S** force of longer range is required to explain all the data, especially the data on de-polarisation. There is, however, some controversy, on this issue. The **L·S** force required phenomenologically is, however, ten times more than that expected on the meson-theory. In practice, one uses an **L·S** force empirically to fit the data. Such a force is, of course, not inconsistent with the general form of the potential as given Eqs. 6.82 to 6.89.

The velocity dependent forces of the type $(\sigma_1 . \mathbf{p}_1)$, $(\sigma_2 . \mathbf{p}_2)$ or $(\sigma_1 . \mathbf{L}_i)$ seem to exist at high energies and have been used for the *p-p* scattering by Breit[11] and later by Tabakin[18], and also by Green[18]. They are, however, not important for most of the nuclear structure problems.

The representation of nucleon-nucleon interaction, through a potential is valid at lower-energies. One can replace it by the dispersion-relation theory using scattering amplitudes rather than a potential. This, however, requires a complete experimental knowledge of the scattering matrix over a whole range of energies.

Recent studies[26] of nucleon-nucleon interaction based on meson-exchange model, have pushed the accuracy of predictions to a higher level. As for example, one of the most spectacular case was the calculation of (*n, p*) capture cross-section at thermal energy, using meson-exchange current correction which yielded an agreement within 1%. This was further improved upon, by using effective field theory[27], to explain accurately the deuteron properties, (*n-p*) $1S^0$ scattering and (*n-p*) radioactive capture process. As a matter of fact, using a 2π exchange contribution to *NN* interaction, it has been possible to calculate using the charge symmetry breaking (CSB) model, due to mass splitting between neutron and proton, the small difference between *p-p* and *n-n* interaction, which agrees with experimental data[28]. Similarly, in phenomenological calculation it has been possible to obtain parameters of optical model for nucleon-deuteron potential explaining and predicting the result of scattering and polarisation at low and intermediate energies. Also models for three-body forces have been explored for predicting the experimental binding energy of say H^3 system[29].

6.6 THE QUARK MODEL

It is now well-known[30] that proton and neutron are the two lightest particles which are members of a group of particles called baryons with masses and iso-spins (I) as given in Table 6.4.

J in J^P represents total angular momentum and P is parity. There are, similarly ten $J^P = 3/2^+$ baryons and nine $J^P = 0^-$ mesons and nine $J^P = 1^-$ mesons. All these baryons and mesons are termed as hadrons; their mutual interactions are governed by strong forces of a similar nature, as are available in nuclei; binding protons and neutrons. All these hadrons are governed by the quantum numbers, Q (charge); I (Isotopic spin), I_3 (z-component of isotopic spin); B, baryon quantum number which is +1 for baryons and 0 for mesons—also called bosons, and S (strangeness) quantum number. The significance of S is explained as follows.

Table 6.4 *The light eight baryons ($J^P = 1/2^+$). Masses in MeV/C^2*

No.	*I*	*Particles*	*Mass*
1	1/2	p	938.28
2		n	939.57
3	0	Λ_0	1115.60
4	1	Σ^+	1189.37
5		Σ^0	1192.46
6		Σ^-	1197.34
7	1/2	Ξ^0	1314.9
8		Ξ^-	1321.32

It was experimentally found,[31] that some particles in the hadron family were copiously produced, as expected in a strong interaction but decay very weakly, *i.e.* with low cross-sections and hence have comparatively long lifetimes. As for example:

$$\pi^- + p \rightarrow K^0 \text{ (Mass = 498 MeV)} + \Lambda^0 \text{ (Mass = 1116 MeV)}$$

$$K^0 \text{ (498 MeV)} \rightarrow \pi^+ + \pi^- \; (\tau = 0.89 \times 10^{-10} \text{ secs})$$

$$\Lambda^0 \text{ (1116 MeV)} \rightarrow p + \pi^- \; (\tau = 2.63 \times 10^{-10} \text{ secs}) \quad ...(6.94)$$

These lifetimes should be compared to the typical times taken by a strong interaction process of 10^{-23} secs. This paradox was resolved by suggesting that a new quantum number called strangeness, and denoted by S should be associated with these hadrons. It was found that in the above case $S = 0$ for π-mesons, $S = -1$ for Λ^0 and $S = +1$ for K^0. For proton and neutron, $S = 0$. So in the above reaction in the production equation; $S = 0$, on both sides; but in the first decay equation; $S = -1$ on the left side of Eq. 6.94, and $S = 0$ on the right side. Hence $\Delta S = +1$. One can also have $\Delta S = -1$ in Eq. 6.94 for the other decay process. So for strong interaction $\Delta S = 0$; and for weak interactions $\Delta S = \pm 1$. One can also have $\Delta S = 0$ for weak interactions as in the case of β-decay.

With these definitions, another quantum number called strong hypercharge (Y) has been defined as:

$$Y \equiv B + S \qquad ...(6.95)$$

where B is the baryon quantum number. One can assign these six quantum numbers, *i.e.* Q, J^P, I, Y, S and B to all the hadrons. Any hadron having quantum numbers Y, B, S, will have an anti-particle with quantum numbers $-Y$, $-B$, and $-S$. Further, the charge Q is empirically related to the third component of Isospin (I_3) as:

$$Q = I_3 + \frac{Y}{2} \qquad ...(6.96)$$

In (1964), GellMann and Zweig[32], separately suggested that all hadrons could be made from spin $J = 1/2$ fermion-substructures called quarks. One could classify all hadrons and bosons, in terms of these quarks. They plotted the hadrons with $J^P = 1/2^+$ in a plot of Y against I_3 called the weight diagram. It was found, that the lightest baryons ($J^P = 1/2^+$) fall in clear muliplets, as an octet. Similarly $J^P = 3/2^+$ baryons fall into a decuplet, and similarly the mesons could be arranged in nontes. It is this classification, which led to the assumption of quarks, assigning to each hadron, with these substructures. Details of this classification led to the following three quarks, with their quantum numbers, as given in Table 6.5. For details *see* Ref. (33).

Table 6.5 *Properties of Quarks*

Quark	*Charge*	I	I_3	Y	S	B
u	+ 2/3 e	1/2	1/2	1/3	0	1/3
d	– 1/3 e	1/2	– 1/2	1/3	0	1/3
S	– 1/3 e	0	0	– 2/3	– 1	1/3

Figure 6.6, shows the classification of eight light baryons and their composition in terms of these quarks. The symbol u stands for 'up'; d for 'down' and S for 'strange'. According to this assignment, proton consists of $u\ u\ d$, *i.e.* 2 'up' quarks and one 'down' quark and neutron consists of $d\ d\ u$, *i.e.* 2 'down' quarks and one 'up' quark. Other baryons are assigned the various quarks as shown in Fig. 6.6. Similarly $J^P = 3/2^+$ baryons and the bosons have been assigned the quarks. All the baryons can be assigned three quarks and all the bosons two quarks; one quark and other anti-quark of different or same type. As for example π^- contains du^-, π^0 contain uu^- and π^+ contains ud^-. Is there any direct evidence of three quarks in a nucleon? In 1967, at SLAC, 20 GeV/c electron beams were used to study the elastic and inelastic scattering of electrons from proton[34], *e.g.*,

$$e^- + p \rightarrow e^- + p \quad \text{(elastic scattering)} \qquad ...(6.97)$$

It was found, that the elastic scattering is diffractive with a forward peak, whose angular size θ is connected to the total momentum p; and p_T, the momentum transverse component of the scattered electrons by the relationship:

$$p_T = p\theta \qquad ...(6.98)$$

By the uncertainty principle,

$$p_T r \approx h \qquad \text{...(6.99)}$$

Hence

$$\theta \approx \frac{h}{rp} \qquad \text{...(6.100)}$$

where r is the radius of say, proton. The diffraction pattern shows, that proton is an extended object having an internal structure.

Also, Panofsky[35] et al. have reported in 1968 experiments on inelastic scattering of electrons from proton targets corresponding to the reaction:

$$e^- + p \rightarrow e^- + \text{hadrons} \qquad \text{...(6.101)}$$

The inelastic distribution shows little change with θ. The detailed analysis showed that the scattering must be taking place from something point like; which were taken as quarks.

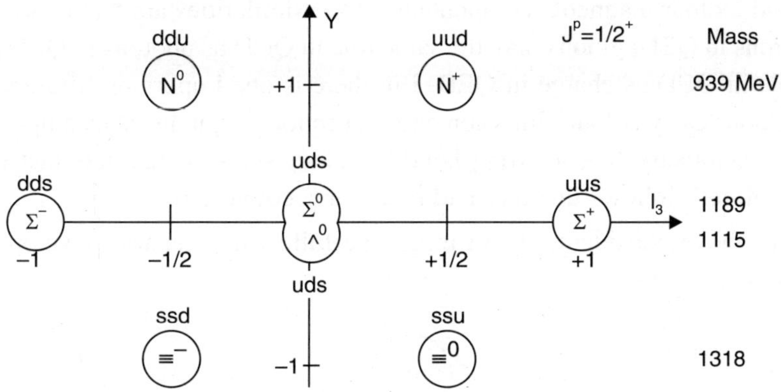

Fig. 6.6 Classification of eight light baryons and their compositions in terms of u (up) and d (down) and strange (s) quarks (Ref. 33).

The forces between nucleons, are taken to be due to interaction of these quarks. The field particles of these forces are called gluons which are massless bosons. We have discussed till now only three species of quarks, *i.e.* u, d and s. Subsequent experiments[36] have shown three more species. Their charges and approximate masses (in MeV/c^2) are given as:

$$Q = \frac{2}{3} \quad u\,(\sim 10) \quad c\,(1300) \quad t\,(> 40{,}000)$$

$$Q = -\frac{1}{3} \quad d\,(\sim 10) \quad s\,(150) \quad b\,(4200) \qquad \text{...(6.102)}$$

The heavier species carry the names charm (c) beauty (b) and top (t). The top quark has been only discovered recently[40].

These six species are called the flavours of quarks. Each quark carries further a quantum number called colour. As for example, each quark has three colours: red (r), blue (b) and yellow (Y). This is required, to achieve the necessary anti-symmetrisation of many-quark-system. One can, therefore, write for the quarks, a wave-function describing internal structure as:

$$\psi_{total} = \text{(space-term)} \times \text{(spin-term)} \times \text{(flavour-term)} \times \text{(colour-term)} \quad ...(6.103)$$

The formal quantum field theory describing the interaction between quarks is called Quantum Chromo-Dynamics (QCD).

The nuclear forces as discussed till now, were explained by using the concept of exchange of π-meson of heavier mesons as developed in Section 6.4, and proposed in pion-exchange model, developed by Yukawa[4]. But after the discovery of quarks as qualitatively discussed above; the quarks assume the role of basic blocks of nuclear matter and hence become the source of nuclear forces.

We explain below qualitatively, how quarks explain the nuclear forces, through the theory of Quantum Chromo Dynamics (QCD). In Fig. 6.7*a*, we have shown an exchange of π^+ meson, between a proton and a neutron, in the Feynman diagram, showing the *n-p* interaction. In Fig. 6.7*b*, this one-pion exchange is represented by the interaction of quarks and gluons. QCD is similar to QED (Quantum Electro Dynamics). The strong field, for which the carriers are gluons, is also a vector field having 'colour-electric' and 'colour magnetic' components. Other similarities are that quarks in QCD play the same role as electrons in QED; gluons play the same role in QCD as photons in QED and finally colour plays the same role in QCD as charge in QED. But there is one important difference. While photons carry no charge, gluons carry colour. Emission and absorption of gluons is accompanied by changes in colour. Also colour symmetry (*i.e.* a strong bond in colour singlets) requires that the total colour is conserved and therefore it follows that the field must carry colour too.

Keeping the above considerations in mind, the following scenario develops for exchange of quarks as shown in Fig. 6.7*b*.

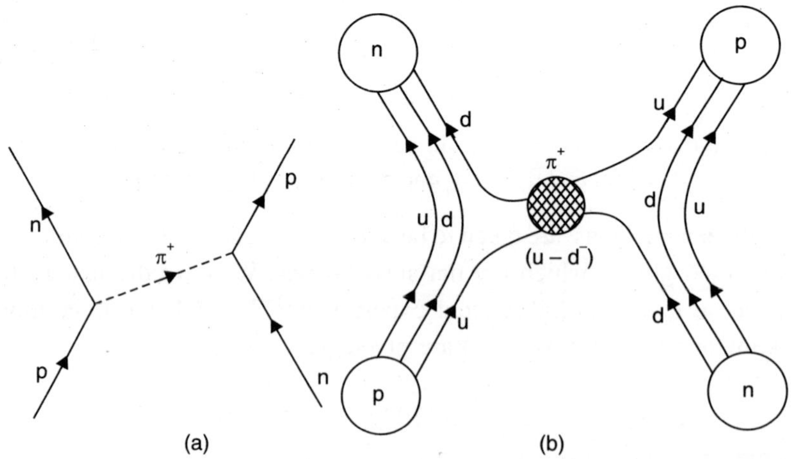

Fig. 6.7 The nuclear force, as seen in the interaction of quarks, representing one-meson change (*a*) Exchange of one-meson (π^+) as discussed in Section (6.4); (*b*) Same represented, by exchange of quarks-antiquarks, as explained in the text (Ref. 37).

Both the proton and neutron contain quarks and gluons, keeping the total colour zero. One of the gluons inside say a proton produces a $d\bar{d}$ pair; the $\bar{d}$ can unite with a *u*-quark to become $u\bar{d}$

(a π^+-meson) represented by shaded area in Fig. 6.7*b*. Then other *d*, along with *u* and *d* in the nucleon gives *udd*, *i.e.* a neutron. Then $\bar{d}$ of $u\,\bar{d}$ joins with *d* of neutron, converting it into a gluon and *u* joins with *ud* of neutron, converting it into *uud*, *i.e.* a proton. In this manner, the quarks-gluon system in the neutron and proton in neutron-proton interaction can create and annihilate a π^+-meson; as envisaged in Yukawa theory of one-pion exchange. It should be realised that $u\bar{d}$ pair is in a state, which is coherent superposition of all $I = I_3 = 1$ mesons, *i.e.* not only π^+, but also ρ^+, A_1^+, A_2^+, etc. In this manner, quark-gluon model, can account for the exchange of other mesons.

The use of heavy quarks, *b, c* and *t*, and production of many mesons, is a somewhat complicated phenomenon encountered at higher energies. We will not go into it, for which *see* Ref. (38).

The above qualitative description of the nuclear forces through an exchange of $u\bar{d}$, applies to the case when the two nucleons are too close. It is analogous to the coming close together of two neutral atoms, which will interact through van der Waal forces. In the case of two nucleons, we have a case of a system of two three-quark system, which interacts through colour exchange of gluons. It is difficult to solve the field equations of QCD in such a complicated situation, though some attempts have been made[39].

One of the recent-most discoveries in this field is the experimental discovery of top quark (*t*) in which many laboratories from many countries all over the globe including India have participated (*see* Ref. 40).

6. Nuclear Forces 2000–2008

In an interesting calculational work, the proton-emulsion interactions at high energies (200–800 GeV), have been analyzed, using the Monte-Carlo procedure where the concept of classical strings is used. Results, very nearly match the theory. This analysis, was done at Panjab University, Chandigarh (India) by S. Dhamija, Kaur and Dahaya [Phy. Rev. 63, 035201 (2001)].

In a theoretically interesting paper authorized by 42 authors, the dissociation of two neutron halo nuclei (He^6, Le^{11}, Be^{14}) are explained, using a technique based on intensity interferometry and Dalitz plots (Numerical methods of analysis). This approach provides the combined treatment of $n + n$ and core-*n* interaction. [Phy. Rev. C. 64, 0601, 301 (R) (2001)].

In another theoretical paper Y.K. Gambhir and R.K. Bhagwat (India) has used Relativistic Mean Field (RMF) approach, and Relativistic Hartree Bouglibouv (RHB) equation, to calculate peripheral factors, in the coordinate space to explain the anti-proton annihilation experiments. The calculated peripheral factors, are closer relatively, to the corresponding values, if one uses RHB equations. [Phy. Rev. C. 66. 034306 (R) (2002)].

In an interesting paper, Gerald Miller from University of Washington, Seattle Washington (USA) [Phy. Rev. C. 68, 0220 (R) (2003)], has calculated the shape of proton using spin-dependent quark densities as the input. For high momentum quarks, with spins parallel to that of proton, the shape of proton result into that of peanut; but with quarks antiparallel, the shape is that of a bagel (doughnut).

In a generalised theoretical paper the authors (3) from US, Germany and Spain, have classified a nuclear force, according to isopin dependence, and discuss the most general isopin structure of three-nucleon force. [Phy. Rev. C. 71, 024001 (2005)].

In a theoretical paper, the author P. Dalekshall Hangary investigated the dependence of the theoretical *p-d* and *n-d* elastic scattering results, on the triplet wave *N-N* interaction, using non-local *N-N* interaction. Finally a non-local interaction is constructed which reproduces simultaneously *N-N* data and most of *n-d* elastic scattering experiments up to 30 MeV [Phy. Rev. C. 77, 034002 (2008)].

REFERENCES

1. Chapters 3, 4, and 5.
2. H. Tyren and P. A Tove, Phy Rev. 96, 773, (1954). B.M. Rustad and S.L. Ruby, Phy. Rev. 97, 991, (1955). Amer. Inst. Phy. Handbook, McGraw-Hill, (1957) P. 8-59.
3. R. Middleton and D.J. Pullen; Nuclear Physics 51; 50, 63, 77, (1964). A. Gallinan et al., Phy. Rev. 138, *B* 560, (1965). Breit G. and E. Feenberg, Phy. Rev. 49, 519, 642 (1936).
4. H. Yukawa, Proc. Math. Soc. Japan, 17, 48, (1935). Proc. J. Phy. 7, 747, (1936); H. Bethe, Phy. Rev. 57, 260, (1940); G. Wentzel, Quantum Theory of Fields, Interscience, New York, (1949).
5. Nuclear Physics—Theory and Experiment, R.R. Roy and B.P. Nigam; New Age International (P) Ltd., Publishers New Delhi, (1986). Theoretical Nuclear Physics, J.M. Blatt and V.F. Weisskopf, John Wiley & Sons, New York (1952). The Quantum Theory of Radiation, W. Heitler, Clarenden Press, Oxford, (1954).
6. W. Heisenberg: Z. Physik, 7, 1, (1932); E. Majorona Z. Physik 82, 137 (1933).
7. Klein-Gordon. Zeitsfur, Phy. 40, 117, (19260; Ibid 40, 121, (1926), Ibid 48, 11 (1928).
8. R.P. Feynman: Quantum Electrodynamics, Benjamin, New York, (1961). Nuclear Interaction: Sergio DeBenedetti, (P. 427) John Wiley & Sons, Inc. New York, (1964). A. Kremner, Poe. Cambridge Phil. Sec. 34, 354, (1938). L. Eisenbud and E. P. Wigner: Proc. Nat., Acad. Science: U. S. 27, 281, (1941). Okubo, S., and R. E. Marshak; Ann. Phys. N. Y. 4, 166 (1958).
9. J. L. Gammel, R. Christians and R.M. Thaler, Phy. Rev. 105, 311 (1951).
10. J.L. Gammel, and R. Thaler, Phy. Rev. 107, 291, (1957). Gartenhaus S. Phy. Rev. 88, 725 (1952).
11. K.E. Lasilla, Hull M.H., Ruppal H.H., Mc Donald F.A and Breit H., Physical Rev. 126, 881, (1962); Breit G., M.H. Hull, K.E. Lassilla and K.D. Pyatt. Jr. Phy. Rev. 120, 2227 th. (1960).
12. T. Hamada and J.T. Johnson, Nuclear Physics 34, 382 (1962).
13. P. Signall and Marshak R.E. Phy. Rev. 109, 1229 (1958).
14. Theory of Nuclear Structure; M.K. Pal, Affiliated East-West Press Pvt. Ltd. (1982).
15. H.P. Srapp, T.J. Ypsilantis, and N. Metropolis, Phy. Rev 105, 302 (1957).
16. M.H. McGregor, Phy Rev. 113, 1559, (1959), L.H. Hulthén and M. Sugawara, Encyclopedia of Physics, ed. S. Flugge. 39, Berlin, Springer (1957).
17. A.N. Mitra and J.H. Naqvi, Nuclear Physics 25, 307, (1961), J. H. Naqvi Nuclear Physics, A 103, 565 (1967).
18. F. Tabakin, Ann. Phy. (N.Y.) 30, 51 (1964).
19. A.M. Green, Nuclear Physics, 33, 218 (1962).
20. S.A. Moszkowski, Phy. Rev. 129, 1901 (1965).
21. S.A. Moszkowski and B.L. Scott, Ann. Phy. (N.Y.), 11, 65 (1960).

22. K. Brueckner and L. Gammel, and H. Weitzner, Phy. Rev. 110, 431, (1938), K. Brueckner and J.L. Gammel; Phy. Rev. 109, 1023 (1958).

23. H.A. Bethe, H. Brandow and A.G. Petschre, Phy. Rev. 129, 225 (1963).

24. K. Brueckner, A.M. Lockett and M. Rotenberg; Phy. Rev. 121, 255, (1961). Mitra A.N. Advances Nuclear Physics 3, 1, (1969). Levinger, J.S., Nuclear Physics, Springer Tracts in Modern Physics V. 71; edited by G. Holer, Springer-Verlag, (1974), P. 88. Perey F. G., and B. Buck; Nuclear Physics, 32, 353 (1962).

25. J. Iwadare; R. Tamagaki and W. Watari Progress. Th. Phy, (Kyoto), Supp. 1, 52, (1956). Hamada J; J. Iwadare. S. Otsaki; R. Tamagaki and W. Watari. Progress Th. Phy. (Kyoto) 22, 566, (1959); 23, 61, (1960). Tabevani M. S. Takamura and H. Sasaki, Progress. Theor. Phy. (Kyoto) 6, 581 (1951).

26. T. S. Park, D.P. Minh and M. Rho, Phy. Rev. letters 74, 153, (1995), Nuclear Physics A 596, 515 (1996).

27. T. Sun-Park, K. Kubodera, Deng Pil-Min and M. Rho. Phy. Rev. C. 58, P 637 (1998).

28. G.Q. Li and R. Machleidt, Phy. Rev. C. 58, P. 1393 (1995).

29. J. Carlson, V.R. Pandharipande the R.B. Wiringa, Nuclear Physics. A 401, (1983). R. Schiavilla, V. R. Pandharipande and R.B. Wiringa Nuclear Physics, *A* 449, 219, (1986) L.D. Knuston and A. Kievsky; Phy. Rev. C 58, P. 49 (1998).

30. Introduction to High Energy Phy. D.H. Perkins, Addison-Wesley Publishing Company, Reading (Mass), (1992).

31. A. Pais Phy. Rev. 86, 663 (1952).

32. Gell-Mann M. Phy. Rev. 92, 833, (1953), Ibid Phy. Letters 8, 214, (1964), Nishijima K. Progress, Th. Phy 13, 285, (1955), Zweig G. CERN Report, 841, 91, Th. 412 (1964).

33. Ref. (28), (26) and (32).

34. G. Weber, Proceeding Instrument Symposium Electron and Photon interactions at high energies, Standford, California, P. 59 (1967).

35. W. Panofsky (data of E. Blooni et al.) Int. Conference Third Energy Phy., Vienna, (1968).

36. S. Glashow L., J. Lliopulis and L. Maini, Phy. Rev. D2, 1285, (1970), Abrams (G. S. et al., Phy. Rev. Letters 33, 145, (1974), Aubert J. J. et al., Phy. Rev. letters 33, 1404, (1974).

37. Concepts of Particle Physics V1, Kurt Goddttfried, and Victor Weisskopf, Clarenden Press, Oxford, (1984).

38. S. W. Herb et al., Phy. Rev. Letters 39, 252, (1977), B.K. Jain IAPS Seminar, P. 51 (1999) (Private communication).

39. A.S. Kronfeld and P.B. Markenzii, Ann. Rev. Nuclear and Particle Science 43, 793, (1993).

40. S. Abachi, ...Paul Granis, ... H.E. Montgomery, ... V. Narsinham, ... J.M. Kohli, ... R. Raja,... R.K. Shivpuri, ... J.B. Singh, ... P.M. Sood, ... A. Zylbersetjns (303 authors); Phy. Rev. Letter V. 74, P. 2632–2637 (1995).

PROBLEMS

1. Though tensor potential may be quite strong, comparable to the central potential in deuteron, the wave-function for tensor forces is much weaker than that for central forces. Explain this anomaly and its implications.

2. Show that if the nuclear forces is charge-independent; then the nucleon-nucleon potential cannot contain terms of the type; τ_{1z} , τ_{2z} and $\tau_{1z} + \tau_{2z}$, where τ_1 and τ_2 are the iso-spins of the two nucleons.

3. Prove that $P_M \psi = (-)^1 \psi$; where P_M is Majorona exchange operator, ψ is the two-nucleon Wave-function and 1 is the orbital angular momentum.

4. Show that Heisenberg exchange potential is equivalent to an ordinary potential, which changes sign according to whether **L** + **S** is even or odd. Here **L** is the total two nucleon orbital angular momentum and **S** is the total spin.

5. The tight binding of He^4, and lack of binding of He^5 and L^5 can be regarded as evidence in favour of exchange forces of Majorona type. Explain.

6. The potential energy between the two nucleons depends on operators.

$$1, \sigma_1 \cdot \sigma_2, \tau_1 \cdot \tau_2, (\sigma_1 \cdot \sigma_2)(\sigma_1 \cdot \sigma_2),$$

$$S_{12} = [3(\sigma_1 \cdot \mathbf{e})(\sigma_2 \cdot \mathbf{e}) - \sigma_1 \cdot \sigma_2] \text{ and } (\tau_1 \cdot \tau_2) S_{12}.$$

Prove that the square and products of any two of these are a linear combination of the above six operators.

7. Show that the following dependence for the potentials are not acceptable for nucleon-nucleon potential.

(*i*) $(\mathbf{r} \times \mathbf{S}) \cdot (\mathbf{r} \cdot \mathbf{S})(\mathbf{S} \cdot \mathbf{S})$ (*ii*) $(\mathbf{L} \cdot \mathbf{S})(\mathbf{L} \cdot \mathbf{L})$

(*iii*) $(\mathbf{r} \cdot \mathbf{p})(\mathbf{L} \cdot \mathbf{p})$ (*iv*) $(\mathbf{r} \times \mathbf{L}) \cdot \mathbf{p}$

(*v*) $(\mathbf{r} \cdot \mathbf{p})(\mathbf{r} \cdot \mathbf{S})$

8. If $V(\mathbf{r}_1, \mathbf{r}_2, \mathbf{r}_1', \mathbf{r}_2') = (R - R')\, V(r, r')$, where $\mathbf{r}_1$ and $\mathbf{r}_2$ are the coordinates of the two particles, before the potential is switched on, and $\mathbf{r}_1'$ and $\mathbf{r}_2'$, after that and $\mathbf{r} = (\mathbf{r}_1 - \mathbf{r}_2)$; and $\mathbf{r}' = (\mathbf{r}_1' - \mathbf{r}_2')$ and $\mathbf{R} = (\mathbf{r}_1 + \mathbf{r}_2)/2$, then show that the potential $V(\mathbf{r}_1, \mathbf{r}_2, \mathbf{r}_1', \mathbf{r}_2')$ satisfies the conservation laws.

9. Prove Eq. 6.57, having the reasons as given for Eq. 6.49.

10. Show that meson-theory, leads not only to the attractive radial depending part, but also to a repulsive core. Consult Nuclear Physics 34, 382, (1962), and Phy, Rev. 126, 881, (1962) for Hamada-Johnson and Yale potentials, and Annual Review of Nuclear Science 10, (291), (1960).

7

CHAPTER

Radiative Transitions

7.1 ENERGETICS AND EXPERIMENTAL

Radiative transitions are encountered in nearly every case, when a nucleus is excited, either through a nuclear reaction or as a result of a radiative decay, corresponding to alpha or beta decay. Figure 7.1(*a*) shows[1], a typical set of excited states, obtained from a daughter nucleus ${}_{26}Fe^{56}$, produced in a beta decay. As shown in the diagram, the excited states decay through radiative transitions (or gamma (γ) transitions), to the lower energy states because the energy of the excited state is not high enough for particle emission. We will discuss, in this section, the various properties of gamma transitions encountered in such cases. Typically in the fifties, inelastic scattering of neutrons was used in 1955, to excite gamma rays in many elements.[1]

Evidently, in gamma transitions, the excitation process results, either from the prominent mode of the oscillation of electric charge, or from the oscillation of the electric current. In both cases, the electromagnetic radiations are emitted. In the former case, oscillation of electric charge produces oscillating electric field, which in turn, produces oscillating magnetic field; while in the latter case, the oscillation of electric current produces, an oscillating magnetic field. In the first case, the emitted electromagnetic radiation is called the electric multiple radiation and is designated as *EL* and in the second case, it is called the magnetic multiple radiation and is designated as *ML*; where *L* is the total angular momentum carried by the radiation. We will discuss this subsequently in detail.

The energetics of gamma ray transitions are straight-forward, as given below:

$$[M_i^* \ (A, Z) - M_j \ (A, Z)]\, c^2 = E_{ij} \qquad ...(7.1)$$

where *i* and *j* denote the levels of excitation, of the initial and final states of the nucleus. In other words, the energy E_{ij} of the gamma transition determines, the energy difference of the two levels of an excited nucleus. If *j* corresponds to the ground state, the value of E_{ij} gives directly the energy of the excitation of the level.

The extra-nuclear electrons of the atom, while orbiting around the nucleus have a finite probability of interacting with the nucleus, through electromagnetic interactions. This is especially true for *K* and *L* electrons for which the orbiting path partly lies within the nucleus. The interaction of these atomic electrons, with the electromagnetic fields present in the nucleus, results in imparting the full energy of transition to these electrons. This phenomenon is called 'internal conversion' as opposed to the external conversion, when the gamma ray after being emitted by the nucleus, interacts with the external electrons either in the same atom or other atoms, say through phtoelectric or compton effects. The internal conversion, is a direct interaction of electromagnetic field of the nucleus, with the extra-nuclear electrons, and not a two-step process of the creation of gamma ray and its interaction with the electron. There does exist, however, a parallel process, where gamma rays from one nucleus, may interact with electrons of another atom, in a two step process, which, leads to the external conversion, on the basis of which converters are used, to detect the gamma rays by say G.M. counters, or ionisation chambers, etc. The energetics of the internal conversion is expressed by the relation:

$$E_e = E_{ij} - B_e \qquad ...(7.2)$$

where E_{ij} is the transition energy between two levels and E_e is the kinetic energy of the emitted electron and B_e is the atomic binding energy of the electron which is emitted.

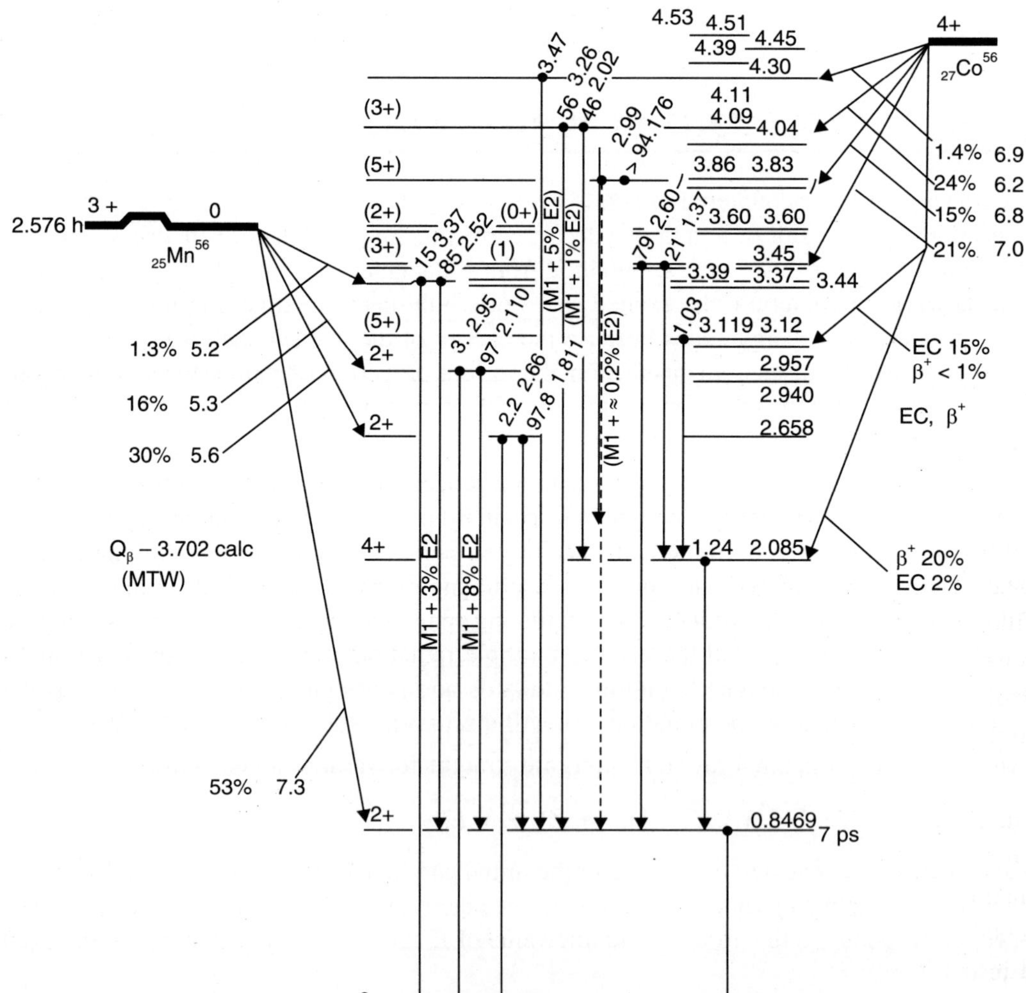

Fig. 7.1 **(a)** The decay of $_{27}Co^{56}$ and $_{25}Mn^{56}$ through $\beta^{\pm}$-decay, to various levels of $_{26}Fe^{56}$ giving rise to various gamma ray transitions (Ref. 1).

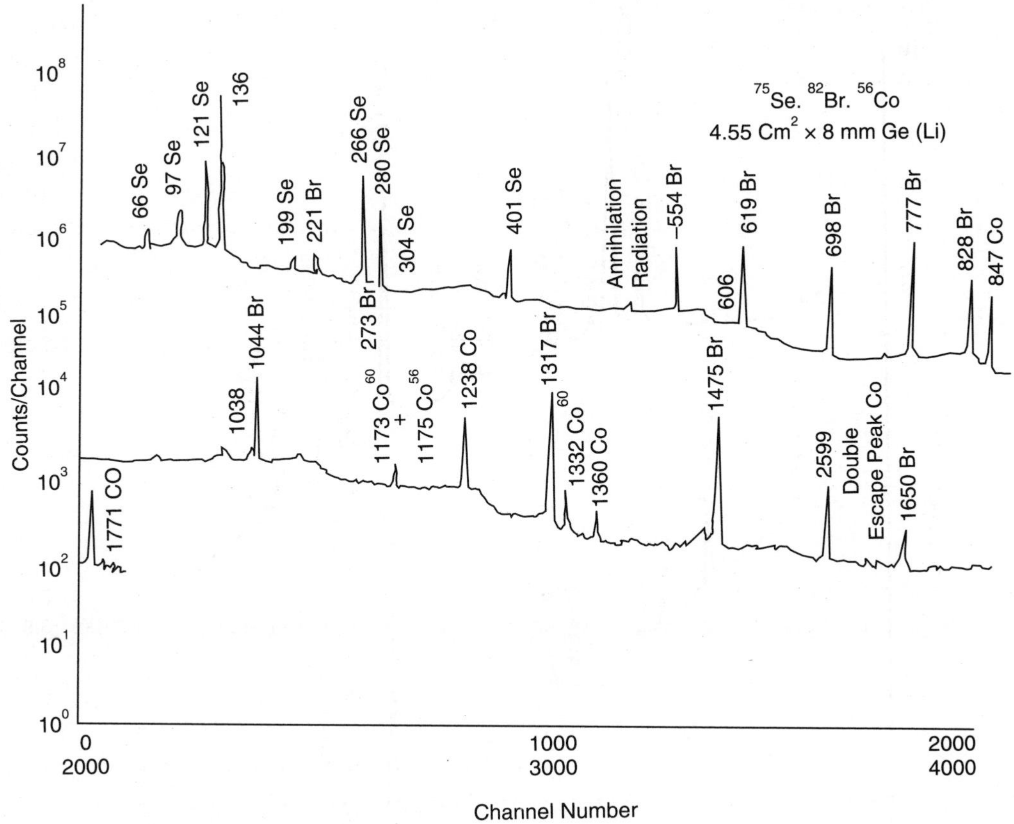

Fig. 7.1 (b) Spectrum of gamma rays, from β^-decay of Co^{56} + Se^{75} + Br^{82} (Ref. 1).

Evidently, the emitted electron leaves the atomic level, which it was occupying, as unoccupied for a short time. This level is, then filled by an electron from higher energy level, and starts a cascade of electromagnetic transitions by electrons of higher energy filling the levels at lower energy, till the last level is filled by surrounding free electrons. This results in X-rays, corresponding to the binding energies of electrons. Figure 7.1 (*b*) demonstrates the gamma rays spectrum of ${}_{26}Fe^{56}$ and some other radioactive nuclei obtained with the most modern techniques using a high resolution Ge (Li) detector. Figure 7.1 (*c*) shows on the other hand, a discrete internal electron spectra in the presence of continuous β-spectrum from Ba^{140} decay, using a mini orange electron spectrometer.[14]

The theory of radiative transitions as developed in the following pages is expected to yield the dependence of transition probabilities of such transitions on the angular momentum and other wave-functional properties of the initial and final levels. Naturally such theories should also explain the experimental values of internal conversion coefficient and their dependence on the various parameters involved in the intial and final states and the energy and the nature of the transitions. Also angular dependence (distributions or correlations) should be described by such a theory.

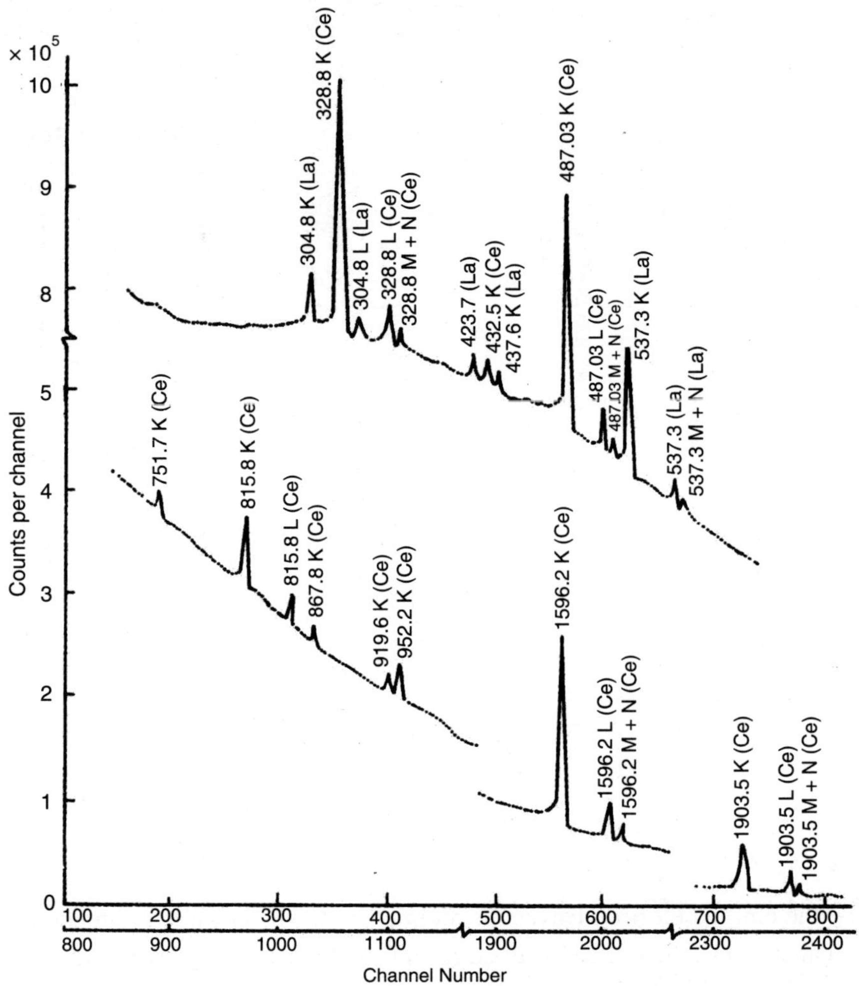

Fig. 7.1 (c) Conversion electron spectrum from Ba^{140} decay, taken with mini-orange electron spectrometer (Ref. 14).

7.2 CLASSICAL THEORY OF RADIATIVE TRANSITIONS[2]

Before, embarking on the quantum mechanical theory of transition probabilities in the next sections, we briefly describe here, the classical theory of the electromagnetic radiations, as emitted by a source corresponding to electric or magnetic multiple transitions. Some of these concepts, may then be easily carried to the quantum mechanical theory.

The source of these electromagnetic oscillations, *i.e.,* the nucleus has a size of the order of 10^{-12} cm, or so; while wavelength of a gamma ray of energy say 1 MeV is $(\lambda/2\pi) \approx 10^{-11}$ cm. It is, therefore, a system where $kr \approx 10^{-1}$, which is very much less than 1 and the theoretical treatment should be based on quantum theory of radiation with $kr << 1$. However, for orientating oneself, one may obtain considerable insight by treating the problem, first as of classical radiation field, emitted by a distribution of charges and currents, which vary with time.

First, we use the classical theory of electromagnetic fields and understand the significance of (*i*) total angular momentum L carried by the radiation; (*ii*) the meaning of electric and magnetic multiple fields and (*iii*) the relationship of parity and the selection rules in the two cases.

We know, that Maxwell equations in a region free of sources of radiation may be written as:

$$\nabla \times \mathbf{E} = -\frac{1}{c}\frac{\partial \mathbf{B}}{dt} \;;\; \nabla \cdot \mathbf{E} = 0 \qquad ...(7.3)$$

$$\nabla \times \mathbf{B} = \left(\frac{\mu \in}{c}\right)\frac{\partial \mathbf{B}}{dt} \;;\; \nabla \cdot \mathbf{B} = 0 \qquad ...(7.4)$$

in C.G.S. system where **E** and **B** are electric and magnetic fields carried by electromagnetic waves, μ is the constant of permeability and $\in$ is the dielectric constant (permitivity). For vacuum both μ_0 and ε_0, are equal to 1. As is evident, from these equations, electric field **E** can be created by an oscillating magnetic field **B**: [Eq. 7.3]; or a magnetic field **B** can be created by an oscillating electric field **E** [Eq. 7.4]. One can, as a matter of fact, assume the form of **E** (r, t) and **B** (r, t) as sinusoidal, *i.e.*,

$$\mathbf{E}(r, t) = R_e\, \mathbf{E}(r)\, e^{-i\omega t} \qquad ...(7.5)$$

and

$$\mathbf{B}(r, t) = R_e\, \mathbf{B}(r)\, e^{-i\omega t} \qquad ...(7.6)$$

Then, one can write an explicit expression of **E** in terms of $\nabla \times \mathbf{B}$ and of **B** in terms of $\nabla \times \mathbf{E}$, by combining Equations 7.3, 7.4, 7.5 and 7.6 appropriately. These expressions then, yield two sets: the first set as:

$$(\nabla^2 + k^2)\,\mathbf{B} = 0;\; \nabla \cdot \mathbf{B} = 0 \qquad ...(7.7\ a)$$

and

$$\mathbf{E} = \frac{i}{\sqrt{\mu \in} k}\, \nabla \times \mathbf{B} \qquad ...(7.8)$$

In this set, we have eliminated **E** between Eqs. 7.3 and 7.4 for Eq. 7.7 *a* and written Eq. 7.8 from Eqs. 7.5 and 7.4. Here we have used the relationships:

$$\nabla \times (\nabla \times \mathbf{B}) = \text{grad div } \mathbf{B} - \nabla^2\mathbf{B} = -\nabla^2\mathbf{B} \qquad ...(7.7\ b)$$

because, grad div $\mathbf{B} = 0$; as $\nabla \cdot \mathbf{B} = 0$ ...(7.7 *c*)

This set, along with Eq. 7.5 corresponds to oscillating electric field.

Here $$k = \frac{2\pi}{\lambda} \;;\; \omega = vk \text{ and } v = \frac{c}{\sqrt{\mu \in}}$$

Similarly one gets another set:

$$(\nabla^2 + k^2)\,\mathbf{E} = 0;\; \nabla \cdot \mathbf{E} = 0 \qquad ...(7.9\ a)$$

and

$$\mathbf{B} = \frac{-i\sqrt{\mu \in}}{k}\, \nabla \times \mathbf{E} \qquad ...(7.9\ b)$$

corresponding to oscillating magnetic fields, where we have eliminated **B** between Eqs. 7.3 and 7.4 for Eq. 7.9*a* and written Eq. 7.9*b* from Eqs. 7.6 and 7.3. These fields are called magnetic because of the oscillating magnetic field, Eq. 7.6. Knowing the experimental fact of transverse nature of E.M. waves, one can write:

$$\mathbf{r} \cdot \mathbf{E} = 0 \quad ...(7.10\ a)$$

and

$$\mathbf{r} \cdot \mathbf{B} = 0 \quad ...(7.10\ b)$$

Combining Eqs. 7.7, 7.8 and 7.10 *b*, we may obtain equations of $\mathbf{E}^E(r)$ and $\mathbf{B}^E(r)$ which are linearly independent respectively of $\mathbf{E}^M(r)$ and $\mathbf{B}^M(r)$, which are, in turn, obtained by combining Eqs. 7.9 and 7.10 *a*. One can, therefore, write, in general:

$$\mathbf{E}(r) = \mathbf{E}^E(r) + \mathbf{E}^M(r) \quad ...(7.11\ a)$$

and

$$\mathbf{B}(r) = \mathbf{B}^E(r) + \mathbf{B}^M(r) \quad ...(7.11\ b)$$

It may be realised that $\mathbf{E}^E(r)$ and $\mathbf{B}^E(r)$ correspond to $\mathbf{r} \cdot \mathbf{B} = 0$; when radial vectors and magnetic fields are transverse to each other. However, in this case, electric multipole has non-vanishing radial components of electric field: hence superscript *E* for these fields. They are also termed as transverse magnetic. Similarly $\mathbf{E}^M(r)$ and $\mathbf{B}^M(r)$ which correspond to $\mathbf{r} \cdot \mathbf{E} = 0$ are also called magnetic (or transverse electric); hence superscript *M* on them. In this case magnetic multipole transition has non-vanishing radial components of magnetic fields.

Equation with Source

The relationship of the electric and magnetic fields, with their sources, can be understood by writing the Maxwell equations, for a case when E.M. waves are generated by a localised charge density $\rho\ (r, t)$ and current density $\mathbf{j}\ (r, t)$ and magnetisation density $\mathbf{M}\ (r, t)$ say due to the intrinsic magnetic moment of the nucleus; each having a time-dependence given by exp-ickt.

The Maxwell equations, then are in the presence of sources:

Ist Set:

$$\nabla \times \mathbf{E} = \frac{ik}{\sqrt{\mu \in}}\ \mathbf{B} = ik \sqrt{\frac{\mu}{\in}}\ (\mathbf{H} + 4\pi\mathbf{M}) \quad ...(7.12)$$

$$\nabla \cdot \mathbf{E} = \frac{4\pi\rho}{\in} \quad ...(7.13)$$

These fields correspond to magnetic character.

2nd Set:

$$\nabla \times \mathbf{H} = \nabla \times \left[\frac{\mathbf{B}}{\mu} - (4\pi)\mathbf{M}\right] = \left[\frac{4\pi}{c}\right]\mathbf{j} - ik\sqrt{\frac{\in}{\mu}}\ \mathbf{E} \quad ...(7.14)$$

$$\nabla \cdot \mathbf{B} = 0 \quad ...(7.15)$$

These fields correspond to electric character. In Eq. 7.14, we have used the modified Eq. 7.8.

We now define a quantity ε, so that:

$$\varepsilon \equiv \mathbf{E} + \frac{i\,[4\,\pi]}{\omega \in}\,\mathbf{j} = \mathbf{E} + \frac{i}{k}\sqrt{\frac{\mu}{\in}} \times \left[\frac{4\pi}{c}\right]\mathbf{j} \quad ...(7.16\ a)$$

We can also use the well-known relationships of continuity equations:

$$\nabla \cdot \mathbf{j} - i\,\omega\rho = 0 \quad ...(7.17)$$

and $$\omega = \frac{[c]}{\sqrt{\mu \in}} k \quad ...(7.18)$$

Then, using expression of ρ from Eq. 7.17, into Eq. 7.13, we can obtain the relationship:

$$\nabla \cdot \varepsilon = 0 \quad ...(7.16\ b)$$

In this manner, both **B** and ε have divergences equal to zero and hence one can express, in the following equations, the equivalent of Equations 7.7 to 7.9, for the case of the presence of sources, from Eqs. 7.12 to 7.18.

Electric-Radiation:

$$\left(\nabla^2 + k^2\right) \mathbf{B}^E = -\left[\frac{4\pi}{c}\right] \mu \left(\nabla \times \mathbf{j} + [c]\, \nabla \times \nabla \times \mathbf{M}\right) \quad ...(7.19\ a)$$

$$\nabla \cdot \mathbf{B}^E = 0 \quad ...(7.19\ b)$$

$$\varepsilon^E = \mathbf{E}^E + \frac{i}{k}\left[\frac{4\pi}{c}\right] \mathbf{j} \quad ...(7.19\ c)$$

or $$\varepsilon^E = \frac{i}{\sqrt{\mu \in k}} \left(\nabla \times \mathbf{B} - [4\pi]\, \mu\, \nabla \times \mathbf{M}\right] \quad ...(7.19\ d)$$

Magnetic-Radiation:

$$(\nabla^2 + k^2)\, \varepsilon^M = (\nabla^2 + k^2) \left(E^M + \frac{i}{k}\sqrt{\frac{\mu}{k}} \left[\frac{4\pi}{c}\right] \mathbf{j}\right)$$

$$= -\left[\frac{4\pi}{c}\right] ik \sqrt{\frac{\mu}{\in}} \left(\frac{1}{k^2} \nabla \times \nabla \times \mathbf{j} + [c^2]\, \nabla \times \mathbf{M}\right) \quad ...(7.20\ a)$$

$$\nabla \cdot \varepsilon^M = \nabla \cdot \left(E^M + \frac{i}{k}\sqrt{\frac{\mu}{k}} \left[\frac{4\pi}{c}\right] \mathbf{j}\right) = 0 \quad ...(7.20\ b)$$

and $$B^M = -\sqrt{\frac{\mu\varepsilon}{k}} \left(i\nabla \times \varepsilon^M + \frac{[4\pi]}{\omega \in} \nabla \times \mathbf{j}\right) \quad ...(7.20\ c)$$

It can be seen from Eq. 7.19 that the fields ε^E and $\mathbf{B}^E$ behave the same way as $\mathbf{E}^E$ and $\mathbf{B}^E$ in Eq. 7.8 and hence they have the electric character, and similarly ε^M and $\mathbf{B}^M$ behave the same way as $\mathbf{E}^M$ and $\mathbf{B}^M$ in Eq. 7.9 and hence they have the magnetic character.

The significance of terms 'electric' and 'magnetic' will be clear further when we calculate the transition probabilities, and will find that the transition probability calculated from $\in^E$ and $\mathbf{B}^E$ will be mainly connected with electric charge oscillator and that calculated from $\in^M$ and $\mathbf{B}^M$, will be mainly connected with electric current (or magnetic field) oscillations.

7.3 ANGULAR MOMENTUM IN CLASSICAL ELECTROMAGNETIC FIELD[2]

(A) We have seen [Eqs. 7.7 and 7.9] that both electric and magnetic fields can be expressed in the form of non-singular scalar Helmholtz equation:

$$\nabla^2 u + k^2 u = 0 \qquad ...(7.21)$$

where u is the amplitude of the field. It can be shown that the solution of Eq. 7.21 can be written in terms of the components of u, *i.e.*, u_{LM} which can be expressed as:

$$u_{LM} \propto j_L (k r) Y_{LM} (\theta, \phi) \qquad ...(7.22\ a)$$

where $j_L(kr)$, are spherical Bessel functions given by:

$$j_L (kr) = \left(\frac{\pi}{2kr}\right)^{1/2} J_{L+\frac{1}{2}} (kr) \qquad ...(7.22\ b)$$

and $$J_{L+\frac{1}{2}} (kr)$$

is an ordinary Bessel function of half integer order. Asymptotic values of $J_1(kr)$ are given by:

$$\underset{kr\to 0}{j_L (kr)} = \frac{(kr)^L}{(2L+1)!}$$

L is a positive integer and $|M| \le L$, so that,

$$u = \sum_{LM} u_{LM} \qquad ...(7.22\ c)$$

Here $Y_{LM}(\theta, \phi)$ are well known normalised spherical harmonics. The functions u_{LM} have the following characteristics:

(*i*) They form a complete orthogonal set of non-singular solutions to the scalar Helmholtz equation.

(*ii*) If one applied to the u_{LM}, the orbital angular momentum operator **L**, *i.e.*,

$$\mathbf{L} \equiv -i\mathbf{r} \times \mathbf{grad} \qquad ...(7.23\ a)$$

Then, the following relations are satisfied:

$$L_z u_{LM} = M u_{LM} \qquad ...(7.23\ b)$$

and $$L^2 u_{LM} = L(L+1) u_{LM} \qquad ...(7.23\ c)$$

i.e., u_{LM} are the eigen functions of orbital angular momentum operators.

(*iii*) u_{LM} also satisfy the equation:

$$u_{LM}(-\mathbf{r}) = (-1)^L u_{LM} (\mathbf{r}) \qquad ...(7.24)$$

Thus u_{LM} will have negative parity, if L is odd integer and positive parity if L is even integer.

(B) Also one can express **E** and **H** in terms of a single parameter **A**, the Vector potential as:

$$\mathbf{E} = -\frac{1}{c}\frac{\partial \mathbf{A}}{\partial t} \qquad \text{...(7.25 } a\text{)}$$

and

$$\mathbf{H} = \nabla \times \mathbf{A} \qquad \text{...(7.25 } b\text{)}$$

which obey, like Eq. 7.21 the vector Helmholtz equation, *i.e.*,

$$\nabla^2\mathbf{A} + k^2\mathbf{A} = 0 \qquad \text{...(7.26 } a\text{)}$$

and

$$\nabla \cdot A = 0 \qquad \text{...(7.26 } b\text{)}$$

It is possible to express the standing wave solutions for the divergent-less vector equation [Eq. 7.26 *a*]; with the boundary condition[2] $\mathbf{A}_{\tan g}(R_0) = 0$; in terms of solution u_{LM}, as expressed in Eq. 7.21.

As a matter of fact, it can be proved, that:

$$\mathbf{A}_{LM}^{E} = kC_L\, \nabla \times \mathbf{L}u_{LM} \qquad \text{...(7.27 } a\text{)}$$

and

$$\mathbf{A}_{LM}^{M} = iC_L\, \mathbf{L}u_{LM} \qquad \text{...(7.27 } b\text{)}$$

as will be shown below.

We have already seen, from Eq. 7.22 *a* that:

$$u_{LM}\, \alpha\, j_L(kr)\, Y_{LM}(\theta, \phi)$$

We will, hence, prove that:

$$\mathbf{A}_{LM}^{E} = \frac{1}{k^2}\nabla \times \{f_{LM}^{E}(kr)\, X_{LM}(\theta, \phi)\} \qquad \text{...(7.28 } a\text{)}$$

and

$$\mathbf{A}_{LM}^{M} = \frac{-i\sqrt{\mu \in}}{k} f_{LM}^{M}(kr)\, X_{LM}(\theta, \phi) \qquad \text{...(7.28 } b\text{)}$$

where

$$\mathbf{X}_{LM} = \frac{1}{\sqrt{L(L+1)}} L\, Y_{LM}(\theta, \phi) \qquad \text{...(7.28 } c\text{)}$$

This leads to Eqs. 7.27 *a* and 7.27 *b*. Because of the similarities of Eqs. 7.21 and 7.26 *a*, one may presume that each component of **E** and **B** is a solution of the scalar Helmholtz wave-equation. It may be seen that though Eq. 7.26 *a* only contains vector potential **A** it is well known that **A** and electric field (**E**) and magnetic field (**B**) are related to each other by:

$$\mathbf{E}(\mathbf{r}, t) = -\frac{1}{[c]}\frac{\partial \mathbf{A}}{\partial t}(\mathbf{r}, t) \qquad \text{...(7.29 } a\text{)}$$

and

$$\mathbf{B}(\mathbf{r}, t) = \nabla \times \mathbf{A}(\mathbf{r}, t) \qquad \text{...(7.29 } b\text{)}$$

which can be written after eliminating the time dependence in Eqs. 7.5 and 7.6:

$$\mathbf{E}(\mathbf{r}) = \frac{ik}{\sqrt{\mu \in}}\mathbf{A}(\mathbf{r}) \qquad \text{...(7.30 } a\text{)}$$

and

$$\mathbf{B}(\mathbf{r}) = \nabla \times \mathbf{A}\;(r) \qquad \text{...(7.30 } b\text{)}$$

Keeping the above arguments, in mind, electric and magnetic fields, can be written from Eqs. 7.25, 7.26 and 7.30 in the following format in terms of spherical Hankel functions multiplied by the spherical, harmonics Y^L_{LM} (θ, ϕ), as:

$$\mathbf{E}^M(\mathbf{r}) = \sum_{L=0}^{\infty} \sum_{M=-L}^{L} \left[\mathbf{e}^{(1)}_{LM}\, h^{(1)}_L(kr) + \mathbf{e}^{(2)}_{LM} h^{(2)}_L(kr)\right] Y_{LM}(\theta, \phi) \qquad \text{...(7.31 } a)$$

$$\mathbf{B}^E(\mathbf{r}) = \sum_{L=0}^{\infty} \sum_{M=-L}^{L} \left[\mathbf{b}^{(1)}_{LM}\, h^{(1)}_{LM}(kr) + \mathbf{b}^{(2)}_{LM} h^{(2)}_L(kr)\right] Y_{LM}(\theta, \phi) \qquad \text{...(7.31 } b)$$

where $$h^{(1)}_L(kr) \equiv j_L(kr) + i\eta_L(kr)$$

and $$h^{(2)}_L(kr) \equiv j_L(kr) - i\eta_L(kr) \qquad \text{...(7.31 } c)$$

where $\mathbf{e}^{(1,2)}_{LM}$ and $\mathbf{b}^{(1,2)}_{LM}$ are vectors to be determined by boundary conditions.

Then, the divergence relations in (7.3) and (7.4) have to be satisfied, *i.e.*,

$$\nabla \cdot \mathbf{E}^M(\mathbf{r}) = \sum_{LM} \sum_{i=1}^{2} \left[\nabla \cdot \left\{\mathbf{e}^{(i)}_L\, h^i_L(kr)\, Y_{LM}(\theta, \phi)\right\}\right] = 0 \qquad \text{...(7.32 } a)$$

and $$\nabla \cdot \mathbf{B}^E(\mathbf{r}) = \sum_{LM} \sum_{i=1}^{2} \left[\nabla \cdot \left\{\mathbf{b}^{(i)}_L\, h^i_L(kr)\, Y_{LM}(\theta, \phi)\right\}\right] = 0 \qquad \text{...(7.32 } b)$$

Also the transversality conditions Eqs. 7.10 *a* and 7.10 *b* have to be obeyed, *i.e.*,

$$\mathbf{r} \cdot \mathbf{E} = \mathbf{r} \cdot \sum_{LM} \mathbf{e}^{(i)}_{LM}\, Y_{LM} = 0 \qquad \text{...(7.33 } a)$$

$$\mathbf{r} \cdot \mathbf{B} = \mathbf{r} \sum_{LM} \mathbf{b}^{(i)}_{LM}\, Y_{LM} = 0 \qquad \text{...(7.33 } b)$$

We now use the definition of orbital angular momentum operator:

$$\mathbf{L} = \frac{1}{\hbar}(\mathbf{r} \times \mathbf{p}) = -i\,(\mathbf{r} \times \nabla)$$

$$\nabla = \mathbf{r}\,\frac{1}{r}\frac{\partial}{\partial r} - \frac{i}{r^2}\,\mathbf{r} \times \mathbf{L} \qquad \text{...(7.34)}$$

Combining (7.34) and (7.32 *b*), we write:

$$\left[\mathbf{r} \cdot \sum_{LM} \frac{\partial h^{(i)}\,\mathrm{L}}{\partial r}\,\mathbf{b}^{(i)}_{LM}\, Y_{LM} - \sum_{LM} \frac{i}{r}\, h^{(1)}_L \mathbf{r} \cdot \mathbf{L} \times \mathbf{b}^{(i)}_{LM}\, Y_{LM}\right] = 0 \qquad \text{...(7.35)}$$

Consequently, from Eqs. 7.32 to 7.34, we write:

$$\mathbf{r} \cdot \left[\mathbf{L} \times \sum_M \mathbf{e}^{(i)}\, Y_{LM}\right] h_L^{(i)}(kr) = 0 \qquad \text{...(7.36 } a\text{)}$$

$$\mathbf{r} \cdot \left[\mathbf{L} \times \sum_M \mathbf{b}_{LM}^{(i)}\, Y_{LM}\right] h_L^{(i)}(kr) = 0 \qquad \text{...(7.36 } b\text{)}$$

From Eqs. 7.33 to 7.36, we can write:

$$\sum_M \mathbf{e}^{(i)}\, Y_{LM} = \sum_M c_{LM}^{(i)}\, \mathbf{L}Y_{LM} \qquad \text{...(7.37 } a\text{)}$$

and

$$\sum_M \mathbf{b}^{(i)}\, Y_{LM} = \sum_M d_{LM}^{(i)}\, \mathbf{L}Y_{LM} \qquad \text{...(7.37 } b\text{)}$$

where $c_{LM}^{(i)}$ and $d_{LM}^{(i)}$ are constants. Eq. 7.37 satisfies Eq. 7.36, where we have used the relationships $\mathbf{r} \cdot \mathbf{L} = 0$ and the commutations relation $\mathbf{L} \times \mathbf{L} = i\mathbf{L}$. It should be realised that Lx and Ly operating on the function Y_{LM} give linear combination of $Y_{L,\,M+1}$ and $Y_{L,\,M-1}$ and hence summing over M is required in Eq. 7.37.

Substituting Eq. 7.37 in Eq. 7.31, we get:

$$\mathbf{E}^M(\mathbf{r}) = \sum_{LM} \mathbf{E}_{LM}^M = \sum_{L,\,M,\,i} c_{LM}^{(i)} h_L^{(i)}\, \mathbf{L}Y_{LM}(\theta, \phi) \qquad \text{...(7.38 } a\text{)}$$

$$\text{and} \quad \mathbf{B}^E(r) = \sum_{LM} \mathbf{B}_{LM}^E = \sum_{L,\,M,\,i} d_{LM}^{(i)} h_L^{(i)}\, \mathbf{L}Y_{LM}(\theta, \phi) \qquad \text{...(7.38 } b\text{)}$$

Equation 7.38 represent the most general solutions for $\mathbf{E}^M$ and $\mathbf{B}^E$.

The corresponding fields $\mathbf{B}^M$ and $\mathbf{E}^E$ are defined from 7.9 *b* and 7.8, respectively.

It is easy to see now, that one can write from Eq. 7.38 for (*E*-type) radiation:

$$\mathbf{B}_L^E(\mathbf{r}) = \sum_M f_{LM}^E(kr)\, \mathbf{X}_{LM}(\theta, \phi) \qquad \text{...(7.39 } a\text{)}$$

$$\mathbf{E}_L^E(\mathbf{r}) = \frac{i}{\sqrt{\mu \in} k} (\nabla \times \mathbf{B}_L^E) \qquad \text{...(7.39 } b\text{)}$$

and for (*M*-type) radiation:

$$\mathbf{E}_L^M(\mathbf{r}) = \sum_M f_{LM}^E(kr)\, \mathbf{X}_{LM}(\theta, \phi) \qquad \text{...(7.40 } a\text{)}$$

and

$$\mathbf{B}_L^M(\mathbf{r}) = \frac{-i\sqrt{\mu \in}}{k} (\nabla \times \mathbf{E}_L^M) \qquad \text{...(7.40 } b\text{)}$$

where $\mathbf{X}_{LM}(\theta, \phi)$ is defined in Eq. 7.28 *c* and $f^E_{LM}(kr)$ and $f^M_{LM}(kr)$ have the same meaning as in Eqs. 7.28 *a* and 7.28 *b* and are defined as:

$$f^{E,M}_{LM}(kr) = a^{E,M(1)}_{LM}\, h^{(1)}_L(kr) + a^{E,M(2)}_{LM}\, h^{(2)}_L(kr) \qquad ...(7.41)$$

where *a*'s in Eq. 7.41 are related to $c^{(i)}_{LM}$ and $d^{(i)}_{LM}$ of Eq. 7.38.

Now using the relationships 7.30 and the above relations, from Eqs. 7.39 to 7.41, it is easy to see the validity of Eq. 7.28.

We have thus proved that *L*, from Eqs. 7.23 and 7.24 have the same meaning in scalar Helmholtz Equation 7.21 ; as in the divergentless vector Equation 7.29 *a*. As a matter of fact, A^{σ}_{LM} [Eq. 7.27] have properties, similar to u_{LM}[Eq. 7.22].

(C) Similar to solution U_{LM} of the scalar Helmholtz equation [Eq. 7.21]; the solution A^{σ}_{LM} of Eq. 7.26 *a* (where σ denotes electric (*E*) or magnetic (*M*) solutions) also are (*i*) complete set of non-singular solutions of vector Helmholtz equation, Eq. 7.26. (*ii*) They are eigen-vectors of angular momentum operators; each of the three components of A^{σ}_{LM} is an eigen function.

There is, however, one important difference between the scalar function u_{LM} and the vector function A^{σ}_{LM}. The vector function A^{σ}_{LM} is an eigen function of total angular momentum; and symbol *L* attached to A^{σ}_{LM} is the total angular momentum, which is generally written as *J*, for a particle. While in a particle:

$$\mathbf{J} = \mathbf{L} + \mathbf{S} \qquad ...(7.42)$$

where **L**, the orbital angular momentum is a part of the space properties; the **S**, the spin angular momentum is due to the internal structure of the particle. On the other hand, the spin angular momentum of the vector electromagnetic field is due to the space properties of the vector function and not due to internal structure of the photon. Hence both **S** and **L** are part of the space properties of the vector function representing the electromagnetic field. So one writes, for the electromagnetic field:

$$\mathbf{J}^2\, A^{\sigma}_{LM} = J(J+1)A^{\sigma}_{(J,M,r)} \equiv L(L+1)\, A^{\sigma}_{LM} \qquad ...(7.43\ a)$$

and

$$J_z A^{\sigma}_{LM} = MA^{\sigma}_{LM} \qquad ...(7.43\ b)$$

Also

$$\mathbf{S}^2\, A^{\sigma}_{LM} = S(S+1)A^{\sigma}_{LM} = 2A^{\sigma}_{LM} \qquad ...(7.43\ c)$$

Further one can write:

$$S_z\mathbf{A} = i\,\hat{\epsilon}_z \times \mathbf{A} \qquad ...(7.44\ a)$$

Here $\hat{\epsilon}_z$ is a unit vector in the *z*-direction while operator S_z corresponds to the component of intrinsic spin in the *z*-direction. Similarly we can write:

$$S_x\mathbf{A} = i\hat{\epsilon}_x \times \mathbf{A} \qquad ...(7.44\ b)$$

and

$$S_y\mathbf{A} = i\hat{\epsilon}_y \times \mathbf{A} \qquad ...(7.44\ c)$$

where S_x and S_y are the spin components in x and y directions and $\hat{\epsilon}_x$ and $\hat{\epsilon}_y$ are unit vectors in the x and y directions, respectively. Conventionally for electromagnetic radiation, the total angular momentum is represented as **L**, the minimum value for which could be one, as the free photon (with no orbital angular momentum) carries a total angular momentum of $1\hbar$. The total angular momenta of photons, of course, can be 1, 2, 3, ..., depending on the value of orbital angular momentum. The orbital angular momentum component is also represented by **L** and obeys the relationship,

$$\mathbf{L} \times \mathbf{L} = i\mathbf{L} \text{ (Here } \mathbf{L} \text{ is orbital angular momentum operator)} \quad \text{...(7.44 } d\text{)}$$

and should be distinguished from the total angular momentum.

In Eq. 7.43 we have designated **L** as the total angular momentum carried by the field A_{LM}; with M as it z-component. These are the traditional nomenclatures.

One can write:

$$\mathbf{A}(r) = \sum_{J=0}^{\infty} \sum_{M=-J}^{J} \mathbf{A}(J, M, r) \quad \text{...(7.45 } a\text{)}$$

where **A**(J, M, r) is given by:

$$\mathbf{A}(J, M, r) = \frac{1}{r}\{f(J, M, r)\,\mathbf{X}_{JM} + g(J, M, r)\,\mathbf{Y}^{M}_{J, J+1} + h(J, M, r)\,\mathbf{Y}^{M}_{J, J-1}\} \quad \text{...(7.45 } b\text{)}$$

where f, g, h are the functions of radial coordinates r only and the three terms correspond to $L = J$; $L = J + 1$, and $L = J - 1$ respectively, where L is orbital angular momentum here, and J is the total angular momentum. The first term in Eq. 7.45 b has angular dependence given by:

$$\mathbf{X}_{JM}\ (\theta, \phi) \equiv \mathbf{Y}^{M}_{J, J=L}\ (\theta, \phi) = \frac{\mathbf{L}\, Y_{JM}(\theta, \phi)}{\sqrt{J\,(J+1)}} \quad \text{...(7.45 } c\text{)}$$

where **L** is the differential operator $-i\mathbf{r} \times \nabla$ and Y_{JM} are scalar spherical harmonics. The parity of the first term, *i.e.* X_{JM} in Eq. 7.45 is $(-1)^J = (-1)^L$ which corresponds to $J \equiv L$. Here L stands for spatial angular momentum; as well as for J. For other two terms, the parity of $Y_{J,J\pm1,1}$ is given by $(-1)^{J\pm1} = -(-1)^J$ and is normally expressed in the nomenclature of L as $-(-1)^L$, where L, here, means total angular momentum. Further from Eq. 7.22; the two solution A^E_{LM} and A^M_{LM} have opposite parity namely $(-1)^{L-1} = -(-1)^L$ and $(-1)^L$ respectively. Hence the first-term in (7.45 b) corresponds to magnetic vector potential A^M_{LM} and other two terms correspond to electric vector potential A^E_{LM}. Further as we will see subsequently, an EL transition carries $(-1)^L$ parity and ML transition carries $-(-1)^L$. In other words, the parity of radiation field itself is opposite to the parity of the vector potential. Also it may be realised that $|J_a - J_b| \le L \le J_a + J_b$, where J_a and J_b are the total angular momenta of the initial and final states of the nucleus, involved in the transition, and L is the total angular momentum carried by the radiation.

7.4 QUANTUM MECHANICAL TREATMENT OF TRANSITION PROBABILITIES[3]

We now, derive the expression for the probability of an electromagnetic transition between two nuclear states i (initial) and f (final) quantum-mechanically. According to the perturbation theory (*see* Quantum

Mechanics by L.I. Schiff, page 189)[3]; the transition probability W_{fi} between any two states i and f in a quantum mechanical system is given by, Ferm's golden rule:

$$W_{fi} = \frac{2\pi}{\hbar} \rho_f \left| \langle \psi_f \mid H' \mid \psi_i \rangle \right|^2 \qquad ...(7.46\ a)$$

where ρ_f = is the density of the final states given by $\rho_f = (dN/dE)$ *i.e.,* the number of states per unit energy, near the final state; H' is the perturbation Hamiltonian, which in our case will be the interaction Hamiltonian due to electromagnetic interaction between charged particles and E.M. field in the nucleus, *i.e.,* protons and the electromagnetic field of the emitted gamma ray.

The non-relativistic interactions Hamiltonian H' for an electromagnetic field of vector potential A and nucleons of charge e_j and magnetic moment μ_j $(e\hbar/2M_pc)$ is given by:

$$H' = -\sum_j \frac{e}{M_pc} \frac{\mathbf{p}_j \cdot \mathbf{A} + \mathbf{A} \cdot \mathbf{p}_j}{2} - \frac{e\hbar}{2M_pc} \sum_j \mu_j \sigma_j \cdot \mathbf{H} \qquad ...(7.46\ b)$$

where M_p is the mass of proton, and $e_j = e$ for proton and zero for neutron; $\mu_j = \mu_p$ for proton, and $\mu_j = \mu_n$ for neutron; **H** is the magnetic field connected with eloctromagnetic radiation, σ is the Pauli Spin Vector, and $\mathbf{p}_j$ is the momentum operator $- i\hbar\,(\text{grad})_j$, of the jth nucleon. Of course **A** is the vector potential operator associated with the electromagnetic radiation. In practice, if only one proton takes part in the transition, then Eq. 7.46 *b* is given by:

$$H' = -\frac{e}{M_pc} \frac{\mathbf{p} \cdot \mathbf{A} + \mathbf{A} \cdot \mathbf{p}}{2} - \mu_p \frac{e\hbar}{2M_pc} \sigma \cdot \mathbf{H} \qquad ...(7.46\ c)$$

Then the matrix element of H' is to be taken for single quantum emission between two states in such a way that the initial state contains a proton in state i and no quantum of e.m. field while the final state f contains a proton and one quantum of e.m. field in the state.

It is possible (in the language of quantum mechanics) to expand the Hermitian operator A, as a linear combination of eigen-solutions of Equations 7.26, which correspond to the orthogonal waves.

Then

$$A = \sum_\lambda \{a_\lambda\, A_\lambda(r) + a_\lambda^*\, A_\lambda^*(r)\} \qquad ...(7.47)$$

where λ denotes the various quantum numbers associated with these waves and a_λ and a_λ^* are the annihilation and creation operators for the photon which decrease or increase the number of photons respectively, in the system by one. It is well known, (*see* Quantum Mechanics by L.I. Schiff, pages 340–347), that the matrix elements of annihilation and creation operators a_λ and a_λ^*, between photon states $| n_1 ..., n_k ....>$ and $| n'_1 ..., n'_k ...>$; where n_k and n'_k specify the number of states k and k' are given by:

$$\langle n'_i ... n'_k ... | a_\lambda^* | n_1 ..., n_k ... \rangle$$

$$= \left(n_k + \frac{1}{2}\right)^{1/2} \delta_{n'_1 n_1} ... \delta_{n'_k\, n_{k+1}} ... \qquad ...(7.48\ a)$$

and $$\langle n'_i \ldots n'_k \ldots \mid a_\lambda \mid n_1 \ldots, n_k \ldots\rangle$$

$$= (n_k)^{1/2}\, \delta_{n'_1 n} \ldots \delta_{n'_k n_{k-1}} \ldots \qquad \text{...(7.48 } b)$$

Hence using Eqs. 7.47 and 7.48 and realising that the field amplitudes a_λ (and a_λ^*) satisfy the harmonic oscillator equation:

$$\frac{d^2 a_\lambda}{dt^2} + \omega_\lambda^2\, a_\lambda = 0 \qquad \text{...(7.48 } c)$$

we can write the matrix element of $H'(A)$ for single quantum emission between final state and one quantum in state λ and initial state and no quanta and hence,

$$\left\langle f, 1_\lambda \mid H'(A) \mid i, 0 \right\rangle = \left(\frac{\hbar}{2\omega}\right)^{1/2} \left\langle f \right| H'(A_\lambda^*) \left| i \right\rangle \qquad \text{...(7.48 } d)$$

where $$\omega = \frac{(E_1 - E_f)}{\hbar}$$

and commutation relation is obeyed, *i.e.*,

$$a_\lambda\, a_\lambda^* - a_{\lambda'}^*\, a_\lambda = -\left(\frac{\hbar}{2\omega_k}\right)\delta_{\lambda\lambda'} \qquad \text{...(7.49)}$$

From the above discussion, we can, therefore, write from Eqs. 7.46 and 7.48:

$$W_{fi} = \frac{\pi}{\omega}\left(\frac{dN}{dE}\right)_f \left|\left\langle f \right| H'\left(A_\lambda^*\right) \left| i \right\rangle\right|^2 \qquad \text{...(7.50 } a)$$

To determine $(dN/dE)_f$; we consider a sphere of radius R. The allowed values of k in the sphere are governed by the condition, that there are standing multiple waves; *i.e.*, $k = N\pi/R$; hence; $dN = (R/\pi)dk$.

From this, it is easy to see that:

$$\frac{dN}{dE} = \frac{R}{\pi\hbar c} \qquad \text{...(7.50 } b)$$

We, now, use the relations in Eq. 7.28 for $\mathbf{A}_{LM}^E$ and $\mathbf{A}_{LM}^M$, keeping in mind that:

$$f_{lM}^E(kr) = C^E\, j_l(kr) \qquad \text{...(7.50 } c)$$

and $$f_{lM}^M(kr) = C^M\, j_l(kr) \qquad \text{...(7.50 } d)$$

where C^E and C^M are determined from the condition that the energy of a photon confined to a sphere of radius R is given by $\hbar\omega = \hbar c R$.

Then $$C^E = \sqrt{\frac{(4\pi)\,\mu c\hbar k^3}{R}} \text{ and } C^M = k^2\sqrt{\frac{(4\pi)\, c\hbar k^3}{\in R}}$$

Hence from Eqs. 7.50, 7.46 and 7.28, we obtain:

$$H'^{E}_{LM} \equiv H'(A^{*E}_{LM}) = -\sqrt{\frac{(4\pi)\mu c\hbar}{kR}} \sum_j \left[\frac{e_j}{M_p c} \mathbf{p}_j \cdot (\nabla \times j_L(kr)\,\mathbf{X}_{LM})^* + \frac{\hbar e k^2}{2M_p c} \mu_j \cdot \sigma_j (j_L(kr)\,X_{LM})^* \right] \quad ...(7.51\ a)$$

and

$$H'^{M}_{LM} \equiv H'(A^{*M}_{LM}) = ik \sqrt{\frac{(4\pi)\mu c\hbar}{kR}} \sum_j \left[\frac{e_j}{M_p c} \mathbf{p}_j \cdot (j_L(kr)\,X_{LM})^* + \frac{\hbar e}{2M_p c} \mu_j \cdot \sigma_j (\nabla \times j_L(kr)\,\mathbf{X}_{LM})^* \right] \quad ...(7.51\ b)$$

Equation 7.51 may be used to obtain W_{fi} [Eq. 7.50 *a*], by realising that for $kr << 1$, and hence,

$$j_1(kr) \approx \frac{(kr)^L}{[(2L+1)!!]} \quad ...(7.52\ a)$$

and

$$\mathbf{X}_{LM} = \frac{1}{\sqrt{L(L+1)}} \mathbf{L}\, Y_{LM}(\theta, \phi) \quad ...(7.52\ b)$$

Keeping in mind Eqs. 7.28 *c* and 7.52 and that:

$$(\nabla \times j_L(kr)\,\mathbf{X}_{LM})^* \approx i\sqrt{\frac{(L+1)}{L}} (\text{grad}\, j_L(kr)\, Y^*_{LM}) \quad ...(7.53)$$

and

$$(\nabla \times \mathbf{L}\, j_L(kr)\, Y_{LM})^* = i[r\,\nabla^2 j_L(kr)\, Y_{LM} - (\mathbf{r} \times \mathbf{grad} + 2) \cdot (\mathbf{grad}\, j_L(kr))\, Y_{LM}]^* \quad ...(7.54)$$

and

$$\left\langle f \right| \mathbf{p}_j \cdot (\mathbf{grad}\, j_L(kr)\, Y_{LM})^* \left| i \right\rangle$$

$$= \frac{-iM_p}{\hbar} \left\langle f \left| \frac{p^2}{2M_p} j_L(kr)\, Y^*_{LM} - j_L(kr)\, Y^*_{LM} \frac{p^2}{2M_p} \right| i \right\rangle$$

$$= -\,iM_p ck \left\langle f \right| j_L(kr)\, Y^*_{LM} \left| i \right\rangle \quad ...(7.55)$$

Realising that $\omega = vk = \dfrac{c}{\sqrt{\mu \in}} k$

we obtain, using Eqs. 7.51 to 7.55:

$$\left\langle f \right| H'\left(A^{E,M}_{L,M}\right) \left| i \right\rangle = C_L \frac{L+1}{(2L+1)!!} \left(\frac{\omega}{c}\right)^L \left\langle f \right| M^{E,M}_{L,M} \left| i \right\rangle \quad ...(7.56\ a)$$

where $$C_L = \left[\frac{8\pi\omega^2}{L(L+1)R}\right]^{1/2} \qquad \text{...(7.56 b)}$$

and $$M_{LM}^E = \sum_j e_j\, r^L Y_{LM}^* - \frac{i\mu_j\, e\hbar k}{2M_p c}(L+1)^{-1}\,(\sigma_j \times \mathbf{r}).[\mathbf{grad}\,(r^L\, Y_{LM}^*)] \qquad \text{...(7.57 a)}$$

$$M_{LM}^M = \sqrt{\mu \in} \sum_j \left\{\frac{e_j\hbar}{M_p c}\frac{1}{(L+1)}\mathbf{L}\cdot\left[\mathbf{grad}\,(r^L\, Y_{LM})\right]^* + \frac{e_j\hbar}{2M_p c}\mu_j\sigma_j\,.\left[\mathbf{grad}\,(r^L\, Y_{LM})\right]^*\right\}$$

...(7.57 b)

Physically M_{LM}^E represents the electric multipole moment operators, corresponding to the sum of the two terms, representing the orbital motion of the charge (ej) (first term) and the contribution arising from the change in the distributions of the intrinsic magnetic moment (μj) classically (second term). As a matter of fact, these term are referred to us Q_{LM} and Q'_{LM} respectively in literature, so that one can represent

$$\left\langle M_{LM}^E\right\rangle \equiv (Q_{LM} + Q'_{LM}) \qquad \text{...(7.58 a)}$$

where $$Q_{LM} \equiv \sum_j e_j \left\langle f \middle| r^L\, Y_{LM}^*\,(\theta,\phi) \middle| i\right\rangle \qquad \text{...(7.58 b)}$$

and $$Q'_{LM} \equiv -\frac{ie\hbar k}{2M_p c(L+1)}\left\{\sum_j \left\langle f \middle| \mu_j\, \sigma_j \times \mathbf{r}_j\,.\left[\mathbf{grad}\,(r^L\, Y_{LM})\right]^* \middle| i\right\rangle\right\} \qquad \text{...(7.58 c)}$$

Similarly, M_{LM}^M represents the magnetic multipole moment operators; the first term coming from orbital motion of charges and second term coming from intrinsic magnetic moment; so that one can write $\left\langle M_{LM}^M\right\rangle$ as:

$$\left\langle M_{LM}^M\right\rangle \equiv (M_{LM} + M'_{LM}) \qquad \text{...(7.59 a)}$$

where $$M_{LM} \equiv \sum_j \frac{e_j\hbar\sqrt{\mu \in}}{M_p c(L+1)}\left\langle f \middle| \mathbf{L}\cdot\left[\mathbf{grad}\,(r^L\, Y_{LM})\right]^* \middle| i\right\rangle \qquad \text{...(7.59 b)}$$

and $$M'_{LM} \equiv \sqrt{\mu \in}\left[\sum_j \frac{e_j\hbar}{2M_p c}\mu_j \left\langle f \middle| \sigma\cdot\left[\mathbf{grad}\,(r^L\, Y_{LM})\right]^* \middle| i\right\rangle\right] \qquad \text{...(7.59 c)}$$

One can, now, write the expression for W_{LM}^E, and W_{LM}^M the transition probabilities for electric and magnetic radiations using Eqs. 7.50, 7.56, 7.58 and 7.59.

For Electric Multipole Radiation:

$$W_{fi}^{E}(L, M) = \frac{8\pi}{\in} \frac{(L+1)}{L[(2L+1)!!]^2} \frac{1}{\hbar}\left(\frac{\omega}{c}\right)^{2L+1} |Q_{LM} + Q'_{LM}|^2 \quad \text{...(7.60 a)}$$

and for Magnetic Multipole Radiation:

$$W_{fi}^{M}(L, M) = (8\pi\mu)\frac{(L+1)}{L[(2L+1)!!]^2} \frac{1}{\hbar}\left(\frac{\omega}{c}\right)^{2L+1} |M_{LM} + M'_{LM}|^2 \quad \text{...(7.60 b)}$$

for vacuum $\in = \mu = 1$.

[Equation 7.60 was derived by Blatt and Weisskopf[2] earlier, in 'Theoretical Nuclear Physics', Chapter XII]

To understand, the practical application of Eqs. 7.57 to 7.60; we calculate the dipole moment ($L = 1$) operators, for electric and magnetic transitions; for the change in orbital motion and distribution of intrinsic magnetic moment of a proton, respectively. Then we do not sum over j but only take one term for $L = 1$. Using μ_p for the nuclear magnetic moment in magnetons and L_{p_z} for the z-projection of the proton orbital angular momentum, we write; for $M = 0$, $L = 1$, from Eqs. 7.58 and 7.59:

$$\left\langle M_{10}^{M}\right\rangle = \sqrt{\frac{3}{4\pi}}\, ez - i\sqrt{\frac{3}{4\pi}}\, \mu_p \frac{e\hbar}{4M_p c}\, \omega\, (\sigma_p \times \mathbf{r})_z \quad \text{...(7.61 a)}$$

and

$$\left\langle M_{10}^{M}\right\rangle = \sqrt{\mu\varepsilon}\sqrt{\frac{3}{4\pi}} \frac{e\hbar}{2M_p c}(L_{p_z} + \mu_p\, \sigma_{pz}) \quad \text{...(7.61 b)}$$

From Eqs. 7.57 to 7.60, it is apparent that the first term in the expressions of $M_{LM}^{(EM)}$ arise from the orbital motion of charges, and the second term arises due to contribution from the change in the distribution of the intrinsic magnetic moment. It should, however, be further noted that the first term in the electric transition is dominant; *i.e.*, the change in the position or the oscillation of the electric charge contributes mainly to transitions. This also explains the name 'electric' transition. Similarly for the magnetic transitions, the second term is dominant, *i.e.*, the change in the distribution of the magnetic moment of proton, contributes mainly in this case, thus explaining the name 'magnetic' transition.

Selection Rules: We have, already discussed the selection rules for the change of parity, on the basis of classical theory of electromagnetic transition. We discuss now, on the basis of the expression of transition probability derived quantum mechanically; [Eqs. 7.57 to 7.60] the comprehensive selection rules of angular momentum and parity, for electric and magnetic transitions.

It is evident from Eq. 7.60; that higher the value of L, the lesser the transition probability, because a crude estimate yields, the magnitude of multipole operators:

$$M_{LM}^{E} \approx [ea^L] \quad \text{...(7.62 a)}$$

and

$$M_{LM}^{M} \approx \left[\left(\frac{e\hbar}{Mca}\right)(ea^L)\right] \quad \text{...(7.62 b)}$$

from which the transition probabilities W_{fi} are given approximately,

$$W_{fi}^{E} \approx \left[\omega\left(\frac{e^2}{\hbar c}\right)\left(\frac{\omega a}{c}\right)^{2L}\right] \qquad \text{...(7.63 } a\text{)}$$

and

$$W_{fi}^{M} \approx \left[\omega\left(\frac{e^2}{\hbar c}\right)\left(\frac{\hbar}{M}\right)\left(\frac{\omega a}{c}\right)^{2L}\right] \qquad \text{...(7.63 } b\text{)}$$

where a is the radius of the nucleus and hence $a^{2L} << 1$.

For given values of j_i and j_f, the permissible values of L are given by:

$$\left| j_i - j_f \right| \le L \le j_i + j_f \,;\; M = m_i - m_f \qquad \text{...(7.64 } a\text{)}$$

However, the electromagnetic radiative transitions between two states of angular momentum $j_i = j_f = 0$; are absolutely forbidden. In such cases, however, the conversion electrons, may create the transition, as we shall see subsequently. Out of the possibilities of L, [Eq. 7.64] smallest value of L, *i.e.*, $L = | j_i - j_f |$ has the maximum probability, for both electric or magnetic tansitions. Also from the properties of M_{LM}^{E} and M_{LM}^{M} [Eq. 7.57], it can be seen that, their values become zero unless the following selection rules is followed:

$$\left| l_i - l_f \right| \le L \le l_i + l_f \qquad \text{...(7.64 } b\text{)}$$

where l_i and l_f are the orbital angular momenta of initial and final states.

Because of the parity restriction and conservation of parity in the electromagnetic interaction, and for fixed spins and parities of the initial and final states, it may be seen that if the lowest order of transitions carrying angular momentum L is say, electric (or magnetic), then the next order of transition carrying angular momentum $L + 1$ is magnetic (or electric). This can be understood if we realise from Eqs. 7.57 and 7.58 and Eq. 7.61, that:

(*a*) In the case of say $M_{L=1,\,M}^{E}$ (Matrix for electric transition for $L = 1$, *i.e.*, for dipole transition), the three electric dipole matrix elements transform among themselves under space rotation and inversion, through the origin in the same way as the three components of a vector; *i.e.*, they behave like a tensor of rank 1 and of odd parity. Also for the general case, the $2L + 1$ components of M_{LM}^{E} (the electric transition matrix for transition carrying L angular momentum) transform among themselves as the components of a tensor of rank L and parity $(-1)^L$. One can, therefore, express the parity change of the electric transition by:

$$\Delta\pi \equiv \frac{\pi_i}{\pi_f} = (-\,\mathrm{i})^L \text{ for electric transitions} \qquad \text{...(7.65 } a\text{)}$$

(*b*) Similarly in the case of say M_{LM}^{M}, the magnetic dipole matrix elements for $L = 1$ transform as the components of a pseudo-vector; *i.e.*, like a tensor of rank 1, and even parity. In general for M_{LM}^{M}, the components transform like a tensor of rank L and parity $-(-1)^L$, respectively. Hence the parity change of magnetic transition is given by:

$$\Delta\pi \equiv \frac{\pi_i}{\pi_f} = -(-1)^L \text{ for magnetic transitions} \qquad \text{...(7.65 } b\text{)}$$

Hence, it is evident, that if L-transition has electric (or magnetic) nature; the next order of transition, *i.e.*, $L + 1$ transition will have magnetic (or electric) nature. The ratio of intensities between these two orders, will, of course, depend on the nature of the configurations of the initial and final states in the transition.

This ratio of the two matrices is called the mixing ratio and is defined as the ratio of reduced matrix elements, which by definition, are supposed to be independent of the projection quantum numbers m' and m, *i.e.*, writing the reduced matrix elements for L' and L as:

$(j_f \| L' \| j_i)$ and $(j_f \| L \| j_i)$; one defines δ as:

$$\delta \equiv \frac{(j_f \| L' \| j_i)}{(j_f \| L \| j_i)} \qquad ...(7.66\ a)$$

where $L' > L$. The value of L' be $L + 1$, in the above discussion for the next order, after L. However, in principle, still higher orders are also possible for which the values of δ may be determined. The reduced matrix element is defined in such a manner, that the reduced matrix elements are chosen to be real; but the sign of δ will depend on the relative phase of the reduced matrix elements. The ratio of the total intensity (*i.e.*, angle integrated), of the L'-pole to that of L-pole is equal to δ^2. Experimentally, one can measure δ^2, as well as δ, with its sign and then compare them with a nuclear model. Alternatively, the mixing ratio is defined experimentally as:

$$\delta = \pm \sqrt{\frac{\text{Intensity of radiation with } L'}{\text{Intensity of radiation with } L}} \qquad ...(7.66\ b)$$

We will see subsequently, that the angular correlation measurements yield δ, alongwith its sign.

The types of mixtures of multipolarities and the change of parities and angular momenta, which one generally encounters in radiative transitions is summarised below in Table (7.1).

Table 7.1 *Mixture of multipolarities, with the changes in parity and total angular momenta L.*

$\Delta\pi$ \ L	*0*			*1*		2		*3*	
+ 1	(E_0)	M_1	(E_2)	M_1	(E_2)	E_2	(M_3)	M_3	(E_4)
– 1		E_1	(M_2)	E_1	(M_2)	M_2	(E_3)	E_3	(M_4)
		$0 \to 0$	$0 \to 0$		$0 \to 1$		$0 \to 2$		$0 \to 3$
Forbidden	$0 \to 0$		$\frac{1}{2} \to \frac{1}{2}$		$0 \leftarrow 1$		$0 \leftarrow 2$		$0 \leftarrow 3$
	$\frac{1}{2} \to 0,$	$1 \to \frac{1}{2}$							

The values of transitions shown in the brackets, correspond to the higher multipolarity, which may mix with lower multipolarity, shown in the same category.

It may be seen from Table 7.1, that transition between certain values of j_i and j_f are forbidden; because (*i*) the minimum angular momentum which a photon carries is 1 and (*ii*) L has to lie between $|j_i - j_f|$ and $|j_i + j_f|$.

7.5 INTERNAL CONVERSION

7.5.1 Theory of Internal Conversion (Semi-Quantitative)

The phenomenon of the emission of internal conversion electrons is accompanied, by gamma rays emission, whenever there is electromagnetic transition between two nuclear states of a nucleus. As a matter of fact in internal conversion, the number of electrons emitted is proportional to the number of gamma rays and the internal conversion coefficient α is defined as:

$$\alpha = \frac{Ne}{N\gamma} = \frac{\text{Number of conversion electrons}}{\text{Number of gamma rays}}$$

where Ne and $N\gamma$ are both taken for the total emission of conversion electron and gamma rays in 4π solid angle. What is the quantum mechanical mechanism for this process ? It is important to recognise that according to quantum electrodynamics, an electromagnetic interaction which takes place via the retarded potential, can be considered to be due to an exchange of a virtual or a real photon. In the present case, photon is real and corresponds to the gamma ray emission and the internal conversion process may be looked upon as a second order process. In this case, the emission of electron from orbital electrons; can be understood from the following picture: One envisages an initial state in which the nucleus is excited, with energy E and the electron (among the orbiting electrons) is in the ground state. In the final state the electron is in the excited (continuum) state with energy W, above the ground state and the nucleus has lost the energy E and is in the ground state. The number of quanta present is zero both in the initial and final states. The interaction involves a set of intermediate states, in which either the nucleus or electron (but not both) have made transitions to the ground state. In this intermediate state a photon of frequency ω exists. The state in which both particles (electron (1) and nucleus (2)) have their final energy is reached by the mechanism of absorption of the photon by particle 1 (or 2) if it was originally emitted by 2 (or 1). Thus there are actually two sets of intermediate states. The final state in the electron emission, depends on the emission of the gamma ray, which corresponds to the final state in gamma rays emission, but forms the first intermediate state in electron emission. Then for an electron, which is at a distance of more than half the wavelength of the gamma rays from the nucleus; the rate of production of conversion electron is expected to be proportional to the rate of gamma-rays production, a certain fraction of which is absorbed by atomic electrons and does not leave the atom.

Using the well-known quantum mechanical rule for transition probabilities, we can, then define: the conversion coefficient:

$$\alpha_{LM}^{M,E} = \frac{(W_{LM}^{M,E})_{\text{electron}}}{(W_{LM}^{M,E})_{\text{gamma-ray}}} = \frac{\frac{2\pi}{\hbar}\left|\langle f|\ H'_{\text{int}}\ |i\rangle\right|^2 \left|\langle\rho_f\rangle_{\text{electron}}\right.}{\frac{2\pi}{\hbar}\left|\langle f|\ H'_{N_\nu}\ |i\rangle\right|^2 \left|(\rho_f)_{\gamma\text{-ray}}\right.} \quad \text{...(7.68)}$$

where $\langle\rho_f\rangle_{\text{electron}}$ is the density of free electron states per unit energy per unit solid angle. Similarly $\langle\rho_f\rangle_{\gamma\text{-ray}}$ is the density of the final states of the nucleus in gamma rays emission.

We can see from Eq. 7.46 *b*, that the matrix element for the interaction. Hamiltonian, in the case of electromagnetic field, is given by:

$$\left\langle f \left| H'_{N_v} \right| i \right\rangle_{\gamma\text{-rays}} = -\sum_j \frac{e_j}{2M_p c} \left\langle \psi_{jf} \left| \mathbf{p}_j . \mathbf{A}_{LM} + \mathbf{A}_{LM} \cdot \mathbf{p}_j \right| \psi_{ji} \right\rangle$$

$$= -\frac{1}{c} \mathbf{J}_N \cdot \mathbf{A} \qquad \text{...(7.69 } a\text{)}$$

where we have neglected the magnetic moment interaction part and ψ_{jf} and ψ_{ji} are the nuclear wave-functions of the final and initial states, of *j*th nucleon, and have used Eqs. 7.46*b* and 7.47, and definition of nuclear current as:

$$J_N(\mathbf{r}_N) = \sum_j \frac{e_j}{M_p} \psi^*_{jf}(\mathbf{r}_N) \mathbf{p}_j \psi_{ji}(\mathbf{r}_N) \qquad \text{...(7.69 } b\text{)}$$

On the other hand, the interaction Hamiltonian between the nucleus and the atomic electron is given by using the retarded potential as:

$$\left\langle f \left| H'_{\text{int}} \right| i \right\rangle_{\text{electrons}} = -4\pi \int d^3 r_N \int d^3 r_e \left(\frac{\mathbf{J}_N \cdot \mathbf{J}_e}{c^2} - \rho_N \rho_e \right) \times \frac{e^{ik|\mathbf{r}_N - \mathbf{r}_e|}}{4\pi \epsilon_0 |\mathbf{r}_N - \mathbf{r}_e|} \qquad \text{...(7.70)}$$

where $k = (E_f - E_i)/\hbar c$, is the wave number of the intermediate gamma rays and the quantities J_N, J_e and ρ_n and ρ_e are the Dirac current and charge densities for the nucleus and the atomic electrons. Physically this interaction corresponds to the interaction between currents and charges in the nucleus and the atomic electrons.

Physically Eq. 7.69*a* corresponds to the interaction energy due to nuclear electric current density, with the vector potential; as has been described earlier, in Sections 7.4, while discussing electromagnetic transitions.

Equation 7.70 is based on the theory of retarded potentials. By definition, if there are two particles (1) (nucleus) and (2) (electron), with charge and current densities $\rho(r_1, t)$, $\rho(r_2, t)$, $\mathbf{i}\,(r_1, t)$ and $\mathbf{i}\,(r_2, t)$; then the retarded potential produced by the particle 1 at distance r_2 is given as:

$$\phi(r_2, t) = \int \frac{1}{r_{12}} \rho\left(\mathbf{r}_1, t - \frac{r_{12}}{c} \right) d\tau_1$$

$$A(r_2, t) = \frac{1}{c} \int \frac{1}{r_{12}} \mathbf{i} \left(\mathbf{r}_1, t; t - \frac{r_{12}}{c} \right) d\tau_1 \qquad \text{...(7.71)}$$

and the retarded interaction between two particles, then is given by:

$$I_{ret} = \iint \frac{\rho\left(\mathbf{r}_1, t - \frac{r_{12}}{c} \right) \rho(\mathbf{r}_2, t)}{r_{12}} d\tau_1\, d\tau_2 - \iint \frac{\mathbf{i}\left(\mathbf{r}_1, t - \frac{r_{12}}{c} \right) \mathbf{i}(\mathbf{r}_2, t)}{c^2\, r_{12}} d\tau_1\, d\tau_2 \qquad \text{...(7.72)}$$

(*see* Quantum Theory of Radiation[3] by Heitler, p. 233)

Equation 7.70 is basically derived from an equation like 7.72; by using the quantities $\mathbf{J}_N$ and $\mathbf{J}_e$ for Dirac currents, and ρ_N and ρ_e as Dirac charge densities as given by:

$$\mathbf{J}_e = -e\,c\,\psi^*_{ef}(\mathbf{r}_e)\,\alpha\,\psi_{ec}(\mathbf{r}_e)$$

$$\rho_e = -e\psi^*_{ef}(\mathbf{r}_e)\,\psi_{ei}(\mathbf{r}_e) \qquad ...(7.73)$$

where α are Dirac matrices, required for relativistic electrons tightly bound to heavy atom. On the other hand for nuclear particles, their velocities are much less than c and hence Schrödinger expressions may be used, *i.e.*,

$$\rho_N(\mathbf{r}_N) = \sum_{j=1}^{A} e_j\,\psi^*_{jf}(\mathbf{r}_N)\,\psi_{ji}(\mathbf{r}_N)$$

and

$$\mathbf{J}_N(\mathbf{r}_N) = \sum_j \frac{e_j}{M}\,\psi^*_{jf}(\mathbf{r}_N)\,\mathbf{p}\,\psi_{ji}(\mathbf{r}_N) \qquad ...(7.74)$$

Leaving the details of the calculation[2, 5], we produce below, the expressions for internal conversion coefficient for electric transition $(\alpha^E_{LM} \equiv \alpha_{LM})$ and magnetic transitions $(\alpha^M_{LM} \equiv \beta_{LM})$ for point nucleus:

$$\alpha_{LM} = \frac{\pi k\,(4\pi)}{\epsilon_0\,cL\,(L+1)}\left|\int d^3r\left[k\,\mathbf{J}_e\cdot\mathbf{r}_e + ic\,\rho_e\frac{\partial}{\partial r}\right]\times\right.$$

$$\left.\left(r\,h^{(1)}_L(kr)\,Y_{LM}(\Omega)\right)\right|^2 (\rho_f)_{\text{electron}} \qquad ...(7.75)$$

and

$$\beta_{LM} = \pi\mu_0 k\left[\frac{4\pi}{c^2}\right]\times\left|\int_{re<R}\mathbf{J}_e\cdot A^M_{>LM}\,d^3r_e - \right.$$

$$\left.\frac{h^{(1)}_L(kr)}{J_L(kr)}\int_{re<R} J_e\cdot A^M_{<LM}\,d^3r_e\right|^2 (\rho_f)_{\text{electrons}} \qquad ...(7.76)$$

where $R = |\,r_n\,|$ is the value of the radial distance of the nucleon on the surface. In Eq. 7.76 $A^M_{>LM}$ corresponds to the solution outside the charge distribution, and $A^M_{<LM}$ for solution inside the charge distribution.

Rose et al[6], have calculated the values of α_L and β_L, *i.e.,* the values summed over all possible values of M, for different values of L and for different energies (k) and the atomic numbers Z (through the values of $\mathbf{r}_e$ for K, L of M shell orbits) and for the subshells of the electrons in question, *e.g.* L_I, L_{II}, L_{III}, etc. It may be realised, that an actual conversion-coefficient α will occur as a sum of the form:

$$\alpha = \sum_L \left(\alpha_k(L) + \alpha_{L_I}(L) + \alpha_{L_{II}}(L) + \alpha_{L_{III}}(L) + \ldots\right)\delta^2_L \qquad ...(7.77)$$

where δ_L^2 has been defined in Eq. 7.65 and is the relative contribution of the multipoles corresponding to L. Also, many times, it is possible to measure K/L ratio, defined as:

$$\frac{K}{L} = \frac{\alpha_K}{\alpha_L} = \frac{\alpha_K}{\alpha(L_I) + \alpha(L_{II}) + \alpha(L_{III})} \quad ...(7.78)$$

Similarly, terms like

$$\frac{L_I}{L_{II}} = \frac{\alpha_{L_I}}{\alpha_{L_{II}}} \text{ or } \frac{L_I}{L_{III}} = \frac{\alpha_{L_I}}{\alpha_{L_{III}}} \quad ...(7.79)$$

are also measured, where

$$\alpha_L = \sum_{Li} \alpha_{Li}(L)\,\delta_L^2 \quad ...(7.80)$$

where δ_L^2 is specific to $\alpha_{L_I}, \alpha_{L_{II}}$ etc.

Theoretical values of these quantities are now available, which when compared to experimental values, help in determining the L value of the transitions and the mixing ratio δ.

Figures 7.2 to 7.4, summarise some of these calculations for $Z = 85$. For the exact values, detailed tables and the discussions about the various corrections, *see* References (7), (8) and (4).

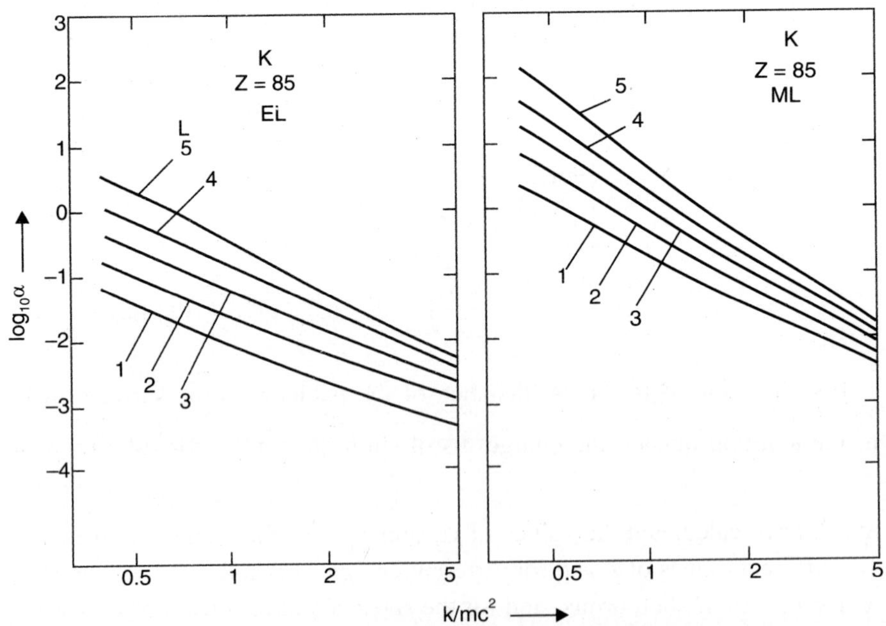

Fig. 7.2 K shell conversion coefficients, for Z = 85, where screening effects have been taken into account (Ref. 8).

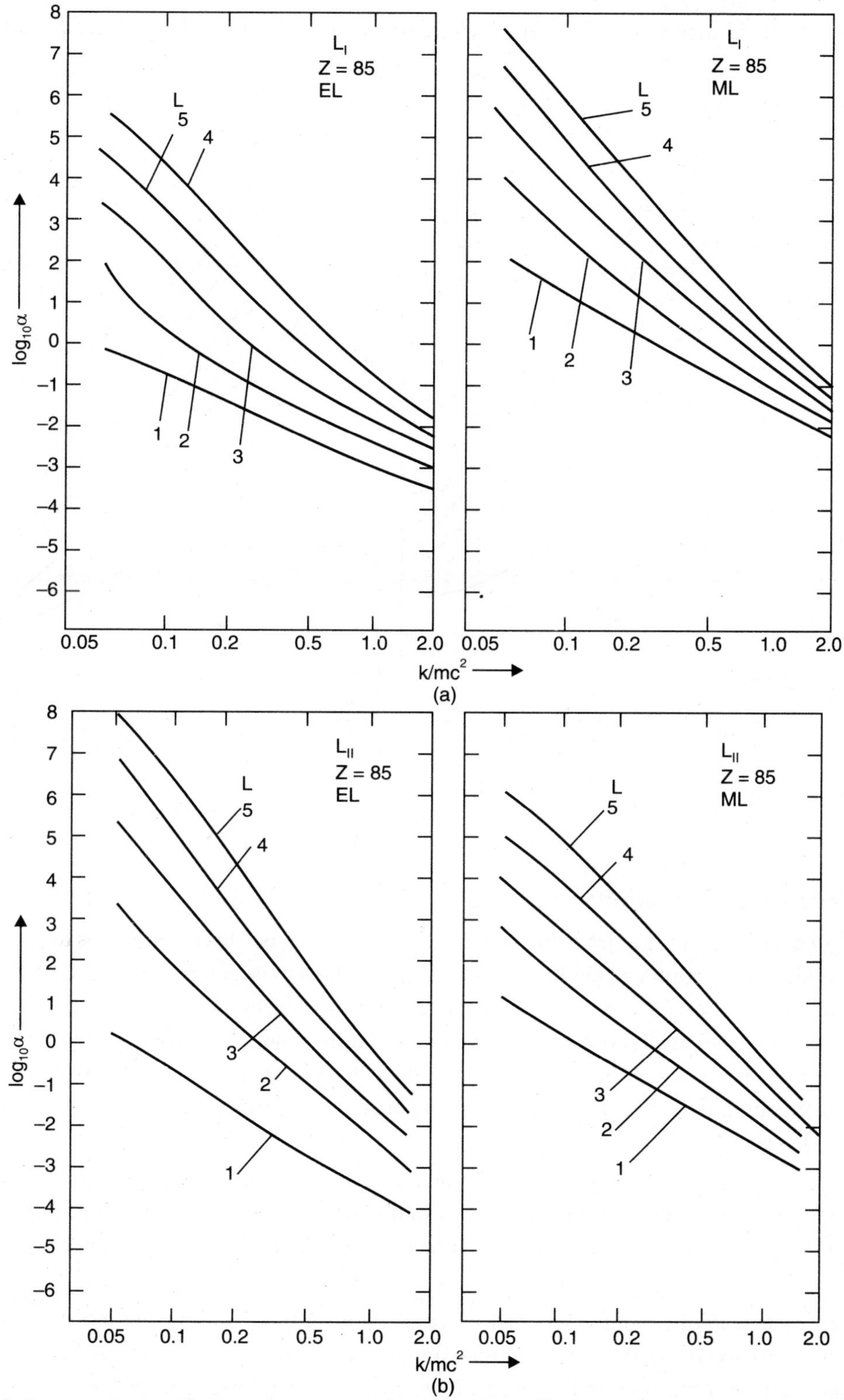

Fig. 7.3 L shell conversion coefficients, for Z = 85, where screening effects have been taken into account (Ref. 8).

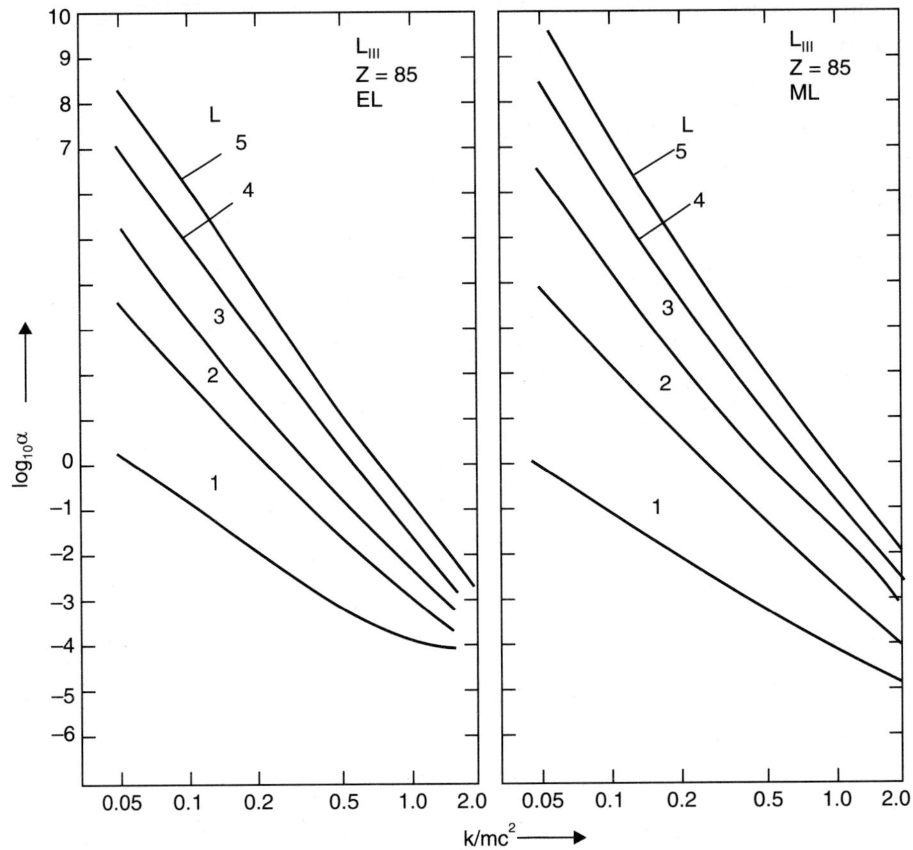

Fig. 7.4 L_{III} shell conversion coefficients, for Z = 85, where screening effects have been taken into account (Ref. 8).

7.5.2 Zero-Zero Transitions

There cannot be any gamma ray transitions, between $0 \to 0$, because this corresponds to $L = 0$; while the intrinsic angular momentum carried by photons is $1\hbar$ and hence no gamma ray can be emitted. But as pointed out, first by Fowler[9], the electromagnetic field created by such a transition, without parity change corresponds classically to that of a radially oscillating spherical charge distribution. Such a field is independent of time outside the sphere, but not inside the charge distribution. This field can transfer energy to an electron, penetrating the nucleus, and hence the electromagnetic transition takes place via conversion electron.

This E_o transition corresponds to the change of the radius of charge distribution of the nucleus say from a monopole of higher radius to a lower radius or vice versa. It is evident, that because of no gamma emission, and because only conversion electrons are emitted; the conversion coefficient will be infinite.

The processes competing, with internal conversion in $0 \to 0$ transition, no parity change; are simultaneous emission of (*i*) Two photons (total angular momentum carried out can become zero; if the two angular momenta are opposed to each other); (*ii*) Two conversion electrons with total angular momentum $L = 0$; (*iii*) One electron and one photon, with opposite angular momenta.

These three processes are, however, very less probable; though they are very much dependent on the nuclear structure.

The case of $0 \rightarrow 0$, with parity change are not possible at all and are totally forbidden. The matrix element in such cases will involve a pseudo-scalar operator, corresponding to oscillating magnetic charge. As free magnetic charges do not exist in nature; such a process is not physical. That will correspond to M_0 transition. In practice, we do not have any M_0 term[9].

7.5.3 Internal Pair Creation

If the transition energy between the two states is more than 2 mc^2; then this energy is large enough to excite a negative energy state (in the negative energy continuum–called 'sea' as expected in Dirac's relativistic theory) into a positive energy state. This creates a positron (vacancy in the negative energy sea) and an electron and hence an electron-positron pair is created.

In the case of internal conversion electrons, the atomic electron interacts, with the electromagnetic field of the nucleus when the transition energy is less than 2 mc^2. On the other hand, the case of internal pair production requires energy greater than 2 mc^2 and interaction is with negative energy states of the electrons, near the nucleus. Of course, in principle, the higher energy transition can interact with positive energy electrons in the atomic orbits also; but this probability is small, especially for light Z (less than 40) and high energy transition (> 2 mc^2).

An interesting case has been reported[10] in the detection of the sharp positron lines, in the conversion spectrum of Ra$\acute{C}$ with transition energy of 1.4 MeV. These have been interpreted as pair conversion in an atom, which is ionised as the result of internal conversion process, preceding the pair-converted transition. In other words, this is a case of comparative probability of interaction, both with positive energy atomic electrons, and negative energy states of electrons, from negative energy sea. Also investigation of the decay of Po^{210} showed that in the case of alpha decay, there is K and L shell ionisation, giving rise to X-rays, over and above those due to internal conversion[10].

The conversion coefficients for pair-production (π_{EL} or π_{ML}) have been calculated by Harton[11] and Rose[12], for low Z and sufficiently high energy, based on the Born approximation, both for magnetic and electric transitions, for $1 \leq L \leq 5$. Figure 7.5*a* shows, these results by Rose[12]. In the range of primary interest, π is of the order of 10^{-4}; which is much smaller than the internal conversion coefficients. We also note that, in contrast to the internal conversion coefficient α, the value of π increases with increasing energy and decreases with increasing Z and also decreases with increasing L.

7.5.4 Internal Bremsstrahlung

Discovered first by G.H. Aston in 1927, (Ref. 12), the internal Bremsstrahlung, arise because of the secondary emission of gamma radiation emitted by changing the dipole moment of the atom when the electronic charge is shifted (as in β-decay) from the nucleus to a region outside by the emission of a β-particle. The energy spectrum of such radiation is continuous. A simple theory by Morrison and Schiff[12], for s electron capture explained many experimental results. B. Saraf[12], investigated the inner Bremsstrahlung experimentally for Cs^{131}, and found disagreement with this theory, which shows that higher order terms should be introduced.

7.5.5 Measurement of Internal Conversion Coefficients and Pair Creation

(i) Internal Conversion Coefficient

The internal conversion coefficients have been measured for a large number of gamma transitions. It, basically, involves the measurement of discrete energy electrons, in the presence of continuous electron spectrum in decay and simultaneous measurement of gamma rays. Another method is to measure the intensity of relevant X-rays, emitted when a gamma rays-transition conveys its energy to say K-electrons, which are, then, emitted and the bound electrons fall into the vacant orbits, resulting in K X-rays emission. The ratio of the emitted K-electrons and gamma rays, or K X-rays and gamma rays yields the value of α_k; as discussed in Section 7.5.1.

We have shown in Fig. 7.2 the discrete electron spectrum[14] (lines) due to conversion electrons in the presence of the continuous spectrum of beta rays, using a magnetic beta spectrometer of the various gamma transitions in the decay scheme of $Ba^{140} \rightarrow La^{140}$.

One can also measure, in this case, the K X-rays along with gamma rays, using modern solid state detectors corresponding to decay scheme of La^{140} and also the X-ray spectrum in coincidence with the subsequent gamma rays from which one can determine the conversion coefficients.

For comprehensive review on conversion coefficient, *see* Slatis and Siegbahn and works by other workers[16].

It is found that conversion coefficients[15] measured by these methods range from 3×10^{-5} in the case of Na^{24}; to infinity in $0 \rightarrow 0$ transition in the decay of Ga^{72} and Ra'C. The probable error in α, in accuracy is 5 to 10 per cent and is good enough to determine the character of the transition and value of *L*.

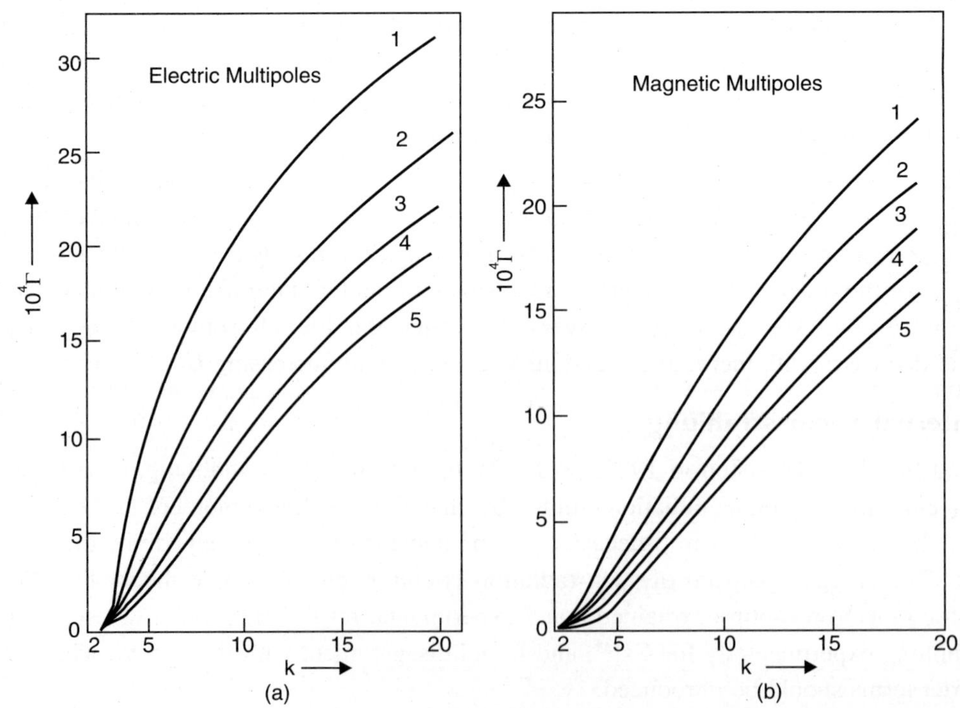

Fig. 7.5 (a) Total internal pair conversion coefficients for (*a*) electric and (*b*) magnetic multipoles for 1 < L < 5. (12), for low Z, and sufficiently high energy (Ref. 12).

(ii) Internal Pair Creation

A review of the entire subject of internal pair creation, has been given by Riou[8]. It has been found that π does not depend markedly on Z, because the Coulomb effects on the electron and positron in pair production largely cancel each other (Fig. 7.5 *b*).

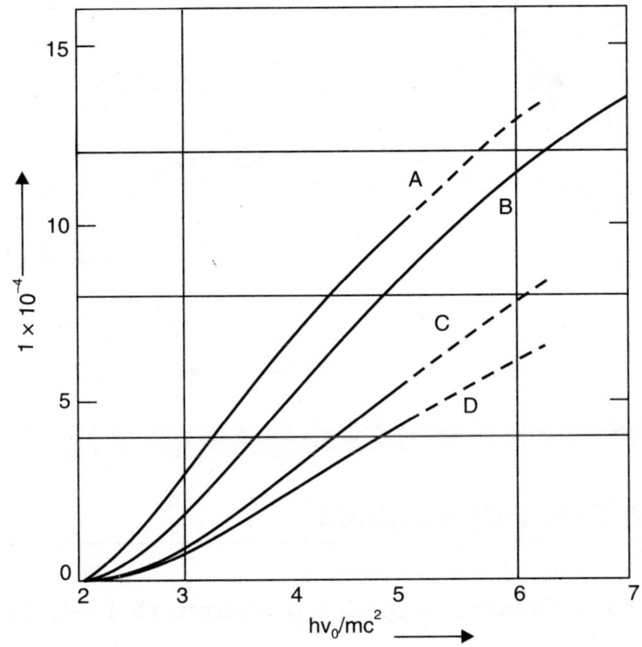

Fig. 7.5 (b) The behaviour of internal pair production for different values of Z and energies, for E_1 transitions. Curve (A), Z = 0, (B) Z = 84; M_1 transitions curve (C), Z = 0, (D) Z = 84 (Ref. 13).

Figure 7.5*a*, gives the behaviour of internal pair conversion coefficient, for E_1 and M_1 transitions. It is interesting to see that the effect of *Z* is comparatively small. However, the spectrum shapes of electrons and positron are such that positron spectrum is placed towards high energies and for electrons it is placed towards low energies, because of Coulomb effect.

Experimentally the coefficient of internal pair production, *i.e.*, $\pi = N\pi/N\gamma$ requires the measurement of electron-positron pairs. This can be accomplished by counting the number of positrons either in a magnetic spectrometer or by coincidence between the annihilation quanta, (0.5 MeV) emitted in opposite directions. Figure 7.6 shows the typical positron spectrum from pair-production in RaĆ. Pair conversion coefficients have been measured in a number of transitions following radioactive decay.

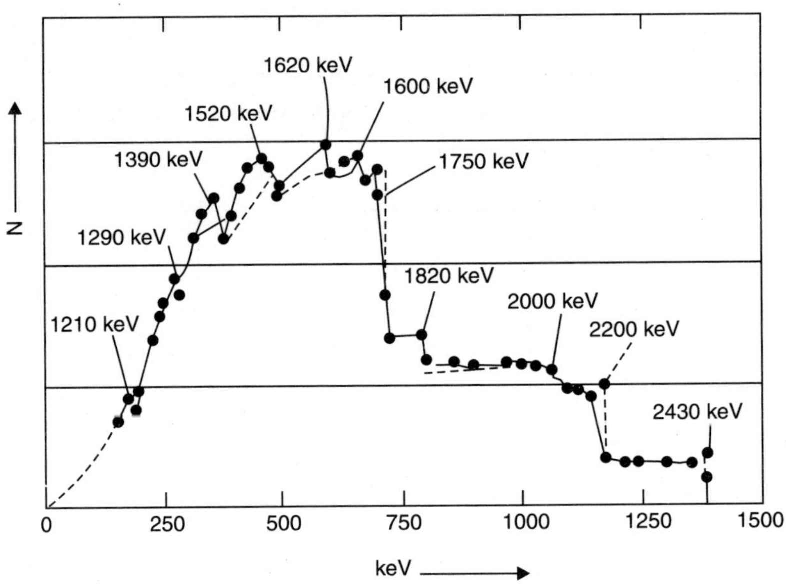

Fig. 7.6 Energy spectrum of the positron from internal pair production from RaĆ (Ref. 13).

7.6 ANGULAR CORRELATIONS

7.6.1 Theory of Gamma-Gamma Angular Correlation (Semi-Quantitative)

The angular correlation studies between two gamma rays γ_1 and γ_2; (called Gamma-Gamma angular correlation) can lead to the determination of either of the angular momenta I_a, I_b and I_c if other two values of angular momentum alongwith the value of mixing ratio δ are known [Fig. 7.7*a*] and/or if conversion coefficients of gamma rays alongwith the angular correlation are available. It is important, therefore, to understand the general theory of the phenomenon of angular correlations, which has proved to be a very effective tool for investigating nuclear structure.

A typical gamma-decay case is shown in Fig. 7.7*a*, where the levels I_a, I_b, and I_c, are shown split into their m-components m_a, m_b, and m_c, (*i.e.*, into $2I_a + 1$, $2I_b + 1$ and $2I_c + 1$, number of components) in the decay sequence of $A \to B \to C$. The experimental arrangement for the study of the angular correlation between gamma rays γ_1 asnd γ_2 is shown in Fig. 7.7*b*. One studies the number of gamma rays γ_2 in coincidence (emitted simultaneously within the resolving time of the electronics system) with γ_1 gamma rays as a function of angle θ between the two gamma rays in a single plane. This measurement corresponds to the summation over all values of m_a, m_b and m_c. The theory, as developed below also assumes, that the states corresponding to m_b of the intermediate state as obtained from the decay of m_a, are not disturbed. In other words, the nuclei exist in free space and there are no magnetic or quadrupole electric fields present, affecting the intermediate state. On the other hand, when the nuclei are embedded in solids or liquids, the value of m_b is disturbed because of the electric or magnetic fields around the nucleus. One, then, determines, the perturbed angular correlation, which we will discuss subsequently.

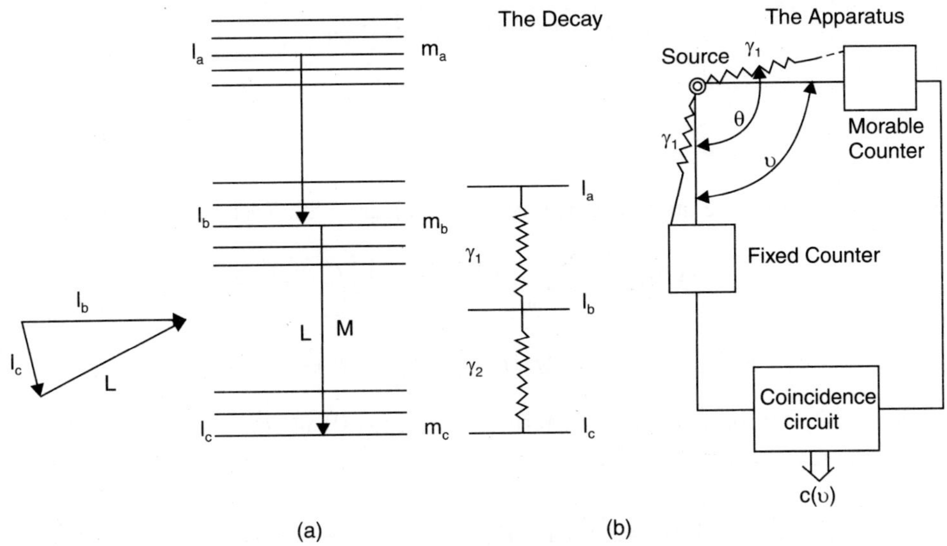

Fig. 7.7 Example of angular correlation sequence and apparatus.

The general theory of angular correlation has been developed by Hamilton[17] and others. We give below the salient features of the theory, emphasising the physical aspects.[8, 17]

Physically, one can get an insight into the angular correlation, by choosing the z-axis, as the direction, of the emission of γ_1; then one can write the angular correlation as, the product of the angular dependence of transition probability of transition from $b \to c$, *i.e.*, F_{L2} (θ) and the transition probability of the transition $a \to b$, *i.e.*, F_{L1} $(\theta = 0)$ in the forward direction of z-axis. Each transition probability depends on three factors:

(*i*) The relative population of each sub-level say $P(m_b)$, for each sub-level m_b ;

(*ii*) The relative transition probability say for each component say from $m_b \to m_c$ as $G(m_b, m_c)$.

(*iii*) The angular distribution of radiation for each transition corresponding to $M = m_b - m_c$, *i.e.*, $F_{L_2}^{M}(\theta)$.

The angular distribution $F_{L_2}(\theta)$ for transition between b and c states with respect to the z-direction of the transition between a and b states is expressed as:

$$\underset{b \to c}{F_{L_2}(\theta)} \approx \sum_{m_b, m_c} P(m_b)\, G(m_b\, m_c)\, F_{L_2}^{M}(\theta) \qquad \text{...(7.81)}$$

where $P(m_b)$ is given by:

$$P(m_b) \approx \sum_{m_a} G(m_a\, m_b)\, F_{L_1}^{\pm}(\theta = 0) \qquad \text{...(7.82 a)}$$

In writing Eq. 7.82 *a*, it is assumed that, the relative probability of population of each sub-level m_a is the same; so that the population $P(m_b)$ for m_b sub-state is given by the transition from m_a to m_b, summed over all sub-states m_a.

Further, the relative transition probability $G(m_b, m_c)$ or $G(m_a, m_b)$ are equal to the square of the Clebsch-Gordon coefficient for the vector addition:

$$\mathbf{I}_b = \mathbf{I}_c + L_1 \quad \text{or} \quad \mathbf{I}_b = \mathbf{I}_a + L_1$$

so that: $$G(m_a\, m_b) = \left(I_b L_1 m_b, \pm 1 \mid I_a\, m_a\right)^2 \qquad ...(7.82\ b)$$

and $$G(m_b\, m_c) = \left(I_c L_2 m_b\, M \mid I_b\, m_b\right)^2 \qquad ...(7.82\ c)$$

One can, then, write the angular correlation, from Eqs. 7.80, 7.81 and 7.82 as:

$$W(\theta) \approx \sum_{m_b, m_c, m_a} (I_b\ L_l\ m_b \pm 1 \mid I_a\ m_a)^2 \times$$

$$F_{L_1}^{\pm 1}(0)\,(I_c\ L_2\ m_c\ M_2 | I_b m_p)^2\ F_{L_2}^{M_2}\,(\theta_2) \qquad ...(7.83)$$

Equation 7.83, can be expressed in terms of P_L (cos θ) or $a^{2\nu} \cos^{2\nu}(\theta)$ for pure dipole and quadrupole gamma-rays; as given by Hamilton, *i.e.*,

$$W(\theta) = \sum_{\nu=0}^{\nu=2} A_{2\nu}\ P_{2\nu}\,(\cos\theta) = 1 + a_2 \cos^2\theta + a_4 \cos^4\theta \qquad ...(7.84)$$

In deriving Eq. 7.83, we have basically derived the angular distribution of the radiation from $b \to c$; with respect to the z-axis, corresponding to the γ_l direction. Such a method of derivation is possible, if $M_1 = \pm 1$ because only angular momentum $\pm \hbar$ can be carried in the direction of motion, along the z-axis. This equation is therefore, useful only, if the transition corresponds to dipole radiation.

For a general case, one requires to develop a theory, where z-axis need not be along γ_1 ray, but is chosen arbitrarily.

For each of the two radiations γ_1 and γ_2, a separate coordinate system coincides with the direction of the emission of the gamma rays. The connection between two radiations and the quantisation along the arbitrary z-axis is established through the transformation properties of the eigen-functions of γ_1 and γ_2 transitions.

While writing the general expression, one should realise that (*i*) we have to sum over all values of $m_{a'}$ m_b and $m_{c'}$ as done earlier in Eq. 7.83 and (*ii*) the radiation γ_1 or γ_2 may not be pure, but may contain a mixture of the two neighbouring multipoles say $E_2 + M_1$; or $M_1 + E_2$, etc.

We give below the detailed discussion (but without derivation) of the angular correlation $W(\Omega_1, \Omega_2)$ between γ_1 and γ_2, for a general case, where Ω_1 and Ω_2 are the solid angles into which γ_1 and γ_2 are emitted. Evidently these solid angles, also contain the directions of these two gamma rays.

Following the original treatment of Fano, Coester and Jauch[18] and as discussed in details by H. Fraunfelder[17] in Beta and Gamma spectroscopy[2]; we give below the final expression of $W(\Omega_1, \Omega_2)$ as:

$$W(\Omega_1, \Omega_2) = \sum (-1)^\nu\,(I_b \parallel L_1 \parallel I_a)(I_b \parallel L_1' \parallel I_a)(I_c \parallel L_2 \parallel I_b)(I_c \parallel L_2' \parallel I_b) \times$$

$$\mathrm{W}\,(I_b\ I_a\ \nu L_1',\, LI_b)\, W\,(I_b\ I_c\ \nu L_2',\, L_2\ I_{b2}) \times$$

$$C_{\nu\tau_1}(L_1'\ L_1)\, C_{\nu\tau_2}\,(L_2'\ L_2)\, D_{T_1T_2}^{\nu}\,(\theta) \qquad ...(7.85)$$

In Eq. 7.85, the summation is carried over L_1, L_1', L_2 L_2', τ_1, τ_2 and ν ; for which the meaning and significance have been earlier defined for L_1 and L'_1. The summation over-z-components of L and L' *i.e.*, L_z and L'_z is also required for getting the reduced matrix elements. Formally $C_{\nu\tau}$ are the eigen functions of the operators for total angular momentum and the z-component of the angular momentum, with eigen values ν and τ. The Racaah coefficients $W(I_b, I_a, \nu, L'_1, L_1, I_b)$ and $W(I_b, I_c, \nu, L'_2, L_2, I_b)$ are well-known, and are required while evaluating the sum over z-components of the product of three Clebsch-Gordon coefficients. For details *see* Ref. (17). $D^{\nu}_{T_1T_2}(\theta)$ is the rotation matrix.

These rotation matrices $D^{\nu}_{T_1T_2}$ are related to the eigen functions belonging to an angular momentum L, so that an eigen function of $LM\,\pi$ with direction and polarisation σ, *i.e.*, $\langle LM\,\pi||\Omega\sigma\rangle$ transforms through these matrices to an eigen function $LM\,\pi$ with direction 0 (z-axis) and polarisations σ, with the following equation:

$$\langle LM\,\pi||\Omega\sigma\rangle = \sum_m D^L_{Mm}(R^{-1})\langle LM\,\pi||0\sigma\rangle \qquad ...(7.86)$$

where m are the z-components of L in the transformed eigen function denoted by τ_1 and τ_2 in Eq. 7.85 and R represents the rotation from the z-axis to the direction of radiation and R^{-1} the reverse of it, *i.e.*, from direction of radiation to z-axis; $D^L_{m_1m_2}(R)^{-1}$; being given by the equation:

$$\sum_{m'} D^L_{m'_1m'_2}\left(R_2^{-1}\right) D^L_{m_1m_2}\left(R_1^{-1}\right)^* = D^L_{m_1m_2}\left(R_2^{-1}\,R_1\right) \equiv D^L_{m_1m_2}(\theta) \qquad ...(7.87)$$

Equation 7.87 corresponds to the case of rotation R_1 followed by rotation R_2^{-1}; so that the final D's refer to the rotations $R_2^{-1}R_1$ which rotates the second rotation into the first (through θ) and hence we call it $D^L_{m_1m_2}(\theta)$.

Some conclusion from the discussion of Eq. 7.85, are given below (without proof):

The Racaah coefficients $W(a; bc\,ef)$ vanishes unless each of the vector triplets (abe), (cdf) and (bdf) forms a triangle. From this restriction, applied to Eq. 7.85; we find the selection rule for ν:

$$0 \le \nu \le \text{Min}\,(2I_a, 2L_1, 2L_2) \qquad ...(7.88)$$

For pure multipole radiations ($L_1 = L'$ and $L_2 = L'_2$); there is only one term in the summation in Eq. 7.85. It can be seen that ν runs from $|L - L'|$ to $|L + L'|$; which in the above case will have $\nu = 0$ to $2L$ and the summation over L_1, L'_1, L_2 and L'_2 degenerate into one term. For mixed radiations, *i.e.*, $L_1 \ne L'_1$ and $L_2 \ne L'_2$; L and L' vary independently over the range of allowed angular momenta. Then the angular correlation will depend on the mixing ratio δ defined for $L' > L$ as:

$$\delta \equiv \frac{(I_b \,\|\, L' \,\|\, I_a)}{(I_b \,\|\, L \,\|\, I_a)} \qquad ...(7.89)$$

The angular correlations are independent of Euler angles χ_1 and χ_2 which denote rotations about the direction of propagation, as only the direction of radiations are observed and τ_1 and τ_2, which are the eigen values of L_z, become zero and hence independent of Euler angles χ_1 and χ_2. It may be seen that $Lx = -i\hbar\,(\partial/\partial_\chi)$. Under these condition, Eq. 7.85, may be expressed as (because for $\tau_1 = \tau_2 = 0$, the functions D^{ν} become Legendre Polynomials).

$$W(\Omega_1, \Omega_2) = W(\theta) = \sum_{\nu_{even}} A_\nu P_\nu (\cos\theta)$$

$$= 1 + A_2 P_2 (\cos\theta) + \ldots A_{\nu_{max}} P_{\nu_{max}} (\cos\theta) \qquad \ldots(7.90)$$

where $\nu_{max} = \text{Min}\,(I_b, 2L_1, 2L_2)$

It can be, now seen, by comparing Eqs. 7.85 and 7.90 that for pure multipole,

$$A_\nu = F_\nu(L_1 I_a I_b)\, F_\nu (L_2 I_c I_b) \qquad \ldots(7.91)$$

where $$F_\nu (L_1 I_a I_b) \equiv W(I_b I_a \nu L_1, L_1 I_b)\, C_{\nu_0} (L_1 L_1) \qquad \ldots(7.92)$$

and $$F_\nu (L_2 I_c I_b) \equiv W(I_b I_c \nu L_2, L_2 I_b)\, C_{\nu_0} (L_2 L_2) \qquad \ldots(7.93)$$

For mixed transitions, in one radiation, it can be seen from Eqs. 7.85 and 7.89; that for cascade $I_a (L, L'_1) I_b (L_2) I_c$, $W(\theta)$ can now be written as:

$$W(\theta) = W_I + \delta^2 W_{II} + 2\delta\, W_{III} \qquad \ldots(7.94)$$

where W_I and W_{II} are angular correlation functions for pure cascades $I_a (L_1) I_b (L_2) I_c$ and $I_a (L'_1) I_b(L_2) I_c$; but W_{III} is interference term, between L_1 and L'_1 and is given by:

$$W_{III} = \sum_{\nu \neq 0_{even}} A_\nu^{III} P_\nu (\cos\theta) \qquad \ldots(7.95)$$

with $$A_\nu^{III} = F_\nu (L_1 L_1' I_a I_b)\, F_\nu (L_2 I_c I_b) \qquad \ldots(7.96)$$

where the new interference term $F_\nu (L_1 L'_1 I_a I_b)$ is given by:

$$F_\nu (L_1 L'_1 I_a I_b) = (-1)^{I_a - I_b} [(2I_b + 1)(2L_1 + 1)(2L_1' + 1)]^{1/2} \times$$
$$G_\nu (L_1 L'_1 I_a I_b) \qquad \ldots(7.97)$$

The functions G_ν have been evaluated by Biedenharn and Rose[8,17], which also tabulates F_ν's.

7.6.2 Experimental Measurement of $\gamma_1 - \gamma_2$ Angular Distribution or Angular Correlations

The Angular distribution is, in general, measured with respect to the incident particles in a reaction, *e.g.* in the case of Coulomb excitation of Cs^{133}, *i.e.* Cs^{133} $(p, p'\gamma)$ Cs^{133}; [a case described in Section (7.7)]; the angular distribution of gamma rays is measured with respect to the incident beam of protons[20]. The angular correlation, on the other hand, is determined by measuring coincidences by the two consecutive gamma rays and determining the angular dependence of such coincidences, keeping one detector in the direction of γ_1-rays and rotating the other which detects γ_2-rays. The decay of Co^{60} is generally used for a standard correlation on which an angular correlation setup is tested. The Co^{60} decays through beta decay to Ni^{60}, which decays through the cascade $4^+ \xrightarrow{2} 2^+ \xrightarrow{2} 0^+$. The theoretical angular correlation is, then, given by:

$$W(\theta) = 1 + 0.102\, P_2(\cos\theta) + 0.009\, P_4(\cos\theta)$$
$$\approx 1 + 0.125 \cos^2(\theta) + 0.0417 \cos^4(\theta) \qquad \ldots(7.98)$$

Figure 7.8 shows, the experimentally measured angular correlation[19] $W(\theta)$; along with the theoretically expected curve. The agreement is excellent. This is a case of pure gamma transitions, with mixing ratio δ being zero both for γ_1 and γ_2.

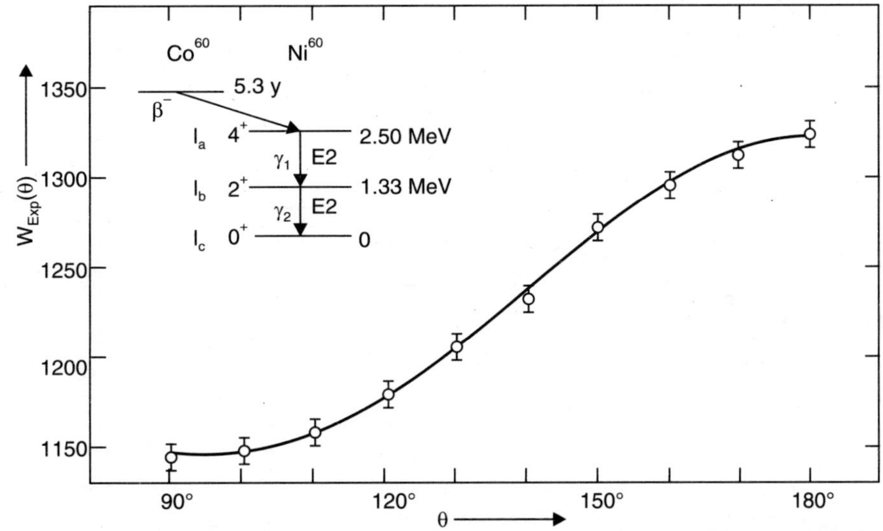

Fig. 7.8 Experimentally measured angular correlation W(θ) for the gamma rays γ_1 (1.17 MeV) and γ_2 (1.33 MeV) of Ni^{60} along with the theoretically expected curve (Ref. 19).

If the mixing ratio has a definite value, then the angular correlation is affected as given earlier in Eq. 7.94. For fixing the value of δ, in a given angular correlation or distribution measurement, one performs a χ^2 test, for the experimental and theoretical yields say for three angles, so that χ^2, defined in the following equation, is minimised.

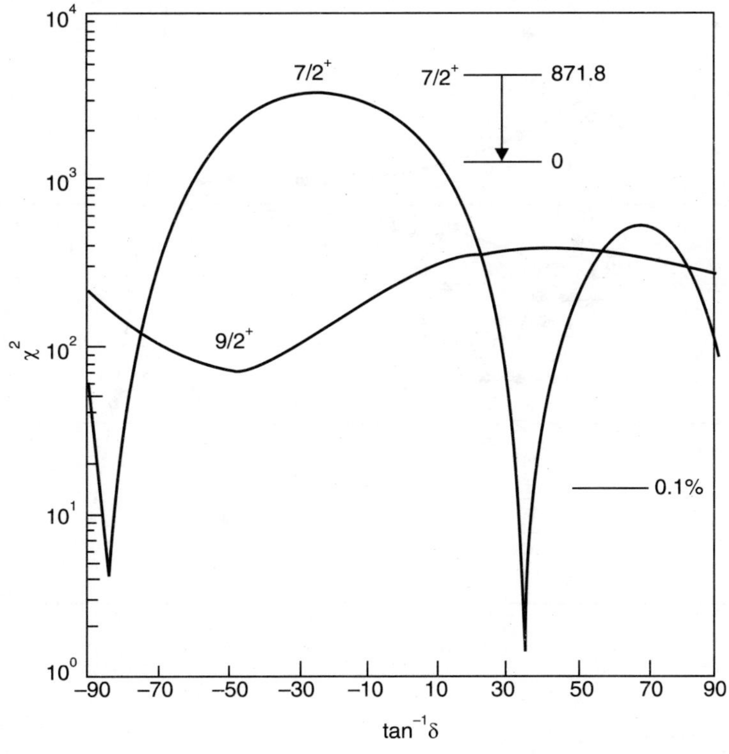

Fig. 7.9 Values of χ^2 as a function of mixing ratios for the 871.8 kV. γ-ray in the decay of Cs^{133} (Ref. 20).

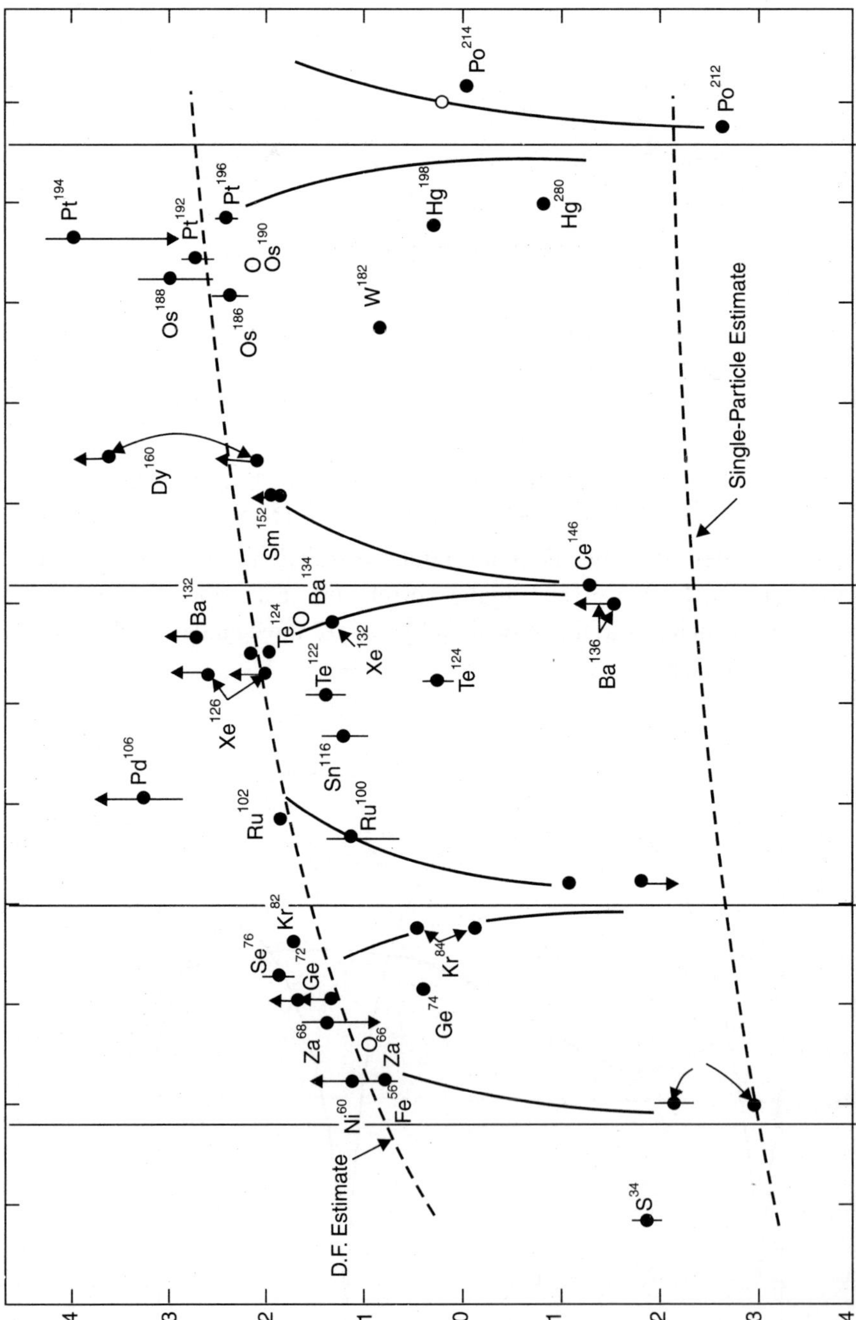

Fig. 7.10 Reduced mixing ratio log $(\delta/E_\gamma)^2$, as a function of neutron number. Most of the data is from the angular correlation experiments(filled circles) (Ref. 21).

$$\chi^2 = \sum_i \left(\frac{W_{\text{th}} - W_{\text{exp}}}{\in_i} \right)^2 \qquad ...(7.99)$$

Here W_{th} is calculated using Eq. 7.94 and $\in_i$ is the error in W_{exp}. We show in Fig. 7.9, a case of mixing ratio δ for the 871.8 keV γ-ray of Cs^{133} shown for (p, p') in Fig. 7.22. A 0.1% confidence limit was fixed to reject the unacceptable spin. As is shown in this figure[20], out of the two spins, *i.e.,* 9/2 and 7/2; it is evident that 7/2 is preferred. Also out of the two possible values of δ; *i.e.* $\delta = 0.7^{+0.05}_{-0.08}$ and -85 ± 2.4, the second value is not physical, hence first value is accepted. This illustrates the typical case of angular distribution, in general, with a certain mixing ratio, δ.

7.6.3 The Systematics of Mixing Ratios

The mixing ratio, basically, indicates the mixture of the angular momenta in the wave functions of the nuclear states involved in the transitions. Extreme situations of the wave functions correspond to pure single particle of excitation. In practice, the wave function will be mixture of a few single particle modes. Potnis and Rao[21] [Fig. 7.10] have systematised the global data on δ's and compared it with the calculated values of δ, on the basis of single particle level estimates given by Moszkowski[22] and collective model estimates as given by Davidov-Filippov[23]. It is interesting to point out that log $(\delta/E_\gamma)^2$ increased as a function of the neutron number, with exceptions at magic numbers 28, 50, 82 and 126. At these neutron numbers, the mixing ratios diminish by a factor of 2 or 3.

Grechukhin[24] has calculated, in great detail, the magnetic transitions of even nuclei which show collective type excitations. He has shown that δ^2 (E_2/M_1) is sensitive to the admixture of single particle excitations in the collective states of nuclei.

7.6.4 Triple $\gamma_1 - \gamma_2 - \gamma_3$ Angular Correlation

The theory of angular correlation of three successive gamma rays was first given by Biedenharn, Arfken and Rose[25], who also applied it to the case, where an emission process was replaced by an absorption process. The complexity and highly tedious nature of the summation over magnetic quantum numbers, even making use of the two gamma rays parameters of radiations, allows only using simple cases to be considered. Ferguson[26] outlined some possible triple angular correlation studies, in which all the three radiations are detected.

Though it is some-what tedious, the interest in triple angular correlation arises due to its advantage in the identification of nuclear states.

Theoretically, one may consider a triple cascade, when the decay is through successive emission of three radiations R_1, R_2 and R_3 in the direction $\mathbf{K}_1$, $\mathbf{K}_2$ and $\mathbf{K}_3$; respectively. Let a, b, c and d be the spin quantum numbers and m_a, m_b, m_c and m_d be the magnetic substates for the triple cascade. Then the triple correlation functions $W(\mathbf{K}_1, \mathbf{K}_2, \mathbf{K}_3)$ can be written as:

$$W(\mathbf{K}_1, \mathbf{K}_2, \mathbf{K}_3) = \sum_{m_b,\, m'_c} \langle m_b \mid \rho(\mathbf{K}_1) \mid m'_b \rangle \langle m_b\, m'_b \mid G(\mathbf{K}_2) \mid m_c\, m'_c \rangle \times$$

$$\langle m'_c \mid \rho(\mathbf{K}_3) \mid m_d \rangle \qquad ...(7.100\ a)$$

where the three matrix elements, on the right hand side of Eq. 7.100*a* are evaluated in Ref. (25, 26).

The matrices $\langle m_b \mid \rho(\mathbf{K}_1) \mid m'_b \rangle$ and $\langle m'_c \mid \rho(\mathbf{K}_3) \mid m_d \rangle$ are density matrices, between the states m_b, m'_b and m'_c, m_d.

While $\langle m_b\, m'_b \mid G(\mathbf{K}_2) \mid m_c\, m¢_c \rangle$ is the coupling matrix, given by:

$$\langle m_b\, m'_b \mid G(\mathbf{K}_2) \mid m_c\, m'_c \rangle = S_2 \langle m_c \mid H_2 \mid m_b \rangle \langle m'_c \mid H_2 \mid m'_b \rangle^* \quad ...(7.100\ b)$$

where H_2 is the Hamiltonian operator which induces transition from intermediate state to final state and S_2 gives summation over unobserved quantities for the second transition.

For a realistic case[27] with three detectors in three direction, with detectors extending solid angles of Ω_1, Ω_2 and Ω_3 respectively; the triple gamma angular correlation function can be written as:

$$W(\Omega_1, \Omega_2, \Omega_3) = \sum_{k_1, k_2, k_3} a_{k_1k_2k_3}\, P_{k_1k_2k_3}(\Omega_1, \Omega_2, \Omega_3) \quad ...(7.101)$$

k_1, k_2 and k_3 will take even values; as the levels have definite parity and the sum will have terms with $(k_1, k_2, k_3) = (0.00)$, (022), (0.44), (2,02), etc. depending on spin quantum numbers and multipolarity of transition. If we put the three detectors in the same plane say that of the table and measure the angles θ_1, θ_2, θ_3 with respect to the detector 1, so that θ_1, is always zero, then θ_2 or θ_3 will be variable, and may be called θ. One measures the angular correlation coefficients of $W(\theta)$ if one takes the triple coincidences of γ_1, γ_2 and γ_3 by varying say θ_3. Then it is possible to write:

$$W(\theta) = 1 + A_2\, P_2(\cos\theta) \pm A_4 P_4(\cos\theta) \quad ...(7.102)$$

The values of A_2 and A_4 are given in references (25), (26) and (27). The values of A_2 and A_4 are related, to the angular momenta of the three levels, involved in the $\gamma_1 - \gamma_2 - \gamma_3$ triple cascade and if the first one is a mixture of dipole and quadrupole with a mixing ratio of δ and the other two are pure radiations, then explicit expression of A_2 and A_4 have been found.

Unambiguous spin assignments and determination of multiple mixing ratios can be done in triple correlation in a better way than in double correlation.

7.6.5 $\gamma - \bar{e}$ and Alpha-Gamma Ray Angular Correlations

Many times it is possible to carry out experimentally, the angular correlation studies involving one charged particle and one gamma ray, *e.g.* an $\gamma - \bar{e}$ involving conversion electrons: or $\alpha - \gamma$ in alpha decay or $\beta - \gamma$ in beta decay. The formalism developed till now can be used for expressing the angular correlation in all these cases.

1. $\gamma - \bar{e}$; involving conversion electrons

As the conversion electrons are created, instead of gamma rays from the electromagnetic interaction, we expect a similar angular correlation as given in Eq. 7.90; except that the coefficients A_2 are now multiplied by a particle parameter $b(\bar{e})$. We can, then write, for $\gamma - \bar{e}$ an angular correlations as:

$$W_{\gamma-\bar{e}\,(\theta)} = 1 + b_2(\bar{e})\, A_2\, P_2(\cos\theta) + b_4(\bar{e})\, A_4\, P_4(\cos\theta) \quad ...(7.103)$$

If both gamma rays are replaced by conversion electrons, each coefficient A_2 is multiplied by two particle parameters, *i.e.*,

$$A_\nu \rightarrow b_\nu(\bar{e}_1)\, b_\nu(\bar{e}_2)\, A_\nu$$

Basically the derivation of $b_2(\bar{e}_i)A_2$ is obtained in the same way as for A_ν, from Eq. 7.91; where now the radiation parameters $C_{\nu T}$ are different. The values of $b_\nu(\bar{e})$ have been calculated by Rose, Biedenharn and Arfken[8]. The coefficients $b_2(\bar{e})$ are, except for dipole radiation, always positive and larger than one. In the high energy limit, the particle parameters approach the value unity; hence $\gamma - \bar{e}$ correlation functions are nearly the same as $\gamma - \gamma$ correlation functions.

The calculations have been extended to the cases, when unconverted gamma ray is mixed by Biedenharn and Rose[17].

2. Alpha-Gamma ($\alpha - \gamma$) angular correlation

The particle parameters $b_\nu(L, \alpha)$ have been calculated for this case by J. Seed and A.P. French[28], using the formalism of Biedenharn and Rose[17]. For the simple case of α-particle carrying the single value of the angular momentum L, then if $b_0 = 1$, $b_\nu(L, \alpha)$ is given by:

$$b_\nu(L, \alpha) = \frac{2L(L+1)}{2L(L+1) - \nu(\nu+1)} \qquad \text{...(7.104)}$$

In case, α-particles carry away more than one angular momentum, then only L, $L + 2$, $L + 4$, etc. can be carried because of parity selections rules. Such calculations have been carried out[28]. If we consider the case of alpha particle carrying, two angular momenta L_1 and $L_1' = L_1 + 2$; then, we denote the intensity ratio of $(L_1'/L_1) = \delta^2$. The $\alpha - \gamma$ angular correlation is obtained as modification of Eq. 7.94 by multiplying A_ν^I of W_I with $b_\nu(L_1, \alpha)$, A_ν^{II} of W_{II} with $b_\nu(L_1', \alpha)$ and A_ν^{III} of W_{III} by $b_\nu(L_1, L_1', \alpha)$.

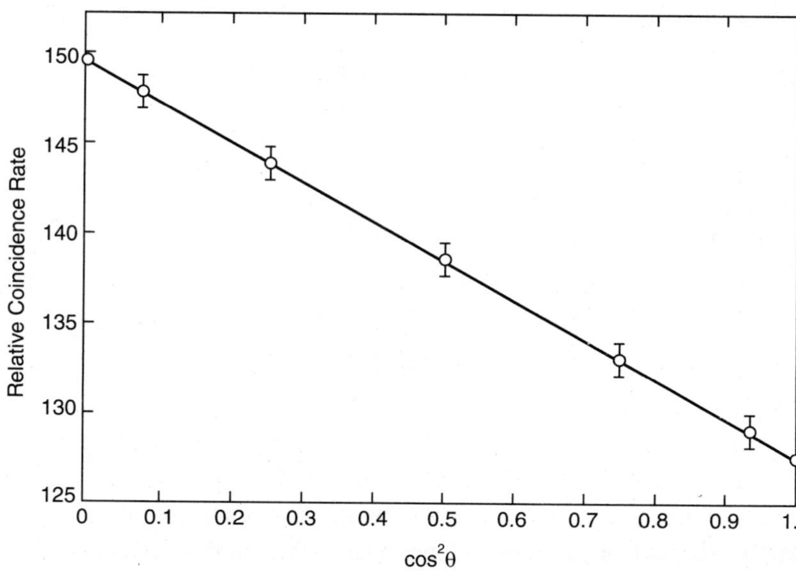

Fig. 7.11 Alpha-gamma directional correlation in Am^{241} embedded in $HClO_4$. The correlation is fitted to $W(\theta) = 1 + A\cos^2(\theta)$ with $A = 0.15$ (Ref. 29).

The values of b_ν's are calculated by S.P. Lloyd[28]. For pure cases of $b_\nu(L_1, \alpha)$ and $b_\nu(L_2, \alpha)$, one can use Eq. 7.104, and for a mixture, b_ν one uses $b(L_1, L_1'; \alpha)$, given by:

$$b_\nu(L_1, L'_1, \alpha) = \frac{2[L(L+1)\,L'(L'+1)]^{1/2}}{L(L+1) + L'(L'+1) - \nu(\nu+1)} \times \cos(\sigma_{L'} - \sigma_L) \qquad ...(7.105)$$

where σ_L is the phase shift due to Coulomb potential. Figure 7.11 shows a typical[29] example.

In general, there is a discrepancy between the experiments and theory, *e.g.* in the case of Th^{228}. This was explained by Abragam and Pound[29], due to interaction of the nuclear quadrupole moment of the emitting nucleus, with the electric field gradients of the surroundings, especially if the material in which source is embedded, is solid or liquid.

7.6.6 β – γ Angular Correlation[17, 30]

Angular correlation of $\beta - \gamma$ is a complex process, because of several new factors entering into the process compared to say $\gamma - \bar{e}$ or other correlations. An important difference with other correlations like $\gamma - \gamma$; or $\gamma - e'$, etc. and $\beta - \gamma$ correlation is, that in this case three particles are emitted, *i.e.*, β^-, antineutrino and γ. That requires us to integrate the angular correlation of $\bar{\beta} - \gamma$ over all directions of electrons (β^-), as antineutrinos are not observed.

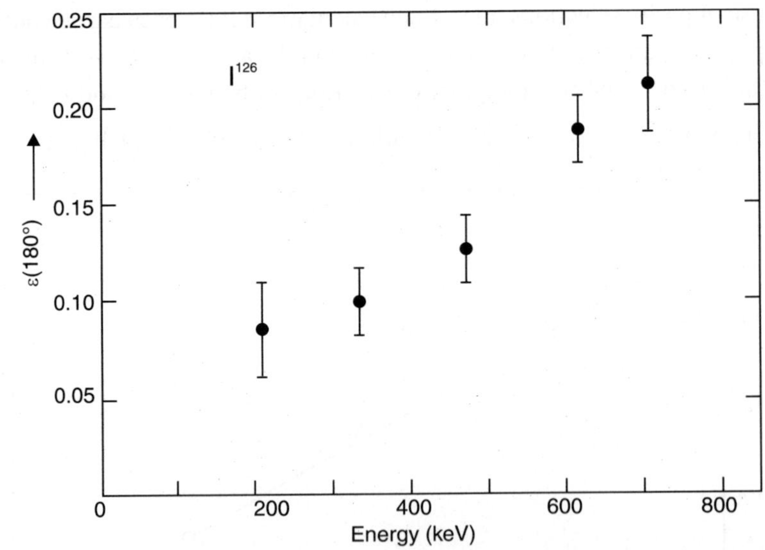

Fig. 7.12 $\beta - \gamma$ Angular correlation of I^{126}, as a function of beta ray energy (Ref. 32).

Instead of a single coupling constant as used in e.m. or nuclear interaction; in the case of $\bar{\beta} - \gamma$ correlation, the interaction must be some combination of the five relativistically invariant pure forms *S*, *V*, *T*, *A* and *P*. We know that, two interactions *V-A*, are responsible for the interaction. Then in the angular correlation function as in the transition probability, there will be interference between different pure interaction terms (*see* Chapter 8). For the same interaction, there may be *L* and *L* + 1 angular momenta carried away by beta rays. Then there will be interference between matrix elements belonging to different values of *L*. It is apparent that this complexity is further compounded by the fact that one is not sure about the coupling which is not generally known uniquely in a given case. The detailed expressions are given[31] in Annual Rev. N. Science 2, (1953). P-129. It is found that if the shape of the energy spectrum is strictly identical with allowed transition, the identical differential correlation $W(\theta, E)$ taken between energy E and $E + dE$ is isotropic. In a forbidden spectrum, differential correlation is anisotropic. The integral correlation $W(\theta)$ corresponds to the observation of β particles of all energies. In general $W(\theta, E)$ has its maximum anisotropy for the maximum β-energy and always becomes isotropic as beta energy approaches zero.

Most of β-cascades which have been investigated so far show an isotropic distribution. However anisotropic directional correlation have been found in cases where differential correlations are measured *e.g.* K^{42}, As^{76}, Rb^{86}, Sb^{124}, I^{126} and Tnl^{170} (Ref. 17). We give in Fig. 7.12 the result of measurement of anisotropy of beta-gamma angular correlation of I^{126}, as a function of beta ray energy; anisotropy ε (180°) being defined as $\varepsilon(180°) = Nc(180) - Nc(90°)/Nc(180°)$. These results are explained on the basis of assuming a given spin $J = 2$ (odd) for excited state and combination of a given set of matrix elements (tensor B_{ij} and $\int\beta\alpha$)(Ref. 32).

7.6.7 Gamma-Gamma ($\gamma_1 - \gamma_2$) Perturbed Angular Correlation, with External Fields—Magnetic or Electric[33, 34]

We have already discussed in Section (7.6.1), that in the case of a radioactive nucleus, which is free, (*e.g.* in a gas); there are no electric or magnetic fields around the nucleus, except that due to extra nuclear electrons of the atom. These fields, due to atomic electrons will, however, have random, directions, because of the random orientation of the nuclear spin in space. Hence in measuring the angular correlation; the population of the intermediate states (m_b) in Fig. 7.7 are either not disturbed or are only randomly oriented, leaving the angular correlation unaffected.

Now if a magnetic field **B** is applied perpendicular to the plane containing two gamma rays γ_1, and γ_2; the angular momentum I of the nucleus processes about the field B, with Larmor angular velocity ω_L' given by:

$$\omega_L = -\frac{B\mu}{\hbar I} = -\frac{g\mu_N B}{\hbar} \quad ...(7.106)$$

where g is the g-factor of the intermediate state, and μ_N is the nuclear magneton ($\mu_N = 5.5 \times 10^{-24}$ ergs/gauss). So if gamma ray γ_1 is emitted say at time $t = 0$ and gamma ray γ_2 after time t, after that; the angular correlation after time t will, now be given by $W(\theta, B, t) = W(\theta - \omega_L t)$, instead of $W(\theta)$. Suppose that the coincidence circuit registers coincidences only if γ_2 is emitted within time interval τ_1 to $\tau_2 = \tau_1 + \tau_0$, after emission of γ_1; then the measured correlation function $W_\perp(\theta, B)$ is given by:

$$W_\perp(\theta, B) \approx \frac{\int_{\tau_1}^{\tau_2} e^{t/\tau_b}\, W(\theta - \omega_L t)\, dt}{\int_{\tau_1}^{\tau_2} e^{-t/\tau_b}\, dt} \quad ...(7.107)$$

where $W_\perp(\theta, B)$ represents angular correlation, when the applied magnetic field **B** is perpendicular to the plane containing two gamma-rays.

If, in the original angular correlation, *i.e.*, $W(\theta, t = 0, B = 0) \equiv W(\theta)$ for $= \nu_{max} = 2$; then we can write, from Eq. 7.90:

$$W(\theta) = 1 + A_2 P_2(\cos\theta) = 1 + B_2 \cos 2\theta$$

where
$$B_2 = \frac{3^2 A_2}{(4 + A_2)} \qquad ...(7.108)$$

In Eq. 7.107, τ_b is the life time of the intermediate state, which is, many times, greater than τ_0; the resolving time of the electronic circuit. Then one can write the time differential correlation pattern about B as obtained from Eq. 7.107, for $\tau_b >> \tau_0$ and $\nu_{max} = 2$, as:

$$W_\perp (\theta, B; t) = 1 + B_2 \cos 2(\theta - \omega_L t)$$
$$= A_{22} P_2 [\cos (\theta - \omega_L t)] \qquad ...(7.109)$$

where decaying term e^{-t/τ_b} has been neglected for $\tau_b >> \tau_0$ and $t \approx \Delta t << \tau_b$. However, if the observations are made in a way, that they correspond to total time–integrated correlation, *i.e.*, $\tau_1 = 0$ and $\tau_2 \to \infty$ then, for $\nu_{max} = 2$.

$$W_\perp (\theta, B, \infty) = \frac{1}{\tau_b} \int_0^\infty e^{-t/T_b} W_\perp (\theta, B, t)\, dt$$

$$= A_{22} \overline{G_{22}(\infty)} P_K [\cos (\theta - \overline{G_{22} (\infty)} \omega_L \tau_b)] \qquad ...(7.110)$$

where
$$\overline{G_{22}(\infty)} = \frac{1}{\tau_b} \int_0^\infty e^{-\lambda_2 t} e^{-t/\tau_b}\, dt = \frac{1}{1 + \lambda_2 \tau_b} \qquad ... (7.111)$$

where λ_2 corresponds to decay constant, for $\nu_{max} = 2$.

The quantity $\overline{G_{22}(\infty)}$ gives an attenuation in the maximum amplitude of the oscillatory pattern of angular correlation and is related to the electric quadrupole interaction of the nuclear electric quadrupole moment and electric field gradient due to the surrounding atoms and electrons.

Apart from attenuation, Eq. 7.110 gives simultaneous rotation in the negative direction. The angular correlation pattern is shifted towards smaller angles by $\Delta\theta$, as shown in Fig. 7.13 for positive value of g. On the other hand, if g is negative, the rotation is in the positive direction. In this manner, the sign of g is determined from the direction of the rotation of the angular correlation pattern.

For a general case, *e.g.* in solids where both magnetic and electrostatic fields may be present, the attenuation coefficient is determined by both the magnetic and electric quadrupole interaction. The shifting of the pattern is, however, not affected by electric quadrupole interaction. The total time integrated perturbed angular correlation [Eq. 7.110] is affected by ω_Q:

$$\omega_Q = -\frac{(eQ V_{zz})}{4I(2I - 1)\hbar} \qquad ...(7.112)$$

where eQ is the electric quadrupole moment of the nucleus in the state connecting gamma rays γ_1 and γ_2 and $V_{zz} \equiv \partial^2 V/\partial Z^2$ is the electric quadrupole field due to the surrounding atoms on the concerned nucleus.

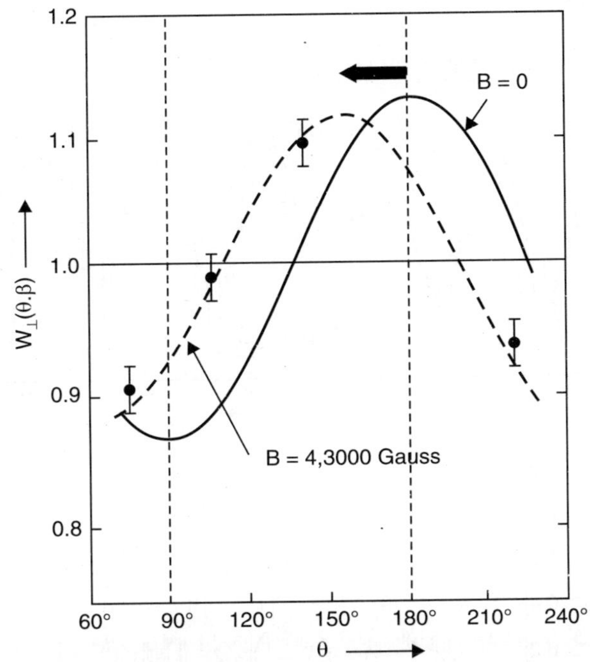

Fig. 7.13 Angular displacement of Pb^{204} $\gamma_1 - (\gamma_2 \cdot \gamma_3)$ cascade angular correlation in a transverse magnetic field (B = 4300 oesterds, g = 0.07) (Ref. 35).

If the resolution of the coincidence circuitry is much less, than the lifetime of the intermediate state, *i.e.*, $\tau_0 << \tau_b$; then the angular correlation pattern is quite complicated, especially when both electrostatic and magnetic fields are present. For details *see* Ref. (34, 36).

The angular correlation can then be measured as a function of *t*, which will show an oscillating pattern, modified in its amplitudes in a complicated way, depending on the relative strengths of electric and magnetic interaction fields.

Figure 7.14 shows a typical angular correlation pattern for Cd^{111} in Cd and Ta^{181} in Pd. The detailed shape depends on the electric quadrupole interaction.[36, 37] While the magnetic field shifts the pattern, for the time-integrated case, [Eq. 7.110]; without attenuation; the electric quadrupole interaction, introduces attenuation. For time-differential case, this corresponds to λ_2 in Eq. 7.111 being positive and non-zero.

Angular correlation or directional studies involving gamma transitions are a powerful tool for measuring the mixing ratios. Recently for Ba^{128}, such studies were carried out, using the reaction Mo^{96} (S^{36}, 4n) Ba^{128} and for Dy^{160}, obtained from spallation of tantalum target with 600 MeV protons. In the case of Ba^{128} the determination of values of some nine states with I^{π} from 13^+ to 21^+ were obtained while in the case of Ho^{160}, ($Ho^{160} \xrightarrow{\beta^+} Dy^{160}$) the values of some 27 states, where the host material Ho^{160} was imbedded in Gd to obtain Ho^{160g+m} Gd. These are the cases, which indicate[37] the possibilities of angular correlation studies in future.

A comparison with the theory of dynamic deformation model[38] leads only to qualitative agreement between the experimental and theoretical values for Dy^{160} levels but it does describe correctly, the mixing ratio sign change.

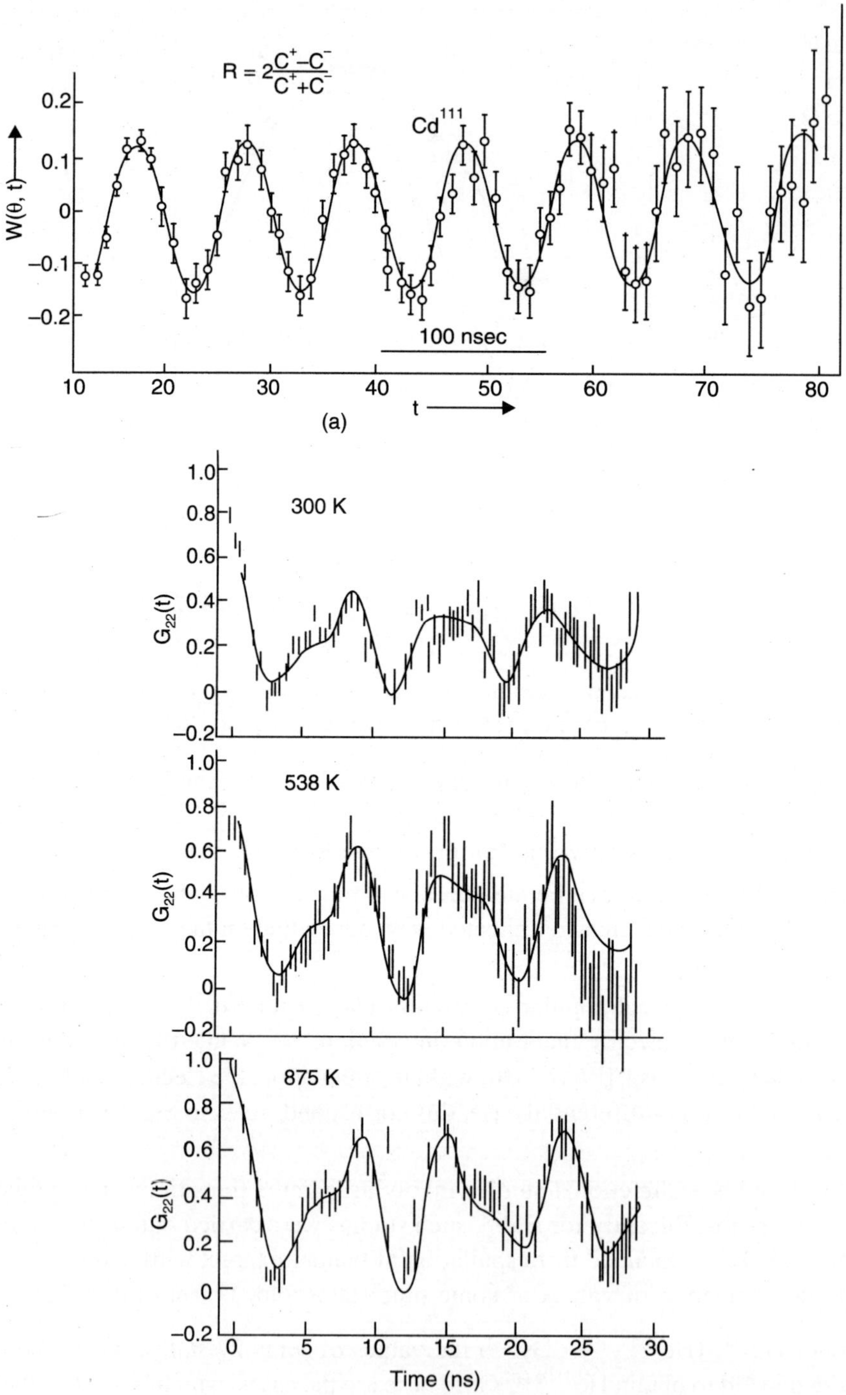

Fig. 7.14 The perturbed angular correlation for time differential measurements, as a function of time, for (*a*) Cd^{111} in Cd; (*b*) Ta^{181} in Pd, at different temperatures (Ref. 36).

As discussed in Section (7.6.7), the magnetic moments of the excited states of nuclei can be obtained, by measuring the perturbed angular correlation in which precession of transition is determined in the presence of a magnetic field. In a recent case[39], the magnetic moments of Pre-Yrast (for meaning of Yrast line, (*see* Chapter 17, Figure 17.32) high spin levels, and first excited levels were determined in neutron deficient $Hf^{162, 163, 164}$ isotopes. The reaction $Ta^{126, 128}$ (Ca^{40}, Xn) Hf^{164} was employed, using 175 MeV Ca beam. The target which consisted of equal mixture of Te^{126} and Te^{128} isotopes, was evaporated on a 3.4 mg/cm^2 Gd ferro-magnetic layer. A considerable fraction of the recoils of each isotope stops in Gd, and experiences the static hyperfine field in addition to the transient magnetic field, experienced by the whole ensemble. Comparison of the measured precession (for perturbed angular correlations studies) of transitions from the first excited longer lived levels with precession from shorter lived high spin transitions yields *g* factor (*see* Eq. 7.106) ratios of the first excited states in three isotopes. Such measurements, in odd isotopes can be instructive, as they provide unique information about nuclear level configuration. Comparison of these measurements with theory, provides this test[39].

7.7 EXPERIMENTAL METHODS AND RESULTS IN GAMMA RAYS SPECTROSCOPY

The detection and energy analysis of gamma-rays is a powerful tool in the nuclear spectroscopy, involving the excited states of nuclei. The detailed properties, of the gamma transitions, *e.g.* angular momenta carried by them, is determined by the methods of measuring the transition probabilities–or lifetimes; conversion coefficients and the angular correlations (or angular distribution in nuclear reactions). We will describe below, in brief, the salient experimental methods and results concerning each of these phenomena.

A. Detection and Energy-Determination

The various experimental methods, used for detection and energy-determination of gamma rays are:

G.M. counters: These are, in general, low efficiency detectors for gamma rays, but can be used for very low energy gamma rays say from 5–100 keV, with larger efficiency, which is hardly 1 per cent for 1 MeV gamma rays. For details *see* work by H. Sanerr and W.M. Good and R.J. Hart[40]. G.M. Counters are only detection devices and are not meant for energy measurement.

Scintillation counters: These detectors, especially Na (T1) are high efficiency (nearly 100%) and medium energy resolution (5 to 20%) detectors, which are used very extensively for work in gamma spectroscopy, for low energies from a few keV to say 10 MeV and are available in many sizes, *i.e.*, from thin wafers for low energy gamma rays to a few hundred C.C. for higher energy gamma rays. For comprehensive review *see* P.R. Bell, in Beta Gamma spectroscopy[41] edited by Siegbahn and other reviews.

Magnetic spectrometers: Basically, a magnetic spectrometer detects and acts as a momentum or energy analyser of a charged particle. For gamma rays detection, one uses the source of secondary electrons from a scatterer (or convector) from which electrons produced by photoelectric effect, can be analysed by a magnetic spectrometer. The efficiencies (1 – 10%) of such detectors, is much lower than scintillation detectors, but energy resolution (0.1 to 1%) is much better.

Double focussing spectrometers based on Svartholm-Siegbahn type spectrometers are commonly used for this purpose[40].

Solid-state detectors: Ge-Li or pure-Ge detectors are the most commonly used detectors these days for gamma rays spectroscopy. They combine the properties of high efficiency and high energy resolution (0.1–0.05%). The gamma rays spectrum shown in Fig. 7.1*b* is a good example of the excellent quality of this detector. Figure 7.15 gives the sketch of a typical array of pure Germanium detectors, with an anti-compton shield used as the detecting system at 15 UD accelerator facility at Nuclear Science Centre at New Delhi[41] India. For review of such detectors, *see* Ref. (41).

Crystal Grating Method: Use of Laue's principle of X-rays diffraction by crystal lattice and its extension by Bragg-Scerttering has given rise to the technique of crystal-grating spectrometers, which are today the maximum resolution instruments in gamma or X-rays spectrometry. The line width of a crystal spectrometer is approximately given by $d\lambda/\lambda \approx \alpha^2/2$, α being half angle of aperture of crystal: its value is much less than 1. So for X-rays of 100 keV, this gives a resolution of 0.1% and for 1 MeV gamma rays, it is 1%.

For comprehensive reviews, one should consult articles by J.W.M. Du Mond, et al. and D.H. Muller[42], et al.

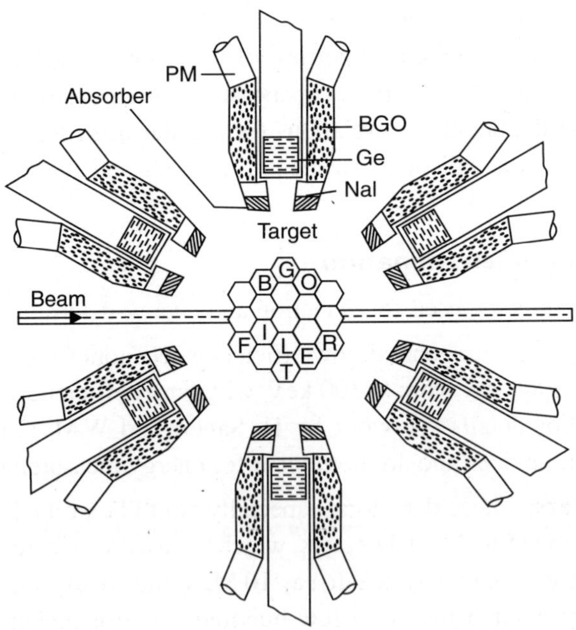

Fig. 7.15 An array of pure gemanium detectors with an anti-compton shield, for gamma rays detection, in the experimental arrangement at 15 UD accelerator at N.S.C. (New Delhi) India (Ref. 41).

B. Measurement of Transition Probabilities: Experimental Methods

We will discuss below the following methods of measuring the transition probabilities of gamma transition or lifetimes of nuclear states. (*i*) Direct lifetime measurement by electronic methods (*ii*) Time of flight method (*iii*) Doppler shift methods (*iv*) Resonance scattering and absorption methods (*v*) Coulomb Excitation.

The lifetimes of the excited state of a nucleus depend on the energy of excitations and its character, *i.e.*, if they have electric character, then lifetime depends on whether they correspond to E_1, E_2, E_3 or E_4 transitions. Similarly for magnetic transitions, the life time depends on whether they are M_1, M_2, M_3 and M_4. We have shown in Figs. 7.16 and 7.17, the measured values of $\log_{10}(T_\gamma A^{2L'/3} E_\gamma^{2L+1})$ for electric transitions versus neutron number and $\log_{10}(T_\gamma A^{\frac{2L-2}{3}} E_\gamma^{2L+1})$ for magnetic transitions, versus neutron number. It is seen, that lifetimes T_γ extend from 10^{-4} secs to 10^{-1} secs for energies near 3 MeV. For higher energies, lifetimes can be as short as 10^{-15} secs or shorter[43].

The horizontal lines in these two diagrams correspond to Weisskopf estimates of lifetimes based on shell model. It is evident from these two diagrams that the observed lifetimes are usually too long for E_3, E_4, M_1, M_2, M_3 and M_4. This only shows that extreme single particle shell model does not hold, and including more than one level, hinders the transition. On the other hand, for E_2 transitions, lifetimes are too short. For spherical nuclei, this is due to strong quadrupole type core-polarisation effect. On the other hand, for deformed nuclei, this is a direct consequence of the charge distribution. Both of these effects are essentially collective effects.

Direct lifetime measurements: Electronic circuits are available, by the use of which one can directly measure the lifetimes from 10^{-4} to 10^{-9} secs, either by the delayed coincidence technique, or using the modern equivalent of time to pulse height converters (TPHC). The time-resolution of the two detectors detecting γ_1 and γ_2, connecting the upper and lower levels to the concerned levels, can be arranged to less than these lifetimes. The fastest detectors are the plastic scintiliators ($\Delta t \approx 10^{-10}$ secs); while Na (T_l) scintillator can be used to give pulses of 10^{-6} secs width and the Ge-Li detectors also have pulse width of 10–15 nano secs and hence are slow. The Ba–F_2 scintillators are fast (pulse widths 10^{-9} secs or so) and also[2] have high efficiency and energy resolution and are quite commonly used for direct measurements of lifetimes of the order of nanoseconds. For details *see* measurement of short lifetimes in excited states by R.E. Bell, in Beta and Gamma Spectroscopy edited by K. Siegbahn. Also *see* A.W. Sunyar and R.E. Bell, and other recent works[44] for recent development.

Recoil method: This method is based on the principle that after mean decay time of the decaying state, the nucleus will recoil before emitting the subsequent gamma rays, because of the preceding reaction or decay process. As for example (p, n, γ) reaction, after the emission of neutron, the residual nucleus will recoil within the conservation laws of momentum and energy. The recoil direction and energy of the recoiling nucleus will depend on the energy of the incident proton, the energy of the emitted neutron and the relative directions of proton and neutron. The residual nucleus after the emission of the neutron will move in a given direction in the excited state and after travelling certain distance will emit the gamma rays. This distance will depend following the laws of probability on the lifetime of the state, so that if lifetime is τ, the probability of emitting the gamma ray will be proportional to $l/\tau = v/d$ where d is the distance of a recoiling nucleus, with recoil velocity v.

Jacobsen[45], has developed a method in which distance d is measured geometrically. [Fig. 7.18] shows the typical arrangement of the target T exposed to beam B. Lead baffles W are arranged, so that the gamma rays can strike the scintillation counter only from recoiling atoms; which have travelled at

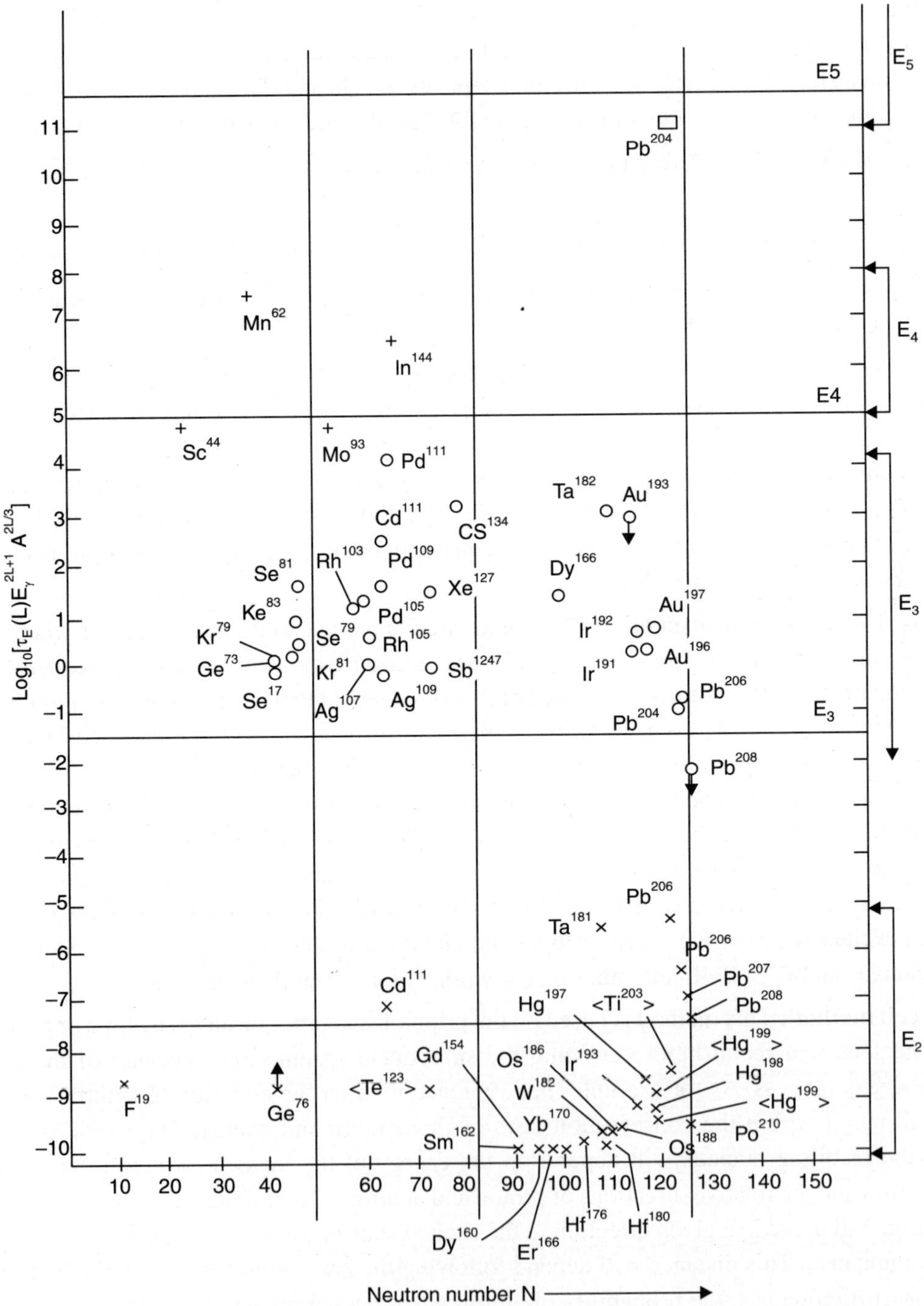

Fig. 7.16 Comparative lifetimes (Experimental) of E_2, E_3, E_4 or E_5 transitions plotted against neutron number N (Ref. 43).

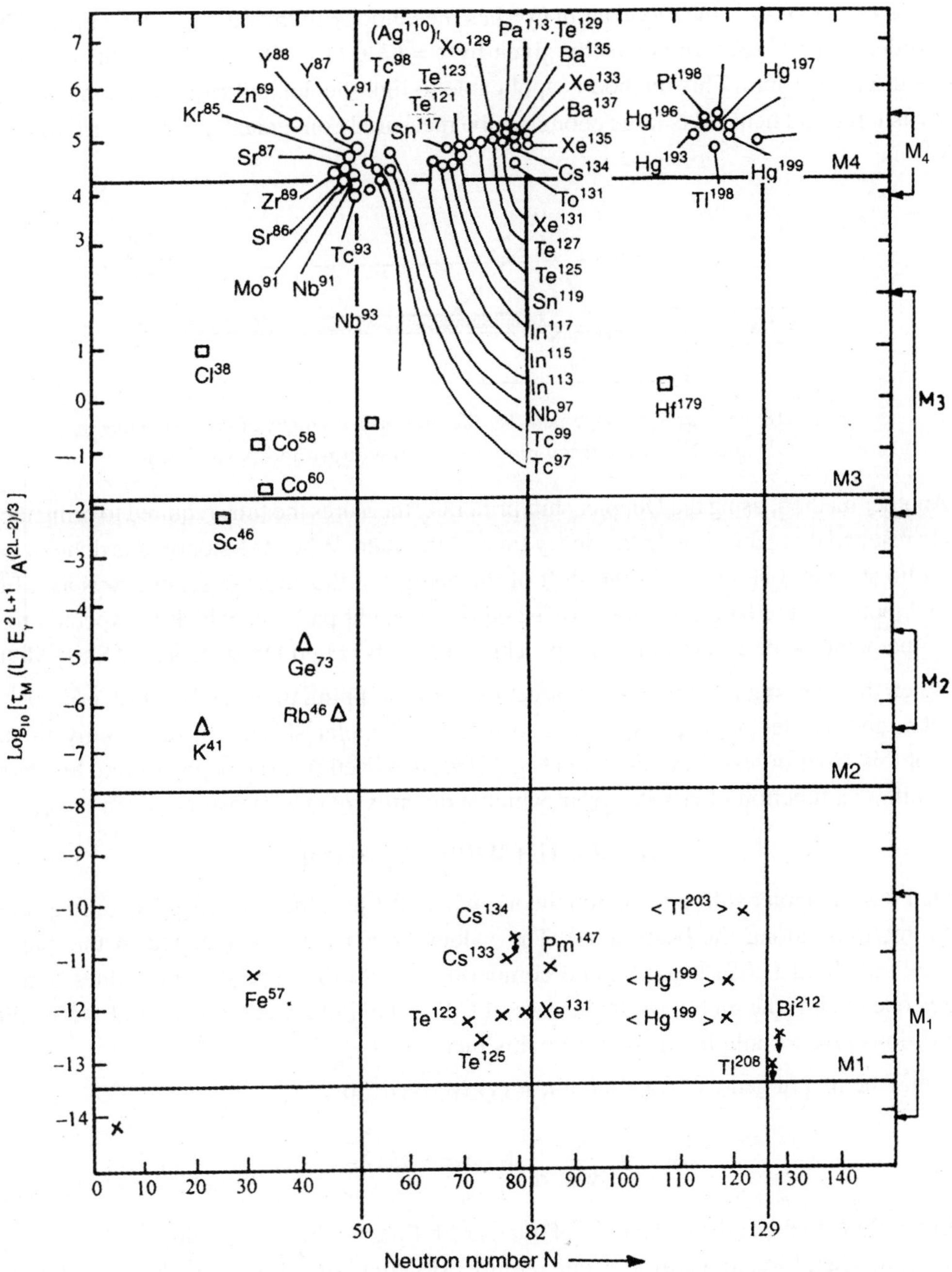

Fig. 7.17 Comparative lifetimes (Experimental) of M_1, M_2, M_3 or M_4 transitions plotted against neutron number N (Ref. 43).

least the distance d from the target to a position, where the counter can be seen through a slit. The counting rate as a function of d measures directly the rate of decay of the excited state in flight. For a reaction like F^{19} (p, α) O^{16}, the recoil atom O^{16} has the velocity in the range of 10^9 cm/sec, the lifetime is of the order of 10^{-10} secs, (it is exactly given by $\tau = (7 \pm 1) \times 10^{-11}$ seconds), then d is only about 0.1 cms and $\Delta d = 0.1$ mm. This method is applicable to light nuclei and large reaction energies, so that velocity v is large and hence d is large enough to be measured conveniently, *e.g.* O^{16} (d, p) O^{17}, and Be^9 (d, n) B^{10}, etc.

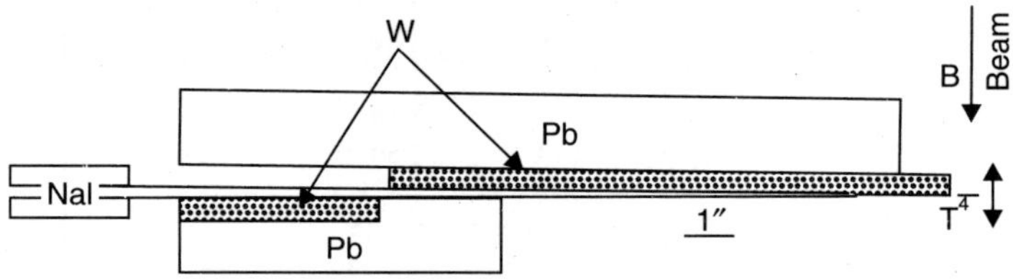

Fig. 7.18 The target arrangement for lifetime measurement of excited states in O^{17} and O^{10} by the recoil method. T is the target and B, the beam.

Another method, using the Doppler shift principle, measures the time required to bring the recoiling nucleus to rest and compares it with the decay time of the state. When the stopping is done in vacuum by another solid stopper, one measures the shift of the energy of the emitted gamma rays as a function of the angle[46], between the emitted gamma rays and the incident particles which is expected to be of the order of 2 per cent of the gamma ray energy. This method is called DSM (Doppler Shift Method).

When the stopping is done in solid target itself the stopping time is of the order of 10^{-13} secs, for light nuclei, and shorter for heavier nuclei, so that the Doppler shifts will occur only in states with lifetimes of this order or less[46]. As shown in Fig. 7.19, such lifetimes can be measured, by observing the Doppler shift as a function of cos (θ); from which a quantity $F(\tau)$ is measured, from the equation:

$$E_\theta = E_{90} \, [1 + \beta \, (0) \, F(\tau) \cos (\theta)] \qquad \text{...(7.113)}$$

as expected from Doppler shift attenuation theory, where $\beta(0)c$ is the velocity of recoiling nuclei, in the forward directions (along the beam axis). The values of $F(\tau)$ were calculated in the framework of Lindhart-Schraff-Shitt (LSS) theory[46], and comparing it with the experimental values obtained from Fig. 7.19*b*, one obtains the lifetime τ. In the case of Cd^{109}, the lifetimes of the order of $(45 - 150) \times 10^{-12}$ secs for various levels could be measured by this method.

It may be mentioned that for vacuum; $F(\tau)$ is given[46] by:

$$F(\tau) = \frac{\lambda}{\beta(0)} \int_0^\infty \beta\,(t)\, e^{-\lambda t}\, dt; \lambda = \frac{1}{\tau} \qquad \text{...(7.114)}$$

Basically $F(\tau)$ depends on the ratio of nuclear lifetime, to the slowing down times α of the nuclei in the material. Therefore, to measure τ, one requires slowing down time; which is based on the theory of slowing of recoil nuclei in materials. For higher energy reaction and thin targets, one can measure the intensity as a function of distance, between the source and a separate stopper and compare the resulting curve with that expected for a given lifetime. These methods are called DSAM (Doppler Shift Attenuation Method), or RDSM (Recoil Doppler Shift Method). (Ref. 46).

Resonance-Methods: The principle of these methods–using scattering or absorption phenomenon is based on the exact matching of the energies of the level in question in the scattering and absorbing nucleus and the incident radiation. The method is applied (*i*) when X-rays or low energy γ-rays are used as incident radiation, *e.g.* in Mössbauer spectroscopy or (*ii*) Doppler shift method of changing the incident energy; *e.g.* by (*a*) mounting the source on rotating system; (*b*) raising the temperature of the source or target, to widen the energy spectrum or (*c*) by using the succeeding gamma rays, *e.g.* in resonance scattering of neutron capture gamma rays, or β-decay, to match the recoil momentum, of the emitting nucleons, by an equal and opposite momentum imparted by the preceding gamma ray emitted in appropriate direction.

All these methods, basically measure the partial level width corresponding to gamma decay which is related to lifetime (if the gamma decay is the mode of decay) by:

$$\frac{1}{\tau_v} = \lambda_\gamma \ (\text{sec}^{-1}) = \frac{\Gamma_v}{\hbar} = 1.52 \times 10^{15} \ \Gamma_\lambda \ (\text{eV}) \qquad ...(7.115)$$

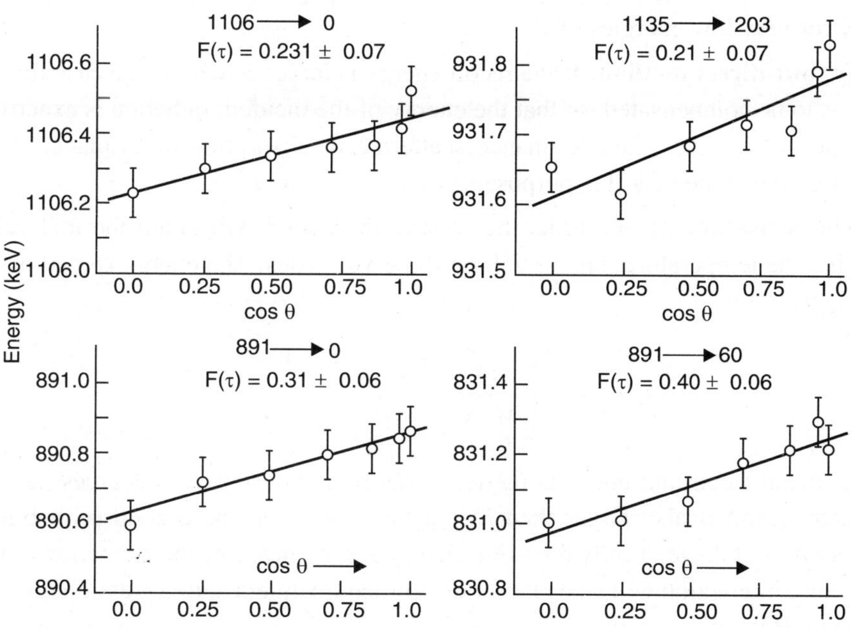

Fig. 7.19 Plots of doppler-shifted energies of gamma rays, for various excited states of recoiling nuclei of Cd^{109} in $49In^{109}$ (p, nγ) $48Cd^{109}$ reaction at 3.4 and 3.9 MeV proton beams from Chandigarh cyclotron (Chandigarh, India) (Ref. 46).

A common situation in all these methods is, that the energy of the scattered radiation is less than the incident radiation, from nuclei not embedded in solids or liquids (*e.g.* in gases or in vacuum). The energy loss Δ*E* recoil is calculated by obtaining the recoil momentum of the recoil nucleus from the application of the law of conservation of linear momentum. Then Δ*E* recoil is given by:

$$\Delta E_{\text{recoil}} = \frac{E_\gamma}{2} \frac{v_{\text{recoil}}}{c} = \frac{1}{2} \frac{E_\gamma^2}{M_p \ Ac^2} \qquad ...(7.116)$$

where E_γ is the resonance energy and M_pA is the mass of the recoiling nucleus. Different methods mentioned above basically provide to the nucleus the energy equivalent to this energy loss, so that the energy of the incident gamma ray falling on the scatterer or absorber is exactly equal to the excitation energy of the level in the scatterer (or absorber), which we want to measure. For comprehensive details of such methods, *see* F.R. Metzger: Progress in Nuclear Physics, edited by O.R. Frisch, V. 7, p. 54, published by Pergamon Press, New York[47].

We give below, briefly, the principles of operations of some of these methods of resonance.

Mössbauer spectroscopy: As described in Chapter 2, the scattered gamma rays from crystals have a certain probability of not losing any energy due to recoil; because in these cases when media are embedded in a crystal; the whole crystal recoils and not only the nucleus. Then in Eq. 7.116, M_pA is replaced by the atomic weight of the whole crystal; which will be of the order of 10^{24} M_pA in a crystal of size of 1 cm^3. Hence E recoil is very negligible. The width of level is determined directly, by using a velocity drive, attached to the source or target, and obtaining the resonantly scattered counts as a function of the velocity of the drive, *see* Fig. 2.18. The level width is, then, obtained directly. This method is applicable only for very low energies of E_γ.

Doppler shift-direct method: If the recoil energy is large, as will be the case for large E_γ, then recoil energy has to be compensated, so that the energy of the incident radiation is exactly the same; as the energy of the excited state, and resonance scattering or absorption takes place. In practice, two methods have been developed for this purpose.

(*a*) **Thermal compensation:** If one heats the source, the nuclei will attain thermal velocities in all directions. Higher the temperature, higher will be these velocities. The energy spread of these thermal energies will be given by:

$$\Delta E_\gamma \approx E_\gamma \left(\frac{2kT}{M_p A} \right)^{1/2} \frac{1}{c} \qquad ...(7.117)$$

where k is Boltzmann's constant and T is the temperature of the source in degrees (kelvin). If ΔE_γ is much larger than Γ_γ and is also larger than E_{recoil}; then there will be a good probability, that some incident gamma rays will have exactly the same energy as the energy of the excitation of the concerned level, and resonance interaction will take place. For detailed treatment and expression for this probability one should consult Ref. (47).

(*b*) The other method of compensation is based on the *use of a rotor of an ultracentrifug*e, as was first done by Moon and others[48]. The source is placed on the rim of the rotor which is made to rotate in such a way, that the target can see the gamma rays from a recoiling nucleus; which is given the compensatory velocity by the ultra centrifuge, so that the incident radiation on the target has exactly the energy required for resonance. The line width of the incident gamma ray is determined by the geometry of the experiment and the shape of the rotor. For details *see* Ref. (48).

(*c*) Another general method of providing compensatory energy to the recoiling nucleus is, by using the *preceding emission of beta ray* or gamma ray, in radioactive decay or in a nuclear reaction. For details, *see* Ref. (49).

One method which has been recently used utilising this principle, is the use of neutron capture gamma rays for the compensatory energy. We will describe somewhat in details the principle of the *resonance scattering of neutron capture gamma rays* for measuring Γ_γ. This method was developed at the Argonne National Lab, using CP_5 reactor[50], by H.S. Hans, G.E. Thomas and L.M. Bollinger; in 1966 for the first time, and is one of the latest methods for measuring Γ_γ. Earlier Fleischman H.H. and F.W. Stanek[50] used the same principle, to demonstrate the feasibility of this method.

Figure 7.20, represents the experimental arrangement. We will illustrate the principle involved, by taking example of the measurement of for the two levels of Sr^{88} at 1.84 MeV and 3.52 MeV. The target used was natural Strontium, which has isotopes of both Sr^{87} (7%) and Sr^{88} (82%). We expect gamma rays of high energies corresponding to 3.52 MeV and 1.84 MeV levels [Fig. 7.21*a*] to be emitted from Sr^{88}, in all directions, giving rise to the emission of 3.52 MeV ± ΔE_1 and 1.84 ± ΔE_2, where ΔE_1 and ΔE_2 will depend on the detailed kinematics of the gamma rays emitted and the recoil energies. Because of all directions into which gamma rays can be emitted, ΔE_1 and ΔE_2 will have a broad width, and hence, when such gamma rays fall on the scatterer, which contains Sr^{88} there will be an energy-overlap, with the levels of Sr^{88}, after taking into account the recoils of the scattering nuclei. On the other hand; if we substitute another scatterer, which has the same electronic properties as Sr; say Zr; we expect the same general spectrum, but without resonances. Differences of two such spectra should yield only the resonance scattering of neutron capture gamma rays from Sr^{88}. Figure 7.21*b* shows, one such spectrum. The spectrum exhibits strong evidence for two resonances—one at 1.84 MeV and the other at 3.5 MeV.

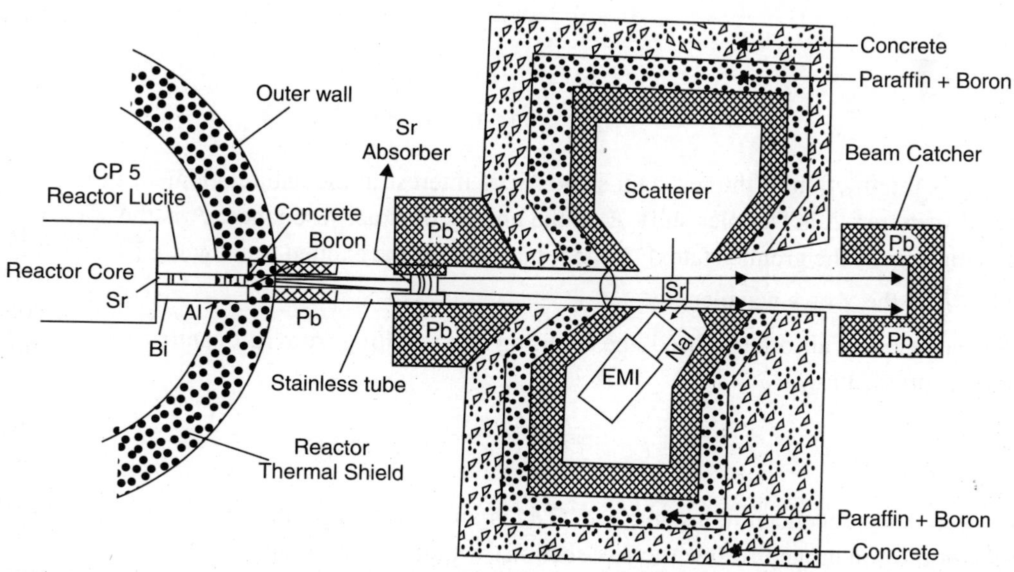

Fig. 7.20 Experimental arrangement for the measurement of radiation width (Γ_γ) of Sr^{88} levels by resonance fluorescence of neutron capture gamma rays (Ref. 50).

Such results have been obtained also for Boron[11] and Se^{78}. Since there is no possibility of knowing the shape of the recoil-broadened gamma ray line, which provides the resonantly scattered radiation, we follow the integral transmission method as follows: We measure spectra for fixed strontium scatterer, but for various thickness of resonant (strontium) absorbers and for equivalent (*i.e.*, same number of electrons/cm^3) non-resonant (zirconium) absorber in the beam. For each thickness, relative intensities were used to determine ratio:

$$R = \frac{I_\gamma}{I_0} \qquad \text{...(7.118)}$$

where I_γ, is the intensity of resonantly scattered gamma rays, from a resonance (strontium) absorber and I_0 is the intensity with non-resonance (Zr) absorber. Assuming that neutrons are absorbed uniformly throughout the target and the probability of detecting resonantly scattered γ-rays is independent of the width within the scatterer, we can write down the intensity within energy E and $E + dE$ of the gamma rays from the target, falling on the scatterer, after passing through a given absorber as:

$$N(E)\, dE = \frac{C}{(\sigma_e + \sigma_r)}[1 - e^{-n_1(\sigma_e+\sigma_r)_1}]\, e^{-n_2(\sigma_e+\sigma_r)_2}\, dE \qquad \text{... (7.119)}$$

where C is a constant, dependent on the geometry and number of neutrons captured; n_1 and n_2 are the thicknesses of the source and absorber, respectively expressed as number of atoms/cm^2, σ_e is the atomic cross-section and σ_r is the resonance absorption cross-section for incident gamma ray. Now σ_r is zero for a non-resonant absorber but for a resonant absorber it is given by:

$$\sigma_r = \sigma_{\max} \exp - \left[\frac{(E - E_r)^2}{\Delta}\right] \qquad \text{...(7.120)}$$

where

$$\sigma_{\max} = \pi ƛ^2 \frac{2J_1 + 1}{2J_0 + 1}\left(\frac{4\Gamma_\gamma}{\Gamma}\right)\left(\frac{\Gamma\pi^{1/2}}{2\Delta}\right) f \qquad \text{... (7.121)}$$

Here f is the fraction of the resonant scatterer of interest in the natural element used for source, absorber and scatterer; Δ is Doppler shift, E_r is the energy of exact resonance, J_1 is the spin of excited state and J_0 the spin of the ground state, Γ is the total width of the excited state, while Γ_λ is the radiation width and $2\piƛ$ is the γ-rays wavelength.

When a resonant absorber is used, the counting rate resulting from resonantly scattered radiation may then be expressed as:

$$I_r = Ke^{-(\sigma_e n)_2} \int_{-\infty}^{\infty} e^{-(\sigma_r n)_2}\, F(E)\, dE \qquad \text{...(7.122)}$$

where K is geometry-dependent constant, and $F(E)$ is an energy dependent function depending on σ_e and σ_r—the non resonant and resonant cross-sections, which was computed.

For a non-resonant absorber; the intensity of scattered counts under the resonance peak is given by:

$$I_0 = Ke^{-(\sigma_e n)_2} \int F(E)\, dE \qquad \text{...(7.123)}$$

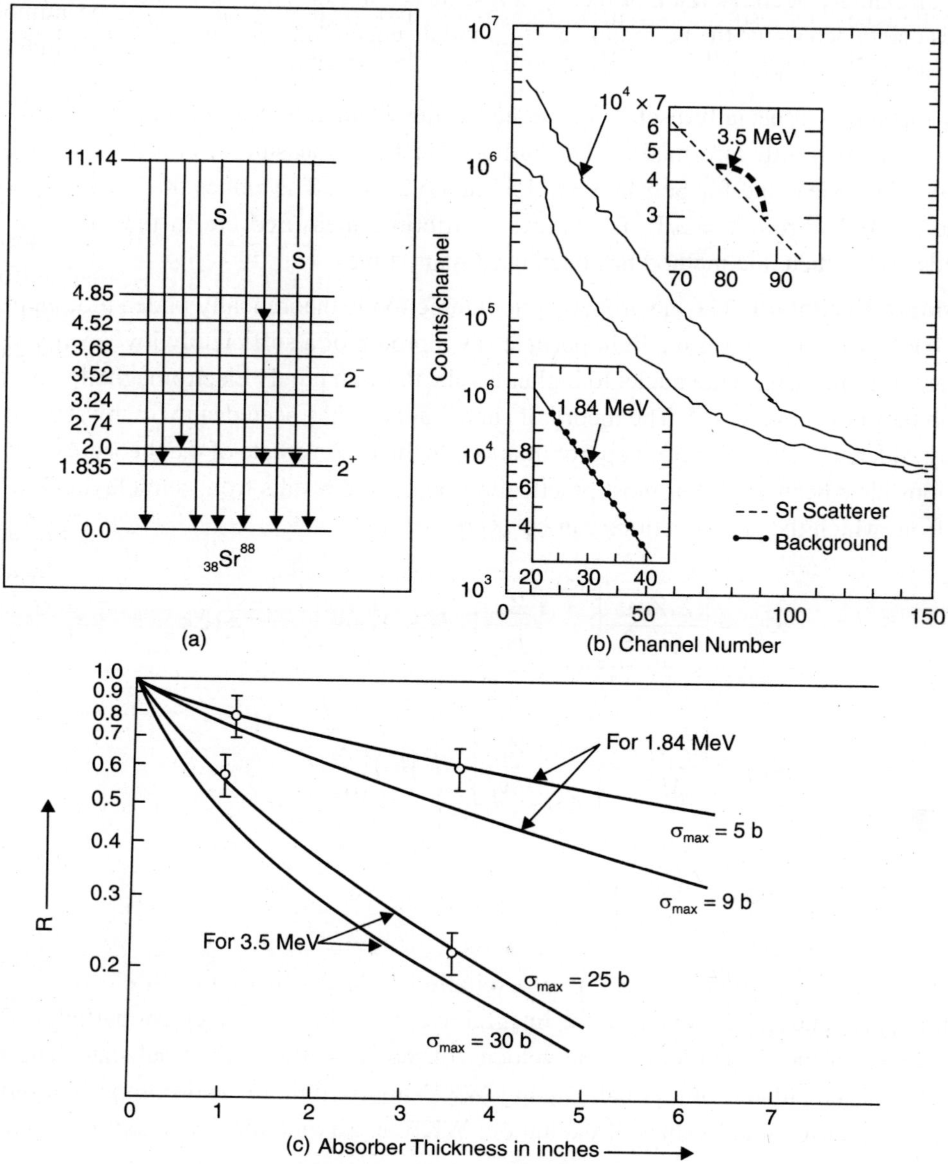

Fig. 7.21 *(a)* The γ-decay scheme of Sr^{88}; *(b)* The original spectra of resonance and non-resonance scattered gamma rays, and *(c)* Absorption curves (Ref. 50).

Therefore, one can write:

$$R = \frac{I_r}{I_0} = \frac{\int_{-\infty}^{\infty} e^{-(\sigma_r\, n)_2}\, F(E)\, dE}{\int_{-\infty}^{\infty} F(E)\, dE} \qquad \text{...(7.124)}$$

The values of R, were computed as a function of $(n\, \sigma_{max})_2$, for a range of (σ_{max}/σ_e) and for various thicknesses of the source and absorber (n_2); for several values of σ_{max} and for known fixed values of σ_e, n_1 and n_2. A comparison of these curves, shown in Fig. 7.21*c* with the experimental value of *R*, enables one to determine the best value of σ_{max} and its uncertainty. From σ_{max}, one can get Γ_γ, if all the other quantities are known.

This method is especially useful for excited states from 2 to 5 MeV range, for which other methods have provided little information. The authors[50] had also measured Γ_γ for three levels of B^{11} at 2.14, 4.16 and 5.03 MeV and for Se^{78} at 3.35, 4.57 and 5.2 MeV. The values of Γ_γ are in the range of 0.002 eV to 1.00 eV. From $\tau = \hbar/\Gamma_v$ the values of lifetime measured are in the range of 10^{-13} to 10^{-16} seconds. Since then, the method has been used many times.

Coulomb Excitation: The electromagnetic interaction is theoretically understood in detail. If a heavy (say ${}_2He^4$, ${}_3Li^7$, etc.) or even a light particle say a proton of a sufficiently low energy is allowed to strike a target; it may excite the nuclei to higher levels, through purely electromagnetic interaction, if the Coulomb barrier is not crossed. The theory of such reactions has been dealt with by several authors, especially by Alder, et al.[51], using Born approximation. In the present state of the theory, exact Coulomb wave-functions have been used. For most practical purposes, the results from semiclassical treatment[51] may apply from which the cross-sections can be expressed as:

$$\sigma_{E\lambda} = \left[\frac{Z_1 Z_2 e^2}{M_p v^2}\right]^{2-2\lambda} \left[\frac{Z_1 e}{\hbar v_i}\right]^2 B(E_\lambda) \left(\frac{v_i}{v_f}\right)^{2\lambda-2} f_{E\lambda}(\xi)$$

and

$$\sigma_{M\lambda} = \left[\frac{Z_1 Z_2 e^2}{M_p v^2}\right]^{2-2\lambda} \left[\frac{Z_1 e}{\hbar v_i}\right]^2 B(M_\lambda) \left(\frac{v_i}{v_f}\right)^{2\lambda-1} f_{M\lambda}(\xi);$$

where

$$\xi = \frac{Z_1 Z_2 e^2}{\hbar}\left(\frac{1}{v_f} - \frac{1}{v_i}\right) = n_f - n_i \qquad \text{...(7.125)}$$

Here $\sigma_{E\lambda}$ and $\sigma_{M\lambda}$ are cross-sections of excitation of levels involving an electric or magnetic transition and $f_{E\lambda}(\xi)$ and $f_{M\lambda}(\xi)$ represent the integral over the orbit of the incident particles where v_i is the initial velocity of the projectile, v_f is the velocity of the projectile in the final state. The values of $f_{EM\lambda}(\xi)$ have been calculated numerically using WKB approximation and comparison with exact calculation show that for small values of excitation; WKB approximation gives correct results.

We obtain $B(E_2)$ from Eq. 7.125, by comparing the experimental values of cross-section, with the theoretically expected values for different values of $B(E_2)$, which is the reduced matrix element, [Eq. 7.56 and Ref. (51)], and can be written as:

$$B(E_\lambda) = \left| \langle f \,|\, H'\left(A_{LM}^{E_\lambda}\right) |\, i\rangle \right|^2 = \frac{1}{2I_i + 1}\left|\langle I_i \,\|\, m(E_\lambda) \,\|\, I_f\rangle\right|^2 \qquad \text{...(7.126)}$$

Which represents the reduced transition probability associated with a radiative transition of multipole of order λ. The value of $B(E_2)$ have been measured by Coulomb excitation, for many excited states, using heavy ions as well as light nuclei. The value of $B(E_2)$ is connected to the transition probability λ_r as:

$$\lambda_r (E_2, I_f \rightarrow I_i) = 1.23 \times 10^{-2} (\Delta E)^5 B(E_2, I_f \rightarrow I_i) \times \frac{2I_i + 1}{2I_f + 1} \text{ sec}^{-1} \qquad ...(7.127)$$

and life-time of the state, *i.e.*, τ is given by $\tau = 1/\lambda_r$ and ΔE corresponds to energy of transition. This method is useful for levels up to excitation of 1 or 1.5 MeV, for light incident particles, where direct Coulomb excitation is possible; but can go to much higher energies for heavy ions incident on the target nuclei, where multiple Coulomb excitation can take place from one excited level to higher excitation. The lifetime of the order of a few nanoseconds to picoseconds can be measured through this method.

Figure 7.22 shows a recently measured case of Coulomb excitation of Cs^{133} by protons 3.2–4.2 MeV, from 160.7 keV to 871.8 keV excitation energies by Singh, et al.[20] The experimental yield can be calculated, by taking into account, the incident flux, the mass numbers of projectile and target and the stopping power of the target. This yield is related to $B(E_2)$ and $f_{E_2}(\xi)$ of Eq. 7.125, which have been tabulated by Alder, et al. (51).

With protons, there is some probability, that compound nucleus is formed, along with Coulomb excitation. Figure 7.22*b*, shows the excitation function both experimentally[20] and expected theoretically from pure Coulomb excitation. It shows that there is less than 1% contribution from compound formation.

Such measurements have been carried out using alpha particles and heavy ions as projectiles– Alphas are most suited, because according to Eq. 7.125, the cross-sections for Coulomb excitation for multipole order E_λ is proportioned to Z_1^2, which means that the larger cross-sections are obtained with the heavier projectiles. So alpha particles are preferable to protons for Coulomb excitation. Also in the case of protons as projectiles, (p, n, γ) reaction is also possible, which complicates the matter. On the other hand for heavier ions, multiple Coulomb excitations is possible, which makes the analysis of pure E_2 transitions more difficult.

The measurement of the lifetimes of the excited states of nuclei has continued to be a topic of interest especially with the availability of modern electronics and computing facilities. Recently mean lifetimes of 22 levels of P^{32} up to an excitation energy of 6.4 MeV, obtained from H^2 (P^{31}, p) P^{32} at bombarding energies of 22—29 MeV of deuterons, of some 20 levels in S^{32} available from H^2 $(P^{31}, n\gamma)$ S^{32}, Si^{28} $(Li^6, pn\gamma)$ S^{32} and P^{31} (p, γ) S^{32} reactions up to an excitation[52] of 8.0 MeV and some six levels of Sn^{115}, up to an excitation of 3.66 MeV, obtainable from In^{113} $(\alpha, 2n\gamma)$ Sb^{115} at bombarding energy of 27.2 MeV have been obtained using Doppler shift attenuation method[52]. The experimental values of the lifetimes of these cases were compared with the theoretically expected values based on the application of extended shell model. For positive parity states, the effective Hamiltonian of Wildenthal was used[53], using an oscillator wave function with $\hbar\omega = (45A^{-1/3} - 25A^{-2/3})$ [*see* chapter on Shell Model]. *Sd* shell mixing or *fp* shell mixing was used for diagonalisation. For negative parity states, another Hamiltonian form (WBHB) was used[54]. The agreements were qualitatively correct, with large quantitative differences. Figure 7.23 shows the comparison. For Sb^{115} the transition probabilities were calculated using interacting Boson Fermion Model[55], with a Hamiltonian of D. Buchurescu[56], et al. Again the agreement is only qualitative.

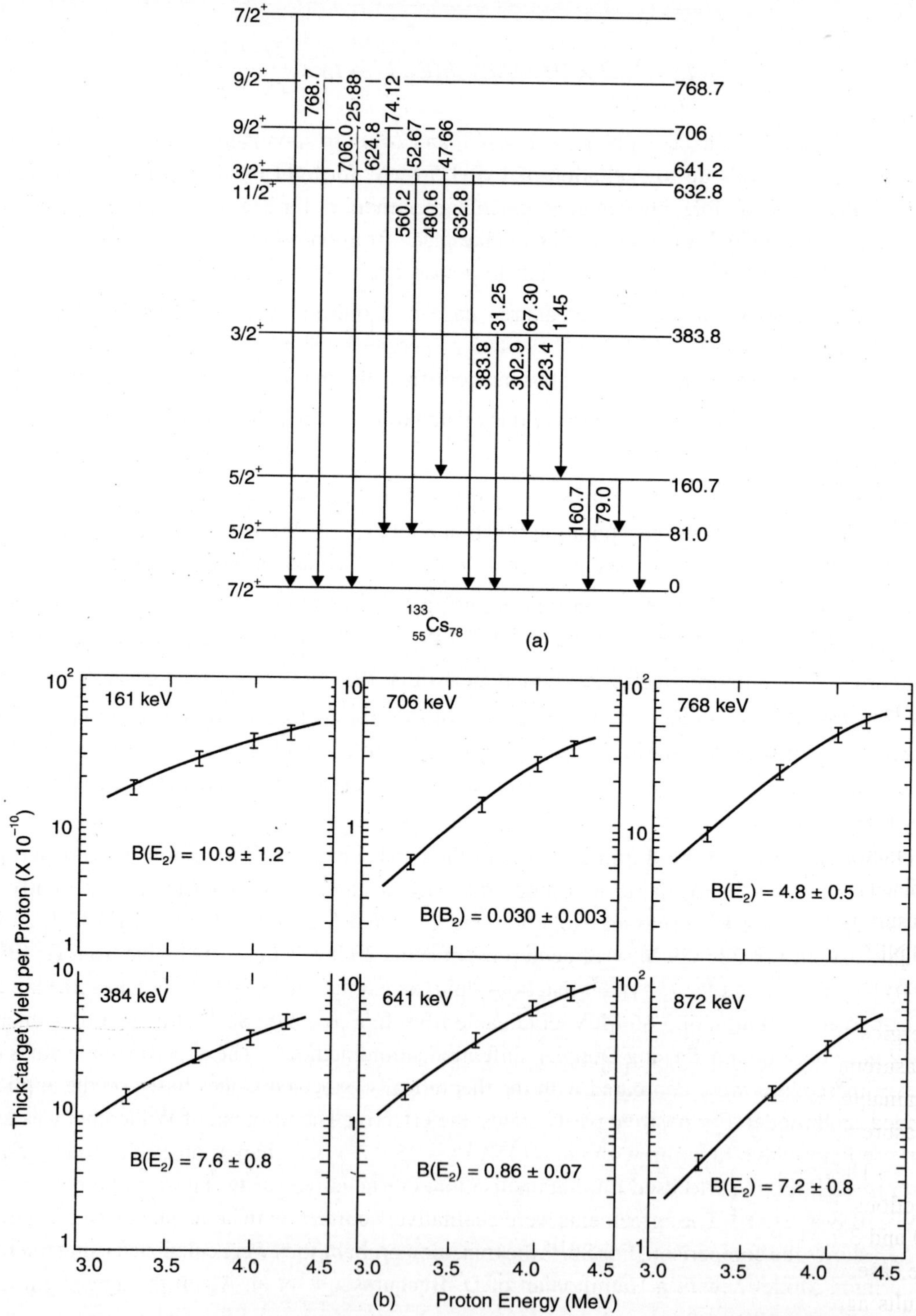

Fig. 7.22 Coulomb excitation of Cs^{133} using protons of 3.4-4.2 MeV. (*a*) The excited levels of Cs^{133}; (*b*) Excitation function for various levels. Points are experimental and curves are derived semi-empirically. Eq. 7.125. B (E_2) values are in units of e^2 cm^4 × 10^{-50} (Ref. 20).

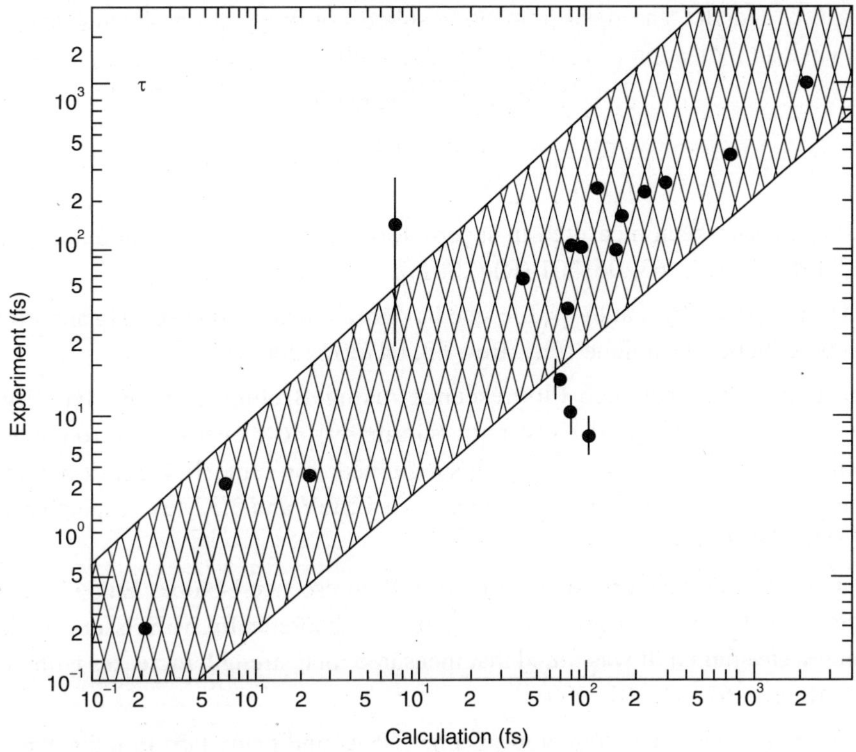

Fig. 7.23 Calculated mean lifetimes compared to those experimentally determined. Experimental upper limits are shown by open triangles. Calculated lifetimes (the shaded area) agree with experiment to within a factor of 5 inside the shaded region (Ref. 52).

The other methods of lifetime measurement which are commonly used these days are: Recoil Distance method, Coulomb Excitation, and Direct methods using electronics. The recoil distance method as discussed earlier in Section (7.7), has been popular, when lifetimes of many states is involved. Recent measurements[57], using this technique has yielded the lifetimes in Nb^{86} and Zr^{86} from Ni^{58} (S^{32}, 4P) Zr^{86} and Ni^{58} (S^{32}, $2pn$) Nb^{86} with 130 MeV projectile energies. Similarly lifetimes[58] of low lying states of $La^{125,\ 127}$ were measured using M_0^{94} (CI^{35}, 2 $p2n$) La^{125} and Cd^{112} (F^{19}, $4n$) La^{127} reaction at the beam energies of 155 MeV of CI^{35} and 84.5 MeV of Ca^{112} using this technique. Another interesting measurement using this technique by I M Govil[59], et al., yielded the lifetimes of ten states in $Xe^{122,\ 124}$ obtainable from Pd^{110} (O^{16}, 4X) Xe^{122} and Pd^{110} (O^{18} $4n$) Xe^{124} at beam energies of 66 MeV. All these measurement could yield lifetimes for 35 levels up to an excitation of 4.4 MeV.

These two methods have become popular methods for lifetime measurement, involving nuclear reactions. In a recent set of papers, Coulomb excitation of Ta^{180} and Cu and Zn was measured[60] using 3.0 and 3.7 MeV protons and 12.20 MeV alpha particles up to excitation energies of several MeV. In one case, evidence of an intermediate state was discovered in Ta^{180}, while in the case of Cu and Zn the results agree with particle phonon interaction model, using the shell model configuration and quasi-particle cluster vibration models[61]. The interesting case of Coulomb excitation, recently measured[62], however, used an exotic beam of Be^{11} at 57–60 MeV/A for Coulomb exciting 320 keV, first excited states in Pb, Au, C and Be. The measurements resulted not only in reproducing $B(E_2)$ values of the transition, but gave an indication of higher order Coulomb and nuclear effects.

7. Radiative Transitions

2000–2008

Lifetimes

(*i*) Lifetimes have been measured in superdeformed states, in Ar^{36} (24 authors) using 80 MeV Ne^{20} [Phy. Rev. C. 63, 061301 (R) (2001)].

(*ii*) In another paper, by 8 authors, lifetime have been measured in Au, Hg and Pb isotopes using photo-activation technique [Phy. Rev. 63, 047307 (2001)].

(*iii*) In a paper of Resonance Flourescence, using Brehmsstrahlung Facility of 4.3 MeV Dynamitron, at Stuttgard University (Germany), low energy electromagnetic excitation strengths in Sb^{121} and Sb^{123}, have been studied; for some 164 transitions in Sb^{121} and 83 transitions in Sb^{123} in the energy range up to 4 MeV and 3.5 MeV respectively [Phy. Rev. C. 65, 024313 (2002)].

(*iv*) In another similar paper, using Resonance Flourescence isotopes $Eu^{151,\,153}$ and Dy^{163} Ho^{165} have been studied by a group of 15 authors. Theoretically M Scissors mode in odd mass nuclei were studied. It was found that measured total strength increases with mass number A: [Phy Rev. C. 67, 034307 (2003)].

(*v*) An international group from Notre-Dame (USA) and many labs in India, used recoil isomer tagging technique to measure the lifetime of 8+ isomer of Z^{98}r, and Mo^{90}, in the microsecond range [Phy. Rev. C.70, 014311 (2004)].

(*vi*) In another case, lifetime measurements of excited states of levels, in Te^{120} were measured in elastic scattering of neutrons–(n, n, γ), from $H^3(p, n)$ He^3 reaction (7, 21 MeV neutrons) using 7 MeV proton beam, resulting in the study of 70 levels, from 666 keV to 3.21 MeV, using Dopplar shift technique for some 138 gamma transitions. [Phy. Rev. C. 69, 064322 (2004)].

Nuclear Moments of Excited States

g factors: Time dependent perturbed angular correlation techniques has been used for measuring *g* factor of Pb^{196} ($\overline{11}$) as 1.04: Pb^{174} ($\overline{11}$) as 1.03, Pb^{196} (9^-), as – 0.037 and Pb^{194} (9^-) as 0.0421. [Phy. Rev. C. 69, 054318 (2004)]. A similar technique has been used in another measurement of *g* factor determination of Yb^{168} (2^+) state at 123.4 keV energy, which has a half life of 0.88 nano-seconds [Phy. Rev. C. 69, 0-34370 (2004)].

In quite a few papers; *g* factor have been studied by various techniques: (*i*) Ion-source laser spectroscopy [Phy. Rev. C. 65, 02435 (2002)]; (*i*) Coulomb excitation [Phy. Rev. C. 65, 024316 (2002)]. (*ii*) Coulomb Excitation in inverse kinematic and transient magnetic fields [Phy. Rev. C. 65, 534308 (2002)].

Quadrupole Moments: In an interesting paper, quadrupole moment in Pb^{194} transition has been measured using double perturbed analysis of a level mixing spectroscopy measurement [Phy. Rev. C. 65, 024320 (2002)].

γ-rays Spectroscopy: This has been a subject of a large number of papers; with big collaborative efforts. In a paper with 27 authors from labs, in Europe. γ-rays spectroscopy of Kr^{73}, ($Z = 36$, $N = 37$) has been studied using $Ca^{40}(Ar^{36}$, 2pn) Kr^{73} and Ca^{40} (Ca^{40}, α2pn) Kr^{73} reactions at 145 and 160 MeV incident energies. Comparison with Cranked Strunsky and Cranked Relativistic Mean Field shows, that they match [Phy. Rev. C. 63, 044331 (2001)]. Some fifteen papers in Phy. Rev. C. 64 (2001). One paper by Govil et al. concerns recoil distance method by measuring lifetimes in Ta^{178}, from Notra Dame and Chandigarh [Phy. Rev. 65, 034303 (2002)]. In an interesting paper, on γ-rays spectroscopy, on Xe^{139}, Be^{140}, Ce^{143} (N = 85) obtained from fission of cm^{258}, the observed excitation in N = 85 are interpreted as being due to quadrupole and oclupole vibrations [Phy. Rev. 64, 044302 (2003)]. In three more papers (*i*) Bi^{193} (64 γ-rays, and Bi^{191} (32 gamma rays), [Phy. Rev. C. 69,064328 (2004)]. (*ii*) $Tc^{105, 107, 109}$ from fission of Cf^{242} and (*iii*) Thermal neutron capture of Ni^{58}, (414 gamma rays in Ni^{57}, 390 in Ni^{59} and 240 in Ni^{60} have been studied [Phy. Rev. C. 70, 044318 (2004)].

In some thirty papers in [Phy. Rev. C. 61(2000)], and Phy. Rev. C. 62 (2000), a massive collaboration is in evidence; where number of authors in each paper ranges from 16 to 19. In a paper in Phy. Rev. C. 74, 064305 (2006), γ-rays from Ne^{26i}, P^{26}, and $Na^{28, 29}$ have been investigated.

In an experimental paper, the authors have carried out gamma rays angular distribution measurement from N_i^{nat} (n, n, γ) reactions, at 1.6 MeV and 1.8 MeV neutron energies. Further, lifetimes have been measured through Doppler shift attenuation method. The results support an enhancement of proton-core excitation and related quadrupole and pairing strengths in the light N_i isotopes, in agreement with mean field and shall model calculation [Phy. Rev. C.77, 064301 (2008)].

REFERENCES

1. C. Michael Lederer, J.M. Hollander and I. Perlman: Table of Isotopes, John Wiley & Sons, New York (1967); M.A. Rothman, H.S. Hans and C.E. Mandeville: Phy. Rev. 100, 83 (1955).

   R.J. Gehrke, J.E. Cline and R.I. Heath: Nuclear Instruments and Methods 91, 349 (1971).

2. Theory of Multipole Radiation; S.A. Moszkowski, Beta and Gamma Spectroscopy; edited by K. Siegbahn, North Holland Publishing Company, Amsterdam (1955). W. Heitler, Prac. Camb. Phil. Soc. 32. p. 112 (1936).

   Theoretical Nuclear Physics, J.M. Blatt and V.F. Weisskopf Chapter XII, John Wiley & Sons, New York, (1952). Nuclear Physics—Theory and Experiment: R.R. Roy and B.P. Nigam; New Age International (P) Ltd., Publishers, New Delhi (1986), W. Franz, Z. Physik, 127, p. 363 (1950).

3. W. Heitler: The Quantum Theory of Radiation, Oxford University Press, London, (1947), Chapters 2 and 3.

   Reference (2); S.A. Moszkowski, Phy. Rev. 89, 474. (1953), Quantum Mechanics, L.I. Schiff, McGraw-Hill Book Company, Inc. (1949).

4. M.E. Rose: Phy. Rev. 76. 678 (1949), M.E. Rose, G.H. Goertzel, B.I. Spinard, J. Harr and P. Strong: Phy. Rev. 83 79 (1951).

   Experimental Nuclear Physics, edited by E. Segre, John Wiley & Sons, p. 341 (1959).

5. Theory of Internal Conversion by M.E. Rose.

   Beta and Gamma Spectroscopy, edited by K. Siegbahn (Ref. 2) (1955).

6. Reference (3).
M.E. Rose: Elementary Theory of Angular Momentum, John Wiley & Sons (1957).
7. L.A. Sliv and I.M. Band, Report 571CC K1, and 581CC K1 University of Illinois; Coefficients of Internal Conversion of Gamma Radiation. Part—I. K-shell & Part—II L-shell. Physico Technical Institute, Academy of Sciences, Leningrad (1958).
8. M.E. Rose, L.C. Beidenharn, and G.B. Arfken: Phy. Rev. 85, 5 (1952); M.E. Rose, 91, 619, (1953): Ibid, 93, 477, (1954), M. Riou, J. Phy. Radium 13,480, (1952). Exp. N. Physics, V.III, edited by E. Segre-M. Deutsch and O. Hanson, p. 344, John Wiley & Sons, New York (1959).
9. B.H. Fowler: Proc. Royal Society (London) 129A, 1, (1930).
E. Segre: Experimental Nuclear Physics, John Wiley & Sons. Inc. New York (1959).
10. L.A. Sliv. Doklady Akad Nauk S.S.R. 64, 321 (1949).
B.P. Singh, H.S. Hans and P.S. Gill, Nuovo Simonto, V. 9, p. 699 (1958).
11. G.K. Harton: Proc. Phy. Soc. (London), 60, 45 (1948).
12. M.E. Rose. Phy. Rec. 76, 678 (1949).
G.H. Aston, Prac. Cam: Phil. Soc. V. 10, p. 935 (1927).
P. Morrison and L. Schiff, Phy. Rev. 56, 24(1940); B. Saraf: Bulletin of A.P.S. (1954).
13. D.R. Inglis: Revs. Mod. Physics. 25, 390 (1953); 27, 76 (1955).
14. B. Chand, J. Goswamy, Divinder Mehta: Nirmal Singh and P.N. Trehen: C and J of Physics 69, p. 90 (1991).
15. H. Slatis and K. Siegbahn: Arkv Fysik 4, 485 (1952).
A.W. Sunyar, J.W. Mihelich, et al., Phy. Rev. 86, 1023 (1952).
F.R. Metzger and H.C. Amacher: Phy. Rev. 88, 147 (1952).
16. H. Slatis and K. Siegbahn, Arkv Fysik 4, 485, (1952). G.H. Nijgh and A.H. Wapstra: Nuclear Physics 1, 245 (1956).
17. D.R. Hamilton, Phy. Rev. 58 (1940), 122; G. Racaah, Phy. Rev. 84, 91, (1951) G. Goetzel, Phy. Rev. 70 (1946). 897, K. Alder, Helv. Phy. Acta, 25, 235 (1952).
L.C. Biedenharn, and M.E. Rose, Rev. Mod. Physics. 25, 729 (1955). Angular Distribution of Nuclear Radiation by H. Frauenfelder (p. 530).
Gamma Spectroscopy in Ref. (2) (1955) P. Loyd: Phy. Rev. 85, 904 (1952).
18. V. Fano, Phy. Rev. 90, (1953). 577, F. Coester and J.M. Jauch, Helv. Physica Acta, 26, 3 (1953).
19. R.M. Steffen, Advances in Physics 4, 293, (1955); Brady E.L. and M. Deutsch Phy. Rev. 78, 558 (1950).
20. K.P. Singh, D.C. Tayal, B.K. Arora, T.S. Cheema and H.S. Hans: Canad J. Physics, V. 63, p. 483 (1984).
21. Potnis V.R., and Rao. G.N., Nuclear Physics, 42, 620 (1963).
22. Moszkowski S.A., Theory of Multipole Radiation: α-B-spectroscopy. V. 2, 9.753 North Holland Publishing Co., Amsterdam (1965).
23. Davidov A.S. and Philipov. G.F.: Nuclear Phy. 8, 237(1958).
24. Grechukin D.P., Nuclear Phy, 40, p. 422 (1963).
25. L.C. Biedenharn, G.B. Arfken and M.E. Rose: Phy. Rev. 83, 586 (1951).
26. A.J. Ferguson, Angular Correlation Methods in Gamma Rays Spectroscopy: North Holland Publishing Co., Amsterdam (1965).
27. U.S. Pande and B.P. Singh: Nuclear Instruments and Methods, 97, 123-130. (1971): B.P. Singh and M.S. Dahaya: Phy. Rev. C. 6, 1789 (1972). B.P. Singh, M.S. Dahaya and U.S. Pande: Phy. Rev. C4, 1510 (1971).

28. J. Seed and H.P. French. Phy. Rev. 88 1007, (1952).

S.P. Lloyd, Phy. Rev. 83; 7(6), (1951); L.C. Biedenharn and M.E. Rose., Rev. Modern Physics 25, 729 (1953).

29. T.B. Novey, Phy. Rev. 96, 547, (1954); J.K. Balling, J. Battey, L. Madensky and F. Rassetti. Phy. Rev. 89. 182, (1953). A. Abragam and R.V. Pound. Phy. Rev. 89, 1306 (1953).

30. Reference (14); D.L. Falkoff and G.E. Uhlenbeck. Phy. Rev. 79 (1950) 334.

31. H. Fraundfelder, Ann. Rev. Nuclear Science 2, 129, (1953).

32. M. Monta and M. Tamadal: Progress of Th. Phy. 8, (1952), 449, 10 (1953), III: 10, (1953) 641, D.T. Stevenson and M. Deutsch; Phy. Rev. 84, 1071, (1951). Ibid Phy. Rev. 83. 1202, (1951) E.K. Darby and W. Opechowskin Phy. Rev. 83, 676 (1951).

33. G. Goertzel, Phy. Rev. 70(1948), 897; K. Alder, Helv. Phy. Acta 25, (1952) 235.

34. Perturbed Angular Correlation: Karlson F., Mathiase E. and Siegbahn K. North Holland Publishing Co. Amsterdam (1964).

35. K. Alder; Phy. Rev. 84 (1951). 389; H. Fraundfelder, J.S. Lawson, and W. Jentschke, Phy. Rev. 93, 1126 (1954).

36. M. Salomon, L. Bostron et al.; (p. 204) in Ref. (34) (1964), A.K. Bhatti, N. Aggarwal, S.C. Bedi and H.S. Hans: Hyperfine Interactions. 23, 125–132 (1985).

37. I. Wiedenhöver, et al. (23 authors), Phy. Rev. 58, p. 721 (1998).

T.I. Trackicova, M. Finger, M. Kremer, K. Kumar, A. Jamata, N.A. Lebedev, V.N. Povlov, and E. Simekova: Phy. Rev. 58, 1987 (1998).

38. K. Kumar, Hyperfine Interactions 75, 43, (1992): Nuclear Models and Search for Unity in Nuclear Physics, (Universte Forlarget, Bergen (1984). K. Kumar and M. Baranger, Nuclear Physics A, 110) (1968).

39. L. Weissman M. Hass, and C. Broude: Phy. Rev. C. 57. 621 (1998).

40. H. Sanerr: Helv. Phy. Acta 23, 381 (1950). H. Slatis: Arkiv Mat Fysik 36A, 17 (1948).

W.M. Good, D. Peasles, and M. Deutsch: Phy. Rev. 69, 313 (1946).

R.J. Hart, K. Russel & R.M. Steffan.

Phy. Rev. 81, 460, (1951).

41. P.R. Bell: The Scintillation Method, Beta Gamma Spectroscopy, edited by Kai Siegbahn, (1955); ref. (2), M. Medadjenovic and A. Hedgram: Physica 18, 1242 (1952); S.C. Pancholi and R.K. Bhowmick, Ind. Journal of Pure and Applied Physics 27, 660 (1989). Private Communication A.P. Patro and, G.K. Mehta. N.S.C., Delhi (1996).

42. J.W.M. Du Mond: Rev. Sc. lnstr. 18, 626(1947). D.H. Muller, H.C. Hoyt, D.J. Klein and J.W.M. Du Mond. Phy. Rev. 88, 775 (1952).

43. M. Goldhaber and A.W. Sunyar—Classification of Nuclear Isomers: $\beta - \gamma$ spectroscopy by K. Sieghbahan p. 453 (Ref. 2); (1955). A.W. Sunyar: Phy. Rev. 95, 626, (1954) M. Goldhaber and A.W. Sunyar, Phy. Rec. 83, 906 (1951).

44. R.E. Bell and H.E. Petch: Phy. Rev. 76, 1409 (1949).

A.W. Sunyar: Phy. Rev. 98, 653 (1955).

I.M. Govil, C.S. Khurana and H.S. Hans: Nuclear Physics 45, 60–64 (1963).

45. Jacobsen, J.C. Phil.: Mag. 47, 23, (1924) Devons S. and Hereward H.G.: Nature 164. p. 586 (1949).

46. D.K. Avasthi, V.K. Mittal and I.M. Govil: Phy. Rev. C 26, 1310 (1982).

I.M. Govil and H.S. Hans: Proceeding of Ind. Acad. Sc. (Engineering section), V.3 p. 237, (1980).

E.K. Warburton, D.E. Alburger and D.H. Wilkinson: Phy. Rev. 129, p. 1280 (1963).

A.E. Litherland, M.J.L. Yates, B.M. Hinds, D. Eccleshal: Nuclear Physics, 44, 220–255 (1963).

K.C. Chan, B.L. Cohen, L. Shabason: Phy. Rev. C. 11, 1064, (1975).

A.E. Blaugrund: Nuclear Physics, 88, 501 (1966).

47. F.R. Metzger: Progress in Nuclear Physics, edited by O.R. Frisch, V. 7, p. 54, Pergamon Press, New York, (1959).
48. P.B. Moon: Proc. Phy. Society (London) 64A, 76, (1951).

    P.B. Moon and A. Storeusste, Proc. Phy. Soc. 66A, 585, (1953).
49. F.R. Metzger, and W.B. Todd: Phy. Rev. 95, 627, (1954), Ref. (47).
50. H.S. Hans, G.E. Thomas and L.M. Bollinger: International Conference on the Study of Nuclear Structure with Neutons edited by M. Neve, de Nevergries, P. Van Assche and J. Vernier, North Holland Publishing Co., Amsterdam, p. 514, (1966); Fleischman, H.H.: Ann. Phy. 12, 133 (1963) and F.W. Stanek Z. Physik 175, p. 172 (1963a); B. Arad, G. Ben David. Rev, of Modern Physics r-45, p. 230 (1973).
51. K. Alder, A. Bohr, T. Huss, B. Mottleson and A. Winther: Rev. of Mod. Phy. 28, 4321 (1956), A Winther and J. De Bow: Coulomb Excitation edited by K. Alder and A. Winther, [Academic Press, New York, (1966)].
52. A. Kangasmaki, P. Tikkenen, J. Keinonen, W.E. Ormand and S. Raman: Phy. Rev. C.V. 55, p. 169 (1997), A. Kangsmaki, P. Tikkenen, J. Keinonen, W.E. Ormand, S. Raman–ZS, Fulop, Z. Kiss and E. Sainorjai: Phy. Rev. C.V. 58. p. 699, (1998), Yu. N. Lobach and D. Bucurescu: Phy. Rev. C.V. 57, p. 2880 (1998).
53. B.A. Brown and B.H. Wildenthal: Nuclear Physics, A 474, 290, (1987) Ibid.: Annual Rev. Nuclear Particle Science 38, 29 (1998).
54. E.K. Warburton, J.A. Becker and B.A. Brown: Phy. Rev. C. 41, 1147 (1990).
55. F. Ichello and O. Scholten, Phy. Rev. letter 43, 679 (1979).
56. D. Bucurescu, I Cata Danil, G. Ilas, M. Ivascu, L. Strole and C.A. Ure: Phy. Rev. C. 52, 616 (1995).
57. R.A. Kaye, J.B. Adams, A. Hale, C. Smith, G.Z. Solomon and S.L. Torber and G. Gracia Bermudez, M.A. Cardenna, A. Filevich and L. Szyhisz: Phy. Rev. C.V. 57, p. 2189 (1998).
58. K. Starosta, Ch. Droste, M. Mon and J. Srebrny, D.B. Fosson, S. Gundel, J.M. Sears, I. Thorslund, P. Vaska and M.P. Waring, S.G. Rohozinski and M. Satula, and U. Garg, S. Naguleswaran and J.C. Walpe, Phy. Rev. C. 55, p. 2794 (1997).
59. I.M. Govil, A. Kumar, H. Iyer, H. Li., U. Garg, S.S. Ghughre, T. Johnson, R. Kaczarcowski, B. Khanuja, S. Naguleswaran and J.C. Walpe: Phy. Rev. C. 57, p. 632 (1998).
60. M. Schumann and F. Kappller: Phy. Rev. CV. 58, p. 1790 (1998).

    K.P. Singh, D.C. Tayal and H.S. Hans: Phy. Rev. C.V. 58, 1980 (1998).
61. L. de. Jaeger and E. Boekor: Nuclear Physics A 106, 3930 (1972), A 216. 349 (1973).
62. M. Fanerbach, M.J. Chtromik, T. Glasmacher, P.G. Hamsen, R.W. Ibbotson. D.J. Morrisay, H. Scheit, P. Thirolf and M. Thoennessen. Phy. Rev. C. 56, p. R, 1 (1997).

PROBLEMS

1. Classically,

(*i*) What is the physical significance of 'electric' or magnetic' transitions? Equations 7.19 and 7.20.

(*ii*) If the scalar Helmholtz Equation 7.21, represents, an equation for orbital angular momenta based function; u_{LM}; the solutions to vector Helmholtz equation. Equation 7.27 represent total angular momenta dependent function. Why? Explain, physically and mathematically.

2. Prove Eqs. 7.58 and 7.59, classically using Theoretical Nuclear Physics, Blatt and Weisskopf; John Wiley & Sons, New York (1952).

3. Compare Eqs. 7.57*a*, 7.58*a*, Eqs. 7.58*b* and 7.59*a*.
4. Using Eqs. 7.60*a*, 7.60*b* and 7.60*c*, calculate the transition probability of M_1 transitions of 160 keV ($5/2^+ \rightarrow 7/2^+$), 383 keV ($3/2^+ \rightarrow 7/2^+$) and 871.8 keV. ($7/2^+ \rightarrow 7/2^+$) of Cs^{133} transitions of Fig. 7.19.
5. Why are gamma rays transitions of E_0, M_1 and E_2 forbidden in $0 \rightarrow 0$ transitions; though they could be allowed say in say $1 \rightarrow 1$ transitions?
6. Obtain the ratio $\frac{\alpha_K}{\alpha_L}$ for $L = 4$ for the transitions of 1 MeV. Explain physically why is $\alpha_K >> \alpha_L$ say for Z = 85 [Fig. 7.2]. Compare Eqs. 7.78 and 7.79 and discuss the factors, which make the difference and explain these physically.
7. List the possible experimental ways of determining the multipole character of a transition and give a careful limitation of the accuracy of theoretical calculations of each effect.
8. (*a*) A nucleus, with Z = 90 decays by emission of a 0.4 MeV gamma ray from a 7/2 state to 1/2 state. What is the ratio of internal conversion to gamma ray emission? What is *k*/*L* ratio; what is L_I/L_{II} or L_I/L_{III} ratio? What would these ratios be if the initial state were $5/2^+$?

   (*b*) What is conversion coefficient for 0.2 MeV, E_5 transition in a nucleus with A = 65? [Figs. 7.3–7.5].
9. Obtain the angular correlation coefficient for the pure transitions for:

   (*i*) $1/2^+ \xrightarrow{E_2} 5/2^+ \xrightarrow{M_1} 7/2^+$ (*ii*) $3/2^+ \xrightarrow{M_1} 5/2^+ \xrightarrow{M_1} 7/2^+$

   (*iii*) $5/2^+ \xrightarrow{M_1} 7/2^+ \xrightarrow{E_2} 3/2^+$ (*iv*) $3/2^+ \xrightarrow{E_2} 7/2^+ \xrightarrow{M_1} 5/2^+$
10. Derive Eq. 7.57 from Eqs. 7.51 to 7.55.

8 CHAPTER

Beta Decay

8.1 EXPERIMENTAL FACTS

Beta decay corresponds to the decay of a parent radioactive nucleus by the emission of e^- and anti-neutrino ($\bar{\nu}$) or e^+ and a neutrino (ν), to any of the levels of the daughter nucleus through weak interaction, and may be expressed as:

$$_{Z}A^{N} \xrightarrow{\bar{\beta},\,\bar{\nu}} {}_{Z+1}A^{N-1} \text{ or } {}_{Z}A^{N} \xrightarrow{\beta^{+},\,\nu} {}_{Z-1}A^{N+1} \qquad ...(8.1)$$

These modes of decay were understood after the discovery that the energy spectrum of $e^{\mp}$ (called beta) was continuous bell-shaped. One typical spectrum is shown[2] in Fig. 8.1 depicting the decay of $_{29}Cu^{64}$, which decays both by negatron decay of e^- to $_{30}Zn^{64}$, and by positron decay e^+ to $_{28}Ni^{64}$ [Fig. 2.23*a*]. Such a continuous spectrum, in the early days led to the surmise that in the case of beta decay, energy may not be conserved. It was, however pointed out by Pauli[1] that the continuous shapes

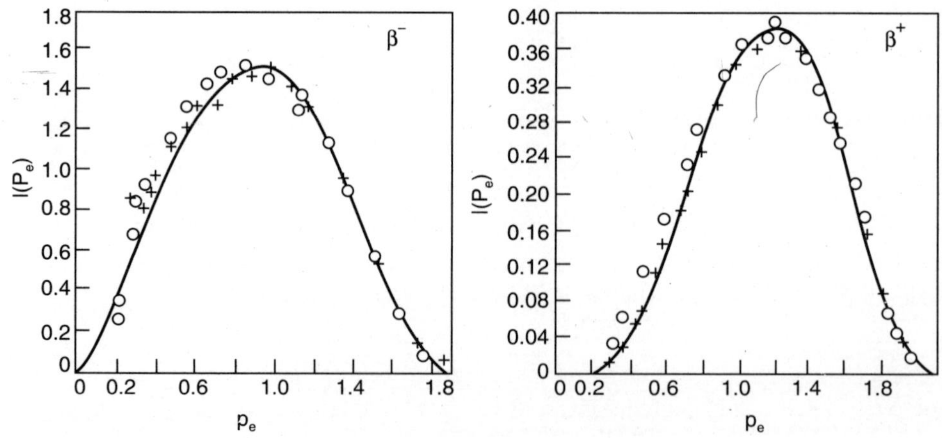

Fig. 8.1 The β-spectra from the decay of $_{29}Cu^{64}$ through the negatron (β^-) decay to $_{30}Zn^{64}$ and through positron (β^+) decay to $_{28}Ni^{64}$ (Ref. 2).

of these electron spectra could be understood, if one assumes the simultaneous emission of a massless particle termed as neutrino or anti-neutrino. The detailed theory of these spectra, involving these ideas was provided by Fermi[1] in 1934.

An alternative mode of decay is the electron-capture, which is energetically and charge-wise equivalent to positron emission. Such a decay can be written as:

$$_{Z}A^{N} + e_{k}^{-} \rightarrow {}_{Z-1}A^{N+1} + \nu \qquad ...(8.2)$$

Here the extra nuclear electrons interact with the nucleus and are captured via the weak interaction of beta decay, emitting the neutrino.

Sometimes, in the observation of continuous spectra, one also sees the sharp lines, representing the direct electrons emitted from radioactive nuclei. These correspond to the internal conversion of gamma-rays, emitting the extra nuclear electrons directly, through an electromagnetic interaction with the nucleus.

There are many naturally occurring $\beta^{\mp}$-emitters, and many thousand beta emitting radioactive nuclei have been produced artificially through nuclear reactions. The study of the characteristics of these spectra include (*i*) the lifetimes of the decaying state, (*ii*) the maximum energy (or energy of end point) of the beta spectrum, (*iii*) the detailed shape of the spectra. Besides these, various angular correlations between betas and neutrinos or betas and γ-rays, etc. have been studied.

While electrons have been detected directly through the use of particle detectors, or magnetic beta spectrometers, neutrino could only be detected indirectly. A very interesting experiment of neutrino detection is based on the detections of the recoil energy of the emitted nucleus. If only electrons were emitted with varying energies, without the neutrino emission; the recoiling nucleus also will have a continuous energy spectrum. On the other hand, if the neutrinos were emitted say, in the case of electron capture, the recoiling nuclei will have sharp recoiling energy. Such an experiment was conducted by Smith and Allen[3], using the radioactive nucleus Be^7. Others have conducted such experiments using Ar^{37} and Cd^{107} nuclei[3].

Ar^{37} and Be^7 decay by electron capture is given by:

$$Ar^{37} + \bar{e}_{K,L} \rightarrow Cl^{37} + \nu\,;\; Be^{7} + \bar{e}_{K,L} \rightarrow Li^{7} + \nu \qquad ...(8.3)$$

The decay of Be^7 has a half-life of 53.5 days. The daughter nucleus Li^7 is formed 90% times in the ground state and 10% time in the excited state, which then, decays in the emission of a γ-ray. The recoil energy spectrum of daughter nucleus corresponding to the ground state of Li^7 will be sharp, and will give a line spectrum, corresponding to 57.3 ± 0.5 eV computed from $Be^7 \rightarrow Li^7$ mass difference. The recoil spectrum corresponding to the excited state of Li^7 will, however, be continuous because of the sharing of energy between neutrino and γ-rays. Figure 8.2*a* shows the energy spectrum of such recoiling nuclei, in the form of counts versus the retarding potential required to stop the nuclei. There is an indication of peak around 50 eV for this source, supporting the emission of a single energy neutrino. A similar case of $Ar^{37} \rightarrow Cl^{37}$ has been studied by Rodeback and Allen[3] [Fig. 8.2*b*].

Experiments have also been performed by Cowen, Reines et al.[4] using the reaction:

$$\bar{\nu} + p \rightarrow e^{+} + n \qquad ...(8.4)$$

The anti-neutrinos were obtained from the reactor from the neutron decay. *i.e.*,

$$n \rightarrow p + \bar{e} + \bar{\nu} \qquad \text{...(8.5)}$$

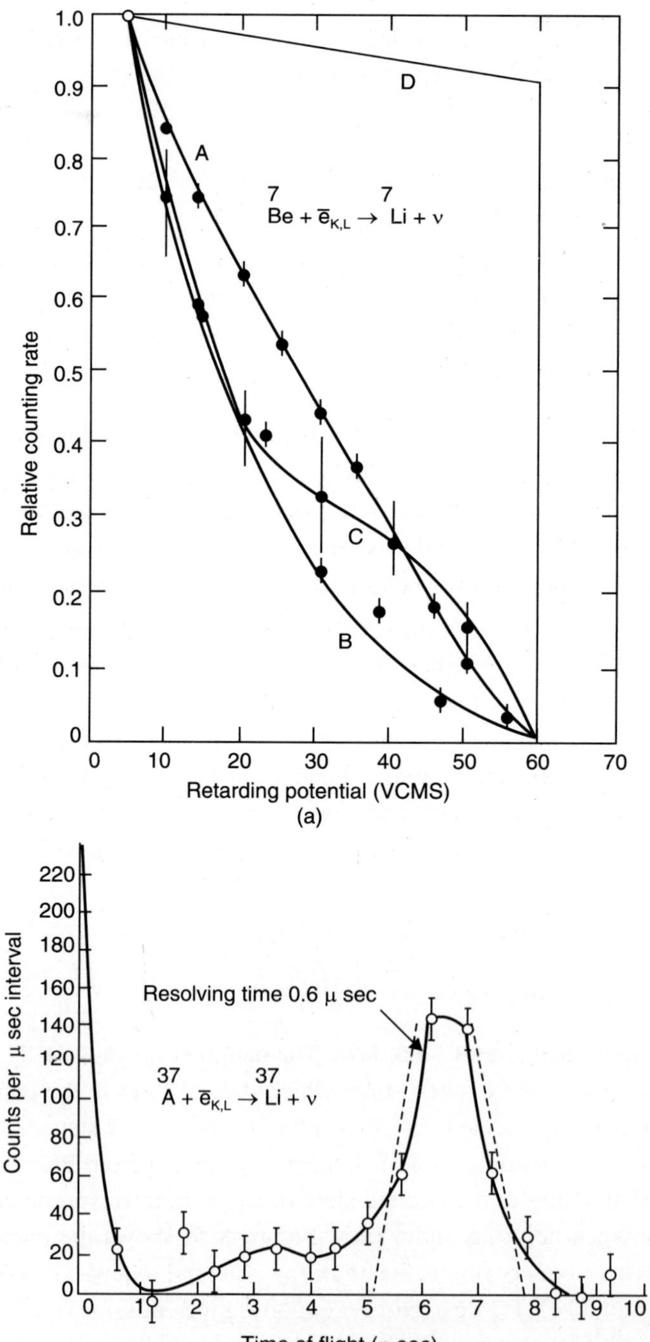

Fig. 8.2 (*a*) Energy spectrum of recoiling nuclei from the decay of $Be^7 + \bar{e}_{K,L} \rightarrow Li^7 + \bar{\nu}$ (*b*) Spectrum of Cl^{37} recoil velocities, as a function of time of flight, following electron capture transitions in A^{37} (Ref. 3, 4).

The anti-neutrinos were allowed to fall on a water-target containing dissolved Cd—Cl_2. Anti-antrinos, interacting with protons produced a positron and neutron. The positrons combined with electrons produced two annihilation γ-rays of 0.5 MeV; which can be detected in coincidence. Further the neutron can be delay-captured by Cd, emitting the characteristic γ-rays of Cd-which can be studied in delayed-coincidence with neutrons. Such experiments[4] yielded positive result, giving a cross-section of events with $\sigma = (1.10 \pm 0.25) \times 10^{-43}$ cm^2, in agreement with the predicted value from two component neutrino theory (*see* Sections 8.4 and 8.9).

Beta decay arises out of weak interaction, compared to the electromagnetic interaction in gamma decay, and nuclear plus electromagnetic interaction in alpha decay. These three interactions—weak, electromagnetic and nuclear—represent the three basic interactions, out of the four; the fourth being gravitational. The relative strengths of these interactions are generally represented by:

Gravitational $$\frac{Gm^2}{\hbar c} \approx 10^{-45}$$

Weak $$\left[\frac{\dfrac{g_\beta\, m_\pi^2\, c^2}{\hbar^2}}{\hbar c}\right]^2 = \left[\frac{\dfrac{\left(g_\beta c^2\right) m_\pi^2}{\hbar^2}}{\hbar c}\right]^2 \cong 10^{-13}$$

Electromagnetic $$\frac{e^2}{\hbar c} \approx 10^{-2}$$

Strong (Nuclear) $$\frac{g_N^2}{\hbar c} \approx 1 \qquad \text{...(8.6)}$$

As is evident from this equation, the weak interaction is much weaker than electromagnetic and nuclear. One of the results of this weak interaction is comparatively longer lifetimes of beta decaying radioactive nuclei. The lifetimes, which one encounters in beta decay, range from 4.4×10^{10} years for Re^{187} to 3×10^{-7} secs for Po^{212} though for double beta decay ($\beta^- \beta^-$), lifetimes of $> 10^{15}$ years for Pt^{198} and $> 10^{14}$ years for Os^{192} have been observed[2]. They are comparable to lifetimes in alpha decay, where tunneling effect predominates but are much slower than lifetimes in gamma decay.

The basic mechanism in any beta decay in a nucleus is the decay of a neutron (in the nucleus) or a proton (in the nucleus). While free neutron decays with a lifetime of ≈ 12 minutes; the free proton is for all practical purposes stable. In nucleus, however, in its dynamic interaction with other nucleons, it can find itself in an unstable condition, decaying as:

$$p \rightarrow n + e^+ + \nu \qquad \text{...(8.7)}$$

while neutron decays as in Eq. 8.5.

Experimentally[4], it has also been possible to observe the converse interaction, *i.e.*,

$$p + \bar{e} \rightleftharpoons n + \nu \quad ...(8.8)$$

and

$$p + \bar{\nu} \rightleftharpoons n + e^{+} \quad ...(8.9)$$

How does one know, that neutrino and anti-neutrinos are different particles? Some specific experiments have been performed to understand this problem. An interesting experiment was performed by Davis[3], who placed one thousand gallons of carbon tetrachloride near a reactor which is a copious source of anti-neutrinos from the decay of neutrons, Eq. 8.5. He looked for the reaction, which should not be observed, if ν and $\bar{\nu}$ are different, *i.e.*,

$$_{17}\mathrm{Cl}_{20}^{37} + \nu \rightarrow {}_{18}\mathrm{Ar}_{19}^{37} + \bar{e} \quad ...(8.10)$$

This corresponds to the case of a neutron converting into proton, *i.e.*,

$$n + \nu \rightarrow p + \bar{e} \quad ...(8.11)$$

Experimentally, they looked for *X*-rays from the electron capture decay of Ar^{37}, and found only (0.3 ± 3.4) Ar^{37} counts per day, which could be taken as zero, considering a small expected background of Ar^{37} from cosmic ray interactions. The experiment of Reins and Cowan[3, 4] as described earlier, had, however, shown that reactor emits anti-neutrinos. We, therefore, conclude that neutrinos and anti-neutrinos are different particles. One can, of course, perform the experiment, corresponding to the capture of neutrinos by Cl^{37}, by exposing Cl^{37} to a positron emitter. The artificially produced positron emitters are not strong enough for such an experiment. Sun, however, emits neutrinos copiously, but the experiment has not been performed, because of the experimental difficulties arising out of very low flux of such neutrinos.

Energetics

As mentioned earlier, beta decay occurs through three processes (*i*) e^--emission, (*ii*) e^+-emission and (*iii*) electron capture.

The energy relationships of the masses of neighbouring nuclei (isobars), decide which decay process, out of the above will take place.

(*i*) A negatron emission (β^-) in beta decay requires that the mass differences of the parent and daughter nuclei is more than the mass of the electrons and anti-neutrinos. Assuming the mass of anti-neutrino to be zero, one can write the energy equation of e^--emission as:

$$[M'(A, Z) - M'(A, Z+1)]\, c^2 = \Delta \quad ...(8.12)$$

where Δ is the total energy of the decay-products, *i.e.* e^- and anti-neutrino. It is, therefore, the sum of their masses and kinetic energies, if no gamma rays are emitted subsequently. Here $M'(A, Z)$ and $M'(A, Z+1)$ are the nuclear masses corresponding to atomic weight A, and charges Z and $(Z+1)$. The maximum kinetic energy of electrons corresponds to the zero kinetic energy of anti-neutrinos. One can easily measure the maximum kinetic energy of the electrons. One may, therefore, write Δ as:

$$\Delta = E_{\max} + (m_e + m_{\bar{\nu}})\, c^2 \quad ...(8.13)$$

If $m_{\bar{\nu}} = 0$; then, $\Delta = E_{\max} + m_e c^2$

One can then write:

$$[M'(A, Z) - M'(A, Z+1)]\,c^2 = m_e c^2 + E_{\beta^- \max} \qquad ...(8.14)$$

or

$$M'(A, Z)\,c^2 = M'(A, Z+1)\,c^2 + m_e c^2 + E_{\beta^- \max} \qquad ...(8.15)$$

One can convert the above equation into one for atomic masses by adding $Z m_e c^2$ on both sides, so that:

$$M'(A, Z)\,c^2 + Zm_e c^2 = M'(A, Z+1)\,c^2 + (Z+1)\,m_e c^2 + E_{\beta^- \max} \qquad ...(8.16a)$$

or

$$M(A, Z)\,c^2 = M(A, Z+1)\,c^2 + E_{\beta^- \max} \qquad ...(8.16b)$$

where $M(A, Z)$ and $M(A, Z+1)$ are neutral atomic masses.

If the beta decay takes place to a level from where a gamma ray of energy E_γ is emitted, then one can write Eq. 8.16*b* as:

$$Q_{\beta^-} \equiv M(A, Z)\,c^2 - M(A, Z+1)\,c^2 \qquad ...(8.17)$$

$$= E_{\beta^- \max} + E_\gamma$$

The quantity Q_{β^-} is defined as the mass difference between the parent and the daughter atoms in beta decays.

(*ii*) An e^+-emission or β^+ decay can be similarly treated. Similar to Eq. 8.14, we can relate the nuclear masses involved in β^+ decay as:

$$M'(A, Z)\,c^2 - M'(A, Z-1)\,c^2 = m_e c^2 + E_{\beta^+ \max} \qquad ...(8.18)$$

or

$$M'(A, Z)\,c^2 + Zm_e c^2 - M'(A, Z-1)\,c^2 - (Z-1)\,m_e c^2 - m_e c^2$$

$$= m_e c^2 + E_{\beta^+ \max} \qquad ...(8.19)$$

or

$$Q_{\beta^+} = M(A, Z)\,c^2 - M(A, Z-1)\,c^2 = 2m_e c^2 + E_{\beta^+ \max} \qquad ...(8.20)$$

In the above equations, $E_{\beta^+ \max}$ corresponds to the decay to the ground state, so that no subsequent gamma rays is emitted. If β^+ decay takes place to an excited state of the daughter nucleus so that a gamma ray is subsequently emitted, one can write the energy equation as:

$$Q_{\beta^+} = M(A, Z)\,c^2 - M(A, Z-1)\,c^2$$

$$= 2m_e c^2 + E_{\beta^+ \max} + E_\gamma \qquad ...(8.21)$$

In Fig. 2.23 in Chapter 2 we have represented diagrammatically, the energy relations given in Eqs. 8.19 and 8.21. The $2\,m_e c^2$ term in Eq. 8.21 is represented by a vertical line[5].

(*iii*) Electron capture is energetically equivalent to β^+ decay, and connects the same parent daughter nuclei as β^+ decay. One can write, in terms of nuclear masses, the energy relationship for electron capture as:

$$M'(A, Z)\, c^2 + m_e\, c^2 = M'(A, Z-1)\, c^2 + E_\nu + E_{M'} + E_\gamma \qquad ...(8.22)$$

The above equation represents energetically the physical phenomenon when in electron capture, a nucleus with energy $M'(A, Z^2)\, c^2$ captures an electron ($m_e c^2$) resulting in a nucleus with energy of $M'(A, Z-1)\, c^2$ emitting a neutrino with energy E_ν and gamma rays with energy E_γ, with a recoil energy of E'_M. If one neglects the recoil energy, then Eq. 8.22 can be written, in terms of atomic masses by adding $(Z-1)\, m_e\, c^2$, on both sides, as:

$$M(A, Z)\, c^2 = M(A, Z-1)\, c^2 + E_\nu + E_\gamma \qquad ...(8.23)$$

or

$$Q_{E.C} = M(A, Z)\, c^2 - M(A, Z-1)\, c^2 = E_\nu + E_\gamma \qquad ...(8.24)$$

It may be realised that in electron capture, the electron is captured from a definite atomic orbit, giving rise to transitions in the atomic orbitals. This gives rise to X-rays, whose nature depends on the orbital in which the electron was captured. *K*-electron capture is most probable and K-X-rays are generally observed, along with the L and M X-rays.

In electron capture, X-rays arise out of the subsequent filling of vacancies arising in the electronic orbits, by electrons from higher atomic orbits. The electron itself can be assumed to be at rest with respect to the nucleus at the time of capture and no kinetic energy is associate with it. As in electron capture the total energy of transition has a definite value; this leads to a definite value of $E_\nu + E_\gamma$, where E_γ is the energy of subsequent γ-ray. As the energy of this γ-ray is also fixed, this gives a definite value, for E_ν. It is on this basis of $E_\gamma = 0$, and E_ν being fixed, that one can have a unique energy spectrum for recoil nuclei as discussed earlier for the case of Be^7. Then $Q_{E.C.}$ can be written as in Eq. 8.24.

One can show diagrammatically, the energy relations in the electron capture transition. As mentioned earlier, the electron capture and β^+ decay are energetically equivalent. As a matter of fact, the same nucleus can decay through both the modes.

The case of the decay of Cu^{64}, is a good example of what is called the dual decay. Here Cu^{64}, decays to $_{30}Zn^{64}$ via β^- decay, along with the decay to $_{28}Ni^{64}$ via β^+ or E.C. It is one of those odd Z-even *A* cases ($_{29}Cu^{64}$) which have stable neighbouring isobars, both for $Z-1$ and $Z+1$. Figure 2.23 in Chapter 2 illustrates this case. There are many such cases in nature.

8.2 ELEMENTARY THEORY OF β DECAY

We have seen, that in beta decay, either an electron (e^-) and an anti-neutrino ($\bar{\nu}$) are emitted or a positron (e^+) and a neutrino (ν) are emitted. We give below Fermi theory of beta decay, based on the simple assumption about the nature of interaction.

Quantum mechanically, the transition probability or rate, of any interaction between two quantum states, is given by Fermi's golden rule:

$$T_{if} = \frac{2\pi}{\hbar} \left| M_{if} \right|^2 \frac{dn_f}{dE_f} \qquad ...(8.26)$$

where dn_f/dE_f are the number of final states per unit energy and, of course $|M_{if}|$ is the matrix element for the transition, given by:

$$M_{if} = \int \Psi_f^* \,|H_\beta|\, \Psi_{in}\, d\tau \qquad ...(8.27)$$

In this case, Ψ_{in} corresponds to the initial state of the nucleus before going through the beta transition, while Ψ_f represents the simultaneous emission (e^-) and ($\overline{\nu}$) and existence of a daughter nuclear state ψ_f, *i.e.*,

$$\Psi_f = \psi_f\, \phi_{\overline{\nu}}\, \phi_{e^-} \qquad ...(8.28)$$

Emission of an anti-neutrino is equivalent to the absorption of neutrino; hence one can write Eq. 8.27 as:

$$M_{if} = \int \psi_f^*\, \phi_{e^-}^* \,|H_\beta|\, \psi_{in}\, \phi_\nu\, d\tau \qquad ...(8.29)$$

where, for symmetry, we have used Ψ and ψ_{in} on the right side.

We will discuss the energy effect of $|M^2|$ in Eq. 8.26 later on. If $|M|$ is energy-independent, then the shape of beta spectrum depends on the behaviour of dn_f/dE_0; where E_0 is the total energy of the transition, *i.e.*,

$$E_0 = E_{e^-} + E_{\overline{\nu}} + E_{\text{rec}} \qquad ...(8.30)$$

where E_{e^-}, $E_{\overline{\nu}}$, E_{rec} are the energies of electron, anti-neutrino and recoil of the nucleus respectively. As the recoil energy is very small, one can neglect it, and can, therefore, write Eq. 8.30 as:

$$E_0 = E_{e^-} + E_{\overline{\nu}} \qquad ...(8.31)$$

It should be realised that E_{e^-} is the total energy of electron. The kinetic energy T_0 (in MeV) of electron may be obtained by subtracting the mass of the electron, *i.e.*,

$$T_0 \equiv E_{e^-} - 0.511 \qquad ...(8.32)$$

To find dn_f; we calculate the number of states in a momentum space in the space between two spheres of radius p and $p + dp$. The momentum volume in the momentum space is $4\pi p^2\, dp$. Hence for electrons:

$$(dn_f)_{e^-} \;\alpha\; \frac{4\pi}{h^3}\, p_{e^-}^2\, dp_{e^-} \qquad ...(8.33)$$

where h^3 is the volume of a unit cell in momentum space. Similarly for anti-neutrinos:

$$(dn_f)_{\overline{\nu}} \;\alpha\; \frac{4\pi}{h^3}\, p_{\overline{\nu}}^2\, dp_{\overline{\nu}} \qquad ...(8.34)$$

The total number of final states dn_f is given by:

$$dn_f = (dn_f)_{e^-}\,(dn_f)_{\bar{\nu}} = \frac{16\pi^2}{h^6}\left(p_{e^-}^2\,dp_{e^-}\right)\left(p_{\bar{\nu}}^2\right)dp_{\bar{\nu}} \qquad ...(8.35)$$

We want to write Eq. 8.35 in terms of the electron energy, which is the observed quantity. We know that relativistically,

$$p_{\bar{\nu}} = \frac{E_{\bar{\nu}}}{c} = \frac{E_0 - E_e}{c} \qquad ...(8.36a)$$

Therefore,

$$dp_{\bar{\nu}} = \frac{dE_0}{c} \qquad ...(8.36b)$$

Hence

$$\frac{dn_f}{dE_0} = \frac{16\pi^2}{h^6 c^3}(E_0 - E_e)^2\,p_{e^-}^2\,dp_{e^-} \qquad ...(8.36c)$$

Collecting these factors, we can write:

$$T_{if} = \left(\frac{64\pi^3}{h^7 c^3}\right)|\,M_{if}\,|^2\,(E_0 - E_e)^2\,p_{e^-}^2\,dp_{e^-} \qquad ...(8.37)$$

We have not taken, into account the Coulomb effect on the emission of electrons. Without going into details, we will depict it as $F(Z, E_e)$, and, therefore, finally write T_{if} as:

$$T_{if} = n(p_{e^-})\,dp_{e^-}$$

$$= \left(\frac{64\pi^3}{h^7 c^3}\right)|\,M_{if}\,|^2\,F(Z, E_{e^-})(E_0 - E_{e^-})^2\,p_{e^-}^2\,dp_{e^-} \qquad ...(8.38)$$

Till now we have calculated the rate of decay, in the energy range corresponding to p_e and $p_e + dp_e$. In practice, the total rate of decay, which determines the lifetime is given by:

$$\lambda \equiv \frac{1}{\tau} = \left(\frac{64\pi^3}{h^7 c^3}\right)\int_0^{p_0} |\,M_{if}\,|^2\,F(Z, E_e)(E_0 - E_{e^-})^2\,p_e^2\,dp_e \qquad ...(8.39)$$

where $E_0^2 = \left(m_{e^-}c^2\right)^2 + (p_0\,c)^2$, where p_0 is the maximum momentum of the electrons.

The shape of the spectrum is given by Eq. 8.38. Many times, the beta-spectrum is plotted as $n_p/p^2\,F(Z, E_e)^{1/2}$ versus E_e. It is apparent from Eq. 8.38 that this should give a straight line, if $|\,M_{if}\,|^2$ is independent of energy. This is called Kurie plot.

In Eq. 8.39 if we define:

$$f(Z, E_0) = \int_0^{p_0} p_{e^-}^2\,(E_0 - E_{e^-})^2\,F(Z, E_e)\,d\,p_e \qquad ...(8.40)$$

then Eq. 8.39 can be written as:

$$f\tau = \frac{1}{G\,|\,M_{if}\,|^2} \qquad \text{...(8.41)}$$

where $$G = \frac{64\,\pi^3}{h^7 c^3} = \text{Const.}$$

Normally log ($T_{1/2}$) is taken as measure of the strength of matrix element $|\,M_{if}|$. Smallest values of log $f\,T_{1/2}$ between 3 to 5 correspond to allowed and super-allowed cases. Larger values correspond to different levels of forbiddenness.

In beta decay literature, $f\,(Z, E_0)\;T_{1/2}$ is written as *ft* which, we will follow subsequently.

Figure 8.3 gives the Kurie plot for the spectrum from $H^3 \xrightarrow{\bar{\beta},\,\bar{\nu}} H^3$, with a half-life of ≈ 12 yrs. This represents one of the very accurately done experiments in beta decay and was performed to determine if anti-neutrino has non-zero mass. As shown in the figure the straight line fits the Kurie-plot very well and detailed analysis showed that the rest mass of the anti-neutrino is less than 250 eV; or:

$$m_\nu < \frac{me^-}{2000} \qquad \text{...(8.42)}$$

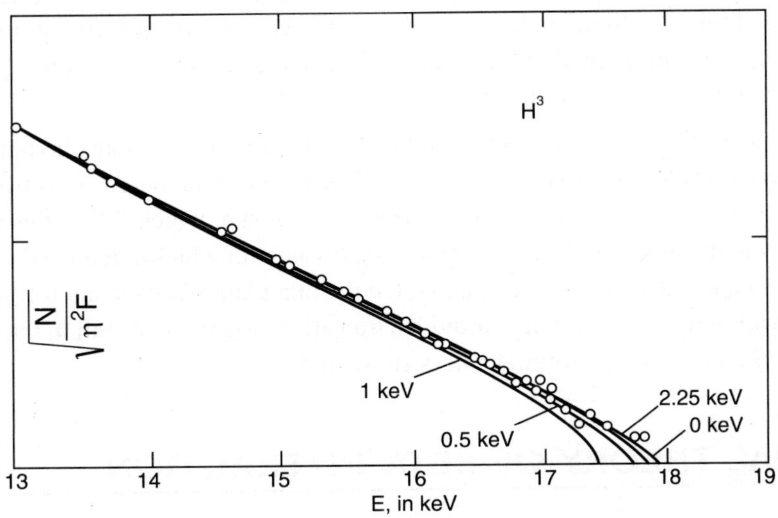

Fig. 8.3 The Kurie plot of β⁻-rays near the end of beta spectrum from H³ (Ref. 6).

Such Kurie plots have been experimentally drawn for a large number of similar cases, where $|\,M_{if}\,|^2$ can be taken as a constant and all of them fall on a straight line, *e.g.*, ${}_{16}S^{35} \xrightarrow{\beta^-} {}_{17}Cl^{35}$ or ${}_{32}Ge^{75} \xrightarrow{\beta^-} {}_{33}As^{75}$, etc. These are cases of 'allowed' beta decay, for which log ft falls between 3 and 5.

It was further found that many cases, for which log ft falls between 6 and 23, do not have a straight line plot, *e.g.*, $_{28}Ni^{65} \xrightarrow{\beta^-} {}_{20}Cu^{65}$ (1-forbidden) (log ft 6.6), $_{56}Cs^{137} \xrightarrow{\beta^-} {}_{56}Ba^{137}$ (First forbidden, log ft (6.7–9.6), and many other cases. Figure 8.4 depicts such cases over the whole periodic table.

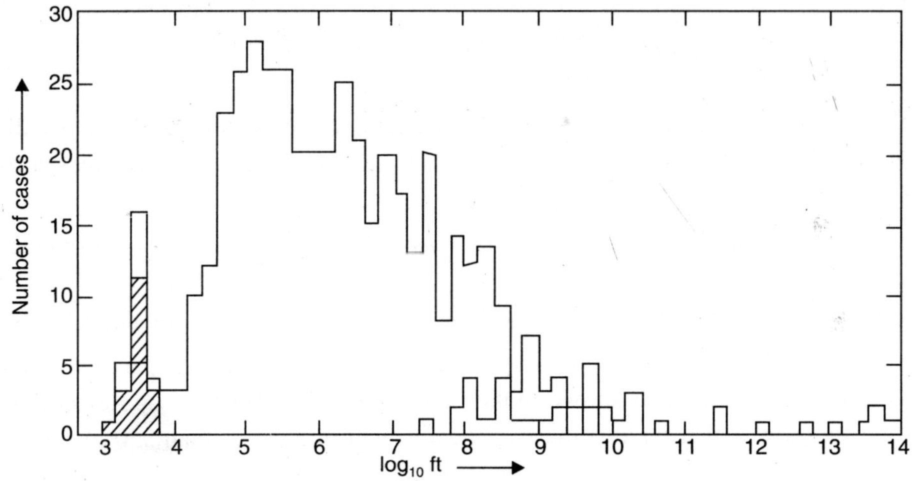

Fig. 8.4 Frequency histogram of log ft, for combined odd-even A nuclei (Ref. 7).

What do we learn from the distribution of log ft? It may be realised that the lifetimes represented vary in a certain small range of light nuclei from about 10^{-2} secs (N^{12}) to 6×10^{10} yr (Rb^{87}) a factor of 10^{23}. The spread in ft is however, smaller than this by about a factor of 10^5, and hence log ft varies by a factor of 5, *i.e.* from log ft $\approx 3 - 4$ to lot ft $\approx 9 - 10$.

The detailed analysis of Fig. 8.4 shows that the group of log ft $\approx$ 4 is most favoured and hence is referred to, as super-allowed transitions. Most of the allowed but unfavoured transitions have their ft values in the interval $4 \leq$ log ft ≤ 7. A few values near 9 are known, which fall in this category. In the high end, they mix with the first forbidden transitions, log ft values of which extend to the neighbourhood of 10, ending with unique-shape transitions mainly. On the other hand, for second forbidden transition, log ft values spread around 13 and for third forbidden transitions log ft values fall in the neighbourhood of 18. We will discuss the theory of forbiddenness subsequently.

8.3 FORMAL THEORY OF BETA INTERACTION

We have discussed above, the elements of Fermi theory of beta decay; which uses the simplest possible form of Hamiltonian, *i.e.* Scalar, as we have used $|M_{if}|^2$ as constant, independent of energy. As we saw in the previous discussion, the Fermi theory is quite satisfactory in explaining the properties of the specra of emitted electrons. Fermi's theory, however, assumes the conservation of parity; which is now known to be not conserved in beta decay. One of the consequences of this is the polarisation of neutrinos. To understand, the phenomena of non-conservation of parity and all its varied consequences, we should develop the basic theory of beta interaction; so that the non-conservation of parity flows out of it as a special case of the general theory.

As has been pointed out earlier, the theory of beta interaction should be based on relativistic quantum mechanics; because of the high velocity of neutrinos and betas. Further we assume the interaction to be a contact interaction; in the absence of any better knowledge about the space-dependence of the interaction, except that it has very short range.

From the very general principles of the quantum mechanical treatment of time dependent phenomena in the microscopic system; the transition probability between two quantum mechanical states is given by Eqs. 8.26 to 8.29 (*See* Q. Mechanics by Schiff p. 193), as described earlier, where dn_f/dE_0 is the density of states in the final system and $|M_{if}|^2$ is the appropriate matrix element between the initial state i and the final f.

8.3.1 General[8] Discussion about $|M_{if}|^2$

The behaviour of $|M_{if}|^2$ is evidently dependent on the properties of nuclear structure embodied in Ψ_f and Ψ_{in} and on the beta interaction operator H_β. Apriori, one can write down the expression for H_β on the general principles, that it should be a scalar or a pseudoscalar quantity or a mixture of the two. As explained earlier, we will further assume that its spatial dependence is given by a delta function $\delta(\mathbf{r}_n - \mathbf{r}_L)$ where $\mathbf{r}_n$ and $\mathbf{r}_L$ are the nucleon and lepton position vectors; so that $(\mathbf{r}_n - \mathbf{r}_L)$ denotes the vector distance between, say, the initial nucleon, and the leptons, *i.e.* emitted anti-neutrino and electron. The assumption which implies a contact interaction is convenient to handle in the absence of any other detailed information about spatial dependence. This is in conformity with the short range nature of weak forces.

Of what factors should M_{if} be made of, apart from the spatial part? An mentioned earlier, one should use the relativistic four component wave functions obtained from Dirac's relativistic equations, *i.e.*,

$$(E + c\,\alpha\,.\,\mathbf{p} + \beta\, m\, c^2)\, \psi = 0 \qquad \text{...(8.43}a\text{)}$$

or

$$\left(i\hbar \frac{\partial}{\partial t} - i\hbar c \alpha\,.\,\mathbf{grad} + \beta m c^2 \right) \psi = 0 \qquad \text{...(8.43}b\text{)}$$

leading to the four component wave-functions

$$\psi_{in,\,f} = \begin{pmatrix} \psi_1 \\ \psi_2 \\ \psi_3 \\ \psi_f \end{pmatrix}_{in,\,f} \quad \text{and } \phi_{\nu,\,e} = \begin{pmatrix} \phi_1 \\ \phi_2 \\ \phi_3 \\ \phi_4 \end{pmatrix} \qquad \text{...(8.44)}$$

which corresponds to two-spin states and two energy states (+ve and –ve). In dealing with Dirac's relativistic formalism[9]; we have used ψ in Eq. 8.43 in the generic sense, standing both for nuclear and laplonic wave functions. As the wave function $\psi^* = \psi^\dagger$; the Hermitian adjoint wave-function may be written as:

$$\psi^* = \psi^\dagger = \overbrace{\underbrace{\psi_1^*\ \psi_2^*\ \psi_3^*\ \psi_4^*}} \qquad \text{...(8.45)}$$

Another wave-function $\overline{\psi} = \psi^+ \beta$ is also sometimes used and has the expression:

$$\overline{\psi} = \overbrace{\underbrace{\psi_1^*\ \psi_2^* - \psi_3^* - \psi_4^*}} \qquad \text{...(8.46)}$$

Where ψ's are again used in the generic sense.

The same definitions apply to the wave functions ϕ also. From this it follows that the operator H_β should also be built out of the Dirac operators, in a manner that the theory of transition probability is invariant under Lorentz transformation. A general probability expression for operator H_β, based on the above ideas, may be written as:

$$H_\beta = \left(O_n\, O_L\, \tau_+^n\, \tau_+^L + O_n^+\, O_L^+\, \tau_-^n\, \tau_-^L\right) \delta\, (\mathbf{r}_n - \mathbf{r}_L) \qquad \text{...(8.47)}$$

where the meaning of isospin operators $\tau_\pm^i$'s has been already discussed in Chapter 6. The second term containing $O_n^+\, O_L^+\, \tau_-^n\, \tau_-^L$ in Eq. 8.47 is called the Hermitian conjugate (*h.c*) of the first term. We can use isospin formalism for leptons also, because positron and electrons have the same interaction. The operator O_n, operates on the nuclear wave-function $\psi_{i,\,f}$, and O_L operates on the lepton wave-function $\phi_{\nu,\,e}$. Both these operators should be built out of the Dirac matrices. The operator τ_+^n is the creation operator for protons from neutrons and τ_+^L is the creation operator for β^--decay. Similarly O_n^+ and O_L^+ are the complex conjugate of O_n and O_L to ensure the reality of the Hamiltonian and τ_-^n is the creation operator for the creation of neutrons from protons and τ_-^L is the creation operator for the creation of β^+ decay (or electron capture). The operator τ_-^L converts a negatron (e^-) into a neutrino, so that $\tau_-^n\, \tau_-^L$ has matrix elements, which operate between states, in which initial state contains a proton and a negatron (e^-) either bound in atom (+ ve energy) for *K*-capture or in the negative energy, so that there is a hole in the negative energy, for positron emission and final state contains a neutron and a neutrino. The operators τ_+ and τ_- are given by (as earlier explained in Chapter 6):

$$\tau_+ = \tau_1 + i\tau_2\,;\ \tau_- = \tau_1 - i\tau_2$$

where
$$\tau_1 = \begin{pmatrix} 0 & 1 \\ 1 & 0 \end{pmatrix};\ \text{and}\ \tau_2 = \begin{pmatrix} 0 & -i \\ i & 0 \end{pmatrix} \qquad \text{...(8.48)}$$

so that
$$\tau_+ = 2\begin{pmatrix} 0 & 1 \\ 0 & 0 \end{pmatrix};\ \text{and}\ \tau_- = \begin{pmatrix} 0 & 0 \\ 1 & 0 \end{pmatrix} \qquad \text{...(8.49)}$$

The proton and neutron wave-function are represented in isotopic spin space as:

$$\gamma = \begin{pmatrix} 1 \\ 0 \end{pmatrix}\ \text{for proton; and}\ \delta = \begin{pmatrix} 0 \\ 1 \end{pmatrix}\ \text{for neutron}$$

Then
$$\tau_+^n\, \gamma = 0;\ \tau_+^n\, \delta = 2\gamma$$

$$\tau_-^n\, \gamma = 2\delta;\ \tau_-^n\, \delta = 0 \qquad \text{...(8.50)}$$

So that τ_+^n operating upon a neutron, creates a proton—a case when β^- is emitted; similarly τ_+^n operating upon proton, creates a neutron—a case when β^+ is emitted or electron is captured. The same arguments can be given for leptons, if one uses the isotopic spin formalism to write their wave functions also. The τ-operators are useful, if one includes in the wave function the isotopic spin. Further, one should realise that in an actual case, either β^+-decay (or electron capture) takes place or β^--decays takes place, but seldom both and therefore, one will in general, either use the first part or the second but seldom both (though in some cases both cases may occur).

8.3.2 Dirac Matrice

The operators O_n and O_L and their Hermitian conjugates, may be built out of the Dirac matrices, which are given as follows:

(i)
$$\alpha = \left(\begin{array}{c|c} 0 & \sigma \\ \hline \sigma & 0 \end{array}\right) \qquad \text{...(8.51)}$$

with components

$$\alpha_1 = \left(\begin{array}{cc|cc} 0 & 0 & 0 & 1 \\ 0 & 0 & 1 & 0 \\ \hline 0 & 0 & 0 & -i \\ 0 & 0 & i & 0 \end{array}\right) \qquad \text{...(8.52)}$$

$$\alpha_2 = \left(\begin{array}{cc|cc} 0 & 0 & 0 & -i \\ 0 & 0 & i & 0 \\ \hline 0 & i & 0 & 0 \\ i & 0 & 0 & 0 \end{array}\right) \qquad \text{...(8.53)}$$

$$\alpha_3 = \left(\begin{array}{cc|cc} 0 & 0 & 1 & 0 \\ 0 & 0 & 0 & -1 \\ \hline 1 & 0 & 0 & 0 \\ 0 & -1 & 0 & 0 \end{array}\right) \qquad \text{...(8.54)}$$

(ii)
$$\beta = \left(\begin{array}{c|c} +1 & 0 \\ \hline 0 & -1 \end{array}\right) = \left(\begin{array}{cc|cc} 1 & 0 & 0 & 0 \\ 0 & 1 & 0 & 0 \\ \hline 0 & 0 & -1 & 0 \\ 0 & 0 & 0 & -1 \end{array}\right) \qquad \text{...(8.55)}$$

(iii)
$$\sigma' = \left(\begin{array}{c|c} \sigma & 0 \\ \hline 0 & \sigma \end{array}\right) \qquad \text{...(8.56)}$$

with components

$$\sigma'_1 = \left(\begin{array}{cc|cc} 0 & 1 & 0 & 0 \\ 1 & 0 & 0 & 0 \\ \hline 0 & 0 & 0 & 1 \\ 1 & 0 & 1 & 0 \end{array}\right)$$

$$\sigma_2' = \left(\begin{array}{cc|cc} 0 & -i & 0 & 0 \\ i & 0 & 0 & 0 \\ \hline 0 & 0 & 0 & -i \\ 0 & 0 & 0 & 0 \end{array}\right)$$

$$\sigma_2' = \left(\begin{array}{cc|cc} 1 & 0 & 0 & 0 \\ 0 & -1 & 0 & 0 \\ \hline 0 & 0 & 1 & 0 \\ 0 & 0 & 0 & -1 \end{array}\right) \qquad \text{...(8.57)}$$

and

(*iv*)
$$1 = \left(\begin{array}{c|c} 1 & 0 \\ \hline 0 & 1 \end{array}\right) = \left(\begin{array}{cc|cc} 1 & 0 & 0 & 0 \\ 0 & 1 & 0 & 0 \\ \hline 0 & 0 & 1 & 0 \\ 0 & 0 & 0 & 1 \end{array}\right) \qquad \text{...(8.58)}$$

The Dirac matrices have the following important properties, which are required for various operations:

$$\alpha_1^2 + \alpha_2^2 + \alpha_3^2 = \beta^2 = 1 \qquad \text{...(8.59)}$$

$$\alpha_1\alpha_2 + \alpha_2\alpha_1 = \alpha_2\alpha_3 + \alpha_3\alpha_2 = \alpha_3\alpha_1 + \alpha_1\alpha_3 = 0$$

and
$$\alpha_1\beta + \beta\alpha_1 = \alpha_2\beta + \beta\alpha_2 = \alpha_3\beta + \beta\alpha_3 = 0 \qquad \text{...(8.60)}$$

or
$$\{\alpha_i, \alpha_j\} = \alpha_i\alpha_j + \alpha_j\alpha_i = 2\delta_{ij}$$

and
$$\{\alpha_i, \beta\} = 0 \qquad \text{...(8.61)}$$

An alternative way of writing the Dirac equation is in terms of γ-matrices related to α-matrices as follows:

$$\gamma_k \equiv -i\alpha_k\beta = -i\alpha\gamma_4 \quad \text{where} \quad k = 1, 2, 3.$$

$$\gamma_4 \equiv \beta \text{ ; and } \gamma_5 \equiv -i\alpha_1\alpha_2\alpha_3 = \gamma_1\gamma_2\gamma_3\gamma_4 \qquad \text{...(8.62)}$$

So that
$$\gamma_k = i\begin{pmatrix} 0 & \sigma_k \\ -\sigma_k & 0 \end{pmatrix} \qquad \text{...(8.63}a\text{)}$$

and
$$\gamma_5 = \begin{pmatrix} 0 & 1 \\ 1 & 0 \end{pmatrix} \qquad \text{...(8.63}b\text{)}$$

A few useful relationships governing γ-matrices are:

(*i*) $\{\gamma_\mu, \gamma_\nu\} \equiv \gamma_\mu\gamma_\nu + \gamma_\nu\gamma_\mu = 2\delta_{\mu\nu}$; $\mu, \nu = 1, 2, 3$

(*ii*) $\gamma_1^2 = \gamma_2^2 = \gamma_3^2 = \gamma_4^2 = 1$

(*iii*) $\gamma_5^2 = 1$

(*iv*) $\{\gamma_5, \alpha_k\} = \gamma_5\,\gamma_\mu + \gamma_\mu\,\gamma_5 = 0\,;\ \mu = 1, 2, 3$ and 4

(*v*) $[\gamma_5, \alpha_k] = \gamma_5\,\gamma_\mu + \gamma_\mu\,\gamma_5 = 0,\ k = 1, 2, 3$

(*vi*) $\alpha = \sigma'\,\gamma_5$; and $\sigma' = \alpha\gamma_5$

(*vii*) $[\sigma', \gamma_5] = \sigma'\gamma_5 - \gamma_5\,\sigma' = 0$ ...(8.64)

The Dirac equation which is generally expressed as:

$$(E + c\alpha \cdot \mathbf{p} + \beta\, mc^2)\,\psi = 0 \qquad \text{...(8.43}a\text{)}$$

can be written alternatively in terms of γ-matrices as:

$$\left(\sum_{\mu=1}^{4} \gamma_4 \hbar \frac{\partial}{\partial x_\mu} - mc^2\right)\psi = 0 \qquad \text{...(8.65)}$$

or

$$(\gamma \cdot \mathbf{p} + \gamma_4\, p_4 - mc^2)\,\psi = 0 \qquad \text{...(8.66)}$$

where

$$\gamma \cdot \mathbf{p} = \sum_{k=1}^{3} \gamma_k\, p_k \qquad \text{...(8.67)}$$

All γ's are Hermitian and so are α, σ' and β.

8.3.3 Discussion about Operator $\mathcal{H}_\beta$

The operators O_n and O_L built out of the above mentioned Dirac matrices should, normally, be scalar as is required for any interaction operator on general considerations of symmetry and conservation laws. This means that the theory should be invariant under Lorentz transformation and also on translation, rotation or reflection of the axes of coordinates. This would be true if the laws of conservation of angular momentum and parity hold good. As has been described earlier, the experiments by C. S. Wu[10] on Co^{60} have proved that the parity is not conserved in beta decay. This requires, therefore, that an operator $\mathcal{H}_\beta$ should be defined in such a way, that a mixture of scalar and pseudoscalar terms can be written as:

$$\mathcal{H}_\beta = H_\beta + H'_\beta \qquad \text{...(8.68)}$$

where H_β is a scalar and H'_β is pseudoscalar. Such a choice of $\mathcal{H}_\beta$ assures that the transition probability which is proportional to $|M_{if}|^2$ and is given by:

$$T_{if}\, \alpha\, |M_{if}|^2$$

$$\alpha \left| \left\langle f\,|\,H_\beta\,|\,i \right\rangle + \left\langle f\,|\,H'_\beta\,|\,i \right\rangle \right|^2 \qquad \text{...(8.69)}$$

will contain, apart from the square of the matrix elements which are scalar, product of a scalar and a pseudoscalar—which is pseudoscalar. In this way, the transition probability will depend on the handedness of the frame of reference—a requirement of the non-conservation of parity. To achieve such an interaction, the operators O_n and O_L will be chosen to obtain H_β such that their product is scalar while to obtain H'_β, the product of the operator O'_n and O'_L should be pseudoscalar.

It so happen that the Dirac matrices mentioned above have the following properties on the reflection of the coordinate system under relativistic conditions, and may be classified under the following types:

(*i*) I, β and γ_4—Scalar (S). Each matrix has one component.

(*ii*) γ_1, γ_2, γ_3 and γ_4—Vector (V). These are the four components of four vector.

(*iii*) α—Vector (V). This has three components.

(*iv*) $\sum_{\mu} \sum_{\lambda} \gamma_\mu \gamma_\lambda$ —Tensor (T). This has six components.

(*v*) $\beta\alpha$—Vector (V). This has three components.

(*vi*) σ'—Axial Vector (A). This has three components.

(*vii*) $\sum_{\mu=1}^{4} \gamma_\mu \gamma_5$ —Axial Four Vector (A). This has four components.

(*viii*) $\gamma_5 = \gamma_1 \gamma_2 \gamma_3 \gamma_4$—Pseudoscalar ($P$). It has only one component.

One may now build H_β and H_β' out of these matrices, in such a manner that there are different types of interactions depending on which type of matrices are used to build the scalar or pseudoscalar matrices. As for example, they may be built out of scalar (S), Vector (V), Tensor (T), Axial Vector (A) and Pseudoscalar (P) type matrices. One may, therefore, write in general,

$$\mathcal{H}_\beta = \sum_i \mathcal{H}_{\beta i} \qquad ...(8.70)$$

where $\mathcal{H}_{\beta i} = H_{\beta i} + H_{\beta i}'$; subscript i denotes S, V, T, A and P as described above.

Then one may write:

$$H_{\beta i} = (O_i)_n (O_i)_L \tau_+^n \tau_+^L \delta(\mathbf{r}_n - \mathbf{r}_L) + h \cdot c \qquad ...(8.71)$$

$$H_{\beta i}' = (O'_i)_n (O'_i)_L \tau_+^n \tau_+^L \delta(\mathbf{r}_n - \mathbf{r}_L) + h \cdot c \qquad ...(8.72)$$

Different types of matrices may now be built as follows:

One should note that properties of γ-matrices form the basis for the nomenclature of types of interactions. As for example, for the case of scalar interaction, one can write:

$$H_{\beta s} = (O\, s)_n (O\, s)_L \tau_+^n \tau_+^L \delta(\mathbf{r}_n - \mathbf{r}_L)$$

and
$$H_{\beta s}' = (O' s)_n (O' s)_L \tau_+^n \tau_+^L \delta(\mathbf{r}_n - \mathbf{r}_L) \qquad ...(8.73a)$$

where
$$(O\, s)_n (O\, s)_L = g_s (\gamma_4)_n (\gamma_4)_L = g_s (\beta)_n (\beta)_L$$

and
$$(O's)_n (O's)_L = g_s' (\gamma_4)_n (\gamma_4 \gamma_5)_L = g_s' (\beta)_n (\beta\gamma_5)_L \qquad ...(8.73b)$$

8.3.4 Matrix Elements: Non-Conservation of Parity[11, 12]

From the above, the matrix element corresponding to the scalar type interaction is given by:

$$\mathcal{M}_s = M_s + M_s' = g_s \int \left(\psi_f^* (\gamma_4)_n \tau_+^n \delta(\mathbf{r}_n - \mathbf{r}_L) \psi_{in}\right)\left(\phi_e^* (\gamma_4)_L \tau_+^L\right) \phi_\nu \, d\tau$$

$$+ g_s' \int \left(\psi_f^* (\gamma_4)_n \tau_+^n \delta(\mathbf{r}_n - \mathbf{r}_L) \psi_{in}\right)\left(\phi_e^* (\gamma_4 \gamma_5)_L \tau_+^L\right) \phi_\nu \, d\tau \quad \text{...(8.74)}$$

and similar expression for M_s in terms of α, β, σ' and γ.

In a very similar manner, one can write the other matrix elements corresponding to (V), (T), (A) and (P) type interactions, where only the operators $(O_i)_n$, $(O_i)_L$ and $(O'_i)_n$ $(O'_i)_L$ change from case to case. We give below the five possible values of the operators $(O_i)_n$ $(O_i)_L$ in the two notations. One should note that one may obtain the dashed pseudoscalar term of the matrix element by replacing $(O_i)_n$ by $(O_i')_n$ and $(O_i')_L$ by $(O_i \gamma_5)_L$ and g_i by g_c'.

Table 8.1 *Five types of β-interactions* $(O_i)_n (O_i)_L \approx g_i (\Gamma_i)_n (\Gamma_i)_L$

Type i	*γ-notation* $[(O_i)_n (O_i)_L]$	*(α, β, σ′) notation* $[(O_i)_n (O_i)_L]$
S	$g_s (\gamma_4)_n (\gamma_4)_L \equiv g_s (\Gamma_s)_n (\Gamma_s)_L$	$g_s (\beta)_n (\beta)_L$
V	$g_v \sum_\mu (\gamma_4 \gamma_\mu)_n (\gamma_4 \gamma_\mu)_L \equiv g_v (\Gamma_v)_n (\Gamma_v)_L$	$g_v [(I)_n (I)_L - (\alpha)_n \cdot (\alpha)_L]$
T	$-g_T \sum_\mu \sum_\nu (\gamma_4 \gamma_\mu \gamma_\nu)_n (\gamma_4 \gamma_\mu \gamma_\nu)_L = g_T (\Gamma_T)_n (\Gamma_T)_L$	$g_T [(\beta\sigma')_n \cdot (\beta\sigma')_L + (\beta\alpha)_n \cdot (\beta\alpha)_L]$
A	$-g_A \sum_\mu (\gamma_4 \gamma_\mu \gamma_5)_n (\gamma_4 \gamma_\mu \gamma_5)_L \equiv g_A (\Gamma_A)_n (\Gamma_A)_L$	$g_A [(\sigma')_n \cdot (\sigma')_L - (\gamma_5)_n (\gamma_5)_L]$
P	$g_p (\gamma_4 \gamma_5)_n (\gamma_4 \gamma_5)_L \equiv g_p (\Gamma_p)_n (\Gamma_p)_L$	$g_p [(\beta\gamma_5)_n (\beta\gamma_5)_L]$

The negative sign before the tensor and axial vector interactions are needed to make sure that Ψ_p^* $O_n \Psi_p$ and $\psi^* O_L \psi$ are anti-Hermitian in contrast to other matrices, which may be Hermitian. This is required to ensure that $|M_T|$ and $|M_A|$ and similarly $|M_T'|$ and $|M_A'|$ in Eqs. 8.76*b* and 8.76*c* are positive. In general, a transition may involve any or all the above mentioned interactions, so that the value of $|M_{if}|^2$ may, be written:

$$|M_{if}|^2 = \left| \int \left\{ \psi_f^* \phi_f^* \left| \delta(\mathbf{r}_n - \mathbf{r}_L) \sum_i \left[(O_i)_n (O_i)_L + (O_i')_n (O_i')_L\right] \times \tau_+^n \tau_+^L + h \cdot c \right| \psi_{\text{in}} \phi_{\text{in}} \right\} d\tau \right|^2$$

...(8.75)

It may be expressed alternatively in line with the form of Eq. 8.74 as:

$$|M_{if}|^2 = \left| \sum_i \int \left(\psi_f^* (O_i)_n \tau_+^n \psi_{in} \right) \delta(\mathbf{r}_n - \mathbf{r}_L) \right.$$

$$\left. \times \left(\phi_f^* \left[(O_i)_L + (O_i \gamma_5)_L \right] \tau_+^L \phi_{in} + h \cdot c \right) d\tau \right|^2$$

$$= |M + M'|^2 \qquad \text{...(8.76a)}$$

because $O_n = O_n'$

where

$$M = \sum_i \int \left(\psi_f^* \left[(O_i)_n \tau_+^n \right] \psi_{in} \, \delta(\mathbf{r}_n - \mathbf{r}_L) \right.$$

$$\left. \phi_f^* \left[(O_i)_L \tau_+^L \right] \phi_{in} + h \cdot c \right\} d\tau \qquad \text{...(8.76b)}$$

and

$$M' = \sum_i \int \left(\psi_f^* \left[(O_i)_n \tau_+^n \right] \psi_{in} \, \delta(\mathbf{r}_n - \mathbf{r}_L) \right.$$

$$\left. \phi_f^* \left[(O_i \gamma_5)_L \tau_+^L \right] \phi_{in} + h \cdot c \right\} d\tau \qquad \text{...(8.76c)}$$

8.3.5 Calculation of Matrix Elements

In calculating the value of $|M_{if}|^2$ in Eq. 8.76, the various points may be kept in mind. The effect of contact-function is taken into account by releasing that the wave-function of the leptons at the contact point of $\mathbf{r}_n - \mathbf{r}_L = 0$ may be taken to be wave function of leptons. So we take $\delta(\mathbf{r}_n - \mathbf{r}_L) = 1$, in Eqs. 8.75 to 8.76. Equation 8.47 represents the general expression for the Hamiltonian, for weak interaction responsible for beta decay. The first term produces β^--decay and the second term produces β^+-decay, and electron capture. As we have seen before, τ_-^n converts a proton into a neutron, and τ_-^L converts a negatron into a neutrino, hence $\tau_-^n \tau_-^L$ contained in matrix elements $\left\langle \psi_f^* \phi_f | H_\beta | \psi_i \phi_i \right\rangle$ represent β^+-decay in which initial state contains a proton (ψ_i) and negatron (ϕ_i); while the final state has a neutron (ψ_f) and neutrino (ϕ_f). In the case of *k*-capture ψ_i is proton; but ϕ_i represents a negatron bound in the atom which is converted to final state which is neutrino. ψ_f is, of course, neutron. If ϕ_i represents a negative energy negatron, then in the final state ϕ_f there is a hole in the negative energy sea, corresponding to positron emission. On the other hand, in the case of β^--decay, only the first term in Eq. 8.47 is effective; where τ_+^n converts a neutron into proton and τ_+^L converts a neutrino into electron. Rest of the arguments follow the complementary logic compared to the one given in the previous para.

The wave functions ψ_1, ψ_2, ψ_3 and ψ_4 of Eq. 8.44 and similarly ϕ_1, ϕ_2, ϕ_3 and ϕ_4, have to be written under conditions of the validity of special theory of relativity; so that the theory must by invariant under Lorentz transformation, including translation and rotation of the axes. As described earlier, under these conditions there are only five essentially different choices for combination of O_n and O_L. These have been described earlier in terms of γ-matrices or (α, β, σ′) matrices of special theory of relativity (Table 8.1). An example of writing H_β and H_β', would be (for scalar interaction):

$$(i)\ O_n = \gamma_4\ ;\ O_L = C\gamma_4 + C'\gamma_4\ ;\ \text{and}\ H_\beta = (O_n\ O_L\ \tau_+^n\ \tau_+^L + h \cdot c)\,\delta(\tau_n - \tau_L)$$

$$(ii)\ O_n' = \gamma_4 = O_n\ \text{and}\ O_L' = C\,\gamma_4\,\gamma_5\ ;\ \text{and}\ H_\beta' = (O_n'\ O_L'\ \tau_+^n\ \tau_+^L + h \cdot c)\,\delta(\gamma_n - \gamma_L)$$

First form is called the scalar interactions, because the nucleon matrix element is a scalar; similarly the second form is called the pseudoscalar; because the lepton matrix element is pseudoscalar. In both forms, the final matrix element would be a mixture of scalar and pseudoscalar. The symbols C and C' represent the coupling constants, which determine the strengths of interaction included in g_i and g_i' in Table 8.1. The notation has been expressed in such a manner, that the terms in the matrix elements containing C are scalars; while those with C' are pseudoscalars.

The most general form of interaction matrix element for β^--decay may be formally written as an explicit form of Eqs. 8.69 and 8.70, as:

$$T_{if}\ \alpha\, |M_{if}|^2\ \alpha \left| \left\langle \psi_p^*\ \phi_e^* \left| \sum_i [O_i]_n\ \{(g_i + g_i'\ \gamma_5)\}_L + h \cdot c \right| \psi_n\ \phi_\nu \right\rangle \right|^2 \qquad ...(8.76d)$$

The subscript L indicates, that these operators act between the lepton wave functions. The subscript i takes on five values S, V, T, A and P, corresponding to scalar, vector, tensor, axialvator and pseudoscalar terms. The coupling constants are different for each type of interaction. The term $[O]_n\ \{O\}_L$ is symbolic and is to be interpreted either as an ordinary product or scalar product depending on the interaction. If matrix element is only scalar; or only pseudoscalar; the square of the matrix element will be scalar and the interaction will be independent of left or right-handed system, *i.e.* parity in the interaction will be conserved. If on, the other hand; the matrix element is a mixture of scalar and pseudoscalar terms, the square of the matrix element will contain scalar as well as pseudoscalar terms and hence the interaction will acquire the property of distinguishing left from right, which means, in other words, that parity of interaction is not conserved.

If we use the notation of Γ_i's instead of O_i's for the interaction matrices; we may write for β^--decay the general form of the four fermion (neutron, proton, electron, neutrino) interaction for a given type of interaction as:

$$\mathcal{H}_i(x) = \frac{1}{\sqrt{2}} \left(\overline{\psi}_p\ \Gamma_i\ \psi_i\right)\left[g_i\left(\overline{\phi}_e\ \Gamma_i\ \phi_\nu\right) + g_i'\left(\overline{\phi}_e\ \Gamma_i\ \gamma_5\ \phi_5\right)\right] + h \cdot c \qquad ...(8.76e)$$

The factor $1/\sqrt{2}$ is inserted in order to conform to the usual conventions. Also the argument (x) of the spinors is not explicitly shown. The operators O's or Γ_i's which correspond basically to products of Dirac matrices, give sixteen linearly independent elements as given in Table 8.1, and are classified into five groups as discussed earlier. These matrices are so defined that under a Lorentz transformation, bilinear quantities formed with the help of Dirac spinors $\overline{\psi}(x) = \psi^*(x)\ \gamma_4$ and $\psi(x)$ [where $\overline{\psi}(x)$ is Hermitian conjugate of $\psi(x)$], have definite transformation properties. Thus,

$$\overline{\psi}(x)\,\Gamma_i\,\psi(x) \quad i = 1, 2, 3, 4 \text{ and } 5$$

$$\text{for } S, V, T, A \text{ and } P, \text{ respectively} \qquad \ldots(8.76f)$$

transform like a (1) scalar (S); (2) a four vector (V); (3) an anti-symmetric tensor of the second rank (T); (4) a tensor of the third rank anti-symmetric in all three indices: hence axial vector (A), (5) Pseudoscalar (P). The most general form of the interaction for beta decay, is given by a linear combination of the five interactions:

$$\mathcal{H}_\beta(x) = \sum_i \mathcal{H}_i(x) \; ; i = S, V, T, A \text{ and } P \qquad \ldots(8.76g)$$

There will be twenty coupling constants in the general beta interaction Hamiltonian, *i.e.* ten complex constants g_i and g_i' (five for scalar and five for pseudoscalar) and each having a real and complex component. But if one imposes the condition of invariance, with respect to space reflection (P); time reversal (T) and charge conjugation (C), there will be restrictions of these constants. As for example, it can be shown that:

(*i*) Conservation of interaction after space-reflections (P) requires that $g_i = 0$. But this means parity is conserved. If parity is not conserved, as in the case in beta decay, then $g_i' \neq 0$.

(*ii*) Charge conjugation C (according to which a particle gets transformed into an anti-particle with equal mass, but opposite signs of the electric charge or lepton number or strangeness or third component I_3 of isotopic spin), then $g_i = g_i^*$ and $g_i' = -g_i^*$. However, in beta decay, charge conjugation does not hold good and hence the above mentioned restriction is not valid.

(*iii*) Time reversal requires,

$$g_i = g_i^* \text{ and } g_i' = g_i'^*, \text{ etc.} \qquad \ldots(8.77)$$

In the present theory of beta decay, since time reversal is assumed to be valid (but not space reflection and charge conjugation separately), the only restriction imposed on these coupling constants is that $g_i = g_i^*$, $g_i' = g_i'^*$.

8.4 NON-CONSERVATION OF PARITY IN BETA DECAY

8.4.1 Experimental Evidence

(A) *Correlation between nuclear spin and electron-momentum* in beta decay, is observed by a very interesting experiment. Parity conservation requires that all terms of an equation representing transition probability, must have the same behaviour under reflection, *i.e.* it should be scalar and should have no pseudoscalar component. If the parity is not conserved in β-decay and both scalar and pseudoscalar terms exist in the square of matrix element; as discussed in Section 8.3.5; (and as will be shown subsequently), we expect the emission of beta rays to depend on the angle between the direction of beta emission and spin direction of the nucleus. This means that if the direction of beta emission is parallel to spin direction of the nucleus, the transition probability is different from the case, when they are anti-parallel.

As a matter of fact as we shall see later, a mixture of scalar and pseudoscalar terms in transition probability; Eq. 8.76; will give rise an expression for transition probability[11, 12] as:

$$T_{if} \propto A\,(1 + \alpha \cos \theta_{jp}) \qquad \text{...(8.78a)}$$

where θ_{jp} is the angle between the direction of total spin **j** of the emitting nucleus and momentum $\vec{P}$ of the emitted electron in beta decay.

An experiment was performed by Wu, Amblar, Hayward, Hoppes and Hudson[10] at National Bureau of Standards in 1957; to test the theory of Lee and Yang[8]; who proposed for the first time in 1956 and 1957; that in weak interactions in particle physics ($\theta \to 2\pi$, $\tau \to 3\pi$), (called $\theta - \tau$ puzzle) or in β-decay in nuclear physics, *e.g.* ($n \to p + \bar{e} + \bar{\nu}$); the parity is not conserved and hence suggested an experiment of angular distribution of the electrons emitted from polarised nuclei.

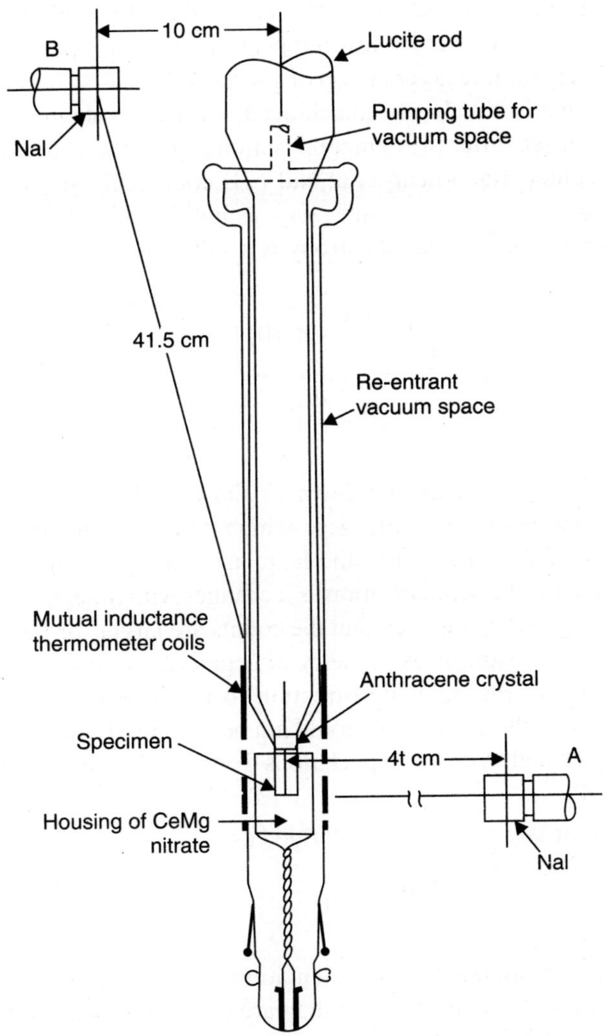

Fig. 8.5 The experimental arrangement used by C.S. Wu and collaborators in their demonstration of asymmetric electron distribution from polarised nuclei (Ref. 10).

For this purpose, experiments were performed[10] using sources of Co^{60} and Co^{58}, by Wu and her collaborators at National Bureau of Standards, Washington, D.C. (U.S.A.). These sources were selected as it is possible to prepare polarised samples of these isotopes, by applying the magnetic fields at low temperatures. Figure 8.5 shows the experimental arrangement.

Co^{60} nuclei are polarised by Rose-Gorter[10] method of demagnetisation; in which one applies a magnetic field at low temperatures. For beta-ray detection, an anthracene scintillator placed inside the cryostate, is used as detector, whose scintillations are seen through a lucite pipe by a photo-multiplier outside the cryostate, for detecting electrons of β-decay; and the degree of polarisation is measured by means of the asymmetry of γ-rays, detected by two NaI (Tl) detectors, placed outside the cryostate, nearly at right angles to each other.

The experiment requires basically two measurements (*i*) The degree of polarisation of Co^{60} nuclei, *i.e.* the relative number of Co^{60} nuclei, which have been oriented and (*ii*) The beta asymmetry with the polarisation direction of nuclei.

While the cooling of the specimen, say Co^{60}, is carried out by demagnetisation, the polarisation is carried out by applying a magnetic field at low temperature. In this experiment, the source consists of a thin layer of Co grown as crystalline layer on top of good single crystals of cerium magnesium nitrate. When the cooling by demagnetisation has been achieved, a vertical solenoid is raised around the lower part of the cryostat within 20 sec, after the demagnetisation so that the warming starts and the counting in the equator NaI (Tl) counter *A* and near-axial NaI (Tl) counter *B* for γ-rays starts as a function of time. After say 8 minutes or so the polarisation is zero, as indicated by the same number of counts in *A* and *B*, as shown in Fig. 8.6*a*. The gamma anisotropy is ε_γ calculated for *A* and *B*, *i.e.*,

$$\varepsilon_\gamma = \frac{\left[W\left(\frac{\pi}{2}\right) - W(\theta)\right]}{W\left(\frac{\pi}{2}\right)}$$

for polarising magnetic fields pointing up and down, as shown in Fig. 8.6*b*. The beta asymmetry, with polarising fields up and down, is shown in Fig. 8.6*c* which indicated, that the intensity of beta rays for $H\uparrow$ is clearly different from that, for $H\downarrow$. But the large asymmetry, as shown in these figures, shows that the effect is closely related to the depolarisation as it changes with time, as the salt (cerium magnesium nitrate) is heated up. From Fig. 8.6*c*, it is seen that the counts are larger for field down, and less for field up. However it is seen, that the sign of asymmetry is negative, *i.e.* the emission of beta particles is favoured for electrons going in the opposite direction to the nuclear spin orientation. In the early measurement, the exact value of the asymmetry could not be measured. But later measurements showed, that they were in general agreement with the expectations from the ideas following maximum polarisation of electrons with helicity of – 1.

Helicity of electron (or any particle) *h* is defined as:

$$h = \frac{\boldsymbol{\sigma}.\mathbf{p}_e}{p_e} = \pm 1 \qquad \text{...(8.79)}$$

Physically $h = +1$, corresponds to spin orientation and direction of motion of electrons being along the same direction and $h = -1$ means, that they are opposite to each other. Also neutrinos have helicity $h = -1$, while anti-neutrinos have helicity $h = +1$. This is an experimental fact and is in accord with the theory of beta decay, which requires electrons in β^--decay to have helicity of – 1 and positrons

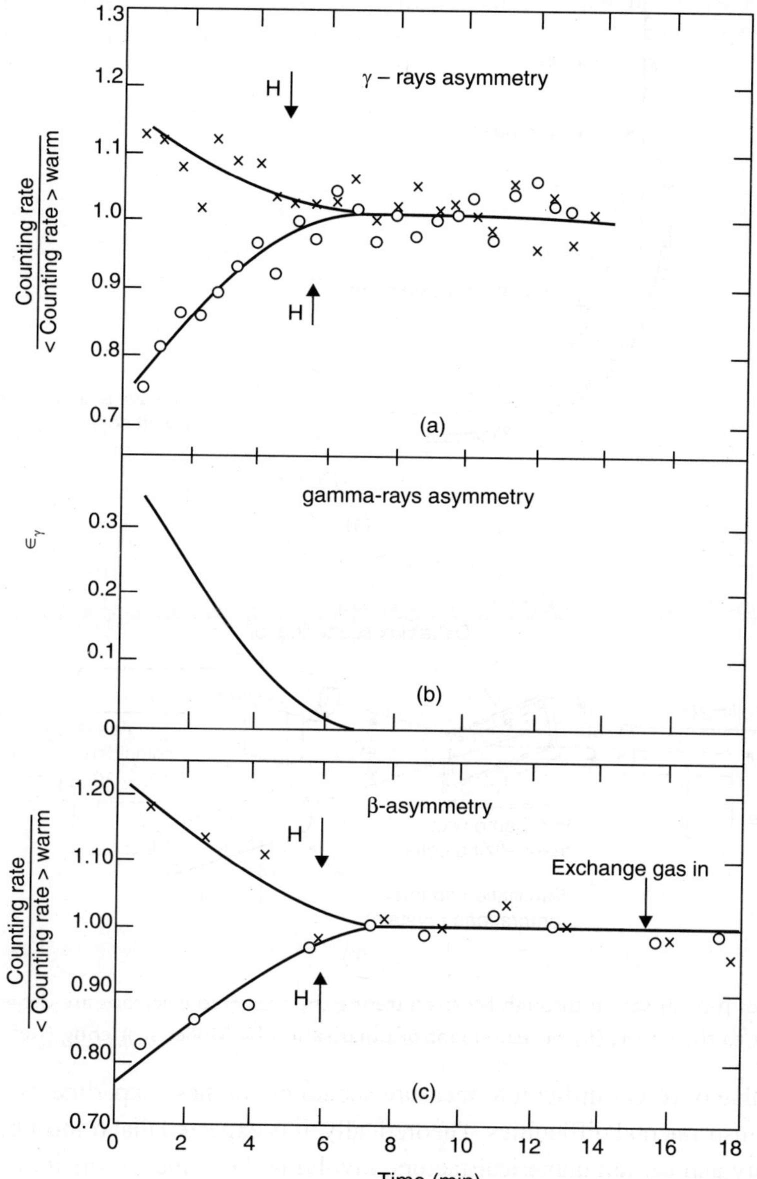

Fig. 8.6 The results on beta-asymmetry obtained with the instruments shown in Fig. 8.5 (Ref. 10).

in β^+-decay to have helicity of + 1. Both of these facts correspond to a definite handedness of the weak interaction. As a matter of fact, it was found that in the experiment by C.S. Wu, the intensity of electrons follows the general law given in Eq. 8.78*a* using $H\uparrow$ corresponding to $\cos\theta_{JP} = 1$ and $H\downarrow$ corresponding to $\cos\theta_{JP} = -1$, with:

$$\alpha < 0 \quad \text{for } \beta^- \ (Co^{60}); \quad \alpha > 0 \quad \text{for } \beta^+ \ (Co^{58}) \qquad \ldots(8.78b)$$

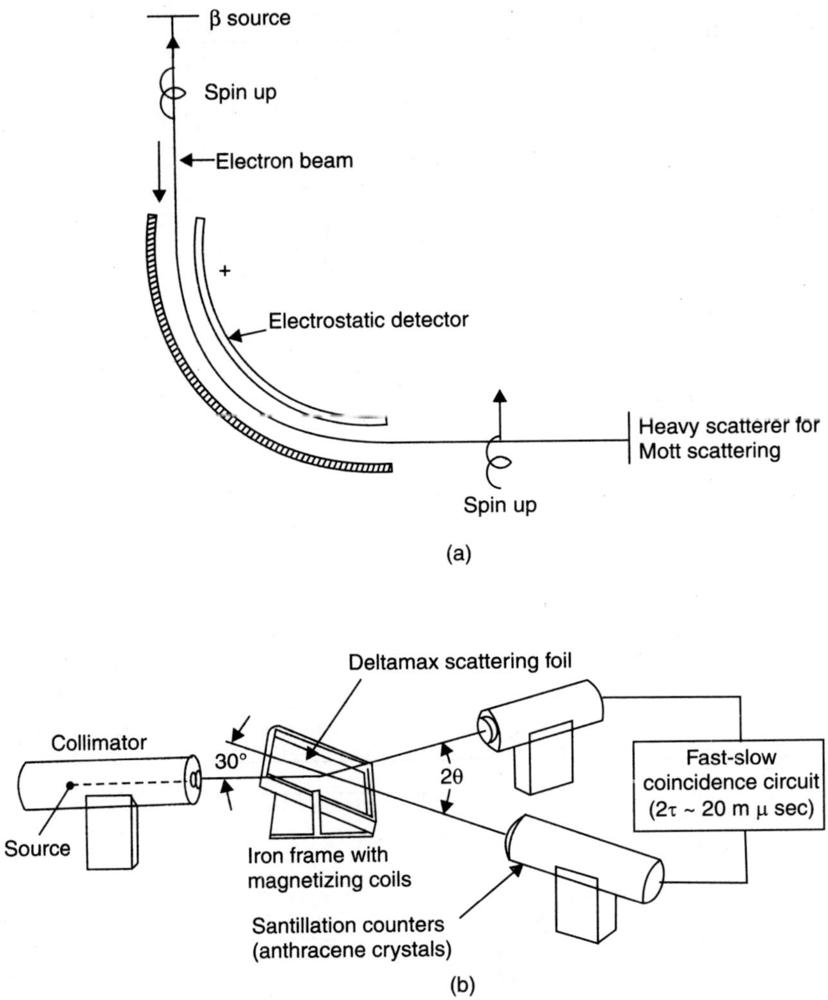

Fig. 8.7 (*a*) Study of β-polarisation through Mott scattering the scattered electrons are detected in a plane perpendicular to the paper; (*b*) Measurement of polarisation by Moller scattering (Ref. 13, 14).

The actual value of α was difficult to measure accurately in these experiments because of back-scattering and other instrumental difficulties. Theoretically, it is expected that α must be proportional to the electron's velocity and certain numerical factors, involving the values of the initial and final spins (*see* Eq. 8.135*b* derived subsequently). When the experimental results are compared to the theory; they all indicate that α is always as large as theoretically possible.

Hence, the probability of the emission of electrons in a given direction depends on the orientation of spin of nucleus, with the direction of the electron emission and is also related to the helicity of the emitted electrons, as we will see subsequently.

Parity violation in β-decay is not a small correction to the parity conserving theories; the effects are as large as they can possibly be.

(*B*) *Polarisation of electrons in* β-*decay*: The other consequence of non-conservation of parity in beta decay is that the electrons emitted are polarised, *i.e.* the spins of negatrons (e^-) are preferentially oriented anti-parallel to the direction of their momentum and spins of positrons (e^+) are oriented parallel to the direction of momentum; as we shall show theoretically in a subsequent section.

Such an experiment was described by Konopinski[13] in which the beta rays are first deflected electrostatically [Fig. 8.7*a*], in order to transform the longitudinal polarisation into a transverse one and then scattered through Mott scattering, in the electrostatic field of a nucleus. Then, the polarisation is measured from the left-right asymmetry of the Mott scattering.

Alternatively, we can study the longitudinal polarisation directly by scattering from a target of electrons, polarised along the direction of incidence (Moller scattering from magnetised iron), as shown in Fig. 8.7*b*. In this case, electrons from the source hit a magnetised foil (Deltamax foil) and both the scattered and the recoil electrons are detected in the scintillation counters and number of coincidences due to Moller scattering are detected with field parallel and anti-parallel to the electron momentum.

The other independent experiments for the same purpose are:

(*i*) Circular polarisation of internal bremsstrahlung; which depends on the direction of the spin of the radiating electrons (Ref. 15).

(*ii*) One can also study the helicity of positrons by means of observations on their annihilation. In this case it is seen that the number of two quantum (singlet) annihilation events in magnetised iron depends on the direction of magnetisation, *i.e.* spin direction of iron, relative to the incoming positrons. For this, *see* References (16) and (17). All these experiments concerning the polarisation of electron (or positrons) concur in showing that electrons and positrons are partially polarised and that the polarisation P_e is given by:

$$\left.\begin{aligned} P_e &= \mp \frac{v}{c} \\ &= \mp \beta \end{aligned}\right\} \begin{aligned} &-\text{(left handed) } \mathbf{P}_e \text{ and } \sigma \text{ anti-parallel for electrons.} \\ &+\text{(right handed) } \mathbf{P}_e \text{ and } \sigma \text{ parallel for positrons} \end{aligned} \quad ...(8.80)$$

These experiments point towards the non-conservation of parity in beta decay as we will see below in theoretical discussion.

8.4.2 Theoretical Treatment of Transition Probability and Polarisation Phenomena[18]

A. We can start, with the discussion of the general case of transition probability, where we consider all the five interactions; Equation 8.76*e*; and write the $|M_{if}|^2$ as:

$$\left|\langle f | \mathcal{H}_\beta | i \rangle\right|^2 = \left|\int d^3 x \Psi_f^*(x)\, \mathcal{H}_\beta(x)\, \Psi_i(x)\right|^2 \quad ...(8.81)$$

where $\Psi_f^*(x)$ and $\Psi_i(x)$ are total wave functions and

$$\mathcal{H}_\beta(x) = \sum_i \mathcal{H}_i(x),\ i = S, V, T, A \text{ and } P \quad ...(8.82)$$

so that

$$W_{fi} = \frac{2\pi}{\hbar}\left|\langle f | \mathcal{H}_\beta | i \rangle\right|^2 \rho_f; \left[\rho_f = \frac{dn_f}{dE_f}\right] \quad ...(8.83)$$

where ρ_f is the density of the states in the final system. As discussed in Section (8.2), vide Eqs. 8.35 and 8.36.

$$\rho_f \, dE_f = V^2 \frac{p_e^2 \, dp_e \, d\Omega_e}{(2\pi\hbar)^3} \frac{p_\nu^2 \, dp_\nu \, d\Omega_\nu}{(2\pi\hbar)^3} \qquad ...(8.84)$$

where E_f denotes the energy of the final state and is given by:

$$E_f = \sqrt{M_f^2 \, c^4 + c^2 \, (\mathbf{p}_e + \mathbf{p}_\nu)^2} + \sqrt{m^2 \, c^4 + c^2 \, p_e^2} + c \, p_\nu$$

$$= E_i = M_i \, c^2 \qquad ...(8.85)$$

where M_i and M_f are the masses of the initial and final nuclei. From Eq. 8.84 we obtain, after neglecting the terms of the order of $(p_\nu / M_f c)$:

$$\rho_f = V^2 \frac{p_e^2 \, dp_e \, d\Omega_e \, p_\nu^2 \, d\Omega_\nu}{c \, (2\pi\hbar)^6} = V^2 \frac{p_e \, E_e \, d\, E_e \, d\Omega_e \, p_\nu^2 \, d\Omega_\nu}{c^3 \, (2\pi\hbar)^6} \qquad ...(8.86)$$

where we have used

$$E_e^2 = m_e^2 \, c^4 + c^2 \, p_e^2 \qquad ...(8.87)$$

Proceeding from Eq. 8.81 and realising that any nucleon in the nucleus can undergo beta decay and hence we introduce isospin-operator [as shown earlier in Eqs. 8.76*a* and 8.76*b*] τ_n^+ in the matrix, *i.e.* we write for a nucleus of *A* nucleons, for the nuclear part of wave function *A*.

$$\overline{\psi}_f \, (x) \, \Gamma_i \, \psi_i \, (x) \rightarrow \sum_{n=1} \overline{\psi}_f \, (x) \, \Gamma_i \, \tau_n^+ \, \psi_i \, (x) \qquad ...(8.88)$$

Because of non-relativistic energies involved in beta decay for nucleons (a few MeV), the nucleus and the nucleons can be treated in non-relativistic approximation. This means that operators Γ_i between the states of the nucleus can be replaced by their large components $\Gamma_i^{(L)}$, (Ref. 19). [(*L*) in $\Gamma_i^{(L)}$ stands for large].

We write

$$\Gamma_S \rightarrow \Gamma_{(S)}^{(L)} = 1$$

$$\Gamma_V \rightarrow \Gamma_{(V)}^{(L)} = \beta$$

$$\Gamma_T \rightarrow \Gamma_{(T)}^{(L)} = \sigma = \sigma_k = \sigma_{ij}$$

$$\Gamma_A \rightarrow \Gamma_{(A)}^{(L)} = - \beta \, \sigma$$

The wave function of electron and anti-neutrino which are emitted with momentum $\mathbf{p}_e$ and $\mathbf{p}_\nu$ respectively can be written as plane waves neglecting the Coulomb interaction of the electron with the nucleus, *i.e.*,

$$\overline{\psi}_e (x) = \overline{U}_e^{(+)} (\mathbf{p}_e) e^{-\mathbf{x}\cdot\mathbf{p}_e} \text{ and } \psi_\nu (x) = U_\nu^{(-)} (-\mathbf{p}_\nu) e^{-i\mathbf{x}\cdot\mathbf{p}_\nu} \quad ...(8.90)$$

where $\overline{U}_e^+$ corresponds to electron and U_ν^- represents anti-neutrino.

$\overline{U}_e^+$ and U_ν^- correspond to the solution of

$$(i\gamma p_\mu + m)\, U(\mathbf{p}) = (i\gamma_k p_k - \gamma_4 E + m)\, U(\mathbf{p}) = 0 \quad ...(8.91)$$

which is Dirac equation in momentum representation.

Further, the exponential factor, *i.e.* $\exp\left(-i\,\mathbf{x}.(\mathbf{p}_e + \mathbf{p}_\nu)/\hbar\right)$ can be taken to be approximately as one if $|x| < \hbar/p_e$ and $\hbar/p_\nu$ as long as $|x|$ is less than the compton wavelength $\hbar/p$. This is true as $|x|$ may be taken as the nuclear size, *i.e.* 10^{-13} cm and $\lambda\!\!\!^{-}_e$ and $\lambda\!\!\!^{-}_\nu$ come out to be 10^{-11} cm and hence $|x|$ is about hundred times smaller than the compton wavelength of electron and neutrino. Keeping the above factors in mind, we can write:

$$\left\langle f \,|\, H_\beta \,|\, i \right\rangle = \frac{1}{V}\sum_{n=1}^{A}\sum_{i=1}^{5} g_i \int d^3x \left[\overline{\psi}_f(x)\,\Gamma_i\,\tau_n^+\,\psi_i(x)\right]\times$$

$$\left(\overline{U}_e^{(+)}(\mathbf{p}_e)\,\Gamma_i\,U_\nu^{(-)}(-\mathbf{p}_\nu)\right)\times e\,\frac{-\mathbf{x}\cdot[\mathbf{p}_e+\mathbf{p}_\nu]}{\hbar}$$

$$= \frac{1}{V}\sum_{i=1}^{5} g_i \left\langle \psi_f \,\middle|\, \Gamma_i^{(L)} \,\middle|\, \psi_i \right\rangle \left(\overline{U}_e^{(+)}(\mathbf{p}_e)\,\Gamma_i\,U_\nu^{(-)}(-\mathbf{p}_\nu)\right) \quad ...(8.92a)$$

where
$$\left\langle \psi_f \,\middle|\, \Gamma_i^{(L)} \,\middle|\, \psi_i \right\rangle = \sum_{n=1}^{A} \int d^3x \left(\overline{\psi}_f(x)\,\Gamma_i\,\tau_n^+\,\psi_i(x)\right) \quad ...(8.92b)$$

Here $1/V$ is used as normalising factor, to cancel V^2 in [Eq. 8.88]. We can, then, write the transition probability per unit time W_{fi}, [Eq. 8.92] as:

$$W_{fi} = \frac{1}{(2\pi)^5 c^5 \hbar^7}\, p_e E_e (E_0 - E_e)^2 \times$$

$$\left|\sum_{i=1}^{5} g_i \left\langle \psi_f \,\middle|\, \Gamma_i^{(L)} \,\middle|\, \psi_i \right\rangle \overline{U}_e^{(+)}(\mathbf{p}_e)\,\Gamma_i\,U_\nu^{(-)}(-\mathbf{p}_\nu)\right|^2 dE_e\, d\Omega_e\, d\Omega_\nu \quad ...(8.93)$$

where we have used $cp_\nu = E_0 - E_e$ in Eq. 8.86.

B. Unpolarised Initial and Final Nucleus: In the normal observation of beta decay when the source is unpolarised so that the polarisation of the final nucleus is not observed, we should develop the expressions for matrix element where we must average over the different orientations of the initial nucleus and sum over the magnetic quantum numbers of the final nucleus. For averaging, we multiply by operator:

$$\left(\frac{1}{2J_i+1}\sum_{mi}\right)$$

which sums over the magnetic numbers mi of the initial states and divide by the total number of magnetic number, *i.e.* $2J_i + 1$. Similarly, we obtain Σ_{mf} of the expression. Keeping these facts in mind, we obtain the squared matrix element as follows, from Eq. 8.92 as:

$$\left|\left\langle f \left| H_\beta \right| i\right\rangle\right|^2 = \frac{1}{V^2\,(2J_i+1)}\sum_{\substack{mi\\mf}}\left|\sum_{i=1}^{5} g_i \left\langle \psi_f \left| \Gamma_i^{(L)} \right| \psi_i\right\rangle \times \left[\overline{U}_e^{(+)}(\mathbf{p}_e)\,\Gamma_i\,U_\nu^{(-)}(-\mathbf{p}_\nu)\right]\right|^2$$

$$= \frac{1}{V^2\,(2J_i+1)}\sum_{\substack{mi\\mf}}\sum_{i=1}^{5} g_i\, g_i^* \left\langle \psi_f \left| \Gamma_i^{(L)} \right| \psi_i\right\rangle \left\langle \psi_f \left| \Gamma_i^{(L)} \right| \psi_i\right\rangle^* \times$$

$$\left[\overline{U}_e^{(+)}(\mathbf{p}_e)\,\Gamma_i\,U_\nu^{(-)}(-\mathbf{p}_\nu)\right]\left[\overline{U}_e^{(+)}(\mathbf{p}_e)\,\Gamma_i\,U_\nu^{(-)}(-\mathbf{p}_\nu)\right]^* \qquad \text{...(8.94)}$$

C. Nuclear Matrix Elements (Unpolarised Nuclei): We now realise that the nuclear part of Eq. 8.94 can be written as:

$$|M_{\text{Nuclear}}|^2_{\text{av}} = \frac{1}{2J_i+1}\sum_{\substack{mi\\mf}} \left\langle \psi_f \left| \Gamma_i^{(L)} \right| \psi_i\right\rangle \left\langle \psi_f \left| \Gamma_i^{(L)} \right| \psi_i\right\rangle^*$$

$$= \left(\left\langle \psi_f \left| \Gamma_i^{L} \right| \psi_i\right\rangle \left\langle \psi_f \left| \Gamma_i^{L} \right| \psi_i\right\rangle^*\right)_{\text{av}}$$

$$= \left[\left(\sum_{n=1}^{A}\int d^3x\,\psi_f(x)\,\Gamma_i^{(L)}\,\tau_n^+\,\psi_i(x)\right)\left(\sum_{n'=1}^{A}\int d^3x'\,\psi_f(x)\,\Gamma_{i'}^{(L)}\,\tau_n^+\,\psi_i(x')\right)^*\right]_{\text{av}} \qquad \text{...(8.95)}$$

We have, now, the following possibilities for different types of interactions:

(*i*) for $i, i' = S, V;\ \Gamma_i^{(L)} = \Gamma_{i'}^{(L)} = 1$

$$\left(\left\langle \psi_f \,|1|\, \psi_i\right\rangle \left\langle \Psi_f \,|1|\, \Psi_i\right\rangle^*\right)_{\text{av}} = \left|\sum_{n=1}^{A}\int d^3x\,\psi_f^*(x)\,\tau_n^+\,\psi_i(x)\right|^2 = |M_F|^2 \qquad \text{...(8.96)}$$

We have assumed in deriving Eq. 8.96 that each magnetic quantum number m_i is equally probable and hence $\sum_{m_i}$ cancels $2J_i + 1$ in the denominator.

(*ii*) For $i = S$ or V (and T, A)

$$i' = T, A \text{ (and } S \text{ or } V\text{), so that } \Gamma_i^{(L)} = 1 \text{ (and } \pm\sigma), \; \Gamma_{i'}^{(L)} = \pm\sigma \text{ (and 1)}$$

Then
$$\left(\langle\psi_f \,|\, \mathbf{1} \,|\, \psi_i\rangle \langle\psi_f \,|\, \sigma \,|\, \psi_i\rangle^*\right)_{\text{av}} = 0 \qquad \text{...(8.97)}$$

because the average of a single σ-operator for an unpolarised nucleus is zero.

(*iii*) $i, i' = T$ and A; $\Gamma_i^{(L)} = \pm\sigma$; $\Gamma_{i'}^{(L)} = \pm\sigma$ (lower sign for A)

(*a*) For $i = i'$

$$\left(\langle\psi_f \,|\, \pm\sigma_k \,|\, \psi_i\rangle \langle\psi_f \,|\, \pm\sigma_{k'} \,|\, \psi_i\rangle^*\right)_{\text{av}}$$

$$= \left(\langle\psi_f \,|\, \sigma_k \,|\, \psi_i\rangle \langle\psi_i \,|\, \sigma_{k'}^* \,|\, \psi_f\rangle\right)_{\text{av}}$$

$$= \delta k k' \left(\langle\psi_f \,|\, \sigma_k \,|\, \psi_i\rangle \langle\psi_i \,|\, \sigma_{k'}^* \,|\, \psi_f\rangle\right)_{\text{av}} \qquad \text{...(8.98}a\text{)}$$

The multiplication shows, that since σ_x and σ_y change the magnetic quantum number of the state over which it operates, k and k' must be the same to get a non-zero result.

Also since $\sigma_x^2 = \sigma_y^2 = \sigma_z^2 = \frac{1}{3}\sum_{k=1}^{3} \sigma_k^3 = \frac{1}{3}(\sigma \cdot \sigma)$

we can rewrite Eq. 8.98*a* as:

$$\left(\langle\psi_f \,|\, \pm\sigma_k \,|\, \psi_i\rangle \langle\psi_f \,|\, \pm\sigma_{k'}^* \,|\, \psi_i\rangle\right)_{\text{av}} = \frac{1}{3}\delta\, k\, k' \left(\left|\langle\psi_f \,|\, \sigma \,|\, \psi_i\rangle\right|^2\right)_{\text{av}}$$

$$\equiv \frac{1}{3}\delta\, k\, k' \,|\, M_{GT} \,|^2 \qquad \text{...(8.98}b\text{)}$$

where
$$|\, M_{GT} \,|^2 = \left(\sum_{k=1}^{3} \left|\sum_{n=1}^{A} \int d^3 x\, \psi_f^*(x)\, \sigma_k\, \tau_n^+\, \psi_i(x)\right|^2\right)_{\text{av}}$$

(*b*) For $i \neq i'$

$$\left(\langle\psi_f \,|\, \pm\sigma_k \,|\, \psi_i\rangle \langle\psi_f \,|\, \mp\sigma_{k'} \,|\, \psi_i\rangle^*\right)_{\text{av}} = -\frac{1}{3}\delta\, k\, k' |M_{GT}|^2 \qquad \text{...(8.98}c\text{)}$$

The nuclear matrix element M_F and M_{GT} defined by Eqs. 8.96, 8.96*b* and 8.96*c* are referred to as Fermi and Gamow-Teller elements, respectively. The matrix element $|\, M_F \,|$ is many times written as $\int \mathbf{I}$, denoting that $\Gamma_i^{(L)} = \Gamma_{i'}^{(L)} = 1$. Similarly $| M_{GT} |$ is referred to as $\int \boldsymbol{\sigma}$, again indicating that $\Gamma_i^{(L)} = \pm \sigma$ and $\Gamma_{i'}^{(L)} = \mp \sigma$. The two selection rules, *i.e.* Fermi selection rules and Gamow-Teller selections rules, depend on the properties of $\int \mathbf{I}$ and $\int \boldsymbol{\sigma}$, respectively as discussed subsequently.

8.5 SELECTION RULES AND SHAPES OF SPECTRA

A. The Gamow-Teller Rules: They emerge when the intrinsic spin of the transforming nucleon is introduced into the Hamiltonian as shown above. This couples the electron-neutrino spin directly to the nucleonic spin. The Fermi rules are applicable, when this is neglected. Fermi rules imply for allowed transitions, that the electron-neutrino pairs are emitted with anti-parallel spins (singlet state) while in Gamow-Teller rules, they are emitted with parallel spin (triplet state). At present, Gamow-Teller rules appear to describe most cases of beta decay; but there are a few cases where Fermi rules or a mixture of Fermi and Gamow-Teller rules are required to describe the observations.

We have discussed in the previous section the theoretical derivations of $| M_F |^2$ and $| M_{GT} |^2$. Physically, it may be seen that if electron and neutrino are emitted with their spins opposite to each other, then the only angular momentum carried by them is **L**; the orbital angular momentum, so that the relationship of the final and initial angular momentum is given by:

$$\mathbf{I}_f = \mathbf{I}_i + \mathbf{L}_i \qquad ...(8.99)$$

This corresponds to Fermi transition

On the other hand, the electron and anti-neutrino spin may be parallel to each other. Then total spin change $\Delta S = 1$, and the relationship of $\mathbf{I}_f$ and $\mathbf{I}_i$ is given by:

$$\mathbf{I}_f = \mathbf{I}_i + \mathbf{L}_i + \Delta\, \mathbf{S} \qquad ...(8.100)$$

This corresponds to Gamow-Teller transition.

The parity change is, of course, given by, in both cases by:

$$\pi_i = \pi_f (-1)^L \qquad ...(8.101)$$

and is independent for Fermi or Gamow-Teller nature of the beta transitions.

B. Allowed Beta Decay-Selection Rules: It will be seen subsequently from Eqs. 8.121 and 8.122 that the value of transition probability depends upon ξ, which, in turn, is dependent on the structure of $\int \mathbf{I} = | M_F |$ and $\int \sigma = | M_{GT} |$. The matrix elements $| M |$ for allowed transitions, where only large (L) components are used, are given in Eqs. 8.96 and 8.98 according to which the allowed transitions depend on these two integrals. From these equations, we can see that for allowed transitions, involving S and V interactions, the integral $\int \mathbf{I}$ is involved; while for transitions involving T and A interactions; the integral $\int \sigma$ is involved. In allowed transition P-interaction does not play any role. Making use of Eqs. 8.96 and 8.98, we can derive the selection rules for the change of angular momentum and parity for the various matrix-elements. The matrix-elements $| M_S |$, $| M_V |$ and $| M'_S |$ and $| M'_V |$ involve integral $\int \mathbf{I}$ and the products of the kind $(\overline{U}_e \mathbf{I} U_\nu)$, $(\overline{U}_e \gamma_4 U_\nu)$ for undashed matrix elements and $(\overline{U}_e \gamma_5 U_\nu)$ and $(\overline{U}_e \gamma_5 \gamma_4 U_\nu)$ for dashed matrix elements. The integral $\int \mathbf{I}$ is given by:

$$\int \mathbf{I} = i\int \overline{\Psi}_f \left(\sum_{n=1}^{A} \tau_+^n\right) \Psi_{inc}\, d\tau$$

Remembering that τ_+, either converts the wave function to zero, or to another wave function, replacing neutron to proton, without any change of angular momentum, it is apparent that $\int \mathbf{I}$ is maximum when there is maximum overlap between the wave function of the final nucleus, and the initial nucleus with a change of neutron to proton, but no change of angular momentum or parity. Hence for maximum $\int \mathbf{I}$; $\Delta I = 0$ and $\Delta \pi = 0$, gives the maximum overlap. Also for the lepton part, *i.e.* for $\overline{U}_e\ \mathbf{I} U_\nu$ and $U_e \gamma_4 U_\nu$ to be non-zero; it is required that both wave functions $\overline{U}_e$ and U_ν correspond to the same spin direction, where $\overline{U}_e$ corresponds to emitted electron and U_ν to the absorbed neutrino. As in practice; it is the anti-neutrino which is emitted; this means, that electron and anti-neutrino should be emitted with opposite spin-directions or in other words, there should be no change in the angular momentum due to the emission of leptons. The same logic applies to M'_s and M'_ν because γ_5 also does not involve any spin-change. Hence for Fermi-interactions, the rules are:

$$\Delta \pi = 0 \text{ for } S, S';\ \Delta \pi = 0 \text{ for } V, V'\ 0 \to 0 \text{ Allowed} \qquad \text{...(8.102)}$$

The condition $0 \to 0$ is specifically mentioned, because though it is generally permissible for Fermi interaction to give $\Delta I = 0$ under the rules of vector addition; yet as we will see later, it may not be generally permissible for Gamow-Teller interaction. Hence it is mentioned specifically. The same arguments hold good for primed matrix elements M'_s and M'_ν; as the operator γ_5 does not affect the spin part of the wave function.

Similarly for T and A interactions the $\int \sigma$ plays the same role as $\int \mathbf{I}$ for S and V interaction. By the very definition:

$$\int \sigma \equiv \int \psi_f^* \left(\sum_{n=1}^{A} \tau_+^n\, \sigma\right) \psi_{in}\, d\tau \qquad \text{...(8.103)}$$

We now realise that the martix $\int \sigma$ is pseudo vector which has three components, but does not change sign under space reflection; $\int \sigma$ is maximum, when ψ_i and ψ_f are coupled by a unit vector and the two states have the same parity. This change of quantum numbers is also reflected in the vector product involving leptons, *i.e.* $(\overline{U}_e\, \sigma\, U_\nu)$. This product is maximum, when again the emitted electron and the absorbed neutrino are coupled by a unit pseudovector, *i.e.* the spins of emitted electron and absorbed neutrino are opposite to each other; or the spins of emitted electron and emitted anti-neutrino is in the same direction. The same arguments apply to the primed matrix elements M_T' and M_A'. From the above consideration, the selection rules for the Gamow-Teller interaction can be written as:

$\Delta I = 0, \pm 1\ 0 \to 0$ forbidden

for T, T' |

$\Delta \pi = 0$ and A, A' | interactions ...(8.104)

The condition $0 \rightarrow 0$ forbidden; corresponds to the fact that when vectorially two functions are connected by a unit vector; $0 \rightarrow 0$ is not possible vectorially. The relative spin direction of the emitted electrons and anti-neutrinos in Fermi and Gamow-Teller interaction are shown in Fig. 8.8 for right-handed anti-neutrinos.

Gamow-Teller β-transitions given in Eqs. 8.98 and 8.103 which correspond to change of spin and especially 1-forbidden (*i.e.* $\Delta\ 1 \neq 0$) transitions are a sensitive probe of the possible influence of extra nucleonic effects, such as meson—current exchange, etc. on the description of low-lying nuclear states. These transitions are special in that ordinary nuclear effects have only very small transition matrix elements and extra nucleonic effects like meson—current exchange become more visible.

Recently two such cases have been studied experimentally—especially their ft values and branching ratios, in the decay of K^{37} and Li^7 and Be^8 from which one can obtain the Gamow-Teller β-decay transition strengths M_{GT} [Eqs. 8.41 and 8.98]. Comparing these values in the case of K^{37} with the theoretically expected values for which the wave function ψ's were written, say for shell model in which residual interaction of Wildenthal type of Brown or Towner-Khanna Model[21], was used, it was possible to calculate the choice of *S-d* shell wave function in the β-decay of $d_{3/2}$ of K^{37} to $S_{1/2}$ of Ar^{37}. Using the effective coupling constants, explicitly evaluated from core polarisation and meson exchange process, the experimental results could be reproduced. In the case of Li^7 and Be^8 again, log ft values were compared with the expected theoretical values[22], which predicted the narrowing of pf and sd shell in the two cases. A case of the study of Gamow-Teller strength from Se^{79} (*n, p*) As^{76} reaction is of special interest[23], at the incident energy of 198 MeV of neutrons. It has been found that between energies 200 MeV and 300 MeV, $\Delta L = 0$, $\Delta J^{\pi} = 1^+$, holds good, so that Spin-flip isovector excitation (*i.e.* Gamow-Teller) interaction dominates over, $\Delta J^{\pi} = 0$; non-flip Fermi interaction. The cross-section for (*n, p*) at 0°, extrapolated to zero momentum are directly proportional[24] to $\int \sigma$. Hence measurement of the cross-section as a function of excitation energy, gives information about the strength of Gamow-Teller transition. This method which provides an alternative method of measuring the Gamow-Teller strength, gave an indication about the strength of β β 2 ν transition of Ge^{76} and check on the ft value of Ge^{76}.

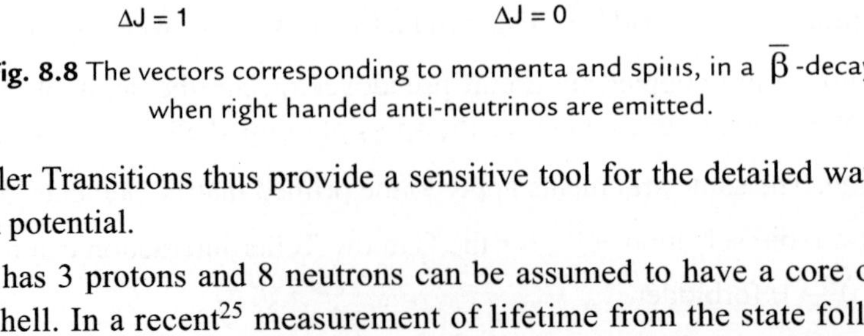

Fig. 8.8 The vectors corresponding to momenta and spins, in a $\bar{\beta}$-decay, when right handed anti-neutrinos are emitted.

Gamow-Teller Transitions thus provide a sensitive tool for the detailed wave function and the detailed interaction potential.

Li^{11}, which has 3 protons and 8 neutrons can be assumed to have a core of He^4, and valence particles in the *p*-shell. In a recent[25] measurement of lifetime from the state following β-decay, for $Li^{11} \rightarrow Be^{11}$ (½⁻, 320 keV) transition, the value of lifetime and the Gamow-Teller Strength obtained from it has been compared with a shell model calculation, assuming that Li^{11} ground state halo wave function, contains about 50% S-wave neutron component and for the resulting Li^{11}, the halo wave function for 7 nucleons has the following structure for the valance nucleons.

$$\Phi = \alpha\left[(0p_{3/2})^{\pi}, (0p_{3/2})^{4\nu}\,(0p_{1/2})^{2\nu}\right] + \beta\left[(0p_{3/2})^{\pi}, (0p_{3/2}\;p_{1/2})^{n_1\nu}\,(IS_{1/2}\;0d_{5/2})^{n_2\nu}\right]$$

where $\qquad n_1 + n_2 = 6$

This reproduces the experimental results.

Another interesting study[26] of β^+-decay from Sb^{108}, Sb^{107}, Sb^{106} and Sb^{105} has been carried out from $Ni^{58} + Cr^{50}$, $Ni^{58} + Cr^{52}$ and $Ni^{58} + Ni^{58}$, reactions, which yield these nuclei through ion induced fusion—evaporation. Out of the resulting isotopes, Sb^{107} and Sb^{108} are neutron deficient. The values of lifetime measured, in the experiment, were compared with RPA approximation calculations which gave qualitative agreement.

C. Allowed Beta Decay—Shapes of Spectra: A very good example[27] of Fermi Transitions is the decay of O^{14}, which decays by the emission of a positron from the state 0^+ of O'^{14} to 0^+ of N^{14}. Similarly the decay of 2 He^6 to 3 L_i^6 by electron emission from $0^+ \to 1^+$ is a good example of a pure Gamow-Teller transition. On the other hand, many transitions from $I_1 \to I_2$ involving both I_1 and I_2 different from zero, will have mixed characteristics[28]. Cases of neutron decay; $H^3 \xrightarrow{\beta} He^3$ and $S^{35} \xrightarrow{\beta^-} Cl^{35}$ are other examples of pure allowed Fermi decay.

Equation 8.38 which was derived in the previous Section, gives the shape of the spectrum of emitted electrons for allowed β^- transitions, for which the various conditions have been stated in the previous Section.

It can be seen from this equation, that if one plots $(n\,(p)/F\,(Z, E)\,p_e\,E_e)^{1/2}$ versus E_e, one expects a straight line if the Fierz constant 'b' as given in Eq. 8.122 is equal to zero and hence $|\,M\,|^2$ is independent of energy but one expects a curvature in the line if $b \neq 0$. Shapes of many beta spectra for allowed transitions have been carefully studied experimentally and it has been found that these lines called Fermi-Kurie plots are very nearly straight lines, with an upper limit of 'b' of about 0.05 to 0.1. One generally assumes that $b = 0$. The shapes of Fermi-Kurie plots of many such cases[22] are shown in Fig. 8.9.

D. Forbidden Beta Decay-Selection Rules: What happens, if $\Delta I = \pm 1$ or 0, and/or $\Delta \pi \neq 0$ but the beta transition is energetically possible? According to Eqs. 8.96 and 8.98, transition probability is zero. These equations along with earlier discussion (Section 8.3.4) were, however, written under the conditions when the higher terms in the expansion of the leptonic wave functions were neglected and the wave function was taken to be the same throughout the nucleus (*i.e.* at $r = 0$). But with the vanishing of zero-order terms in nuclear matrix elements; the higher order terms become important. Hence we should consider these higher terms—which give rise to forbidden transitions of different order.

For the calculation of higher terms, we may, expand $\phi_{\bar{e},\bar{\nu}}$ as,

$$\phi_{\bar{e},\bar{\nu}} = \phi_{\bar{e},\bar{\nu}}\,(0)\,\exp\,\frac{-i\mathbf{p}_{\bar{e},\bar{\nu}}\cdot(\mathbf{r}_n - \mathbf{r}_n)}{\hbar}$$

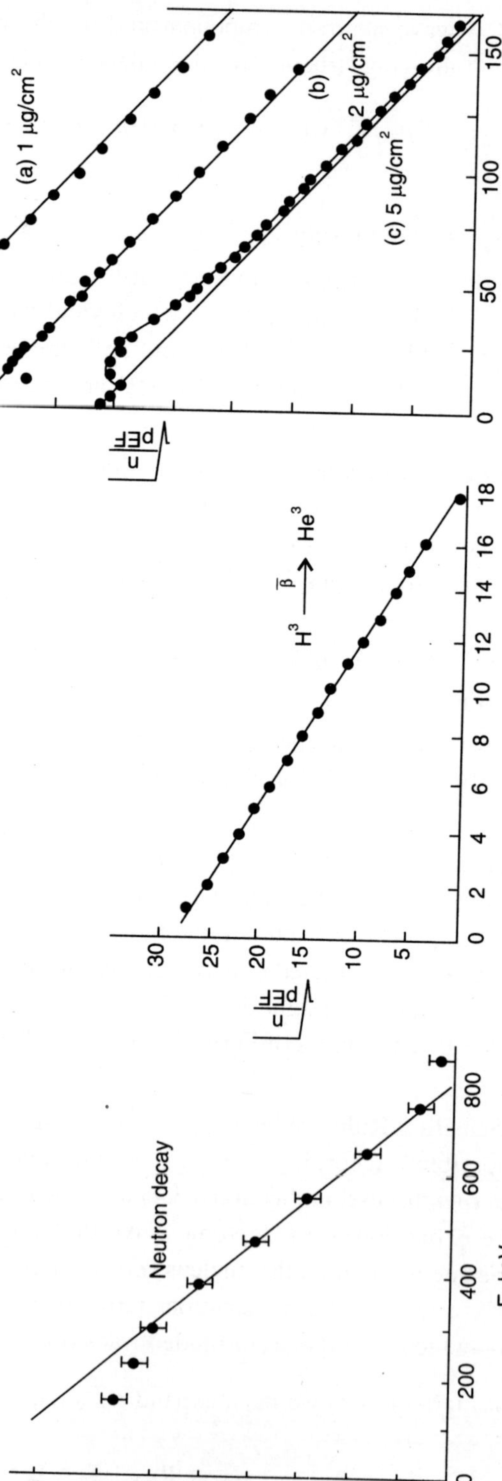

Fig. 8.9 Some allowed Fermi-Kurie plots. The curves of S^{35} decay demonstrate the effect of energy loss due to source thickness [Ref. 29].

$$= \phi_{\bar{e},\bar{\nu}}(0)\left[\frac{1 - i\mathbf{p}_{\bar{e},\bar{\nu}}\cdot(\mathbf{r}_n - \mathbf{r}_L)}{\hbar} + \left[\frac{i\mathbf{p}_{\bar{e},\bar{\nu}}\cdot(\mathbf{r}_n - \mathbf{r}_L)}{\hbar}\right]^2 + \ldots\right] \qquad \ldots(8.105)$$

Denoting $\mathbf{p}_e \to \mathbf{p}$ and $\mathbf{p}_{\bar{\nu}} \to \mathbf{q}$ and $\mathbf{r}_n - \mathbf{r}_L = \mathbf{r}$ and realising from Eq. 8.76 *d* that $|M|^2$ involves, the product of ϕ_e and $\phi_{\bar{\nu}}$, it is easy to see from Eq. 8.76; that the leptonic part of the $|M|^2$ will now involve,

$$\phi_e\,\phi_{\bar{\nu}} = \phi_e(0)\,\phi_{\bar{\nu}}(0)\exp\frac{-i(\mathbf{p}_e + \mathbf{p}_{\bar{\nu}})\cdot(\mathbf{r}_n - \mathbf{r}_L)}{\hbar} \qquad \ldots(8.106)$$

In general, one may write the expansion of exponential terms as:

$$\exp\frac{-i(\mathbf{p}+\mathbf{q})\cdot\mathbf{r}}{\hbar} = \frac{4\pi}{\hbar}\sum_{m,1} i^1\,J_1(|\mathbf{p}+\mathbf{q}|r)\,Y_1^m(\Omega)\,Y_1^{*m}(\Omega') \qquad \ldots(8.107)$$

where Ω denotes the polar angle of **r** and Ω' stands for the polar angles of **p** + **q**. In the case of allowed spectra; for only 1 = 0, j_0 was used for Bessel function and was taken to be unity. To use higher terms corresponding to 1 > 0; if the term J_0 corresponding to 1 = 0 gives zero nuclear matrix, we may use J_1 for 1 > 0; for non-zero results.

Here one may realise that for small x, $J_1(x)$ is of the order of x. Hence the ratio of successive terms in the expansion is of the order of $|\mathbf{p} + \mathbf{q}|\,r \approx 10^{-2}$. It is evident, therefore, that the successive terms are important only if the proceeding terms give zero nuclear matrix. The transition probability corresponding to 1 > 0, corresponds to 1-forbidden transitions of different order, *e.g.* for 1 = 1, it is first forbidden, for 1 = 2, second forbidden, etc. From Eq. 8.107, it can be seen that in the lowest approximation Bessel function $J_0 = 1$ for 1 = 0 and higher terms belonging to higher values of 1 can be neglected. If this term gives a zero nuclear matrix element, the terms for higher values of 1 can be considered. The detailed properties of the transitions and the selection rules will depend on the properties of the nuclear and leptonic wave functions. As discussed in Section 8.2 the plot of $(n(E_e)/P_eE_eF(Z, E_e))$ versus E_e determines the shape of the spectrum. In allowed spectra, where $|M|^2$ is independent of energy (*i.e.* $b = 0$) we get the straight Fermi-Kurie plot.

In forbidden transition, the nuclear matrix elements cannot be given by $\int \mathbf{I}$ and $\int \boldsymbol{\sigma}$, but by expressions like Eq. 8.76 for any order the other similar expression with appropriate operators as given in Table 8.1. Also the leptonic wave functions cannot be taken out of the integral, as they are not constants, but are functions of space-coordinates. As a matter of fact, the matrix element $|M|^2$ in Eq. 8.76 may now be written as for only scalar interaction:

$$|M|^2 = \left|\int \psi_f^+ \left|\sum_{i=1}^{A}\left[\phi_f^*(\mathbf{r}_i)\,O_L\,\phi_i(\mathbf{r}_i)\right]O_n\right|\psi_i\,d\tau\right|^2 \qquad \ldots(8.108)$$

where ϕ_f^* is the electron wave function and ϕ_i the neutrino wave function. For taking into account both the Coulomb effect on electronic wave function and the higher terms in the neutrino wave function, this will evidently introduce the energy dependent terms, even if $b = 0$. These energy dependent terms[30], will give rise to the shape factors. We may, then, write Eq. 8.37 or 8.38 as:

$$N(E)\,dE = G^2 F(Z, E_e)\, S_n\, p_e\, E_e\, (W_0 - E_e)^2\, dE$$
$$[W_0 \equiv E_0 \text{ of Eq. (8.38)}] \quad ...(8.109)$$

where S_n gives the shape-factor for nth forbidden transition, arising from the energy dependence of the marix element $|M|^2$. It is expressed as:

$$S_n = S_n^{(n)} + S_n^{(n+1)} + \delta_{n,1}\, S_1^{(0)} \quad ...(8.110)$$

where $S_n^{(n)}$ is due to tensors of rank n, $S_n^{(n+1)}$ is due to those of rank $n + 1$. $S_1^{(0)}$ corresponds to the contribution from tensor of rank 0, if $n = 1$ and only for this case the last term exists. For the derivation of $S_n^{(n)}$, $S_n^{(n+1)}$ and $S_1^{(0)}$ and their expressions *see* Reference [24]. Suffice to say that they are energy-dependent and hence contribute to the shape of the spectrum. From Eq. 8.109, we can write, in general only for the shape of the spectrum of electrons as:

$$\sqrt{\frac{N(E)}{p_e E_e\, F(Z, E_e)}} = \sqrt{\frac{S_n\, q^2}{2\pi^3}} = (W_0 - E_e)\left(\frac{S_n}{2\pi^3}\right)^{\frac{1}{2}} \quad ...(8.111)$$

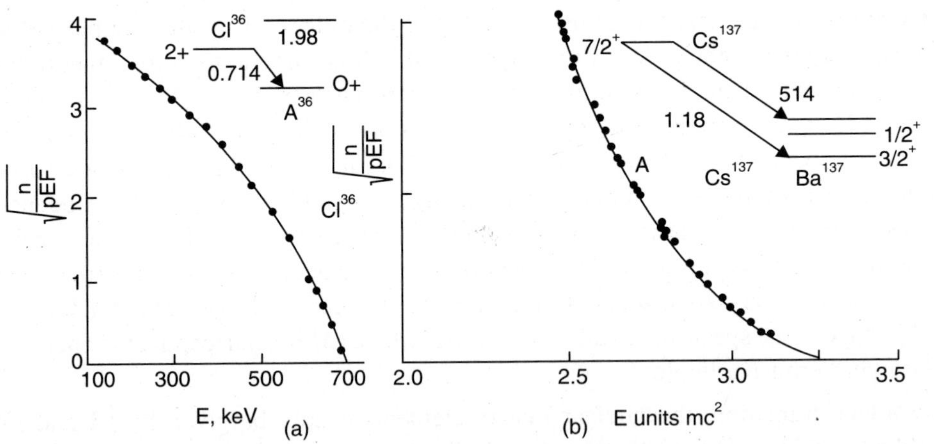

Fig. 8.10a and b (*a*) Conventional Fermi plot for Cl^{36} beta spectrum, I = 2, No, Non-unique (Ref. 32); (*b*) Conventional Fermi plot for the high energy group of Cs^{137} β-spectrum, of non-unique transition, I = 2, No. (Ref. 32).

For non-unique transitions, $S_n^{(n)}$ determines the shape. In general, they may not be straight plots. They can have curvatures of opposite sign, depending on the detailed properties of matrix elements. Figure 8.10 shows the Fermi plots of Cl^{36} and Cs^{137} spectra, showing the opposite curvatures.

For allowed transitions, S_n is constant and, we have the equation for *a*-Fermi-Kurie plot, which for a massless neutrino is a straight line. If neutrino has a mass, then we can write, near the high energy end point by replacing q^2 by $q\, E_\nu$ and using

$$\sqrt{\frac{N(E_e)}{p_e E_e\, F(Z, E_e)}} = \sqrt{\frac{S_n\, q\, E_\nu}{2\pi^3}}$$

$$= \sqrt{\frac{S_n}{2\pi^3}}\,(W_0 - E_e)^{\frac{1}{2}}\left[(W_e - E_e)^2 - m_\nu^2\right]^{\frac{1}{4}} \qquad ...(8.112)$$

and the shape of the spectrum near the end of the spectrum is disturbed. The experiments limit the upper limit of neutrino mass as:

$$m_\nu \le \frac{m_e}{2000}$$

The shapes of the electron-spectra from transitions of different forbiddenness, depend on the properties of S_n. We discuss the properties of S_n and the relevant selection rules. In the first forbidden spectra, it is found experimentally that for $\Delta I = 0$ or 1 and $\Delta\pi =$ Yes, the Fermi-Kurie plots are straight lines; though one expects an energy dependence of

$$S_I^{(I)} = S_1 + S_1^{(0)}$$

and hence these plots are not in general expected to be straight. But the energy dependence of S_1 happens to be very slight; except in a rare case when nuclear matrix elements belonging to different operators may cancel. In fact, in the first forbidden spectra, one can write (Ref. 31):

$$\frac{\sqrt{\dfrac{n(E_e)}{p_e\, E_e\, F(Z, E_e)}}}{(W_0 - E_e)} = \frac{-k}{1 + k\,W_0}\,E_e \qquad ...(8.113)$$

where k is very small, given by $k = 0.058$ for Au^{198} and $k = -0.06$ for OS^{186}.

For $\Delta I = 2$; $\Delta\pi =$ Yes, however, these plots are not straight lines. Their shapes depend on the characteristics of $S_1^{(2)}$ which happens to be proportional to $(q^2 + p^2)$. If one plots

$$\left[\frac{n(E_e)}{\{p E_e\, F_0\,(p^2 + q^2)\}}\right]^{1/2}$$

versus E_e, one gets a straight line, where F_0 is called Fermi function, and takes into account the Coulomb effect[31]. On the other hand, the standard Fermi-Kurie plots will be curved and their exact shape will depend on the end energies. Decay of Y^{91} is a good example of this mode of decay (Ref. 32).

These transitions are called unique transitions because in these cases, the shapes of the spectra entirely depend on the value of nuclear matrix elements. Although shapes of spectra do not uniquely fix the beta decay interaction; it is found that they are in accord with *V-A* theory. Though it is generally assumed that pseudoscalar interaction is not present or is negligible, it can be shown that the selection rules and magnitude of the pseudoscalar matrix elements are such, that this interaction can be noticed only if $\Delta I = 0$, yes, transition is in interference, with axial vector interactions. Several cases *e.g.* Pd^{144}, Ho^{166} and Tl^{207} have been studied and can be satisfactorily explained with no pseudoscalar interaction[33].

E. Double β-Decay: Double β-decay in which the transition of beta decay is single isobaric jump of two units in atomic numbers has been observed[34] in many nuclei, though it is a rare second order weak transition. Such a transition corresponds to:

$$(A, Z) \xrightarrow{\beta\beta 2\nu} (A, Z+2) + 2\overline{e} + 2\overline{\nu}_e \qquad ...(8.114)$$

Such a decay has been observed with its two accompanying neutrinos in nuclei, for which single β-decay is strongly inhibited or energetically forbidden. A variation of the decay as given in Eq. 8.114, corresponds to:

$$(A, Z) \xrightarrow{\beta\beta o\nu} (A, Z+2) + 2e^- \qquad ...(8.115)$$

Search for this mode is also going on in various laboratories. If such a search gives positive results, this will lead to a deviation from, the standard model of universal force, and will also require that neutrinos may have mass[34].

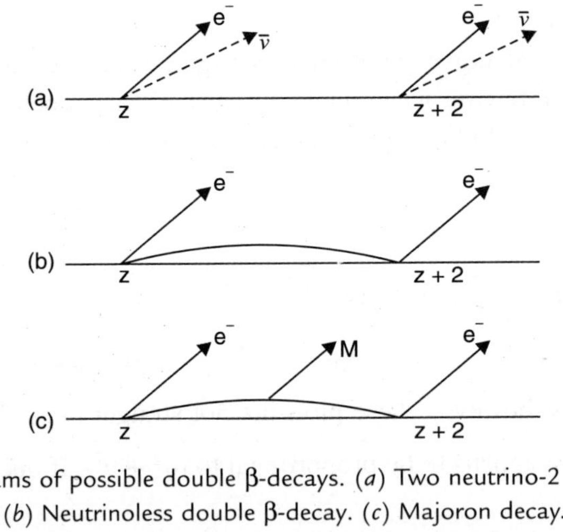

Fig. 8.10c Diagrams of possible double β-decays. (*a*) Two neutrino-2 $\overline{e}$ double β-decay. (*b*) Neutrinoless double β-decay. (*c*) Majoron decay.

See Fig. 8.10*c* for these two models. The majoron (*M*) is a hypothetical boson, coupling to the neutron with sufficient strength to make a significant contribution to β β decay rate.

However, β-transition as given in Eq. 8.114 is more commonly found, experimentally and provides a test for the theory and calculation for both the modes of double β-decay. Already Ge^{76}, Se^{82}, Me^{100}, Cd^{116} and Nd^{150} have been investigated and half lifetime measured which ranges from 1.42×10^{21} years for Ge^{76} to 6.75×10^{18} years for Nd^{150}. Most recent[35] measurement which have been carried out, for Mo^{100} and Nd^{150} are perhaps the most detailed measurements so far, where not only ββ2ν, has been investigated and measured but also ββoν. Attempts for measurements of neutrinoless transitions, where only limits of lifetimes have been obtained[35], have centered around Ca^{48}, Ge^{76}, Se^{82}, Mo^{100}, Te^{130} and Xe^{136} and limits of lifetimes range from 9.5×10^{21} years for Ca^{48} to 3.7×10^{23} years for Xe^{136}. The positive results for these neutrinoless transitions lead to an upper limit of the mass of neutrino (called Majoron Neutrino) which comes out to be 9.3 eV, though some measurements lead to the mass limits of 1.0 eV to 5.0 eV[36].

F. Electron Capture: Equation 8.76, which was discussed for $\beta^{\pm}$-decay is, also applicable to electron capture, which is an alternative mode to β^{+}-decay. The wave function ϕ_i now describes the electron in parent atom, and ϕ_f represents the outgoing neutrino. The only difference between this case and positron emissions is, that in $\beta^{\pm}$-decay, energy continuum is normalised to a particle of any energy, whereas atomic states are of sharp energy. Hence the decay probability does not contain the statistical factor $p_e^2\, dp_e$, [Eq. 8.33 and 8.38], and for an allowed transition, the decay constants λ_x, from any atomic shell X becomes:

$$\lambda_x = \frac{1}{4\pi^2} q^2 \left(g_{kx}^2 + f_{kx}^2\right)\left(g_V^2 \left|\int \mathbf{I}\right|^2 + g_A^2 \left|\int \sigma\right|^2\right) \qquad \text{...(8.116}a\text{)}$$

In this expression g_{kx} and f_{kx} are the two radial functions, which appear in the relativistic expression of the electron in shell X and are evaluated at some average position in the nuclei. For details, *see* References (37) and (38).

8.6 LEPTONIC MATRIX ELEMENTS

For lepton part of the matrix squared; we write from Eqs. 8.93, 8.94 and 8.92*b*.

$$\left|\mathcal{M}_{e\nu}^{*i,i'}\right|^2 = \left(\overline{U}_e^{(r)(+)}(\mathbf{p}_e) g_i\, \Gamma_i\, U_\nu^{(S)(-)}(-\mathbf{p}_\nu)\right)\left(\left(\overline{U}_e^{(r)(+)}(\mathbf{p}_e)\, g_{i'}\, \Gamma_{i'} \times U_\nu^{(S)(-)}(-\mathbf{p}_\nu)\right)^*\right.$$

with $\qquad i, i' = S.\ V.\ T$ and A $\qquad$ ...(8.117*a*)

with the subscripts r and S indicating the spin states of electrons and neutrino respectively and i and i' do not include the value P, since in the non-relativistic approximation for the nucleus, the contribution form pseudoscalar interaction can be neglected.

If we are not interested in the polarisation directions of the electron and antineutrino, we must sum over their spin states: then we should replace Eq. 8.117*a* by:

$$\left|\mathcal{M}_{e\nu}^{i,i'}\right|^2 = \sum_{r,S=1}^{2}\left(\overline{U}_e^{(r)(+)}(\mathbf{p}_e)\, g_i\, \Gamma_i\, U_\nu^{(S)(-)}(-\mathbf{p}_\nu)\right)\times$$

$$\left(\overline{U}_\nu^{(S)(-)}(-\mathbf{p}_\nu)\, \gamma_4\, g_{i'}^*\, \Gamma_{i'}^*\, \gamma_4\, U_i^{(r)(+)}(\mathbf{p}_e)\right) \qquad \text{...(8.117}b\text{)}$$

This was the matrix element for given i and i'. We should now sum overall the interaction types, at least for *S, V, T* and *A* (*P* can be neglected) interactions. Also we have to sum over the spins. Finally, to write the general case, we have to include the parity non-conserving terms in the interaction, as earlier shown in Eqs. 8.74 and 8.76. For this purpose, we should make the following replacements in Eq. 8.117*b*, *i.e.*,

$$g_i\, \Gamma_i \rightarrow \Gamma_i\, (g_i + g_i'\, \gamma_5)$$

and
$$g_{i'}^*\, \Gamma_{i'} \rightarrow \left(g_{i'}^* + g_{i'}^{'*}\, \gamma_5\right) \Gamma_{i'} \qquad \text{...(8.118)}$$

we will not go through the details of summing of spins, for which one may refer to the literature[11].

8.7 TOTAL MATRIX ELEMENTS FOR BETA DECAY FOR UNPOLARISED CASE

We, therefore, write the following result of the total beta matrix element squared, on the basis of the above discussion for the general case:

$$\left|\langle f \,|\, H_\beta \,|\, i \rangle\right|^2_{\text{unpolarised}} = \frac{1}{2V^2}\left[\left| M_{ev}^{SS*}\right|^2 + \left| M_{ev}^{VV*}\right|^2 + 2R_e \left| M_{ev}^{SV*}\right|^2\right] |M_F|^2 +$$

$$\frac{1}{6V^2}\left[\left| M_{ev}^{TT*}\right|^2 + \left| M_{ev}^{AA*}\right|^2 - 2R_e \left| M_{ev}^{TA*}\right|^2\right] |M_{GT}|^2 \qquad \text{...(8.119)}$$

where $\mathcal{M}_{ev}^{i,i'}$ is replaced by $M_{ev}^{i,i'}$, which contains not only the summation over spins of leptons, but also parity non-conservation, and nuclear wave functions.

After calculating the various lepton matrix elements, we finally write the square of the matrix element for beta decay for unpolarised case[17, 38]:

$$\left|\langle f \,|\, H_\beta \,|\, i \rangle\right|^2_{\text{unpolarised}} = \frac{1}{2V^2}\left[\left\{|g_s|^2 + |g_s'|^2\right\}\left(1 - \frac{\mathbf{p}_e \cdot \mathbf{p}_\nu\, c^2}{E_e\, E_\nu}\right) + \right.$$

$$\left\{|g_\nu|^2 + |g_\nu'|^2\right\}\left(1 + \frac{\mathbf{p}_e \cdot \mathbf{p}_\nu\, c^2}{E_e\, E_\nu}\right) +$$

$$\left.\frac{2mc^2}{E_e} R_e \left\{g_S\, g_V^* + g_S'\, g_V^{'*}\right\}\right] |M_F|^2 + \frac{1}{2V^2}\left[\left\{|g_T|^2 + |g_T'|^2\right\}\left(1 + \frac{1}{3}\frac{\mathbf{p}_e \cdot \mathbf{p}_\nu\, c^2}{E_e\, E_\nu}\right) + \right.$$

$$\left.\left\{|g_A|^2 + |g_A'|^2\right\}\left(1 - \frac{1}{3}\frac{\mathbf{p}_e \cdot \mathbf{p}_\nu\, c^2}{E_e\, E_\nu}\right) + \frac{2mc^2}{E_e} R_e \left\{g_T\, g_A^* + g_T'\, g_A^{'*}\right\}\right] |M_{GT}|^2 \qquad \text{...(8.120)}$$

So, combining Eqs. 8.93, 8.94 and 8.120, we can write the transition probability per unit time W_{fi}, by integrating over solid angles over which neutrinos are emitted. Denoting by θ the angle between electron momentum and the neutrino momentum, and using $v_e = c^2\, p_e / E_e$ as the velocity of electron, we can express W_{fi} as follows:

$$W_{fi} = \frac{\xi}{4\pi^3} \frac{p_e E_e}{c^5 \hbar^7} (E_{\max} - E_e)^2 \times \left(1 + a \frac{v_e}{c} \cos\theta + b \frac{mc^2}{E_e}\right) \sin\theta \, d\theta \qquad ...(8.121)$$

where
$$\xi = \frac{1}{2}\{|g_S|^2 + |g_S'|^2 + |g_V|^2 + |g_V'|^2\}|M_F|^2 +$$

$$\frac{1}{2}\{|g_T|^2 + |g_T'|^2 + |g_A|^2 + |g_A'|^2\}|M_{GT}|^2 \qquad ...(8.122a)$$

$$a\xi = \frac{1}{2}\{|g_V|^2 + |g_V'|^2 - |g_S|^2 - |g_S'|^2\}|M_F|^2 +$$

$$\frac{1}{6}\{|g_T|^2 + |g_T'|^2 - |g_A|^2 - |g_A'|^2\}|M_{GT}|^2 \qquad ...(8.122b)$$

$$b\xi = \frac{1}{2} R_e \{g_S \, g_V^* + g_S' \, g_V'^*\}|M_F|^2 + \frac{1}{2} R_e \{g_T \, g_A^* + g_T' \, g_A'^*\}|M_{GT}|^2 \qquad ...(8.122c)$$

The above theoretical result can, now, be compared with experimental ones from which one can determine the various coupling constant g_i and g_i', as follows:

8.8 COMPARISON WITH EXPERIMENTS

(*i*) **$0^+ \rightarrow 0^+$ Decay:** The nucleus O^{14} decays by β^+ from ground state of 0^+ for O^{14} to 0^+ state of N^{14} at 2.31 MeV excited state, *i.e.* this is $0^+ \xrightarrow{\beta^+} 0^+$ case and hence corresponds to pure Fermi transition. This case will, therefore, not involve $|M_{GT}|^2$ and hence we write ξ for this case as:

$$\xi = \frac{1}{2}\{|g_V|^2 + |g_V'|^2 + |g_S|^2 + |g_S'|^2\}|M_F|^2$$

and
$$b\xi = \frac{1}{2} R_e \{g_S \, g_V^* + g_S' \, g_V'^* + \}|M_F|^2 \qquad ...(8.123)$$

Hence b now involves only g_V^*, $g_V'^*$, g_S and $g_S'^*$.

For pure Fermi beta decay; the beta energy spectrum is obtained by integrating Eq. 8.121, over θ:

$$N(E_e)\, dE_e = \frac{1}{2\pi^3 c^5 \hbar^7} p_e \, E_e \, (E_0 - E_e)^2 \times F(Z, E_e)\, \xi \left(1 + \frac{2mc^2 \, b}{E_e}\right) d\,E_e \qquad ...(8.124)$$

where $F(Z, E_e)$ takes into account the Coulomb effect, as mentioned earlier. b is called Fierz constant. The integration over energies of electrons in beta decay, of Eq. 8.107 gives:

$$2\pi^3 c^5 \hbar^7 (\text{ft})^{-1} \log 2 = \xi + b\,(2\,mc^2)\,\xi \left\langle E_e^{-1} \right\rangle \quad ...(8.125a)$$

where
$$\text{ft} = \int_{mc^2}^{E_{\max}} F(Z, E_e)\, p_e E_e\, (E_{\max} - E_e)^2\, dE_e \quad ...(8.125b)$$

and
$$\left\langle E_e^{-1} \right\rangle = (\text{ft})^{-1} \int_{mc^2}^{E_{\max}} F(Z, E_e)\, p_e\, (E_{\max} - E_e)^2\, dE_e \quad ...(8.125c)$$

Experimentally it has been found[39] for $O^{14} \xrightarrow{\beta^+} N^{14}$ decay, that the Fermi plot is a straight line. Therefore, the Fierz constant 'b' is equal to zero, for pure Fermi allowed beta decay case. Then, combining Eqs. 8.123 and 8.125a, we can write:

$$2\,\pi^3 c^5 \hbar^7 (\text{ft})^{-1} \ln 2$$

$$= \xi = \frac{1}{2}\left\{|g_S|^2 + |g_S'|^2 + |g_V|^2 + |g_V'|^2\right\} |M_F|^2 \quad ...(8.126a)$$

Also $b = 0$, leads to (from Eq. 8.122c),

$$R_e\left(g_S\, g_V^* + g_S'\, g_V'^*\right) = 0 \quad ...(8.126b)$$

This can yield either

$$g_S = g_S' = 0 \text{ or } g_V = g_V' = 0 \text{ or } g_S = g_S' = g_V = -\,g_V' \quad ...(8.126c)$$

Experiment for ft value for $0^+ \xrightarrow{\beta^+} 0^+$ for $O^{14} \xrightarrow{\beta^+} N^{14}$ was performed by Bardin R.K. et al.[39] and Butler J.W. and R.O. Bondelid[39], (1962) and (1961). C.S. Wu has measured $Al^{*26} \xrightarrow{\beta^+} Mg^{26}$; Freeman et al. have measured $Cl^{34} \xrightarrow{\beta^+} S^{34}$ and Janeke, and Freeman et al. have measured the β^+ decay of $V^{46} \longrightarrow Ti^{46}$ with $0^+ \xrightarrow{\beta^+} 0^+$ decays (Ref. 40).

We describe below, the interesting case of the measurement of ft value of $O^+ \rightarrow O^+$ for $O^{14} \xrightarrow{\beta^+} N^{14}$; by Bardin[39] et al. It is seen from Eqs. 8.39 and 8.40, that ft measurement requires the maximum energy of β^+ decay and the lifetimes τ.

For obtaining $E_{\max}$ (β^+), the ground state of O^{14}, and first excited state of N^{14} were obtained from the reactions C^{12} (He^3, n) O^{14} and C^{12} $(He^3, p)^*$ N^{14}. The Q-values of these reactions were measured as: $Q = -\,1148.8 \pm 0.6$ keV and $Q = 2468 \pm 1.0$ keV respectively. Then $E_{\max}$ (β^+) was obtained from $C^{12} + He^3 \rightarrow O^{14} + n + Q_n$ and $C^{12} + He^3 \rightarrow N^{14} + p + Q_p$.

Hence
$$E_{\max}(\beta^+) = (O^{14} - N^{*14})\,c^2 - 2m_e\,c^2$$

$$= (Q_n - Q_p) + (M_p - M_n)\, c^2 - 2m_e\, c^2$$
$$= 1812.6 \pm 1.4 \text{ keV}$$

For lifetime measurement of β^+-decay, one measured the delayed gamma ray yield of the reaction:

$$C^{12}\ (He^3, n)\ O^{14}\ (\beta^+, \nu)\ N^{*14}\ (\gamma)\ N^{14}$$

Finally the non-radiative and radiative corrections were made to the experimental value of ft = 3066 ± 10 secs [*see* References (39) and (40)]; which also took into consideration screening, nuclear electromagnetic form-factor and *K*-capture competitions. The final ft value was, then, obtained as 3076 ± 10 secs, after corrections (40).

The above description is typical of the type of experimental details of other cases of $0^+ \to 0^+$ cases mentioned earlier, all of which require a certain nuclear reaction. As for example, Al^{26} is produced from Mg^{26} (*p, n*); Mg^{25} (*d, n*) or Si^{28} (*d, a*); Cl^{34} is produced from p^{31} (*d, n*) and V^{46} is produced from Ti^{46} (*p, n*). [References (39, 40)].

From Eqs. 8.124, 8.125, 8.126 and 8.127 (giving $g_V = g_V'$), we get:

$$2\,\pi^3\, c^5\, \hbar^7\, (\text{ft})^{-1} \ln 2 = \left[|\, g_V\,|^2\, |\, M_F\,|^2\right]$$

using ft values of 3075 ± 10 secs for $O^{14} \xrightarrow{\beta^+} N^{14}$ and $|\, M_F\,|^2 = 2$; Bardin[39] found:

$$g_V = (1.4025 \pm 0.0022) \times 10^{-49} \text{ ergs–cm}^3$$

Other workers have found similar values.

From experiments of $Al^{26} \xrightarrow{\beta^+} Mg^{26}$, $Cl^{34} \xrightarrow{\beta^+} S^{34}$ and $V^{46} \xrightarrow{\beta^+} Ti^{46}$ (Ref. 40) and after introducing the screening effects, Wu[32], found an average value of ft as; ft = 3125 ± 10 secs and hence the value of g_v is obtained as:

$$g_V = (1.4029 \pm 0.0022) \times 10^{-49} \text{ ergs–cm}^3$$

(*ii*) Another experiment on *polarisation of electrons* defined as the ratio of the number of electrons with spins along positive *z*-axis minus the number with spins in the opposite direction, to the total number of electrons, requires [Eq. 8.80], that $P_e = \mp\, \beta$. Theoretically it can be proved (for which the reader is referred to the literature[41]) that

$$P_e = \mp \frac{2R_e\left(g_V^* \, g_V'\right)|\, M_F\,|^2 + 2R_e\left(g_A^* \, g_A'\right)|\, M_{GT}\,|^2}{\left(|g_V|^2 + |g_V'|^2\right)|\, M_F\,|^2 + \left(|g_A|^2 + |g_A'|^2\right)|\, M_{GT}\,|^2}\, v_e \qquad \text{...(8.127)}$$

$$\begin{Bmatrix} -\text{ for electrons} \\ +\text{ for positrons} \end{Bmatrix} \begin{matrix} \text{for pure Fermi transition } [M_{GT} = 0,\ g_V = g_V'] \\ \text{for pure Gamow-Teller transition } [M_F = 0,\ g_A = g_A'] \end{matrix}$$

which when compared with the experiments, resulted in going, for a pure G. T. transition, *e.g.* decay of

P^{32} (spin change $1^+ \xrightarrow{\beta^-} 0^+$)

$P_e = (-0.990 \pm 0.009)\,\beta$, where $\beta = 0.891$ for 616 keV electrons

$P_e = (-1.00 \pm 0.02)\,\beta$, where electron energies are between 600 keV and 990 keV and

$P_e = (-1.02 \pm 0.03)\,\beta$, for electron energy 340 keV. ...(8.128)

This is consistent with *V-A* theory (Ref. 37, 13 and 14).

(*iii*) **Electron-Neutrino Angular Correlation:** Basically electron-neutrino angular correlations are carried out, by the measurement of energy spectra of the recoil ions. Figure 8.11, shows the spectra of recoil nuclei for A^{35}, Ne^{23} and He^6, as measured by Allen, Burman and their coworkers[42], at University of Illinois or at Argonne National Lab. (U.S.A.) in 1959. The experiment was designed to measure the energy spectrum of the recoil ions, without detection of the associated beta particles. The experimentally determined recoil spectra were analysed by comparison with the theoretical spectra predicted from various assumed values of the angular correlation coefficients, with the result:

Nucleus	*Interaction type*	*Angular correlation coefficient 'a'*
He^6	G.T.	-0.39 ± 0.05
Ne^{23}	G.T.	-0.37 ± 0.04
Ar^{35}	Mostly Fermi	0.97 ± 0.14

As we shall see [Section 8.8 and 8.9, Eqs. 8.138 and 8.131], β-interaction can be described by *V-A* theory. These measurements were carried out by using only rare gases because the recoil energies of the ions are small and therefore, the binding energies of the atoms in a solid source may lead to erroneous results. This technique has been developed by Allen[42] and coworkers and involves the use of a double spectrometer. The effective β-source is contained inside an artificially created cone. While the betas are detected, by a β-monitor outside the β-source volume; the recoil ions are passed through the double spectrometer so that the linear momentum is measured. An ion-detector especially designed detects the ions, after they are accelerated somewhat by an accelerator, after emerging from the magnetic spectrometer.

The shapes of the spectra of the recoil ions are compared with that expected for different values of *a*, [Eq. 8.12], after properly integrating overall the energies, and angles of neutrinos, and for specific angle of electrons. It is evident that the recoil energy is larger, if the electron neutrino pair are emitted parallel, rather them anti-parallel to one another.

To select proper possibilities out of those given in Eq. 8.126*c* and 8.127, one uses electron-neutrino angular correlation experimental results, which yield the coefficient '*a*' as given in Eq. 8.122, which experimentally comes out to be 0.97 ± 0.14 for Fermi transitions as discussed earlier. From Eqs. 8.122*a* and 8.122*b*, we get for Fermi transitions:

$$a = \frac{|g_V|^2 + |g_V'|^2 - |g_S|^2 - |g_S'|^2}{|g_V|^2 + |g_V'|^2 + |g_S|^2 + |g_S'|^2} \approx 1 \qquad ...(8.129)$$

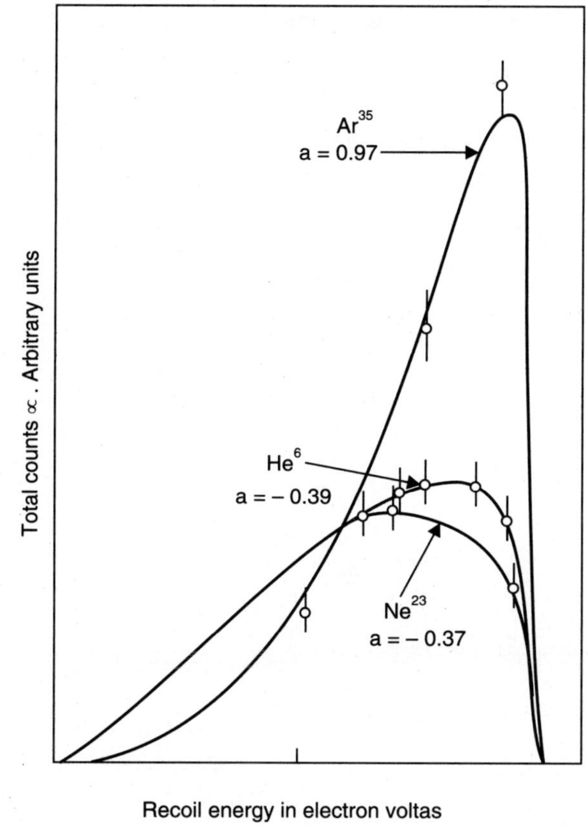

Fig. 8.11 The results of Ar^{35}, Ne^{23} and He^6 experiments, for the energy spectra of recoil ions in their $\beta^{\pm}$ decay (Ref. 42).

Equation 8.129, requires that $|g_S|^2 + |g'_S|^2 = 0$ and therefore, in Eq. 8.126*c*; the condition $g_S = g'_S = 0$ is satisfied. Therefore Eq. 8.126*a* reduces to:

$$2\pi^3 c^5 \hbar^7 (ft)^{-1} \ln 2 = \frac{1}{2}\{|g_V|^2 + |g'_V|^2\} |M_F|^2 \quad ...(8.130)$$

Another electron-neutrino angular correlation experiment has been performed for Ne^{23} (Ref. 42), which is a pure Gamow-Teller transition. The experimental value obtained is:

$$a = -0.35 \begin{bmatrix} +0.033 \\ -0.053 \end{bmatrix} \quad ...(8.131a)$$

Theoretically, one can write the expression for '*a*' for pure *G-T* case as from Eqs. 8.122*a* and 8.122*b*:

$$a = \frac{1}{3}\frac{|g_T|^2 + |g'_T|^2 - |g_A|^2 - |g'_A|^2}{|g_T|^2 + |g'_T|^2 + |g_A|^2 + |g'_A|^2} \quad ...(8.131b)$$

comparing Eqs. 8.131*a* and 8.131*b*; it is seen that:

$$a = \frac{1}{3}; \text{ for } g_A = g_A' = 0 \text{ and } a = -\frac{1}{3} \text{ for } g_T = g_T' = 0$$

showing unambiguously, that, the second option is valid. Then we express:

$$g_A = g_A' \neq 0 \qquad \text{...(8.131}c\text{)}$$

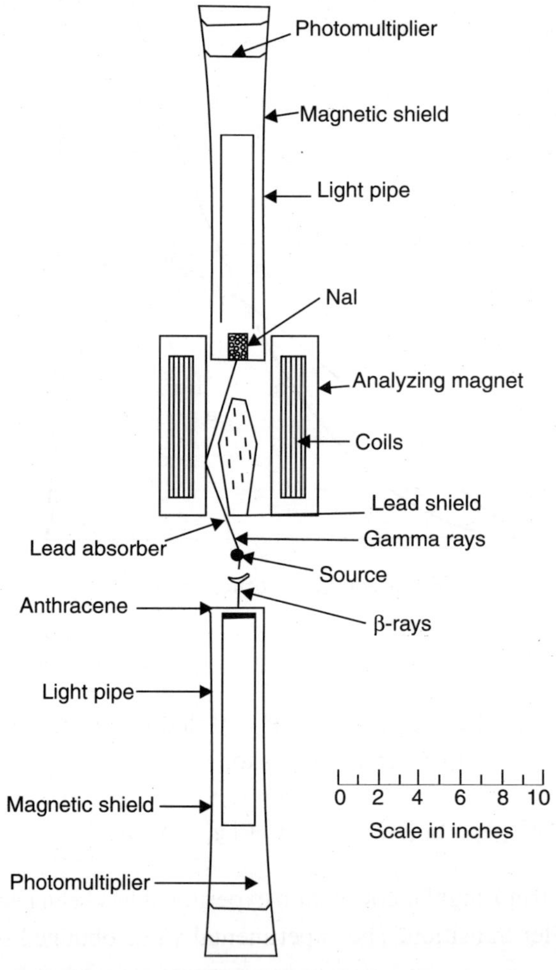

Fig. 8.12 Bohem and Wapstra's arrangement for measurement of beta-gamma circular polarisation correlations (Ref. 43).

(*iv*) **Beta-Gamma (circularly polarised) Experiment:** An interesting experiment connected with beta decay, requires the beta-gamma circularly polarised correlation measurement in which for a definite angle between the beta and gamma rays, the circular polarisation of the latter is observed, by utilising the dependence of the compton cross-section on the spin of the electrons and the circular polarisation of gamma rays [Fig. 8.12]. The electrons (β-rays) emitted from the source are oriented in a given direction; depending upon the direction of the magnetisation of the analysing magnet. The gamma rays following beta decay are scattered from the inside of the magnetised iron of the analysing magnet at 52° and detected in the NaI (Tl) detector. The beta rays are detected on the other side, by anthracene detector.

Direct gamma rays are suppressed by means of a lead shield. The average angle between beta and gamma rays is 148°. One observes coincidences between β and γ-rays for fields up and down. It is possible to calculate the amount of circular polarisation which when compared in details with the theoretical expectation, yields the ratios of nuclear matrix elements and spin changes in nuclear decay. They add to the evidence that in beta decay parity is not conserved[43].

(*v*) **Measurement of Helicity of Neutrinos:** Another experiment, which had a direct bearing on non-conservation of parity was carried out by Goldhaber, Grodzins and Sunyar[44]. Actually, the experiment belongs among the recoil experiments, but it utilises circular polarisation of gamma rays following *K*-capture, selecting definite recoils and direction of emitted neutrinos. It then follows that if the neutrino and gamma rays are emitted in approximately opposite direction, the helicity of gamma rays is the same as that of the neutrino. In this way the helicity of the neutrino in *K*-capture (*i.e.* in β^+-decay) is found. It is concluded that neutrinos in β^+-decay are left handed, or have their spins anti-parallel to momentum.

The experimental arrangement requires a Eu^{152} source which emits neutrinos followed by gamma rays, which are resonantly scattered from a Sm_2O_3 scatterer. The recoil which may put the gamma ray energy off the resonance, is compensated by neutrino emission in appropriate direction, so that the resultant Doppler shift puts the gamma ray energy at the resonance energy. As shown below, the decay is of type:

$$\bar{0} \xrightarrow{K} \bar{1} \xrightarrow{\gamma} 0^+$$

It then follows that, if the neutrinos and the gamma rays are emitted in approximate opposite direction, so that the effective Doppler shift is nearly zero and resonance scattering can take place; then the helicity of the gamma ray is the same, as that of the neutrino. By counting the scattered gamma rays, with magnetic field up and down and using the energy interval corresponding to resonance, the helicity of gamma rays in coincidence with 180° neutrino was measured. The results give a circular polarisation of $86 \pm 14\%$; which should be compared with 75% for the expected value for helicity of -1, for neutrino.

(*vi*) **Evidence Concerning Time-Reversal Invariance:** The evidence about time-reversal in β-decay should really come from the theory of weak coupling constants being complex and its comparison with experiments. However, one could look at the problem physically; according to which one should assume that the transition probability must not include terms that change sign, when the direction of the time-axis is inverted. This means, that terms like $\vec{J} \cdot \vec{p} \times \vec{q}$ should not enter into the formula of the transition probability, since it contains an odd number of factors, which change sign under time-reversal.

Experiments were performed on the neutrons polarised[45] along the *z*-axis. If time-reversal invariance does not hold good; then the transition probability should be different for electrons in the *x*-direction and neutrinos in the *y*-direction compared to the mirror situation of electrons in $-x$-direction and neutrinos on $-y$-direction. The experiments, however, showed no such difference. This shows, that the weak interactions are invariant under time-reversal.

According to the general field-theoretical consideration, any Lorentz invariance theory of interacting field, must be invariant under the product of three operators *PTC* [Parity (*P*); time (*T*) and *C* (charge conjugations)]. Hence weak interactions are:

(*i*) Invariant under *PTC*

(*ii*) Invariant under *T*

(*iii*) Invariant under *PC*, but

(*iv*) Not invariant under *P* and *C* separately. This means, that although β-decay is not invariant under reflection, the decay of charge conjugated Co^{60} is the mirror-image of Co^{60}-decay. Thus if the electrons from Co^{60} decay are left-handed; the positrons from anti-Co^{60}, must be right handed. No such experiments have, however, been conducted.

8.9 THEORETICAL EXPRESSION FOR ANGULAR DISTRIBUTION FOR BETA DECAY FROM ORIENTED POLARISED NUCLEI

A. Nuclear Matrix Elements: We have already discussed the matrix elements—nuclear and leptonic—for unpolarised nuclei. We will now, consider the case of initial nucleus polarised—as done in the famous experiment of Wu and her coworkers[10]. We denote by $W(m_i)$, the weight factors, which specify the relative population m_i in the initial nucleus.

$$\sum_{m_i} W(m_i) = 1 \text{ and } W(m_i) = \frac{1}{2J_1 + 1}$$

for an unpolarised nucleus, the value of $W(m_i)$ being different for different states of polarisation. Then for three cases, *i.e.*,

(*i*) $i, i' = S$ or V., *i.e.*, pure Fermi interaction;

(*ii*) $i = S$ or V (and T, A) and $i' = T$, A (and S or V), and

(*iii*) $i, i' = T, A$.

We discuss these cases, one by one.

(*i*) $i, i' = S$ or V

$$\left(\left\langle \psi_f \middle| \Gamma_i^{(L)} \middle| \psi_i \right\rangle \left\langle \psi_f \middle| \Gamma_{i'}^{(L)} \middle| \psi_i \right\rangle^*\right)_{av} = |M_F|^2 \qquad ...(8.132a)$$

(*ii*) $i = S$ or V (and T, A) and $i' = T$, A (and S or V)

then

$$\left(\left\langle \psi_f \middle| \Gamma_i^{(L)} \middle| \psi_i \right\rangle \left\langle \psi_f \middle| \Gamma_{i'}^{(L)} \middle| \psi_i \right\rangle^*\right)_{av}$$

$$= \sum_{mi,\, mf} W(m_i) \left\langle \psi_f | \mathbf{I} | \psi_i \right\rangle \left\langle \psi_i \middle| \mp \sigma_K \middle| \psi_i \right\rangle^*$$

$$= \sum_{mi,\, mf} W(m_i) \left(\sum_{n=1}^{A} \int d^3 x\, \psi_f^*(x)\, \tau_n^{(+)}\, \psi_i(x) \right)$$

$$\left(\sum_{n=1}^{A} \int d^3 x \, \psi_f^* (x) (\mp \sigma_K) \tau_n^{(+)} \psi_i (x) \right)^* \quad ...(8.132b)$$

(*iii*) *i*, *i′* = *T*, *A*

$$\left(\left\langle \psi_f \left| \Gamma_i^{(L)} \right| \psi_i \right\rangle \left\langle \psi_f \left| \Gamma_{i'}^{(L)} \right| \psi_i \right\rangle^* \right)_{av}$$

$$= \pm \sum_{mi, mf} W(m_i) \left\langle \psi_f \left| \pm \sigma_K \right| \psi_i \right\rangle \left\langle \psi_f \left| \mp \sigma_K \right| \psi_i \right\rangle^*$$

$$= \pm \sum_{mi, mf} W(m_i) \left(\sum_{n=1}^{A} \int d^3 x \, \psi_f^* (x) \sigma_K \tau_n^{(+)} \psi_i (x) \right) \times$$

$$\left(\sum_{n=1}^{A} \int d^3 x \, \psi_f^* (x) \sigma_{K'} \tau_n^{(+)} \psi_i (x) \right)^* \quad ...(8.132c)$$

where the plus sign is for $i = i'$, and minus sign for $i \neq i'$.

B. Leptonic Matrix Elements (polarised nuclei): If we restrict ourselves to vector (V) and axial vector (A) interactions, then only

$$\left| M_{ev}^{VV*} \right|^2 \text{ and } \left| M_{ev}^{AV*} \right|^2 \text{ and } \left| M_{ev}^{AA*} \right|^2$$

are effective. The values of these matrix elements squared are the same whether the beta emitting nuclei are polarised or not; especially when we are not interested in the polarised directions of electron and neutrino, and have summed over their spin states.

C. Beta Decay from Oriented Nuclei: Now combining Eq. 8.132 with Eq. 8.117 under conditions given in '*A*' we can write:

$$\left| \left\langle f \right| H_\beta \left| i \right\rangle \right|^2_{\text{polarised nuclei}} = \frac{1}{2V^2} \Big[| M_F |^2 | M_{ev}^{VV*} |^2 + \left(\left\langle \psi_f | \sigma_{K'} | \psi_i \right\rangle^* \right) | M_{ev}^{AA*} |^2$$

$$- \left\langle \psi_f | \mathbf{I} | \psi_i \right\rangle \left\langle \psi_f | \sigma_K | \psi_i \right\rangle 2R_e \left| M_{ev}^{AV*} \right|^2 \quad ...(8.134)$$

where the nuclear matrix elements can be now calculated from Eq. 8.132, and leptonic matrix elements from Eq. 8.117.

Finally, realising the

$$\sum_{mi} m_i \, W(m_i) = 0 \text{ for unpolarised nuclei}$$

$= J_i$ for nucleus completely polarised along z-axis,

so that for $m_i = j_i$ $W(j) = 1$, and $W(m_i \neq j_i) = 0$

we finally write[8, 11] the equation as:

$$\int \left| \left\langle f \left| H_\beta \right| i \right\rangle \right|^2_{\text{Polarised nuclei}} d\Omega$$

$$= \frac{4\pi}{V^2} \Bigg\{ |g_V|^2 |M_F|^2 + |g_A|^2 \left[\langle \psi_f | \sigma_K | \psi_i \rangle \langle \psi_f | \sigma_{K'} | \psi_i \rangle^*_{av} \right] \times \left[\delta_{KK'} + i\,\varepsilon_{KK'1} \frac{(p_e)_1}{E_e} \right] -$$

$$2R_e\, g_V\, g_A^* \left[\langle \psi_f | \mathbf{I} | \psi_i \rangle \langle \psi_f | \sigma_K | \psi_i \rangle^*_{av} \right] \left(\frac{(p_e)_k}{E_e} \right) \Bigg\} \qquad \text{...(8.135}a\text{)}$$

where $\varepsilon_{k\,k',\,1} = +1$, for k, k', 1 cyclic, and $\varepsilon_{k\,k',\,1} = -1$ for non-cyclic case. This leads[8,11] to:

$$\int \left| \left\langle f \left| H_\beta \right| i \right\rangle \right|^2_{\text{Polarised nuclei}} d\Omega = \frac{4\pi\xi}{V^2} \left[1 + A \frac{\hat{\mathbf{z}} \cdot \mathbf{p}_e}{E_e} \frac{1}{j_i} \sum_{m_i} m_i\, W(m_i) \right] \qquad \text{...(8.135}b\text{)}$$

where ξ is given by Eq. 8.122a, with $g_S = g'_S = 0$, $g_T = g'_T = 0$; $g_V = g'_V$ and $g_A = g'_A$ and

$$\xi A = -|g_A|^2 |M_{GT}|^2 \frac{2 + J_i(J_i + 1) - J_f(J_f + 1)}{2(J_i + 1)} -$$

$$2R_e \left(g_V\, g_A^*\, M_F\, M_{GT}^* \right) \sqrt{\frac{J_i}{J_i + 1}}\, \delta_{J_i J_f} \qquad \text{...(8.135}c\text{)}$$

and $\hat{\mathbf{z}} \dfrac{1}{J_i} \sum_{m_i} W(m_i)$ is the polarisation vector of the nucleus.

An experiment was performed by Burgy et al.[45], by observing the results of the beta decay of the polarised (free) neutrons which will corresponds to $j_i = j_f = 1/2$; $M = 1$; $M_F = 1$; $M_{GT} = \sqrt{3}$.

This is equivalent of Wu's experiment[10], with known matrix elements. This gives from Eqs. 8.135 and 8.122a:

$$\xi = |g_V|^2 + 3|g_A|^2 \qquad \text{...(8.136}a\text{)}$$

and hence

$$\xi A = -2|g_A|^2 - 2R_e \left| g_V\, g_A^* \right| = -2|g_A|^2 - |g_V||g_A| \cos\phi \qquad \text{...(8.136}b\text{)}$$

where ϕ is the phase difference (complex) between the couplings g_V and g_A.

Therefore,

$$A = -\frac{2|g_A|^2 + 2|g_V||g_A|\cos\phi}{|g_V|^2 + 3|g_A|^2}$$

$$= \frac{1}{2}(1 + \cos\phi) \text{ for } |g_A| = |g_V|$$

$$= \begin{cases} -1 & \text{for } \phi = 0 \\ 0 & \text{for } \phi = \pi \end{cases} \qquad ...(8.136c)$$

The experimental value by various authors[46, 47] for polarised neutron decay has yielded $A = -0.11 \pm 0.02$, which favours relative phase difference of $\phi = \pi$ between g_V and g_A, with $g_A/g_V = -1.24$.

There are two more types of experiments: (*i*) Decay of unpolarised, but free neutrons. In this experiment conducted by Robson[48] J. M., a beam of neutrons, is brought out of the reactor and using a magnetic beta ray spectrometer on one side of the neutron beam, for detecting electrons and a proton counter on the other side of the beam one detects the coincidences between the recoil protons in the proton counter and the decay electrons detected in beta ray spectrometer, as expected from Eq. 8.11. In this way momentum spectrum of decay electrons is measured with recoils emitted into the direction of the recoil counter. The results of such a momentum spectrum is shown in Fig. 8.13. The results are fitted to an angular distribution between angle electrons and recoil protons, *i.e.*,

$$P(\theta) = 1 + a\beta\cos\theta \qquad ...(8.137)$$

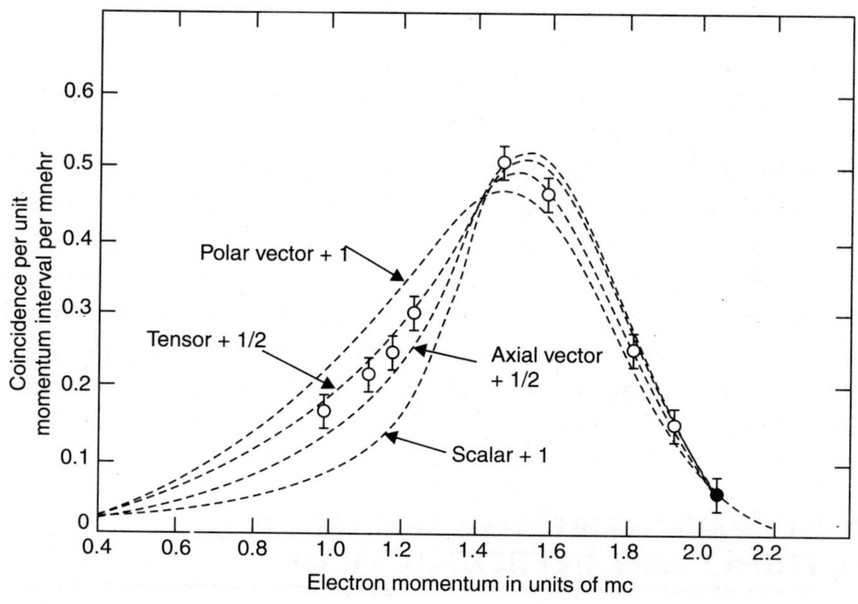

Fig. 8.13 Momentum spectrum of electrons from the decay of free unpolarised neutrons, in coincidence with recoil protons. Comparison with different interactions is also shown (Ref. 48).

where θ is given by $\sin \phi = (q/r) \sin \theta$, where ϕ is the angle between electron and recoil proton, and q in momentum transfer, to the electron. Experimentally a was found out to be:

$$a = + 0.089 \pm 0.108 \qquad ...(8.138a)$$

in the earlier experiments. It is interesting to see that these early experiments did not give clear results, in favour of *V-A* theory—later experiments[49] showed that $a = - 0.39 \pm 0.02$. This is in better agreement with the interaction *V-A*.

(*ii*) A more recent experiment was conducted by Burgy[45] et al., using a beam of polarised neutrons. Polarisation was created by allowing the neutrons from a reactor at Argonee National lab, to pass through a narrow collinator (85 3/4 inch long, 8 inch high and 1/4 inch wide). After leaving the collinator, the beam was polarised by reflection from a mirror of 95% cobalt, and 5% Fe alloy magnetised in a vertical direction by a field of roughly 250 oesterds. The beam from the reactor struck the mirror at a grazing angle of about 8 min and then entered a vacuum chamber, from where the decay was observed. This reflected beam contained a total of about $\approx 5 \times 10$ neutrons/sec spread out over an area about 8″ high and 1/4″ wide. The beam was found to be polarised, 87% with an uncertainty of 7%. Detector arrangement was, in principle, similar to the one described earlier for unpolarised neutrons. The experimental results yielded the value of A, from the correlation between the momentum of the beta particle and the neutron spins—it resulted in

$$A = - 0.114 \pm 0.0019 \qquad ...(8.138b)$$

which is nearly zero, and hence corresponds to $\phi = \pi$, of Eq. 8.136*c*.

Both the coupling constants, *i.e.* g_V and g_A have, thus been found from the analysis of these experiments, with the absolute value of g_V as:

$$g_V = (2.87 \pm 0.11) \times 10^{-12} \text{ natural units}$$

$$= (1.35 \pm 0.05) \times 10^{-49} \text{ ergs-cm}^3 \qquad ...(8.138c)$$

In summary, the experiments on (*i*) Angular distribution of electrons from polarised nuclei, as observed in free polarised neutron decay; (*ii*) The longitudinal polarisation of electrons emitted from unpolarised nuclei; (*iii*) The Fermi-curie plot of a pure Fermi beta decay (O^{14}); (*iv*) The electron-neutrino angular correlation experiment, and (*v*) The experiment on circular polarisation of gamma rays emitted following beta decay and the comparison of these results with theoretical expressions based on general consideration of non-conservation of parity, but involving all interactions, *i.e.*, *S*, *V*, *A* and *T*, shows very clearly that parity, is not conserved in weak interaction, and parity violation is not a small fraction to the parity conserving theories. As a matter of fact, the effects are as large as they can possibly be. Also it shows that only g_V and g_A are non-zero, so that *V-A* interaction for β-decay explains all experiments.

8.10 TWO-COMPONENT THEORY OF NEUTRINO—AN INTRODUCTION

We are familiar with the four component theory of electron, obeying Dirac equation conditions of invariance under Lorentz transformation. Physically this means that electrons have their spin up and down and similarly positrons have their spins up and down; thus giving a four component wave function as given in Eq. 8.44. But we assumed there, that neutrinos have also similarly a four component wave functions. Now, invariance under Lorentz transformation also requires invariance under reflection of space coordinates, *i.e.* the conservation of parity. But we have seen that in beta decay; parity is not conserved. One of the consequences of this was, as proved in the circular polarisation of gamma rays after emission of neutrino and electron; that neutrinos have negative helicity, *i.e.* their spins are anti-parallel to momentum direction, and anti-neutrinos have positive helicity. These are the only two states of neutrinos and anti-neutrinos.

The two-component theory of the neutrino with zero mass and spin in 1/2 $\hbar$ was first proposed by H. Weyl before the experimental discovery by Wu et al.[10, 51] as suggested by Lee and Yang's theory[50] of the non-conservation of parity. It has been later on reviewed by Landau and Salam[18,50]. It is basically a modification of Dirac equation for zero mass, which can be written as:

$$\gamma_4 \frac{\partial}{\partial x_\mu} \psi(x) = 0 \qquad \text{...(8.139}a\text{)}$$

or in Hamiltonian form as:

$$\gamma_4\, \gamma_k \frac{\partial}{\partial x_\mu} \psi(x) = i \frac{\partial \psi}{\partial t} \quad (k = 1, 2, 3) \qquad \text{...(8.139}b\text{)}$$

or as earlier expressed in Eq. 8.43*a*, we can express it in momentum representation as:

$$(E_\nu - i\gamma_4\, \gamma_k\, p_{\nu k})\, U = 0 \qquad \text{...(8.139}c\text{)}$$

where

$$U = \begin{pmatrix} \chi \\ \phi \end{pmatrix}, \gamma_k = \begin{pmatrix} 0 & -i\sigma_k \\ i\sigma_k & 0 \end{pmatrix}$$

$$\gamma_4 = \begin{pmatrix} 1 & 0 \\ 0 & -1 \end{pmatrix} \text{ and } \gamma_5 = \begin{pmatrix} 0 & 1 \\ 1 & 0 \end{pmatrix} \qquad \text{...(8.139}d\text{)}$$

These are 2 × 2 matrices, corresponding to two component spinors x and $\phi \cdot \sigma_k$ are 2 × 2 Pauli matrices. One can then, write Eq. 8.139*c* as:

$$E_\nu\, \phi + \sigma \cdot \mathbf{p}_\nu\, \chi = 0 \text{ and } E_\nu\, \chi + \sigma \cdot \mathbf{p}_\nu\, \phi = 0 \qquad \text{...(8.140}a\text{)}$$

We now define two wave function ψ_+ and ψ_- as:

$$\psi_+ = \frac{1}{2} \begin{pmatrix} \phi + \chi \\ \phi + \chi \end{pmatrix}$$

and

$$\psi_- = \frac{1}{2} \begin{pmatrix} -(\phi - \chi) \\ (\phi - \chi) \end{pmatrix} \qquad \text{...(8.140}b\text{)}$$

Equation 8.140*b* is equivalent to:

$$\psi_\pm = \frac{1}{2} (1 \pm \gamma_5)\, \psi \qquad \text{...(8.140}c\text{)}$$

Multiplying Eq. 8.140*c*, by γ_5, we get:

$$\gamma_5\,\psi_\pm = \frac{1}{2}\,(\gamma_5 \pm 1)\,\psi = \pm\,\psi_\pm \text{ or } (1 \pm \gamma_5)\,\psi_\pm = 0 \qquad ...(8.140d)$$

This means that ψ_+ and ψ_- are eigen functions of γ_5, with eigenvalues + 1 and – 1. These are the two components of wave function for neutrino-one corresponding to left-handed neutrino and other of right handed anti-neutering.

It can be seen from Eq. 8.140*a*, that

$$\phi = -\frac{(\sigma\cdot\mathbf{p})}{p_\nu}\chi\,;\,\chi = -\frac{(\sigma\cdot\mathbf{p})}{p_\nu}\phi \qquad ...(8.141)$$

and $$E_\nu = p_\nu$$

Using Eq. 8.141, and Eq. 8.140*b* we can write:

$$\psi_+ = \frac{1}{2}\begin{pmatrix}(1-h)\,\phi\\(1-h)\,\phi\end{pmatrix} \text{ and } \psi_- = \frac{1}{2}\begin{pmatrix}-(1+h)\,\phi\\(1+h)\,\phi\end{pmatrix} \qquad ...(8.140e)$$

with the helicity $h = \dfrac{\sigma\cdot\mathbf{p}}{p_\nu}$

We write from Eqs. 8.140*a* and 8.140*e*:

$$\psi_+ = \phi + \chi = (1-h)\,\chi = (1-h)\,\phi = \frac{1}{2}\,(1-h)\,(\phi+\chi)$$

$$= \frac{1}{2}\,(1-h)\,\psi_+ \text{ or } \psi_+ = -\,h\psi_+$$

Similarly $$\psi_- = +\,h\,\psi_- \qquad ...(8.142)$$

Physically Eq. 8.142 indicates, that ψ_+ and ψ_- are fields with only two independent components; ψ_+ corresponding to a particle with negative helicity (a neutrino, with spins anti-parallel to the momentum) and ψ_- with a positive helicity (spin parallel to the momentum).

This is two-component model of neutrino; indicating that neutrino-wave function can be written as a two-component column matrix. The two-component model leads to two distinguishable states of a neutrino, *i.e.* left handed neutrino and right handed anti-neutrino. The law of conservation of neutrinos required the fact that lepton number is to be conserved. The lepton number of left handed neutrino is + 1 and of right handed anti-neutrino is – 1. The reaction $\bar{\nu}$ (p, e) n has been studied by Cowan et al.[4], as shown earlier, where the cross-section for this reaction has been calculated, and it is in accord with the cross-section[52] from the two-component neutrino theory.

8.11 CONSERVATION OF VECTOR CURRENT (CVC) THEORY (QUALITATIVE)

This law of conservation holds in beta decay and in weak interactions in particle physics and electroweak model.

V-A interaction in beta decay is expected to be applicable to weak interaction in problems of particle physics also, *e.g.* in muon decay, where we can write:

$$\mu^{\pm} \rightarrow e^{\pm} + \nu_{\mu,e} + \bar{\nu}_{\mu,e} \qquad ...(8.143)$$

Evidently one expects the coupling constants of weak interaction based on *V-A* theory to play their role here. Interestingly, it was found experimentally that coupling constants $g_{V\beta}$ in β-decay and $g_{V\mu}$ in muon decay are practically the same, showing that *V-A* theory holds for both cases. However, it was found, experimentally that while in μ-meson decay, axial vector constant is the same as vector constant, it is 20% large in β-decay, as shown earlier. However, it has been shown that the two component theory[53] requires $g_A / g_V = -1.00$. We now answer the question; why do we expect $|g_A| = |g_V|$ in two component theory in both μ-decay and beta decay and how do we explain 20% difference found experimentally in beta decay? The answer is: In beta decay, weak interaction takes place in the presence of strong interaction. It is, therefore, expected that even if fundamental coupling constants are the same for all weak interactions, the coupling constants for the particles, that also have strong interaction, should be screened off because of 'renormalisation effects', in the presence of strong interaction.

Gell-Mann and Feynmann[53] have proposed a conservation of current vector theory (CVC), which explains the phenomenon[54].

Most recently weak interactions have been understood as a special case of electroweak interaction as described in Glashow-Salam-Weinberg (GSW) model[55]. This model unifies the electromagnetic and weak interaction theories. (This theory is also called the standard model). One of the predictions of the model was the existence of massive gauge bosons $W\pm$ and Z°. These are the massive particles with $M(W^{\pm}) \approx 85$ GeV and $M(Z^{\circ}) \approx 95$ GeV. They were experimentally discovered in 1983, at super proton synchrotron at CERN, in January 1983, and May 1983 respectively[56]. They are the carriers of weak-interaction.

The standard theory naturally leads to *V-A* interaction for weak interaction, and two component theory for neutrinos and also to CVC theory.

8. Beta Decay
2000–2008

Among some half a dozen experiments in β-decay in the Phy. Rev. C. 61 (2000), one experiment on β-decay of Co^{60}, Co^{68} and Co^{67} has been authored by 24 authors, in Europe. This experiment has used a 30 MeV proton + U^{238} (Target) for creation of fission, and a laser ionization separator [Phy. Rev. C. 61, 054308, (2000)].

In many experimental and theoretical papers double beta decay (*i*) 0νββ or (*ii*) 2νββ *i.e.* neutrino less or two neutrino emission are involved.

In an interesting theoretical discussion involving sub-ev, neutrino masses from $0\nu\beta\beta$ decay to an excited of state of Mo^{96}, it has been suggested that first excited O^+ state in Mo^{96} can be described as a monopole-vibrational state; and neutrino mass *i.e.* Majorona neutrino mass, $\simeq 8 \times 10^{-2}$ eV; can be determined in modern underground experiments using huge active-mass detectors [Phy. Rev. C. 62, 04250 (*R*) (2000)]. In an experiment for measuring mass of neutrino majori, in Cd^{116} double beta decay, they have obtained neutrino mass ≤ 2.6 (1.4) eV. [Phy. Rev. C. 62, 045501 (2000)].

Many cases of double beta decay have been studied (*i*) $\beta\beta 2\nu$ decay in Ca^{48} (*ii*) Neutrinoless double beta decay calculations of matrix elements, within second quasi-random phase approximation; (*iii*) calculation of limits on majora neutrino mass, and right handed weak currents by neutrinoless double beta decay of Mo^{100} (11 authors) (Phy. Rev. 63, 065501 (2001). Two authors (Raina and Dhiman from Shimla (India) have calculated using microscopic vibrational model, the systematic of promising β-β decaying medium mass nuclei Mo^{100}, Ru^{104}, Pd^{110}, cd^{114}, cd^{116}, Sn^{124}, Te^{128} and Te^{130}. [Phy. Rev. 64, 024310 (2001)].

An interesting paper on Fermi super-allowed β-decay with $T = 1$, ground states of heavy odd-odd $N = Z$ nuclei is authorised by 31 authors, from U.K., France, Sweden, Berkley etc. for the case of Rb^{74}, Y^{78}, Nb^{82} and $Tc^{83} \underset{\beta^+}{\rightarrow}$ decay [Phy. Rev. C. 63, 044307, (2001)].

In a paper authored by 21 authors from laboratories in USA and Europe; β-decay studies of $Sn^{135-137}$, have been carried out using relative resonance laser ionization technique. Neutron rich nuclei were produced at CERN, by spallation, of UC_2 target with IGeV protons. β-decayed neutrons were counted. Neutron decay rate; γ-rays singles and γ-γ concidences data was collected as a function of time. Shell model calculations are consistent with observed Sb_{135} level structure. [Phy. Rev. C. 65, 034313 (2002)].

In a paper, authored by a group of 16 authors from Japan and Germany, [Phy. Rev. C. 67, 064312 (2003)] the comparative study of G.T. strength for $Sc^{26} \rightarrow Al^{26}$, have been compared with GT strength of Al^{26}, using the reaction Mg^{26} (*p,n*) Al^{26} at Ep = 135 MeV. This comparison is interpreted by shell model, and particle rotor model assuming correlation of proton and neutron pair around a Mg^{24} core.

In a theoretical survey study carried out by P.C. Sood et al. for log ft values of 500 transitions; of $A > 228$ nuclei, has interestingly shown, that the first forbidden transition ($\Delta\pi$ = Yes), there is a mis-match between the global (for the whole periodic table); and authors' evaluation of log ft values, indicating the effect of second order elements. [Phy. Rev. C. 69. 057303, (2004)].

In an experiment conducted by 21 authors; plus ISLODE collaborations, the GT strength distribution of transitions in Kr^{74} of some eighteen excited states from ground state to 978 keV, up to 3.00 MeV the experimental data was compared with theory, which showed that Kr^{74} is neither oblate nor prostate [Phy. Rev. C. 69, 034307 (2004)].

In a long paper of 16 pages with 11 authors in Phy. Rev. *C*. 74, 0.54309, (2006), β-decay proportion of ${}_{28}Ni^{72}$ and ${}_{29}Cu^{67}$, have been investigated at CERN ISOLDE facility. These neutron rich nuclei have been produced in the proton induced fission of U^{258}. Comparison is made with schematic shell model picture of Cu^{72}, and with large scale shell model calculation.

In this paper [Phy. Rev. C. 77, 064303 (2008)] studies on the double β-decay using (d, He^4), charge exchange reaction on the double decay (ββ) nucleus 30 Zn^{64} [30 Zn^{64}_{34} + $d \rightarrow$ 31 $Ga^{66}_{35} \rightarrow$ 29 Cu^{64+2}_{35} + 2 He] have been studied at an incident energy of 1.83 MeV. The two proton in IS_O (indicated by 2 He) were both momentum analyzed and detected simultaneously by BBS magnetic spectrometer and position-sensitive detector. 2 He spectra with a resolution of about 115 keV, have been in the residual intermediate nucleus 29 Cu^{64}_{35}. With the nuclear matrix elements of ββ decay of Zn^{64}, the $GT_{\pm}$ distribution are compared with shell model calculation.

REFERENCES

1. W. Pauli: Rapports de septieme Council de Physique Solvey Brussels, (1933), Ganthier-Villars and cie, Paris (1934).

   E. Fermi: Z. Physik, 88, 161 (1934).

   G. Gamow and E. Teller: Phy. Rev. 49, 895 (1936).

2. J.R. Reitz: Phy. Rev. 77, 10 (1950).

   C.M. Lederer, J. Hollander, I. Perlman: Table of Isotopes, John Wiley & Sons, New York (1967): J. Shirley (1986).

3. P.B. Smith and J. S. Allen: Phy. Rev. 81, 381 (1951).

   R. Davis: Phy. Rev. 86, 976 (1952).

   G.W. Rodeback and J. S. Allen: Phy. Rev. 446 (1952). (for Cl^{37}).

4. C.L. Cowan, F. Reines, F.B. Harrison, H.W. Kruse and A.D. McGuire: Science 124, 103 (1956).

   Reines, and F. Annals: Rev. Nuclear Sciences 10, 1 (1960).

5. J.C. Dickens, F. G. Perey and R.J. Silva: Phy. Rev. 132, 1190 (1963); Ibid. Physics Letters 6, 53 (1963).

6. L.M. Langer and R.J.D. Moffat: Phy. Rev. 88, 689 (1952).

7. E. Fernberg and K.G. Hammack: Phy. Rev. 75, 1877 (1949).

   A.M. Feingold: Rev. Mod. Phy. 23, 10 (1951).

   E.J. Konopinski: Rev. Mod. Physics 15, 209 (1943).

   E. Fernberg and G.T. Trigg: Rev. of Mod. Physics 22, 3 (1950).

8. T.D. Lee and C.N. Yang: Phy. Rev. 104, 254 (1956).

   Ibid: Phy. Rev. 105, 1671 (1957); T.H.R. Skyrme: Progress in Nuclear Physics 1, 115 (1950).

   E.J. Konopinski: Annual Rev. of Nuclear Science. 9, 99 (1959).

   W. Pauli: Encyclopedia of Physics, ed. S. Flügee, Vol. VII, Springer-Verlag, Berlin (1958).

9. P.A.M. Dirac: The Principles of Quantum Mechanics, 2nd ed. Section 20, 21 Chapter XII, Oxford, New York (1953); Proc. Royal Society A 117, 610 (1928) Quantum Mechanics L.I. Schiff, McGraw-Hill Book Company, Inc., New York (1949).

10. E. Ambler, R.W. Hayward, D.D. Hoppes, R.P. Hudson and C.S. Wu: Phy. Rev. 106, 1361 (1957).

11. R.R. Roy and B.P. Nigam: Nuclear Physics—Theory and Experiment, New Age International (P) Ltd., New Delhi (1986).

    M.A. Preston: Physics of the Nucleus.

    Addison-Wesley Publishing Company, Inc. Reading Massachusetts (1968).

12. M.T. Burgy, V.E. Krohn, T. B. Novey, G.R. Ringo and V.L. Telegei: Phy. Rev. 120, 1829 (1960); J.M. Robson: Phy. Rev. 100, 933 (1955).

H. Postma, W.J. Huiskamp, A.R. Miedema, J. Steenland, H.A. Tolhock and C.J. Gorter: Physica 23 (259) (1957).

13. E.J. Konopinski: Annual Rev. of Nuclear Sciences, 9, 99 (1959).

14. H. Frauenfelder, R. Bobone, E. Von. Goeler, N. Levine, H.R. Lewis, R.N. Deacock, A. Rossi and G. De Pasquali: Phy. Rev. 106, 386 (1957).

15. M. Goldhaber, L. Grodzins and A.W. Sunyar: Phy. Rev. 106, 826 (1957).

16. S.S. Handa and R.S. Preston: Phy. Rev. 106, 1363 (1957); 108, 160 (1957).

17. L.A. Page and M. Heinberg: Phy. Rev. 106, page 1220 (1957).

18. L. Landau: Nuclear Phy. 3, 127 (1957); A. Salam: Nuovo Cimento 5, 299 (1957), Ref. (8), J.D. Jackson, S.B. Treiman and H.W. Wyld: Phy. Rev. 106, 517 (1957).

19. R.F. Feynman: Quantum Electrodynamics Benjamin, New York (1961); M.E. Rose: Relativistic Electron Theory, Wiley, New York (1961); Quantum Mechanics by L.I. Schiff, McGraw-Hill Book Company, Inc. (1949).

20. E. Hagberg, I.S. Towner, T.K. Alexander, G.C. Ball, J.S. Forester, J.C. Hardy, J.G. Hykawy, V.T. Koslowsky, T.R. Leslie, H.B. Mark; I. Nelsen and G. Savard: Phy. Rev. C. 56, p. 135 (1997); Toshio Suzuki and Takahasu Otsuka: Phy. Rev. C.V. 56, p. 847 (1997).

21. B.H. Wildenthal: Progress, Nuclear Physics 11, 5 (1984); B.A. Brown and B.H. Wildenthal: Phy. Rev. 28, 2397 (1983); I.S. Towner and F.C. Khanna: Nuclear Physics A 399, 334 (1983), Ibid; Physics Rev. Letter 42, 51 (1979).

22. N. Aoi et al.: Nuclear Physics A 616, 181, (1997), F. Ajzenberg: Selov Nuclear Physics A 433, 1 (1985), Ibid A 506, 1 (1990).

23. R.L. Helner et al. (16 authors): Phy. Rev. C. 55, p. 280 (1997).

24. T.N. Taddeucei, C.A. Goulding, T.A Carey, R.C. Byod, C.D. Goodman, C. Gaarde, J. Larson, D. Horen, J. Rapahorn and E. Sugerbaker: Nuclear Physics A 469, 125 (1987).

25. M.J.G. Borge et al. (17 authors): Phy. Rev. C 55, p. R 8 (1997).

26. M. Shibater et al. (18 authors): Phy. Rev. C. 55, p. 1715 (1997).

27. G. Frick, A. Gallmann, D.E. Alburger, D.H. Wilkinson and J.D. Coffin: Phy. Rev. 132, 2169 (1963).

28. E.J. Konopinski and L.M. Langer: Ann. Rev. Nuclear Sciences 2, 261 (1953).

29. J.M. Robson: Phy. Rev. 83, 349 (1951); O.S.Curran, J. Angus and A. Cockcroft: Phil Mag. 40, 53 (1949); D.R. Albert: Phy. Rev. 74, 847 (1948); C.S. Wu and R.D. Albert: Phy. Rev. 75, 315 (1949).

30. N. Dismuke, M.E. Rose, C. L. Perry and P.R. Bell: U.S Atomic Energy Commission Report OR NL-1222, (1952), Tables for the Analysis of Beta Spectra, U.S. National Bureau of Standards, Applied Math ser. No. 13, Washington D.C. (1952); A.H. Wapstra, G.J. Nijgh, and R. Van, Llichout: Nuclear Spectroscopy Tables, North Holland Publishing Co. Amsterdam (1959); K. Siegbahn; Beta and Gamma Rays Spectroscopy, North Holland Publishing Co. Amsterdam (1958).

W. Pauli: Encyclopedia of Physics ed. S. Flügge; Vol. VC/1, Springer Verlag, Berlin (1958).

31. G.E. Lee-Whiting: Cand. J. of Physics, 36, 1199 (1958), T. Ahrens and E. Feenberg: Phy. Rev. 86, 64 (1952): M.E. Rose and R.K. Osborne: Phy. Rev. 93, 315 and 1326 (1954), M.E. Rose and D.K. Holmes: Phy. Rev. 83, 190 (1951).

A.H. Wapstra: Nuclear Physics, 9, 519 (1958–59). A.H. Wapstra, G.H. Nijgh and R. Van Leishout: Nuclear Spectroscopy Tables, North Holland Publishing Co., Amsterdam (1959).

F.T. Porter, M.S. Freedman, T.B. Novey and F. Wagner: Phy. Rev. 103, 921 (1956); L.M. Langer and H.C. Price: Phy. Rev. 75, 1109 (1949).

32. L. Feldman and C.S. Wu: Phy. Rev. 87, 1091 (1952).

C.S. Wu: Rev. Mod Phy. 22, 386 (1950).

C.S. Wu and L. Feldman: Phy. Rev. 76, 693 (1949).

L.M. Langer and R.D. Moffat: Phy. Rev. 82, 635 (1951).

33. M.E. Rose and R.K. Osborne: Phy. Rev. 93, 1315–1326 (1954).

R.L. Graham, J.S. Gigger and T.A. Eastwood: Cand. J. of Physics, 36, 1084 (1958).

F.T. Porter and P.D. Day: Phy. Rev. 114, 1286 (1959); J.M. Robson Cand: J. of Physics 38, 148 (1960). F.T. Porter, M.S. Freedman, T.B. Novey and F. Wagner Jr.: Phy. Rev. 103, 921 (1956).

34. M. Moe and P. Vogel: Annual Rev. Nuclear, Particle Sciences, 44, 247 (1994), B. Kayser: Nuclear Physics A 546, 399 C, (1992), Z. Berezhiani, A. Somisnov and J.Valle.: Phy. Letters B 291, 99 (1992), C. Burges and J. Cline: Phy. Letter B 298, 141 (1993). Phy. Rev. D 49, 59 (1994).

35. M. Alston Garnjost, B.L. Dougherty, R.W. Kenny, R.D. Tripp, J.M. Kirivicich, H.W. Nicholson and C.S. Sutten: B.D. Dieterle, S.D. Foltz, C.P. Leavit, R.A. Reeder, J.D. Baker and A.J. Caffery: Phy. Rev. C.V. 55, p. 474 (1997), A. De. Silver, M.K. More, M.A. Nelson and M.A. Vrient: Phy. Rev. C.V. 56, p. 245 (1997).

36. M.K. More: Int. J. Mod. Physics E2, 507 (1993).

37. L.A. Mikaclyan, I.E. Kuteskov and V.F. Apelin: Seviet Physics JETP 12, 1027 (1961).

J.D. Ullman; H. Fraunfelder, H.J. Lipkin and A. Rossi, H.R. Bobone, E. Von Goeler, N. Levine, H.R. Lewis, R.N. Peacock, A. Rossi and G. De. Pasquali: Phy. Rev. 106, 386, 1957; Spivak, P.E., L.A. Mikaclyan, et al: Nuclear Physics, 20, 475 (1960).

38. K. Alder, B. Stech and A. Winther: Phy. Rev. 107, 728 (1957); B.T. Feld: Phy. Rev. 107, 797 (1957).

39. R.K. Bardin, C.A. Barues, W.A. Fowler and P.A. Seeger: Phy. Rev. 127, 853 (1962); J.W. Butler and R.O. Bondelid: Phy. Rev. 121, 1770 (1961). M.E. Rose: Phy. Rev. 49, 727 (1936).

40. C.S. Wu: Rev. Modern Physics, 36, 618 (1964); J.W.Freeman, J.H. Montague D. Kerst and R.E. White: Physics Letter 3, 126 (1962) J. Jeneke: Physics Letters 6, 66 (1963) and J.W. Freeman et al.: Physics Letters 8, 115 (1964).

41. J.D. Jackson, S.D. Tremous and H.W. Wyld Jr.: Phy. Rev. 106, 517 (1957); K. Alder, B. Stech and A. Winther: Phy. Rev. 107, 728 (1957) B.T. Feld: Phy. Rev. 107, 797 (1957). Nuclear Physics, R.R. Roy and B.P. Nigam; New Age International (P) Ltd., (1967): (Ref. 28).

42. J.S. Allen, R.L. Burman, W.B. Aemansfeld, T. Stahelin, T.H. Brand: Phy. Rev. 116, 124 (1958).

B.W. Ridley: Nuclear Physics 25, 483 (1961).

M.A. Clark, J.M. Robson and R. Nathens: Phy. Rev. Letter 1, 100 (1958).

43. F. Bohem and A.H. Wapstra: Phy. Rev. 106, 1364 (1957); 107, 1202 (1957); 107, 1202 (1957).

44. M. Goldhaber, L. Grodzins and A.W. Sunyar: Phy. Rev. 109, 1015 (1958); 106, 826 (1957).

45. M.T. Burgy, V.E. Krohn, T.B., G.R. Ringo and V.L. Telegdi: Phy. Rev. 110, 1214 (1958); Phy. Letters 1, 324 (1958); Ref. (32); Nuclear Interactions; S. De Benedetti, John Wiley & Sons, New York (1964).

46. G. Luders: Ann. Physics 2, 1 (1957).

47. A.N. Sosnovshi, P.E. Spivak, Yu. A. Prokofev, V.V. Vladminiskii, V.K. Grigovev and V.A. Ergakov, JETP 36, 931 (1959).

48. J.M. Robson: Phy. Rev. 100, 933 (1955).

49. W.B. Berrmannsfeldt, R.L. Burman, T. Stahelin and J. Allen: Phy. Rev. 107, 641 (1957).

50. L. Landau: Nuclear Physics. Refer (8) 3, 127 (1957); Salam A, Nuovo Cimento 5, 299 (1957); T.D. Lee and C.N. Yang: Phy. Rev. 119, 1410 (1960).

51. H.Z. Weyl: Physik. 56, 330 (1929).

52. F. Reines: Ann. Rev. Nuclear Sciences 10, 1 (1960); R.E. Carter, F. Reines, J.J. Wagner and M.E. Wyman: Phy. Rev. 113, 280 (1959).

53. R.P. Feynmann and M. Gell Mann: Phy. Rev. 109, 193 (1958).

54. M. Gell Mann and S.M. Berman: Phy. Rev. Letters 3, 99 (1959)

M. Gell-Mann: Phy. Rev. 111, 362 (1958); Mayer-Kuchuk, T and C.M. Michel: Phy. Rev. 127, 545 (1962); Lee Y.K., L.W. Mo and C.S. Wu: Phy. Rev. Letters 10, 253 (1963).

55. S.L. Glashow: Nuclear Physics, 22, 579, (1961); S. Weinberg: Phy. Rev. Letters 19, 264, (1967); A. Salam: Elementary Particle Theory: edited by N. Svartholm, E. Almquist and Wilksells, Stockholm. (1968); H. Fritzsch, M. Gell-Mann and H. Loutwayler: Phy. Letters B47, 365, (1973); S. Weinberg: Phy. Rev. 31, 1494 (1973).

56. G. Armison et al.: Phy. Letters, 122B, 103 (1983); Phy. Letters, 126B, 398, (1983); M. Banner et al.: Phy. Letters 122B, 476 (1983).

PROBLEMS

1. Equation 8.38 has been developed for zero neutrino mass. Write the new form of this equation if neutrino mass is small but not zero.

2. (*i*) Prove relations in Eqs. 8.59 to 8.65; by actual substitution of the matrices.

(*ii*) Substitute the actual values of α and β matrices in Eqs. 8.34*a* and 8.34*b*; write the individual equations and interpret each one physically.

(*iii*) Show from (*ii*), that for $v < c$, the four relations for given momentum are given by

$$\psi_s = U_s \bar{e}\,(Et - \mathbf{p}\cdot\mathbf{x}),\; S = 1, 2, 3, 4 \text{ as } U_s \text{ satisfies}$$

$$\underset{\text{For } E>0}{\begin{pmatrix} \xi \\ \dfrac{\sigma\cdot\mathbf{p}}{|E|+m}\xi \end{pmatrix}};\; \underset{\text{For } E>0}{\begin{pmatrix} \dfrac{-\sigma\cdot\mathbf{p}}{|E|+m}\xi \\ \xi \end{pmatrix}};\; \xi = \begin{pmatrix}1\\0\end{pmatrix} \text{ or } \begin{pmatrix}0\\1\end{pmatrix} \text{ for } E>0 \text{ or } E<0.$$

3. It is often said that the theory of β-decay is a generalisation of the theory of electromagnetic radiation. Show their similarities. In particular show how their selection rules arise, and have similarities.

4. Show that a simple nuclear model gives the following values for β-decay matrices:

$$\left|\int \sigma\right|^2 = 3, \frac{1}{3}, \frac{7}{5}, \frac{3}{5} \text{ and } \frac{9}{7}$$

for H^3, O^{15}, F^{17}, Ca^{39} and Si^{41} respectively. Also show that $|\int \mathbf{1}|^2 = 1$ for all these nuclei.

5. From the consideration of spin orientation, show Gamow-Teller selection rule $0^+ \to 0^+$ is forbidden.
6. If (*i*) He^3 $(1/2^+) \to He^3$ $(1/2^+)$; log ft = 3.03,

   and $| M_{GT} |^2 = 3$ and $(M_F)^2 = 1$

   and (*ii*) Cl^{34} (O^+), log ft = 3.49 $| M_F |^2 = 2$,

   Determine the ratio g_A/g_V.
7. What would be the relationship between g_i and g'_i in expression for β-decay Hamiltonian H_β, if Wu's experiment had found that H_β was non-variant under charge conjugation and time reversal?
8. Assuming the first experimental information about the helicity of neutrino and $g_A/g_V = -1$; prove that two component theory follows automatically and also the non-conservation of parity.
9. Show that with vector interaction, the transitions $\Delta\, 1 = 1$, $\Delta\, \pi =$ yes, $\Delta\, J = 0$ are first forbidden.
10. Using reference (27), show why for non-unique transitions, of Cl^{36} and Cs^{137}; the curvatures of Fermi plots are opposite to each other.

9

CHAPTER

Alpha and Charged Particle Decay

9.1 ENERGETICS AND EXPERIMENTS

Alpha (α) decay was discovered in the initial stages of the discoveries of radioactivity from naturally occurring radioactive chains, *e.g.* thorium series[1]. Some of these series are shown in Figs. 9.1 to 9.4. Alpha decay corresponds to the decay of radioactive nucleus, emitting the nucleus of a $_2He^4$ atom, leading to the decay, *e.g.*,

$$_Z A^N \xrightarrow{\;_2He^4\;} {}_{Z-2}(A-4)^{N-2} \qquad ...(9.1)$$

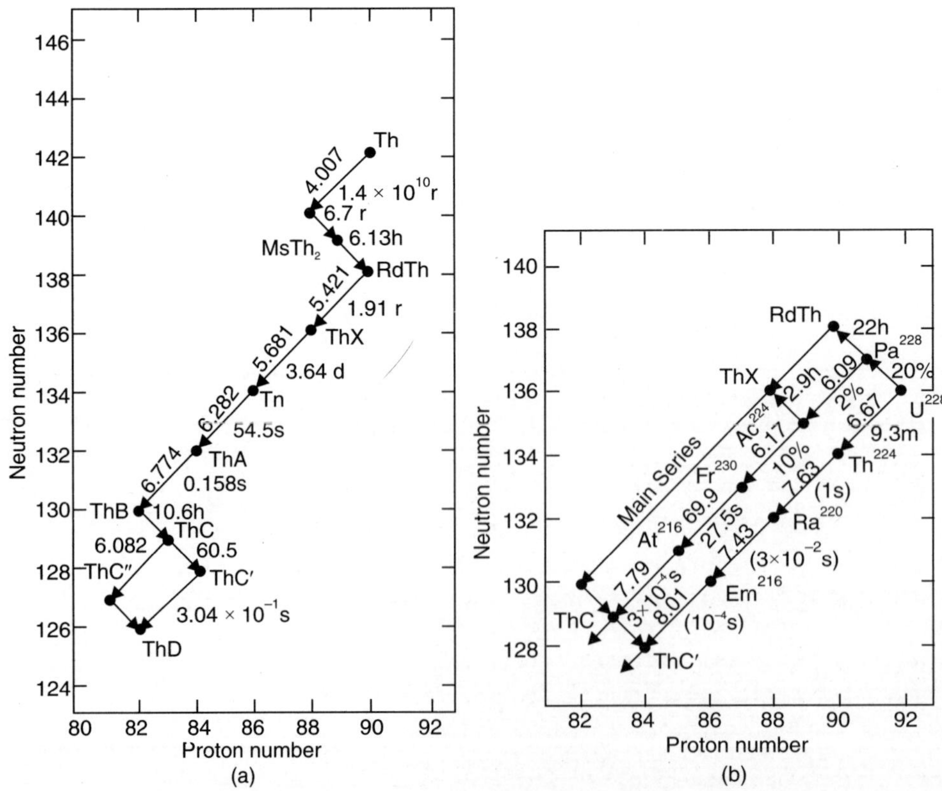

Fig. 9.1 4n (Thorium) series (*a*) Main series and (*b*) Collateral series (Ref. 1).

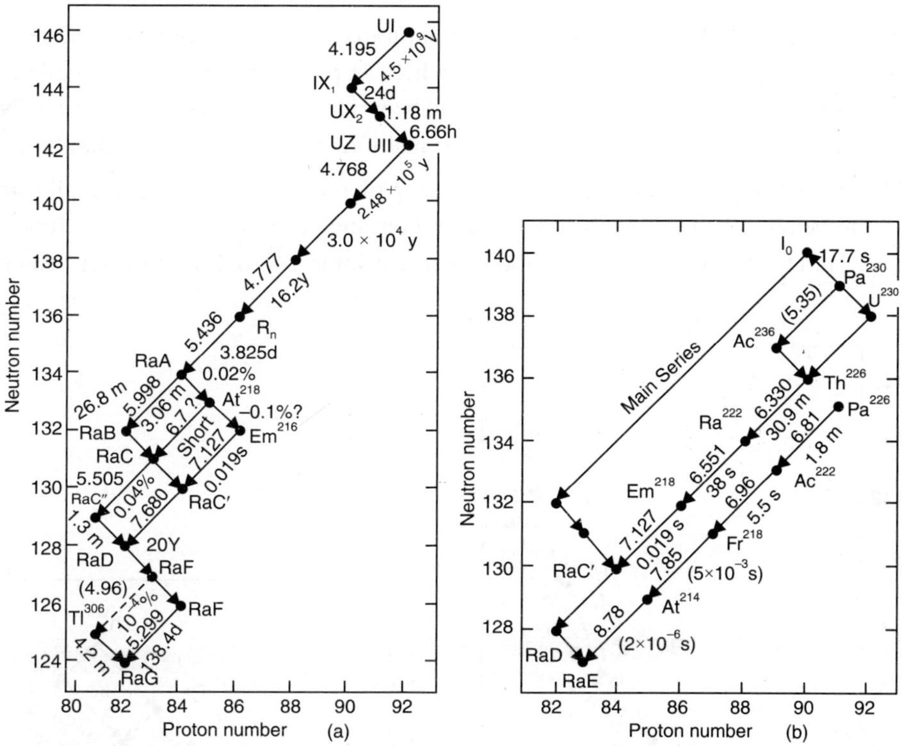

Fig. 9.2 4n + 2 (Uranium) series (*a*) Main series and (*b*) Collateral series (Ref. 1).

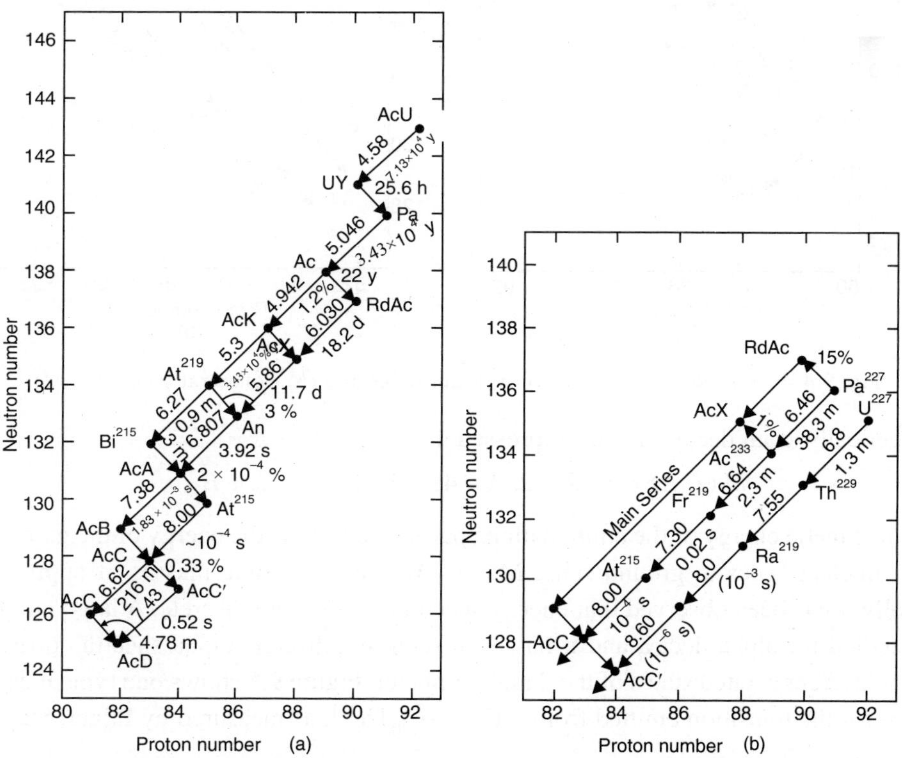

Fig. 9.3 4n + 3 (Actinium) series (*a*) Main series and (*b*) Collateral series (Ref. 1).

There are some thirty α-emitters in the three naturally occurring radioactive series[1]. Out of the four available series, there is one series, (4*n* + 1), which is produced artificially—called Neptunium series[2]. The number of nucleons in these four series can be written as: 4*n* (Thorium), 4*n* + 1 (Neptunium), 4*n* + 2 (Uranium), 4*n* + 3 (Actinium). There are no transitions from one series to another.

In addition to the above series[3], a large number of nuclei produced artificially are α-active. Many of these are heavier than Pb, but also include lighter nuclei[4] especially some isotopes of Sn, Cm, Hf, Nd, Eu, Gd, Tb, Dy, Au and Hg. All together, there are some 140 α-active nuclei, which have been studied.

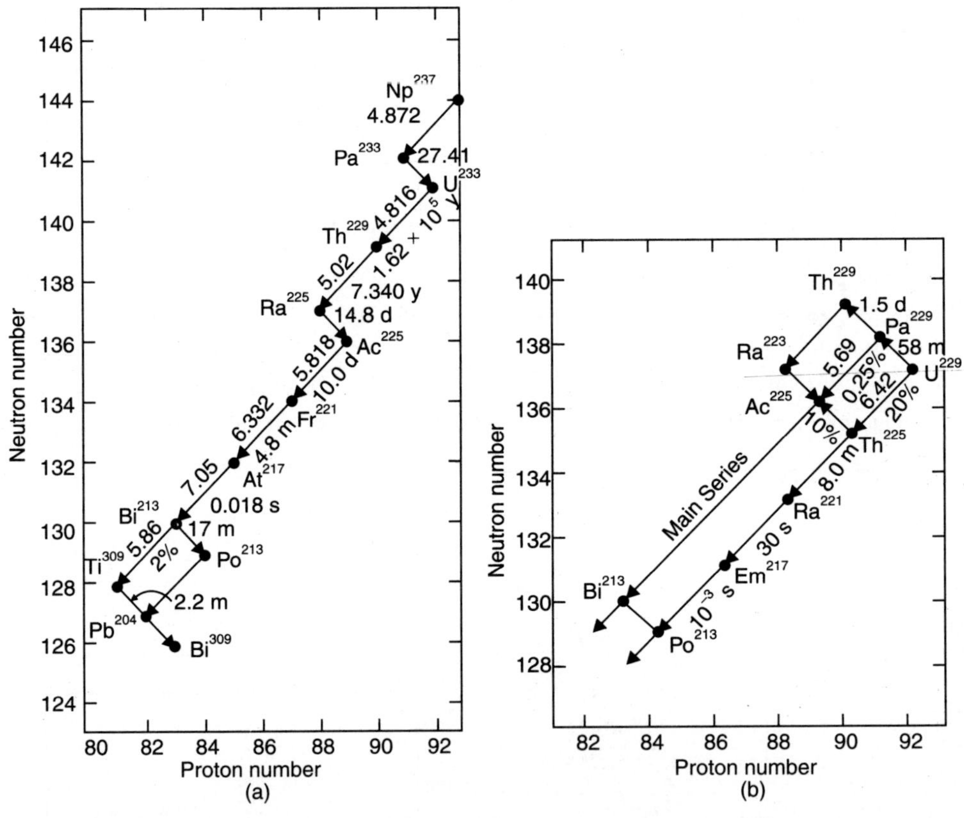

Fig. 9.4 4n + 1 (Neptunium series) (*a*) Main series and (*b*) Collateral series (Ref. 2).

Energetically, each decay may be expressed as:

$$M\,(Z, A) = M\,(Z - 2, A - 4) + M\,(\text{He}^4) + E_0 + E_\gamma \qquad \text{...(9.2)}$$

where E_0 is the kinetic energy of the emitted alpha particle, and E_γ is the energy of the gamma ray, when the daughter nucleus, goes to ground state. Masses M_i are the atomic masses of appropriate atoms. Experimentally it has been observed that energy spectra of γ-rays are discrete. A large number of cases have been studied for alpha decay, and it has been seen that discrete alpha spectra correspond to the transitions to different excited states of the daughter nuclei. Figure 9.5 shows one typical case of energy level diagram for the transition emitted from ${}_{92}U^{235} \rightarrow {}_{90}Th^{231}$, as measured by Pilger, Stephens, Asaro and Perlman[5], and others with magnetic spectrometer. The energy of alpha particles in the most cases of decay ranges between 4 to 8 MeV.

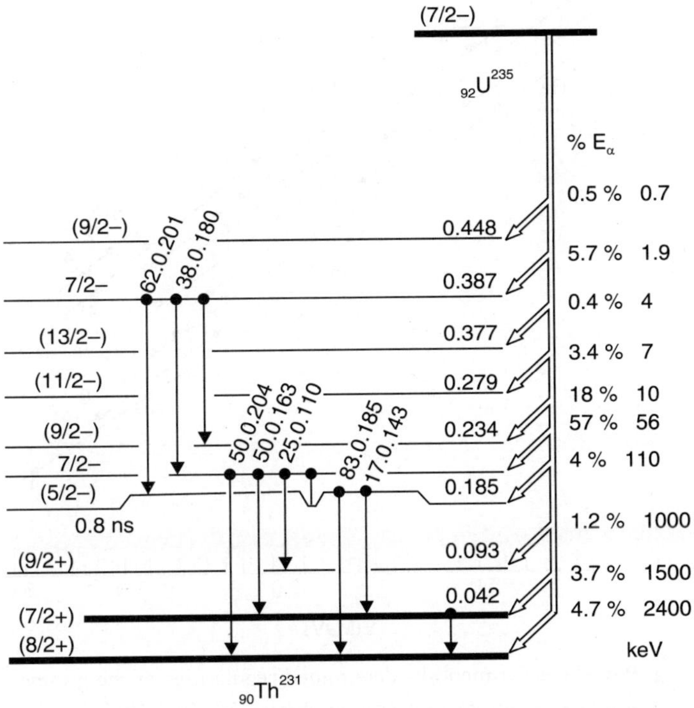

Fig. 9.5 Energy level of $_{92}U^{235} \rightarrow {}_{90}Th^{231}$, indicating energies of α's and percentage fractions of α's. Also are indicated gamma ray transitions in Th^{231} (Ref. 5).

9.2 EMPIRICAL ALPHA DECAY LAW

The life-times of alpha emitters has a large range, illustrated by the shortest half life[6] of 3.04×10^{-7} secs (0.964×10^{-14} years) for Po^{212} emitting 8.776 MeV alphas, and the longest half life of 1.39×10^{-10} years, for Th^{232}, emitting 4.607 MeV alphas[7]. This example also illustrates the empirical Geiger-Nuttal Law[8], according to which the energy of the emitted alpha particle and its half life are related by:

$$\log \lambda = a + b\,\bar{v}^{-n} \left(\lambda = \frac{1}{\tau}\right)$$

where $n \approx 1$ or 2 ...(9.3)

and a and b are constants, and v is the velocity of the alpha particles, related to the energies E by $v^2 = 2\,E/m$. Figure 9.6 illustrates the law, in terms of half life $T_{1/2}$ and energy E, *i.e.*,

$$\log T_{1/2} = A + BE^{-1/2} \quad \text{...(9.4)}$$

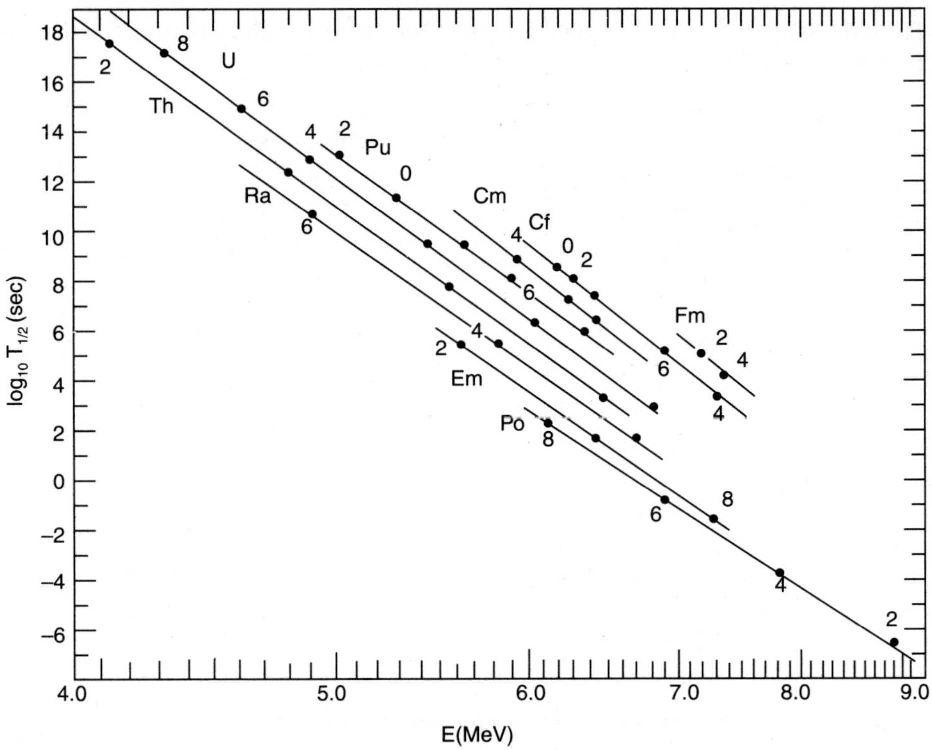

Fig. 9.6 The experimentally determined regularities of the ground state α-transitions of even-even nuclei (Eq. 9.4) (Ref. 8).

Very low energy alphas cannot be emitted because of the Coulomb barrier. The high energy limit is governed by the relative binding energies of nuclei, connectecd through alpha decay. From the proper correlation between different energies of emitted alphas, one can infer the energies of the excited states of the daughter nuclei. The relationships (observed through coincidence studies), with accompanying gamma-rays, further confirm these studies), Fig. 9.7 illustrates[5] the systematics of energy release as a function of the mass and charge number of alpha-emitters. The regularities in these systematics may be summarised as: (*i*) For a given mass number, E_α (max) increase as Z increases. (*ii*) For a given element E_α (max) increases as A decreases. (*iii*) For small A, (for neutrons less than 128); there is a sharp reversal of the trend of E_α (max) as given in (*ii*) but the trend is resumed for smaller A. Corresponding to discontinuity for $N \leq 128$, there is also a discontinuity for $Z \leq 84$.

These systematics are related to binding energies.

Why does a nucleus decay through alpha-emission, and not by emission of say proton, neutron, deuteron or He^3 ? The answer lies in the large binding energy of ≈ 8 MeV, for alpha particle. Further, emission of an alpha particle reduces the relative charge of the nucleus and hence results in a low-energy system. As a matter of fact, the systematics of alpha decay may be understood from the considerations of Weiszäcker's mass-formula, based on liquid-drop model. Because of slow decrease of binding energy per nucleon for heavy nuclei, all heavy nuclei with $A \geq 150$ are expected to be alpha-unstable, with an energy release, that is smoothly varying function of mass and charge. The mass number limit of alpha-stability is expected to be somewhat indefinite, and depends on the sensitivity of detection methods.

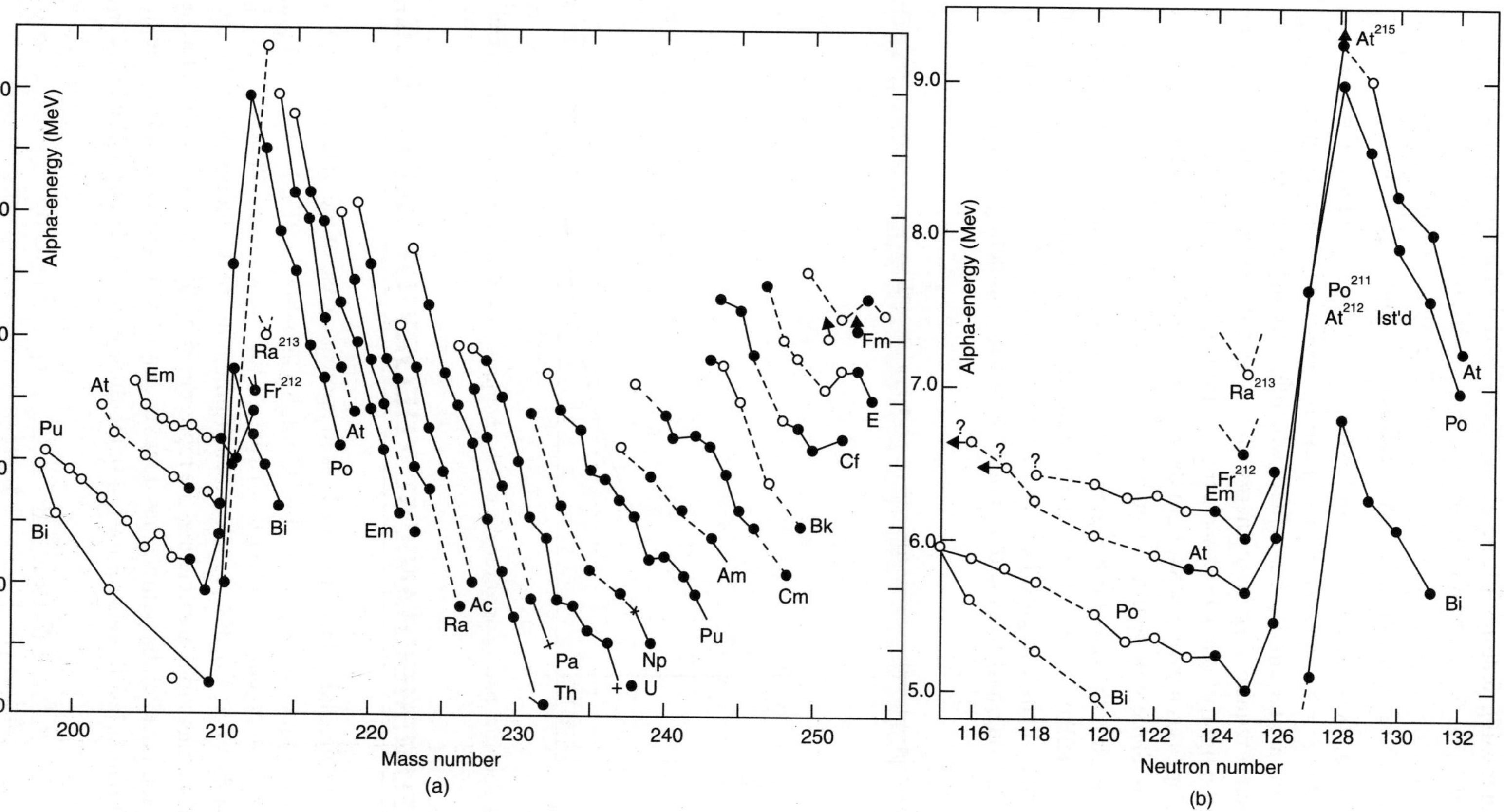

Fig. 9.7 (*a*) Energy release in alpha decay in heavy elements as a function of Z and A. (*b*) Effect of 126-neutron shell (Ref. 5).

One of the interesting features of the alpha decay is the effect of magic numbers. We have already seen, in Fig. 9.7 that for neutron number 128, there is discontinuity in the energies released in the alpha decay. There is similar discontinuity at $Z \leq 84$. The alpha energies of the Bismuth isotopes ($Z = 83$ for Bi) are considerably lower than expected from the trend for $Z \geq 84$.

As a matter of fact, these are the effects of having the closure of shells at $Z = 82$ and $N = 126$. An interesting fact that Pb ($Z = 82$) and Tl ($Z = 81$) are not radioactive is, because of this effect. For α-emission, two protons are required to be emitted, which in both these cases are very tightly bound — at the closed shell, hence require very large energy of separation; resulting in no α-emission. Bismuth ($Z = 83$) has one proton tightly bound and hence, there is tremendous drop in the energy of the alpha, as shown in Fig. 9.7*b*. For cases above $Z = 84$, both the protons are loosely bound and there is increase in the energy of emission. Same will be true for nuclei, with $N \geq 128$.

A region of interest is the rare-earth region, where there is no α-radioactivity, basically because of the effect of closed shell at $N = 82$. Because of the tightly bound last two neutrons; the nuclei with $N = 84$, will require large alpha energies. For other neutron numbers in this region, energies of emitted alphas are smaller with long half lives and a poorer chance of competing with decay due to electron capture.

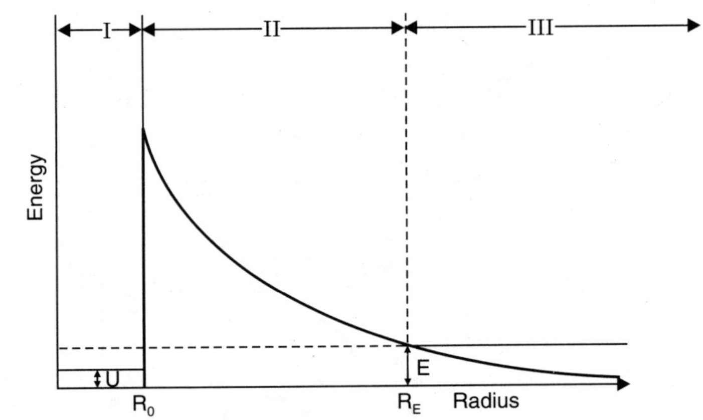

Fig. 9.8 The potential barrier used for the one-body model calculations, carried subsequently.

9.3 QUANTUM MECHANICAL THEORY OF ALPHA DECAY

Quantum mechanically, the alpha decay may be looked upon as a case of the transmission of a one-body wave-function, representing α -particle tunnelling through a potential barrier, presented by the Coulomb potential created by the interaciton of electric charges of the nucleus and alpha particle. The energy relationship between the energy of the particle, nuclear potential and Coulomb potential is represented in Fig. 9.8. It is evident, that alpha particle may be represented by $\Psi = \phi / r$ where Ψ is the radial, time-independent wave-function of alpha particle; has three regions in which its value and its radial derivative has to match at the boundaries. In general for $L \neq 0$, the differential equation obeyed by ϕ is given by:

$$\frac{d^2\phi}{dr^2} + \frac{2M}{\hbar}\left[E - \frac{L(L+1)\hbar^2}{2Mr^2} - V(r)\right]\phi(r) = 0 \qquad \text{...(9.5)}$$

where M, the reduced mass of the alpha-particle plus residual nucleus is given by $M = M_\alpha M_r / M_\alpha + M_r$ where M_r is the mass of the residual nucleus (which also recoils after emission of alpha), E is the energy of alpha particle plus that of recoil nucleus and is given by:

$$E = \frac{M}{2}(v_r + v_\alpha)^2 = \frac{Mv^2}{2} \qquad \text{...(9.6)}$$

Essentially E is the total energy of the α-decaying system.

We consider below only the case for $L = 0$, then the wave-function φ, obeys the equation:

$$\frac{d^2\phi}{dr^2} + \frac{2M}{\hbar^2}(E - V(r))\,\phi(r) = 0 \qquad \text{...(9.7)}$$

Keeping in mind (Fig. 9.8), that there are three distinct regions, we consider the solution in each region separately:

Region I:

$$0 < r < R;\ V(r) = U\ (< E)$$

In this region,

$$\phi_1(r) = A_1 \exp \pm ik_1 r \qquad \text{...(9.8)}$$

where $k_1 = \frac{1}{\hbar}[2M(E-U)]^{\frac{1}{2}}$

Region II:

$$R < r < R_E,\ V(r) = \frac{zZe^2}{r}\ (> E)$$

Then from Quantum Mechanics[9] (Schiff L.I., page 180), for WKB approximation solutions, One can write:

$$\phi_\pm = A_2\, k_2^{-1/2}(r) \exp \pm \int_r^R k_2(r)\,dr = A_2\, k_2^{-1/2}(r) \exp[\pm K_2(r)] \qquad \text{...(9.9)}$$

where $k_2(r) = \frac{[2M(V(r)-E)]^{\frac{1}{2}}}{\hbar}$ and $[\pm K_2(r)] \equiv \pm \int_r^{R_E} k_2(r)\,dr$

Region III:

$$R_E < r;\ V(r) = \frac{zZe^2}{r}\ (< E)$$

Again from (Schiff[8] p. 180), we get the solution:

$$\phi_{\pm} = A_3\, k_3^{-1/2} \text{ exp. } \pm i \int_{R_E}^{r} k_3 (r)\, dr = A_3 k_3^{-1/2} \exp \pm iK_3 (r) \qquad \text{...(9.10)}$$

where $k_3 (r) = \dfrac{[2\, M\, (E - V\, (r))]^{\frac{1}{2}}}{\hbar}$ and $K_3 (r) \equiv \int_{R_E}^{r} k_3 (r)\, dr$

As $r \to \infty$; the solution in region (III) should represent a pure outgoing wave physically; we may therefore, write ϕ (r) in this region as:

$$\underset{r \to \infty}{\phi_3} = A_3\, e^{ikr},\ k = \frac{(2\, M\, E)^{1/2}}{\hbar} \qquad \text{...(9.11)}$$

Now we use the WKB connection formula between Region (II) and the asymptotic formula (*see* Schiff, page 184) Region (III), which gives near $r = R$, in Region (II) a solution:

$$\phi\, (r) = A_2\, k_2^{-1/2}\, (r) \exp\, [+\, K_2\, (r)] = A_3 \exp + i\, k\, r \qquad \text{...(9.12)}$$

giving $$A_2 \left(\frac{1+i}{2\,k}\right)^{\frac{1}{2}} = A_3$$

where we have neglected the term in exp. $[-\, K_2\, (r)]$ in ϕ (r); because K_2 (r) is very large near R.

At $r = 0$; ϕ (r) must vanish, and hence the solution in region (I) is

$\phi_1\, (r) = A_1 \sin k_1\, r$

At the boundary (I) and (II), we match the value and derivative of the wave-functions;

So matching ϕ_1 and ϕ_2:

$$A_1 \sin k_1\, r = A_2\, k_2^{-1} \exp\, (+\, K_2\, R)$$

and matching

$$\left(\frac{d\,\phi_1}{d\,r}\right)_{r=R} = \left(\frac{d\,\phi_2}{d\,r}\right)_{r=R}$$

we get $$A_1\, k_1 \cos k_1\, R = -\, [A_2\, k_2^{1/2}\, (R) \exp\, (+\, K_2\, (R))]\, [1 + \gamma] \qquad \text{...(9.13a)}$$

where $$\gamma = \frac{1}{2\, k_2^2\, (R)} \left.\frac{d\, k_2\, (r)}{d\, r}\right|_R = \frac{1}{4\, k\, R} \frac{(E\,/\,V\, (R))^{1/2}}{(1 - E\,/\,V\, (R))^{3/2}} \qquad \text{...(9.13b)}$$

so that $$k_1 \cot\, (k_1\, R) = -\, (1 + \gamma)\, k_2\, R \qquad \text{...(9.14a)}$$

As k_2 $(R) >> k_1$; we get $$\frac{-\cot k_1\, R}{1 + \gamma} >> k_1 \qquad \text{...(9.14b)}$$

since $\gamma \approx 0.017$ and $k_1 R = \pi$

therefore, from Eq. 9.13

$$A_1 \approx A_2\, k_1^{-1}\, k_2^{1/2}(R) \exp(+K_2(R)) \qquad \text{...(9.15)}$$

Normalising the wave function ϕ_1 in the region 0 to R, we get

$$4\pi A_1^2 \int_0^R \sin^2(k_1 r)\, dr = 2\pi R A_1^2 = 1 \qquad \text{...(9.16)}$$

The number of alpha particles leaving the nucleus per unit time, *i.e.* λ is now given by:

$$\lambda = 4\pi v |A_3|^2$$

Hence using Eqs. 9.12 and 9.15 and 9.16 we can write:

$$\lambda = 4\pi v |A_3|^2 = 4\pi v |A_2|^2 \left[\frac{(1+i)(1-i)}{2k}\right]$$

$$= \frac{4\pi v}{2k} |A_2|^2 \times 2 = \frac{2\pi v}{k} k_1^2 \frac{|A_1|^2 \exp[-2K_2(R)]}{k_2(R)}$$

$$= \frac{2\pi v \times 2}{k \times 2\pi R} \frac{k_1^2}{k_2(R)} \exp[-2K_2(R)]$$

$$= \frac{2v}{R} \frac{k_1^2}{k\, k_2(R)} \exp[-2K_2(R)]$$

$$= \left[\frac{2v_i}{R}\right]\left(\frac{(E-U)^{1/2}}{(B-E)^{1/2}}\right) \exp(-2C) \qquad \text{...(9.17a)}$$

where $v_i \equiv v\left(1 - \frac{U}{E}\right)^{1/2}$

$$B \equiv V(r) \text{ (barrier-height)}$$

$$C \equiv K_2(R) = \frac{2BR}{\hbar v}(\alpha_0 - \sin\alpha_0 \cos\alpha_0)$$

and

$$\cos^2\alpha_0 \equiv \frac{E}{B} \qquad \text{...(9.17b)}$$

An alternative form is

$$\lambda = \frac{2v}{R} \frac{\mu^2}{\tan\alpha_0} \exp(-2C)$$

where $\mu^2 = \left(1 - \frac{U}{E}\right)$ ...(9.18)

More exact calculations have been carried out by Winslow and Simpson[10] and later on by Preston[11] who has given a rigorous, treatment avoiding WKB approximate methods.

According to Preston[11]

$$\lambda = \frac{2v}{R}\frac{\mu^2 \tan \alpha_0}{\mu^2 + \tan^2 \alpha_0} \exp(-2C) \quad ...(9.19a)$$

where $\mu = -\tan \alpha_0 \tan \mu K R$

The quantities α_0 and μ^2 have been defined earlier.

Comparing with Eq. 9.18; it is apparent that Preston's formula has a similar form, as obtained earlier by WKB approximation.

One can interpret Eq. 9.17 in a physical manner. The term ($v/2R$) corresponds to the striking frequently of alpha particle within the nuclear potential wall, $(E - U)^{1/2} / (B - E)^{1/2}$ is the reflection coefficient due to discontinuity at R and expt. $(-2C)$ is the barrier penetrability or tunnelling probability. In other words, the scenario which emerges is, that alpha particles strike the potential walls many times; every time they strike, a fraction is reflected and the rest is transmitted into the Coulomb barrier, which further allows some fraction given by e^{-2C} to tunnel through the Coulomb barrier. It is a beautiful example of quantum mechanical tunnelling. The three expressions, *i.e.* Eqs. 9.17, 9.18 and 9.19 differ only slightly, in the final determination of λ. As for example, the coefficient of e^{-2C} in the three expression differ from each other by not more 15–20%.

We have neglected, in the above discussion some factors, which are significant.

(*i*) We have assumed the nuclear potential to be square well, which realistically has a diffused boundary. For heavy nuclei, deformations may set in, and the boundary may not even be spherically symmetrical. This, however, introduces an error of only a few per cent.

(*ii*) We have only considered $L = 0$. If we include, the higher values of L's, the value of λ_0 compared to λ_L could be higher by 75% for $L = 1$; 98% for $L = 2$; 75% for $L = 3$, etc.

(*iii*) The Geiger-Nuttal[8] law is written as:

$$\log \lambda = a + bv^{-n} \ (n = 1 \text{ or } 2) \quad ...(9.20)$$

It is explained quite well by the theoretical expressions; Eqs. 9.17–9.19. It is, however, interesting that while the experimental curves, Fig. 9.6, show that straight lines can be drawn (more or less) through species of constant isotopic number (A–$2Z$), while the theory predicts that a and b will be constant only for constant Z.

(*iv*) It is easy to see from Eqs. 9.17 to 9.19, that U, the value of the potential depth and R, the radius of the potential play an important role in the determination of the values of λ, the transition probability of alpha decay. But the shape of the potential-well may not be square and U and R will have to be defined, there, in a particular manner.

In practice, one determines experimentally the values of E and λ and calculates the effective values of U and R. It can be seen that the dominant term in the expression for λ is the barrier penetration term (e^{-2C}), which is independent of U and hence λ is less sensitive to U and more sensitive to R. Hence large errors in U, can be tolerated. Also the calculated value of R from the experimental value of λ, gives the radius of the product nucleus, ignoring the alpha particle radius. Therefore, the value of R obtained semi-empirically may be considered the sum of the values for residual nucleus and alpha particles.

While discussing the above theory, we have neglected two aspects (*i*) the theory of the formulation of the alphas in the nucleus, and (*ii*) the effect of deformation in the nuclear potential as expected in deformed nuclei.

The mechanism for alpha particle formation inside the nucleus, was extensively dealt with, by Tolhock and Brussaard[12], in 1955. They considered the alpha particles as formed from nucleons in outer orbits, with the inner part acting only as the origin of potential-well, without exchanging energy with the alpha particle. They calculated the probability P_α, of 2 neutrons and 2 protons, combining together to form an alpha particle by taking the component of the wave-function, which represents the wave function of an alpha particle with the same total energy as total kinetic energy of 4 nucleons, *i.e.*, $E = 2E_p + 2E_n + E_x$, where E_x is the binding energy of the alpha particle inside the nucleus. Then the probability P_α was estimated by considering that the alpha particle is formed when four nucleons are within the alpha particle radius r_α. With the rough assumption that the nucleon wave functions are constant over the nuclear volume (a sphere of radius R); the final value used for the probability of alpha formation was then taken to be $n_\alpha P_\alpha$, with

$$P_\alpha = 64\left(\frac{r_\alpha}{R}\right)^9 \qquad \text{...(9.21}a\text{)}$$

where n_α is the number of ways in which an alpha particle can be formed from all the nucleons in outer orbits. The value of n_α was estimated to be 3. A value of $P_\alpha = 1.4 \times 10^{-4}$ has been semi-empirically accepted from the decay data and electron scattering from Po^{214}; with $r_\alpha = 1.6 \times 10^{-13}$ cm (Ref. 13).

The effect of deformation has been calculated by many authors since fifties: Rasmussan[14] (1953, 56), Preston[15] (1958), Fröman[16] (1957). We will not go in details of the theory for which the reader should see the original references mentioned above, and reference (17). We only mention that now we use a potential $V(r)$ which takes into account the asymmetry of the nucleus so that,

$$V(r, \theta) = V_0(r) + \sum_{\lambda=2}^{\infty} V_\lambda(r)\, Y_{1,0}(\theta)$$

$$\text{where} \quad V_0(r) = \frac{Z_1 Z_2 e^2}{r^2} \qquad \text{...(9.21}b\text{)}$$

It should be noted that $\lambda = 0$ corresponds to, no deformation and odd values of λ, which correspond to asymmetry of reflection, is not permissible in nuclei.

One can define the deformation parameters β_λ, so that for small values of β_λ, and uniformly charges nucleus; $V_\lambda(r)$ can be related to β_λ as:

$$V_\lambda(r) = \frac{Z_1 Z_2}{r^{\lambda+1}} \times \frac{3 R_0^\lambda \beta_\lambda}{2\lambda + 1} \qquad \text{...(9.21}c\text{)}$$

for $\lambda = 2$,

$$V_2(r) = \left(\frac{\pi}{5}\right)^{1/2} \frac{Z_1 e^2}{r^3} Q_0$$

where Q_0 is the intrinsic quadrupole moment of the concerned nucleus. Equation 2.126 relates β with the intrinsic quadrupole[17] moment.

Typical values of β_λ, form various experiments of scattering and α-decay have been found to be $\beta_2 = 0.26$; $\beta_4 = 0.041$ for thorium; $\beta_2 = 0.26$ and $\beta_4 = 0.029$ for uranium, etc. Even higher values of β_6 and β_8 may be determined by precise comparisons.

Though the theory of alpha decay is well understood, it requires further development if one has to include the effect of nuclear deformation in the theory. Recently[18] a microscopic description of alpha decay has been developed, using the framework of Hartee Fock-Bogoliubov approximations for super-deformed nuclei. This requires the solution of the coupled system of equation describing the motion of alpha particles in the deformed potential. Such an equation is written as:

$$\left\{-\frac{\hbar^2}{2M}\frac{d^2}{dr^2} + \frac{\hbar^2}{2}\frac{l(l+1)}{M_\alpha r^2}\right\} g_1(r) + \sum_{1'} V_{11'}\, g_{1'}(r) = E_\alpha\, g_1(r) \qquad \text{...(9.22)}$$

where $V_{11'}(r)$ is the matrix element connecting the channels 1 and 1′. It seems WKB approximation method for deformation larger than $\beta = 0.3$, are not valid. So Hartree-Fock-Bogoliubov (HFB) method was employed. One result of this calculation is that alpha-decay which forms the head of a super deformed band in Pb^{192} to the corresponding states in Ag^{188} is 14 order of magnitude larger than the corresponding probability from ground state to ground state. This, therefore, predicts the decay of alpha-decay transition from super deformed bands.

On the experimental side, there have been many measurements[19] of lifetime of Po, Ra, Rh and Th nuclei with $N \geq 134$ from which a systematics of reduced widths (δ^2) for various neutron numbers could be built. Figure 9.9 shows the plot of alpha decay reduced widths. A magic effect is evident. [For the definition of reduced width of a level see Chapter 13, Eqs. 13.14 and 13.15. The quantity δ^2 has the same meaning as Γ_i^r in Chapter 13].

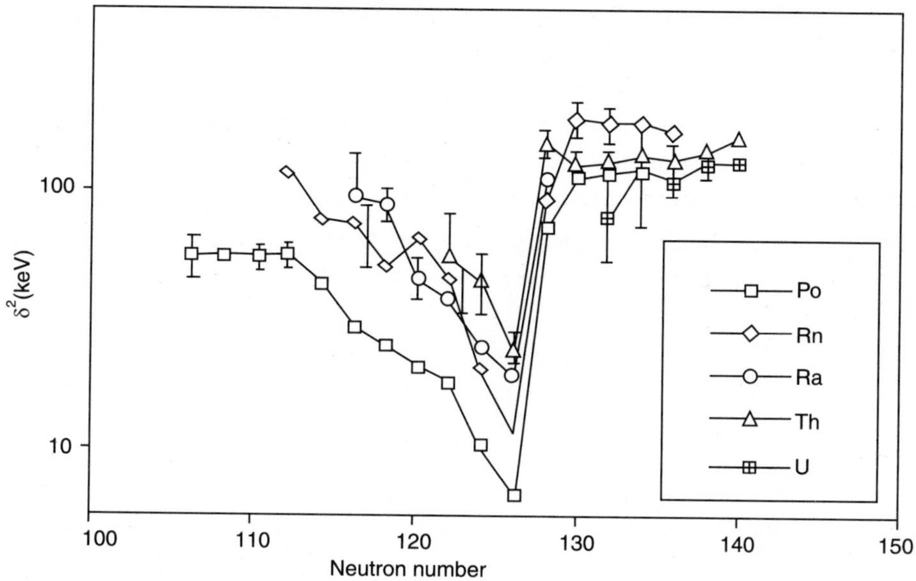

Fig. 9.9 Plot of α-decay reduced widths (δ^2) versus neutron number for even-even Po, Ra, Rn, Th nuclei with N ≤ 134 (Ref. 19).

9.4 CLUSTER DECAY

We, generally, consider radioactive decay of nuclei, througth alpha and beta decay, with subsequent emission of gamma rays in many cases. Also it is well known since 1939, that many radioactive nuclei decay through spontaneous fission. However, only very recently, [20, 22] it has been experimentally found that many high Z nuclei decay through the emission of particles heavier than alphas—say C^{14}, Ne^{24}, $Mg^{30}$and Si^{34}, with the lifetimes of the order of some years to 10^3 years to 10^{19} years.

The first experimental identification of a case of radioactive decay, through the emission of such comparatively heavy fragments, was accomplished by Rose and Jones[21] from Oxford University in 1984. They observed the radioactive decay of Ra^{223} by C^{14} emission with a half life of $T_{1/2} = 3.7 \pm 1.1$ years.

However, the first communication in a presentation in a conference was as early as 1975–77, by a group from Brazil, [20, 22] where in the spontaneous nuclear disintegration of U^{238} by the emission of large clusters in the region from neon to nickel was observed. Since 1984, many cases of such decays by heavy clusters have been observed and identified. The nuclear detectors used for such measurements have ranged from Polycarbonate track recording films, loaded with different radioactive heavy nuclei, to solid state $\Delta E - E$ telescopes and magnetic spectrometers. Table 9.1 gives a summary of the present status of the experimentally observed and identified cases of radioactive nuclei decaying through various clusters. It is interesting to note, that all these decays, end up with nuclei, which have either protons in the closed shell, *i.e.* Z-82 corresponding to lead isotopes or neutrons in closed shell, *i.e.* $N = 126$ corresponding to say Hg^{206} ($Z = 80$, $N = 126$), or Tl^{207} ($Z = 81$, $N = 126$). The comparison of half lives due to cluster decay with alpha decay from the same parent nuclei, shows a factor of $10^{-10} - 10^{-15}$, of the probability of cluster decay compared to alpha decays. This is a case of high experimental skills, in detecting the clusters. The number of events, corresponding to the emission of a specific cluster, is very small, ranging from say three or four to 20–30, over a period of several days. It is due to increased sensitivity of the mehods to measure the ratio of charge and mass of the emitted particles, that has made these experiments possible in recent[28], times; since the first experiments[22] in 1975–77.

Table 9.1*a* *Summary of the observed decay modes of cluster emission and experimental details*

S. No.	*Decay Mode*	*Detector*	*Reference*
1.	${}_{87}Fr^{21} \xrightarrow{C^{14}} {}_{81}Tl^{207}$	Film	Price et al.[23]
2.	${}_{88}Ra^{221} \xrightarrow{C^{14}} {}_{82}Pb^{207}$	Film	Same as above[23]
3.	${}_{88}Ra^{222} \xrightarrow{C^{14}} {}_{82}Pb^{208}$	Film	Same as above[23]
4.	${}_{88}Ra^{224} \xrightarrow{C^{14}} {}_{82}Pb^{210}$	Mag Sp.	Same as above[23]
5.	${}_{88}Ra^{226} \xrightarrow{C^{14}} {}_{82}Pb^{212}$	Mag Sp.	Hourani et al.[24]
6.	${}_{90}Th^{230} \xrightarrow{Ne^{24}} {}_{80}Hg^{206}$	Film	Tretyakova et al.[25]
7.	${}_{91}Pa^{231} \xrightarrow{Ne^{24}} {}_{81}Tl^{207}$	Film	Samdulescu et al.[26]
8.	${}_{92}U^{232} \xrightarrow{Ne^{24}} {}_{82}Pb^{208}$	Film	Barwick et al.[27]
9.	${}_{92}U^{233} \xrightarrow{Ne^{25,\,28}} {}_{82}Pb^{208,\,209}$	Film	Tretyakova et al.[28]
10.	${}_{93}Np^{237} \xrightarrow{Mg^{30}} {}_{81}Tl^{207}$	Film	Tretyakova[28]

Table 9.1*b*

Decay	*Experimental log $T_{1/2}$*	*Calculated half lives (log $T_{1/2}$)*			
		Poenaru[30]	*De Carvillo*[31]	*Bleadowske*[32]	*Gupta*[33]
$Ra^{222} \xrightarrow{C^{14}} Pb^{208}$	10.9–11.1	12.6	12.4	11.0	11.2
$Ra^{223} \xrightarrow{C^{14}} Pb^{209}$	14.9–15.5	14.8	14.5	15.2	14.1
$Ra^{224} \xrightarrow{C^{14}} Pb^{210}$	15.8–16.0	17.4	17.1	15.9	15.0
$U^{232} \xrightarrow{Ne^{24}} Pb^{208}$	21.3–21.5	20.4	–	–	16.5

9.5 THEORY OF CLUSTER DECAY

There are two possible theoretical approaches for understanding the disintegration of radioactive nuclei through the emission of clusters. (*i*) Super asymmetric fission as a dynamical mass fragmentation process and (*ii*) Emission of heavy cluster, through the Coulomb barrier, similar to Gamow's theory of alpha decay; including the theory of the formation of the cluster in the nucleus. It is interesting, that already in1980, Samdulescu[29] et al. had predicted such cluster decays, using both the above approaches. Since then, this group (Poenaru, Greiner, Ivasco and Yi-Jin-Shi et al.[30]) have calculated the half lives for Ra^{223}, Ra^{276}, Ac^{227}, via Ne^{24} emission, and for U^{238} via Si^{34} emission, and for Cf^{252} via Ar^{46} emission. Also some other authors (De Carvillo[31], Bleadowske[32], and Gupta[33]) have carried out similar calculations using one or other modes of cluster decay.

We give, in brief, the theory of these two approaches:

Model of Cluster Formation and Decay: This model is very similar to Gamow's theory of alpha decay, except that the calculations of pre-formation of the clusters in the nucleus, is calculated somewhat differently. We look at cluster decay as a two-step mechanism of (*i*) formation of the fragment (the cluster and the daughter nucleus), in their ground state with probability P_0 and (*ii*) Impinging on the confining nuclear interaction barrier with frequency ν and tunnelling through it with probability *P*. Then we define the decay constant as:

$$\lambda = P_0 \; \nu P \qquad ...(9.23)$$

We will calculate these three factors now, following De Carvillo[31] et al. for ν and *P* and Gupta[33] et al. and Maruhn and Greiner[34] for P_0.

(*i*) *Frequency Factor* ν: The calculation of ν can be accomplished in the same manner as we did for alpha decay, *i.e.*,

$$\nu = \frac{v}{2R} \qquad ...(9.24)$$

For the case of cluster decay, *v* should be the relative velocity between the cluster and the daughter nucleus, and can be written as:

$$v = v_1 + v_2$$

where v_1 and v_2, are the velocities of the cluster and daughter nucleus respectively, in the C.M. system. Similarly, let $R = R_1 + R_2$, where R_1 and R_2 are the radii of the cluster and daughter nucleus.

We may then, write:

$$\nu = \frac{|v_1 + v_2|}{2(R_1 + R_2)} = \frac{(2Q/\mu)^{1/2}}{2(R_1 + R_2)} \qquad ...(9.25)$$

where *Q* is the *Q*-value of the reaction, and μ is the reduced mass given by:

$$\mu = \frac{M_1 \, M_2}{M_1 + M_2}$$

where M_1 and M_2 are masses of the cluster and daughter nucleus.

The value of v comes out to be $10^{21} - 10^{22}$/sec.

(*ii*) *The Penetrability or Tunnelling Probability P:* Again referring to alpha decay, P is calculated similary and is expressed similar to Eqs. 9.17 and 9.18, as:

$$P = \exp(-G)$$

where $$G = \frac{2}{\hbar}\int_c^d \{2\mu[V(r) - Q]\}^{\frac{1}{2}}\, dr \qquad ...(9.26)$$

is called the Gamow factor; $C = R_1 + R_2$ is the inner turning point and d corresponds to R, in $d = Z_1 Z_2 e^2/R$, and represents outer turning point. The symbol, e, of course, stands for the electronic charge. The Gamow factor G is similar to $2C$ or Eqs. 9.17 and 9.19. In actual practice, the potential $V(r)$ is written as (for the inner turning point):

$$V(r) = \frac{Z_1 Z_2 e^2}{r} + V_p \quad \text{for } r \geq C \qquad ...(9.27)$$

where V_p is called the proximity potential; which physically incorporates the nuclear effects, due to nuclear surface tension. Figure 9.10 gives semi-empirically[18] determined shape of $V(r)$. In practice, the expression for P is obtained in two steps, *i.e.*,

$$\exp(-G) = P = P_i P_b \qquad ...(9.28)$$

where $$P_i = \exp\left(\frac{-2}{\hbar}\right)\int_c^{R_i} \{2\mu[V(r) - V(R_i)]\}^{\frac{1}{2}}\, dr$$

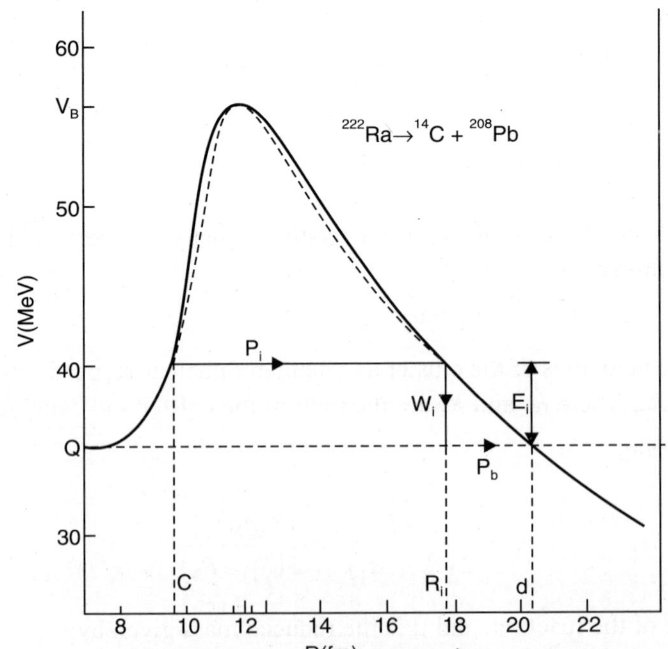

Fig. 9.10 Empirically determined shape of V (η) (Ref. 18, 23).

and $$P_b = \exp\left(\frac{-2}{\hbar}\right)\int_{R_i}^{d}\{2\mu[V(r)-Q]\}^{\frac{1}{2}}\,dr \qquad ...(9.29)$$

The Gamow factor G, as given in Eq. 9.28, has been calculated both for alphas and cluster decay. The typical values of the ratio, of Gamow factors, for cluster decay to alpha decays varies from 10^{-11} to 10^{-13}.

(*iii*) *Pre-formation Probability* P_o: This quantity has been calculated by many authors, [32, 33, 35] using different approaches. We, however, follow the approach by Gupta and co-workers[33].

We define the pre-formation probability of clusters in a nucleus as a quantum-mechanical probability of finding the fragments A_1 and A_2 (with fixed charges Z_1 and Z_2 respectively) at a point of the relative motion. For this purpose, we use the mass and charges asymmetry coordinates, defined as:

$$\eta \equiv \frac{A_1 - A_2}{A_1 + A_2} \text{ and } \eta_Z \equiv \frac{Z_1 - Z_2}{Z}\,;\; Z = Z_1 + Z_2 \qquad ...(9.30)$$

we, then solve the Schrödinger equation in η; at fixed η_Z and $r = R$, *i.e.*,

$$\left[\frac{-\hbar^2}{2\sqrt{B_{\eta\eta}}}\frac{\partial}{\partial\eta}\frac{1}{\sqrt{B_{\eta\eta}}}\frac{\partial}{\partial\eta} + V(\eta)\right]\psi^{(v)}_{R_{\eta Z}}(\eta) = E^{(v)}_{R}\,\psi_{R_{\eta Z}}(\eta) \qquad ...(9.31)$$

Approximately, Eq. 9.31 determines the wave function of mass parameter in terms of mass asymmetry. In Eq. 9.31, the mass parameter $B_{\eta\eta}$ behaves like mass and is formally defined according to BCS formalism[33]. We will not go in details about the derivation of this parameter, for which one should consult literature [33, 34, 38]. Then after proper scaling and normalising the solution of Eq. 9.31, we can write the cluster-formation probability P_o as:

$$P_o = \left|\psi^{(0)}_{R_{\eta Z}}(\eta)\right|^2 \sqrt{B_{\eta\eta}(\eta)}\,\frac{4}{A} \qquad ...(9.32)$$

In Eq. 9.31, $V(\eta)$ is the fragmentation potential and is defined as the sum of experimental binding energies, $B_i(A_i, Z_i)$, Coulomb interaction and proximity potential V_p, *i.e.*,

$$V(\eta, R) = \sum_{i=1}^{2} B_i(A_i, Z_i) + \frac{Z_1 Z_2 e^2}{r} + V_p \qquad ...(9.33)$$

In Eq. 9.33, the charge Z are fixed by minimising the sum of two binding energies in charge asymmetry. We have not used the angular momentum dependent term $E_1 = \hbar^2\, l\,(l+1)\,/\,2\,\mu R$ since l-values involved are small ($\approx 5\hbar$) whose contribution to life times have been shown to be small as in alpha decay. The proximity potential V_p is given in literature[34, 36]. The values of mass parameters $B_{\eta\eta}$ have been calculated by Kroger and Scheid[37] and closed expressions for the same have been found. The ratio of Pre-formation P_o for clusters and alphas has been calculated by many authors and has been found to vary from 10^{-7} to 10^{-13}. For a typical case for Ra^{223} decaying through alpha and C^{12} decay, the value of P_o for alphas is 10^{-8} and for C^{14} is 4×10^{-16}; so that the ratio P_o (cluster)/P_o (alphas) $\approx 5 \times 10^{8}$.

An alternative method of cluster decay has been developed, on the lines of the theories of spontaneous fission, which have been discussed since fifties, and been recently formulated properly to give quantitative results, which agree with the experiments reasonably. The Analytical Super-Asymmetrical Fission Model (ASAFM) is based on these later formulations. This model gives an analytical expression for half life, calculated as the WKB penetration probability through a barrier $E(R)$, approximated by a second order polynomial in separation distance R, for the overlap of two spheres from the parent nucleus (Z, A) to the touching sphere. For detailed discussion *see* Ref. (30) and (41).

9.6 SPONTANEOUS FISSION

Nuclear fission of uranium was first discovered experimentally by O. Hahn and F. Strassmann[42], in 1939 and was explained by Bohr and Wheeler[39] in the same year by using a model of deformation of the nucleus—considering it like a liquid-drop, which under conditions of extreme instability becomes like a dumb-bell; as shown in Fig. 9.11*c*. Then still greater extension will occur, with electrostatic energy being set free faster than the consumption of energy in the increase of the surface. The movement thus accelerates till the nucleus breaks into two or more parts, *i.e.* the spontaneous fission has taken place.

However, a detailed first unified picture of the experimental and theoretical explanation of the phenomenon of spontaneous fission was given by D.L. Hill and J.A. Wheeler[39] in 1953, based on the collective motion of the nuclear fluid and a hydrodynamic model of the mass asymmetry and its variation with energy was proposed.

Two important experimental facts have been known since fifties, *i.e.* fragment mass-distribution for thermal neutrons[43] and spontaneous fission half-life[44] against Z^2; as shown in Figs. 9.11*a* and 9.11*b*. The analogy of Fig. 9.11*b* with α -decay case, Fig. 9.8, is quite evident showing that penetrability factor in both cases is based on similar considerations.

Theory of Spontaneous Fission: Figures 9.11*a* and 9.11*b* form the basis of barrier penetration theory of Bohr and Wheeler[39] and later Hill and Wheeler[39]. Basically a deformed or a spherical nucleus, belonging to heavy nucleus, say $A > 100$, has a binding energy per nucleon B/A, which decreases with A. [*see* Fig. 2.4]. For $A = 238$, say for U^{238}; $B/A = 7.6$ MeV. If such a nucleus is divided into two halves of $A \approx 119$; $B/A = 8.5$ MeV, and some 0.9 MeV/A is released, making it 210 MeV for the total fission.

The basic equation, which can be used to understand the energetics in the mechanism of spontaneous fission is the binding energy Equation 2.15; based on liquid drop model. If energy-wise; fission releases the extra energy as described earlier, the following scenario can develop, for spontaneous fission : A nucleus in the ground state, has two energy terms [Eq. 2.15], which are distance-dependent, *i.e.* (*i*) E_C for Coulomb energy. Equation 2.17 and (*ii*) E_S for surface energy

Adding them gives

$$E = E_C + E_S = \frac{3}{5}\frac{e^2}{r_0}\left(\frac{Z^2}{A^{1/3}}\right) + a_2\, A^{2/3} \qquad ...(9.34)$$

where a_2 is surface tension coefficient for the nucleus. Writing $a_2 = 4\pi\, r_0^2\, S$, where S is surface tension per unit area; one can write Eq. 9.34 as:

$$E = 4\pi\, r_0^2\, SA^{2/3} + \frac{3}{5}\frac{(Ze)^2}{\left(r_0\, A^{1/3}\right)} \qquad ...(9.35)$$

In the case of spontaneous fission, a spherical nucleus of radius r_0, goes through various stages of deformation and fission as shown in Fig. 9.11*a*; due to oscillations induced by the internal motion of nucleons. Such oscillations can also be induced by reaction with thermal neutrons. The shape of distorted surface of a spheroidal nucleus be described in term of two coordinates r and θ as:

$$r = R\left[1 + a_2\, P_2\,(\cos\theta) + a_3\, P_3\,(\cos\theta) + \cdots\right] \qquad ...(9.36)$$

where a's are small numbers, which determine the distortion. Expanding the Legendre Polynomials, and replacing r_0 in Eq. 9.35 by r and by integration of the mutual Coulomb energy of all pairs of charge elements contained within the distorted surface, Bohr and Wheeler, obtained the expression of Coulomb plus surface energy as:

$$E = 4\pi\, r_0^2\, A^{2/3}\left(1 + \frac{2a_2^2}{5} + \frac{5a_3^2}{7} + \cdots\right)S$$

$$+ \left[\frac{3\,(Ze)^2}{5r_0\, A^{1/3}}\right]^2 \left[1 - \frac{a_2^2}{5} - 10\,\frac{a_3^2}{49} + \cdots\right] \qquad ...(9.37)$$

One should realise that, as the nucleus becomes ellipsoidal from spherical, Coulomb energy decreases (because charges are further apart on the average) only slowly because of the long range nature of Coulomb forces; but the surface energy increases, because of the increase in surface area. When the changes in attractive surface energy and repulsive Coulomb energy are equal, are balanced or $\Delta E_S = \Delta E$ coulomb.

i.e.,
$$\frac{3}{5}\frac{Z^2\, e^2}{r_0\, A^{1/3}}\left(\frac{1}{5}\, a_2^2\right) = S\, A^{2/3}\left(4\pi\, r_0^2\right)\left(\frac{2}{5}\, a_2^2\right)$$

or
$$\left(\frac{Z^2}{A}\right) \text{limit} = 2 \times \frac{S \times 4\pi\, r_0^3}{\frac{3}{5}e^2} = 45 \qquad ...(9.38)$$

then, these two forces are balanced and nucleus is stable for small deformation. But for large deformation the long range repulsive Coulomb force has a greater advantage over the short-range attractive forces due to surface tension. Hence for $Z^2 / A \le 44$ if the nucleus is nudged enough, it goes on becoming more and more ellipsoidal, till the fission takes place. For spontaneous fission, this nudging is provided by the internal clusters tunnelling through the Coulomb potential of each other. For $Z^2 / A \ge 45$ the nucleus is instantaneously fissile. Figure 9.12*a* gives experimental cases of spontaneous fission between $Z^2/A = 35$ and $Z^2 / A = 40$.

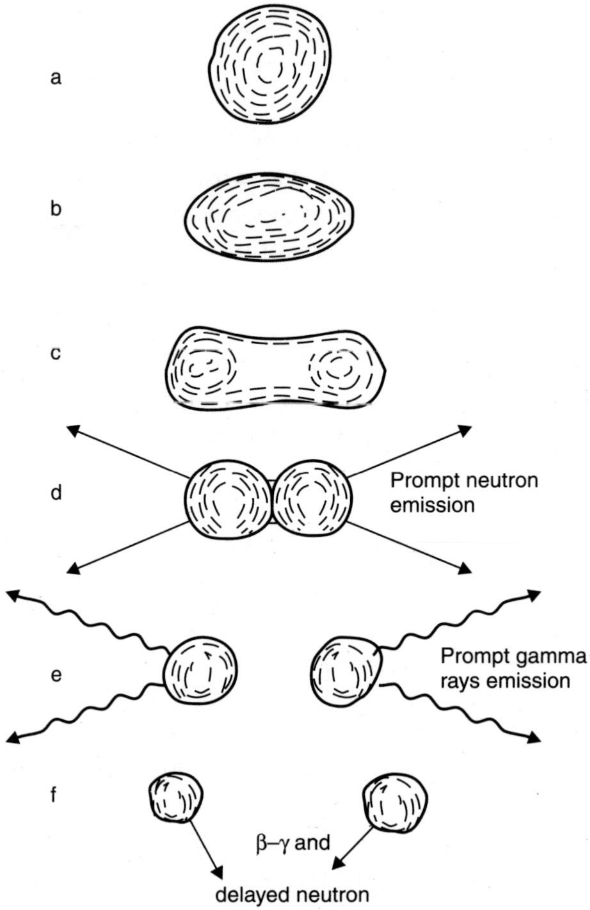

Fig. 9.11a Different stages of nucleus, undergoing fission—stage (*a*) (a sphere); (*b*) (a spheroid); (*c*) (a dumb-bell); (*d*) (near separation); (*e*) (separated); (*f*) (further apart).

Hill and Wheeler Model: Till now, we have basically discussed the energetics leading to fission, based on the classical analysis of liquid drop model; as was originally suggested by Bohr and Wheeler[39] in 1939. However, the fissioning of a nucleus requires, the quantum mechanical treatment, in which the nucleus, after being disturbed from its equilibriums position, say as a sphere, either due to internal motions of nucleons; which will be the case of spontaneous fission; or due to say the incidence of a thermal neutron, as will be the case of induced fission. The disturbed nucleus goes on progressing through various stages of distortion; as shown in Figs. 9.11*a* and 9.11*b*. Hill and Wheeler, developed a quantum mechanical theory, in which, the probability of leakage of any one of the fragments, through Coulomb potential can be calculated, by writing out an appropriate Schrödinger equations for such a case.

Assuming one-dimensional inverted harmonic oscillator, one writes the Schrödinger equation as:

$$H\psi = E\psi \qquad ...(9.39)$$

where

$$H = -\frac{\hbar^2}{2B}\frac{\partial^2}{\partial\alpha_1^2} - \frac{1}{2}B\omega_1^2\,\alpha_1^2 + E_f \qquad ...(9.40)$$

where α_1 is the distortion-parameter, connecting the radius R_0 of the undeformed spherical drop, to

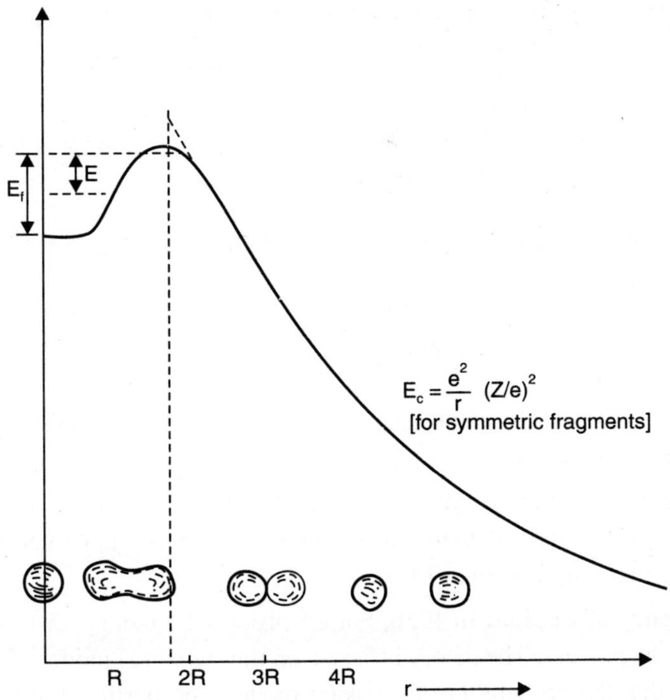

Fig. 9.11b Potential energy barrier, opposing the spontaneous fission. Different stages of evolution of fission at different radial positions of potential energy curve, given approximately.

the radius of deformed nucleus R, as:

$$R(\theta) = R_0 \left[1 + \sum_{1=0}^{\infty} \alpha_1 \, P_1 (\cos \theta)\right] \quad \text{...(9.41)}$$

In Eq. 9.40, B is an inertia factor for the undistorted nucleus, so that the first-term in Eq. 9.40 corresponds to the kinetic energy term due to shape-distortion and second term corresponds to potential due to distortion as in a harmonic oscillator equation and E_f is square well barrier. Solving the Schrödinger Equation 9.39, one obtains the penetration factor p, in a somewhat similar manner, as in alpha decay, Eq. 9.18 or cluster decay, Eq. 9.26. They obtained the penetration factor P as:

$$P = \left\{1 + \exp\left[\frac{-2\pi (E - E_f)}{\hbar \, \omega_L}\right]\right\}^{-1} \quad \text{...(9.42)}$$

One can calculate the decay lifetime, for spontaneous fission from this, which have been calculated by Frankel and Metropolis[45] as:

$$t_0 = 10^{-21} \times 10^{7.85 E_f} \text{ secs}. \quad \text{...(9.43)}$$

The lifetimes of spontaneous fission, which have been found experimentally[1, 44], and some of which are given in Fig. 9.12 *a* are:

U^{235} (≈ 2 × 10^{17} yrs); U^{234}, U^{238}, Pu^{239}, U^{236} ($10^4 - 10^9$ yrs), Pu^{240} (10^{11} yrs), Cm^{248} (≈ 4.7 × 10^5 yrs), Cm^{50} (≈ 10^5 yrs), Cf^{252} (≈ 2.65 yrs), Cf^{254} (55 days) and Fm^{256} (2.7 hrs).

Theoretical values of lifetimes, obtained, say from Eq. 9.42, do not match with the experimental values, if E_f is calculated from classical theory of liquid drop; according to which

$$E_f = 4\pi\, r_0^2\, S\, A^{2/3}\, f(x) \quad ...(9.44)$$

where $f(x)$ is a semi-empirical function depending on x, which is given by:

$$x = \frac{Z^2/A}{\left(Z^2/A\right)_{\text{limit}}} \quad ...(9.45)$$

When $x < 1$, the nucleus is stable against spontaneous fission, and for $x > 1$, the nucleus is unstable against spontaneous fission.

The disagreement of experimental lifetimes with theory of Hill and Wheeler as used by many authors, shows, that E_f has to be calculated differently. Also theory does not predict the asymmetric fission, as observed experimentally, Fig. 9.12*b*.

Statistical Theory of Fission: In 1956, Fong[46] proposed a theory for fission which predicts the asymmetric nature of the process. The theory is based on the statistical model of the compound nucleus, with the assumption, that the probability of a fission mode is proportional to the density of quantum states. We give below, a semi-qualitative description of theory of Fong.

As will be discussed in Chapter 13 on compound nucleus model, the level density W_o (E) of nucleus of the mass number A, excited by energy E, can be written as:

$$W_o(E) = C \exp 2\sqrt{a\,E} \quad ...(9.46)$$

For excitation of the nucleus in fission model, let the two fragments have energies of excitation E_1 and E_2. Hence the density of quantum states of the two fragment-nucleus is given by:

$$N = C_1 \exp\left(2\sqrt{a_1\,E_1}\right).\, C_2 \exp\left(2\sqrt{a_2\,E_2}\right) \quad ...(9.47)$$

As the two fragments are in contact, they will have the same temperature. Then

$$E_1 : E_2 = a_1 T^2 : a_2 T^2 = a_1 : a_2 \quad ...(9.48)$$

writing $E = E_1 + E_2$, we can write:

$$N = C_1\, C_2 \exp\left\{2\sqrt{(a_1 + a_2)\,E}\right\} \quad ...(9.49)$$

For symmetric fission, $a_1 = a_2 = a_0$. One can, then write the expression for the ratio of probabilities of asymmetric to symmetric fission P/P_0; as:

$$\frac{P}{P_0} = \frac{N}{N_0} = \frac{C_1\,C_2}{C_0\,C_0} \exp\left[2\sqrt{2a_0}\left(\sqrt{E} - \sqrt{E_0}\right)\right] \quad ...(9.50)$$

where C_0, a_0 and E_0 correspond to symmetric fission and C_1 and C_2 and E correspond to the asymmetric fission.

Similarly the ratio of probabilities for two different modes A and B for fragments can be written as:

$$\frac{P_A}{P_B} = \frac{(C_1\, C_2)_A}{(C_1\, C_2)_B} \exp\left[2\sqrt{2a_0}\left(\sqrt{E_A} - \sqrt{E_B}\right)\right] \quad \text{...(9.51)}$$

$$\approx \exp\left[2\sqrt{2a_0}\left(\sqrt{E_A} - \sqrt{E_B}\right)\right] \quad \text{...(9.52)}$$

because C_1 and C_2 are slowly varying factors and can be taken to be the same for two fragments. Now E, the excitation energy of a given fragment can be written as:

$$E = M^* (A, Z) - M (A_1, Z_1) - M (A_2, Z_2) - K - D \quad \text{...(9.53)}$$

where M^* (A, Z) is the mass of the excited compound nucleus undergoing fission, M (A_1, Z_1) and M (A_2, Z_2) are the masses of two fragments, in their ground state, K is the total kinetic energy of the fragments and D is the total deformation energy of the fragments. It can be seen from Eq. 9.53 that the excitation energy is different for different target nuclei for different incident particles and for different incident-energies. One can calculate the values of E, for different fragments from this formula.

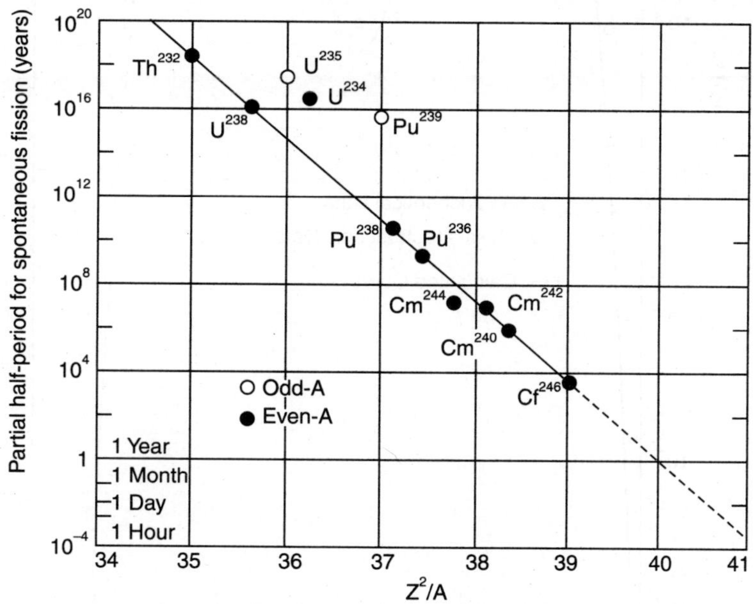

Fig. 9.12a Half life periods of nuclei, which undergo spontaneous fission, plotted on log scale, versus Z^2/A (Ref. 44).

Fong[46] has derived the mass distribution curve for U^{235} + ${}_0n^1$ fission and compared it, with the experimental result. The agreement is good. However results are not so fitting the experimental values when incident neutrons are in the energy range of a few MeV; as shown by Perring and Story[47] for Pu^{239}. Theory gave four peaks, instead of the experimental two peaks. Figure 9.12*b* gives the comparison for thermal incident neutrons for U^{235}, which is extremely good.

Recently Greiner and his colleagues[33] have developed the so-called fragmentation theory; allowing the calculation of the mass-distribution in Uranium isotopes[43]. Similarly many workers have explained the asymmetry of fission fragments in the mass-distribution curve, using the shell effects[40], Maruhn and

Greiner[34] have, afterwards, calculated the mass asymmetry in fission, using mass-asymmetry vibrations in the final stage of the fission process, with an approximate treatment of the coupling to relative masses, somewhat akin to the treatment by Gupta, Scheid and Greiner earlier[33] for cluster decay.

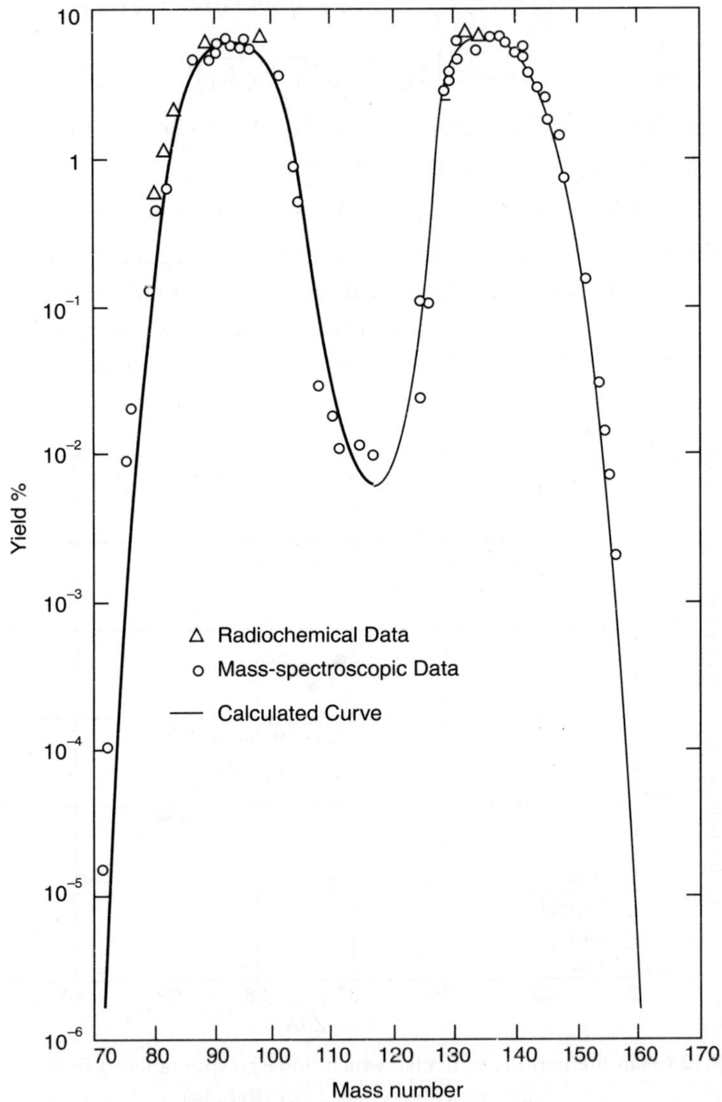

Fig. 9.12b Calculated mass-distribution curve of fission products in the case of U^{235} + thermal neutrons compared with experimental data (Ref. 46).

Fragmentation Theory of Fission: We give below briefly, the theory of spontaneous fission based on the work of Greiner[33], Gupta[34] and their co-workers. We start with a similar equation as in Eq. 9.31; for the collective motion of two fragments A_1 and A_2 (following Maruhn and Greiner[32]) and write:

$$\left[-\frac{\hbar^2}{2\sqrt{B}}\frac{\partial}{\partial Z}\frac{1}{\sqrt{B}}\frac{\partial}{\partial Z}+V(\xi,\lambda)\right]\phi_\nu = E_\nu(\lambda)\,\phi_\nu(\xi) \qquad \text{...(9.54)}$$

where B is approximation of $B_{\xi\xi}$, and ξ is the same as η of Eq. 9.31, *i.e.*,

$$\xi = \frac{A_1 - A_2}{A_1 + A_2} \qquad ...(9.55)$$

and λ is an elongation parameter to be found, by comparison with an experiment. It is defined as the total length of the deformed nucleus in the units of diameter R_0 of a spherical nucleus of equal mass, *i.e.* total length of the deformed nucleus is written as $2\,R_0\,\lambda$, Fig. 9.13*a*. Our purpose, now, is to find out the distribution of mass or charge. For this purpose, one wants to calculate $B_{\xi\xi}$ versus ξ, which for a proper value of λ, will give proper mass distributions. To calculate $B_{\lambda\lambda}$, expectation value of Hamiltonian H are calculated and optimised. One writes the expected value of H as:

$$\langle H \rangle_{av} = \frac{1}{2} \langle B_{\lambda\lambda} \rangle\, \lambda^2 + \sum_{v} | a_v(\lambda) |^2\, E_v(\lambda) \qquad ...(9.56)$$

where the first term is the kinetic energy, and the second term is the energy both at the ground and excitation level, representing the potential energy. Starting with a given initial set of λ and ξ, one calculates the $B_{\lambda\lambda}$ from Eq. 9.54 and Eq. 9.56 and also $B_{\xi\xi}$ from Eq. 9.31. For details see Ref. (31). Figure 9.13*b* shows the mass parameter $B_{\xi\xi}$ for asymmetry oscillations for different values of λ. It is seen that for $\lambda = 1.80$, it very much resembles the mass-distribution curves, as found experimentally. for fission.

$$\xi = \frac{A_1 - A_2}{A_1 + A_2}$$

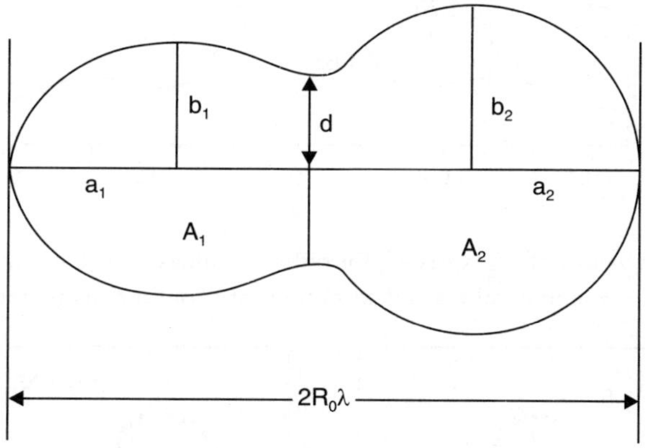

Fig. 9.13a The bell shape configuration of the two fragments in fission. Various symbols are explained in the text (Ref. 34).

In this model, the penetration probability is given by[34]:

$$P = \exp\left[-\frac{2\,m}{\hbar} \int \sum \left(|E - V|\, B_{x_i x_j}\, X_i\, X_j\right)^{1/2} dt\right] \qquad ...(9.57)$$

with $x - x(t)$, as the variable parameter of a path in the space of N coordinates. By varying the path; one may search for that one, which has highest probability and take that probability as multi-dimensional one. This gives a good approximation to a real multi-dimensional WKB theory[48]. Calculations for a large number of cases and comparison with experimental lifetimes, Fig. 9.12*a*, has to be still carried out.

In Fig. 9.13*c* are shown, the results of such calculations, where theoretical fission product-yields for U^{236}, for $\lambda = 1.80$ and 1.85, are plotted for different masses of fission fragments, and compared with experimental results[43]. The agreement seems to be qualitatively reasonable. The theoretical curves are for different excitation energies, which is related to temperatures. For fission from the ground state, temperature $\theta = 0$; (solid curve) the humps of theoretical distribution are somewhat narrower and the valley is too deep, compared to the experimental data. When higher excitations are taken into account, the theoretical curves have the valley filled up, and there is a general flattering of the distribution, making the agreement with experiment, more quantitative. In these calculations there was no fitting of the parameters for the position and height of the peaks. They were theoretical values, automatically gave proper heights, position and width of peaks.

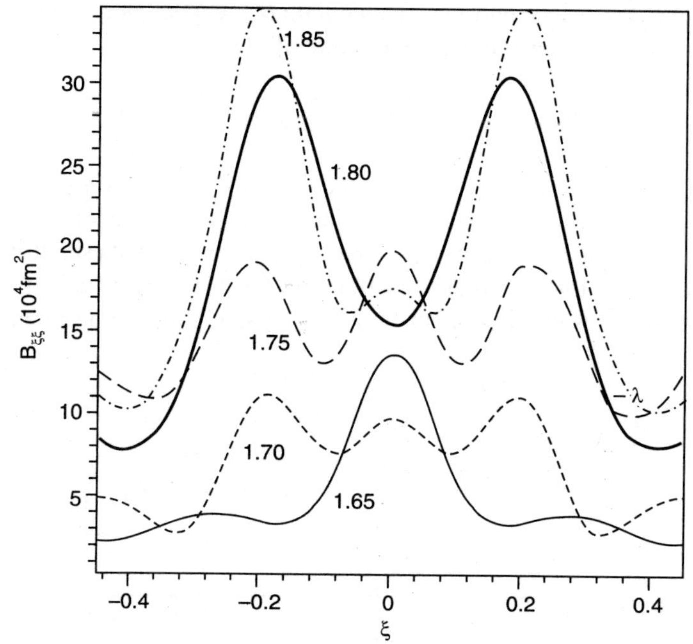

Fig. 9.13b The calculated values of $B_{\xi\xi}$, versus ξ for different values of λ. One value of λ = 1.80 gives a similar shape as experimental mass distribution, of spontaneous fission (Ref. 34).

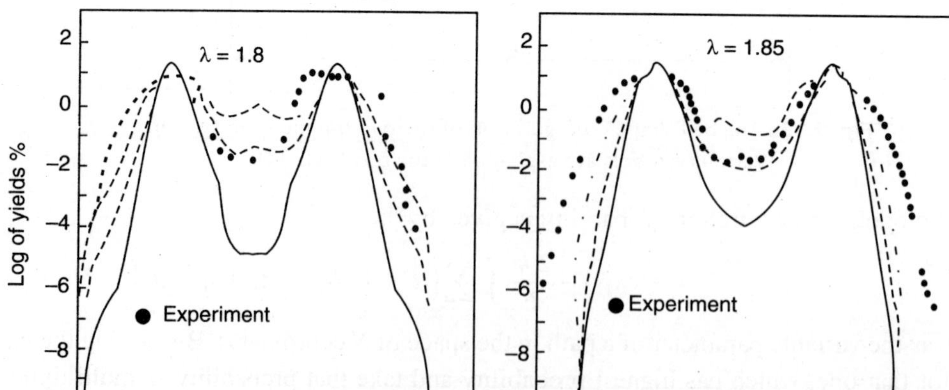

Fig. 9.13c The comparison of experimental and theoretical fission yields of U^{236}, for λ = 1.8 and 1.85. The theoretical dashed, dot-dash-dot and continuous curves correspond to different excitation energies (Ref. 34).

Emission of prompt gamma rays in the thermal neutron fission of U^{238} , and the study of their angular distribution has proved to be useful, for observing the effect of angular momenta of fragments on emission of gamma rays[49] as found experimentally by Rammanna and his group. Theory and experiments on fission-decay have entered into a phase of sophistication, only available recently. In a very interesting theoretical work by Bency John and S.K. Kataria[49], they have calculated the angular distribution of fission fragments induced in $O^{16} + Cm^{248}$, $O^{18} + Th^{232}$, $F^{19} + Pb^{208}$, $O^{18} + Bi^{209}$, $O^{16} + Pb^{208}$, $O^{18} + U^{238}$, $S^{32} + Pb^{208}$, $S^{32} + Au^{197}$, $Si^{28} + Pb^{208}$, $Mg^{24} + Pb^{208}$ and $He^4 + U^{238}$ reactions, in the energy ranges of 90 to 280 MeV of projectile except for He^4 for which the energy range was 30 to 120 MeV. These reactions have been studied for fission by many workers[50]. The authors have compared the anisotropies W (O)/W (90°), as obtained experimentally and theoretical calculation based on the assumption of a statistical equilibrium for the tilting mode at the precession point. They have included, in calculating, the population probability, the effect of the wriggling and twisting mode, on the tilting mode angular momentum.

In another theoretical study M. Mirea[51], have studied, the effect of avoided level crossing of nuclear levels, created out of Nilsson orbits for two nuclei approaching each other. One writes the wave function keeping this effect in mind while writing the expression for lifetime. Such fine effects have been compared as in Fig. 9.14 with the experimental results of the lifetimes of clusters (say alphas). This effect is called Landau-Zener effect. Figure 9.14 show clearly that Landau-Zener effect is quantitatively reproduced.

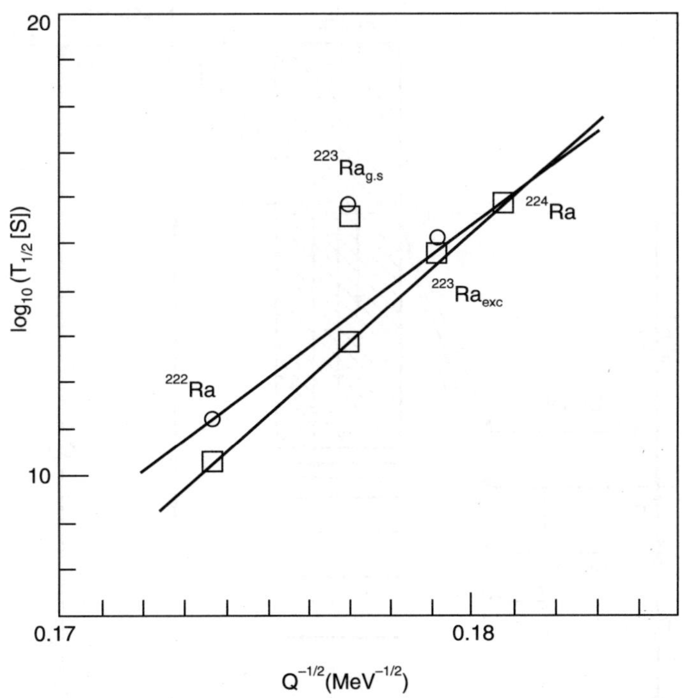

Fig. 9.14 Landau-Zener effect, using Nilsson orbits in calculating lifetimes. The experimental points (rectangles and circles) are compared, with theoretical calculations, shown as solid lines (Ref. 51).

On the experimental side[52], an interesting study has been made of the charge of fission products in 24 MeV proton-induced fission of U^{238}. The fission products were identified by γ-ray energies and lifetimes of 143 fission products, belonging to 40 mass chains, available from the measurements by authors. One could measure the yield from γ-ray intensities and determine the most probable of fission products. It was found that the most probable charge, Z_p, mainly lies on the proton rich side in the light fragment mass region and on the proton deficient side in the heavy mass region. This means that charge polarisation occurs in the fission process and the most probable charge is determined before separation.

9.7 BETA DELAYED AND SELF DELAYED, PROTON DECAY

A. Beta Delayed Proton Decay: This topic only marginally, belongs to the category of charged particle decay; in which Coulomb barrier plays an important role in determining the decay probabilities or the lifetimes, as in the case of α-decay, cluster decay or spontaneous fission, discussed in the previous sections. What is, generally, termed as proton-decay is really a case of beta-delayed proton decay, where the initial nucleus called 'precursor' decays to the immediate neighbour nucleus-called 'emitter' via electron capture or β^+ -emission with a high Q-value. If the energy E of the emitter state is greater that B_p the binding energy of proton; then these levels can undergo proton transition (emission) to the final nucleus. The emitted protons will seem delayed, because of the decay lifetime of the beta decay.

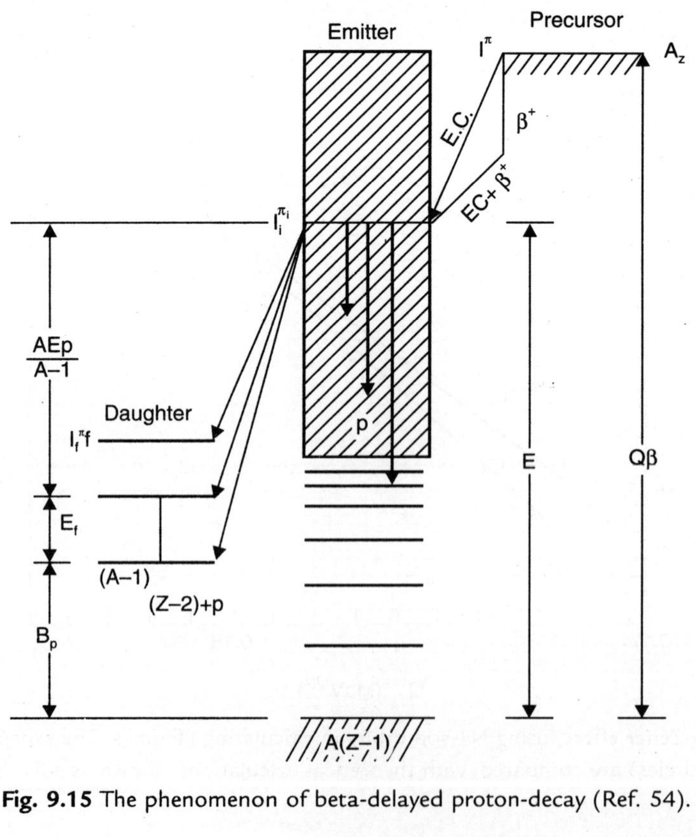

Fig. 9.15 The phenomenon of beta-delayed proton-decay (Ref. 54).

Such proton-decay cases have been observed since fifties, when exotic nuclei, *i.e.* nuclei far from beta stability line (*see* Fig. 2.3) started being produced[53]. These nuclei are, in general 3 or 4 boxes away from the line of stability and have excessive proton/neutron ratio and hence decay by β^+ emission. We show in Fig. 9.15, the phenomenon of beta delayed proton emission[53].

Apparently, the lifetime of proton decay will be the lifetime of beta decay because protons are emitted, in general, from level of the emitter nucleus for which energy is much higher than the Coulomb barrier of the proton and emitter nucleus and, therefore, have very much shorter lifetimes for proton decay from these excited states. Hence it is the properties of the levels of the emitter nucleus and the daughter nucleus-like the level widths, level densities and the statistical consideration, which determine the shape and cross-sections, etc. of the emitted protons; along with the properties of beta-decay. Apparently, this is not a case of penetration of Coulomb barrier. Examples of such cases are: Kr^{73}, Sn^{101}, Sn^{103}, Te^{111}, Ba^{117}, Ba^{119}, Ba^{121}, $Xe^{115,117}$ and $Hg^{179,181,183}$. These are the examples[53, 55] of cases for $A > 70$. There is another set of nuclei, for which proton decay takes place. They belong to $38 < A < 70$. *e.g.* Ti^{41} or to $A < 38$, *e.g.* Ar^{33}.

There are also cases of beta-delayed neutron decay and alpha decay[55] and fission. For the first category, the precursors are nuclei like Ba^{121} and I^{137}, *i.e.* from the neutron rich nuclei, on the other side of beta-stability line, compared to the cases of proton decay[55]. For β-delayed fission only very heavy nuclei are involved. For these cases, *see* Reference (55).

B. Self-delayed Proton-Decay: These cases involve the low energy protons, because the proton emitting level may be at a low excitation energy and the long lifetime is required because of the Coulomb-barrier and not because of beta decay. Such cases have a meta-stable state in the precursor nucleus.

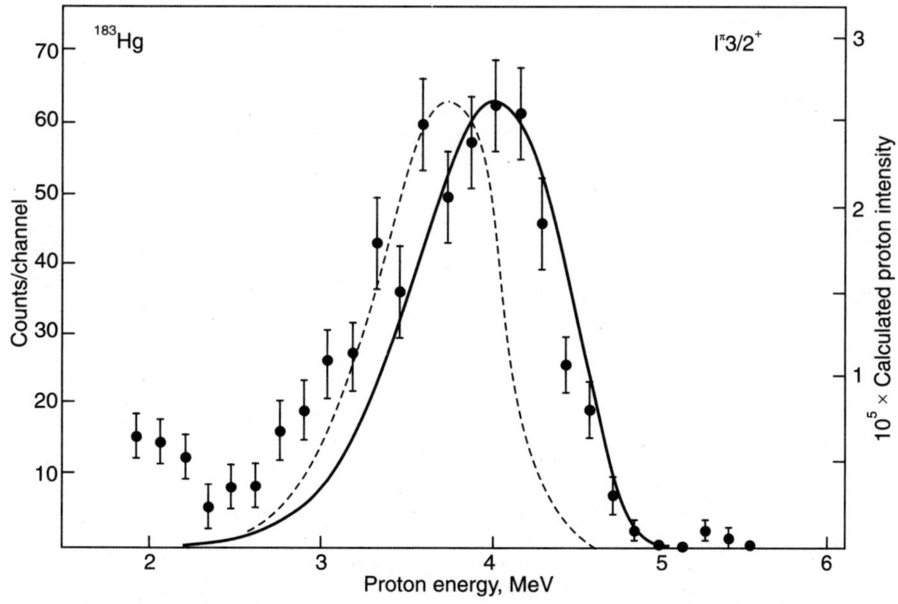

Fig. 9.16a The measured β-delayed proton spectra of Hg^{193}. Experimental points are compared with the theory, Eq. 9.60 [Pb, (p, 3p, xn) reaction] using 600 MeV protons-assuming $I^{\pi} = 3/2$. The dotted line and solid line show different values of Q and the binding energy E_p (Ref. 54).

A very good example is the case of Co^{53m} produced from heavy[57] ion-reactions like Ca^{40} (O^{16}, $2np$) Co^{53m} or Fe^{54} (p, $2n$) Co^{53m}. The fact that protons from Co^{53m} are self-delayed was established by introducing a plastic scintillator to detect electrons, and no β-rays were observed to be in coincidence, with protons. Also the energy spectrum of these protons, has only a single peak, compared to a broad spectrum from beta delayed prections. Figure 9.16*b* shows the typical shape the proton spectra from such cases.

In the case of self-delayed proton spectra; the proton peak is very sharp because now only proton evaporation plays role without the intensity-spectrum of beta rays coming into play.

C. Theory: (*i*) *Beta delayed proton emission:* This is not a case of penetration of Coulomb barrier; because the level of the emitter from which protons are emitted, is generally in the high excitation region, which for higher *A*, corresponds to the region where levels have $\Gamma >> D$, and hence continuous energies of excitations are available. This is indicated by the bell-shaped spectral distribution of protons as shown in Fig. 9.16*a*.

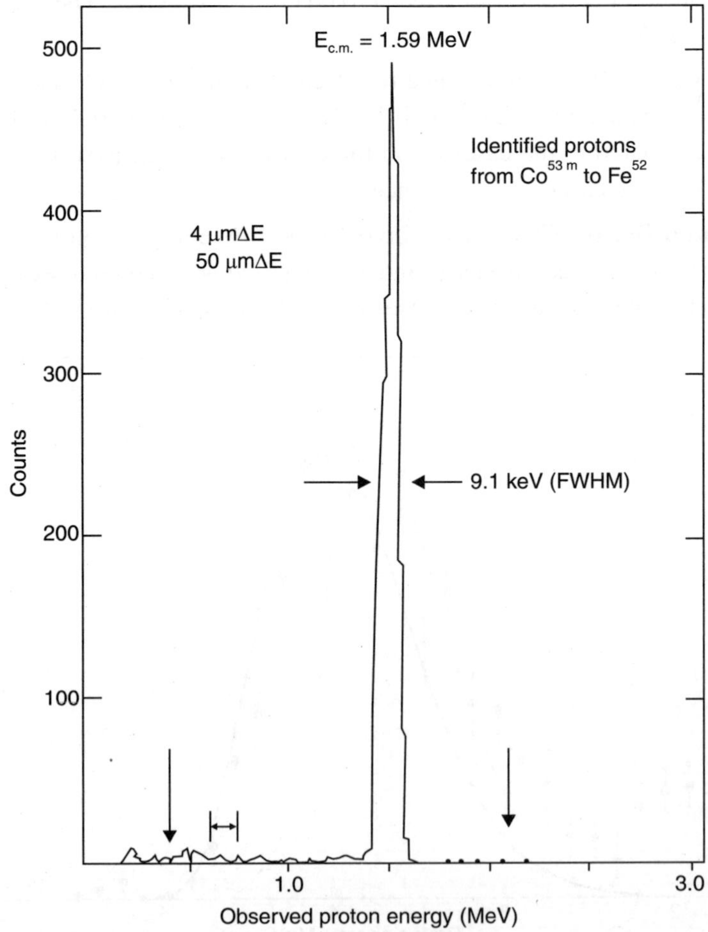

Fig. 9.16b Self-delayed proton spectrum from the decay of Co^{53m} produced in Fe^{54} (p, 2n) Co^{53m} reaction (Ref. 55, 57).

The theory of the shape of the proton spectrum and the decay probability as a function of proton energy E_p are described here, in brief. The proton intensity I_p as a function of E_p is determined by two factors of (*i*) behaviour of the intensity of betas, I_β, as a function of E_p and (*ii*) the behaviour of Γ_p/Γ, as a function of E_p where Γ_p is the level width for proton decay and Γ is the total level width given by $\Gamma = \Gamma_\gamma + \Gamma_p$. The quantity Γ_p/Γ is the relative branching ratio for protons. This means:

$$I_p\,(E_p)\,\alpha\, I_\beta \frac{\Gamma_p}{\Gamma} \qquad ...(9.58)$$

I_β decreases with E_p; because as E_p increases, the beta energy decreases and hence its decay probability. On the other hand Γ_p / Γ increase with energy E_p, because higher E_p means higher excitation energy for which the probability of proton-emission will increase. It is well known that level widths increase with excitation energy. Hence it can be seen qualitatively that $I_p\,(E_p)$ can have a bell-shaped spectrum.

For detailed quantitative theory, see work by Hornshoj et al. (54). In brief, following the above arguments, the average β-intensity per unit energy interval to the compound levels $I_i^{\pi_i}$ at energy E can be expressed as product of the total β-intensity $I_\beta(E)$ and a weight factor $W\,(I, I_i)$, denoting the fraction of the β -decays leading to compound levels with spin and parity $I_i^{\pi_i}$. The intensity from transitions $i \to f$, can, then, be obtained from the compound nucleus expression, *i.e.*,

$$I_p^{if}\,(E_p) = W\,(I, I_i)\, I_\beta\,(E) \frac{\Gamma_p^{if}\,(E_p)}{\Gamma^i} \qquad ...(9.59a)$$

where
$$\Gamma^i = \Gamma_\gamma^i + \sum_{f'} \Gamma_p^{if}\,(E_p) \qquad ...(9.59b)$$

and
$$E = B_p + E_f + \frac{A}{A-1}\,E_p \qquad ...(9.59c)$$

where B_p is the binding energy of proton, E_f is the end energy of betas and the last term is the energy of proton keeping in mind the recoil of the emitting reactions.

Therefore the total intensity of protons with energy E_p is given by:

$$I_p\,(E_p) = \sum_f \sum_i W\,(I, I_i)\, I_\beta\,(E) \frac{\Gamma_p^{if}\,(E_p)}{\Gamma^i} \qquad ...(9.60)$$

We will not go into details of calculating $I_\beta\,(E)$, except to say that it is given by:

$$I_\beta\,(E) = \frac{S_\beta\,(E)\, f\,(Z, Q, E)}{\int_0^Q S_\beta\,(E)\, f\,(Z, Q, E)} \qquad ...(9.61)$$

where f is the statistical weight function and $S_\beta\,(E)$ is called the β-strength function involving the parameters of beta decay theory. For details *see* Ref. (57).

The statistical weight factor $W\,(I, I_i)$ is given by:

$$W\,(I, I_i) = \frac{2\,I_i + 1}{3\,(2I + 1)} \qquad ...(9.62)$$

In Fig. 9.16*a*, the theoretical curve corresponds to Eq. 9.60.

(*ii*) *Self-delayed proton decay:* In this case, the factor I_β (*E*) does not exist and therefore, for a single level decay of proton of 1.5 MeV from level 11.49 MeV of Co^{53} to 10.74 MeV of Fe^{52} + *p*, a single proton peak is observed as shown in Fig. 9.16*b*.

J. Cerny et al.[57] have experimentally plotted yields for Fe^{54} (*p*, *pn*) Fe^{53m} and Fe^{54} (*p*, 2*n*) Co^{53m} versus incident energy E_p and compared them with that predicted from the spin dependent cluster evaporation model of Grover and Gilat[58], using optical model for populations and for transmission coefficients, including Coulomb part for charged particles emission. Figure 9.17*a* shows, such a comparison. Figure 9.17*b* shows the concerned decay scheme relations. In the case of Fe^{54} (*p*, *pn*) Fe^{53m}, the penetrability factor, due to Coulomb barrier plays the role. We conclude that in self-delayed proton decay; the penetration of an optical potential explains, the behaviour of these protons. The decay of Co^{53m} by protons to Fe^{52}, requires penetration of Coulomb barrier.

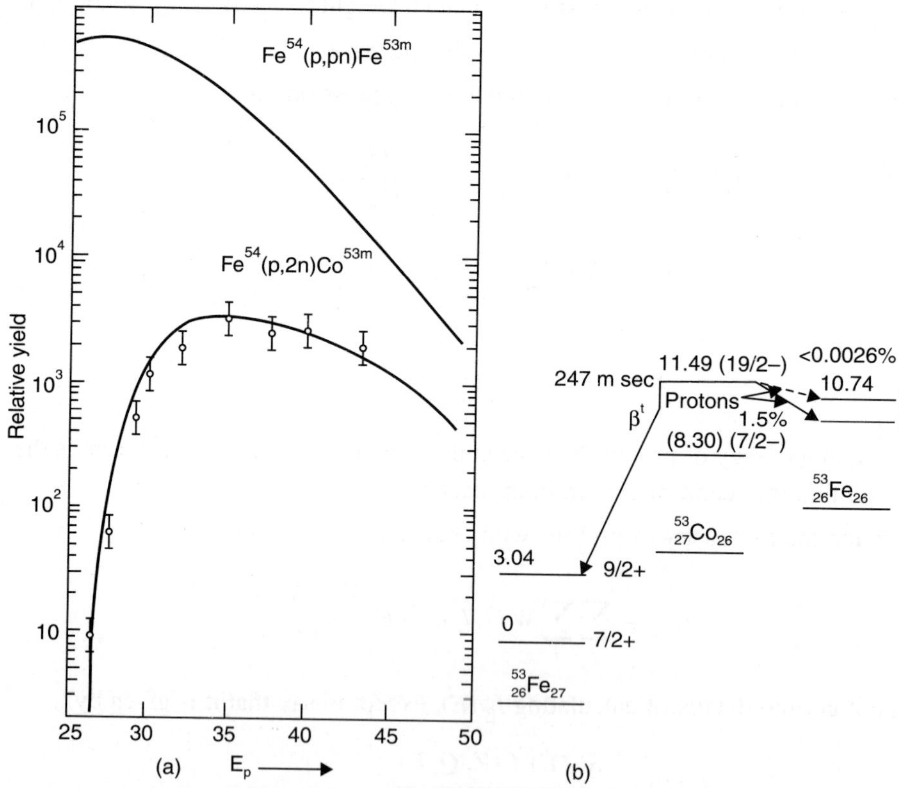

Fig. 9.17 (*a*) Theoretical predictions for the excitation function of Fe^{53m} and Co^{53m}, employing the approach of Grover and Gillet (Ref. 57); (*b*) The decay scheme of Co^{53m} to Fe^{53} by β^+ -decay and by protons to Fe^{52} (Ref. 57).

Finally, we conclude that delayed proton decay is a special case of a charged particle emission, delayed either due to beta decay or by transmission barrier due to the properties of the potential including the Coulomb part.

More and more delayed proton and neutron decay cases are being investigated, using various reactions, by either using heavy ion projectiles or radioactive projectiles. A few cases have recently been investigated. As for example, the proton decay[59] of N^{11}, using a radioactive beam of O^{18} of 80 MeV/N energy impinging on Be^9 target, causing fragmentation out of which 40 MeV/N, N^{12} beam was obtained using mass separation techniques. N^{12}, then impinged further on a secondary Be^9 target giving N^{11} by Be^9 (N^{12}, N^{11}) Be^{10}. The β-delayed decay of N^{11} to $C^{10} + p$, was then studied and proton spectrum observed. Another[60] light nucleus case is the β-delayed neutron decay of Be^{14}, which was produced by a similar technique. A 100 MeV/N, O^{18} beam was allowed to fall on Be^9 target and Be^{14} ions were produced by projectile fragmentation. Its delayed neutron decay was studied by time-of flight method. The detailed studies in both these cases ended up in giving information about new states of N^{11}, and B^{14}.

Also $Ir^{165, 166, 167}$ and Au^{171} and Kr^{69} have been recently produced using heavy ion reactions. Theses isotopes decay by delayed proton emission, which have been studied in details[61].

Theoretically, of course, shell model[62] with WKB approximation could be used to compare with experiments. The importance of the work lies in the experimental methods and new techniques and new proton radioactive nuclei.

9. Alpha and Charged Particle Decay

2000–2008

The charged particle studies are quite popular in the last ten years. In 2000, some one dozen papers were published on (*i*) α-radioactivity (*ii*) Fusion Fragmentation, (*iii*) Proton Decay (*iv*) Heavy ion emission etc.

In one paper, on helium break-up in Be^{10} and Be^{12}, made available in (p, C^{12}) Be^{12}, He^6, He^5 and (p, C^{12}) Be^{12}, He^4, - using 378 MeV^{12} beam of protons has been studied by 34 authors from U.K., France and Sweden [Phy. Rev. 63, 034301) (2001)]. In another paper 14 authors from Finland have participated where α-decay studies of Rh^{195} and Ru^{196}, were undertaken (Phy. Rev. C. 63, 044315 (2001)).

In a theoretical study; fusion barriers of super-heavy nuclei has been studied by a Russian group involving 112^{286}; 114^{292}, 116^{296} which were produced in $Ca^{48} + U^{238}$; $Ca^{48} + Pu^{244}$; and $Ca^{48} + Cu^{249}$ resulting in $3n$ and $4n$ evaporation channels. From the calculations of survivability probability they obtained for fission barrier heights as: for $112^{283–286}$ (5.5 MeV), for $116^{282–286}$ (6.7 MeV); and for $116^{292, 296}$ (6.40 MeV) [Phy. Rev. C. 65, 044602 (2002)].

In another, interesting theoretical study R.K. Gupta and co-workers, from Panjab University, Chandigarh have studied the decay of ${}_{28}Na^{56}$ formed in ${}_{16}S^{32} + {}_{12}Mg^{24}$ at $E_{c.m} = 51.6$ MeV and 60 MeV [experimentally heavier than mass 12 and for total kinetic energies for only favoured α-nucleus fragments]. The comparison between theory and experiment has brought about the difference in measured mass spectra and calculated ones, requiring the inclusion of I-dependent potential in the calculations [Phy. Rev. C. 68, 014610) (2003)].

Fission studies have been carried out both by using light particles, *i.e.* neutrons as projectiles, as well as by using heavy ion projectiles.

In one of such studies, neutron-involved fission in Bi^{209}, Pb^{nat}, Pb^{208}, Au^{197}, W^{ant}, and Ta^{181} targets were studied to determine the cross section ratios for Pb^{nat}/Bi^{209}, Pb^{208} / Bi^{209}, Aa^{197} / B^{207}, W^{nat} / Bi^{209} and Bi^{209} / U^{238} from 30–180 MeV neutron energy range [Phy. Rev. 70, 054603 (2004)]. The comparison with theory showed consistency.

In heavy induced fission, many cases were studied *e.g.* $C^{12} + Pb^{nat}$, $C^{12} + Bi^{209}$, $C^{12} + U^{235}$ and $B^{14} + Bi^{209}$ [Phy. Rev. C. 70, 051602 (R) (2004)]. The purpose of these studies was to determine the variance of fragment-mass distribution σ_A^2 at E_{lab} = 76 MeV. Many other such studies were undertaken in 2004; One of them using 15 UD at -N.Sc. New Delhi [Phy. Rev. C. 69, 03160 (R) (2004)].

Decay of the excited nuclear states by proton-Decay is a well known phenomenon and has been observed in fusion evaporated in the case of Tb^{135}, produced by Cr^{50}, + Mo^{92}, reaction, with a beam of 310 MeV of Cr^{50} [Phy. Rev. 69, 051302 (2004)].

In an interesting calculational paper, in [Phy. Rev. C. 74, 044311, (2006)] the authors have calculated the overlap of the mean field of Slater determinant with one containing pure Gauassian and perfect spin and iso-spin symmetry, optimizing the overlap by varying α-particle positions and radii, for nuclei from Be^8 through Br^{36} as well as S^{32}, Ar^{36} and Ca^{40}. They find quite large overlap for some of the higher systems that diminish for nuclei above Ne^{20}, but again strong clustering in Ai^{36}.

In an experimental paper; with 18 authors, they have measured C^{12} (O^{16}, O^{14}) C^{14} reaction at 234 MeV with O^{14} as ejectible particles, detected at forward angles. The decay of C^{14} in the backward angles was observed, so that C^{14} decays into Be^{10} + α, and alpha decay are candidates for three-body molecular cluster [Phy. Rev. C. 78, 014317 (2008)].

REFERENCES

1. J.M. Hollander, J. Perlman and G.T. Seaborg: Rev. Mod. Phys. 25, 469 (1953).
2. Asaro F. and I. Perlman: Phy. Rev. 107, 318 (1957).
3. F. Heggemann, L.I. Katzin, M.H. Studies, G.T. Seaborg and A. Ghiorso: Phy. Rev. 79, 435 (1950);

   A.C. English, T.E. Cranshaw, P. Demers, J.A. Harvey, E.P. Hinks, J.V. Jelly and A.N. May: Phy. Rev. 72, 523 (1947).
4. W. Porshen and W. Rielser: N. Naturforsch 119, 143 (1956); K. H. Sun, F. A. Pecjack, B. Jennings and A.J. Allan: Phy. Rev. 85, (726) (1952).
5. R. Pilger, F.S. Stephens, F. Asaro and I. Perlman, reported in the Nuclear Properties of the Heavy Elements. Vol. II, Detailed Radioactivity Properties: Prentice-Hall, Inc. Englewood Cliffe; New Jersey; edited by E.K. Hyde and G.T. Seaborg (1964).

   Michael Lederer et al.: Table of Isotopes: John Wiley & Sons, Inc., New York, p. 421 (1967).
   I. Perlman, A. Ghiorso and G.T. Seaborg: Phy. Rev. 77, 26 (1956).
6. D.E. Bunyanm, A. Kundley and D. Walker: Proc. Phys. Soc. (London) 62A, 253, (1949). Table of Isotopes, p. 131 (1967).
7. Thomas A. Farley: C and J. of Physics, 38, 1059 (1960).
8. H. Geiger and J.M. Nuttal: Phil. Mag. 613 (1911), Ibid; 23, 439 (1912); 24, 647 (1912).

   I. Perlman and F. Asaro. Ann. Rev. Nuclear Science 4, 157 (1954).
9. L.I. Schiff, Quantum Mechanics: McGraw-Hill Book Company, Inc. (1949).
10. G.H. Winslow and O.C. Simpson, ANL 4841 (1952).
11. M.A. Preston: Phy. Rev. 71, 865 (1947).
12. H.A. Tolhock and P.J. Brussard: Physica 21, 449 (1955).
13. R. Hofstadter: Rev. Mod. Physics 28, 214 (1956).
14. J.O. Rasmussen and B. Segall: Phy. Rev. 103, 1298 (1956).

15. E.M. Pennington and M.A. Preston: Cand. J. Physics 36, 944 (1958).

16. P.O. Fröman; Kgl, Danske Vidauskab, Selskab, Natfys, skr. 1, 3 (1957).

17. E. Segre: Experimental Nuclear Physics, John Wiley & Sons, Inc., (New York); (1959). R. F. Christy: Phy. Rev. 98, 1205 (1955).

18. D.S. Dellon and R.J. Liotta: Phy. Rev. C. 58, 2073 (1998).

19. J.C. Batcheldor et al. (10 authors): Phy. Rev. C. V. 55, p. 2142 (1997)

J.C. Batcheldor, K.S. Toth, E.G. Zganjor, D.M. Molz, C.R. Bingham, D.J. Lowell,

T.J. Ogariboene, and M.W. Rowe: Phy. Rev. C. 54, 949, (1996), A.W, Quint et al. (12 authors): J. Phy. A 346, 119 (1993).

20. Raj Kumar Gupta: Private Communication (1994).

21. Rose H.J. and G.A. Jones: Nature 307, 245 (1984).

22. R.K. Gupta: Private Communication, Ref. (31) and (32).

23. P.B., Price, J.D. Stevenson and S.W. Barwick: Phy. Rev. Letters 54, 297 (1985).

24. E. Hourani, Hussnnois M., Stab I; L. Billard S. Gales and J.P. Schapira: Phy. Letters, 160 B, 375 (1985).

25. S.P. Tretyakova: Samdulescu A., Zanystenin Yus. Korotkin Yus. and Mikheev V.L., JI NR report 7.85, (1985); Ibid JINR report no. 13–85 (1985).

26. A. Samdulescu, Zanstenin Yus., I.A. Lebedev, B.F. Mysasbedov, S.P. Tretyakova, and D.JINR Hasegen report 5–84 (1984).

27. S.W. Barwick, P.B. Price and J.D. Stevenson: Phy. Rev. C. 31, 1984 (1985).

28. Same as Ref. (25) and Ref. (33).

29. A. Samdulescu, D.N. Poenaru and W. Greiner: Soviet J. of Particle and Nucleus 11, 528 (1980).

30. D.N. Poenaru, W. Greinere, M. Ivascu, and A. Samdulescu: Phy Rev. C. 32, 2195 (1985); D.N. Poenaru, M. Ivascu, A. Samdulescu, and W.J. Greiner: Phy. G, Nuclear Physics V. 10. L. 183 (1984), Shi-Yi-Jin and W. Swiatecki: Phy. Rev. Letters, 54, 300 (1985); Ibid, Nuclear Physics A 438, 450 (1985).

31. H.G. De Carvillo, J.B. Martins, and Tavares G.A.P.: Phy. Rev. C. 34, 2261 (1985).

32. R. Bleadowske, T. Fliesshack, and H. Wallster: Nuclear Physics C. 34, 2261 (1985).

33. R.K. Gupta, W. Scheid, and W. Greiner: Phy. Rev. Letters 35, 353 (1975); R.K. Gupta, D.R. Saroha and N. Malhotra, J. De: Phy. Rev. Letters. 45, C. 6–477 (1984); R.K. Gupta, S. Gulahati, S.S. Malik and R.J. Sultana: Phys. G. Nuclear Physics 13, 127 (1987).

34. J.A. Maruhn and W. Greiner: Phy. Rev. C. 13, 2404 (1976). J.A. Maruhn and W. Greiner: Phy. Rev. Letters 32, 548 (1974); R.K. Gupta; J. Sov: Particle Nuclear 8 (4). July–August, p. 289–309 (1977).

35. H. Iriondo, D. Jerrestio, and R.J. Liotta: Nuclear Physics A. 454, p. 252 (1986). References (18), (29), (30), (31) and S. Landowne and C.H. Dasso: Phy. Rev. C. 33, 387 (1986).

36. J. Blocki, J. Randrup, W.J. Swiatecki, and C.F. Tsang: Ann Phys. (NY) 105, 427 (1977).

37. H. Kroger and W. Scheid, J.G. Physics: Nuclear Physics 5, L. 85 (1980).

38. Same as Ref. (32) and (33).

39. N. Bohr, J.A. Wheeler: Phy. Rev. 56, 426 (1939); D.L. Hill and J.A. Wheeler: Phy. Rev. 89. 1102 (1953).

40. R. Ramanna: Physics Letters, V-10. p. 321–323 (1964), R. Ramanna and V.S. Rama Murthy: Physics Letters V. 21, p. 437 (1966). V.M. Strutinski: Nuclear Physics A. 95, 420 (1967), A, 122, 1 (1968); M. Back, J. Damgaad; A.S. Jensen: H.C. Pauli: V.M. Strutinski, R.C.Y. Wong: Rev. of Mod. Physics,

44, 320 (1972); J.A. Maruhn, and W. Greiner, Phy. Rev. V. 13, p. 2404 (1976); Vanderboch R. and J.R. Huizenga; Nuclear Fission, Academic, New York, (1973), P. Moller and S.G. Nilsson, Phy. Letters, 31 B, 283 (1970); V.V. Pashkevish, Nuclear Physics A 169, 275 (1971); B.D. Wilkins, E.P. Steinberg, and R.R.Chasman, Phy. Rev. C. 14, 1832 (1976); J.R. Nix, Annual Rev. Nuclear Science 22, 65 (1972).

41. D.N. Poenaru, Ivascu M., Sandulescu A., and Greiner W.: Phy. Rev. C. 32, 572 (1985) and earlier references there-in.

42. O. Hahn and F. Strassman: Naturiss 27, 11 (1939).

43. Plutonium Project Report, Rev. Mod. Physics, 18, 513 (1946); J. American Chemical Soc. 68, 2411 (1946); A.S. Newton: Phy. Rev. 75, 17 (1949); V.K. Rao, V.K. Bhargava, S.G. Marathe, S.M. Shakundu, and R.H. Iyre: Phy. Rev. C.V. 19, p. 1372 (1979); H. Farrar, H.R. Fickel and R.H. Tomlinson, Cand. J. Physics, V. 40, 1017, (1962); Ibid, Nuclear Physics 34, 367 (1962).

44. G.T. Seaborg: Phy. Rev. 85, 157L (1952), A. Ghiorso, G.H. Higgens, A.E. Larsh, G.T. Seaborg and S.G. Thompson: Phy. Rev. 87, 163 (1952).

45. S. Frankel and N. Metropolis: Phy. Rev. 72, 914 (1947).

46. P. Fong: Phy. Rev. 89, 332 (1953), Phy. Rev. 102, 434 (1956).

47. J.K. Perring and J.S. Story: Phy. Rev. 98, 1525 (1955).

48. H.J. Fink, J. Marahn, W. Scheid and W. Greiner; J. Phys. 268, 321 (1974); W.H. Miller, J. Ch. Phy. 53, (1949); 3578 (1970).

49. S.S. Kapoor, R. Ramanna and P. N. Rama Rao: Phy. Rev. 131, 283 (1963); S.S. Kapoor and R. Ramanna: Phy. Rev. 133, B 598 (1964); Bency John and S.K. Kataria; Phy. Rev. C. 57, p. 1337 (1998).

50. B.B. Back, R.R. Betts, J.E. Ginder, B.D. Wilkins, S. Saini, M.B. Tang, C.K. Gelbke, W.G. Lynch, M.A. McMohan and P.A. Baisden: Phy. Rev. C. 32, 195, (1985); D.J. Hinde, C.R. Morton, M. Dasgupta, J.R. Leigh, J.C. Mein and H. Timmers: Nuclear Physics A 592, 271 (1995); S.S. Kapoor, H. Baba and S.G. Thompson: Phy. Rev. 149, 965 (1966); J. Hinde, K. Mahaboub, J. Richert and K. Remorski: J. Phy. A. 354, 59 (1996).

51. M. Mirea: Phy. Rev. C. 57, p. 2484 (1998).

52. H. Kudo, M. Maruyama and M. Tanikawa: T. Shinozuka and M. Fujioka. Phy. Rev. C. 57. 178 (1998).

53. V.I. Goldansky: Annual Rev. of Nuclear Science 16, 1 (1996).

54. P. Hornshoj and K. Wilksky, P.G. Hanson and B. Johnson and O.B. Nielson: Nuclear Physics A-187. p. 609 (1972); Ibid, Phy. Letters 34B, 591 (1971).

55. J. Hardy: Nuclear Spectroscopy and Reactions, V. VIII, edited by J. Cerney; Academic Press, New York (1974).

56. C.L. Duke, P.G. Hansen, O.B. Nielson and Rudstan: Nuclear Physics, A151, 609 (1970).

57. J. Cerney, R.A. Ghouch, R.G. Sextro and John E. Estrel: Nuclear Physics A 188, 666 (1972).

58. J.R. Grover and J. Gilat: Phy. Rev. 157, 802 (1967).

59. A. Azhari, T. Baumun, J.A. Brown, M. Hellstrom, J.H. Kelley, R.A. Kryger, D.J. Millener, H. Modani, E. Ramakrishnan, D.E. Russ, T. Suomijari, M. Thoennessen and S. Yokoyama: Phy. Rev. C. 57, p. 678 (1998).

60. M. Belbet et al. (18 authors): Phy. Rev. C. 56, 3038 (1997).

61. X.J. Xu, W.X. Huang, R.C. Mai, Z.D. Gu, Y.F. Yang, W.W. Wang, C.F. Deng and L.L. Xu: Phy. Rev. C.55, p. 553 (1997).

62. C.N. David et al. (18 authors): Phy. Rev. C. 55, 2255 (1997); R.D. Lawson: Theory of Nuclear Shell Model (Clarendon, Oxford) (1980) pp. 76 and 440.

PROBLEMS

1. If the potential depth is $V_o = 50$ MeV, E between 0 and 9 MeV, and width at the base as 10 Fermis and the height of the potential barrier 10 MeV; Calculate the penetration probability P versus E; for alpha decay [Fig. 9.8].

2. Take the case of highly excited state of $\overline{9}/2$ in Pb^{209}, decaying to $\overline{5}$ state Pb^{208}, by neutron emission of 3 MeV, calculate the barrier penetration probability of protons of energies between 1 to 10 MeV from Pb^{208}.

3. (*i*) The ground state of Pb^{208} is stable; then why does a highly excited state of Pb^{208} decay by protons ?

   (*ii*) Can we find an excited state of Pb^{208}, which decays by α-emission ?

   (*iii*) Calculate its penetration probability for alphas and compare it with protons.

4. Calculate the lifetimes from Eq. 9.17 for alpha decay, for U, Th, Ra, and Po for 6 MeV energy of alphas. Compare them with the values in Fig. 9.7.

5. Estimate transmission and reflection coefficients for alpha particles and proton beam of current *i*, and energy from 1 MeV to 20 MeV; from say U^{238} nucleus.

6. Taking the pre-formation values P_o to be 10^{-8} for alphas, and 4×10^{-16} for C^{14}; compare the transition probabilities of alpha decay and cluster decay for

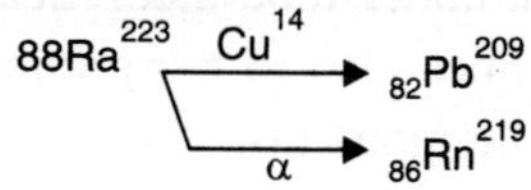

   Using Eq. 9.23

7. Calculate the above lifetimes, using Eqs. 9.26 and 9.32. Compare these results with the above calculations.

8. (*i*) Why has I^{131} beta delayed neutron-decay, while B^{121} has beta delayed proton decay ?

   (*ii*) Why are there less beta-delayed alpha-decay cases ?

   (*iii*) Why is Co^{53m} self delayed proton decay case ?

9. Search for fission cases, and find out the cases for beta delayed fission ? Are there many cases also prone to be beta-delayed-alpha emission ? If no, then why ?

10. Discuss, qualitatively, how shell effect are responsible for the asymmetry of mass distribution in fission [consult reference (38)].

10

CHAPTER

Shell Model

10.1 GENERAL

Nuclear properties of energy levels of excited states, and their transition probabilities through alpha, beta and gamma transitions can only be understood theoretically by treating the nucleus as a quantum mechanical system. Of course, this applies also to the ground state properties of nuclei like binding energies, spins, magnetic and quadrupole moments.

We have seen, in the last few chapters that nuclear forces are very strong and have short range. So, in principle, one has to solve a many-body problem, of many nucleons in a nucleus, with very strong and short range forces. Such attempts, in general, are expected to be quite difficult; though recently some success in this direction has been achieved in a limited manner. However, it was realised in the very beginning that because of the strong forces; the nuclei may act like a fluid so that all the nucleons move collectively. We have already seen, that one can explain the binding energies of nuclei, through, Weizsäcker's mass formula based on the concepts of a fluid. As is now well-known, this could not explain, the spins or magnetic moments of nuclei.

The first nuclear structure model, which could, explain such detailed nuclear properties was the shell model[1]. Some properties of the excited states, especially the systematics in energies could, however, be explained only in terms of collective motion of nucleons, *e.g.* their rotational and vibrational behaviour. This has given rise to Bohr-Mottelson collective model and rotational model[2]. The coupling of single particle excitations with collective model, as initially calculated by Mottelson[3] and Nilsson[4] brings in collective and particle excitation together, and has become the basis of modern nuclear structure calculations. Excitation by heavy ions has yielded the highly excited states of rotational model, with very high spin states, and also has resulted in super-deformed nuclei. These collective phenomena have been discussed, subsequently.

Magic Numbers

Many phenomena, in a set of nuclei, show a greater stability than expected from a smooth behaviour. A large amount of data is available, which shows that for Z or N = 2, 8, 20, 28, 50, 82 and 126—the so-called magic number nuclei—show a behaviour corresponding to extra stability. Some interesting examples are:

(*i*) As shown in Fig. 2.22, the quadrupole moments[5] of the ground state of nuclei, when plotted as a function of odd nucleons, indicate zero quadrupole moments for Z or $N = 8, 20, 28, 50, 82$ and 126 showing that these nuclei have spherical shapes, indicating no loose nucleons, and hence a closed structure.

(*ii*) If we plot the experimentally determined energies of the first excited states[6] of even-even nuclei versus the number of neutrons in a nucleus, it is interesting to note that for the magic number, the energies are extraordinarily high showing that nuclei with these neutrons are very stable and it requires higher energy than normal to excite them, to their first level.

(*iii*) An interesting piece of systematics is the behaviour of neutron cross-section for thermal neutrons[7] plotted as a function of neutron number of target nucleus, as shown in Fig. 10.1. It is interesting to note that for $N = 50$, 82 and 126, the cross-sections are exceptionally low, showing the special nature of these numbers of neutrons, for which the nuclei are especially stable and hence the probability (or cross-sections) of neutron-capture is exceptionally low, due to reasons, all connected with the stability of the target nucleus.

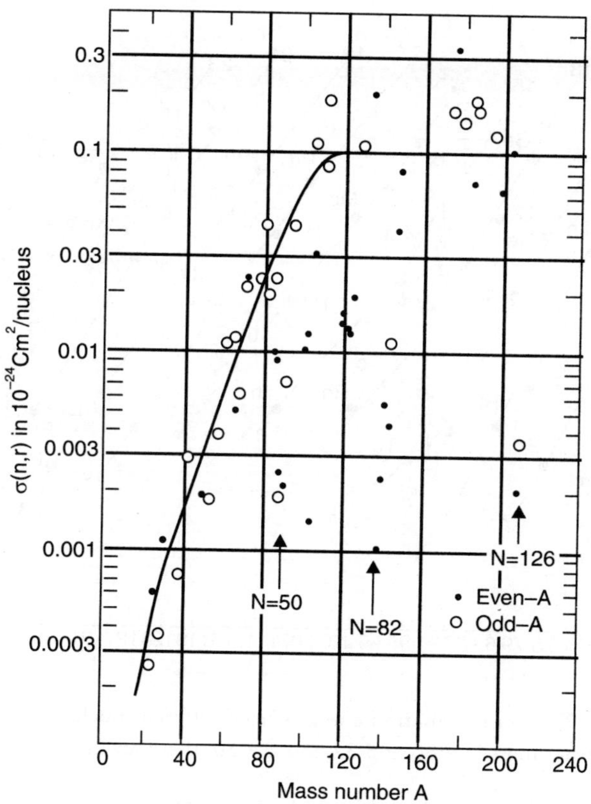

Fig. 10.1 Thermal cross-sections for (n, γ) as a function of mass number A. At magic number N-50, 82 and 126, the cross-sections are abnormally low (Ref. 7).

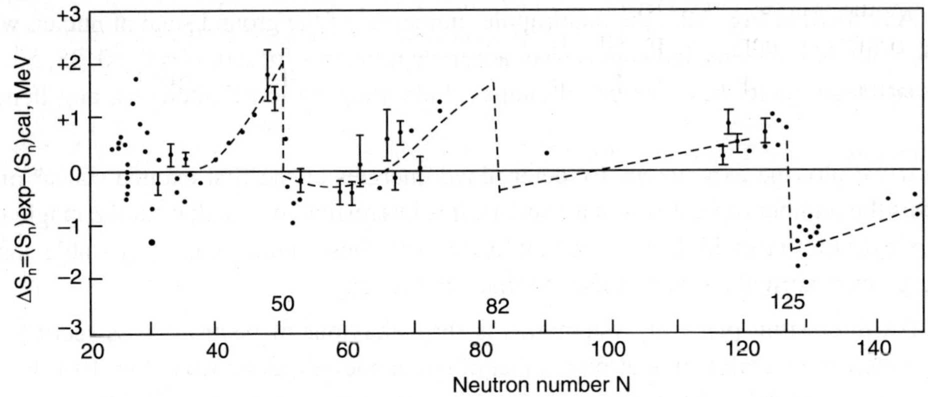

Fig. 10.2 Observed neutron separation energies (S_n) Exp − $(S_n)_{cal}$ compared with the smooth behaviour of (S_n) cal. Expected from semi-empirical mass formula. At magic number N = 50, 82 and 126, there are clear breaks (Ref. 8).

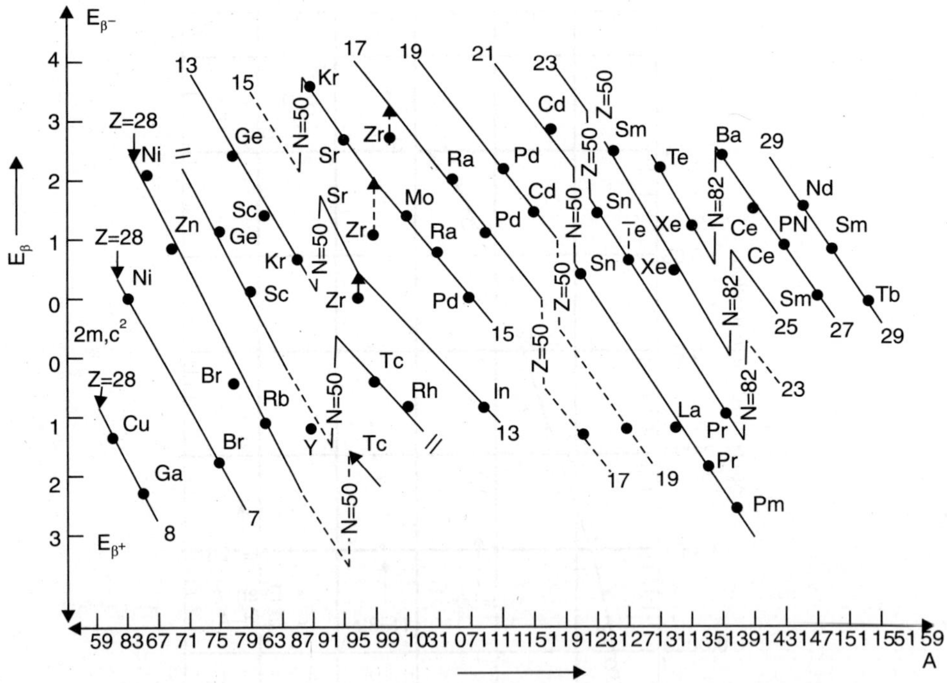

Fig. 10.3 Beta transformation energies from different nuclei plotted versus A. At magic numbers N = 50 and 82 and Z = 50, there are clear breaks (Ref. 9).

(*iv*) A fact, which brings out the effect of magic number very strongly is, the deviation of experimental separation energies[8] and the ones obtained from Weizsäcker's formula uncorrected for shell effects. Figure 10.2 shows this difference, between these masses, as a function of number of neutrons. It is very interesting that for $N = 28$, 50, 82 and 126, there are very significant dips. The extra stability exceeds 2 MeV for 126 neutrons. Similar effects are seen for $Z = 28$, 50 and 82, if one plots the difference of proton separation energies as a function of the number of protons in nuclei.

(*v*) Another interesting experimental fact, indicating strongly the presence of magic numbers is, that β-transformation energies representing β^+-emission *i.e.* converting an odd neutron to proton; or proton into neutron (β^-), show very clear breaks, when $E_{\beta^\pm}$ is plotted against A; where $E_{\beta^\pm}$ represents the maximum kinetic energy of electron or positron; (Total $E_{\beta^\pm} = (K \cdot E)_e + 2mc^2$). These transformations may be represented by:

$$ {}_{Z}X^{N} \xrightarrow{\beta^-} {}_{Z+1}X^{N-1} \text{ and } {}_{Z}X^{N} \xrightarrow{\beta^+} {}_{Z-1}X^{N+1} \qquad \text{...(10.1)}$$

The breaks at Z and $N = 50$; $N = 82$, represent the start of a new shell; Fig. 10.3 (Ref. 9).

There are many other physical phenomena, which show that nuclei with magic number of Z or N are especially stable. The phenomenon of magic numbers was explained by Mayer and independently by Haxel, Jensen and Suess[1], by assuming, that each odd nucleon in a nucleus, moves in the average nuclear potential of the even-even core of the rest of nucleus, to which should be added a potential proportional to **s . l** where **s** represents the spin of the nucleon and **l** represents the orbital angular momentum of the same nucleon. This assumes that spin-orbit coupling as strong enough that it produces a definite total angular momentum **j**, and the angular momenta of various nucleons are coupled through *jj* coupling. The average potential $V(r)$ under which a nucleon moves in a nucleus is, then assumed to be:

$$V(r) = V_o(r) + U(r)\,(\mathbf{s.l}) \qquad \text{...(10.2}a\text{)}$$

The average potential $V_o(r)$ is assumed to have Woods-Saxon potential shape (*see* Section 2.1.2), expressed as:

$$V_o(r) = \frac{U_o}{1 + \exp\left[\dfrac{(r-R)}{a}\right]} \qquad \text{...(10.2}b\text{)}$$

where U_o is a constant and the significance of R and 'a' have been discussed earlier in Eq. 2.30 in Section 2.1.2 and $U(r)$ is taken to be proportional to $1/r\, \partial V_o/\partial r$ and is frequently taken to be of the Thomas form[10]:

$$U(r) = \frac{-\lambda\hbar^2}{4M^2c^2}\left(\frac{1}{r}\frac{\partial V_o(r)}{\partial r}\right) \qquad \text{...(10.3)}$$

where λ is a dimensionless constant and is taken as $\lambda = 40$. Though L.H. Thomas[10] wrote this form for the atomic case; it is adapted here in nuclear-structure, but here $U(r)$ is negative and hence leads to levels with $j = 1 + 1/2$, being depressed and those with $j = 1 - 1/2$, being raised in energy. This is opposite to the atomic case. As a matter of fact; the genesis of spin-orbit coupling lies, in the basic nature of two-body nuclear force, discussed in Chapter 6, [Equations 6.79 to 6.90, and Table 6.3].

10.2 COMMON POTENTIAL V (r) IN SHELL MODEL

It is pertinent to understand, the significance of the potential used in Eq. 10.2*a*. First of all, it should be realised that the even-even core mentioned above, can be looked upon, as the one built on the principles of the shell model itself.

While calculating the values of $V_o(r)$, and hence U_o, as we go on adding more nucleons; it may be realised that these values will change for different additions of nucleons. Let us illustrate.

(*i*) The accepted values of the various constants[11] in the expression for $V_o(r)$ in Eq. 10.2 are:

$$U_o \approx 57 \text{ MeV} + \text{correction } C$$
$$R \approx 1.25\, A^{1/3} \times 10^{-13} \text{ cm}$$
$$a \approx 0.6 \times 10^{-13} \text{ cm} \qquad \text{...(10.4)}$$

The correction *C*, arises due to the asymmetry energy, as has been discussed in Weizsäcker formula, and is due to asymmetry between neutron and proton. It is given by:

$$C = \pm 27 \text{ MeV} \times \frac{N-Z}{A} \left[\begin{array}{l} - \text{ for neutrons} \\ + \text{ for protons} \end{array} \right. \qquad \text{...(10.5)}$$

(*ii*) Further $V_o(r)$ in Eq. 10.2*a* should be corrected for the repulsive Coulomb potential. It is well known, that the field $E(r)$, due to the Coulomb charge at distance *r* from the centre of a charged sphere is given by (as given by a well-known theorem of electrostatics):

$$E(r) = \frac{\left(\frac{4\pi}{3} r_o^3 \rho\right) e}{4\pi \in_o r_o^2} = \frac{r}{R_c} E(R_c) \text{ for } r < R_c \qquad \text{...(10.6a)}$$

where
$$\rho = \frac{Ze}{4\pi/3\, R_c^3}$$

for a nucleus of charge *Z* and R_c is a distance from centre, where the electric charge density has sharp edge, so that, since all the charge is inside the radius R_c:

$$E(r) = \frac{Ze^2}{4\pi \in_o r^2} \text{ for } r > R_c \qquad \text{...(10.6b)}$$

and
$$E(R_c) = \frac{Ze^2}{4\pi \in_o R_c^2} \text{ for } r = R_c \qquad \text{...(10.6c)}$$

Then,
$$V_{oc}(r) = \int_\infty^r E(r)\, dr = \int_\infty^{R_c} \frac{Ze^2\, dr}{4\pi\in_o r^2} + \int_{R_c}^r \frac{Ze^2}{4\pi\in_o R_c^2} \frac{r}{R_c} dr$$

$$= \frac{Ze^2}{4\pi\epsilon_o R_c}\left\{1+\frac{1}{2}\left[1-\left(\frac{r}{R}\right)^2\right]\right\};\ r < R_c$$

and
$$V_{oc}(r) = \frac{Ze^2}{4\pi\epsilon_o r} \text{ for } r > R_c \qquad \text{...(10.7)}$$

This Coulomb potential should be added to nuclear potential $V(r)$ in Eq. 10.2.

(*iii*) The potential should also be corrected for second-order velocity dependence of the nucleon-nucleon force. This effect has been expressed in terms[12] of the expansions of $V_o(r)$:

$$V_o(r) = V_{oo}(r) + \alpha p^2 + \beta p^2 + ... \qquad \text{...(10.8)}$$

For the lowest order term, we consider the expression for energy as:

$$\frac{p^2}{2M} - V_o(r) = E \quad \text{or} \quad \frac{p^2}{2M} - V_{oo}(r) - \alpha p^2 = E \qquad \text{...(10.9a)}$$

This can be re-written as:

$$\frac{p^2}{2M^*} - V_{oo}(r) = E \qquad \text{...(10.9b)}$$

where effective mass M^* is defined as:

$$\frac{1}{2M^*} \equiv \frac{1}{2M} - \alpha \qquad \text{...(10.9c)}$$

where α is a constant determined semi-empirically.

(*iv*) Another correction[13] is required, because the total potential for protons is shallower, than for neutrons, because of Coulomb forces. For the same total energy E, the kinetic energy is, therefore, lower for protons; hence velocity dependent term is different for protons and neutrons.

A typical V_o given for protons in the excitation energy range between 9 and 22 MeV, is expressed semi-empirically as:

$$V_o\,(\text{MeV}) = 53.0 - 0.55\,(\text{MeV}) + \frac{0.4Z}{A^{1/3}} + 27\frac{N-Z}{A} \qquad \text{...(10.10)}$$

One should, therefore, express $V_o(r)$ in Eq. 10.2 by combining all the above terms (Ref. 13), for different values of N and Z.

As discussed in Section (2.2) in Chapter 2 the nuclear shape is very similar to a Woods-Saxon potential, showing the shape of nuclear matter. Therefore, a realistic potential shape (similar to the one in Eq. 10.2*b*) is written as:

$$V(r) = -\frac{V_o}{[1+\exp\mu(r-R)]} \qquad \text{...(10.11)}$$

where μ and R are constants, determined from scattering experiments discussed in Chapter 2. Accepted values of these constants[14] are: $\mu^{-1} \approx 0.5 \times 10^{-13}$ cm and $R \approx 1.33\, A^{1/3} \times 10^{-13}$ cm, A being the mass number of the nucleus and V_o is about 50–60 MeV; $V(r) \to -V_o$ for $r < R$ for $\mu \to \infty$ and $V(r) = 0$ for $r > R$, when the potential corresponds to square well.

10.3 THE WAVE-FUNCTION AND NUCLEAR POTENTIAL

It is not possible to write analytically, the wave-function for the potential as given in Eqs. 10.2–10.11. However, for getting the first orientation, one can approximate the potential by a harmonic oscillator-type potential:

$$V = -V_o + \frac{1}{2} M\omega^2 r^2 \qquad \text{...(10.12)}$$

where V_o and ω are constants which are adjusted to experimental data. Then, for the oddeth particle, the wave-function ψ, can be expressed in the Schrödinger equation:

$$\left[\left(\frac{-\hbar^2}{2M}\right)\nabla^2 - V_o + \frac{1}{2} M\omega^2 r^2\right]\psi = E\,\psi \qquad \text{...(10.13)}$$

Then, one can express for a three dimensional spherical polar coordinates, and for a spherically symmetrical potential:

$$\psi_{n_1 m_1}(r, \theta, \phi) = \frac{R_{n,1}(r)}{r} Y_{1m_1}(\theta, \phi) \qquad \text{...(10.14)}$$

Where $R_{n,1}(r)$ are the radial functions and may be expressed in terms of Associated Laugerre polynomials $v_{n,1}(r)$ as:

$$R_{n,1}(r) = N_{n,1} \exp\left(-\frac{1}{2} \nu r^2\right) r^{1+1}\, v_{n,1}(r) \qquad \text{...(10.15)}$$

In Eq. 10.15, n signifies the number of nodes in $R_{n,1}(r)$ including the one at the origin and $\nu \equiv (M\omega)/\hbar$ and $v_{n,1}(r)$ is the Associated Laugerre polynomial. Nodal surfaces, associated with these wave-functions are planes containing the z-axis (n in number); cones of constant θ, ($1 - m_1$ in number) and spheres ($(n-1)/2$ in number). The functions Y_{1m_1} are, of course, familiar spherical harmonic functions, where l corresponds to orbital quantum number and m_l is the projection of l on z-axis.

The eigen values of energies above V_o corresponding to the eigen function $\psi_{n,1,m_1}$ is given by:

$$E_{n1} = \hbar\omega\left(2n + l - \frac{1}{2}\right) = \hbar\omega\left(\eta + \frac{3}{2}\right) = E_\eta \qquad \text{...(10.16a)}$$

with $n = 1, 2, 3, \ldots; \quad l = 0, 1, 2, \ldots$

and $\eta = 2n + l - 2$ ...(10.17)

Because of the two spin states of each particle, the total degeneracy for each l value is $2(2l+1)$. Also there is degeneracy attached to each value of η. Since $2n = \eta - l + 2 =$ even, for a given value of η, the degenerate eigen states of each η are:

$$(n, l) = \left(\frac{\eta+2}{2}, 0\right), \left(\frac{\eta}{2}, 2\right) \ldots (2, \eta-2), (1, \eta) \text{ for even } \eta$$

$$(n, l) = \left(\frac{\eta+1}{2}, 1\right), \left(\frac{\eta-1}{2}, 3\right) \ldots (2, \eta-2), (1, \eta) \text{ for odd } \eta \qquad \ldots(10.18)$$

putting $l = 2k$ for even η and $l = 2k+1$ for odd η, we can write the expression for the number of neutrons or protons, *i.e.* the degeneracy for odd or even η as:

$$N_\eta = \sum_{k=0}^{\eta/2} 2\,[2\,(2k) + 1] \text{ for even } \eta$$

$$N_\eta = \sum_{k=0}^{(\eta-1)/2} 2\,[2\,(2k+1) + 1] \text{ for odd } \eta$$

both of which can be written as:

$$N_\eta = (\eta + 1)(\eta + 2) \text{ for both even or odd } \eta \qquad \ldots(10.19)$$

The accumulated total number of particles for all levels up to η is:

$$\sum_\eta N_\eta = \frac{1}{3}(\eta+1)(\eta+2)(\eta+3) \qquad \ldots(10.20)$$

Further assuming an equal number of neutrons and protons, *i.e.* $A = 2Z$, we can write:

$$A = \sum_{\eta=0}^{\eta_o} 2N_\eta = \frac{2}{3}(\eta_0+1)(\eta_0+2)(\eta_0+3)$$

$$= \frac{2}{3}\left(\eta_0^3 + 6\eta_0^2 + 11\eta_0 + 6\right)$$

$$= \frac{2}{3}(\eta_0+2)^3 - (\eta+2) \approx \frac{2}{3}\left(\eta_0 + \frac{3}{2}\right)^3 \qquad \ldots(10.21)$$

Now we develop an expression for the harmonic oscillator frequency ω, as a function of A. Realising that in a harmonic oscillator potential, the expectation value of the kinetic energy in any state is equal to the expectation value of the potential energy; we can write for each single particle energy; the total single particle energy as:

$$\langle E \rangle_{s.p} \approx 2 \times \left(\frac{1}{2} M\omega^2 \langle r^2 \rangle \right) \approx M\omega^2 \langle r^2 \rangle \qquad ...(10.22a)$$

Factor 2 comes from the fact, that the potential energy and kinetic energy are assumed to the equal, hence total energy is twice the kinetic energy. Hence total energy E is given by:

$$E \approx MA\omega^2 \langle r^2 \rangle \qquad ...(10.22b)$$

Also from Eq. 10.16a, one can write,

$$E \approx \hbar\omega A \left(\eta + \frac{3}{2} \right) \qquad ...(10.16b)$$

Combining Eqs. 10.22b and 10.16b, one can write:

$$\langle r^2 \rangle = \frac{\hbar}{M\omega} \left(\eta + \frac{3}{2} \right) \qquad ..(10.22c)$$

Hence $\sum_p \langle r^2 \rangle$ for both protons and neutrons is given by:

$$\sum_p \langle r^2 \rangle = \frac{\hbar}{M\omega} \sum_{\eta < \eta_o} (\eta + 1)(\eta + 2) \left(\eta + \frac{3}{2} \right)$$

$$\approx \frac{\hbar}{M\omega} \sum_{\eta} \left(\eta_o + \frac{3}{2} \right)^3 \approx \frac{\hbar}{M\omega} \int \left(\eta_o + \frac{3}{2} \right)^3 d\eta$$

$$= \frac{\hbar}{M\omega} \frac{1}{4} \left(\eta_o + \frac{3}{2} \right)^4 \qquad ...(10.22d)$$

Further the proton number Z is related to the oscillator quantum number η by the relation [Eq. 10.19],

$$\frac{A}{2} = Z = \sum_{0 < \eta < \eta_o} (\eta + 1)(\eta + 2)$$

$$= \int \left(\eta_o + \frac{3}{2} \right)^2 d\eta \approx \frac{1}{3} \left(\eta_o + \frac{3}{2} \right)^3 \qquad ...(10.22e)$$

Hence the mean square radius is given by from Eqs. 10.21 and 10.22b:

$$\overline{\langle r^2 \rangle} = \frac{\sum_p \langle r^2 \rangle}{Z} = \frac{\hbar}{M\omega} \frac{3}{4} \left(\eta_o + \frac{3}{2} \right) \qquad ...(10.22f)$$

From which we can write, from Eqs. 10.16*b* and 10.22,

$$\frac{E}{\hbar\omega} = A\left(\eta_o + \frac{3}{2}\right) \approx \text{Constant } A^{4/3} \qquad ...(10.22g)$$

As shown in Chapter 2, Eq. 2.29, the mean square radius $\langle r^2 \rangle$ can be estimated from the relation:

$$\langle r^2 \rangle \approx \frac{3}{5} R_c^2 \qquad ...(10.22h)$$

where R_c is the Coulomb radius, determined experimentally [*see* Chapter 2]; as $R_c = 1.2\ A^{1/3} \times 10^{-13}$ cms. Then, from Eqs. 10.22*b*, 10.22*c*, 10.22*d* and 10.22*e*, we can write, calculating the various constants, the relationship:

$$\hbar\omega = \text{Constant } A^{-1/3} \approx 40\ A^{-1/3} \qquad ...(10.23)$$

The accepted value of $\hbar\omega$, keeping in mind that Z is always less than 1/2 A; is then given as

$$\hbar\omega = 41\ A^{-1/3} \qquad ...(10.24)$$

The treatment given above gives an order of magnitude of the properties of the harmonic oscillator potential.

10.4 THE ROLE OF SPIN-ORBIT COUPLING AND EXTREME SINGLE PARTICLE SHELL MODEL

In extreme single particle shell model, it is envisaged that the oddeth nucleon moves in the common potential $V(r)$, [Eq. 10.2*a*]. We will first consider the situation when spin-orbit-interaction term is absent.

While calculating the energy levels of such a single particle in a common potential as given in Eq. 10.2*a*, without **s . l** term, but with a realistic Woods-Saxon potential as given in Eq. 10.11, it is not possible to write the analytical expression for eigen-functions or eigen-values of energies. One has to solve it only numerically.

However, one can approximate the Woods-Saxon potential by the one given in Eq. 10.12, for which the solutions has been discussed in Section 10.3, without spin-orbit coupling. The other possible potential, for which solutions can be found analytically is the square well potential, *i.e.*,

$$V(r) = -V_o \quad r < R$$
$$= 0 \quad r > R \qquad ...(10.25)$$

The solution of such a potential have been discussed in Section (3.2), for the deuteron problem. For the case of square well for infinite depth, however, the solutions of Schrödinger equation:

$$\left[\nabla^2 + \frac{2M}{\hbar^2}(E - V(r))\right]\psi(r) = 0 \qquad ...(10.26a)$$

can be given in the form:

$$\psi_{n,\,l,\,m}(r, \theta, \phi) = U_{n,\,1}(r)\, Y_{l,\,m}(\theta, \phi) \qquad ...(10.26b)$$

where $Y_{l,\,m}(\theta, \phi)$ are the spherical harmonics and $U_{n,\,1}(r)$ is the radial solution. The exact solutions for $U_{n,\,1}(r)$ is given by, (for a square-well potential):

$$U_{n,\,1}(r) = j_i(kr) = \frac{J_{1+1/2}(kr)}{r^{1/2}} \qquad ...(10.26c)$$

These solutions are such that for $r = 0$, $U_{n,\,1}(r) = 0$ and $j_i(kr)$ are the spherical Bessel functions; $J_{l+1/2}$ are ordinary Bessel function of half integral order, and k is given by:

$$k = \sqrt{\frac{2ME}{\hbar^2}} \qquad ...(10.26d)$$

The bound eigen-solutions for $r < R$, are obtained by the boundary conditions at $r = R$, *i.e.*,

$$j_l(k_{n,\,1}r) = 0 \qquad ...(10.26e)$$

which corresponds to the solution disappearing at the boundary $r = R$, if there is infinite potential barrier. Energy eigen-values are determined by the nth zero of the spherical Bessel function j_l. Such levels have orbital momenta l and radial quantum number n.

One can now fill different levels in such a case by particles of one type (proton or neutron) using Pauli Exclusion Principle. Each level with angular momentum l is degenerate by 2 $(2l + 1)$, and hence will have 2 $(2l + 1)$ particles (the factor 2 being due to two spin states and $2l + 1$ are the m-states of 1). So for $l = 0$ (s-state) there will be 2 particles and this closes the first shell. For $l = 1$ (p-state) we have six particles, so that for, $(2 + 6 = 8)$ particles, we have a closed shell. In this manner for $l = 2$ (d-state) we have 10 particles and have a total of 18 particles. To create a closed shell for 20 particles, we require again s-state but for higher radial quantum number $n = 2$, *i.e.* 2s state [For eigen-energy states, *see* Quantum Mechanics; L.I. Schiff; p. 36]. Higher values of l give occupancy number 34, 40, 58, etc. They do not reproduce magic numbers.

However, if we assume a simple harmonic potential, as written in Eq. 10.12, it can be seen (Quantum Mechanics; L.I. Schiff, p. 60), that the eigen-values of energies for three dimensional harmonic oscillator, are given by as shown in Eq. 10.16a:

$$E_{n,\,l} = \hbar\omega\left(2n + l - \frac{1}{2}\right); \; n = 1, 2, 3, ...; \; l = 0, 1, 2, ...$$

$$= \hbar\omega\left(\eta + \frac{3}{2}\right) = E_\eta \qquad ...(10.27)$$

where $\qquad \eta = 2n + l - 2$

Again, as in the square well case, the degeneracy for each l is 2 $(2l + 1)$; but the eigen states corresponding to the same value of $2n + 1$, (*i.e.* the same η) are also degenerate. So for different value of η, the eigen-values of energies are given by Eq. 10.16. But for each η one can have different combinations of n and l. Hence for $\eta = 0$, and $n = 1$, only possible value of l is zero and hence we have the configuration $1s$. Similarly for $\eta = 1$, $n = 1$; $l = 1$ is the only possibility and hence lp configuration. For $\eta = 2$, $n = 1$, $l = 2$ and $n = 2$, $l = 0$ are two possibilities and hence $1d$, $2s$ configurations and so on.

The number of particles for each value of l is given $2\,(2l + 1)$ and hence we have 2 particles for $\eta = 0$, and $1s$ state; then 6 particles for $\eta = 1$ and $1p$ state; next 12 particles and $1p$ and $2s$ states, and so on. This sequence ends up giving 2, 8, 20, 40, 70, 112 and 168 particles as the number of particles of one type *i.e.* neutrons or protons. Evidently they do not reproduce the magic numbers.

It was at this stage of logic, that spin-orbit interaction, proportional to $\mathbf{l} \cdot \mathbf{s}$, was introduced. The spin-orbit energy for a state can be written as:

$$E_{\mathbf{l}\cdot\mathbf{s}} = U\,(r)\,\langle(\mathbf{l}\cdot\mathbf{s})\rangle$$

where

$$\langle \mathbf{l}\cdot\mathbf{s}\rangle = \frac{1}{2}\left\{\langle j^2\rangle - \langle l^2\rangle - \langle s^2\rangle\right\}$$

$$= \frac{j\,(j+1) - l\,(l+1) - s\,(s+1)}{2}$$

$$= \begin{cases} \dfrac{1}{2}\,l & \text{for } j = l + \dfrac{1}{2} \\ -\dfrac{1}{2}\,(l+1) & \text{for } j = l - \dfrac{1}{2} \end{cases} \qquad \text{...(10.28)}$$

so that if $U\,(r)$ is negative, then the energy due to spin-orbit coupling will be depressed for $j = l + 1/2$, and will be enhanced for $j = l - 1/2$, as has been briefly mentioned earlier. This is the reverse of atomic case, where the energy is enhanced for $j = l + 1/2$ and depressed for $j = l - 1/2$. This is so because in atomic level, spin-orbit-coupling constant $U\,(r)$ arises essentially out of Coulomb-interaction and is positive. On the other hand spin-coupling constant $U\,(r)$ in nuclear interaction is part of nuclear interaction arising out of meson-currents and it turns out to be negative (*see* Chapters 5 and 6 for details).

In this manner, every energy level (as discussed earlier for a common potential) is split into two levels—one corresponding to $J = l + 1/2$ which is depressed, and the other for $j = l - 1/2$, for which energy is enhanced. Such a level pattern for an odd proton is shown in Fig. 10.4. Such a diagram is not only the energy diagram of a single particle in a potential; but also gives occupancy of each level, as we build the sequence of angular momenta and energies by adding nucleons; to the nucleus. While building such occupancy diagram, one makes semi-empirical assumptions, which are now experimentally confirmed.

(*i*) Even number of protons, or even number of neutrons couple their angular momenta to give a total angular momentum of zero.

(*ii*) The total angular momentum $\mathbf{I}$ in an odd-even nuclear system is given by the angular momentum $\mathbf{I} = \mathbf{l} + \mathbf{s}$; where $\mathbf{l}$ is the orbital angular momentum and $\mathbf{s}$ is the spin of the oddeth particle. Apparently, as described earlier, $I = l + 1/2$ or $I = l - 1/2$.

(*iii*) The total angular momentum **I**, of an odd-odd system, whose oddeth nucleon may be in identical states, is equal to double the angular momenta of a nucleon. As for example, if the two nucleons—one a proton and the other a neutron are in a $j = 5/2\ \hbar$ level, the total angular momentum will be $5\hbar$. This means that their individual total spins **j** are parallel to each other.

(*iv*) The energy of a level with a given *n*, increase, with *l*; where *n* is the quantum number of the radial wave-function of the nucleons in the level. It may be noted that *n* signifies the number of nodes in the radial wave-function including the one at the origin.

When the spin-orbit coupling [Eq. 10.28] is introduced, we obtain the following quantitative situation. To each level-energy; one adds or subtracts an energy-term, depending on **l** . **s** and [Eqs. 10.2*a* and 10.3], thus splitting each level of a given 1, to two levels with $j = l + 1/2$ and $j = l - 1/2$. The energy shift of each level from central values are:

$$\Delta E_{nl}\left(j = l + \frac{1}{2}\right) = +\frac{1}{2}l\int dr\,|\,\psi_{nl}(r)\,|^2\;U(r) = -\frac{1}{2}F_{nl} \qquad \text{...(10.29}a\text{)}$$

and

$$\Delta E_{nl}\left(j = l - \frac{1}{2}\right) = -\frac{1}{2}l\int dr\,|\,\psi_{nl}(r)\,|^2\;U(r) = \frac{1}{2}(l+1)\,F_{n,l} \qquad \text{...(10.29}b\text{)}$$

where

$$F_{nl} \equiv -\int U(r)\,|\,\psi_{nl}(r)\,|^2\,dr$$

The total spin-orbit, energy splitting is then given by:

$$\Delta(E_{nl}) = \Delta E_{nl}\left(j = l - \frac{1}{2}\right) - \Delta E_{nl}\left(j = l + \frac{1}{2}\right)$$

$$= -\left(l + \frac{1}{2}\right)\int dr\,|\,\psi_{nl}(r)\,|^2\,U(r) \qquad \text{...(10.29}c\text{)}$$

Because $U(r)$ is negative, this energy splitting is positive. One then builds the sequence of levels, with the sequence as shown in Fig. 10.4. It is apparent from Eq. 10.4, that for higher level of *l*, *e.g.* for $l = 3, 4, 5, 6$, which in Fig. 10.4 correspond to *lf, lg, lh* and *li*, the spin orbit coupling results in lowering the energy of the level with $I = l + 1/2$ by ΔE_{nl} which is proportional to *l*. The number *l*, before the symbols *f, g, h, i* representing different values of *l*; corresponds to $n = l$ in Eq. 10.27. Similar is the case for higher values of *n*.

This results in splitting levels like *lg, lh* and *li.* in such a manner, that $lg_{9/2}$ comes close to $2p_{1/2}$; $lh_{11/2}$ comes close to $3s_{1/2}$, and $li_{13/2}$ comes close to $3p_{1/2}$, as shown in Fig. 10.4, leading to closed shells at 50, 82 and 126. Of course *lp, ld* and *lf* levels also split, but with lesser amount. As seen in Fig. 10.4, this also leads to closed shells at 8 and 20. This results in final sequence of levels, so that at the magic numbers of total occupancy of 2, 8, 20, 28, 50, 82 and 126; the next energy level has a large gap. This means that nuclei with neutrons or protons of magic number, are stable, and it requires, a large energy to the next excited state. In this manner the properties of extra-stability for magic number nuclei are very naturally explained.

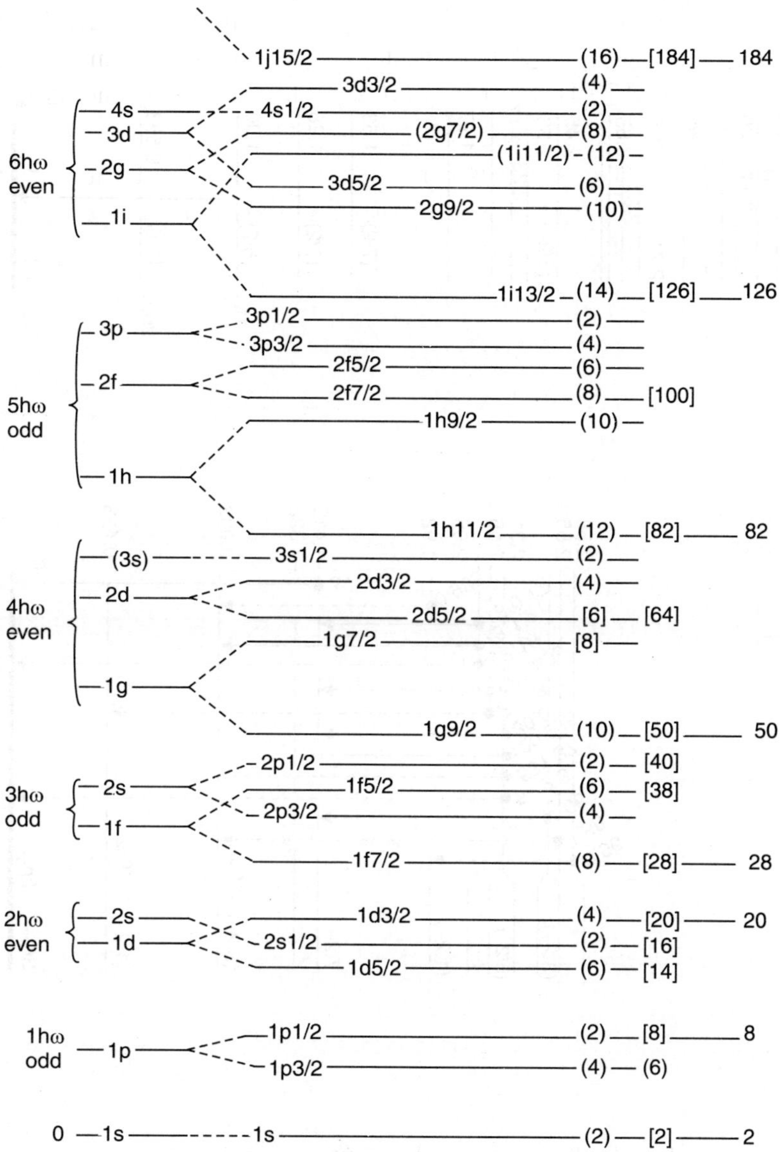

Fig. 10.4 Approximate level pattern for protons. On the extreme left are harmonic oscillator energies, based on quantum numbers n [Eq. 10.16]. Next are levels based on n and l quantum numbers. The next are levels, in which spin-orbit coupling has been included. The numbers on the right are particles allowed in each level. Next are number, which are sum of the occupied number up to that level. The numbers on the extreme right are the numbers up to the closed shell (Ref. 1).

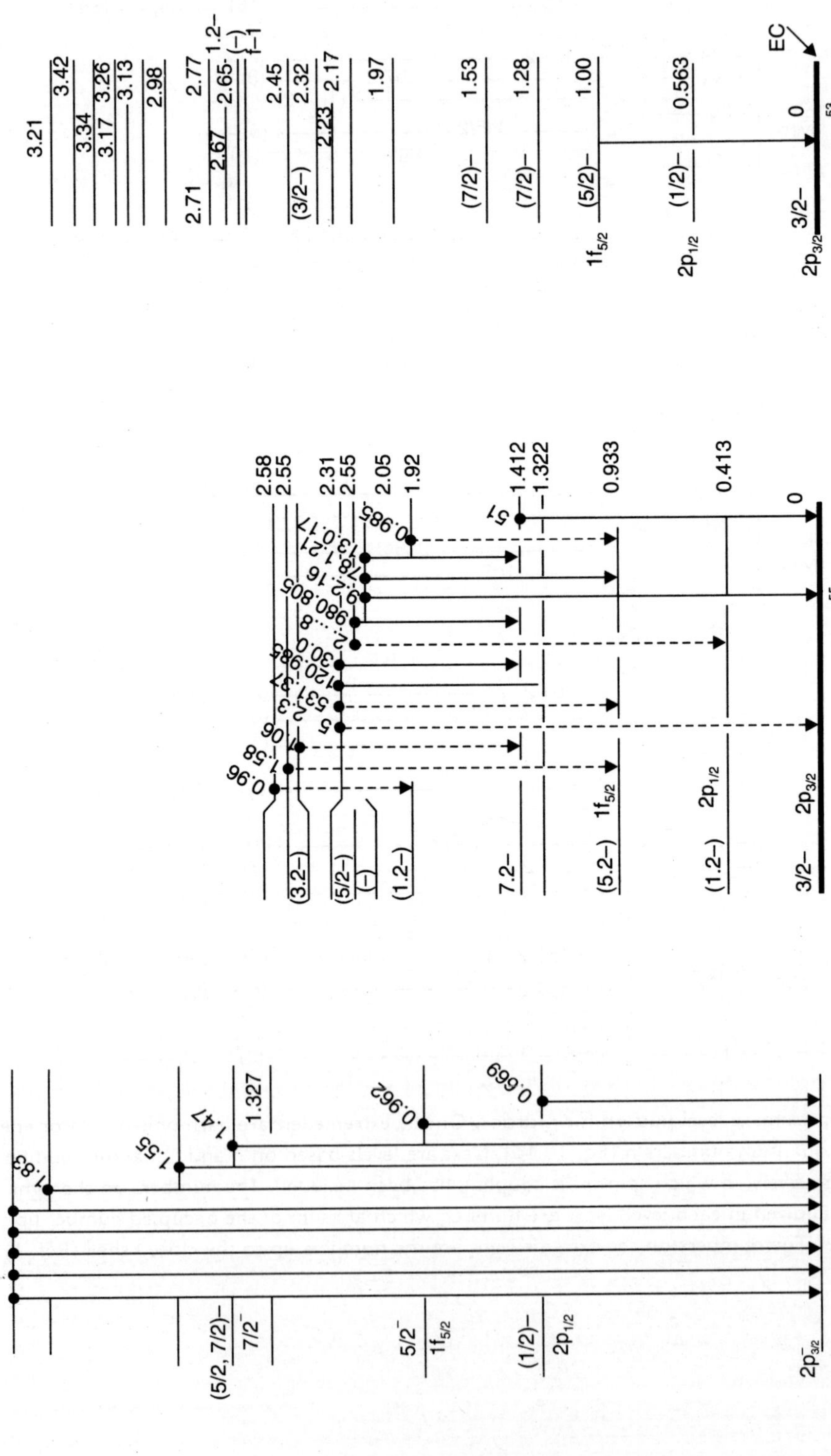

Fig. 10.5 The spins and parities of the first few excited levels of different nuclei with 29 protons or neutrons. The spins and parities of the first few excited levels is the same on all these cases (Ref. 9).

The ground state spins and parities are more easily explained by the direct application of Fig. 10.4. This is especially so, up to $A = 33$. As for example, for ${}_{16}S^{33}{}_{17}$, the experimentally determined spin and parity is $I^{\pi} = 3/2^{+}$. The single particle model will give the configuration of 17th neutron as $(1s_{1/2})^2$ $(1p_{3/2})^4$ $(1p_{1/2})^2$ $(1d_{5/2})^6$ $(2S_{1/2})^2$ $(1d_{3/2})^1$; so that; the first 16 neutrons (and also 16 protons) fill the levels up to $2s_{1/2}$ and 17th neutron goes into the 1 $d_{3/2}$ level.

On the other hand, in case like ${}_{28}\text{Ni}^{61}$, the first 28 protons and 28 neutrons, fill up the levels up to $1f_{7/2}$ and hence while the extreme single particle model predicts the last five neutron to be in $(2p_{3/2})^4$ $(1f_{5/2})^1$ configuration and hence should have $I^{\pi} = \frac{5}{2}^{-}$; the experimental value is $I^{\pi} = \frac{3}{2}^{-}$. How do we explain it? This can be explained by putting 2 particles in $1f_{5/2}$ level which pair to give zero spin and three particles in $1p_{3/2}$ level, yield $I^{\pi} = \frac{3}{2}^{-}$.

The exact ordering of the level, also may change slightly from the one shown in Fig. 10.4, because the magnitude of splitting, may not be exactly the same as assumed.

In Fig. 10.5, we have shown the experimental properties of the ground state and the excited states of three nuclei, *i.e.* ${}_{29}\text{Cu}_{23}{}^{63}$, ${}_{26}\text{Fe}_{29}{}^{55}$ and ${}_{24}\text{Cr}_{29}{}^{53}$. They have all 29 odd nucleons (29 protons in Cu^{63} and 29 neutrons in Fe^{54} and Cr^{53}), *i.e.* one nucleon more than the magic shell number of 28. All of them have $I^{\pi} = \frac{3}{2}^{-}$ for the ground state, corresponding to $2p_{3/2}$, as can be seen in Fig. 10.4. The first excited state for all the three nuclei have $I^{\pi} = \frac{1}{2}^{-}$ corresponding to $2p_{1/2}$, *i.e.* the next excited state after $2p_{3/2}$ is $2p_{1/2}$. The second excited state is $I = 5/2$, which is expected to correspond to $1\,f_{5/2}$. Comparison with Fig. 10.4 shows that the level $If_{5/2}$ has moved up, giving more splitting of $If_{5/2}$ level than given there. So these three states correspond to the configuration of an even-even case $(28p + 24n)$ in Cu^{63}, $(26p + 28n)$ in Fe^{55} and $(24p + 28n)$ in Cr^{53} and 29th nucleon going to excited states corresponding to extreme single particle shell model. For higher states core-excitation and other configuration mixing can take place.

10.5 TWO PARTICLES OUTSIDE A CLOSED SHELL

If the number of nucleons outside the closed shell are more than one, then complications arise. The general case of n nucleons outside the closed shell is, of course, too complicated, for the scope of this book. For that *see* References (14) and (16), though we will discuss it briefly later.

But a few theoretical concepts, which have been developed for two loose particles say 2 protons or 2 neutrons, or a neutron and a proton outside the close shell are very useful for solving even n particle system outside the closed shell. We will discuss these concepts one by one.

It should be immediately realised that these loose particles will interact not only with the core through a common potential $V_o\,(r)$ but also there will be an 'effective' residual two-body interaction, *i.e.* if only two body residual interactions are considered; the shell model Hamiltonian may be written for two nucleons outside the closed shell, as:

$$H = H_o + V_{ij} \quad ...(10.30)$$

where V_{ij} is the residual nucleon-nucleon potential.

This residual nucleon-nucleon interaction is taken to be charge independent. This is, however, only partially correct. Henley[16] (1969) has found that V_{np} is about 2% stronger than V_{pp} or V_{nn}. One, however, assumes, for the sake of first orientation, that concepts of invariance of iso-spin is valid and H commutes with T^2 the iso-spin operator as shown in Eq. 6.13, in Chapter 6. Then the most general residual two-body interaction is written as:

$$V_{12} = V_{\mathbf{w}} (\mathbf{r}_1, \mathbf{r}_2) + V_H (\mathbf{r}_1, \mathbf{r}_2) P_H + V_B (\mathbf{r}_1, \mathbf{r}_2) P_B + V_M (\mathbf{r}_1, \mathbf{r}_2) P_M \quad ...(10.31)$$

where the tensor forces and velocity dependent forces are neglected. The exchange operators P_H, P_B and P_M have been defined in Chapter 6.

The radial dependence of these potentials representing residual interaction are generally taken to be one of the following type; *i.e.*,

(*i*) Delta type potential, *i.e.*,

$$V_{ij} = -4\pi V_o \delta (\mathbf{r}_i - \mathbf{r}_j) \quad ...(10.32)$$

This is based on the assumption, that the residual nucleon-nucleon interaction potential is surface interaction. Other radial dependent functions which are assured many times for nucleon-nucleon radial potential are:

(*ii*) Yukawa type, *i.e.*,

$$V_{ij} = \frac{e^{-\mu |\mathbf{r}_i - \mathbf{r}_j|}}{|\mathbf{r}_i - \mathbf{r}_j|} \quad ...(10.33)$$

where $\mu = 0.7 \text{ Fm}^{-1}$.

(*iii*) Gaussian potential:

$$V_G (\mathbf{r}_i, \mathbf{r}_j) = e^{-\mu |\mathbf{r}_i - \mathbf{r}_j|^2} \quad ...(10.34)$$

where $\mu = 0.6 \text{ Fm}^{-1}$.

10.5.1 A. The Wave-Function

Let us assume that the two particles are in the same orbit; so that $j_1 = j_2 = j$; then an allowable (anti-symmetric) state of say two neutron system would be:

$$\psi_{IM} (j\,j) = 2^{-1/2} B \sum_{m_1 m_2} (j\,j\,m_1\,m_2 \,|\, I\,M) \times \{\Phi_{jm_1} (1)\, \Phi_{jm_2} (2) - \Phi_{jm_1} (2)\, \Phi_{jm_2} (1)\} \quad ...(10.35)$$

where B is determined from the normalisation condition. Keeping in mind, that anti-symmetrised wave-function can be expressed as:

$$|\,\Phi_{jm_1}\, \Phi_{jm_2}\,| = |\, a^{\dagger}_{jm_1}\, a^{\dagger}_{jm_2}\, |0\rangle \quad ...(10.36)$$

where $a^{\dagger}$ are creation operators (L.I. Schiff, p. 343)[12], we can write Eq. 10.35 as:

$$\psi_{IM}(j\,j) = 2^{-1/2}\, B \sum_{m_1 m_2} (j\,j\,m_1\,m_2 \mid IM) \times a^{\dagger}_{jm_1}\, a^{\dagger}_{jm_2} \left|0\right\rangle \qquad ...(10.37)$$

where $|0\rangle$ corresponds to the ground state belonging to even-even core, and

$$\sum_{m_1 m_2} (j\,j\,m_1\,m_2 \mid I\,M)$$

are Clebsch-Gorden Coefficients[12].

B. Energies and Angular Momenta

To get the energies, of the various states in a two-nucleon system outside the closed shell, we write the expectation value of H as:

$$\langle \psi_{IM}(jj) \mid H \mid \psi_{IM}(jj) \rangle = 2\varepsilon_j + \left\langle \psi_{IM}(jj) \middle| V_{jj} \middle| \psi_{IM}(jj) \right\rangle$$

$$= 2\varepsilon_j + E_f\,(jj,\, jj) \qquad ...(10.38)$$

An example of such as case will be ${}_{22}Ti^{50}_{28}$. In this case ${}_{20}Ca^{48}_{28}$ acts as an inert core, and the low lying states should be those corresponding to two protons in $1f_{7/2}$, which can couple to 7^+, 6^+, 5^+, 4^+, 3^+, 2^+, and 1^+, 0^+. Experimentally, the levels are found at 0^+, 2^+, 4^+ and 6^+ for ${}_{22}Ti^{50}_{28}$ while for ${}_{21}Sc^{42}_{21}$, *i.e.* for neutron plus one proton outside the core of ${}_{20}Ca^{40}_{20}$, one has observed 0^+, 1^+, 3^+, 5^+ and 7^+ levels below 3 MeV. Hence all the possible combination of angular momenta have been observed, experimentally.

For a bit more general case, when two different states are involved for one possibility (j_1, j_2) to give ψ_{IM} and two other states are involved for another possibility $(j_3\, j_4)$ for producing ψ_{IM}; then we can write:

$$H\,(\alpha_1\, \psi_{IM}(j_1\, j_2) + \alpha_2\, \psi_{IM}(j_3\, j_4))$$

$$= \tilde{E}\left[\alpha_1\, \psi_{IM}\,(j_1 j_2) + \alpha_2\, \psi_{IM}\,(j_3 j_4)\right] \qquad ...(10.39)$$

where α_1 and α_2 are normalising constants,

Multiplying both sides by $\psi_{IM}(j_1\, j_2)$ on the left and integrating over all the space, one gets:

$$\alpha_1 \left\{\left(\varepsilon_{j_1} + \varepsilon_{j_2} + E_I\,(j_1\, j_2,\, j_1\, j_2)\right)\right\} + \alpha_2\, \dot{E}_1\,(j_1\, j_2,\, j_3\, j_4) = \alpha_1\, \tilde{E} \qquad ...(10.40)$$

Similarly multiplying Eq. 10.39 by $\psi_{IM}(j_3\, j_4)$, on the left, we get

$$\alpha_1\, E_I\,(j_3\, j_4,\, j_1\, j_2) + \alpha_2 \left\{\varepsilon_{j_3} + \varepsilon_{j_4} + E_I\,(j_3\, j_4,\, j_3\, j_4)\right\} = \alpha_2\, \tilde{E} \qquad ...(10.41)$$

These two equations, *i.e.* 10.40 and 10.41, can be written as:

$$\sum_k \left\{\langle H\rangle_{ik} - \tilde{E}\delta_{ik}\right\} = 0,\; i = 1, 2 \qquad ...(10.42a)$$

(*i* corresponds to two configurations or two levels)

where $$\langle H \rangle_{11} = \varepsilon_{j_1} + \varepsilon_{j_2} + E_I\,(j_1\,j_2, j_1\,j_2) \qquad ...(10.42b)$$

with $$E_I\,(j_1\,j_2, j_1\,j_2) \equiv \langle \psi_{IM}\,(j_1\,j_2) | V_{12} | \psi_{IM}\,(j_1\,j_2) \rangle \qquad ...(10.42c)$$

and $$\langle H \rangle_{22} = \varepsilon_{j_3} + \varepsilon_{j_4} + E_I\,(j_3\,j_4, j_3\,j_4) \qquad ...(10.43a)$$

where $$E_I\,(j_3\,j_4, j_3\,j_4) \equiv \langle \psi_{IM}\,(j_3\,j_4) | V_{12} | \psi_{IM}\,(j_3\,j_4) \rangle \qquad ...(10.43b)$$

$$\langle H \rangle_{12} = \langle H \rangle_{21} = E_I\,(j_1\,j_2, j_3\,j_4) \qquad ...(10.43c)$$

and $$E_I\,(j_1\,j_2, j_3\,j_4) \equiv \langle \psi_{IM}\,(j_1\,j_2) | V_{12} | \psi_{IM}\,(j_3\,j_4) \rangle \qquad ...(10.43d)$$

In writing the above equations, we have assumed that V_{12} is Hermitian. We can solve these equations either through perturbation theory or through other methods, if E is large. A typical simple case following the above situation is ${}_8O^{18}_{10}$; where we can consider ${}_8O^{16}_8$ as the inert core and outer two neutrons in $1d_{5/2}$ and $1s_{1/2}$ orbitals. Experimentally,[14, 16] one has ground state with o^+; the first excited state at 1.98 MeV with 2^+ and the second excited state at 3 MeV, with 4^+. It is easy to see that these two states can produce the following spins.

$$(d_{5/2})^2_I,\ I = 0, 2, 4;\ (d_{5/2}, s_{1/2})_I\,,\ I = 2, 3;\ (S_{1/2})^2_I,\ I = 0 \qquad ...(10.44)$$

One can, therefore, write the wave-functions ψ_{IM} as:

$$\psi_{00} = a_o\,(d_{5/2})^2_{0,\,0} + b_0\,(s_{1/2})^2_{0,\,0} \text{ for } I = 0$$

and $$\psi_{2M} = a_2\,(d_{5/2})^2_{2M} + b_2\,(d_{5/2}\,s_{1/2})_{2M} \text{ for } I = 2$$

$$\psi_{4M} = a_3\,(d_{5/2})^2{}_{4M} \text{ and } \psi_{3M} = a_4\,(d_{5/2}\,s_{1/2})_{3M}$$

for $I = 4$ for $I = 3$ ...(10.45)

Here $j_1 = j_2 = 5/2; j_3 = j_4 = 1/2$. It can be easily seen from Eq. 10.45 that for $I = 3$, and 4, only one possibility exists and the wave-function has only single component in these cases. So the Hamiltonian matrix for these states is (1×1) and hence their energies are given by an expression similar to Eq. 10.38, but with two different values of j.

$$\langle H_{I=3} \rangle = \varepsilon_d + \varepsilon_S$$

This corresponds to one particle in $d_{5/2}$ state, and the other in $s_{1/2}$ state:

$$\langle H_{I=4} \rangle = 2\varepsilon_d - \frac{2}{7} V_o\,R_o \qquad ...(10.46)$$

This corresponds to both the particles in $d_{5/2}$. The quantity $2/7\ V_o\,R_o$ corresponds to interactions between the two particles, *i.e.*, $E_1\,(j_3 j_3, j_3 j_3)$, as in Eq. 10.38, where V_o comes from the delta function assumed for the V_{ij} potential, which will be the case for nucleons on the surface, for which we write:

$$V_{ij} = -4\pi\,V_o\,\delta\,(\mathbf{r}_i - \mathbf{r}_j) \qquad ...(10.47)$$

and R_o comes from the expression of radial wave-function integral given by:

$$\overline{R} = \int R_{j_1}(r)\, R_{j_2}(r)\, R_{j_3}(r)\, R_{j_4}(r)\, dr = (-1)^{n_1+n_2+n_3+n_4} R_o \quad \text{...(10.48)}$$

where n_i = the number of radial nodes for the state j_i, and R_o is a positive number and factor 2/7 in Eq. 10.46 comes from the properties of C.G.F. coefficients. Using the wave-function for $I = 0$ and $I = 2$ [Eq. 10.45], we can write (Ref. 16):

$$\text{For } I = 0 \quad \langle H_{I=0}\rangle = \begin{pmatrix} 2\varepsilon_d - 3V_o R_o & -\sqrt{3}\,V_o R_o \\ -\sqrt{3}\,V_o R_o & 2\varepsilon_S - V_o R_o \end{pmatrix}$$

$$\text{and for } \quad I = 2 \quad \langle H_{I=2}\rangle = \begin{pmatrix} 2\varepsilon_d - \frac{24}{35}V_o R_o & -\frac{12\sqrt{7}}{35}V_o R_o \\ -\frac{12\sqrt{7}}{35}V_o R_o & \varepsilon_d + \varepsilon_S - \frac{6}{5}V_o R_o \end{pmatrix} \quad \text{...(10.49)}$$

The exact matrix elements are obtained by calculating the eigen-functions ψ's and eigen values E's, calculating in details the value of $E_I\,(j_1 j_2 j_3 j_4)$ for two identical particles.

Diagonalisation process of matrix in Eq. 10.49 gives the energies, corresponding to $(d_{5/2})^2$ and $(S_{1/2})^2$ configurations with

$$\varepsilon_d = \text{B.E.}\left(8O_9^{17}\right) - \text{B.E.}\left(8O_8^{16}\right) = -4.143 \text{ MeV}$$

where B.E. is the negative of the total binding energy of the nucleus, found experimentally. Since $s_{1/2}$ state lies at an excitation energy of 871 keV for ${}_8O_9^{17}$, one can write:

$$\varepsilon_s = -4.143 + 0.871 = -3.272 \text{ MeV}$$

On the other hand, the value of $V_o R_o$ may be fixed by requiring that the observed excitation energy of one of the states be reproduced or by requiring that the experimental binding energy of ${}_8O_{10}^{18}$, relatives to ${}_8O_8^{16}$ is correctly obtained *i.e.* lower of the eigen value E_o of $<H_{I=0}>$ should be:

$$E_o = \text{B.E.}\left({}^{18}_{8}O_{10}\right) - \text{B.E.}\left({}^{16}_{8}O_8\right) = -12.189 \text{ MeV}$$

Then from $<H_{I=0}>$ [Eq. 10.49]

$$V_o R_o = 1.057 \text{ MeV}$$

One can, similarly calculate the energy of the next higher state for $I = 2$, and from $\langle H_{I=2}\rangle$, one can calculate the energies of the first two 2^+ states, and from $\langle H_{I=3}\rangle$, and $\langle H_{I=4}\rangle$, the energies of 3^+ and 4^+ states and compare them with the experimental values. Figure 10.6 shows the comparison of energy levels of ${}_8O_8^{16}$ between theory and experiment. Except the experimental energy level at 3.63 MeV (o^+) level, all other levels have one to one correspondence. The quality of agreement between theory and experiment can be somewhat improved by taking of finite range spin dependent residual two body interaction.

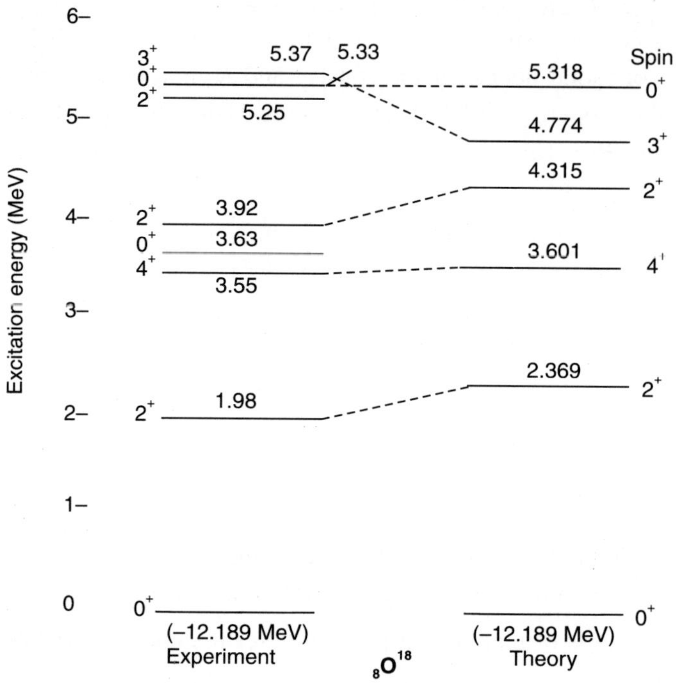

Fig. 10.6 Comparison of experimental and theoretical energies of energy levels [Eqs. 10.44 to 10.49 of $_8O^{18}_{10}$ (Ref. 14, 16)].

Even, the first o^+ excited state at 3.63 MeV is explained by using an oscillator wave-function, and finite range potential. A 3 MeV level, with predominant $(s_{1/2})^2$ configuration is predicted, as was done by Elliot and Flowers[15] in 1955. As a matter of fact, later detailed calculations showed that this corresponds to the core excitation, due to four-particle two-hole component.

C. Eigen-Functions

The process followed for obtaining eigenvalues of $\langle H_{I=0}\rangle$ and $\langle H_{I=2}\rangle$ can also be used to obtain eigen-functions.

The results are:

$$\psi_{00} = 0.929(d_{5/2})^2_{00} + 0.371\,(s_{1/2})^2_{00}$$

and

$$\psi_{2M} = 0.764\,(d_{5/2})_{2M} + 0.645\,(d_{5/2}\,s_{1/2})_{2M} \quad ...(10.45b)$$

D. Angular Momenta

First of all, we write the nuclear configuration of a shell, with two nucleons outside the closed shell. It can be written as:

$$(n_1\, l_1\, j_1)^{2(2j_1+1)}\,(n_2\, l_2\, j_2)^{2(2j_2+2)} \ldots (n_i\, l_i\, j_i)^{2(2j_i+1)}\,(n, l, j)^{k_1}\,(n, l, j)^{k_2} \quad ...(10.50)$$

where $k_1 = 1$ and $k_2 = 1$. The closed shells up to $(n_i\ l_i\ j_i)^{2(2j_i+1)}$ yield a resultant angular momenta zero, and acts as an inert core for the first few excited levels.

Let two particles be in the single particle states $j_{k_1=1}$ and $j_{k_2=1}$. Then, the maximum angular momentum possible is

$$I_{\max} = j_{k_1=1} + j_{k_2=1} \quad \text{...(10.51}a\text{)}$$

other values of I are given by:

$$\left| j_{k_1=1} - j_{k_2=1} \right| \leq I \leq \left| j_{k_1=1} + j_{k_2=1} \right| \quad \text{...(10.51}b\text{)}$$

Two neutrons or two protons, in the same single particle orbit j (j = half integral) can only couple their spins to even values of I, *i.e.*,

$$I = 0, 2, 4, \ldots (2j - 1)$$

This can be seen, by writing the wave-function ψ_{IM} for two indentical particles:

$$\psi_{IM}(1, 2) = \frac{B}{\sqrt{2}} \sum_{m_1 m_2} (j_1\ j_2\ m_1\ m_2 \mid I\ M)$$

$$\left[\phi_{j_1 m_1}(1)\, \phi_{j_2 m_2}(2) - \phi_{j_1 m_1}(2)\, \phi_{j_2 m_2}(1)\right] \quad \text{...(10.52)}$$

where B is determined from normalisation condition.

It can be seen, that

$$\langle \psi_{IM}(1,2)\, \psi_{IM}(1,2) \rangle = B^2 \sum_{m_1 m_2 m_1 m_2} \sum (j_1\ j_2\ m_1\ m_2 \mid IM)(j_1\ j_2\ m_1'\ m_2' \mid IM)$$

$$\times \left\{\delta_{m_1 m_1'}\, \delta_{m_2 m_2'} - \delta_{j_1 j_2}\, \delta_{m_1 m_2'}\, \delta_{m_2 m_1'}\right\}$$

$$= B^2 \left\{1 - \delta_{j_1 j_2} (-1)^{j_1 + j_2 - 1}\right\} \sum_{m_1 m_2} (j_1\ j_2\ m_1\ m_2 \mid IM)^2$$

$$= B^2 \left\{1 - \delta_{j_1 j_2} (-1)^{j_1 + j_2 - 1}\right\} \quad \text{...(10.53)}$$

where summation properties of Clebsch-Gorden Coefficients have been used. So, if $j_1 = j_2$, it is impossible to write a normalised antisymmetric wave-function when I is odd. Hence only even values of I are possible. Experimentally, the excited energies of ${}_{22}\mathrm{Tc}_{28}^{50}$ as shown in Fig. 10.7 bear out this theorem.

10.6 THREE OR MORE PARTICLES OUTSIDE A CLOSED SHELL (SINGLE PARTICLE-MODEL)

The example of two particles outside the closed shell, discussed in the last section was the simplest example of the applications of the so-called individual particle model which, in principle, can be used for any number of the particles outside an inert core; so long as the core itself is not excited. We will, first discuss the general case and then apply it to the three nucleons outside the close shell inert-core.

10.6.1 Wave-Function

(*i*) We have already discussed the wave-function (anti-symmetrised) for two particles outside the closed shell. If there are n particles outside closed shell, the anti-symmetrised wave-function for nucleons outside the closed shell is given by Slater determinant:

$$\psi_{IM}^{n}(r) = (n!)^{-1/2} \begin{vmatrix} \phi_{j_1m_1}(1) & \phi_{j_2m_2}(1) \ldots \phi_{j_nm_n}(1) \\ \phi_{j_1m_1}(2) & \phi_{j_2m_2}(2) \ldots \phi_{j_nm_n}(2) \\ \vdots & \\ \phi_{j_1m_1}(n) & \phi_{j_2m_2}(n) \ldots \phi_{j_nm_n}(n) \end{vmatrix}$$

$$\equiv \left| \phi_{j_1m_1} \quad \phi_{j_2m_2} \ldots \phi_{j_nm_n} \right| \qquad \ldots(10.54)$$

which is anti-symmetric (from the properties of the determinant), to the interchange of any two particles and vanishes when two particles occupy the same quantum state. Due to this condition, the only possible M state, for the configuration $I = (j)^{2I+1}$ (when particles completely fill the orbital), is

$$M = \sum_i m_i = 0 \qquad \ldots(10.55a)$$

where $\qquad m_i \equiv j_{m_i}$.

Without proof, for which *see* Ref. (16), we write down a few more useful theorems for $n < 2j + 1$ particles outside a closed shell.

(*i*) The maximum possible angular momentum that can arise in the configuration j^n is

$$I_M = n\left[j - \frac{n-1}{2}\right] \qquad \ldots(10.55b)$$

(*ii*) There is no state of j^n, with $I = I_M - 1$ $\qquad \ldots(10.55c)$

(*iii*) In the configuration j^n, there is one state with

$$I = I_M - 2 = n\left\{j - \frac{n-1}{2}\right\}^{-2} \qquad \ldots(10.55d)$$

where M corresponds to maximum M-value, *i.e.* $I_M = M_{\max}$.

Mayor and Jenson[1] have given a comprehensive table of total spin I for various configuration J^n, from which we give some results, to illustrate the above theorems (*see* Table 10.1).

Table 10.1

j = 3/2
$n = 1, I = 3/2$
$n = 2, I = 0, 2$
j = 5/2
$n = 1, I = 5/2$
$n = 2, I = 0, 2, 4$
$n = 3, I = 3/2, 5/2, 9/2$
j = 7/2
$n = 1, I = 7/2$
$n = 2, I = 0, 2, 4, 6$
$n = 3, I = 3/2, 5/2, 7/2, 9/2, 11/2, 15/2, ...$
$n = 4$, $I = 0$, 2 (twice), 4 (twice), 5, 6, 8
j = 9/2
$n = 1, I = 9/2$
$n = 2, I = 0, 2, 4, 6, 8$
$n = 3$, $I = 3/2$, 5/2, 7/2 (twice), 11/2, 15/2, 17/2, 21/2
$n = 4$, $I = 0$ (twice), 2 (twice), 3, 4 (3 times) 5, 6 (3 times), 7, 8, 9, 10, 12, etc.

The above values of I, for different values of n obey the theorems mentioned above. Sometimes one can obtain the same I by different possible combination of m_1 belonging to the orthogonal states. The number of such combinations is given in brackets for cases where this is applicable.

10.6.2 Nordheim's Rules for Total Angular Momenta[18]

Apart from the rules, which have been discussed for obtaining the total angular momenta for n nucleons outside the closed shell; we give below Nordheim's coupling rules, without proving them; but which are based on detailed consideration as above and empirical data from β-decay. These are basically semi-empirical in nature.

If we have k_n neutrons of the configurations $(n_n, l_m, j_l)^{kn}$ coupled to j_1 and similarly k_p protons of the configuration $(n_p, l_p, j_2)^{kp}$ coupled to j_2; then apparently, the resultant J of j_1 and j_2 will obey the law:

$$|j_1 - j_2| \leq J \leq |j_1 + j_2|$$

Nordheim proposed two empirical coupling rules, that predict the ground state spin J of odd-odd nuclei; these are:

(*i*) Strong rule: which is obeyed frequently:

$$J = |J_1 - J_2| \text{ for } j_1 = l_1 \pm \frac{1}{2} \text{ and } j_2 = l_2 \pm \frac{1}{2}$$

(*ii*) Weak rule: which is obeyed less frequently:

$$|J_1 - J_2| \leq J \leq J_1 + J_2 \text{ for } j_1 = l_1 \pm \frac{1}{2}; j_2 = l_2 \pm \frac{1}{2}$$

Here j_1 l_1 and j_2 l_2 are single particle angular momenta while J_1 and J_2 are the total angular momenta of adjacent odd-*A* nuclei involved in β-decay.

(*iii*) Modified Nordheim's rules by Brennan and Brenstein[19] are further given as:

$$(a)\ J = |J_1 - J_2| \text{ for } j_1 = l_1 \pm \frac{1}{2} \text{ and } j_2 = l_2 \pm \frac{1}{2}$$

$$(b)\ J = |J_1 \pm J_2| \text{ for } j_1 = l_1 \pm \frac{1}{2} \text{ and } j_2 = l_2 \pm \frac{1}{2}$$

$$(c)\ J = J_1 \pm J_2 - 1$$

The above rules, basically indicate the tendency of spins of the protons and neutrons to line up parallel or anti-parallel and are essentially empirical.

10.6.3 A. Coefficients of Fractional Parentage (C.F.P.) for *n*-Particles[20]

In writing the expectation values, like total energy of *n* number of nucleons outside the closed shell; one finds it convenient to express the Slater determinant, Eq. 10.54, in such a way, that the single particle wave-function of one particle (say *n*th) is explicitly separated and the Slater integral is expanded in terms of the minor of the last row, excluding the wave-function of the *n*th particle. As for example, one can write the wave-function of *n* particles as:

$$\psi_{IM}(1 \ldots n)$$

$$= \sum_{J\beta} \left\langle j^{n-1} J\beta, j \middle|\right\} j^n I\alpha \left.\right\rangle \left[\Phi_{J\beta}(1 \ldots n-1)\, \phi_j(n)\right]_{IM} \qquad \ldots(10.56)$$

where $$\left\langle j^{n-1} J\beta, j \middle|\right\} j^n I\alpha \left.\right\rangle$$

are called the coefficients of fractional parentage (c.f.p.) which are so chosen that the wave-functions is anti-symmetric to the interchange of any two particles and are real quantities in the representation chosen above.

It may be seen that the wave-function has been divided between the $\Phi_{J\beta}$ (1 ... n – 1), which will be written in the form of Slater determinant of the terms of minors of the last row and ϕ_j (n) the single particle wave-function of *n*th particle. In writing Eq. 10.56; *J* is the total angular momentum for (n – 1) particles so that

$$\mathbf{J} + \mathbf{j} = \mathbf{I}$$

and α and β are the additional quantum numbers, in case the angular momentum is not sufficient to specify completely the state of many-particle systems.

It has been shown that (Ref. 16, 20):

$$\left\langle j^{n-1}\, J\beta, j \,\middle|\middle\}\, j^n\, I\alpha \right\rangle = \frac{(-1)^{n-1}\left\langle \psi_{I\alpha} \,\|\, \alpha_j^+ \,\|\, \Phi_{J\beta} \right\rangle}{\sqrt{n}} \qquad ...(10.57)$$

These coefficients have been tabulated in References (20), (21) and (22).

We now, illustrate the use of c.f.p. for a case; say $(1f_{7/2})^3_{I=5/2}$ in the next section.

B. Coefficients of Fractional Parentage for a Three Particle Configuration $(\mathbf{If}_{7/2})^3$

When the number of 'loose' particle are three or more, one developes a certain coupling scheme for the calculations of the matrix elements for

$$V = \sum_{i<j} V_{ij}$$

Let us consider the case of 3 particle configuration of $(1f_{7/2})^3$. This will be the case of say ${}_{20}Ca^{43}_{23}$, where 3 neutrons will be in the orbit of $1f_{7/2}$. Theoretically, it is possible to prove, that V has non-vanishing matrix elements, only for $I = I'$ in the reduced matrix element $\left\langle \Psi_{I'} \,\|\, Q_\lambda \,\|\, \Psi_I \right\rangle$; [Eq. 10.87]; and is independent of M. Thus one requires to evaluate V for $I = 15/2$, $M = 15/2$, *i.e.* the value of $\psi_{15/2,\,15/2}$ in terms of Slater determinant, involving single particle wave-functions $\phi_{7/2,\,7/2}$, $\phi_{7/2,\,5/2}$ and $\phi_{7/2,\,3/2}$ for the three particles. One, then, writes:

$$\psi_{15/2,\,15/2} = (3!)^{-1/2} \begin{vmatrix} \phi_{7/2,7/2}(1) & \phi_{7/2,5/2}(1) & \phi_{7/2,3/2}(1) \\ \phi_{7/2,7/2}(2) & \phi_{7/2,5/2}(2) & \phi_{7/2,3/2}(2) \\ \phi_{7/2,7/2}(3) & \phi_{7/2,5/2}(3) & \phi_{7/2,3/2}(3) \end{vmatrix}$$

$$\equiv \left| \phi_{7/2,7/2} \quad \phi_{7/2,5/2} \quad \phi_{7/2,3/2} \right| \qquad ...(10.58)$$

Then, the matrix element of V becomes a sum of three terms, where

$$V = \sum_{i<j} V_{ij}$$

and

$$\langle V \rangle = \left\langle \psi_{15/2,15/2} \,\middle|\, \sum_{i<j} V_{ij} \,\middle|\, \psi_{15/2,15/2} \right\rangle$$

$$= \left\langle \psi_{15/2,15/2} \,\middle|\, V_{12} + V_{13} + V_{23} \,\middle|\, \psi_{15/2,15/2} \right\rangle \qquad ...(10.59)$$

To evaluate each term, say V_{12}, it is convenient to rewrite the wave-function in a form, in which particle 1 and 2 are explicitly separated from particle 3. This is done by expanding the Slater determinant in terms of the minor of the last row. Thus we get:

$$\psi_{15/2,\,15/2} = \frac{1}{\sqrt{3}}\Big\{\Big|\,\phi_{7/2,5/2}\,\phi_{7/2,3/2}\Big|\,\phi_{7/2,7/2}\,(3) - \Big|\,\phi_{7/2,7/2}\,\phi_{7/2,3/2}\,\Big|\,\phi_{7/2,5/2}\,(3)$$

$$+ \Big|\,\phi_{7/2,7/2}\,\phi_{7/2,5/2}\,\Big|\,\phi_{7/2,3/2}\,(3)\Big\} \quad ...(10.60)$$

Now keeping in mind Eqs. 10.52 and 10.53 we can write:

$$\Phi_{JM}\,(1, 2) = \frac{1}{2}\sum_{mm'}(7/2, 7/2\;mm' \mid IM)$$

$$\times \{\phi_{7/2}\,(1)_m\,\phi_{7/2}\,(2)_{m'} - \phi_{7/2}\,(2)_m\,\phi_{7/2}\,(1)_{m'}\}$$

$$= \frac{1}{\sqrt{2}}\sum_{mm'}(7/2\;7/2\;mm' \mid JM)\Big|\,\phi_{7/2m}\,\phi_{7/2m'}\,\Big| \quad ...(10.61)$$

Further, using completeness relationship for Clebsch-Gorden Coefficients, [Ref. (22)], one finds:

$$\Big|\,\phi_{7/2m}\,(1)\,\phi_{7/2m'}\,(2)\,\Big| = \sqrt{2}\sum_{JM}(7/2, 7/2\;mm' \mid JM)\,\Phi_{JM}\,(1, 2) \quad ...(10.62)$$

Using the table of Clebsch-Gorden Coefficients[22], we write:

$$\psi_{15/2,\,15/2} = \frac{1}{\sqrt{3}}\Bigg[-\sqrt{\frac{15}{22}}\,\phi_{44}\,(1, 2)\,\phi_{7/2\,7/2}\,(3)$$

$$+ \sqrt{\frac{51}{22}}\Bigg\{\sqrt{\frac{22}{51}}\,\Phi_{66}\,(1, 2)\,\phi_{7/2\,3/2}\,(3) - \sqrt{\frac{22}{51}}\,\Phi_{65}\,(1, 2)\,\phi_{7/2\,5/2}\,(3)$$

$$+ \sqrt{\frac{7}{51}}\,\Phi_{64}\,(1, 2)\,\phi_{7/2\,7/2}\,(3)\Bigg\}\Bigg] \quad ...(10.63)$$

The coefficients $\sqrt{22/51}$, $-\sqrt{22/51}$ and $\sqrt{7/51}$ are precisely the Clebsch-Gorden Coefficients, (C.G.C.), that ensure that spins 6 and 7/2 couple to $I = 15/2$, $M = 15/2$. Also C.G.C. (4, 7/2, 4, 7/2 | 15/2, 15/2) = 1. This gives the result for $\psi_{15/2,\,15/2}$ in terms of $\Phi_6\,(1, 2)$, and $\Phi_4\,(1, 2)$, where ϕ's are the wave-functions for 1 and 2 particles coupled to obtain spin 6 and 4, respectively. Then

$$\psi_{15/2,\,15/2} = \sqrt{\frac{17}{22}}\,[\Phi_6\,(1, 2)\,\phi_{7/2}\,(3)]_{15/2,\,15/2} - \sqrt{\frac{5}{22}}\,[\,\Phi_4\,(1, 2)\,\phi_{7/2}\,(3)]_{15/2,\,15/2} \quad ...(10.64)$$

It should be realised that ϕ's are single particle wave-function while Φ's are for two particles 1 and 2. Then the matrix element for V_{12} may be written, using the above equation and keeping in mind that V_{12} is independent of M; nor does it operate on particle 3. Hence the m value in $\phi_{7/2}(3)_m$ is unchanged. Then,

$$\langle \psi_{15/2,15/2} | V_{12} | \psi_{15/2,15/2} \rangle = \frac{17}{22} \sum_{mM} (67/2\ Mm | 15/2\ 15/2)^2$$

$$\times \langle \Phi_{6M} | V_{12} | \Phi_{6M} \rangle + \frac{5}{22} \sum_{mM} (47/2\ M\, m | 15/2\ 15/2)^2 \langle \Phi_{4M} | V_{12} | \Phi_{4M} \rangle$$

$$= \frac{17}{22} E_6 \left(\frac{7}{2}\frac{7}{2}, \frac{7}{2}\frac{7}{2} \right) + \frac{5}{22} E_4 \left(\frac{7}{2}\frac{7}{2}, \frac{7}{2}\frac{7}{2} \right) \qquad \text{...(10.65)}$$

where E_6 and E_4 are energies corresponding to total spin 6 and 4; from Eq. 10.64 and $3\varepsilon_{7/2}$ in Eq. 10.66 is the eigen value for H_o. It should be kept in mind, that eigen-energies of H_{13} and H_{23} have the same coefficients as for V_{12} and hence the multiplication of 3. Finally, one can write the matrix element of V as the sum of the $V_{12} + V_{13} + V_{23} + H_o$ and one gets:

$$\langle \psi_{15/2,15/2} | H_o + V_{12} + V_{13} + V_{23} | \psi_{15/2,15/2} \rangle$$

$$= 3\varepsilon_{7/2} + 3 \left[\frac{17}{22} E_6 \left(\frac{7}{2}\frac{7}{2}, \frac{7}{2}\frac{7}{2} \right) + \frac{5}{22} E_4 \left(\frac{7}{2}\frac{7}{2}, \frac{7}{2}\frac{7}{2} \right) \right] \qquad \text{...(10.66)}$$

where E_6 and E_4 are defined like E_f in Eq. 10.38.

The number $\sqrt{17/22}$ and $-\sqrt{5/22}$ in Eq. 10.64 are the coefficients of fractional parentage (c.f.p.). They are basically numbers whose values give the probability that, in the anti-symmetric three particle $(1f_{7/2})^3$ system, one will find the configuration $[\Phi_6\,(1, 2) \times \phi_{7/2}\,(3)]_{15/2,\,15/2}$ and $[\Phi_4\,(1, 2) \times \phi_{7/2}\,(3)]_{15/2,\,15/2}$ respectively.

They are apparently useful in writing the matrix elements of V as shown in Eq. 10.66.

10.6.4 Seniority[22]

Another useful concept in dealing with n 'loose' nucleons outside the close shell is that of seniority quantum number, for the identical particle configuration j^n. We will see that seniority quantum number is a 'good' quantum number and is very useful in describing a given configuration.

When there are n identical nucleons, outside a closed shell, it is self-evident, that some nucleons may pair themselves, to give zero angular momentum, while others may be unpaired. It can be proved that total interaction energy of n nucleons becomes less attractive as I increases. So minimum possible I gives stable state. Hence there will be a tendency to pair, and only minimum possible number of nucleons will be unpaired. The normalised wave-function describing a paired state may be written as:

$$\psi_{00}\,(j^2) \equiv (j\,j\,m, -m \,|\, 00)\ \phi_{jm}\ \phi_{j,\,-m}$$

where
$$(j\,j\,m\,m' \,|\, 00) = (-1)^{j-m}\,(2j + 1)\ \delta_{m,\,-m'} \qquad \text{...(10.67)}$$

and can be expressed in terms of creation and destruction operator $a^{\dagger}_{jm}$ and a_{jm} as:

$$\psi_{00}(j^2) = \sqrt{\frac{2}{(2j+1)}}\, S_+(j)\,|\,0\rangle \qquad ...(10.68)$$

where
$$S_+(j) = \sum_{m>0} (-1)^{j-m}\, a^{\dagger}_{jm}\, a^{\dagger}_{j,-m} \qquad ...(10.69)$$

Hence $S_+(j)\,|\,0\rangle$ corresponds to a case of two particles oppositely oriented. It is convenient to write:

$$S_+(j)\,|\,0\rangle \;\Rightarrow\; \uparrow\downarrow \qquad ...(10.70)$$

On the other hand, the two particle creation operator $A_{IM}^{+}(j, j\,|\,0)$ which does not pair in spin, can be pictured as:

$$A^{\dagger}_{IM}(j\,j\,|\,0\rangle \;\Rightarrow\; \uparrow\uparrow \qquad ...(10.71)$$

Then the following relationships follow:

(*i*) For even *n* particles, the lowest energy state, would correspond to all nucleons having their spins paired and the eigenfunction may written as:

$$\{S_+(j)\}^{n/2}\,|\,0\rangle \;\Rightarrow\; \uparrow\downarrow\uparrow\downarrow\ldots\uparrow\downarrow \qquad ...(10.72)$$

(*ii*) When this system is excited to say 1.5–3.0 MeV, the next energy state may be described by wave-functions having two unpaired particles: *i.e.*,

$$A^{\dagger}_{IM}(j\,j)\,\{S_+(j)\}^{\frac{n-2}{2}}\,|0\rangle \Rightarrow \uparrow\uparrow\uparrow\downarrow\ldots\uparrow\downarrow \qquad ...(10.73a)$$

$A^{\dagger}_{IM}$ stands for two particle creation operator, given by:

$$A^{\dagger}_{IM}(j\,j) = \sum_{\substack{m_1\\ m_2}} (j\,j\,m_1\,m_2\,|\,IM)\, a^{*}_{jm_1}\, a_{jm_2} \qquad ...(10.73b)$$

(*iii*) The next excited states, will, then, be with four unpaired nucleons and may be written as:

$$B^{\dagger}_{IM}\,\{S_+(j)\}^{\frac{n-4}{2}}\,|0\rangle \;\Rightarrow\; \uparrow\uparrow\uparrow\uparrow\uparrow\downarrow\ldots\uparrow\downarrow \qquad ...(10.74)$$

where $B^{\dagger}_{IM}$ stands for a four particles creation operator, in which no particles are paired.

It is evident, that the number of unpaired particles can be used as a device for labelling eigen-function for the configuration of j^n. We then, define a seniority quantum number ν of nuclear state as the number of unpaired nucleons in the eigenfunction describing the state, so that Eq. 10.70 represents a state of zero seniority; Eq. 10.73 a state of seniority two and Eq. 10.74 a state of seniority 4, and so on.

For an odd *n*, it may be easily seen, that the lowest state will have seniority one and other possible states will correspond to seniority 3, 5, ... *n*. The ground state of this system may be written as

$$a_{jm}^{\dagger}\,\{S_+(j)\}^{\frac{n-1}{2}}\,|0\rangle \;\Rightarrow\; \uparrow\uparrow\downarrow\uparrow\downarrow\ldots\uparrow\downarrow \qquad \text{...(10.75)}$$

The seniority quantum number is a good quantum number for the identical particle configuration j^n, if the residual interaction is a delta function potential. For medium and large A; for which $R \approx 1.2 \times A^{1/3} \times 10^{-13}$ cm and hence much larger than the π-meson compton wavelength, which is of the range of residual interaction; the δ-function potential may be assumed and hence the seniority quantum number is a good number for medium and large A nuclei[23].

We now show that seniority quantum number ν is a good quantum number. For this we realise that for two nucleons coupled to $I = 0$, their spins are paired and the normalised wave-function can be described as:

$$\psi_{00}\,(j\,j) = \frac{1}{\sqrt{2}}\sum_m (j\;j\;m\;m' \,|\, 00)\,\phi_{jm}\,(1)\,\phi_{j,-m}\,(2)$$

$$= \left[2\,(2j+1)\right]^{-1/2}\sum_m (-1)^{j-m}\,a_{jm}^{\dagger}\,a_{j,-m}^{\dagger}\,|0\rangle \qquad \text{...(10.76)}$$

[Because $(j\;j\;m\;m' \,|\, 00) = (-1)^{j-m}\,a_{jm}^{\dagger}\,a_{j,-m}^{\dagger}\,|0\rangle \times \phi_{jm}\,(1)\,\phi_{j,-m}^{(2)}$

or
$$\psi_{00}\,(j,j) = \left[2\,(2j+1)\right]^{-1/2}\,S_+\,(j)\,|00\rangle\,] \qquad \text{...(10.77)}$$

where
$$S_+\,(j) = \sum_{m>0}(-1)^{j-m}\,a_{jm}^{\dagger}\,a_{j,-m}^{\dagger}$$

Physically $S_+\,(j)$ is an operator, which creates a pair with quantum numbers j and m and j and $-m$, *i.e.* two oppositely oriented particles.

Similarly we define another operator $S_-\,(j)$ as:

$$S_-\,(j) \equiv \sum_{m>0}(-1)^{j-m}\,a_{j-m}\,a_{jm} \qquad \text{...(10.78)}$$

Physically operator $S_-\,(j)$ destroys a pair, which is coupled to give zero spin. We can, therefore define N, the number operator as:

$$N = \sum_m a_{j,m}^{\dagger}\,a_{j,-m} \qquad \text{...(10.79)}$$

Operation N has the property that when it operates on many-particle wave-function constructed say from ϕ_{jm}, it gives back the wave-function again; multiplied by the number of particles in the state j. Physically N destroys a particle with j, $-m$ and creates a particle with j, m. We further define:

$$\Omega_j = \frac{2j+1}{2} \qquad \text{...(10.80)}$$

So that Ω_j gives the pair-degeneracy of the state j, *i.e.* the number of pairs existing in the same state:

Then we can define an operator S_z, so that

$$S_Z(j) = \frac{1}{2}(N_j - \Omega_j) \qquad ...(10.81)$$

Physically $S_z(j)$ corresponds to pair degeneracy left in the state, after operation of S_z. It has been proved that following commutation relation (Kerman, 1961)[23] holds good, *i.e.*,

$$[S_k - S_1] = S_k S_1 - S_1 S_k = iS_m \text{ } (k, l, m \text{ cyclic}) \qquad ...(10.82)$$

where $S_{k, l, m}$ are called the quasi-spin operators. Equation 10.82 shows that quasi-spin operators obey the usual angular-momentum commutation relationships.

Now seniority quantum number ν, is defined in such a way, that a ν-particle state with seniority ν has zero coupled pairs in the makeup, of ν-particle state. Thus, if

$$(j^{\nu})_{IM\nu} = \Phi_{IM\nu} |0\rangle \qquad ...(10.83a)$$

is a ν-particle wave-function with seniority ν, *i.e.,* then,

$$S_{-} \Phi_{IM\nu} |0\rangle = 0 \qquad ...(10.84)$$

i.e. S_- does not destroy any pair in ν-particle wave-function, because there are no pairs in it. For a general expression for the normalised n-particle, ν-seniority wave-function (*see* Ref. 22, p. 78); it is given by:

$$(j^n)_{IM\nu} \equiv \psi_{IM\nu}(j^n) = \left\{\frac{2^{n-\nu}(2j+1-\nu-n)!!}{(n-\nu)!!(2j+1-2\nu)!!}\right\}^{\frac{1}{2}} \Phi_{IM}\{S_+(j)\}^{\frac{(n-\nu)}{2}} |0\rangle \qquad ...(10.83b)$$

From Eq. 10.77, and Eqs. 10.81, 10.82 and 10.84, it can be proved [Ref. (23)] that

$$S^2 = \frac{1}{2}(S_+ S_- + S_- S_+) + S_Z^2 = S_+ S_- + S_Z(S_Z - 1) \qquad ...(10.85)$$

It has been shown[22] from Eqs. 10.81 and 10.85, that

$$S_Z \Phi_{IM\nu} |0\rangle = -\frac{1}{2}(\Omega - \nu)\Phi_{IM\nu}|0\rangle \qquad ...(10.86a)$$

and

$$S^2 \Phi_{IM\nu}|0\rangle = S_Z(S_Z - 1)\Phi_{IM\nu}|0\rangle$$

$$= \left\{\frac{1}{2}(\Omega - \nu)\right\}\left\{\frac{1}{2}(\Omega - \nu) + 1\right\}\Phi_{IM\nu}|0\rangle \qquad ...(10.86b)$$

Thus a ν-particle state, with seniority ν is an eigenfunction of the quasi-spin operator S^2 and S_z with eigenvalues, $S = 1/2\ (\Omega - \nu)$ and $S_z = -\ 1/2\ (\Omega - \nu)$, and these eigenvalues are independent of any other quantum number of the state.

Without going into the detailed wave-functions, we can readily deduce some interesting selection rules from the 'arrow' diagrams, as given in Eqs. 10.70 to 10.75. We start with a general form of any operator, say any irreducible tensor operator of rank λ, which may be written as the sum of the single particle operators. Such an operator may be written as:

$$Q_{\lambda\mu} = \sum_i (Q_{\lambda\mu})_i = \sum_{jm}\sum_{j'm'} \langle \phi_{j'} \| Q_\lambda \| \phi_j \rangle \langle j\lambda\, m\mu \mid j'\, m' \rangle \qquad \text{...(10.87)}$$

where $\langle \phi_j' \| Q_\lambda \| \phi_j \rangle$ is the single particle reduced matrix element. The commonly encountered single particle reduced matrix elements have been evaluated in Appendix (2) of Reference (22). The single particle operator $Q_{\lambda\mu}$, corresponds to the destructions of a particle in the state (j, m), followed by its recreation in the state (j, m'). This single particle operator $Q_{\lambda\mu}$, operating on a seniority ν eigen-function gives:

For $\Delta\, v = 0$ $\quad Q_{\lambda\mu} \uparrow\uparrow \ldots \uparrow\uparrow\uparrow\downarrow \ldots \uparrow\downarrow \Rightarrow \uparrow\uparrow \ldots \uparrow\uparrow\uparrow\downarrow \ldots \uparrow\downarrow$

For $\Delta\, v = 2$ $\quad Q_{\lambda\mu} \uparrow\uparrow \ldots \uparrow\uparrow\uparrow\downarrow \ldots \uparrow\downarrow \Rightarrow \uparrow\uparrow \ldots \uparrow\uparrow\uparrow\uparrow \ldots \uparrow\downarrow$

For $\Delta\, v = -2$ $\quad Q_{\lambda\mu} \uparrow\uparrow \ldots \uparrow\uparrow\uparrow\downarrow \ldots \uparrow\downarrow \Rightarrow \uparrow\uparrow \ldots \uparrow\downarrow\uparrow\downarrow \ldots \uparrow\downarrow$ ...(10.88)

Therefore, a selection-rule emerges: Any operator which is the sum of the single-particle operators has non-vanishing elements only between states that differ in seniority by $\Delta v = 0$ or ± 2 units because single particle operators will either operate on the last of the closed shell ($\Delta v = 2$) or the first of the open particle ($\Delta v = -2$). As for example ${}_{24}Cr^{52}_{28}$ has seniority $v = 0$ for ground state, and $v = 4$ for excited state of 3.96 MeV, as shown by R.D. Lawson[22], and Horoshko et al.[24] and hence there will be no gamma ray transition between 2^+ state at 3.96 MeV ($v = 4$) and the ground state ($v = 0$), which is experimentally found.

This rule, also has been found to be useful in predicting ${}_{20}Ca^{43}_{23}$ (d, p) ${}_{20}Ca^{44}_{24}$, and ${}_{20}Ca^{42}_{22}$ (t, p) ${}_{20}Ca^{44}_{24}$, direct reaction probabilities for different levels. It predicts that 3.044 MeV state in ${}_{20}Ca^{44}_{24}$, which is more strongly populated than 2.83 MeV state has predominantly $v = 2$, because the target ${}_{20}Ca^{42}_{22}$ has $v = 0$ seniority and the reaction has $\Delta\, v = 2$. On the other hand in ${}_{20}Ca^{43}_{23}$ (d, p) ${}_{20}Ca^{44}_{24}$ case, the target ground state is $(vf_{7/2})^3_{I=7/2}$ and has seniority one, and hence transfer of one particle will correspond to transfer of seniority one of the target to create either seniority zero or seniority 2. Hence only $v = 0$ or $v = 2$ states of ${}_{20}Ca^{44}_{24}$ can be reached. Similarly in (t, p) reaction, because two particles are transferred, so $v = 0$ or $v = 2$ can be transferred again reaching $v = 0$ or $v = 2$ states of ${}_{20}Ca^{44}_{24}$. So the concept of seniority transfer or conservation is a useful tool for nuclear structure studies (Ref. 24).

10.7 MORE ASPECTS OF SHELL MODEL

We have, in the previous sections, discussed the basic assumptions, procedures and results based on the extreme single particle model, and its extended version, where more than one particle is occupying the orbitals outside the closed shell.

A few interesting and useful aspects are still required to be discussed, which we do below briefly:

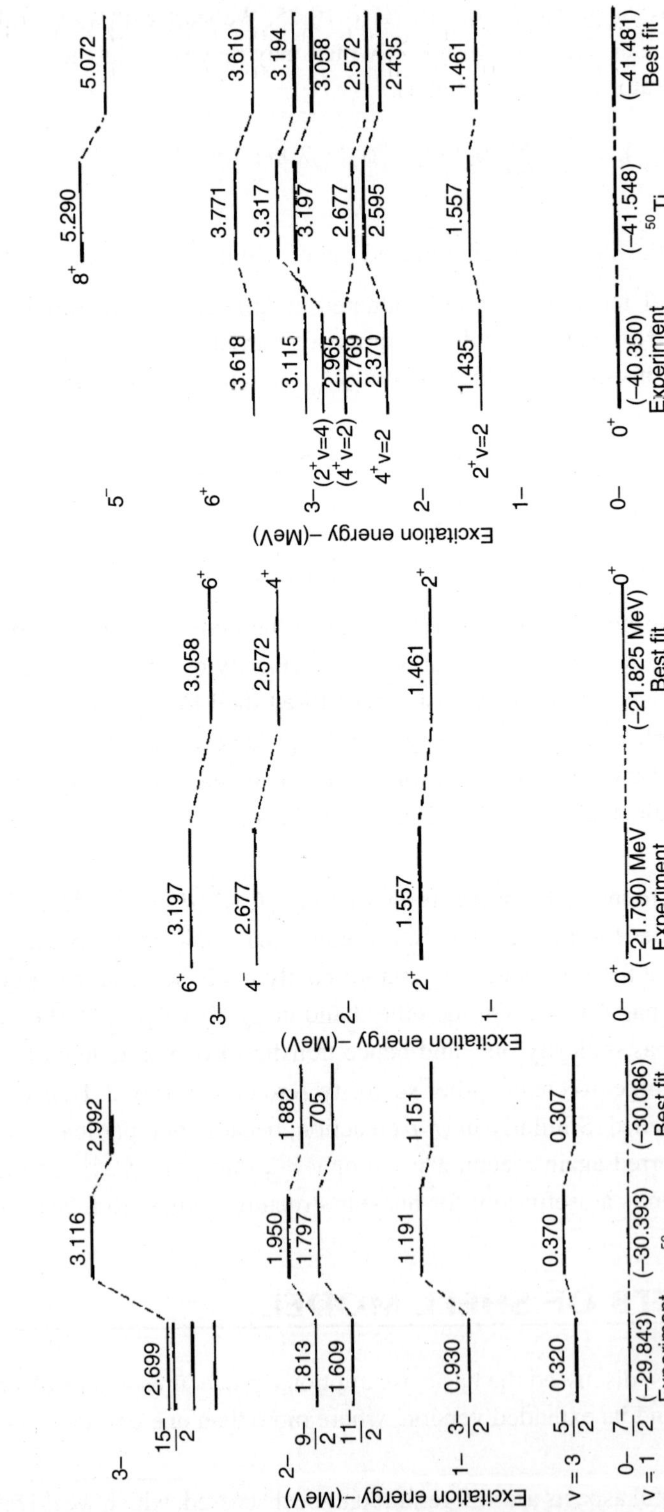

Fig. 10.7 The energy levels of $_{22}Ti^{50}_{28}$, $_{23}V^{51}_{28}$ and $_{24}Cr^{52}_{28}$; experimental and theoretical. The calculated spectra are based on considering $_{20}Ca^{48}_{28}$ as core and two nucleons (Ti^{50}), three nucleons (V^{51}) and four nucleons (Cr^{52}) (Ref. 17).

10.7.1 Configuration Mixing

This aspect, corresponds to valence particles being in more than one orbital. We have discussed it for two valence nucleons in Eqs. 10.39 to 10.51. However, there are many cases, where nuclear configuration may be represented by:

$$(n_1\ l_1\ j_1)^{2(2j_1+1)}\ (n_2\ l_2\ j_2)^{2(2j_2+1)}\ \ldots$$

$$(n_{i1}\ l_{i1}\ j_{i1})^{k_{i1}}\ (n_{i2}\ l_{i2}\ j_{i2})^{k_{i2}}\ (n_{i3}\ l_{i3}\ j_{i3})^{k_{i3}} \qquad \ldots(10.92a)$$

A simple example is that of ${}_9F_9^{18}$, where one proton and one neutron may be in $1d_{5/2}$, $2s_{1/2}$ and $1d_{3/2}$ orbitals. Similarly, in ${}_8O^{18}$ two neutrons will be in these orbitals. We have the following possibilities:

$T = 0, I = 1\ ({}_9F_9^{18})$	$T = 1, I = 0\ ({}_8O_{10}^{18})$
Possible states	**Possible states**
$(d_{5/2})^2\ (d_{5/2}\ d_{3/2})$	$(d_{5/2})^2,\ (d_{3/2})^2$
$(s_{1/2})^2\ (d_{3/2})^2\ (d_{3/2}\ s_{1/2})$	and $(s_{1/2})^2$ ...(10.92b)

The ground states of F^{18} and O^{18} are expected to be (T, I) = (0, 1) and (1, 0) respectively. Diagonalisation of the matrix elements, determines the wave-function $\psi\,(T, I)$ and eigenvalues as follows:

$$\underset{\left[\text{for } {}_9F_9^{18}\right]}{\psi\,(0,1)} = 0.732\ \psi\,(d_{5/2})^2 + 0.477\ (d_{5/2,\ 3/2})$$

$$+\ 0.464\ \psi\,(s_{1/2})^2 - 0.131\ \psi\,(d_{3/2})^2 - 0.009\psi\,(d_{3/2}\ s_{1/2})$$

and

$$\underset{\left[\text{for } {}_8O_{10}^{18}\right]}{\psi\,(1,0)} = 0.895\ (d_{5/2})^2 + 0.37\ (s_{1/2})^2 + 0.243\ (d_{3/2})^2 \qquad \ldots(10.93b)$$

The above expression and coefficients have been obtained from the work of Redlich[29].

Such calculations have been carried out for nuclei ranging from Si^{21} to Ca^{40}, by Glavdenans, Wiechers and Brussard[25]. The diagonalisations involve a large number of parameters; say 17 parameters were involved in the case of these nuclei, considering the loose nucleons outside the core of Si^{28}. There have been further comparisons of the experimental and theoretical energies of many nuclei[25]. S.P. Pandya and co-workers have carried out many calculations, for nuclei, with many particles outside the closed shell like ${}_{28}Ni^{62}$ and nuclei with A = 38, 40 nuclei, using Pandya theorems, which connect particle-particle interaction energy to particle-hole interaction energy[30].

Magnetic Moments and Configuration Mixing: We have already seen above that the concept of configuration mixing which explains the energies of many nuclei, can be conveniently supported by the single shell model picture. As for example, if the single particle configuration $(1\ s_{1/2})^k$ is accepted for D^2, H^3 and He^3; the experimental and theoretical magnetic moments match quite well. Similarly if Li^6, Li^7, Be^9, B^{10}, ground states are taken as $He^4\ (1p_{3/2})^k$ configuration, the comparison is good. This goes on till Mg^{25}. For higher A, one requires various mixtures of orbitals, to explain the experimental values of the magnetic moments. For details *see* Reference (26).

10.7.2 Individual (Independent)—Particle Model[27]

We have discussed in the previous sections, the various aspects of the shell model; a common feature of all of them was the existence of an inert core of even-even nucleons in all these cases, the inert core belonged to magic number particles and it was assumed that the core is not excited. But what happens, if there is no inert core and all particles are excited? It is in this case, that the individual particle model is invoked. The wave-function $\psi^{\nu}(r)$ for configuration of the nucleus is a Slater determinant like Eq. 10.54 of the single particle wave-functions ψ^{ν_i} for all A particles, where ν_i specifies the quantum state. But now there is no central common potential and ψ^{ν} are the solutions of Schrödinger equation:

$$H_1 \psi^{\nu} = E_1 \psi^{\nu} \qquad ...(10.94a)$$

where

$$H_1 = \sum_{i=1}^{A} \left[\frac{-\hbar^2}{2M} \nabla_i^2 + V_1(r_i) \right] \qquad ...(10.94b)$$

where $V_1(r_i)$ is suitably chosen single particle potential in which each nucleon moves and is obtained by certain self-consistent procedure—say Hartree Fock method or the wave-function is chosen by an insight into the physical situation. We will not go into the details of this method. However, by this process a set of wave-functions ψ^{ν_i}, which is complete and orthogonal, is obtained. Next we set up the energy matrix, that is the matrix formed by evaluating $(\psi^{\nu} \mid H \mid \psi^{\nu})$, where

$$H = -\frac{\hbar^2}{2M} \sum_{i=1}^{A} \nabla_i^2 + \sum_{i<j=2}^{A} V_{ij}(r_{ij}) \qquad ...(10.95)$$

Here $V_{ij}(r_{ij})$ is the two-nucleon interaction between nucleons i and j. This matrix is then diagonalised. The diagonal elements are the eigenvalues E_k of the energy of the actual system. In this manner we get the eigen-functions and eigen-energies of Schrödinger equation:

$$H\psi = E\psi \qquad ...(10.96a)$$

$$\psi = \sum_{\nu} a_{\nu} \psi^{\nu} \qquad ...(10.96b)$$

For a realistic case, the number of elements ψ^{ν} are very large or infinite, though in practice only a few values of ν are sufficient to determine ψ.

One chooses either L-S coupling or so-called Russel-Saunders coupling while writing the wave-functions (which is generally applicable for light nuclei), or j-j coupling for intermediate nuclei.

10.7.3 Hartree-Fock Method (Qualitative)

We have mentioned in the previous section, that in individual particle model, one requires a potential, $V_1(r_i)$ obtained by a self-consistent method like Hartree-Fock procedure. The physical idea behind this approach is that for some systems, it will prove adequate to suppose that each nucleon experiences a potential, which is constructed by averaging its interaction, with all other nucleons over their orbitals. It may then be possible to account for the properties of at least some nuclei in this average potential.

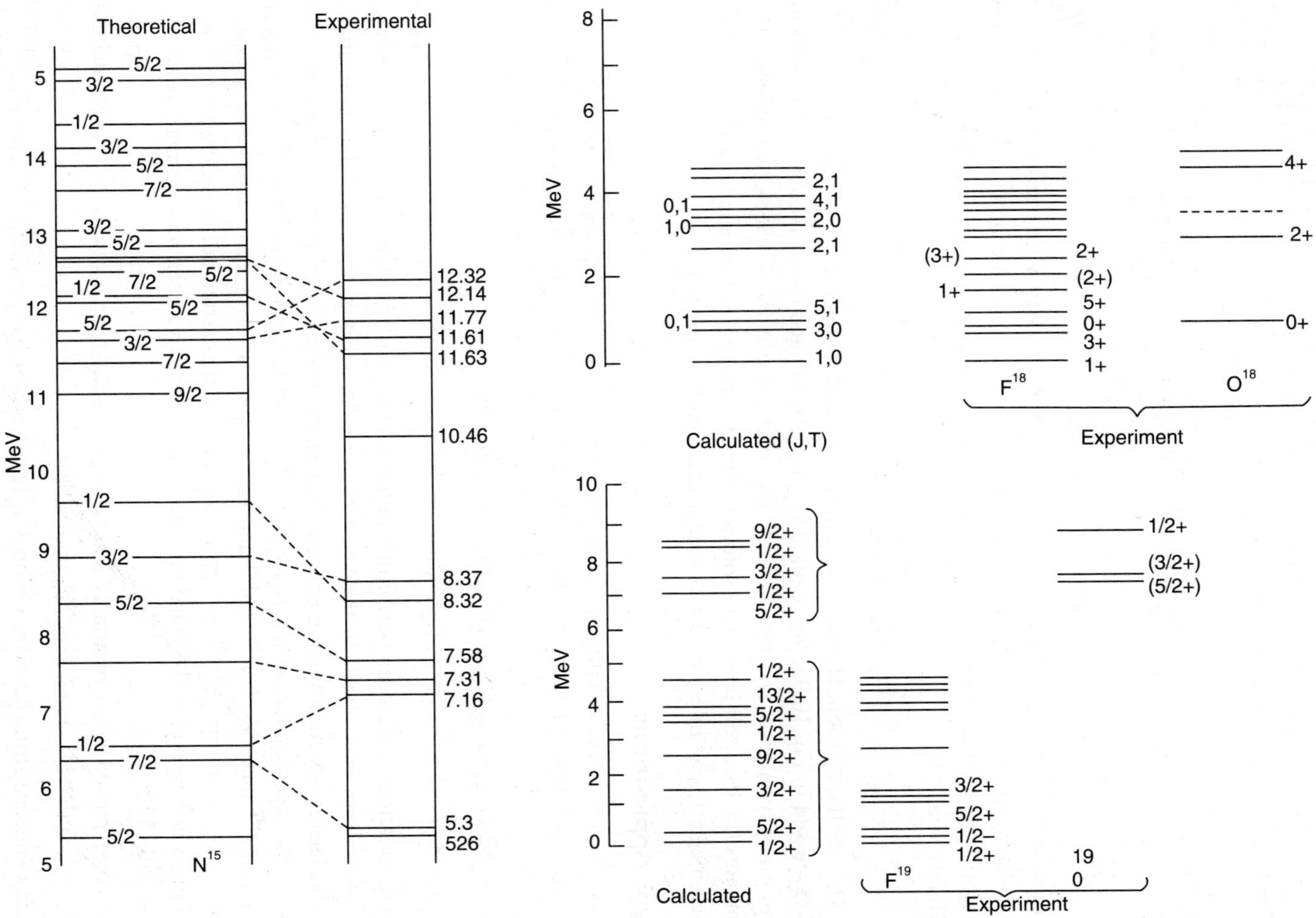

Fig. 10.8 Individual particle calculations for (*a*) $_7N_8{}^{15}$ (*b*) $_9F_9{}^{18}$, $_8O_{10}^{18}$ and (*c*) $_9F_{10}{}^{19}$, $_8O_{11}{}^{19}$, compared to experimental levels (Ref. 27).

We write the total Hamiltonian for A particles and also write the Schrödinger equation for these systems. Then we assume the many particle wave-function ψ^{ν} as denoted in the previous section (10.7.2). Equation 10.94*a* represents a function of A single particle wave-functions ϕ (or ψ^{ν} as written in the section (10.7.2); where each particle obeys a Schrödinger equation involving potential due to all A particles in a self-consistent manner, *e.g.*,

$$\left[\frac{-\hbar^2}{2\,M_k}\nabla_k^2 + \sum_{j\neq k}\left(\phi_j(r_j)\,V_{jk}(r_j, r_k)\,\phi_j(r_j)\,d\,r\right)\right]\phi_k(r_k)$$

$$\approx g_k\,\phi_k(r_k) \qquad \text{...(10.97)}$$

This equation is solved by a variational principle, where start with trial wave-functions ϕ's. Then we solve the Schrödinger equation by a re-iterative process. One proceeds with successive approximations, till one arrives at a set of $\phi_j(r, j)$ which do not change on further interaction. Then we write H, the Hamiltonian as:

$$\mathrm{H} = \mathrm{H}_{sc} + \mathrm{H}_{res} \qquad \text{...(10.98)}$$

Where H_{sc} is self consistent Hamiltonian and H_{res} is residual interaction, which can be two-body interaction as expressed in Eq. 10.95. We assume H_{res} to be small and then first obtain the solution for H_{sc}; as a starting approximation and then include H_{res} and solve it by re-interactive process. In this manner, one obtains finally a self consistent solution giving ψ and H. For details *see* Ref. (28). The use of Hartree-Fock method is an integral part of independent particle model.

10.7.4 Recent Applications

Detailed and exact-shell model calculations have been very successful in predicting or explaining the properties of the excited states, up to an excitation of several (4 – 5) MeV, for nuclei near magic numbers, *e.g.*,

(*i*) For $Z \approx 82$ *i.e.* $_{79}$Au, $_{80}$Hg, $_{81}$Th, $_{82}$Pb, $_{83}$Bi and $_{87}$Fr.

$N \approx 82$ *i.e.* Dy^{149}, Tb^{149}, Dy^{150}, Ho^{151}, Ho^{152} and Tm^{153}

Several calculations have been performed in the lead region[31–33] for which phenomenological two-body potentials have been used. In the latest calculations[33] Bonn potential[34] has been used for $Pb^{204,\,205,\,206}$. The energy spectra, binding energies and electromagnetic properties are calculated with good agreement with experiments. Similarly for $N \approx 82$, which correspond to cases of $148 < A < 152$, *e.g.* $Er^{150;\,152}$ and Dy^{150} and then Dy^{149}, Tb^{149} and Ho^{152}, again a lot of shell model calculations[35–37] have been carried out, assuming $_{64}Gd^{146}$ as a core, the latest being carried out in Reference (35). Other cases in these ranges correspond to Bi^{210} and Fr^{216} for $Z = 82$ and have been most recently dealt with in References (38) and (39).

(*ii*) The other closed shell region is near $Z \approx 50$, which corresponds to nuclei like Te^{120}, $Xe^{114,\,116,\,120,\,124,\,128}$, $Ba^{133-134}$, and $Ce^{126-148}$, and Nd^{143}.

The study of nuclear structure of the isotopes of Te to Cerium, has provided[40, 41] a lot of information about the particle excitation. This is a region where both collective and particle excitations coexist. As a matter of fact one can study, the variation of shape from spherical to quite well deformed nuclei. In this region cranked Hartree Fock-Bogoliubov (HFB) formalism has been extensively used, to study the interplay of single particle and collective aspects of nuclear motion. Recently region of Xe-Be has been investigated for the properties of the ground states and 2^+ state using Hartree–Fock–Bogoliubov method. Related to these studies in the region of $N \approx 50$ are Nb^{91}, Te^{93}, Rh^{95}, and nuclei like Zr^{90}, Zr^{91}, Mo^{92}, Pd^{96}, etc. A comprehensive list of such calculations is given in Ref. (42). Recently[43] energy level in Tc^{93}, Ru^{94}, Rh^{95}, Nb^{41} and Mo^{92} and Pd^{96}, have been calculated allowing single particle excitation from $p_{1/2}$ shell into $d_{5/2}, s_{1/2}, d_{3/2}$ and $g_{7/2}$ shell making use of Sr^{88} as a core with active protons in $p_{1/2}$ and $g_{9/2}$ shells. Agreement with experiment was quite good.

Another interesting calculation[44] of reduced E_2 transition probabilities has been carried out for even mass Xe nuclei $Xe^{114, 116, 120, 124, 128}$ by using HFB technique and quadrupole-quadrupole plus pairing model of the two body interaction. Comparison with experiments yields very good agreement.

(iii) $N, Z \approx 20$ or 28, *e.g.* Ca^{47}, Sc^{47}, Ti^{47} and V^{47}, and Ca^{48}, Sc^{49}, Ti^{49}, V^{49} and Cr^{49} and on the lighter side, with holes, we have Cl^{38}, Ar^{38}, Ar^{39}, Ar^{40}, Ar^{41}, Ar^{42}, K^{42} and K^{43}.

These nuclei have been handled by using Ca^{48} as a core by many workers. As for example[45], complete diagonalisation in *pf* major shell outside this core has been used for calculating the properties of Ca^{47}, Ti^{49}, V^{49} and Cr^{49} and Mn^{49}. Kuo-Brown interaction potential has been used. A part from the energy, the dipole magnetic moments, the transition probabilities $B(E_2)$ and $B(M_1)$ and quadrupole moments have been calculated for nuclei from $Z = 20$ up to $Z = 28$ over the atomic weight range of $A = 46$ up to $A = 56$. Comparison with experiments is reasonable. Similar calculations along with experimental results for energies for K^{42} and K^{43} have given results[45] in good agreement with experimental values.

(iv) $8 \le Z \le 20$ or $8 \le N \le 20$, *e.g.* Al, Mg, Na, Ne, F, and O, etc. Typically[46], the valence space is defined, so that the space for *Z*-8 protons corresponds to full *sd* shell space, and for *N*-20 neutrons, it is *pf* full space. The effective interaction was such, that its main parts, were Hamiltonian of Wildenthal for *sd* shell, and modified Kuo–Brown interaction energies for pf shell. Predictions for separation energies could be made for these neutrons-rich nuclei whose neutron number was varied from 14–30. Also $B(E_2)$ and $\langle Q \rangle$ was predicted and compared with experiments. Also shell model calculations[47] of magnetic moments in odd–odd ($N = Z$) nuclei have been carried out using the shell model wave-function ψ_{core} $(J = 0, T = 0)$, $\psi_{sp}(J, T = 0)$ where ψ_{core} describe the $(Z - 1, N - 1)$ even-even core and ψ_{sp}, the shell model wave-function of odd neutron and odd proton occupying the same single particle orbit (n, l) with a total angular momentum *J* and iso-spin $T = 0$. The results are very much in agreement with experimental.

(v) For $A = 7 - 11$, which corresponds to:

He^7, Le^7, Be^7, $(A = 7)$, He^8, Li^8, Be^8, $(A = 8)$,

He^9, Li^9, Be^9, B^9 and C^9, $(A = 9)$,

He^{10}, Li^{10}, Be^{10}, B^{10}, C^{10}, $(A = 10)$

and Li^{11} and Be^{11}, $(A = 11)$. Some of these (Li^{11}, Be^{11}) are halo nuclei.

These nuclei have been, theoretically dealt with, through large basis[48], no core shell-model calculation. One starts with the one plus two body, Hamiltonian for *A*-nucleon system, *i.e.*,

$$H^{\Omega} = \sum_{i=1}^{A} \frac{p_i^2}{2M} + \sum_{i<j}^{A} V_N(\mathbf{r_i} - \mathbf{r_j}) + \frac{1}{2} H_m \Omega^2 \mathbf{R}^2 \qquad ...(10.99)$$

where the first term represents the kinetic energy, the second term a nucleon-nucleus potential and the last term is the centre of mass harmonic oscillator potential. The symbol Ω is the oscillator number and

$$\mathbf{R} = \left(1/A \sum_{i=1}^{A} \mathbf{r}_i \right)$$

Using this Hamiltonian and a complete $N \hbar \omega$ where N was taken to be $N = 4$ for Be^{11}, Be^{10}, Li^{10}, B^9, Be^9 and C^9 and He^9, Li^8, B^{10}, He^8 and Be^8 and $N = 6$ for B^7, Li^7, He^7, and Be^7, just for convenience for computer capabilities. The result for binding energies E_B, magnetic moments (μ), quadrupole moments (Q) and rootmean square of radius, of these nuclei for ground as well as the excited states were calculated. An effective interaction

$$\sum_{i<j}^{A} V_N (\mathbf{r}_i - \mathbf{r}_j)$$

was derived by a special procedure, *see* Reference (48).

For $Li^{6,\,7}$, a shell model structure with $\alpha + d$ in a two body description and $\alpha + p + n$ in a three body description for Li^6, and $\alpha + t$ for Li^7, have been used. Shell model calculations[49] have been carried out, involving space range from conventional $o \hbar \omega$ space to (0 + 2 + 4 +) $\hbar\omega$ space. Again using different types of interaction potentials, the values for r_{rms}, μ and Q and cross-sections for elastic and in elastic scattering for p-Li^6 and p-Li^7 have been obtained, and compared with experiments with good agreements.

10. Shell Model

2000–2008

Many studies have been undertaken for specific necessity of understanding the nuclear structure near the closed shell nuclei.

An interesting and a massive collaborative effort was undertaken, by a group of 30 authors from half a dozen laboratories, and Universities of Europe and USA, in which excitation up to high spin of 20 h have been studied in $_{48}Cd_{52}^{100}$ which is two– proton-hole and two neutrons away from magic number $Z = N = 50$, using T_i^{46} (N_i^{58}, p2n) Cd^{100} at 215 MeV of N_i^{58} beam [Phy. Rev. C.61, 044311 (2000)].

In a paper, authored by 36 authors from Hungary, France, Russia, Denmark, Romania and U.K., the structure of $S^{40,\,42,\,44}$, obtained from Be^9 (target) and Ca^{48} (19^+) beam of 60.3 MeV from fragmentation of $Ca^{48} \rightarrow S^{40,\,42,\,44}$ has been studied through in-beam γ-rays spectroscopy. The results were interpreted by the use of microscopic collective model and large scale shell model. Both models suggest an erosion of $N = 28$ shell structure, closure at $N = 8$, and suggest a deformed ground state for $S^{40,\,42}$ and a spherical deformed mixed configuration for S^{44} [Phy. Rev. C. 66, 054302, (2002)].

In a theoretical paper, where calculations are carried out ab-initio no core, shell model (NCSM) for a realistic three body interaction for Li^{6-7}, He^6, $Be^{7,\,8,\,10}$, $B^{10,\,11,\,12}$, N^{12}, and $C^{10,\,11,\,12,\,13}$, using Argonne V^8 and Thomas Mebourne three nucleon interaction, the authors have obtained the correct state spin for $B^{10,\,11,\,12}$, with three body potential [Phy. Rev. C. 68, 034305 (2003)].

In a massive collaborative effort, involving 30 authors from China, UK, Germany, Spain, Brazil, Poland, and Romania, an experiment on deep inelastic scattering of Se^{82} beam, at 460 MeV for $_{37}Rb^{87}_{50}$, $_{35}Br^{85}_{50}$, $_{34}Se^{84}_{50}$, and $_{32}Ge^{82}_{50}$, targets - all having $N = 50$, has been conducted; exciting these nuclei up to 8.822 MeV in Rb^{87}; 4.34 MeV in Br^{85}, 4.406 MeV in Se^{84} and 3.68; 4.34 MeV in Br^{85}; 4.406 MeV in Se^{84}. The detailed calculations based on shell model; including neutron configuration showed the importance of neutron-core; and was indicative of the persistence of $N = 50$ shall up to $Z = 32$ [Phy. Rev. C. 701, 024301, (2004)].

In an interesting treatment of shell model, a Hamilton is written for p-shell nuclei; which properly takes into account spin-isopin interaction to obtain the cross-section of neutrino - C 12 reactions; induced by the deacy at rest nutrition as well as supernuono neutrinos Branching ratio, to various decay channels are calculated using Hauser - Feshbach theory. Also neutrino- He^4 cross-sections are calculated. These cross-section for both He^4 and C^{12} are compared to previous calculations. One of the results of these calculations is the possible enhancement of yields of light elements Li^7 and B^{11}, during supernova explosions. [Phy. Rev. C. 74, 0344307 (2006)].

In a calculational paper, from authors from Japan, they have correlated the spectra and quadrupole moments close to isomeric states 7^+ and 21^+ states in $_{47}Ag^{94}_{47}$ (N = Z) nucleus, using $g_{9/2}$ – shell nuclei, nuclear structure. It is found that 7^+ state is oblately deformed and is suggestive to be shape-isomer in nature. On the other hand 21^+ state is isomeric because of general inversion of 19^+ and 21^+ due to core polarization of 21^+ state. [Phy. Rev. C. 77, 064304, (2008)].

REFERENCES

1. M.G. Mayer: Phy. Rev. 78, 19 (1950); H.E. Suess and J.H.D. Jensen; Arkiv F. Fysik (Stockholm), 3 (1951), 577; J.H.D. Jensen in Beta-Gamma Spectroscopy, edited by K. Siegbahn, North-Holland Publishing Co., Amsterdam, (1995); A de Shalit and I. Talmi: Nuclear Shell Theory, Academic Press (1963); R.D. Lawson: Theory of Nuclear Shell Model, Clarenden Press, Oxford (1980); Mayer M.G. and J.H.D. Jensen and H.E. Suess: Phy. Rev. 75, 1766 (1949); Mayer M.G.: Phy. Rev. 75, 1969 (1949).
2. A. Bohr and B.R. Mottleson: Dan Mat. Physik Medd. 27 (1953) No. 16; Ibid in Beta-Gamma Spectroscopy, edited by K. Siegbahn (1955).
3. A. Bohr and B.R. Mottleson: Phy. Rev. 89, 316 (1953), Alga G., B.K. Adair, and B. Mottelson, Dam-Fys Medd 30, 29, 9 (1956).
4. S.G. Nilsson, K. Danske Vidensk selsk mat-fys. Medd. 29 No. 16 (1955); B.R. Mottelson and S.G. Nilsson; K. Danske Vidensk selsk Mat-fys skr. 1, No. 8 (1959).
5. E. Segre: Nuclei and Particles, p. 252, W.A. Benjamin, New York (1964); C.H. Townes: Determination of Nuclear Quadrupole Moments, Handbuch der, Physik, XXXVIII/ip. 377, Springer-Verlag (1958); C.H. Townes, H.M. Foley, and W. Low: Phy. Rev. 76, 1415L (1949).
6. O. Nathen and S.G. Nilsson; in K. Siegbahn (ed) Beta and Gamma Spectroscopy; North Holland Publishing Co., Amsterdam (1966); B.L. Cohen: Concepts of Nuclear Physics, McGraw-Hill Book Company, New York; p. 115 (1971).
7. D.J. Hughes: Pile Neutron Research, Addison-Wesley Publishing Company, Cambridge, Mass. (1953); D.J. Hughes R.C Garth and J.S. Levin: Phy. Rev. 91, 1423 (1953).

8. J.A. Harvey: Phy. Rev. 81, 53 (1951).
9. L. Kowarshi: Phy. Rev. 78, (1950); 477; H.E. Suess: Phy. Rev. 81 (1951), 10, 477.
10. L.H. Thomas: Nature, London, 117, 514 (1926).
11. B.J. Malonka: Phy. Rev. 86, 68 (1952); Levinger J.S. and D.C. Kent: Phy. Rev. 95, 418 (1954).
12. B.L. Cohen: Concepts of Nuclear Physics, McGraw-Hill Book Company (1971); Quantum Mechanics, L.I. Schiff: McGraw-Hill Book Company, (New York) 1971; J.M. Blatt: Theoretical Nuclear Physics and V.F. Weisskopf, John Wiley & Sons, New York (1952).
13. P.G. Perey: Phy. Rev. 131, 745 (1963).
14. A. Moszkowski: Models of Nuclear Structure, Handbuchder Physik, V. 39, 411, S. Flügge ed. Springer-Verlag, Berlin (1957).

    G. Racah: Phy. Rev. 63, 367 (1943); Flowers B.H., Process Royal Socicty, (London), A 212, 248 (1952).
15. J.P. Elliot and B.H. Flowers: Process Royal Society A 229, 536 (1955).

    F. Ajzenberg Selov: Nuclear Physics A 190, 1, 1972, A 281 (1977).
16. R.D. Lawson: Theory of Nuclear Shell Structure, Clarenden Press, Oxford, p. 33 (1978); E.M. Henley: Isospin in Nuclear Physics (ed. D.H. Wilkinson) North Holland, Amsterdam (1969).
17. T. Nomura, H. Gil C. Saito, T. Yamazaki and M. Ishihara: Phy. Rev. Letters 25, 1342 (1970); Horosko, R.N., Cline D. and Lesser P.M–S., Nuclear Physics A 149, 562 (1970); M.S. Freedman, E.Jr. Wagner, F.T. Porter, and H.H. Boloton: Phy. Rev. 146, 291 (1966).
18. L.W. Nordheim: Phy. Rev. 78, 294, (1950); Rev. of Modern Phy. 23, 322 (1951).
19. M.H. Brennan and A.M. Brenstein: Phy. Rev. 120, 927 (1960).
20. R.F. Bacher and S. Goudsmit: Phy. Rev. 46, 948 (1934).

    G. Racah: Phy. Rev. 63, 367 (1943).

    B.F. Bayman and A. Lande: Nuclear Phy. 77, 1 (1966).
21. I.S. Towner and J.C. Hardy: Adv. Phy. 18, 74, 401 (1969*a*).
22. R.D. Lawson: Theory of Nuclear Shell Model; Clarenden Press, Oxford (1978); M. Rotenberg, R. Bivens, N. Metropolis and J.K. Wooten Jr.: The 3*j* and 6*j* Symbols, Technology Press, MIT, Cambridge, Mass. (1959).
23. G. Racah, Phy. Rev. 76, 1352 (1949).

    G. Racah, and Talmi I., Physica 18, 1097 (1952); C. Schwarty and De shalit A., Phy. Rev. 94, 1257 (1954); Kerman A.K., R.D. Lawson and Macfarlane M.H.: Phy. Rev. 124, 162 (1961).
24. C. Gil, H. Saito, T. Nomura, T. Yamazaki and I. Shihara: Phy. Rev. Letters 25, 1342 (1970); R.N. Horoshko, D. Cline, and P.M.S. Lesser: Nuclear Physics A 149, 562 (1970); M.S. Freedman, E.Jr. Wagner, F.T. Porter and H.H. Bolotin: Phy. Rev. 146, 791 (1966); J.H. Bjerregaard, O. Hansen, O. Nathan, R. Chapman, S. Hinds and R. Middle: Nuclear Physics A 103, 33 (1967).
25. P.W.M. Glavdenens, G. Wiechers and P.J. Brussard: Nuclear Physics 56, 529 (1964), 548 (1964).
26. L. Landau, and Ya Smorodinsky: Lecture on Nuclear Theory, Plenum Press Inc., New York (1959); H. Horie and K. Sugimoto: J. Phy, Soc. Japan, Supp. 34, 1 (1973).
27. S.P. Elloit and A.M. Lane: Encyclopedia of Physics, ed. S. Filügge, V. 39, p. 24, Springer-Verlag (1957);

    E.C. Halbert and J.B. French: Phy. Rev. 105, 1563 (1957).
28. G. Ripika: Advance in Physic V. 1, (Eds M. Baranger and E. Vogt), Plenum Press, New York (1968).

    K.A. Brueckner, T. Soda, P.W. Anderson, and P. Morel: Phy. Rev. 118, 1442 (1960); K.A. Brueckner and D.T. Goldman: Phy: Rev. 116, 424 (1959).

A.K. Kerman. J.P. Svenne, and F.M.H. Villars, Phy. Rev. 147, 710 (1966); Microscopic Theory of the Nuclear Physics, V. 3, (Nuclear Theory), Eisenberg J.M. and W. Greiner; North Holland Publishing (1979).

29. M.G. Redlich: Phy. Rev. 99, 1421 (1955).

30. S.P. Pandya: Phy. Rev. 103, 959 (1956); Nuclear Physics 43, 636 (1963); Cohen S. Lawson R.D. Mcfarlane M.H., Pandya S.P. and Soga M.: Phy. Rev. 160, 903 (1967).

31. J.B. McGory and T.T.S. K Nuclear Physics A 247, 283 (1975).

32. C.A. Ceneviva, L. Losano, N. Tesuya and H. Dias: Nuclear Physics A. 169, 129 (1997).

33. L. Coraggio, A. Covello, A. Gargano, N. Itace and T.T.S. Kuo: Physics Rev. C. 58, p. 3346 (1948).

34. R. Machleidt, K. Hollinde, and Ch. Elster: Phy. Rev. C. 49, 1 (1987).

35. K.S. Toth et al. (10 authors): Physics Rev. C. 32, 342 (1985).

36. D. Horn, I.S. Towner, O. Hausser, D. Ward and H.R. Andrews, M.A. Lone, J.F.S. Schafer, N. Rud and P. Taras.: Nuclear Physics Rev. A 441, 344 (1985).

37. Chang-Hua-Zhang, Shun-Jin-Wang and In-Van Gir: Phy. Rev. C. 58. p. 851 (1998).

38. P. Allxa. Jan Kvasli, Nhugen Viet Minh and Raymand K. Sheline: Phy. Rev. C. 55, 179 (1997); Ibid, Phy. Rev. C. 55 p. 2395 (1997).

Phy. Rev. 56, 3087 (1997); J.N. Gu, A-V. [illegible]n, C.H. Zhang, P. Guazzoni, L. Zeta G. Graw, M. Jaskola and G. Staudt: Phy. Rev. C. 5[illegible], p. 2395 (1997); R.K. Sheline, C.H. Liang, P. Paris and A. Grizon: Phy. Rev. C. 55, p. 1162 (1997).

39. J.R. Hughes, D.B. Fosson, D. Larosse, Y. Liang, P. Yaska, and M.P. Waring: Phy. Rev. C. 44. 2390 (1991); R. Goswami, B. Sethi, P. Banerjee and R.K. Chattopadhye: Phy. Rev. C. 47. 1013 (1993); Y. Liang. D.B. Fossan et al (10 Authors): Phy. Rev C. 45, 1041 (1992).

40. M. Saha Sirkar, A. Goswami. S. Bhattacharya, B. Dasannacharya, P. Bhattacharya, and S. Sen: J. of Physics G 23, 169 (1992).

41. M. Saha Sarker and S. Sen: Phy. Rev. C. 56. p. 3140 (1997).

42. H.A. Roth, S.E. Arnel, D. Foltescu, O. Skeppstedt, T. Kuloyangi, S. Mitarai and J. Nyberg.: Phy. Rev. C 50, 1330 (1994).

43. I.P. Johnstone and L.D. Skouras: Phy. Rev. C. 55, p. 1227 (1997).

44. Rani Devi, S.P. Sarwat, Arun Bharti, and S.K. Khosa: Phy. Rev. C.55, p. 2433 (1997).

45. G. Martinez-Pinedo, A.P. Zuker, A. Pones and E. Courier, Phy. Rev. C. 55, p. 187 (1997); A. Kav, M.S. Sarker. J.M.G. Gomez, V.R. Manfredi and L. Salasmich: Phy. Rev. C.55, p. 1260 (1997); M. Moralles, P. Janker et al. (10 authors): Phy. Rev. C. 58, p. 739 (1998).

46. B.H. Wildenthal: Progress of Particle and Nuclear, Physics, 11, 5 (1984); A. Povos and A. Zuker: Phy. Reports, 70, 4 (1981).

47. Yigal Ronen and Shalom Shalomo: Phy. Rev. C.V. 58, p. 884 (1998).

48. D.C. Zheng. B.R. Barret, L. Jaqua, J.P. Vary and R.J. McCarthy: Phy. Rev. 48, 1083 (1993), Phy. Rev. C. 52, 2488 (1995); P. Navaratil and B. Barret: Phy. Rev. C. 54, 2986 (1996). P. Navaratil and B.R. Barret and W.E. Ormand: Phy. Rev. 56, 2542 (1997); P. Navaratil and B.R. Barret: Phy. Rev. C. 57, p. 3119 (1998).

49. S. Karataglidis, B.A. Brown, K. Amos and P.J. Dortman: Phy. Rev. C. p. 2826 (1997).

PROBLEMS

1. Find the energies of possible states, formed from 1, 2, 3 and 4 particles in p 3/2 level of shell model, when Majorona potential acts between pairs of particles.
2. Write down the angular momenta, and parities predicted by extreme single particle shell model of $_6C^{12}$, $_5B^{11}$, $_8O^{17}$ and $_{57}La^{139}$, $_8O^{18}$, $_{10}Ne^{20}$, $_{12}Mg^{25}$, $_{30}Zn^{67}$, $_{43}Tc^{99}$ and $_{57}La^{139}$. Compare them with experimental values and write out their shell configurations.
3. From the first three levels of $_{29}Cu^{63}$, $_{28}Fe^{55}$ and $_{24}Cr^{53}$, find out the value of R_{n1} [Eq. 10.15] and spin-orbit coupling and hence the value of λ, for each case.
4. Prove Eqs. 10.22 and 10.23.
5. Using the properties of Clebsch-Gorden Coefficients, prove that for two particles and odd J

   $\psi_J^M(r_1, r_2) = 0$.
6. The lowest levels of $5B_5^{10}$ are:

E (MeV)	G.S	0.77	1.74	3.58
J^π	3^+	1^+	0^+	2^+

   (*i*) What are the iso-spins (T) and seniorities (S) of these levels?

   (*ii*) What are the values of T and S for Be^{10} and C^{10}?
7. Taking the clue from Eq. 10.64, write down the value of coefficients of fractional parentage for $j = 5/2$, $n = 3$ and $I = 9/2$ [$_8O^{19}$].

   [Consult the Table of Clebsch-Gorden Coefficients].
8. For the following nuclides, the most likely experimental ground state spins and parities are given below in parenthesis. State the single particle configurations

$$He^3\left(\frac{1^+}{2}\right), N^{14}(1^+), Al^{27}\left(\frac{5^+}{2}\right), Cl^{35}\left(\frac{3^+}{2}\right), K^{40}(4^-), Ca^{43}\left(\frac{7^-}{2}\right)$$

$$Cr^{53}\left(\frac{7^-}{2}\right), Cu^{65}\left(\frac{3^-}{2}\right), Mo^{97}\left(\frac{5^+}{2}\right), Rh^{106}(1^-), Te^{130}(1^+)$$

   and Bi^{250} (1^-).
9. Show that the r.m.s. radius of a particle in the oscillator state, with quantum numbers n, l, m, is α^{-2} [2 (n –1) + 1 + 3/2].

   Where $\alpha = (m\omega / \hbar)^{1/2}$. Find the radius of the whole nucleus for Ca^{40}, K^{41} and Ni^{60}.
10. If V_{00} is 110 MeV and V_0 is 57 MeV, near $E = 0$, calculate M^* [Eqs. 10.8 and 10.9].

    Also calculate M^* from Eq. 10.10, if V_{00} = 100 MeV. Compare and comment on these values.

11

CHAPTER

Collective Model

We have already seen, that the behaviour of binding energies as a function of atomic number may be understood, on the basis of liquid drop model of the nucleus (*see* Section 2.1.1.2 of Chapter 2). This was expected, on the basis of very strong nuclear forces.

As a matter of fact the nuclear shell model was a surprise; because it assumes that the 'loose' nucleon outside a core can move in the average potential of the core, without interacting with the individual nucleons in the core. Now, of course, it is understood, that this independent particles model holds good because of the Pauli exclusion principle. Because all the ground state energy levels available to the nucleons in the core are filled and therefore, the loose nucleon, inspite of strong nuclear forces, only gets scattered elastically from the nucleons, in the core, and behaves as if it was an 'independent' particle.

Also if the interaction of the 'loose' nucleons with the core is not very strong, the shape of the core is expected to be spherical. It was because of this reason, that we assumed in the shell model, the shape of the potential to be spherical, Eq. 10.2.

However, there are a lot of indications, that extreme single particle model with a spherical potential does not hold good all the time. As for example, it is known that many nuclei have quadrupole moments in their ground state; [*see* Fig. 2.22]. This means that they are spheroidal in shape. Such nuclei, when excited will either have collective mode of excitation, like rotational or vibration, or if the individual particles are excited in the spheroidal potential; the energies of the excited levels of individual 'loose' nucleons will be different from the ones discussed in the previous chapter.

11.1 ROTATIONAL MODE

11.1.1 Semi-Empirical

The behaviour of the excitation of even-even nuclei in the rare earth region provides an indication of the collective nature of the motion of nucleons in the nucleus, especially at lower energies of excitation.

We show in Fig. 11.1*a* the experimental excitation level diagrams of a few typical even-even nuclei[1, 2]. It is interesting to note that empirically the energies, E_I of many of the excited levels are related to their angular momenta I by the approximate relationship.

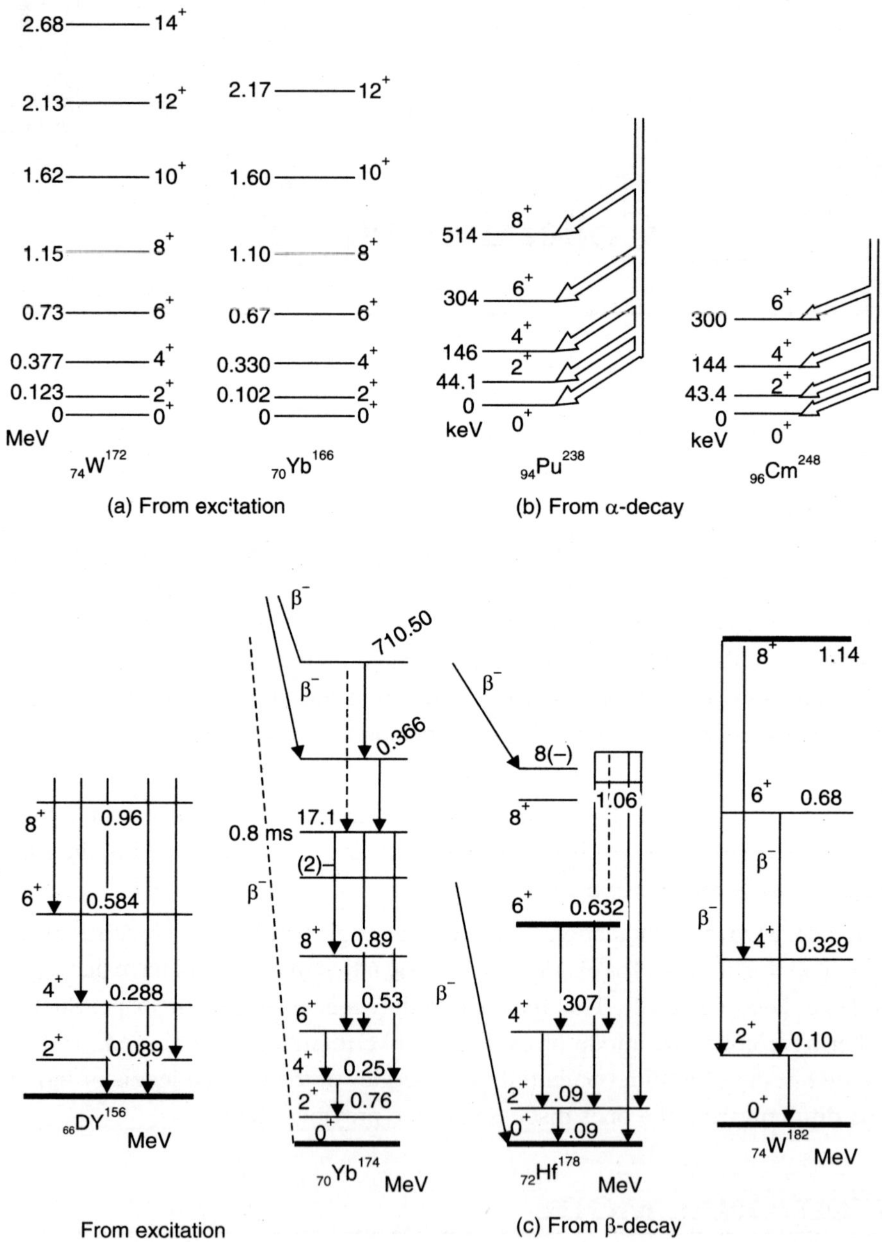

Fig. 11.1a Experimental energy level diagrams of some spheroidal nuclei, along with their spins and parities (*a*) From excitation experiments; (*b*) From α-decay; (*c*) From β-decay experiments (Ref. 1, 2).

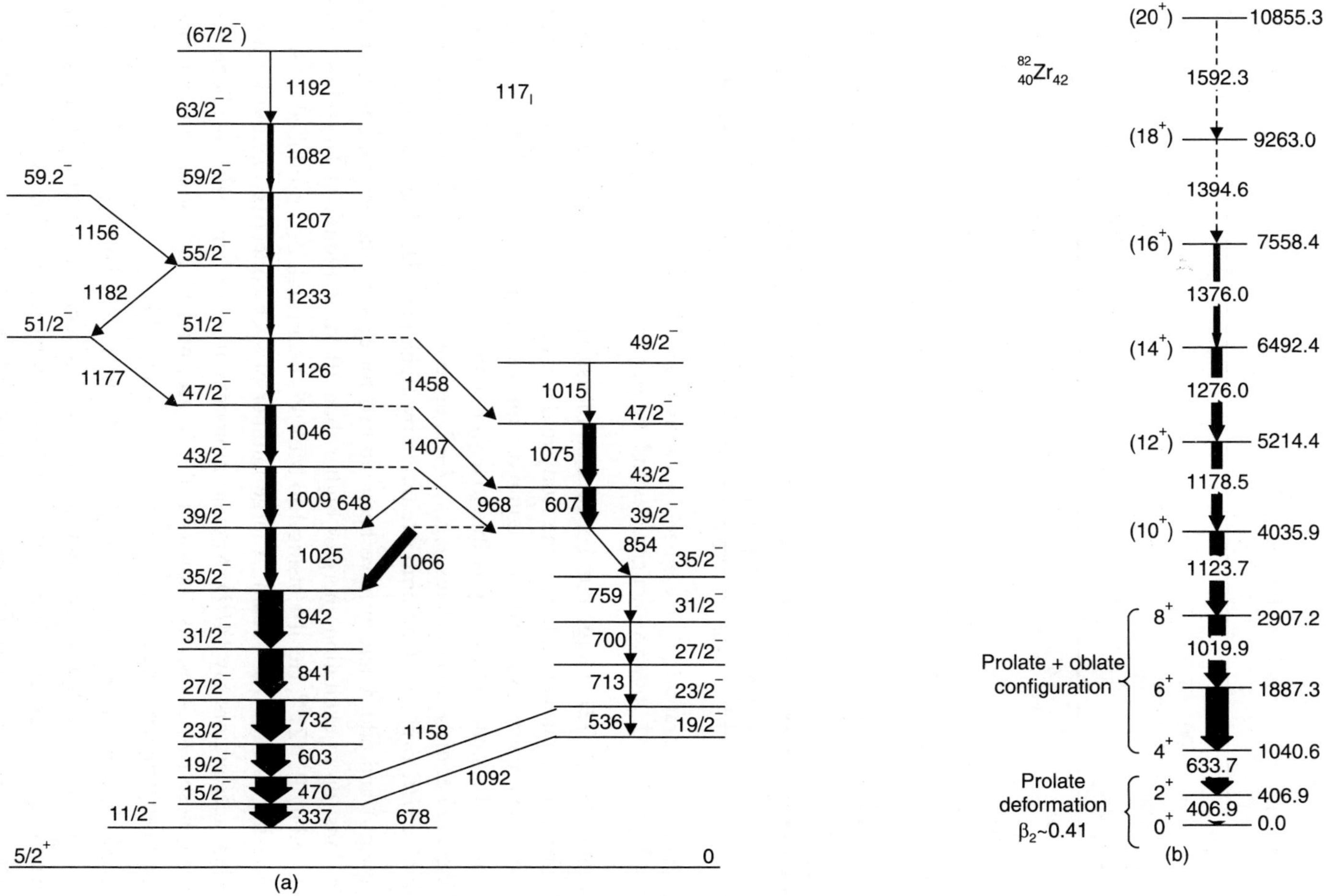

Fig. 11.1b The rotational states built over the ground or near ground states in (*a*) an odd A nucleus I^{117} and (*b*) an even A nucleus ${}_{48}Z^{82}_{42}$, by heavy ion induced reactions. The quadrupole transitions downward are shown (Ref. 1, 2).

$$E_I = \frac{\hbar^2}{2\mathcal{I}} I(I+1) \qquad \text{...(11.1)}$$

where $\mathcal{I}$ is a constant for a given nucleus, [Fig. 11.2].

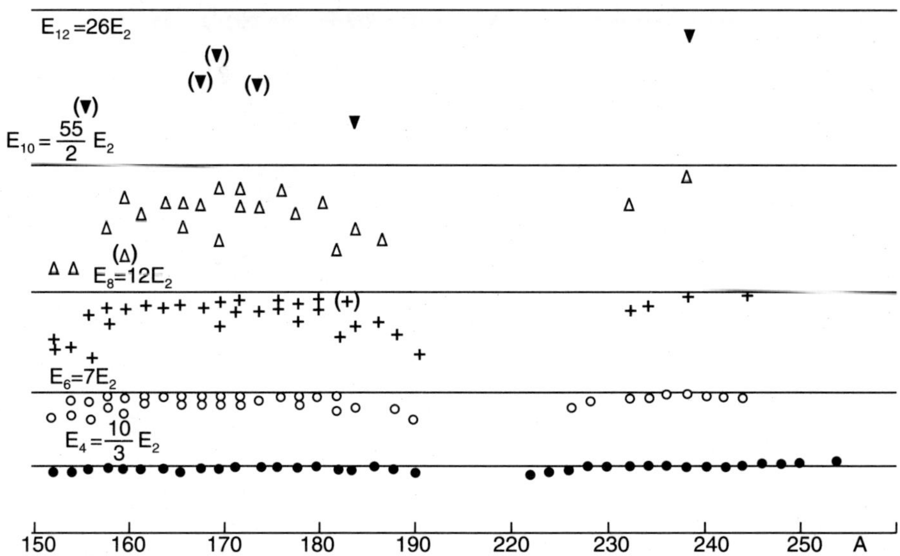

Fig. 11.2 The rotational energies ratio E_1/E_2 in even-even nuclei as a function of A; where E_1 (I = 4, 6, 8, 10, 12) is a obtained from Eq. 11.1, shown as straight lines and the experimental values as points (Ref. 2).

In the seventies (1971–72), a large number of such measurements were made, from Stockholm and subsequently from other places; including the study in details of the weak transitions at the top of the ground state rotational bands. Such studies required coincidence spectra, angular distribution and excitation functions. In this manner, it was possible to assign, uniquely each observed line to a definite transition involving angular momenta of states, *e.g.* I, $I + 2$, up to as high as $I = 22$ in some cases. These measurements involve (*HI*, *xny*), any Coulomb excitation by alpha particles [Fig. 11.1*a*].

Recently even *A* and odd *A* nuclei have been studied for quadrupole excitations from ground state, using heavy ions; up to excitation energies higher than 10 MeV. Figure 11.1*b* shows an odd *A* nucleus, I^{117} produced by P^{31} (Zr^{90}, He^4)I^{117} using 150 MeV P^{31} beam and, Zr^{82} produced by Al^{29} (Ni^{58}, *p*2*n*) Zr^{82} reaction using 92 MeV Al^{27} beam. A large number of such collective excitations are being studied using tandem accelerators say 14*UD*. Angular momenta of $\geq 30\hbar$ and $20\hbar$ have been added to the ground states in these two cases. As shown, in the figure; nucleus Zr^{82} goes through rotational states; as the shape of the nucleus changes from prolate to prolate-oblate.

Some of these nuclei (*e.g.* $64Gd^{156}$, $70Yb^{166}$, $70Yb^{174}$, Hf^{178}, $74W^{182}$), etc. are known to be spheroidal, which gives rises to their permanent quadrupole moments—as discussed earlier. In such cases, for even-even nuclei the ground states have zero total angular momentum; because both protons and neutrons pair to give total spin zero. The angular momenta of the excited states, due to rotation, for such nuclei, may be understood if we realise that classically,

$$\mathbf{R} = \mathcal{I}\omega = \mathbf{I}$$

$$E = \frac{1}{2}\mathscr{I}\omega^2 = \frac{1}{2}I\omega$$

Therefore, $$\frac{\partial E}{\partial I} = \frac{\omega}{2} \propto \omega \quad \text{and} \quad \frac{\partial E}{\partial \mathscr{I}} = \frac{1}{2}\omega^2 \propto \omega^2 \qquad ...(11.2)$$

where **R** is the angular momentum due to rotation and $\mathscr{I}$ is the effective moment of inertia and ω is the angular velocity.

Then one can write:

$$E = \frac{1}{2}\mathscr{I}\omega^2 = \frac{R^2}{2\mathscr{I}} \qquad ...(11.3)$$

Quantum mechanically, if **R** is the only angular momentum, *i.e.* if $\mathbf{R} = \mathbf{I}\hbar$, then one can write, from Eq. 11.3.

$$E = \frac{I(I+1)}{2\mathscr{I}}\hbar^2 \qquad ...(11.4)$$

The moment of inertia $\mathscr{I}$ used in Eqs. 11.2 to 11.4 corresponds to rotation around an axis perpendicular to the axis of symmetry of the nucleus because Eq. 11.2 is based on this assumption. In Fig. 11.8*a*, we have represented the angular momentum due to rotation as **R**, perpendicular to three-axis, for a nucleus with axial symmetry. The angular velocity ω corresponds to rotation perpendicular to three-axis. However, for a general situation for a spheroidal nucleus, if $\mathscr{I}_3$ is the moment of inertia for rotation about symmetry three-axis of the nucleus and $\mathscr{I}_1 = \mathscr{I}_2 = \mathscr{I}$ is the moment of inertia around an axis, perpendicular to the three-axis; then Hamiltonian for the energy of rotation can be written as (assuming I_3 exists),

$$H = \sum_{i=1}^{3}\frac{\hbar^2}{2\mathscr{I}_i}I_i^2 = \frac{\hbar^2}{2\mathscr{I}}(I^2 - I_3^2) + \frac{\hbar^2}{2\mathscr{I}_3}I_3^2 \qquad ...(11.5)$$

If K is quantum number for the projected component of total angular momentum corresponding to I_3 along 3-axis, *i.e.* axis of symmetry of the nucleus; then the energy of rotation can be written as:

$$E_{I,K} = \frac{\hbar^2}{2\mathscr{I}}[I(I+1) - K^2] + \frac{\hbar^2 K^2}{2\mathscr{I}_3} \qquad ...(11.6)$$

We will discuss this further in Section 11.1.6.

We, however, can see from Fig. 11.2 that in general the experimental points do not obey Eq. 11.4. The experimental points indicate that the moment of inertia increases with increasing spin. As a matter of fact the actual moment of inertia are found to be lower by factors of two to ten than for the spherical body-value $\mathscr{I}_{\text{rigid}} = 2/5\, MR^2$. This shows that in nuclear rotation, the whole nucleus does not rotate like a classical rotor. Further, the increase of moment of inertia with spin shows, that the internal behaviour of the nucleus is affecting the moment of inertia. This is now attributed to the effect of pairing of nucleons in the nucleus. We will discuss this more in details in Section 11.1.5.

11.1.2 Phenomenological and VMI Model

We now discuss the empirical results of the rotational energy levels as systematised and evaluated by various workers.

One of the earliest systematisation was done in 1930, by Thibund[3], who predicted that the rotational spectra should follow, Eq. 11.1 on the partial analysis of the excited levels of U^{238}. Also many authors[4] like Niels Bohr and Frenkel, later worked on the basics of this model in 1939. However, it was the work of Bohr A; Rainwater, and Mottleson and Jensen in fifties and sixties which led to the present understanding of the model.[4] Later, the detailed analysis of many spectra, led Diamond, Stephens and Swiatecki[5] to the idea of a spinning nucleus being stretched out under the influence of centrifugal force—the so called Beta stretching model—and the energy of the rotational states in unit of $\hbar^2$ was expressed as:

$$E_I(\beta) = \frac{1}{2}C(\beta - \beta_0)^2 + \frac{I(I+1)}{2\mathcal{J}(\beta)} \qquad ...(11.7)$$

where $\mathcal{J}(\beta)$ is the moment of inertia in terms of β ; β_0 is the ground state deformation parameter, $\beta - \beta_0$ is the deviation from ground state deformation parameter, and C is the stiffness parameter. Equation 11.7 represents a semi classical model, of a stretchable liquid drop and therefore it was assumed that $\mathcal{J}(\beta)$ is proportional to β^2. The equation explained the rotational spectra of strongly deformed neutron deficient nuclei quite well.

This expression, was the fore-runner to variable moment of inertia, (VMI) model of the rotational levels proposed by Scharff Goldhaber G[6], who re-wrote Eq. 11.7 which presupposes axial symmetry, in a more general form in which β of Eq. 11.7 was replaced by a general variable t, which might include the effects not only of deformation but also of the effective pairing energy. Then $\mathcal{J} = t^n$ was used with $n = 1, 2, 3$ and $t \approx \beta$ for ground state bands, ranging from $3.33 > E4/E2 > 2.34$. Fits were obtained for $n = 1$. Then one can write:

$$E(\mathcal{J}) = \frac{C}{2}(\mathcal{J} - \mathcal{J}_0)^2 + \frac{I(I+1)}{2\mathcal{J}} \qquad ...(11.8)$$

where $\mathcal{J}_0$ is a constant, corresponding to the moment of inertia of the ground state.

The equilibrium condition

$$\frac{\partial E(\mathcal{J})}{\partial \mathcal{J}} = 0 \qquad ...(11.9)$$

determines the moment of inertia called variable moment of inertia by minimising energy.

From Eqs. 11.8 and 11.9, one obtains

$$\mathcal{J}_I^3 - \mathcal{J}_0\,\mathcal{J}_I^2 = \frac{I(I+1)}{2C} \qquad ...(11.10)$$

Equations 11.9 and 11.10 represent VMI model. They have a real root for $\mathcal{J}_I$ for a given $\mathcal{J}_0$ and C and were fitted to a large number of rotational bands[7, 8] (88 bands ranging from $A = 108$ to $A = 248$), for E_I ($I = 0, 2, 4, ...$). For each band the quantities C, $\mathcal{J}_0$, $\mathcal{J}_I$ and σ were obtained, where

$$\sigma = \left(\mathcal{I}_I^{-1} \frac{\partial \mathcal{I}_I}{\mathcal{I} I} \right)_{I=0} = (2\, C\, \mathcal{I}_0^3)^{-1} \qquad ...(11.11)$$

Figure 11.3 shows the behaviour of C as a function of A and σ vs R_4 where $R_4 = E_4/E_2$. Essentially R_4 represents the deviation from the assumption of pure rotational behaviour of the nuclei. Physically σ-parameter is termed softness parameter and C is termed stiffness parameter—both implying the deviation from rigid behaviour of $\mathcal{I}_{\text{rigid}}$ which is independent of I as mentioned earlier. The stiffness parameter C decreases by approximately five orders of magnitude between the nucleus C^{12} and heavy actinides ($A > 200$), which decay spontaneously[9, 10]. So less C means, that peripheral nucleons are not tightly bound.

The variation of moment of inertia $\mathcal{I}$ of deformed nuclei as a function of angular momentum I, has been dealt with semi-classically from the early days as discussed in the beginning of this chapter. Even recently[11], continuation of this interest has brought some interesting results. The concept of nuclear softness as discussed earlier in Eq. 11.11 was first introduced by Morinaga[12] in 1966. Based on these concepts, Mariscotti et al. [Ref. (7)] developed a model of variable moment of inertia (VMI) and its latter-version[7]—the generalised VMI, (GMVI). In recent work, the softness parameter σ and stiffness parameter C Eq. 11.11, have been calculated by making different assumptions about the equilibrium condition. As for example, one assumption[13] (J.B. Gupta et al.) corresponds to Eq. 11.9 as given earlier, *i.e.* $\dfrac{\partial E(\mathcal{I})}{\partial \mathcal{I}} = 0$ and other (R.K. Gupta et al.) corresponds to the expansion of $\mathcal{I}$ by Taylor series. The two approaches yield some what different results of σ and C as a function of deformation parameter, or the neutron number n. However,[14] these approaches are only qualitative. A more quantitative approach for the variation of $\mathcal{I}$ with I has been most recently[15], made by dividing the moments of inertia into kinematic and dynamic parts, given by:

$$\frac{\mathcal{I}^{(1)}}{\hbar^2} = \mathcal{I}_0\, [1 + bI\,(I+1)]^{1/2} \qquad \text{(Kinematic Part);}$$

$$\frac{\mathcal{I}^{(2)}}{\hbar^2} = \mathcal{I}_0\, [1 + bI\,(I+1)]^{3/2} \qquad \text{(Dynamic Part)} \qquad ...(11.12a)$$

where a and b is derived from the energy expression obtained from Bohr-Hamiltonian for a well deformed nucleus and $\mathcal{I}_0 = \hbar^2/ab$ is referred to a band moment of inertia. This formulation has been applied to super deformed (SD) band in $A \approx 190$, region for 70 SD bands for which spins could be predicted.

It is interesting how these semi-empirical approaches are useful in the range of collective motion.

11.1.3 High Spin-States

Till now, we have confined our discussion to comparatively low energy rotational states, as shown in Fig. 11.1, where we can reach up to $8\hbar$ rotational states through β or α -decay and up to $16\hbar$ through excitation by light particles say α-particles. If we want to impart higher angular momentum directly to the target; one can employ Coulomb excitation methods using heavy projectiles. The angular momentum imparted to the target nucleus, can be expressed by:

$$\mathbf{l}\hbar = \langle\, \mathbf{r} \times \mathbf{p}\, \rangle$$

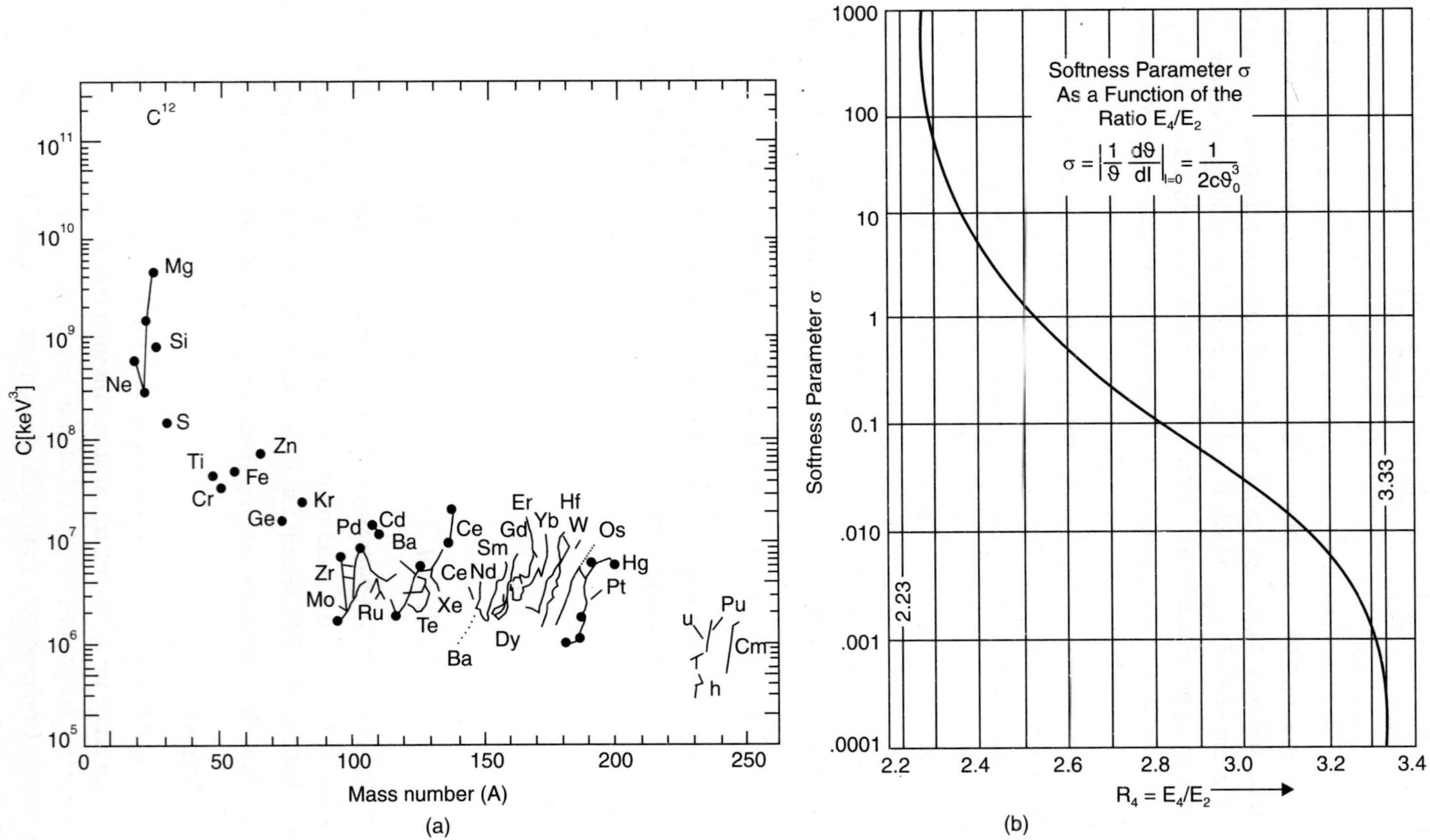

Fig. 11.3 (*a*) Stiffness parameter C as a function of A. For heavy nuclei, C is less, than for lighter nuclei. For isotopes of the same element, C is highest for the most stable nucleus [Ref. (9, 10)]; (*b*) Softness parameter σ vs R_4 where $R_4 \equiv E_4/E_2$ (Ref. 9. 10).

where **l** is orbital angular momentum imparted, **r** will be of the order of the nuclear radius of the target and **P** is the linear momentum of the projectile, related to energy $E \equiv p^2/2\,M$. So if M is large; for the same E; p will be large (for a heavy ion) and therefore a large value of angular momentum can be imparted.

Experiments in recent years, by heavy ion projectiles have yielded high angular momenta[16]. As for example, projectiles with high charge (heavy ions) were found to excite successively a number of rotational transitions ($\Delta\, I = 2$) in a single collision. In 1977 Fuchs et al.[16] observed the excitation of a state with $30\hbar$ in U^{238} using Pb^{208} projectiles. Another more effective way is fusion, in which all the angular momentum of the initial system is retained.[17] In this process, one can bring in angular momenta, up to $100\hbar$ into the compound system say by Ar^{40}. In practice, an angular momentum of $38\hbar$, for the highest value been achieved in 1978 by Khoo et al.[18] and has been extended by Beck et al.[19] The limits of the high angular momentum are determined by different criteria in different regions of A. For light nuclei, this will correspond to the maximum angular momentum that can be generated by valence nucleons say $4\hbar$ for p-shell nucleus Be^4; $8\hbar$ for s-d shell nucleus Ne^{20}. For mass around $A \approx 170$ the limit exceeds $100\hbar$, or so, most generally affected by instability against fission[20]. Between these regions, for $40 < A < 100$, the highest angular momentum that can be conveniently studied is limited by what can survive the particle evaporation cascade that follows the production of compound nucleus. We are talking, in the above discussion about the transfer of angular momenta, which affects, the nuclear behaviour like emission of measurable gamma rays. But, in principle, a target and projectile nucleus may be in contact momentarily—say for 10^{-20} secs or less—with as much as $500\hbar$ in the system. But this may not transfer any internal degree of freedom to the internal structure of the nucleus.

In Fig. 11.4, we show the theoretical results of Cohen et al.[21, 22] for the comparison with the angular momentum values for a given mass number A, where fission barrier just vanishes, or where

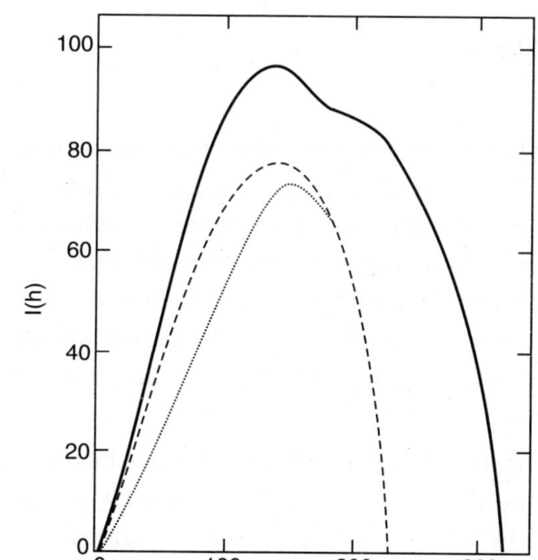

Fig. 11.4 Theoretical curves for the values of angular momentum beyond which fission starts, (the solid line). For the cases, when fission just vanishes, is shown by dashed line, and dotted line is an estimate of the boundary between particle evaporation (above) and γ-ray emission (below) estimated from data by Newton et al. (1977) (Ref. 21, 22)

fission takes place spontaneously. It is interesting to see that the maximum angular momentum that a 'cold' idealised nucleus could contain, is about $100\hbar$ for $A \approx 130$. It is lower, for higher mass number, because of higher Coulomb barrier and also is lower for lower mass number, because of lower surface energy and hence higher rotational frequency required by smaller moments of inertia and this gives higher barrier due to centrifugal forces. For further details *see* Reference (23). For some cases of high spins excitation, *see* Reference (24).

11.1.4 Physical Significance of $\mathcal{I}$; Phenomenon of Back-Bending

We have seen in Eqs. 11.8 and 11.9, that $\mathcal{I}_0$, the moment of inertia of the ground states of rotating nuclei, is dependent on β_0, and for higher states due to rotation, depends on β, the deformation parameter of the excited states. In other words, the moment of inertia depends on the excitation energy of the states. An interesting advance was made by Harris et al.[27]; who introduced the dependence of $\mathcal{I}$ on ω^2; ω being the angular velocity due to rotation, associated with the states. Physically this can be understood from the fact that $\mathcal{I}$ arises from the nucleons outside the closed shell in an interactive manner, and can be dependent on the I, C and ω. As a matter of fact, Harris[27] wrote the following two possibilities.

$$E = \frac{I(I+1)}{2\mathcal{I}} + \frac{C}{2}(\mathcal{I} - \mathcal{I}_0)^2 \qquad ...(11.8)$$

$$\left(\frac{\partial E}{\partial \mathcal{I}}\right)_I = 0 \text{ (Equilibrium condition for VMI model)} \qquad ...(11.9)$$

and

$$\frac{dE}{dI} = \omega \qquad ...(11.12b)$$

$$I(I+1) = (\mathcal{I}\omega)^2 \qquad ...(11.12c)$$

and

$$\mathcal{I} = \mathcal{I}_0 + C\omega^2 \qquad ...(11.12d)$$

Equations 11.12 are written intuitively using the classical concepts of Eq. 11.2. Eqs. 11.8 and 11.9 correspond to VMI model[26] and Eqs. 11.12, represent Harris Cranking model[27]. It can be seen that they are somewhat equivalent. Harris found excellent agreement for his model, with energies in ground state bands of deformed nuclei.[29] As a matter of fact, one can plot $\mathcal{I}$ versus ω^2. One expects from Eq. 11.12*d*, a straight line which was obtained for a large number of cases and a reasonable agreement was found. The intercept of these curves at $\omega^2 = 0$ is 2 and the initial slope is $1/C$. These results have been compiled from Se^{72} to Pu^{292} (Ref. 30). Figure 11.5*a* reproduces some of these results. The following points emerge from these curves:

As magic number limit is approached, the slope becomes increasingly steeper. Apart from this gradual increase, there is an ultimate flattening, and even downwards decrease; as shown for Er^{168}, W^{182} and Hf^{180}. Some cases show a back-bending *i.e.* a triple valued curve in ω^2 as shown for Pd^{104}, Er^{158}, Dy^{158}, Er^{160}, Er^{162} and Yb^{166}. As a matter of fact for Pd^{104}, a double back-bending is observed[31]. These back-bending breaks occur at $I = 14 - 16$. The values of $B(E_2)$ for each transition up to the top of the band at spin 18, are of the order of enhanced rotational collective rates; but for $I = 14 - 12$ transitions rate $B(E_2)$ is reduced by 20–40%. This indicates that the nucleus is changing its structure.

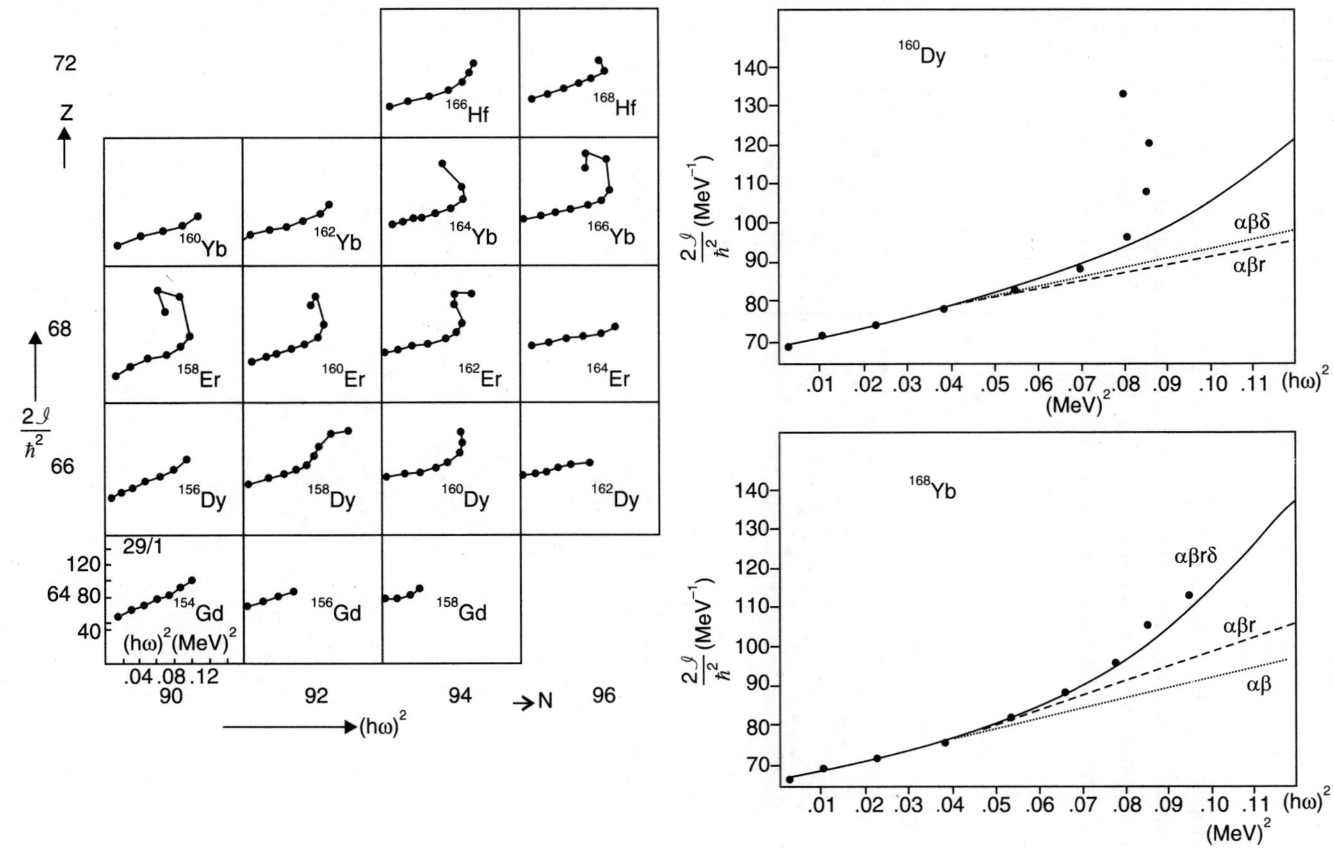

Fig. 11.5a The curves (Experimental) representing $2\mathscr{I}/\hbar^2$ versus ω^2 showing back-bending phenomenon (Ref. 30), of the excited states of many spheroidal nuclei in the rare earth region (Ref. 30, 24, 25, 28).

Harris[27] has expressed VMI model, by expansion of $E\,(I)$ and $\mathscr{I}$ in terms of an angular velocity ω, *i.e.*,

$$E\,(I) = \alpha\,\omega^2 + \beta\,\omega^4 + \gamma\,\omega^6 + \delta\,\omega^8 \qquad \text{...(11.13}a\text{)}$$

and

$$\mathscr{I} = 2\alpha + \frac{4}{3}\beta\omega^2 + \frac{6}{5}\gamma\,\omega^4 + \frac{8}{7}\delta\,\omega^6 \qquad \text{...(11.13}b\text{)}$$

The angular velocity expansion Eq. 11.13, gives very good fits, with four parameters to the energies of deformed nuclei, all the way up to the start of back-bending region and for many cases for the moments of inertia as a function of ω^2 as shown in Fig. 11.5*a* for Yb^{168}. But for cases like Dy^{160} these parameters do not reproduce the experimental parameters, and of course, they do not reproduce the back-bending[26, 30].

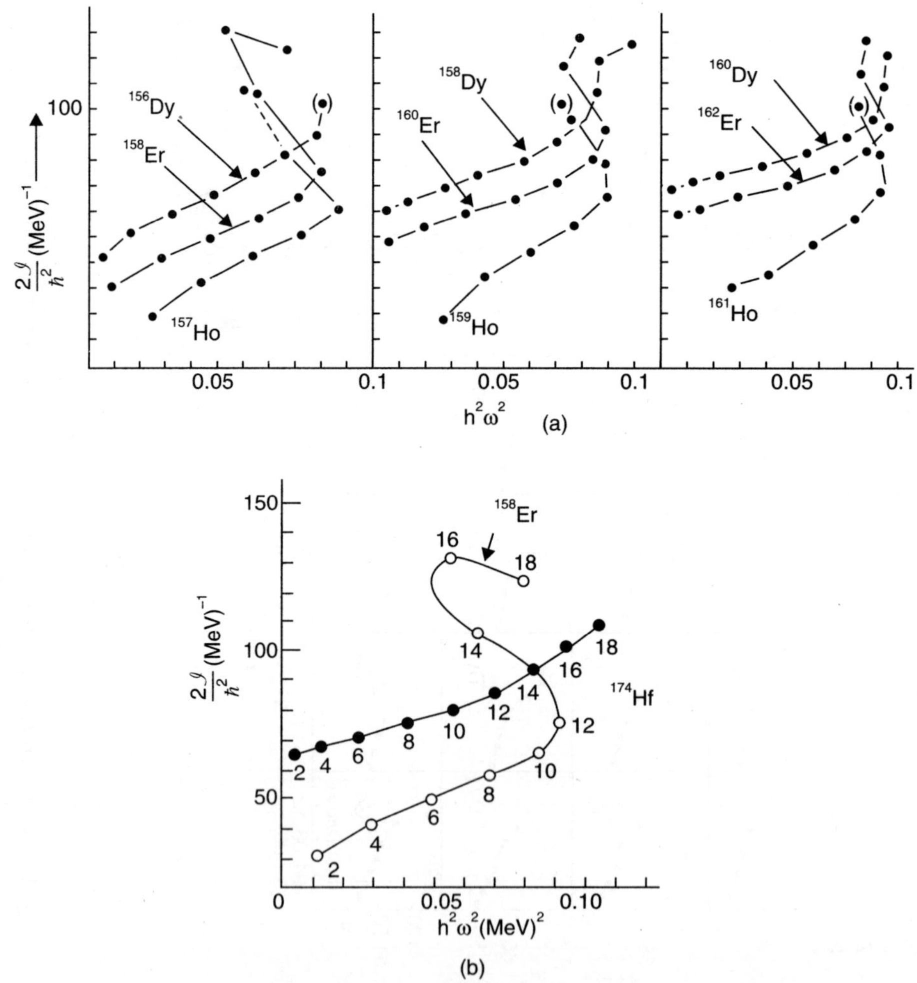

Fig. 11.5b (*a*) Comparison of back-bending and its absence in Er^{158} and Hf^{174}; (*b*) Comparison of back-bending plots of even and odd mass nuclei (Ref. 43, 61).

Pure rotational cases, discussed above are applicable when nucleus in the ground state is strongly deformed and does not get disturbed by the incidence of exciting energy. On the other hand, if the deformation is weak or for even-even nuclei near closed shell, vibrational mode of collective motion takes over. This is shown in Fig. 11.6, where we have plotted E_2/E_1 as function of neutron number, $E_2/E_1 \approx 2.2$ corresponds to vibration state and $E_2/E_1 \approx 3.33$ to rotational mode. We will discuss the vibrational mode in the subsequent section.

The phenomenon of back-bending, shown in Fig. 11.5*a* for Er^{158}, Er^{160}, Er^{162} and Yb^{166}, as measured by Harris[27] et al., has been observed afterwards for many deformed nuclei.[43, 61] Figure 11.5*b* shows the relationship of $\frac{2\mathscr{I}}{\hbar^2}$ for different values of $\hbar^2\,\omega^2$ for various nuclei, giving a comparison of odd mass and even nuclei, and showing indications of back-bending, in the range of $I = 12 - 18$, especially shown for Er^{158} in comparison to Hf^{174}, for which the curve is monotonously smooth.

As we will see later for the ideal vibrator; the ratio $E_2/E_1 = 2$. The experimental value of 2.2 is understood on the basis of the removal of the degeneracy of the triplet state, using the shell-model calculations. As we see in Fig. 11.6, where we have plotted the experimental values of E_2/E_1, as a function of neutron number N, there is an abrupt transition from the vibrational to the rotational pattern between $N = 88$ and 90 and $Z = 86$ and 88. These seem to arise due to $\hbar_{11/2}$ shell breaking up as a function of deformation. For more details *see* Reference (34).

11.1.5 Qualitative Theoretical Explanation of the Variation of $\mathscr{I}$ with ω^2

The observed energy spacing of 0^+, 2^+, 4^+ etc., which belongs to the first ground state band, increase less rapidly than $E_I = \hbar^2\, I\,(I + 1)\,/\,2\,\mathscr{I}$ expected for a constant moment of inertia (Fig. 11.2). As has been discussed earlier this was explained by VMI model, which assumes some sort of centrifugal stretching causing the moment of inertia to increase smoothly with angular momentum right up to the highest spins [Fig. 11.5]. Equations 11.12*d* and 11.13*b* explain this semi-empirically. It has been pointed out by many authors[32, 35], that the intrinsic moment of inertia and the nuclear softness parameter are correlated with change in nuclear deformation.

The inclusion of third and fourth parameter; (γ and δ) does improve the fit but is not correlated with deformation. The nuclear structure calculations[36] of the moment of inertia which have been based on the Cranking model, and employ such characteristics, as the strength of pairing force, the nuclear deformation, the characteristic of single particle orbitals, etc. seem to agree with experimental data within an accuracy of 10 to 20%. In this model the occurrence of the higher order terms in ω^2 expansion are understood to be primarily due to the nuclear coriolis, anti-pairing effect[36], especially developed by Mottleson and Valatin[39].

An elementary presentation, of the cranking model requires, the derivation of moment of inertia, by comparing it with 'Pushing Model' formula for the mass, step by step[37]. The basic assumption of the simplest form of the cranking model is that the nuclear many-body system in its rotation, can be described in terms of independent non-interacting particles contained in an external potential well, which is rotating[38].

Then one solves the Schrödinger equation:

$$H\,(t)\;\psi = i\;\hbar\;\dot{\psi} \qquad ...(11.14a)$$

where $H\,(t) = P^2/2m + V\,(t)$; where $V\,(t)$ is the external potential of fixed shape moving in space with a constant angular velocity. It is time dependent. For the cranking case,

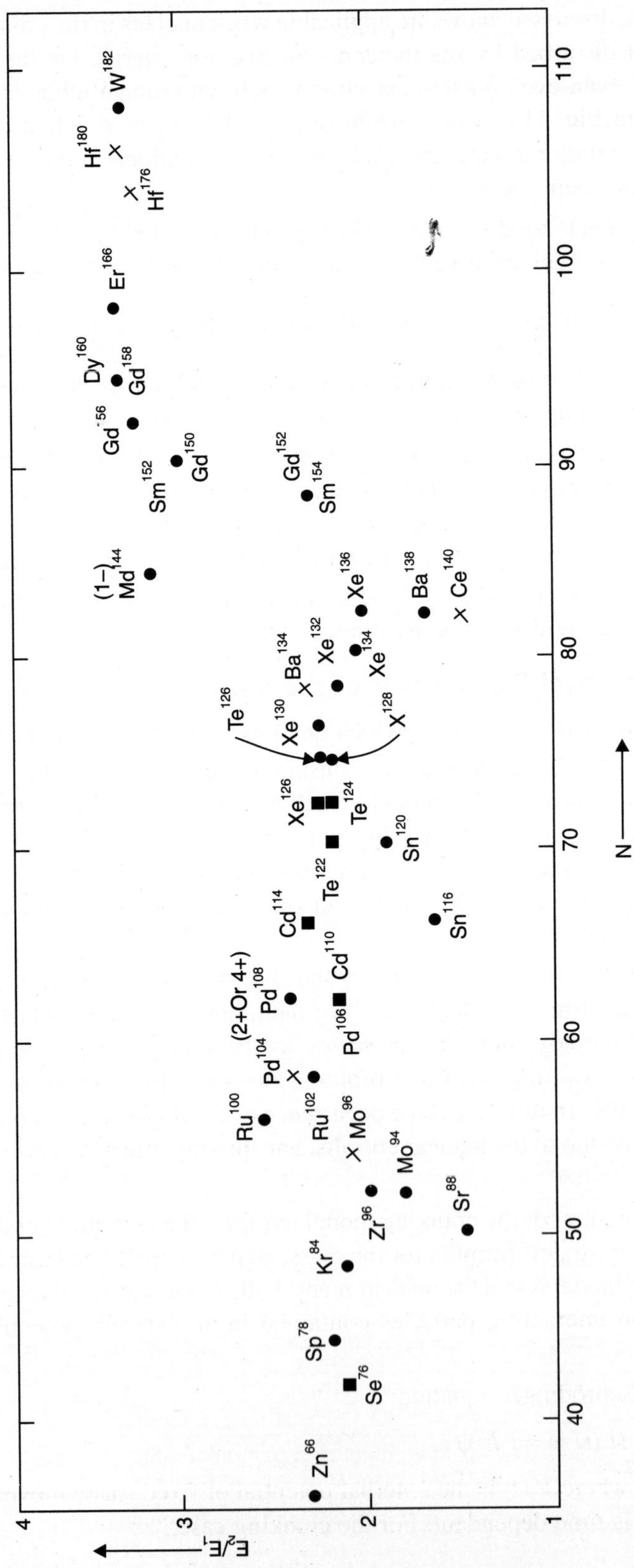

Fig. 11.6 The ratios of E_2/E_1 as a function of neutron number, N. The on-set of rotational states at about N ≈ 90 is evident (Ref. 33)

$$V(t) = h(r, \theta, \phi - \omega t) \qquad \text{...(11.14b)}$$

where $t = 0$ potential-shape must have dependence on ϕ, *i.e.* it must be a deformed non-spherical potential. One then calculates the energies to be associated with rotational inertia, *i.e.*,

$$E = \langle \psi | H(t) | \psi \rangle$$

$$= E_0^0 + \frac{\hbar\omega^2 \left| \langle i | L_z | 0 \rangle \right|^2}{(E_1^0 - E_0^0)} = E_0^0 + \frac{1}{2}\mathcal{I}\omega^2 \qquad \text{...(11.15a)}$$

Hence
$$\mathcal{I} = \frac{2\hbar \left| \langle i | L_z | 0 \rangle \right|^2}{(E_1^0 - E_0^0)} \qquad \text{...(11.15b)}$$

where L_z is the spin, and can be replaced by I_z. This is cranking formula for moment of inertia.

The above model can be used to explain the parabolic portions of the curves of Fig. 11.5 by including different higher orders. But it cannot explain the sudden increase of $\mathcal{I}$ with ω^2, giving rise to back-bending.

Calculations for high spin states have been carried out by many authors[39]. In Fig. 11.7*a* we show the calculations by BCS approximation, where energy E consists of the rotational part involving the moment of inertia $\mathcal{I}$ and an intrinsic part which depends on pairing and quadrupole deformation. It seems that stronger the pairing force relative to the critical value at which pairing just becomes effective, the stronger is the back-bending tendency. So the cranking model of the nucleus, with particles moving in a rotating deformed potential and having pair interactions may be able to give a satisfactory description of the moment of inertia including back-bending.

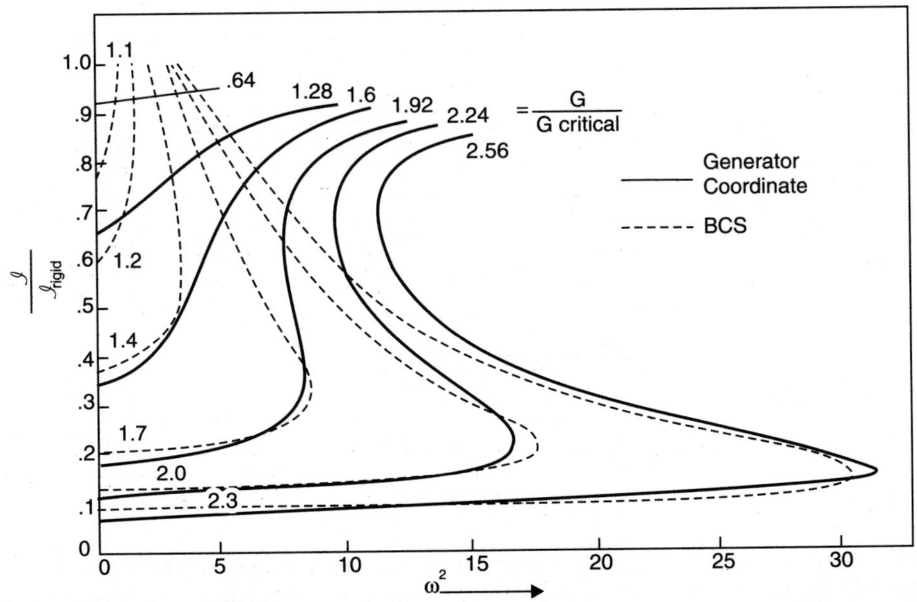

Fig. 11.7a Curves of $\mathcal{I}/\mathcal{I}_{\text{rigid}}$ versus ω^2, as function of the strength of pairing force for the two level model. Critical is the minimum pairing force strength at which there is pairing at $\omega = 0$, in the BCS approximation. The solid curves are with an improved pairing approximation (Ref. 39).

Summarising, the rotational states to high spin give a unique opportunity for the study of the response of a nucleus to continuously variable disturbance, over a wide range of strength, namely the coriolis and centrifugal forces. These are explained to a large extent by VMI model. However, experimentally, the phenomenon of sudden change in the spectrum at a certain values of I (back-bending) shows that a new theoretical approach is required. The origin of this effect is now understood as coming from competition between the pairing force, which prefers pairs of like particles near the Fermi surface to have opposite angular momentum (j^2) $I = 0$ and coriolis force, which breaks the time reversal j m, j-m, degeneracy of the single particle states, thus reducing the pairing and prefers particles near the Fermi surface to be aligned with their angular momenta pointed to the rotation direction. The singularity represents the start of the process of such a breaking of pairs. BCS calculations, by Chan and Valatin, Mottleson Valatin, Kreunlinde and Sorensen[39], have yielded qualitatively satisfactory results. Essentially the effect of residual interaction has been taken into account by using Hartree-Fock-Bogoliubov[40] (HFB) functions in BCS approximation, using two level cranking model. Figure 11.7*a* shows the results of calculations by Sorensen[39]. This is called coriolis decoupling model.

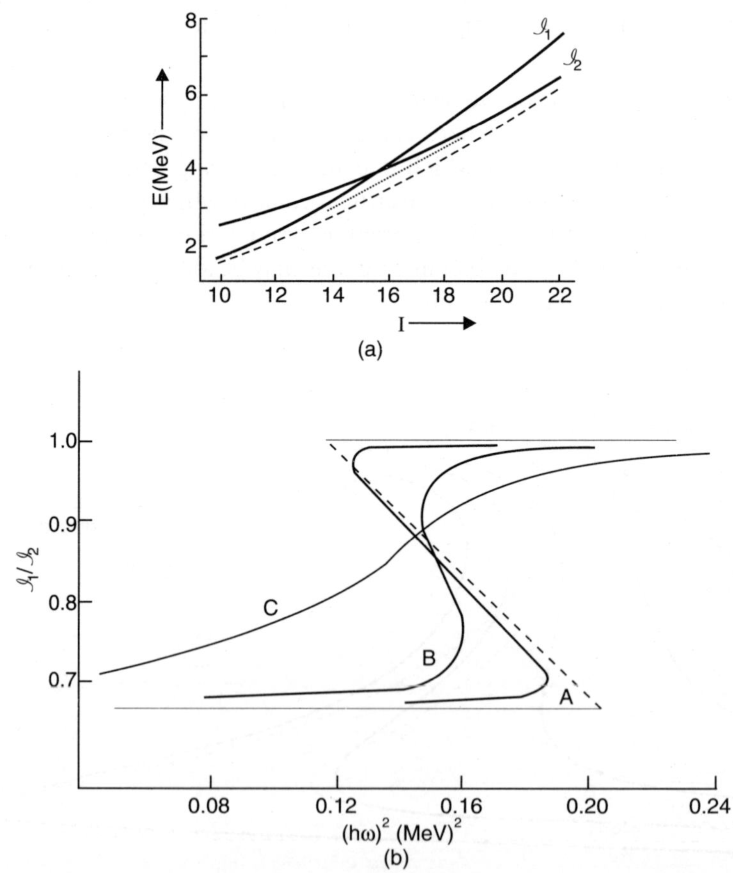

Fig. 11.7b (*a*) It illustrates the crossing of two rotational bands, with moments of inertia $\mathscr{I}_1$ and $\mathscr{I}_2$ at $I_c = 16$; (*b*) Back-bending plot for Yrast band as expected from crossing of bands, shown in (*a*) above (39, 40, 41).

One plausible explanation of this back-bending or phase transition is, that crossing of two rotational bands—namely one ground state band and the other excited band—occurs and hence there is a sudden jump in deformation[41].

We reproduce in Fig. 11.7*b*, a back-bending plot for Yrast bands as per idealised situation of crossing of two bands, with moments of inertia $\mathcal{I}_1$ and $\mathcal{I}_2$ at $I_c = 16$; where I_c is the value of angular momentum where the two bands cross.

The coriolis decoupling model has been extended, and has been applied to cases where departure from axial symmetry is significant[74], for different cases.

Theoretical discussion have been extended to very high spin $I \approx 60$–$70\hbar$ where the rotation is nearly classical, and the nuclear rotation may give rise to an appearance of oblate shape. A giant back-bending is expected at such high spins[75].

11.1.6 Wave-Functions

Even-Even Nuclei

The wave-function for the rotational motion, in general, corresponds to a rotating spheroid which corresponds to $\mathcal{I}_1 = \mathcal{I}_2 = \mathcal{I}$ and $\mathcal{I}_3$ has a different value. So that it is invariant with respect to rotation by π, about any axis passing through the centre. It is required that under rotational operator:

$$R\ (\theta \to \pi + \theta;\ \phi \to \pi - \phi;\ \text{and}\ \psi \to \psi)$$

the wave-function is invariant, where θ, ϕ and ψ are Euler angles.

If we consider the collective rotational motion of nuclei, which have axial symmetry, *e.g.* a symmetric top, there will be two sets of orthogonal system of axes (*i*) body-fixed reference frame giving rise to 1, 2, 3-axes and (*ii*) laboratory fixed reference frame; with *x*, *y*, *z*-axes. In body-fixed reference frame; the 3-axis is generally used as the axis of symmetry, for nuclei with axial symmetry. As described earlier, $\mathcal{I}_3$ is taken as moment of inertia for rotation about symmetry—3-axis and $\mathcal{I}_1 = \mathcal{I}_2 = \mathcal{I}$ is taken around an axis (1-axis or 2-axis), perpendicular to it. I is, then, the total angular momentum operator, with components I_1, I_2 and I_3 along the body-fixed axes. The eigen functions which are the *D*-functions, should be eigen function of I^2 and I_3 and I_z where I_z is the angular momentum of the body around a fixed space axis [Fig. 11.8]. Such wave-functions will correspond to pure rotation, which pertain to even-even nuclei, with ground state $I = 0$ and higher excited states of $I = 2, 4$, etc. As shown in Fig. 11.8*a*, in an axially symmetric nucleus, this rotational angular momentum corresponds to **R**; whose projection on z'-axis is zero; so that *K*, [Eq. 11.6] for pure rotation is zero. *D*-functions which represent such wave-function; have the property of transformation of spherical harmonics under finite rotations. We can write the *D*-function such that:

$$Y_{l\,m}(\Omega) = \sum_{m'=-l}^{l} Y_{l\,m'}(\Omega')\ D^{l}_{mm'}(\theta, \phi, \psi) \qquad \text{...(11.16)}$$

In Eq. 11.16, l represent an angular momentum due to motion in space, and hence can represent rotational motion. So total rotational angular momentum represented by I, can be used instead of l. Then *D*-function transforms, through rotation of Euler angles θ, ϕ and ψ through a counterclock rotation of θ $(0 \le \theta \le 2\pi)$ about *z*-axis; followed by a rotation of ϕ $(0 \le \phi \le \pi)$ about the *y*-axis and a rotation of ψ $(0 \le \psi \le 2\pi)$ about the *x*-axis. The solid angles Ω and Ω' correspond to initial and final polar angles (θ, ϕ) and (θ', ϕ') and m and m' take any of values of $-I, -I+1, \ldots 0, \ldots I-1, I$. For more details about the D-functions, *see* appendix and for the properties of D-matrices, *see* References (42 and 43). If such a wave-function is eigen-function of I^2, I_z and I_3, then it satisfies the following:

$$I^2\, D^I_{MK} = I\,(I+1)\; D^I_{MK}$$

$$I_z\, D^I_{MK} = M\; D^I_{MK}$$

$$I_3\, D^I_{MK} = K\; D^I_{MK} \qquad \text{...(11.17)}$$

and the normalised wave-function is, then, written as:

$$\psi^I_{MK} = I\,M\,K> = \left(\frac{2I+1}{8\pi^2}\right)^{1/2} D^I_{MK}\;(\theta,\phi,\psi) \qquad \text{...(11.18)}$$

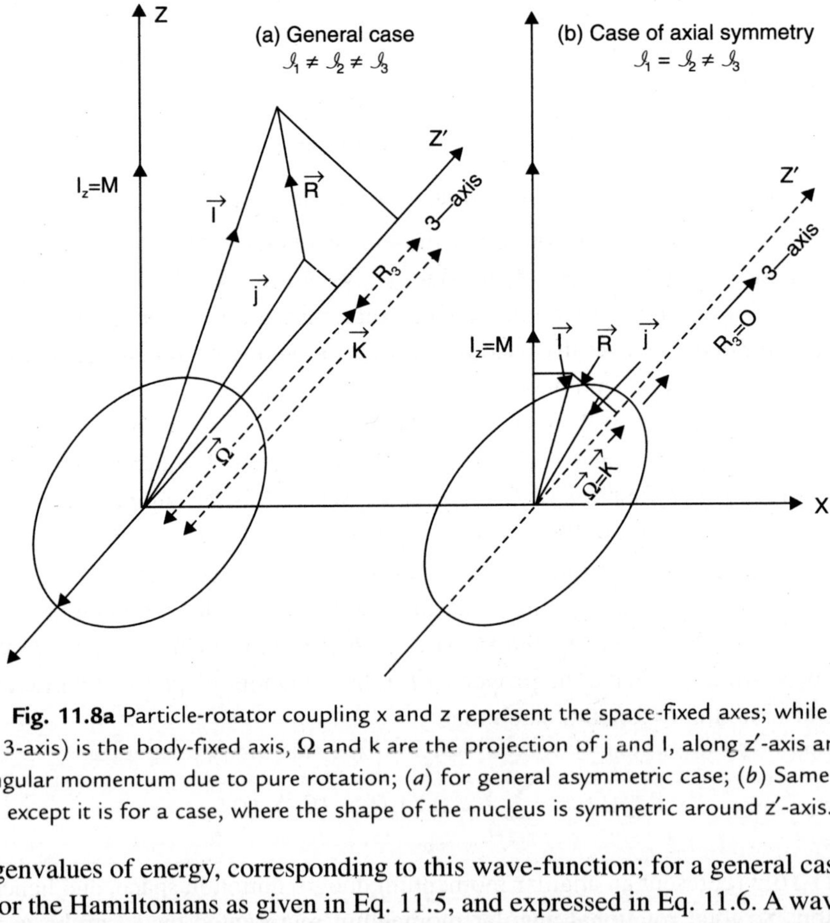

Fig. 11.8a Particle-rotator coupling x and z represent the space-fixed axes; while z′ (or 3-axis) is the body-fixed axis, Ω and k are the projection of j and I, along z′-axis and R is the angular momentum due to pure rotation; (*a*) for general asymmetric case; (*b*) Same as (*a*), except it is for a case, where the shape of the nucleus is symmetric around z′-axis.

The eigenvalues of energy, corresponding to this wave-function; for a general case; when $K \neq 0$, can be given for the Hamiltonians as given in Eq. 11.5, and expressed in Eq. 11.6. A wave-function like given in Eq. 11.18, is invariant with respect to rotation of π, about any axis passing through the centre.

For even-even nuclei which are axially symmetric, $K = 0$. This can be physically understood from the fact the even-even spherical nuclei do not show experimentally any rotational spectra and hence indicate that the wave-function of rotation are such that the angular momentum about any symmetry axis vanishes. Hence $K = 0$. This has been shown in Fig. 11.8*a*, by drawing **R**, as perpendicular to z'-axis. The wave-function for the general symmetric case, can, then, be written as:

$$\psi^I_{M0} \equiv | I\, M\, 0 \rangle = i^{-M} \left[\frac{(2I+1)}{8\pi^2} \right]^{1/2} D^I_{M0}\,(\theta, \phi) = Y_{IM}\,(\theta, \phi)$$

and eigenvalues of energy are:

$$E_I = \frac{\hbar^2}{2\mathcal{I}}\, I\,(I+1) \qquad \text{...(11.19)}$$

as expressed earlier in Eq. 11.4 classically. In Eq. 11.19, the factor i^{-M} has been inserted in the wave-function of ψ^I_{MO}; so that the wave-function corresponds to Y_{IM}.

A general form of the wave-function for rotation satisfying the condition of invariance under R_1, requires that the normalised wave-function can be written as:

$$\psi^I_{MK} = | IMK \rangle = \left[\frac{2I+1}{8\pi^2} \right]^{1/2} \times \frac{1}{\sqrt{2}}\,(1 + R_1)\, D^I_{MK}\,(\theta, \phi, \psi)$$

$$= \left[\frac{2I+1}{16\pi^2} \right]^{1/2} [D^I_{MK} + (-1)^{I+K}\, D^I_{M,-K}] \qquad \text{...(11.20}a\text{)}$$

and if we include the spin-rotation wave-function χ_i then due to additional internal degrees of freedom the wave-function can be written as:

$$\psi^I_{MK} \equiv | IMK \rangle$$

$$= \left[\frac{2I+1}{16\pi^2} \right]^{1/2} [D^I_{MK}\, \chi_i + (-1)^{I+K} D^I_{M,-K}\, R_1\, \chi_i] \qquad \text{...(11.20}b\text{)}$$

where R_1 is rotation operator, expressed as:

$R_1\,(\theta \to \pi + \theta;\ \phi \to \pi - \phi;\ \psi \to \psi)$. A wave-function of a spheroid is invariant under rotations R_1. For relationship (11.20*a*) and (11.20*b*), one has used the relations:

$$R_1\, D^I_{MK}\,(\theta\,,\, \phi,\, \psi) \to e^{i\pi(I+K)}\, D^I_{M,-K}\,(\theta, \phi, \psi)$$

$$R_1^2 = 1;\ \text{and}\ R_1\, \psi^I_{MK} = \psi^I_{MK} \qquad \text{...(11.20}c\text{)}$$

If $K = 0$, then, only for $I = 0, 2, 4$; the value of $\psi^I_{MK} \neq 0$, so that for $K = 0$, only even values of I are permitted. Of course, in writings (11.20*b*), we have assumed that the total wave-function of space rotation and spin rotation is the product of two wave-functions.

Odd A Nuclei

For odd nuclei, the ground state angular momentum is expected to be different from zero due to the odd particles. The total Hamiltonian for such a case will be:

$$H = H_{\text{rot}} + \frac{p^2}{2\,M} + V\,(r) \qquad \text{...(11.21)}$$

Now $\mathbf{I} = \mathbf{j} + \mathbf{R}$ [Fig. 11.8*a*]

Hence writing,

$$H_{rot} = \sum_{i=1}^{3} \frac{\hbar^2}{2\mathcal{I}_i} R_i^2 = \frac{\hbar^2}{2\mathcal{I}}(R^2 - R_3^2) + \frac{\hbar^2}{2\mathcal{I}_3} R_3^2 \quad ...(11.22)$$

where we have used $\mathcal{I}_1 = \mathcal{I}_2 = \mathcal{I}$. For $R_3 = 0$, we write:

$$H_{rot} = \frac{\hbar^2}{2\mathcal{I}} R^2 = \frac{\hbar^2}{2\mathcal{I}} (\mathbf{I} - \mathbf{j})^2 \quad ...(11.23)$$

In Eq. 11.23, we have assumed $R_3 = 0$; which corresponds to the physical fact that rotation around the nuclear symmetry axis does not give rise to any angular momentum because this correspond to no change of wave-function in space [*see* Fig. 11.8*b*].

The wave-function, which corresponds to quantum numbers *I*, *M* and *K*, like Eq. 11.20 *b*, can now be written as:

$$\psi^I_{\Omega, MK} = | IMK \rangle$$

$$= \left(\frac{2I+1}{8\pi^2}\right)^{1/2} \chi_\Omega (r)\, D^I_{MK} (\theta, \phi, \psi) \quad ...(11.24)$$

where Ω is the eigenvalue of the operator j_3 and $\chi_\Omega (r)$ is the solution of non-rotational part of the Hamiltonian corresponding to individual particle angular momenta j and j_3, its component along 3-axis. Then Eq. 11.21 represents the Hamiltonian H for this wave-function.

(*i*) Now remembering that R_3, the angular momentum of the rotor about its symmetry axis is zero, we can write:

$$K = I_3 = R_3 + j_3 \rightarrow j_3 = \Omega \text{ (for } R_3 = 0) \quad ...(11.25)$$

and hence writing Eq. 11.21, with the help of Eq. 11.22, we get:

$$H = \frac{\hbar^2}{2\mathcal{I}}[(\mathbf{I} - \mathbf{j})^2 - R_3^2] + \frac{\hbar^2}{2\mathcal{I}_3} R_3^2 + \frac{p^2}{2M} + V(r) \quad ...(11.26)$$

$$\approx \frac{\hbar^2}{2\mathcal{I}} [\mathbf{I} - \mathbf{j}]^2 + \frac{p^2}{2M} + V(r) \text{ for } R_3 = 0$$

We can rewrite it, then, as:

$$H = H_0 + \frac{\hbar^2}{2\mathcal{I}}[I(I+1) - 2I_3 j_3] = H_0 + \frac{\hbar^2}{2\mathcal{I}}[I(I+1) - 2K^2] \quad ...(11.27)$$

where

$$H_0 = \frac{p^2}{2M} + V(r) + \frac{\hbar^\theta}{2\mathcal{I}} j^2 \quad ...(11.28)$$

where we have used $R_3 = 0$, and $I_3 = j_3 = K$ ...(11.29) (*ii*) If $R_3 \neq 0$ which is true for a general non-symmetric nucleus [Fig. 11.8*a*]; then $K \neq j_3$ and the Hamilton H, is, written, from Eq. 11.21 and $R_3 = K - j_3$ from Eq. 11.26, then

$$H = H_0' + \frac{\hbar^2}{2\mathcal{I}}[I(I+1) + j^2 - 2\mathbf{I}\cdot\mathbf{j} - (K-\Omega)^2] + \frac{\hbar^2}{2\mathcal{I}_3}(K-\Omega)^2 \qquad \text{...(11.30)}$$

where we have used $j_3 = \Omega$, and $H_0' = \dfrac{p^2}{2M} + V(r)$

Hence
$$H = \frac{\hbar^2}{2\mathcal{I}}[I(I+1) - K^2 - \Omega^2] + \frac{\hbar^2}{2\mathcal{I}_3}(K-\Omega)^2 + H_0 + R.P.C. \qquad \text{...(11.31}a\text{)}$$

where
$$R.P.C. = \frac{\hbar^2}{2\mathcal{I}}[-2\mathbf{I}\cdot\mathbf{j} - 2K\Omega] = -\frac{\hbar^2}{2\mathcal{I}}[2I_1 j_1 + 2I_2 j_2]$$

$$= -\frac{\hbar^2}{2\mathcal{I}}[(I_1 + iI_2)(j_1 - ij_2) + (I_1 - iI_2)(j_1 + ij_2)]$$

$$= -\frac{\hbar^2}{2\mathcal{I}}[I_+ j_- + I_- j_+] \qquad \text{...(11.31}b\text{)}$$

And I_1, I_2, I_3 and j_1, and j_2 and j_3 are the components of the concerned angular momenta. As for example, if angular momenta J have components J_x, J_y and J_z along the space fixed axes; then $J_\pm = J_x \pm i J_y$ and commutation rules are $[J_x, J_y] = iJ_z$ and cyclic permutations.

Also
$$[J, J_z] = 0 \text{ and } [J_z, J\pm] = \pm J\pm \qquad \text{...(11.31}c\text{)}$$

Then,
$$H\,\psi^I_{\Omega MK} = E\,\psi^I_{\Omega MK} \qquad \text{...(11.32)}$$

yields the eigen-function $\psi^I_{\Omega MK}$. If we denote the solution of H_o by $\chi_\Omega(r)$, where Ω is the eigenvalue of the operator j_3; a normalized and anti-symmetric eigen-function can be written, so that it represents a function which is invariant under operation R_1 and R_3. The total general wave-function, then, becomes:

$$\psi^I_{\Omega MK} = |\,IMK\rangle$$

$$= \left[\frac{2I+1}{16\pi^2}\right]\left[D^I_{MK}|\chi_\Omega\rangle + (-1)^{I+K+\Omega} D^I_{M,-K} \times \sum_j (-)^j C_{j\Omega}\,|\chi_j, -\Omega\rangle\right] \qquad \text{...(11.33}a\text{)}$$

where
$$|\chi_\Omega\rangle = \sum_j C_{j\Omega}\,|\chi_{j\Omega}\rangle \qquad \text{...(11.33}b\text{)}$$

and
$$C_{j\Omega} = \sum_{1\Lambda}\left(1\frac{1}{2}\Lambda\,\Sigma\,|\,j\Omega\right)a_{1\Lambda} \qquad \text{...(11.33}c\text{)}$$

where $a_{1\Lambda}$ are the coefficients for single particle functions $\chi_{1\Lambda\Omega}$; (Λ is 3-component of 1 and Σ is 3-component of S); *i.e.*,

$$\chi_\Omega(r) = \sum_{1\Lambda} a_{1\Lambda\Omega}\, \chi_{1\Lambda\Omega}(r) \qquad ...(11.33d)$$

Further, because of axial symmetry, the wave-function given by Eqs. 11.33*a*, 11.33*b* and 11.33*d* are invariant to the rotations of the body-fixed frame about the symmetry axis, by a small amount α. The operator of this rotation is R_3 ($\theta \to \theta$, $\phi \to \phi$, $\psi \to \psi + \alpha$) and hence:

$$R_3\, D^I_{MK} = e^{iK\alpha}\, D^I_{MK} \qquad ...(11.34a)$$

and

$$R_3\, \chi_\Omega = e^{-i\Omega\alpha}\, \chi_\alpha \qquad ...(11.34b)$$

The exponents of Eqs. 11.34*a* and 11.34*b* have opposite sign because D is the wave-function of the body system, with respect to lab. system; while χ is the wave-function with respect to body system. Hence we satisfy the requirement,

$$R_3\, \psi^I_{\Omega MK} = \psi^I_{\Omega MK} \qquad ...(11.35)$$

if we should put $\Omega = K$, *i.e.* for a spheroidal nucleus.

Further there is invariance with respect to the operator R_1-rotation by π about 1-axis. The effect of R_1 on D^I_{MK} has been given by Eq. 11.20*c*. For transformation of χ under R_1, we expand in terms of wave-functions $\chi_{j\Omega}$, which are the eigen-function of angular momentum j, *i.e.*,

$$\chi_\Omega = \sum_j \alpha_{j\Omega}\, \chi_{j\Omega} \qquad ...(11.36)$$

Also χ in body system is related to the wave-function $\chi'_{j\Omega}$, in lab. system $j\Omega$, by the relations:

$$\chi_{j\Omega} = \sum_{\Omega'} \chi'_{j\Omega'}\; {}^*D^j_{\Omega'\Omega} \qquad ...(11.37)$$

Hence, R_1-operation yields,

$$R_1\, \chi_{j\Omega} = \sum_{\Omega'} \chi' R_1\; {}^*D^j_{\Omega'\Omega} = e^{-i\pi(j+\Omega)} \sum_{\Omega'} \chi'_{j\Omega'}\; {}^*D^j_{\Omega'\Omega}$$

$$= e^{-i\pi(j+\Omega)}\, \chi_{j,-\Omega} \qquad ...(11.38)$$

Cases of the wave function for (*i*) $K = \Omega \neq 0$ and (*ii*) $K = \Omega = 0$ can be derived from [Eq. 11.33]. It can be seen, for example, that,

(*i*) For $K = \Omega \neq 0$

$$\psi^I_{MK} \equiv |\, IMK\rangle = \left[\frac{2I+1}{16\pi^2}\right]^{1/2} \left[\chi_K\, D^I_{MK} + (-)^{I-j}\, \chi_{-K}\, D^I_{M,-K}\right] \qquad ...(11.39)$$

where term $(-)^{I-j}$ acts separately on each j-component for χ_Ω.

(*ii*) For $K = \Omega = 0$

$$\psi^I_{MO} \equiv | IM_o \rangle = \frac{1}{\sqrt{2\pi}} \chi_o Y_{LM} (\theta, \phi) \qquad ...(11.40)$$

It can further, be proved that parity operator P operating on ψ^I_{MK} is the same as the parity operator on the particle wave-function χ_{jK} in the lab. system, *i.e.*,

$$P \psi^I_{MK} = P\chi'_{j,\, K} \qquad ...(11.41)$$

We have, thus, been able to write the wave-function of rotational states under different conditions. These can be used for calculating, for instance, the transition probabilities and electric quadrupole moments, or magnetic dipole moments, of rotational states.

11.1.7 Energy Eigenvalues

We have already discussed in the previous sections the empirical behaviour of the energies of rotational excited states of even-even spheroidal nuclei [Eqs. 11.1, 11.3, 11.6, 11.9 and 11.13*a*]. These were pure rotational cases; where in Eqs. 11.9 and 11.13 the effect of the variation of the moment of intertia were taken into account. One of the well-known relationship, to explain the experimental points in Fig. 11.2 is:

$$E_I = \frac{\hbar^2}{2\mathcal{I}} [I (I + 1)] - B [I (I + 1)]^2 \qquad ...(11.42)$$

where $\hbar^2/2\mathcal{I}$ turns[43] out to be 7.371 keV.

The value B is found to be positive and constant. Intrinsically the second term arises out of the stretching of the liquid drop as it rotates. Theoretically attempts have been made to understand B in terms of the coupling between the rotation and the β-vibrations. There do arise cases, when the odd particle is tightly bound to the rotator so that, the energy separation of H_o in Eq. 11.28 are large compared to rotational energies, *e.g.*, after the closed shell, the next level is quite apart and also $HRPC = 0$. Then it turns out, that the eigenvalues E_I of $H_o + H^o_{\text{rot}}$ [Eqs. 11.27 and 11.28], are given by the rotation spectrum.

$$(E_I)_{\text{rot}} = E_o + \frac{\hbar^2}{2\mathcal{I}} [I (I + 1) - 2 K^2] \qquad ...(11.43)$$

For calculating the energy eigenvalues for say $K = 1/2$ we have to keep in mind the contribution of *RPC* (Rotation-Particle Coupling) or coroilis force. It can be seen, that *RPC* connects states with K values differing by one unit. Since the nuclear states are degenerate with respect to $\pm K$; *RPC* term contributes only for the case $K = 1/2$. For a given K, the expectation value of *RPC* part of the energy, has been calculated in the next chapter, for the case of axial symmetry.

Another interesting case is that of the collective motion in nuclei without axial symmetry. Here neither K nor Ω are constants of the motion, and coriolis force is important in all states. Now we have three unequal moments of inertia, and hence coriolis force does not have simple *RPC*-form. For some odd-*A* nuclei *e.g.*, HO^{165} and W^{183} such a situation can arise. Davidov and Fillipov have calculated the energy spectra of rotational states of such axially non-symmetric[44, 45] nuclei.

11.1.8 Some Examples

Many prolate or oblate nuclei, in the range of atomic weight $A \approx 80$ mass region in the even-even category (Kr^{74}, Sr^{78} and Zr^{82}) and in the even-odd category in $A \approx 100$ region (Ag^{107}) and $A \approx 130$ region (Nd^{133}) have been studied recently for the properties of rotational bands excited in appropriate reaction. As for example:

(*i*) For Kr^{74}, Sr^{78} and Zr^{82}, the reaction Ni^{58} (Si^{28}, 3α) Kr^{74}, Ni^{58} (Si^{28}, 2α) Sr^{78}, and Ni^{58} (Si^{28}, 2*p*2*n*) Zr^{82} have been used[46], at 130 MeV beam energy of Ni^{58} ions.

(*ii*) In odd-even casc, Nd^{133} has been obtained[49] using Pd^{104} (S^{32}, 2*pn*) at 135 MeV of Pd^{104} projectiles.

(*iii*) Similarly for Ag^{107}, the reaction Mo^{100} (B^{11}, 2*pn*) Ag^{107} has been used[47], at 39 MeV beam energy.

(*iv*) Another nucleus in this mass region *i.e.* Sn^{105} has been obtained[48] using Cr^{50} (Ni^{58}, 2*pn*) Sn^{105}, at 210 MeV of projectile energy. It may be seen, that energy range of projectiles which lie between 1–5 MeV/N, is a little above Coulomb barrier. With the availability of heavy ions in many laboratories, such experiments on excitation of collective mode in nuclei have become one of the major activities.

These four cases represent variation of the rotational mode of excitation. In even-even nuclei, *i.e.* Se^{70}, Kr^{71}, Sr^{78} and Zr^{82} in the mass range of $A \approx 80$, the nuclear structure is governed by a large shell gap of 2 MeV, at oblate shape (*i.e.* $\beta \approx -0.3$, particle numbers 34, 36) and at prolate shape (*i.e.* $\beta \approx +0.4$, particle numbers 38, 40). From these considerations, Y, Rb and Sr isotopes are predicted to be very elongated[50] with shape mixing of prolate-oblate shape coexistence, as well as island of superdeformed prolate ($\beta \approx 0.55$), was predicted for N = 44 in Sr to Zr. Because of the low density of single particle energy levels, the nuclear shapes are strongly dependent on configuration. Hence both positive parity and negative parity are found in the rotational excitation.

In the case of Sn^{105}, Ag^{107}, Nd^{133}, which are even-odd or odd-even cases, the rotational band heads, have particle plus collective configurations. Figures 11.8*b* and 11.8*c*, for Nd^{133} represents one of the latest cases, where both positive and negative parity levels coexist, as well as normally deformed (*ND*) shapes *i.e.* ($\beta = 0.2 - 0.25$) as well as highly deformed (*HD*) ($\beta = 0.35 - 0.40$) shapes[51]. In the mass region of $A = 80$, 150, 190 we have super deformed (*S D*) nuclei ($\beta = 0.5 - 0.6$).

The study of super deformed (*SD*) nuclei has been at the forefront of nuclear structure studies during the last decade. Apart from $A \approx 80$ region, $A \approx 60$ region also has been found to have (*SD*) nuclei[52], Good examples are Zr^{86}, Nb^{87}, Sr^{80} etc. Apart from their *SD* nature, some of these nuclei have triaxial shapes; as for example, for Z^{86}, $\gamma \approx 20°$. One determines the triaxial nature of the nuclei, by calculating the potential energy surfaces by plotting $X = \varepsilon_2 \cos(\gamma + 30°)$ and $Y = \varepsilon_2 \sin(\gamma + 30°)$, where ε_2 is deformation parameter. In the mass region of $A \approx 190$, we have Hg, Te, Db, Bi and Po isotopes[53], showing the (*SD*) shapes. An interesting case of (*SD*) states of Hg^{194} has been studied by M. Kaci[54] *et al.* who obtained this nucleus, through $O^{16} + W^{186}$ reaction at 110 MeV, corresponding to an incomplete fusion reaction.

The case of Sn^{105} is an interesting case of magnetic rotation. Such rotations were first discovered[55] in Pb^{200}. The dominant transition in these cases are $\Delta I = I$, M_1 and E_2 transitions. There are only weak

E_2 transitions. $B(M_1)$ values are very large (several units of μ_N^2). In these nuclei, one observes regular rotation-like sequence in nuclei, with very low deformation. These cases correspond to magnetic rotation due to magnetic dipole that rotates about the angular momentum vector in contrast to the normal electrical charges distribution[56]. In the *Pb* region the angular momentum increase along the band is generated by the simultaneous re-orientation of the spins of proton particles ($\hbar_{9/2}$ or $i_{13/2}$). The angular momentum vector aligns along the total angular momentum of the nucleus, for increasing rotational frequency, through shear mechanism. The appearing of such bands have also been predicted in $A \approx 100$ mass region, like In^{114} and Cd^{112}. The case of Sn^{105} also falls in this category. In these cases, the neutron particles in $\hbar$ 11/2 orbital and proton hole in $\hbar$ 9/2 orbital, combine to give the rotating dipole. In the case of Sn^{105}, the magnetic rotational levels have been found, extending from $5/2^+$ in the ground state to $43/2^+$ in the excited state at 10.29 MeV.

11.2 VIBRATIONAL MODE

Spherical nuclei, do not have any rotational energies because the wave-function is invariant under rotation. Such nuclei and also the spheroidal nuclei, however, can have collective excitation, in vibrational mode. Such a mode requires, that while the deep core continues, to have the shape of the ground state, a few nucleons outside this core take part in the surface oscillation, giving rise to the vibrational spectra. In Fig. 11.9, we have shown the evolution of a spherical shape to spheroidal or octupole shapes; or vibration of neutron sphere against a proton sphere.

It may be mentioned here that both in rotational and vibrational modes, only the external nucleons take part. In the rotational mode, the rotation is not that of rigid rotor, but the low energy rotational spectrum depends upon the so-called irrotational moment of inertia, as has been described in the previous section. Also, there will be cases, where a large number of external 'loose' nucleons interacting strongly with the core will give rise to a deformation of the nucleus, which is large compared to its zero point vibration. This corresponds to strong coupling limit. When there are only a few particles in the outermost unfilled *j*-subshell, weak coupling limit turns out to be appropriate approximation. Then there is no static deformation and hence pure vibration spectra is observed. In permanently deformed nuclei, however, both rotational and vibrational modes are possible. We discuss first, the pure vibrational mode of collective excitation over a spherical core of even-even nuclei. Figure 11.10 depicts the actual energy levels due to vibrational mode of some spherical even-even nuclei.

The vibrational mode corresponds to the surface oscillations of protons and neutrons, on the nuclear surface. As is well known in the classical analysis of the surface of liquid drop model, since nineteenth century, when Lord Rayleigh[59] developed the theory; these oscillatory modes are proportional to the associated Legendre polynomials $P_{\lambda\mu}$. Therefore for $\lambda = 0$, *i.e.* $P_{oo} = 1$ mode corresponds to no θ-dependence and no oscillation and no shape-change. For $\lambda = 1$, $\mu = 0$ and $P_{10} = \cos\theta$ as shown in Fig. 11.9*a*, it is not an oscillation but only a movement of the centre of mass of the drop. But for $\lambda = 1$, $\mu = \pm 1$, in a nucleus, neutron and proton spheres move in opposite directions. Such a change cannot take place by the internal forces in the nucleus, and hence is unphysical. For $\lambda = 2$, $\mu = 0$, $P_{20} = 3\cos_2\theta/4$,

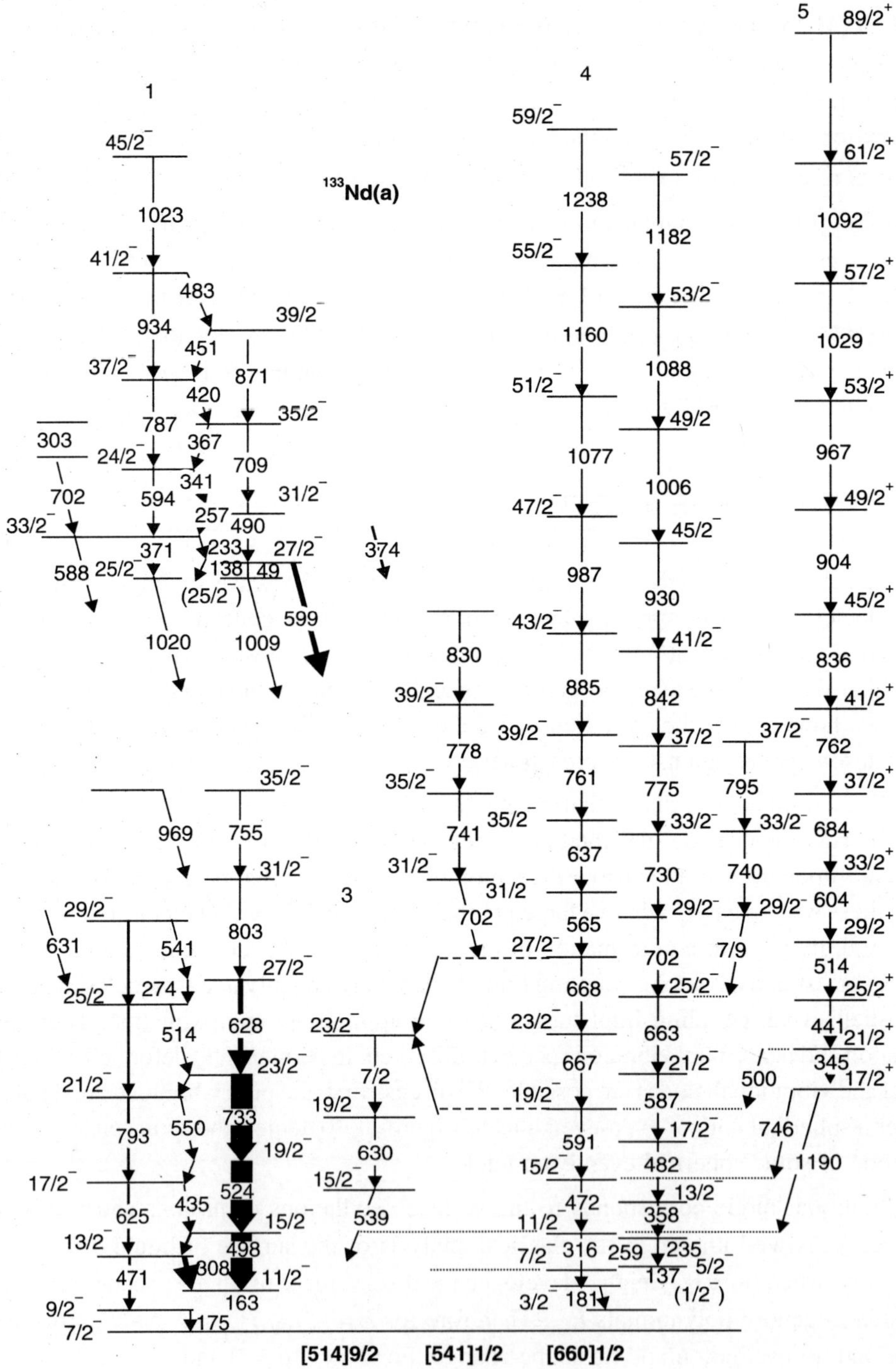

Fig. 11.8 (b) Five rotational bands, in the excited states of Nd^{133}, obtained using Pd^{104} (S^{32}, 2pn) Nd^{133} at 135 MeV of Pd^{104} (Ref. 49).

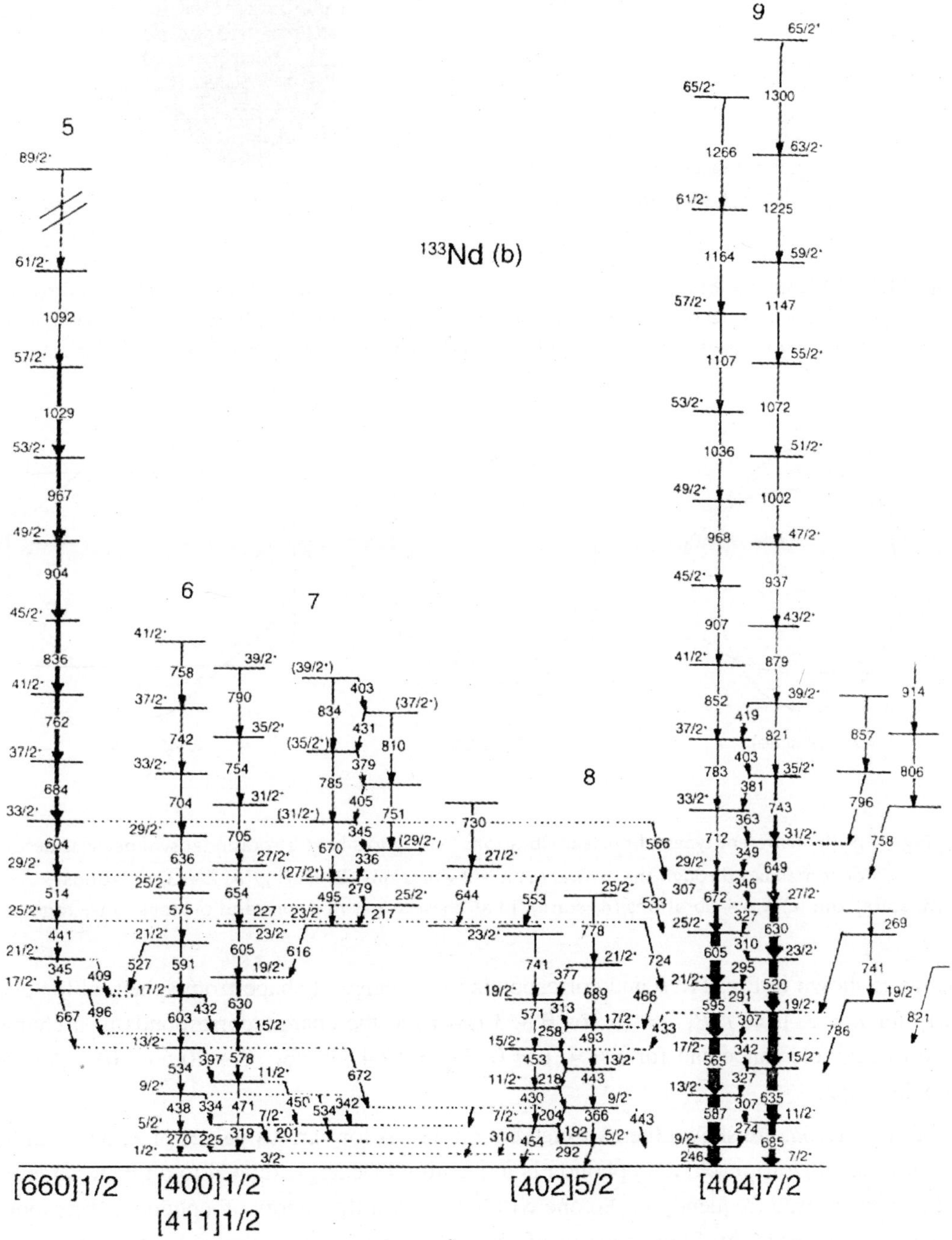

Fig. 11.8 (c) Continuation of the presentation of the rotational bands in the excited states of Nd^{133}, from the previous figure (Ref. 49).

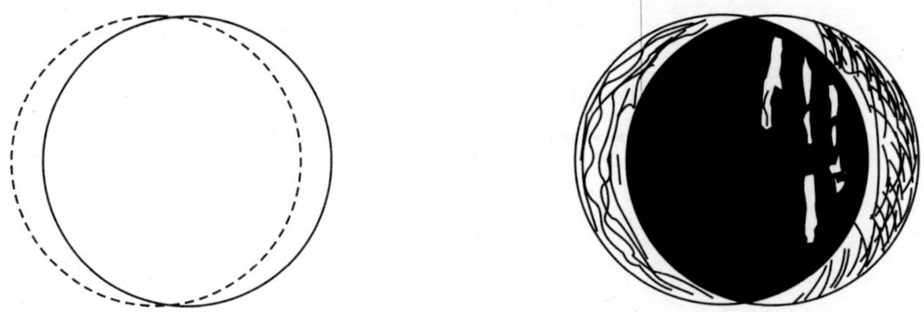

$P_{10}(\lambda=1, \mu=0)$
(a)

$P_{11}(\lambda=1, \mu=1)$
(b)

Fig. 11.9 (*a*) Vibrations for λ = 1 μ = 0 correspond to only displacement; (*b*) λ = 1; μ = ± 1 correspond to vibrations, where neutrons and protons move in opposite direction, (1^- state). However such a case cannot take place by internal forces and such a change is, therefore, unphysical.

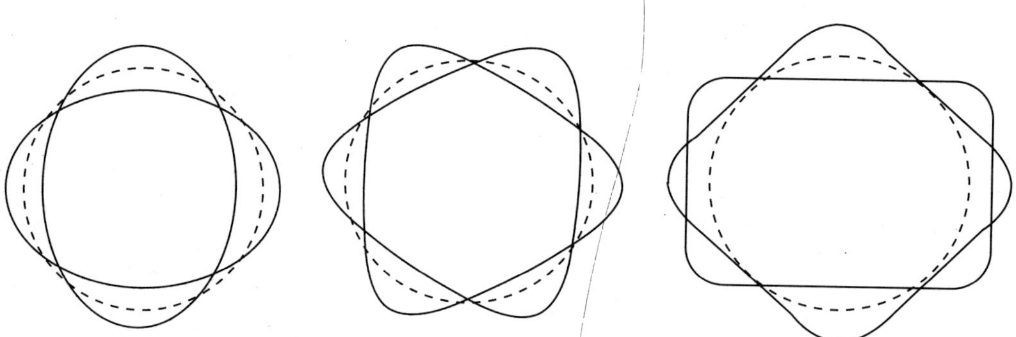

λ=2(2^+state)
(a)

λ=3(3^-state)
(b)

λ=4(4^+state)
(c)

Fig. 11.9 (b) Different types of nuclear vibrations. All figures have a rotational symmetry about a horizontal axis through the centre. (*a*) λ = 2 vibrations (2^+ state); (*b*) λ = $\overline{3}$ vibrations (3^- state) and (*c*) λ = 4 vibrations (4^+ state). In all these states neutrons and protons move together.

the change is shown in Fig. 11.9*b* and corresponds to a change of shape from spherical to spheroidal. Similarly for $\lambda = 3$, $\mu = 0$, $P_{3,\,0} = (5 \cos \theta + 3 \cos \theta)/8$; the change corresponds to a change from spherical to octupole shape and for $\lambda = 4$, $\mu = 0$, $P_{40} = 1/64\ (35 \cos_4 \theta + 20 \cos_2 \theta)$, gives rise to a hexadupole shape.

In Fig. 11.10*a*, are shown the expected energy levels for various values of λ, but with different number of phonons *n* for each, λ. A phonon is a quantum of energy, related to a given value of λ; for which there will exist a frequency ν_λ. So one will have $h\nu$ for the energy of a phonon, corresponding to one phonon. The value of this frequency *v* is, of course, determined by the consideration of the properties of oscillations which classically are related to surface tension. We will discuss the quantum mechanical treatment subsequently, to determine the value of ν_λ. It is, however, easy to understand, that the lowest energy vibration states, have energies $h\nu_2$, $2h\nu_2$, $h\nu_3$, $h\nu_2 + h\nu_3$, $2h\nu_3$, $h\nu_4$, etc. Each phonon of vibration

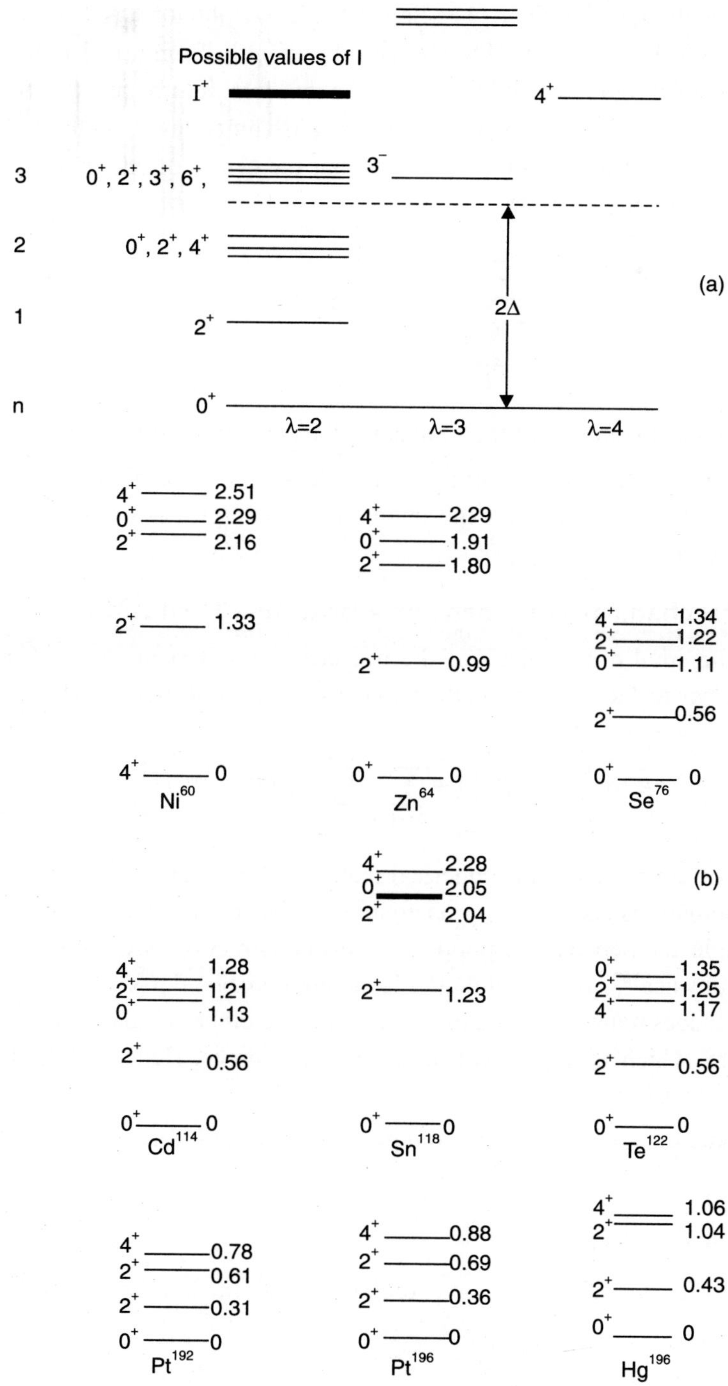

Fig. 11.10 The low energy states of various spherical even-even nuclei due to vibrational mode. (*a*) Expected energies if the nucleus could be considered a liquid drop and (*b*) The experimental levels. The n in (*a*) corresponds to number of phonons (Ref. 45, 57, 58).

carries an angular momentum, and parity $(-1)^{\lambda}$. So for $\lambda = 2$, one phonon has 2^+ state; while 2 phonons can have angular momenta $0^+, 2^+, 4^+$ as will be shown subsequently. Similarly for phonons of a given λ, one can calculate the angular momenta and parity (*see* Table 11.1). It turns out, that for any given λ and n, the levels are aggregated, which shows that m–splitting are much less than the level differences for different values of n. As a matter of fact, it turns out that the centre of energy of the three levels for $\lambda = 2$, $n = 2$; calculated from the relation:

$$\overline{E}_1 = \frac{\sum_I (2I+1)\,\overline{E}_I}{\sum_I (2I+1)} \qquad \text{...(11.44)}$$

turns out to be about twice the energy of the first state. This is typical of the vibrational spectra.

For two successive shells, the m-splitting is much less than the level differences for the adjacent shells. Such shell structures are available experimentally for energies of orbits in $n = 5$ and $n = 6$ for various values of A [Ref. (39)].

11.2.1 Quantum Mechanical Treatment of Vibrational Excitation

Let us assume, that the equilibrium shape of the nucleus is a spherical shape. The time dependent fluctuations in the nuclear surface vector, *i.e.* the radial coordinates of such a surface may be written as:

$$R(\theta, \phi, t) = R_0 \left[1 + \sum_{\lambda=0}^{\infty} \sum_{\mu=-\lambda}^{\lambda} \alpha_{\lambda\mu}(t)\, Y_{\lambda}^{\mu}(\theta, t) \right] \qquad \text{...(11.45)}$$

where R_0 is the equilibrium radius of the spherical nucleus. For spherical shape for the core nucleus, over which surface oscillations take place $\alpha_{\lambda\mu}$ is assumed to be small. As discussed earlier, $\lambda = 0$ terms are independent of angle and hence correspond to uniform contraction or expansion of the nucleus. At low energy this is not physical[62], as nuclear matter is incompressible. For $\lambda = 1$ the centre of mass of the nucleus, shifts—which does not correspond to oscillation. The other alternative is that the neutron and proton sphere separate, which is unphysical under internal forces. Hence the first physical case of vibrations corresponds to $\lambda = 2$.

We now calculate the kinetic energy T and potential energy V for such a vibrating system.

Kinetic Energy

We make the following assumptions:

The nuclear density is ρ which may be expressed so that the kinetic energy T can be written as:

$$T = \frac{1}{2}\rho \int v^2 \, dT \qquad \text{...(11.46)}$$

where $\mathbf{v}(r, \theta, \phi)$ is the velocity of the fluid at the point (r, θ, ϕ).

The flow of nuclear fluid is irrotational *i.e.*, it is vortex free and therefore, the velocity $\mathbf{v}$, obeys the relation:

$$\nabla \times \mathbf{v} = 0 \qquad \text{...(11.47)}$$

An expression of **v**, like

$$\mathbf{v} \approx R_0 \sum_{\lambda\mu} \frac{1}{\lambda}\, \dot{\alpha}_{\lambda\mu} \text{ grad} \left\{ \left(\frac{r}{R} \right)^{\lambda} Y_{\lambda\mu} (\theta, \phi) \right\} \qquad \text{...(11.48)}$$

satisfies the Eq. 11.47 because

$$\nabla \times \nabla \left[\frac{r}{R} \right]^{\lambda} Y_{\lambda\mu} (\theta, \phi) = 0 \qquad \text{...(11.49)}$$

If we define a velocity potential Ψ, so that

$$\mathbf{v} = \nabla \Psi \qquad \text{...(11.50)}$$

then, because the fluid is incompressible, it is required that

$$\nabla \cdot \mathbf{v} = \nabla^2 \Psi = 0$$

So that, it can be seen that Ψ turns out to be

$$\Psi = R_0 \sum_{\lambda\mu} \frac{1}{\lambda} \left(\frac{r}{R} \right)^{\lambda} \alpha_{\lambda\mu} Y_{\lambda\mu} (\theta, \phi) \qquad \text{...(11.51)}$$

One can, then, write:

$$T = \frac{\rho}{2} \int | \nabla \Psi |^2 \, d\tau$$

$$= \frac{\rho}{2} R_0^5 \sum_{\lambda\mu} \frac{1}{\lambda} \, | \dot{\alpha}_{\lambda\mu} |^2 \qquad \text{...(11.52)}$$

$$= \frac{1}{2} \sum_{\lambda\mu} B_\lambda \, | \dot{\alpha}_{\lambda\mu} |^2 ; \text{ where } B_\lambda = \frac{\rho R_0^5}{\lambda} = \frac{3}{4\pi} \frac{MAR_0^2}{\lambda} \qquad \text{...(11.53)}$$

B_λ, physically, corresponds to the moment of inertia of the nucleus with respect to the changes in deformation.

The Potential Energy and Total Hamiltonian

The potential energy arises in vibrational mode, due to two sources; (*i*) Surface deformation and (*ii*) Electrostatic interaction.

Surface Deformation: This comes from the change in the spherical surface area which is $4\pi R^2$. Hence, if σ is surface energy per unit area, then the energy changes due to deformation may be written as:

$$V_s = \sigma (S - 4\pi R_0^2) = \sigma \int_S dS - (4\pi R_0^2)\, \sigma \qquad \text{...(11.54)}$$

where S is the area of the deformed (due to vibration) surface.

It can be shown[45] that

$$\int dS = \int R^2 \sin\theta \, d\theta \, d\phi \left[1 + \left(\frac{1}{R}\frac{dR}{d\theta}\right)^2 + \left(\frac{1}{R\sin\theta}\frac{\partial R}{\partial \phi}\right)^2\right]^{\frac{1}{2}} \quad ...(11.55a)$$

Putting $\xi = \sum_{\lambda,\mu} \alpha^*_{\lambda\mu} \, Y^{\mu}_{\lambda} \, (\theta, \phi)$ and $R = R_0 \, (1 + \xi \, (\theta, \phi))$ ...(11.55b)

and $d\omega = \sin\theta \, d\theta \, d\phi$

We can write Eq. 11.55*a*

$$\int dS = \int R_0^2 \, d\omega \, (1 + 2\xi + \xi^2) \left[1 + (1-\xi)^{-2} + \left\{\left(\frac{\partial \xi}{\partial \theta}\right)^2 + \frac{1}{\sin\theta}\left(\frac{\partial \xi}{\partial \phi}\right)^2\right\}\right]^{\frac{1}{2}}$$

$$= R_0^2 \int d\omega \left[1 + 2\xi + \xi^2 + \frac{1}{2}\left(\frac{\partial \xi}{\partial \theta}\right)^2 + \frac{1}{2}\frac{1}{\sin^2\theta}\left(\frac{\partial \xi}{\partial \phi}\right)^2\right] \quad ...(11.56)$$

Hence

$$\sigma \int dS = \sigma \, R_0^2 \int d\omega \left[(1+\xi)^2 + \frac{1}{2}\left\{\left(\frac{\partial \xi}{\partial \theta}\right)^2 + \text{cosec}^2\theta \left(\frac{\partial \xi}{\partial \phi}\right)^2\right\}\right] \quad ...(11.57)$$

Using the relations of Eq. 11.55, the above intergral comes out to be:

$$\sigma \int dS = E_{S_1} + E_{S_2} + E_{S_3}$$

where $E_{S_1} = 4\pi \, R_0^2 \, \sigma$. This is the surface energy of a sphere of radius R_0.

$$E_{S_2} = - R_0^2 \sigma \sum_{\lambda\mu} \left| \alpha_{\lambda\mu} \right|^2$$

and

$$E_{S_3} = \sigma R_0^2 \sum_{\lambda\mu} \sum_{\lambda'\mu'} \alpha_{\lambda\mu} \, \alpha_{\lambda'\mu'} \times \int d\omega \left[\left(\frac{\partial Y^{\lambda}_{\mu}}{\partial \theta}\right)\left(\frac{\partial Y^{\lambda'}_{\mu'}}{\partial \theta}\right) + \text{cosec}^2\theta \left(\frac{\partial Y^{\lambda}_{\mu}}{\partial \phi}\right)\left(\frac{\partial Y^{\lambda'}_{\mu}}{\partial \phi}\right)\right]$$

$$= \frac{1}{2} \sigma \, R_0^2 \sum_{\lambda\mu} \lambda(\lambda + 1) \left| \alpha_{\lambda\mu} \right|^2 \quad ...(11.58)$$

Then one can write V_S from Eq. 11.54 as:

$$V_S = \frac{1}{2} \sum_{\lambda,\mu} C_{\lambda} \, (S) \left| \alpha_{\lambda\mu} \right|^2 \quad ...(11.59)$$

where $C_{\lambda} \, (S) = R_0^2 \sigma \, (\lambda - 1)(\lambda + 2)$ ...(11.60)

Electrostatic Interaction: The Coulomb interaction energy V'_C between the charges located between the volume elements of, $d\tau_1 = r_1^2\, dr_1\, d\omega_1$; and $d\tau_2 = r_2^2\, dr_2\, d\omega_2$; is given by:

$$V'_C = \frac{1}{2}\rho_0^2 \int d\omega_1 \int d\omega_2 \int_0^{R_0(1+\xi_1)} r_1^2\, dr_1 \int_0^{R_0(1+\xi_2)} r_2^2\, dr_2 \frac{1}{|\mathbf{r}_1 - \mathbf{r}_2|} \quad \text{...(11.61)}$$

We insert extra factor of 1/2, to cancel double counting. Writing[61]:

$$\frac{1}{|\mathbf{r}_1 - \mathbf{r}_2|} = \sum_{K=0}^{\infty} \frac{r<K}{r>K+1} P_K(\cos\omega_{12})$$

$$= \sum_{K=0}^{\infty} \frac{r<K}{r>K+1} \frac{4\pi}{2K+1} \sum_{q=-K}^{K} Y_q^{*K}(\theta_1, \phi_1)\, Y_q^K(\theta_2, \phi_2) \quad \text{...(11.62)}$$

It can be proved that, extra potential energy[61] over and above, the potential energy of a sphere, can be written as:

$$V_C = 4\pi\, \rho_0^2\, R_0^5 \int d\omega_1 \int d\omega_2 \times \sum_{K,q} \frac{1}{2K+1}$$

$$\left[\frac{\xi_1}{K+3} + \frac{1-K}{2(K+3)}\xi_1^2 + \frac{1}{2}\xi_1\xi_2\right] \times Y_q^{*K}(\theta_1, \phi_1)\, Y_q^K(\theta_2, \phi_2) \quad \text{...(11.63)}$$

which leads to:

$$V_C = 4\pi\, \rho_0^2\, R_0^5 \sum_{\lambda\mu} |\alpha_{\lambda\mu}|^2 \times \frac{1}{2}\left(\frac{1}{2\lambda+1} - \frac{1}{3}\right) \quad \text{...(11.64a)}$$

or

$$V_C = -\frac{3}{4\pi}\frac{(Ze)^2}{R_0} \sum_{\lambda\mu} \frac{\lambda-1}{2\lambda+1} |\alpha_{\lambda\mu}|^2 \quad \text{...(11.64b)}$$

where we have used the relation:

$$\frac{4\pi}{3}\rho\,(R_0)^3 = Ze \quad \text{...(11.65)}$$

For $\lambda = 1$, $V_C = 0$ where for all higher λ's, V_C has a negative sign, showing that Coulomb energy decreases, when the sphere gets deformed. We rewrite Eq. 11.64*b* in the form:

$$V_C = \frac{1}{2}\sum_{\lambda\mu} C_\lambda\ (V)\, |\alpha_{\lambda\mu}|^2 \quad \text{...(11.66)}$$

where

$$C_\lambda\,(V) = -\frac{3}{2\pi}\frac{Z^2e^2}{R_0}\frac{\lambda-1}{2\lambda+2} \quad \text{...(11.67)}$$

Hence, one can write the total potential energy as:

$$V_{\text{Total}} = \frac{1}{2}\sum_{\lambda\mu} C_\lambda \,|\, \alpha_{\lambda\mu} \,|^2 \qquad \text{...(11.68)}$$

where $$C_\lambda = C_\lambda(S) + C_\lambda(V)$$

Combining the kinetic energy T in Eq. 11.53 and total potential energy in Eq. 11.68; the Hamiltonian of spheroidal liquid drop, whose surface is undergoing shape vibration has the form:

$$H = \sum_{\lambda\mu} \frac{1}{2} B_\lambda \,|\, \dot{\alpha}_{\lambda\mu} \,|^2 + \frac{1}{2} C_\lambda \,|\, \alpha_{\lambda\mu} \,|^2 \qquad \text{...(11.69)}$$

Defining the conjugate momentum $\pi_{\lambda\mu}$ as:

$$\pi_{\lambda\mu} \equiv B_\lambda \,\dot{\alpha}^*_{\lambda\mu}$$

so that B_λ plays the role of mass, and the total Hamiltonian can be written as:

$$H_{\lambda\mu} = \frac{1}{2B_\lambda} \,|\, \pi_{\lambda\mu} \,|^2 + \frac{1}{2} C_\lambda \,|\, \alpha_{\lambda\mu} \,|^2 \qquad \text{...(11.70)}$$

Hence, the energy eigenvalues (levels) are harmonic oscillator energies given by:

$$E = E_o + \sum_{\lambda\mu}\left(n_{\lambda\mu} + \frac{1}{2}\right)\hbar\omega_\lambda \qquad \text{...(11.71)}$$

where E_o is the energy of the nucleus for a spherically symmetric shape and $n_{\lambda\mu}$ is the number of oscillators or phonons in $\lambda\mu$ mode of oscillation.

The wave-function for the lowest $E_{\lambda\mu}$ *i.e.* ground states, corresponding to $n = 0$ is given by:

$$\psi_o = N \exp\left[-\sum_{\lambda\mu} \frac{B_\lambda \omega_L}{2\hbar} (\alpha_{\lambda\mu})^2\right] \qquad \text{...(11.72)}$$

where N is normalisation constant. For one phonon state of mode of energy $E_{\lambda\mu}$, the wave-function is given by:

$$\psi_1 = N_1 \,\alpha^*_{\lambda\mu} \psi_o \qquad \text{...(11.73)}$$

Further from Eqs. 11.70 and 11.71, the classical frequency of oscillation ω_λ is given by:

$$\omega_\lambda = \sqrt{\frac{C_\lambda}{B_\lambda}} \qquad \text{...(11.74)}$$

and hence $\hbar\,\omega_\lambda \approx 13\,\lambda^{3/2} A^{-1/2}$ MeV ...(11.75)

The excitation energy of the collective mode from Eq. 11.72 is, then, given by:

$$E = \sum_{\lambda\mu}\left(n_{\lambda\mu} + \frac{1}{2}\right)\hbar\omega_\lambda \qquad \text{...(11.76)}$$

It is seen from Eq. 11.75, that the energy of emitted gamma rays should decrease as a function of mass number which is in agreement with experimental observations. Further, the electric quadrupole transition E_2, between the first excited state and ground state is expected to be much stronger because of the collective motion of many nucleons involved; than one would expect from a single particle model. Experimentally E_2 is more often, at least one order of magnitude, larger than the prediction of a single particle model (*see* Fig. 7.16).

The angular momentum of a phonon in the state $\lambda\mu$; is λ and μ is z component and parity is $(-1)^{\lambda}$. In writing the energies of various excited states, frequency of emitted radiation ω_3 is found to be about twice the value of ω_2 and $\omega_4 \approx 3\omega_2$. The first excited state corresponds to $\lambda = 2$.

It is interesting to see, that for both $\lambda = 0$ and $\lambda = 1$, $\omega_1 = 0$ as has been explained earlier and hence possible frequencies correspond to $\lambda \geq 2$.

As stated earlier the surface vibrations are characterised by the number of phonons n, each having an angular momentum λ. The phonons are said to be quadrupole type if $\lambda^{\pi} = 2^+$; octupole if $\lambda^{\pi} = 3^-$ and hexadecapole if $\lambda^{\pi} = 4^+$. Figure 11.9 represents the vibrations for these values of λ. A phonon, being a 'particle' with integral spin is a boson; hence many bosons can be in the same state. There is no Pauli Exclusion principle, here, unlike for the fermions.

We have stated earlier, that for $\lambda = 2$, the ground state is 0^+, then there is 2^+ state corresponding to one phonon; next there are three states close by, with 0^+, 2^+ and 4^+ corresponding to 2 phonons. How do we arrive at these angular momenta and parity for these phonon numbers? The 2 phonon cases could have maximum $J^{\pi} = 4^+$, with $m = 2$ for each phonon. Also one phonon could have $m_1 = 2$, and the other $m_2 = 1$, then $M = m_1 + m_2 = 3$. But this corresponds to $J^{\pi} = 4^+$. It could also be $m_1 = 2$ and $m_2 = 0$, with $M = 2$ this belongs to $J = 2^+$, as well as to 4^+. If it is $m_1 = 1$ and $m_2 = 1$ with $M = 2$, it again corresponds to $J^{\pi} = 2^+$. If it is $m_1 = 1$ and $m_2 = 0$, with $M = 1$, it again corresponds to $J^{\pi} = 2^+$. If $m_1 = 1$ and $m_2 = -1$, or $m_1 = 0$, $m_2 = 0$, it corresponds to $J^{\pi} = 0^+$ or 2^+ or 4^+. Hence the only possible values of J^{π} are 0^+, 2^+, 4^+. Similarly for three phonons, M varies from 6 to 0; which corresponds to $J^{\pi} = 0^+, 2^+, 3^+, 4^+$ and 6^+. The value of $J^{\pi} = 5^+$ is not allowed because $M = 5$, can belong to $J = 6^+$. Also $M = 1$, can be obtained from $m_1 = -1$, $m_3 = 2$. Similarly it could be seen, that:

$$m_1 = -1,\ m_2 = 1,\ m_3 = 1,\ (M = 1)$$

or

$$m_1 = 0,\ m_2 = 0,\ m_3 = 1,\ (M = 1)$$

or

$$m_1 = -2,\ m_2 = 1,\ m_3 = 2,\ (M = 1)$$

These are the only possible combinations of three phonons. All of them give $M = 1$, which corresponds to $J^{\pi} = 2^+$. As it is not possible to obtain $M = -1$, it shows that $J = 1$ is not allowed. Following this method, one can obtain the positive values of J^{π} for $\lambda^{\pi} = 2^+$, $\lambda^{\pi} = 3^-$. and $\lambda^{\pi} = 4^+$ for any number of phonons. We give in Table 11.1, the possible values of J^{π} for $n = 0, 1, 2$ and 3.

Table 11.1 *Allowed J^π states for $\lambda = 2^+, 3^-$ and 4^+ and other values of J^π for $n = 0, 1, 2$ and 3 (Ref. 63)*

Number of phonons	$\lambda = 2^+$	$\lambda = 3^-$	$\lambda = 4^+$
0	0^+		
1	2^+	3^-	4^+
2	$0^+, 2^+, 4^+$	$0^+, 2^+, 4^+, 6^+$	$0^+, 2^+, 4^+, 6^+, 8^+$
3	$0^+, 2^+, 3^+, 4^+, 6^+$	$1^-, 3^-, 4^-, 6^-, 7^-, 9^-$	$0^+, 2^+, 3^+, 4^+, 4^+, 4^+, 5^+, 6^+, 6^+, 7^+, 8^+, 9^+, 10^+, 11^+, 12^+$

11.3 β AND γ VIBRATIONS

Till now, we have considered the case of vibrations in an even-even spherical nucleus. However, there may be spheroidal nuclei *e.g.* the rare earth nuclei like *Hf, Gd*, etc. which have quite a few 'loose' nucleons outside the closed shell which deform the nucleus to a spheroidal shape. Let the deformation correspond to $\lambda = 2$, in Eq. 11.48. Then the surface of such an ellipsoid may be described by $R(\theta', \phi')$ and the deformation as $\delta R(\theta', \phi')$ may be expressed as:

$$\delta R(\theta', \phi') = R(\theta', \phi') - R_0. \qquad \text{...(11.77)}$$

where R_0 is the radius of the equivalent (of the same volume) sphere. From Eq. 11.45 it is evident, that

$$\delta R(\theta', \phi') = R_0 \sum_{\mu=-2}^{\mu=2} \alpha_{2\mu}\, Y_{2\mu}(\theta', \phi') \qquad \text{...(11.78)}$$

where the deformation parameter $\alpha_{2\mu}$ and angles θ', ϕ' are given with respect to the rotation system. As the deformation δR is real; it requires that

$$\alpha_{2,-\mu} = (-1)^\mu (\alpha_{2\mu})^* \qquad \text{...(11.79)}$$

Equation 11.79 implies, that deformation constant $\alpha_{2\mu}$($\mu = -2$ to $\mu = +2$) has five independent values. These are related to two values corresponding to change in shape (*i.e.*, values of major and minor axes); and three values of angles corresponding to orientation (θ', ϕ', ψ').

We now define new variable '$a_{\lambda\mu}$' related to $\alpha_{\lambda\mu}$ as:

$$a_{\lambda\nu} = \sum_{\mu} \alpha_{\lambda\mu}\, D^{\lambda\lambda\nu}_{\nu\mu}\,(\theta, \phi, \psi) \qquad \text{...(11.80)}$$

where $D^{\lambda}_{\nu\mu}$ are the rotation matrices and θ, ϕ and ψ are the Euler's angles, and signify the rotation of body-fixed coordinate system around the space fixed axes. This means, that $\alpha_{\lambda\mu}$ are the variables in the rotating frame while $a_{\lambda\nu}$ is in the lab frame. Then for $\lambda = 2$:

$$a_{2\nu} = \sum_{\mu} \alpha_{2\mu}\, D^2_{\nu,\mu}\,(\theta, \phi, \psi) \qquad \text{...(11.81)}$$

and

$$\alpha_{2\mu} = \sum_{\nu} a_{2\nu}\, D^{*2}_{2\nu}\,(\theta, \phi, \psi) \qquad \text{...(11.82)}$$

From the reflection properties of D-matrices, it can be seen that

$$a_{21} = a_{2,-1} \text{ and } a_{2,2} = a_{2,-2} \qquad ...(11.83)$$

We now define

$$a_{2,0} \equiv \beta \cos \gamma \qquad ...(11.84a)$$

and

$$a_{22} \equiv \frac{1}{\sqrt{2}} \beta \sin \gamma \qquad ...(11.84b)$$

It can, then, be easily seen that

$$\sum_{\mu} |\alpha_{2\mu}|^2 = \sum_{\mu} |a_{2\mu}|^2 = \beta^2 \qquad ...(11.84c)$$

Then from Eq. 11.78 it may be easily seen, that for $\lambda = 2$

$$\delta R = R - R_0 = \sqrt{\frac{5}{16}} R_0 \beta \left[\cos \gamma (3 \cos^2 \theta' - 1) + \sqrt{3} \sin \gamma \sin^2 \theta' \cos 2\phi'\right] \qquad ...(11.85a)$$

Physically, one can now determine the values of δR along the principal axes. Replacing θ' by θ and ϕ' by ϕ, one can denote the principal axes as:

1-axis for $\theta = \frac{\pi}{2}$, $\phi = 0$ is along R_1

2-axis for $\theta = \frac{\pi}{2}$, $\phi = \frac{\pi}{2}$ is along R_2

3-axis for $\theta = 0$, $\phi = \pi$ is along R_3 ...(11.85b)

One gets from Eq. 11.85

$$\delta R_1 = \sqrt{\frac{5}{4\pi}} \beta R_0 \cos\left(\gamma - \frac{2\pi}{3}\right) \qquad ...(11.86)$$

$$\delta R_2 = \sqrt{\frac{5}{4\pi}} \beta R_0 \left(\cos\gamma - \frac{4\pi}{3}\right) \qquad ...(11.87)$$

$$\delta R_3 = \sqrt{\frac{5}{4\pi}} \beta R_0 (\cos \gamma) \qquad ...(11.88)$$

or, in general

$$\delta R_k = \sqrt{\frac{5}{4\pi}} \beta R_0 \cos\left(\gamma - \frac{2\pi k}{3}\right), k = 1, 2, 3 \qquad ...(11.89)$$

It may be seen that physically β corresponds to deviation from sphericity, and we can think of β varying, in time, in β-vibrations; while γ-stays fixed. The nucleus preserves its symmetry axis, but alters the eccentricity of the ellipse, which is the cross-section of ellipsoid. And γ corresponds to the asymmetry of the nucleus so that in γ-vibrations, the nucleus loses its axial symmetry and as γ changes, the direction of axial symmetry for some values of γ changes; while for others it has no axis of symmetry at all. For $\gamma = 0$ the body fixed 3-axis or z'-axis is the axis of symmetry and the nucleus is prolate (or cigar-shaped), for $\gamma = 2\pi/3$ it is prolate spheroid with x'-axis as the axis of symmetry and for $\gamma = 4\pi/3$, the axis of symmetry is y'-axis. Similarly for $\gamma = \pi$, $\pi/3$ and $5\pi/3$, the nucleus is oblate spheroid. If γ is not a multiple of $\pi/3$, then the nucleus has an ellipsoidal shape with three unequal axes. The nuclear shapes[64,65] for different values of γ are given in Table (11.2).

Table 11.2 *Different shapes for various values of* γ

γ	*Shapes*	*Symmetry axis*
0	spheroid (prolate)	3-axis
π	spheroid (oblate)	3-axis
$2n\pi/3$	spheroid (prolate)	1 or 2-axis
$(2n + 1)\,\pi/3$	spheroid (oblate)	
$\neq n\pi/3$	ellipsoid	three axes along principal axes

11.3.1 Hamiltonian and Wave-function

For vibrational mode, the kinetic energy T is written from Eqs. 11.82 to 11.84 and 11.53 as:

$$T_{\lambda\mu} = \frac{1}{2} B_2 \sum_{\mu} |\dot{\alpha}_{2\mu}|^2 \qquad \text{...(11.90)}$$

or

$$T_{\lambda=2} = \frac{1}{2} B_2 (\dot{\beta}^2 + \beta^2 \dot{\gamma}^2) + \frac{1}{2} \sum_{k=1}^{3} \mathcal{I}_k \, \omega_k^2 \qquad \text{...(11.91)}$$

where ω_k is the rotational angular velocity of the body fixed coordinate system with respect to the space-fixed one and can be written in terms of θ, ϕ and ψ, and their derivatives, [Goldstein][61]. Also it is found that, the effective moment of inertia $\mathcal{I}_k$, turns out to be:

$$\mathcal{I}_k = 4\, B_2\, \beta^2 \sin^2\left(\gamma - \frac{2\pi}{3}k\right) \qquad \text{...(11.92)}$$

or

$$\mathcal{I}_k = \frac{15}{4\pi} \mathcal{I}_{\text{rigid}}\, \beta^2 \sin^2\left(\gamma - \frac{2\pi}{3}k\right) \qquad \text{...(11.93)}$$

where B_2 is the mass parameter of collective quadrupole oscillations, given by:

$$B_2 = \frac{3}{4\pi}\frac{MAR^2}{2}; \quad \mathcal{I}_{\text{rigid}} = \frac{8\pi\rho R_0^5}{15} = \frac{16\pi}{15} B_2 \qquad \text{...(11.94)}$$

where ρ is the density of nuclear matter. The last term in Eq. 11.91, corresponds to rotational kinetic energy, though its value is small for small values of β. Also for $\gamma = 0$, π and $k = 3$; $\mathcal{I}_3 = 0$ and $\mathcal{I}_1 = \mathcal{I}_2 = 3B_2\beta^2$, which means that 3-axis is an-axis of symmetry and that for spheroidal nuclei, the rotation around the axis of symmetry does not produce any kinetic energy due to rotation. In sphere $\mathcal{I}_1 = \mathcal{I}_2 = \mathcal{I}_3$. Hence in an irrotational case, no rotational kinetic energy for spherical nuclei. Both these phenomena (for spheroidal and spherical nuclei), are experimentally observed.

The potential energy V in a vibrational case is given by:

$$V = \frac{1}{2} C \mid \alpha_{2\mu} \mid^2 = \frac{1}{2} C \beta^2 \qquad ...(11.95)$$

from Eqs. 11.78, 11.83 and 11.84. Then for a vibrating nucleus, the total Hamiltonian H is given by (Eqs. 11.91 and 11.95):

$$H = T + V = \frac{1}{2} B_2 \dot{\beta}^2 + \frac{1}{2} B_2 \beta^2 \dot{\gamma}^2 + \frac{1}{2} \sum_{k=1}^{3} \mathcal{I}_k \mathcal{I}\omega_k^2 + \frac{1}{2} C \beta^2$$

$$= H_\beta + H_\lambda + \frac{1}{2} C \beta^2 + \frac{1}{2} \sum_{k=1}^{3} \mathcal{I}_k \omega_k^2$$

$$= H_0 + \sum_{k=1}^{3} \frac{L_k^2}{2\mathcal{I}_k} \qquad ...(11.96)$$

where $$H_\beta \equiv \frac{1}{2} B_2 \dot{\beta}^2 ; H_\gamma \equiv \frac{1}{2} B_2 \beta^2 \dot{\gamma}^2$$

and $$H_0 = H_\beta + H_\gamma + \frac{1}{2} C\beta^2 ; L_k \equiv \mathcal{I}_K \omega_k \qquad ...(11.97)$$

The quantities $\mathbf{L}_k$ are referred to moving axis (body axes) and have some-what different commutation properties than the angular momenta along the fixed axes (space axes).

Now

$$[L_1, L_2] = -i L_3; [L_x, L_y] = -i L_z \qquad ...(11.98)$$

Though L^2 and L_z are constants of motion; L_k are, in general, not constant of motion; except that L_3 is a constant of motion; if the nucleus has an axis of symmetry. Then, for a nucleus with axis of symmetry.

$$L^2 \psi = L(L+1) \psi$$
$$L_z \psi = M \psi$$
$$L_3 \psi = K \psi \qquad ...(11.99)$$

So, physically, the Hamiltonian H in Eq. 11.96 represents a vibrational plus rotational modes of motion. The quantity β is connected with the intrinsic quadrupole moment Q_0, which can be seen as follows: for $\gamma = 0$, *i.e.*, for symmetrical ellipsoid $a_{21} = a_{2,-1} = 0$ and $a_{22} = a_{2,-2} = 0$, so only a_{20} survives out of the five possible $a_{2\mu}$. Hence from Eq. 11.84 a, only $a_{20}^2 = \beta^2$. One can, then write from Eqs. 11.45 and 11.78 to 11.85, as:

$$R = R_0 (1 + \alpha_{2,0} Y_{2,0} (\theta, \phi))$$
$$= R_0 (1 + a_{2,0} Y_{2,0} (\theta, \phi))$$
$$= R_0 (1 + \beta Y_{20} (\theta, \phi)) \quad \text{...(11.100}a\text{)}$$

Then one can write the expression for quadrupole moment Q_0 as:

$$Q_0 = e\int d^3 r\, \rho(r) \langle IM \mid (3z^2 - r^2) \mid IM \rangle_{M=I}$$

$$= \frac{3}{\sqrt{5\pi}} Ze R_0^2\, \beta (1 + 0.36\beta + ...) \quad \text{...(11.100}b\text{)}$$

[*see* Eqs. 2.115 to 2.126].

Further, the deformation parameter β can be related to the radial difference from [Eq. 11.100*a*] as:

$$\beta = \frac{(\Delta R/R_0)}{Y_{2,0}(\theta,\phi)} = \frac{4}{3}\sqrt{\frac{\pi}{5}}\frac{\Delta R}{R_0} \quad \text{...(11.100}c\text{)}$$

$$= 1.06 \times \frac{\Delta R}{R_0} \quad [\textit{see Eq. 2.125}].$$

Wave-function: We can, now write the wave-function for the total collective motion, for spheroidal nuclei, as:

$$\psi = f_{I\tau v} (\beta) \sum_{K=-I}^{I} g_K^{I\pi} (\gamma)\, D_{MK}^I (\theta, \phi, \psi) \quad \text{...(11.101}a\text{)}$$

where D_{MK}^I are the functions of Euler angles, as discussed earlier for rotational transformation and $f_{I\tau v}(\beta)$ corresponds to vibration where τ stands for quantum numbers (two) associated with γ-vibration and v stands for one quantum number associated with β-vibrations. The β-vibrations have the same character as the radial vibrations of a three-dimensional harmonic oscillator. The function $g(\gamma)$ is single valued in the quantities $\alpha\mu$ which refer to space fixed axes, but various choices of body axes are possible which give the same $\alpha\mu$, but different Euler angles θ, ϕ, ψ and γ, as in

$$\sum_{\mu} |\alpha_2\mu|^2 = \sum_{\mu} |a_2\mu|^2 = \beta^2 \quad \text{...(11.101}b\text{)}$$

the value of β is independent of the choice of the axes. Equation 11.101 *a* represents a wave-function, which must be invarient under any change of the body-axes, which does not alter $\alpha\mu$'s. These requirements lead to certain relationships of $g_K(\gamma)$ with $g_{-K}(\gamma)$ or $g_K(-\gamma)$ or $g_K(\gamma = 2\pi/3)$ for integral I. One of these relationships is, that $g_K = 0$ if K is odd and that only even values of K exist. Hence, one can write the collective wave-functions as:

$$\psi_{I,M} = f_{IT v} (\beta) \sum_{K=0}^{I\,,} g_K^{I,T} (\gamma) \mid IMK \rangle \quad \text{...(11.102}a\text{)}$$

where $\mid IMK \rangle$ is normalised wave-function given by:

$$| IMK \rangle = \left[\frac{2I+1}{16\pi^2(1+\delta_{K,0})} \right]^{\frac{1}{2}} \sum_K{}' (D^I_{MK} + (-1)^I D^I_{M\bar{K}}) \qquad \text{...(11.102}b\text{)}$$

and prime over the summation indicates that it is restricted to even values of K. For detailed properties of $f_{I\tau v}(\beta)$ and $g_K^{IT}(\gamma)$, *see* References (44) and (45), and appendix.

For the case of axial symmetry; where I_3 is a constant of motion; then,

$$\psi = | IMK; \gamma\, v \rangle$$

$$= f_{IK,\lambda v}(\beta)\, g_K^{I,\lambda}(\gamma)\, | IMK \rangle \qquad \text{...(11.102}c\text{)}$$

where λ and v are the two vibrational quantum numbers. In Eq. 11.102*c*, the collective motion is separated into a vibrational part $f(\beta)\, g(\gamma)$ and the rotational part $| IMK \rangle$. One can obtain, the energy eigenvalues due to rotation only, from using the rotational wave-function *i.e.*,

$$\left\langle IMK \left| \sum_K \frac{L_K^2}{2\mathcal{J}_k} \right| IMK \right\rangle$$

$$= \hbar^2 \left\langle \frac{L^2 - L_3^2}{2\mathcal{J}_1} + \frac{L_3^2}{2\mathcal{J}_3} \right\rangle$$

$$= \hbar^2 \frac{\{I(I+1) - K^2\}}{2\mathcal{J}_1} + \frac{\hbar^2 K^2}{2\mathcal{J}_3} \qquad \text{...(11.103)}$$

It is the same as given Eqs. 11.26 to 11.30 which has been discussed earlier. For discussion, for pure rotational case, for even-even nuclei, therefore, one should refer to the previous section. It is, however, interesting to note here, that in Eq. 11.96; the term

$$\frac{1}{2} \sum_K \mathcal{J}_K \omega_K^2 = \sum_K \frac{L_K^2}{2\mathcal{J}},$$

comes out automatically as a part of kinetic energy of the vibrations of a deformed nucleus.

Other terms in Eq. 11.102*c*, *i.e.* $f_{IK\lambda v}(\beta)\, g_K^{I,\lambda}(\gamma)$ are[60, 61] responsible for vibration; as discussed briefly earlier, corresponding to variations of β and γ. For each of these vibrational modes; one has associated quantum numbers λ and v. Generally the rotational states are built when nuclei have larger distortion, *i.e.* the expectation value of β_0 is large; but $\langle(\beta - \beta_0)^2\rangle$ is comparatively small and expectation value of γ is zero. Many nuclei have such conditions in their ground states; which have axially symmetric distortions, *i.e.* they are spheroidal in shape, and have large quadrupole moments. On such states, one builds a rotational band. There will be β and γ-vibrations associated with the head (the lowest state) of such rotational states. Many such rotational bands may exist in the excitation of a deformed nucleus. There may be cases, where an axially symmetric state in the ground state, may be excited by a γ-vibration, to any unsymmetric shape; giving rise to 3^+ and 5^+ states in such even-even nuclei. There may, however, be unsymmetric ground state, then such states are possible among their rotational levels. One should, by now, realize that for even-even nuclei not far from the magic numbers, the departure from spherical shape is small, so that their equilibrium shape is nearly spherical, and collective oscillations about spherical shape may be well approximated by quadrupole deformation.

The sum $\sum_{\mu} \alpha_{\mu}$ is the predominant deformation parameter of displacement and $\sum_{\mu} \dot{\alpha}_{\mu}$ that of velocity. Under such condition, the energy eigenvalue of the state containing phonons is given by Eqs. 11.70 and 11.71 where the value of β^2 can be calculated from Eq. 11.84 for $\lambda = 2$, if B_2 and C_2, *i.e.* Coulomb energy parameters and surface energy parameters, are known. The values of these parameters can be calculated from the energy differences of vibrational levels and as we shall see subsequently from electric quadrupole transition rates between the first excited state and the ground state. These energy levels have uniform spacing between non-degenerate levels, and have much larger spacing between two levels than the average spacing between rotational levels in a given situation. The spacing between the vibrational levels is large for near spherical nuclei near the magic number ($N = 82$, 126) and small for deformed nuclei in the intermediate region. A few points of systematics for energy emerge from the analysis of these diagrams of relationships of energy levels. Each rotational band based on $K = 0$ represents a case of the head of the band corresponding to a nucleus with spheroidal space having an axis of symmetry, and in general, $K = 0$ remains constant for all the states of the rotational band. At a higher excited state a new rotational band may start, with say $K = 2$ which again may stay constant for the rotational band built on this. This band, however, may have β or γ-vibrations and hence will have n_{β} and $n_{\gamma} \neq 0$. When $n_{\beta} = n_{\gamma} = 0$, then the nuclear shape has an axis of symmetry. As the shape is very much deformed, only rotational states are created by its excitation. If shape is only slightly deformed from the spherical shape, then vibrational modes of excitation are physically possible, with either β-unstable or γ-unstable nuclei. As a matter of fact, if one starts from spherical nuclei with magic number of neutrons (say 82); first we have only vibrational mode for spherical nuclei; then for a little deformed nuclei β-instability starts and one has β-vibrations; till the deformation is large enough and only rotational modes occur. For further deformation, γ-instability starts and γ-vibrations are found till one reaches the other end of spherical nuclei (say $N = 126$); where again only vibrational modes are available [Fig. 11.11]. In the region of A where both vibrational and rotational modes are available, it is fascinating to see the relationship of the collective properties of levels in one region, with the neighbouring regions. It may be further seen, as has also been shown theoretically by Sheline[66]; that in a deformed nucleus with γ-vibrations and a rotational band, the spin sequence is 2^+, 3^+, 4^+, 5^+ etc. and for a β-vibration over which a rotational band rides, the spins sequence is 0^+, 2^+, 4^+, 6^+ etc. and $K = 0$, generally at about 1 MeV excitation. In the γ-vibration also K is nearly a constant of motion which is given by $K = 2$.

Then there are the octupole vibrations at lower energy than β and γ-vibrations which are shown for some nuclei in Fig. 11.12*a*. These vibrations can carry 0 to 3 units of angular momentum parallel to 3-axis. Angular momentum of this vibrational mode with 3-component v is coupled with the angular momentum of rotator, with 3-component zero, gives a resultant $K = v$. In general, the states have $I = K$, $K + 1$, etc. but if $K = 0$, then only even I, with even parity are allowed because of the symmetry requirements. The same considerations of symmetry results in negative parity $K = 0$ states, to $\overline{1}$, $\overline{3}$, $\overline{5}$ etc.

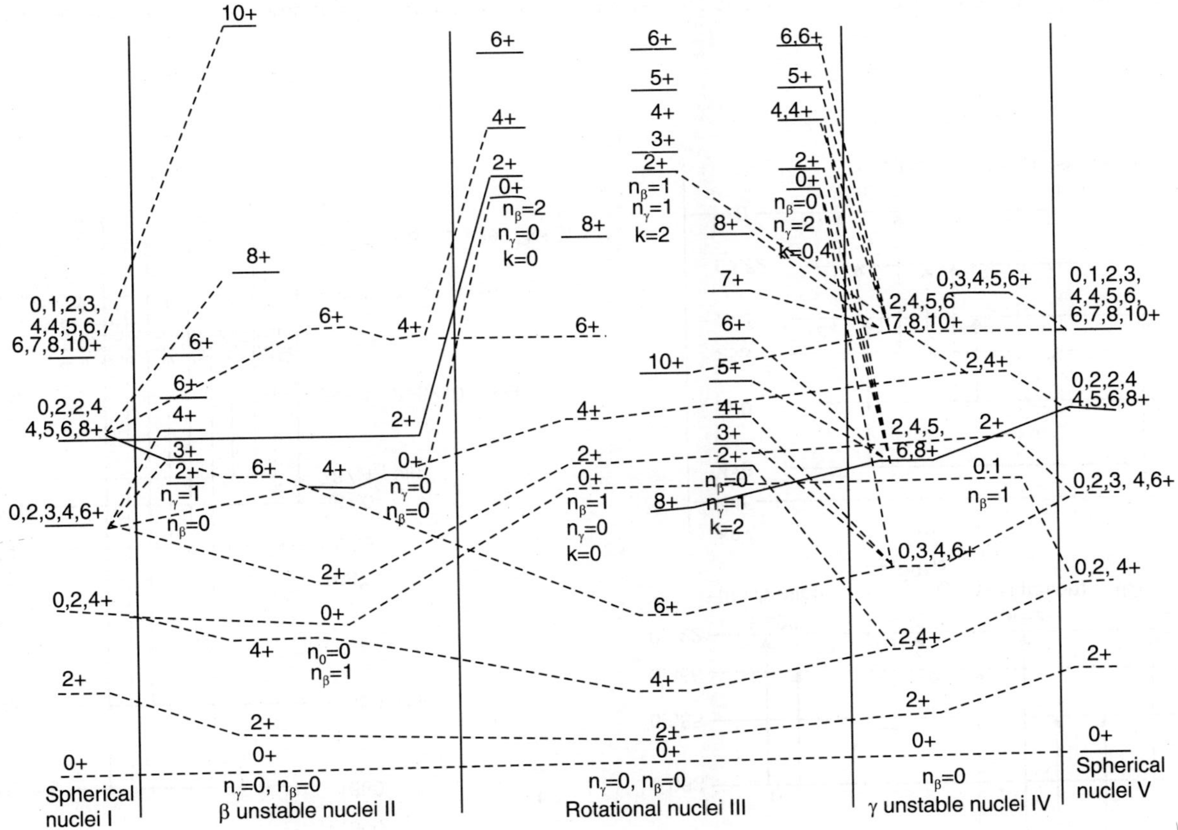

Fig. 11.11 Relationship between the levels in spherical, deformed β-unstable, and γ-unstable nuclei. Related states are connected by dashed lines (Ref. 66)

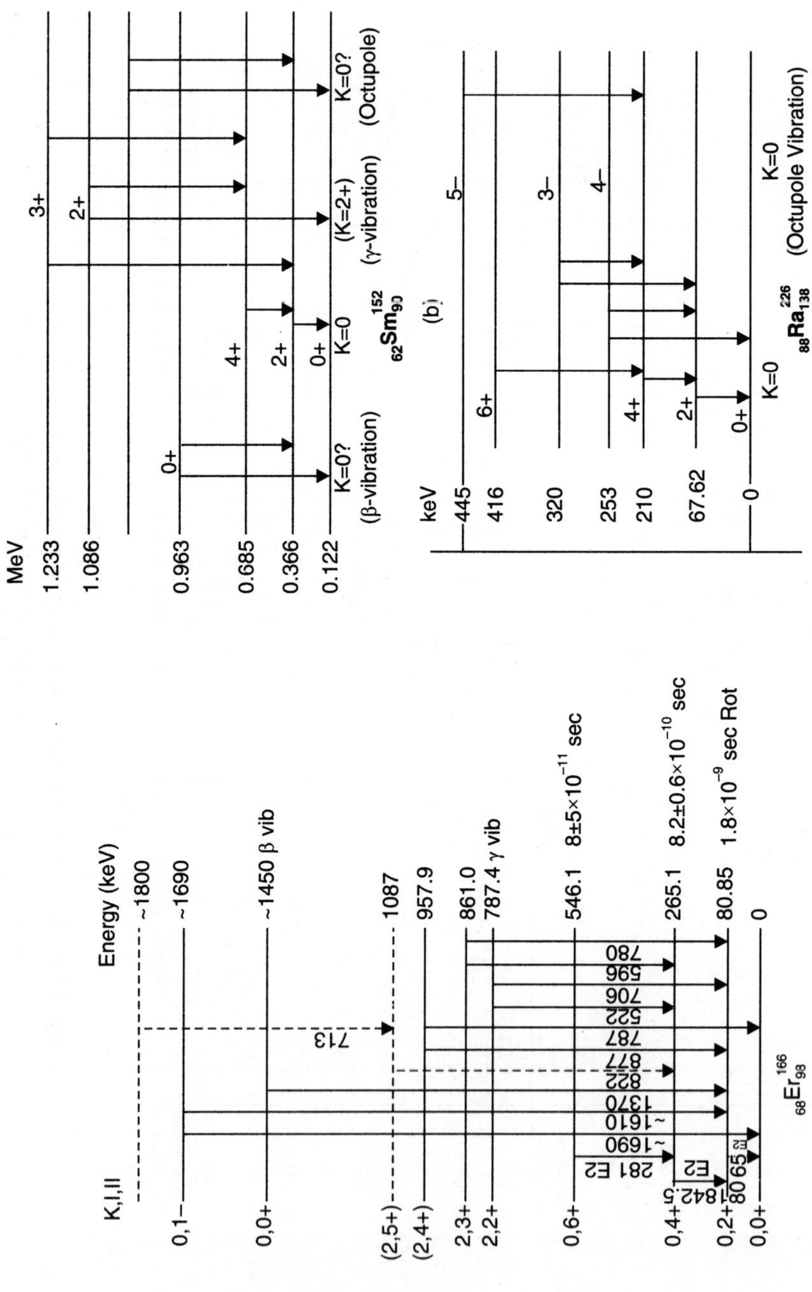

Fig. 11.12 Assignment of K to rotational and vibrational states in even-even nuclei $_{68}Er^{166}$, $_{62}Sm_{90}^{152}$ and $_{88}Ra_{138}^{226}$ (Ref. 66, 65).

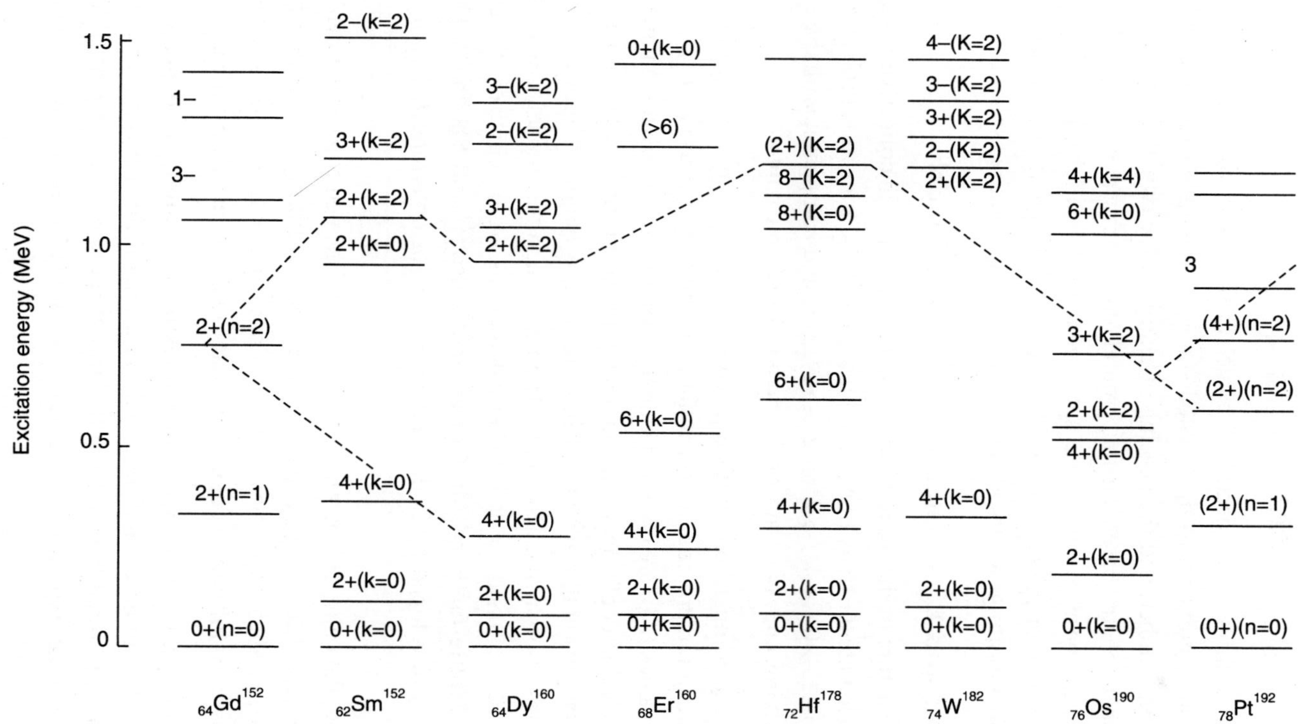

Fig. 11.13 Spectra of nuclei, along with the assignments of I, π and n and K in the range of Gd^{152}—Pt^{192} (Ref. 69).

In many cases like Ra^{226} and Sm^{152}, such negative states have been recognised from which it may be deduced that the lowest energy vibrational state has $v = 0$, *i.e.* it is the vibration with axial symmetry. Because of the form of Legendre polynomial P_3 (cos θ), these vibrations may be described as pear shaped, with the bulge alternating between one end of the nucleus and the other. These vibrations are available in Er^{166} and also in Sm^{152} and Ra^{226} as shown in Fig. 11.12.

In Fig. 11.13, we have shown the spectra of nuclei in the range of Gd^{152} to Pt^{192}, with the assignments of various quantum numbers to the excited states, as obtained experimentally. From the assignments of n, *i.e.*, phonon number, and the spacing of energies, the two extreme nuclei, *i.e.* Gd^{152} and Pt^{192} have the vibrational spectra based on the ground state, corresponding to different values of n; other nuclei, in between these extremes are spheroidal nuclei, all having rotational spectra, either riding over the ground state or the excited vibrational states. We have seen already [Fig. 11.11], how the nuclei evolve from vibrating spherical nuclei to vibrating-rotating spheroidal nuclei.

11.3.2 Nuclear Moments and E_2-transitions in Collective Model

One method of exploring the nature of the wave-function of various states in collective model-involving rotation and vibration is to calculate the transition probabilities say $B(E_2)$ transition probability involving an E_2 transition say between $2^+ \rightarrow 0^+$ or $4^+ \rightarrow 2^+$ or $2^+ \rightarrow 2^+$, etc., and compare it with the experimental values; or to measure the mixing ratio δ, as in Eqs. 7.66*a* and 7.66*b* in Chapter 7.

Such calculations and comparisons with experiments have been carried out extensively. Experimentally Coulomb excitation provides a very reliable method of measure $B(E_2)$: as discussed in section (7.7). Of course the values of δ^2 are obtained in many experiments involving gamma transitions, using the techniques of conversion coefficients and angular correlation (or distribution) as discussed in Chapter 7. Theoretically, Davydov-Fillipov[8] model has provided a major attempt to calculate these quantities. Figure 7.10 is a result of such calculations. The nucleus, in this model, deviated from axial symmetry within the limits of $\gamma = 0$ and $\gamma = \pi/3$. Other calculations are by Grechekin[67] and Davidson and Davidson[68]; Bohr and Mottleson[69] and Mc Gowan and Stelson[70] and Sheline[66]. The theory of Coulomb excitation has been developed by Adler and Winther[71] for obtaining $B(E_2)$ and compared with experiments.[73]

As shown in Fig. 7.10, the value of δ increases as function of the neutron number with exceptions at magic numbers, 28, 50, 82 and 126. At these neutron numbers, the mixing ratio diminishes by factor of two or three from the values predicted by Davydov and Fillipov[8] model. It also shows that experimental values are very close to Davydov-Fillipov model, compared to single particle estimates.

Transition probabilities provide another source of testing the validity of the collective model. As for example, the ratio of Coulomb excitation transition probabilities for the excitation of lowest 2^+ states in the ground and vibrational bands, from 0^+ ground state, have been plotted as a function of C, the stiffness parameter. The experimental values fit very well with the theoretically expected values[72].

Appendices

(A)

The D-functions, used in Eqs. 11.16 to 11.38 and Eqs. 11.101 to 11.102c in Chapter 11 and Eqs. 12.36 to 12.41 in Chapter 12.; basically are meant to relate the spherical harmonics of the unrotated frame of reference *i.e.*, $Y_m^1(\Omega)$ to the spherical harmonics of the rotated frame, *i.e.* $Y_{m'}^1(\Omega')$ with the relationship.

$$Y_m^1(\Omega) = \sum_{m'=-1}^{1} Y_{m'}^1(\Omega')\, D_{mm'}^1(\theta, \phi, \psi) \qquad \text{...(A.1)}$$

In other words $D_{mm}^1(\theta, \phi, \psi)$ is a rotation matrix. An infinitesimal rotation about any axis $\hat{n}$ is produced by an infinitesimal rotation operator $R_n(\varepsilon)$ as:

$$R_{\hat{n}}(\varepsilon) = 1 - i\,\varepsilon\hat{n}.\mathbf{L} \qquad \text{...(A.2)}$$

where $\mathbf{L}$ is the angular momentum operator. Then a finite rotation can be generated by applying $R_n(\varepsilon)$ operator N times in succession, so that if $\alpha = \varepsilon N$, one can, then write:

$$R_n(\alpha) = \underset{N\to\infty}{\text{Lt}}\left(1 - \frac{i\alpha}{N}\hat{n}\cdot\mathbf{L}\right) = e^{-i\alpha\,\hat{n}\cdot\mathbf{L}} \qquad \text{...(A.3)}$$

Then using this definition, the spherical harmonic $Y_m^1(\theta, \phi)$ can be related to $Y_m^1(\theta', \phi')$, by successive Eularian rotations α, β, γ, to get:

$$\begin{aligned} Y_{m'}^1(\theta', \phi') &= R_n\, Y_m^1(\theta, \phi) \\ &= \exp(-i\alpha L_X)\exp(i\beta L_Y)\exp(-i\gamma L_Z)\, Y_m^1(\theta, \phi) \end{aligned} \qquad \text{...(A.4)}$$

The successive rotation around x, y and z are due to the properties of Eulers angles. Using the completeness relation of the spherical harmonics one can write:

$$\begin{aligned} Y_m^1(\theta', \phi') &= \sum_{1'm'} Y_{m'}^{1'}(\theta, \phi) \int_0^{2\pi} d\phi \int_{-1}^{+1} d(\cos\theta)\, Y_{m'}^{1'}(\theta, \phi)\, R\, Y_m^1(\theta, \phi) \\ &= \sum_{1'm'} Y_{m'}^{1'}(\theta, \phi)\, \langle 1'\, m' \mid R \mid 1\, m \rangle \\ &= \sum_{m'} Y_{m'}^{1}(\theta, \phi)\, \langle 1\, m' \mid R \mid 1\, m \rangle \\ &= \sum_{m'} D_{mm'}^{*1}(\alpha, \beta, \gamma)\, Y_{m'}^1(\theta, \phi) \end{aligned} \qquad \text{...(A.5}a\text{)}$$

where
$$D_{mm'}^{*1}(\alpha, \beta, \gamma) \equiv \langle 1m' \mid R \mid 1\, m \rangle \qquad \text{...(A.5}b\text{)}$$

In writing (A.5a), we have omitted the sum on l', because the operator L_Y and L_Z contained in R are diagonal in quantum number l.

For a general case when **L** is replaced everywhere by **J** and $Y_m^1(\theta, \phi)$ by $Y_m^j(\theta, \phi)$ and hence $|\, l\ m \rangle$ by $|\, j\ m \rangle$; then we can express this state after rotation as $|\, j\ m \rangle_R$ which can be written as:

$$|\, j\ m \rangle_R = R\,|\, jm \rangle = \sum_{m'} D^j_{mm'}(\alpha, \beta, \gamma)\,|\, j\ m' \rangle \qquad \text{...(A.6)}$$

where
$$R = \exp(-i\alpha J_z)\exp(-i\beta J_y)\exp(-i\gamma J_z) \qquad \text{...(A.7)}$$

and
$$D^j_{mm'}(\alpha, \beta, \gamma) = \langle j\ m' \,|\, R \,|\, j\ m \rangle$$

$$= \langle jm' \,|\, \exp(-i\,\alpha\,J_z)\exp(-i\,\beta\,J_y)\exp(-i\,\gamma\,J_z)\,|\, jm \rangle \qquad \text{...(A.8)}$$

$$= e^{-i\,m'\alpha}\, d^j_{mm'}(\beta)\, e^{-i\,m\gamma} \qquad \text{...(A.9)}$$

where
$$d^j_{mm'}(\beta) = \langle jm' \,|\, \exp(-i\,\beta\,J_y)\,|\, jm \rangle$$

Expression for $d^j_{m'm}(\beta)$ has been derived by Wigner[61] as

$$d^j_{mm'}(\beta) = [(j+m)!\,(j-m)!\,(j+m')!\,(j-m')!]^{1/2} \times$$

$$\sum_r \frac{(-)^{r+m-m'}\cos\left(\frac{1}{2}\beta\right)^{2j-2r+m'-m}\sin\left(\frac{1}{2}\beta\right)^{2r+m-m'}}{r!(j+m'-r)!(j-m-r)!(r+m-m')!} \qquad \text{...(A.10)}$$

A few properties of *D*-functions:

(*i*) $\overset{j}{D}{}^{*}_{mm'}(\alpha, \beta, \gamma) = (-1)^{m-m'} D^j_{-m,-m'}(\alpha, \beta, \gamma)$ ...(A.11)

(*ii*) $D^1_{mo}(\alpha, \beta, \gamma) = \sqrt{\frac{4\pi}{[1]}}\, Y^1_m(\beta, \alpha)$ ...(A.12)

(*iii*) $D^1_{o,m}(\alpha, \beta, \gamma) = (-1)^m \sqrt{\frac{4\pi}{[1]}}\, Y^1_m(\beta, \gamma)$ ...(A.13)

(*iv*) $D^1_{oo}(\alpha, \beta, \gamma) = \sqrt{\frac{4\pi}{[1]}}\, Y^1_0(\beta) = (P_1\cos\beta)$ ...(A.14)

(B)

We derive the physical significance of $f(\beta)$ and $g^I_K(\gamma)$ as used in Eqs. 11.101 to 11.102 in Chapter 11 and Eqs. 12.36 to 12.39 in Chapter 12.

The kinetic energy in a collective motion consists of rotational and vibrational motions given by Eqs. 11.91 and 11.97, respectively; *i.e.*,

$$T_{\text{rot}} = \frac{1}{2}\sum_k \mathcal{I}_k\, \omega_k^2 = \frac{1}{2}\sum_k \mathcal{I}_k \sum_{jj'} q_{kj}\, q_{kj'}\, \dot{\Theta}_j \dot{\Theta}_{j'}$$

$$= \sum_K \frac{R_k^2}{2\mathcal{I}_k} \qquad \text{...(B.1)}$$

and
$$T_{\text{vib}} = \frac{1}{2}B(\dot{\beta}^2 + \beta^2\dot{\gamma}^2) \qquad \text{...(B.2)}$$

As proved in (Ref. 45), Chapter 11.

$$H\psi = E\psi$$

or
$$[(T_{\text{rot}} + T_{\text{vib}}) + V]\,\psi = E\psi \qquad \text{...(B.3)}$$

or
$$\left[-\frac{\hbar^2}{2B}\left(\frac{1}{\beta 4}\frac{\partial}{\partial\beta}\beta^4\frac{\partial}{\partial\beta} + \frac{1}{\beta^2}\frac{1}{\sin 3\gamma}\frac{\partial}{\partial\gamma}\sin 3\gamma\frac{\partial}{\partial\gamma}\right) + \right.$$

$$\left.\sum_k \frac{\hbar^2}{2\mathcal{I}_k} R_k^2 + \frac{1}{2}C\beta^2\right]\psi(\beta, \gamma, \theta_1, \theta_2, \theta_3)$$

$$= E\,\psi(\beta, \gamma, \theta_1, \theta_2, \theta_3) \qquad \text{...(B.4)}$$

Using the expressions for $\mathcal{I}_k$ in terms of β and γ, [Ref. (45)], and Eq. 11.92

$$\mathcal{I}_k = 4\,B\,\beta^2 \sin^2\left(\gamma - k\frac{2\pi}{3}\right) \qquad \text{...(B.5)}$$

and multiplying Eq. B.4 throughout by β^2, we find that the term containing $\partial / \partial\gamma$ and rotational terms are independent of β. This means that the wave-function is separable in coordinates and we can write:

$$\psi(\beta, \gamma, \theta_1, \theta_2, \theta_3) = f(\beta)\,\Phi(\gamma, \theta_1, \theta_2, \theta_3) \qquad \text{...(B.6)}$$

Substituting this value of ψ and $\mathcal{I}_K$ from Eq. B.5 in Eq. B.4, and multiplying from the left by $\beta^2[\psi(\beta, \gamma, \theta_1, \theta_2, \theta_3)]^{-1}$, we get:

$$-\frac{\hbar^2}{2B}\left[\frac{1}{f(\beta)}\frac{1}{\beta^2}\frac{d}{d\beta}\beta^4\frac{d}{d\beta}f(\beta) + \frac{1}{\sin 3\gamma}\frac{1}{\Phi}\frac{\partial}{\partial\gamma}\sin 3\gamma\frac{\partial}{\partial\gamma}\Phi\right]$$

$$+\frac{1}{2}C\beta^2 - E\beta^2 + \sum_k \frac{\hbar^2}{8\beta}\frac{1}{\Phi}\frac{R_k^2}{\sin^2\left(\gamma - k\frac{2\pi}{3}\right)}\Phi = 0 \qquad \text{...(B.7)}$$

or separating the terms containing β and γ we write:

$$-\frac{\hbar^2}{2B}\frac{1}{f(\beta)}\left[\frac{1}{\beta^2}\frac{d}{d\beta}\beta^4\frac{d}{d\beta}f(\beta)\right] + \frac{1}{2}C\beta^2 - E\beta^2$$

$$= -\left[\sum_k \frac{\hbar^2}{8\beta}\frac{1}{\Phi}\frac{R_k^2}{\sin^2\left(\gamma - k\frac{2\pi}{3}\right)}\Phi\right]$$

$$-\frac{\hbar^2}{2B}\frac{1}{\sin 3\gamma}\frac{1}{\Phi}\frac{\partial}{\partial\gamma}\sin 3\gamma\frac{\partial}{\partial\gamma}\Phi \qquad \text{...(B.8)}$$

In Eq. B.7, use has been made of the fact, that operator R_k^2, acts only on a function of θ_1, θ_2 and θ_3 and not on a function of β. Now we separate the variables, and by the argument of separation of variable method, we equate each side by a constant, say

$$\left[-\frac{\hbar^2}{2B}\Lambda\right]$$

Then from the left side we get:

$$\left(-\frac{\hbar^2}{2B}\frac{1}{\beta^4}\frac{d}{d\beta}\beta^4\frac{d}{d\beta} + \frac{\Lambda\hbar^2}{2B}\frac{1}{\beta^2}\right) = Ef(\beta) \qquad \text{...(B.9)}$$

Similarly, the right side of Eq. B.8 leads to a coupled equation for rotation and γ-motion, *i.e.*,

$$\left[-\frac{1}{\sin 3\gamma}\frac{\partial}{\partial\gamma}\sin 3\gamma\frac{\partial}{\partial\gamma} + \frac{1}{4}\sum_K \frac{R_K^2}{\sin^2(\gamma - k\,2\pi/3)}\right]$$

$$= \Lambda\,\Phi\,(\gamma, \theta_1, \theta_2, \theta_3) \qquad \text{...(B.10)}$$

Equation B.9 for $f(\beta)$ can be solved by standard numerical methods. Physically $f(\beta)$ is the amplitude of vibrational motion, in which β is changed *i.e.*, the ellipsoidal shape changes. On the other hand, Eq. B.10 corresponds to vibrational motion and rotation and their coupling. The first term corresponds to γ-change and second term to rotation so that Φ contains both the effects. If nucleus is rigid against γ-vibration, the first term in Eq. B.10 can be taken to be zero. On the other hand, if only γ-motion exists, then $\Phi_{I=0}(\gamma) = P_\gamma \cos(3\gamma)$, corresponds to pure vibration. In general, from Eq. B.10:

$$\Phi_M^I\,(\gamma, \theta_1, \theta_2, \theta_3) = \sum_K g_K^I(\gamma)\, D_{MK}^I(\theta_1, \theta_2, \theta_3) \qquad \text{...(B.11)}$$

The function $g_K^I(\gamma)$, thus corresponds to a coupling of rotation and γ-vibrations.

(C)

TRI-AXIAL DEFORMATION

For $\lambda = 2$, we have defined β and γ, [Eqs. 11.85 and Eqs. 11.86 to 11.89 and Table 11.2], from which it may be inferred that nuclei with $\beta \neq 0$ and $\gamma \neq 0 \neq \frac{n\pi}{3}$ are tri-axial with three axes along principle axes.

Both theoretical and experimental developments since seventies have made it possible to investigate a large number of deformed tri-axial nuclei and assign the shapes to the ground and excited levels.

1. Theory

Theoretically the Strutinsky[76]—theory has been developed since 1967 to describe the rotational problem of nuclei, with γ-degree of freedom. Basic parameters arise from a derivation, which relates the Strutinsky theory for the rotational problem, to the constrained Hartree Fock or Hartree-Fock-Bogoliubov approach to the deformation energy problem. We follow here the approach of Anderson[77], et al., and Bengtsson & Frauendorf[78].

We start with,

$$H^{\omega} = H^{o} - \hbar\,\omega\, I_x = \sum_i h_i^{\omega} = \sum_i (h_i^{o} - \hbar\omega\, j_{x_i}) \qquad \text{...(C.1)}$$

where H^{ω} is the Hamiltonian for higher angular momentum $I = \langle I_x \rangle$ in the rotating system, rotating with angular frequency ω and H^o is the single particle Hamiltonian for the zero angular momentum given by:

$$H^{o} = \sum_i h_i^{o} \qquad \text{...(C.2)}$$

and the static single particle Hamiltonian, $h^o = T + V$, for which for modified harmonic oscillator (MHO), we write V as:

$$V = \frac{1}{2}\hbar\omega_o \rho^2 \left[1 - \frac{2}{3}\beta_2\sqrt{\frac{4\pi}{5}}\cos(\gamma)\,Y_{2,0} + \frac{2}{3}\beta_2\sqrt{\frac{4\pi}{4}}(\sin\gamma)\,(Y_{2,2} \pm Y_{2,-2})\right]$$

$$- \hbar\omega_o^o\,(2\text{K}\,\vec{l}_t\cdot\vec{s}_t + \mu\left(\vec{l}_t^{\,2} - \left\langle \vec{l}_t^{\,2}\right\rangle\right)) \qquad \text{...(C.3)}$$

Here the first term correspondents to space-terms of V and the second term to the intrinsic spin-orbit and orbital interaction terms of potential. The deformation is contained only in the first term. The term $\hbar\,\omega\, I_x$ in Eq. (C.1) is a Lagrange multiplier, so that H^{ω} acts like a Routhian operator for a configuration of a rotating system. Two symmetries of the rotating case, *i.e.* parity π and rotation of an angle π around *x*-axis, have to be added in Eq. (C.2). The latter symmetry is designed by α, the signature of the wave function, so one has (π, α) as the new quantum numbers of χ_i^{ω} for which, one diagonalises the equation.

$$\hbar\chi_i^{\omega} = e_i^{\omega}\chi_i^{\omega} \qquad \text{...(C.4)}$$

and obtains modified eigen-functions χ_i^{ω}. The corresponding real single particle energies are the expectation value of

$$\langle e_i \rangle = \left\langle \chi_i^{\omega} \middle| h^o \middle| \chi_i^{\omega} \right\rangle = e_i^{\omega} + \hbar\omega\langle m_i \rangle \qquad \text{...(C.5)}$$

And the total single particle energy, as the sum over single particle energies is given by:

$$E_{s\cdot p} = \sum_i \langle e_i \rangle = \sum_i e_i^{\omega} + \hbar\omega\, I \qquad \text{...(C.6)}$$

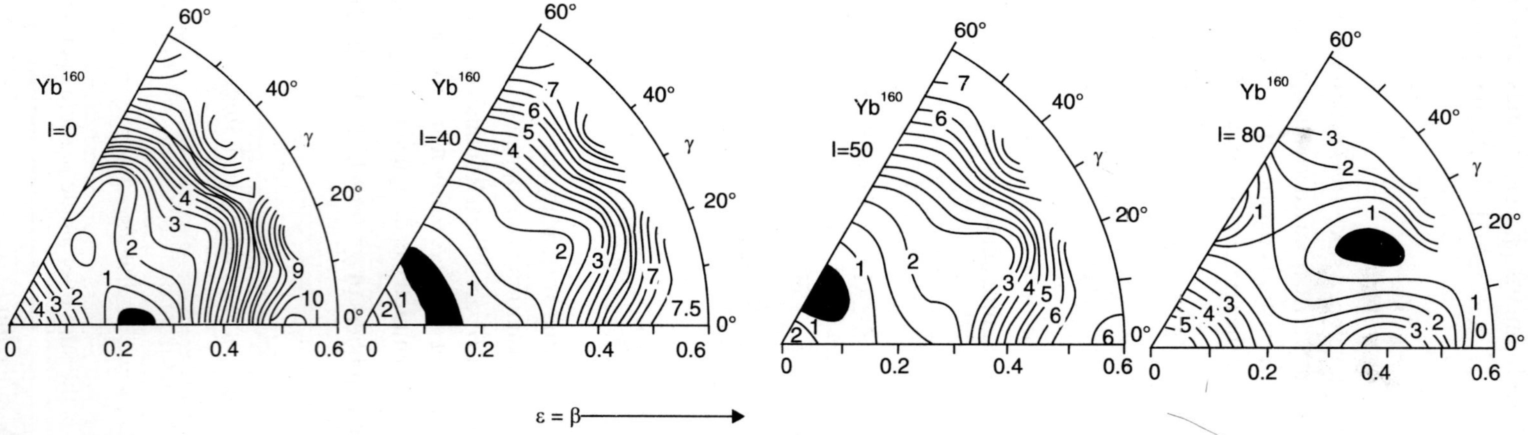

Fig. 11.14 Potential energy (Eq. C.12) surfaces in (ε, γ) plane, with the inclusion of shell energy for Yb^{160}, as a function of angular momentum I. For the surface, coulomb, and macroscopic rotation energy terms, a minimisation is performed with respect to ε_4; but keeping $\varepsilon_4 = 0$ for E_{shell} (77, 80).

Here

$$I = \sum_i m_i = \sum_i \left\langle \chi_i^{\omega} \,|\, j_x \,|\chi_i^{\omega} \right\rangle \qquad \text{...(C.7)}$$

It may be seen that it is convenient to add explicitly, the particle number operator A and corresponding Langrange multiplier λ, so that

$$e_i^{\omega} - \lambda = e_i - \hbar\,\omega\, m_i - \lambda \qquad \text{...(C.8)}$$

We have neglected here, the pairing field $P+$, for which see Bengtsson, et al., (Ref. 78). The energies e_i^{ω} are called, Ruthians. For the total single particle energy $E_{s.p}$, [Eq. (C.6)]; the summation is carried over A-states, for which the values of $e_i^{\omega} - \lambda$ are, the lowest possible, say negative so that $e_i = \langle \hbar\omega m_i + \lambda\rangle$. From classical rigid rotation problem [Eqs. 11.2 and 11.4], it can be seen that,

$$\hbar\omega = \frac{\partial E}{\partial I} \qquad \text{...(C.9)}$$

and

$$\frac{\hbar^2}{2\mathcal{I}} = \frac{\partial E}{\partial I^2} \qquad \text{...(C.10)}$$

It may be mentioned here, that ω_o and ω_o^o are related to single particle energies and are converted to the well depth of MHO, while ω is the rotational angular momentum connected with a rotational band.

In actual practice a normalisation procedure called Strutinsky smearing is followed by defining Strutinsky density functions for level density $g_1\,(e^{\omega})$, and spin density $g_2\,(e^{\omega})$ function. The energy, which is mapped, in practice, is the total energy given by:

$$E_{\text{total}}\,(\beta, \gamma) = E_{\text{macroscopic}}\,(\beta, \gamma, \min\beta_4) + \delta\, E_{\text{shell}}\,(\beta_2, \gamma) \qquad \text{...(C.11)}$$

where $\delta\, E_{\text{shell}}$ is the difference between the fluctuating $E_{s\,p}$ and the smooth function $\tilde{E}$ and $E_{\text{macroscopic}}$ is the sum of Coulomb, surface and rotational energy, for which the liquid drop parameters are chosen according the Myers and Swiatecki (79).

In literature (77, 78), Eq. (C.11) is alternatively written as:

$$E_{\text{total}}\,(\varepsilon, \gamma) = E_{\text{macroscopic}}\,(\varepsilon, \gamma, \min\, \varepsilon_4) + \delta\, E_{\text{shell}}\,(\varepsilon, \gamma) \qquad \text{...(C.12)}$$

where

$$\delta\, E_{\text{shell}} = \sum_{\substack{\text{protons}\\ \text{neutrons}}} \left(\sum_{\varepsilon_i < \lambda} \varepsilon_i - \int_{-\infty}^{\lambda} \varepsilon\, \tilde{g}(\varepsilon)\, d\varepsilon \right) \qquad \text{...(C.13)}$$

where ε_i denotes the eigenvalues of the single particle operator.

While mapping $E_{\text{total}}\,(\varepsilon, \gamma)$, one chooses, $\varepsilon_4 = 0$ for E_{shell}. One plots potential energy surface, *i.e.* $E_{\text{total}}\,(\varepsilon, \gamma)$ for different values of ε, 1 and $\varepsilon_4 = 0$ for E_{shell} and minima is obtained by choosing ε_4 to given minimum of $E + E_{\text{shell}}\,(\varepsilon_4 = 0)$ for a grid of deformation points in (ε and γ). In E_{macro}, ε_4 corresponds to β_4, as mentioned in the beginning and is contained in the expression for E_{macro} as given in Ref. (79).

Figure 11.14 shows [77, 80] the potential energy surfaces in (ε, γ) plane, with inclusion of shell energy for Yb^{160} as a function of I. For the surface, Coulomb and macroscopic rotation energy terms, a minimisation is performed with respect to ε_4. To create a mapping like Fig. 11.14, requires a large numerical computer programme to create a mesh of (β, γ) points.

In Fig. 11.14, the coordinate system plots deformation ε, radially outward and γ as an angle from the horizon; $\gamma = 60$ corresponds to an oblate shape rotating about its symmetry axis. For low spins, however, Yb^{160} is prolate, as shown for $I = 0$, in Fig. 11.14. At higher spin, *i.e.* $I = 40$, it is tri-axial; at $I = 50$ and 60, it is oblate and fissions for $I = 80$.

2. Experimental

In recent years, there have been extensive experimental and theoretical studies,[53, 54, 82, 83] involving rotational levels of tri-axial nuclei.

In a typical case in 1989, investigating high spin states in Kr^{75}, produced in the reaction Ti^{46} (S^{32}, $2pn$) Kr^{75}, beam of 97 MeV of S^{32} was used. One observes gamma rays spectrum, by using proper gates, to obtain parity and negative parity transition, from which a decay scheme of Kr^{75} in built, showing rotational bands. Using cranked shell model analysis[81, 82] to the data, one obtains, $e_i = e_i^{\omega}$ [Eq. C.8], as a function of ω/ω_o. Experimental values of (é) are obtained from the relationship, developed in (1981), *i.e.*,

$$\acute{e}\,(\omega) = \acute{E}\,(\omega) + \frac{1}{2}\,\mathcal{I}_1\omega^2 + \frac{1}{4}\omega^2 - \frac{\hbar^2}{2\mathcal{I}} + \Delta \qquad \text{...(C.14)}$$

where

$$\acute{E}(\omega) = E\,(\omega) - \omega I_x \qquad \text{...(C.15)}$$

$$\omega\,(I) = \frac{E(I+1) - E(I-1)}{I_x(I+1) - I_x(I-1)} \qquad \text{...(C.16)}$$

and

$$I_x = \sqrt{\left(I + \frac{1}{2}\right)^2 - K^2} \qquad \text{...(C.17)}$$

is the x-component of the angular momentum and $\mathcal{I}_0$ and $\mathcal{I}_1$ are the two moment of intertia (Eq. 11.13a) are related to I_x^g, the reference angular momentum of the ground state of the nucleus, as:

$$I_x^g = \mathcal{I}_0\,\omega + \mathcal{I}_1\omega^3 \qquad \text{...(C.18)}$$

for which

$$i_x\,(\omega) = I_x\,(\omega) - I_x^g\,(\omega) \qquad \text{...(C.19)}$$

or

$$I_x\,(\omega) = \mathcal{I}_0\,(\omega) + \mathcal{I}_1\omega^3 + i_x\,(\omega) \qquad \text{...(C.20)}$$

The corresponding effective moment of intertia is, then, obtained from

$$\mathcal{I}_{eff}(\omega) = \frac{I_x}{\omega} = \mathcal{I}_0 + \mathcal{I}_1\omega^2 + i_x\,(\omega)/\omega \qquad \text{...(C.21)}$$

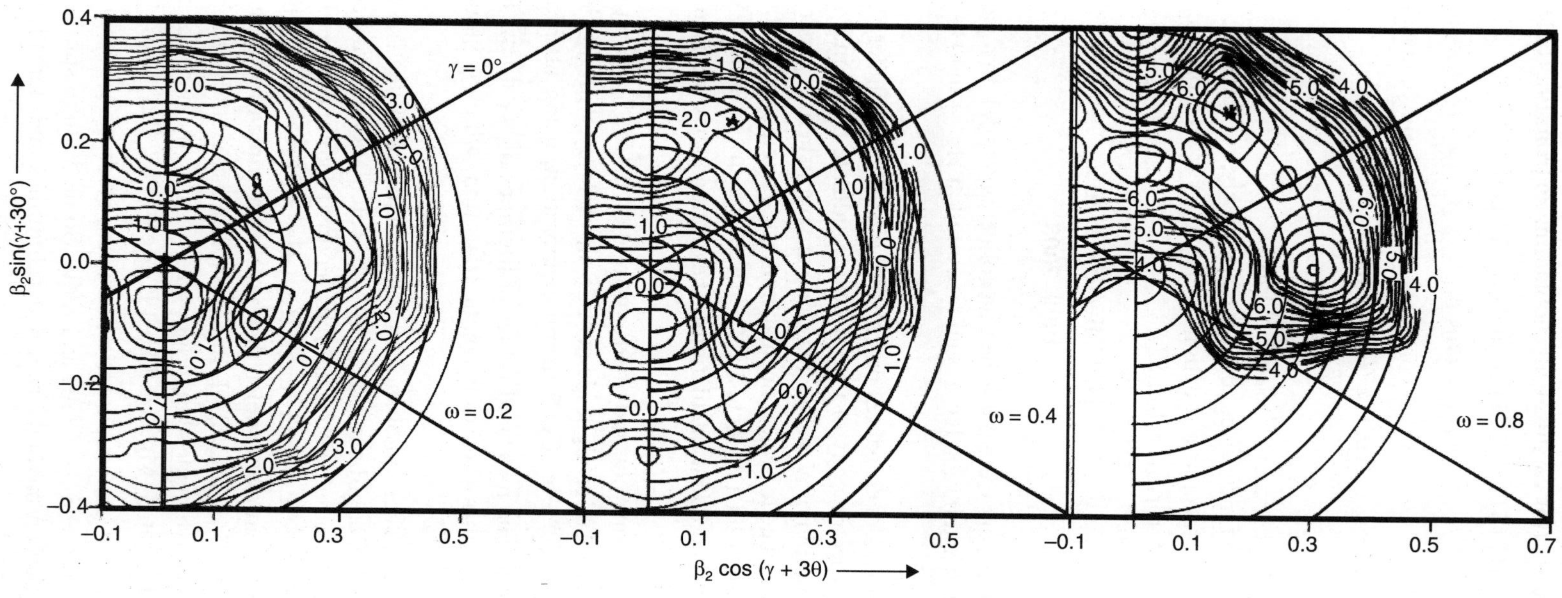

Fig. 11.15 Total Routhian (E_{total}) surface for Br^{76}. The deformation parameter β_4 was chosen to minimise the Routhian at each point, on the grid measured in 1996 and 1997 (Ref. 82).

One can then plot $\mathcal{I}_{eff}(\omega)$ versus ω, after obtaining ω from Eq. (C.16)

One may combine (C.15) to (C.21) suitably; and obtain:

(*i*) É versus ω; (*ii*) é versus ω (*iii*) I_{eff} versus ω

(*iv*) I_x versus ω (*v*) Δ constant (*vi*) I_o constant

(*vii*) $\mathcal{I}_1$ (constant) and (*viii*) $\mathcal{I}_x(\omega)$ versus ω.

Also one obtains

$$E_{\text{total}} = E_{\text{macro}} + \delta E_{\text{shell}} \quad \text{...(C.22)}$$

Using Ref. 84, and 76 for $E_{L.D.}$ and δE_{shell} respectively, for different values of β_2 and γ, one can, then plot the E_{total} energy surface, as a functions of $\beta_2 \sin(\gamma + 30°)$ on *y*-axis, and $\beta_2 \cos(\gamma + 30°)$ on *x*-axis as shown in Fig. 11.15 for Br^{76} (Ref.82).

11. Collective Model

2000–2008

In 2000, some ten papers were published on collective model. In a paper, authored by 30 authors octupole and quasi-particle excitation have been measured in Hg^{178} and Hg^{180} [Phy. Rev. C.62, 044305 (2000)].

In 2001, some half a dozen papers on collective excitation, under condition of super deformed bands of Hg^{194}; detailed spectroscopy of Chiral–twin candidate bands in Pm^{136}, anti-magnetic rotation in Pd^{100}; and rotational motion of $N = Z$ nucleus Kr^{72} are published. A paper on possible contribution of a doubly excited collective state to proton-nuclear-multi step interaction at 300 MeV, is worth mentioning [Phy. Rev. C.64, 01101 (R) (2001)]. Inclusive C^{12} (*p*, *p*) spectra at 300 MeV were measured in a wide range and compared with theoretical calculations. The discrepancy was discussed in terms of a collective excitation process.

In a paper on 'shape transition and titled axis rotation in Ce^{136}, a group of 6 authors from Roorkee, Mumbai and Amritsar, have used an eight-element CS clover array, with .14 elements NaI (*Tl*) multiplying filter, in a reaction Sn^{124} (O^{16}, 4n) Ce^{136} fusion evaporation, using 80 MeV O^{16} beam, from 14 UD pellotron at TIFR Mumbai. The measured values of lifetimes of $1\overline{8}$ to $2\overline{2}$ levels, were compared with theory indicating a transition for PAC to magnetic transition, which is induced by shape change [Phy. Rev. C.66, 041303 (R), (2002)].

Some half a dozen papers published in 2002, involved collective excitation. In a theoretical paper, using Hartree-Fock Bouglibouv (HFB) model, calculations have been made for all even-even nuclei from proton dip line to the neutron dip line, having proton numbers $Z = 2...4...$ 108; and neutron numbers $N = 3, 4...$ 188. These nuclei range from He^4 to Pb^{182}. It has been found that there exist numerous particles bound even-even nuclei, that have at the same time negative two-neutron separation energies [Phy. Rev. C.68, 054312 (2003)]. In a comprehensive study, 19 authors from Argentina and Italy have studied the collective mode of high spin states of Ir^{178}, by means of in-beam γ-rays spectroscopy using Tb^{159} (Mg^{24}, 5n) Ir^{178} at E (Mg^{24}) = 131–141 MeV [Phy. Rev. C.67, 024308 (2003)].

An interesting case of collection excitation is the study of Y^{80} obtained from Fe^{54} (Si^{28}, pn) at 90 MeV by 16 authors from U.S.A., China and India. Some 30 high spin levels in Y^{80} were measured, using Doppler shift-attestation method, from which lifetimes and transitional quadrupole moments *Q*, are measured. This shows an abrupt decrease of Q, with spin which is attributed to shape change [Phy. Rev. C. 69, 10643 (4) (2004)].

In a massive collaborative effort of 76 authors from Europe, U.S.A. and India, spin states in Ba^{124} were experimentally investigated using Ni^{64}, (Ni^{64} 4n) Be^{124} [Phy. Rev. C.70, 14304 (2006)]. Through γ-rays spectroscopy it was proved that one of the bands showed a transition from collective to non-collective behaviour.

In a paper, on search for multiple step Coulomb excitations, and inelastic neutron scattering of Sn^{152}, it is explored to find 685 keV excitation energy of the first excited O^+ states in Sn^{152} to explore the expected two phonon excitation at low energy. Multiple-step-Coulomb excitation and inelastic scattering studies of Sm^{152} are used to probe E_2 collectivity of excited 2+ states in this 'soft' nucleus and the results are compared with model predictions. No candidate for two phonon K^π = o+ quadrupole vibrational state are found. A 2^+, $K = 2$ state with strong $E2$ decay to the first excited $K^\pi = 0^+$ band and a probable 3+ band members are established [Phy. Rev. C. 77, 061301 (R) (2008)].

REFERENCES

1. Table of Isotopes sixth edition; C.M. Lederer; J.M. Hollander, I. Perlman, John Wiley & Sons, New York (1961), S. Törmänen, et al.: Nuclear Physics A575, 417 (1994); E.S. Paul et al.: Phy. Rev. C51, R28571 (1995).
2. O. Nathan and S. G. Nilsson: Collective Nuclear Motion and the Unified Model, Alpha, Beta and Gamma Ray Spectroscopy V.I., p. 601, North Holland Publishing Co., Amsterdam (1965); S. Mitarai et al.: 2. Physik A344, 405 (1993).
3. J. Thibund C.R. Acad. Science, Paris, 1911. 778-89 (1930).
4. Niels Bohr: Nature, 137, p. 344-48 (1939); Frenkel J. Phy. Rev. 55. 987 (1939); Bohr A., Don Mat Fys. Medd 26, No.14 (1952); Bohr A. and B.R. Mottleson: Dan Mat. Fyz. Medd. 27, No. 16 (1953); Rainwater J., Phy. Rev. 79, 432 (1950).

   H.D. Jensen: Zur Geschichte der Theories des Atomkerns, pp. 153-64 les prix Noble en 1963, Stockholm, Norstedt, (1963).
5. R.M. Diamond, F.S. Stephen, W.J. Swiatecki: Phy. Rev. Letters 11, 315-318 (1964).
6. G. Scharff-Goldhaber: Proc. Int. Conf. Nuclear Structure, Tokyo, pp. 150-59 (1967).
7. M.A.J. Marriscoti, G. Scharff-Goldhaber B. Buck: Phy. Rev. 178, 1864-1887 (1969).
8. Same as Ref. (7), A.S. Davydov, G.F. Fillippov: Nuclear Phy. 8, 237-49 (1958).
9. J.D. Garret, G. Scharff-Goldhaber, J.P. Vary: Bull. Am. Phy. Soc. 19, p. 59 (1970).
10. A. Johnson, Z. Szymanski: Phy. Rep. C. 7, 181 (1973).

    R.A. Sorensen: Rev. Mod. Phy. 45, p. 353 (1973).
11. J.S. Batra and R.K. Gupta: Phy. Rev. C. 43, p. 1725 (1991).
12. M. Morinaga: Nuclear Physics 75, 385 (1966).
13. J.B. Gupta, A.K. Kavathekar and Y.P. Sabharwal: Phy. Rev. C P 3417 (1997).
14. A. Klein: Nuclear Phy. A 347, 3 (1980): Phy. Letter 93B11 (1980); D. Bonatros and A. Klein: Phy. Rev. C.29, 1879 (1984) Atomic Data, Nuclear Data Table, 30, 27 (1984).
15. S.X. Liu and J.Y. Zheng: Phy. Rev. C. 58, P. 3266 (1998).

16. P. Fuchs, Emling L., Folk-man F., Greese E.: Schwalm Jeharbericht, Darmstadt. Germany G.S.I. (1977).

17. F.S. Stephens, D. Ward and J.O. Newton: Am. J. Phy. Suppl. 27, 164 (1968).

18. T.L. Khoo, R.K. Smither, B. Hass, O. Hausser, H.R. Andrews, D. Horn, D. Ward: Phy. Rev. Letters 41, 1027 (1978).

19. F. Beck, K.H. Müller and H.S. Köhler: Phy. Rev. Letter 40, 837 (1978)

20. R.M. Diamond and F.S. Stephens: Annual Rev. of Nuclear and Particle Physics, V.30, pp. 85-157; (1980).

21. S. Cohen, F. Plasil, W.J. Swiatecki: Ann. Phy. 82, p. 557 (1974).

22. Same as Ref. (16), J.O. Newton, I.Y. Lee, R.S. Simmon, M.M. Aleonard, El.Y. Mastri, F.S. Stephens, R.M. Diamond: Phy. Rev. Letter 38, 810 (1977).

23. A. Bohr, B.R. Mottleson: Nuclear Structure, V.2, Reading, Mass Banjamin (1975).

J. Blocki, J. Rendrup , W.J. Swiatecki and C.F. Tsang: Annual Phy. 105, 427 (1977).

24. S.A. Seethre, A. Johnson , S. Jagare, H. Ryde and Z. Szymanski: Nuclear Physics A 207, 486-518 (1973).

25. Same as Ref. (19), P. Taras, W. Dehnhardt, S.J. Mills, M. Veggian, J.C. Hardinger, U. Neumann and B. Povh: Phy. Letters B, 41B, 295 (1972).

R. A. Sorensan: Rev. of Mod. Physics, V. 45, pp. 353-377 (1973).

26. M.A.J. Marriscotti, G. Scharff-Goldhaber and B. Back: Phy. Rev. 178, 1864-87 (1969).

27. S.H. Harris: Phy. Rev. B, 138, 509-13 (1965).

28. References (19) and (20)., R.O. Sayer, J.S. Smith, W.T. Milner: Atomic and Nuclear Data Table 15, 85-110 (1975).

29. G. Scharff-Goldhaber and A.S. Goldhaber: Phy. Rev. Letter. 24, 1349 (1970).

30. Same as Ref. (23)., P. Thieberger, A.W. Sunyar, P.C. Rogers, N. Lark, O.E. Kistner, E. der Mateosian, S. Cochavi, E.H. Auerbach: Phy. Rev. Letter 28, 972-74 (1972).

31. S. Cochavi, O.E. Kistner, M. Mc Gowen: G.J. Scharff-Goldhaber: Phy. France 33, 102 (1972).

32. R.K. Gupta: Phy. Letters B, 36B, 173 (1971); L.E.H. Trainer and R.K. Gupta: Cand. J. Physics 49, 133 (1971)., M. Satpathy and L. Satpathy: Phy. Letters B, 34B, 377 (1971).

33. G. Scharff-Goldhaber and J. Weneser: Phy. Rev. 98, 212 (1955).

34. B.R. Mottleson, J.G. Nilsson: Phy. Rev. 99, 1615-17 (1955).

35. Same as Reference (27)., S.G. Nilsson and O. Prior: Mat. F.S. Medd. Dan. V.d. Selsk. 32 No. 16 (1961).

36. J. Kreunlinde: Nuclear Physics A 121, p. 306, (1968); A160, p. 471 (1971).

D.R. Bes, S. Landowns and H.A.J. Mariscotti: Phy. Rev. 166, p. 1045 (1968).

37. D.R. Inglis: Phy. Rev. 96, 1059 (1954); Villars F., Annual Rev. Nuclear Sciences, 7, p. 211 (1951).

38. D. Bes, S. Landowne and M.A.J. Mariscotti: Phy. Rev. 166, 1045 (1968). Griffin J.J. and M. Rich: Phy. Rev. 118, p. 850 (1960).

S.G. Nilsson and O. Prior: K. Dan Vidensk, Seisk Mat-Phys. Medd. 32, (16) (1961).

39. R.A. Sorenson: Rev. of Mod. Physics V. 45, 333-377 (1973).

K.C. Chan and J.G. Valatin: Nuclear Physics V. 82, p. 222 (1966); Ibid, Phy. Letters, 11, 304 (1964). B.R. Mottleson and J.G. Valatin: Phy. Rev. Letters, 5, p. 511 (1960); F.S. Stephens: Rev. Mod. Physics, 47, p. 58 (1975); A. Foessler, K.F.A. (Jälich) Preprint, XI International School of Nuclear Physics, Predeal, Romania (1976); J. Kreunlinde: Nuclear Physics A 121, 306 (1968); Ibid. Nuclear Phy. A. 160, 471,

(1971); J. Kreunlinde and Z. Szymenski: Phy. Rev. letters B, 36, 157 (1971); Ibid. Phys. Letters B, 36B, 471 (1972); B.L. Cohen, Phy. Letters. 27N, 271 (1968).

40. J. Meyer, J. Speth and J.H. Vogler: Nuclear Physics A, 193-60 (1972); E.R. Marshall: Phy. Rev. 139, B 770 (1965).

41. A. Foessler, Lecture given at XIth International School on Nuclear Physics Aug. 26-Sept. 4, Predcal, Romania (1976); Y.R. Waghmare: Introduction to Nuclear Physics, Oxford and IBH Publishing Co., New Delhi (1981).

42. J. Rainwater: Phy. Rev. 79, 432 (1950).

43. R.R. Roy and B.P. Nigam: Nuclear Physics, Theory and Experiment, New Age International (P) Ltd., New Delhi (1980).

    M.A. Preston: Physics of Nucleus, Addison-Wesley Publishing Co. Inc., Reading Massachusetts, London (1962).

44. A.S. Davydov and G.F. Fillipov: Nuclear Physics 8.237 (1958); Ibid, 12, 58 (1959); A.S. Davydov, Soviet Physics. JETP V. 36, p. 1103 (1959); zh. Eksper Leo. Fiz. 36, p. 1555 (1959).

45. De Mille et al.: Canad. J. Physics 37, 1036 (1959).

46. D. Rudolph et al. (17 authors): Phy. Rev. C.V. 56, p. 98 (1997).

47. F.R. Epinoza-Quinones, et al. (10 authors): Phy. Rev. C. 55, p. 1548 (1997)

48. A. Gader et al. (31 authors): Phy. Rev. C. 55, Rl (1997).

49. D. Bazzco et al. (15 authors): Phy. Rev. C.V. 58, 2002 (1998).

50. I. Ragnarsson, S.S. Nilsson and R.K. Sheline, Phy. Reports, 45, 1 (1978); P. Bonche et al.: Nuclear Phy. A443, 39 (1985), A Petrovici, K.W. Schmid and A Faessler: Nuclear Physics A 605, 290 (1996); H. Dejbakhsh et al.: Phy. Letter B, 249, 195 (1990).

51. R. Wyss, J. Nyberg, A. Johnson, R. Bengtsson and W. Nazarewitz.: Phy. Rev. letter B, 195, 53 (1987).

52. D.G. Sarantities et al. (15 authors): Phy. Rev. C.R. 1 (1998); C. Svensson et al.: Phy. Rev. leter 79, 1223 (1997).

53. C. Baktash, B. Hass and W. Nazarwitz, Annual Rev. Nuclear, Particle Sci. 45, 484 (1995); C. Schuck et al. (19 authors): Phy. Rev. C.R. 1667 (1997); D.P. Mc Nabb et al. (22 authors): Phy. Rev. C. 55, 148 (1997); D.F. Winchell, L. Welner, J.X. Saladin, M.S. Kaplan, E. Landulpho and A. Apsohasmian: Phy. Rev. C 55, 111 (1997).

54. M. Kaci et al. (24 authors): Phy. Rev. C. 565, R. 600 (1997).

55. G. Baldsiefen et al.: Nuclear Physics A 574, 521 (1994), M. Neffgev et al.: Nuclear Physics A 595, 499 (1995).

56. S. Frauendorf: Nuclear Physics A., 557, 2590 (1993).

57. A. Bohr, B.R. Mottleson: Nuclear Spectroscopy, Part B, ed . F. Ajzenberg-selove, Acad. Press, New York (1960), p. 1029.

    R. McHerson and J.C. Hardy: Cand. J. Physics 43, 1 (1965).

58. L. Cohen Bernard: Concepts of Nuclear Physics, McGraw-Hill Company, New York (1971).

    Table of Isotopes: Michael Lederer, M. Hollander, Isadore Perlman, John Wiley & Sons, Inc. (1967).

59. Rayleigh Lord, The Theory of Sound; V.2, ectro 364, McMillan and Co., London (1977).

60. A. K. Bohr, Danake Vidensk.: Selsk, Mat-fys. Medd, 26, No. 14 (1952); A. Bohr and B. Mottleson: Ibid 27, No. (16) (1953).

61. M.K. Pal: Theory of Nuclear Structure, Affiliated East-West Press Pvt. Ltd., New Delhi and Chennai, India. (1982); pp. 383-402., H. Goldstein: Classical Mechanics, Addison-Wesley Publishing Co., p. 134 (1959).

62. Same as References (38) and (43) (1980).

63. Reference (38), A. Bohr, K. Danske: Vidensk. Selsk. Mat. Fgs. Medd. 26, No. 14 (1952).

64. Same as Ref. (41): Also Ref. (45) and Ref. (38).

65. M.A. Preston: Physics of the Nucleus Addison-Wesley Publishing Co. Inc. Reading, Massachusettes, London, 247 (1962).

66. R.K. Sheline: Rev. Mod. Phy. 32, 1 (1960).

67. D.P. Grechekhin: Nuclear Physics 40, p. 422 (1963).

68. J.P. Davidson and M.G. Davidson: Phy. Rev. 138 B, 316 (1965).

69. A. Bohr and B.R. Mottleson: Dan. Mat-Fgs. Medd. 27, No.16 (1953); Ibid, Nuclear Spectroscopy Part B, edited by F. Ajzenberg: Salove, Academic Press, New York (1960), p. 1028.

70. F.K. Mc Gowan, and P.H. Stelson: Phy. Rev. 122, p. 1274 (1961).

71. K. Alder and A. Winther: Phy. Rev. 91, p. 1578 (1953).

72. R. Graetzer and E.H. Bernstein: Phy. Rev. 129, 1772 (1963).

73. K. Alder and A. Winther: Dan Mat-Fys. Medd 32, No. 8 (1960); Alder K. Proc. Conference Reactions complex Nuclei 3d, California, p. 253, University of California Press, Berkley (1963).

74. J. Meharter Vehn: Nuclear Physics A249, 111, 141 (1975)., H. Toki and A. Faessler: Nuclear Physics A253, 231 (1975); Phys. Letter 59B, 211 (1975).

75. S. Chandershekhar, Ellipsoidal Figures of Equilibrium, Yale Press, Yale, Connecticut (1969); 'Johnson A. and Z. Szymanski: Phy. Rev. 7, 181 (1973); G.B. Broda R. Hage-mann, B. Herskind, M. Ishihara, S. Ogaza, and H. Ryde: Nuclear Physics A 245, 166 (1972).

76. V.M. Strutinsky: Nuclear Phy. A95, 420 (1967); yad-Fiz. 3-614 (1966); Transl. Sov. J. Nucl. Physic, 3.449 (1966).

77. G. Anderson, S.E. Larsson, G. Leander, P. Moller, S.G. Nilsson, I. Rangarsson, S. Berg, R. Bengtsson, J. Dudek, B. Nerlo-Pomoska, K. Pomorski and Z. Szymariski: Nuclear Phy. A 268, 205 (1976).

78. K. Neergard, V.V. Pashkavich and S. Frauendorf: Nuclear Phy. A.262, 61 (1976); R. Bengtsson, & S. Franaudorf, N. Physics A. 314, 27 (1979); Ibid, N. Physics, 327, 139 (1979).

79. W.D. Myers and W.J. Swiateki, Ark. Fys. 36, 343 (1967).

80. R.M. Diamond, and F.J. Stephons: Ann. Rev. Nuclear. Part, Science, 85 (1980).

81. J.A. Pinston, R. Bengtsson, E. Monnand, F. Schussler and D. Barneond: N. Physics. A. 361, 464 (1981).

82. M.S. Kaplan, J.X. Saladim; L. Taro, D.F. Winchell, H. Takail, C.N. Knatt: Phy. Letters, B215, 251 (1988), Ibid Phy. Rev. 40, 2672 (1989); Ibid, Phy. Rev. C.55, 111 (1997); Ibid Phy. Rev. C.54 p. 626 (1996).

83. J.W. Holcomb, J.D. Johnson, P.C. Womble, S.L. Tabor, F.E. Dursham and S.G. Bucino: Phy. Rev. C 43, 470 (1991); S.D. Paul, H.C. Jain, S. Chattopadhyay, M.L. Shingran and J.A. Sheikh: Phy. Rev. C51, 2959, (1995).

84. Mollen and J.R. Nix: Nuclear Physics, A 361, 117 (1981).

PROBLEMS

1. Assume the following energy spin sequence.

(*i*) U^{234}	E	G.S	0.044	0.143	0.296	0.499		
	J^π	0^+	2^+	4^+	6^+	8^+		
(*ii*) W^{72}	E	G.S	0.1229	0.377	0.727	1.147	1.616	2.129
	J^π	0^+	2^+	4^+	6^+	8^+	10^+	12^+
(*iii*) Hf^{177}	E	G.S.	0.249	0.5913	1.0177			
	J^π	$7^-/2$	$11^-/2$	$15^-/2$	$19^-/2$			
(*iv*) Pu^{239}	E	G.S.	0.0078	0.057	0.076	0.164		
	J^π	$1/2^+$	$3^+/2$	$5^+/2$	$7^+/2$	$9/2^+$		

Fit these energies and spins to the systematics they follow and describe the nuclear structure, to which they belong.

2. A nucleus of $A = 180$, is ellipsoidal in shape, with $\beta = 0.15$; $\gamma = 15°$. What are the lengths of three principle axes ?

3. Consult Reference (7) and using, $\mathscr{I}_{\text{rigid}}$ and $C = 5 \times 10^6$ (keV)^3, plot $\mathscr{I}$ versus $I(I + 1)$, form Eq. 11.10, and compare with experimental results of cases in Problem (1). Explain the difference, between the experimental results and theoretical results.

4. Derive Eq. (11.82), from the original definition of Eqs. 11.80 and 11.45.

5. Show that, if the body axes are the principal axes; the products of inertia are zero in the expansion of nuclear surface *i.e.*,

$$R = R_0\left[1 + \sum_{\mu} a_{2\mu}\, Y_2^{\mu}(\theta, \phi)\right]$$

and hence

$$a_{2,1} = a_{2,-1} \text{ and } a_{2,2} = a_{2,-2}$$

6. Show that, if

$$T = \sum_{\mu\nu} G_{\mu\nu}\, \dot{\beta}_\mu \dot{\beta}_\nu$$

and
$$H\psi = (T + V)\,\psi = \left(-\frac{1}{2}\hbar^2 \sum_{\mu\nu} |G|^{-1/2} \frac{\partial}{\partial\beta_\mu} |G|^{1/2}(G^{-1}) \times \frac{\partial}{\partial\beta_\nu} + V\right)$$

(*i*) Prove that the wave-function ψ is separable in β-coordinates, *i.e.*,

$$\psi(\beta, \gamma, \theta_1, \theta_2, \theta_3) = f(\beta)\,\phi(\gamma, \theta_1, \theta_2, \theta_3)$$

(*ii*) Derive the equation for $f(\beta)$ and prove that:

$$\left(-\frac{\hbar^2}{2B}\frac{1}{\beta^2}\frac{d}{d\beta}\beta^4\frac{\alpha}{\alpha\beta} + \frac{1}{2}C\beta^2 + \frac{\lambda\hbar^2}{2B}\frac{1}{\beta^2}\right) f(\beta) = Ef(\beta)$$

where
$$|G| = 4\, B^5\, \beta^8 \sin^2\theta_2 \sin^2 3\gamma.$$

7. If forces between nucleons are taken to be charge-symmetric; but not charge-independent, what relationship would be expected between states of Li^7 and Be^7 or between C^{14}, N^{14} and O^{14}?

8. In Fig. 11.12, allot the nature of each level (rotational or vibrational) in all nuclei. What is the significance of negative parity levels appearing for the same K, for many cases?

9. Prove Eq. 11.31.

Given $\langle I M K | I_{\pm} | I M K \pm 1 \rangle = \sqrt{(I \mp K)(I \pm K + 1)}$

and $(j \Omega | j^{\mp} | j, \Omega \pm 1) = \sqrt{(j \mp \Omega)(j \pm \Omega + 1)}$

keeping in mind that I corresponds to space-axes and j to the body axes.

10. What will be the configuration of K^{41} states, which are isobaric analogues of the low lying states of $_{18}A^{41}_{23}$?

12

CHAPTER

Particle States and Collective Motion in Nuclei

12.1 PARTICLE STATES IN NON-SPHERICAL NUCLEI

We have seen in the previous chapter, that there are many nuclei, which have spheroidal shape in their ground states. The rotational and vibrational mode of excitation for such nuclei has been discussed. If we combine the two collective modes, *i.e.* vibrational and rotational, then each rotational band rides over a vibrational mode. We have discussed this in the previous chapter and will again discuss in the next section. But over and above these collective motions, individual nucleons can be excited to higher states, without inducing collective modes of excitation. The energies of these excitations, are determined by the energy levels available to the nucleons moving in a non-spherical nucleus. The average shell model potential, for such non-spherical nuclei, was first written by Nilsson[1], and has the shape of an oscillator potential with axial symmetry, and a strong spin-orbit coupling, *i.e.*,

$$V(r) = \frac{1}{2} M [\omega^2 (x^2 + y^2) + \omega_z^2 z^2] + C\,\mathbf{l} \cdot \mathbf{s} + Dl^2 \qquad \text{...(12.1)}$$

where $$\omega^2 = \omega_x^2 = \omega_y^2 = \omega_o^2 (\delta) \left(1 + \frac{2\delta}{3}\right); \ \omega_z^2 = \left(1 - \frac{4}{3}\delta\right) \omega_o^2 (\delta)$$

$\delta \equiv \Delta R / R$ (the deformation parameter) and C, D, and ω_o are constants. Apparently, the potential strength is dependent on δ. Many times symbols like ω_1, ω_2 and ω_3 are used for angular velocities with the relations $\omega_1 \equiv \omega_X$, $\omega_2 \equiv \omega_Y$ and $\omega_3 \equiv \omega_Z$. While spin-orbit coupling term $C\,\mathbf{l} \cdot \mathbf{s}$ is expected from shell model considerations; the term Dl^2 was introduced by Nilsson, in order to obtain detailed characteristics of heavy, strongly deformed nuclei. It may be realised that ω_i is the angular velocity of the single particle around the axis represented by i. The energy levels of a single particle in a deformed potential of Eq. 12.1 are given in Fig. 12.1. Evidently they involve new quantum numbers N, 1, Λ and Σ oscillator quanta where N corresponds to the total numbers of oscillator quanta, while 1, Λ and Σ are the quantum numbers, corresponding to operators l, l_z and s_z respectively. It may be pointed out here that:

$$\omega_x\, \omega_y\, \omega_z = \text{Constant} \qquad \text{...(12.2)}$$

which is the condition of the constant volume of the nucleus and $\omega_o(\delta)$ is the angular velocity associated with a nucleus of deformations δ. We now write:

$$\omega_z = \omega_o(\delta)\left(1-\frac{4}{3}\delta\right)^{1/2}$$

$$\omega_x = \omega_y = \omega_\perp = \omega_o(\delta)\left(1+\frac{2}{3}\delta\right)^{1/2} \quad ...(12.3a)$$

Hence Eq. 12.2 yields

$$\omega_o^3(\delta)\left(1+\frac{2}{3}\delta\right)\left(1-\frac{4}{3}\delta\right)^{1/2} = \text{Constant} = (\tilde{\omega}_o)^3$$

so that

$$\omega_o(\delta) = \tilde{\omega}_o\left[\left(1+\frac{2}{3}\delta\right)^2\left(1-\frac{4}{3}\delta\right)\right]^{-1/6}$$

$$= \tilde{\omega}_o\left[1-\frac{4}{3}\delta^2-\frac{16}{27}\delta^3\right]^{-1/6} \quad ...(12.3b)$$

The quantity $\tilde{\omega}_o$ is related to the size of the nucleus. The relation comes out to be $\tilde{\omega}_o \,\alpha A^{-1/3}$ [Eq. 10.24]. Smaller the size of the nucleus, the larger is the value of the angular velocity.

The quantities C and D in Eq. 12.1 are to reproduce the energies of the shell model for spherical nuclei. The value of C is assumed to be dependent on 'N' the neutron number[3]. Two parameters $K = -1/2\ C/\hbar\omega_o(o)$ and $\mu = 2\,D/C$, have been evaluated by Nilsson[1], Bhatt and Bishop[6], giving $K = 0.7$ to 0.1 and $\mu \approx 0$ to 0.33 and 0.5.

The quantity $\tilde{\omega}_o = \omega_o(o)$ corresponds to oscillator frequency of a sphere, with $\delta = 0$.

To understand, the significance of the quantum numbers, $N\ 1\ \Lambda\ \Sigma$ in depth: we write the Hamiltonian from Eq. 12.1 as:

$$H = H_o + H_\delta + C\,\mathbf{l}\cdot\mathbf{s} + Dl^2 \quad ...(12.4)$$

where

$$H_o = -\frac{1}{2}\hbar\,\omega_o\nabla^2 + \frac{1}{2}M\omega_o^2 r^2 \quad ...(12.5)$$

and

$$H_\delta = -\frac{4}{3}\sqrt{\frac{\pi}{5}}\,\delta\,\omega_o r^2 Y_{2o} \quad ...(12.6)$$

In writing the above, we have assumed that

$$r = r_o(1+\beta Y_{2,0}) \text{ and } \beta = \frac{4}{3}\sqrt{\frac{\pi}{5}}\,\delta \quad ...(12.7)$$

It leads to $r = r_o(1+\beta^2/4\pi)$, if we assume the volume of ellipsoid to be the same as that of the sphere with the radius r_o.

The eigenvalues of H_o are obtained by the representation in which H_o is diagonal together with operators l^2, l_z and s_z, which all commute with H_o. But these do not commute, with the total Hamiltonian H, with which only $I_z = l_z + s_z$ commutes. Denoting eigenvalues of l_z by Λ and S_z by Σ, as mentioned earlier, the eigenvalues equation, for the unperturbed Hamiltonian is given by:

$$H_o \mid N 1 \Lambda \Sigma \rangle = \left(N + \frac{3}{2} \right) \hbar \omega_o (\delta) \mid N 1 \Lambda \Sigma \rangle \qquad ...(12.8)$$

In other words, energy eigenvalues for H_o are given by:

$$E_N = \left(N + \frac{3}{2} \right) \hbar \omega_o (\delta) \qquad ...(12.9a)$$

These eigenvalues correspond to vibrational motion only, as mentioned in the previous chapter. For the total Hamiltonian H, the energy eigenvalues which are obtained after proper diagonalization, are:

$$E_{N\Omega} = \left(N + \frac{3}{2} \right) \hbar \omega_o (\delta) + k \hbar \omega_o^o \, r^{N\Omega} \qquad ...(12.9b)$$

We now, describe how the quantum numbers, N, n_3, Λ and Σ, and $\Omega = \Lambda + \Sigma$ and spin and parity (Ω^π) are assigned to single particle scheme obtained in Nilsson Mode [Ref. (1)] for the Hamiltonian H, in Eq. 12.4.

In this model, $I_z = \Omega$ is a good quantum number, because we are assuming an axial symmetry for spheroidal potential. In Fig. 12.1, each state is represented by the quantum number: $I_z = \Omega$ and parity π and the triad $[N\, n_3\, \Lambda]$.

Each of the five numbers are constants of motion for the state, I_z or Ω is a constant for a symmetrical ellipsoid. Parity (π) of course, is constant for any nuclear system, $N = n_1 + n_2 + n_3$ is the total number of oscillator quanta, and is a sum of three quantum numbers; which are oscillator numbers along three body axes. The quantum number n_3 is not a constant of motion in the spherical representation, but it becomes very nearly so for very large distortions, for spheroidal nuclei, with an axis of symmetry. The energies of levels are a function of the distortion δ, which is connected to three distortions, for spheroidal nuclei:

$$\delta_1 = \delta_2 = \frac{1}{3}\delta = \frac{\Delta x}{x} = \frac{\Delta y}{y} \qquad ...(12.10a)$$

and

$$\delta_3 = -\frac{2}{3}\delta = \frac{\Delta z}{z} \qquad ...(12.10b)$$

so that

$$\delta = \frac{\Delta R}{R} \doteq \sqrt{\frac{(\Delta x)^2 + (\Delta y)^2 + (\Delta z)^2}{x^2 + y^2 + z^2}} \qquad ...(12.10c)$$

The value of δ is also related to β as [Eq. 12.7]:

$$\delta = \frac{3}{2}\sqrt{\frac{5}{4\pi}}\,\beta = 0.95\,\beta \qquad ...(12.10d)$$

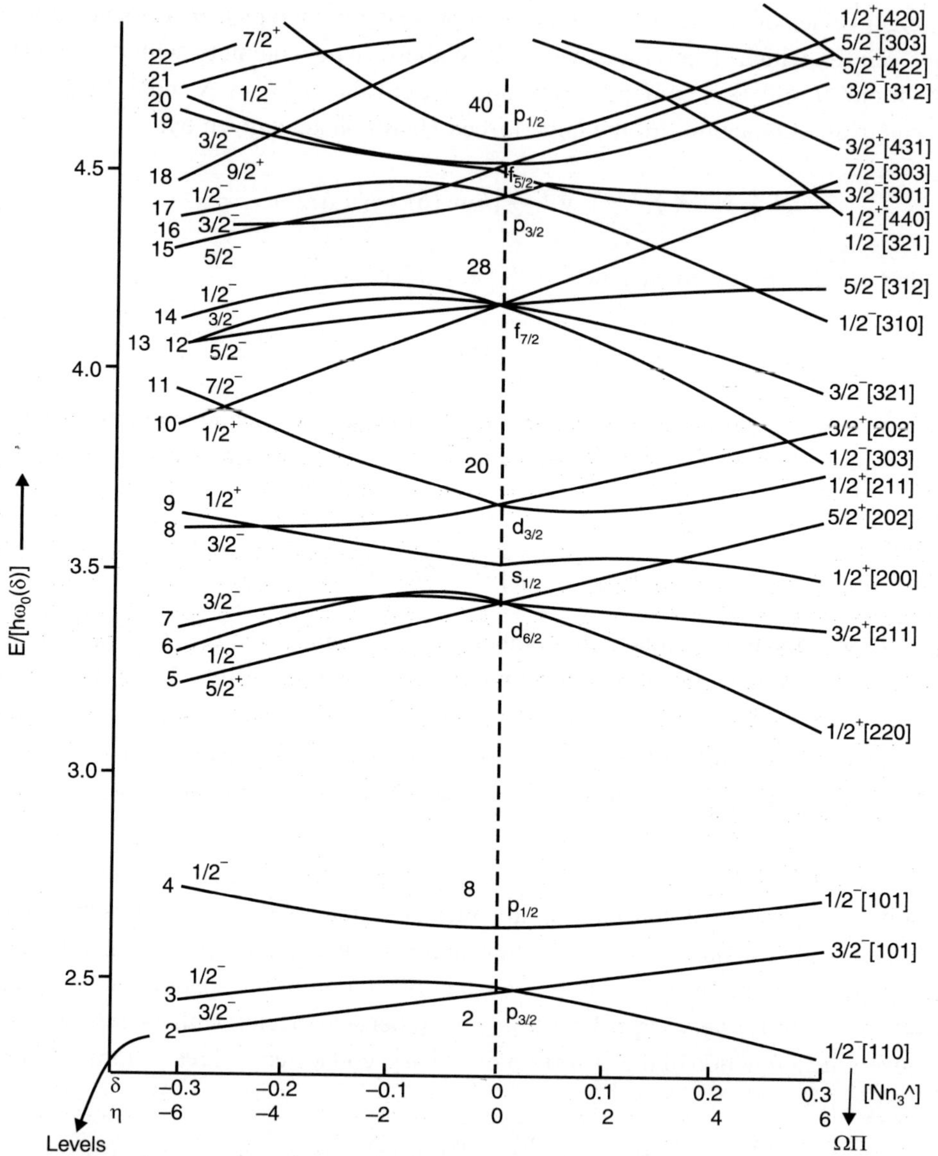

Fig. 12.1a Nilsson levels calculated as a function of deformation parameters and for the first few major shells, for magic numbers 8 to 40 (Ref. 4).

For each orbital, [as developed already for shell model for spherical potential, Fig. 10.5], the energy is plotted in Fig. 12.1 as a function of the distortion, δ or η of the nucleus from spherical shape to a spheroidal shape where the quantity δ is related to η as:

$$\eta = \frac{2\delta\hbar\omega_o(\delta)}{C} \qquad ...(12.11)$$

where C is the same constant, which appears in front of $\mathbf{l}\cdot\mathbf{s}$ term in Eq. 12.1 or Eq. 12.4. The Nilsson's wave-function ψ_α which should be eigen-function of Hamiltonian H of Eq. 12.4, with eigenvalues E_α can be written as:

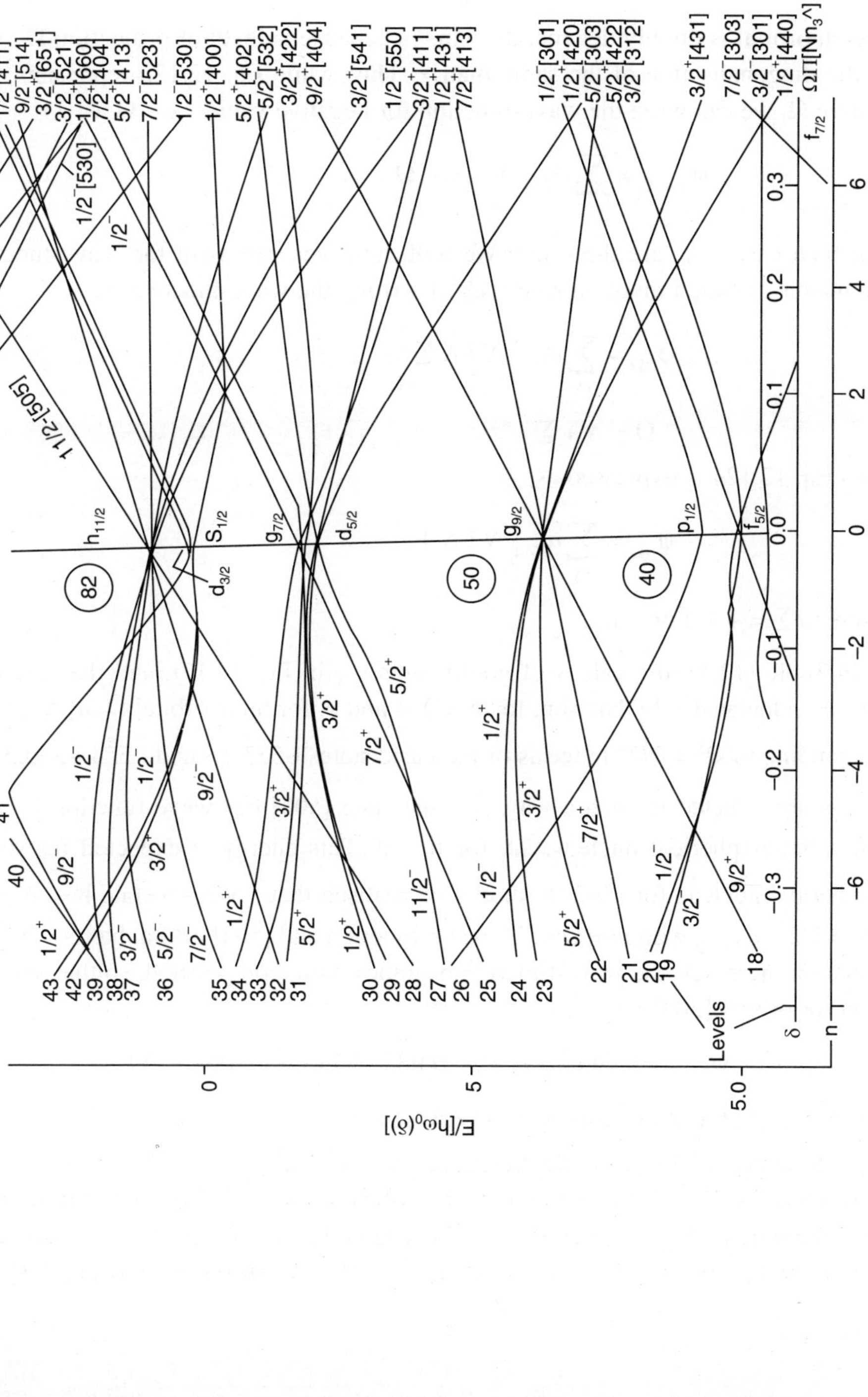

Fig. 12.1b Nilsson level diagram as a function of deformation parameter for shells between magic numbers 40 and 82 (Ref. 4).

$$H\,\psi_\alpha = E_\alpha\,\psi_\alpha \qquad \text{...(12.12}a\text{)}$$

where ψ_α were written by Nilsson in terms of the expansion coefficients such that; (α represents any quantum numbers, beyond the ones specified below):

$$\psi_{\alpha\Omega} = \sum_{l\Lambda} a_{l\Lambda}\,|\,N\,l\,\Lambda\,\Omega\,\rangle \qquad \text{...(12.12}b\text{)}$$

Each Nilsson orbital is doubly degenerate. This is because, a level of projection Ω and that of $-\Omega$ result from the diagonalisation of the same matrix. Thus while Eq. 12.12*b* represents the wave-function for positive Ω, we can write the wave-function for negative value, *i.e.*– Ω as:

$$\psi_{\alpha,-\Omega} = \sum_{l\Lambda} a_{l\Lambda}\,|\,N\,l\,\Lambda, -\Omega\,\rangle \qquad \text{...(12.12}c\text{)}$$

So that in both cases, $a_{l\Lambda}$ are the same. We will, however, represent the wave-function as in Eq. 12.12*b*. As a matter of fact, a more popular way of writing the wave-function is:

$$\psi_{\alpha\Omega} = \sum_{l\Lambda} a_{l\Lambda}\,|\,N\,l\,\Lambda\,\Sigma\,\rangle \qquad \text{...(12.12}d\text{)}$$

where

$$\Omega = \Lambda + \Sigma \qquad \text{...(12.12}e\text{)}$$

Sometimes Eq. 12.12*d* is expressed as,

$$\psi_{\alpha\Omega} = \sum_{l\Lambda} a_{l\Lambda}\,|\,N\,l\,\Lambda\,\pm\,\rangle \qquad \text{...(12.12}f\text{)}$$

where ± sign refers to $\Sigma = +\,1/2$ or $-\,1/2$.

We give in Table (12.1), the values of coefficients $a_{l\Lambda}$ in Eq. 12.12*f*, and the eigenvalues of energy E_α (MeV), as calculated[1,5] by Nilsson, for $\eta = 0$, 4 and 6, for three orbital ($N\,n_3\,\Lambda$) [503], [514] and [523], corresponding to $\Omega^\pi = 7/2^-$ in terms of the basic states $|\,533\,+\rangle$ and $|\,554\,-\rangle$ and $|\,553\,+\rangle$. These basic vectors are selected in such a way, that the state [503] has wave-function $|\,533\,+\rangle$ and energy –11.4 MeV for a spherical nucleus, *i.e.* for $\eta = 0$. This energy is expected for shell model calculation for spherical nucleus for $f\,7/2$. It should be realised that $|\,533\,+\rangle$ state has, $N = 5$, $l = 3$, $\Lambda = m_l = 3$ and $\Sigma = 1/2$, *i.e.*, $f_{7/2}$ with $m = \Omega = 7/2$ wave-function and [503] orbital has $N = 5$, $n_3 = 0$ and $\Lambda = 3$. As deformation increases, the wave-function becomes more complicated, so that say for $\eta = 4$; the wave-function for orbital [503] is:

$$\psi_{5,\,7/2} = 0.144\,|\,533+\rangle - 0.987\,|\,533+\rangle - 0.067\,|\,554-\rangle \qquad \text{...(12.13}a\text{)}$$

Subscripts 5 and 7/2 for ψ, correspond to $N = 5$, $\Omega = 7/2$.

It is still predominantly $|\,533\,+\rangle$ wave-function; but now, the only valid quantum numbers are N and $\Omega = \Lambda + \Sigma$, which is 3 + 1/2 = 7/2 in the first two terms and 4 – 1/2 = 7/2 in the last term. As $\eta \to \infty$, the last term corresponding to $\Lambda = 4$, which is never very large for the state [503], tends to be zero and one is left only with the first two terms, which describe $n_3 = 0$ state. Hence it leads to [503].

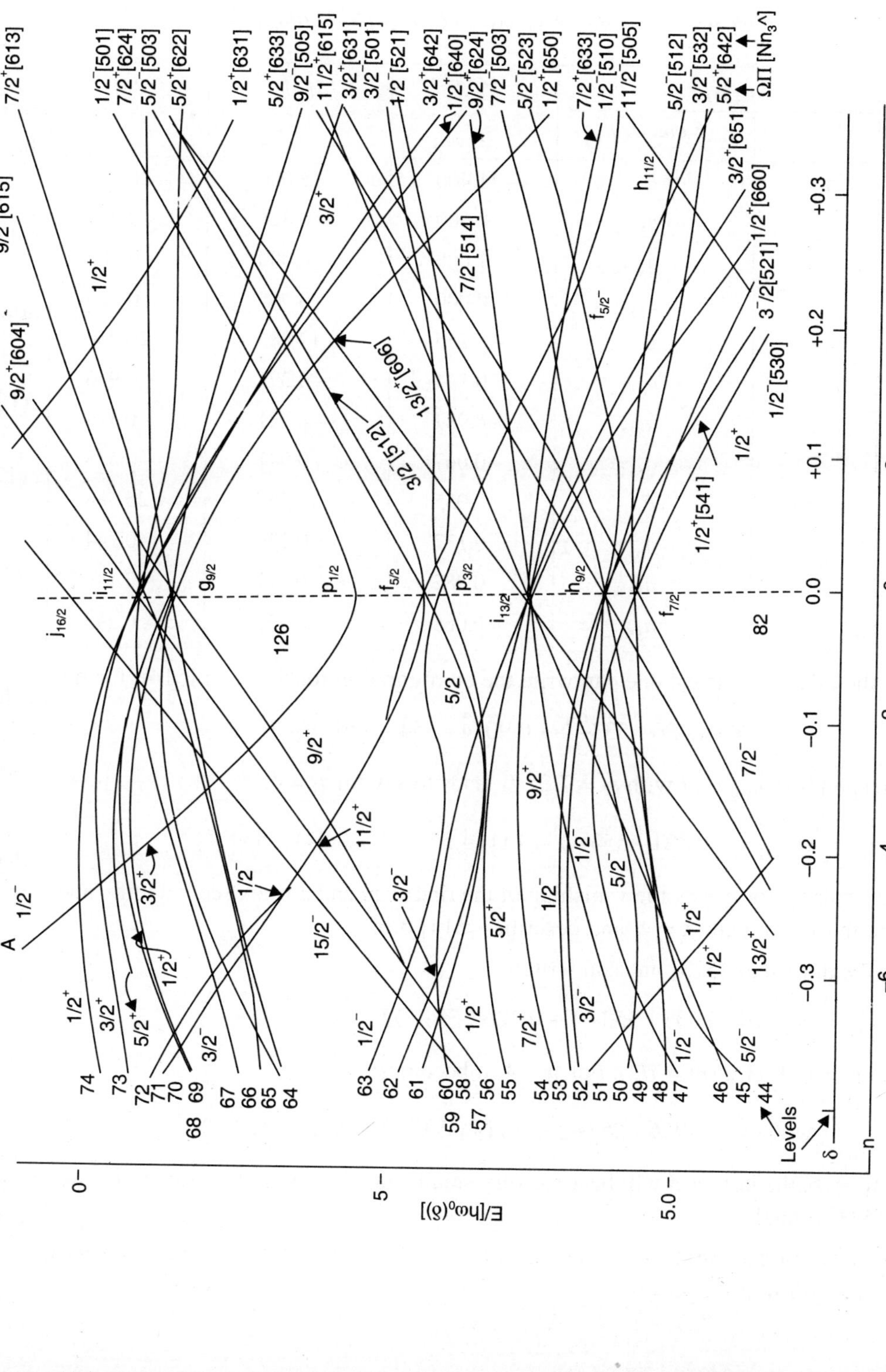

Fig. 12.1c Nilsson levels, as a calculated as a function of deformation, for shells between magic numbers 82 and 126, and beyond (Ref. 4).

Table 12.1 *The values of $a_{l\Lambda}$ and E_{α} for Nilsson wave-functions, corresponding to $N = 5$, $\Omega^{\pi} = 7/2^{-}$ with basic vectors $|\,553+\rangle$, $|\,533+\rangle$, $|\,554-\rangle$ for $\eta = 0$, 4 and 6 representing $|\,N\,l\,\Lambda\,\Sigma\rangle$. The first number in the first row in each state gives energies E in units of $\hbar\omega_o$ and other three rows give the values of $a_{N\,l\,\Lambda\,\pm}$ of Eq. 12.12f for three basic vectors for three states [503], [514] and [523].*

[Nn₃ Λ]	*E and* $a_{N\,l\,\Lambda\,\pm}$	*[503]*	*[514]*	*[523]*
$\eta = 0$	E	– 11.400	– 15.000	– 26.000
	a_{533+}	0.000	– 0.426	0.905
	a_{533+}	1.000	0.000	0.000
	a_{554-}	0.000	0.905	0.426
$\eta = 4$	E	– 5.255	– 12.885	– 26.260
	a_{553+}	0.144	– 0.321	– 0.936
	a_{533+}	– 0.987	– 0.111	– 0.114
	a_{554-}	– 0.067	0.941	– 0.333
$\eta = 6$	E	– 2.055	– 11.757	– 26.588
	a_{553+}	0.177	– 0.279	– 0.944
	a_{533+}	– 0.982	– 0.117	– 0.149
	a_{554-}	– 0.068	0.953	– 0.295

For the other two states, one can write the configuration for the state [514], for $\eta = 0$.

$$\psi_{5,\,7/2} = -\,0.426\,|\,533+\,\rangle + 0.9054\,|\,554-\,\rangle \text{ for } \eta = 0 \qquad \text{...(12.13}b\text{)}$$

which corresponds to $h_{9/2}$ (but with $m = 7/2$). It leads to say for $\eta = 4$,

$$\psi_{5,\,7/2} = -\,0.321\,|\,553+\,\rangle - 0.111\,|\,533+\,\rangle + 0.941\,|\,554-\,\rangle \qquad \text{...(12.14)}$$

This leads to, for $\eta = \infty$, the vanishing of the first term and a week second term, and corresponds to dominant third term with $\Lambda = 4$, and describes [514] orbit.

For the third state [523] one can write:

$$\psi_{5,\,7/2} = 0.905\,|\,553+\,\rangle + 0.426\,|\,554-\,\rangle \text{ for } \eta = 0$$

and corresponds to h 11/2 ($m = 7/2$). For $\eta = 4$, it becomes:

$$\psi_{5,\,7/2} = -\,0.936\,|\,553+\,\rangle - 0.114\,|\,533+\,\rangle + 0.333\,|\,554-\,\rangle \qquad \text{...(12.15)}$$

As $\eta \to \infty$, the last term will become very small, leaving behind first two terms with $\Lambda = 3$, and describes [523] orbital.

It may be pointed out, that for these orbitals, in Figs. 12.1*a* to 12.1*c*, one can determine N, n_3, Λ and Ω and π as follows:

1. N are the harmonic oscillator numbers, as used for spherical potential. Same values of N are allotted to Nilsson orbitals for $\eta = 0$ and then value stays the same, for different values of η.

2. From the fact, that for positive deformation ($+\eta$); the value of ω_3 is less than ω_1 and also the δ-dependence term in ω_3 and ω_2 is negative and in ω_1 and ω_2 it is positive [*see* Eqs. 12.1 and 12.3]; it is evident that $n_3\,\omega_3\,(\delta)$ term should be minimum to have the maximum energy. Hence, for different orbitals emanating from a configuration for spherical potential, (*i.e.*, $\eta = 0$); the orbital corresponding to highest energy will have $n_3 = 0$. As the energy of the orbitals for positive δ or η decrease for other Ω^{π} states, the value of n_3 increases for the same N, for these states.

3. The value of Ω for each orbital can be obtained for a given configuration say $f\,7/2$; by realising that the highest energy orbital has $m_j = \Omega = 7/2$; the next lower energy orbital has $m_j = \Omega = 5/2$ and so on till lowest energy orbital has $\Omega = 1/2$. The parity of all the orbitals, belonging to the same shell model configuration is the same, *e.g.* for $f\,7/2$, it is –ve; because for $f\,7/2$, $1 = 3$, and parity is given by $(-1)^3 = -1$.

4. The value of Λ is obtained from the relationship:

$$\Omega = \Lambda + \Sigma \qquad ...(12.16)$$

The value of Σ for each orbital is given by the manner, in which it enters the total angular momentum of the shell model configuration. As for example, in $f\,7/2$ configuration, the value of 7/2 can be taken as maximum of m_j; the value of $1 = 3$ for $f\,7/2$, may be taken as the maximum of m_l then $\Sigma \equiv m_s$ is given by:

$$m_j = m_l + m_s = \Lambda + \Sigma;\ \frac{7}{2} = 3 + \frac{1}{2}$$

So for all orbits of $f\,7/2$; Σ is +1/2. This is expressed as +sign in the expression of basic vector; $a_{l\Lambda}$ $|\,N\,l\,\Lambda+\rangle$ as discussed earlier. Similarly for $f5/2$; $\Sigma = -1/2$ and the basic vector is written as $a_{l\Lambda}\,|\,N\,l\,\Lambda-\rangle$. Of course, there can be contributions from higher values of say $1 = 4$ or 5, then for the same $\Omega = m_j$; there could be different values of Λ and Σ. Another method of determining Λ is based on the fact that the oscillator motion is such that Λ is even or odd as $N - n_3$ is even or odd. This determines Λ out of the two choices from Eq. 12.12*e*, for $\Sigma = \pm 1/2$.

It is interesting to the note from Fig. 12.1, that, while for extreme values of Ω, the orbitals are very asymmetrical for $\pm\,\delta$; for intermediate value of Ω they are comparatively less asymmetrical. This may be understood, if one realises that $+\delta$ corresponds to prolate state and $-\delta$ to oblate shape and extreme values of Ω corresponds to the direction of the angular momentum along the axis of symmetry. So naturally the energies will be quite different in the two cases, and extreme asymmetry in energy results.

Besides Nilsson tables for the wave-function, other authors[6] have used alternative quantum numbers, especially Rassey and Newton[7] have prepared tables similar to those of Nilsson; but using the basis vectors of $|\,N\,n_3\,\Lambda\,\rangle$, instead of $N\,l\,\Lambda\,\Sigma\,\rangle$ of Nilsson and also using various values of β and γ. Without going into details, we may discuss some of the results of Nilsson and Rassey (6, 7). They have calculated the equilibrium shapes of the deformed nuclei, determining the total energy of the average single particle in a deformed potential for various values of β and γ and determine those values which minimise the energy. The calculations have many uncertainties, but they give an average picture of the distortion of the potential. Deformation for nuclei, under equilibrium condition, have been calculated for many nuclides by Mottleson and Nilsson[i].

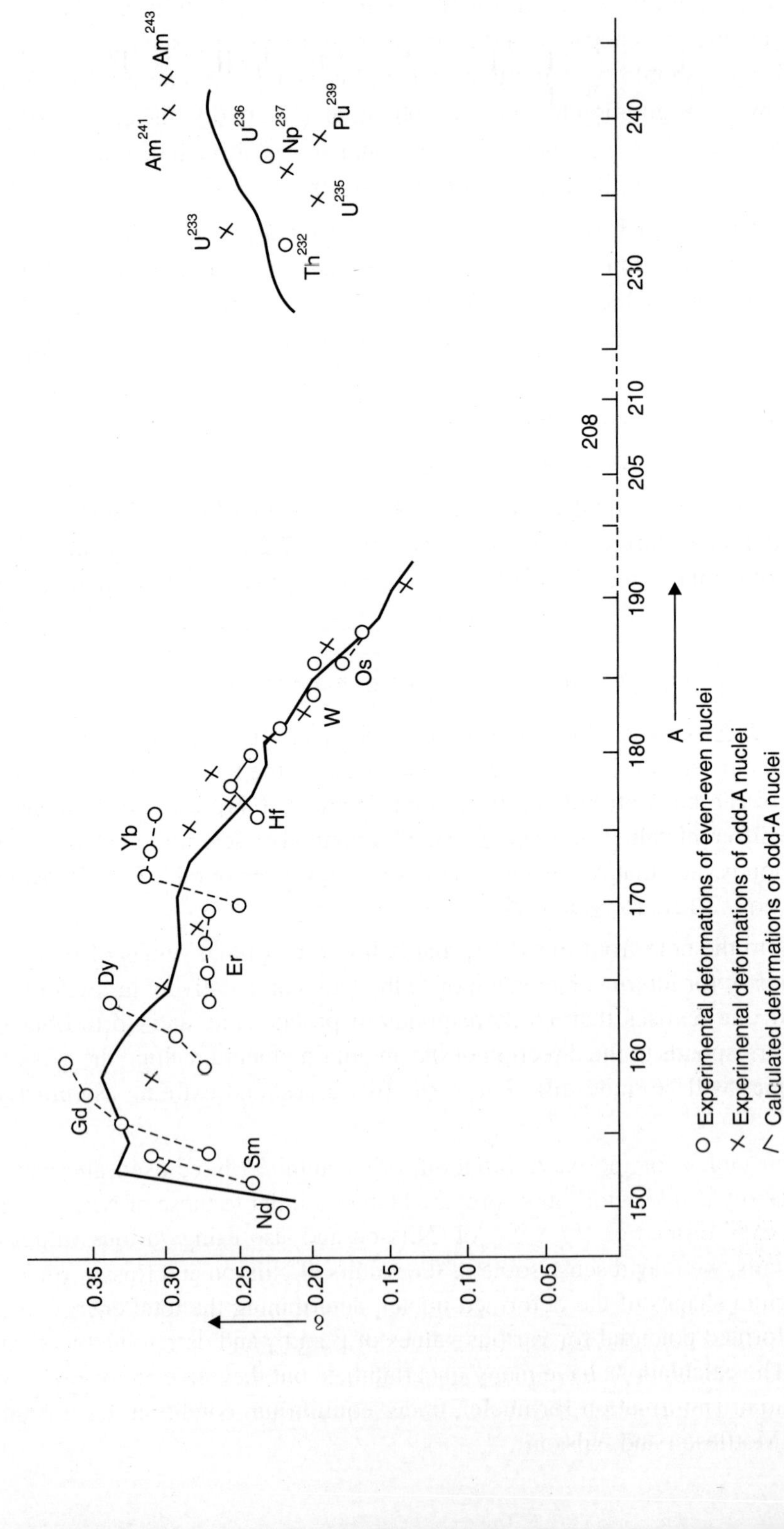

Fig. 12.2 Comparison of equilibrium deformation δ of rare earth nuclei, obtained from Nilsson model with the measured values (Ref. 1).

These results are shown in Fig. 12.2, where the nuclear deformation is plotted as a function of A. Experimental values follow the general trend of theoretical solid line.

12.2 UNIFIED MODEL (COUPLING OF PARTICLE AND COLLECTIVE MOTION)

In the last two chapters we have discussed (*i*) the motion of individual particles in a spherical nuclear potential, giving rise to shell model; (*ii*) the collective motions of the nuclei, *i.e.* rotational and vibrational, and finally in this chapter we consider (*iii*) the motion of individual particles in spheroidal nuclei. In practice, however, one is likely to come across many nuclei, which when excited, will undergo both the collective modes of excitation as well as the particle excitation; in spherical or spheroidal nuclear potentials, depending on the nucleus under consideration.

We, therefore, write the Hamiltonian for such a general unified case. We have partly done so already, when we combined the rotational and particle motion in Eqs. 11.24, 11.27 and 11.31 to 11.38, for rotational case. We will now bring in the vibrational motion as well as the particle motion in deformed nuclei.

We write, the general Hamiltonian, containing both the collective and particle contributions as:

$$H = H_c + H_p \qquad ...(12.17)$$

where H_c represents collective motion and H_p, represents the particle motion. For a general case for spherical or spheroidal potential, we can write H_p as:

$$H_p = \sum_{i=1}^{k} [T + V(\beta, \gamma_i ; r_i, l_i, s_i)] \qquad ...(12.18)$$

where the average of body potential V is a function of the nuclear shape. Assuming that the potential at a point in the nucleus is closely related to the nuclear density, and hence equi-potentials are also the surface of constant nuclear density.

If we write:

$$r = r_o\, f(\beta, \gamma, \theta, \phi) \qquad ...(12.19)$$

then r_o represents a surface of constant nuclear density, and one can write:

$$V = V_o \left(\frac{r}{f(\beta, \gamma, \theta, \phi)}, \mathbf{l} \cdot \mathbf{s} \right) \qquad ...(12.20)$$

where $f(o, o, \theta, \phi) = 1$, and V_o $(\mathbf{r}, \mathbf{l}, \mathbf{s})$ is the shell model potential of a spherical nucleus. The exact expression for $V(\beta, \gamma, r_i, \mathbf{l_i}, \mathbf{s_i})$ will depend on the shape of the nucleus.

As mentioned earlier; the particle motion may be coupled to the collective motion in two ways, *i.e.* (*i*) 'Weak coupling' when β is small, and the changes in the potential shape are slow and small and particle excitations are much larger. Then the particles move in an effective spherical potential; which in itself may undergo small vibrational changes, *i.e.* the coupling between particle and collective motion is weak. This corresponds to vibrational motion and applies to even-even spherical nuclei, with very small number of particles as 'loose' nuclei. (*ii*) On the other hand, 'strong coupling' corresponds to the case, where a large number of 'loose' nucleons move around a closed shell 'core' and deform the core. In other words, their individual motion is strongly coupled to the core, which, then, may undergo a resultant rotational motion due to the collective effect of the 'loose' particles. We will treat these two cases separately.

12.2.1 Weak Coupling

As mentioned above, we can expand, in this case, the potential $V(r)$, around a spherical shape $V_o(r)$.

If $\alpha_{\lambda\mu}$ is small, the particle is, then, taken to move only in an average $V_o(r)$ potential. Then H_C may be written as described earlier [Eq. 11.69], as:

$$H_C = \sum_{\lambda\mu}\left(\frac{1}{2} B_\lambda |\dot{\alpha}_{\lambda\mu}|^2 + \frac{1}{2} C_\lambda |\alpha_{\lambda\mu}|^2\right) \qquad ...(12.21)$$

and H_p may be taken for zero-approximation only, around the spherical nucleus *i.e.* $H_p = H_p^o$ and may be written as:

$$H_p^o = \sum_i T_i + V_o(\mathbf{r_i}, \mathbf{l_i}, \mathbf{s_i}) \qquad ...(12.22)$$

Then under the assumption of weak coupling, one can write the total Hamiltonian H as:

$$H = H_C + H_p^o + H_{int} \qquad ...(12.23)$$

where H_{int} is the effect of the second term in Eq. 12.21 and may be written as:

$$H_{int} = -\sum_i k(r_i) \sum_{\lambda\mu} \alpha_{\lambda\mu} Y_\lambda^\mu (\theta_i, \phi_i) \qquad ...(12.24)$$

where $k(r_i)$ is the radial function of the ith nucleon, that determines the strength of coupling of the core with the nucleon. It is assumed to be, independent of the state of the nucleon, *i.e.* whether particle frequency is comparable or much larger than collective frequency. As a matter of fact, $k(r)$ is of the form rdV_o/dr and, therefore, is a surface phenomenon, since it is appreciable only in tail, where nuclear density is changing, *i.e.* rdV_o/dr is large. It can be written as:

$$k(r) = R_o V_o^{(o)} \delta(r - R_o) \qquad ...(12.25)$$

where V_o^o and R_o are the depth and range of the shell model potential which is taken as square well in shape. The matrix elements of $k(r)$ are obtained from the relation:

$$\langle n\,l \,|\, k(r) \,|\, n'\,l' \rangle = \int U_{nl}(r)\, k(r)\, U_{n'l'}(r)\, r^2\, dr$$

$$= V_o^o R_0^3\, U_{nl}(R_o)\, U_{n'l'}(R_o) \qquad ...(12.26)$$

Where $U_{nl}(R_o)$ and $U_{n'l'}(R_o)$ are the radial parts of wave-functions at $r = R_o$. Hence the energy of a state will be a sum of the vibrational energy, given by Eq. 12.21, plus particle energy, obtained from shell model calculation, say Eq. 12.4, plus interaction energy from Eqs. 12.24 and 12.26 of the motion of the particle and the vibrational mode. This interaction energy represented by the last term in Eq. 12.23 is roughly independent of n and 1 and is of the order of 40 MeV. It is generally treated as a constant.

It may be realised that the weak coupling requires that the parameter, which determines the coupling between particle and collective excitation is, essentially the ratio of the matrix of $k(r)$ to the vibrational quantity $(\hbar\omega_2 C_2)^{1/2}$. So the criterion for perturbation treatment for particle motion is that:

$$\frac{kr}{(\hbar\omega_2 C_2)^{1/2}} << 1 \qquad ...(12.27)$$

Weak coupling model is applicable to nuclei where one nucleon may be moving around even-even core. Then one can use the standard perturbation theory. States corresponding to H_C to H_p^o form the core states. The combined wave-function can be written, by coupling the angular momentum of the particle and the collective (R) angular momentum. Then H_{int} acts as a perturbation. The total wave-function is then written, after taking into account, H_{int}, *i.e.*,

$$\psi\,(J=j, M) = |\; j, o\, o, J\,M\rangle - \sum_{j'}\sum_{\lambda} |\; jl\;\Lambda, J\,M\rangle \times \frac{\langle\, nl\,|\,k\,|\,n'l'\,\rangle\langle\, j\,\|\,Y_1\,\|\,j'\,\rangle}{(\hbar\,\omega_o + E_{j'} - E_j)} \times \frac{(\hbar\omega)^{1/2}}{2C_\lambda} \qquad ...(12.28)$$

where j and j' belong to the wave-functions of the two neighbouring states, which are connected to each other by Y_l. In general $\Lambda = 2$ is dominant and then only states with $(j - j') \leq 2$ and with same parity are connected[8].

Complete diagonalization calculations with particle vibration coupling model have been carried out by many authors. The necessary formulas were first derived by Chowdhary[9] and later on by Ford and Levinson[9]. Later, Lande and Brown[10], and then Sen[11] et al. applied the model including both the quadrupole and octupole phonons. For odd mass isotopes of Sb, In, Sr, Kr, Zr, Te, Rb and Y; calculations have been especially carried out by Sen[11], who obtained not only the energy levels, but also the electromagnetic transition strengths and spectroscopic factors of various levels.

12.2.2 Strong Coupling

Under the assumption of strong coupling, between the particle motion and the collective motion, the expansion of the potential term in powers of β is no longer attempted but instead one solves the single particle Schrödinger equation:

$$H_p\,\psi_\alpha = [T + V\,(\beta, \gamma, \mathbf{r}, \mathbf{l}, \mathbf{s})]\;\psi_\alpha = E_\alpha\,(\beta, \gamma)\;\psi_\alpha \qquad ...(12.29)$$

where β and γ are parameters, to be determined from the solution of this equation and subscript α represents the various quantum numbers. One determines, for a given number of particles, the values of β and γ, so that the total energy is minimised. This is basically a case of self-consistency problem; which satisfies two inter-dependent conditions (*i*) the wave-functions ψ_α for a given β and γ, give a proper level density and (*ii*) such level density and particle motion in turn, yield the same potential, with which we started. This is accomplished by using the so-called restricted Hartree-Fock approach[12]. In this approach, one guesses which orbits have to be included, (by making a proper guess about the Hamiltonian in Eq. 12.29). Then one calculates the Hamiltonian matrix, which is diagonalised. The diagonalised matrix should yield the same parameter, say β or γ, with which we started. If not, one re-iterates the process. With a good guess, in a few reiterations one should be able to obtain a self-consistent result. It can be proved that, if deformations are admitted to the generation of nuclear Hartree-Fock solutions, then certain nuclei with a large number of nucleons outside a closed shell exhibit a preference for a permanently deformed nuclear state. In this case, one assumes an inert core and carriers out the Hartree-Fock procedure for those nucleons which are outside the core. Thus the deformation is carried only by the particles in the last non-closed shell. These methods are best suited for application in Sd shell nuclei, where large deformations are encountered but sufficiently few particles are involved to make microscopic calculations tractable.

It is, then, found that it is possible to separat collective energy and the single particle energies due to certain extra core nucleons. The number of these nucleons, depends on how complicated a model we wish to use. Ordinarily we take either one nucleon or all the nucleons, beyond a major shell. The collective energy, in the lowest approximation is taken to be the rotational energy to which can be added the vibrational energy in the next approximation.

We have already seen in Eq. 12.21, that the Hamiltonian H for the collective motion contains the vibrational and rotational motion. Assuming only the rotational motion to exist as a collective motion, we write the complete wave-equation for the nucleus as:

$$H_T \psi = E_T \psi \qquad ...(12.30)$$

Where

$$H_T = T_{rot} + \sum_p H_p \qquad ...(12.31)$$

(where H_T = Total Hamiltonians, and E_T is the total energy) ...(12.32)

and

$$T_{rot} = \sum_i \hbar^2 \frac{R_i^2}{2\mathcal{I}} \qquad ...(12.33)$$

and H_p is the Hamiltonian due to particle motion. Now following Eq. 11.31, we can write, for p 'loose' nucleons in the valence shells outside the core,

$$H_T = H + \frac{\hbar^2}{2\mathcal{I}} \left(\sum_p j_p \right)^2 \qquad ...(12.34)$$

where H has been taken from Eq. 11.31a, except that now

$$H_o = \sum_i \frac{p^2}{2M} + V_i(r) \qquad ...(12.35)$$

so that the term of $\frac{\hbar^2}{2\mathcal{I}} j^2$ as given Eq. 11.28 is not included here. It is included separately, after summing it over p as in Eq. 12.33. The rest of the treatment is similar to the one given in Chapter 11, after Eq. 11.31.

We may then get the wave-function and energies, as we have obtained earlier by Eqs. 11.26 to 11.37 and the individual motion of the particles is described similarly by Nilsson model, *i.e.* from Eqs. 12.1 to 12.18. The composite wave-function of a particle and collective rotational motion may be, then expressed as:

$$\psi = D^{I}_{MK}\ \chi_{\Omega} \qquad \text{...(12.36)}$$

as the wave-functions for odd mass nucleus, with an axially symmetric case. If we take into account the vibrations also, it may be written as:

$$\psi = \sum_{K} g^{I}_{K}\ (\gamma)\ D^{I}_{MK}\ (\theta_1, \theta_2, \theta_3)\ \chi_{\Omega} \qquad \text{...(12.37)}$$

The transformation properties of $\chi_p\ (\mathbf{r}_p)$ and $g^{I}_{K}\ (\gamma)\ D^{I}_{MK}$ are such [*see* Ref. (4)] that one can determine $g^{I}_{K}\ (\gamma)$ for γ from $0 \to 360°$, if $g^{I}_{K}\ (\gamma)$ is known from $0 \to 60°$. As for example $\gamma = 0 \to 60°$ enables us to go $\gamma \to -60°$ from the relation (Ref. (13)),

$$g^{I}_{K}\ (\gamma) = i g^{I}_{K}\ (\gamma) = (-1)^{K}\ g^{I}_{K}\ (\gamma)$$

Then starting from the relation,

$$g^{I}_{K}\ (\gamma) = g^{I}_{-K}\ (\gamma)\ (-1)^{I-K} \qquad \text{...(12.38)}$$

with γ in the interval of –60° to 60°, we can go to the values between 60° and 180°. Again using Eq. 12.38, one can go between 180° to 300°, which is equivalent to – 60. Hence, in this manner, we cover the whole range. While the general solution $g^{I}_{K}\ (\gamma)$ is quite complicated, for special cases, the value of g^{I}_{K} can be determined from the condition of normalisation of the total wave-function, when g^{I}_{K} is independent of γ so that γ-dependence can be ignored. So writing the anti-symmetrised wave-function of a spheroidal body as:

$$\Phi^{I}_{MK}\ (\theta_1, \theta_2, \theta_3) = g^{I}_{K}\ [D^{I}_{MK}(\theta_1, \theta_2, \theta_3) + (-1)\ D^{I}_{M,-K}\ (\theta_1, \theta_2, \theta_3)] \qquad \text{...(12.39}a\text{)}$$

(K = even integers only)

Then the normalisation condition can be written as:

$$\int_0^{2\pi} d\theta_1 \int_0^{\pi} \sin\theta_2\, d\theta_2 \int_0^{2\pi} d\theta_3\, \Phi^{*I}_{MK}\ (\theta_1, \theta_2, \theta_3)\ \Phi^{I}_{MK}\ (\theta_1, \theta_2, \theta_3) = 1 \qquad \text{...(12.39}b\text{)}$$

Using Eq. 12.39*a* in Eq. 12.39*b*, and using properties of D^{I}_{MK}, it can be seen, that for a case of g^{I}_{K} being independent of γ, one obtains g^{I}_{K} as:

$$g^{I}_{K} = \sqrt{\frac{2I+1}{16\pi^2}} \qquad \text{...(12.39}c\text{)}$$

So that, we can, now, write the wave-function Eq. 12.39*a* as:

$$\Phi^I_{MK}(\theta_1, \theta_2, \theta_3) = \left(\frac{2I+1}{16\pi^2}\right)^{1/2} [D^I_{MK}(\theta_1, \theta_2, \theta_3) + (-1)^I D^I_{M,-K}(\theta_1, \theta_2, \theta_3)] \quad ...(12.39d)$$

For $K = 0$, both terms in (12.39*a*) become D^I_{M0} and realising that only even I is possible, we can write:

$$\Phi^I_{MK}(\theta_1, \theta_2, \theta_3) = g^I_0 \, D^I_{M0}(\theta_1, \theta_2, \theta_3)$$

$$= \sqrt{\frac{2I+1}{8\pi^2}} \, D^I_{M0}(\theta_1, \theta_2, \theta_3) \quad ...(12.39e)$$

We can now similarly express the transformation properties of the combined wave-function $\chi_p \, D^I_{MK}$, as required for the actual wave-function of an ellipsoid. We conclude that for K = even integer, in the case of axial symmetry, ϕ being arbitrary,

$$K - \Omega = 0 \text{ (for axial symmetry) or } K = \Omega \quad ...(12.40)$$

We use Eq. 12.39 and require that in Eqs. 12.36 and 12.37 the value of the wave-function remains unaltered under R_1; then the anti-symmetrised wave-function ψ^I_{MK} for particle rotation can be written as:

$$\psi^I_{MK}(\theta_1, \theta_2, \theta_3, \mathbf{r}_p) = \sqrt{\frac{2I+1}{16\pi^2}} [D^I_{MK}(\theta_1, \theta_2, \theta_3) \chi_K(\mathbf{r}_p) +$$

$$(-1)^{I-j} D^I_{M-K}(\theta_1, \theta_2, \theta_3) \chi_{-K}(\mathbf{r}_p)] \quad ...(12.41)$$

which is the same as Eq. 11.33*a* discussed earlier.

This formula is good for odd mass nuclei and therefore; the case for $K = 0$ does not arise. Here I = half integer, the minimum value being $I = 1/2$, and hence K is also half integer.

For a given band of rotational levels, of various angular momenta I, belonging to a given band quantum number K; the permitted values of I in a band must necessarily be $I \geq K$. For all the levels of a rotational band, the particle wave-function χ_K remains the same; the rotational part D^I_{MK} however, acquires different values of I, as shown in Fig. 12.3, for the excited[14.6] states of Al^{25}.

Apart from the rotational energy and the particle excitation energy in a deformed potential; we will also have a rotational particle coupling (RPC), as given in Eq. 11.31*b*.

The contribution from H_p to the energy of a level remains constant for all the levels of a rotational band; which may be denoted by E_o (*K*). So the total energy of the level E (*I*, *K*) consists of (*i*) the rotational energy E_o (*K*) (*ii*) particle energy and (*iii*) the rotational particle coupling energy (RPC), as shown in Eqs. 11.31*b* and 11.41. The lowest of change in energy due to the RPC term is given by:

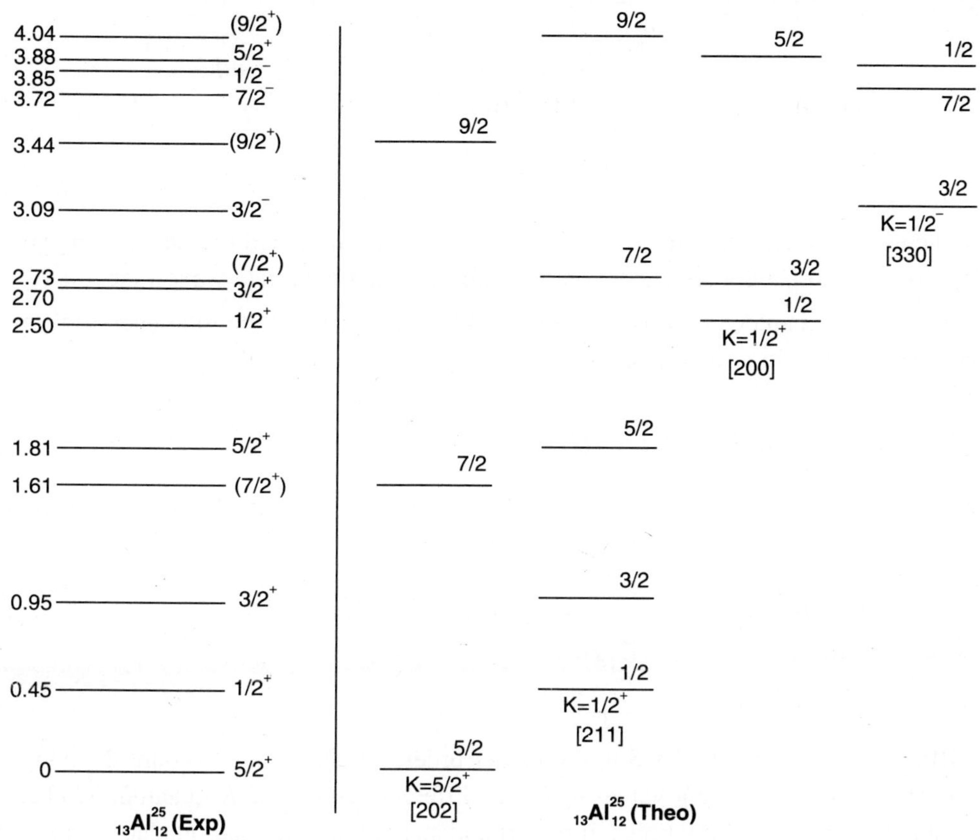

Fig. 12.3 Rotation spectra of Al[25], with a ground state of $5/2^+$. Rotation bands on K = $1/2^+$ and K = $1/2^-$ are explicitly shown to the right of vertical line, for different particle energies (Ref. 6, 14).

$$\langle \text{R.P.C.} \rangle = \Delta E(I, K) = \frac{-\hbar^2}{2\mathcal{I}} \langle \psi^I_{MK} \mid I_+ j_- + I_- j_+ \mid \psi^I_{MK} \rangle \qquad ...(12.42)$$

where ψ^I_{MK} is given by Eq. 12.41.

Since $I_\pm$ changes K to $K = \mp 1$, we conclude that a non-vanishing contribution to this matrix element can arise from the cross-terms, between $D^I_{M,K}$ and $D^I_{M,-K}$ and that requires $K = 1/2$. For this value of K; $-K = -1/2 = 1/2 - 1 = K - 1$.

For any other value of K; such a condition is not fulfilled. It can, then, be seen, that Eq. 12.42 reduces to, (using definitions of Eq. 11.33):

$$\langle \text{R.P.C.} \rangle = \Delta E(I, K) = +\,\delta_{K,1/2} (-1)^{I+1/2} \left(I + \frac{1}{2} \right) a \qquad ...(12.43a)$$

where $$a = \frac{\hbar^2}{2\mathcal{I}} \left\langle \chi_{1/2} \mid j_+ \mid \chi_{-1/2} \right\rangle (-1)^{j-1/2}$$

$$= \frac{\hbar^2}{2\mathcal{J}} \sum_{nlj} \left| C_{nlj,1/2} \right|^2 \left\langle nlj, +\frac{1}{2} \middle| j_+ \middle| nlj, -\frac{1}{2} \right\rangle$$

$$= \frac{\hbar^2}{2\mathcal{J}} \sum_{nlj} \left(j + \frac{1}{2} \right) \left| C_{nlj,1/2} \right|^2 \quad \text{...(12.43}b\text{)}$$

It is interesting to note that in Eq. 12.43*b*, the parameter '*a*' called the de-coupling parameter, does not depend on *I* and is only dependent on the deformed single particle wave-function $\chi_{1/2}$.

As discussed in Chapter 11, [Eqs. 11.33 to 11.43], we can write, the total energy of level with tal angular momentum *I*, in a rotational band as:

$$E(I, K) = \frac{\hbar^2}{2\mathcal{J}} [I(I+1) - K^2] + E_0(K) + \text{R.P.C.} \quad \text{...(12.44}a\text{)}$$

where $E_0(K)$ is the particle energy; so that $E(I, K)$ from Eqs. 12.42, 12.43 and 12.44, can, now, be expressed as:

$$E(I, K) = \varepsilon_0(K) + \frac{\hbar^2}{2\mathcal{J}} I(I+1) + \delta_{K,1/2}\, a(-1)^{I+1/2} \left(I + \frac{1}{2} \right) \quad \text{...(12.44}b\text{)}$$

where $$\varepsilon_0(K) = E_0(K) - \frac{\hbar^2}{2\mathcal{J}} K^2$$

The RPC term may be calculated to the higher order, by taking into account the non-diagonal matrix elements connecting different values of *K*, *i.e.* different bands. The *K*-quantum numbers mixed in this way, can differ by ± 1. Many times, if the ground state of an odd-mass nucleus has $I_o = K_o$; the RPC term of the higher order can mix the ground state band with a band of $K = K_o + 1$. Each of the higher level of the ground state band $I = I_o + 1, I_o + 2$, finds a state of the same angular momentum in the excited band, $K = K_o + 1$, with which it mixes through RPC of higher order. The resultant states for a given *I* can be obtained by diagonalising a 2 × 2 matrix of the Hamiltonian, using the states, $|I, K_o>$ and $|I, K = K_o + 1>$, as the basis.

Similarly, one can calculate $E(I, K)$ for higher order RPC, for which one should consult Reference (4), page 443, for details. This term, renormalises the inertia parameter $\hbar^2/2\mathcal{J}$ to $\hbar^2/2\mathcal{J}'$. These normalised moments of inertia have been calculated by J.J. Griffin and M. Rich[15] who used the principle of long range correlations using B.C.S. theory. Comparison of experimental values of $\mathcal{J}'$ as a function of δ; with theoretical values based on this model gave a very good fit, showing that strong coupling model holds good here.

12.3 COMPARISON OF EXPERIMENTAL LEVELS OF ODD NUCLEI WITH COLLECTIVE MODELS

We have, earlier, discussed briefly the case of ${}_{13}\text{Al}_{12}^{25}$ in Fig. 12.3, an odd mass nucleus with $K = 1/2$. According to Litherland et al.[6], the quadrupole moment in this region suggests $\delta \approx 0.3$, hence one can get from Fig. 12.1, for this distortion, Ω^{π} $[N\, n_3\, \Lambda]$ = 5/2⁺ [202] for $d_{5/2}$ for 12 particles and close by of orbital 1/2⁺ [211] for $d_{3/2}$ configuration for 13 particles. Experimentally, the ground state of Al^{25} is 5/2⁺ and an excited state next to it is 1/2⁺ at 450 keV. So we have a rotational band over 5/2⁺ and another rotational band over 1/2⁺. The rotational head at 1/2⁺ [200], and 1/2⁺ [211] and 1/2⁻ [330] are the possibilities obtainable from Fig. 12.1. The energy levels, however, can only be understood, if we use Eq. 12.44, and values of δ may be different for the three K = 1/2 bands. Of course $K = \Omega$ has been used for all cases. Besides Al^{25}, other nuclei which have been studies in this region are: Mg^{25}, F^{19}, Mg^{24}, Al^{28}, Si^{29} and P^{31} [Ref. 16].

Another interesting case is that of various rotational bands in $74W^{183}_{109}$ and $75Re^{183}_{108}$ as shown[8] in Fig. 12.4. For 109th neutron in Fig. 12.1, the configuration for $\delta \geq 0.2$ is 1/2⁻ [510]; which is consistent with the ground state assignment for W^{183} and arises from $p3/2$ configuration for $\delta = 0.3$. There is a rotational band based on this state up to I = 9/2. Then another band starts at the excitation energy of 209 keV, which corresponds to 3/2– [512] and arises from f 5/2. Two other states of its band appear,

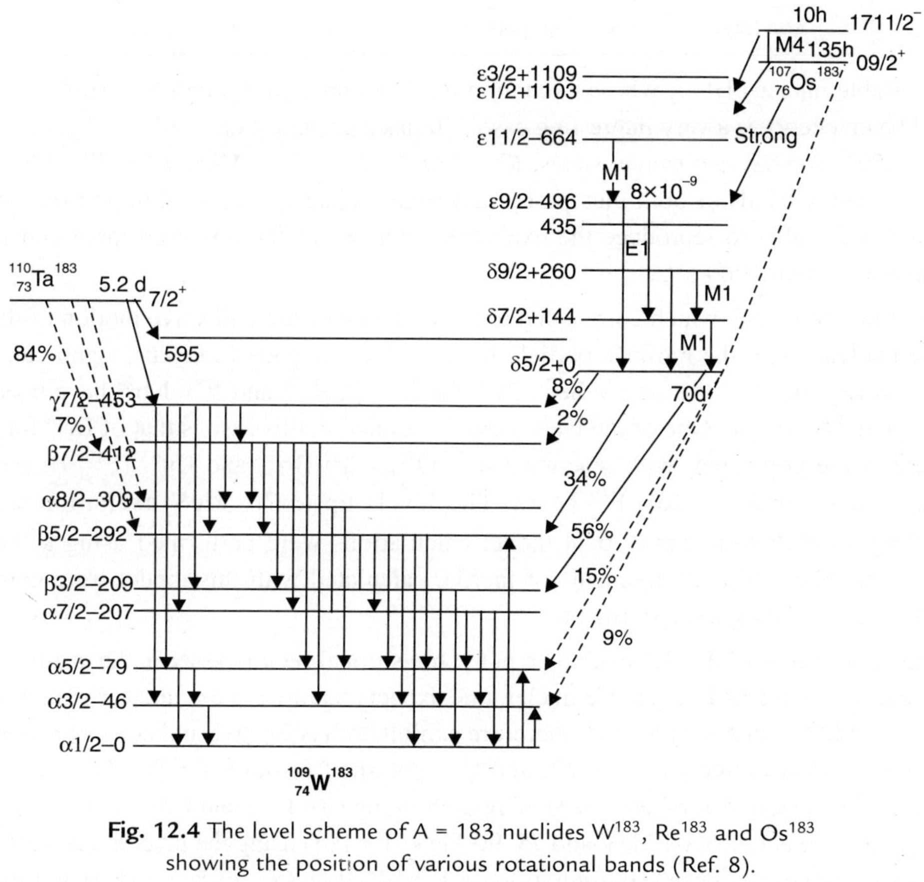

Fig. 12.4 The level scheme of A = 183 nuclides W^{183}, Re^{183} and Os^{183} showing the position of various rotational bands (Ref. 8).

which correspond J (J + 1) law. The 7/2 state at 453 keV, seems to be 7/2⁻ [503], belonging to f 7/2 configuration. Similarly for Re^{183}, the ground state of Re^{183} belongs to 5/2⁺ [642]; arising from i 13/2 configuration over which a rotational band is built up to 260 keV excitation. Another band starts at

496 keV, belonging to $9/2^-$ [505] state arising from configuration h 9/2 at 496 keV excitation energy of Re^{183}, or it could be $5/2^-$ [514] state of h $11/2^-$, if it is the oddeth proton.

Such cases of rotational bands are available for many elements for $A \geq 150$. A few interesting cases[17] are that of Pu^{239}, Th^{228}, U^{234}, and Pu^{238}. These nuclei, not only contain the rotational band based on ground states, but also the β and γ vibrations.

Many nuclei, both through nuclear reaction and radioactive decay, have been studied recently, corresponding to even-odd and odd-even proton-neutron configuration. We give below a partial list:

Mass-range	*Nuclei*	*Method of production*
$A \approx 70$	Kr^{71}, Br^{71}	Radioactive[18]
$A \approx 100$	Ru^{97-111}, $Pd^{104-111}$	Fission fragments[19]
$A \approx 132$	Sn^{131}, Sn^{133}, In^{133}, Sb^{133}	Fission fragments[20]
$A \approx 150$	Eu^{153}	Reactions[21]
$A \approx 180$	Au^{187} and Hf^{177}	Radioactive and reactions[22]

The stable nucleus Hf^{177}, whose excitation has been studied through Y^{176} (Be^9, α, $4n$) Hf^{177} at 70 MeV of beam energy is a very interesting case[23]. It has a ground state of $K^\pi = 7/2^-$, corresponding to $[N, n_3, \Lambda] = [503]$ and has two isomer states; $K^\pi = 23/2^+$ ($T_{1/2} = 1.13$ Min), $K^\pi = 37/2$ ($T_{1/2} = 51$ Min). It has nearly twenty bands of both plus parity and negative parity states. Multiparticle quasi-particle calculations[24] were able to reproduce the excitation energies of the observed three and more quasi-particle states to within 100 keV.

The odd-odd nuclei, which correspond to the coupling of the collective motion of the deformed case of the nucleus, with the intrinsic particle motion of the unpaired nucleus, again can be put into different mass regions. As for example Se^{74}, Br^{76}, Br^{77}, Br^{78}, Kr^{79} and Y^{82}, Nb^{86} have been studied in the mass region[25] of $A \approx 80$. One of the latest cases[26], studied in this case is that of Br^{76} for which high energy states were populated, with reaction Cu^{63} (O^{16}, $n2p$) Br^{76}, and Cu^{63} (F^{19}, α, pn) Br^{76} with 69 MeV O^{16} beams and 67 MeV F^{19} beam. The levels up to 7.6 MeV excitation were excited corresponding to $19^+\hbar$ spin. Theoretical model calculations were performed using a Woods-Saxon potential[27]. The kinematics and moments of inertia, calculated with this model were compared with experimental value, with good agreement.

In the mass range of $A \approx 120$ which lie in the transitional region between the primarily spherical Sn nuclei and well deformed La and Ce nuclei, one expects features in the nuclear structure from both sides. The odd-odd nuclei $A = 113 - 125$, therefore exhibit both collective and non-collective features[28]. The recent-most measurements[29], on I^{120} and I^{122} obtained through $Pd^{108} + O^{16}$ and $C^{114} + B^{11}$ fusion-evaporation reaction at 84 and 60 MeV respective by (for I^{120}) and $Cd^{116} + B^{11}$ and $Pd^{110} + O^{16}$ fusion-evaporation reaction at 64 MeV and 75–90 MeV (for I^{122}) using the tandem accelerator at Delhi. New bands were obtained, for the configuration for I^{120} based on (π_g $7/2 \otimes \nu$ h 11/2) and for I^{122}, based on π ($g7/2/d5/2$) $\otimes$ ν h 11/2 and π $g_{9/2}^{-1}$ $\otimes$ ν h 11/2 configuration.

In the higher mass region, doubly odd Pm-nuclei with $Z = 61$ and $N < 80$ have been recently studied[30], by in-beam γ-spectroscopy using Cs^{133} (C^{13}, $4n$) Pm^{142} at $E = 63$ MeV. Levels up to 19^- $\hbar$, were excited. The highest mass region of $A \approx 180$, has also been explored recently[31], for odd-odd nuclei by experimentally studying Ir^{186}, using Hf^{180} (B^{11}, $5n$) Ir^{186} at 65 MeV and Ta^{180} by using Y^{176} (B^{11}, $3n$) Ta^{180} and 65 MeV and Yb^{176} (Li^7, $3n$) Ta^{180}. Basically the results are explainable by multi-quasi-particle states. As a matter of fact, eight new two quasi-particle states could by identified for Ta^{180} in the case of an aligned proton and a neutron pseudo-spin-doubles and could be used to explain the properties of the band.

In summary, odd-odd nuclei represent an interesting field of interaction of multi-quasi-particle configurations.

12. Particle States and Collective Motion in Nuclei 2000–2008

In about eight papers, where high spin states spectroscopy has been reported, for odd even, or odd-odd nuclei, an interesting paper of smooth band transformation concerns odd-mass nuclei La^{127}, La^{129} and La^{131} authored by 26 Workers, using Mo^{100} (S^{32}, $p4n$) La^{127} etc. at $E = 100$ MeV. [Phy. Rev. C. 62, 034315 (2000)].

Among half a dozen papers, on odd-even or odd nuclei, in year 2001; there is one authored by 20 authors, where there has been the first observation of excited states in neutron deficient isotones ($N = 86$) Ta^{159}_{86} and W^{168}_{86} + Ni^{58} was used as a beam at 286 MeV and 291 MeV, and 298 MeV at cd^{186} target [Phy. Rev. C. 63, 064309 (2001)].

In a paper on lifetime measurements, using Doppler shift technique, Au^{183} was studied by a group from Panjab University (I.M. Govil and A. Kumar et al.) and from Delhi (N.Sc.) and Notre Dame; using the reaction Tb^{159} (Si^{28}, $4n$)^{183} Au with Beam energy of 140 MeV; using 15 UD at N.Sc. (Delhi) [Phy. Rev. C. 66, 044306 (2002)].

In 2001-2002, some twenty papers were published on collective motion and particle states. In a paper authored by 26 authors, with Rmayya et al., level structure of Ba^{141} and Xe^{139} have been studied, obtained from Cf^{252} fission. It finally resulted in understanding $N = 85$ even odd isones [Phy. Rev. 66, 014305 (2002)].

An interesting case of odd-odd nucleus Tb^{146} was investigated by a group of 26 authors from India, USA, using In^{115} (S^{34}, $3n$)^{146} using a 140 MeV incident S^{34} beam using 15 UD petraction facility at N.Sc. (Delhi). Level scheme has even extended up to 10 MeV excitation energy, and 30 units of angular momentum. This is a case of dominane of single particle nature and shell model. [Phy. Rev. C.70, 0443151 (2004)].

In a paper [Phy. Rev. C. 73, 064304, (2006)], the high spin structure of Er^{157} has been expanded, using Gamma-sphere spectroscopy to investigate Cd^{114} (Ca^{48}, $5n$) E^{157} reaction at 215 MeV. States up to 40 h have been excited. Cranked Nilsson collective configuration that break $Z = 64$ semi-magic core explain the excited states.

In a paper, authored by 27 authors, from Europe and Canada, excited states of extremely neutron deficient isotopes with $N = 111, 113, 115$ in $Rn^{197, 199, 201}$ have been studied, using Sm^{150} (Cr^{52}, $3n$) Rn^{199}, Sn^{118} (Kr^{82}, $5n$), Rn^{197}, Sn^{120} (Kr^{82}, $3n$), Rn^{199}, and Sn^{122} (Kr^{82} , $3n$) Rn^{201}, using accelerated laboratory of the University of Jynoskyla (Finland), γ-spectroscopy procedures showed that in excitation of 17/2+ levels built on 13/2 + level, a transition from anharmonic vibrational structure towards a rotational structure at low spin in these nuclei takes place. It is an interesting paper about the evolution of the shape of the nucleus on excitation [Phy. Rev. C. 77, 054303 (2008)].

REFERENCES

1. S.G. Nilsson: Kgl. Danske Videnskab, Selskab, Mat. Fys. Medd. 29, No. 16, 68, 9 (1955); B.R. Mottleson, and S.G. Nilsson: Kgl. Danske, Videnskab, Mat. Fys Skrifter, 1, 8, 76 (1959); Nuclear Physics 13, 281 (1959).
2. R.H. Lammer: Phy. Rev. 117, 1515 (1960).
3. Reference (1); T.D. Newton: Canad. J. Phy. 38, 700 (1960); Atomic Energy of Canada Limited, Report, No. CRT–886, (1960), Also Canad. J. Physics 37, 944 (1959).
4. Reference (1); Mottleson and Nilsson: M.K. Paul: Theory of Nuclear Structure, Affiliated East West Press Ltd., New Delhi (1982), pp. 452–53.
5. Reference (1); M.A. Preston: Physics of the Nucleus, Addison-Wesley Publishing Company Inc., Reading (Mass.) (1962).
6. A.E.H. Litherland, E.B. McManus, D.A. Bromley Paul, and H.E. Gove: Canad. J. of Physics, 36, 378, (1958); E.B. Paul, Phil Mag (8), 2, 311 (1957); G.R. Bishop: Nuclear Physics 14, 376, 1959/60 K.H. Bhatt: Nuclear Physics 39, 375 (1962).
7. Reference (3); A.J. Rassey, Phy. Rev. 109, 949 (1958).
8. M.A. Preston: Physics of the Nucleus, Addison-Wesley Publishing Company Inc., Reading (Mass.) p. 251 (1962).
9. D.D. Chowdhary: Kgl. Danske Videnskab Mat. Fys. Medd. 28 No. 4 (1954); Nuclear Physics *A*93, 300 (1967), Phy. Rev C3, 1619 (1971); K.W. Ford and C. Levinson: Phy. Rev. 100.1 (1955).
10. A. Lande and G.E. Brown: Nuclear Physics, 75, 344 (1996).
11. S. Sen et al.: Nuclear Physics A, 157, 497 (1970): Ibid A 191, 29 (1972); Ibid *a* 220, 580, (1974), S. Sen et al.: Phy. Rev. C 13, 20, 55, (1976); Ibid C14, 758 (1976); S. Sen et al.: J. Phy. G. Nuclear Physics 1, 286 (1975).
12. K.A. Brueckner, T. Soda, P.W. Anderson, and P.l. Morel, Phy. Rev. 118, 1442 (1960); A.K: Kerman, K. Danske Vidensk, Selsk, Mat. Phys. Medd. 30, 15 (1956); A.S. Davydov: Soviet Phy. JEPT, 36, 1103 (1959); Zh.eksper.teor. Fiz 36, 1555 (1959); Ibid Nuclear Physics 12, 58 (1959); Nuclear Physics 8, 237 (1958).
13. Reference (4); M.K. Pal. p. 419.
14. Reference (6); A. Bohr and B.R. Mottleson: Nuclear Spectroscopy, Part B, edited by F. Ajzenberg-Selov, Academic Press, New York, (1029) (1960).
15. J.J. Griffin and M. Rich: Phy. Rev. 118, 850 (1960).
16. E.B. Paul, Phil. Mag. 2, 311 (1957); (for F19); R. Batchelor, A.J. Ferguson, H.E. Gove and A.E. Litherland: Nuclear Physics 16, 38 (1960); (for Mg^{24}); R.K. Sheline: Nuclear Physics 382 (1956); (for Al^{28}). D.A. Bromley, H.E. Gove and A.E. Litherland: Cand. J. of Physics 37, 53 (1959); (for Si^{29}); A.E. Litherland, E.B. Paul, G.A. Bartholomew and H.E. Gove: Cand. J. of Physics 37, 53 (1959).

17. R.K. Sheline: Rev. of Mod. Physics 32, 1 (1960).

18. P. Urkedal and I. Hamamoto: Physics Rev. C.V. 58, p. R. 1889 (1998).

19. F. Fotiades et al. (14 authors): Phy. Rev. V. 58, p. 1997 (1998); T. Kutsarova, et al. (17 authors): Phy. Rev. C.58, p. 1996 (1998).

20. Jing-Ye-Zheng, Yang Sun, Mike Guidny, L.I. Ridinger and G.A. Lalazissis: Phy. Rev. C. 58, p. 2663 (1988).

21. S. Basu et al. (9 authors): Phy. Rev. C. 56, p. 1756 (1997).

22. D. Rupnik, E.F. Zganjar, J.L. Wood, P.B. Semmes and P.F. Mantica: Phy, Rev. C. 58, p. 771 (1998).

23. S.M. Mulins, A.P. Byrne, G.D. Dracaulis, T.R. McGoram and W.A. Seals: Phy. Rev. C. 58, p. 831 (1988).

24. F.G. Kondev, G.D. Dracoulis, A.P. Byrne, T. Kibedi and S. Bayer: Nuclear Physics A617, 91 (1997); Kiran Jain, P.M. Walker, and N. Rowley: Phy, Letter B2, 322, 27 (1994).

25. F.D. Cottle, J.W. Holcomb, T.D. Johnson, K.A. Stuckey, S.L. Tabore, P.C. Womble, S.G. Buccim and F. Durham: Phy. Rev. C. 42, 1254 (1990); G.N. Sylvan et al. (10 authors): Phy. Rev. C. 48, 2252 (1993); E. Landulfo et al. (12 authors): Phy. Rev. C. 54, 628 (1996) and R. Schwengner, J. Doring, L. Finke, G. Winter, A. Johnson and W. Nazarewicz: Nuclear Physics A. 509, 550 (1990); S.L. Tabor et al. (18 months): Phy. Rev. C.56, 142 (1997).

26. D.F. Winchell, L. Welner, J.X. Saladin, M.S. Kaplan and E. Landulfo: Phy. Rev. C.55, p. 111 (1997).

27. W. Nazarewicz, K.J. Dudek, R. Bengtsson, T. Bengtsson, and I. Ragnasson: Nuclear Physics A.435, 397 (1985), D.F. Winchell, M.S. Kaplan, J.X. Saladin, H. Takai, J.J. Kolata and K.J. Dude: Phy. Rev. C.40, 2672 (1989).

28. M. Waring et al. (25 authors): Phy. Rev. C.51, 2427 (1995); E.S. Paul, R.M. Clake, B.A. Forbs, D.B. Fossan et al.: (13 authors): J. Physics G. 18, 837 (1992); Z.G. Kostora, W. Andrejtscheff, L.K. Kostora, F. Donnaus, L. Kaubler, H. Prade and H. Rotter: Nuclear Physics A. 485, 31 (1988).

29. Harjeet Kaur, Jagbir Singh, A. Sharma, D. Mehta, Nirmal Singh, P.N. Trehen, H.C. Jain, S.D. Paul, E.S. Paul and R.K. Bhowmik: Phy. Rev. C.55, p. 2234 (1997), H. Kaur et al. (9 authors): Phy. Rev. C.55, 512 (1997).

30. G.de. Angelis, S. Wnardi, D. Bazzacco, J. Rico, T. Tessasi, M. Muporinen, A. Atac and J. Nyberg: Z. Phy. A. 347, 93 (1993), C.W. Beausang et al.: Phy. Rev. C.36, 1810, (1987), V-Datta Pramanik, et al.: Nuclear Physics A.632, 307 (1998); Sarmishtha Bhattacharya et al. (8 authors): Phy. Rev. C.P. 2998 (1998).

31. C. Schlegel et al. (9 authors): Phy. Rev. C.50, 2198 (1994); G.D. Dracoulis et al. (9 authors): Phy. Rev. C.58, 1444 (1998); M.A. Cardona et al. (25 authors): Phy. Rev. C.55, 144 (1997).

PROBLEMS

1. Write down the quantum numbers Ω^π $[N\, n_3, \Lambda]$, for the ground states of the following Nuclei: Nd^{147}, Sm^{145}, Gd^{153}, Dy^{163}, Hf^{177} and Os^{183} and W^{183}, by using Figs. 12.1*a* to 12.1*c*.

2. Using Eq. 12.33, derive Eq. 12.39.

3. From Ref. (1), Chapter 11 obtain level scheme of U^{234} and Np^{237}. Write down the character of each state. At what energy would there be 10^+ state in U^{234} and an $13/2^+$ in Np^{237}?

4. Derive Eq. 12.28, using Eqs. 12.24 and 12.26. Which aspect of the weak coupling has been used in writing Eq. 12.28?

5. What will be the configuration of K^{41} states, which are isobaric analogues of the low lying states of $_{18}Ar_{23}^{41}$?

6. Prove Eq. 12.25; using Eqs. 11.81 to 11.86 and also derive the relationship between deformation parameter β, and $\Delta R/R$.

7. Derive

$$T = \frac{1}{2}\beta(\dot{\beta}^2 + \beta^2\dot{\gamma}^2) + \frac{1}{2}\sum_{K=1}^{3} \mathcal{I}_K \omega_K^2.$$

8. According to Nilsson, 1/2⁻ [510], level has the following wave-functions:

$$\eta = 4;\ \psi = 1.000\,|\,500+\rangle - 2.730\,|\,530+\rangle + 2.440\,|\,510\rangle - 0.606\,|\,551-\rangle + 0.780\,|\,531\rangle + 1.431\,|\,511-\rangle$$

$$\eta = 6;\ \psi = 1.000\,|\,550+\rangle - 2.359\,|\,530+\rangle + 1.981\,|\,510\rangle - 0.469\,|\,551-\rangle + 0.421\,|\,531-\rangle + 0.993\,|\,511-\rangle$$

Find the decoupling constant for $\eta = 4$ and $\eta = 6$ and compare the results with the experimental values of energies for W^{183}.

9. The lowest level of B^{10} are:

E (MeV)	G.S.	0.77	1.74	3.58
J^π	3^+	1^+	0^+	2^+

Write down:

(*i*) Iso-spins and seniorities of these levels.

(*ii*) Your comments about the properties and comparison of Be^{10} and B^{10}.

10. Show that the allowed total angular momenta for two quadrupole phonons are 0^+, 2^+ and 4^+, and there are no odd angular momenta.

13

CHAPTER

Compound Nucleus Model

13.1 INTRODUCTION

We have discussed in the previous chapters the problems of nuclear structure, *i.e.* the properties of the various nuclear states including ground and excited states. The various models gave us an insight into the motion of nucleons inside the nucleus below the Fermi energies so that the nucleons stay within the nucleus. However, when the excitation energy takes a nucleon to an energy much higher than the Fermi energy and above the binding energy using an incident projectile, the excited nucleus may emit a particle and itself may be left in either a low excitation energy or the ground state. Description of such phenomenon requires models of reaction mechanism, *i.e.* models describing the relationship of the incident projectile to the target, the composite nucleus, the residual nucleus and the ejectile, for their energies, momenta, angular momenta and other quantum mechanical properties like the probability of transition and hence cross-sections.

The study of the cross-sections of the nuclear reactions is important for two reasons: (*i*) For empirical knowledge for various applied purposes, *e.g.* in fission or fusion studies required for designing reactors or for various fission devices for purposes of energy production or for understanding the evolution of stars in which nuclear reactions play the most important role or some other nuclear science problems of applied nature. (*ii*) For understanding the reaction mechanism. At microscopic level, any reaction corresponds to the interaction of the projectiles with nucleons in the target-nucleus and hence should be understood theoretically, through the application of many-body interaction techniques or theories. This procedure, however, is somewhat difficult because of strong nature of nucleon-nucleon interaction and hence it is not easy to apply the various appropriate approximation techniques. One, therefore, takes course to various reaction-mechanism models, for explaining the absolute values of cross-sections as a function of incident energy or the energy of the residual nucleus and angles θ, ϕ with respect to the incident direction in which the emitted particles are observed. These models basically represent the alternative ways of looking at the reaction in which the many-body process of nucleon-nucleon interaction is replaced by an approximate two body interaction, *i.e.* the incoming projectile and the whole nucleus as a target.

We briefly discuss the four reaction mechanism models, *i.e.* (*i*) compound nucleus model, (*ii*) direct reaction model, (*iii*) optical model and, (*iv*) pre-compound or pre-equilibrium model. We will also discuss at the end, the heavy ion induced reactions, which involves many reaction mechanisms. In the above description, the three models. *i.e.* the compound nucleus formation, the direct reaction and the pre-compound reaction model, correspond to different mechanisms for calculating the probabilities of a projectile to interact with the target nucleus. If the projectile enters the nucleus and interacts with one nucleon (or a cluster) and both of them are ejected after a single interaction—this is called direct reaction. If the projectile interacts with a limited number say 2 to 5 nucleons, before the striking projectile and or the ejectiles come out, this is a case of pre-compound reaction mechanism. On the other hand, the incident projectile may interact with a large number say thousands or more of the nucleons, inside the nucleus in some sequence and the energy of the incident particle is shared by a very large number of nucleons, so that only after a lot of interactions including many reflections from the surface, some nucleons (or clusters) find enough energy and get emitted. This scenario corresponds to the compound nucleus formation.

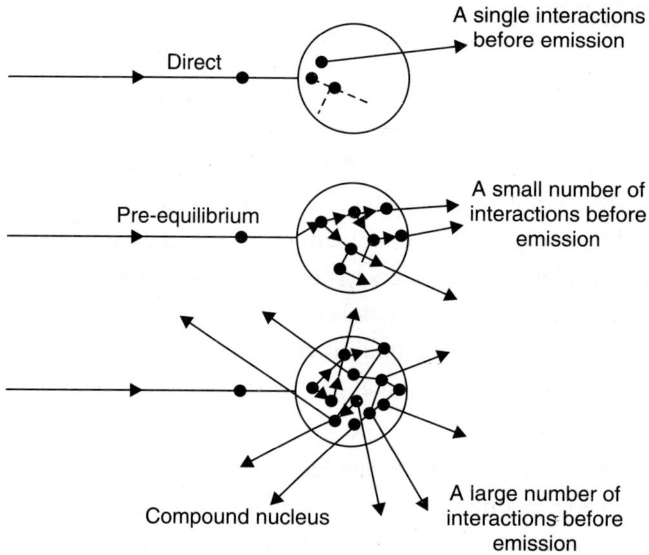

Fig. 13.1 Three modes of reaction-mechanism. In the direct mode, there is only one interaction; in pre-compound 2–4 or 5 and in compound several thousand interactions before emission.

Figure 13.1 gives a graphical picture of these three mechanisms. The optical model is basically a mathematical model, which replaces the physical nucleus by a potential, which has both real and imaginary parts. The imaginary part corresponds to the interaction with the nucleus by all three processes mentioned above.

The heavy ion induced reactions are somewhat complex because of the complexity of the incident projectile. Because of the heavy mass of the incident heavy ion, a lot of energy and angular momentum can be imparted to the target nucleus giving rise to new phenomenon, *e.g.* fusion-fission, etc.

We describe these cases in the subsequent chapters, starting from this chapter, which describes the compound nucleus model.

13.2 COMPOUND NUCLEUS MODEL

13.2.1 General

Because of the very strong nature of the nucleon-nucleon interaction, it is assumed in this model[1] that after a projectile enters the nucleus, it interacts very strongly with nucleons in the nucleus and the projectile loses its initial parameters and a new nuclear configuration, *i.e.* a combination of the projectile and the target nucleus (completely merged) is formed. This is called the compound nucleus. From the conservation of energy, it is apparent that the compound nucleus will be formed at a highly excited state, from which it has to decay to its own ground state by emitting gamma-rays or to another neighbouring residual nucleus in the ground or excited state by emitting a particle.

The behaviour of the cross-sections is explained in many cases by this model[2]. Of course, some time lapses between the entry of the projectile and the formation of the complete compound nucleus. In this period, the projectile may travel back and forth in between the surface points or may break up or pick up nucleons and go through a complicated energy mixing with the nucleons of the nucleus before setling down to the equilibrium state of the compound nucleus. The experimental lifetime of such a compound state is of the order of $T_c \approx 10^{-12} - 10^{-16}$ secs compared to the small time required for single traversal of the nucleus, which is of the order of 10^{-19}–10^{-22} sec. From experimental values of T_c for the compound state, it can be seen that the number of traversals in a typical case before a compound nucleus is formed is of the order of 10^3 to 10^7–10^{10}.

The theory of compound nucleus of nuclear cross-sections is based on two assumptions: (*i*) The incident particle after entering the nucleus, forms a compound nucleus, in such a way that the final equilibrium state of the compound nucleus does not 'remember' how it was formed. In other words, the compound state may be formed by various methods but, the properties of the compound state will be independent of the methods of formation of the compound state as long as the compound nuclear state is the same. (*ii*) The decay of this compound state depends on the properties of the compound state and not on how it was formed. This was experimentally demonstrated by the famous experiment of S.N. Ghoshal[3], where he produced the same compound nucleus ${}_{30}Zn^{*64}$ with the bombardment of ${}_{28}Ni^{60}$ by alpha particles and of ${}_{29}Cu^{63}$ by protons with appropriate energies to reach the same excitation energy as shown in Fig. 13.2. The reactions observed were:

1. ${}_{28}Ni^{60}$ (α, n) ${}_{30}Zn^{63}$
2. ${}_{28}Ni^{60}$ $(\alpha, 2n)$ ${}_{30}Zn^{62}$
3. ${}_{28}Ni^{60}$ (α, pn) ${}_{29}Cu^{62}$
4. ${}_{29}Cu^{63}$ (p, n) ${}_{30}Zn^{63}$
5. ${}_{29}Cu^{63}$ $(p, 2n)$ ${}_{30}Zn^{62}$
6. ${}_{29}Cu^{63}$ (p, pn) ${}_{29}Cu^{62}$ ...(13.1)

These reactions correspond to the formation of the same compound state ${}_{30}Zn^{*64}$ but, decay into different outgoing channels.

Now we assume that ${}_{30}Zn^{*64}$ after formation, will decay in such a way that for every case we can write:

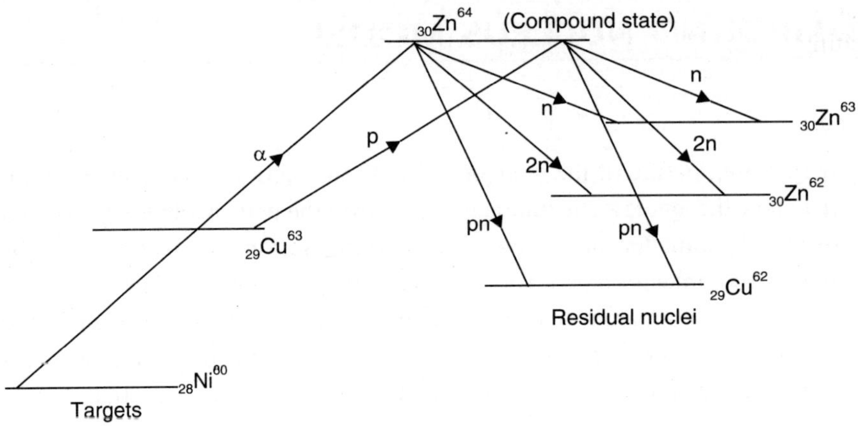

Fig. 13.2 Diagram depicting the common compound state formed in Eq. 13.1, formed from various incoming channels and decaying into various outgoing channels.

$$x + X \rightarrow C \rightarrow y + Y \qquad ...(13.2)$$

and

$$\sigma\,(x, y) = \sigma_c\,(x)\,G_c\,(y) \qquad ...(13.3)$$

where x is the incident particle, X is the target, y is the outgoing ejectile and C is the compound state and, Y is the residual nucleus. Then the cross-section for a given reaction (x, y) is a two step process, *i.e.* the formation of the compound state for which the cross-section is given by $\sigma_c\,(x)$; and $G_c\,(y)$, the probability that compound state C decays with emission of y with a residual nucleus Y. Equation 13.3 assumes that the decay probability is dependent only on the property of the compound state and is independent of the way it was formed. Hence if there is a reaction,

$$x + A \rightarrow C \rightarrow z + Z \qquad ...(13.4a)$$

then we can write:

$$\sigma\,(x, z) = \sigma_c\,(x, A)\,G_c\,(z) \qquad ...(13.4b)$$

Similarly for a reaction

$$y + B \rightarrow C \rightarrow z + Z \qquad ...(13.5a)$$

One can write, then

$$\sigma\,(y, z) = \sigma_c\,(y, B)\,G_c\,(z) \qquad ...(13.5b)$$

It is easy to see that according to the hypothesis of compound nucleus, one should get from reactions 13.1 and Eqs. 13.4 and 13.5

$$\sigma\,(p, n)\!:\,\sigma\,(p, 2n)\!:\,\sigma\,(p, p\,n) = \sigma\,(\alpha, n)\!:\,\sigma\,(\alpha, 2n)\!:\,\sigma\,(\alpha, p\,n) \qquad ...(13.6)$$

This is based on the assumption that $G_C\,(z)$ is the same for proton induced or α-induced reaction of the same ejectiles (z), if the compound state C is the same. Ghoshal's experiment[3] basically proved this relation empirically. Similar experiments done subsequently have confirmed the above assumptions.

From the uncertainty principle, it is well-known that

$$\Delta E\,\Delta t \approx \hbar;\ \text{or}\ \Delta E \approx \frac{\hbar}{\Delta t} \qquad ...(13.7)$$

If we put $\Delta t = \tau_c$, then ΔE has the meaning of the energy width of the level and is referred to as Γ_c. Further, writing Γ_c as:

$$\Gamma_c = \frac{\hbar}{\tau_c} = \hbar\lambda_c \qquad \text{...(13.8)}$$

it is easy to see that Γ_c, the level width of the compound state is proportional to λ_c, the probability per unit time of the decay of the level. If the widths are less than the energy-difference between successive states D, *i.e.* $\Gamma_c << D$ then, the cross-sections for excitation as a function of incident energy will have discrete structure as shown in Fig. 13.3*a*. On the other hand, if $\Gamma_c >> D$, the cross-section becomes a smoothly varying function of energy as shown in Fig. 13.3*b*. The quantity D is normally referred to level spacing. It may be realised that Γ_c in the above discussion is the total width. But as we will see later, this consists of many partial widths, corresponding to different types of emitted particles and different energy channels.

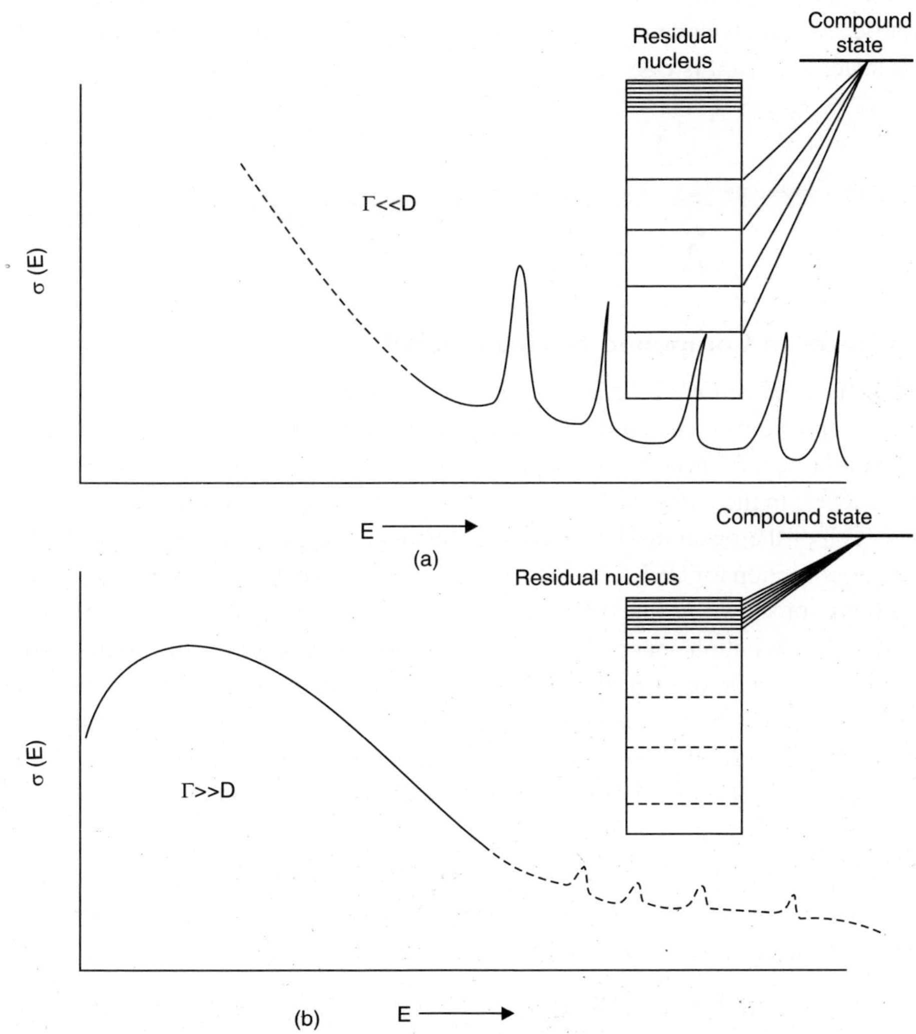

Fig. 13.3 The emitted spectra (solid lines) (*a*) when the levels in the residual nucleus are far apart, i.e. Γ << D and (*b*) when the levels are very close by, i.e. Γ >> D.

We now discuss the case of the resonances when $\Gamma_c << D$. Let us take the case of interaction of neutrons with a nucleus. The major modes of reaction will be (*i*) elastic scattering (*n*, *n*) and (*ii*) reaction, *i.e.* (*n*, *n′*) and (*n*, *R*), where (*n*, *n′*) corresponds to inelastic scattering and (*n*, *R*) corresponds to the reaction proper. According to Eq. 13.8, the total width Γ_c is then given by:

$$\Gamma_c = \Gamma_n + \Gamma_R \qquad ...(13.9)$$

where Γ_n represents width for elastic scattering and *R* represents the reaction including inelastic scattering. We now define a quantity Γ_i^r, the reduced width in such a way that the overall rate of decay of state *C* to a state *i*, can be written as:

$$\frac{\lambda_i}{\lambda_c} = \Gamma_i^r \qquad ...(13.10)$$

where λ_c is the total probability per unit time of the decay of compound state and λ_i is the partial probability for decay in a given channel *i*. It is being assumed here that the compound state decays to many intrinsic states *i*. Then it is easy to see from Eqs. 13.8 and 13.10 that,

$$\Gamma_i^r = \frac{\Gamma_i}{\Gamma_c} \qquad ...(13.11)$$

and

$$\sum_i \Gamma_i^r = \sum_i \frac{\Gamma_i}{\Gamma_c} = 1 \qquad ...(13.12)$$

13.2.2 Resonances in Compound Nucleus Model

While discussing the implication of the level width Γ_c of a level, we showed in Fig. 13.3 that the higher end of the energy spectrum of the emitted particles will have imprint of discrete levels indicated by peaks in the spectrum which showed the properties of the states in the residual nucleus. To obtain the properties of the levels in the compound nucleus, however, we require to study the excitation function, *i.e.* the cross-section of the reaction as a function of incident energy as shown in Fig. 13.4 for the case of total neutron cross-section for U^{238}. The energetics of the levels in the compound nucleus and expected excitation functions are shown in Fig. 13.5. As the levels in the compound nucleus in such a reaction will correspond to the excitation energy $E_x = E_{\text{inc}} + Q$, which in the case of (*n*, γ) will be equal to E_{inc} + Binding energy, *i.e.* of the order of 8 MeV, when the incident neutrons are of very low energy say in electron-volts. At such high excited energies, the levels width of the levels are generally in the range of a few electron volts. So if the incident beam energy resolution is less than Γ_c and $\Gamma_c << D$, then one observes resonance peaks as shown in Fig. 13.5 and are actually observed in an experiments shown in Fig. 13.4.

The value of total level width as measured in an experiment like given in Fig. 13.4, corresponds to decay to all possible channels and from the principle that the total probability λ_c is the sum of partial probabilities λ_i, one can write from Eq. 13.8:

$$\lambda_c = \sum_i \lambda_i \qquad ...(13.13a)$$

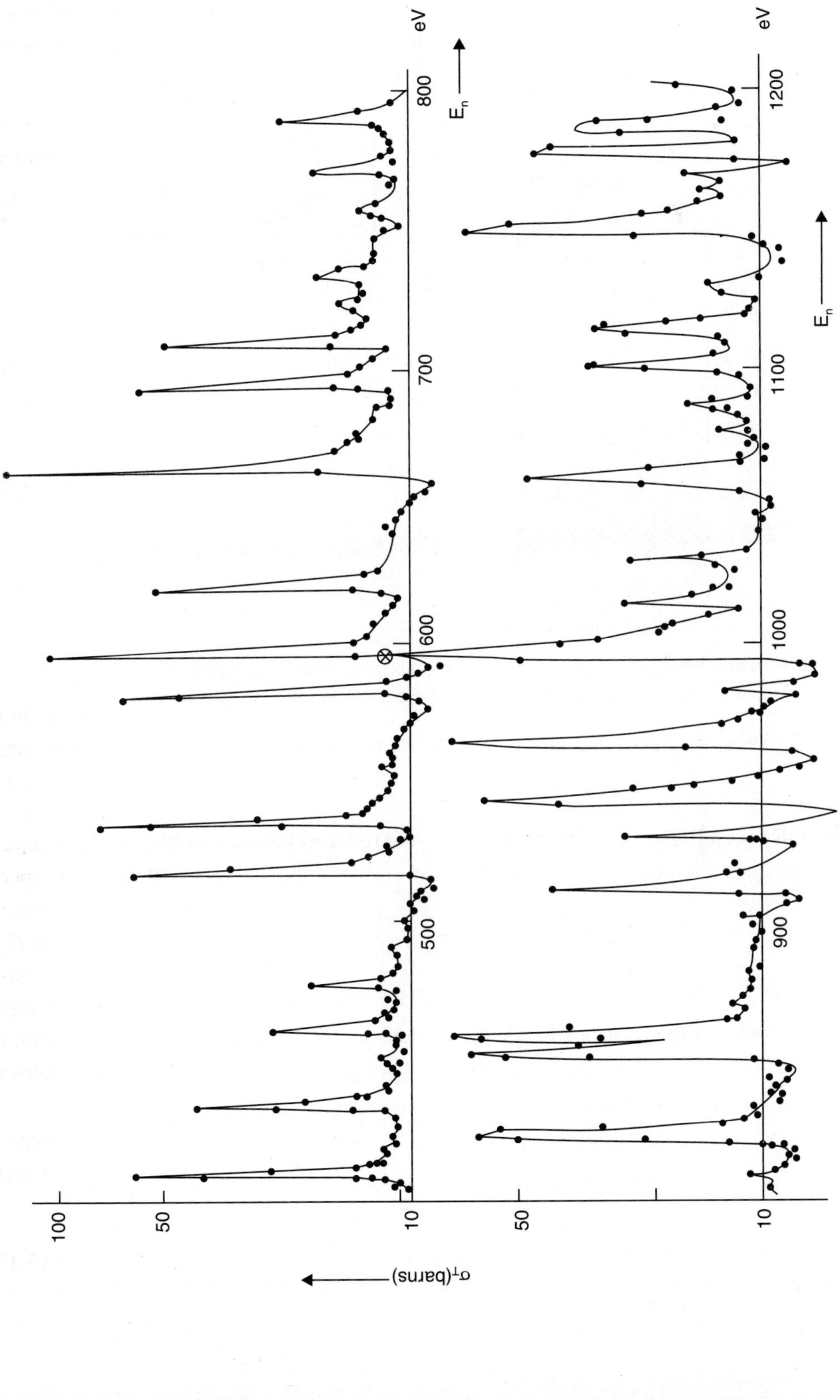

Fig. 13.4 Resonances in the total cross-sections for interaction of slow neutrons with U^{238} from 400 eV–1200 eV [after Firk et al. (Ref. 4)]. The minimum in the cross-sections are due to the interference effects [Eq. 13.32]

and hence, $$\Gamma_c = \sum_i \Gamma_i \qquad ...(13.13b)$$

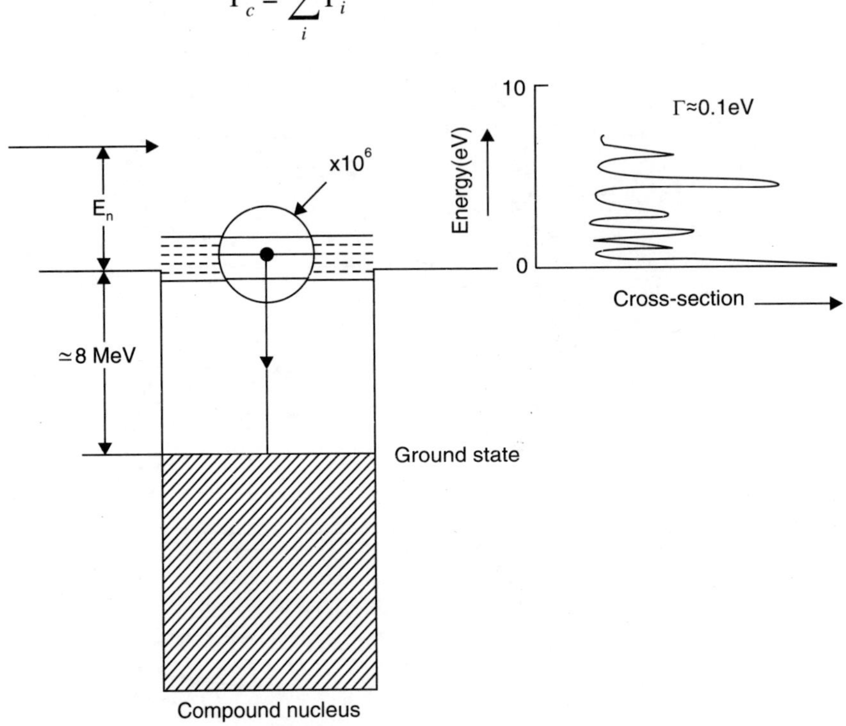

Fig. 13.5 Relationship of energies of ground and excited states of compound nucleus, incoming energy E_n and resonances in the excitation function (Ref. 4).

From Eq. 13.13*b*, it is evident that any Γ_i say Γ_n is a fraction of Γ_c from which one can write:

$$\Gamma_n = \Gamma_n^r \Gamma_c \qquad ...(13.14)$$

where Γ_n is the width for emission of neutrons and Γ_n^r is the reduced neutron width, *i.e.* the fraction of the total width which goes into neutron emission. One can further express Eq. 13.14 using Eq. 13.8 as:

$$\Gamma_n = \Gamma_n^r \hbar\lambda_c = \Gamma_n^r \hbar \frac{v_e}{2R} \times \frac{1}{N} \qquad ...(13.15)$$

where N = Total number of traversals in the nucleus before decay.

Where v_e = velocity of the neutrons inside the nucleus and R = Radius of the target nucleus. We now proceed to show that Γ_n^r is proportional to $1/v$, *i.e.* it obeys $1/v$ law, where v is the velocity of incident particles. This is a well-known law in neutron cross-sections and hence its importance.

It is easy to see that, one can write:

$$1/\lambda_c = \tau_c = 2R/v_e \times \text{Total number of traversals, before compound state decays} \qquad ...(13.16)$$

The total number of traversals are proportional to $K/4k$ as shown below for $k << K$. The wave-function $\psi(x)$ of the incident projectile ($x < 0$) inside the nucleus and ($x > 0$) outside of the potential can be written as:

$$\psi(x) = A \exp e^{ikx} + B \exp e^{-ikx} \quad x < 0$$
$$\psi(x) = C e^{iKx} \qquad x > 0 \qquad \text{...(13.17)}$$

where
$$k = \frac{(2ME)^{1/2}}{\hbar} \quad x < 0 \qquad \text{...(13.18a)}$$

and
$$K = \frac{[2M(E + V_o)]^{1/2}}{\hbar} \quad x > 0 \; [V = -V_o] \qquad \text{...(13.18b)}$$

where V_o is the depth of a nuclear potential, *e.g.* a square well potential from which transmission coefficient T is given by (*see* Schiff[5]):

$$T = \frac{|A|^2 - |B|^2}{|A|^2} \qquad \text{...(13.19)}$$

By equating the value and derivative of $\psi(x)$ at $x = 0$, we obtain:

$$T = \frac{4kK}{(K + k)^2} \qquad \text{...(13.20a)}$$

which for $k << K$, *i.e.* at low energies becomes:

$$T = \frac{4k}{K} \qquad \text{...(13.20b)}$$

The probabilities for particles to be in some range of x values in the incident, reflected and transmitted waves, are proportional to $|A|^2$, $|B|^2$ and $|A|^2 - |B|^2$. These probabilities are proportional to the currents, *i.e.* the number of particles per second passing each x value times, the length of time they spend in the range of x value which is inversely proportional to their velocity. Hence I/T which corresponds to the inverse of probability of transmission will signify the number of reflections in this period. One can, then write from Eq. 13.16 as:

$$\tau_c = \frac{2Rk}{V_e \hbar} = \frac{2R}{V_e} \frac{V_c}{4V_i} \cong \frac{R}{2V_i} \qquad \text{...(13.21a)}$$

where V_i is the velocity of the particle incident on the nucleus and V_e is the velocity of the particle inside the nucleus. However, Eq. 13.21*a* is only approximate. If we use the correct expression of T, *i.e.* Eq. 13.20*a* and take into account the fact that for getting out, a neutron must reach a radius considerably larger than R, Eq. 13.21*a* changes to a more accurate (but still not exact), expression:

$$\tau_c = \frac{2}{V_i} = \frac{1}{\lambda_c} \qquad \text{...(13.21b)}$$

Then from Eq. 13.15, we can write:

$$\Gamma_n^r = \frac{\Gamma_n}{\hbar \lambda_c} = \frac{R\Gamma_n}{\hbar V_i} = \frac{RM}{2\hbar} \frac{\Gamma_n}{E^{1/2}} \qquad \text{...(13.22)}$$

From definition 13.14, the reduced width Γ_n^r is the fraction of total width, which goes into neutron emission, hence, the cross-section for formation of any individual state by neutron is given by:

$$\sigma\,(n, n) = \frac{\Gamma_n}{E^{1/2}} \propto \frac{\Gamma_n}{V} \qquad ...(13.23)$$

for very low energies. This is well-known $1/V$ law.

Resonance Cross-sections

We have already seen in Figs. 13.3 and 13.4, that the phenomenon of resonances in the excitation function is observed in many reactions. This occurs basically because when the excitation energy is exactly equal to the energy of a level in the compound nucleus, there is an exact overlap between the wave-functions of incident particle and the target nucleus giving rise to the phenomenon of resonance in the interaction cross-section. As a matter of fact, it is the properties of the wave-function of neutron of the incident particle just inside the nuclear surface of target plus incident-particle, which decides the resonance behaviour of the interaction. We will discuss this in details in the next section. The phenomenon, however, resembles very much with the emission of light in the resonance regions discussed in details in Classical Electrodynamics by J.D. Jackson[5] (p. 601). The line-shape of such an emission of light in optics and neutrons in nuclear physics is given by Lorentz shape:

$$\sigma\,(n, R) \propto \left[(E - E_r)^2 + \left(\frac{\Gamma}{2}\right)^2\right]^{-1} \qquad ...(13.24)$$

where E_r is the energy at the centre of resonance and E is the energy of the incident neutrons. The expression in Eq. 13.24 has a maximum at $E = E_r$ and falls to half of that value at $E - E_r = \Gamma/2$ which are the expected properties of a resonance.

We can now write the cross-section of any reaction in the region of resonance. Classically, the maximum cross-section of any reaction corresponds to the situation when it is as large as the actual area of the target within which the absorption is strong, *i.e.* about πR^2. This means that every particle on the nuclear surface will form a compound nucleus. Then, one can write for incident neutrons:

$$\sigma_c\,(n) = \pi R^2 \qquad ...(13.25a)$$

If we take into account, the physical extension of the incident particle and consider it as of the order of de-Broglie wavelength of the particle, Eq. 13.25*a* may be modified to:

$$\sigma_c\,(n) = \pi\,(R + \lambda\!\!\!^{-})^2 \qquad ...(13.25b)$$

which for large values of $\lambda\!\!\!^{-}$, *i.e.* for small energies becomes:

$$\sigma_c\,(n) = \pi\,\lambda\!\!\!^{-2} \qquad ...(13.25c)$$

As shown in Eq. 13.23 for $\sigma\,(n, n)$, it can also be seen that for reaction cross-section also, that

$$\sigma\,(n, R) \propto \frac{\Gamma_n}{E^{1/2}} = \frac{f\,\Gamma_n}{E^{1/2}} \qquad ...(13.26)$$

This can be seen from the argument that for any reaction, which corresponds to the emission in a particular mode, incident neutron has to enter the same mode of compound state as for neutron emission hence, the $\Gamma_n / E^{1/2}$ dependence still holds as in Eq. 13.23. Further, realising that λ_i is the probability of decay per unit time in a particular channel i, one can write from Eq. 13.12 and 13.13:

$$\sigma(n, R) \propto \frac{\lambda_R}{\lambda_R + \lambda_n} \propto \frac{\Gamma_R}{\Gamma_R + \Gamma_n} \propto \frac{\Gamma_R}{\Gamma} \quad ...(13.27)$$

Now combining Eqs. 13.24 to 13.27, one can write for a resonance condition for low energies, the reaction cross-section as:

$$\sigma(n, R) = \pi \lambda\!\!\!^{-2} \frac{\Gamma_n \Gamma_R}{(E - E_r)^2 + \Gamma^2 / 4} \frac{f}{\Gamma E^{1/2}} \quad ...(13.28a)$$

It is easily seen from Eq. 13.28a that for $E = E_r$ and $\Gamma_n \approx \Gamma_R = \Gamma / 2$, the value of $\sigma(n, R)$ is maximum and is given by:

$$\sigma(n, R)_{max} = \pi \lambda\!\!\!^{-2} \frac{f}{\Gamma E^{1/2}} = \pi \lambda\!\!\!^{-2} \quad ...(13.28b)$$

Hence, $$f = \Gamma E^{1/2} \quad ...(13.28c)$$

and $$\sigma(n, R) = \pi \lambda^2 \frac{\Gamma_n \Gamma_R}{(E - E_r)^2 + \Gamma^2 / 4} \quad ...(13.28d)$$

which, for a particular reaction channel x, can be written as:

$$\sigma(n, x) = \pi \lambda^2 \frac{\Gamma_n \Gamma_x}{(E - E_r)^2 + \Gamma^2 / 4} \quad ...(13.28e)$$

The width of a channel is connected to total reaction by, $\Gamma_R = \sum_x \Gamma_x$ and Eq. 13.28e assumes that the reaction cross-section in a given channel x is proportional to Γ_x, Eq. 13.28d represents the well known Breit-Wigner one level formula[6] for very low energy incident neutrons say with 1 = 0. For elastic scattering, *i.e.* for $\sigma(n, n)$, one then writes:

$$\sigma(n, n) = \pi \lambda\!\!\!^{-2} \frac{\Gamma_n^2}{(E - E_r)^2 + \Gamma^2/4} \quad ...(13.28f)$$

For $l \neq 0$, we again look at the problem semi-classically and express the angular momentum 1 as:

$$r_1 p = \sqrt{l(l+1)}\, \hbar \approx (l + 1)\, \hbar \quad ...(13.29a)$$

or $$r_1 \approx (l + 1) \frac{\hbar}{p} \approx (l + 1)\, \lambda\!\!\!^{-} \quad ...(13.29b)$$

This means, that a nucleon at a distance r_1 between $l\, \lambda\!\!\!^{-}$ and $(l + 1)\, \lambda\!\!\!^{-}$ has an angular momentum $1\hbar$. The area enclosed between these r-values then, gives an estimate, of the cross-section $\sigma_{R,\, l}$ for an incident projectile of angular momentum $1\hbar$ as:

$$\sigma_{R,\, l} \approx \pi (l + 1)^2 \lambda\!\!\!^{-2} - \pi l^2 \lambda\!\!\!^{-2} \approx (2l + 1)\, \pi \lambda\!\!\!^{-2} \quad ...(13.30)$$

This is the maximum cross-section for an incident projectiles with an orbital angular momentum $1\hbar$. Hence Eq. 13.28 should be modified for angular momentum $1\hbar$ as:

$$\Gamma(n, R) = \pi \lambda\!\!\!^{-2} (2l + 1) \frac{\Gamma_n \Gamma_R}{(E - E_r)^2 + \Gamma^2/4} \quad \text{...(13.31}a\text{)}$$

$$\Gamma(n, x) = \pi \lambda\!\!\!^{-2} (2l + 1) \frac{\Gamma_n \Gamma_x}{(E - E_r)^2 + \Gamma^2/4} \quad \text{...(13.31}b\text{)}$$

and

$$\sigma(n, n) = \pi \lambda\!\!\!^{-2} (2l + 1) \frac{\Gamma_n^2}{(E - E_r)^2 + \Gamma^2/4} \quad \text{...(13.31}c\text{)}$$

These are Breit-Wigner formula for resonances for $1 \neq 0$.

Equations 13.28 and 13.31 hold good strictly in the region of resonances. What happens in the regions away from resonances, *e.g.* in the energy range between two resonances, especially in the elastic scattering say (α, α) or (n, n) reactions? There will be interference effects between the re-emission of the incident particle by the compound nucleus and the scattering from the nuclear surface or from the potential outside the nucleus. We will discuss this subsequently in details. Enough to say here that because of this expected interference, we write the scattering cross-sections off the resonance as the square of the sum of amplitudes corresponding to resonance (coming from the re-emission from the compound nucleus) and off-resonance (from the nuclear surface or the potential outside the surface). This leads to the expression say for neutrons as:

$$\sigma_{sc,\,l} = (2l + 1)\, \pi \lambda\!\!\!^{-2} \left| A_{res}^{l} + A_{pot}^{l} \right|^2 \quad \text{...(13.32}a\text{)}$$

where

$$A_{res}^{l} = \frac{i\Gamma_n}{(E_n - E_\gamma) + (1/2)\, i\Gamma} \quad \text{...(13.32}b\text{)}$$

and

$$A_{pot}^{l} = \exp(-2i\,\xi_l) - 1 \quad \text{...(13.33}a\text{)}$$

which reduces to, for $1 = 0$, as:

$$A_{pot}^{0} = \exp(2\, i\, K\, R) - 1 \quad \text{...(13.33}b\text{)}$$

The meaning of ξ will be discussed subsequently in Section 13.2.3.

When $E \neq E_r$ and $E - E_r >> \Gamma$, we can neglect A_{res} in Eq. 13.32*a* and write σ_{sc}^{l} as:

$$\sigma_{sc}^{l} = (2l + 1)\, \pi \lambda\!\!\!^{-2} \left| A_{pot}^{l} \right|^2 \quad \text{...(13.34)}$$

Also we will derive the expression for A_{pot}^{0} in the next section. Experimentally Eq. 13.32 are applicable only for scattering. We have already seen in Fig. 13.4, the effects of interference of A_{res} and A_{pot}, for (n, n) reaction. The minimum in the cross-section are due to this effect. Similarly while such minima have been observed for (α, α) for (p, p) but no such minima for (α, n) or (p, n) as shown in Figs. 13.6*a* and 13.6*b*.

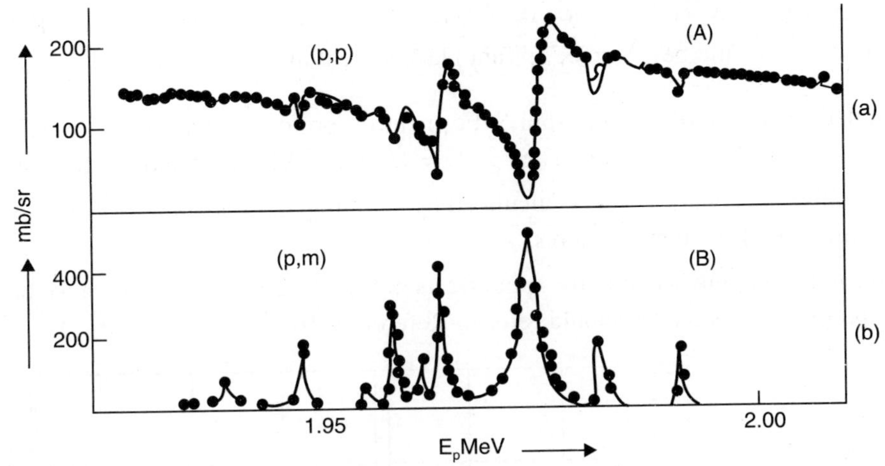

Fig. 13.6a (*a*) The cross-section vs energy for protons on Ca^{48} for elastic scattering (p, p) showing interference between resonance and potential scattering. (*b*) Reaction (p, n) leading to final nucleus Sc^{48} in the ground state. It shows no such interference (Ref. 7).

A special case of resonance is the capture cross-section say σ (*n*, γ) or σ (*p*, γ) near a resonance with *l* = 0. We then can write the reaction cross-section as:

$$\sigma_{r,0} = g\,(S)\,\pi\,\lambda\!\!\!^{-2}\,\frac{\Gamma_n \Gamma_{\text{rad}}}{(E - E_r)^2 + (\Gamma/2)^2} \qquad \text{...(13.35}a\text{)}$$

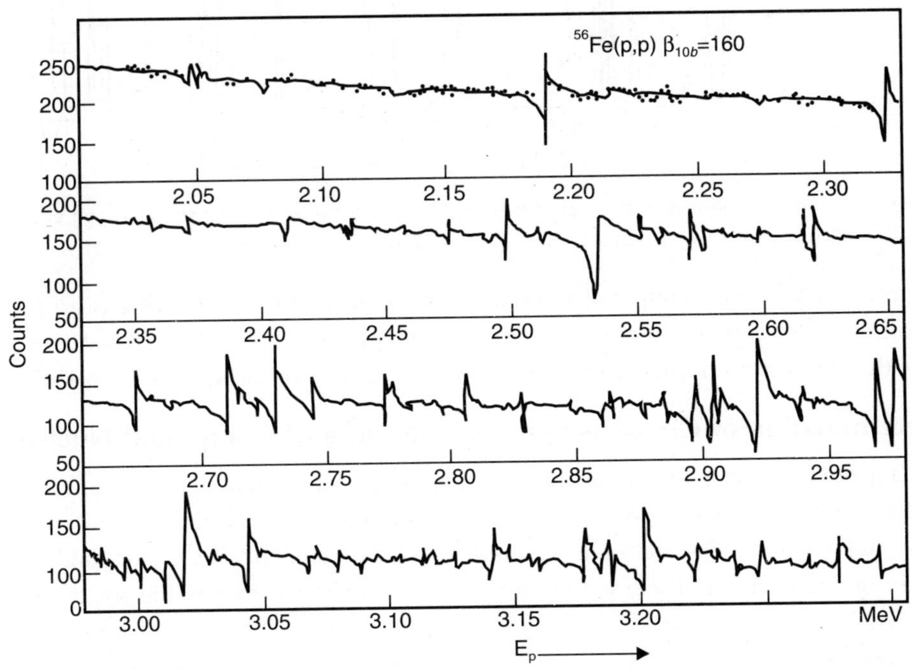

Fig. 13.6b Differential cross-sections at 16° for the process of elastic scattering in Fe^{58} (p, p). Very strong interference exists between the Coulomb scattering, [i.e. potential scattering] and scattering through resonant states of Co^{59} (D.P. Lindstron et al.) (Ref. 8).

where $$g(S) = \frac{2S+1}{(2s+1)(2I+1)} \tag{13.35b}$$

$g(S)$ is the statistical weight of channel spin S. For neutron-proton scattering $s = I = 1/2$ and S can be either 0 (single-state) or 1 (triplet state). Then $g(S) = 1/4$ for singlet state and 3/4 for the triplet-state.

Basically, the statistical weight is calculated by summing over m-states of S and averaging over the incoherent mixture of incident channels.

We have not taken into account, the spin effects in Eqs. 13.28 to 13.34, which we have now done for neutron capture. For this reader should refer to Reference (9).

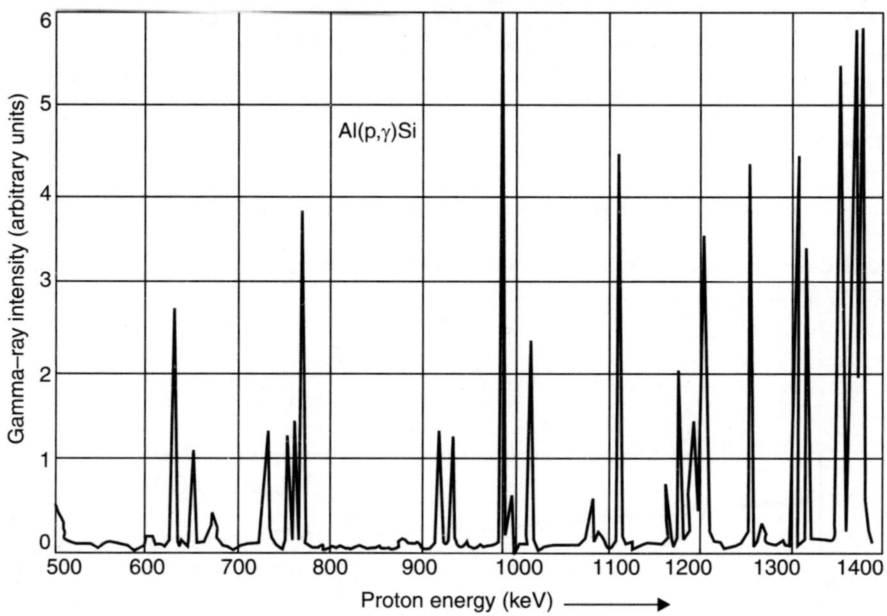

Fig. 13.7 The resonances in the yield of Al (p, γ) Si reaction as a function of proton energy; (Bronström K.G., T. Huus and B. Tangen, Phy. Rev. 82, 190 (1951) (Ref. 10).

Figure 13.7 shows an example of resonances in (p, γ) reactions, following Eq. 13.35. It may be seen that there are no interference minima as were found in (p, p) reaction in Fig. 13.6. These experimental results show the correctness of the various resonances equations from Eqs. 13.32 to 13.35.

13.2.3 Determination of Cross-sections on the Basis of Compound Nucleus Model

A. General Theory: We have already dealt in Chapter 4, the basic quantum mechanical procedures of scattering of nucleons from a nucleus. We proceed in a somewhat similar manner to derive expressions for scattering and reaction cross-section in a general case and then apply it to compound nucleus theory. Proceeding in the manner, we discussed the Eq. 4.22, we write the expression for the incident plane wave as:

$$e^{ikz} \approx \sum_l \frac{i^{l+1}}{2kr}(2l+1)\left[e^{-i(kr-l\pi/2)} - e^{i(kr-l\pi/2)}\right] P_l(\cos\theta) \qquad \text{...(13.36)}$$

We now introduce a factor η_l which is a complex amplitude for the lth partial wave and is related to the phase shift δ_l by $\eta_l = |\eta_l| e^{2i\delta_l}$. We write the wave-functions $\psi(r)$ describing the outgoing total wave after interaction as:

$$\psi(r) \approx \sum_l \frac{i^{l+1}}{2kr}(2l+1)$$

$$\left\{\exp -i\left(kr - \frac{l\pi}{2}\right) - \eta_l \exp i\left(kr - \frac{l\pi}{2}\right)\right\} P_1(\cos\theta) \quad ...(13.37)$$

It is apparent, that η_l is also connected with f_1 of Eq. 4.28. We are however going to use η_l in Eqs. 13.37 and 13.38 because, the symbol f_1 will now be used in a different manner as discussed subsequently. It is evident from Eq. 13.37 that η_l signifies the change in the amplitude and phase of the outgoing part of the spherical wave. Then ψ_{sc} can be expressed as:

$$\psi_{sc} = \psi(r) - e^{ikz}$$

$$= \sum_l \frac{i^{l+1}}{2kr}(2l+1)(1-\eta_l)\exp\left[i\left(kr - \frac{l\pi}{2}\right)\right] P_1(\cos\theta) \quad ...(13.38)$$

Then one can write (as described earlier in Eq. 4.26*a*)

$$N_{sc}(\theta)\, d\Omega = \frac{\hbar}{2iM}\left|\left(\frac{\partial\psi_{sc}}{\partial r}\psi_{sc}^* - \frac{\partial\psi_{sc}^*}{\partial r}\psi_{sc}\right)\right|_{r=r_o} r^2\, d\Omega$$

$$= \frac{\hbar k}{M}|\psi_{sc}(r,\theta)|^2 r_o^2\, d\Omega \quad ...(13.39)$$

Writing the cross-section of scattering σ_{sc} as:

$$\sigma_{sc}(\theta) = \frac{N_{sc}(\theta)\, d\Omega}{v\, d\Omega} \quad ...(13.40)$$

where $v = \hbar k/M$,

It can be seen from Eqs. 13.38 and 13.39, that

$$\sigma_{sc}(\theta)\, d\Omega = \frac{\pi}{k^2}\left|\sum_{l=0}^{\infty}\sqrt{2l+1}(1-\eta_l)\, Y_{l,0}(\theta)\right|^2 \quad ...(13.41)$$

Realising that

$$N_{sc} = \int N_{sc}(\theta)\sin\theta\, d\theta\, d\phi$$

We can write:

$$\sigma_{sc}^l = \int \sigma_{sc}(\theta)\, d\Omega = \frac{\pi}{k^2}(2l+1)|1-\eta_l|^2 \quad ...(13.42)$$

In these equations N_{sc}, corresponds to the number of particles being scattered per second from the target in the whole space.

Similarly for writing the expression for reaction cross-section, we realise that it corresponds to the net flux entering the sphere with radius r, *i.e.*, it is the flux absorbed and hence corresponds to the reaction. The total wave-function including the effects of the target nucleus as has been given in Eq. 13.37 corresponds to this situation. Hence, one can write for the reaction in analogy to Eq. 13.39, an expression for N_r as:

$$N_r = -\frac{\hbar}{2\,i\,M}\int\left(\frac{\partial\psi}{\partial r}\psi^* - \frac{\partial\psi^*}{\partial r}\psi\right) r_o^2 \sin\theta\, d\theta\, d\phi \qquad ...(13.43)$$

It may be seen that the angular dependence of N_r does not make any sense as it is a case of absorption in the nucleus and not any emission. Different reaction products going out correspond to the division of the total reaction cross-section into different channels. Also we have used total wave-function ψ in Eq. 13.43, which corresponds to $\psi(r) = e^{ikz} + \psi_{sc}$, because N_r represents the total net flux into the sphere.

One then obtains the reaction cross-section as:

$$\sigma_r = \frac{N_r}{V} = \pi\,\lambda\!\!\!^{-2}\,(2l+1)\left[1 - |\eta_l|^2\right] \qquad ...(13.44)$$

It is easy to see that for $\eta_l = -1$

$$\sigma_{sc,\,l} = 4\,(2l+1)\,\pi\,\lambda\!\!\!^{-2} \qquad ...(13.45a)$$

and

$$\sigma_r = 0 \qquad ...(13.45b)$$

It is interesting to see that for this situation, ($\sigma_{sc,\,l}$) is four times the maximum physical cross-section offered to the incident beam.

Again for $\eta_l = 0$

$$\sigma^l_{reaction} = \sigma^l_{sc} = \pi\lambda\!\!\!^{-2}\,(2l+1) \qquad ...(13.45c)$$

For $\eta_l = 1$, on the other hand,

$$\sigma^l_{reaction} = \sigma^l_{sc} = 0 \qquad ...(13.45d)$$

The above three equations show that it is impossible to have only reaction cross-section without scattering cross-section. But it is possible to have only scattering cross-section without any reaction cross-section. Also maximum value of reaction cross-section is equal to the scattering cross-section and hence half of the total cross-section. For a black nucleus for which, all the projectiles falling on it are absorbed, the maximum 1 value can be given by:

$$l_{max}\,\hbar = \frac{R\hbar}{\lambda\!\!\!^-} \text{ or } l_{max} = \frac{R}{\lambda\!\!\!^-} \qquad ...(13.45e)$$

For this case of l_{max}:

$$\sigma_{reaction} = \sigma_{sc} = \sum_{l=0} R/\lambda\!\!\!^-\,(2l+1)\,\pi\lambda\!\!\!^{-2} = \pi R^2 \qquad ...(13.45f)$$

Hence $$\sigma_{total} = \sigma_{reaction} + \sigma_{sc} = 2\pi R^2 \quad ...(13.45g)$$

i.e., the total cross-section is twice the geometrical cross-section. Both Eqs. 13.45*a* and 13.45*g* correspond to the shadow scattering due to quantum mechanical effects. It means physically, that because of the wave-nature of the incident projectiles, there is diffraction in the forward direction, which accounts for these anomalies. The shadow behind the nucleus is not perfectly sharp and at far away distance say L from the nucleus in the region of Fraunhoffer diffraction, the edge of the shadow is blurred over its full extent and the shadow itself has disappeared. The distance L is of the order of $R/\lambda\!\!\!^{-}$ or larger. The scattering angles of shadow scattering are small. Particles scattered beyond L in practice are counted and give rise to the extra cross-section.

B. Calculation of Cross-sections: Making use of the general theory described in the last section, we now proceed to discuss the actual mathematical steps required to derive the various cross-sections. To start with, we only discuss $l = 0$ case, *i.e.* only very low energy projectiles. Further we assume the incident particles to be neutrons and also neglect their spins. For such projectiles, the wave-function at $r > R$ (where $R = R_a + R_x$ is the sum of radii of incident and target nuclei) may be written as:

$$\frac{d^2\psi_0}{dr^2} + k^2\psi_0 = 0 \quad ...(13.46a)$$

Writing $\psi_0(r) = U_0(r)/r$, Eq. 13.46*a* becomes:

$$\frac{d^2U_0(r)}{dr^2} + k^2 U_0 = 0 \quad ...(13.46b)$$

The solution of Eq. 13.46*b* may be expressed as:

$$U_0(r) = a\exp(-ikr) + b\exp(ikr) \quad ...(13.47)$$

Comparing this with Eq. 13.37 for $l = 0$, it is easily seen, that one can write:

$$a = \frac{i\sqrt{\pi}}{k};\ b = \eta_0 a \quad ...(13.48)$$

The quantities a and b and hence η_0 may be determined from the continuity conditions at the surface for which one defines a logarithm derivative f_0 at $r = R$, *i.e.*,

$$f_0 \equiv R\left[\frac{dU_0/dr}{U_0}\right]_{r=R} \quad ...(13.49)$$

Substituting the value of η_0 from Eq. 13.48 into Eq. 13.49, it is evident that

$$\eta_0 = \frac{f_0 + ikR}{f_0 - ikR}\exp(-2ikR) \quad ...(13.50)$$

Using this value of η_0, it is easy to see from Eq. 13.42 for $l = 0$ and Eq. 13.50 that

$$\sigma_{sc,0} = \pi\lambda\!\!\!^{-2}\left|A_{res} + A_{pot}\right|^2$$

Where $$A_{res} = \frac{-2\,i\,k\,R}{f_0 - i\,k\,R};\; A_{pot} = \exp(2\,i\,k\,R) - 1 \qquad ...(13.51)$$

The meaning of the two terms A_{res} and A_{pot} can be seen from Eq. 13.32 and (13.33) of the last section. The term A_{res} is dependent on 'f_0' which contains information about internal nuclear structure. Hence, only this term can give rise to the resonances. On the other hand, the term 'A_{pot}' only depends on k and R and hence represents interaction only with the surface and corresponds to the scattering from the potential surface. In the extreme case of hard sphere, $U_0 = 0$ at $r \leq R$, then f_0 becomes infinite and hence, A_{res} become zero and only A_{pot} exists.

In a similar manner, one can write the expression for reaction cross-section from Eq. 13.44 by substituting the value of η_0 from Eq. 13.50 and obtain:

$$\sigma_{\text{reaction, }0} = \pi \lambda\!\!\!^{-2} \frac{-4\,k\,R\,I_m\,f_0}{(\text{Re}\,f_0)^2 + (\text{Im}\,f_0 - k\,R)^2} \qquad ...(13.52)$$

The division of f_0 into imaginary f_0 and real f_0 corresponds to reaction or absorption component and elastic scattering component respectively. This can be seen from the definition of f_0 in Eqs. 13.49 and 13.50 and the fact that an imaginary f_0 corresponds to an absorption for outgoing wave in Eq. 13.47, while real f_0 will mean an outgoing wave with a phase change, but no absorption.

As $\sigma_{r,0}$ has to be positive, this requires that,

$$I_m\,(f_0) \leq 0 \qquad ...(13.53)$$

which also follows from the fact that it is required that,

$$|\eta_0|^2 \leq 1 \qquad ...(13.54)$$

For the charged particles as projectiles one has to introduce the change in the potential, due to Coulomb interaction. For these details, *see* Nuclear Physics by Blatt and Weisskopf[2], p. 330.

It may be further realised that $\sigma^l_{\text{reactions}}$ is the total reaction cross-section, *i.e.* it does not tell us about the division into various channels. Nor does it describe the angular distribution of different channels. Of course, the angular distribution of the total reaction cross-section does not make any sense.

For individual channels say $\alpha \to \beta$, however, we can write the wave-function ψ in the incident channel as a plane wave, *i.e.*,

$$\psi_{inc}(\alpha) = e^{i\,k\,z}\,(v_\alpha)^{-1/2}\,\chi_a \text{ (A plane wave in channel } \alpha) \qquad ...(13.55a)$$

For outgoing wave in channel β

$$\psi_{outgoing}(\beta) = q(\theta)\,\frac{\exp(i\,k\,R)}{r}\,(v_\beta)^{-1/2}\,\chi_\beta \qquad ...(13.55b)$$

where θ is the angle between the outgoing particle and the incident particle a. Equation 13.55*b* represent a spherical wave of β channel over which a θ dependent function $q(\theta)$ called the reaction ampltitude of $\alpha \to \beta$ reaction is superimposed. The function χ_α and χ_β are the wave-functions involving the specification of the direction of the spins of the particles.

The differential reaction cross-section for (α, β) reaction may, then, be written as:

$$d\,\sigma = |\,q(\theta)\,|^2\,d\,\omega \qquad ...(13.56a)$$

If we take into account the spin of particles *a* and *b* and if the incident beam is unpolarised and if the detector is not sensitive to spin direction, the observed reaction cross-section is obtained by averaging over the spin directions of *a* and target *X* and summing over the spin directions of outgoing particle and residual nucleus *Y*. We call the spin of the incident particle *s* and of the target nucleus *i*, then

$$d\,\sigma = \frac{1}{2s+1} \times \frac{1}{2i+1} \sum_{\rho} q_\rho\,(\theta, \phi)^2\,d\omega \qquad ...(13.56b)$$

where ρ indicates the four spin indices of incident particle, the target nucleus and the emerging particle and the residual nucleus. The derivative of $|\,q_\rho\,(\theta, \phi)^2\,|$ is however, involved and will not be discussed here.

C. Decay (or Emission) Rates of the Compound Nucleus: We have, till now, discussed, the theory of the compound nucleus formation, which corresponds to the total reaction cross-section. Now we discuss the decay of the compound nuclear state into various reaction channels.

Because of the assumptions of the compound nucleus theory, it is quite easy to understand that a reaction cross-section (*a*, *b*) may be written as (it has been discussed earlier, while explaining Ghoshal's experiment):

$$\sigma\,(a, b) = \sigma_c\,(a)\,G_c\,(b) \qquad ...(13.57)$$

where $\sigma_c\,(a)$ is the cross-section for the formation of compound nucleus by an incident projectile *a* and $G_c\,(b)$ is the probability of the compound state to decay into channel *b*. It is assumed in Eq. 13.57, that the two processes are independent of each other. If we sum over all the outgoing channels then of course, we get:

$$\sum_b G_c\,(b) = 1 \qquad ...(13.58)$$

This corresponds to the assumption that total reaction cross-section is equal to the cross-section for compound states formation. Equation 13.58 for incoming channel α and outgoing channel β may be expressed as:

$$\sigma\,(\alpha, \beta) = \sigma_c\,(\alpha)\,G_c\,(\beta) \qquad ...(13.59)$$

We have already seen in Eq. 13.8 that

$$\Gamma\,(E_c) = \frac{\hbar}{\tau(E_c)} \qquad ...(13.60)$$

where E_c represents the compound state energy. We have also discussed that the level width of any state is proportional to the probability of decay. $\Gamma\,(E_c)$ represents the total probability of decay and therefore it may be expressed as:

$$\Gamma\,(E_c) = \sum_\beta \Gamma_\beta\,(E_c) \qquad ...(13.61)$$

where β represents different channels. It is easy to see then, that the relative probability of decay $G_c(\beta)$ into a given channel β may be expressed as:

$$G_c(\beta) = \frac{\Gamma_\beta}{\Gamma} \quad ...(13.62)$$

It is well known in the formal theory of cross-sections (*see* Theoretical Nuclear Physics, Blatt and Weisskopf [2]) that reciprocity theorem holds good in all nuclear reactions.

This theorem states that the cross-sections σ (α, β) and reciprocal reaction σ (β, α) are related to each other by the relationship:

$$\frac{\sigma(\alpha, \beta)}{\lambda_\alpha^2} = \frac{\sigma(\beta, \alpha)}{\lambda_\beta^2} \quad ...(13.63)$$

Where λ_α and λ_β are the channel wavelengths. Now using Eqs. 13.63 and 13.59, it can be seen that

$$\frac{\sigma_c(\alpha)\, G_c(\beta)}{\lambda_\alpha^2} = \frac{\sigma_c(\beta)\, G_c(\alpha)}{\lambda_\beta^2} \quad ...(13.64)$$

Using Eq. 13.62, we can write Eq. 13.64 as:

$$\frac{\sigma_c(\alpha)\, \Gamma_\beta}{\Gamma \lambda_\alpha^2} = \frac{\sigma_c(\beta)\, \Gamma_\alpha}{\Gamma \lambda_\beta^2} \quad ...(13.65)$$

or

$$\frac{\sigma_c(\alpha)}{\Gamma_\alpha \lambda_\alpha^2} = \frac{\sigma_c(\beta)}{\Gamma_\beta \lambda_\beta^2} = \text{Constant} \quad ...(13.66)$$

Therefore,

$$\Gamma_\beta = \text{Const}\, \frac{\sigma_c(\beta)}{\lambda_\beta^2} \quad ...(13.67)$$

Hence, one can write from Eqs. 13.62 and 13.67,

$$G_c(\beta) = \frac{\Gamma_\beta}{\Gamma} = \frac{k_\beta^2\, \sigma_c(\beta)}{\sum_\gamma k_\gamma^2\, \sigma_c(\gamma)} \quad ...(13.68)$$

where $k = 1/\lambda$.

We are now ready to write an expression for the shape of the energy spectrum of the emitted particles on the basis of the above assumptions.

Let a particle of energy E_α be incident in channel α on the target and let E_β be the energy of the emitted particle, then they are energy related to each other by:

$$E_\beta = E_\alpha + Q_{\alpha\beta} \quad ...(13.69)$$

where $Q_{\alpha\beta}$ is the Q-value from channel α to β. Translating this into energy of the incident channel α and outgoing particle b, it can be proved that

$$E_{\beta,0} = E_{bY} = E_{\alpha,0} + Q \qquad \text{...(13.70)}$$

where $E_{\beta,0}$ is the kinetic energy in channel β, leaving the residual nucleus in the ground state and hence, it corresponds to the E_{bY}, *i.e.* the maximum energy of the outgoing particle b and the residual nucleus Y in the ground state. $E_{\alpha,0}$, is of course, the energy of the incident particle in channel α, when the target is in the ground state. It is easy to see then that for an excited state of the residual nucleus with an excitation energy of ε_β, energy in the outgoing channel E_β will be given by:

$$E_\beta = E_{bY} - \varepsilon_\beta \qquad \text{...(13.71)}$$

We have given in Fig. 13.3 a typical energy spectrum of the outgoing particles, along with the excitation levels of the residual nucleus. It is expected that the higher energy end of the spectrum contains evidence of discrete levels, while at low energy end there will be a continuous spectrum because of the high density of excited states of the residual nucleus. We give now the theory, which reproduces the continuous shape of such a spectrum.

The shape will evidently be governed by the shape of the decay function $G(E)\,dE$ given by:

$$G(E)\,dE = \sum_{E < E_\beta < E + dE} G_c\,(E_\beta) \qquad \text{...(13.72)}$$

Let the number of terms in the above sum be given by the number of levels of the residual nucleus Y with an excitation energy between ε and $\varepsilon - d\varepsilon$. If $W(E)$ is the level density, then the number of terms will be $W(E)\,dE$. We now insert the expression of $G_c\,(\beta)$ from Eq. 13.68 and assume that the denominator may be taken as constant, then

$$G(E)\,d\,E = \text{Constant}\ \sigma_c\,(\beta)\ k_\beta^2\ W_Y\,(\varepsilon)\,d\,\varepsilon \qquad \text{...(13.73)}$$

Remembering that k_β^2 is proportional to the incident energy E, we can write the shape of the spectrum from Eqs. 13.73 and 13.71 and Eqs. 13.59 and 13.68 as:

$$I_b\,(E)\,d\,E = \text{Constant}\ \sigma_c\,(\beta)\ E\ W_Y\,(E_{bY} - E)\,d\,E \qquad \text{...(13.74)}$$

Because of the use of $W(E)\,dE$ for the number of levels in the energy range of E and $E + d\,E$, this theory corresponds to continuum of energy levels. This theory is therefore called continuum theory or statistical model of nuclear reactions, where we have used $|\,d\varepsilon\,| = |\,dE\,|$ and $\varepsilon = E_{bY} - E$ from Eq. 13.71.

Here $\sigma_c\,(\beta)$ is the compound nucleus cross-section, if the incident particle had the energy of the outgoing channel, which is a function of channel energy $E = E_\beta$. Hence $E\,\sigma_c\,(\beta)$ is an increasing function of E. $W_Y\,(E_{bY} - E)$ is easily seen to be decreasing function of E, because $W_Y\,(\varepsilon)$, the level density increases with excitation energy ε; hence $W_Y\,(E_{bY} - E)$ decreases with E. Because of these two opposing trends, the function $I_b\,(E)$ has a maximum in the middle of the energy range.

We now define a quantity T, the nuclear temperature as:

$$\frac{1}{T} = \frac{d\,S(E)}{d\,E} \qquad \text{...(13.75)}$$

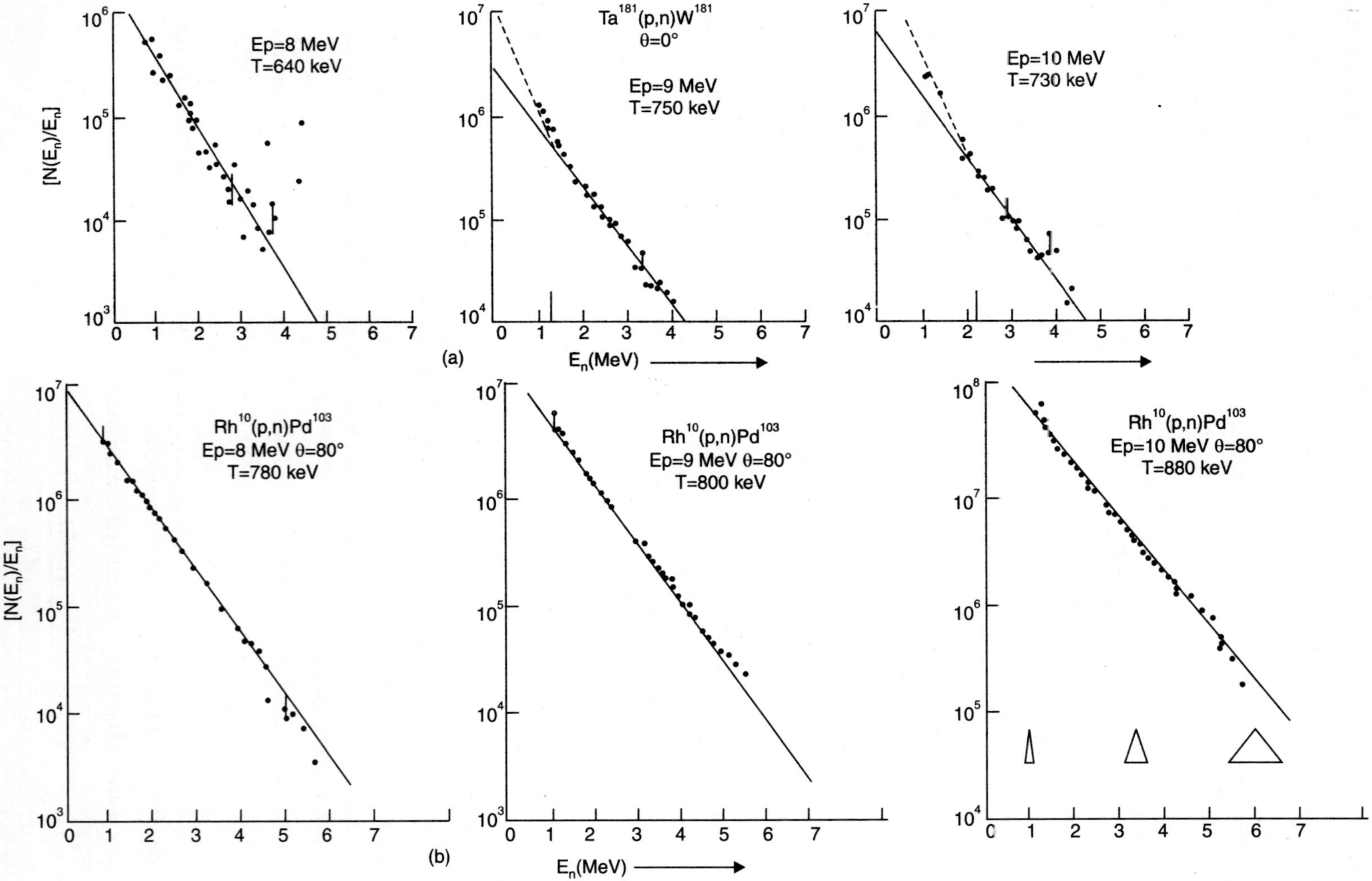

Fig. 13.8a The spectra of neutrons, plotted as log $(N(E_n)/E_n)$ versus E_n (MeV), from (p, n) section for Ta^{181} and Rh^{103} targets for 8, 9 and 10 MeV protons, Ref. (11).

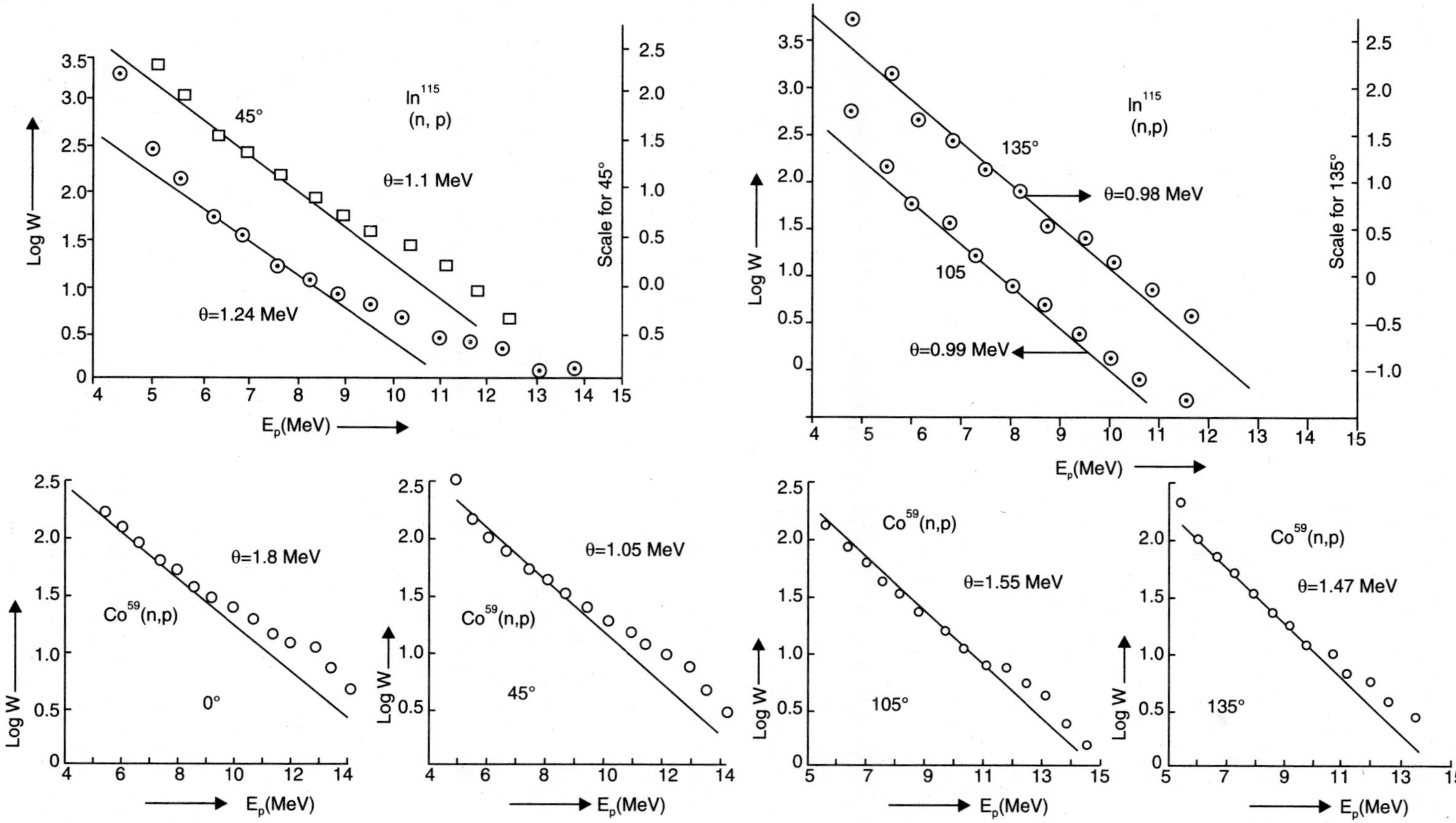

Fig. 13.8b The spectra of protons from (n, p) reaction for In^{115} and Co^{59} targets for 14 MeV neutrons at various angles. The spectra are plotted as log W versus E_p, where $W = I_p/\sigma_c(E_p)\,E_p$ (Ref. 12).

where $$S(E) = \log W(E) \qquad ...(13.76)$$

Let $S(E)$ which is entropy in classical thermodynamics be expanded through Taylor series, *i.e.*,

$$S(E_{bY} - E) = S(E_{bY}) - E\left(\frac{\partial S}{\partial E}\right)_{E = E_{bY}} + ... \qquad ...(13.77)$$

Then from Eqs. 13.75, 13.76 and 13.77 we can write:

$$W(\varepsilon) = \exp S(E_{bY}) \exp - E\frac{dS}{d\varepsilon}$$

$$= \text{Const} \exp\left(-E\frac{dS}{d\varepsilon}\right) - \text{Const} \exp\left(\frac{-E}{T(E_{bY})}\right) \qquad ...(13.78)$$

One may, therefore, write Eq. 13.74 as:

$$I_b(E)\, dE = \text{Const}\, E\, \sigma_c(E) \times \exp\left(\frac{-E}{T(E_{bY})}\right) dE \qquad ...(13.79)$$

If $T(E_{bY})$ in the above equation is taken as constant then,

$$\log\left(\frac{I_b(E)}{E\,\sigma_c(E)}\right) = -\left(\frac{1}{T}\right)E \qquad ...(13.80)$$

Hence a plot between $\log(I_b(E)/E\,\sigma_c(E))$ versus E will yield a straight line (as shown in Figs. 13.8*a* and 13.8*b*, representing actual experimental spectra), which will determine the nuclear temperature.

The spectra of emitted particles, as shown in Fig. 13.8*a* for (p, n) reactions and in Fig. 13.8*b* for (n, p) reaction shows that Eq. 13.80 holds good for low energy part of the emitted spectrum but, for high energy, the (n, p) experimental cross-sections are larger than expected from Eq. 13.80. These are due to pre-equilibrium effects, which we have discussed earlier and which will be discussed in Chapter 16 in details.

D. Statistical Theory of Level Densities: The concept of nuclear level density is applicable only for higher excitation energies, when $\Gamma >> D$. This is especially true for medium and heavy nuclei for energies much above the Fermi energies, *i.e.* above 8 MeV excitation energy as has been shown in Fig. 13.3. Though the degrees of freedom available to the system is limited because, the number of particles (nucleons) is not very large and at these energies, the degrees of freedom available to each particle is also limited obeying Fermi gas model of single particles, one still applies the principles of statistical model as a first approximation to calculate the nuclear level densities as first introduced in nuclear physics by Bethe[18] in (1937). We reproduce below the derivation of the formula for nuclear level density $W(E)$ showing its dependence on the excitation energy (E) and single particle level density a, which in turn again depends on A, the atomic weight of the nucleus concerned and excitation energy ε.

For the derivation, we follow Gilbert and Cameron[19], who have incorporated the basic elements of Bethe's treatment of the nucleus as a gas of two types of fermions, *i.e.* of protons and neutrons.

We start with a thermodynamic system, which obeys the well-known law of entropy S, *i.e.*,

$$S = k \ln \Omega \qquad ...(13.81a)$$

where k is Boltzmann constant and Ω is the total number of quantum states available to the system. We will now show that for a quantum system of N neutrons and Z protons, *i.e.* two Fermi gases, it is possible to write a similar law, *i.e.*,

$$S(\varepsilon) = \ln \rho(\varepsilon) \qquad ...(13.81b)$$

where $\rho(\varepsilon)$ is the density of states, so that $\rho(\varepsilon)\, d\varepsilon$ is the total number of nuclear states in the nucleus within the excitation energy range.

We start with the well-known definition of grand partition function P expressed in terms of thermodynamic free energy F and β where $\beta = 1/T$ and T is the thermodynamic temperature. Physically P is defined in such a way that the probability p_N^K that a system exists which has N particles and has the energy level ε_N^K can be written as:

$$p_N^K = W_N \exp - \beta\varepsilon_N^K \qquad ...(13.82a)$$

and

$$P \equiv e^{-\beta F} = \sum_{N,K} W_N \exp - \beta\, \varepsilon_N^K \qquad ...(13.82b)$$

In a nucleus of Z protons and N neutrons let neutrons; occupy single particle levels in energy with occupation number $n_S a_S$ and magnetic quantum number m_{1S}. Similarly for proton single particle levels, these quantities be designated as bs, Z_S and m_{2S}. Then a single state of the whole nucleus is defined by four constants of motion.

$$N = \sum_S n_S \qquad Z = \sum_S z_S$$

$$\varepsilon = \sum_S n_S\, a_S + \sum_S z_S\, b_S; \qquad M = \sum_S n_S\, m_{1S} + \sum_S z_S\, m_{2S}$$

Also the total angular momentum is a constant of motion. Then the grand Partition function P can be written as:

$$P = e^{-\beta F} = \sum_{N',Z',M',\varepsilon'} \exp[\beta(\mu_1 N' + \mu_2 Z' + \mu_3 M' - \varepsilon')] \qquad ...(13.83)$$

where μ_1 and μ_2 are the energies per neutron and per proton occupation numbers respectively and μ_3 is the energy per magnetic quantum number. One can, integrate Eq. 13.83, over energies by introducing the level-density $\rho(\varepsilon' N' Z' M')$, so that Eq. 13.83 becomes:

$$P \equiv e^{\lambda} = e^{-\beta F} = \sum_{N',Z',M'} \int d'\varepsilon \rho(\varepsilon' N' Z' M') \exp[\beta(\mu_1 N' + \mu_2 Z' + \mu_3 M' - \varepsilon')] \qquad ...(13.84)$$

One can then obtain ρ (*E N Z M*) by a succession of inverse Laplace transform from Eq. 13.84.

The calculation in Eq. 13.84 requires us to replace the sum in Eq. 13.83 by integrals over the energy ε′ introducing for this purpose the single particle level densities $g\ (\varepsilon, m_1)$ and $g\ (\varepsilon, m_2)$ for neutrons and protons, respectively. Though these single particle level densities are theoretically dependent on the excitation energy, the value of ρ (*E, N, Z, M*) have been calculated for the case, where all derivatives of $g_1\ (\varepsilon, m_1)$ and $g_2\ (\varepsilon, m_2)$ are taken to be zero. Then defining

$$g_1 = \sum g_1\,(\varepsilon_1, m_1);\; g_2 = \sum g_2\,(\varepsilon_2, m_2);\; g = g_1 + g_2 \qquad ...(13.85)$$

it can be seen that from inverse Laplace transform of Eq. 13.84, one can obtain:

$$\rho\,(\varepsilon') = \frac{1}{2\pi_i}\int_{\gamma - i\infty}^{\gamma + i\infty} d\beta\; e^{\beta\,(\varepsilon - F)} \quad [F = \text{Thermodynamic free energy}]$$

This integral can be calculated by using the method of steepest descent, (*i.e.*, saddle point method). The integrand has a saddle point at β = β (ε) given by:

$$\frac{d}{d\beta}\,(\beta F) = \varepsilon \qquad ...(13.86)$$

where ε has been treated as an independent variable.

We can see from Eq. 13.86 that from the properties of saddle point:

$$\frac{d}{d\varepsilon}\,[\beta(\varepsilon - F)] = \frac{d}{d\varepsilon}\,S\,(\varepsilon) = \beta = \frac{1}{T} \qquad ...(13.87)$$

where $S\,(\varepsilon) \equiv \beta\,(\varepsilon - F)$, by definition of entropy *S* (ε). After integrating Eq. 13.84, we get:

$$\rho\,(\varepsilon) = \frac{1}{\sqrt{-2\pi\partial\varepsilon/\partial\beta}}\,e^S \qquad ...(13.88)$$

Hence,
$$\frac{d}{d\varepsilon}\ln\rho\,(\varepsilon) = \frac{\partial S}{\partial\varepsilon} - \frac{1}{2}\frac{d}{\partial\varepsilon}\ln\left(-\frac{\partial\varepsilon}{\partial\beta}\right) = \frac{1}{T} - \frac{1}{2}\frac{d}{d\varepsilon}\left(T^2\,\frac{\partial\varepsilon}{dT}\right) \qquad ...(13.89)$$

Neglecting the last term, we can write:

$$\frac{d}{d\varepsilon}\ln\rho\,(\varepsilon) = \frac{dS}{d\varepsilon} = \frac{1}{T} \text{ or } S\,(\varepsilon) = \ln\rho\,(\varepsilon) \qquad ...(13.90)$$

Alternatively, using the consideration of Nernst theorem, that the specific heat vanishes at $T = 0$, we can write:

$$\left(\frac{d\varepsilon}{dT}\right)_{T=0} = 0 \qquad ...(13.91)$$

Hence the first term in the expansion of ε will be proportional to T^2, so that we may write:

$$\varepsilon = a\,T^2 \qquad ...(13.92)$$

Keeping in mind that $U = \varepsilon - U_0$, we can also write:

$$U = a\,T^2 \qquad \text{...(13.93)}$$

Then one can write from Eqs. 13.89 and 13.92:

$$\left(\frac{d}{d\varepsilon}\right) \ln \rho\,(\varepsilon) = \sqrt{\frac{a}{\varepsilon}} \qquad \text{...(13.94)}$$

As a matter of fact, Gilbert and Cameron[19] derived through a more detailed procedure from Eq. 13.89:

$$\rho\,(U) = \sum_M \rho\,(U\,N\,Z\,M) \approx \int dM\,\rho\,(U\,N\,Z\,M)$$

$$= \frac{\sqrt{\pi}}{12}\,\frac{\exp(2\sqrt{aU})}{a^{1/4}U^{3/4}} \qquad \text{...(13.95}a\text{)}$$

where $$a = \frac{\pi^2 g}{6}; \text{ and } U = \varepsilon - U_0 \qquad \text{...(13.95}b\text{)}$$

U_0 is the energy of fully degenerate system, which is constant.

In deriving Eq. 13.95*a*; *M* has been taken as a constant of motion. However, Lang and LeCouteur[20] have used an alternative dependence of *U*, which fits the experimental data better; *i.e.*,

$$U = a\,T^2 - T \qquad \text{...(13.97}a\text{)}$$

and do not include *M* as a constant of motion. They finally obtain for the level density an expression.

$$\rho_{\text{L.C.}}\,(U) = \frac{\sqrt{\pi}}{12}\,\frac{\exp(2\sqrt{aU})}{a^{1/4}(U+T)^{5/4}} \qquad \text{...(13.97}b\text{)}$$

A slightly different form of the level density formula is due to Newton[15] according to which

$$U = a\,T^2 - \frac{3}{2}T \qquad \text{...(13.98}a\text{)}$$

$$\rho_{\text{Newton}}\,(U) = \frac{\sqrt{\pi}}{12}\,\frac{\exp(2\sqrt{aU})}{a^{1/4}(U+3/2T)^{5/4}} \qquad \text{...(13.98}b\text{)}$$

In all these formulations

$$a \approx \text{Constant } A \qquad \text{...(13.98c)}$$

Figure 13.9 shows, the dependence of '*a*' determined empirically as a function of *A*. It is interesting to see that while the general trend of '*a*' does follow Eq. 13.98*c*, it has a strong dip at around $A \approx 208$ which corresponds to $Z = 82$ and $N = 126$. Also there is strong spike at $A_1 \approx 90$ corresponding to $N \approx 50$, $Z \approx 40$. Both these effects seem to be connected with the magic numbers.

Further, *g* has been taken to be independent of ε. We will see later, in Chapter 15 on pre-equilibrium that this may not be the case.

The theory of compound nucleus decay is well known as given in Section (13.2.3). However, the level density formula has adjustable parameters. Recently[13], decay of the compound nucleus of Sn^{112} has been studied through the reaction $F^{19} + Nb^{93} \rightarrow Sn^{*\,112}$ at projectile energies of 73 and 95 MeV. The elastic scattering; evaporation residues, their excitation function and angular distribution of protons or alpha from this reaction were studied using statistical model code and it was found that the experimental results are better produced if one uses Kataria-Ramamurty-Kapoor (KRK) level density formula[14] rather that Gilbert-Cameron (GC) formula[19]. This is basically because of the realistic way by which the KRK formula treats the shell corrections. A new formula[14] for level density was developed in 1982, using Nilsson Hamiltonian. It compared well with those by H.A. Bethe (Ref. 18) for U^{236}, Pu^{240} and Pb^{208}.

In another interesting work[16], authors have determined detailed excitation function with high resolution for Al^{27} (*d, n*), Sc^{28} (*n, p*) and Si^{28} (*n, d*). They have been studied over an excitation energy range of 3 to 22 MeV. After a lot of analysis, it was found that the conventional Fermi gas form of level density seems to be valid while using the Ericson analysis.

In a recent work[17], particle emission as a probe for deformation has been reported, where $Si^{28} + V^{51} \rightarrow$ fission + α emission at 140 MeV incident energy has been studied and explained by the statistical calculation using rotating liquid drop model values of the moment of inertia. This indicates that in the symmetric system, probably the collisions in the early stages of the equilibration excite the nucleons near the surface of the nucleus resulting in the unexpectedly low emission barrier and thus low average particle energies.

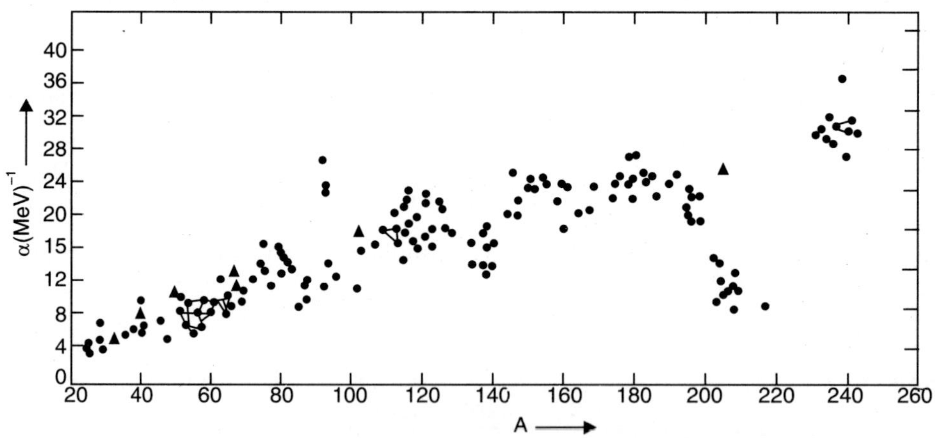

Fig. 13.9 The variation of 'a' with A as found experimentally[21].

E. Experimental Results for Illustrating Compound Nucleus Model: The theory of compound nucleus formation and the decay-process involving statistical model has been tested in many experiments. In actual experiments, compound states, *i.e.* the excited states formed, in the Compound nucleus are created by different types of reactions. As for example, the compound states just-above the Fermi energy can be created by thermal neutron-capture, while reactions like (p, γ), (p, n), (p, α), etc. or similar reactions induced by neutrons say (n, γ), (n, p), (n, α) or induced by alphas like (α, p), (α, n) and (α, $2n$) etc. can be used to search the higher energy compound states.

One wants to understand the behaviour of these compound states as a function of the excitation energy to bring out their properties to throw light about their formation. We have already seen in Figs. 13.4 and 13.6 the behaviour of total cross-section for neutrons and elastic scattering cross-sections σ (p, p) as a function of incident energy showing the cases of interference between resonance scattering and potential scattering as expected from Eqs. 13.29, 13.50 and 13.56 and lack of it—in σ (p, n).

Similarly we have shown in Figs. 13.8*a* and 13.8*b*, the validity of the concept of temperature T as expected from Eq. 13.94 and also some deviations.

There are some other physical phenomena which correspond to compound nucleus formation. As for example, we expect long time of interaction within nucleus and a large number of close-by nuclear states, around the energy of excitation corresponding to compound nucleus formation.

Physical Phenomena: One can get insight into various theoretical assumption of compound nucleus formation from the following physical quantities:

(*i*) The lifetimes or level widths of excited levels,

(*ii*) The energy distribution of emitted particles, and

(*iii*) The angular distribution of emitted particles.

We will discuss next, the various experiments concerned with the above properties of excited nuclear states in the compound nucleus and bring out the characteristics having a bearing on the compound nucleus model.

(i) Lifetime or Level Widths of Compound Nucleus Levels: (*a*) One of the most well-known methods of measuring the level widths is the measurements of resonance widths in the elastic scattering of thermal neutrons. A thermal beam of neutrons is generally obtained by allowing the neutrons say from a reactor or from a target, where neutrons are produced from reactions say (p, n) or (α, n), etc. using an accelerator to pass through a medium, generally containing deuterons or other low-Z elements in solid form so that the emerging neutrons are in thermal equilibrium of the medium which is kept at room temperature. Generally the medium is paraffin, heavy water or graphite. The neutrons emerging out of this thermalising medium have an energy distribution which is Maxwellian with an epithermal tail ($1/E$ flux distribution) extending into the intermediate energy range.

To measure the scattering or total cross-sections of neutrons as a function of energy in the range of electron-volts, one uses monochromators, which give a beam of monoenergic neutrons in the energy range of 0.001eV up to 10,000 eV. Table 13.1 gives[26] the characteristics of some such monochromators used for neutron beams from reactions and accelerators.

Table 13.1

No.	*Device*	*Usable range (eV)*	*Resolution (eV)*
1.	Thermal chopper	0.002 – 0.2	0.005 – 0.05
2.	Fast chopper	0.001 – 5000	0.001 – 0.05, 0.1–1.7, 40 – 1000
3.	Columbia modulated cyclotron	001 – 10,000	0.001 – 0.05, 0.1 – 1.2, 30 – 1000
4.	Crystal spectrometer	002 – 100	0015 – 0.01, 0.2 – 5

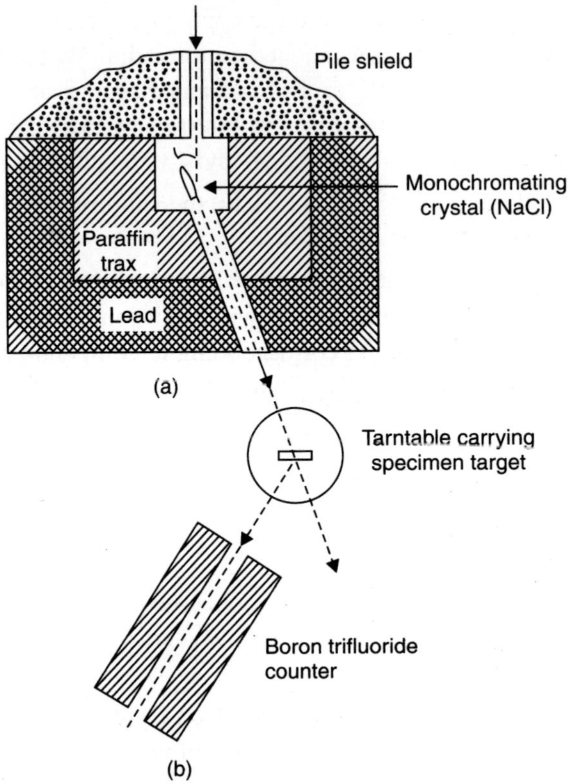

Fig. 13.10 (*a*) The crystal spectrometer[24, 25] for obtaining the monoenergetic slow neutrons from a reactor, (*b*) Experimental arrangement of the target and the detector for determining the cross-sections and neutron-widths.

For thermal or epithermal neutrons from reactors, one uses mechanical velocity selectors or choppers or crystal spectrometer to obtain the monochromatic neutrons. The mechanical monochromators called choppers as first developed by Dunning, Pegram, Fink, Mitchell and Segre[22] are based on the principle used by Fizeau in his measurements of the velocity of light. In this device, two cadmium disks were mounted, one at each end of a shaft. Both disks had a series of uniformly spaced radial slits. The disks could be displaced by an arbitrary angle with respect to each other. In this configuration, thermal neutrons moving along the direction of the shaft, which pass through the first set of slits were absorbed in the second disk, provided shaft remained stationery. If the shaft was rotating with such an angular velocity that the neutrons passing through the slits of first disk passed through the slits of second disk, then the neutrons will be detected. This requires a certain relationship of the angular velocity of the shaft, the velocity of neutrons (or their energy) and the distance between the slits on the disks. Thus, by varying the velocity of rotation or the angular displacement, it was possible to select neutrons of predetermined velocity. Later on Fermi, Marshall and Marshall[23] restructured an improved version of these choppers.

The principle of crystal spectrometer is illustrated in Fig. 13.10. The thermal neutrons are allowed to fall on a crystal at an angle θ with crystal planes. Those neutrons will be preferentially reflected at an angle of reflection, whose wavelengths satisfy the Bragg condition:

$$m\,\lambda = 2\,d\sin\theta$$

where $$\lambda = \frac{\hbar}{(2ME)^{1/2}} \quad ...(13.99)$$

where m, an integer is the order of reflection and d is the distance between crystal planes. Of course, E is the kinetic energy of neutrons and M is the neutron mass and λ is de Brgolie wavelength. These crystal spectrometers require high intensity of neutrons available from reactors. These neutrons with Maxwellian distribution of energies are allowed to impinge on a crystal at an angle θ with the crystal planes and after reflection, are made to fall on the target for which the cross-section is to be determined. They are finally detected by a detector say a BF_2 counter[24].

(*b*) With accelerators, the variability of energies of neutrons is achieved either by varying the energies of charged particles or by causing the beam to impinge intermittently on the target by producing the beam by modulation of ion-source or the acceleration mechanism or by deflecting a continuous beam into target at stated intervals. In this manner, one produces the bursts of low energy neutrons using an endoergic reaction. Then time of flight method is used to select neutrons of a specific energy. For obtaining the variability of energy of the charged particles, the accelerator energies are directly varied. For approaching the compound states just above the Fermi-energies of the target + projectile system, study of neutron capture reactions like (n, γ) or charged particle reactions like (p, γ) or (α, γ) are very useful. We have shown in Fig. 13.5, the energy relationship of compound nucleus model. It is seen that on entering the nucleus, a slow neutron can excite any of the closely packed levels and hence an excitation function is expected to show resonances as shown in Fig. 13.5. Figure 13.4 gives a total cross-section for interaction of slow neutrons with U^{238} where the presence of resonances is clearly shown. The level widths and level spacing D can be directly measured. Similarly in Fig.13.7, we show[10] the excitation function for (p, γ) showing resonances.

(*c*) To measure directly lifetimes of the order of $10^{-16} - 10^{-19}$ secs which are the expected values of the lifetimes of the compound states, very few methods are available[26].

The values of Γ_n^S, and Γ_γ^S at the resonance energies E, are given in Table 13.2 as defined in Eqs. 13.8 to 13.24. Evidently these levels are just above the Fermi energy of the compound states. These are cases where $\Gamma << D$. A few points emerge from this table. For heavy nuclei, radiation width Γ_γ^S is always near 0.1 eV, *e.g.* for In, Sm, Gd, Au, etc. As a matter of fact it was found from actual measurements that

$$0.03 \text{ eV} < \Gamma_r^S < 0.2\text{eV (For heavy nuclei)} \quad ...(13.100)$$

We also find from this table that experimentally

$$\Gamma_r^S << \Gamma_n^S \quad ...(13.101)$$

This means from Eqs. 13.28 and 13.31 that cross-section for the emission of neutrons is much larger than for the emission of gamma rays from a compound state if the energy requirements permit. This is expected, because neutron emission is determined by nuclear interaction (or forces) compared to electromagnetic interaction, which determines the gamma rays emission. The experimental ratio of the capture cross-section to total cross-section is about 10^{-3}, *i.e.* $\Gamma_r/\Gamma \approx \Gamma_r/\Gamma_n \approx 10^{-3}$. As the neutron width Γ_n are of the orders of 5 to 20 keV, this gives the values of Γ_r to be order of 3 to 30 eV for light and intermediate nuclei. For heavy nuclei, $\Gamma_r \approx 0.1$ eV as mentioned earlier.

Table 13.2 *Some cases of neutron widths*[27] , Γ_n^S *and radiation widths* Γ_r^S *and level spacing D as measured experimentally:*

Target nucleus	*Resonance energies*	*Neutron width* (Γ_n^S)	*Radiation width* (Γ_γ^S)	*Spacing D*
Al^{27}	155 keV	10 keV		300 keV
S^{32}	115 keV	25 keV		300 keV
Mn^{55}	345 keV	13 keV		> 1000 eV
Co^{59}	120 keV	2.6 keV		≈1000 eV
Ag^{109}	5.1 eV	11 keV	0.17 eV	~ 40 eV
I^{127}	20 eV	2.4 eV		~ 15 eV
Au^{197}	4.87eV	21.1 eV	0.17 eV	> 100 eV

From the simple relation of Eq. 13.8, *i.e.*,

$$\Gamma = \hbar/\tau_c \qquad ...(13.102)$$

If we use, $\Gamma = 10$ keV, it gives $\tau_c = 10^{-16}$—10^{-17} seconds. A neutron striking the nucleus even with low velocities, will attain in the nucleus, the energies of the order of 8 MeV for which the velocity is of the order of 10^4-10^5 cm/sec. With nuclear radius of 10^{-12} cms, the time for one traversal is of the order of 10^{-17} secs, so that about 10 traversersals take place, before the emission of the neutron. On the other hand, if $\Gamma = 1$ eV, the number of traversals become, 10^6 or so. So in a compound nucleus formation, a large number of traversals take place, giving rise to the mixing of nuclear configurations. At the end of such a mixing; the effect of the configuration of initial channel is lost. Hence decay becomes independent of the channel of formation as assumed in Eq. 13.74.

As discussed in Section 13.2.3(*E*), one of the important activities in the field of compound state formation, is the determination of level-widths of neutron capture resonance. This work is continuing even right now. As for example, recently neutron resonance spectroscopy of Pd^{106} and Pd^{108} was carried out from 20 to 2000 eV, using time of flight technique[28]. The epithermal neutrons were obtained by using an 800 MeV pulsed proton beam, from a linear accelerator, which are allowed to hit a tungsten target, creating the neutrons through spallation. The neutrons are then moderated to epithermal energies by a water moderator[29]. A total of 28 resonances in Pd^{106} and 32 resonances in Pd^{108} were studied. The value of $g\,\Gamma_n$ and Γ_r were determined for these resonances. Apart from adding to the data parity non-conservation studies, were measured in *p*-wave resonances of Pd^{106} and Pd^{108} by using longitudinal polarized epithermal neutrons, to measure the asymmetries in resonance shapes for *p*-wave resonance in the energy range of 0.5 eV – 2 keV. The asymmetries are a measure of the mixing of *s*-wave and *p*-waves, which corresponds to parity non-conserving effects. These are technically very involved experiments.

Similar measurements[30] of parity non-conservation were made for 24 *p*-wave neutron resonances in U^{238} from 10 to 300 eV and 24 *p*-wave resonances from 8-100 eV in Thorium, by a similar experimental arrangement as described above for Pd^{106} and Pd^{108}.

Some set of authors have also measured the neutron resonances in Ag^{107} and Ag^{109}, by the same technique[31] and observed 28 previously unreported resonances and the parity violation for a number of previously unreported resonances.

In another example[32] of the case of neutron-capture resonances in Nd^{142} and Nd^{144}, the neutrons were obtained from Li^7 (p, n) Be^7 reaction using pulsed proton beam from 3.75 MeV Van de graaff accelerator. The capture yields and cross-sections were measured for 52 resonances in Nd^{142} and for 78 resonances in Nd^{144} using time-of-flight technique in the energy range of 2.5 keV to 20 keV.

(*d*) Isomeric ratios, defined as the ratio of the cross-sections of the production of an isomer state in a given thermal neutron capture reaction (n, γ) were investigated by Mateosian and Goldhaber[27] in fifties, and later on the work was extended by Sehgal, Hans and Gill[33] in sixties. It was shown that the isomer state with spin close to that of the compound nucleus is favoured. This shows the dependence of Γ_γ on the spin difference between the initial and final states.

***(ii)* Secondary Emission:** An interesting example of the application of the compound nucleus model is the calculation of the cross-section of a secondary reaction say (a, b, c). Such cases are $\sigma(\alpha, 2n)$ or $\sigma(p, 2n)$ or $\sigma(\alpha, np)$ or $\sigma(n, 2n)$, etc. We will now discuss the cases of $\sigma(a, n)$ and $\sigma(a, 2n)$ in some details. Generally a reaction $\sigma(a, b)$ like $\sigma(a, n)$; means that after neutron emission a γ-ray or another particle may be emitted, *e.g.* (a, n, γ) or (a, n, n) etc. or in general terms $\sigma(ab) = \sigma^*$ (a, b) where σ^* (a, b) represents the cross-section when particle has been emitted leaving behind an excited residual nucleus, which may emit only a gamma ray and come to the ground level. On the other hand, if another neutron or proton is emitted following the first neutron, one can write:

$$\sigma(a, b) = Y^* \quad \Rightarrow \quad \sigma_z^*(a, b) + \sum_c \sigma(a;\ b, c) \qquad \text{...(13.103}a\text{)}$$

Following the logic of Eq. 13.57, one can write (if compound nucleus model is applicable for the emission of both the particles):

$$\sigma(a, b, c) = \sigma_c(a) \sum_\beta{}' G_c(\beta)\, G_{z\beta'}(c) \qquad \text{...(13.103}b\text{)}$$

where prime on the sum indicates that it is extended only over those channels β, in which residual nucleus z is excited high enough to emit a particle. The term $G_c(\beta)$ is the relative probability that the compound nucleus will decay into channel β through emission of particle b, while $G_{z\beta}(c)$ is the relative probability that the nucleus z in β'-state will emit the particle c. A secondary particle c is emitted, if we are above a certain energy. Referring to Fig. 13.11, we define E_{sec} and ε_{sec}, in such a way that physically E_{sec} represents the energy of the emitted particle above which only gamma rays are emitted and below which particles, are emitted then ε_{sec} represents the energy of excitation above $S_{min}(Z)$; so we define:

$$E_{sec} = E_{bz} - S_{min}(Z) \qquad \text{...(13.104}a\text{)}$$

Then we define quantity ε_{sec} so that

$$\varepsilon_{sec} \equiv [E_{bz} - S_{min}(Z)] - E_{sec} \qquad \text{...(13.104}b\text{)}$$

where E_{bz} is the energy of transition between compound state and ground state of the nucleus to which compound state decays and $S_{min}(Z)$ is the smallest separation energy of a particle from the nucleus Z. The energy ε_{sec}, becomes the maximum energy available for particle b from the compound nucleus for which a particle c can be emitted from Z^* (*see* Fig. 13.11).

We can, now write the expressions for both (*i*) $\sigma^* (a, b)_2$ which corresponds to the emission of *b* particle along with the capacity of secondary particle emission. *i.e.* $\sigma (a, b, c)$ where particle *c* is emitted and (*ii*) $\sigma^* (a, b)_1$ where *b* is emitted plus only gamma rays. Under these conditions,

$$\sigma^* (a, b)_1 = \sigma (a, b) \text{ for } \varepsilon_{sec} < 0 \text{ [This corresponds to option (ii)]}$$

and

$$\sigma^* (a, b)_2 = \sigma (a, b) \times \frac{\int_0^{E_{sec}} I_b (E)\, dE}{\int_0^{E_{bz}} I_b (E)\, dE} \text{ for } \varepsilon_{sec} > 0 \text{ [This correspond to option (i)]} \quad ...(13.105)$$

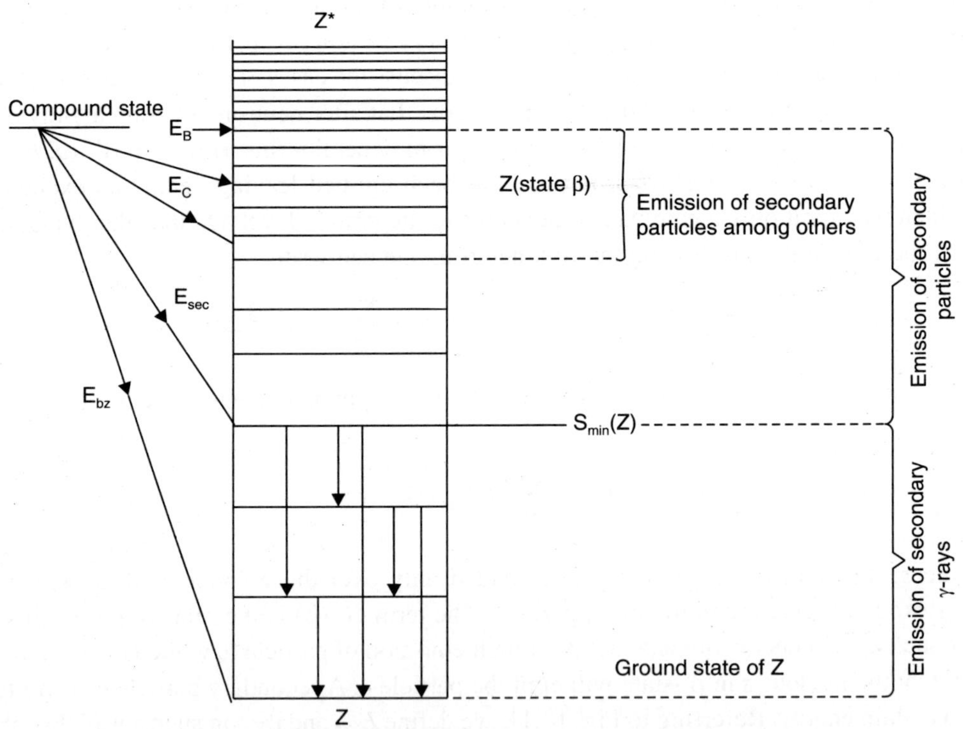

Fig. 13.11 The energy-relationship of emitted particles from the compound state and the excited states of residual nucleus Y from where the secondary particles are emitted.

Physically $\sigma^* (a, b)_1$, for $\varepsilon_{sec} > 0$ corresponds to the emission of *b* particles, where no secondary particles are emitted but gamma rays may be omitted and $\sigma^* (a, b)_2$ for $\varepsilon_{sec} > 0$ means, both gamma ray and secondary particle emission are possible. In Eq. 13.79, $I_b (E)\, dE$ is the relative probability of emission of particles *b* with energies between *E* and $E + dE$. Using the expression from Eq. 13.79, we can write the expression for $\sigma^* (a, b)_2$ for $\varepsilon_{sec} > 0$ as:

$$\sigma^* (a, b)_2 \approx \sigma (a, b) \times \left(1 + \frac{E_{sec}}{T}\right) \exp\left(-\frac{E_{sec}}{T}\right) \quad ...(13.106)$$

where T is the temperature $T(E_{bz})$ governing the emission of neutrons. Continuing this argument and using Eq. 13.103 b and defining E_c as the threshold of (a, n, c) reaction (given in Eq. 13.109), we can write the cross-section for the emission of particles c as:

$$\sigma(a, b, c) \approx \sigma(a, b) \frac{\int_0^{E_c} E \exp(-E/T)\, G_{z\beta'}(c)\, dE}{\int_0^{E_{bz}} E \exp(-E/T)\, dE} \qquad \text{...(13.107)}$$

where $G_{z\beta'}(c)$ is given in Eq. 13.103b.

If both b and c particles are neutrons (n), we can write Eq, 13.107 as:

$$\sigma(a, 2n) \approx \sigma(a, n)\left[1-\left(\frac{1+E_c}{T}\right)\exp\left(-\frac{E_c}{T}\right)\right] \qquad \text{...(13.108)}$$

Here we have assumed that the neutron emission is predominant mode of decay, if the energetic permit, $G_{z\beta'}(n) \approx 1$. This only shows that $\Gamma_n \approx \Gamma$. In Eq. 13.107, E_c is the excess energy over threshold of $(a, 2n)$ reaction and is given by (*see* Fig. 13.11):

$$E_c = E_{bz} - S_c(Z) \qquad \text{...(13.109)}$$

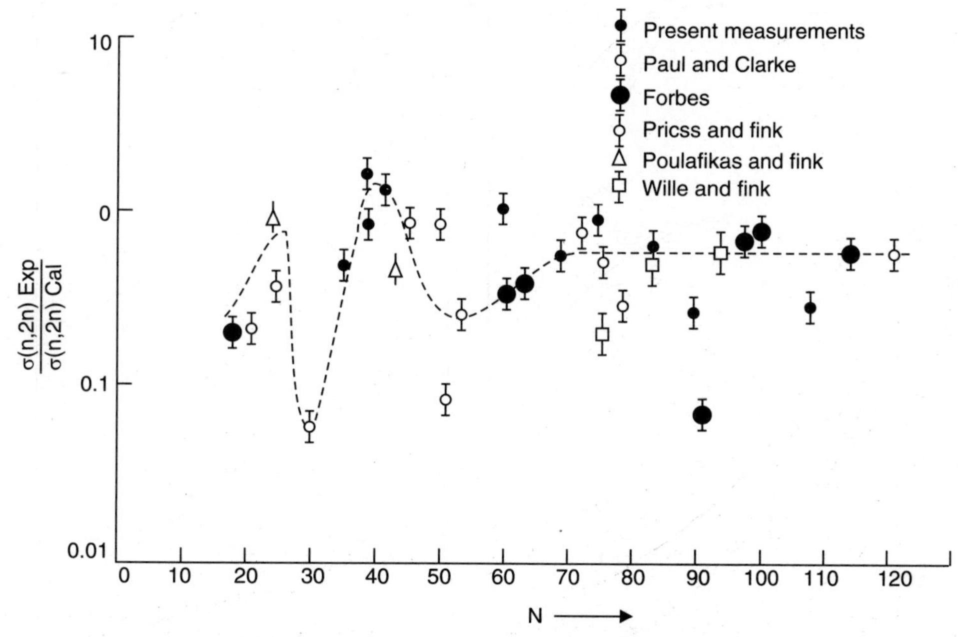

Fig. 13.12 The ratio of observed and calculated values of σ (n, 2n) based on Eq. 13.108 is plotted against neutron number N. The dotted line is an arbitrary line to show the trend (Ref. 33).

where $S_c(Z) = S_n(Z)$ = separation energy of neutron from Z nucleus. Also T is the temperature of the intermediate residual nucleus Z for the emission energy E_{bz}. Further, the cross-section $\sigma(a, n)$ may be taken equal to the cross-section for the compound nucleus formation of the state from where the first neutron is emitted, *i.e.* $\sigma_c(a)$, whose variation with energy E is ignored.

Experimental Results from σ (n, 2n): In 1961, a large number of measurements were made of σ (*n*, 2*n*) at 14.8 MeV by Khurana and Hans and compared with the calculated values expected from Eq. 13.108. Figure 13.12 gives the ratio. Except for a structure around the region of $N = 20$, 28 and 50, which is due to shell effects, the other points lie close to the value, a little less than unity. Weisskopf has suggested that the most probable cause of such a discrepancy may be that the compound nucleus may some times, decay before thermal equilibrium between the various nucleons in the nucleus has been achieved. The neutrons emitted under these conditions will generally have higher energy, than those after the thermal equilibrium has been reached. The calculated values based on the assumption of the compound nucleus will, therefore tend to be higher than the measured values and hence ratio will tend to be smaller than unity. This is a case of the effects of pre-equilibrium discussed in Chapter 16 in detail.

***(iii)* The Energy and Angular Distribution of Emitted Particles:** We have already discussed the emission spectra of (*p*, *n*) and (*n*, *p*) reactions in terms of the nuclear temperature *T*. It seems from Figs. 13.8*a* and 13.8*b*, that while the spectra do fit Eq. 13.79 over a large portion of energies, some portions of the spectrum do not fall on the straight line in Fig. 13.8*b* showing that there may be deviations from compound nucleus model.

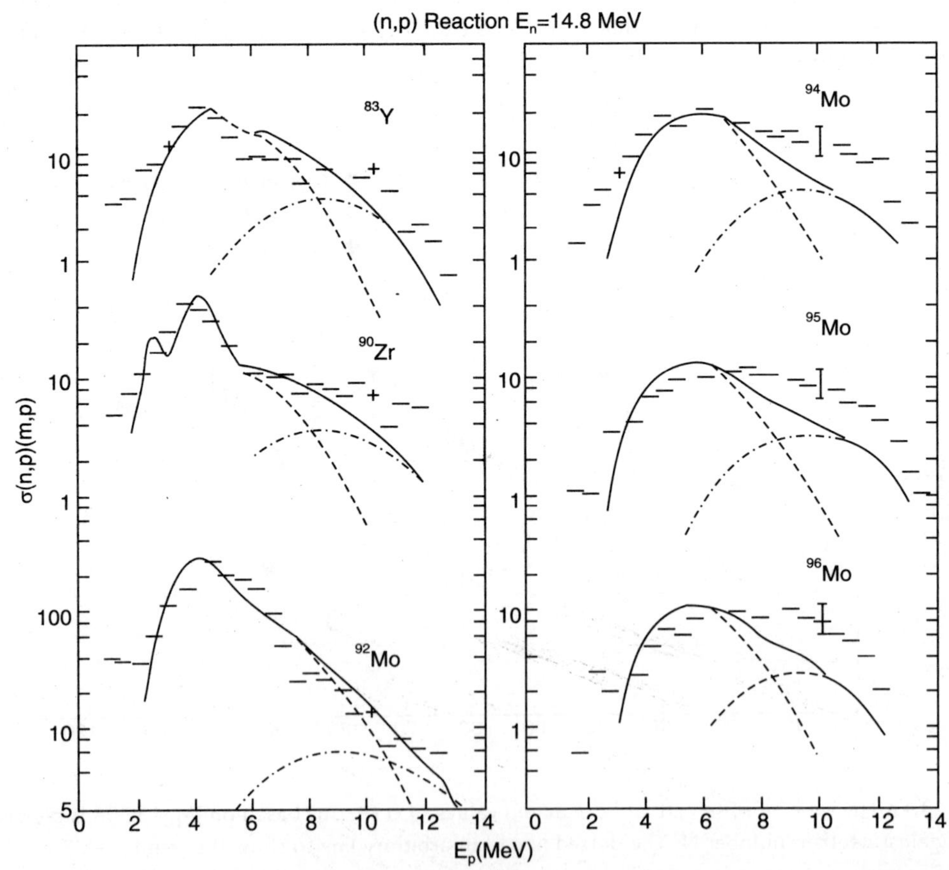

Fig. 13.13 Angle integrated proton spectra for emitted protons for reactions induced by 14.8 MeV neutrons on targets of Y^{89}, Zr^{90}, $Mo^{92, 94, 95, 96}$. The horizontal lines represent a spread of 500 keV for each point. The solid line represents multi-stage Hauser Feshbach calculations based on compound nucleus model, while dot-dash lines represent pre-equilibrium contribution calculated on the basis of Hybrid model (Ref. 34).

A large number of spectra of emitted particles has been analysed[34] in the recent years for reactions like (p, n), (p, α) or (α, p), (α, n) or (n, p), (n, α), etc. at energies from say 4 or 5 MeV up to 100 MeV. We show in Fig. 13.13, a few cases of (n, p) reactions for the incident neutron energy of 14.8 MeV, which have been theoretically analysed, using both the Compound nucleus model-based calculation of Hauser and Feshbach[35] and the pre-equilibrium reaction mechanism using Hybrid model. It is seen from these comparisons that the lower end of the spectra in all the cases was explained, more or less satisfactorily by the compound nucleus model, the higher energy end could be understood only by pre-equilibrium model. This combination of compound nucleus formation and pre-equilibrium, basically belongs to the same broad reaction mechanism as described in the beginning of this chapter. After the incident particle enters the nucleus, a series of interactions will start. The compound nucleus formation corresponds to a large number of such interactions before the particles are emitted. The outgoing particles in this case have lower energies because, the energy of the incident channel has been shared by a large number of interactions of particles and will follow a statistical behaviour. The bell-shape behaviour of the emitted particles is then expected to follow [Eq. 13.79].

On the other hand, if the emitted particles come out only after a few interactions say 4 to 5, then they will have higher energies. The detailed theory for this case, *i.e.* pre-equilibrium emissions will be discussed in Chapter 16.

One interesting situation which arises in the experiments, when the energy resolution of the incident beam is of the same order as the total width Γ_i of the compound nucleus level, one observes in the energy spectrum of emitted particles, fluctuations known as Ericson[36] fluctuations. According to Ericson, for large intervals of incident energy, the fluctuation in the cross-section for a reaction like Si^{28} (n, α) Mg^{25} is roughly given by:

$$\overline{(\sigma - \overline{\sigma})^2} = \frac{2\,\overline{\sigma^2}}{n} \qquad \text{...(13.110)}$$

where $\overline{\sigma}$ is the mean value of the cross-section over the energy interval considered and n is the number of final states. When Eq. 13.110 is applied to the spectra, this relationship is found to hold good. These cross-sections versus neutron energy have been measured by Colli et al., from 12.15 MeV to 18.4 MeV for Si^{28} (n, α) Mg^{25}, which clearly brings out physically the concept of Ericson fluctuations (Ref. 37).

The Angular Distribution of Emitted Particles: The angular distribution of the emitted particles, in general, depends on the reaction-mechanism, (*e.g.* compound, direct or pre-equilibrium) and the angular momenta—both orbital and spin—of the two states connecting the emitted particle and, of course, on the angular momenta of the emitted particle itself.

We now, discuss the angular distribution of the emitted particles b in a nuclear reaction $a + X \rightarrow Y + b$, which takes place via compound nucleus formation, *i.e.* $a + X \rightarrow C^* \rightarrow Y^* + b$, where C^* is the compound nucleus in a highly excited state and Y^* is the residual nucleus after the emission of particle b as discussed in the previous section. If Y^* is excited to an energy so that only γ-rays are emitted after emission of particle b, then the angular distribution of particle b is of interest. The differential reaction cross-section of such a reaction (a, b) may be written as (as explained from Eqs. 13.56*a* and 13.56*b*):

$$d\sigma \,\alpha \sum_{\rho} \left| q_{\rho}(\theta) \right|^2 d\omega \qquad \text{...(13.111)}$$

where ρ represents angular momentum quantum numbers, and $|q_\rho(\theta)|$ has similar meaning of $|f_0|$ of Eq. 13.49 or $|\eta_0|$ of Eq. 13.48, connected through Eq. 13.50. However till now, we only considered all the reactions together, so that Eqs. 13.44, 13.45 and 13.52, correspond to absorption cross-section for which angular distribution has no significance. In contrast, we are now considering only one channel of reaction for which angular distribution has a clear meaning. Now assuming the compound nucleus formation, we can split the cross-section σ(*a*, *b*) into two factors:

$$d\sigma_{a_k, b_\lambda}(\theta) = \sigma_c(a_k)\, G_{b_\lambda, M_J}(\omega)\, d\omega \qquad \text{...(13.112)}$$

where $\sigma_c(a_k)$ is the cross-section of the formation from channel a_k of a definite compound state with an angular momentum *J*, whose *z*-component is M_J. The symbol *k* stands for definite spin index of the incoming channel. $G_{b_\lambda, M_J}(\omega)\, d\omega$ is the branching probability that this compound state decays into a channel $b_\lambda\, M_J$ from a compound state whose *M*-component is M_J. Evidently $\sigma_c(a_k)$ has no angular dependence as it does not know in what direction it will disintegrate, when it was formed.

In writing Eq. 13.112, we have assumed a single state of the compound state with a specific *J* and *M*. In practice, when the incident beam is unpolarised, we require to sum over the spin directions of the outgoing channel and averaging over the spin direction of the incoming channel:

$$d\sigma_{ab}(\omega) = \frac{1}{2s+1} \times \frac{1}{2i+1} \sum_k \sigma_c \left[\sum_k G_{b_\lambda M_J}(\omega)\, d\omega \right] \qquad \text{...(13.113)}$$

This equation is similar to Eq. 13.56*b*,

where we have assumed that the compound nucleus formation has no angle-dependence. Eqs. 13.112 and 13.113 are based on the excitation of a single compound state level with a given *J*. This is true under conditions of monoenergetic incident beams with Δ*E*, the energy spread of the beam being less than Γ, the level width of the state. Also it assumes that $\Gamma/D << 1$. However, if many levels are excited in the compound state and when $\Gamma/D > 1$ and/or when the incident beam energy has a large energy spread, then the expression in Eq. 13.113 has to be modified to include all the levels involved. It has been proved by Blatt and Weisskopf[2], that if only one level is involved, the angular distribution in Eq. 13.113 has an angular distribution, no more complicated than $(\cos\theta)^2$ both for integral or half integral values of *J*. However, if many compound levels are involved and the angular distribution is a result of their interference, it becomes much more complicated. It is no longer true that outgoing beam has a definite parity hence, odd powers of cos θ enter and the angular distribution can be as complicated as $(\cos\theta)^{2J_{\max}}$, where $J_{\max}$ is the highest value of *J* of the various compound states, which contribute to the reaction.

While the exact shape of such an averaged angular distribution will vary for different cases, depending on the strength of the contribution and values of *J* involved, it will be in general, symmetric about 90. This is true for compound nucleus model, which is excited at somewhat lower energies. In Fig. 13.14*b*, we have plotted the experimental angular distribution of (*d*, α) reaction after summing over all the outgoing channels for different incident energies[39]. The angular distributions tend to be symmetrical around 90° but, not exactly so. But if we sum over all the incident energies from 5.6 MeV to 6.8 MeV, the angular distribution is quite symmetrical around 90°.

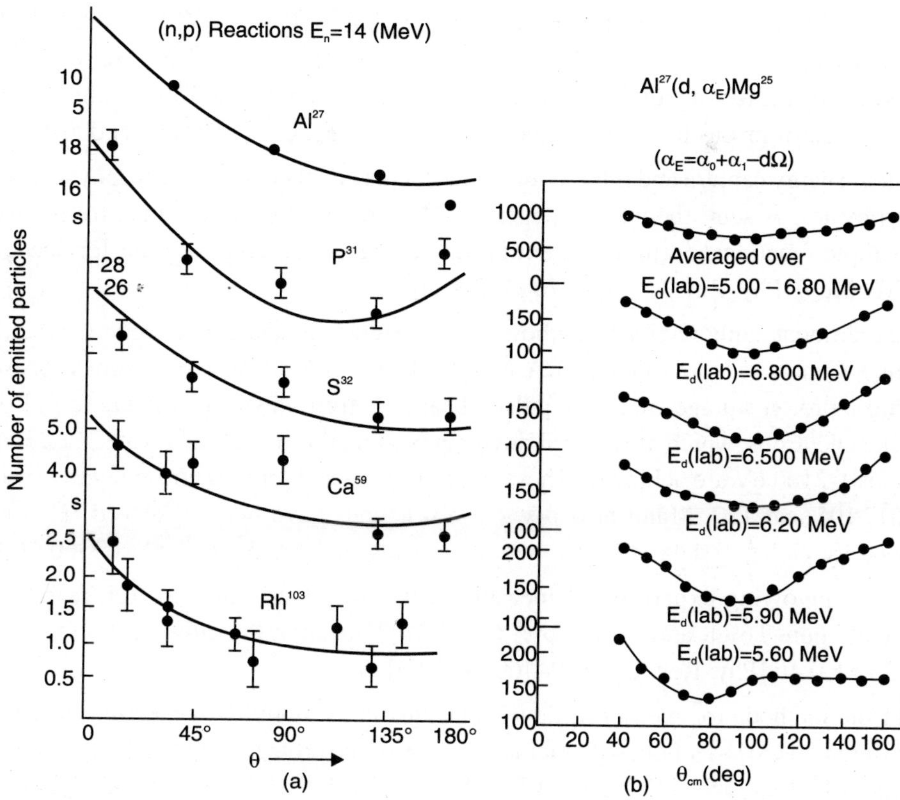

Fig. 13.14 The angular distributions of the emitted particles for two reactions, i.e. for (*a*) protons in (n, p) sections at E_n = 14 MeV averaged over all the energies of outgoing protons (*b*) Alphas in (d, α) reaction at E = 5, 6 – 6.8 MeV, where averaging has been done for outgoing channels[39, 38].

An interesting case of angular distribution is plotted[38] in Fig. 13.14*a*, where the angular distribution of protons of all outgoing channels is plotted for (*n*, *p*) reaction at 14 MeV incident neutrons, which, in general, have an energy spread covering many levels in the Compound nucleus. In all the cases, there seems to be two components of angular distribution, (*i*) Symmetric around 90°, which is due to compound nucleus formation, (*ii*) A forward peaked angular distribution assumed to be due to pre-equilibrium reaction mechanism. As a matter of fact the angular distribution was analysed using a comprehensive theory of F.K.K[40] (Feshbash, Kerman and Koonin) and the data could be satisfactorily explained, where both the compound nucleus model and pre-equilibrium play role. The solid continuous lines for various targets represent the results of calculations based on such a model.

13. Compound Nucleus Model 2000–2008

There is enough interest in problems, involving compound nucleus formation *e.g.* (*i*) Dispersion relation for (n, n'), (n, p) and (n, a) reaction for K^{39} and K^{40} at $E_n = 4.5$, 5.5 and 6.5 MeV etc. (*ii*) Neutron resonance spectroscopy of Ti^{99} from 3 eV to 150 keV in which 120 resonance were recorded. This has been accomplished by measuring total neutron cross-sections, using time of flight technique with pulsed white neutron source from Electron linear accelerator. [Phy. Rev. C. 69, 054608 (2000)].

Many low energy capture and interaction experiments and theory have astrophysical significance as well as in physics. A semi-theoretical paper by Majumbdar, Deb and Amar from Australia, has developed a simple functional form, representing the total reaction cross-section for scattering in the range of 20-300 MeV [Phy. Rev. C. 64, 027603 (2001)].

In an experiment, authored by 14 authors from USA, Japan and Russia conducted at Los Almos Neutron Science Centre at New Maxico (LA NSC) a beam of 800 MeV H^- from LANSC Linac is transported into a proton storage ring. The pulsed beam H^+ from this ring is made to fall on tungsten target, giving fast neutrons, which are slowed down by passing through chilled water moderator. Finally neutron energies 1-2100 eV are achieved. These neutrons are used for measuring neutron resonances in Palladium 101, 105 and 110. Many new *p*-wave resonances have been observed. This has helped, obtaining the parity violating data, due to *p*-nature resonances. [Phy. Rev. C. 65, 024607 (2002)].

In a paper involving experimental measurement, using Pakistan Research reactor, and using spectrographically pure Tb_4O_7 and using Tb^{159} (n, α) Tb^{160} reaction the cross-section for reaction was measured as $23.6 \pm 0.4b$ [Phy. Rev. C. 68, 044608 (2003)].

One of the methods for measuring level densities is by studying the Eriscon fluctuation [The variation in differential cross-section, with incident energy of the projectile] using the compound reaction. In a recent paper Al^{22} (p, n_0) Si^{27}, Se^{45} (p, n_4) Ti^{45}; Sc^{45} (p, n_5) Ti^{45} and V^{51} (p, n_0) Cr^{51}, and Co^{59} (p, n_0) Ni^{59} have been measured giving level densities as: Si^{28} $(a = 5.5)$; Ti^{46} $(a = 4.8)$; and Cr^{52} $(a = -4.8)$; [Phy. Rev. C. 70, 024311 (2004)].

In a collaborative effort from India, with 21 authors from Delhi, Mumbai, Guwahati, Kolkata, the elastic scattering and fusion cross-section of Be^7, $Li^7 + Al^{27}$ system has been experimentally studied at 17, 11 and 22 MeV in the angular range of 12°–43° using 15 MV Pelletron at New Delhi. The fusion cross-sections were consistent with the calculations of coupled channels compound nucleus model and optical model were used to explain the fusion and elastic scattering data [Phy. Rev. C. 73, 024609 (2006)].

In a paper authored by 12 authors, the $(n, 2n)$ cross-sections have been measured from 7.6 MeV to 14.5 MeV, for the radioscope Am^{241} $(T_{1/2} = 4326Y)$. The induced γ-rays activity of Am^{240} activity was measured. Very good agreement is found with the END-*B* VII evaluation; where JENDL-3.3 evaluation is in fair agreement. These are compound nucleus model programmes [Phy. Rev. C. 77, 054610 (2008)].

REFERENCES

1. N. Bohr: Nature 137, 344. (1936); N. Bohr and F. Kalcher. Kgl. Denske: Videnskab Mat. Fys. (1937).
2. John M. Blatt and Victor F. Weisskopf: Theoretical Nuclear Physics, John Wiley & Sons, New York (1952).
3. S.N. Ghoshal: Phy. Rev. 80, 939, (1950).
4. F.W.K. Firk, J.E. Lynn, and M.C. Moxon: Proceedings International Conference on Nuclear Structure, Kingston, Canada, Ontorio University of Toronto (1960); Nuclear Physics, W.E. Burcham, Longmans, London (1963).
5. J.D. Jackson: Classical Electrodynamics, p. 601, Wiley, New York (1962), L.I. Schiff, Quantum Mechanics, p. 250, McGraw-Hill, New York (1949).
6. G. Breit and E.P. Wigner: Phy. Rev. 49, 519, 642 (1936).
7. P. Wilhjelm et al. Phy. Rev. 177, 1553 (1969).
8. D.P. Lindstron et al. : Nuclear Physics A, 168, 37 (1971).
9. R.C. Mobley and R.A. Laubenstein: Phy. Rev. 80, 309 (1950).
10. K.G. Bronstrom, T. Huus, and R. Tangen: Phy. Rev. 82, 190 (1951).
11. C.H. Holbrow and H.H. Barschall: Nuclear Physics 42, 269 (1963).
12. R.K. Mohindra and H.S. Hans: Nuclear Physics 44, p. 597 (1963).
    H.S. Hans and R.K. Mohindra: Nuclear Physics 47, 473 (1963).
13. Bency John and S.K. Kataria, B.S. Tomar, A. Goswami, G.K. Gubbi and S.B. Manohar: Phy. Rev. C.V. 56 p. 2582 (1997).
14. S.K. Kataria, V.S. Ramamurthy and S.S. Kapoor: Phy. Rev. C 18, 549 (1978); M. Rajasekaran and V. Devananthan: Phy. Letters, V. 113 B, No. 6, 433 (1982).
15. T.D. Newton: Canad J. of Phy. 34, 804 (1956).
16. F.B. Bateman, S.M. Grimes, N. Boukharouba, V. Mishra, C.E. Brient, R.S. Pedroni, T.N. Massey and R.C. Haight: Phy. Rev. C. 55 p. 1336 (1997).
17. I.M. Govil, R. Singh, A. Kumar, A.K. Sinha, N. Madhavan, D.O. Kataria, P. Sugathan, S.K. Kataria, K. Kumar, Bencyjohn and G.V. Ravi Prasad: Phy. Rev. C. 57, p. 1269 (1998).
18. H.A. Bethe: Rev. Mod. Phy. 9, 69 (1937).
19. A. Gilbert and A.G.W. Cameron: Canad J. Physics 43, 1446 (1965).
20. J.M.B. Lang and K.J. Le Couteur: Proc. Phy. Society, 67 A, 586 (1954); Newton T.D.: Canad J. Physics 34, 804 (1956).
21. U. Fachini: Proc. Conf. Direct Interactions. Nuclear Reaction Mechanism (1962) p. 243, Gorden and Breach Science Publishers, Inc., New York (1963).
22. J.R. Dunning, G.B. Pegram; G.A. Fink, D.P. Mitchell and E. Segre: Phy. Rev. 48, 704 (1935).
23. E. Fermi, J. Marshall and L. Marshall: Phy. Rev. 72, 193 (1947).
24. E.O. Wollan and C.G. Shull: Phy. Rev. 73, 830 (1948); C.G. Shull, E.Q. Swallen, G.A. Morton and W.L. Davidson: Phy. Rev. 73, 842 (1948); R.B. Sawyer, E.O. Wollen, S. Bernstein and K.C. Peterson, Phy. Rev. 72, 109 (1947).
25. E. Fermi and L. Marshall: Phy. Rev. 75, 666 (1947).

26. E. Segre, B.T. Feld (Ed.): Experimental Nuclear Physics V-II, p. 429, John Wiley & Sons, Inc., New York (1953).
27. T. Teichmann: Phy. Rev. 77, 506 (1950); Ref. (2); p. 474; E. Le Metosian and G. Caedhaber: Phy. Rev. 108, 786 (1951).
28. B.E. Crawford et al. (16 authors): Phy. Rev. C. 58, p. 729 (1998).
29. D.W. Lisowski, C.D. Bowman, Er. J. Russel and S.A. Wender: Nuclear Science Engineering, 106, 208 (1990).
30. B.E. Crawford et al. (18 authors): Phy. Rec. C. 58, p. 1228 (1998), S.L. Stephenson (18 authors): Phy. Rev. C. 58, 1236 (1998)
31. L.Y. Lowie et al. (17 authors): Phy. Rev. C. 56, p. 90 (1997).
32. K. Wisshak, F. Voss and E. Kappaler: Phy. Rev. C. 57, p. 3452 (1998).
33. C.S. Khurana and H.S. Hans: Nuclear Physics, V. 28, p. 560 – 569 (1961); M.L. Sehgal, H.S. Hans and P.S. Gill: Nuclear Physics V-12, 264 (1954): V. 20, 183 (1960).
34. S.M. Grimes, R.C. Haight and J.D. Anderson: Phy. Rec. C. 17, 508 (1978); Ibid Phy. Rev. C.19, 2127 (1979); R.C. Haright, S.M. Grimes, P.G. Johnson and H.H. Barschall: Phy. Rev. C. 23, 700 (1981).
35. W. Hauser and H. Feshbach, Rev. Modern Physics 46, 1 (1974).
36. T. Ericson: Advances Physics 9, 425 (1960).
37. L. Colli, I. Lori, M.G. Marcazzan and M. Milazzo: Proceedings Conference Direct Interactions, p. 387, Gordon and Beach Sciences Publishers Inc., New York (1963).
38. Gulzar Singh, H.S. Hans, T.S. Cheema, K.P. Singh, D.C. Tayal, Jahan Singh and Sudip Ghosh: Phy. Rev. C. 49, p. 1066 (1994).
39. I.M. Naqib, R. Gleyvod and N.P. Heydenberg: Nuclear Physics 66, 129 (1965) (adapted form).
40. H. Feshbach, A. Kerman and S. Koonin: Annals Physics, New York, 125, 429 (1980).

PROBLEMS

1. The total cross-section for neutron absorption in $_{42}In^{115}$ ($I = 9/2$) target shows a large peak at $E = 1.44$ eV, which has a width of 0.085 eV and the peak cross-section is observed to be 26400 barns. Obtain an approximate value of neutron-width. For thermal neutron (0.025 eV), it gives an absorption cross-section of 17 barns. Explain it is consistent with the above resonance data.

2. Extending the scope of equations 13.51 and 13.52, calculate the reaction and scattering cross-sections using plane wave Born approximation for calculating $U_l^{(+)}$ (R) and $U_l^{(-)}$ (R) with the potential of the form

$$V = v_o\, \delta\, (\mathbf{r}_a - \mathbf{r}_b)\, (\mathbf{r}_a - \mathbf{R})$$

for the process $a + x \rightarrow y + b$, for $1 \neq 0$ [*see* Ref. 2].

3. If 10 MeV neutrons induce nuclear reactions on Cu^{63}, what is the largest 1-value with which they have a good chance of entering the nucleus? What is the mean free path of nucleons of 1, 10 and 30 MeV inside a nucleus say with $A = 120$?

4. Prove, using the assumptions of statistical model that

$$\frac{\Gamma_p}{\Gamma_n} = \frac{\int \varepsilon_p \, \sigma(\varepsilon_p) \, \rho(E_p) \, d\varepsilon_p}{\int \varepsilon_n \, \sigma(\varepsilon_n) \, \rho(E_n) \, d\varepsilon_n}$$

where ε_i's are energies of the emitted particles, E_i's are the excitation energies and Γ_i are level-widths for $i = p, n$, *i.e.* for protons and neutrons, respectively.

5. A proton of energy T collides with an electron at rest. Neglecting Coulomb forces, find the maximum energy of the electron and the minimum energy? What would be the difference if electron is replaced by a proton or an alpha particle?

6. From Eq. 13.26 and Eqs. 13.28 and 13.31, draw the curves for cross-section with respect to E and find out the physical meaning of Γ_R.

7. From Eqs. 13.56*b* and 13.113, prove that in a compound nuclear reaction, the angular distribution of emitted particles with respect to the direction of incidence is symmetric about 90°.

8. Show that,

$$\frac{1}{\rho(U)} = D(U) = T\left(2\pi \frac{dU}{dT}\right)^{\frac{1}{2}} e^{-s}.$$

9. Using Gilbert-Cameron [Eq. 13.95*b*] Lang-le-Couteur [Eq. 13.97*b*] and Newton formulae [Eq. 13.98*b*], find out the values of temperatures for Al^{27}, Cu^{63}, and U^{238} at the excitation energies U of 8 MeV and 20 MeV.

10. In a mechanical chopper for getting mono energetic neutrons, develop the relationship between the angular velocity of the shaft, the velocity (and hence energy) of neutrons and distance between the slits on the disks.

14

CHAPTER

Direct Reactions

14.1 INTRODUCTION—ELEMENTARY THEORY

In many cases of reactions, the target nucleus has one or a few very loosely bound nucleons, *e.g.* in $_{83}Bi^{209}$. Alternatively the projectile itself may have loosely bound nucleons, say in a deuteron (H^2) or helium^3 (He^3) or tritium (H^3). In such cases, much before the projectile can interact with many nucleons of the nucleus; the loose particle in the target nucleus or projectile in the very first encounter acquires enough energy to get separated from the target or projectile and is emitted. Also a projectile may interact with the target through Coulomb excitation leaving the target nucleus excited, while itself having less energy than the incident energy. This is inelastic scattering through Coulomb direct interaction. Also a (p, n) or (p, p') or (α, p) or (α, n) section may be through direct knockout reaction. Further break-up reaction like $Mg^{24} + C^{12} \rightarrow C^{12} + (O^{16}, Be^8)$ is also a direct reaction. These are special cases of direct reaction. Good examples are:

$_1H^2 + {_1H^2} \rightarrow {_2He^3} + {_0n^1}$	Stripping
$_1H^2 + {_1H^3} \rightarrow {_2He^4} + {_0n^1}$	
$_1H^2 + {_6C^{14}} \rightarrow {_1H^3} + {_6C^{13}}$	Pickup
$_1H^1 + {_3Li^7} \rightarrow {_1H^2} + {_3Li^6}$	
$_1H^2 + {_{15}P^{31}} \rightarrow {_{15}P^{32}} + {_1p^1}$	Stripping
$_1H^2 + {_{83}Bi^{209}} \rightarrow {_{84}Po^{210}} + {_0n^1}$	
$_2He^4 + Er^{168} \rightarrow Er^{*168} + {_2He'^4} \rightarrow$	Inelastic scattering
$Mg^{24} + C^{12} \rightarrow C^{12} + (O^{16} + Be^8)$	Break-up
$P^{31} + He^4 \rightarrow S^{34} + {_1p^1}$	Knockout ...(14.1)

All these reactions have a positive value of Q and outgoing projectiles are emitted with comparatively large energies.

The indications of the direct reaction mechanism are: (*i*) the forward peaking in the angular distributions of the outgoing particles and (*ii*) the short lifetimes of the composite system. As a matter of fact, while in the compound nucleus formation, many traversals take place, before the outgoing particle is emitted; in the direct reaction, there is only one traversal across the nucleus. So, if say, a 10 MeV nucleon enters the nucleus, which has a potential depth of –50 MeV; the nucleon will attain a kinetic energy of 60 MeV. The time required to cross one diameter of the nucleus will be of the order of only 10^{-22} seconds, compared to 10^{-14} to 10^{-16} seconds in a typical compound formation.

The first set of experiments, corresponding to direct reactions were performed by Lawrence E.O., E. McMillan and R.L. Thornton[1] in 1935 who observed proton and neutron spectra from (*d*, *p*) and (*d*, *n*) reactions and found that (*d*, *p*) reactions were more frequent than (*d*, *n*) sections. This is not expected in the compound nucleus model, because Coulomb barrier should make the emission of proton less frequent. Oppenheimer and Phillips[2] were the first one to recognise that a new type of reaction is taking place, not included in the scenario of the compound nucleus formation. They explained the reaction induced by deuterons by stating that the deuteron is a loosely bound system and when it approaches the target; the proton is detached from the deuteron-system due to Coulomb field and moves forward, while the neutron is captured. This is the stripping process as described in Eq. 14.1. At lower energies, (*d*, *p*) is more probable than (*d*, *n*), because Coulomb field helps in detaching proton. At higher energies; it is the nuclear interaction with the nucleons in the nucleus, which is effective for both protons and neutrons. Hence (*d*, *p*) and (*d*, *n*) reactions are equally probable at higher energies. These are the examples of stripping reactions. However, it is the angular distributions of emitted particles, which distinguishes, the compound nucleus reaction mechanism from the direct reactions. In the later case; the angular distribution is peaked in the forward direction. It was the work of Butler[3], which explained this forward-peaking. Qualitatively as we explained earlier, the uncaptured nucleon, proceeds in the forward direction, giving a forward peak. Such forward-peaking was observed in the fifties in Na^{23} (*d*, *p*) Na^{24}, P^{31} (α, *p*) S^{34}, Li^7 (*p*, *d*) Li^6, and ${}_6C^{14}$ (*p*, *d*) ${}_6C^{13}$ and many other cases[4]. These are the examples of stripping and pickup reactions in the first two and second two cases respectively. P^{31} (α, *p*) S^{34} represents a case of multi-nucleon transfer or a knockout process. These direct reactions have become one of the important tools to obtain the information about the orbital angular moment transferred to the states of target-nucleus. The angular distribution with forward-peaking, can be understood semi-classically, by assuming, that in the case of, say a knockout or stripping process, for an incoming projectile; momentum $\mathbf{p}_i$ is shared vectorially between the emitted particle with momentum $\mathbf{p}_f$ and that which is transferred to the residual nucleus, *i.e.* $\mathbf{p}_t$. Then from Fig. 14.1*a*:

$$p_t^2 = p_i^2 + p_f^2 - 2\, p_i\, p_f \cos\theta \qquad ...(14.2)$$

where θ, is the angle between $\mathbf{p}_i$ and $\mathbf{p}_f$. If we write $\mathbf{p}_i \equiv \mathbf{p}$, and $\mathbf{p}_f = \mathbf{p} - \delta$; Eq. 14.2 may be expressed as:

$$p_t^2 = 2p^2\left[(1-\cos\theta)\left(1-\frac{\delta}{p}\right)\right] + \delta^2$$

$$\approx p^2\left[\theta^2\left(1-\frac{\delta}{p}\right)+\left(\frac{\delta}{p}\right)^2\right] \qquad ...(14.3)$$

where we have expanded cos (θ) for small angles and have taken only the first term.

Solving Eq. 14.3 for θ^2, we get:

$$\theta^2 = \frac{(p_t/p)^2 - (\delta/p)^2}{1 - \delta/p} \qquad ...(14.4)$$

Realising that semi-classically

$$\mathbf{r} \times \mathbf{p}_t \approx \hbar\sqrt{l_t\,(l_t + 1)} \qquad ...(14.5a)$$

and further realising that $\mathbf{r} \times \mathbf{p}_t$ must be equal or less than $\mathbf{R}' \times \mathbf{p}_t$, where $\mathbf{R}_1'$ is the radius at which most of the reaction takes place, minimum value of p_t can be written as:

$$p_t \geq \frac{\hbar\sqrt{l_t\,(l_t + 1)}}{R'} \qquad ...(14.5b)$$

Then one can write from Eq. 14.4, as:

$$\theta^2 \geq \frac{\left(\frac{\lambda\!\!\!^{-}}{R'}\right)^2 (l_t + 1) - \left(\frac{\delta}{p}\right)^2}{1 - \frac{\delta}{p}} \qquad ...(14.6)$$

where $\lambda\!\!\!^{-} = \frac{\hbar}{p}$

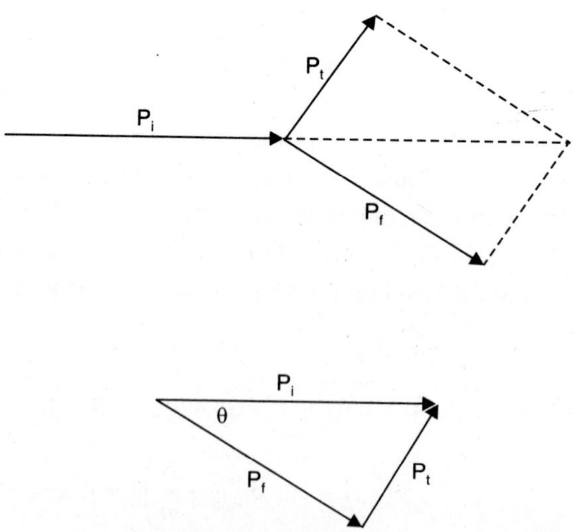

Fig. 14.1a Linear momentum relationship, as used in Eqs. 14.2 and 14.3.

One can also express the orbital angular momentum l_t in terms of angle β, between **r** and $\mathbf{p}_t$, (as shown in Fig. 14.1b), classically as:

$$\mathbf{l} = \mathbf{r} \times p_t \quad \text{or} \quad l = p_t \, r \sin \beta \qquad ...(14.7)$$

Equation 14.7, shows that for a given value of l, the direct interaction points say P lie over a surface of a cylinder of radius $r \sin \beta = l/p_t$, with the momentum transfer lying in the direction of $\mathbf{p}_t$ or $\mathbf{q}_t$ as shown in Fig. 14.1*b*. However, the reaction mechanism of direct reaction requires a single interaction and therefore the reactions from inside the cylinder will give rise to the compound nucleus, because there will be internal scattering and reflection. So for direct reaction in a given direction, only the two spherical caps at the end of the cylinder are effective. Interference from the waves originating at these two areas, leads to the maxima and minima in the angular distribution.

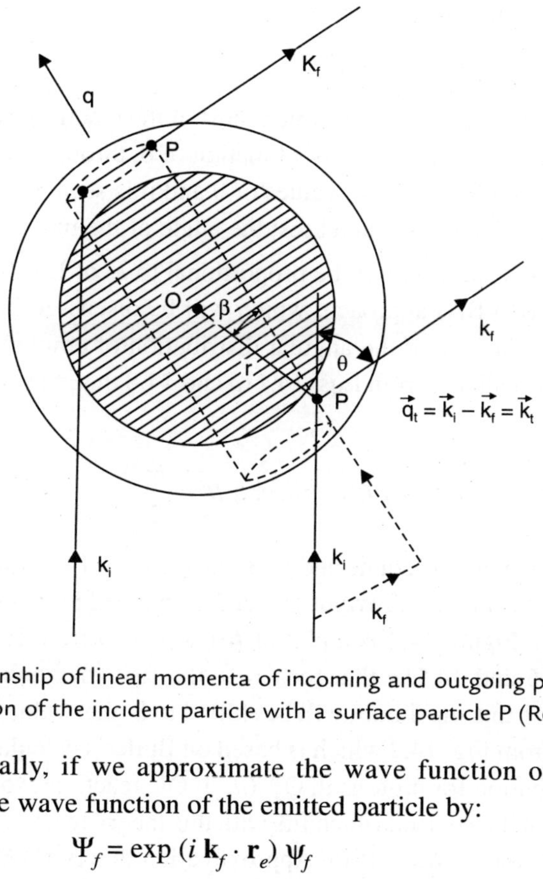

Fig. 14.1b The relationship of linear momenta of incoming and outgoing particles, involving the collision of the incident particle with a surface particle P (Ref. 5).

Quantum mechanically, if we approximate the wave function of the incident particle by $\Psi_i = \exp(i\mathbf{k}_i \cdot \mathbf{r}_i)\, \psi_i$, and the wave function of the emitted particle by:

$$\Psi_f = \exp(i\, \mathbf{k}_f \cdot \mathbf{r}_e)\, \psi_f$$

then, the transition probability λ_i between these two states is given by the Fermi's Golden rule:

$$\lambda_i = \frac{2\pi}{\hbar^3} \left| T_{if} \right|^2 \frac{dn_f}{dE_f} \qquad ...(14.8)$$

where $\frac{dn_f}{dE_f} \, \alpha \, p_f^2 \frac{dp_f}{dE_f}$ and $T_{if} = \int \Psi_f^* \, V \Psi_i \, d\tau$

We assume V as a zero-range potential and write it as:

$$V = V_0 \, \delta(\mathbf{r}_i - \mathbf{r}_e) \, \delta(\mathbf{r}_i - \mathbf{R}) \qquad ...(14.9)$$

where $\mathbf{r}_i$ corresponds to incident particle and $\mathbf{r}_e$ corresponds to the emitted particle, and R is the radius of the potential well.

Therefore,

$$T_{if} = V_0 \int \exp[i\mathbf{q}\cdot\mathbf{R}]\,\psi_f^*\,\psi_i\,d\tau \quad \text{where} \quad \mathbf{q} = \mathbf{k}_i - \mathbf{k}_f = \frac{\mathbf{p}_t}{\hbar}$$

writing,
$$\exp[i(\mathbf{q}\cdot\mathbf{R})] = \sum_l{}' \, i\sqrt{4\pi(2l+1)}\; J_l(qR')\; Y_{l,0}(\theta) \quad ...(14.10)$$

One can express the relationship

$$T_{if}\,\alpha\, J_l(qR') \quad ...(14.11a)$$

where R' is the radius (as explained earlier), at which most of the reaction takes place, J_l is the Spherical Bessel function of order 1 and $q = p_t/\hbar$, which is a function of θ, through Eq. 14.3. The values of $q\,R'$ at the first maximum can be found for different values of θ and it is seen, that the first maximum of $l = 0$ occurs at the smallest value of θ and as l increases; the angle of the maximum also increases.

The direct reaction cross-section will be proportional to λ_i of Eq. 14.8, as expected from Plane Wave Born Approximation (PWBA) approach, which will be discussed in details in the next section. The semi-quantitative understanding of the angular distribution of emitted particles in direct reaction, however, can be obtained by realising from Eqs. 14.11*a* and 14.8 that, the angular distribution $I(\theta)$, may be written as:

$$I(\theta)\,\alpha[J_1(q\,R')]^2 \approx (q\,R')^{-2}\sin^2\left(q\,R' - \frac{l\pi}{2}\right) \quad ...(14.11b)$$

As q is equal to $p_t/\hbar$, which is connected to θ, through Eq. 14.4; the dependence of $I(\theta)$ on θ is evident. The maximum for the angular distribution can be obtained from Eq. 14.11*b*, for the angles for which $J_1(q\,R)$ is maximum. Figure 14.2 is a plot of $I(\theta)$ versus θ from Eq. 14.11*b*. One can also see, qualitatively from Eqs. 14.6 and 14.11*a* that for $l = 0$, the first maximum will occur for very small angles (nearly $l = 0$); for $l = 1$ this maximum will be shifted to higher value of θ; for $l = 2$, to still higher value and so on, as is seen from Fig. 14.2 which is based on Butler's calculations. A comparison with an experimental angular distribution for protons in O^{16} (d, p) O^{17} reaction[5], for $l = 2$ is also shown; giving a very satisfactory fit. A lot of data exists, which fits with this theory to a large extent, but not completely. As we shall see later on, Distorted Wave Born Approximation theory (DWBA) gives better results[6].

Apart from stripping and pick up reactions in which direct reaction is involved, the other reactions which fall in the category of direct reactions are given below:

(*i*) Knock out reactions, like (p, n); or (n, p), (α, p) or (p, α) or (γ, p) where the incident particle knocks out in a single encounter a particle or a cluster from the target nucleus.

(*ii*) Inelastic scattering, through direct reaction in a single encounter. Examples are (p, p'), (α, α') or (d, d'). In the direct reaction mechanism applicable to these cases; the incident particle is subject to the optical potential of the target nucleus; losing some of the energy in either imparting to the nucleus a collective mode of excitation or a particle excitation and then escaping from the nucleus.

(*iii*) Coulomb excitation is another mode of direct reaction mechanism induced interaction.

(*iv*) Break-up reaction[30], in which in a single interaction, either the projectile or the target breaks up into two fragments, *e.g.* $Mg^{24} + C^{12} \rightarrow C^{12} + (O^{16} + Be^{8})$. This however, requires higher energies, say 170 MeV in this case.

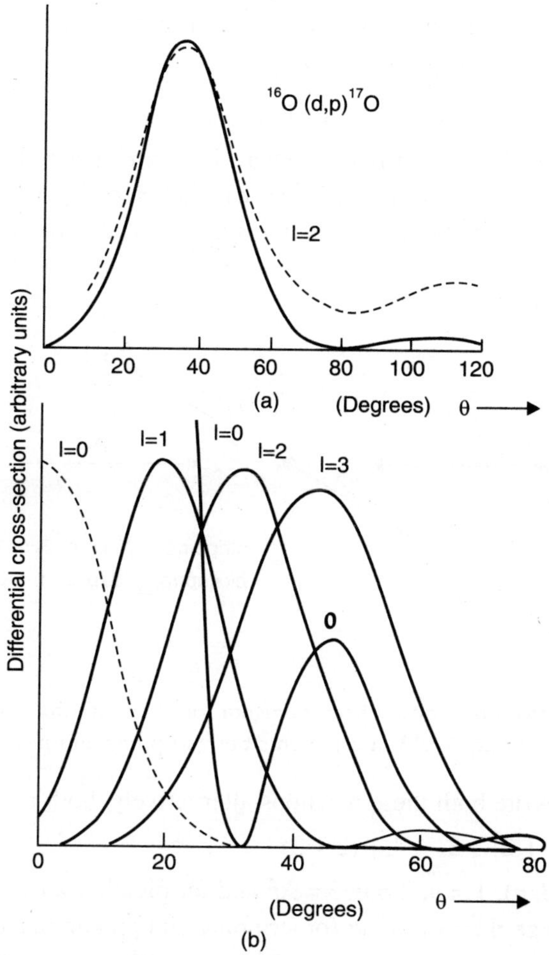

Fig. 14.2 (*a*) Angular distribution of protons in O^{16} (d, p) O^{17}. Angles are in C.M. system. The solid line is experimental, the dashed line is theoretical (PWBA) for 8 MeV deuterons; (*b*) Theoretical angular distribution for l = 0, 1, 2, 3, (PWBA) [Ref. 5].

Gamma-rays emitted in transitions between states along with energy and angular distribution of emitted particles provide a strong tool of nuclear spectroscopy.

14.2 PLANE-WAVE THEORY OF DIRECT REACTIONS

We may now describe in some details, the theory of direct reactions as developed first by Oppenheimer and Phillips[2] and later in details by Butler[3]. We have already mentioned the examples of stripping [(*d*, *p*), (*t*, *p*)] and pickup [(*p*, *t*), (*p*, *d*)] which fall into this category. Also reactions like (*p*, *p*′) and (*p*, *n*), when they take place only by involving the particles in a single interaction at the surface of the target nuclei, will fall in the category of direct reaction as inelastic scattering and knock out direct reactions. The theory of direct reactions should cover all these cases. It also covers, the break-up mode of direct reactions; some details for which will be discussed later.

One can represent these different processes of direct reactions as:

(*i*) $(1 + 2) + 3 \rightarrow (2 + 3) + 1$ Stripping

(*ii*) $1 + (2 + 3) \rightarrow (1 + 3) + 1$ Pickup ...(14.12)

Examples are X^A (*d*, *p*) X^{A+1} for stripping, where $1 + 2 = d$; $1 = p$, $2 = n$ and 3 is the X^A nucleus, so that (2 + 3) is X^{A+1}. For pickup again say in X^{A+1} (*p*, *d*) X^A; $1 = p$; $(2 + 3) = X^{A+1}$, $3 = X^A$ and $1 + 2 = d$, *i.e.* $2 = n$. Hence by writing in the above manner, the symbols 1, 2 and 3 retain their meaning in both cases.

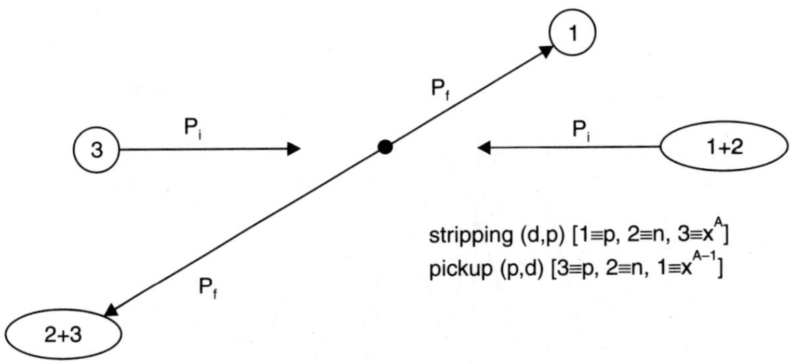

Fig. 14.3 The directions of various linear momenta for the scattering in the centre of mass system; following Eq. 14.13. It represents both stripping and pickup reactions.

One may, however, write both these reactions alternatively, both for stripping and pickup:

$$(1 + 2) + 3 \rightarrow 1 + (2 + 3) \qquad ...(14.13)$$

Then for stripping (*d*, *p*); $1 = p$, $2 = n$, $3 = X^A$ and for pickup (*p*, *d*); $3 = p$; $1 = X^{A-1}$, $2 = n$. Here the symbols 1, 2 and 3 change their meaning for stripping and pickup reactions. Another example is a multi-nucleon pickup reaction, say (*p*, α); *i.e.* $p + X^A \rightarrow \alpha + X^{A-3}$, in which case $1 = X^{A-3}$, $2 = H^3$ and $3 \equiv p$.

We will use the notation of Eq. 14.13 for the development of the subject of direct reactions for both types.

Following the treatment by Roy and Nigam[7], we see that for stripping and pickup reactions, Fig. 14.3 has to be interpreted properly. In the case of say (*d*, *p*) reaction, corresponding to stripping; the incident particle deuteron, *i.e.* (*p* + *n*) is (1 + 2), with the linear momentum $-\mathbf{p}_i$ while the target nucleus X^A is 3, with linear momentum $\mathbf{p}_i$. On the other hand for pick up, say (*p*, *d*) reaction, the incident particle proton (*p*) is 3, and the target is $X^{A-1} + n$, *i.e.* (1 + 2), with linear momentum of $\mathbf{p}_i$ and $-\mathbf{p}_i$, respectively. For both cases, the sharing of linear momentum between particle 1 and 2 is given by:

Particle 1 Particle 2

$$-\frac{M_1}{M_1+M_2}\,\mathbf{p}_i \qquad\qquad -\frac{M_2}{M_1+M_2}\,\mathbf{p}_i \qquad ...(14.14)$$

Similarly for both stripping ($p = 3, n = 2, X^{A-1} = 1$) and pickup ($n = 2, X^A = 3, p = 1$); the sharing of linear momentum between particle 2 and 3 is given by:

Particle 2 Particle 3

$$-\frac{M_2}{M_2+M_3}\,\mathbf{p}_f \qquad\qquad -\frac{M_3}{M_2+M_3}\,\mathbf{p}_f \qquad ...(14.15)$$

Interpreting the meaning of particles differently for the two cases of stripping and pickup, we can write:

(*i*) Momentum transfers $\mathbf{q}_k$ ($k = 1$, 2 and 3) corresponding to transfers to particles 1, 2 and 3, after the reaction are:

$$\mathbf{q}_1 = \mathbf{p}_f + \frac{M_1}{M_1+M_2}\,\mathbf{p}_i$$

$$\mathbf{q}_2 = -\frac{M_2}{M_2+M_3}\,\mathbf{p}_f + \frac{M_2}{M_1+M_2}\,\mathbf{p}_i$$

$$\mathbf{q}_3 = -\frac{M_3}{M_2+M_3}\,\mathbf{p}_f - \mathbf{p}_i \qquad ...(14.16)$$

(*ii*) Reduced Masses μ_i and μ_f are given by:

$$\frac{1}{\mu_i} = \frac{1}{M_1+M_2} + \frac{1}{M_3}\,;\ \frac{1}{\mu_f} = \frac{1}{M_1} + \frac{1}{M_2+M_3}$$

and reduced masses of the system (1 + 2) (μ_{12}) and (2 + 3) (μ_{23}) are given by:

$$\frac{1}{\mu_{12}} = \frac{1}{M_1} + \frac{1}{M_2}\,;\ \frac{1}{\mu_{23}} = \frac{1}{M_2} + \frac{1}{M_3} \qquad ...(14.17)$$

(*iii*) Total Energy E, can be written as:

$$E = \frac{p_i^2}{2\mu_1} - \varepsilon_{12} = \frac{p_f^2}{2\mu_f} - \varepsilon_{23} \qquad ...(14.18)$$

where ε_{12} and ε_{23} are the binding energies of the initial and final bound systems:

(*iv*) The centre of masses $\mathbf{R}_{12}$ and $\mathbf{R}_{23}$ of (1 + 2) system and (2 + 3) system can be easily related to $\mathbf{r}_1$, $\mathbf{r}_2$ and $\mathbf{r}_3$ as:

$$\mathbf{R}_{12} = \frac{M_1\mathbf{r}_1 + M_2\mathbf{r}_2}{M_1+M_2}\,;\ \mathbf{R}_{23} = \frac{M_1\mathbf{r}_2 + M_3\mathbf{r}_3}{M_2+M_3} \qquad ...(14.19)$$

(*v*) The relative coordinates $\mathbf{r}_{12}$ and $\mathbf{r}_{23}$ are given by:

$$\mathbf{r}_{12} = \mathbf{r}_1 - \mathbf{r}_2;\ \mathbf{r}_{23} = \mathbf{r}_2 - \mathbf{r}_3$$

Then it is easy to see that

$$\mathbf{r}_i = \mathbf{r}_3 - \mathbf{R}_{12} \quad \text{and} \quad \mathbf{r}_f = \mathbf{r}_1 - \mathbf{R}_{23} \qquad ...(14.20)$$

Of course, for all these relations:

$$M_1\mathbf{r}_1 + M_2\mathbf{r}_2 + M_3\mathbf{r}_3 = 0 \qquad ...(14.21)$$

which means, that the centre of mass of the whole system is at rest.

(*vi*) $\mathbf{p}_{12}$ and $\mathbf{p}_{23}$ are the momenta canonically conjugate to $\mathbf{r}_{12}$ and $\mathbf{r}_{23}$ and $\mathbf{p}_i$ and $\mathbf{p}_f$ are, of course canonically conjugate to $\mathbf{r}_i$ and $\mathbf{r}_f$ respectively. They are, therefore expressed as (with $\hbar = 1$);

$$\mathbf{p}_{12} = \frac{1}{i}\frac{\partial}{\partial \mathbf{r}_{12}} = \sum_{k=1,2} \mathbf{p}_k \frac{\partial \mathbf{r}_k}{\partial \mathbf{r}_{12}} = \mu_{12}\left(\frac{\mathbf{p}_1}{M_1} - \frac{\mathbf{p}_2}{M_2}\right) \qquad ...(14.22a)$$

$$\mathbf{p}_{23} = \frac{1}{i}\frac{\partial}{\partial \mathbf{r}_{23}} = \sum_{k=2,3} \mathbf{p}_k \frac{\partial \mathbf{r}_k}{\partial \mathbf{r}_{23}} = \mu_{23}\left(\frac{\mathbf{p}_2}{M_2} - \frac{\mathbf{p}_3}{M_3}\right) \qquad ...(14.22b)$$

[It should be realised that $\mathbf{p}_1$, $\mathbf{p}_2$ and $\mathbf{p}_3$ are canonical conjugate to $\mathbf{r}_1$, $\mathbf{r}_2$ and $\mathbf{r}_3$ respectively]

$$\mathbf{p}_i = \frac{1}{i}\frac{\partial}{\partial \mathbf{r}_i} = \sum_k \mathbf{p}_k \frac{\partial \mathbf{r}_k}{\partial \mathbf{r}_i} = \mu_i\left(-\frac{\mathbf{p}_1 + \mathbf{p}_2}{M_1 + M_2} + \frac{\mathbf{p}_3}{M_3}\right) \qquad ...(14.22c)$$

$$\mathbf{p}_f = \frac{1}{i}\frac{\partial}{\mathbf{r}_f} = \sum_k \mathbf{p}_k \frac{\partial \mathbf{r}_k}{\partial \mathbf{r}_f} = \mu_f\left(\frac{\mathbf{p}_1}{M_1} - \frac{\mathbf{p}_2 + \mathbf{p}_3}{M_2 + M_3}\right) \qquad ...(14.22d)$$

(*vii*) In Eq. 14.22, we have obtained the explicit expressions for $\mathbf{p}_{12}$, $\mathbf{p}_{23}$, $\mathbf{p}_i$ and $\mathbf{p}_f$. We can now write from these expressions, the kinetic energy operator T_C in the centre of mass system as:

$$T_C = \frac{\mathbf{p}_{12}^2}{2\mu_{12}} + \frac{\mathbf{p}_i^2}{2\mu_i} = \frac{\mathbf{p}_{23}^2}{2\mu_{23}} + \frac{\mathbf{p}_f^2}{2\mu_f} \qquad ...(14.23)$$

Using the relationships developed from Eqs. 14.14 to 14.23; we now develop the simplified version of Plane-Wave-Born-Approximation (PWBA) model of direct sections.

14.3 GENERAL THEORY

In this theory, we assume that the total wave function is obtained from

$$H\psi = E\psi \qquad ...(14.24)$$

where $$H = T_C + V_{12} + V_{13} + V_{23}$$

and ψ, itself, may be expressed as a product of the wave function of the relative motion of particles of the system and the free particle, and V_{ij}'s are, of course, the potentials between the particles *i* and *j*. If ϕ_{12}

and ϕ_{23} are the wave functions of the (1 + 2) and (2 + 3) system, respectivley, then the Schrödinger equations satisfied by ϕ_{12} and ϕ_{23} may be written as:

$$\left(\frac{\mathbf{p}_{12}^2}{2\mu_{12}} + V_{12}\right)\phi_{12}(\mathbf{r}_{12}) = -\varepsilon_{12}\,\phi_{12}(\mathbf{r}_{12})$$

and
$$\left(\frac{\mathbf{p}_{23}^2}{2\mu_{23}} + V_{23}\right)\phi_{23}(\mathbf{r}_{23}) = -\varepsilon_{23}\,\phi_{23}(\mathbf{r}_{23}) \qquad \text{...(14.25)}$$

Denoting $\chi_i(\mathbf{r}_i)$ and $\chi_f(\mathbf{r}_f)$ as the wave functions of particle 3, with respect to the centre of mass of (1 + 2) in the initial state and of 1-particle with respect to the centre of mass of (2 + 3) in the final state respectively, then one can write:

$$\psi(\mathbf{r}_{12}, \mathbf{r}_i) = \sum_{\alpha} \phi_\alpha(\mathbf{r}_{12})\,\chi_\alpha(\mathbf{r}_i) \qquad \text{...(14.26)}$$

for the initial state, and

$$\psi(\mathbf{r}_{23}, \mathbf{r}_f) = \sum_{\alpha} \phi_\alpha(\mathbf{r}_{23})\,\chi_\alpha(\mathbf{r}_f) \qquad \text{...(14.27)}$$

for the final state.

Here summation over α, corresponds to complete set of states of the bound state and the free particle. One can, then, write an expression for χ's as:

$$\chi_i(\mathbf{r}_i) = \int \phi_{12}^*(\mathbf{r}_{12})\,\psi(\mathbf{r}_{12}, \mathbf{r}_i)\,d^3\,\mathbf{r}_{12}$$

for the initial state, and

$$\chi_f(\mathbf{r}_f) = \int \phi_{23}^*(\mathbf{r}_{23})\,\psi(\mathbf{r}_{23}, \mathbf{r}_i)\,d^3\,\mathbf{r}_{23} \qquad \text{...(14.28)}$$

for the final state.

Then we use Eqs. 14.24 and 14.27 and write:

$$H\psi(\mathbf{r}_{23}, \mathbf{r}_f) = \left(\frac{\mathbf{p}_{23}^2}{2\mu_{23}} + \frac{\mathbf{p}_f^2}{2\mu_f} + V_{12} + V_{13} + V_{23}\right)\psi(\mathbf{r}_{23}, \mathbf{r}_f)$$

$$= E\psi(\mathbf{r}_{23}, \mathbf{r}_f) \qquad \text{...(14.29)}$$

where
$$E \equiv \left(\frac{\mathbf{p}_f^2}{2\mu_f} - \varepsilon_{23}\right)$$

from Eqs. 14.18, 14.23 and 14.24.

Realising that $\mathbf{p}_f^2$ on, the left side of Eq. 14.29 is an operator and can be written as $-\nabla_f^2$ and p_f^2 on the right side, included in the expression for E is a number; we can write Eq. 14.29, as:

$$\frac{1}{2\mu_f}\left(\nabla_f^2 + p_f^2\right)\psi\,(\mathbf{r}_{23}, \mathbf{r}_f)$$

$$= \left(\frac{\mathbf{p}_{23}^2}{2\mu_{23}} + V_{12} + V_{13} + V_{23} + \varepsilon_{23}\right)\psi\,(\mathbf{r}_{23}, \mathbf{r}_f) \qquad ...(14.30)$$

From Eq. 14.27, one applies operator $(\nabla_f^2 + p_f^2)$ to $\chi(\mathbf{r}_f)$ and using Eqs. 14.28, 14.29 and 14.25, we obtain:

$$\left(\nabla^2 + p_f^2\right)\chi_f\,(\mathbf{r}_f) = 2\mu_f \int \phi_{23}^*\,(\mathbf{r}_{23})\left[V_{12} + V_{13}\right]\psi\,(\mathbf{r}_{23}, \mathbf{r}_f)\, d^3\,\mathbf{r}_{23} \qquad ...(14.31)$$

Using the outgoing Green function, *i.e.*,

$$-\frac{\exp i\,\mathbf{p}_f\left|\mathbf{r}_f - \mathbf{r}\right|}{4\pi\left|\mathbf{r}_f - \mathbf{r}\right|}$$

we can write the solution for Eq. 14.31 as:

$$\chi_f\,(\mathbf{r}_f) = \frac{-\mu_f}{2\pi}\int \frac{\exp i\,p_f\,|\mathbf{r}_f - \mathbf{r}|}{|\mathbf{r}_f - \mathbf{r}|} \times \phi_{23}^*\,(\mathbf{r}_{23})\,[V_{12} + V_{13}]\,\psi\, d^3\,r_{23}\, d^3\,r \qquad ...(14.32)$$

As we saw earlier, in Chapter 13, Section 13.2.2, one can write from physical considerations, the outgoing wave as a spherical wave, which may be expressed as:

$$\chi_f\,(\mathbf{r}_f) \underset{r_f \to \infty}{=} f\,(\theta, \phi)\,\frac{e^{i\,p_f\,r_f}}{r_f} \qquad ...(14.33)$$

where $f\,(\theta, \phi)$ is the reaction amplitude. Comparing Eq. 14.32 with Eq. 14.33 it may be seen that

$$f\,(\theta, \phi) = -\frac{\mu_f}{2\pi}\int \phi_{23}^*\,(\mathbf{r}_{23})\,e^{-i\mathbf{p}_f\cdot\mathbf{r}_f} \times [V_{12} + V_{13}]\,\psi\, d^3\,r_{23}\, d^3\,r_f \qquad ...(14.34)$$

where ψ describes the reaction process $(1 + 2) + 3 \rightarrow 1 + (2 + 3)$.

As has been explained in Chapter 13 in Eq. 13.39, the differential cross-sections may be expressed as the outgoing current divided by the incident current density p_i/μ_i and hence from Eq. 14.34, one expresses $\sigma(\theta, \phi)$ as:

$$\sigma(\theta, \phi) = \underset{R\to\infty}{\mathrm{Lt}}\,\frac{1}{2\,i\,\mu_f}\left(\chi_f^*\frac{\partial\chi_f}{\partial r_f} - \frac{\partial\chi_f^*}{\partial r_f}\chi_f\right)_{r_f = R} \times \frac{R^2}{p_i/2\mu_i}$$

$$= \frac{\mu_i\left|\mathbf{p}_f\right|}{\mu_f\left|\mathbf{p}_i\right|}\left|\,f(\theta, \phi)\right|^2 \qquad ...(14.35)$$

14.3.1 Born-Approximation (Plane Wave)—Butler's Theory

We apply the Born approximation here, because the incident energy is assumed to be much larger than potentials $V_{12} + V_{13}$. Then, according to the well known theory of Born approximation (*see* Quantum Mechanics by L.I. Schiff, p. 161), one can express the function ψ of the incident channel as [using Eqs. 14.19 and 14.20].

$$\psi \approx \phi_{12}(r_{12}) \exp(-i\, \mathbf{p}_i \,.\, \mathbf{R}_{12} + i\, \mathbf{p}_i \,.\, \mathbf{r}_3)$$
$$\approx \phi_{12}(r_{12}) \exp i\, \mathbf{p}_i \,.\, \mathbf{r}_i \qquad \text{...(14.36)}$$

which is a product of the wave function of (1 + 2) target system and the incident plane wave. One can, then, write, from Eq. 14.34:

$$f(\theta, \phi) \approx -\frac{\mu_f}{2\pi} \int \phi_{23}^*(\mathbf{r}_{23})\, e^{i(\mathbf{q}_3 \cdot \mathbf{r}_{23} - \mathbf{q}_1 \cdot \mathbf{r}_{12})} \times [V_{12} + V_{13}]\, \phi_{12}(\mathbf{r}_{12})\, d^3 r_{12}\, d^3 r_{23} \qquad \text{...(14.37)}$$

where the relations,

$$-\mathbf{p}_f \cdot \mathbf{r}_f + \mathbf{p}_i \cdot \mathbf{r}_i = \mathbf{q}_3 \cdot \mathbf{r}_{23} - \mathbf{q}_1 \cdot \mathbf{r}_{12}$$

and

$$d^3 r_{23}\, d^3 r_f = d^3 r_{23}\, d^3 r_{12}$$

have been used. One can simplify Eq. 14.37 by neglecting the interaction V_{13}. This can be understood by realising that both in stripping, and pickup sections, V_{13} represents the potential of the particle with the nucleus, to which it is not bound. On the other hand, V_{12} equals the binding energy in both of these reactions, *i.e.* for stripping and pickup. In stripping reaction say (d, p), V_{12} represents interaction of n and p in deuteron, where they are bound. In pickup reaction say (p, d), V_{12} represents the interaction between X^{A-1} ($\equiv 1$) and n ($\equiv 2$); which are bound in the target nucleus X^A. So V_{12} represents an interaction leading to bound states in both cases, and hence is expected to be much stronger than V_{13}. Essentially V_{12} equals V_{np} in both stripping and pickup reactions as in deuteron in stripping case, and in $(n + X^{A-1})$ in the pickup case, because (p, d) reaction is inverse to (d, p) reaction:

Alternatively it can be seen that in the reaction $(1 + 2) + 3 \rightarrow 1 + (2 + 3)$; particles 1 and 3 never appear in the bound state and hence V_{13} is, in general weak and gives a non-vanishing result in pickup reaction only if the final state corresponds to the excitation of core. On the other hand, V_{12} contributes a non-vanishing results in pickup reaction provided the final state contains components of the core, *i.e.* the target nucleus, is left in the ground state.

Hence, in pure direct reaction, where there is no core-excitation, $V_{13} << V_{12}$. One, then expresses from Eqs. 14.34, 14.35 and 14.37,

$$\sigma(\theta, \phi) = \frac{\mu_i\, \mu_f}{(2\pi)^2} \frac{p_f}{p_i} \times \left| \int \phi_{23}^*(\mathbf{r}_{23})\, e^{i(\mathbf{q}_3 \cdot \mathbf{r}_{23} - \mathbf{q}_1 \cdot \mathbf{r}_{12})} \times \left[V_{12}(\mathbf{r}_{12})\, \phi_{12}(\mathbf{r}_{12})\, d^3 r_{12}\, d^3 r_{23} \right] \right|^2 \qquad \text{...(14.38)}$$

where we have neglected V_{13}.

Without trying to solve Eq. 14.38, in details, for which the reader is referred to the original papers of Butler, we give only the final results where, we have used the following basic assumptions:

(*i*) $\phi_{12}(\mathbf{r}_{12})$ is a solution of the Equation: [*see* Eq. 14.25].

$$\left[-\frac{\nabla^2}{2\mu_{12}} + V_{12}(\mathbf{r}_{12}) + \varepsilon_{12} \right] \phi_{12}(\mathbf{r}_{12}) \qquad \text{...(14.39a)}$$

A similar equation for $\phi_{23}\left(\vec{r}_{23}\right)$ can be written from Eq. 14.25:

(*ii*) ϕ_{12} ($\mathbf{r}_{12}$) and ϕ_{23} ($\mathbf{r}_{23}$) can be expressed as:

$$\phi_{12}(\mathbf{r}_{12}) = \phi_{12}(r_{12})\, Y_{l_{12},m_{12}}(\hat{r}_{12})$$

$$\phi_{23}(\mathbf{r}_{23}) = \phi_{23}(r_{23})\, Y_{l_{23},m_{23}}(\hat{r}_{23}) \quad \text{...(14.39}b\text{)}$$

where m_{12} and m_{23} are the magnetic quantum numbers of the orbital quantum numbers, l_{12} belonging to (1 + 2) bound system and l_{23} belonging to (2 + 3) bound system respectively. Finally, evaluating the cross-section by averaging over the initial quantum numbers m_{12} and summing over the final magnetic quantum numbers m_{23}, we obtain the expression for cross-section $\sigma(\theta, \phi)$ as:

$$\sigma(\theta, \phi) = 4\,\mu_i\,\mu_f\,\frac{p_f}{p_i}\left(\varepsilon_{12} + \frac{q_{12}^2}{2\,\mu_{12}}\right)^2 \times (2l_{23}+1)\, R_1^2(q_1)\, R_3^2(q_3) \quad \text{...(14.40}a\text{)}$$

where $$R_1(q_1) = \int_0^\infty r_{12}^2\, \phi_{12}(r_{12})\, J_{l_{12}}(q_1\, r_{12})\, dr_{12}$$

and $$R_3(q_3) = \int_0^\infty r_{23}^2\, \phi_{23}(r_{23})\, J_{l_{23}}(q_3\, r_{23})\, dr_{23} \quad \text{...(14.40}b\text{)}$$

As we have seen earlier, the function $J_l(q\, r)$ depends on the angle θ, between the vectors $\mathbf{p}_i$ and $\mathbf{p}_f$. In a typical (*d*, *p*) reaction; the experimentally observed quantities are momentum, $\mathbf{p}_i$ of the lighter incident particle $d = (2 + 3)$, and $\mathbf{p}_f$, the momentum of lighter particle $p \equiv 1$.

Figure 14.4*b* shows, the comparison of the experimental data on the angular distribution of protons in Sn^{116} (*d*, *p*) Sn^{117} reaction[8] and the comparison with the theoretical expectations based on Eq. 14.40 for different values of l_{23} = 1. It is evident that the angular distribution of such reactions are very sensitive to 1. Evidently, the shape and absolute value of the cross-sections are obtained from Eqs. 14.39 and 14.40 which contain R_1 (q_1) and R_3 (q_3). The wave function ϕ_{12} will be a solution of Eq. 14.25 for appropriate l value. To evaluate R_3 (q_3), on the other hand, use is made of specific approximation of direct reaction model, which assumes that the interaction takes place between the incident particle and target nucleus 3 in the outer region of the nucleus 3. Therefore limits of integration in Eq. 14.40*b* should extend from R_0 to ∞. Then V_{23} may be dropped, if the range of potential V_{23} is less than R_0, for determining ϕ_{23} (r_{23}) from Eq. 14.25.

Then ϕ_{23} in Eq. 14.40*b* obtained from properly written Eq. 14.25 is a solution of the equation:

$$\left[\frac{d^2}{dr^2} - \frac{l(l+1)}{r^2} - 2\,\mu\,\varepsilon\right] r\phi_1(r) = 0 \quad \text{for } r > R_0 \quad \text{...(14.41)}$$

where $l = l_{23}$; $\varepsilon = \varepsilon_{23}$ and $r = r_{23}$.

One determines R_3 (q_3) from Eqs. 14.40*b* and 14.41, and determine $\sigma(\theta, \phi)$ from Eq. 14.40*a*.

14.3.2 Some Applications of Butler Model and Experiments

(*i*) **One-Nucleon Transfer Reactions:** This corresponds to both stripping and pickup, *e.g.* Sn^{116} (*d, p*) Sn^{117} or O^{16} (*p, d*) O^{15} respectively. Both these types of reactions have been extensively used for the investigation of bound nuclear states [Figs. 14.4*b* and 14.4*c*; Ref. 8, 9].

In (*d, p*) reaction, one inserts a neutron into a nucleus, so that the neutron occupies one of the available quantum states in the target nucleus say, l_t, the orbital quantum number of the target state. The outgoing particle, proton, carries information concerning this state. This is a case of stripping. Experimentally, we can measure the angular and energy distribution of the particle proceeding in the forward direction. An accurate measurement of the energy of the emerging proton determines, the energy of the level, in which neutron has been captured. Figure 14.4*a* shows the proton spectra from the reaction Ni^{58} (*d, p*) Ni^{59} in the forward direction of $\theta = 29°$. The numbers, attached to peaks are excitation energies (in MeV) of corresponding states, in Ni^{59}.

The orbital angular distribution for each peak can be determined from angular distribution. The highest energy peak, represents the case where Ni^{59} is left in the ground state. The other peaks represent the various excited states of the Ni^{59}. The numbers in parenthesis correspond to l value of the orbits into which neutron is captured. In Ni^{58}, there are 30 neutrons; out of which 28 neutrons fill the orbit corresponding to $lf_{7/2}$ in $n = 3$ shell and two neutrons either are in $lf_{5/2}$ orbit for $l = 3$ in $n = 4$ shell, or for $l = 4$, in $1g_{9/2}$ orbit or for $l = 1$, in $2p_{3/2}$ or $2p_{1/2}$ orbit. In $l = 2$ cases it is probably $2d_{5/2}$ orbit, and for $l =$ 0, it is $3s_{1/2}$ orbit. At higher energies neutron goes into higher energy levels of $n = 5$ shell. Evidently, simply determination of l value of the orbit into which neutron enters is not good enough to determine j. One has to use the sequence of shell model levels. Sometimes detailed analysis of DWBA, especially at the backward angles, can help determine j, as has been carried[11] out for Fe^{54} (*d, p*) Fe^{55}. Figure 14.4*b* shows the angular distribution of protons from Sn^{116} (*d, p*) Sn^{117} revealing orbital angular momentum of capturing level. It may be seen, that $l = 0$ curve peaks very much in the forward direction while for $l = 2$, and $l = 4$, peaks occur at higher angles, as expected from Eq. 14.40.

Figure 14.4*c*, on the other hand, shows the angular distribution of a pickup reaction[8] from O^{16} (*p, d*) O^{15} at 18.5 MeV proton energy in which plane-wave Butler theory is used for $l = 1$, $r_0 = 5.2 \times 10^{-13}$ cm, to give it a good fit in the forward direction (solid line). This pickup reaction can be looked up as reverse reaction to the stripping section.

(*ii*) **Two-Nucleon Transfer Reactions:** Again the transfer may be in stripping mode, *e.g.* Mg^{26} (*t, p*) Mg^{28} or Fe^{56} (*t, p*) Fe^{58} where two neutrons are stripped[12]. Such measurements have been conducted, by Middleton and Cohen et al. For pickup mode in the two nucleon-transfer reaction; a typical case is Y^{89} (*d*, α) Sr^{87}, where a neutron + proton are picked up. This reaction has been studied by Micheletti and Mead[13].

The case of Mg^{26} (*t, p*) has been studied experimentally by Middleton[11] and compared with an approximate expression for the double stripping process as given by Newns[14], *i.e.*,

$$\sigma(\theta) \approx \sum_L \left| A(L)\, J_L\,(k\,R) \right|^2 \qquad \text{...(14.42)}$$

This expression, which is somewhat similar to Eq. 14.40 is based on the assumption that the two nucleons are captured as one unit and is similar to calculations by Bhatia, Huang, Huby, and Newns[15] for deuteron stripping. It has been possible to measure the angular distribution of the ground state and first

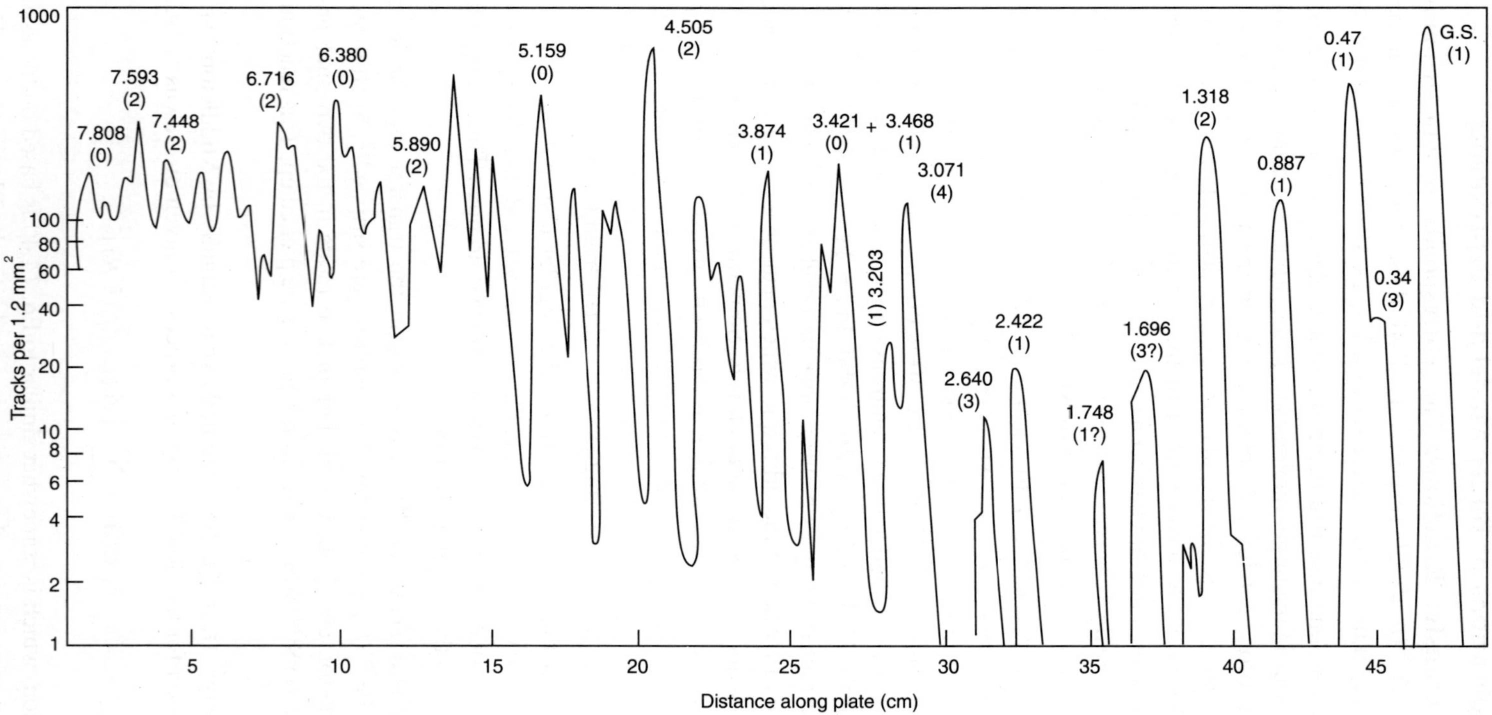

Fig. 14.4a Proton spectrum from the reaction Ni^{58} (d, p) Ni^{59}. The numbers on each peak represent energies of Ni^{59} levels to which the neutron is captured, as shown in the energy levels of Ni^{59}. The numbers in the brackets are the 1 value associated, with the captured neutron, from which the angular momentum of the level is determined (Ref. 10).

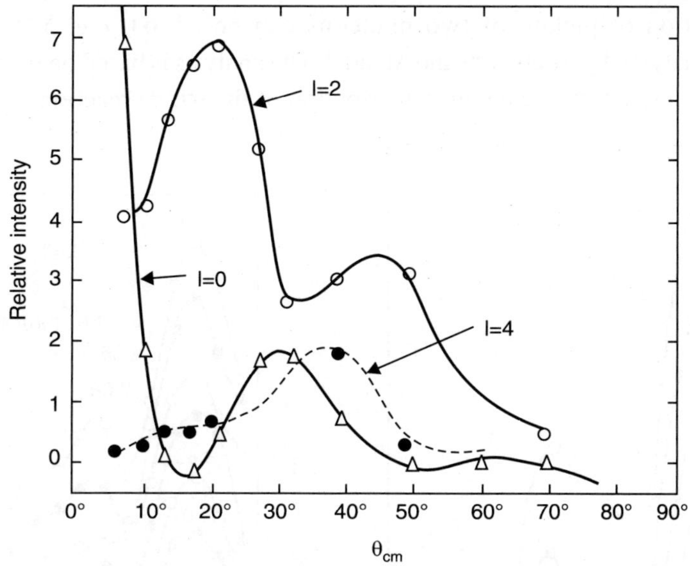

Fig. 14.4b Angular distribution of protons from Sn^{116} (d, p) Sn^{117} showing the dependence of angular distribution on the transfer of orbital angular momentum on stripping reaction (Ref. 8).

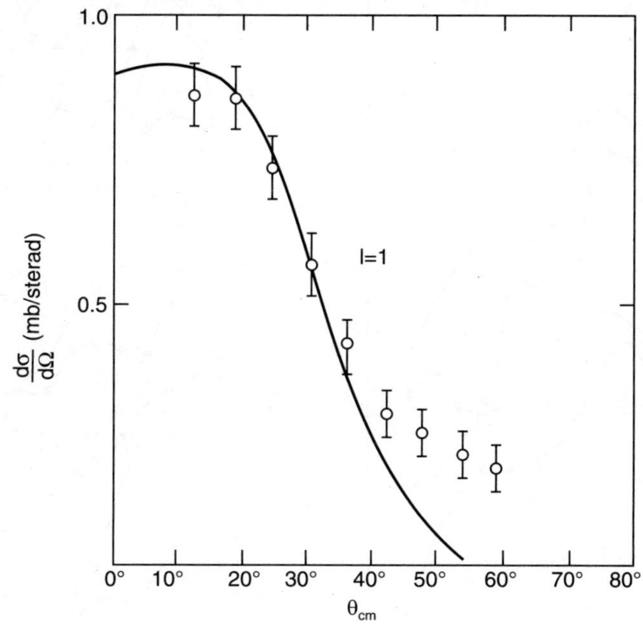

Fig. 14.4c Angular distribution of deuterons from O^{16} (p, d) O^{15} ground state, illustrating a pickup reaction. The solid line is a plane wave Butler curve with l = 1. The energy of proton was 18.5 MeV (Ref. 8).

excited state proton-groups from Mg^{26} (t, p) Mg^{28} reaction at 10 MeV triton energy. The theoretical, curves fit for $l = 0$ and $l = 2$ for these two states; using Eq. 14.42. Similarly, we show in Fig. 14.5*a* the angular distribution of protons from 12 MeV tritons induced Fe^{56} (t, p) Fe^{58} reactions[12].

Reverse reaction of pickup of two nucleons can be (d, α), *e.g.* Y^{89} (d, α) Sr^{87}, studied experimentally and analysed by Michaletti and Mead[13]. The analysis is based on theory of Glendenning[16], who has given an expression for the angular distribution of the (α, d) reaction in the plane wave Born

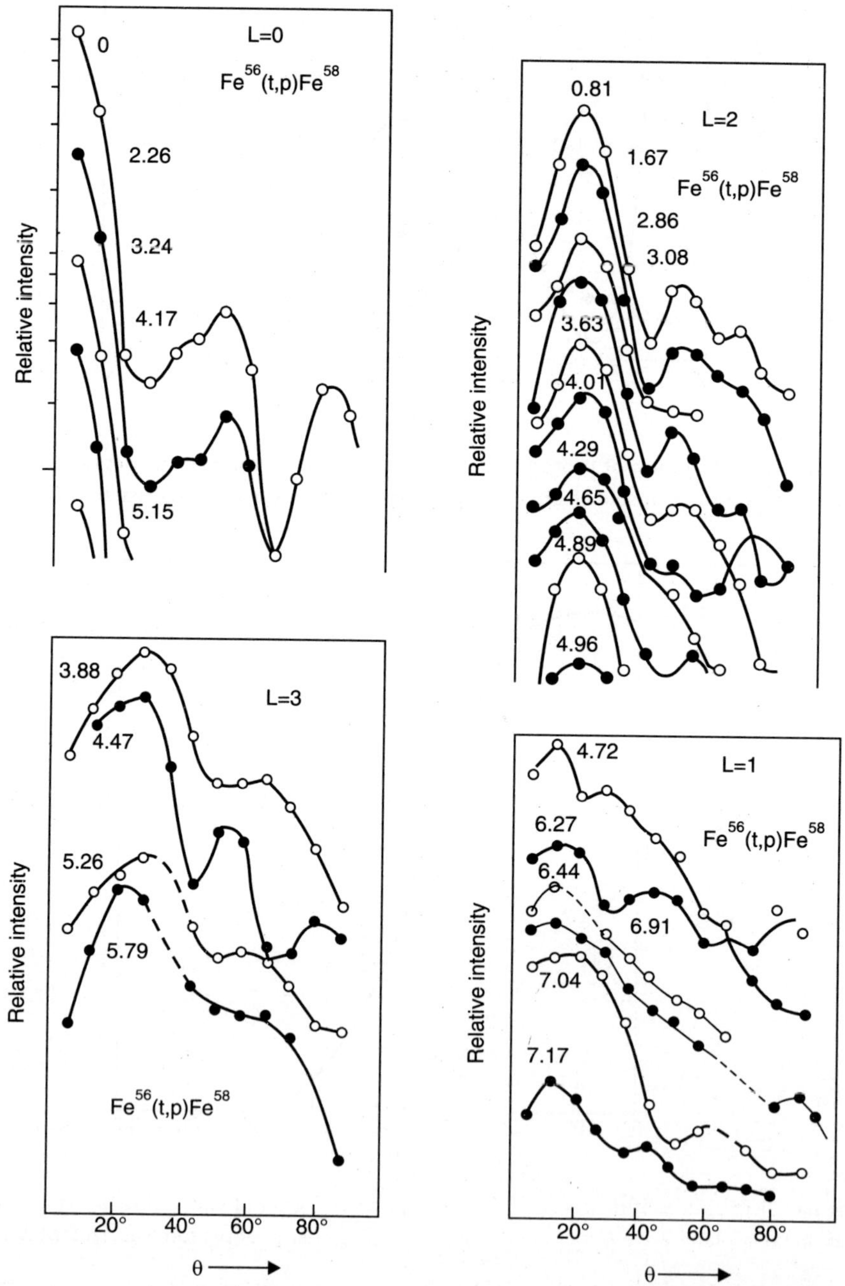

Fig. 14.5a An example of two-nucleon transfer case of Fe^{56} (t, p) Fe^{58}; showing the angular distribution of protons from 13 MeV triton induced reaction on Fe^{56}; leading to selected states of Fe^{58}. We have grouped the angular distributions of a given l-transfer together (Ref. 12).

approximation of the form:

$$\frac{d\sigma}{d\Omega} \propto \exp\left[\frac{-K^2}{8\gamma^2}\right]\sum_L \frac{1}{2L+1} C_L \,|\, B_L(Q)\,|^2 \qquad ...(14.43)$$

where

$$\mathbf{K} = \mathbf{K}_d - \frac{1}{2}\,\mathbf{K}_\alpha$$

is the linear momentum transferred to the outgoing deuteron, and

$$\mathbf{Q} = \mathbf{K}_\alpha - \frac{M_i}{M_f}\,\mathbf{K}_\alpha \qquad ...(14.44)$$

is the linear momentum carried into the nucleus by the stripping pair. The term $B_L(Q)$ is proportional to $J_L(q R_0)$ for a point α-particle. The quantity γ in Eq. 14.43 represents the r.m.s. radius of the alpha-particle charge density.

The two-nucleon transfer reactions follow the selection rules for double stripping sections:

$S = 1, T = 0$ for (t, n), (He^3, p) and (d, α)

$S = 0, T = 0$ for (t, p), (He^3, p) and (He^3, n); and

$\Delta\pi = (-1)^L$, $\mathbf{L} = \mathbf{L}_n + \mathbf{L}_p$; $\mathbf{J}_i = \mathbf{J}_f + \mathbf{J}$

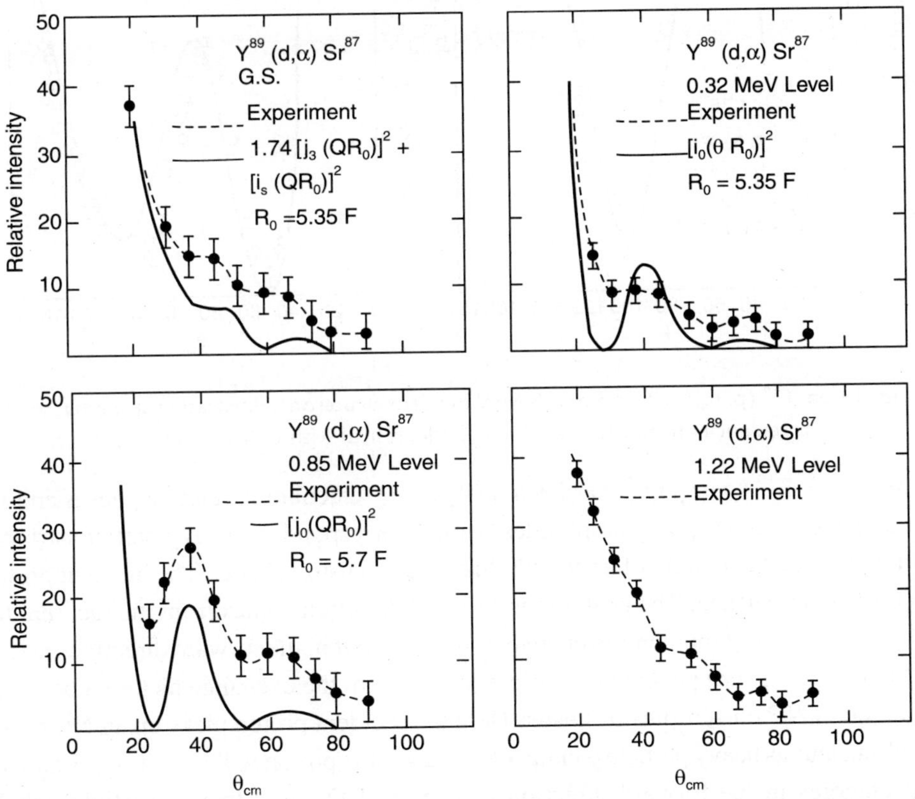

Fig. 14.5b Angular distribution of alpha particles from the formation of four states of Sr^{87} from the reaction $Y^{89}(d, \alpha) Sr^{87}$, illustrating a case of pickup, with two-nucleon transfer. The solid curves are based on plan-wave theory and lack of fits shows the need of more detailed theory (Ref. 13).

We show in Fig. 14.5*b* the angular distribution for alpha particles from the formation of the four levels of Sr^{87} in the reaction Y^{89} (*d*, α) Sr^{87}. It is interesting to note that the fits are only qualitative, and especially amplitudes do not match; showing that plane theory is inadequate.

(*ii*) **Multi-Nucleon Transfer Reactions:** Examples of transfer to more than two nucleons are[17, 18]: F^{19} (p, α_0) O^{16} and F^{19} (d, Li^6) N^{15}. The first case is three nucleons pickup one proton and two neutrons and the second case is pickup of 4 particles, *i.e.* 2 protons and 2 neutrons—which is an alpha particle. Figures 14.6*a* and 14.6*b* show the angular distributions of emitted particles in these cases of (*p*, α) and (d, Li^6). Because of large electric charge involved in the transfer, the plane wave Born theory does not give good fit and only Distorted Wave Born Approximation (DWBA) gives a reasonable fit.

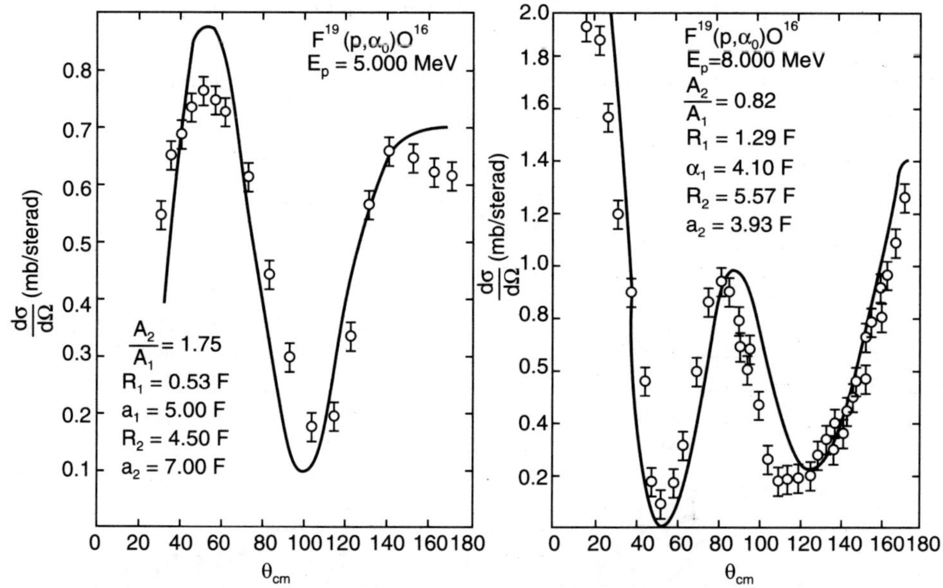

Fig. 14.6a F^{19} (p, α_0) O^{16} at 5 and 8 MeV incident deuterons, illustrating stripping, with three nucleon transfer. Solid line is theoretical with DWBA (Ref. 17).

The reaction[17] F^{19} (p, α_0) O^{16}, has been analysed by calculating relative phases and magnitudes of the direct and exchange stripping amplitudes, which corresponds to heavy particle stripping. In this pickup reaction, F^{19} can be thought of as an O^{16} cluster plus a triton cluster. The incident proton picks up the triton to form an α-particle. The peak in the forward direction comes from the fact that the incident proton will have the smallest momentum change, if it comes out in the forward directions as a constituent of particle. Also the backward peaking of particle results from the exchange pickup model in which F^{19} can be considered as N^{15} cluster plus α-cluster. Here the incident proton picks up an N^{15} cluster to form the ejected O^{16} nucleus as heavy particle pickup. Once again, the proton will have its minimum momentum change, if it emerges in the forward direction as a part of O^{16} and then α-particle will be emitted predominantly in the backward direction. So the total reaction is a sum of direct (proton + triton cluster) and exchange ($F^{19} = N^{15} + \alpha$). The relative proportion of the two processes will depend on the relative probabilities of F^{19} in the respective clusters.

The case[18] of F^{19} (*d*, Li^6) N^{15} corresponds to the four-nucleon transfer and has been studied by W.W. Dahenick and L.J. Dennes[18], for 15 MeV deuterons, and has been analysed by using DWBA calculations.

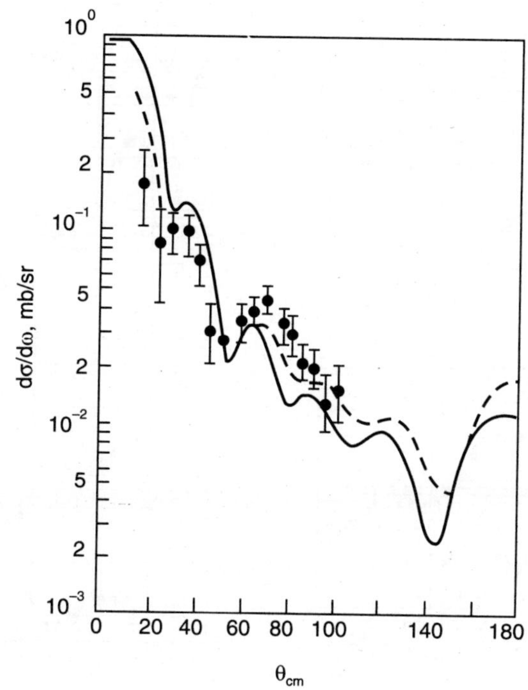

Fig. 14.6b F^{19} (d, Li^6) N^{15}, for the ground state of N^{15}, for 15 MeV deuterons. Solid curves are DWBA calculations for l = 1, for three nucleon transfer (Ref. 18).

14.4 DISTORTED-WAVE BORN APPROXIMATION (DWBA)

In discussing the theory of direct reactions in the previous section, we have neglected (*i*) the Coulomb interaction and (*ii*) short range nuclear interactions V_{13}; while writing Eqs. 14.39*a* and 14.39*b* for ϕ (12) and ϕ (23). Also the wave function ψ in Eq. 14.36 was replaced by ϕ (12) exp $\mathbf{p}_i \cdot \mathbf{r}_i$; which corresponds to the product of the incident plane wave and the wave functions of the target. They do not take into account these two factors.

A more accurate treatment takes these factors into account and for any two body wave function ϕ_{ij}, one solves the following equation numerically, instead of Eq. 14.41:

$$\left[-\frac{d^2}{d\,r_{ij}^2}+\frac{l(l+1)}{r_{ij}^2}+\frac{2\mu_{ij}}{\hbar^2}\left\{V_{ij}\left(r_{ij}\right)+\frac{Z_i\,Z_j\,e^2}{r_{ij}}\right\}-k_{ij}^2\right]\times r_{ij}\,\phi_{ijl}\,(k_{ij},\,r_{ij})=0 \quad ...(14.45)$$

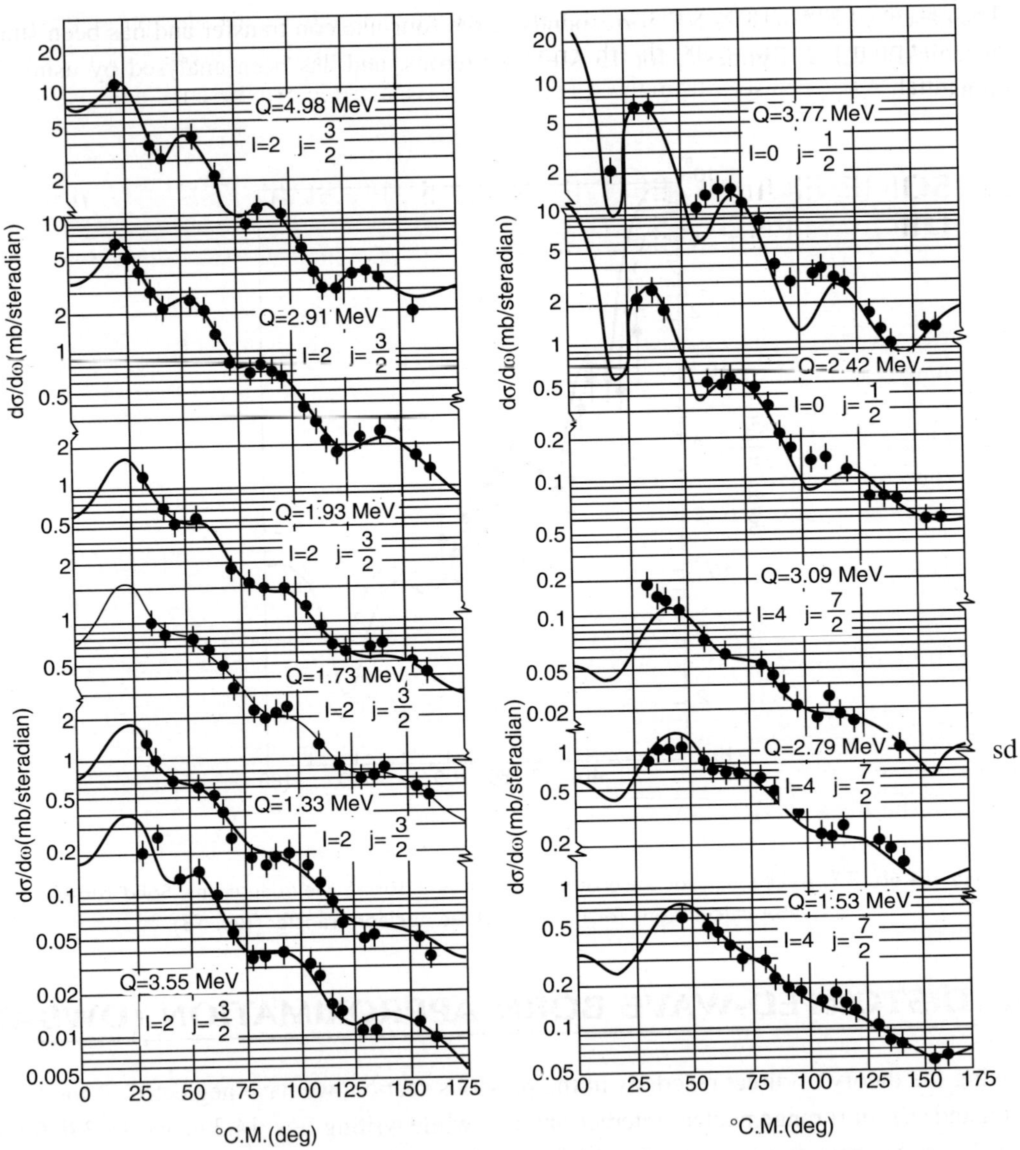

Fig. 14.7 Differential cross-sections for Zr^{90} (d, p) Zr^{91} with 12 MeV deuterons, leading to different states. The solid lines represent calculated curves using Distorted Wave Born Approximation (DWBA) theory with non-local optical potential using Eq. 14.45 (Ref. 19).

The potential V_{ij} may be either taken as corresponding to zero-range force, *i.e.*,

$$V_{ij}(r_{ij})\,\phi_{ij}(r_{ij}) = -V_0\, r^{2/3}\,\delta(r_{ij}) \qquad \text{...(14.46)}$$

or may be taken as an optical potential. Equation 14.45 which is now used for ϕ_{12} or ϕ_{23}, is solved numerically. In Eq. 14.45, μ_{ij} and k_{ij} are the reduced mass and wave number respectively and V_{ij} is the nuclear interaction potential. Solution of Eq. 14.45 involves still more numerical calculations for finite range nuclear potential.

Figure 14.7 shows a comparison[19] for the experimental angular distribution of protons in Z^{90} (d, p) Zr^{91} (stripping reaction) and the theoretical calculations based on DWBA, using a non-local optical potential. As can be seen from the fit, the theory explains the experimental data extremely well.

14.5 SOME EXAMPLES OF INELASTIC SCATTERING AS DIRECT PROCESS

An inelastic scattering reaction like (α, α') or (p, p'), etc. can take place through two channels: (*i*) Compound nucleus formation (*ii*) Direct process. In the case of compound nucleus model; the compound state decays into many energetically possible channels; one of them may be the inelastic scattering. Its angular distribution is expected to be symmetric around 90° in the centre of mass system. On the other hand, there may be a surface reaction or direct process, by which the incident particle may either knockout a particle of its own kind, in a scattering process, leaving the residual nucleus excited and emitted particle has less energy than the incident particle. Also the incident particle, may be directly scattered from the whole nucleus from its surface; leaving the nucleus excited in say a collective mode of vibration or rotation.

Evidently these processes are similar to direct reactions. As discussed in the previous chapter; the reaction time in a direct reaction is about 10^{-22} sec, *i.e.* the time taken by the incident particle to travel the nucleus dimensions, compared to say 10^{-24} secs involved in a compound nucleons process.

For inelastic scattering, a large amount of experimental data and its analysis exists in literature, for which the direct surface interaction theory of Austern, Butler and McManus[20] has been used for comparison: Experiments[21] performed with 31.5 MeV alphas on C^{12} and Mg^{24} have been especially studied, for (α, α') reaction. The actual experimental angular distribution and comparison with theory are shown in Fig. 14.8 for Mg^{24} (α, α') Mg^{*24} for the excited states of Mg^{24} at 1.37 MeV and 4.12 MeV.

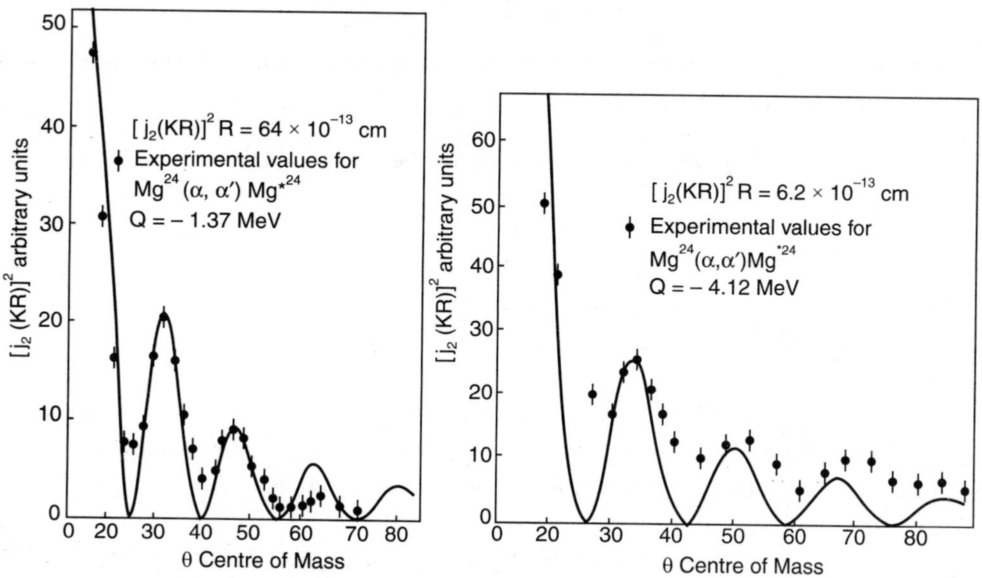

Fig. 14.8 Experimental points for interaction Mg^{24} (α, α') Mg^{*24} for (*a*) Q = – 1.37 MeV and (*b*) Q = – 4.12 MeV, with theoretical curves (solid lines) (Ref. 21).

The theory for inelastic scattering is similar to the theory of direct reaction developed earlier. We consider, as before, the reaction, say

$$(1 + 2) + 3 \rightarrow 1 + (2 + 3) \qquad ...(14.47)$$

Equation 14.47 represents (p, n) reaction, so that the incident particle 3 (proton) falls on a target say X^A = (neutron + X^{A-1}) represented by $1 \equiv$ neutron and $2 \equiv X^{A-1}$ emitting particle 1 (neutron), leaving behind an excited nucleus Y^A represented by:

$$Y^A \equiv X^{A-1} + \text{proton} \equiv (2 + 3)$$

where Y^A is a nucleus with the same number A of nucleons; but with a proton replacing a neutron in the target X^A. The emitted neutron can either be a result of knock out, or exchange process on the surface. Theoretically one can describe the direct reaction by Eqs. 14.35 and 14.37, where the interaction V_{13} $(= V_{np})$ contributes much more than the interaction V_{12} (V_{nA}). It is the shape of potential $V_{13} + V_{12}$ which will decide what type of process, *i.e.* (knock out or exchange type) gives rise to the emission. The inelastic scattering process say (p, p'), is similar to this process of knock out or exchange process, hence the following logic is applicable to inelastic scattering also. In such a case, the amplitude of the inelastic scattering is given (following Eq. 14.37 and neglecting V_{12}), as:

$$f(\theta, \phi) \approx -\frac{\mu_f}{2\pi} \int \phi^*_{23}(\mathbf{r}_{23})\, e^{i(\mathbf{q}_3 \cdot \mathbf{r}_{23} - \mathbf{q}_1 \cdot \mathbf{r}_{12})} \times V_{13}\, \phi_{12}(\mathbf{r}_{12})\, d^3 r_{23}\, d^3 r_f \qquad ...(14.48)$$

For, the zero-range interaction for $V_{13}(\mathbf{r}_{13})$, *i.e.*

$$V_{13}(\mathbf{r}_{13}) = V_0\, \delta(\mathbf{r}_{13}) \qquad ...(14.49)$$

we write Eq. 14.48 as:

$$f(\theta, \phi) = -\frac{\mu_f}{2\pi} V_0 \int \phi^*_{23}(\mathbf{r}_{23})\, \phi_{12}(\mathbf{r}_{12})\, e^{i\mathbf{Q} \cdot \mathbf{r}_{12}}\, d^3 r_{12} \qquad ...(14.50)$$

where
$$\mathbf{Q} = \frac{M_2}{M_1 + M_2} \mathbf{p}_i - \frac{M_2}{M_2 + M_3} \mathbf{p}_f \qquad ...(14.51)$$

If we want to take into account, the angular momentum dependence of the scattering amplitude, we expand $e^{i\,\mathbf{Q} \cdot \mathbf{r}}$ into spherical harmonics; *i.e.*,

$$e^{i\,\mathbf{Q} \cdot \mathbf{r}} = 4\pi \sum_{1, n_1} i^1\, J_l(Qr_2)\, Y_l^m(\hat{r}_{12})\, Y^*_{l_m}(\hat{Q}) \qquad ...(14.52)$$

and using the formula:

$$\int Y_{l_1 m_1}(\Omega)\, Y^*_{l_2 m_2}(\Omega)\, d\Omega = \left[\frac{(2l_1 + 1)(2l_2 + 1)}{4\pi(2l_3 + 1)}\right]^{1/2}$$

$$\langle l_1\, l_2\, m_1\, m_2 \,|\, l_3\, m_3 \rangle \times \langle l_1\, l_2\, 00 \,|\, l_3\, 0 \rangle \qquad ...(14.53)$$

[*see* Ref. (7)], and writing $\phi_{23}(\mathbf{r}_{23})$ and $\phi_{12}(\mathbf{r}_{12})$, in terms of spins ($s_1$, s_2 and s_3) and orbital (l_{23}, l_{12}) and total angular momenta (j_{23} and j_{12}); it is possible to write, Eq. 14.50 as:

$$f(\theta, \phi) = -\frac{\mu_f}{2\pi} V_0 \sum \int dr_{12}\, r_{12}^2\, \phi^*_{23}(r_{23}, l_{23})\, \phi_{12}(r_{12}, l_{12})\, j_1(Qr_{12}) \times$$

$$C(s_2, M_{23}, M_{12}, m_3\, l_{23}\, l_{12}\, j_{23}, j_{12}\, l) \qquad ...(14.54)$$

where $C(s_2, M_{23}, M_{12}, m_3\ l_{23}\ l_{12}\ j_{23}, j_{12}\ l)$ is a function calculated by Roy and Nigam[7].

The angular momenta l_{12}, l, l_{23} are related to each other as

$$l = l_{23} - l_{12} \quad ...(14.55)$$

which gives rise to the various selection rules.

Using the relationships and the expression for the differential cross-sections from Eqs. 14.35 to 14.54, one can analyse the experimental angular distribution of say (α, α′) or even the angular distribution of polarisaters P (θ), in terms of surface reactions of direct reactions. In earlier literature[21], angular distribution for (α, α′) reaction for Li^6, C^{12} and Mg^{24}, at 31.5 MeV alphas were studied experimentally and compared with the theoretical expectation based on Eqs. 14.35 and 14.42. Apparently, the angular distribution is expected to be proportional to $| J_l (KR) |^2$ from these equations. In Fig. 14.8*a*, we show this comparison for Mg^{24} (α, α′) Mg^{*24} for 1.37 MeV excited state for which $J = 2^+$; while for the ground state of Mg^{24}, $J = 0^+$. Incoming alphas have to transfer $l = 2$ to raise Mg^{24} to $J = 2^+$. Hence the angular distribution is expected to be proportional to $[j_2 (KR)]^2$ which is the case in Fig. 14.8. Same is the case for 4.12 MeV excited state in Mg^{24}, as shown in Fig. 14.8. Similarly we have shown such comparisons for Li^6 (Q = – 2.19 MeV) and (Q = – 4.5 MeV) and for C^{12} (Q = – 4.43 MeV) and (Q = – 7.65 MeV) in Fig. 14.9*a* and Fig. 14.9*b*.

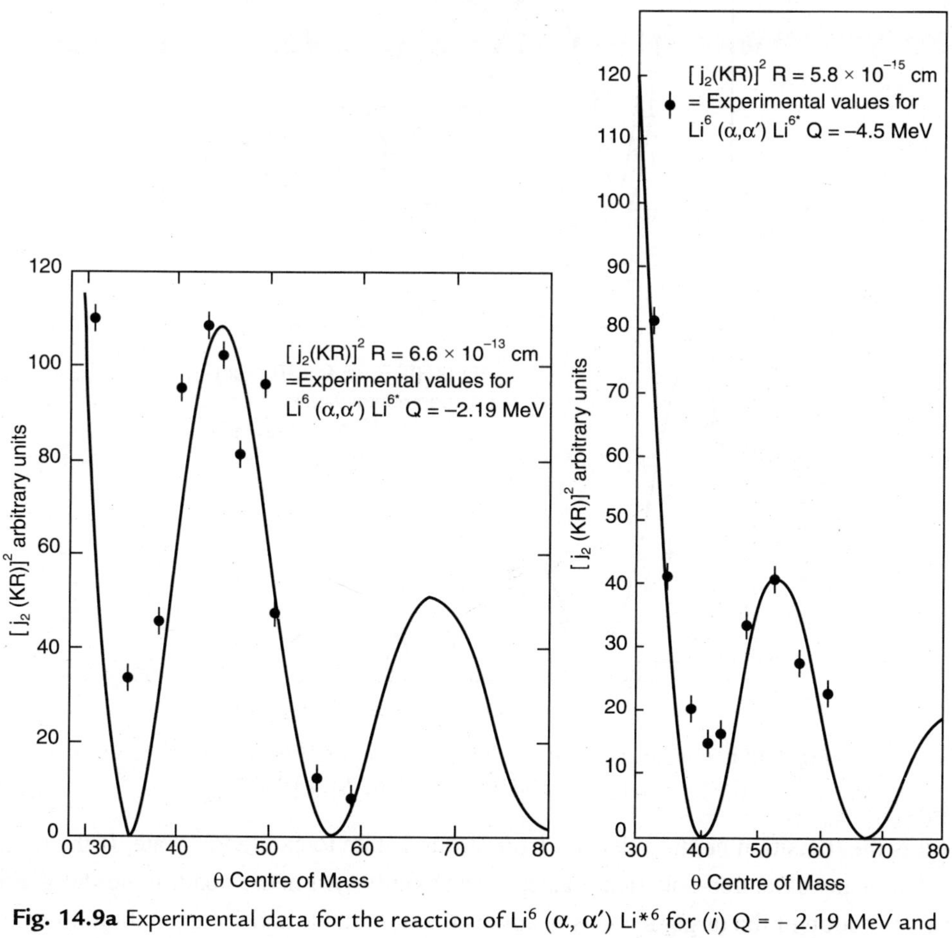

Fig. 14.9a Experimental data for the reaction of Li^6 (α, α′) Li^{*6} for (*i*) Q = – 2.19 MeV and (*ii*) Q = – 4.5 MeV compared with theoretical angular correlation (Ref. 21).

An interesting case of measurement[29] of angular distribution of Sr^{88} (α, α′) Sr^{*88} (2^+, 1.84 MeV), has been reported at $E_\alpha = 50$ MeV, which on comparison with a microscopic DWBA calculation, yielded neutron transition multiple moment M_n, defined as:

$$M_n = \int \rho_n (r)\, r^4\, dr$$

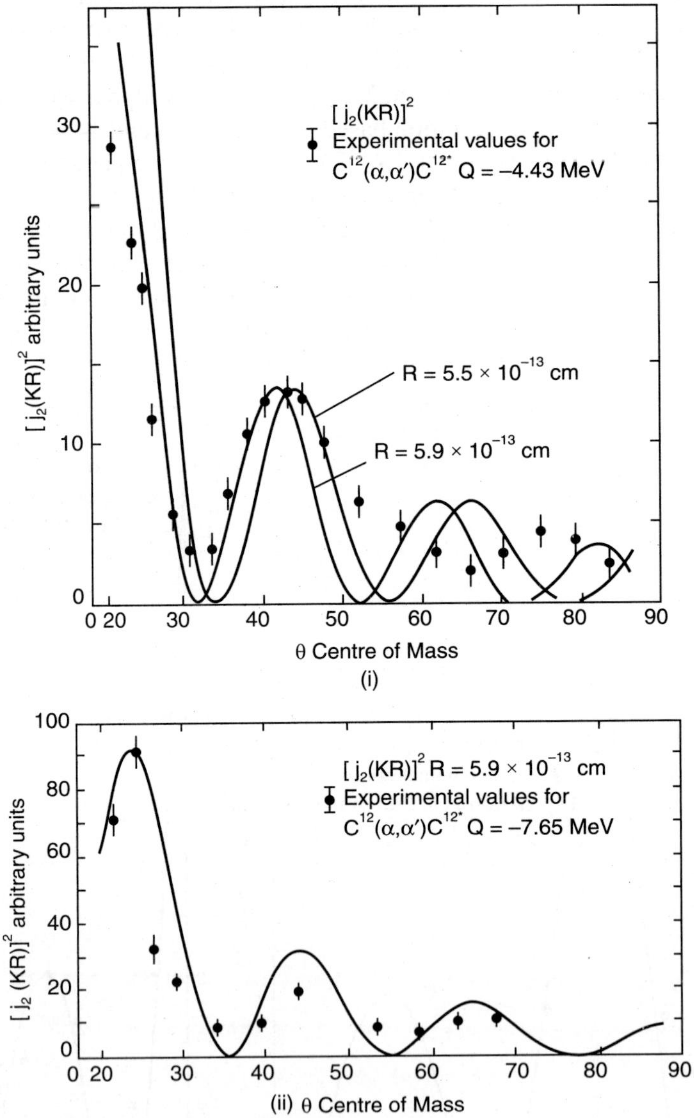

Fig. 14.9b Experimental data for the reaction of C^{12} (α, α′) C^{*12}, for (*i*) Q = – 4.43 MeV and (*ii*) Q = – 7.65 MeV (Ref. 21).

where $\rho_n(r)$ is the transition neutron density from ground state to excited 2^+ state. If the target nucleus is a spheroid or of any other non-spherical shape, say an axially symmetric quadrupole-deformed nucleus; then collective motion of the target nucleus becomes the dominant mode of excitation, in contrast to the case of Mg^{24} (α, α') Mg^{*24} where the shell model single particle configuration were excited. Such cases are; Er^{168} (α, α') Er^{*168} and Y^{172} (α, α') Yb^{*172} recently investigated experimentally[22] by Govil et al. by using incident alphas of 36 MeV. Both the target nuclei are spheroidal; with positive quadrupole moment, *i.e.* they are prolate in shape. It is well known from a large number of experiments on these nuclei, that here the states excited near the ground state include all the members of the ground state band with $K = 0^+$, and some members of band-heads at $K = 2^+$, $K = 1^-$, 2^- and 0^-; on which are built, the rotational states. This shows, that excited states involving vibrational and rotational modes of collective motion should be expected in their excitation in the inelastic scattering of alphas.

We will describe the case of Er^{168} (α, α') Er^{*168} in some details[22], especially the excitation of all the members of the ground state band starting from the ground state of $K^\pi = 0^+$ and going into the excitation of 2^+, 4^+, 6^+, 8^+ and of 10^+ rotational states as shown in Figs. 14.10 and 14.11.

The experimental angular distribution of alphas corresponding to elastic scattering, *i.e.* leading to 0^+ state, and inelastic scattering, *i.e.* leading to 2^+, 4^+, 6^+ and 8^+ states are given in Figs. 14.10 and 14.11.

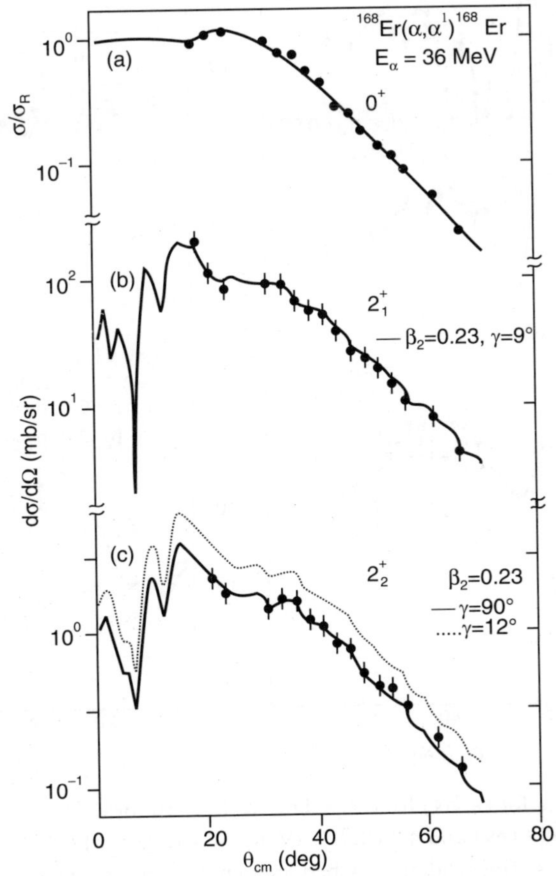

Fig. 14.10 Experimental data and asymmetric rotor model results for (α, α') at 36 MeV, for three states of Er^{168}, i.e. (*a*) 0^+ g.s, (*b*) 2_1^+ (79 keV) and (*c*) 2_2^+ (821 keV) states (Ref. 22).

The thick lines passing through the points, correspond to the theoretically calculated values, in which the collective states are coupled together, so that the wave-function was written according to Davydov and Fillipov model, and the nuclear states were expanded in terms of axially symmetric eigen-functions, *i.e.*,

$$|I\,M,\alpha\rangle = \sum_{K} A^{I}_{\alpha\,K}\,|I\,M\,K\rangle \qquad ...(14.56)$$

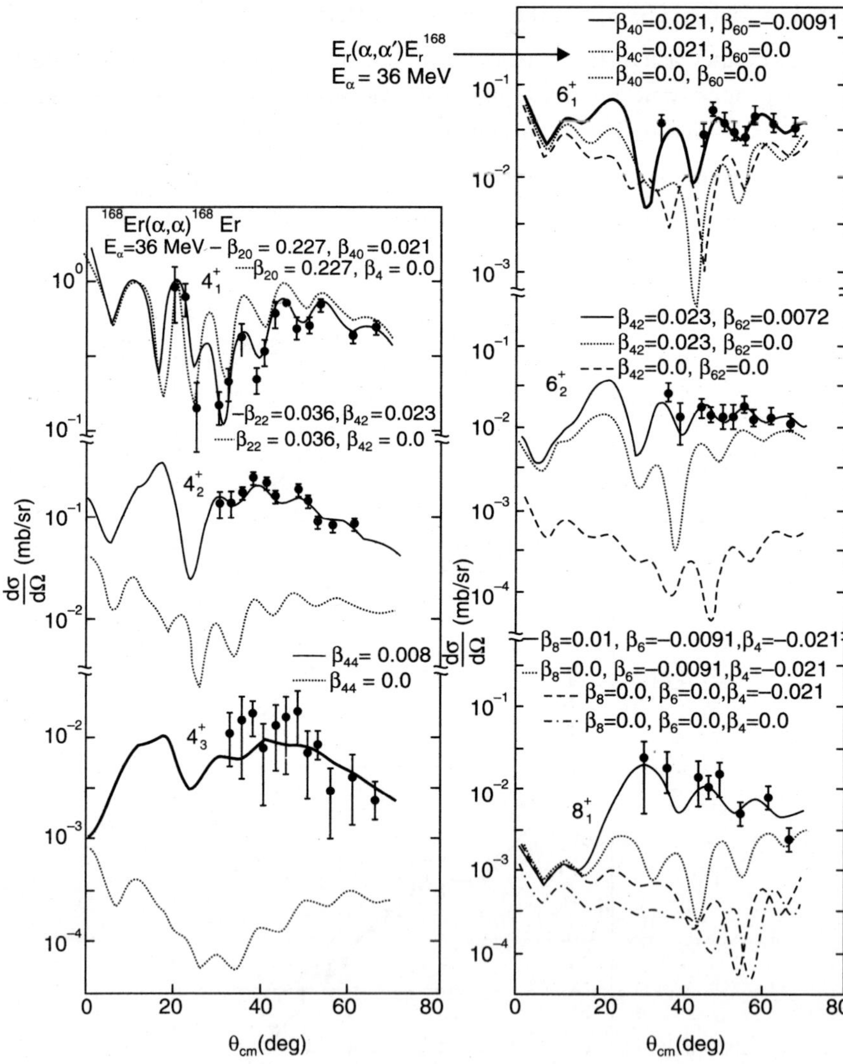

Fig. 14.11 Experimental data for (a, a¢) for target Er^{168} at E_{α} = 36 MeV, for different energy states of Er^{168}. (*a*) 4_1^+ (264 keV), 4_2^+ (994 keV) and 4_3^+ (2055 keV, K = 4^+) states; (*b*) 6_1^+ (548 keV), 6_2^+ (1263 keV) and 8_1^+ (928 keV) states. The solid and dashed lines correspond to theoretically calculated values based on asymmetric rotor model, coupled channel results, except for 8_1^+, which corresponds to symmetric rotor model (Ref. 22).

where $$|I\,M\,K\rangle = \left|\frac{2I+1}{16\pi^2\,(1+\delta_{K,O})}\right|^{1/2} \left[D^I_{M,K} + (-1)^I\, D^I_{M,-K}\right] \qquad ...(14.57)$$

is the wave function of a state with a given K and $|I\,M,\alpha\rangle$ which contains the band mixing coefficients $A^I_{\alpha K}$, representing the coupling between various bands. These are determined by using the eigen-equation:

$$\sum_{K'} \langle I\,M\,K'|\,H_{\text{rot}}\,|I\,M\,K\rangle\, A^I_{a\,K'} = E^I_a\, A^I_{a\,K}$$

where $$|I\,M;\alpha_a\rangle = \sum_{K'} A^I_{a\,K'}\,|I\,M\,K\rangle \qquad ...(14.58)$$

and the rotational Hamiltonian is:

$$H_{\text{rot}} = \sum_{i=1}^{3} \frac{J_i^2}{2\,\vartheta i} \qquad ...(14.59)$$

The three moments of inertia ϑi were calculated to best-produce the level energies of 2^+ states of the ground band and γ-band in Er^{168}, starting at 821 keV [Figs. 14.10 and 14.11].

The band mixing coefficients $A^I_{\alpha K}$ determined from Eq. 14.58 are included in the coupled channel computer code called ECIS, used for calculating the differential cross-sections for (α, α') from an axially symmetric quadrupole shaped nucleus, say Er^{168}. The potential between the incident particle and the target [as was used in Eq. 14.48] for V_{13} was taken as the complex optical model with Woods-Saxon form with volume and surface absorption, to which Coulomb potential is added, *i.e.*,

$$V(r, \theta', \phi') = (V + i\,W)\,(1+e)^{-1} - 4\,i\,W_S\,(1+e)^{-2} + V_{\text{Coulomb}} \qquad ...(14.60)$$

with $$e \equiv \exp\left\{\frac{[r - R(\theta', \phi')]}{a}\right\} \qquad ...(14.61)$$

and Woods-Saxon potential shape say for V was taken as:

$$V = -V_0\left[1 + \frac{r-R}{\alpha_0}\right]^{-1}$$

The prime attached to the angles corresponds to the fact that the arguments are polar angles referred to the body-fixed coordinate system. The function $R(\theta', \phi')$ in Eq. 14.61 contains the rotational expression used by Govil[22] et al. and are given as:

$$R(\theta', \phi') = R_0\left[1 + \beta_2 \cos(\gamma)\, Y'_{2,0} + \frac{1}{\sqrt{2}}\,\beta_2 \sin(\gamma)\, Y'_{2,\pm 2}\right.$$

$$\left. + \beta_{40}\, Y'_{4,0} + \frac{1}{\sqrt{2}}\,\beta_{42}\, Y'_{4,\pm 2} + \frac{1}{\sqrt{2}}\,\beta_{44}\, Y'_{4,\pm 4} + \right.$$

$$\left. \beta_{60} Y'_{6,0} + \frac{1}{\sqrt{2}} \beta_{62} Y'_{6,\pm 2} + \beta_{80} Y'_{8,0} \right] \quad ...(14.62)$$

where Y' have arguments θ' and ϕ'. As can be seen from Figs. 14.10 and 14.11 the fits between theory and experiments are very good; the parameters, in Eq. 14.62, which fit into this comparison are:

$$\beta_2 = 0.23 \pm 0.01; \quad \gamma = 9°$$

$$\beta_{40} = -0.021 \, {}^{+0.025}_{-0.009}; \quad \beta_{42} = +0.023 \pm 0.002$$

$$\beta_{44} = \pm 0.008 \pm 0.002$$

$$\beta_{60} = -0.0091 \pm 0.0012; \quad \beta_{62} = +0.0072 \pm 0.0008 \quad ...(14.63)$$

Similarly the values of V_0, r_0 and a_0 for deformed volume; potential (real) and V_w, r_w and a_w, for deformed volume (imaginary) potential and similarly parameters for deformed, surface (imaginary) and deformed Coulomb and deformed spin-orbit potential were obtained.

Physically, β_2 corresponds to quadrupole deformation, connected with the quadrupole moment as given in Eq. 11.84 in Chapter 11 and β_{40}, β_{42} and β_{44} correspond to $\lambda = 4$ electric moments for different m-components. Similarly β_{60} and β_{62} correspond to still higher pole moments for $\lambda = 6$ [*see* Chapter 11].

14.6 NUCLEAR SPECTROSCOPY FROM DIRECT REACTIONS

Direct reactions are very convenient and unambiguous source of information for the angular momentum and wave-functional properties of the ground and excited states of the residual nucleus. As for example, (d, p) reaction is essentially a fancy way of inserting a neutron with a specific orbital momentum $\mathbf{l}_i$ into an orbit of the nucleus; resulting in total angular momentum $\mathbf{j}_i$, as the vector sum of $\mathbf{l}_i$ and spins $\mathbf{s}_i$. We take a specific example of the target of Ni^{58} on which a beam of deuterons in incident, giving rise to Ni^{58} (d, p) Ni^{59}, so that Ni^{58} captures a neutron in a l_i, j_i orbit. Then the configuration of the resultant nucleus is:

$$Ni^{58}\text{ (G.S.)} + n(l_i, j_i) \quad ...(14.64)$$

which for $l = 1$, $j = 3/2$ state can be written as:

$$Ni^{58}\text{ (G.S.)} + n(p_{3/2}) \quad ...(14.65)$$

As Ni^{58} has 28 protons (closed shell) and 30 neutrons, 28 of which are in the closed shell of, $1f_{7/2}$ and 2 neutrons in $2p_{3/2}$ shell, they are coupled to give total angular momentum zero. Hence the neutron in (d, p) reaction will be captured in $p_{3/2}$ state which can have four particles of the same type; so l_i of the captured state is 1. How do we write the wave function of the state, representing Eq. 14.65? If we assume, that the neutron is captured in $3/2^-$ state, without disturbing the configuration in the target nucleus and if the only possibility of the neutron getting captured is in $p_{3/2}$ state, then one can write the wave function of configuration of Eq. 14.65 as:

$$\psi_{3/2}(p_{3/2}) = C_{3/2}(i)\, \psi\, [Ni^{58}\text{ (G.S.)}] \times \psi(p_{3/2}) \quad ...(14.66a)$$

However, the neutron may not go into $p_{3/2}$ orbit with 100% probability; but may go into some other orbits like $f_{5/2}$, or $p_{1/2}$, etc. so it goes into $p_{3/2}$ orbit only with a partial probability; then one can write the wave function $\psi_{3/2}(i)$ for a state i as given by:

$$\left[C_j^2\ (i) \equiv S_j\ (i)\right]$$

$$\psi_{3/2}\ (i) = \left[\left\{S_j\ (p_{3/2})\right\}^{1/2}\ \psi\left[\text{Ni}^{58}\ (\text{G.S.})\right] \times \psi\ (p_{3/2})\right]$$

$$+ \left[\left\{S_j\ (f_{5/2})\right\}^{1/2}\ \psi\left[\text{Ni}^{58}\ (\text{G.S.})\right] \times \psi\ (f_{5/2})\right] + \ldots \quad \text{...(14.66}b\text{)}$$

where $S_j\ (i)$ is called the spectroscopic factor. As a matter of factor, the spectroscopic factor may be formally described for (*d*, *p*) direct reaction by the relationship,

$$\sigma_{\text{exp}} = \sigma_{\text{DWBA}}\ S_j \quad \text{...(14.67)}$$

where σ_{DWBA} is absolute theoretically calculated cross-section, under the assumption that the transferred neutron enters one of the orbits of shell model without otherwise disturbing the nucleus. Then the spectroscopic factor S_j describes the degree to which the model used in the calculations of DWBA, deviates from the actual experimental situation. S_j depends on the *j*-value of the state involved.

We have already seen from Fig. 14.4*a* to 14.9; that one can determine the *l*-value of the transferred particle; from the angular distribution of the emitted particle. It has been shown by Lee and Schiffer[23] that the angular distribution for two *p*-states, *i.e.* $p_{3/2}$ when **l** and **s** are parallel and for $p_{1/2}$, when they are anti-parallel, are different-especially at the backward angles, for Fe^{54} (*d*, *p*) Fe^{55}. For the *j*-values for each state of the residual nucleus, the comparison of the absolute experimental cross-section, for exciting it, with DWBA calculations, leads to the determination of S_j.

A quantity closely related to S_j, is the occupation number V^2 which indicates how fully occupied is the orbit of a given value of *n*, *l*, *j*. We will give an example of O^{18}. The O^{18} ground state wave function[24] corresponds to 8 protons and 8 neutrons being in the closed shell and two neutrons being in the ($1d_{5/2}$, $2s_{1/2}$, $1d_{3/2}$) shell; coupled to spin zero, *i.e.*,

$$\psi(\text{O}^{18}) = a_1(d_{5/2})_0^2 + b_i\ (s_{1/2})_0^2 + c_1\ (d_{3/2})_0^2 \quad \text{...(14.68)}$$

where a^2, b^2, and c^2, are the probabilities of the two neutrons being in $d_{5/2}$, $S_{1/2}$ or $d_{3/2}$ orbits. Physically one may say that the two neutrons are in $d_{5/2}$ orbits, a fraction of time given by $a_1^{\,2}$. Because these are one-third full; since $d_{5/2}$ can accommodate. $2j + 1 = 6$ neutrons, we define:

$$V_{5/2}^2 = \frac{1}{3} a_1^2 \quad \text{...(14.69)}$$

Similarly,

$$V_{1/2}^2 = b_1^2 \quad \text{...(14.70)}$$

and

$$V_{3/2}^2 = \frac{1}{2} c_1^2 \quad \text{...(14.71)}$$

From this example, we can readily see, that in general, the wave function of a single quasi-particle (SQP) state of nucleus with *N* neutrons can be written as:

$$\psi_{N,j}\ (\text{SQP}) = \sqrt{1 - V_j^2}\ \psi_{N-1}\ \psi(j) + V_j\ \psi_{N+1}\ \psi(j^{-1}) \quad \text{...(14.72)}$$

Single quasi-particle (SQP) states are the states formed by adding a particle or a hole in the nearest even-even nucleus. Physically, Eq. 14.72 says that the wave function of the SQP state is the sum of two terms, the ground state of even-even nucleus ($N-1$), plus a particle in the state j and the ground state of the even-even nucleus ($N+1$) plus a hole in the state j. The fraction of the time it is in a hole state is V_j^2 and the fractions of time it is a particle state is $1-V_j^2$, which is the degree to which the state j is empty. The above definitions and meaning of V_j^2 explain Eq. 14.72. Therefore, from Eqs. 14.67 to 14.72 it can be seen that if the target is even-even and if the state excited is a pure SQP particle state, we get:

$$S_j\,(\text{SQP}) = 1 - V_j^2 \qquad \text{...(14.73)}$$

If the SQP state is mixed among several particle states; each state is excited with an S_j-value of $(1-V_j^2)$ times the fraction of the SQP state, it contains. In such situations, the fractions appearing in all nuclear states must add up to unity. From this it is evident that the sum of the S_j-values of these states will be $1-V_j^2$, so that one can write:

$$\sum_i S_j\,(i) = 1 - V_j^2 \text{ (for } (d, p) \text{ on even-even target)} \qquad \text{...(14.74)}$$

Equation 14.74 represents a sum rule which is valid separately for each neutron orbit j. One can plot V_j^2 versus N, the neutron number, from the experimental data. As for example for Ni^{58} as target, the values of V_j^2 have been found to be 0.69, 0.91, 0.91 and 0.99 for $2p_{3/2}$, $1f_{5/2}$, $2p_{1/2}$ and $1g_{9/2}$, respectively. As each state contains 2 neutrons, this means that on an average, the ground state Ni^{58} contains, $(2 \times 3/2 + 1) \times (1 - 0.69) = 1.24$, $2p_{3/2}$ neutrons. Similarly we have 0.54, $1f_{5/2}$ neutron, 0.8, $2p_{1/2}$ neutrons and 0.1, $1g_{9/2}$ neutrons. As these numbers correspond to two neutrons at any one time in the orbit, to give zero angular momentum; hence this means that each neutron must spend 62% of their time in $2p_{3/2}$ orbit; 27% time is $1f_{5/2}$ ortbit, 9% in $2p_{1/2}$orbit and 0.05% time in $1g_{9/2}$ orbit. This interesting information comes from the analysis of angular distribution data of (d, p) reaction and its analysis through DWBA from which the values of S_j are obtained.

One can thus experimentally determine S_j for many nuclei, for various excited states, from which one can determine the centre of gravity[25] of these levels, defined as:

$$\left(E_j\right)_{\text{C.G.}} = \frac{\sum_i S_i\, E_i}{\sum_i S_i} \qquad \text{...(14.75)}$$

One can plot $(E_j)_{\text{C.G.}}$ as a function of a number of neutrons, and thus get the energies of the various shell and sub-shells in the shell model diagram.

Plots of energy centroids versus target mass number A have been drawn to yield useful information about interaction parameter for which sufficient experimental data are available[26].

Such a data has been analysed, to obtain average two-body interaction parameters, using the formalism of sum rules, as developed by Bansal and French and their co-workers[27]. For formalism and detailed discussion *see* R.D. Lawson[28].

14.7 OTHER MODES OF DIRECT REACTION

Apart from the normal transfer reactions, like (*d*, *p*) in Pb^{207} (*d*, *p*) Pb^{208}, or (He^3, *p*), *e.g.* in Ni^{62} (He^3, *p*) Cu^{64} or (*d*, He^3), *e.g.* in Mg^{25} (*d*, He^3) Na^{24}, which have been recently studied both experimentally and analysed[30] and compared with DWBA predictions, two very interesting and somewhat exotic modes of direct reactions and associated nuclear spectroscopy have been studied. These are the cases of two neutron transfer[31] as in F^{19} (α, He^2) F^{21} and direct emission[32] as in Os^{192} ($p, pn\gamma$) Os^{191}.

The reaction F^{19} (α, He^2) F^{21} was investigated at 55 MeV of incident energy. The target was a self-supporting Ca^{48} F_2 target, and the unbound reaction product He^2 was detected by the two breakup protons in coincidence. The two protons are focused in one direction, due to kinematics and hence one can determine differential cross-section in the centre of mass[33]. Of course, DWBA technique was used to obtain the transferred angular momentum of prominent transitions. Theoretically[34] the results were compared with shell model calculation, with the unified *sd* space interaction with good reproduction of the excitation energies and spectroscopic factors.

Similarly in the case of Os^{192} ($p, pn\gamma$) Os^{191} reaction at 18.6, 20.8, 24.2, 27.2 and 31.1 MeV incident energies, the experiment included, excitation function, angular distribution and coincidence measurements. This is similar to a single nucleon transfer (emission), and therefore is like (*d*, *t*) for which special theoretical programmes[35] were used. The results gave the deformation parameters ε and γ as $\varepsilon = 0.175$ and $\gamma = 24°$.

Breakup Reactions

Some good examples[36] of a breakup reaction are C^{12} (Ne^{20}, O^{16} Be^8) Be^8 and C^{12} (Mg^{24}, O^{16} Be^8) C^{12} reactions which follow the sequence:

$$C^{12} + Ne^{20} \rightarrow Mg^{*24} + Be^{8}$$
$$\qquad\qquad \hookrightarrow O^{16} + Be^{8}$$

or

$$C^{12} + Mg^{24} \rightarrow Mg^{*24} + C^{12}$$
$$\qquad\qquad \hookrightarrow O^{16} + Be^{8}$$

Both of these reactions have been recently studied at 160 MeV using Ne^{20} beam on C^{12} for the first reaction and 170 MeV. Mg^{24} beam for the second reactions. The detection of O^{16} was carried through a heavy ion telescope ($\Delta E - E_1 - E_2$) and detection of Be^8 was carried out by detecting the two alpha particles from the breakup of Be^8 by using a 1000 mm thick silicon strip detectors. Excitation function and angular correlation measurements, were compared with Cranked[37] cluster model and Hartree[38]-Fock calculations, which resulted in the proper spin assignment.

An interesting case of a theoretical study of B^8 has been carried out by R. Shyam and K. Lenske[39], in the framework of post-form DWBA. B^8 is an interesting nucleus having the last proton with a binding energy of only 137 keV[40]. It is suspected to have a proton halo structure. The theoretical study and comparison with elastic and inelastic cross-section of $B^8 + Si^{28} \rightarrow Be^7 + X$ showed that most of the contribution to the breakup cross-section comes from distances far beyond the nuclear surface, which favours elastic break-up mode.

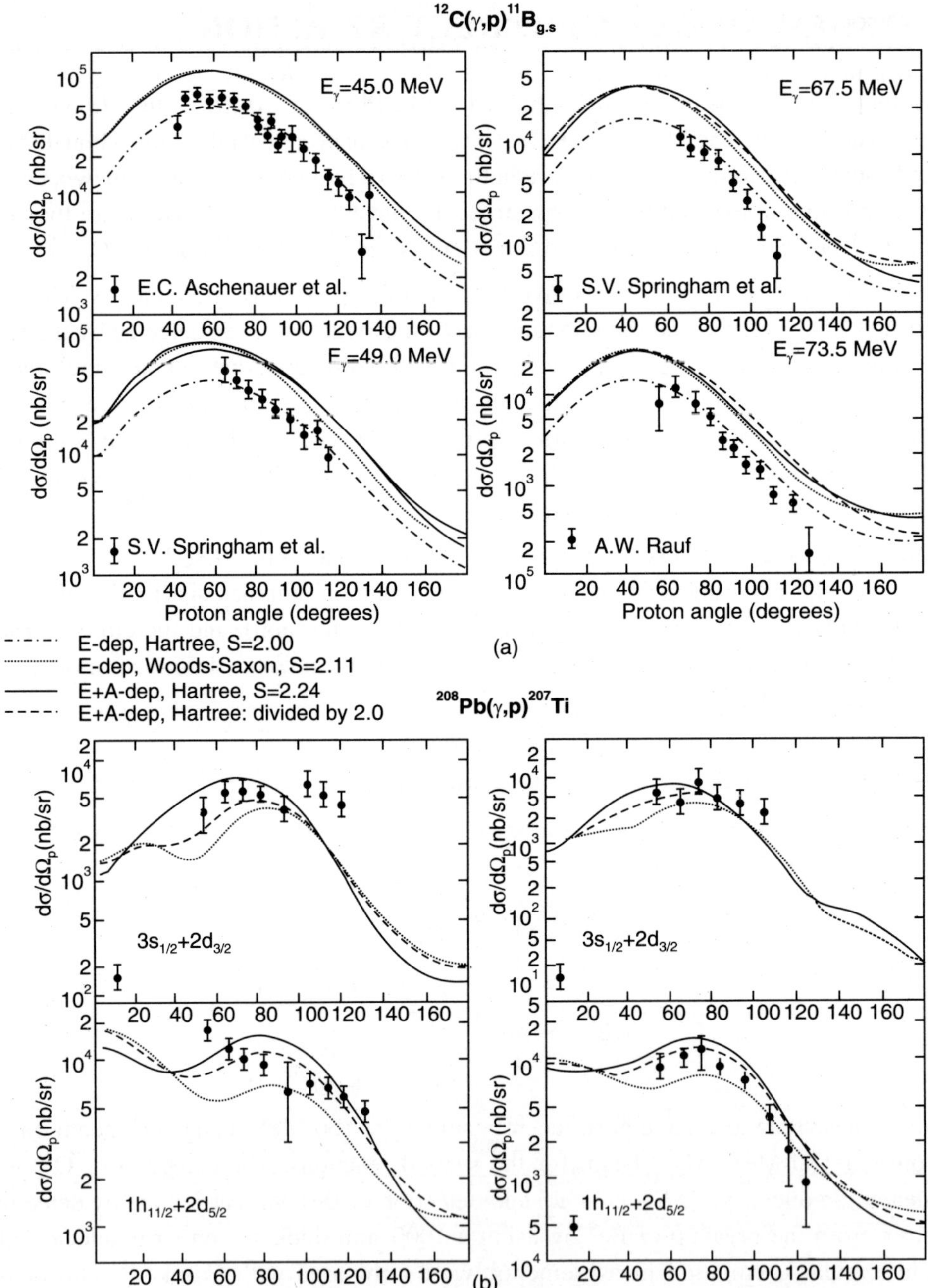

Fig. 14.12 (*a*) Angular distribution for two different photon energies for a knock out of $1p_{3/2}$ protons from a C^{12} target, leading to B^{11} g.s. state and (*b*) Angular distribution of protons, from two levels in a Pb^{208} target at two energies (Ref. 42).

Knock out Processes

In any nuclear interaction of a particle with a nucleus, knock out reaction forms a part of direct-reaction, within the broad framework of pre-compound nucleus model. Kalbach[41] calculated the knock out reaction component; assuming the single knock out process in competition with single-step (direct) inelastic scattering of the projectiles. Such knock out ejectiles are not observed separately and are a part of the overall pre-compound process of the emission of particles, as in an (α, n) reaction on Au^{197} shown in Fig. (16.9) in Chapter 16.

However (γ, p) reactions present a case of direct observation of a knock out process studied recently[42] on targets of C^{12}, B^{10}, B^{11}, O^{16} and Pb^{208}. The angular distribution of directly knocked out protons have been studied for the incident photon energies ranging from 45 MeV to 78.5 MeV obtained from tagged photon facilities of an 185 MeV electron microtron accelerator with mode of operation, using an aluminium radiator foil, using (e, e' γ) Bremsstrahlung reaction and a system of collimators.

Figure 14.12 shows[42] the comparison with relativistic Hartree calculations[43] for a knock out process. The forward peaking of protons is obvious. The discrepancy of a factor of 2 seems to be due to an underestimate of meson exchange current (MEC).

14. Direct Reactions
2000–2008

In a paper, authored by 10 authors, asymptotic normalization coefficient for $N^{14} \rightarrow C^{13} + p$ from C^{13} (He^3, pn)N^{14} reaction at 26.3 MeV have been reported in Phy. Rev. C. 62, 024320 (2000). This is among half a dozen papers on direct reaction, reported in [Phy. Rev. C. 62 (2000)].

In 2001, in an interesting experiment, properties of Cd^{112} have been extracted from ($n, n'\gamma$) reaction especially the level densities, using 7 MeV Van de Graaf, using H^3 (p, n) He^3 reaction, giving 2.5 MeV neutrons. A total of 375γ rays were placed in a level scheme comprising 200 levels. The level density parameter extracted were found to be a sensitive function of the maximum energy used in the fit [Phy. Rev. C. 64, 02316 (2001)].

In a multi institutional experiment, involving 12 institutions from USA, and Poland, and 16 authors, an interesting multi-step process in striping reaction has been carried out in involving 0.0 MeV (0^+), 6.3 MeV (3^-), 6.92 MeV (2^+), 8.87 MeV (2^-) and 10.35 MeV (3^+), states of O^{16} at the bombarding energies of 34 MeV and 50 MeV. These results suggest that these are significant multi-step contributions to transfer mechanism in Direct Reaction [Phy. Rev. C. 67, 044604 (2003)].

In a collaborative effort, 21 authors from India, France, UK, USA and Australia have measured in He^6, + $Ca^{63, 65}$ reaction at 19.5 MeV and 30 MeV beam energies of He^6, transfer reaction cross-section, by using the characteristic γ-rays. Similar measurements were presented for $He^{4, 6} + Os^{188, 192}$ at 30 MeV $He^{4,6}$ beam obtained at 14UD pelletron at Mumbai and CIME Cyclotron at Ganil in France. This work has helped in showing the importance of identifying and delineating the mechanism of residual formation for understanding fusion [Phy. Rev. C. 70, 044601 (2004)].

In a paper, multi nucleon transfer reactions have been studied in reaction Ca^{40} + Zr^{96} and $Zr^{90} + Pb^{208}$, at energies close to Coulomb barrier in a high resolutions γ-particle coincidence experiment. Specific transitions in Zr^{95} are populated in one particle transfer, and are discussed in terms of particle phonon coupling. The γ-rays from excited states of Ca^{42}, in the exciton energy region expected from pairing vibration are also observed [Phy. Rev. C. 76.024604 (2007)].

In a paper, directly concerned with Direct Reaction Mechanism, the structure of excited states in Mg^{21} was studied in one neutron knock out reaction Be^9 (Mg^{22}, Mg^{21}, γ) Be^{10} using Be^9 beam of 74 MeV Nuclear projectile energy. Spectroscopic factors for one neutron removal from Mg^{22} to Mg^{21} are extracted and compared to shell model calculations, using WBP effective interaction. The proposed excitation schemes are in agreement with mirror-nucleus and shell model calculations [Phy. Rev. C. 77, 064309 (2008)].

REFERENCES

1. E.O. Lawrence, E. McMillan and R.I. Thornton: Phy. Rev. 48, p. 493 (1935).
2. J.R. Oppenheimer, and H. Phillips: Phy. Rev. 48, 500 (1935).
3. S.T. Butler: Phy. Rev. 80, 1095 (1950); Nature 166, 709 (1950); Proceedings of Roy. Society (London) 208A, 559 (1951).
4. S. Edwards: Proc. Conference Direct Interactions Nuclear Reactions Mechanism, p. 46, 9 (1962); Gorden and Breach Science Publishers Inc., New York (1963).
5. W.E. Burcham: Nuclear Physics, p. 561 Longmans, Green and Co. Ltd, London (1963); also Reference (6), Butler S., Proc. Royal Society A208, 559 (1951).
6. W. Tobacman: Phy. Rev. 94, 655 (1954), R. Huby, M.Y. Refai and G.R. Satchler: Nuclear Phy. 9.94 (1958).
7. R.R. Roy and B.P. Nigam: Nuclear Physics; New Age International (P) Ltd., New Delhi, (1986).
8. B.L. Cohen and R.E. Price: Phy. Rev. 21, 1441 (1961).
9. J.C. Legg: Phy. Rev. 129, 272 (1963).
10. B.L. Cohen, R.H. Fulmer and A.L. McCarthy: Phy. Rev. 126, 638 (1962).
11. R. Middleton, Proc. Conference Direct Interactions, Nuclear Reaction Mechanism, Gorden and Breach Science Publishers, New York (1963).
12. Reference (4) and (11).
13. S. Michaletti and J.B. Mead: Nuclear Physics, 37, 201 (1962).
14. H.C. Newns: Proc. Phy. Soc. 76, 489 (1960), El-Nadi, M. Phy. Rev. 119, 242 (1960); K. Seth: Nuclear Physics 25, 169 (1961).
15. A.B. Bhatia, K. Huang, R. Huby and H.C. Newns: Phil. Mag. 73, 43, 485 (1952).
16. N.K. Glendenning: Nuclear Physics; 29, 109 (1962).
17. K.L. Warsh, G.M. Tremmer and H.R. Bieden (Experimental Results quoted in) Edwards S. Proc. Conf. Direct Interactions Nuclear Reactions Mech. (1962); p. 469, Gorden Breach Science Publishers Inc., New York (1963).
18. W.W. Dahenick and L.J. Dennes: Phy. Rev. 136, B1325 (1964).
19. References (11) and (17).
20. N. Austern, S.T. Butler and H. McManus: Phy. Rev. 92, 350 (1953).
21. H.J. Walters: Phy. Rev. 103, 1763 (1956).
22. I.M. Govil, H.W. Fulbright, D. Cline, E. Wesolowski, B. Kootlinski, A. Backlin and K. Grindev: Phy. Rev. C. 33, 793 (1986); I.M. Govil, H.W. Fulbright and D. Cline: Phy. Rev. C. 36, 1442 (1987).
23. L.L. Lee, and J.P. Schiffer: Phy. Rev. Letters, 12, 108 (1964).
24. B.L. Cohen: Concepts of Nuclear Physics, McGraw-Hill Company, (NY) (1971).

25. Reference (10), B.L. Cohen, and O.V. Chubinsky: Phy. Rev. 131, 2184 (1963); B. Gott-Schalk and K. Strauch: Phy. Rev. 120, 1005 (1960); Grooding, T.J. and J.J. Pugh: Nuclear Physics 18, 46 (1960); P. Hillman; K. Tyrene, and Th. A.J. Maris: Phy. Rev. Letters 5, 107 (1960).

26. A. Pfeiffer, Mairle G. Knöpfle K.T., Kihm T. Seegert G., Grambmayr P., Wagner G.J., Beshchold V., and Freidrich L.: Nuclear Physics A. 455, 381 (1986); John R. Wienandes U., Wenzel D., and Von Neumann-Cosel P.: Phy. Letters, B150, 331 (1985).

27. R.K. Bansal and J.B. French: Phy. Letters 11, 145 (1964); Ibid Phy. Letters 19, 223, (1965); R.K. Bansal and Ashwini Kumar, Pramana—M J.: Physics. V. 32, 4, 341, (1989); R.K. Bansal, H. Sharada, Ashwini Kumar: Phy. Letters B, 386, p. 17–22 (1996).

28. R.D. Lawson: Theory of Shell Model; Clarenden Press, Oxford (1980); B.L. Cohen, J.B. Moorhead and R.A. Moyer: Phy. Rev. 161, 1257 (1967).

29. S.K. Datta, Subiniti Ray, H. Majumdar, S.K. Ghosh, S. Ray and Das Gupta, S.N. Chintalpudi and S.R. Bannerjee: Phy. Rev. C. 39, p. 1281 (1989).

30. M. Schramum et al. (18 authors): Phy. Rev. C. 56, p. 1320 (1997); A.K. Basak, M.A. Basher, A.S. Mondel, M.A. Uddin, S. Bhattacharya, A Hussain, S.K. Dass Masudul Haque, and H.M. Sen Gupta: Phy. Rev. C. 56, p. 1983 (1997); J. Vernotte, G. Berrier-Ronsin, S. Fortier, E. Hourani, J. Kalifo, J.M. Kalifa, J.M. Maison, L.H. Rosier, G. Rothard and B.H. Wildenthal: Phy. Rev. C. 57, p. 1256 (1998).

31. P. Von Neumann Cosel, R. Jahn, V. Fister, T.K. Jolie, J. Kern, H. Lehmann, S.J. Mannanal and N. Warr: Phy. Rev. C. 58, p. 3734 (1998).

32. P.E. Gasset, D.G. Burke, M. Deleze, S. Drissi, J. Jolie, J. Kem, H. Lahman, S.J. Mannanal, and N. Warr: Phy. Rev. C. 58, p. 3734 (1998).

33. V. Fisher, R. Jahn, P. Von Neumann, Cosel, P. Schank, T.K. Trelle, D. Wenzel, and U. Wienands: Phy. Rev. C. 42, 2375 (1990).

34. B.A. Brown, and B.H. Wildenthal: Ann. Rev. Nuclear Particle Science 38, 29 (1988).

35. S.E. Larsson, G. Leander and I. Ragnarsson: Nuclear Physics A. 307, 189 (1978); T. Bengisson and I. Ragnarson: Nuclear Physics A 436, 14 (1985).

36. J.T. Murgatroyd et al. (14 authors): Phy. Rev. C. 58, p. 1569 (1998).

37. S. Marsh and W.D.M Rae.: Phy. Letter B. 180, 185 (1986).

38. A.S. Umar, M.R. Strayer, R.Y. Cusson, P.G. Reinhard, and D.A. Bromley: Phy. Rev. C. 32, 172 (1985); M.R. Strayer, R.Y. Cusson, J.A. Maruhn, D.A. Bromley and W. Greiner: Phy. Rev. C. 28, 228 (1983).

39. R. Shyam and H. Kenske: Phy. Rev. 57, p. 2427 (1998).

40. K. Rusager: Rev. Mod. Phys. 66, 1105 (1994).

41. C. Kalbach: Phy. Rev. C. 23, 123, (1981), Ibid, C. 24, 819 (1981); P.K. Sarkar, T. Bandopadhaye and G. Muthukrishnan: Phy. Rev. C. 43, 1855 (1991).

42. J.T. Johansson and H.S. Sciuff: Phy. Rev. C. 56, 378 (1997), S.V. Springham et al.: Nuclear Phy. A 517, 93 (1990), L. Bodeldijk, et al.: Phy. letters B 356 (1995).

43. P.G. Blunden and M.J. Iqbal: Phy. Letters. B. 196, 295 (1987); E.D. Cooper, S. Hama, B.C. Clark, and R.I. Mercer: Phy. Rev. C. 47, 2044 (1995).

PROBLEMS

1. Find out the mean-free path of a neutron in a nucleus say Fe for energies of 1 MeV, 10 MeV, 30 MeV and 100 MeV; taking the *n-p* cross-sections from Fig. 4.1 in Chapter 4. Explain from these results, why direct reaction is more probable for higher energies.

2. For the process $d + X^A \rightarrow X^{A+1} + p$; evaluate the reaction cross-section, using Eq. 14.35, with a potential of the form $V = v_o\,\delta\,(\mathbf{r}_i - \mathbf{r}_e) \times (\mathbf{r}_i - \mathbf{R})$ as given in Eq. 14.9.

3. Some of the higher energy proton groups observed when Ni^{58} is bombarded with deuteron of say 15 MeV energy, are strongly peaked in the forward direction. What is the angular momentum and parity of the corresponding levels? In which nucleus are these level formed?

4. Consult Phy. Rev. 156, 1315 (1967); find out the values of l in the reactions Sn^{117} (t, p) Sn^{119} (G.S.) and Sn^{119} (t, p) Sn^{121} (G.S.).

5. Plot $[J_l\,(q\,R')]^2$, for a given l (say $l = 0, 1, 2, 3$, and 4) for different values of θ, using Eqs. 14.42 and 14.11b, for the function $J_l\,(q\,R')$, for the case of Nickel target, for 40 MeV proton-induced (p, p') reaction.

6. If the $d_{5/2}$ and $d_{3/2}$ single quasi-particle states in Sn^{116} are three-fourths and one-fourth full; how sensitive is the ratio of spectroscopic factors $S\,(d, p)/S\,(p, d)$, for distinguishing between $3/2^+$ and $5/2^+$ states.

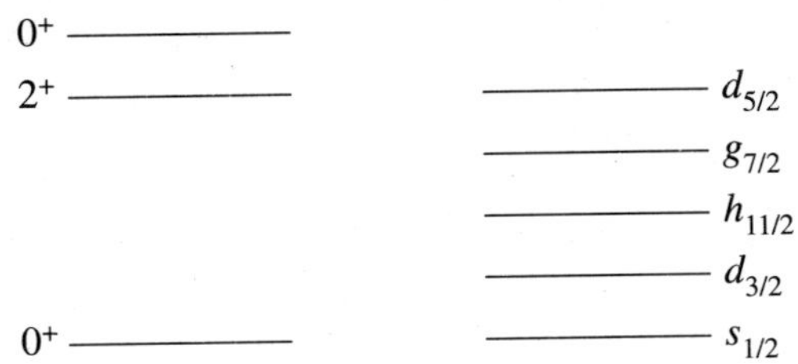

Sn^{116} (Exp.) Single quasi-particle states in Sn^{116}

7. Assuming the Q-value of the reaction Ca^{44} (d, p) Ca^{45} as 3.3 MeV and energy of the incident deuteron to be 7.0 MeV; calculate the momentum transfer and cross-sections using Eq. 14.38.

8. Calculate the ratio of spectroscopic factors for analogue states.

9. Estimate the angles of the first maxima in the angular distributions for various l values from Eq. 14.6; for any target and compare it with the actual experimental values in literature.

10. What is the minimum energy needed to excite the lowest energy 2^+ states in Cd^{114} and Pt^{192} by Coulomb Excitation? At what proton energies do nuclear reactions become important in these nuclei instead of only Coulomb excitation?

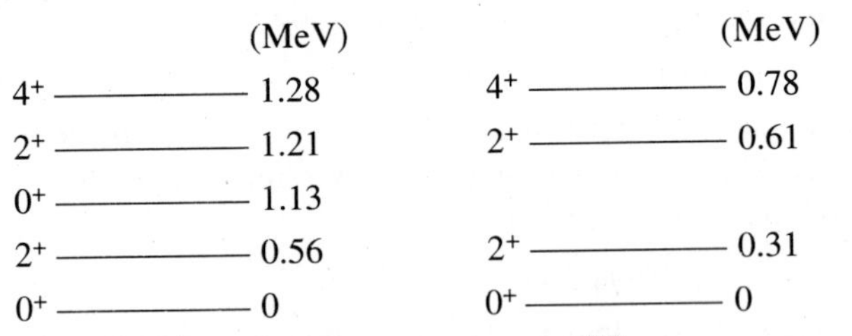

$Cd^{1/4}$ Pt^{192}

11. Using, $I(\theta) \propto [J_{li}\,(q\,R')]^2$, calculate the angle expected for the second maxima for $l = 0$ and $l = 1$ angular distributions for (d, p) reaction on Ni^{58}.

15

CHAPTER

Optical Model

15.1 INTRODUCTION

The interaction between the projectile and the target may, in general, be not describable by the two extreme models of the compound nucleus formation and direct reaction as described in the last two chapters. The interaction may be described (in most of the cases) by an intermediate process, where both the processes occur. This can be understood either by devising methods to measure separately the cross-sections corresponding to compound nucleus formation and the direct reaction and also the pre-equilibrium component. Theoretical formulation of such problems is slowly taking place through these models of reaction mechanism as described in Chapters 13, 14 and 16.

One way, however, take into account all these aspects through a model called optical model in which the target nucleus is replaced by a potential called the optical potential written generally in the form[1, 2]:

$$U(r) = -V(r) - iW(r) + \left(\frac{\hbar^2}{\mu c^2}\right)^2 \left(\frac{1}{r}\frac{df(r)}{dr}\right)(V_{S.O} + iW_{S.O})\,\mathbf{l}\cdot\mathbf{S} \qquad ...(15.1)$$

The first two terms describe the combined nuclear and Coulomb effects and the last term describes the spin-orbit coupling mostly contributed by the outer surfaces of the interacting nuclei. The quantities $V(r)$ and $W(r)$ are expected to have the Woods-Saxon shape[3] and $V_{S.O} + iW_{S.O}$ are taken as constant for a given incident energy, while $1/r\ df/dr$ represents a surface potential of the Thomas type expressing the fact that the spin-orbit coupling is a surface phenomenon.

Equation 15.1 represents a target, which has the property of pure scattering from whole nuclear body $[-V(r)]$ and from the surface, through

$$\frac{1}{r}\frac{df}{dr}V_{S.O}$$

and of inelastic scattering and reaction—corresponding to the absorption of the incoming wave—represented by $iW(r)$ from the whole nucleus and from the surface by:

$$\left(iW_{S.O} \frac{1}{r} \frac{df}{dr} \right)$$

The physical reasons for representing the target with such a potential is based on the physical fact, that an incoming wave representing the projectile is expected to be partially scattered (representing elastic scattering) and partially absorbed (representing inelastic scattering and reactions).

The various parameters used in Eq. 15.1 can be determined by comparing the experimental differential cross-sections of elastic scattering or the total cross-section with the theoretically expected values using the optical potential for which the theory is given in the next section.

15.2 THEORETICAL CROSS-SECTIONS WITH OPTICAL MODEL

One can use the optical potential model for calculating the elastic scattering and reaction cross-sections following the treatment[4], which resulted in Eq. 13.42 for elastic scattering cross-section and in Eq. 13.44 for reaction cross-sections. We reproduce those results here:

$$\sigma_{sc} = \frac{\pi}{k^2} \sum_l (2l+1) |1 - \eta_l|^2 \qquad \text{...(15.2)}$$

and

$$\sigma_{re} = \frac{\pi}{k^2} \sum_l (2l+1) \left[1 - |\eta_l|^2\right] \qquad \text{...(15.3)}$$

Of course, the differential cross-section for elastic scattering according to Eq. 13.41 is given as:

$$\frac{d\sigma_{el}}{d\Omega} = \frac{\pi}{k^2} \left| \sum_l (2l+1)^{1/2} (1 - \eta_l)\, Y_{l0}(\theta) \right|^2 \qquad \text{...(15.4)}$$

It is easy to see then that $\sigma_{\text{total}} = \sigma_{sc} + \sigma_{re} = \frac{\pi}{k^2} (2l+1)\,[2R_e\,(1 - \eta_l)]$ ...(15.5)

For replacing the target nucleus with a potential as given in Eq. 15.1, the detailed structural properties of the target nucleus cannot be considered and hence only the average behaviour of the cross-sections can be obtained. This requires that η_l used in the above equation should correspond to the average properties of the nucleus, which is obtained when $\Gamma >> D$ and levels overlap. This condition can be simulated by mathematically carrying out the averaging over a certain range of energy which corresponds to the overlapping of many levels, *e.g.* over an energy ΔE such that $D << \Delta E \leq E$ or carrying out the experiment when the incident energy has a large spread or/and the detecting system has poor resolution so that, many levels are involved, simultaneously in the measurement. Under these conditions, we carry out the average of the cross-sections in the above expressions and write:

$$\langle \sigma_{sc} \rangle = \pi \lambda\!\!\!^{-2} \sum_l (2l+1) \langle |1 - \eta_l|^2 \rangle$$

$$= \pi \lambda\!\!\!^{-2} \sum (2l+1) \left[1 - \langle \eta_l \rangle - \langle \eta_l^* \rangle + \langle |\eta_l^2| \rangle\right]$$

$$= \pi \lambda\!\!\!^{-2} \sum_l (2l+1) \left| 1 - \langle \eta_l \rangle \right|^2 + \pi \lambda\!\!\!^{-2} \sum_l (2l+1) \left[\left\langle \left| \eta_l^2 \right| \right\rangle - \left| \langle n_l \rangle \right|^2 \right]$$

$$= \sum_l \langle \sigma_{se}^l \rangle + \sum_l \langle \sigma_{ce}^l \rangle \qquad \text{...(15.6)}$$

where we have used $1/k^2 = \lambda\!\!\!^{-2}$ and

$$\langle \sigma_{se}^l \rangle \equiv \pi \lambda\!\!\!^{-2} (2l+1) \left| 1 - \langle \eta_l \rangle \right|^2 \qquad \text{...(15.7)}$$

is called the shape elastic scattering and

$$\langle \sigma_{ce}^l \rangle \equiv \pi \lambda\!\!\!^{-2} (2l+1) \left[\langle | \eta_l |^2 \rangle - \left| \langle \eta_l \rangle \right|^2 \right] \qquad \text{...(15.8)}$$

is called the compound elastic scattering. Similarly one can write:

$$\langle \sigma_r \rangle = \pi \lambda\!\!\!^{-2} \sum_l (2l+1) \ \langle 1 - | \eta_l |^2 \rangle$$

$$= \pi \lambda\!\!\!^{-2} \sum_l (2l+1) \left[1 - \left| \langle \eta_l \rangle \right|^2 - \left(\langle | \eta_l |^2 \rangle - | \langle \eta_l \rangle |^2 \right) \right]$$

$$= \sum_l \langle \sigma_c^l \rangle - \sum_l \langle \sigma_{ce}^l \rangle \qquad \text{...(15.9)}$$

where $$\langle \sigma_c^l \rangle \equiv \pi \lambda\!\!\!^{-2} (2l+1) \left(1 - \left| \langle \eta_l \rangle \right|^2 \right) \qquad \text{...(15.10)}$$

is called the cross-section for the formation of the compound nucleus. Physically, Eqs. 15.7 and 15.9 indicate that the elastic scattering is the sum of scattering from the surface (σ_{se}), (the shape elastic) and from within the volume after forming the compound nucleus (σ_{ce}). Equation 15.8 corresponds to difference between

$$\left\langle | \eta_l |^2 \right\rangle - \left| \langle \eta_l \rangle \right|^2$$

which approaches zero as incident particle energy increases. Then $\left\langle \sigma_{ce}^l \right\rangle \to 0$. Similarly the reaction cross-section $\left\langle \sigma_{re} \right\rangle$ as given in Eq. 15.9 corresponds to the total cross-section for compound formation, out of which the elastic part is subtracted.

It is of course, easy to see that, the total cross-section $\left\langle \sigma_T \right\rangle$ is given by:

$$\left\langle \sigma_T \right\rangle = \left\langle \sigma_{se} \right\rangle + \left\langle \sigma_c \right\rangle \qquad \text{...(15.11)}$$

The averaging process in the above equations has brought out the role of shape elastic scattering and compound elastic which are meaningful especially at comparatively lower energies. At energies say greater than 10 MeV for medium to heavy nuclei:

$$\sigma_{ce} << \sigma_{se} \text{ and } \sigma_{ce} << \sigma_c. \text{ Then } \left\langle \sigma_{el} \right\rangle \approx \left\langle \sigma_{se} \right\rangle$$

At these high energies $\Gamma >> D$, the value of η_l varies slowly with energy, hence $\left|\langle\eta_l\rangle\right|^2 \approx \left\langle\left|\eta_l\right|^2\right\rangle$ and hence $\langle\sigma_{ce}\rangle \to 0$. At energies lower than 10 MeV, $\langle\sigma_{ce}\rangle$ should be taken into account.

15.3 COMPARISON WITH EXPERIMENTS

As described in sections 15.1 and 15.2, one can theoretically calculate the differential cross-sections for elastic scattering using different parameters and compare with experimental results.

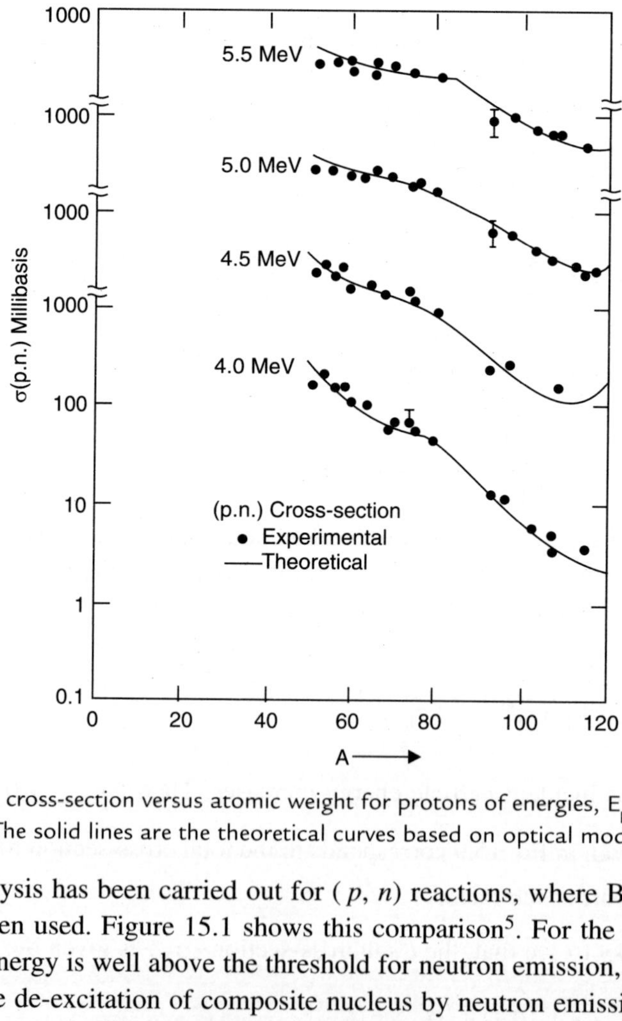

Fig. 15.1 The (p, n) cross-section versus atomic weight for protons of energies, E_p = 5.5, 5.0, 4.5 and 4.0 MeV. The solid lines are the theoretical curves based on optical model (Ref. 5).

(*i*) Such an analysis has been carried out for (p, n) reactions, where Bjorklund and Fernbach[2] optical potential has been used. Figure 15.1 shows this comparison[5]. For the energies of 4.0 MeV to 5.5 MeV, the incident energy is well above the threshold for neutron emission, but below the height of Coulomb barrier. So the de-excitation of composite nucleus by neutron emission is favoured over the charged particle emission for which Coulomb barrier is high. So (p, n) reaction dominates over the competing compound elastic (p, p) or inelastic (p, p') scattering processes. (p, n) reaction will be due to imaginary term of the optical model.

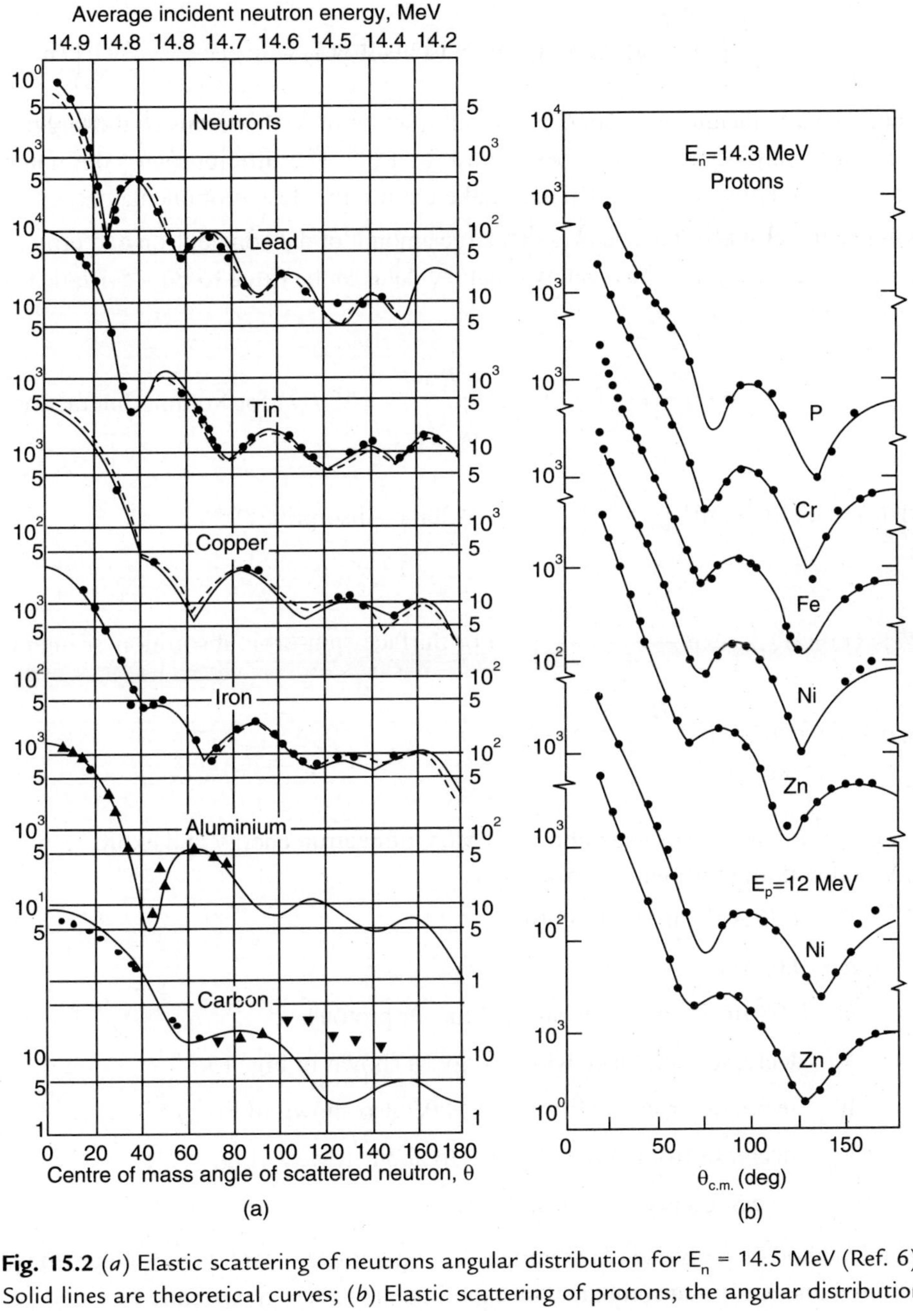

Fig. 15.2 (*a*) Elastic scattering of neutrons angular distribution for E_n = 14.5 MeV (Ref. 6). Solid lines are theoretical curves; (*b*) Elastic scattering of protons, the angular distribution for E_p = 14.3 MeV (Ref. 9). Solid lines are theoretical curves.

(*ii*) A large amount of data exists of scattering of 14 MeV neutrons[6] as well as at lower energies[2, 7]. Figure 15.2*a* gives some typical curves and the comparison with theory. The potential of optical model makes the target behave like a cloudy crystal ball, which makes the angular distributions say of neutrons look like a diffraction pattern. As in a diffraction pattern[8] the minima of scattering from opaque disc of radius *R* occur whenever,

$$\frac{2R}{\lambda}\sin\left(\frac{\theta}{2}\right) = \text{zero of Bessel Function } J_1 \qquad ...(15.12)$$

At about 14 MeV, the quantity $2R/\lambda$ is a small fraction of R, the radius of the target nucleus and three to five minima may occur as one goes from 8 to 180°. The differences of the potential due to Woods-Saxon shape and absorption coefficient make the minima less prominent.

Bjorklund and Fernbach[2] have analysed a large amount of data on neutron and proton-scattering. Through this intensive survey they have shown that the data can be fitted to Eq. 15.1*a*, if $V(r)$ and $W(r)$ are written as below:

$$V(r) = V_v\left[1+\exp\left(\frac{r \quad R}{d}\right)\right]^{-1} = V_v f(r) \text{ for Volume Interaction}$$

and

$$(a)\ W(r) = W_v\left[1+\exp\left(\frac{r-R}{d}\right)\right]^{-1} \quad \text{For Volume Absorption}$$

$$(b)\ W_{S.O.}(r) = W_{S.O.}\left[\exp\frac{-(r-R)^2}{b^2}\right] \quad \text{For Surface spin-orbit absorption or interaction}$$

$$V_{S.O.}(r) = V_{S.O.}\left[\exp-\frac{(r-R)^2}{b^2}\right] \qquad ...(15.13a)$$

where expression (*a*) for $W(r)$ was used above 50 MeV of neutron energy and expression (*b*) was used below 50 MeV and with the following constants:

$R = 1.25\,A^{1/3}$ fm, $d = 0.65$ fm

$d = 0.65$ fm

$b = 0.98$ fm for neutrons and 1.2 fm for protons

V_v decreases smoothly with energy as shown in Fig. 15.3

W_v increases from 3 MeV to 20 MeV also shown in Fig. 15.3

V_{SO} decrease from about 10 MeV to about 1 MeV [Fig. 15.4]

W_{SO} is essentially zero [Fig. 15.4].

The usual method of analysis is to assume a set of parameters and then calculate the phase shifts and cross-sections numerically and compare with experimental values. Systematic variation of parameters using computers ends in the best fit with the experimental data.

In Fig. 15.2*b*, we have shown the angular distribution of protons of energy 14.3 MeV. The pattern is similar to the neutron angular distribution with some changes due to Coulomb interaction. In this case $V(r)$ is written as:

$$V(r) = V_c(r) + V_N(r)$$

where the Coulomb potential $V_c(r)$ is given by:

$$V_c(r) = \frac{Ze^2}{2R_c}\left(3 - \frac{r^2}{R_c^2}\right) \text{ for } r \le R_c$$

$$= \frac{Ze^2}{2R_c} \text{ for } r \ge R_c \qquad \text{...(15.13}b\text{)}$$

and R_c is the charge radius and $V_N(r)$ is the nuclear part $V(r)$ as given in Eq. 15.1.

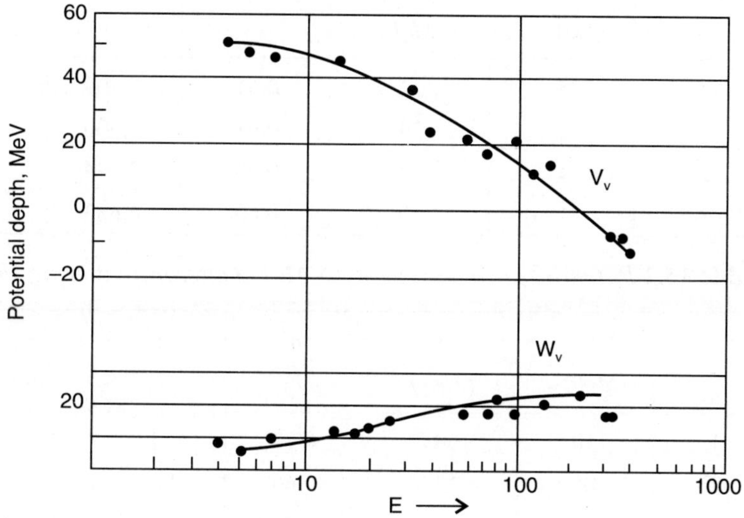

Fig. 15.3 The variation of V_v and W_v as a function of energy from the analysis of Bjorklund and Fernbach (Ref. 2).

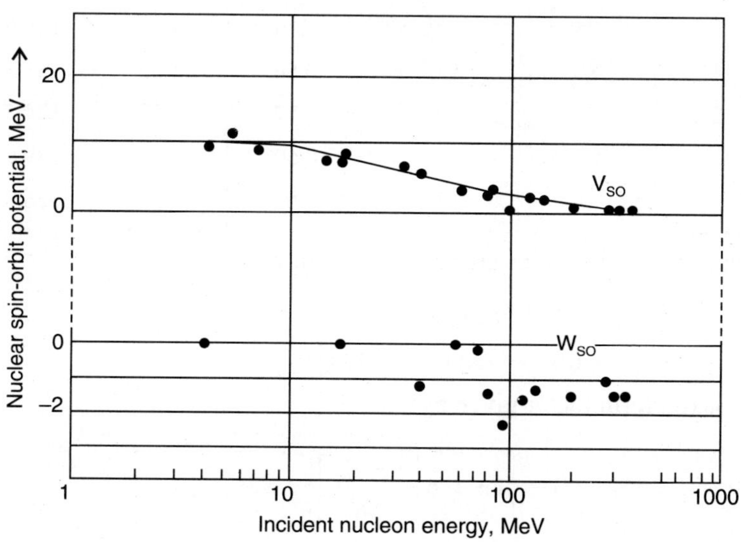

Fig. 15.4 The variation of spin-orbit constants V_{SO} and W_{SO} according to Ref. 2.

(*iii*) In Fig. 15.5, we show the experimental angular distribution of scattering of (*a*) alphas[10] at 40 MeV from Ag and (*b*) of deuterons at 15 MeV from many targets along with the theoretically calculated curves based on optical model.

An interesting feature of both these cases is that, for each target there can be more than one set of optical model parameters, which fit the experimental data. Tables 15.1 and 15.2 shows, these sets of parameters for deuterons and alphas respectively.

Table 15.1 *Potential parameters for 40 MeV alpha scattering (Ref. 10)*

E	*Set No.*	*V* (*MeV*)	*W* (*MeV*)	*a(F)*	$R_0(F)$	X^2
40	1	35	22	0.61	7.709	3.74
	2	50	29	0.61	7.500	3.81
	3	75	37	0.60	7.244	4.07
	4	1.50	70	0.60	6.683	3.43

Table 15.2 *Potential parameters for 15 MeV deuteron scattering (Ref. 11)*

Target element	*Set No.*	*V* (*MeV*)	*W* (*MeV*)	*a* (*F*)	R_0 (*F*)	σ_R (*mb*)	X^2
Ag^{107}	1	110	68.7	0.70	1.17	136.4	0.69
	2	85	13	0.60	1.33	130.6	0.66
	3	50	20	0.58	1.6	176.2	0.68
Er^{168}	1	100	55	0.78	1.17	132.5	0.43
	2	79	11	0.64	1.33	130.2	0.70
	3	35	9	0.56	1.60	164.0	0.30
Y^{89}	1	80	12	0.56	1.33	131.5	1.42
	2	55	9	0.57	1.33	124.5	1.40
Mo^{96}	1	85	14	0.60	1.33	139.9	1.90
Nb^{93}	1	85	12	0.56	1.30	126.0	0.86

The parameters used in the tables are related as follows:

$$V = V_N(r) + V_c(r) \qquad \text{...(15.14a)}$$

$$V_N(r) = -V_0(1 + iW)f(r) \qquad \text{...(15.14b)}$$

and $f(r)$ is a Woods-Saxon form factor given by:

$$f(r) = \left\{1 + \exp\left[\frac{r - R_0}{a}\right]\right\}^{-1} \qquad \text{...(15.14c)}$$

$V_c(r)$ is the Coulomb potential as given in Eq. 15.13*b*.

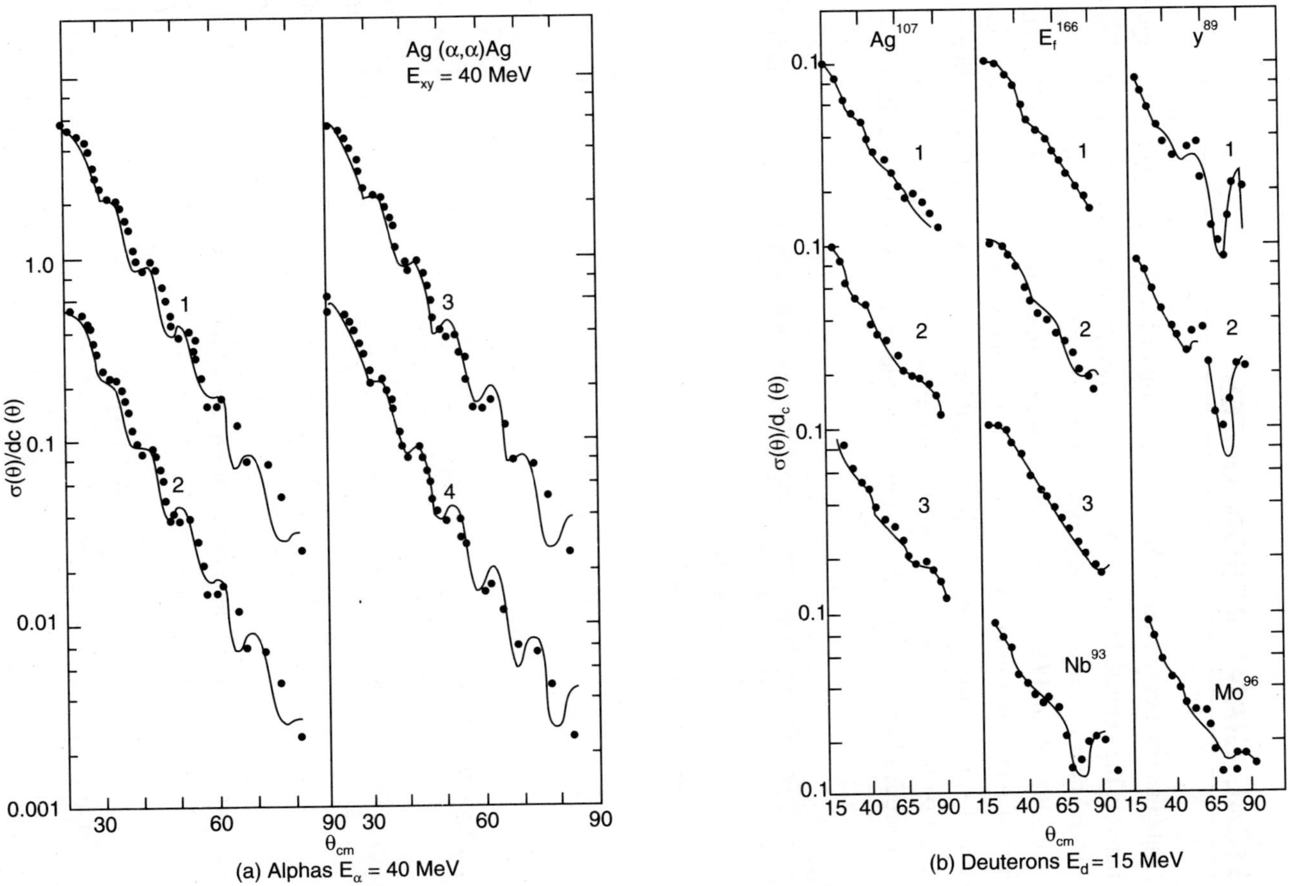

Fig. 15.5 Some typical examples of the comparison of experimental and theoretical angular distribution of elastic scattering of (*a*) alphas[10] (E_α = 40 MeV) and (*b*) deuterons[11] (E_d = 15 MeV) [Ref. (10, 11)]. The numbers on curves correspond to various sets of parameters of Table 15.2.

It has been found by actual variation of the parameters, that the shape and magnitude of the angular distributions are insensitive to large changes in V and W, but the shape is very sensitive to the diffuseness parameter 'a'. As R_0 increases, an increase in the amplitude of oscillations is observed.

15.4 OPTICAL GIANT RESONANCES

A large amount of data has been collected for $\overline{\sigma}_T$ as a function of incident energy[12] for a large number of nuclei and compared to the expected values obtained from the optical model calculations. Figure 15.6 gives the calculated total neutron cross-sections based on the optical model, which has an indication of resonances for $S(l = 0)$, $P(l = 1)$ and $D(l = 2)$ as is observed experimentally by Barschall[12] and his group in the measurements of total neutron cross-sections from thermal to several MeV. These resonances have spacing typical of single particle energies in the real part of the potential and widths of these maxima were found to be directly connected with the complex part of the optical potential $W(r)$. We treat this problem theoretically for $l = 0$ and will only give the results of general values of l without proof.

(*i*) $l = 0$, case.

We write the optical potential in this case as:

$$V = V_0 + iW_0 \qquad \text{...(15.15)}$$

so that the Schrödinger equation can be expressed as:

$$\left[-\frac{\hbar^2}{2M}\frac{d^2}{dr^2} + (V_0 + i\,W_0)\right] U(r) = E\,U(r) \qquad \text{...(15.16a)}$$

where $$U(r) = r\,\psi(r)$$

Then Eq. 15.15 yields, for $r < R$

$$U(r) \approx \sin Kr \qquad \text{...(15.16b)}$$

where $$K = K_1 + iK_2$$

$$= \left[\frac{2M}{\hbar^2}(E + V_0 + i\,W_0)\right]^{1/2} \qquad \text{...(15.16c)}$$

Then the logarithmic derivative f_0 as defined in Eq. 13.49 has the form:

$$f_0 = X \cot X \qquad \text{...(15.17a)}$$

where $$X = X_1 + i\,X_2 = (K_1 + i\,K_2)\,R \qquad \text{...(15.17b)}$$

Using Eq. 13.50, it can be easily seen that

$$\langle \eta_0 \rangle = e^{-2iKR}\left[1 - \frac{2iKR}{i(KR - \operatorname{Im} f_0) - \operatorname{Re} f_0}\right] \qquad \text{...(15.18)}$$

If we write $W_0 \equiv \xi\, V_0$ for $r \le R$ and assuming $\xi << 1$.

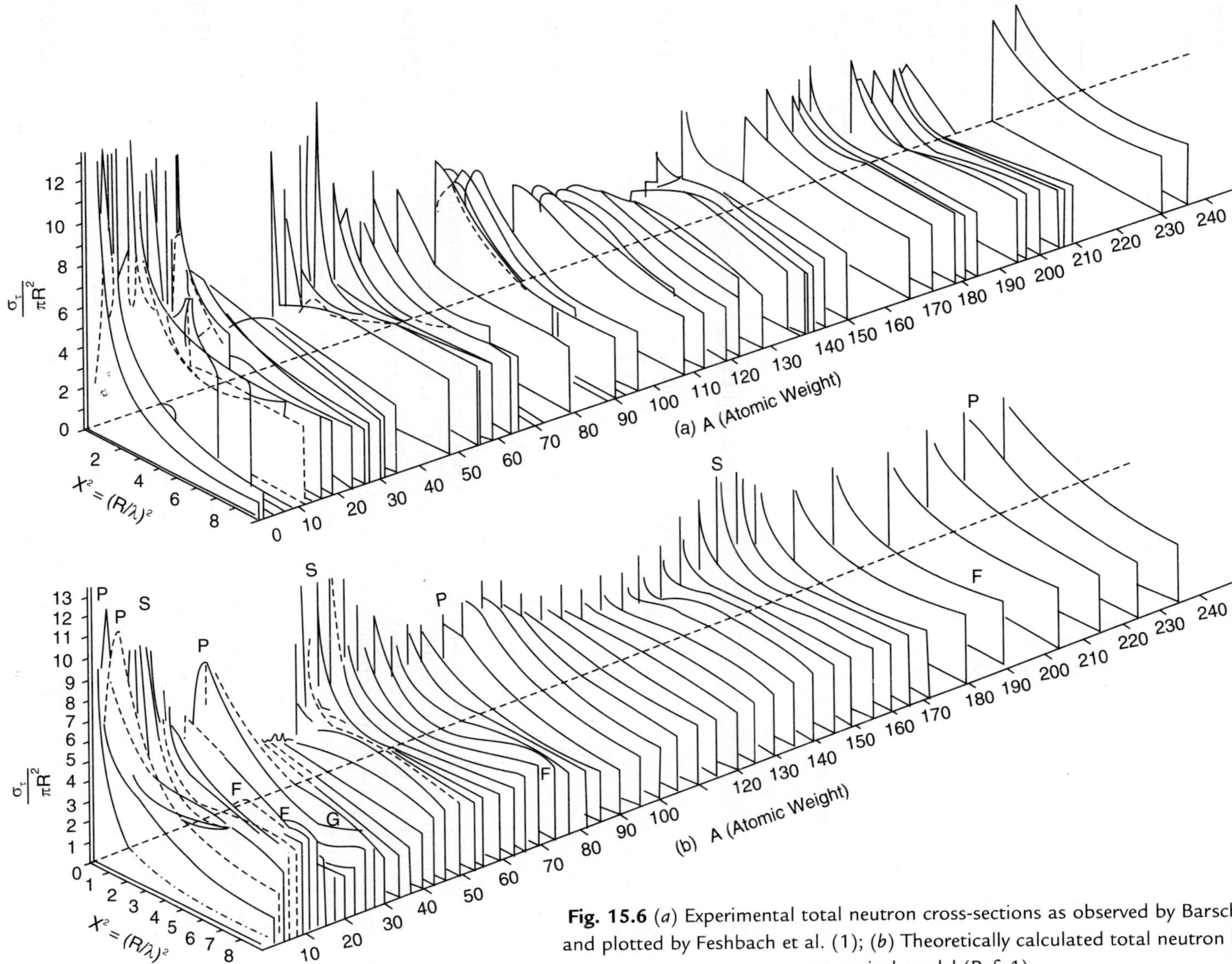

Fig. 15.6 (*a*) Experimental total neutron cross-sections as observed by Barschall[12] and plotted by Feshbach et al. (1); (*b*) Theoretically calculated total neutron based on optical model (Ref. 1).

We get $$\langle \eta_0 \rangle = e^{-2iKR} \left[1 - \frac{2iKR}{iKR - iX_1 \left(\cot X_1 - \frac{X_1}{\sin^2 X_1} \right) - X_1 \cot X_1} \right] \qquad ...(15.19)$$

Defining $N_0 \equiv X_1 \cot X_1 = \text{Re} f_0$

and $M_0 \equiv KR - \text{Im} f_0$ ...(15.20)

The giant resonance occurs in Eq. 15.19, if $N_0 = 0$, *i.e.*,

$$X_1 \cot X_1 = 0 \text{ or } X_1 = \left(n + \frac{1}{2} \right) \pi \qquad ...(15.21)$$

where n is an integer. Expanding N_0 around $E = E_s$, and dividing both the numerator and the denominator by $dN_0(E)/dE$ one obtains:

$$\langle \eta_0 \rangle = e^{-2KR} \left(1 - \frac{i\Gamma^0}{E - E_s + 1/2\, i\Gamma_s} \right) \qquad ...(15.22)$$

where $$\Gamma^0 \equiv \left. \frac{-2KR}{(dN_0/dE)} \right|_{E=E_s} = \frac{2\hbar K}{M_0 R} \qquad ...(15.23)$$

and $$\Gamma^s \equiv \Gamma^0 + \left. \frac{2 \text{ Im } f_0}{(dN_0/dE)} \right|_{E=E_s} = \Gamma^0 + 2W_0 \qquad ...(15.24)$$

From Eq. 15.24, it can be inferred, that Γ^0 is physically the width of resonance due to potential scattering (it only depends on radius R and energies in K) and is called the single particle width. The second term in Eq. 15.24 corresponds to the width due to absorption and is of the order of a few MeV. The quantities M_0 and N_0 are defined in Eq. 15.20.

We can now write the expression for $< \sigma_T >$ as:

$$\langle \sigma_T \rangle_{l=0} = \langle \sigma_{sc} \rangle_{l=0} + \langle \sigma_c \rangle_{l=0} \qquad ...(15.25)$$

Using Eqs. 15.7, 15.10 and 15.22 one obtains:

$$\langle \sigma_{se} \rangle_{l=0} = \frac{\pi}{k^2} \sin^2 KR + \frac{KR}{M_0^2 + N_0^2}$$

$$\times (KR - M_0 + M_0 \cos 2KR - N_0 \sin KR) \qquad ...(15.26a)$$

and $$\langle \sigma_c \rangle_{l=0} = \frac{4\pi}{k^2} \frac{KR(-\text{I}_\text{m} f_0)}{M_0^2 + N_0^2} = \frac{4\pi}{k^2} \frac{KR(M_0 - KR)}{M_0^2 + N_0^2} \qquad ...(15.26b)$$

From the above two equations, one can write for $< \sigma_T >_{l=0}$ as:

$$\langle \sigma_T \rangle_{l=0} = \frac{4\pi}{k^2}\left(\sin^2 KR + KR\frac{M_0 \cos 2KR - N_0 \sin 2KR}{M_0^2 + N_0^2}\right) \quad ...(15.26c)$$

To get average cross-section from an expression like Eq. 13.50

$$\left[\eta_0 = \exp(-2ikR)\frac{f_0 + ikR}{f_0 - ikR}\right] \quad ...(13.50)$$

$$\langle \sigma_c \rangle = \frac{1}{\Delta E}\int \sigma_c \, dE \quad ...(15.27)$$

where σ_c may be obtained from Eq. 13.50 for a resonance conditions. Then one obtains:

$$\langle \sigma_c(\alpha) \rangle = \frac{1}{\Delta E}\int_{E-\Delta E/2}^{E+\Delta E/2} \frac{\pi}{k^2}\sum_S \frac{\Gamma^S \Gamma_\alpha}{(E-E_S)^2 + (\Gamma^S/2)^2} dE \quad ...(15.28)$$

or

$$\langle \sigma_c(\alpha) \rangle = \frac{2\pi}{k^2}\frac{\pi}{\Delta E}\sum_S \Gamma_S^\alpha = \frac{2\pi}{k^2}\pi\left(\frac{\sum_S \Gamma_S^\alpha}{dE}\right)$$

$$= \frac{2\pi}{k^2}\pi\left(\frac{\overline{\Gamma}_\alpha}{D}\right) \quad ...(15.29)$$

where α corresponds to incident channel.

Equating Eq. 15.29 with Eq. 15.26*b* one obtains:

$$\frac{\overline{\Gamma}_\alpha}{D} = \frac{2KR(M_0 - KR)}{\pi\left(M_0^2 + N_0^2\right)} \quad ...(15.30)$$

Further, if we write:

$$\langle \sigma_c(\alpha) \rangle = \frac{\pi}{k^2}(4\pi KRS) \quad ...(15.31a)$$

Then equating Eq. 15.31*a* with Eq. 15.29, we get:

$$\frac{\overline{\Gamma}_\alpha}{D} = 2KRS \quad ...(15.31b)$$

where S is called the strength function which is a measure of the average ratio of the energy level width to level spacing.

Keeping in mind, from Eqs. 13.24 and 13.28, that

$$\sigma\,(n, R) = \sigma_c\,(\alpha)\,\frac{\Gamma_R}{\Gamma} \qquad ...(15.32)$$

and using from Eq. 13.49; $f_0 = -\,iKR$ and from Eq. 13.52, one gets:

$$\sigma_c\,(\alpha) \approx \sigma_{\text{reaction}} = \frac{\pi}{k^2}\,\frac{4\,k\,K}{(k+K)^2} \qquad ...(15.33)$$

and hence from Eq. 15.32:

$$\sigma\,(n, R) = \frac{\pi}{k^2}\,\frac{4\,k\,K}{(k+K)^2}\,\frac{\Gamma_R}{\Gamma} \qquad ...(15.34)$$

Comparing Eqs. 15.29 and 15.33, one obtains:

$$\frac{\overline{\Gamma}_\alpha}{D} = \frac{1}{2\,\pi}\,\frac{4\,k\,K}{(k+K)^2} \approx \frac{2\,k}{\pi\,K}$$

Hence the quantity $(\overline{\Gamma}_\alpha\,/\,D)$ depends on the energy of the incident particle. One defines a quantity called reduced strength function $(\overline{\Gamma}^0_\alpha\,/\,D)$, which is independent of energy as:

$$\left(\frac{\overline{\Gamma}^0_\alpha}{D}\right) = \left(\frac{E_0}{E}\right)^{1/2}\left(\frac{\overline{\Gamma}_\alpha}{D}\right) \qquad ...(15.35)$$

where E_0 is an arbitrarily chosen energy: $E_0 = 1$ eV. The reduced strength function can be obtained from actual experimental data. Figure 15.7 depicts the measured values of reduced strength function for $l = 0$ for neutrons as a function of atomic weight A. The solid line corresponds to the calculated values of $\overline{\Gamma}^0_n/D$ based on the optical model. It is interesting to note that the effect of increasing the imaginary central potential is to decrease the height of the giant resonances. Typical values of the parameters used by Feshbach[1] et al. for getting the giant resonances are:

$$V = V_0 + iW_0$$

$$V_0 = \begin{bmatrix} -U_0 & r \le R \\ 0 & r \ge R \end{bmatrix}$$

$$W_0 = \xi\, U_0$$

$$\xi = 0.03;\ R = 1.45 \times 10^{-13}\, A^{1/3}$$

$$U_0 = 42 \text{ MeV} \qquad ...(15.36)$$

For a detailed and quantitative agreement, a Woods-Saxon potential is used, *i.e.*,

$$V_0 = -\frac{U_0}{1+\exp\left(\dfrac{r-R}{d}\right)} \qquad ...(15.37a)$$

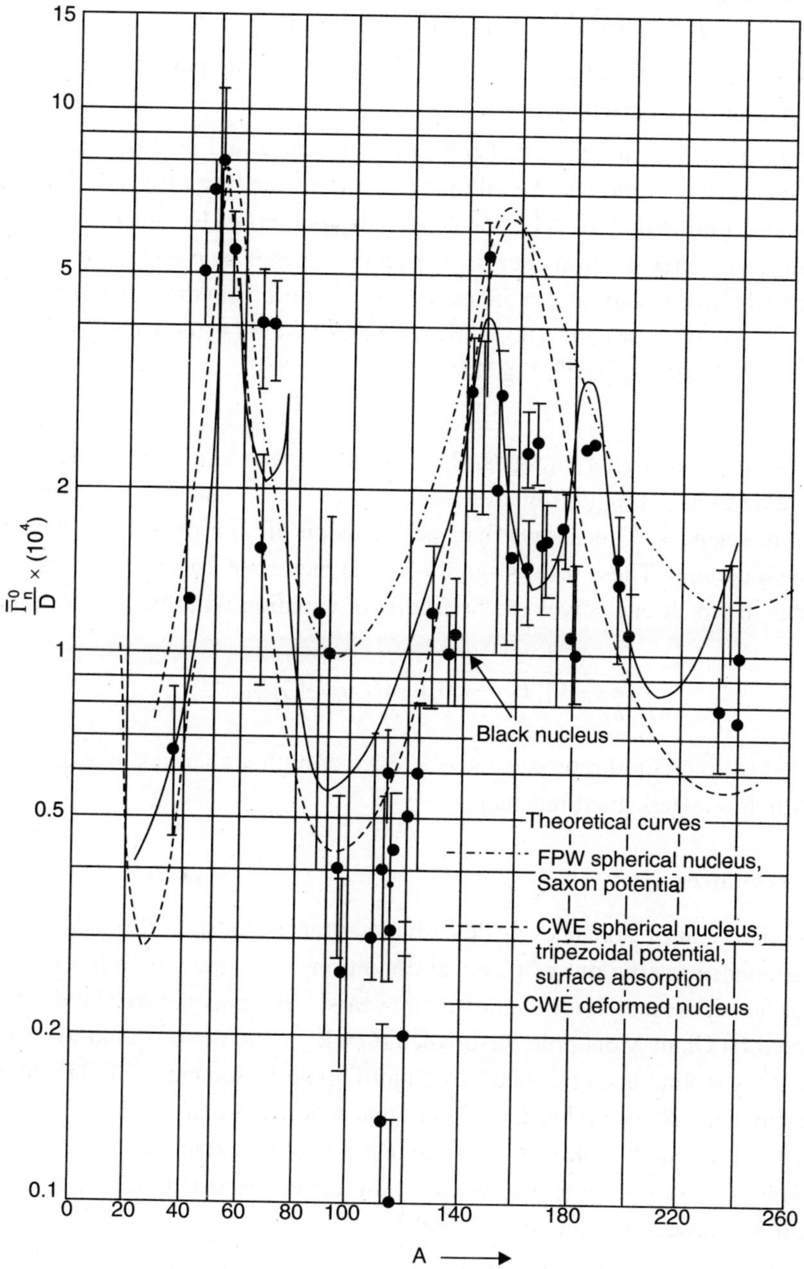

Fig. 15.7 Experimental S Wave (l = 0) neutron reduced strength-function values for different values of A are compared to the theoretical expected values or the basis of optical potential (solid line) (Ref. 13).

where R and d are parameters. Typical values of these constants as given by Campbell, Feshbach, Porter and Weisskopf[14] are: $U_0 = 52$ MeV, $W_0 = 3.1$ MeV and

$$W = \frac{-W_0}{1 + \exp\left(\frac{r - R}{d}\right)} \qquad \text{...(15.37}b\text{)}$$

$$R = (1.15\, A^{1/3} + 0.4) \times 10^{-13} \text{ cms and } d = 0.52 \text{ cms.}$$

Such giant resonance curves have also been experimentally supported by the experimental points by Hughes, Zimmerman and Chrien[15].

Such strength function curves have been drawn recently and analysed for spherical and non-spherical nuclei[16] by Feshbach, Porter, Weisskopf[1] and are denoted by FPW for spherical potential and by Chase, Wilets and Edmonds[16] (CWE) for spheroidal potential. In general, the strength function reaches a maximum, where the neutron energy (essentially zero) is near an $l = 0$ of the optical potential. These maxima actually occur at about $A = 55$ and $A = 155$. Apart from these major resonances, there are minor resonances at $A = 100$, 140 and 180, corresponding to shell effects for S-wave neutrons, which are reproduced by CWE calculations[16], Fig. 15.7.

(*ii*) $l \neq 0$

One can also plot for higher energies, the strength function corresponding to p-wave neutrons ($l = 1$). For p wave neutrons, strength function peaks are at $A = 91$ and $A = 109$, corresponding to $3p_{3/2}$ and $3p_{1/2}$ neutron resonances. This shows that the inclusion of spin-orbit force is clearly essential to produce such a double hump. These results agree with the picture of Lane, Thomas and Wigner[17] model of giant resonances in which one solves the Schrödinger equation with the complex potential, *i.e.*,

$$\frac{\hbar^2}{2M}\frac{d^2\phi}{dr^2} + \left[\left(E_\mu - i\,W\right) - (U - i\,W)\right]\phi_\mu = 0 \qquad ...(15.38)$$

so that η calculated from it could reproduce $< \eta >$, even at higher energies, *e.g.* corresponding to $l = 2$ (D-waves), for which one gets giant resonances.

Shape Giant Resonances

Apart from the optical giant resonances as described earlier, one also comes across the excitation and decay of giant resonance arising out of physical conditions of shape which has been a topic of great interest in nuclear physics research for the last many years[18]. In general, three kinds of giant resonances (GRS) are observed (*i*) Giant Monopole Resonance (GMR), which correspond to collective breathing motion of the nucleus around a single pole, (*ii*) Giant Dipole Resonance (GDR) which corresponds to two poles oscillating against each other (*iii*) Giant Quadrupole Resonance (GQR), which correspond to a deformed nucleus with a quadrupole shape, so that in a collective excitation we get four poles oscillating in the nucleus. The microscopic structure based on the collective small-amplitude nuclear response is explained by random phase approximation model[19].

The isoscalar giant monopole resonance is of particular interest because its energy is directly related to the compressibility of nuclear matter[20]. Experimentally GMR has been observed in only a few nuclei with $A < 90$ (Si^{28}, Ca^{40}, Ni^{58}) by studying the inelastic scattering as a function of incident energy up to an excitation of many MeV's. As for example, GMR was detected in Si^{28} by inelastic scattering of 240 MeV α-particles[21] at small angles including 0° with energy of excitation from 12 MeV to 35 MeV. It had not only the GMR (E_0) but also GQR (E_2) with $Ex = 21.5 \pm 0.3$ MeV and $Ex = 19.0 \pm 2$ MeV for E_2. Strength wise, E_0 strength corresponds to $54 \pm 6\%$ and E_2 corresponds to $32 + 5\ \%$. Experimentally, one measures the spectrum of inelastically scattered alphas as well as the angular distribution. Similarly GMR was discovered[22] in Ca^{40} using 240 MeV α-particles, at $Ex = 17.5 \pm 0.4$ MeV and at 129 and 240 MeV α-paritcles[23] for Ni^{58}. For $A > 190$, the giant monopole resonance has been observed[24] in Pb^{208} using 129 MeV alpha particles. Theoretically these giant resonances are understood in term of Woods-Saxon potential and the deformed potential and folding models. Apart from alpha particles, heavy ion projectiles also have been used[25] for observing giant resonances, *e.g.* in Ni^{60} by O^{17} scattering.

An interesting case of giant resonance exists involving (p, n) reaction at intermediate energies in which there is a role of giant-resonance transition operator describing both the β-decay in nuclei and the reaction mechanism of charge exchange reaction at small momentum transfer. Such giant-resonance are called Gamow-Teller (GT) giant resonances. One such case was studied[26] in Nb^{90} by observing the reaction Zr^{90} (Li^6, He^6) Nb^{60} at 156 MeV bombarding energy. The giant resonance was observed between 8 MeV and 12 MeV excitation energy. Apart from measuring the energy spectrum of (Li^6, He^6), it was also possible to observe the proton decay of Nb^{60} in which Zr^{90} (Li^6, He^6) Nb^{60} was observed in coincidence of decay protons. The decay characteristics showed a dominant statistical damping. On the other hand, the proton decay[27] in giant quadrupole resonance in Ca^{40} has been observed in Ca^{40} + Ca^{40} reaction at 50 MeV/ N, at 14 MeV and 17.5 MeV, excitation energy and corresponds to both statistical mode as well as pre-equilibrium mode.

15.5 THE OPTICAL MODEL PARAMETERS

We have discussed in the previous sections, the experiments concerning total cross-sections, angular distributions and the strength functions and their relationship with the various optical parameters as used in the optical model calculations as given in Eq. 15.1. A perspective about these parameters has finally emerged, which may be summarised.

For protons and neutrons Figs. 15.3 and 15.4 summarise the data on V_v, W_v, V_{so} and W_s of Eqs. 15.13*a* and 15.1. Also are given the values of *R*, *b* and *d* of Eq. 15.13*a* as written below this equation. These parameters are based on elastic scattering of protons and neutrons. However, (p, n) data[5] as given in Fig. 15.1 has yielded for E_p= 5.5, 5.0, 4.5 and 4.0 MeV, the values of $V_v \approx 45$ MeV and W_v = 7.0–8.7 MeV. It is interesting to see from Fig. 15.3, that V_v decreases from around 55 MeV for $E_n \approx 3$ MeV to about 10 MeV for $E_p \approx 500$ MeV, while W_v increases from about 7 MeV to 25 MeV in this range of energies. Increase in W_v is physically understood as an increase in nuclear reaction as incident energy increases, while decrease in V_v corresponds to decrease in the share of elastic scattering compared to the nuclear reaction.

Another analysis about the trend of the values of V_v as a function of incident energy has been carried out by Perey[9], who have plotted the real-well depth after correcting for Coulomb effect and symmetry energy. The slope is about 0.55 *E*, which is higher than 0.3 *E* used to correct for Coulomb potential effect only. Also it was found by Perey[9] that real well depth increased as a function of *Z* at a given incident energy, plotted as a function of $Z/A^{1/3}$. It is found that the slope at each energy is unity, which means that the increase is too large for the variation in well depth due to Coulomb effect. It is found that if one assumes for the real well-depth energy variation of 0.3*E*, then the correction for increased *Z* is 0.3 times the average Coulomb potential inside the well.

The radial dependence of $V(r)$ and $f(r)$ in Eqs. 15.1 and 15.14*c*, have the Woods-Saxon shape as given in Eq. 15.13*a*. This basically represents the shape of the nuclear-matter in the nucleus as we saw in the electron scattering in Chapter 2.

The radial dependence of the imaginary part of the potential $W(r)$ can be calculated using the Thomas-Fermi approximation, which assumes that a Fermi energy can be defined as a function of the nuclear density. Because of the exclusion principle, the imaginary part of the potential is not proportional to nuclear density but decreases more slowly as given in Eq. 15.13*a*.

However, various experiments on scattering say (*d*, *d*) scattering or neutron scattering have shown, that the imaginary part of the potential is expressible differently for volume absorption and surface absorption. As for example; Halbert, Bassel and Satchler[28] have carried out an analysis of the *d-d* scattering data at 11 MeV[29] and 11.8 MeV[30] and have concluded, that the data on Ni, Zr, Ag and Sn can be fitted to the optical model if one assumes that the potential—real plus imaginary—can be written as:

(*i*) For volume absorption:

$$V\left[1+\exp\left(\frac{r-r_0\,A^{1/3}}{a}\right)\right]^{-1}+iW\left[\frac{\left(1+\exp\left(r-r_W\,A^{1/3}\right)\right)}{a_W}\right]^{-1} \quad ...(15.39a)$$

and

(*ii*) For surface absorption:

$$V\left[1+\exp\left(\frac{r-r_0\,A^{1/3}}{a}\right)\right]^{-1}-i\,a_W\,W_0\,\frac{d}{d\,r}\left[1+\frac{\exp\left(r-r_W\,A^{1/3}\right)}{a_W}\right]^{-1} \quad ...(15.39b)$$

For the Coulomb potential part uniform charge distribution is assumed as given in Eq. 15.13*b*.

Similarly for neutron scattering say by Zaffiratos, Oliphant, Levin and Cranberg[31] analysis has yielded a potential of the following type:

$$V\left[1+\exp\left(\frac{r-r_i\,A^{1/3}}{a_0}\right)\right]^{-1}+iW\frac{4\exp\left[\frac{(r-r_i\,A^{1/3})}{a_i}\right]}{\left[1+\exp\left\{\frac{r-r_i\,A^{1/3}}{a_i}\right\}\right]^2}$$

$$+\frac{U\,\mathbf{l.S}}{r_s}\left(\frac{\hbar}{2\,M_p c}\right)^{1/2}\frac{d}{d\,r}\left[1+\exp\left(\frac{r-r_s\,A^{1/3}}{a_s}\right)\right]^{-1} \quad ...(15.40)$$

with the following values of parameters for neutron scattering: V = 45.93 MeV, r_o = 1.23 F, W = 5.16 MeV, $r_i = 1.34 \times 10^{-13}$ cms, $a_i = 0.48 \times 10^{-13}$ cms, U = 995.1 MeV, $r_s = 1.23 \times 10^{-13}$ cms and $a_s = 0.69 \times 10^{-13}$ cms.

Though, the radial variation of the spin-orbit term is introduced in analogy with the Thomas term in atomic physics and is chosen proportional to the derivative of the real potential, its origin lies in the meson interaction and hence the spin-orbit term is much stronger and is of opposite sense compared to the atomic case.

As shown in Tables 15.1 and 15.2 for deuteron and alpha-scattering, the same experimental data can be fitted using different sets of parameters. This is only empirical and shows that optical model has its limitations. Hodgson[32], however, has shown that the values of radius parameter and the diffuseness parameters a and b can be fixed for all nuclei at all energies. The value of radius parameter r_0 according to Hodgson is: $r_0 = 1.25 \times 10^{-13}$ cms for deuterons; $r_0 = 1.60 \times 10^{-13}$ cms for tritons and He^3 and $r_0 = 1.70 \times 10^{-13}$ cms for alphas.

The values of diffuseness parameter for all particles are $a = 0.65 \times 10^{-13}$ cms and $b = 0.98 \times 10^{-13}$ cms.

The potentials mentioned above are local potentials, *i.e.* their radial dependence is unique, so that for a given value of r, θ and ϕ, the potential has a single value. As the potential has a spherical symmetry in spherical nuclei, we have expressed the potential as $V(r)$. However, Perey and Buck[33] have shown that the study of elastic scattering of 7 MeV neutrons in the lead region can determine non-local optical potential parameters. What is a non-local potential? It is a potential in which the energy of the particle at point **r** depends not only on **r** but also on the wave-function at $r' \neq r$. The Schrödinger equation in non-local potential takes the form:

$$\frac{-\hbar^2}{2M}\nabla^2 \psi(r) + \int d r' \, V(r, r')\, \psi(r') = E\, \psi(r) \qquad \text{...(15.41)}$$

Evidently, the non-local potential takes into account, that the particle at a point **r** is associated with the properties of the wave-function, which has a spread over the whole nucleus.

The non-local potential, used by Perey and Buck[33] for explaining the scattering of neutrons by nuclei is given by:

$$V\psi(r) = \int V(r, r')\, \psi(r')\, d r'$$

$$V(r, r') = V(r', r)$$

$$V(r, r') = U\left(\frac{1}{2}|r + r'|\right) H(|r - r'|)$$

and

$$H(|r - r'|) = \frac{\exp\left[-\left(\frac{r - r'}{\beta}\right)^2\right]}{\pi^{3/2}\,\beta^3} \qquad \text{...(15.42)}$$

where U is the complex optical potential and β is the range of non-locality. With a single set of parameters and non-locality range β of 0.85×10^{-13} cms, it was possible to explain neutron elastic scattering up to 24 MeV. They used the following set of parameters

$V = 45.31$ MeV, $r_0 = 1.25 \times 10^{-13}$ cm, $a_0 = 0.65 \times 10^{-13}$ cm

$W = 6.57$ MeV, $r_i = 1.25 \times 10^{-13}$ cm, $a_i = 0.47 \times 10^{-13}$ cm

$U = 12.02$ MeV, $r_s = 1.25 \times 10^{-13}$ cm and

$a_s = 0.65 \times 50^{-13}$ cm

[*see* Eqs. 15.40 and 15.43 for the meaning of various quantities above]

Afterwards, these calculations[34] were extended to 35 angular distributions[35] in various elements at different proton energies.

Various basic approaches have also been attempted to understand the relationship of optical potential with nucleon nuclear effective potential using the *R*-matrix formalism. In essence, this formalism developed initially by Lane, Thomas and Wigner[17] who assumed that single particle states is distributed over many true nuclear states and an *R*-matrix is built out of these states and the optical model can be built out of *R*-matrix. This theory is known as Wigner and Eisenbud theory[36], who formalised it in the useful form. This theory was preceded by a proposal by Kapur and Pierls[37], which laid the basis of these theories.

New systematisation of the parameters of the optical potential model both from new experimental data and theoretical analysis have been recently carried out. For some low energies say at 35 MeV and for low atomic weights, *i.e* $17 < A < 48$, a detailed systematics of optical potential parameters have been recently[38] developed for (*p*, *n*) reaction with $O^{17,18}$, Ne^{22}, $Mg^{25,26}$, Ae^{27}, Si^{30}, S^{32}, $Ar^{38,40}$ and $Ca^{42,44,48}$ targets.

These are *sd* and *fp* shell nuclei, so one can study the differential cross-section for isobaric analog $\Delta J^{\pi} = 0^{+}$ (Fermi Type) transition as well as from mixed components. Finally one obtains, the *A*-dependence of real and imaginary potentials by adjusting the best fit for parameter for each target.

Table 15.3 *Best-fit parameters of isovector potential for each nucleus*[38]

Reaction	E_{exc} *of IAS* (*MeV*)	V_1 (*MeV*)	W_1 (*MeV*)	r_1 (*fm*)	a_1 (*fm*)
^{17}O (*p*, *n*) ^{17}F	0.1	12.00 ± 1.25	6.00 ± 0.80	1.750 ± 0.050	0.450 ± 0.050
^{18}O (*p*, *n*) ^{18}F	1.041	11.26 ± 1.08	5.76 ± 0.60	1.560 ± 0.039	0.451 ± 0.053
^{22}Ne (*p*, *n*) ^{22}Na	0.657	12.19 ± 0.73	5.59 ± 0.63	1.600 ± 0.054	0.450 ± 0.055
^{25}Mg (*p*, *n*) ^{25}Al	0.0	12.06 ± 1.56	6.24 ± 0.87	1.584 ± 0.065	0.500 ± 0.065
^{26}Mg (*p*, *n*) ^{26}Al	0.228	12.66 ± 0.58	6.27 ± 0.40	1.543 ± 0.034	0.533 ± 0.036
^{27}Al (*p*, *n*) ^{27}Si	0.0	11.00 ± 1.50	5.89 ± 0.98	1.403 ± 0.061	0.500 ± 0.070
^{30}Si (*p*, *n*) ^{30}P	0.677	11.84 ± 1.22	6.50 ± 0.32	1.493 ± 0.032	0.528 ± 0.034
^{34}S (*p*, *n*) ^{34}Cl	0.0	12.90 ± 0.78	6.20 ± 0.59	1.425 ± 0.033	0.580 ± 0.047
^{38}Ar (*p*, *n*) ^{38}K	0.130	13.27 ± 0.87	6.50 ± 0.48	1.404 ± 0.023	0.668 ± 0.032
^{40}Ar (*p*, *n*) ^{40}K	4.384	13.97 ± 0.55	5.90 ± 0.30	1.373 ± 0.024	0.722 ± 0.029
^{42}Ca (*p*, *n*) ^{42}Sc	0.0	13.20 ± 0.50	6.69 ± 0.50	1.414 ± 0.029	0.704 ± 0.029
^{44}Ca (*p*, *n*) ^{44}Sc	2.783	15.55 ± 0.97	7.69 ± 1.02	1.414 ± 0.050	0.699 ± 0.050
^{48}Ca (*p*, *n*) ^{48}Sc	6.677	13.86 ± 0.93	6.38 ± 0.80	1.400 ± 0.029	0.720 ± 0.028

For this purpose, Lane optical potential[17] was used, for the quasi-scattering, *i.e.*,

$$U(r) = -U_0(r) + \frac{4}{A} U_1(r)\,\mathbf{t}\cdot\mathbf{T} + U_{s.o}(r) + \left(\frac{1}{2} - t_z\right) V_c(r)$$

where $$U_1(r) = -V_1 f(x_R) - 4ia_I W_I \frac{d}{dx} f(x_I);\; f(x) = (1 + e^x)^{-1}$$

parameters $$x = \frac{(r - R_i)}{a_i},\; R_i = r_i A^{1/3}\; (i = 0 \text{ or } I)$$

These parameters are given in Table 15.3 (Ref. 38).

Similarly a detailed study[39] of Sm^{144}-α optical potential was undertaken at astrophysically relevant energy of $E_{c.m} = 9.5$ MeV which correspond to $E_{lab} = 20$ MeV of alphas. The importance of this reaction comes from the theory of Woosley and Howard[40] about nucleosynthesis process in type-II supernovae, requiring the production of samarium isotope Sm^{144}. The cross-section for Sm^{144} (α, γ) Gd^{148} from which one obtains the cross-section of the reverse reaction Gd^{148} (γ, α) Sm^{144} is required in nucleon synthesis theory. Hence the detailed and precise measurements.

A detailed analysis of the existing three hundred data-set has been most recently[41] carried out for energies from 1 to 200 MeV and for targets from $A = 40$ up to $A = 209$ in terms of nucleon-nucleus Optical Model Potential (OMP) built from the nuclear matter approach of Jeukenne, Lejune and Mahaux (JLM). The global, OMP built in this manner produces[41] a good overall description of the neutron and proton scattering and reaction measurements available up to 200 MeV. This analysis has yielded a detailed dependence of volume integral for complex central term, the volume integral for complex spin orbit term and the root mean square radii for complex central term from P + Ca^{40} up to P + Pb^{208} reactions.

The optical potential approach has, therefore, become a very important semi-empirical tool for nuclear reactions over a wide range of energies and target nuclei.

15. Optical Model
2000–2008

In about 9 papers, in 2000, nuclear reaction or nuclear structure experiments have been reported where an interaction potential is determined. In an interesting paper; global analysis of proton nucleus interaction cross-section has been reported for $9 \le A \le 238$; and $6 \le E \le 800$ MeV [Phy. Rev. C. 62, 064612 (2000)]. Among a set of four papers; involving optical model an unusual near threshold behaviour is witnessed for weakly bound nucleus. Be^9, in elastic scattering for Bi^{209} using 40–48 MeV Be^9. The values of V_0 has been obtained as 49.4 MeV; for $E_{lab} = 40.0$ MeV and 115.0 MeV for $E_{lab} = 480$ MeV [Phy. Rev. C. 61, 061603 (2000)].

About a dozen papers were published in 2001 on studies of reaction for determining the interaction potential. In one such paper $C^{12} + C^{12}$ reaction at 32.0 MeV to 126.7 MeV was studied [Phy. Rev. C. 63, 0584607 (2001)]. In another experiment, Mo^{92} (α, α) Mo^{92} scattering was carried out at $E_{cm} = 13.0$ MeV, 16.0 MeV, and 19.0 MeV by 12 authors. The real and imaginary part of the optical potential was derived. [Phy. Rev. C. 64, 065805 (2001)].

In one interesting theoretical paper, 7 authors, have derived nucleus-volume integrals of the real potentials from proton-scattering studied from 5 MeV to 100 MeV incident energy. The derived isospin potential is compared with those obtained for proton and neutron scattering in the previous investigation [Phy. Rev. C. 66, 06405 (2002)].

In an interesting calculational paper, authors have calculated volume integrals of potentials, derived from elastic scattering studies of deuterons, tritons He^3 and α-particles; at E/Ap from 5 MeV – 1000 MeV. Both real and imaginary parts of these integrals have been calculated using potential parameters in literature [Phy. Rev. C. 68, 014613 (2003)].

In a paper in 2004; the giant resonance region from 10 MeV $\leq Ex \leq$ 55 MeV, has been studied for E_0, E_1,E_2 and E_3 transitions for Sn^{116}, Sm^{144}, Sm^{154} and Pb^{208}, using inelastic scattering of 240 MeV α-particle at Texas A and M, College station (USA) by a group with youngblood et al. They have found nearly all of iso-scaler E_0, E_1, E_2 and E_3 giant resonance strength was located in Sn^{116}, Sm^{144}, Sm^{154} and Rb^{208} [(Phy. Rev. C. 69, 034315) C. (2004)].

In a theoretical paper; the authors have presented a global theoretical optical model for nucleons, with incident energies up to 200 MeV containing dispersive terms and a total energy approximation. The optical model is able to reproduce scattering date as well as bound single-particle state for neutron and protons. However, for the scattering of protons, the situation is not as satisfactory as for neutron [Phy. Rev. C. 76, 044601 (2007)].

In a paper involving the study of fusion excitation of 120 reactions, the fusion (capture) cross-sections have been well described by a modified Woods-Saxon potential, for a unified description of the entrance channel fusion barrier and the fission energy, based on Skyrme energy density function approach. Incorporating a statistical model (HI, VAP), the cross-sections of 51 fusion-fission reactions have been systematically investigated [Phy. Rev. C. 77, 014603 (2008)].

REFERENCES

1. H. Feshbach, C. Porter and V.F. Weisskopf: Phy. Rev. 96, 448 (1954).
2. F. Bjorklund and S. Fernbach: Phy. Rev. 109, 1295 (1985): H.S. Hans and S.C. Snowdon, Phy. Rev. 108, 1028 (1957).
3. R.D. Woods and O.S. Saxon: Phy. Rev. 95, 577 (1954).
4. J.M. Blatt and V.F. Weisskopf: Theoretical Nuclear Physics, John Wiley & Sons, New York (1952).
5. R.D. Albert: Phy. Rev. 115, 925 (1959).
6. J.H. Coon, R.W. Davis, H.E. Felthauser and D.B. Nicodemus: Phy. Rev. 111, 250 (1958).
7. P.J. Wyatt, J.G. Wills and A.E.S. Green: Phy. Rev. 119, 1031 (1959).
8. F.A. Jenkins and H.E. White: Fundamentals, Optics, p. 320–375, McGraw-Hill Book Company, Inc., New York (1950).
9. F.G. Perey: Proceedings of Conference on Direct Interaction Nuclear Reaction Mechanism, p. 125, Gorden and Breach Science Publishers, Inc., New York (1962).
10. M. El-Nadi and F. Riad: Nuclear Physics 65, 99 (1965).
11. El-Nadi and A. Rabie: Nuclear Physics 65, 90 (1965).
12. H.H. Barschall: Phy. Rev. 86, 431 (1952); M. Walt and H.H Barschall: Phy. Rev. 93, 1062 (1954)

13. J.A. Harvey: Proceedings of the International Conference on Structure, Kingston, p. 670, Toronto, University of Toronto Press (1960).
14. E.J. Campbell, H. Feshbach, C.H. Porter and V.F. Weisskopf: Laboratory for Nuclear Science Tech. Report No. 73, p. 132, M.I.T. (1960).
15. D.J. Hughes, R.L. Zimmerman and R.E. Chrien: Phy. Rev. Letters, 1, 465 (1958).
16. D.M. Chase, L. Wilets and A.R. Edmonds: Phy. Rev. 110, 1080 (1958).
17. A.M. Lane, R.G. Thomas and E. Wigner: Phy. Rev. 98, 698 (1955).
18. A. Yander Woude: Progress Particle Nuclear Physics 18, 217 (1987); G.F. Bertsch, P.F. Bortingron and R.A. Broglia: Rev. Mod. Physics 55, 287 (1983); L.S. Cardman: Nuclear Physics A 354, 173 (1981).
19. G.A. Rinker and J. Speth: Nuclear Physics A 306, 360 (1978).
20. J.P. Blaizot: Phy. Rep. 64, 171 (1980).
21. D.H. Youngblood, H.C. Clark and Y.W. Lui: Phy. Rev. C. 57, p. 1134 (1998).
22. D.H. Youngblood, Y.W. Lui and H.C. Clark: Phy. Rev. C. 55, p. 2811 (1997).
23. G.R. Satchler and D.T. Khoa: Phy. Rev. C. 55, p. 285 (1997).
24. D.H. Youngblood, P. Bogucki, J.D. Brodson, U. Garg, Y.W. Lui and C.H. Roza: Phy. Rev. C. 23, 1997 (1981); D.H. Youngblood: Phy. Rev. C. 55, 950 (1997).
25. D.J. Horen, J.R. Beene and G.R. Satchlar: Phy. Rev. C. 52, 1554 (1995).
26. M. Moosburger, E. Aschenauer, H. Dennert, W. Eyrich, H. Lehmann, N. Sohalz, H.Worth, H.J. Gils, H. Rebel and S. Zagromski: Phy. Rev. C. 57, p. 602 (1998).
27. C.A.P. Ceneviva, N. Teneja, H. Dias and M.S. Hussein: Phy. Rev. C. 55, p. 1246 (1997).
28. E.C. Halbert, R.H. Bassel and G.R Satchler: Process Conference Direct Interaction Nuclear Reaction Mechanics, p. 167, Gorden and Breach Science Publishers, Inc., New York (1962).
29. M.J. Takeda: Phy. Soc. Japan, 15, p. 557 (1960).
30. G. Igo, W. Lorenz and V. Schmidt Rohr: Phy. Rev. 124, 832 (1961).
31. C.D. Zaffiratos, T.A. Oliphant, J.S. Levin and L. Cranberg: Phy. Rev. Letters. 14, 913 (1965).
32. P.E. Hodgsen: Proc. Direct Interactions Nuclear Reaction Mechanism, p. 103, Gorden and Breach Science Publishers Inc., New York (1962).
33. F.G. Perey and B. Buck: Nuclear Physics 32, 353 (1962).
34. F.G. Perey: Proc. Conference, Direct Interaction Nuclear Reaction Mechanism 125, Gorden and Breach Science Publishers Inc., New York (1962).
35. G.W. Green-less, L.G. Kuo and M. Petraine: Proc. Roy. Soc. A. 243, 206 (1957); Dayton I.E., and G. Schvank: Phy. Rev. 101, 1356 (1956); C.B. Fulmer, Phy. Rev. 125, 631 (1962).
36. E.P. Wigner and L. Eisnebud: Phy. Rev. 72, 29 (1947).
37. P.L. Kapur and R.E. Pierls: Proc. Roy. Soc., (London), A 166, 277 (1938).
38. C.C. Jon, H., Orihara, T. Niizeki, M. Oura, K. Ishii, A. Terakawa, M. Hosaka, K. Itoh, C.C. Yum, Y. Fujii, T. Nakagawa, K. Miura and H. Ohnuma: Phy. Rev. C. 56, p. 900 (1997).
39. P. Mohr, T. Rauscher, H. Oberhimmer, Z. Mate, Zs. Fulop, E. Som Orjai, M. Jalger and G. Staudi: Phy. Rev. C. 55, p. 1523 (1997).
40. S.E. Woosley and W.H. Howard: Astrophy. J. Suppli. 36, 285 (1978).
41. E. Bauge, J.P. Dalaroche and M. Girod: Phy. Rev. C. 58, p. 118 (1998).

PROBLEMS

1. Consider the scattering of neutrons by a complex potential of the form

$$V(r) = V_0 (1 + i\xi) \quad r \leq R$$
$$= 0 \qquad r > R$$

Compute the differential cross-sections for 14 MeV neutrons, for

$$V_0 = 50 \text{ MeV}, \xi = 0.05, 0.2, R = \text{[illegible]}$$

2. In optical model, the target nucleus is referred to as cloudy cry[illegible]ll with complex refractive index $N = n + iT$. Relate n and T with V and W of optical model potential.

3. Solve Eq. 13.42 for $l = 0$ and $l = 1$ with the scattering potential (*i*) $V(r)$ and $V(r) + i\,W(r)$ and obtain the $\sigma_{sc}^{\,l}$ and σ^l_{ce} from Eqs. 13.36 and 13.37 for $l = 1$ and Eqs. 13.50 and 13.52 for $l = 0$.

4. Obtain from Eq. 15.26, the shape of giant resonance in cross-section σ_T and from Eq. 15.30, the shape of $\overline{\Gamma}_\alpha / D$ for different values of M_0, KR and N_0.

5. Keeping in mind, the effect of Coulomb forces, calculate in (p, n) reaction on a Cu target, the compound elastic scattering σ_{ce} for incident protons of 3 MeV, 4 MeV and 6 MeV and 15 MeV from Eqs. 15.7 and 15.8 the definition of η_l from Chapter 13.

6. Calculate the scattering amplitude $< \eta_0 >$, from Eq. 15.22 using Woods-Saxon potential for low energies and show its dependence on the nuclear radius.

7. Explain, that direct reaction mechanism corresponds to interaction for V and W in Eq. 15.1, while compound nucleus formation corresponds to a very large value of W.

8. Neutron can also be scattered in the Coulomb field of a nucleus, because of its magnetic moment. What is the interaction Hamiltonian ? Calculate the spin averaged differential cross-section in Born approximation.

9. Find out, the scattering cross-section for a given V_v and W_v but for different values of V_{so} and W_{so} in Eq. 15.1, taking the values of a and b from constants of Eq. 15.13.

10. Find an expression for inelastic scattering cross-section assuming a uniform-sphere model, which corresponds to a constant refractive index within the nuclear sphere and unity outside.

16

CHAPTER

Pre-Equilibrium Model

16.1 GENERAL

We have already discussed in Chapters 13 and 14, the cases of compound nucleus and direct reaction models of nuclear reactions. We discuss in this chapter, the pre-equilibrium model of nuclear reaction, which is intermediate between the other two models in the mode of interaction.

It is evident, that an intermediate case is likely to exist for many actual cases, where the incident projectile interacts with a limited number of nucleons in the nucleus and the emission of the particles takes place after the energy is shared only by a few nucleons (say 2–5). Such a case of pre-equilibrium reaction mechanism was historically first treated theoretically by Serber[1] (1947) for high energy interactions and later on by Goldberger[2] (1948) and Metropolis[2] (1958). Experimental comparisons with this model has been conducted by Bertini[3] et al. in 1974. However, great fillip about the understanding of pre-equilibrium phenomena was given by the work of J.J. Griffin[4] in 1966, when the exciton pre-equilibrium statistical process was introduced by him through Exciton model to explain semi-quantitatively the proton and neutron spectra from targets bombarded by protons and α-particles. This model was then modified by Cline and Blann[5] in 1968–72 to include the emission of complex particles like deuteron, triton and alphas, etc. Simultaneously or even earlier, Harp-Miller-Berne (HMB) model[6] was proposed (1968), which is somewhat similar to the Exciton model, but with different method of grouping of the excited states. The common feature of these models is the use of semi-classical scenario in the evolution of the composite nuclear states in the first few encounters.

As a matter of fact, the first model developed by Serber[1] called—the Cascade model—uses the Monte Carlo technique to explain the scattering from 10 MeV to 300 MeV incident energy for both the cross-sections and angular distributions of the emitted particles. The model assumes a degenerate Fermi gas of nucleons confined in a nuclear potential. As the incident particle enters the nucleus, a three dimensional cascade is created. The kinematics of each collision is calculated relativistically. The angles are selected from appropriate distributions. Similarly mean free paths calculated from the known cross-section above a certain cut-off energy gave the distance traveled between the two encounters. In this manner, three dimensional nuclear cascades were worked out using Monte-Carlo technique, finally yielding the shapes of spectral and angular distributions. Exciton model on the other hand, develops an evolutionary scenario of excited particles and holes and gives the energy distribution after a certain number of such excitations. Similarly Harp-Miller-Berne[6] (HMB) model divides the phase-space of particles below Fermi energy into groups or bins and calculates the occupation probability of an average configuration in the *i*th bin as a function of time, which is short-compared to N-N collision time. Another model—called hybrid model[7] combines the above two models and in a modified form called Geometry Dependent-Hybrid (GDH) model, takes into account the surface effects.

These models—especially the excitons model—have been extended by Kalbach[9] and others to include quantum-mechanical theory of Multiple Statistical Direct (MSD) and Multiple Statistical Compound (MSC) contributions as initially enunciated by Feshbach, Kerman and Koonin[8] (FKK) in 1980. Essentially the difference with the exciton model of Cline and Blann[5] lies in counting of levels and the quantum mechanical treatment. We discuss the exciton model in details, while describing the other models only briefly. The M.S.D. and M.S.C. models based on quantum mechanical treatment and their applications are also given next in somewhat details. A good example is provided by Fe^{54} (p, p') $^*Fe^{54}$ reaction[10] at 38.7 MeV energy of the projectile, when one looks at the energy spectrum of outgoing protons as shown in Fig. 16.1 taken from literature[10]. The spectrum has apparently three components. At the lowest energy end, is a bell shaped spectrum as predicted by Eq. 13.78 based on the compound nucleus statistical model. As was discussed in Chapter 13, the compound nucleus is formed at excitation energies where the nuclear levels are very close to each other and $\Gamma >> D$, so that one may consider the levels to be in continuum, hence the shape of the spectrum is continuous with no sharp structure. On the other hand, at the highest energy end of the spectrum of outgoing particle there are clear indications of discrete levels, which is possible under conditions of direct reaction. At the intermediate energies, the spectrum has a structured shape, though it does not indicate the discrete levels. This portion of the spectrum corresponds to the pre-compound nucleus model of nuclear reaction.

In the case of direct reactions, the time of the decay of the composite nucleus (projectile + Target) is generally of the order of the time taken by the projectile to traverse the nucleus once. This is approximately given by $\approx 2\, R/v$, where R is the radius of the target nucleus and v is the velocity of the projectiles. This can be easily seen to be of the order of $\beta^{-1} A^{1/3} \times 10^{-23}$ secs, where $\beta = v/c$. As stated earlier in the compound nucleus formation, the projectile may interact with the nucleons inside the nucleus many times before some particle is emitted. Therefore, if L is the mean free path of the projectile in the nucleus, the NL/v gives the time to form the compound nucleus formation where N is of the order of 10^3 to 10^6. The mean life time of the compound nucleus with $L \approx 10^{-13}$ cms comes out to be $10^{-16} - 10^{-19}$ secs. In pre-equilibrium $N = 2 - 5$ and the time in which such a reaction takes place $\approx 10^{-22}$ secs.

An interesting feature of the reaction mechanism is, the difference between angular distributions expected in different types of reactions. We have already seen in the case of direct reactions, that the angular distribution of the outgoing particles are l-dependent. In the case of the compound nucleus on the other hand, they are expected to be either isotropic or symmetric. In the case of pre-compound nucleus, because of certain amount of sharing of the momentum of the projectile, by the emitted particles, a certain amount of forward-peaking is expected to occur. This is borne out both by experiments and theoretical models.

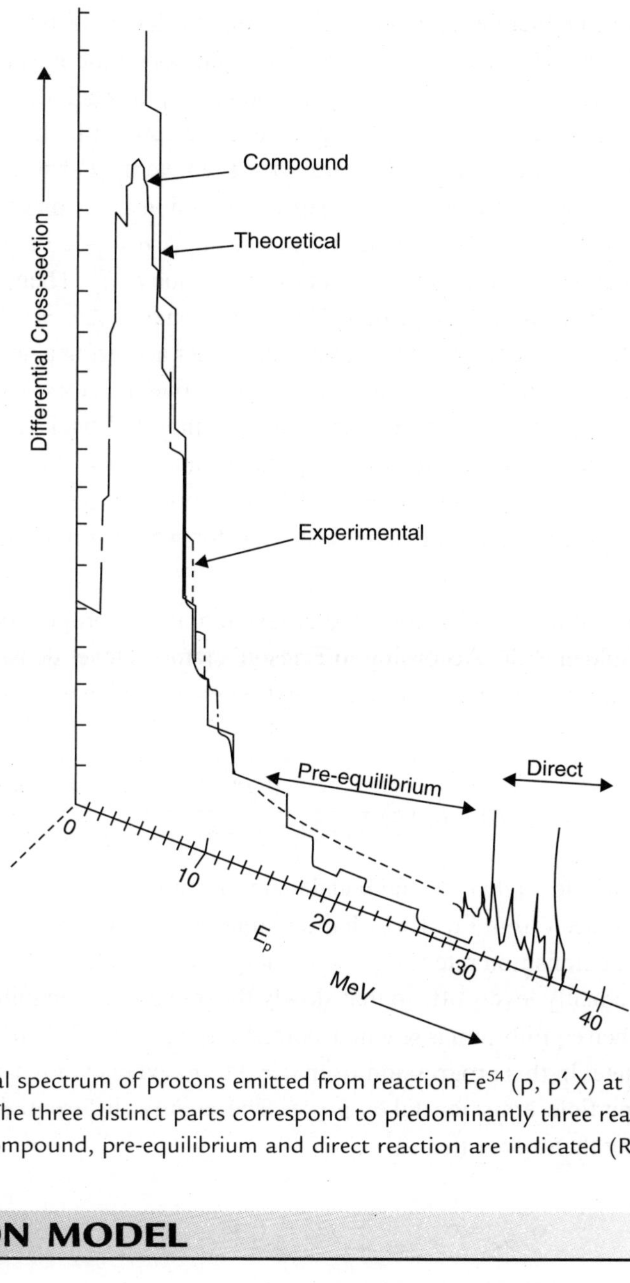

Fig. 16.1 A typical spectrum of protons emitted from reaction Fe^{54} (p, p′ X) at 38.7 MeV incident protons at 60°. The three distinct parts correspond to predominantly three reaction mechanisms, i.e. compound, pre-equilibrium and direct reaction are indicated (Ref. 10)

16.2 EXCITON MODEL

Figure 16.2, expresses[5] the central idea of the exciton model. We have already discussed some of its features in the last section. We assume that the target nucleus is represented by a potential in which there are equally spaced single particle levels, *i.e.*, levels whose occupancy is 0 or 1. Initially all the levels below the Fermi energy E_f are filled. This is the ground state of the target nucleus.

The projectile of energy E above Fermi energy E_f enters the nucleus as shown in Fig. 16.2. In the language of excitons, we start with $n = 1p + 0h = 1$, when the particle just enters where p represents particles above the Fermi energy and h represents the hole below the Fermi energy. It is still in the entrance channel. Here it can either be scattered back, leading to shape-elastic scattering or can proceed into the nucleus and interacts with one nucleon lifting it from below the Fermi energy filled levels to one of the unfilled levels above the Fermi energy. If the energy of excitation of the second particle above separation energy B is U, then the projectile will be left with the energy $\varepsilon = E - B - U = E - (B + U)$ and the struck particle will have energy $B + U$ above the Fermi energy, This corresponds to $n = 2p + 1h = 3$ exciton state. Now either one particle emission takes place and particle remaining behind may fall into the hole. This corresponds to $n = 1$, *i.e.* $\Delta n = -2$ or one of the two particles interacts with a particle below the Fermi energy and excites it above E_f. Then we have 3 particles and 2 holes and hence $n = 5 = 3p + 2h$ state is formed. This corresponds to $\Delta n = +2$. It is also possible that the two nucleons above the Fermi energy interact with each other acquiring a new configuration without change of exciton number n. Then $\Delta n = 0$. Hence in this mode, one has $\Delta n = \pm 2, 0$. Similarly for $n = 5$, either it goes to $n = 7$, *i.e.* $\Delta n = +2$ or one of the particles may fall back in the hole and $n = 3$, *i.e.* $\Delta n = -2$ or there is only scattering between three particles and $\Delta n = 0$ and so on, so forth till the equilibrium is reached for $\Delta n = +2$ and $\Delta n = -2$. Then exciton number becomes constant at $n = \bar{n}$. This is the beginning of the compound nucleus formation for which n may be many times the value of $\bar{n}$.

The transition rates of these excitations and de-excitations are proportional to the level density, as required by Fermi's Golden Rule. According to Ericson's famous level density formula[5, 11] the level density $\rho_n(E)$ at a given exciton number n at the excitation energy for p excited particles and h hole is given by:

$$\rho_{p,h}(E) = \rho_n(E) = \frac{g^n E^{n-1}}{p!h!(p+h-1)!} = \frac{g^{p+h} E^{p+h-1}}{p!h!(p+h-1)!} \qquad \text{...(16.1)}$$

where g is the single particle level density and p and h are the numbers of excited particles and holes as described earlier. It can be seen from Eq. 16.1, that for small n, the level density is a rapidly increasing function of n and hence the transition rate for $\Delta n = 2$ is larger, than for $\Delta n = -2$. But as n increases, the particles level density gradually levels off, so that slowly the transition probability for $n = +2$ becomes the same as for $n = -2$. Then equilibrium is reached, corresponding to $n = \bar{n}$. This is the state approaching the compound nuclear state. In this progression from $n = 3$ to $\bar{n}$, whenever a particle in a given exciton state is in continuum, particle emission takes place. The energy differential cross-section for pre-equilibrium emission may then be written as:

$$\sigma_{\text{PEQ}} = \sigma_{abs} \sum_{\substack{n=n_o \\ \Delta n = \pm 2, 0}}^{\bar{n}} D_n P_n(\varepsilon) \qquad \text{...(16.2)}$$

where σ_{abs} is the absorption cross-section of the projectile by the target and is obtained experimentally, D_n is the probability of reaching the n exciton state, without prior emission and is called the depletion factor and $P_n(\varepsilon)$ is the emission probability from n exciton state of the projectile with energy ε. The summation in Eq. 16.2 is from $n = n_0$, which is taken to be 3, if the projectile is a nucleon and is taken to be the number of nucleons in the projectile plus 2 (1 excited particle + 1 hole), if the projectile is a cluster of nucleons say an alpha particle, etc. Equation 16.2 is the basic equation of the cross-section for emitted particles in exciton model of pre-compound emission.

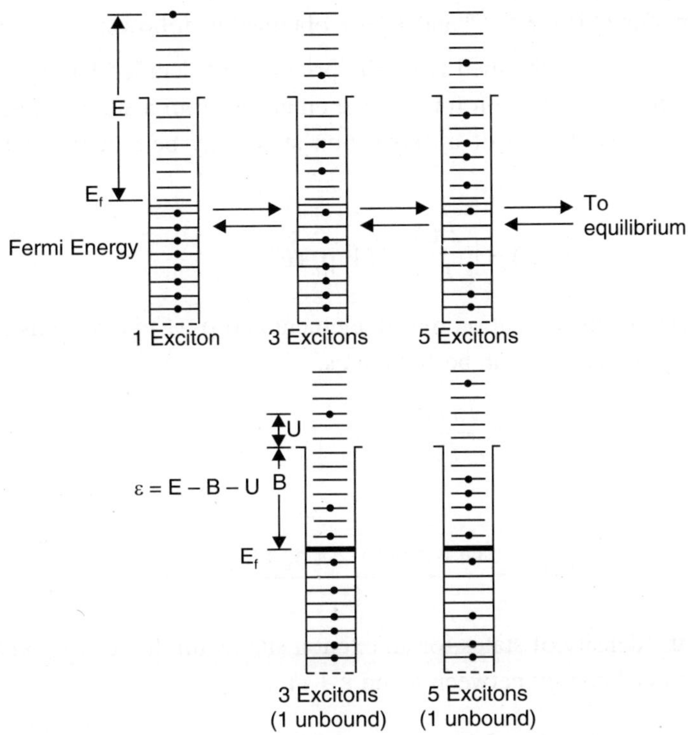

Fig. 16.2 The diagram shows the behaviour of particles and holes in the excitation model. The horizontal lines indicate equally spaced single particle states in the potential well. The particles are shown as solid circles. The holes are represented by states below Fermi energy without any particle within (Ref. 5).

It may be easily seen, that the depletion factor D_n is related to P_n (ε) as:

$$D_n = \prod_{\substack{n' = n_o \\ \Delta n' = \pm 2}}^{n} \left[1 - \int d\,\varepsilon\, P_{n'}(\varepsilon)\right] \quad ...(16.3)$$

Physically the term in the bracket in Eq. 16.3, subtracts the emission probability at each exciton number $n' < n$ from the total probability of reaching the exciton level, thus giving the net probability of reaching n exciton state after taking into account the depletion of the state due to prior emission.

The emission probability $P_n(\varepsilon)$ is defined as the ratio of the rate of emission from n exciton state in a particular channel to the sum of rates of transitions to all states including emission to all channels. If $\lambda_c^{\,n}(\varepsilon)$ is the emission rate from exciton state n to a certain channel c with the energy ε and λ_+^n, λ_-^n and $\lambda_0^{\,n}$ be the emission rates for $n = +2, -2$ and 0 transitions respectively, then one can define $P_n(\varepsilon)$ from the above discussion as:

$$P_n(\varepsilon) = \frac{\lambda_c^n(\varepsilon)}{\lambda_+^n + \lambda_-^n + \lambda_0^n + \int d\,\varepsilon\, \lambda_c^n(\varepsilon)} \quad ...(16.4)$$

The expressions for $\lambda_c^n(\varepsilon)$, λ_+^n, λ_-^n and λ_0^n are obtained as follows:

(*i*) The value of $\lambda_c^n(\varepsilon)$ is obtained from the principle of detailed balance by Blann M. and his coworkers. We start with the expression for particle emission from a state characterised by $n = p + h$. The decay probability per unit time of an n-exciton state may be calculated from the well-known perturbation theory as:

$$\omega_n(E) = \left(\frac{2\pi}{\hbar}\right) |M|^2 \rho_n(E) \qquad \text{...(16.5}a\text{)}$$

where $|M|^2$ is the square of the transition matrix element and $\rho_n(E)$ is the density of n-exciton states. Realising that the decay probability can be written as:

$$\omega = \frac{v\sigma(v)}{\Omega} \qquad \text{...(16.5}b\text{)}$$

We write $|M|^2$

$$|M|^2 = \frac{(\hbar\, v\, \sigma(v) / 2\, \pi\, \rho_n(E))}{\Omega} \qquad \text{...(16.5}c\text{)}$$

We now write the density of states for an exciton state with the restriction that one exciton is in the continuum with channel energy between ε and $\varepsilon + d\varepsilon$ as:

$$\rho_{n-1}(U) = \frac{4\,\pi\, p^2\, d\, p\, d\,\Omega}{(2\,\pi\,\hbar)^3\, d\,\varepsilon} \qquad \text{...(16.5}d\text{)}$$

so that if Eq. 16.5*d* is substituted for $\rho_n(E)$ in Eq. 16.5*a* and also one uses $|M|^2$ of Eq. 16.5*c* in this equation, one then obtains:

$$\omega_n(\varepsilon)\, d\varepsilon = \frac{m\,\varepsilon\,\sigma_{\text{inv}}(\varepsilon)}{\pi^2\,\hbar^3}\,\frac{\rho_{n-1}(U)}{\rho_n(E)}\, d\varepsilon \qquad \text{...(16.5}e\text{)}$$

where ε is the channel energy, m is the reduced mass of the emitted particle and $\sigma_{\text{in v}}(\varepsilon)$ is the inverse cross-section. $\rho_{n-1}(U)$ is the level density of the residual nucleus, $\rho_n(E)$ is the level density of the composite state (not of an equilibrium ensemble, but the one formed by incident particle of energy E and the target nucleus) of particles and holes. In Eq. 16.5*e*, the relationship of $\sigma_{\text{in v}}(\varepsilon)$ and $\sigma(v)$ have been taken into account[7], and $v\,\sigma(\varepsilon)$ has been equated to total number of states in volume $p^2\, dp$ of phase space multiplied by $\sigma_{\text{in v}}(\varepsilon)$, the cross-section for inverse channel.

We generalise Eq. 16.5*a* by considering the case of the emission of particles with ν nucleons, so that $\rho_{n-1}(U)$ is replaced by $\rho_{n-\nu}(U) = \rho'_n(U)$. Also Eq. 16.5*a* must be multiplied by the statistical, spin degeneracy of the emitted particle, *i.e.* by $(2S + 1)$. Then, realising that $\lambda_c^n(\varepsilon)$ has the same meaning as $\omega_n(\varepsilon)\, d\varepsilon$, *i.e.* the emission rate to a channel between ε and $\varepsilon + d\varepsilon$ for the particle denoted by c, we can express for a general situation from Eq. 16.5*e*:

$$\lambda_c^n(\varepsilon) = \frac{\rho_{n'}(U)}{\rho_n(E)} \times \frac{(2S+1)\, m\varepsilon\sigma_{\text{in}}(\varepsilon)}{\pi^2\hbar^3} \qquad \text{...(16.5}f\text{)}$$

where n' is the excitation number after emission of ejectile with ν nucleons, so that $n' = n - \nu$ and U is the residual excitation energy given by $U = E - B - \varepsilon$ as defined earlier. It is, thus, possible to calculate, the cross-section for pre-equilibrium from Eq. 16.2 using the subsequent equations.

(*ii*) Using Fermi's rule of the transition rate, we write the expression for λ_+^n, λ_-^n and λ_0^n as:

$$\lambda_+^n = \frac{2\pi}{\hbar}|M_+|^2 \rho_{n+2}$$

$$\lambda_-^n = \frac{2\pi}{\hbar}|M_-|^2 \rho_{n-2}$$

$$\lambda_0^n = \frac{2\pi}{\hbar}|M_0|^2 \rho_0 \qquad \text{...(16.5g)}$$

where M_0 and $M_\pm$ are the matrix elements of the respective transitions and ρ_{n+2}, ρ_{n-2} and ρ_0 are the level densities of states in the $n + 2$, $n - 2$ and n exciton states after $\Delta n = +2, -2$ and 0 transitions respectively.

In principle, one can calculate the matrix elements empirically by a global fit of the calculations with experimental data. William Jr[12] has calculated these transitions using $M_0 = M_- = M_+ = M$ and

$$\rho_{n+2} = \frac{g^3 E_c^2}{2(2+1)}; \quad \rho_{n-2} = \frac{gph(n-2)}{2}$$

and

$$\rho_0 = \frac{g^2 E_c (3n-2)}{4} \qquad \text{...(16.5h)}$$

when $\lambda_+{}^n = \lambda_-{}^n$, $n = \bar{n}$, then from Eq. 16.5*h*, $\bar{n} = \sqrt{2g E_c}$.

It was also found by the authors, that $|M|^2 = (K/E_c A^3)$ MeV^2, $K = 190$ (± 32) MeV^3 and E_c is in MeV. A review of the values of $|M|^2$ is given by Kalbach and coworkers[9]. Alternatively, as one only requires $\lambda_t{}^n = \lambda_+{}^n + \lambda_-{}^n + \lambda_0{}^n$ in Eq. 16.4, one can write $\lambda_t{}^n$ as: $\lambda_t{}^n = v/L\ \sqrt{2E/m}$, where L is mean free path given by $L = 1/\rho_\sigma$, where ρ = nuclear matter density and σ = nucleon-nucleon cross-section.

The Master Equation

While Eq. 16.2, yields an energy spectrum of the emitted particles on the basis of pre-equilibrium emission, it is also possible to develop a time-evolution equation, so that one obtains a time dependent spectrum as n increases. One obtains $P(n, t)$ of the exciton state n at time t from time dependent Master equation as developed by Williams F.C. Jr[12] and Cline and Blann[5] (1971), *i.e.*,

$$\frac{d}{dt} P(n, t) = \lambda_+{}^{n-2} P(n-2, t) + \lambda_-{}^{n+2} P(n+2, t) - \left[\lambda_+^n + \lambda_-^n + \int d\varepsilon\, \lambda_c^n(\varepsilon)\right] P(n, t) \quad \text{...(16.6)}$$

This Master equation resembles the equation for radioactive decay. The first two terms give the growth states of n exciton states by creation ($\lambda_+{}^{n-2}$) and annihilation ($\lambda_-{}^{n+2}$) from ($n - 2$) exciton and ($n + 2$) exciton states, respectively. The terms in the square bracket give the decay rate of the n exciton state by $\Delta n = \pm 2$ transitions and the particles emission. The various terms in the equations are, of course, obtained from the relations of exciton model as discussed earlier. Of course, $P(n-2, t)$, $P(n+2, t)$ and $P(n, t)$ represent emission rates from n state to $n - 2$, $n + 2$ and n state, respectively.

The Master equation is solved numerically for $P(n, t)$ for different values of times. Figure 16.3 shows the shapes of energy spectra at 24 MeV and 96 MeV[5] calculated from Eq. 16.6. Numerical techniques are used to solve the set of coupled differential equations as expressed in Eq. 16.6. The energy spectrum of the emitted particles is described by:

$$I_\nu(\varepsilon, t)\, d\varepsilon = \sum_{\substack{n \\ \Delta n = \pm 2}} P(n, t)\, \omega_\nu(n, \varepsilon)\, d\varepsilon \qquad ...(16.7)$$

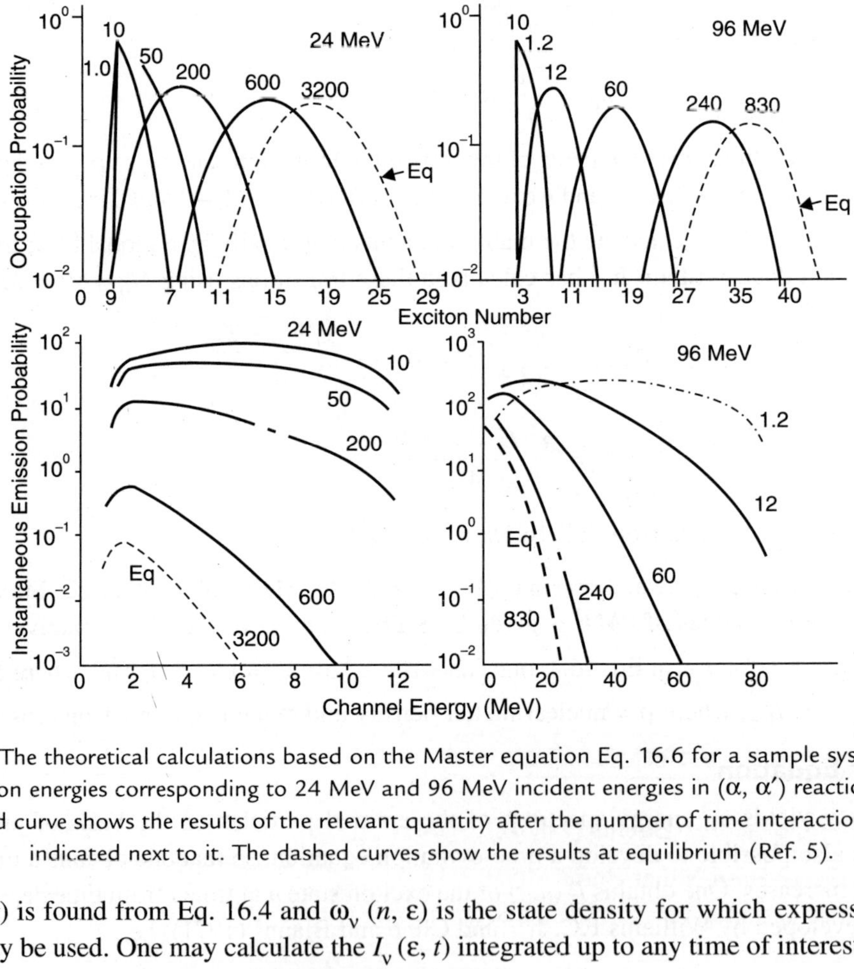

Fig. 16.3 The theoretical calculations based on the Master equation Eq. 16.6 for a sample system at two excitation energies corresponding to 24 MeV and 96 MeV incident energies in (α, α′) reaction. Each solid curve shows the results of the relevant quantity after the number of time interactions as indicated next to it. The dashed curves show the results at equilibrium (Ref. 5).

where $P(n, t)$ is found from Eq. 16.4 and $\omega_\nu(n, \varepsilon)$ is the state density for which expression similar to Eq. 16.5*e* may be used. One may calculate the $I_\nu(\varepsilon, t)$ integrated up to any time of interest say T. Hence denoting such spectra as $S_\nu(\varepsilon, t)$, one defines them as:

$$S_\nu(\varepsilon, t) = \int_0^T I_\nu(\varepsilon, t)\, d\varepsilon\, dt \qquad ...(16.8)$$

The mean life time τ_n of the n exciton state is, then obtained by:

$$\tau_n = \int_0^\infty dt\, P(n, t) = \int_0^{t_{eq}} dt\, P(n, t) \qquad ...(16.9)$$

where t_{eq} is the time taken for the transitions to reach equilibrium.

In the above figures, we have plotted the instantaneous emission probabilities $P(n, t)$ and occupation probability based on Eq. 16.6. The number denoted over the various curves corresponds to the time-iterations carried out in the calculations, *i.e.* on a certain time scale, a certain time increment Δt is used for the calculation of each spectrum, *i.e.* $I_v(\varepsilon, t)\,\Delta t$ and the number indicated is the number of times which this re-iteration was used. It seems, that a large number of iterations are required to achieve the equilibrium position. It is also to be seen, that the higher energy component corresponds to very small number of iterations and hence belongs to pre-equilibrium part of the mechanism.

In Fig. 16.4, a comparison is shown between the experimental and theoretical spectra based on this theory for P_t^{nat} (α, p) at Au at $E = 40$ MeV. The comparison is quite favourable[5].

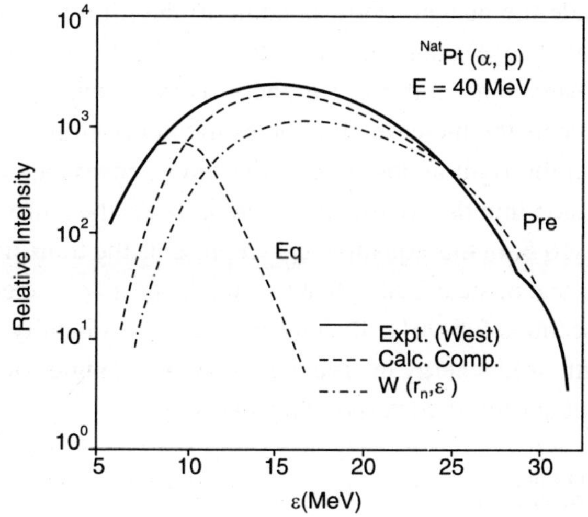

Fig. 16.4 Experimental proton spectrum for $P_t^{nat} + \alpha$ reaction system at 40 MeV of incident energy compared with the calculated pre-equilibrium and equilibrium spectra (Ref. 5).

16.3 OTHER SEMI-CLASSICAL MODELS

16.3.1 The Harp-Miller-Berne (HMB) Model[6]

As stated earlier, the Harp-Miller-Berne Model, which was proposed before the exciton model by Griffin, divided the phase-space of particles below Fermi energy into groups or bins, whose size ΔE is chosen to be of some convenient dimension. This is in contrast to the exciton model, where all energy partition between particles and holes in a given exciton state occur with equal probability. In H.M.B. model, one calculates the occupation probability of an average state in the *i*th bin as a function of time. Figure 16.5 illustrates this scenario. At time $\tau = \tau_0$ all the particles and levels are below the Fermi-energy and the incident particle is in excited state giving certain group occupation probability. Two-body interactions then, lead to re-distribution of probabilities. This goes on till an equilibrium is reached. At each time during the equilibration process, the energy spectrum of emitted nucleons are calculated and final spectrum is obtained.

The following Master equation for a single particle gas of nucleons illustrates this scenario[5, 6] on HMB model as:

$$\frac{d}{dt}(n_i\, g_i) = \sum_{j,k,l} \omega_{kl,ij}\, g_k\, n_k\, g_l\, n_l\,(1-n_i)(1-n_j)\, g_i\, g_j$$

$$-\sum_{j,k,l} \omega_{ij,kl}\, g_i\, n_i\, g_j\, n_j\,(1-n_k)(1-n_l)\, g_k\, g_l - n_i\, g_i\, \lambda_c\,(i') \qquad \text{...(16.10)}$$

where i' is the energy outside the nucleus corresponding to the ith bin in the nucleus. Physically this equation gives the net rate of change of number of nucleons in the ith bin with g_i as the number of single particle states/MeV in one MeV bin-width for ΔE and n_i as the number of particles in each state. The first term on the right represents the increase of particles in bin i from the inter-nucleon *N-N* scattering process, the second term on the right as loss due to the same process and the third term as a loss of particles from i due to emission into the continuum where $\lambda_c\,(i')$ is the emission rate into the continuum of a particle as given by Eq. 16.5. In this equation $\omega_{kl,\,ij}$ represents the transition rate for a nucleon in one of the state i to collide with one in state j, such that two nucleons go to the energy conserving state k, l. Free *N-N* scattering cross-sections for 90° collisions are used. The quantity $ni\ gi$ or $gj\ nj$, etc. gives the number of nucleons in the ith or jth energy interval with g_i or g_j as single particle state/MeV, $\omega_{ij,\,kl}$ again represents the transition probability of corresponding states.

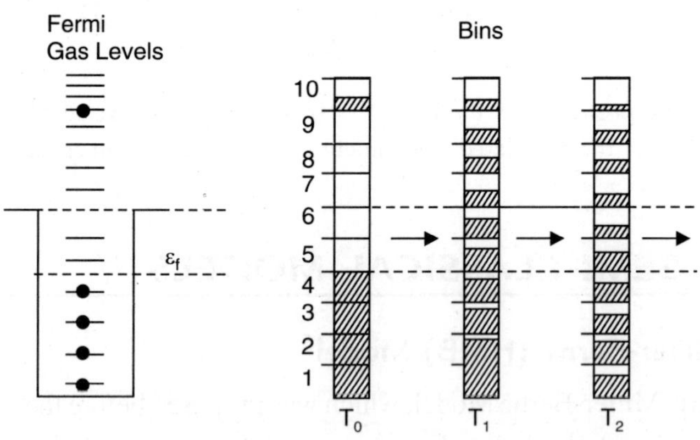

Fig. 16.5 Illustration of equilibrium process as formulated by the Master equation of HMB model [Eq. 16.10]. The shaded areas represent the occupied fraction of each bin with occupation changing after each time interval.

16.3.2 The Hybrid Model

This model[7] combines HMB model and exciton model in such a manner that, while the basic scenario of the evolution of the reaction process inside the nucleus follows the exciton procedure, the counting of levels is done, by using HMB model.

So we write in this model, the emission cross-section of a nucleon of type (neutron or proton) as:

$$\sigma_{PEQ}(\varepsilon) = \sigma_{abs} \sum_{\substack{n=n_0 \\ \Delta n = \pm 2, 0}} D_n \, P_n^x\,(\varepsilon) \qquad \text{...(16.11)}$$

Equation 16.11 is very similar to Eq. 16.2 of exciton model, but $P_n^x\,(\varepsilon)$ has different method of evaluation. It is of the form:

$$P_n^x\,(\varepsilon)\, d\,\varepsilon = f_n^x \left[\frac{g\rho_{n-1}\,(U)\, d\,\varepsilon}{\rho_n\,(E_c)}\right] \frac{\lambda_x^n\,(\varepsilon)}{\lambda_t^n + \lambda_x^n\,(\varepsilon)} \qquad \text{...(16.12)}$$

where f_n^x is the number of x type of exciton particles in the n exciton state and $\lambda_x^n\,(\varepsilon)$ is the emission rate of x from nth exciton state. It may be noted that the calculation of $\lambda_x^n\,(\varepsilon)$ will be done in a different manner than given in Eq. 16.5. The ratio $g\,\rho_{n-1}\,(U)\, d\varepsilon/\rho_n\,(E_c)$ gives the probability of a nucleon having an energy ε in the nth exciton unbound state. How ? The hybrid model describes the n-exciton state as made of two systems, a real system corresponding to the residual nucleus with exciton number $n - 1$, but with excitation energy U, $[U = E - (B + \varepsilon)]$ characterised by the level density $\rho_{n-1}\,(U)$ and a virtual system of nucleons, (which will later be emitted) in an unbound state characterised by the single particle level density, g. Each unbound particle will have $g\, d\varepsilon$ levels available to it between energies ε and $\varepsilon + d\varepsilon$. For each of these levels, the level density is $\rho_{n-1}\,(U)$. Hence the total level density available to the outgoing particles is $g\,\rho_{n-1}\,(U)\, d\,\varepsilon$. Hence, the probability that the n exciton state with level density $\rho_n\,(E_c)$ can be partitioned into two systems, one for the particle going out and the other, the rest of it, is given by:

$$\left[\frac{g\,\rho_{n-1}\,(U)\, d\,\varepsilon}{\rho_{n(E_c)}}\right] \qquad \text{...(16.13)}$$

The factor $\lambda_x^n\,(\varepsilon)$ is again calculated using the principle of detailed[7] balance.

Thus, the emission rate is written as similar to Eq. 16.5*a* as:

$$\lambda_x^n\,(\varepsilon) = \frac{\sigma_{inv}\, v\, p_c}{g_v\, V} \qquad \text{...(16.14)}$$

where σ_{inv} is the inverse cross-section, v is the velocity of the particle having a density of states in the continuum, g_v is the single-particle density in the nucleus and V is the arbitrary volume cancelled by the same value in ρ_c. The factor λ_t^n is calculated similarly as before [Eq. 16.5*g*] and corresponds to the transition rate of the particle of interest by nucleon-nucleon scattering below the Fermi energy to give a state with an additional excited particle plus hole. Now g_v is made geometry-dependent and is expressed semi-empirically as:

$$g_v = g_v\,(R) = \left[\frac{40}{V\,(r)}\right] \frac{A}{28} \qquad \text{...(16.15)}$$

where $V(R)$ is the nuclear potential and can be written as:

$$V\,(r) = 40 \left[\frac{\langle d\,(R)\rangle}{\bar{d}}\right]^{2/3} \text{MeV} \qquad \text{...(16.16)}$$

where 40 MeV is taken as the depth of the potential in the centre of the nucleus and $d(R)$ is the nuclear density at radius R and $\overline{\mathbf{d}}$ is nuclear density in nuclear matter. Thus we introduce a geometry-dependent factor.

Figure 16.6 illustrates a case of comparison of this model with experimental data.

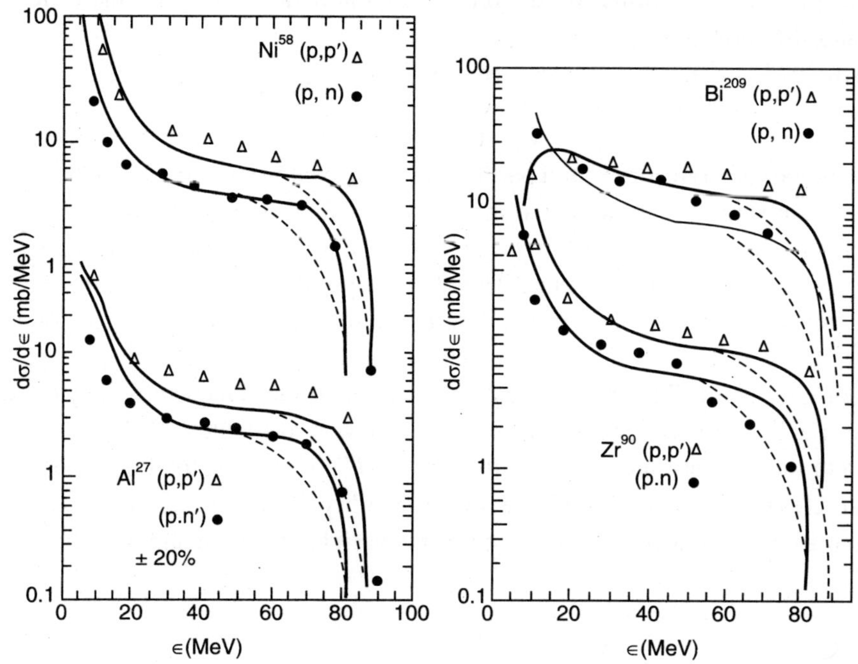

Fig. 16.6 Calculated and experimental shapes of emitted spectra from Al^{27} (p, p′), Al^{27} (p, n) $_{209}Ni^{58}$ (p, p′) and Ni^{58} (p, n) and Zr^{90} (p, p′) Zr^{90} (p, n) and Bi^{209} (p, p′) and Bi^{209} (p, n). The thin solid lines are GDH + evaporation. Dashed lines are hybrid + evaporation (Ref. 7).

16.3.3 Inter-Nucleon Cascade Model

The Cascade Model, originally developed for energies above 100 MeV by Serber[1] has been recently used for lower energies for explaining the pre-equilibrium spectra. The first interaction may send a target nucleon above the Fermi sea. Both of these particles travel further interacting with other nucleons–thus developing a cascade in three dimensions. The trajectory of an excited particle is followed until it reaches the nuclear surface or its energy goes below a certain level governed by Pauli Exclusion Principle. Every particle which reaches the surface above a certain energy—required to cross the surface—is assumed to belong to an emitted particle. Then one starts with another impact parameter and goes through the same calculations—till one covers all the values of impact parameters. This gives the cross-section of the emitted particles of different emitted energies and their angles (Ref. 6, 10, 1, 14).

Figure 16.7 obtained from the work of Porile et al. (Ref. 14) shows the comparison of the experimental and theoretical results for the cross-section of the reaction Ga^{69} (p, xn) at $E_p = 46.5$ MeV for various values of x ranging from 1 to 4. It is evident that for lower values of x, only the cascade evaporation calculations are compatible with the experiment showing that for initial stages, the Cascade model is quite effective.

Detailed Monte-Carlo Calculations on Inter-nuclear Cascade model have been carried out earlier by N. Metropolies et al.[14], who compared these calculations with experimental data on angular distributions and excitation form-function of (p, p') and $(p, 2p)$ and (p, n), etc.at proton energies from 10 MeV to 300 MeV or so and quite good fits were obtained.

Detailed comparisons of differential yields at 39 MeV and 62 MeV proton energies using the Cascade theory for many targets—C^{12}, Fe^{54} and Bi^{209} have been carried out by Bertini et al.[3] and it was found that the major fraction of pre-equilibrium reaction cross-section has the angular distributions consistent with the predictions of Cascade model, except at the backward angles, where there is a disagreement by a factor of ten.

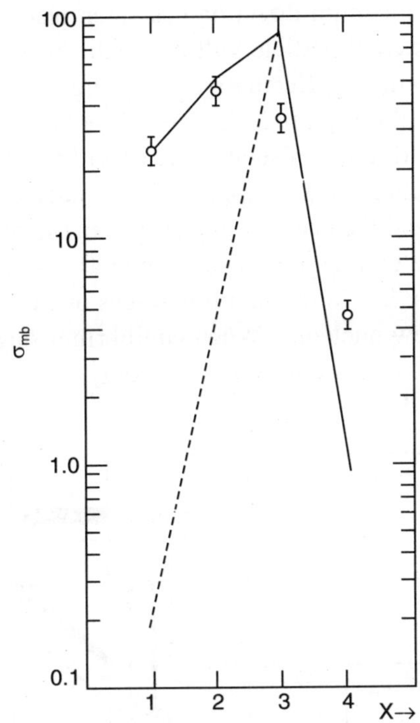

Fig. 16.7 Experimental and calculated Ga^{69} (p, xn) cross-sections for E_p = 46.5 MeV. The straight line shows cascade evaporation calculations, and the dashed line shows compound nucleus evaporation calculations (Ref. 14).

16.4 QUANTUM MECHANICAL SEMI-EMPIRICAL MODELS FOR PRE-COMPOUND EMISSION

16.4.1 Multi-Step Direct (M.S.D.) and Multi-Step Compound (M.S.C.) Model

This model was developed by Feshbach, Kerman and Koonin[8] (FKK) for pre-equilibrium reactions in 1980 and has been extended by Kalbach[9] (1981–88). Essential features of the scenario of excitons are somewhat similar to exciton model, but counting of levels and particles are calculated quantum mechanically. M.S.D. model is based on applying the concepts of excitons to initial stages of the development of the excitation process, through two-body interaction, when only a few degrees of freedom are excited in the cascade of two body interactions. At every excitation stage there is expected to be at least one particle in the continuum and hence there is a finite probability of particle emission at each stage. A theory, which describes these initial configurations and calculates the probability of emission on the basis of each excitation stage having at least one particle in the continuum stage is called the multi step direct (MSD) model. Evidently, because MSD applies to the first few interactions, after the projectile enters the nucleus, the particles emitted in this process will share the momentum direction of the projectile and hence one expects these particles to be peaked in the forward direction in their angular distribution.

After a certain number of two-body interactions which will be large enough so that the complete sharing of its initial energy and momentum-direction has taken place, the number of particles excited will be large, but the energy per excited particle will be small. At this stage, in general, one may not expect to be in continuum, but statistical fluctuations in energy distribution may send a particle in continuum. The emission direction of this particle, however is not expected to be related to the direction of the projectile and the angular distribution of such particles will be either isotropic or symmetric about 90. This scenario corresponds to the multi-step compound (MSC) nucleus formation, where an equilibrium in energy-sharing has taken place and the energy per particles in the bound states is low. The energy of the particle, which is excited to the continuum, once a while as a statistical fluctuation (MSC Model) will be governed by statistical laws. But in the first few stages of excitations (MSD Model) the energy is shared kinematically among a few nucleons. When equilibrium stage is just reached, then particles emitted as a result of statistical fluctuations have higher energy, than those emitted much later, when a compound nucleus is finally formed.

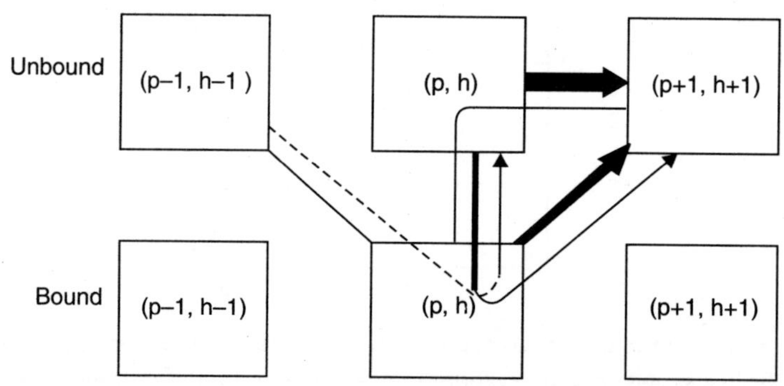

Fig. 16.8 Schematic diagram of two-body interactions-producing strength in unbound states specified by p + 1, h + 1 and E. The boxes represent different classes of states and the heaviness of the lines indicates roughly the amount of strength involved. All strength shown denote ending of (p, h) and (p – 1, h – 1) states in unbound states and are assumed to pass by creation to the final (p + 1, h + 1) state (Ref. 9).

In this manner, the FKK theory of pre-equilibrium visualises the energy sharing process to proceed through two non-interfering chains called *P*-chain and *Q*-chain. In *P*-chain, there is at least one particle at each stage which is in continuum. In the *Q*-chain, all the particles are in bound states and only statistical fluctuation sends a particle in the continuum. After the initial two-body interaction, the reaction may proceed along *P*-chain route or *Q*-chain route. This is shown in Fig. 16.8.

In developing the quantum mechanical FKK theory of pre-compound emission, one keeps in mind, the following considerations based on physical assumptions:

(*i*) *P*-chain and *Q*-chain develop independently as shown in Fig. 16.8. Each chain goes through various successive configurations without any interference from the neighboring chain.

(*ii*) Transitions from *Q*-chain to the *P*-chain take place only through statistical fluctuations.

(*iii*) At low projectile energies, the *Q*-chain interactions dominate giving MSC emissions with angular distributions that are symmetric around 90°. As the energy increases, *P*-chain interactions become increasingly important until finally they are responsible for all the cross-section giving forward peaked MSD emission.

(*iv*) Because of the above consideration, one can write:

$$\sigma_{\text{PEQ}} = \sigma_{\text{MSD}}(\varepsilon) + \sigma_{\text{MSC}}(\varepsilon) \qquad ...(16.17)$$

Equation 16.17 assumes no interference between the two processes. The values of $\sigma_{\text{MSD}}(\varepsilon)$ and $\sigma_{\text{MSC}}(\varepsilon)$ are now calculated quantum-mechanically on the basis of FKK Model. This model is a sort of quantum mechanical extension of exciton model. The exciton model-based quantum mechanical calculation, lead to $\sigma_{\text{PEQ}}(\varepsilon)$ from the following equation based on exciton model:

$$\sigma_{\text{PEQ}} = \sigma_{abs} \sum_{\substack{n=n_o \\ \Delta n=\pm 2,0}}^{\bar{n}} D_n \lambda_c^n(\varepsilon)\, T_U(p, h) \qquad ...(16.18a)$$

$$= \sigma_{abs} \sum D_n P_n(\varepsilon) \qquad ...(16.18b)$$

where

$$P_n(\varepsilon) = \lambda_c^n(\varepsilon)\, T_U(p, h) \qquad ...(16.19)$$

has been used as the expression of $P_n(\varepsilon)$, which was used in Eq. 16.4 of exciton model. In Eqs. 16.18 and 16.19, $\lambda_c^n(\varepsilon)$ and D_n have the same meaning as in the exciton model, though calculated differently as will be seen subsequently. $T_u(p, h)$ is given by:

$$T_u(p, h) = \frac{1}{\lambda_+^n + \lambda_-^n + \lambda_0^n + \int d\varepsilon\, \lambda_c^n(\varepsilon)} \qquad ...(16.20)$$

and hence physically it corresponds to the mean life of the *n*-exciton state with a configuration of *p* particles and *h* holes. The calculations of D_n now requires, that we take into account the two chains of interaction, *P*-chain and *Q*-chain. We therefore replace the symbol D_n by $S(p, h)$ and calculate D_n in such a way, that $S(p, h)$, represents the probability of reaching an *n*-state for different configurations of *p* and *h*, *e.g.* $n = 5 = 3p + 2h = 4p + 1h$. Because the values of *S* will be different for *P*-chain and *Q*-chain, we write:

$$D_n = S_U(p, h) + S_b(p, h) \qquad ...(16.21)$$

where $S_U(p, h)$ is the probability of having a (p, h) configuration with at least one particle in the continuum both from *P*-chain directly and through statistical fluctuation from *Q*-chain and $S_b(p, h)$ corresponds to the probability of having all particles in the bound state—which do not lead to emission. Similarly $\lambda_c^n(\varepsilon)$ is replaced by $\lambda_c^U(\varepsilon)$. $\lambda_c^n(\varepsilon)$ of Eq. 16.5 was calculated by considering $\rho_n(E_c)$ in the denominator as the level density of all bound and unbound states. We replace it by $\rho^\nu(p, h, E_c)$, so that this quantity represents the level density of all the unbound states only in the (p, h) configuration. Writing $\rho'_n(U) = \rho(p - \nu, h, U)$, where ν is the number of nucleons in the ejectile system, we write:

$$\lambda_c^U (p, h) = \frac{\rho (p - \nu; h, U)}{\rho^U (p, h, E_c)} \qquad ...(16.22)$$

then, Equation 16.18 is replaced by:

$$\sigma_{\text{PEQ}} = \sigma_{abs} \sum_{p = p_o}^{\bar{p}} S_U (p, h)\, T_U (p, h)\, \lambda_c^U (p, h, \varepsilon) \qquad ...(16.23)$$

where the summation is now over p (particles), with $\bar{\mathbf{p}}$, as the number of excited particles in the equilibrated compound state.

We now calculate σ_{MSD} and then σ_{MSC} can be obtained from Eqs. 16.23 and 16.17. The value of σ_{MSD} is calculated in a similar manner as in Eq. 16.23, except that now we replace $S_U (p, h)$ by $S_d (p, h)$ which is the probability of formation of (p, h) configuration with at least one unbound particle so that the state has evolved from configurations, which all had atleast one particle in the continuum. This ensures that the system has always been in the P-chain prior to emission and hence contributes to MSD. Then

$$\sigma_{\text{MSD}} = \sigma_{abs} \sum_{p = p_o}^{\bar{p}} S_d (p, h)\, T_U (p, h)\, \lambda_c^U (p, h, \varepsilon) \qquad ...(16.24)$$

The values of $S_U (p, h)$, $S_d (p, h)$, $T_U (p, h)$ and $\rho^U (p, h, E)$ have been calculated quantum-mechanically by Kalbach[19]. His results are given below. Initially, [*i.e.* for (p_o, h_o) state], it is assumed, that the unbound states are populated in proportion to their relative state densities and hence:

$$S_U (p_o, h_o) = S_d (p_o, h_o) = \frac{\rho^U (p_o, h_o, E)}{\rho (p_o, h_o, E)} \qquad ...(16.25)$$

where $\rho^U (p_o, h_o, E)$ represents density of unbound states and $\rho (p_o, h_o, E)$ is the total density of states at the excitation energy E. Hence, the expression for $S_b (p_o, h_o)$ for the bound state is given by:

$$S_b (p_o, h_o) = 1 - S_U (p_o, h_o) \qquad ...(16.26)$$

The evolution from (p_o, h_o) to $(\bar{p}, \bar{h})$ via intermediate steps is then obtained from various recursion relationships obtained in Ref. (6). It is assumed in this model, that the strength of the system is imagined to pass subsequently through configuration of increasing complexity, *e.g.*:

$$(p_o, h_o) \rightarrow (p_o + 1, h_o + 1) \rightarrow (p_o + 2, h_o + 2) \ldots \rightarrow (\bar{p} - h, \bar{h} - 1) \rightarrow (\bar{p}, \bar{h})$$

$\bar{p}$ and $\bar{h}$ represent the states at the equilibrium. In this process, we see from Eq. 16.25, that we have to get the expression for $\rho^U (p, h, E)$ in general to obtain $S_d (p, h, E)$. From Ref. (9, 18 and 19) we write this expression as:

$$\rho^U (p, h, E) = \frac{g_o^n (E - A_{p-1, h} - S)^{n-1}}{p!\,h!\,(n-1)!} \qquad ...(16.27)$$

where $$A_{p,h} = \frac{p_m^2}{g_o} - \frac{p^2 + h^2 + n}{4g_o}$$ ...(16.28)

and g_o is the density of equally spaced single particle states and S is the minimum excitation energy which a particle must have in order to be unbound. This equation, however, has to be corrected for the depth of the potential V for which *see* Ref. (13, 18 and 19). Further the density of bound states specified by p, h and E is given by:

$$\rho^{(b)}(p, h, E) = \rho(p, h, E) - \rho^{(U)}(p, h, E)$$ ...(16.29)

The movement from (p, h) or (p_o, h_o) to say $(p + 1, h + 1)$ and the relevant relations are given in Ref. (13), (19).

If the hypothesis of equal occupation probabilities for all states of a given particle-hole class is valid, then for each class of states, the ratio $S_U(p, h) / S(p, h)$ where $S(p, h) = S_U(p, h) + S_b(p, h)$ should be given by:

$$\frac{S_U(p,h)}{S(p,h)} = \frac{\rho^{(U)}(p,h)}{\rho(p,h)}$$ ...(16.30)

This is, however, not the actual case, which is proved by actual calculations obtained by the evolution from p_o and h_o to higher values of p and h. The expressions for λ's are derived from the expression in Eq. 16.5g where the matrix elements $|M|^2$ are calculated quantum-mechanically. The expression for $\rho(p, h, E)$ is obtained from the general expression for the density of states[5, 7] as given in Eq. 16.1. Then one obtains the expression for $\rho^U(p, h, E)$ by integration between proper units, so that atleast one particle is unbound or is in continuum. Then the density of bound state for (p, h, E) configuration is given by Eq. 16.29.

Figure 16.9 gives a comparison of the shape of the experimental spectrum from $_{79}Au^{197}$ (α, n) $_{81}Tl^{200}$ reaction[15] with the pre-equilibrium calculations discussed above.

16.4.2 Kalbach Model

Basically making use of FKK model and using the empirical data, Kalbach and Mann[9] have been able to write the double differential cross-section in terms of MSD and MSC processes as:

$$\frac{d^2\sigma}{d\varepsilon\, d\Omega}(E_\alpha, \varepsilon, \theta) = a_o\,(\text{MSD}) \sum_{l=0}^{l_{\max}} b_1\, p_1(\cos\theta) + a_o\,(\text{MSC}) \sum_{l=0}^{l_{\max}} b_1\, p_1(\cos\theta)$$ ...(16.31)

where $$a_o\,(\text{MSD}) = \frac{1}{4\pi}\left[\sigma_{\text{MSD}}(E_\alpha, \varepsilon) + \sigma_N^{\bullet}(E_\alpha, \varepsilon) + \sigma_{K.o}(E_\alpha, \varepsilon)\right]$$ ...(16.32a)

and $$a_o\,(\text{MSC}) = \frac{1}{4\pi}\left[\sigma_{\text{MSC}}(E_\alpha, \varepsilon) + \sigma_{\text{evap}}(E_\alpha, \varepsilon)\right]$$ ...(16.32b)

where E_α is the incident energy of the projectile α, ε is the energy of the emitted nucleon and coefficients b_1 are the functions of the projectile energy. $\sigma_N(E_\alpha, \varepsilon)$ is the nucleon transfer and $\sigma_{K0}(E_\alpha, \varepsilon)$ is the knock-out term, so that the pre-equilibrium process includes these processes also, which are not included in the MSD term of Eq. 16.24. Similarly in the expression for the compound part, σ_{MSC} as given in

Eq. 16.17 does not include the evaporation part, which is, therefore, separately included in Eq. 16.32*b*. The expressions for σ_N and σ_{kO} are given in the work of Kalbach and Mann[9] and Sarkar et al.[15] and for σ_{evap} the work of Weisskopf and Ewing[16] contains the derivation and the final expression.

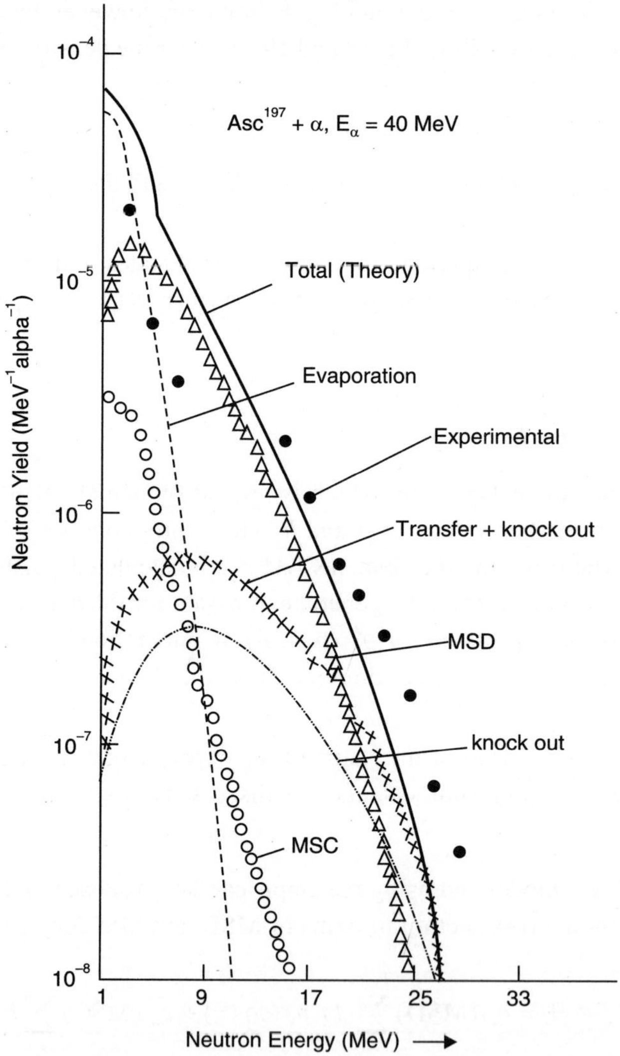

Fig. 16.9 Neutron yield versus the neutron energy of the emitted neutrons, for Au^{197} (α, n) reaction at E_α = 10 MeV. The experimental values are given as solid points (•), while the final theoretical results are shown as solid line. All other spectra are components of calculated spectra. Evaporation (---) MSD (Δ Δ Δ), MSC (o o o), knock out (-·-·-), transfer + knock out (x x x), (Ref. 15).

The coefficients b_1 are assumed to be of the form:

$$b_1 = \frac{(2l+1)}{1+\exp\left[A_1\,(B_1 - \varepsilon - B_e)\right]} \quad ...(16.33)$$

where A_1 and B_1 are free parameters and have been obtained by fitting the observed angular distributions by Kalbach and Mann (KM) and B_e is the ejectile binding energy and ε is the ejectile energy. A_1 and B_1 are assumed to have the form:

$$A_1 = k_1 + k_2\left[l\,(l+1)\right]^{m_1/2} \text{ and } B_1 = k_3 + k_4\left[l\,(l+1)\right]^{m_2/2} \quad \text{...(16.34)}$$

where m_1 and m_2 are assumed to be integer variables and k's are continuous variables. From the semi-empirical analysis, m_2 seems to have unambiguously the value of – 1, while for m_1 values of 1, 2 and 3 are possible. Typical values for (p, p') below 45 MeV are:

$$A_1 = 0.036 \text{ MeV}^{-1} + 0.0039 \text{ MeV}^{-1}\, l\,(l+1)$$

and

$$B_1 = 92 \text{ MeV} - 90\,[l\,(l+1)]^{-1/2} \quad \text{...(16.35)}$$

Figure 16.10 shows the experimentally determined values of A_1 and B_1.

The comparison of some of the experimentally measured angular distribution and the theoretical calculation, based on the above KM model have been made. In general, the fit is quite good showing that empirically arrived systematics for A_1, B_1 and hence b_1 are reliable and are independent of the type of projectile or ejectile. They only depend on 1 and, of course, on the ejectile energy. For higher energy of ejectile, the fit is less satisfactory. Also where deuterons, He^3 and tritium are involved, the degree of forward peaking is slightly underestimated. One of the parameters, which goes in these calculations is the fraction of MSD to the total pre-equilibrium cross-section. Though there is no theoretical model for such a ratio, the expressions for σ_N and $\sigma_{k\,o}$ have been developed in Ref. (15).

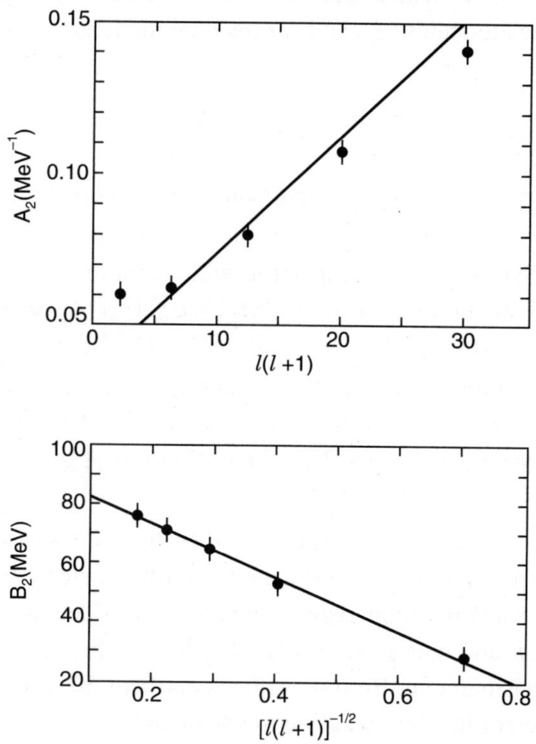

Fig. 16.10 The systematics of A_1 and B_1 values, obtained by least square fitting to (p, p′) data at 62 MeV for many targets (Ref. 9).

16.4.3 Some Experimental Facts

A lot of data[17, 18] on (p, p'), (α, α') and (α, n) at incident energies from 20 to 100 MeV have been collected and analysed on the basis of FKK theory[8] using Kalbach or other equivalent models[19]. We have already shown in Fig. 16.9, one such comparison.

What properties of nuclear structure do we obtain from such studies? As described earlier, the pre-equilibrium phenomenon involves only a few nucleons in the nucleus in the beginning of the reaction. Hence, it is expected that the surface of the target nucleus will play a significant role in it. Also the level densities of the compound nucleus system and the residual nucleus are specifically involved in the theory: [Eqs. 16.1, 16.22, 16.25, 16.27 to 16.30]. The shape and angular distributions will therefore, be quite sensitive to the assumption made in these equations. Also the initial conditions of excitons $n_o = p_o + h_o$ decides, whether the incident particle remained intact—unbroken or got broken, which may happens at higher incident energies. We describe some interesting results based on the analysis of energy spectra or angular distribution and the systematic of various variable parameters involved in the theoretical models.

1. Single Particle Level Densities: A large amount of experimental data[20, 21] exists in literature for the energy spectra and angular distribution of the protons emitted from (n, p) reactions at 14 MeV neutron energy. At such energies, a good fraction of the cross-section goes through pre-equilibrium emission. Recently, G. Singh, H.S. Hans et al. obtained from this data the standard normalised angle integrated energy spectra and energy integrated angular distribution of protons and compared them with various models including Kalbach model as mentioned above. Figure 16.11*a* shows such a comparison with energy spectra and Fig. 16.11*b* with angular distribution using the single particle level density as a free parameter for the composite-states (g_c) and for residual nucleus states (g_R). In Fig. 16.12 we have plotted, the values of

$$a_1 = \frac{\pi^2 g_c}{6}; \text{ and } a_2 = \frac{\pi^2 g_R}{6}$$

as a function of A, as obtained from such comparison. Some interesting observations emerge out of these results:

In a recent analysis, based on the comparison with Shlomo's theory[22, 23] it was found that the values g_c agree with theoretical values $g_c^{Th}(\varepsilon) = g_{cp}^{Th}(\varepsilon_p) + g_{cn}^{Th}(\varepsilon_n)$ [ε_p corresponds to excitation energy of protons; and ε_n to excitation energy of neutrons, for $V_o^{\,i} = 45$ MeV].

This agreement of g_c with $g_c^{Th}(\varepsilon)$, at the excitation of the composite nucleus for 14.8 MeV incident neutrons is interpreted as due to the predominant role of multistep direct (M.S.D.) part of pre-equilibrium reaction mechanism for $h \leq 2$, in the evaluation of g_c; as explained by Kalbach [Phy. Rev. 23, p. 124-126, (1981)] in using PRECO-D_2.

On the other hand, the values of g_R are interpreted as, the values of single particle level densities at the Fermi energies of the residual nuclei following the multiple step compound (M.S.C.) part of the pre-equilibrium reaction mechanism. Comparison with Shlomo's model, using for protons), gives $g_R \sim g_R^{Th}$ (ε_F), for all values of A. We found that g_c was about 10% higher than $g_c^{Th}(\varepsilon)$ for most of the cases, and also g_c/g_R was always greater than one. Both these facts confirm the correctness of our interpretation, for the role of M.S.D. in pre-equilibrium reaction mechanism.

The region of $45 \leq A \leq 64$ which includes a region of 27 or 29 neutrons for the composite and residual nuclei, has very high values (spikes) of a_1 and a_2 showing the shell effects. Similarly the second abnormal region occurs around $A \sim 90$, which has dips in the values of a_1 and a_2 for nuclei having 50, 52 and 54 neutron and 42 protons, due to shell effects. This shows, the sensitivity of pre-equilibrium to detailed properties of level densities.[24]

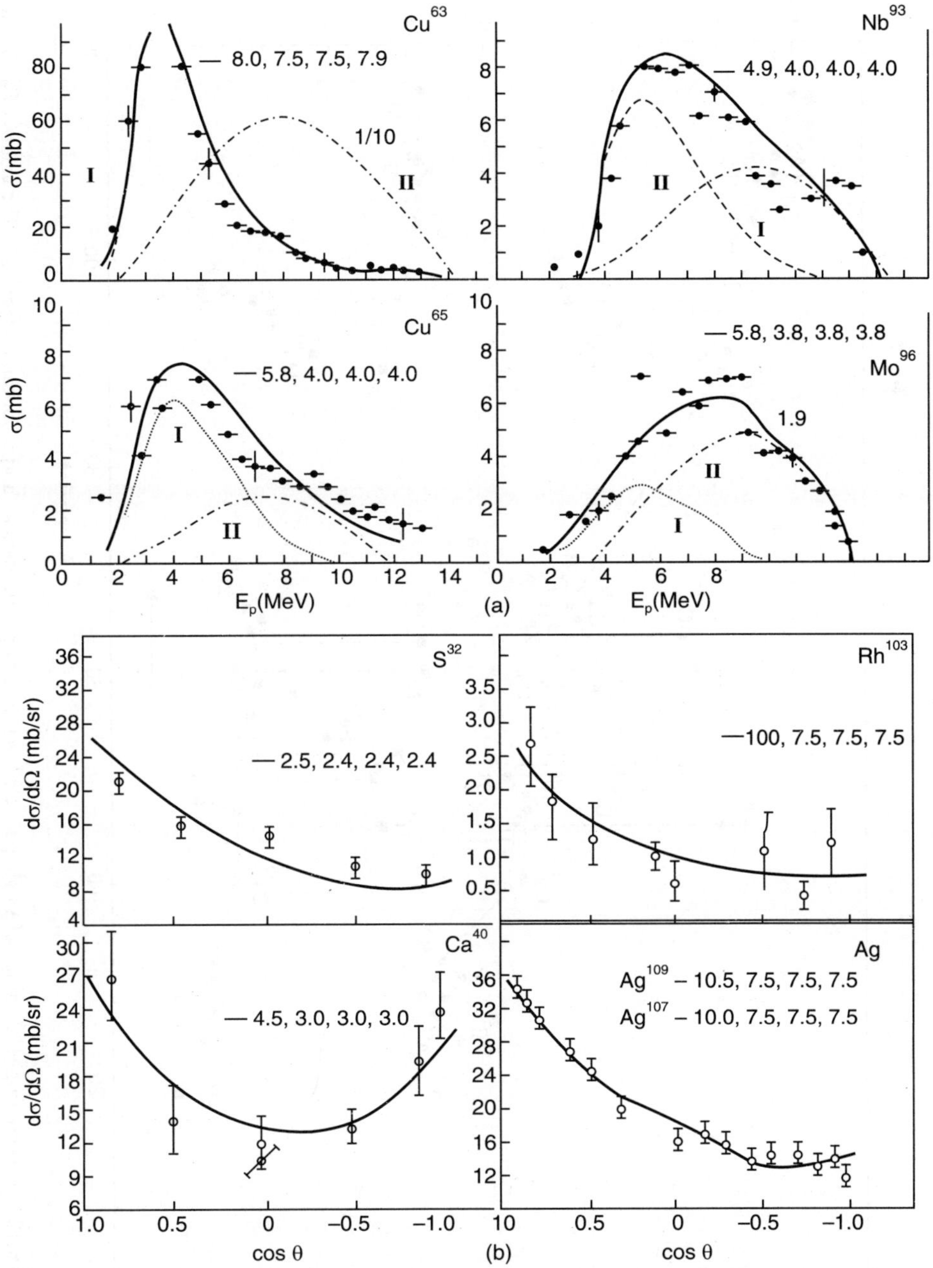

16.11 Typical experimental data on (*a*) energy spectra and (*b*) angular distributions of protons from (n, p) reaction at 14 MeV neutron energy. The solid lines represent the theoretical curves based on Kalbach model. In (*a*), the curves marked as I (- · · · · -) represent compound contribution and II (– – –) represent pre-equilibrium contribution (Ref. 21).

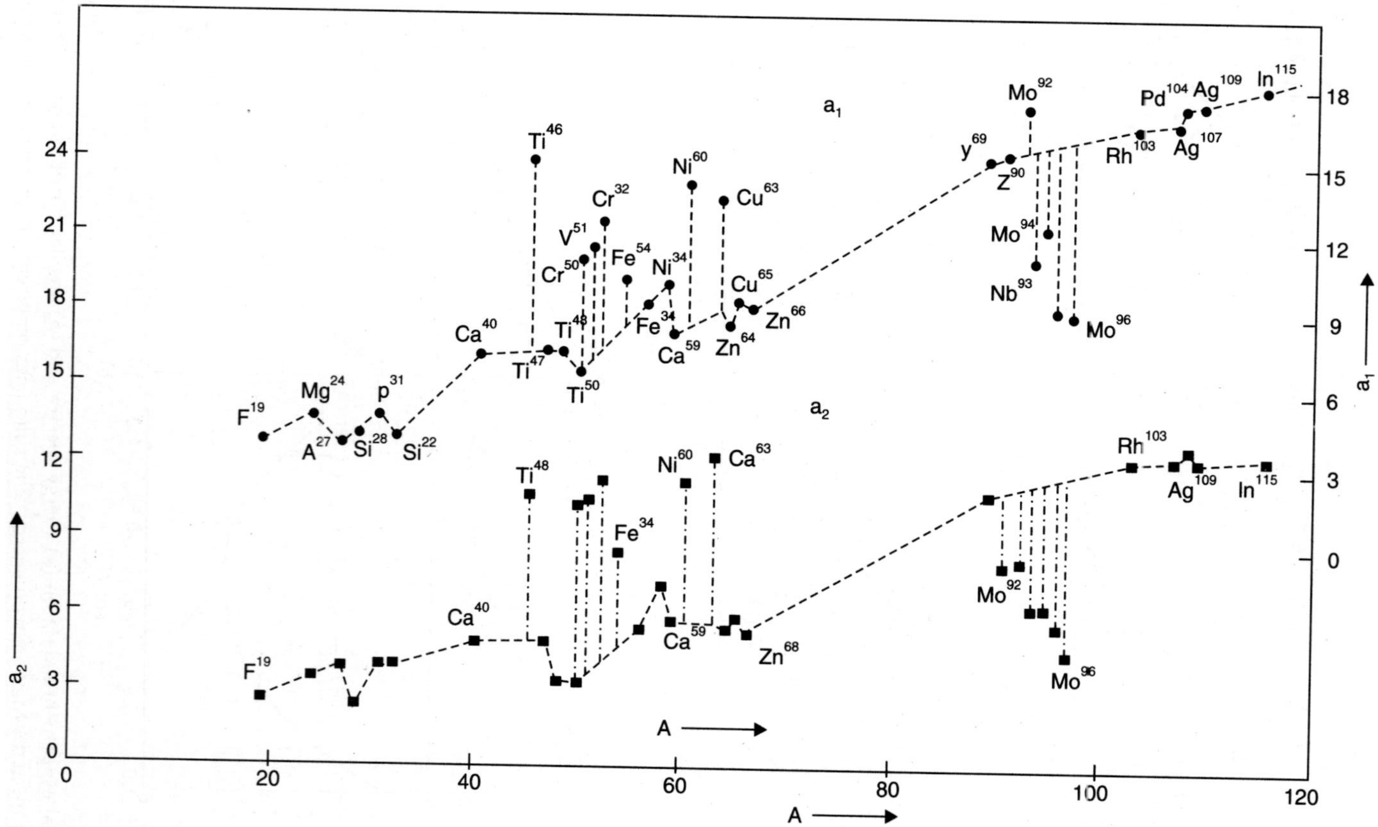

Fig. 16.12 The semi-empirical values of single particle level density parameters a_1 and a_2 obtained from g's. The values of a_1 in the upper curves corresponds to composite nucleus and lower one for a_2 corresponds to the Fermi energy of the residual nucleus (Ref. 21).

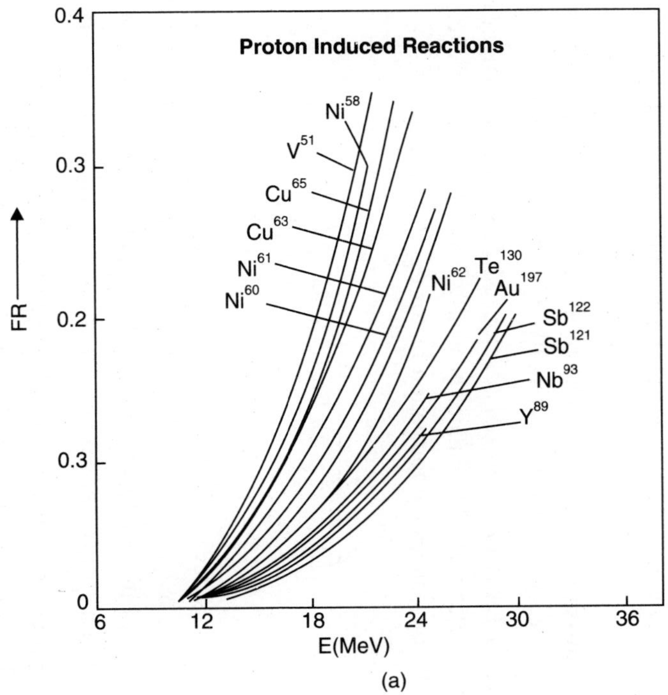

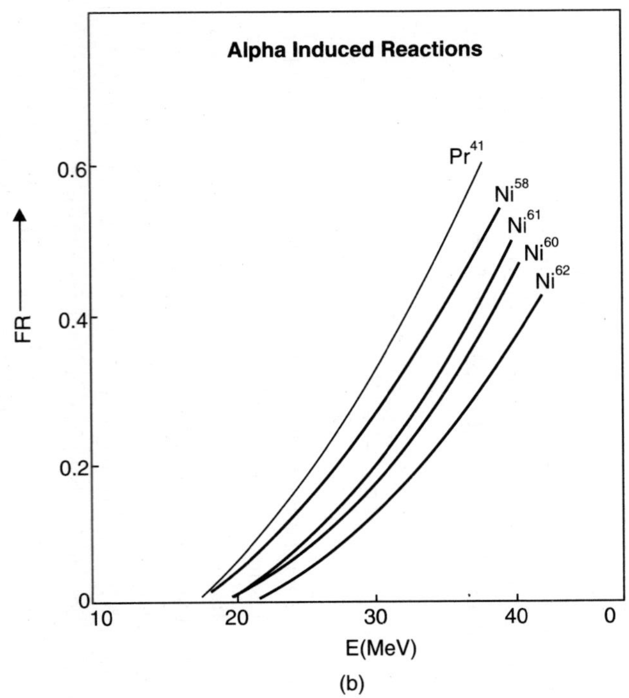

Fig. 16.13 The variation of pre-equilibrium fraction FR as a function of excitation energy (*a*) for protons induced reactions and (*b*) for alpha induced reactions on different targets for incident energies of 30 MeV for proton and 45 MeV for alphas (Ref. 25).

2. Surface-Effects: The pre-equilibrium is comparatively a surface phenomenon, because as explained in the beginning of the chapter, it is concerned with the first few interactions in the nucleus, after the projectile enters the nucleus. In recent experimental studies[25], excitation functions of a large number of cases were experimentally studied for proton and alpha induced reactions in energy ranges of 10–30 MeV and 18–45 MeV, respectively for a large number of targets. The experimental excitation function were compared with theoretical models of compound nucleus and hybrid models of pre-compound nucleus formation. It was possible to obtain from these studies, the pre-equilibrium fraction FR which is defined as:

$$\text{FR} = \frac{\sigma\,(\text{pre} - \text{eqi})}{\sigma\,(\text{Total})} \qquad \text{...(16.37)}$$

The values of FR as a function of the projectile energy was plotted for various targets, as shown in Fig. 16.13. Some interesting points emerged in these plots for proton-induced and alpha-induced reactions. (*i*) In general, the probability of pre-equilibrium emission at a given incident energy is larger for lighter target nuclei than for heavier ones. (*ii*) Also in general, it was found that (FR) for alpha induced reaction was larger by a factor of 2, than for proton induced reactions for the same $E - E_{CB}$ where E_{CB} is the Coulomb barrier for the same target nucleus. The detailed comparison with different values of n_o, the initial exciton number showed that for proton induced reaction $n_o = 3$ and for alpha induced number $n_o = 6$ so the higher values of $(\text{FR})_{\text{alpha}}$ compared to $(\text{FR})_{\text{proton}}$ may be attributed to n_o in the two cases. Also it shows that in alpha induced reactions, alpha breaks up into four constituents, right after the first interaction. The larger value of FR for lighter target nuclei, shows that surface plays the major role in pre-equilibrium because in lighter nuclei the surface is relatively larger, than for heavier nuclei.

3. Cluster Emission: Cluster emission means, emission of light clusters like *d*, He^3, *t* and alpha which fall into one category or heavier clusters like C^{12}, etc. which fall into another category.

Lighter Cluster Emission: A model developed by Iwamoto and Harada[26] calculates the spectra of (p, α), (p, He^3), (p, t), (p, d). The model basically is an extension of exciton model of Blann and Griffin[4, 5]. The model introduces the formation factor $F_{lm}(\varepsilon)$, defined as:

$$F_{lm}(\varepsilon) = \frac{1}{(2\pi\hbar)^3\,(n_x - 1)} \int\limits_{\substack{S \\ |R_x| = R_{res} \\ \overline{P}_x = \text{fixed}}} \pi_{l=0}^{n_x - 1}\, d\,\xi_i\, d\,P_{\xi_i} \qquad \text{...(16.38)}$$

where n_x = number of constituent nucleons to form the outgoing composite particle, so for triton $n_x = 3$; for deuteron $n_x = 2$; for alpha $n_x = 4$, etc. The centre of mass coordinate R_x of the composite particle is set to be the radius R_{res} of the residual nucleus and the centre of mass momentum P_x is fixed, which is connected to the emission energy of light composite particle. The ξ_i and P_{ξ_i} are the internal coordinates (space and momentum coordinates) of the composite particle being emitted. The integration range *S* corresponds to the phase space volume for the intrinsic motion, bounded by the ground state trajectory under the conditions,

$$|\mathbf{P}_i| \text{ for } i = 1 \approx l \text{ and } |\mathbf{P}_i| < |\mathbf{P}_F| \text{ for } i = l + 1 = n_x$$

where $\mathbf{P}_f$ is Fermi-momentum and

$$|\mathbf{r}_i| \le R_o \text{ for } i = 1 \approx n_x \text{ and } R_o = R_{oS} + \Delta R \qquad ...(16.39)$$

is the effective radius of the parent nucleus. One calculates F_{lm} (ε). Then σ_{PEQ} (ε) is calculated by replacing P_n (ε) by F_{lm} (ε) × P_n (ε) in the various expressions of σ_{PEQ} discussed earlier. The values of F_{lm} (ε) for different outgoing particles have been calculated for a full range of energies (0–150 MeV or so) of the emitted particles[26]. Their use for calculating the shapes of deuteron, tritium, He^3 and alphas spectra from proton-induced reactions (62 MeV proton energy) for different targets (say Al, Fe, Ni, Y, Sn, Au) has been demonstrated and found to be reasonably satisfactory, except the high energy end is not fully explained[26].

Heavier Cluster Emission: Another model, which describes the cluster emission is called Exciton Coalescence Model (ECM) as developed by H. Machner and his collaborators[27]. The basis of the model is that complex particle formation occurs from excited nucleons with small transfer momenta. In this model, the cross-section of the emitted composite particles is written by:

$$\frac{d^2\,\sigma_i\,(\varepsilon, \theta)}{d\,\varepsilon\,d\,\Omega} = \sigma_o \sum_{\substack{n = n_o\,(i) \\ \Delta n = 2}}^{\bar{n}} W_i\,(p, h, \varepsilon_i)\,A\,(n, \Omega)\,\tau\,(n, E) \qquad ...(16.40)$$

where σ_o is the total reaction cross-section derived from optical model and W_i (p, h, E) is the emission probability for a particle of type i with energy ε_i form a state with p particles and h holes and is given by:

$$W_i\,(p, h, \varepsilon_i) = \frac{2\,s_i + 1}{\pi^2\,\hbar^3}\,\mu_i\,\varepsilon_i\,\sigma^i_{in}\,(\varepsilon) \times F\,(p, h, p_i, E, t)\,\gamma_i \qquad ...(16.41)$$

where s_i is the spin of the incident particle and F (p, h, p_i, E, t) denotes the probability of having particles with energy $\varepsilon_i + B_i$, (B_i = Binding energy) in unbound states, γ_i denotes the fraction of the states where the p_i particles are coalesced into a cluster of type i and has been calculated to be:

$$\gamma_i = \gamma_{x+y} = \left[\frac{4\pi}{3}\left(\frac{P_o}{\mu c}\right)^3\right]^{x+y-1} \qquad ...(16.42)$$

where P_o is radius of the space around the momentum P_i of the first nucleon, so that if the second nucleon comes within this range it gets coalesced, x (y) are the number of protons (neutrons) in the cluster i, μ is the nucleon mass and c is velocity of light. In Eq. 16.40, A (n, Ω) τ (n, E) is given by:

$$\int_0^{t_{eq}} P\,(n, \Omega)\,d\,t \equiv A\,(n, \Omega)\,\tau\,(n, E) \qquad ...(16.43)$$

where P (n, Ω, t) denotes the occupational probability of states (n, Ω) at time t. One can compare Eq. 16.43 with Eq. 16.19 for exciton model and see the similarities.

For details and experimental comparison, *see* Reference (27).

A good example of the application of this theory was provided by the study of nuclear structure effects in pre-equilibrium reactions in α-induced reaction on Mg^{24}, Al^{27} and Si^{28}. The (α, p) reaction on Mg^{26} and Al^{27} are good examples for coalescence model, according to which following Eqs. 16.40 and 16.41, we can write:

$$\frac{d^2\sigma}{d\varepsilon\, d\Omega} = \sum_{\substack{n=n_o \\ \Delta n = 2}}^{\bar{n}} f(n\,x)\, W_x(n\,E)\, A\,(n - p_x + p_y; \Omega)$$

$$\times\, \theta\,(n - p_x + p_y) \left(\int_0^{teq} P\,(n, t)\, dt \right) \alpha\, \varepsilon_p\, \sigma_{\text{inv}}\, (\varepsilon)\, U^{p_o + h_o - 2} \quad ...(16.44)$$

[For the last step see H. Machner et al., Phy. Rev. C.33, p. 1933, (1986)].

In Eq. 16.44, $f(x)$ is correction for using one Fermi gas for nucleons, instead of different ones for protons and neutrons, W_x (ε, E) is the emission rate per unit time, A is function containing angle information only, θ is a step function, which takes care of having at least one additional nucleon which can carry the recoil momentum, P_x is the number of nucleons in a light nucleus of type x and P_y is the number of nucleons of type y in the projectile.

So that

$$\ln\left[\left(\frac{d^2\sigma}{d\Omega\, d\varepsilon}\right) \times \frac{1}{\varepsilon_p\, \sigma_{\text{inv}}}\right] = (p_o + h_o - 2)\ln U + \text{Const} \quad ...(16.45a)$$

Hence, plotting

$$\ln\left[\frac{(d\sigma / d\varepsilon)}{\varepsilon_p\, \sigma_{\text{inv}}}\right] \text{ versus } \ln \bar{U} \quad ...(16.45b)$$

one gets straight line, as shown in Fig. 16.14, obtained for (α, p) reaction at 100 MeV incident energies for various target nuclei (Mg^{24}, Mg^{25}, Mg^{26}, Al^{27} and Si^{28}). These are called Griffin Plots (Ref. 27).

It is interesting, that $n_o = 5$ fits and not $n_o = 6$ *i.e.* only particles take part and not the hole, *i.e.* $n_o = p\,(_2He^4) + 1p$ and no holes.

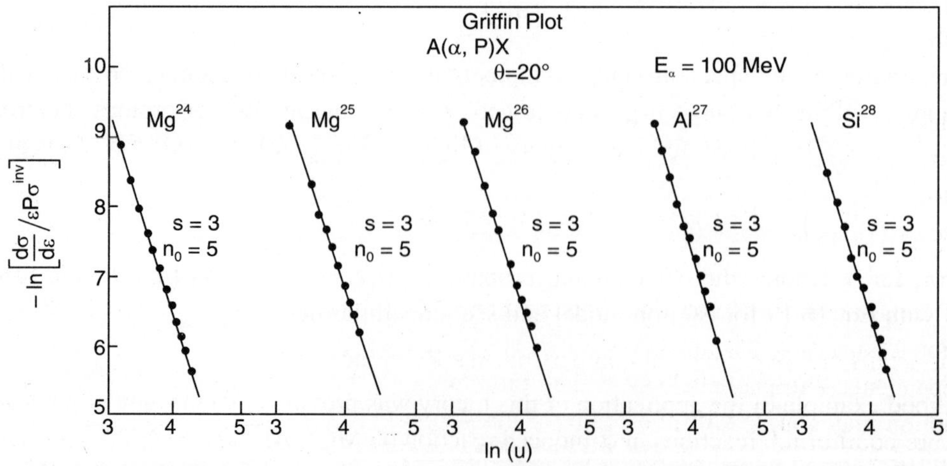

Fig. 16.14 Griffin plots, for (α, p) reactions at 100 MeV incident energies for various target nuclei (Mg^{24}, Mg^{25}, Mg^{26}, Al^{27} and Si^{28}) (Ref. 27).

16.4.4 Pre-equilibrium in Heavy-Ion Induced Reactions

Heavy ion induced reactions, basically differ from the light ion induced interactions in the fact that the heavy ion projectiles have a large number of nucleons in it and hence when a projectile enters the nucleus; there is a diffusion of nucleons to the projectile from the nucleus and from the projectile to the nucleus. This has to be taken into account, over and above the other modes of interaction with the nucleus, *e.g.*,

(*i*) Scattering of particles into the energy interval ΔE_i,

(*ii*) Scattering of particles out of the energy interval ΔE_i,

(*iii*) The emission into the continuum.

Theoretically, this scenario is treated by writing the Master-Equation of the Harper-Miller-Berne Model, as consisting of four terms—the fourth term being due to diffusion–and also including the distribution function of the Fermi ions in the projectile. In heavy ion-induced reactions; the actual coalescence period is expected to be comparable to nucleon-nucleon collision period; so during coalescence there is, significant relaxation and particle emission, leading to diffusion process. Because of the process of collisions, the initial exciton number n_o comes closer to the equilibrium number $\bar{n}$. Also apart from the intrinsic particle excitation, there may exist, in heavy ion induced reactions, the collective modes and nucleon-exchange mode of excitation.

The modified HMB master equation describing the heavy ion reaction for a one-Fermi-ion gas type gas in the projectile and target nuclei is written as:

$$\frac{d\,(n_i\, g_i)}{dt} = \sum_{j,k,1} \omega_{k1,ij}\, g_k\, n_k\, g_1\, n_1\, (1-n_j)\, g_i\, g_j$$

$$- \sum_{j,k,l} \omega_{ij,kl}\, g_i\, n_i\, g_j\, n_j\, (1-n_k)(1-n_1)\, g_k\, g_l - n_i\, g_i\, \omega_{ii'}\, g_{i'} + \frac{d}{dt}(n_i\, g_i)_{\text{fusion}} \quad ...(16.46)$$

where n_i = average occupation number for the *i*th bin above the bottom of nuclear well, each bin being of 1 MeV interval. g_i = number of single particle states per MeV in an energy introduced around *i*th bin. $\omega_{ab,\,cd}$ = the transition probabilities for nucleons in initial state *a* and *b* to scatter into final states *c* and *d*. These are evaluated from free nucleon-nucleon scattering cross-section.

The fractional occupation members g_j $(1 - n_j)$ etc. which multiply the free nucleon-nucleon collision rates give Pauli-Exclusion correction; $\omega_{ii'}$ gives the rate, for particle at energy *i* within the nucleus to go to energy *i*′, outside the nucleus (continuum). The first-two terms in Eq. 16.46 give the rates of scattering particles into and out of the interval *i* by two body collisions and the third term gives the rate of emission into continuum.

If emission into the continuum takes place before an interval, in which equilibrium nucleon distribution is attained, the contribution is part of the pre-compound spectrum. The fourth term represents a time dependent addition of nucleons to the equilibrating system (coalescing system) from the projectile. Fermi motion may couple with the centre of mass motion to give an equal a prior distribution for all degrees of freedom in a manner which conserves energy. The distribution function which is used, is:

$$\mathrm{Ni}\,(U)\,dU = \frac{\left[U^{n_o - 1} - (U - \Delta U)^{n_o - 1}\right]}{E^{n_o - 1}}\, n_t^i \qquad ...(16.47)$$

where n_t^i represents the number of nucleons of type i (neutron or proton) entering the mix during ith time interval; N $(U)\,dU$ represents the number of nucleons of type i in an exciton interval dU centered at U MeV of excitation and E is the compound nucleus excitation, n_o is the initial exciton-number. The fourth term is the time-dependent function representing the time-dependent addition to nucleons to the equilibrating (Coalescing) system from the projectile. This contribution of nucleons has been calculated in a simple manner, by assuming that n_t^i nucleons of type i (n or p) from the projectile enter the reaction zone, when the assumed sharp surface of the projectile crosses the surface of the target nucleus, through a plane normal to the line of centres of target and projectile. The relative velocity of approach is calculated by the centre of mass energy reduced by Coulomb barrier height. Only S-wave-collisions were considered. The coupled differential Equation 16.46 was evaluated on a time scale of 2.1×10^{-23} secs per step. Seven such steps are required to complete fusion process. The energy distribution of these nucleus was taken from Eq. 16.47.

Calculation made earlier[28] for C^{12} induced reactions, showed that below 10–12 MeV/A incident energy the pre-compound decay was negligible.

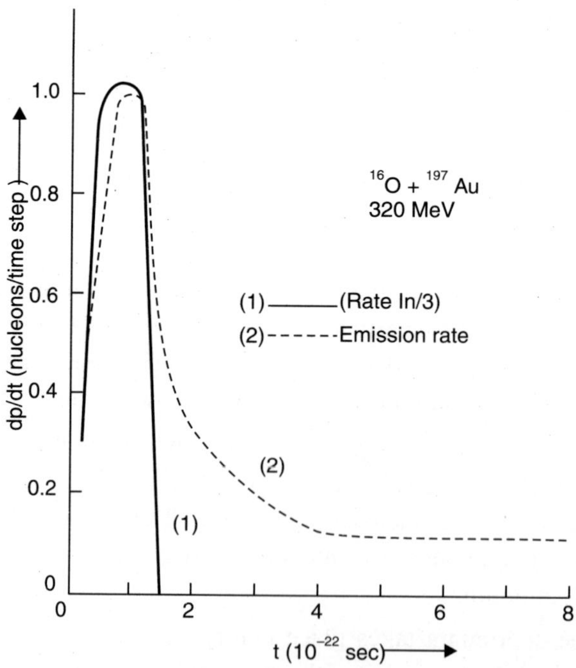

Fig. 16.15 Theoretical curves, based on Eq. 16.46, of the rates of particle insertion and emission for Au^{197} (O^{16}, x) at 320 MeV (lab) energy. The ordinate represents the number of particles 'in' (divided by 3), or 'out' versus time on the abscissa (Ref. 28).

Equation 16.46 is used to obtain the rate of transition probabilities, *i.e.* $d\,(ni\ gi)/\ dt = dP/dt$ as a function of time. One can obtain the rates of particle insertion and emission as a function of time. This will give the internal nucleon distribution of the Coalesced (or Coalescing) nucleus, for different times from the entry time of the projectile.

Figure 16.15 give the rates of particle insertion and emission as predicted by the Master Equation 16.46, for Au^{197} (O^{16}, x) at 320 MeV (lab) energy. The ordinate represents the number of particles inserted in, divided by 3 and emitted out versus time on the abscissa.

The first term and fourth term in Eq. 16.46 correspond to particles going in and the second and third terms correspond to particles going out.

Figure 16.16 shows, the internal nucleon distribution for the fusion of 214 MeV O^{16} with Au^{197}. Evidently in the first step, higher energies particles are there (at 2.0×10^{-23} secs); then high energies particles are further increased up to 1.4×10^{-22} secs; then they get reduced up to 1.3×10^{-21} secs. So only in the first steps, the pre-equilibrium emission takes place directly.

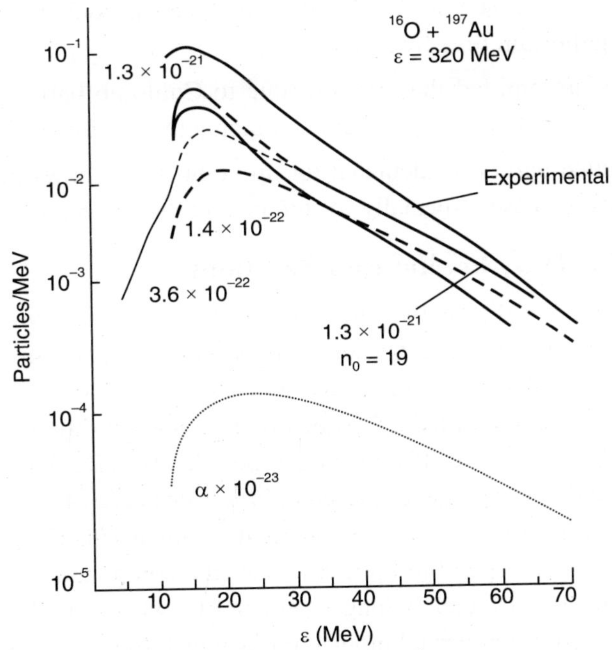

Fig. 16.16 Proton spectra versus time for Au^{197} (O^{16}, p) at 320 MeV, for different times after the insertion of the incident projectile. The first time of Coalescence is taken to be 2×10^{-23} secs and the curve after 1.3×10^{-21} secs; seems to approach the experimental curve, shown as solid curve (Ref. 28).

The analysis of time-evolution and energy evolution of the excitation, neutron and proton-emission and fusion and equilibration process yields the following interesting results for Au^{197} (O^{16}, x) and Ag^{109} (Ar^{40}, x) at 320, 214 and 104 MeV, for the former and at 240 and 320 MeV for the reaction are given as follows:

Reaction Au^{197} (O^{16}, x) calculated

Energy (MeV)	*Time for fusion* (10^{-22} secs)	*Time for equilibrium* (10^{-22} secs)	*Fusion*		*Equilibrium*	
			Nn	*Np*	*Nn*	*Np*
320	1.5	3.6	0.62	0.36	1.08	0.52
214	1.9	4.2	0.35	0.16	0.59	0.53
104	4.7	11	0.061	0.009	0.095	0.012
Ag^{109} (Ar^{40}, x)						
240	4.6		0.092	0.037		
320	3.6	8.2	0.27	0.12	0.72	0.31

This shows that:

(*i*) Higher the energy of projectile, less is the fusion time and equilibrium time. Generally fusion time is much less than equilibrium time by a factor of 2–3.

(*ii*) More neutrons are emitted than protons due to Coulomb barrier effects. Naturally more neutrons and protons are emitted up to equilibration compared to the fusion time.

(*iii*) Spectra of emitted particles calculated for different times shows that as the time increases, higher energy components increase, especially for protons, in the beginning.

16.4.5 Experiments in Heavy Ion Induced Reactions

Heavy ion induced reactions, in the energy range of 10 MeV/A proceed via fusion-fission and light particle emission in a complicated manner, because light particles may be emitted (*i*) in the initial stages of fusion (*ii*) during the fusion-fission process (*iii*) After the compound nucleus is formed and fission takes place and light particles are emitted subsequently or along with the fission. In this manner, the emitted particles are: (*i*) Projectile-like (PL), or target like recoils (TR). The latter case occurs due to deep inelastic scattering and corresponds to pre-equilibrium phase (*ii*) Fusion-fission fragments (FF) which corresponds to one form of decay of the compound nucleus (*iii*) Evaporation residues (ER) or Heavy residues (HR) which correspond to a process after the compound nucleus is formed in which light particles are emitted. But residues belong to either ER, where nearly complete momentum is transferred to evaporated particles or partial momentum is transferred as in HR. The light particles can be emitted in any of these processes. So to understand the mechanism of emission, one has to measure the intensity or multiplicity (number of particles per fragment per MeV-Stredian) in coincidence with the various residual recoil particles.

One such measurement was carried out by E. Holub and coworkers[29], on $Ho^{165} + Ne^{20} \rightarrow A + n$ at $E_{\text{lan}-\text{Ne}}$ for 220, 292 and 402 MeV, where A = evaporation residues or fusion-fission fragments and n is the number of neutrons.

Figure 16.17 shows[29], the neutron-spectra in coincidence with evaporation residue (ER). The pre-equilibrium component is evident in these curves. The experimental data are fitted to an empirical expression from a stationary (compound nucleus) and a moving (pre-equilibrium) source given as:

$$\frac{d^2 M_n(\theta_n)}{dE_n\, d\Omega_n} = \sum \frac{M_i \sqrt{E_n}}{2(\pi T_i)^{3/2}} \exp\left| -\frac{E_n - 2\sqrt{E_n\, \varepsilon_i}\cos(\theta_i - \theta_n) + \varepsilon_i}{T_i} \right| \quad \text{...(16.48)}$$

$i = 1$ for evaporation (EV)

$i = 2$ for Pre-Equilibrium (PE)

for $E = 220$ MeV; $M_{EV} = 6.7 + 0.5$, $T = 2.4 + 0.2$

and $M_{PE} = 0.4 + 0.1$, $T = 4.5 + 0.3$

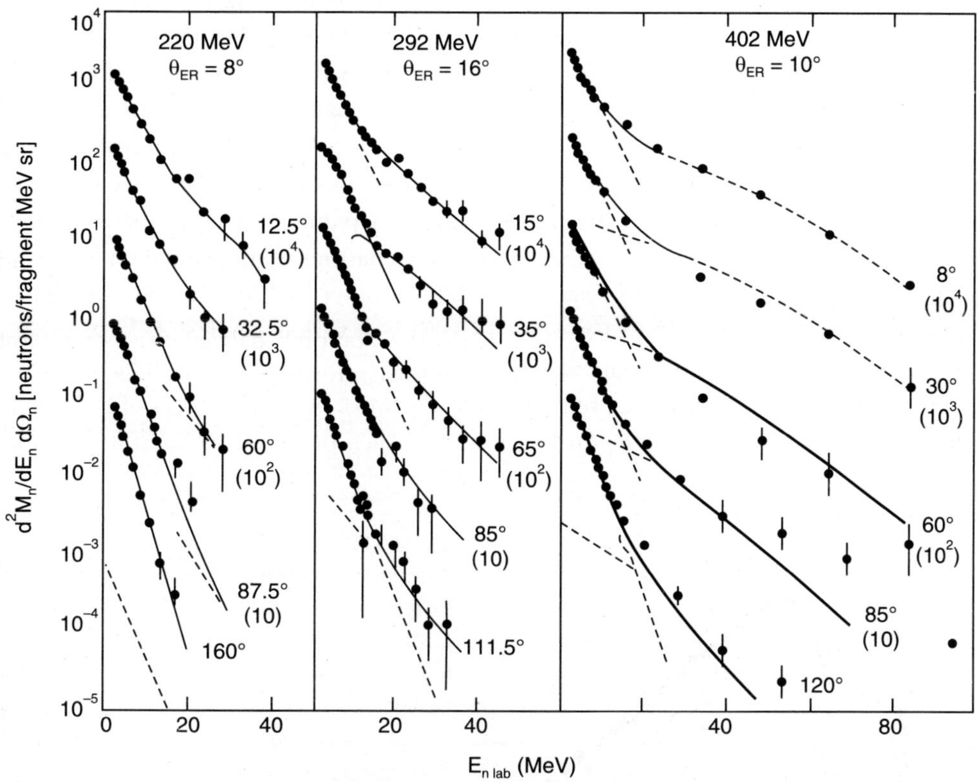

Fig. 16.17 Experimental differential neutron multiplicity in coincidence with evaporation residues at 220, 292 and 402 MeV incident energies. The full curves represent the sum of evaporation (dotted line) (Ref. 29).

The theory by Blann[28] was fitted to the pre-equilibrium part and it was interestingly found that, as the energy increases, n_o of the theory also increases as shown in table below:

		Evaporation (EV)	*Pre-equilibrium (PE)*
$E_{Ne} = 220$ MeV	$n_o = 20$	$T = 2.4 + 0.2$ MeV;	$T = 4.5 + 0.3$ MeV
$E_{Ne} = 292$ MeV	$n_o = 24$	$T = 2.3 + 0.2$ MeV;	$T = 6.3 + 0.3$ MeV
$E_{Ne} = 402$ MeV	$n_o = 28$	$T = 2.8 + 0.2$ MeV;	$T = 8.6 + 0.3$ MeV

Similar data is also available for fusion-fission, which will be discussed in the next chapter.

16.4.6 Recent Developments

The recent developments in pre-equilibrium studies in nuclear reaction, can be divided into two parts: (*i*) The recent experimental data and its analysis, using known pre-equilibrium model studies of neutron induced reaction at 14 MeV incident energies[30] or in the energy range of 9–21 MeV[31] and at 62.7 MeV for many targets for light charged particle emission[32], fall into this category. (*ii*) Somewhat in the same category but for alpha-induced reactions are the measurements[33] and calculations of excitation functions in copper from 8.27 MeV to 50 MeV and the comparison with pre-equilibrium and statistical models. While most of the theoretical calculations reproduce the experimental results well, there has been seen for the first time the effect of collective excitation in Ref. (30). This effect has been very clearly brought out in microscopic two-component multi-step direct theory for continuum nuclear reactions by A.J. Konny and M.B. Chadwick[34]. The multi-step theory, as described earlier was developed by Kalbach and has been now further improved upon by two principal advances. (*a*) A microscopic approach is given for calculating DWBA transition to the continuum. (*b*) A two-component formulation of multi-step direct reaction is given, where neutron and proton excitations are explicitly accounted for, in the evolution of reactions for all orders of scattering. This multi-step direct theory is applied along with the theories of multi-step compound and collective reactions to analyses experimental emission spectra for a range of targets and energies. The energy spectra of Zr^{90} (*n*, *xn*) at 14.1 MeV at various angles and also the same reaction at 18 MeV; then Pb^{208} (*n*, *xn*) at 14.1 MeV at various angles, Zr^{90} (*p*, *xn*) at 25 MeV, 45 MeV and 80 MeV and Zr^{90} (*p*, *xp*) again at 80 MeV and A^{127} (*p*, *xn*) at 90 MeV; have been compared with theory in detail. The agreement is extremely good. The most beautiful aspect of this comparison is the reproductions and its exact fit. Figure 16.18 shows one such fit for 14.1 MeV Pb^{208} (*n*, *xn*) at 90°.

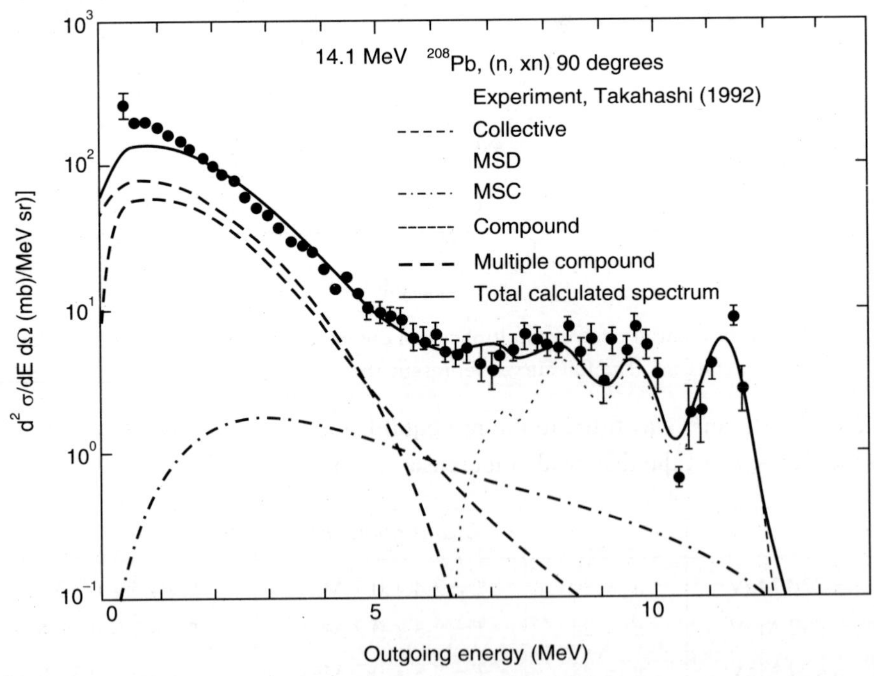

Fig. 16.18 Comparison of fully microscopic calculations, with experimental data for Pb^{208} (n, xn) reaction at 14.1 MeV for angle of emission of 90° (Ref. 34).

A further development of the theory has been carried out by M. Blann and M.B. Chadwick[35]. This is an extension of Monte Carlo sampling method as developed by M. Blann[36] and is called Hybrid Monte Carlo simulation (HMS). This approach follows nucleon-nucleon scattering, such that only two and three quasi-particle scattering distributions are involved. This method has been proved to be consistent, with the kinetic result of nucleon-nucleon scattering in nuclear matter, as shown by M. Blann and coworkers earlier[37]. The actual result for 256 MeV Zr (*p*, *xn*) and Pb (*p*, *xn*) give again an extremely good agreement.

16. Pre-Equilibrium Model 2000–2008

In an interesting paper, evidence for non-equilibrium-emission in a low energy heavy ion reaction has been reported, in a reaction of O^{16} and Al^{27} with 72 MeV beam of O^{16}, giving rise to emission of protons, deuterons, and alphas. Correlation function were measured for *p-p, d*-α and α-α pairs. The observed anticorrelation is stronger at more forward directions [Phy. Rev. C. 61, 024603 (2000)]. An interesting paper by K. Gul, Islamabad, Pakistan, involves calculations based in pre-equilibrium and Hauser Feshbach model [Phy. Rev. C. 62, 067603 (2000)]. In another paper by C. Kalbach, surface and collective effects in pre-equilibrium reactions have been reported for E_{Lab} = 14 – 100 MeV. [Phy Rev. C. 62, 044608 (2002)].

In an paper on nuclear, reactions, Pt^{196} (n, xny, $p\,\gamma$) reaction was studied from E_n = 1 to 250 MeV; by measuring prompt reaction γ-rays. Hauser-Feshach reaction model only partially agreed with the γ-rays yields. Discrepencies appear due to inadequate level information in the calculation describing the pre-equilibrium processes [Phy Rev. C. 64, 054614 (2001)].

In a calculation paper, authors from Poland, France and Germany have calculated nuclear level density parameter in the frame work of relativistic mean filed at temp. between 0 and 4 MeV for 193 spherical and even-even nuclei [(Phy. Rev. C. 66, 051302, (2002)].

In three papers, two from Kolkata (India); and one from Europe, fast neutrons (25–65 MeV), are involved as incident particles, with the production of light particles (Europe); or with the production of 30–50 MeV alphas (Kolkata). Theoretical calculations for Kolkata experiment using Variable Energy Cyclotron (VCE); are based on pre-compound PRE COD2 and Alice (both having pre-equilibrium assumptions) [Phy. Rev. C. 67, 02407 (2003) and Phy. Rev. C. 67, 064611 (2003)].

In a theoretical paper Kalbach, has explained in the (*n*, *xp*) reaction; the continuum spectra of protons, from incident energies of 28 MeV to 63 MeV; neutrons on the basis of surface localization of the reaction by revised version of PRE-COD2 [Phy. Rev. C. 69, 14605 (2004)].

In a paper, on the use of Feshbach Karman-Koonin (FKK) model for calculating the cross reaction for Ti^{48} (n, $n'\gamma$) Ti^{48} and Ti^{48} (n, $2n\gamma$) Fi^{47}, by using the experimentally measured γ-rays spectra from E_n = 1 to 35 MeV: the comparison shows that above 10 MeV; pre-equilibrium model holds good. The detailed comparison support FKK model [Phy. Rev. C.75, 05612 (2007)].

In a paper [Phy. Rev. C. 77, 054605 (2008)], on New calculational method for initial excitron number, nucleon-induced pre-equilibrium reactions, has been discussed.

REFERENCES

1. R. Serber: Phy. Rev. 72, 1114 (1947).
2. M.L. Goldberger: Phy. Rev. 74, 1268 (1948); N. Metropolis et al.: Phy. Rev. 110, 185 (1958); Gadioli-Erba, E. Sona P.G.: Nuclear Physics A 251, 589 (1973).
3. H.W. Bertini: Phy. Rev. V.31, p. 1801 (1963); H.W. Bertini et al.: Phy. Rev. 10, p. 2472 (1974).
4. J.J. Griffin: Phy. Rev. Letters 17, 478 (1966).
5. M. Blann: Annual Reviews of Nuclear Sciences V. 25, 123 (1975).

   G.H. Brega-Marazzan, Gadioli-Erba et al.: Phy. Rev. C6, 1398 (1972), E. Gadioli et al.: Phy. Rev. C29, 76 (1984).

   C.K. Cline and M. Blann: Nuclear Physics A 172, 225 (1971).
6. G.D. Harp, J.M. Miller and Berne B.J.; Phy. Rev. 165, p. 1166 (1968); Ibid, Phy. Rev. C3, 1847 (1971).
7. M. Blann, and H.K. Vanoch: Phy. Rev. C28, 1475 (1983); Blann and H.K. Vanoch: Phy. Rev. Letters V. 27, p. 337 (1971).
8. H. Feshbach, A. Kerman and S. Koonin: Annals Physics, (N.Y.), 125, 429 (1980).
9. C. Kalbach: Phy. Rev. C23, 124 (1981); Phy. Rev. C24, 819 (1981); C. Kalbach: Z. Phy. A 283 401 (1977); C. Kalbach & F.M. Mann: Phy. Rev. C23, p. 112 (1981); C. Kalbach: Phy. Rev. C37, 2350 (1988); T. Tamura, T. Udagawa and H. Lanske: Phy. Rev. 26, 379 (1982); Also Ref. (7); H. Gruppelaar, P. Nagel and P.E. Hodgson: Revista, Nuovo Cemento Cin 9, 1 (1986); C. Kalbach and C.K. Cline: Nuclear Physics A210, p. 590 (1973); C. Kalbach et al.: Z. Phy. A. 275, p. 175 (1975) C. Kalbach: Z., Phy. A287, p. 319 (1970); De A. et al.: J. Phy. 11, L 79 (1985); De A. et al. and S.K. Ghosh: Phy. Rev. C37, p. 2441 (1988).
10. F.E. Bertrand and R.W. Peele: Phy. Rev. C8, 1045 (1973); F.E. Bertrand and R.W. Peele: Phy. Rev. C10, 1028 (1974).
11. T. Ericson: Advances in Physics, 25 (1960); C. Bloch: Phy. Rev. 93, 1094 (1954); I.Kenestrom: Nuclear Physics 83, 332 (1966); M. Blann: Phy. Rev. Letters 21, 1357 (1968); M. Blann and F.M. Lanzafame: Nuclear Physics A. 142, 559 (1970).
12. F.C. Williams Jr.: Phy. Letters 31B, 184 (1970).
13. K. Kikuchi and M. Kawai: Nuclear Matter and Nuclear Reaction, North Holland, Amsterdam, Chapter 2 (1968).
14. N.T. Porile et al.: Nuclear Physics 43, p. 500 (1963); J.D. Jackson: Canadian J. of Physics 43, 767 (1956), N. Metropolis et al., Phy. Rev. V. 110, p. 185 (1958).
15. P.K. Sarkar, T. Bandopadyay, G. Muthukrishnan and Sudip Ghosh: Phy. Rev. C. V 43, 1855 (1991).
16. V.F. Weisskopf and D.G. Ewing: Phy. Rev. 57, 472, 935 (1940).
17. B.L. Cohen, G.R. Rao et al.: Phy. Rev. C7, 331–343 (1973), J.R. Wu, C.C. Chang and H.D. Holmgren, S.M. Grimes, J.D. Anderson: Phy. Ev. V. 19, p. 698 (1979), for (p, n); E. Gadioli, E. Gadioli Erba et al.: Nuclear Physics A 217, 598 (1973), F. E. Bertrand and R.W. Peele: Phy. Rev. C. 8, 1045 (1973); A. De S. Ray and S.K. Ghosh: Phy. Rev. C37, 2441 (1988).
18. For (α, p): F.E. Bertrand: Phy. Rev. C10, 1028 (1974); A. Chavarier, N. Chavarier. A. Demeyer, G. Hollinger, P. Pertosa and Trau Mink Duc: Phy. Rev. C8, 2155 (1973); J.R. Wu, C.C. Chang and H.D Holmgren, Phy. Rev. C19, 659 (1979); For (α, α'): H. Machner: Phy. Rev. C. 33, p. 1931 (1986); M. Blann: Nuclear Physics A., V. 257, p. 15, (1976), For (α, n); Ref. (15).

19. C. Kalbach: Phy. Rev. C. 37, 2350 (1988); A. De et al.: Phy. Rev. C. 37, 2441 (1988); M. Blann & H.K. Vanoch, Phy. Rev. C. 28, 1475 (1983).

20. Murrey D. Goldberg, Victoria M. May and John R. Stehn, National Neutron Cross-sections Centre, Brookhavln National Laboratory, Upton, (N.Y.), Report No. BNL 400, V. 1, 2, (1962–1970) (unpublished), I.A.E.A. OINDAV V. 1.2, (1967–77) Ref. (29).

21. Gulzar Singh, H.S. Hans, T.S. Cheema, K.P. Singh, D.C. Tayal, Jahan Singh and Sudip Ghosh: Phy. Rev. C. 49, p. 1066–1078 (1994).

22. Shalom Shilomo: Nuclear Physics A 539, 17, 36 (1992).

23. Y.A. Bogila, V.M. Kolomietz, A.I. Sorzahur, and S. Shlomo: Phy. Rev. C. 53, 855 (1996).

24. J.R. Huizenga and L.G. Moretto: Annual Rev. Nuclear Sciences 24, 427 (1972).

25. M.M. Mustafa, B.P. Singh, H.G.V. Sankaracharyulu, H.D. Bhardwaj and R. Prasad, Phy. Rev. C. 32, 3174 (1994): Ph. D. Thesis, Aligarh Muslim University, Aligarh, M.M. Mustafa (1997).

26. A. Iwamato and K. Harada: Phy. Rev. C26, 1821 (1921); Sato, A. Iwamato and K. Harada, Phy. Rev. C28, 1527 (1983); H. Machner: Physics Letters 86B, 129 (1979).

27. H. Machner et al.: Phy. Rev. C29, p. 411 (1982); H. Machner et al.: Phy. Rev. C.33, p. 1931 (1986); H. Machner et al.: Phy. Rev. C21, p. 2695 (1980); H. Machner et al.: Phy. Reports 127, 309 (1985).

28. M. Blann: Phy. Rev. C23, 205 (1981); G.D. Harp and J.M. Miller: Phy. Rev. C13, 1847 (1971); M. Blann et al.: Nucleonika 21, 335 (1976); M. Blann et al.: Nucleonika 23, 1 (1978); M. Blann: Nuclear Physics 235, 211 (1974); T. Tamura et al.: Phy. Letters 66B, 109 (1977).

29. E. Holub et al.: Phy. Rev. C28, p. 252 (1983); Phy. Rev. C.33, p. 143 (1986).

30. A. Priller, P. Steirer, A. Pavlik, B. Strohmaier, H. Vonach, A. Wallner, G. Winkler and M.B. Chadwick: Phy. Rev. C.56, p. 1424 (1997).

31. A. Faessler, E. Wathecamps, D.L. Smith and S.M. Quaim: Phy. Rev. C. 58, p. 996 (1998).

32. S. Benck, I. Slypen, J.P. Meulders, V. Carcalciuc, M.B. Chadwick, P.G. Young and A.J. Koning: Phy. Rev. C. 58, p. 155 (1998).

33. K. Gul: Phy. Rev. C. 58, p. 586 (1998), S. Mukherjee, B. Bindu Kumar, M.H. Rashid and S.N. Chintalpudi: Phy. Rev. C. 55, 2556 (1997).

34. A.J. Konny and M.B. Chadwick: Phy. Rev. C. 56, p. 970 (1997).

35. M. Blann and M.B. Chadwick: Phy. Rev. C57, p. 233 (1998).

36. M. Blann: Phy. Rev. C. 54, p. 1341 (1996).

37. M. Blann and H.K. Konach: Phy. Rev. C.28, 1475 (1983); J. Bisplingoff: Phy. Rev. C.33, 1569 (1986).

PROBLEMS

1. For a square well potential of a depth of 40 MeV calculate for $l = 0$, $l = 1$ and $l = 2$, say on shell-model basis, the energy sequence of levels up to ten levels. Plot the energy gaps as a function of energy. What do we learn from this? What will be the number of particles in each level?

2. In a nucleus with say $A = 100$, calculate the thickness near the surface of the nucleus in which the incident particle encounters 3, 5, 10 and 20 particles, if the density of the nucleons in the nucleous is uniform or has Woods-Saxon distribution.

3. Explain physically, how can one write the expression.

$$\omega = \frac{v\,\sigma(v)}{\Omega}$$

4. Starting from the initial exciton number $n_o = p_{o+H_o} = 2 + 1 = 3$, proceed to create excitons at higher energies till $n = 30$, taking the single particle levels from problem 1; when $l = 0$, 1 and 2 are taken. Up to what energy will they go ? Also plot exciton number as a function of excitation energy. What do we learn from this?

5. Using the master Equation 16.6, obtain the expression of the integrated $P(n, t)$, by using the expression 16.5*g* and 16.5*e* for λ_+, λ_- and λ_c.

6. Take an incident proton of 40 MeV and using the differential cross-section of (p, n) reaction from literature, plot the angular distributions of neutrons from first five interaction in a nucleus, taking for each case all possibilities at different angles.

7. Compare the expression of level density, *i.e.* $\rho(E) \approx \exp(2\sqrt{aE})$ obtained on statistical model, where 'a' is a constant and is equal to $\pi^2 g/6$ and $\rho(E)$ as obtained in Eq. 16.1 with $g(\varepsilon) = K^n E^{n-1}$. What is the physical significance of K and n?

8. Compare Griffin's straight line plot with a similar plot of compound model [Fig. 16.14] giving temperature. Discuss the differences in the physical phenomena in the two cases.

9. Compare Eq. 16.10 for light-particle reaction with Eq. 16.46 for heavy ion reaction. How will this change the shape of the emission spectra?

10. Why is empirically measured temperature for pre-equilibrium higher than for evaporation [Eq. 16.48]?

17

CHAPTER

Heavy Ion Induced Nuclear Reactions

17.1 INTRODUCTION

Nuclear reactions discussed till now, generally correspond to the lighter projectile say nucleons (*p* or *n*), alphas (He^4), deuterons (H^2), Tritons (H^3) or Helium 3 (He^3). While the loosely bound projectiles follow, in general, direct reaction-mechanism in their reactions; nucleons and alphas follow either the compound nucleus formation process or pre-compound emission mechanisms. Optical model is, of course, applicable to all these particles in a specified manner in each case. Recently, however a lot of attention has been focussed on nuclear reactions induced by heavy ions. What is a heavy ion? Any nuclide heavier than alpha (He^4) may be called a heavy ion. The word ion is used because in experimental studies, it is a positively charged ion, with many electrons removed from the neutral atom, which is used either say in a tandem electrostatic, accelerator or in a cyclotron. So any ion from He^4 up to Pb^{208} or U^{238} is a heavy ion. Reactions have been studied using a large number of heavy ions as projectiles. Very commonly used heavy ions are : Li^6, C^{12}, N^{14}, O^{16}, Ne^{20}, Mg^{24}, Al^{27}, S^{32}, A^{40}, Ca^{40}, Cr^{52}, Ni^{58}, Sn^{120}, Pb^{208}, etc. (Ref. 1).

Of course, in principle, any heavy ion projectile for which there is a stable nucleus can be used as a heavy ion projectile. However, the ease of creating the ions in the ion-source of the accelerator, varies from an atom to atom and hence some heavy ion projectiles as mentioned above, are more easily available and are commonly used as projectiles. For light nuclei, it is possible to remove all or most of the electrons, by carefully chosen stripping process, while for medium or heavy nuclei, about half or less of the electrons can be removed.

The range of energies of heavy ions, with which we are generally concerned in nuclear reactions related to nuclear structure problems, may be related to the Coulomb barrier, between the heavy ion projectile and the target. Typical Coulomb barrier E_B are given in Table 17.1 along with the typical energy (E_{lab}) of the projectiles required to just overcome the Coulomb barrier.

Table 17.1

	Projectile (*m*)	*Target* (*M*)	*Coulomb barrier* (E_B) (*MeV*)	E_{lab} (*MeV*)
1.	C^{12}	C^{12}	7.9	15.7
2.	C^{12}	Ca^{40}	21.0	27.5
3.	C^{12}	Sn^{120}	41.5	45.7
4.	C^{12}	Pb^{208}	59.9	68.7
5.	Ca^{40}	Ca^{40}	58.5	117.0
6.	Ca^{40}	Sn^{120}	119.7	159.6
7.	Ca^{40}	Pb^{208}	175.5	209.2
8.	Sn^{120}	Sn^{120}	253.4	506.9
9.	Sn^{120}	Rb^{208}	377.6	595.5
10.	Pb^{208}	Pb^{208}	567.4	1134.9

The relationship of E_B and E_{lab} with the charges and masses of the projectiles and target are (Ref. 1):

$$E_B = \frac{Z_1 Z_2 e^2}{R_1 + R_2}; \; E_{\text{lab}} = \frac{M+m}{M} E_B \qquad \text{...(17.1}a\text{)}$$

It may be realised, that at these energies, the wavelength of the projectiles is given by:

$$\lambda\!\!\!^{-} = \frac{\hbar}{p} = \frac{1}{2\pi}\left(\frac{h^2}{2mE_{\text{lab}}}\right)^{\frac{1}{2}} \qquad \text{...(17.1}b\text{)}$$

These wavelengths are of the order of 0.15 – 1.2 Fermis for the energies given in Table 17.1. On the other hand, $R_1 + R_2$ varies from say 17 fm for $Pb^{208} + Pb^{208}$ to 6.59 fm for $C^{12} + C^{12}$. Hence, in some respects, the motion of heavy ions inside the target nuclei are similar to that of classical particles. It can be seen from Table 17.1, that energies of a few MeV per nucleon are required to overcome the Coulomb barrier.

Nuclear reactions, induced by heavy ions below but near Coulomb barrier are useful for exciting the collective modes of excitation of the target nuclei.

Because Coulomb barrier is comparatively high in heavy ion-induced reactions and the projectile energies generally available are of the same order as the Coulomb barrier, the Coulomb barrier energy becomes a very good reference energy to describe the reactions. Further, one can decide the reactions according to the impact parameter (R_1) in a semiclassical manner. It may be noted, that the impact parameter is the closest distance of approach of the projectile as indicated in Fig. 17.1. One, then, has the following possibilities:

(*i*) If $R_{l_1} >> R = R_1 + R_2$, the projectile passes quite away from the nuclear range of interaction of the two nuclei and only Coulomb interaction is possible between the projectile and the target. Then only Rutherford elastic scattering takes place, or the target and/or projectile may be internally excited, through Coulomb excitation giving rise to inelastic scattering of the projectile. This is shown in trajectory (1) in Fig. 17.1.

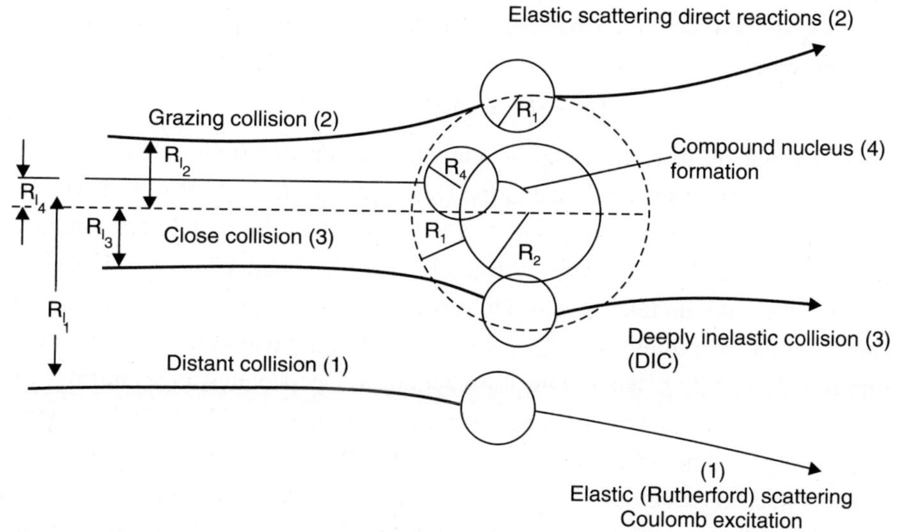

Fig. 17.1 The four possibilities of heavy ion interactions, as seen classically showing the trajectories, corresponding to distant collision (Trajectory 1), grazing (Trajectory 2), close collisions (Trajectory 3) and for compound nucleus formation (Trajectory 4).

(*ii*) If R_{l_2} is $\approx R_1 + R_2$, *i.e.*, the outer surface of the incident projectile just grazes along the surface of the target nucleus, then the edges of the nuclear ranges of the nuclei just touch, so that only the outer portions of the skins of the two nuclei interact. Then only the extremely outer lying nucleus in the two nuclei come within the nuclear range of interaction. This may, lead to one or two nucleon-transfer from one nucleus to the other or only elastic scattering or inelastic scattering may take place when any one of the two nuclei are internally excited, either via Coulomb excitation or nuclear interaction. This is indicated by trajectory (2) in Fig. 17.1.

(*iii*) If $R_{l_3} \le R_1 + R_2$, *i.e.*, the incident projectile just enters the nuclear range of the interaction of the two nuclei. This is a case, where nuclear interaction is prominent and elastic scattering is nearly absent. Then deep inelastic collisions predominate. Here a few nucleons—anything from one to five or six nucleons—can get transferred from one nucleus to the other.

(*iv*) If $R_{l_4} << R_1 + R_2$ and the energy of the incident projectile is high enough, so that it can penetrate the Coulomb barrier, the projectile, then enters the nuclear body much beyond the nuclear 'skin'. Then the interaction of the nucleons in the incident nucleus will be very strong with the nucleus in the target nucleus giving rise to various phenomena, *e.g.*, (1) Fusion: So that the two nuclei completely merge into each other giving rise to the highly excited state of the compound nucleus, which decays through gamma emission. (2) Fusion-Neutron Evaporation: In this case, the compound nucleus may decay to another nuclear species, by the evaporation of many neutrons from the compound nucleus. This may lead to exotic nuclei. (3) Fusion-Fission: So that after the compound nucleus formation, it fission into two less heavy nuclei—according to laws of fission.

These various stages of heavy ion interaction can be expressed not only in terms of the impact parameters and energies, but also in terms of the angular momentum 1 carried by the projectile. Semi-classically, one can write:

$$1\hbar = pR_1 - k\hbar R_1 \qquad \text{...(17.1}c\text{)}$$

We can, now divide the different types of heavy ion nuclear reactions in term of impact parameters and angular momenta for the same incident energy. Defining $R_N = R_1 + R_2$, it can be seen from the above discussion, that if $R_I > R_N$, which corresponds to $I > I_N$, *i.e.*, the orbital angular momentum carried by the projectile is greater than the one given by Eq. 17.1*c*, if we change R_I to R_N. Then only Rutherford scattering or Coulomb excitation takes place. This corresponds to large distance or grazing collision. At a somewhat shorter impact parameter say $R_N > R_I > R_{DIC}$, the orbital angular momentum is also reduced and can be written as $I_N > I > I_{DIC}$, where DIC corresponds to Deep-Inelastic-Scattering, when there will be few nucleon transfer reactions. For still smaller impact parameter and hence orbital angular momenta say I_F, we get the compound nuclear formation and fission. Sometimes I_N is called I_{gr}, *i.e.*, for grazing collisions, I_{DIC} is the maximum value of 1 up to which deep inelastic scattering can take place and hence it is called I_{max} and I_F is called $I_{critical}$, because beyond this, the compound nucleus formation including fusion starts. This is also connected with fission, because of the process of fusion-fission. In the above discussion and referring to Fig. 17.1, R_N is the impact parameter which takes the projectile nucleus in the range of the nuclear forces of the target as shown in trajectory No. (2), *i.e.*, $R_N = R_1 + R_2 \approx R_{l_2}$. Again R_{DIC} is the impact parameter corresponding to deep inelastic collisions as shown in trajectory No. (3), *i.e.*, $R_{DIC} \approx R_3 = R_{l_3}$; $R_4 \approx R_{l_4}$ corresponds to a case, when compound nucleus is formed (fusion), *i.e.*, $R_4 \approx R_F$.

An interesting formulation of this division of reaction-mechanisms can be obtained by defining cross-section semi-classically as:

(*i*) Total cross-section

$$\sigma_T = \pi R^2 = \frac{\pi}{k^2} I^2 = \pi \lambda\!\!\!^{-2} I^2 \qquad \text{...(17.2)}$$

from Eq. 17.1*c*.

(*ii*) Cross-section for Compound Nucleus is given by:

$$\sigma_{CN} = \pi \lambda\!\!\!^{-2} I_F^2 = \pi R_F^2 = \pi R_{l_4}^2 \qquad \text{...(17.3)}$$

(*iii*) σ_{DIC} is defined in such a manner that for $R < R_4$, it is fusion and for $R_3 > R > R_4$, it is deep inelastic collision then, it can be seen that

$$\sigma_{DIC} = \pi\lambda\!\!\!^{-2}\left(I_{DIC}^2 - I_F^2\right) \qquad ...(17.4)$$

Figure 17.2 shows this division in terms of $d\sigma/dl$ as a function of l.

We have, thus described the full range of heavy ion-interactions semi-classically. While upward sloping line in Fig. 17.2 gives the semiclassical value of $d\sigma/dl$ and dotted vertical lines represent the various limits of 1 and the semiclassical values of $d\sigma/dl$, the shaded areas represent the approximate realistic values of $d\sigma/dl$ obtained from more detailed considerations. Here σ_D corresponds to direct reaction, σ_{EL} to elastic scattering, σ_{CE} to Coulomb excitation and σ_{CN} to compound nucleus formation.

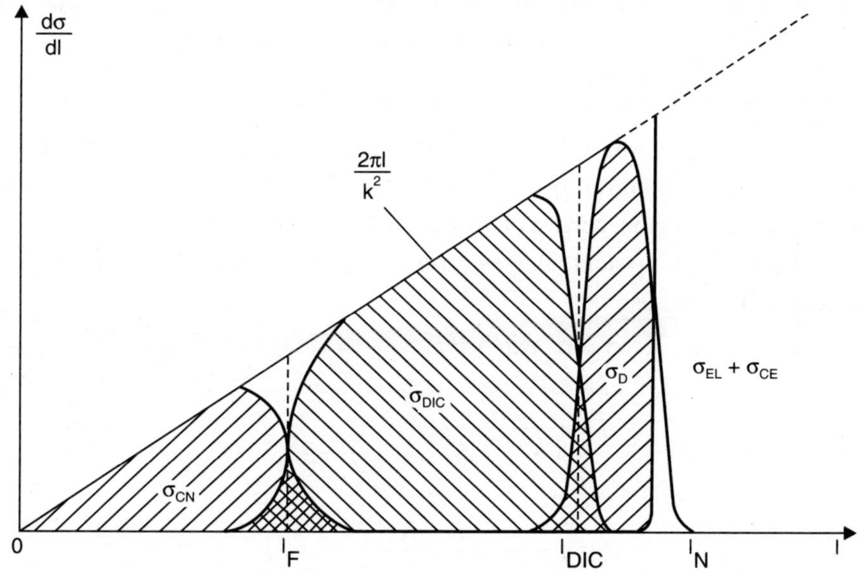

Fig. 17.2 dσ/dl as a function of l, showing different reaction mechanisms (Ref. 2).

17.2 ELASTIC SCATTERING

Some definitions: We have seen, as depicted in Fig. 17.1, that elastic scattering of heavy ions takes place:

(*a*) In pure Coulomb field as in trajectory (1),

(*b*) In Coulomb + nuclear field, as in trajectories (2) and (3) and

(*c*) In pure nuclear field, when one part of the compound nucleus decay is compound elastic scattering. However, the maximum elastic cross-section corresponds to a situation, when only pure Coulomb field is present. This can be seen, if one plots for various heavy ion-reactions, experimentally measured cross-section for elastic differential cross-sections as a function of $d(\theta)$, which is written as impact parameter $D(\theta)$ divided by:

$$\left(A_1^{1/3} + A_2^{1/3}\right)$$

It is seen, that the elastic cross-sections are nearly constant and are given by Rutherford cross-section, *i.e.*,

$$\frac{d\sigma}{d\Omega}\,(\text{Rutherford}) = \left(\frac{Z_1 Z_2 e^2}{4E}\right) \text{cosec}^4\left(\frac{\theta}{2}\right) \qquad ...(17.5)$$

(*see* Classical Mechanics by Goldstein)

for values of impact parameter greater than a certain value given empirically by:

$$R_N = 1.68\left(A_1^{1/3} + A_2^{1/3}\right)$$

But for values lower than this, the cross-section linearly falls to nearly a negligible portion. How do we explain this? [*see* Ref. (3) for details and the curve. Also *see* Fig. 17.7].

Qualitatively this can be understood again in terms of Fig. 17.1. There, $R_N \approx R_{L_2}$ corresponds to the interaction radius corresponding to a very small overlap between the density distribution of the interacting nuclei. The exponential fall-off of cross-sections for impact parameter less than R_N implies that there is absorption in this region, corresponding to nucleon transfer, inelastic scattering, deep inelastic scattering and finally due to compound nucleus formation. The probability of absorption in this region has been empirically expressed as:

$$\sigma = \sigma_0 \exp - Pd \qquad ...(17.6a)$$

where
$$d = \frac{R}{(A_1^{1/3} + A_2^{1/3})} \qquad ...(17.6b)$$

and
$$P = 1 - \exp\left(\frac{R - R_N}{\Delta}\right) \qquad ...(17.7)$$

with $\Delta \approx 0.55$ fm and σ_0 as the Rutherford cross-section for $R \approx R_N$; so for small impact parameters, elastic scattering fraction decreases, as $d(\theta)$ decreases, but above a certain value of $d(\theta)$; elastic scattering is constant.

In describing elastic scattering, a few terms have come in vogue, which have the physical significance in semiclassical treatment and also have been calculated quantum mechanically and have been used in literature extensively. These are (*i*) Critical angle θ_c and critical angular momenta l_c (*ii*) Deflection angle and (*iii*) Rainbow scattering and glory scattering. We describe and define them as below:

(*i*) **Critical angle** (θ_c): It is physically defined as the scattering angle for which the nuclear surfaces just touch. Referring to Fig. 17.1, this will correspond to trajectory (2), which is drawn in a way, so that classically the scattering is Rutherford for $\theta < \theta_c$ and zero for $\theta > \theta_c$. But quantum mechanically, the cross-section falls to zero only gradually for $\theta > \theta_c$. Referring to Fig. 17.3, we can define θ_c from the relationship of closest approach in such a way that θ_c corresponds to (Ref. 1):

$$R = a\left(1 + \text{cosec}\,\frac{\theta_c}{2}\right)$$

where $$R = R_1 + R_2;\ \text{and}\ a = \frac{Z_1 Z_2 e^2}{mv^2} \qquad ...(17.8)$$

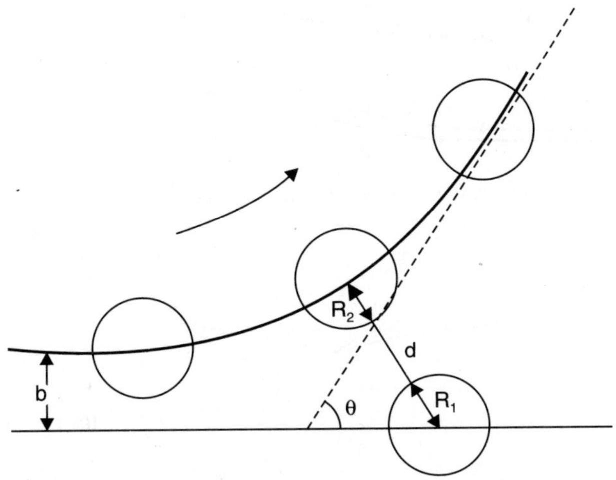

Fig. 17.3 Relation between the impact parameter b, the distance of closest approach R = $R_1 + R_2$ and the scattering angle θ.

Then Eq. 17.8 may be written as:

$$\sin\frac{\theta_c}{2} = \frac{a}{R-a} \qquad ...(17.9)$$

It is easy to see, that for $E_B = Z_1 Z_2 e^2/R$ and $E = 1/2\ mv^2$; $E_B/E \equiv 2a/R$ and for $E >> E_B$ and $a << R$

$$\theta_c = \frac{2a}{R} = \frac{E_B}{E} = \frac{2\eta}{kR} \qquad ...(17.10)$$

where $\eta \equiv Z_1 Z_2 e^2/\hbar v = ka$ is called Coulomb barrier.

Using $\hbar k = mv = p$; critical angular momenta (l_c), from Fig. 17.1 can be written as:

$$l_c \hbar = \mathbf{r} \times \mathbf{p} = bmv_c \qquad ...(17.11a)$$

Hence quantum mechanically

$$l_c (l_c + 1)\ \hbar^2 = b^2 m^2 v_c^2 \qquad ...(17.11b)$$

So the critical angular momenta l_c is connected to the critical angle θ_c, through Eqs. 17.11 and 17.10.

(*ii*) **Deflection Function θ (*b*):** The concept of deflection function is physically based on the fact that for a given θ, there may be many trajectories or for many values of *b*, they may go in the same direction. Some of these trajectories may correspond to θ, some to – θ and some to going around the scattering centre many times. In Fig. 17.4 we have shown independent trajectories *a*, *c* and *d* which are scattered to the same angle.

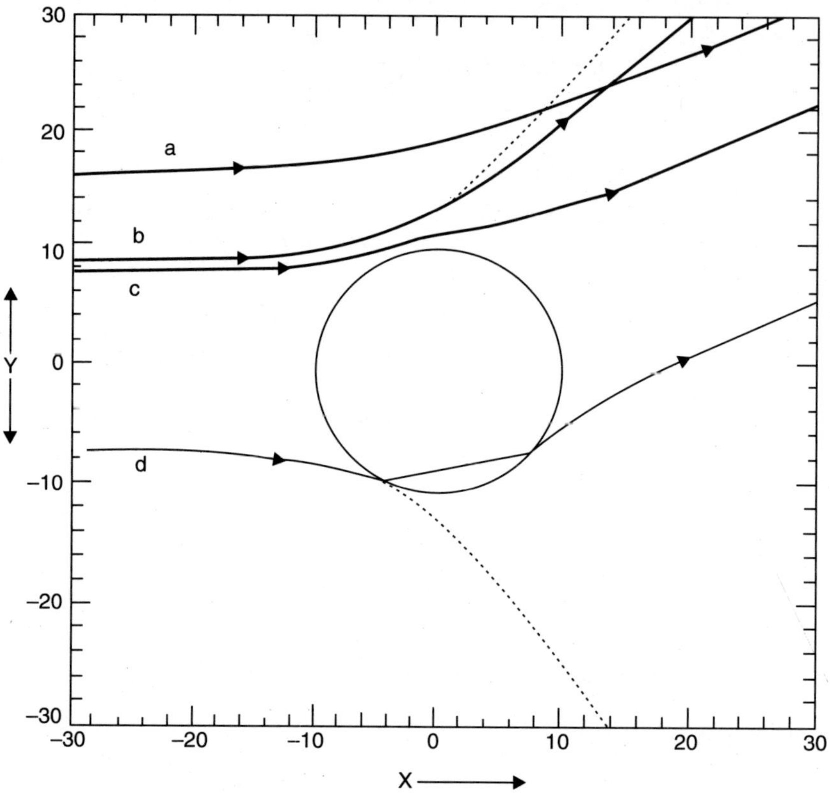

Fig. 17.4 Four classical orbits in the real part of the potential are plotted, three of which, having impact parameters, a, c and – d ($d < c$) scatter to the same angle. The orbit *b* may scatter to a stronger repulsion than *a* and therefore will scatter to a large angle. The circle represents the Woods-Saxon nuclear potential. The orbits are for O^{18} scattered by Sn^{120} at 100 MeV laboratory energy. The dotted curves show pure Coulomb orbits (Ref. 4). Y and X are the two coordinates of the trajectories, which are along Y and X axes of the coordinate system.

This phenomenon as shown in Fig. 17.4 can be described[5] by defining a function θ (*b*) given by:

$$\theta\,(b) \equiv \pi - 2b \int_{r_{\min}}^{\infty} \frac{1}{r^2} \left\{1 - \frac{V(r)}{E} - \frac{b^2}{r^2}\right\}^{1/2} dr \qquad \text{...(17.12)}$$

where $r_{\min}$ is the distance of closest approach at which value, the expression in the brackets in Eq. 17.12 by definition reduces to zero. If *V* (*r*) is positive, *i.e.,* repulsive, the deflection function is also positive, so for Coulomb potential only; θ (*b*) is positive. On the other hand, for an attractive potential like the nuclear potential, θ (*b*) is negative. Hence the character of θ (*b*) will depend on the relative strength of Coulomb and nuclear potentials. Some authors[5] have used the concept of deflection function to understand, in terms of the potential, the difference between several types of orbits.

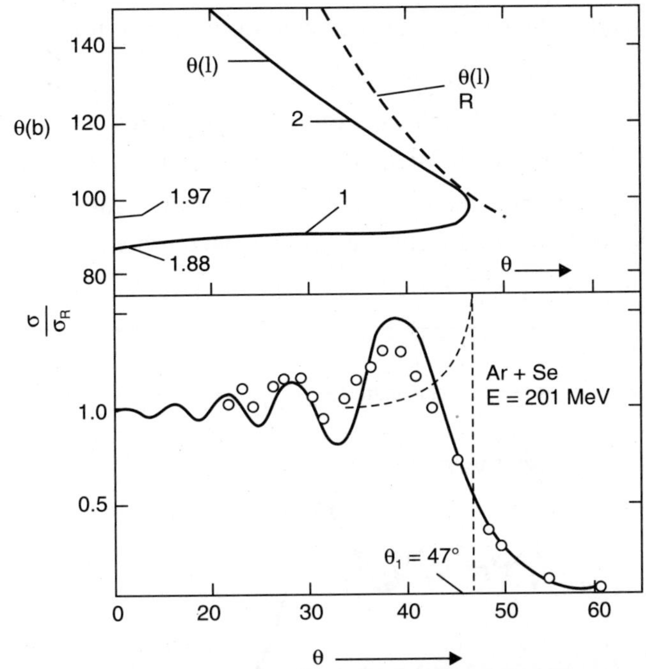

Fig. 17.5 Differential cross-section for the elastic scattering of Ar by Se at 201 MeV compared with calculations using the uniform Rainbow approximation (solid curve) and the pure classical result (broken curve). The upper curve shows the parameterized deflection function (solid curve) and the pure Coulomb deflection function (broken curve) (Ref. 7).

(*iii*) **Rainbow Scattering :** Classically, the scattering cross-section in terms of impact parameter b is given by (Classical Mechanics Goldstein):

$$\frac{d\sigma}{d\Omega} = \frac{b}{\sin\theta}\frac{db}{d\theta} = \frac{b}{\sin\theta}\frac{1}{k(d\theta/dl)} \qquad ...(17.13)$$

$$\left[\because \quad l\hbar = b\hbar k \text{ or } l = b\,k;\right.$$

or $$\left.\frac{dl}{d\theta} = \frac{db}{d\theta}k \text{ or } \frac{db}{d\theta} = \frac{1}{k}\frac{dl}{d\theta} = \frac{1}{k(d\theta/dl)}\right]$$

Therefore, $d\sigma/d\Omega$ will have singularity, when $d\theta/dl = 0$. It has been shown[5], that there are two singularities in the cross-sections at the angles corresponding to the inverse of deflection function having extremes. These angles, correspond to $d\theta/dl = 0$ and $d\theta/dl \approx \infty$. These angles are called the Rainbow angles because the theory, which explains this scattering is based on semiclassical Rainbow scattering mechanism.

Figure 17.5 shows the differential cross-sections[7] for the elastic scattering of Ar by Se at 201 MeV around the Rainbow angles. The calculated values using Rainbow scattering theory of Berry[6] are compared with the experimental points along with the classical theory. The shape of the cross-sections, evidently gives rise to a Rainbow type spectrum, corresponding to the fluctuation of cross-section for different angles.

The deflection function θ (*l*) may be expanded in the region of Rainbow angle as:

$$\theta (l) = \theta_R + q (1 - l_R)^2 \qquad ...(17.14)$$

where l_R is the angular momentum, corresponding to Rainbow angles. The cross-section as calculated by Ford and Wheeler[8] and further improved upon by Berry[6] are based on Rainbow type mechanism.

When θ (*l*) = 0 or π for finite *l*, the scattering is called glory scattering, because this corresponds to large bending in or the scattering in, for small angles and larger intensity. These are the special cases of Rainbow scattering.

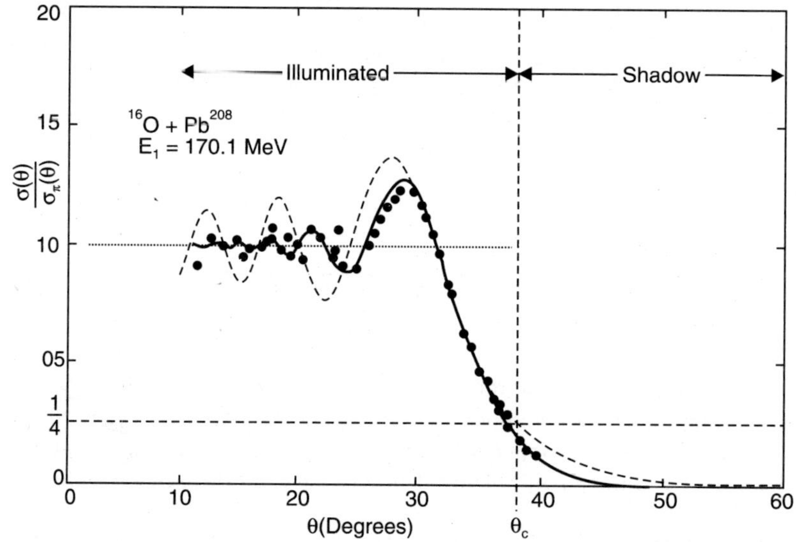

Fig. 17.6 Differential cross-section for the elastic scattering of 170.1 MeV O^{16} by Pb^{208} (Baker and McIntyre 1967) showing a typical Fresnel diffraction pattern. The solid curve is obtained from the strong absorption model (Frahn and Venter 1963) and the broken curve is the Fresnel cross-section (Ref. 9).

(*iv*) **Diffraction in Elastic Scattering:** Figure 17.6 shows[9] the experimentally measured, differential cross-sections for elastic scattering of 170.1 MeV O^{16} from Pb^{208}. A diffraction pattern is evident. A theory for diffraction developed semi-classically can be applied to such cases.

It is well known[10, 11] in Optics, that when both the source of light and the point of observation are situated at very large distances (infinite distances) from the screen, where diffraction pattern is formed, then the Fraunhauffer diffraction is observed. However, if the source is at a finite distance say by using a lense between the source of light and screen, then laws of Fresnel diffraction apply.

In elastic scattering of heavy ions, such conditions can be satisfied, by adjusting the critical angular momentum and critical angle in such a way, that

(*i*) $l_c \sin \theta_c << 1$ for Fraunhauffer Scattering

(*ii*) $l_c \sin \theta_c \geq 1$ for Fresnel Scattering ...(17.15)

How does one assume these conditions ? We have already seen from Eq. 17.10 that

$$\theta_c \approx \frac{2a}{R} = \frac{2\eta}{k\,R} \qquad ...(17.16a)$$

where
$$\eta \equiv \frac{Z_1 Z_2 e^2}{\hbar v} \quad \text{(Coulomb-Parameter)} \qquad ...(17.16b)$$

Putting
$$l_c = (l_c + 1) \approx \left(l_c + \frac{1}{2}\right)^2 = \Lambda^2 \qquad ...(17.17)$$

Then
$$\frac{\eta}{\Lambda} = \frac{\eta}{l_c + 1/2} \approx \frac{\eta}{l_c} \approx \frac{\eta}{kR} \approx \frac{\theta_c}{2} \qquad ...(17.18a)$$

or
$$\tan\frac{\theta_c}{2} \approx \frac{\eta}{\Lambda} \qquad ...(17.18b)$$

As η is an indicator of the Coulomb field strength and Λ indicates the energy of the incident ion and the target nucleus—through kR, it is evident that the general requirement for diffraction is that $l_c >> 1$ from Eq. 17.18 because this corresponds to small angle θ_c, required for diffraction. Then both k and b_c may be large [$l_c \approx kb_c$]. This makes incident energy higher than Coulomb energy and the impact parameter b_c is large enough, so that only Coulomb field is effective. Then condition (*i*) implies that θ_c is small or $\eta << \Lambda$ [Eq. 17.18], and condition (*ii*) means that θ_c is not so small. In this way Coulomb field determines whether scattering is Fraunhauffer (small Coulomb field) or Fresnel (large Coulomb field). Large Coulomb field bends the incident charges particles. They, therefore seem to come from a nearby distance. Thus Coulomb field acts like a lens.

17.3 ELASTIC AND INELASTIC SCATTERING

17.3.1 Nuclear and Coulomb

As discussed earlier, the inelastic scattering takes place mainly in the region where $R_{l_1} >> R >> R_1 + R_2$, where the trajectory is at a large distance from the target nucleus, so only Coulomb field is present. Then through Coulomb excitations, one of the interacting nuclei say target can get excited and the projectile goes out with that much less energy. This picture remains till $R_{l_2} \approx R_1 + R_2$, *i.e.* when the trajectory is such that the projectile just grazes over the boundary of the target nucleus.

In Fig. 17.7 is shown an experimental set of cross-sections of differential elastic and inelastic scattering (corresponding to excitation of 2^+ level in Ni^{58} at 1.45 MeV) for the reaction $Ni^{58} + O^{16}$ from 30 to 60 MeV incident energy of O^{16} at $\theta_{lab} = 60°$. A few features of the excitation function in elastic and inelastic cross-sections emerge.

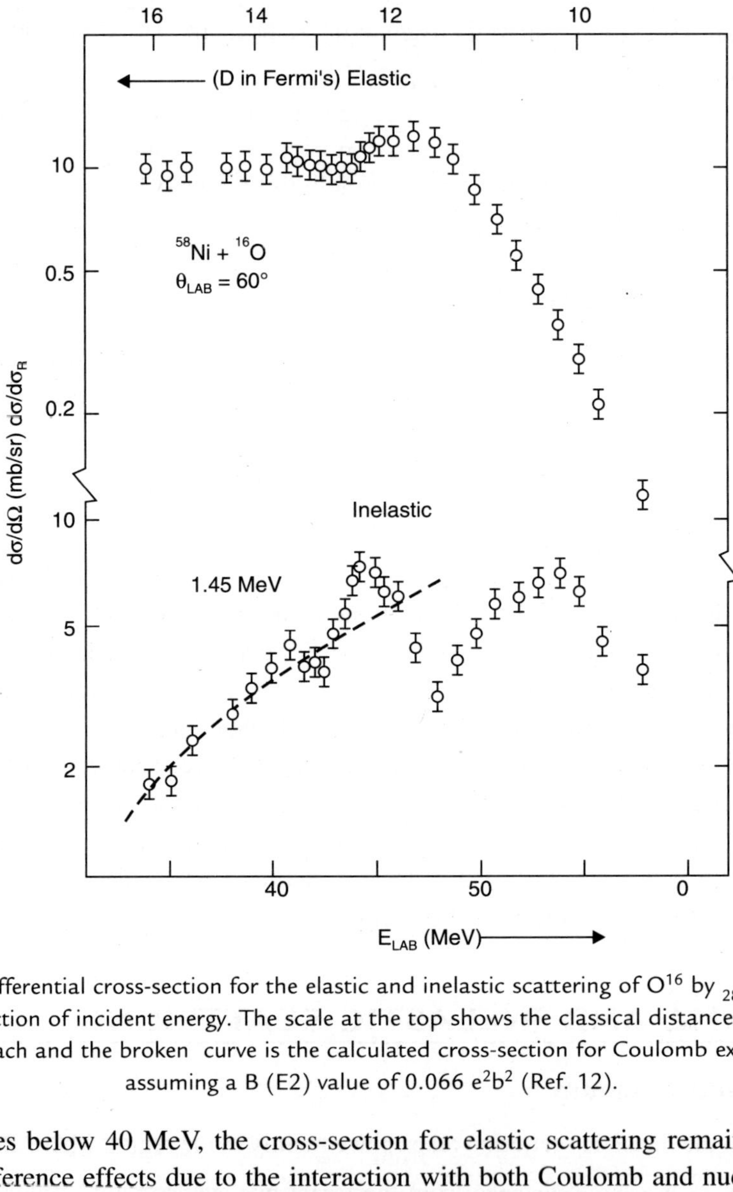

Fig. 17.7 Differential cross-section for the elastic and inelastic scattering of O^{16} by $_{28}Ni^{58}$ at $60°_{lab}$ as a function of incident energy. The scale at the top shows the classical distance of closest approach and the broken curve is the calculated cross-section for Coulomb excitation assuming a B (E2) value of 0.066 e^2b^2 (Ref. 12).

For energies below 40 MeV, the cross-section for elastic scattering remains constant. Above 40 MeV, the interference effects due to the interaction with both Coulomb and nuclear fields become important. This is evident in the figure, in the curve for elastic scattering, from 40 MeV to 50 MeV, beyond which the cross-section decays exponentially in the region, where nuclear absorption dominates. In the case of inelastic scattering, cross-section rises smoothly in the Coulomb interaction region, *i.e.,* below 40 MeV of incident energy. This corresponds to the width of the Coulomb barrier reducing slowly. Then we see again the interference effects between 40 MeV and 55 MeV and finally the exponential decay beyond 55 MeV, where nuclear absorption dominates. The interference phenomenon is a new feature of these observations.

To understand the interference between Coulomb and Nuclear interactions, in elastic and inelastic scattering one should realise that the deflection of an incident ion is not a monotonic function of the impact parameter. In reality, as we saw in Fig. 17.4, the three distinct orbits with different impact parameters, may contribute to the outgoing beam at the same angle. The contribution from these orbits leads to interference.

The qualitative features of this interference can be shown, in the case of elastic scattering by writing the cross-section at a particular angle as coherent superposition of the amplitudes from the turning points dominated by nuclear and Coulomb fields respectively. Then one can write:

$$\sigma_{\text{el}} = \left| (\sigma_N)^{1/2} \exp(i\delta_N) + (\sigma_c)^{1/2} \exp(i\delta_c) \right|^2 \quad \text{...(17.19)}$$

where $(\sigma_N)^{1/2}$ and $(\sigma_c)^{1/2}$ are scattering amplitudes due to the nuclear and Coulomb interactions, written semi-empirically in terms of the nuclear and Coulomb cross-sections σ_N and σ_c. The parameters δ_N and δ_c denote the phase factors. Writing $\sigma_N \approx \sigma_c = \sigma$, one can express, Eq. 17.19 as:

$$\sigma_{\text{el}} = \sigma \left| \exp(i\delta_N) + \exp(i\delta_c) \right|^2$$

$$= 2\sigma\left[1 + \cos(\delta_N + \delta_c)\right] \quad \text{...(17.20)}$$

Similarly, one can write for inelastic scattering cross-section:

$$\sigma_{\text{inel}} = \left| (\sigma_N)^{1/2} a_N^{(1)} \exp(i\delta_N) + (\sigma_c)^{1/2} a_c^{(1)} \exp(i\delta_c) \right|^2 \quad \text{...(17.21)}$$

where $a_N^{(1)}$ and $a_c^{(1)}$ are the parameters, related to the inelastic nucleon scattering.

The total potential $V(r)$ is, of course, the sum of the Coulomb $V_c(r)$ and nuclear potential $V_N(r)$ so that,

$$V(r) = V_c(r) + V_N(r) \quad \text{...(17.22a)}$$

where $V_c(r)$ has the shape of a Coulomb potential of two charged overlapping spheres. In practice, it is assumed, that such a potential can be replaced for sufficient accuracy by the potential between a point charge and a uniform spherical charge distribution of radius R, so that,

$$V_c(r) = \left\{ \frac{Z_1 Z_2 e^2}{2R} \left\{ 3 - \frac{r^2}{R^2} \right\} \right\} \text{ for } r \leq R;\ V_c(r) = \frac{Z_1 Z_2 e^2}{R} \text{ for } r \geq R$$

where,

$$R = r_c \left(A_1^{1/3} + A_2^{1/3} \right) \quad \text{...(17.22b)}$$

Also *see* Ref. (22). $V_N(r)$ is obtained from optical model assumption and is usually expressed to have the form:

$$V_N(r) = U f_V(r) + iW_V f_V(r) + iW_S\, g(r) \quad \text{...(17.22c)}$$

where U, W_V and W_S are the depths of the real volume, imaginary volume and imaginary surface potentials. The radial dependences $f_U(r)$, $f_V(r)$ and $g(r)$ are given by Saxon-Woods form factors, *i.e.*,

$$f_i(r) = \frac{1}{1 + \exp\{(r - R_i)/a_i\}} \quad \text{...(17.23)}$$

$$i = U, V \text{ and } g(r) = df_s/dr \quad \text{...(17.24)}$$

and R_i and a_i are radius and diffuseness parameters as have been explained in Chapter 6.

If only Coulomb potential is effective, then for indistinguishable interacting ions, *e.g.* $O^{16}-O^{16}$ or $Mg^{24}-Mg^{24}$ or $Si^{28}-Si^{28}$, etc.; the differential cross-sections for well below the Coulomb barrier, can be written as:

$$\frac{d\sigma}{d\Omega} = |\, f_c\,(\theta) \pm f_c\,(\pi - \theta)\,|^2 \qquad ...(17.25)$$

where $f_c\,(\theta)$ is classically given as:

$$f_c(\theta) = -\frac{Z^2 e^2}{4E_{cm}} \operatorname{cosec}^2 \frac{\theta}{2}\left(2 \exp\left(2i\sigma_o - 2i\gamma \ln \sin \frac{\theta}{2}\right)\right) \qquad ...(17.26)$$

where σ_o is the *S*-wave Coulomb phase shifts and γ is a constant connected with nuclear surface-tension and the ± sign in Eq. 17.25 refers to systems with symmetric and anti-symmetric spatial wave-functions respectively which is determined by the statistics of interacting particles. The scattering for *I*-spin particles is given by Mottscattering with the inclusion of spin effect.

At energies above or near Coulomb barrier, where both Coulomb and nuclear fields are present, one writes different expressions for particles of different spins. For particles of zero spin, the scattering amplitudes for θ and $\pi - \theta$ act coherently, so that the differential cross-section is given by:

$$\frac{d\sigma}{d\Omega} = |f\,(\theta) + f\,(\pi - \theta)\,|^2 \qquad ...(17.27)$$

Then keeping in mind, that both Coulomb and nuclear scattering amplitudes are involved in Eq. 17.27, one can write the composite cross-section above Coulomb barrier as:

$$\frac{d\sigma}{d\Omega} = 4\left| f_c(\theta) + \frac{1}{2ik} \sum_{L_{even}} (2L+1) \exp 2i\delta_L\,(S_L - 1)\, P_L \cos(\theta) \right|^2 \qquad ...(17.28)$$

where δ_L is phase shift corresponding to a given *L*, due to nuclear part of scattering and S_L is a quantity connected with scattering amplitudes. The scattering is, thus, described by even phase shifts.

For particles of spin 1/2 $\hbar$ the singlet and triplet cross-section add incoherently. So in the case of identical particles with spin 1/2 $\hbar$, the singlet total state (opposite spins), the spin wave-functions is anti-symmetric and vice versa for triplet state. This requires that the space part of the wave-function is symmetric for single state and anti-symmetric for triple state, to make the over-all wave function anti-symmetric. Hence the cross-section for spin 1/2 $\hbar$ particles is given by:

$$\frac{d\sigma}{d\Omega} = \frac{3}{4}|\, f(\theta) - f\,(\pi - \theta)\,|^2 + \frac{1}{4}|\, f(\theta) + f\,(\pi - \theta)\,|^2 \qquad ...(17.29)$$

$f(\theta)$ and $f\,(\pi - \theta)$ are again given by combining Coulomb and nuclear amplitudes as in Eq. 17.28. The cross-section of higher spins can also be similarly described.

17.3.2 Inelastic Scattering

Some Details: The inelastic scattering of heavy ions predominantly excites collective nuclear states and provides one of the most powerful ways of studying them. The collective nuclear states which are excited, may be rotational, vibrational (β, γ) or quadrupole (Quad), octopole (oct) or giant resonances (GR), Fig. 17.8, displays the excitation energies of these modes, excited in ion-ion collision in inelastic scattering.

In Fig. 17.8*a* are shown the result of calculations by Holm[13] et al.; for three typical cases: (*a*) Scattering of two Nd^{148} ions (*b*) Scattering of two Sn^{122} ions and (*c*) Scattering of Gd^{158} on U^{238}, for the time evolution of various collective modes of excitation, *e.g.* quadrupole vibration E_{quad}, octupole vibrations (E_{oct}), giant resonances (E_{GR}), β and γ vibration ($E_{\beta\gamma}$), rotational mode (E_{rot}) and fission (E_f). They used a collective model. Holm et al.[13] derived the coupled equations describing these processes and derived the excitation energy of a given excitation mode as a function of time, the time $t = 0$ corresponding to a point of the closest approach of the ions. It has also been possible (Fig. 17.8*b*), to calculate[14] the probabilities of excitation of the ground state band of rotational states of U^{238}. These results show the phenomenon of collective excitation in inelastic scattering; as theoretically calculated.

Since the excitation energies are usually small compared to the kinetic energies of ions in inelastic scattering, one can assume that ions traverse Coulomb trajectories (corresponding to elastic scattering) in each other's electrostatic field and then one calculates nuclear excitation quantum-mechanically.

Using this semiclassical approach, one can write the expression for the differential cross-section of the inelastic scattering as:

$$\left[\frac{d\sigma}{d\Omega}\right]_{\text{Inel}} = \left[\frac{d\sigma}{d\Omega}\right]_{\text{Rutherford}} \times P_{if} \qquad \text{...(17.30)}$$

where $[d\sigma/d\Omega]_{\text{Ruth}}$ is the differential cross-section for elastic scattering, which has been discussed earlier and P_{if} indicates the transition from the initial state *i* to the final state *f* and can be evaluated by the time dependent perturbation theory, *i.e.*,

$$P_{\text{if}} = \frac{1}{i\hbar}\int_{-\infty}^{\infty} (f \mid H_{\text{int}}(t) \mid i) \times \exp\left[i\frac{E_f - E_i}{\hbar}\right] dt \qquad \text{...(17.31)}$$

where E_i and E_f are the energies of the initial and final states and $H_{\text{int}}(t)$ indicates the electromagnetic interaction between the interaction ions.

The transition probability P_{if} may be calculated under the following assumptions:

If the probability of multiple excitation is large, because of strong Coulomb field acting for a long time, then one solves a set of coupled equations connecting the occupation probabilities of the states of the excited nucleus. Such equations arise from many-level coupled Schrödinger equation, *e.g.*,

$$\sum_{\alpha} P_\alpha(t)\,(n \mid H_{\text{int}} \mid \alpha)\exp\left[-\frac{iE_{\alpha t}}{\hbar}\right] = i\hbar\dot{P}_n(t)\exp\left\{-\frac{iE_{nt}}{\hbar}\right\} \qquad \text{...(17.32)}$$

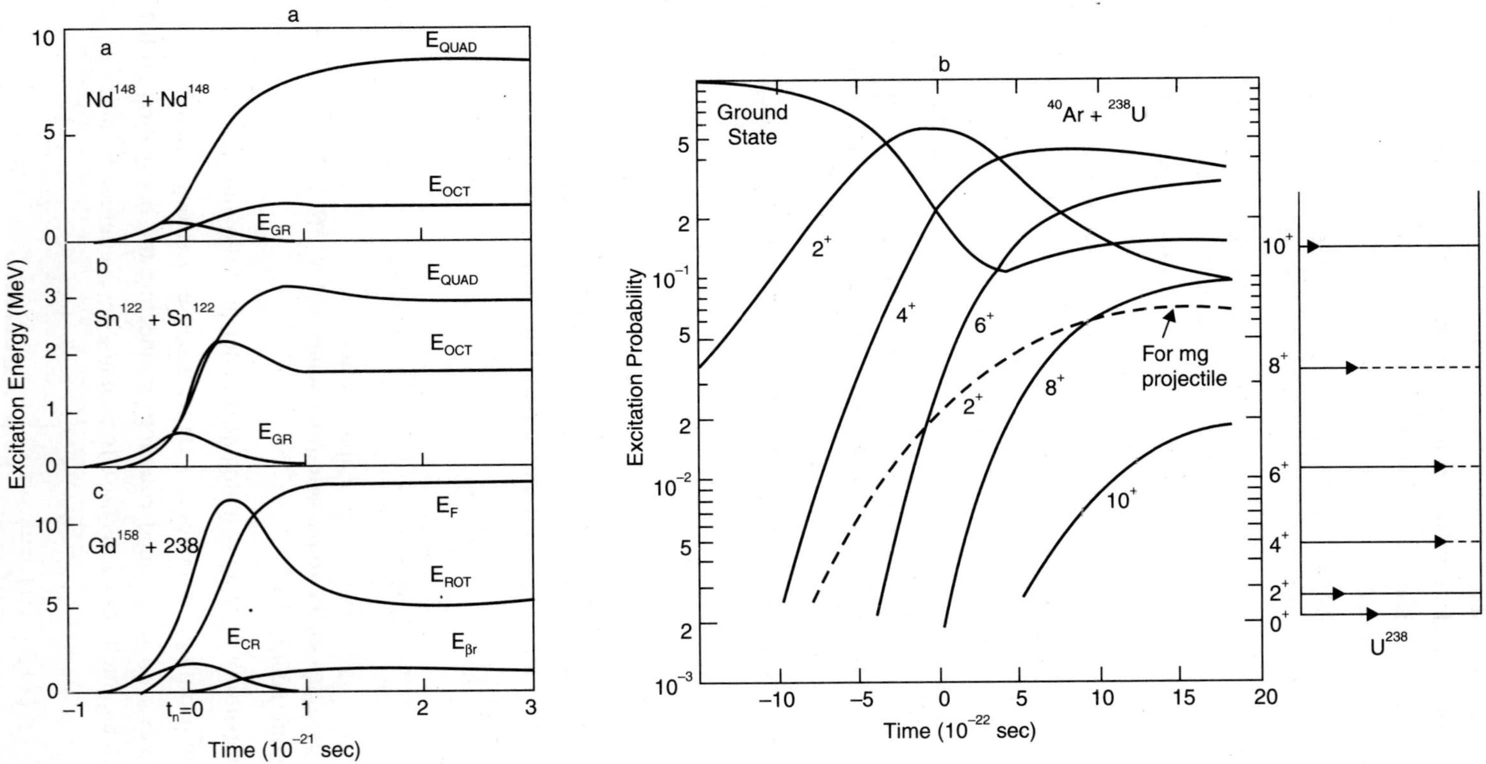

Fig. 17.8 The time evolution of different collective modes of excitation of inelastic scattering: (*a*) Quadrupole vibration, (E_{quad}), Octupole vibration (E_{oct}) giant resonance (E_{GR}) Beta and gamma vibration ($E_{\beta\gamma}$) fission (E_f) and rotation (E_{rot}) as calculated in Ref. (13); (*b*) Excitation of ground state rotational band of U^{238} (Ref. 13, 14).

where a state n is coupled to many states denoted by α. Such coupled equations have been extensively used, by K. Alder and H. Pauli[15] and A. Winther and J. De Boer[16] for calculation of Coulomb excitation probabilities; especially for E_2 transitions involving first excited 2^+ states. Involved computer programs have been written for this purpose.

Coulomb excitation by heavy ions, involving multi-Coulomb excitation, especially at backward scattering angles have been calculated for many cases, using this method.

If the coupling between elastic and inelastic scattering is weak, the transition matrix element is calculated by using the perturbation theory under the DWBA approximation. Then the transition matrix element, (*i.e.,* cross-section for inelastic scattering) is given by, for transferred angular momenta 1 using a DWBA expression[16], by L.C. Briedenhasn and Brussard.

The above formulation has been used also, when the coupling between the elastic and inelastic scattering is strong. Then one uses the coupled channel formalism in which the total wave-function is expressed as the sum of the wave-functions in the elastic and inelastic channels multiplied by the appropriate nuclear structure factors.

The matrix elements or form factors $F_L(r)$ are obtained separately for Coulomb and nuclear forces, so that the total form factor is given by the sum, *i.e.,*

$$F_L(r) = F_L^c(r) + F_L^N(r) \qquad ...(17.33)$$

where subscripts c and N on the right side denote the Coulomb and nuclear parts.

The optical potential for combined nuclear and Coulomb interaction is given by:

$$V(r) = V_c(r) + Uf(r) + iWg(r) \qquad ...(17.34)$$

as earlier discussed in Eq. 17.22. Inclusion of Coulomb potential $V_c(r)$, means that very many partial waves contribute to the interaction and that the radial wave equation has to be integrated to a large radius to secure sufficient accuracy. This is the case of Coulomb excitation, using heavy ions.

17.3.3 Experimental Results

Experimental data is, now available for differential cross-sections and excitation function, both for elastic and inelastic scattering at different incident energies and the comparison with the theories as discussed, earlier. It may be mentioned here, that apart from the various approximations used in theories, there is a choice of the parameters of optical model potentials in the theories, especially at higher energies, where nuclear effects are important.

Figure 17.9 depicts the differential cross-section for elastic scattering of $Si^{28} + Si^{28}$, at $E = 20$ MeV which is well below the Coulomb barrier. The comparison with the Mott scattering formula is exact showing, that the semi-empirical approach of the theory is completely applicable.

Similarly the excitation function for $O^{16} + O^{16}$ elastic scattering at different angles and comparison with optical model calculations with a Saxon-Woods potentials, shows, that agreement is quite good, qualitatively (Ref. 18).

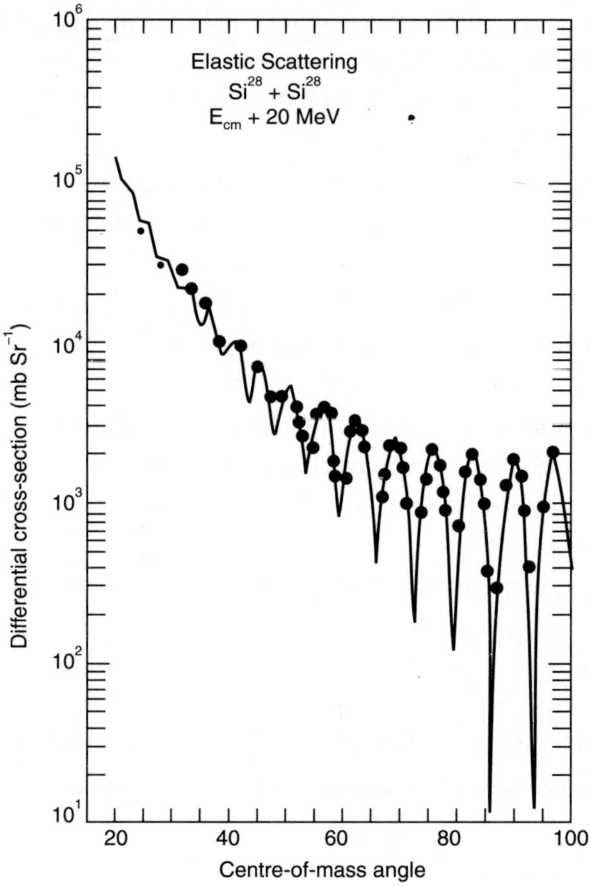

Fig. 17.9 The case of pure Rutherford scattering for Si^{28} + Si^{28} at 20 MeV incident energy. It is much below Coulomb barrier. The theoretical solid line correspond to pure Mott scattering (Ref. 17).

Also the data of differential cross-section for the inelastic scattering of 168 MeV C^{12} on O^{16} with the excitation of 4.43 MeV state of C^{12}, gives a very good fit, when compared with distorted wave calculations (Ref. 19).

Sometimes a coupled channel calculation is more suitable compared to single channel DWBA calculation, because of the presence of nearby two levels. Then one includes couplings of (*i*) direct excitation (*ii*) de-excitation and (*iii*) reorientation. This was especially shown[20] in the case of inelastic scattering of O^{16} by Ni^{58}. While in the elastic scattering; there is not much difference between DWBA and coupled channel calculations, in inelastic scattering, this becomes quite evident, as shown in Fig. 17.10.

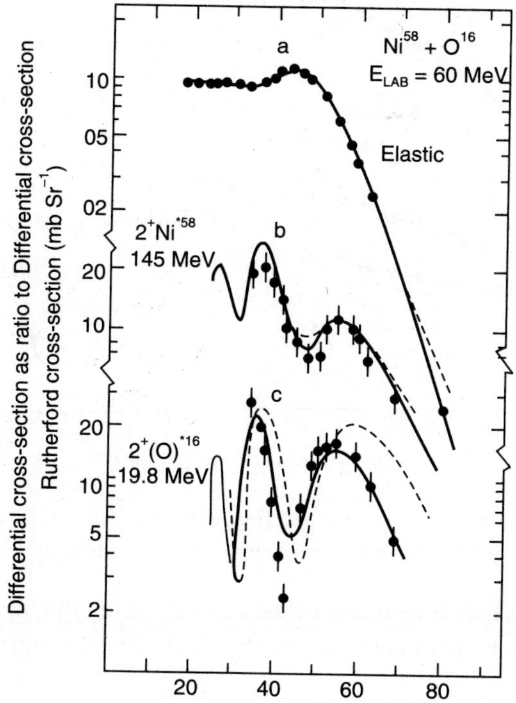

Fig. 17.10 Differential cross-sections for (*a*) the elastic and (*b*) and (*c*) inelastic scattering; of 60 MeV O^{16} ions by Ni^{58} compared with DWBA (broken curves) and coupled channels (solid curves calculations) Ref. 20.

17.4 HEAVY ION POTENTIALS

17.4.1 Optical Potentials

We have already shown that in heavy ion interaction, a modified Coulomb interaction Eq. 17.22*b* and a nuclear optical potential as given in Eqs. 17.22*c*, 17.23 and 17.24 are applicable in understanding both elastic and inelastic scattering. The various parameters of these potentials are generally, obtained semi-empirically.

Figure 17.11 shows[21, 22], the energy dependent behaviour of the real part and imaginary part of these potentials, which fit the data of $O^{16} - O^{16}$ elastic scattering, as given in Ref. 28. It has also been possible to get the energy dependence of $\bar{f}$ in Eqs. 17.23 and 17.24, which is the strength parameter, semi-empirically, of the real and imaginary potential[21]. These values of optical parameters are typical, because the closed shell structure of O^{16} increases the probability of isolating the specific interaction mechanism.

Theoretically, heavy ion optical potentials have two parts: (*i*) Electrostatic (*ii*) Nuclear. We have already discussed these earlier to some extent. The shapes of electrostatic potential[22] [Eq. 17.22*b*] are quite understandable in terms of overlap of the two charged spheres. But the nuclear part, whose shape has also been discussed in Eqs. 17.22 and 17.34, require some discussion about their origin.

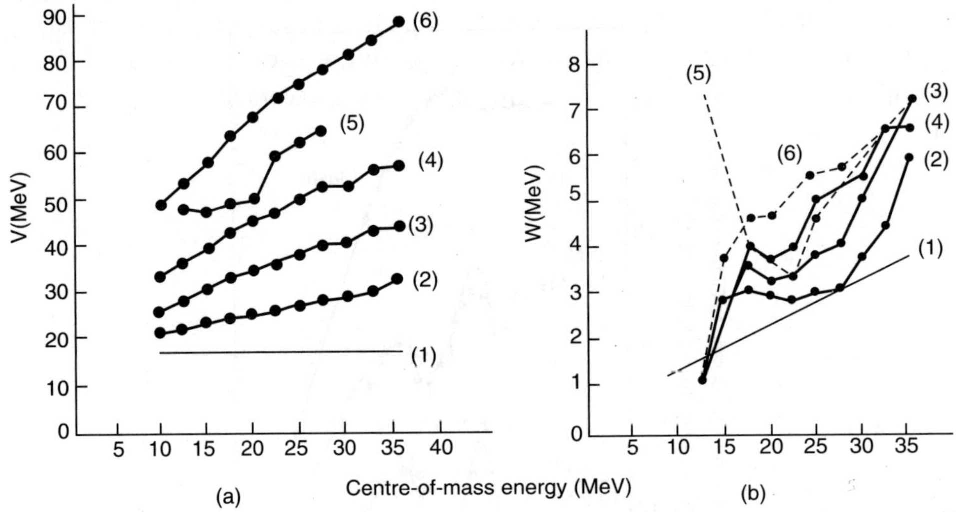

Fig. 17.11 (*a*) Real potential, V_v and (*b*) Imaginary potential W_v, [Eq. 17.22], which fit the data of O^{16}–O^{16} elastic scattering of J.V. Mehr et al. (Ref. 21. 22).

In general, the nuclear part is estimated classically by using the liquid drop theory. This requires that the optical potential may be parameterised by the Saxon-Woods expression:

$$V(r) = \frac{V_o}{1 + \exp[(r - R_o)/a]} \qquad ...(17.35)$$

The three parameters, *i.e.* V_o, R_o and 'a' are determined from the following considerations:

(*i*) The attractive force F at the sum R of the half density radii[21] R_1 and R_2 is given by:

$$F = 4\,\pi\,\gamma\,\frac{R_1\,R_2}{R_1 + R_2} \qquad ...(17.36)$$

$\gamma = 95$ MeV– fm^{-2} is called the surface tension coefficient. F therefore, can be found from Eq. 17.35 by writing:

$$F = -\left(\frac{\partial V}{\partial r}\right)_{r = R_o} = \frac{V_o}{4a} \qquad ...(17.37)$$

Again the potential energy of the system at $r = 0$ is given by:

$$V_o = b_3\left[A_1^{2/3} + A_2^{2/3} - (A_1 + A_2)^{2/3}\right] \qquad ...(17.38)$$

where $b_3 = 17$ MeV, is the surface energy parameter. So combining Eqs. 17.37 and 17.38 one can find V_o and a. The half density radii R_1 and R_2 are given by:

$$R_{1,2} = 1.128\,A_{1,2}^{1/3}\left(1 - 0.786\,A_{1,2}^{-2/3}\right) \qquad ...(17.39)$$

Hence all the parameters of Eq. 17.35 are estimated. The comparison of optical potential obtained in this manner compares favourably with obtained from analysis of scattering data.

One can combine Eqs. 17.36, 17.37 and 17.38 and obtain the expression for the diffusion parameter as:

$$a = \frac{0.356\, R_o \left[A_1^{2/3} + A_2^{2/3} - (A_1 + A_2)^{2/3}\right]}{R_1\, R_2} \qquad \text{...(17.40)}$$

Typical optical potentials (*a*) for heavy ions determined[22, 25] from the liquid drop model or (*b*) from optical model analysis, is given below:

(*i*) $O^{16} + Ca^{40}$: (*a*) $V_o = 57.9$ MeV, $R_o = 6.09$ Fm, $r_o = 1.025$ Fm, $a = 0.825$ Fm
(*b*) $V_o = 59.5$ MeV, $R_o = 6.09$ Fm, $r_o = 1.025$ Fm, $a = 0.846$ Fm

(*ii*) $O^{16} + Sn^{120}$: (*a*) $V_o = 72.0$ MeV, $R_o = 7.87$ Fm, $r_o = 1.056$ Fm, $a = 0.885$ Fm
(*b*) $V_o = 69.1$ MeV, $R_o = 7.87$ Fm, $r_o = 1.056$ Fm, $a = 0.850$ Fm

17.4.2 The Folding Model

This model assumes that the reactions between two nuclei goes so fast that the target nucleus is undisturbed, then the potential between a nucleon of the incoming nucleus and the target nucleus is the sum of all the constituent nucleon-nucleon interactions $V_n(r)$, so that, for a nucleon-nucleus interaction, one writes a folding integral between the nuclear density ρ (*r*) and the nucleon-nucleon interaction:

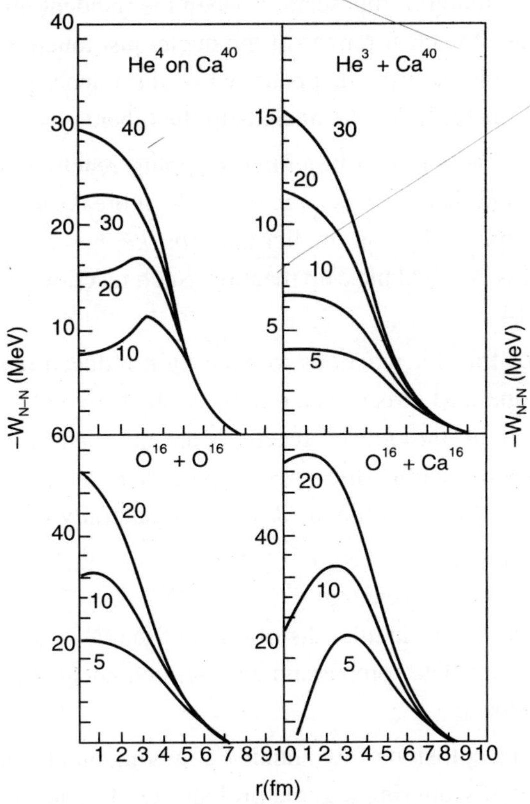

Fig. 17.12 Imaginary optical potentials calculated by folding model for various particles (Ref. 23). The number on the curves represent energies per nucleon of projectile.

$$v_n (|\mathbf{r} - \mathbf{r}_1|)$$

i.e.,

$$V_n (r) = \int \rho (\mathbf{r}_1)\, v_n (\mathbf{r} - \mathbf{r}_1)\, dr_1 \qquad ...(17.41)$$

Then for heavy ion-heavy ion interaction, one writes either as a single folding of the nucleon-nucleus potential of Eq. 17.41, *i.e.* one writes:

$$V_1 (r) = \int \rho (\mathbf{r})\, v_n (\mathbf{r} - \mathbf{r}_1)\, dr_1 \qquad ...(17.42)$$

or as a double folding of the nucleon-nucleon interactions, *i.e.*,

$$V_2 (r) = \int\int \rho_1 (\mathbf{r}_1)\, \rho_2 (\mathbf{r}_2)\, v_n (\mathbf{r}_{12})\, d\mathbf{r}_1\, d\mathbf{r}_2 \qquad ...(17.43)$$

A more detailed folding model of the heavy ion optical potential has been developed by Dover and Vary (Ref. 21). Also *see* Holm and Greiner (Ref. 23) for details. B. Sinha[23] has calculated imaginary optical potential by using folding model given in Fig. 17.12.

17.5 NUCLEON-TRANSFER REACTIONS

17.5.1 Theoretical (One-Nucleon Transfer)

As briefly discussed in the beginning of this section, when the incident projectile just grazes along the surface of the target, the edges of nuclear ranges of the nuclei just touch, so that only the outer portion of the skins of the two nuclei interact (this happens, when the impact parameter $R_L \approx R_1 + R_2$). This leads to one or two nucleon transfer from one nucleus to the other.

Such nucleon-transfer reaction, can be either stripping reactions or pick-up reactions. If the reaction is described as $A(a, b)\, B$, then if 'a' is larger than 'b', it breaks up into $a \to b + x$ and x is added to A to become B. This is stripping. If 'a' is smaller than 'b', *i.e.* $b = a + x$ and x is picked from A and added to 'a' to become 'b'. This is called pick-up reaction. Such reactions have been discussed earlier in Direct Reactions in Chapter 14.

Basically the theory of transfer reactions in heavy ion induced reactions is similar to the one discussed for light projectile-induced direct reaction, except that several mathematical approximations that are commonly made for light ion induced reactions are not made for heavy ion induced transfer reactions. Physically, in heavy ion reactions, the outgoing particle is in the excited state and the transferred particle may not be in S-state in the projectile, as it happens generally in light ion induced reactions.

In the early years of heavy ion interactions (1950–70), one-nucleon transfer below the Coulomb barrier was extensively studied. A typical example[25] is Ca^{40} (N^{14}, N^{13}) Ca^{41} at energies between 25.6 MeV and 29.5 MeV. Figure 17.13 shows the angular distribution of the (N^{14}, N^{13}) reaction. The data includes the transitions to all possible states. The comparison with semiclassical tunneling theory, is quite fitting; which can be expressed as follows:

The transfer process in a collision of two nuclei[26], moving on classical scattering orbits, will be proportional to the probability of scattering and the probability of neutron to tunnel from one nuclear

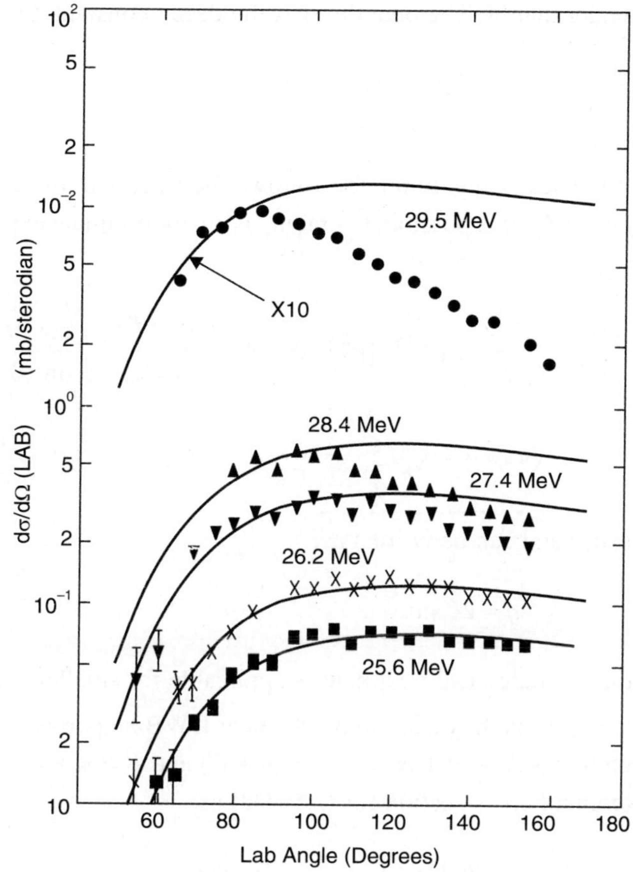

Fig. 17.13 Angular distribution of one nucleon (neutron) transfer reaction[25] Ca^{40} (N^{14}, N^{13}) Ca^{41}, at five different energies. Solid lines are predicted on the basis of semiclassical tunneling theory (Ref. 26).

potential to the other, *i.e.*,

$$\sigma(\theta) = |f_{\text{trans}}(\theta)|^2 = |f_{\text{scatt}}(\theta)|^2 \cdot |f_{\text{tun}}(\theta)|^2 \qquad ...(17.44)$$

The last factor corresponding to an overlap between the initial and final states for a reaction; $a + A \to b + B$ with transfer of particle C, $[a = (b + C), B = (A + C)]$, we obtain, neglecting the difference of binding energy E_B of neutron between the initial and final channel:

$$f_{\text{tun}}(\theta) = \langle \psi_B \psi_b | V | \psi_a \psi_A \rangle$$

$$= \theta^B \theta^b N_B N_b \left(\frac{e^{-\alpha R}}{\alpha R} \right) \qquad ...(17.45)$$

where θ^B and θ^b are spectroscopic amplitudes, (discussed earlier in Chapter 14) for particle C in nuclei B and b; and N_B and N_b are the normalisation factors and $e^{-\alpha R}/\alpha R$ is the shape of the nuclear potential (Yukawa). Equation 17.45 is proportional to the probability of finding the neutron (particle C) outside the nucleus, at distance greater than R. The quantity α is the decay constant, given by:

$$\alpha = \left(\frac{E_B \, 2\,mc}{\hbar^2}\right)^{1/2} \qquad \text{...(17.46)}$$

Remembering, that the scattering amplitude follows the treatment of scattering of protons but without Coulomb part (Chapter 4), it is possible to obtain, as an approximate expression for the neutron transfer cross-section:

$$\frac{d\sigma}{d\theta} \approx \frac{Z_1 Z_2}{(2E^2)^2}\left(\theta^B\right)^2\left(\theta^b\right)^2 N_B^2 N_b^2 \left(\frac{e^{-2\alpha R_{\min}}}{\alpha R_{\min}}\right)\frac{\cos\theta/2}{\sin^3\theta/2}$$

$$\approx S^B S^b N_B^2 N_b^2 \frac{C}{q^4} e^{-2\alpha R_{\min}} \qquad \text{...(17.47}a\text{)}$$

where q = momentum transfer and can be written as:

$$q \approx 2k \sin\frac{\theta}{2} \qquad \text{...(17.47}b\text{)}$$

It is evident, that this semiclassical treatment is applicable for smaller angles.

A more appropriate theory is the quantum mechanical DWBA approach in which the transition amplitude T_{ab} for the reaction $a + A \to b + B$ with $a = (b + C)$ and $B = A + C$ can be written quantum mechanically and the differential cross-section for (j_c = 1/2) as:

$$\frac{d\sigma}{d\theta} = |T_{ab}|^2 = 4\pi \frac{m_i m_f}{\left(2\pi\hbar^2\right)}\frac{k_f}{k_i}\frac{2I_B+1}{(2I_A+1)(2j_2+1)} \times$$

$$\left(O_{l_1 j_1}\right)^2\left(O_{l_2 j_2}\right)^2 N_1^2 N_2^2 |A_{l_1}|^2 \times \sum_{l,m}\left(j_1 \frac{1}{2} l_1 0 \middle| j_2 \frac{1}{2}\right)^2 |T_{lm}(\theta)|^2 \qquad \text{...(17.48)}$$

where $O_{l_1 j_1}$ and $O_{l_2 j_2}$ are spectroscopic amplitudes and N_1 and N_2 are the normalisation factors, as discussed in connection with Eq. 17.47*a*. T_{lm} (θ) gives the angular dependence. For details *see* Ref. (26). $[A_{l_1}]^2$ is a constant, obtained by integrating over $\mathbf{r}_1$ (or $\mathbf{r}_2$).

The other problems, connected with nuclear-transfer are:

(*i*) Transfer reaction at high energies.

(*ii*) Surface effects in nucleon-transfer.

(*iii*) Multi-nucleon transfer reactions and nuclear spectroscopy, *e.g.* two-nucleon, three nucleon or four nucleon or even more than four nucleon transfer.

We will discuss some experimental facts about these problems in the next section without discussing theory.

17.5.2 Selection Rules

As mentioned in Chapter 14, the total angular momentum is conserved, throughout the transfer reactions. This gives rise to several selection rules[27, 28].

Defining $\mathbf{J}_A$, $\mathbf{J}_B$ as the spins of A and B, $\mathbf{S}_a$ and $\mathbf{S}_b$ as spins of a and b, in the reaction $a + A \rightarrow b + B$, where $a = b + X$ and $A + X = B$ and $\mathbf{L}_1$ and $\mathbf{L}_2$ as the orbital angular momenta of X bound to b and A, then the total angular momentum transfer $\mathbf{J}$ is given by:

$$\mathbf{J} = \mathbf{J}_A - \mathbf{J}_B;\ \mathbf{L} = \mathbf{L}_1 - \mathbf{L}_2;\ \mathbf{S} = \mathbf{S}_a - \mathbf{S}_b;\ \mathbf{J} = \mathbf{L} + \mathbf{S} \qquad ...(17.49)$$

If $\mathbf{J}_A$ and $\mathbf{J}_B$ do not change directions which corresponds to the core remaining inert throughout the reaction, then if $\mathbf{J}_1$ and $\mathbf{J}_2$ are the total angular momenta of X in the two bound states and if there is no spin-orbit interaction in the incoming and outgoing channels, then

$$\mathbf{S} = \mathbf{J}_1 \text{ where } \mathbf{J}_1 = \mathbf{L}_1 + \mathbf{S}_X;$$

and

$$\mathbf{J} = \mathbf{J}_2 \text{ where } \mathbf{J}_2 = \mathbf{L}_2 + \mathbf{S}_X \qquad ...(17.50)$$

Further, if a nucleon is transferred from a projectile orbit, specified by quantum numbers $n_1\ l_1\ j_1$ to an orbit $n_2\ l_2\ j_2$ in the residual nucleus, the allowed values of the orbital angular momentum transfer L in the no-recoil formalism are given by:

$$|\ L_1 - L_2\ | \leq L \leq |\ L_1 + L_2\ |$$

$$|\ j_1 - j_2\ | \leq J \leq |\ j_1 + j_2\ |;\ l_1 + l_2 + L = \text{even} \qquad ...(17.51)$$

An interesting feature of these heavy ion reactions is that the largest contribution comes from grazing collisions, in which the ions just touch each other. This increases the surface effect and therefore higher values of L are enhanced and this can lead to J-dependent effects in the cross-sections[27, 28].

17.5.3 Experimental Facts about Nuclear Transfer Reactions

I. One Nucleon Transfer Reactions: In light nuclei, we have come across one nucleon transfer reactions like (d, p), (p, d), (d, n), (α, He^3), etc. In heavy ion reactions, we have similarly (O^{16}, N^{15}), (C^{12}, B^{11}), (N^{14}, C^{13}), reactions which behave like (d, n) stripping reactions where one proton is stripped. Similarly (N^{14}, O^{15}), (N^{15}, C^{14}) are like (n, d) reactions where one proton is picked up. Similarly there are neutron stripping reactions, *e.g.* (N^{14}, N^{13}) similar to (d, p) and one-neutron pick-up reactions, *e.g.* (B^{11}, B^{12}) similar to (p, d) reactions. Many of these reactions have been studied on nuclei with A = 50–110 and have been compared with theories discussed earlier[28, 30].

An interesting feature of these studies is the similarity of spectroscopic factors (S) obtained in (He^3, d) and (O^{16}, N^{15}) for $1f_{7/2}\ 2p_{3/2}$ final states in the Sc isotopes, as shown in Fig. 17.14. This shows, that essentially the basic mechanism of the reaction in the light and heavy ion reactions for the nucleon transfer is very similar. The comparison of experimental cross-sections with the theory with different approximations, brings out the following interesting points.

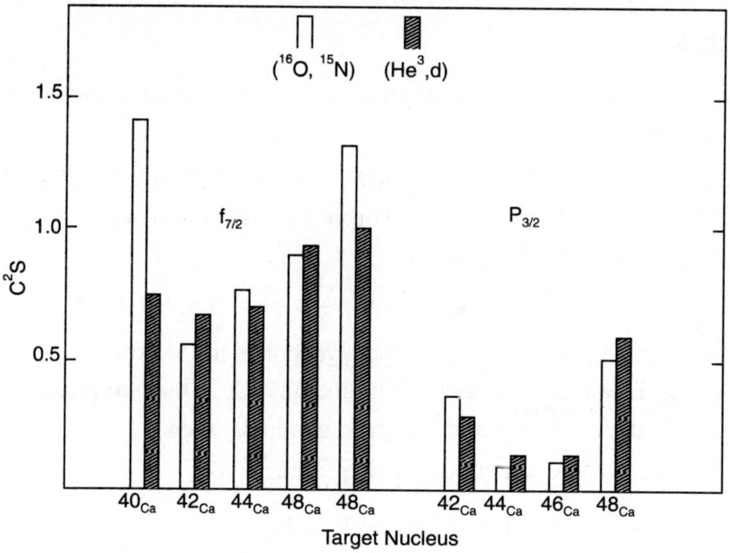

Fig. 17.14 Comparison between (He^3, d) and (O^{16}, N^{15}) spectroscopic factors for 1$f_{7/2}$ and 2$p_{3/2}$ states in the Sc isotopes (Ref. 28).

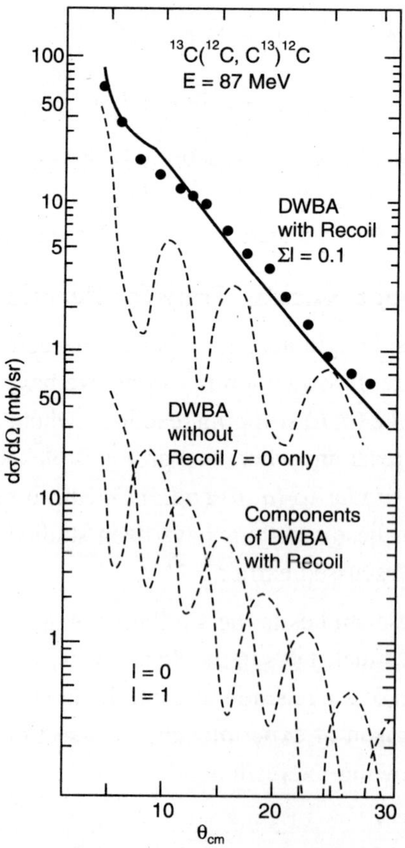

Fig. 17.15 Differential cross-sections for the reaction C^{13} (C^{12}, C^{13}) C^{12} at E_{lab} = 87 MeV (Ref. 29).

The zero range approximation with no recoil assumption is not successful in heavy ion-induced transfer reactions. In general, DWBA calculations with recoil and with finite range for the interaction potential are applicable to heavy ion-data, as shown in Fig. 17.15, where the differential cross-sections of the reaction C^{13} (C^{12}, C^{13}) C^{12}, at $E = 87$ MeV are compared with DWBA with and without recoil. It is evident that DWBA with recoil are the most successful in explaining the data.

Not only are the DWBA calculations with finite range, more appropriate, but they are sensitive to the different parameters of the optical potential used, as is indicated, for Pb^{208} (B^{11}, B^{12}) Pb^{207}, a case of one-neutron pickup reactions. The double folding model potential, as discussed earlier, has also been successfully used to explain the experimental data; Ca^{48} (N^{14}, C^{13}) Sc^{49} and Ca^{40} (C^{13}, C^{12}) Ca^{41}, (Ref. 30).

An interesting case of coherent addition of transition amplitudes of inelastic scattering via direct inelastic scattering and one-nucleon transfer reaction is provided by the study of O^{17} (O^{16}, O'^{16}) O^{17}. Here, apart from the direct inelastic scattering, one can have a one-neutron exchange between O^{17} (target) and O^{16} (projectile) so that: $O^{17} + O^{16} \rightarrow O^{16} + O^{17}$ (*see* Ref. 30).

II. Two Nucleon Transfer Reaction: In general, the two nucleon transfer case is the simplest case of multi-nucleon transfer case.

The wave-functions of the two transferred nucleons, may be calculated as eigenvalues of a Saxon-Woods potential with depth adjusted to give each nucleon one half the separation energy of the pair of nucleons. This theory has been developed under various approximations and formulations.

The first and second order DWBA theory for two nucleon transfer reaction has been developed in the no-recoil approximation by Kamuri[31]. They have been also further developed, to include the recoil efforts. Figure 17.16 shows the comparison of the bell-shaped angular distribution with this theory, for the reactions Mo^{94} (O^{18}, O^{16}) Mo^{96} at three energies. The fit is quite satisfactory. However, a normalisation is required to be applied for absolute cross-sections[32].

An interesting dependence on Q-value of the transfer-reactions has been found for the Nickel isotopes. For Ni^{62} and Ni^{64}, the angular distribution shape is bell-type, while for Ni^{60} and Ni^{58}, it steadily falls with angle. However, for both cases, oscillatory character is also there. This is understood now because of the fact, that the absorption of the waves (outgoing and in-going) is such that the heavier Ni-isotopes can be considered strong absorbers (smallest + Q-values) while there is slightly lower absorption for lighter isotopes (larger + ve Q-values). An oscillatory behaviour of Ni isotopes in Ni^{64} (O^{16}, C^{14}) Zn^{66} reaction can also be obtained at higher energies [Fig. 17.17; Ref. 33].

Comparison with no-recoil formalisation is very good. At higher energies, the shape can be oscillatory and at lower energies a bell shapes, for the same reasoning.

Most of the two-particle transfer experimental cross-sections are higher by a factor of 5 to 1000, compared to the theoretically calculated based on one component of the basis. If one calculates the cross-section by increasing, the basis, as was done by Bang et al.;[34] the cross-sections go up by a factor of 25. The physical reason for this is, the large number of basis states, which are required to describe the tail of the wave-functions, which is the most important region for the interaction.

The problem of normalisation and the lack of reproduction of absolute cross-section, has been solved by proper adjustment of the absorption and still being able to fit the elastic scattering data. Multi-

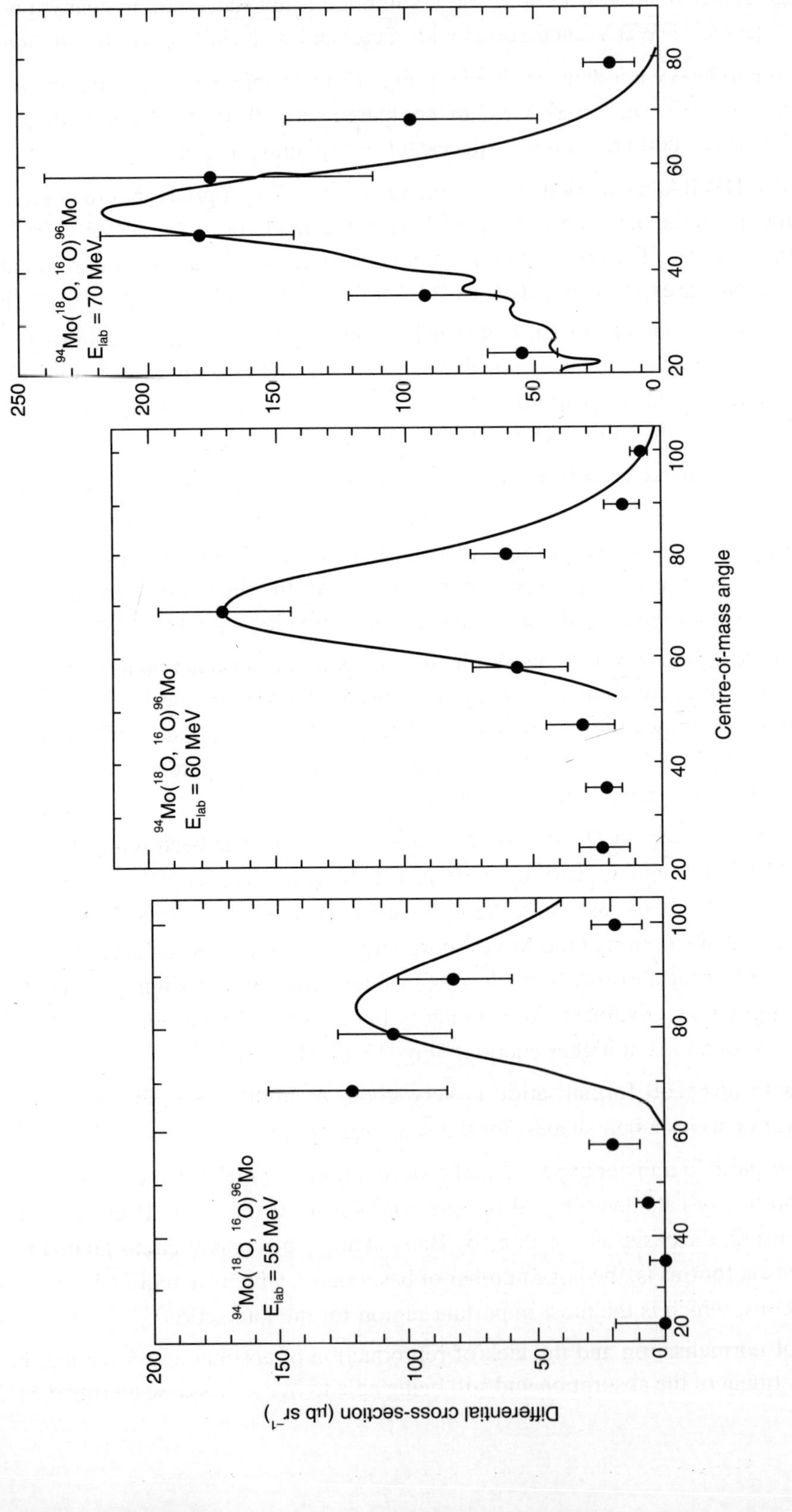

Fig. 17.16 Comparison of bell shaped angular distribution of Mo^{94} (O^{18}, O^{16}) Mo^{96} with theory at three energies (Ref. 32).

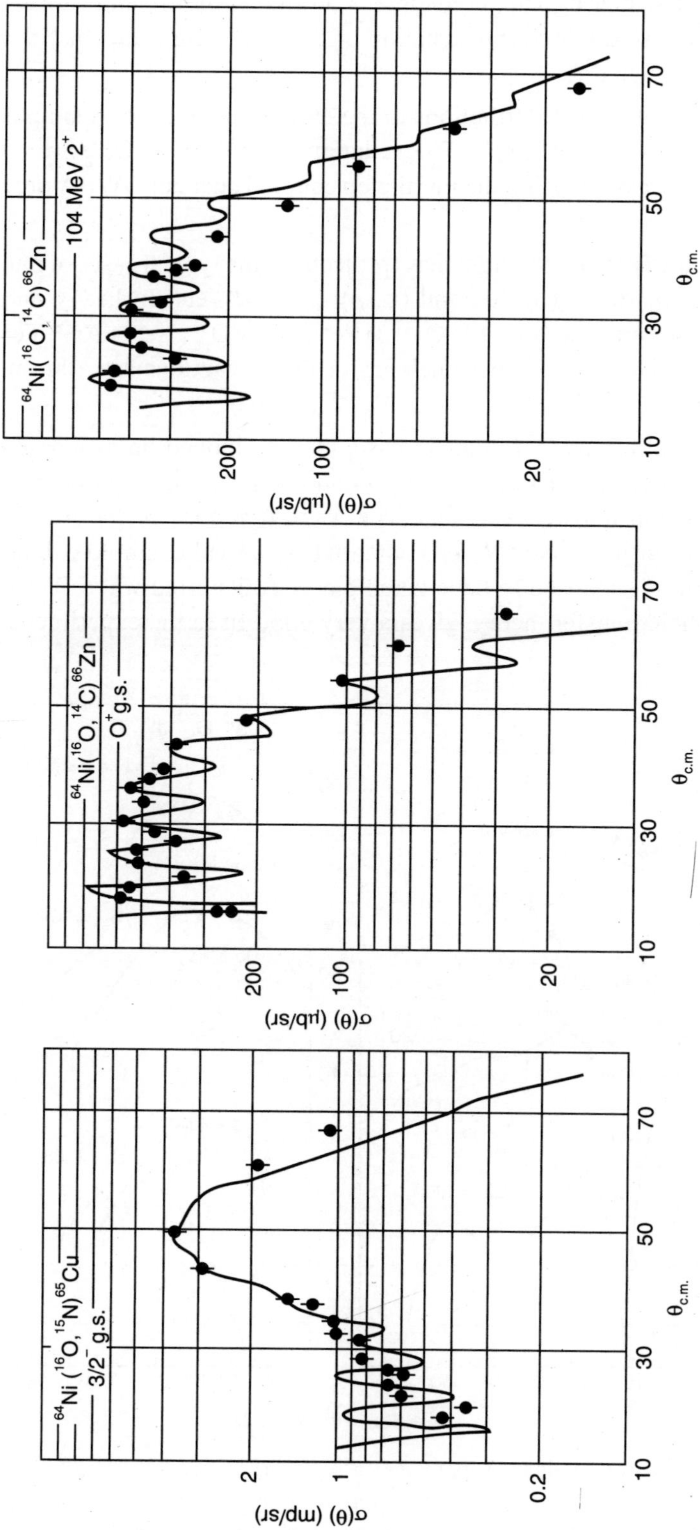

Fig. 17.17 Oscillatory behaviour of Ni^{64} (O^{16}, C^{14}) Zn^{66} and Ni^{64} (O^{16}, N^{15}) Cu^{45} reaction (Ref. 33) at 56 MeV incident energy. (Ref. 33)

step process of inelastic scattering can, sometimes contribute significantly to two nucleon transfer reactions, as shown in the schematic representation of the low lying states of Nd^{142} and the two-step process of their being populated[35].

One extends the multi-step contributions to the transfer amplitude. A simple one-step distorted Wave-Born approximation does not explain 2^+ excitation in Sn^{120} (O^{18}, O^{16}) Sn^{122} but if one includes the additional reaction process, by using the multi-step coupled channel DWBA process, the reaction is explained[36].

III. Multi-Nucleon Transfer: In light mass projectiles, the typical cases of multi-nucleon transfer cases are: (He^3, n) and (n, He^3) or (p, α) and (α, p) or (α, d), etc. In heavy nuclear reactions, the examples of such reactions are: $Ca^{42,44}$ (O^{16}, O^{14}) $Ti^{44, 46}$, Ca^{40} (O^{18}, Ne^{20}) Ar^{38}, Fe^{54} (O^{18}, Ne^{20}) Cr^{52}, Te^{24} (C^{12}, C^{14}) Te^{22}, etc. The energies per nucleon for these reactions are a bit higher than for one-nucleon case.

Theoretically, the multi-nucleon transfer case, may be looked up as a case of cluster-transfer, especially when the transferred particle is tightly bound, *e.g.* an alpha particle. Such cases may be considered like one-nucleon transfer reactions. The wave-functions of the transferred cluster is the eigen function of the cluster in a Saxon-Woods potential. A typical example will be Ca^{40} (Li^6, d) Ti^{44}, where an alpha-stripping is supposed to have taken place. A distorted wave DWBA calculation with finite range, in one particle transfer theory, gives a very good fit to the experimental data as shown in Fig. 17.18 (Ref. 37).

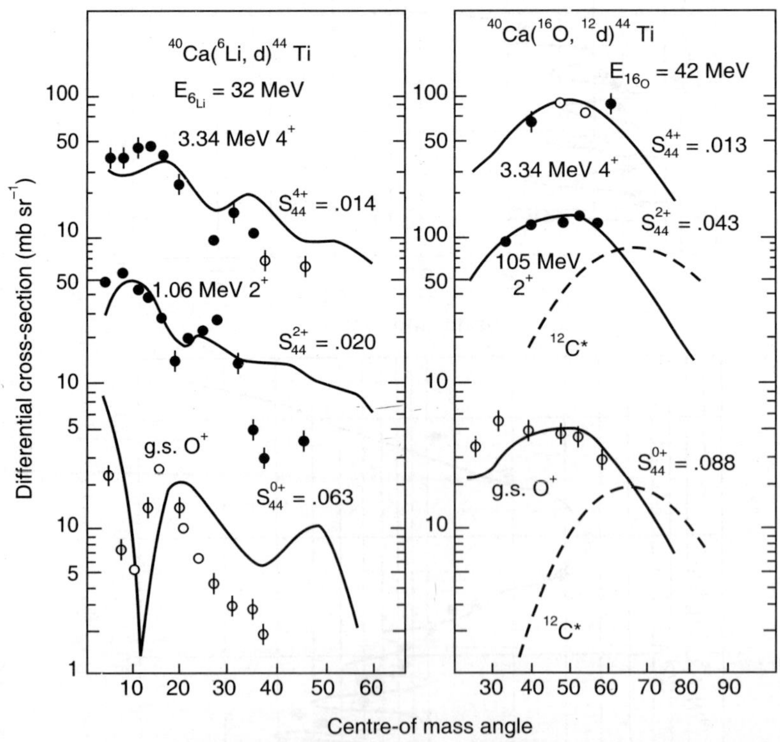

Fig. 17.18 Comparison of alpha stripping in Ca^{40} (Li^6, d) Ti^{44} and Ca^{40} (O^{16}, C^{12}) Ti^{44} with finite range recoil DWBA calculation, supporting the direct reaction mechanism (Ref. 37). S mentioned for each curve is the spectroscopic factor.

Other reactions, when an alpha particle is either picked up or transferred are also very popular for studying transfer reactions. Such transfer reactions are (C^{12}, Be^8), (O^{16}, C^{12}), (Li^7, t^3), (Be^8, α) and (Ne^{20}, O^{16}) reactions.

As stated earlier, such reactions may be treated like one particle transfer reactions. But it is also possible, that a sequential transfer of individual nucleons may take place giving a transfer of alpha particle. Also the alpha emission may take place via compound nucleus decay. So the study of alpha transfer reaction and comparison with different models, can give us information about the probability of various processes. The many-step processes require the appropriate fractional parontage coefficients and nuclear structure wave-function, which are not always available (Ref. 39).

At somewhat lower energies, (*d*, Li^6) reaction can be understood as direct pick-up reaction. It yields at 15 MeV, a forward peaking, explainable with alpha cluster DWBA calculations. At higher energies of 55 MeV, there are indications of alpha pickup plus a weak populations of some unnatural parity states, indicating more complex mechanism. O^{16} (α, Be^8) C^{12} reaction, which is also an alpha pickup reaction, shows compound nucleus-like behaviour at 35.5 – 41.9 MeV.

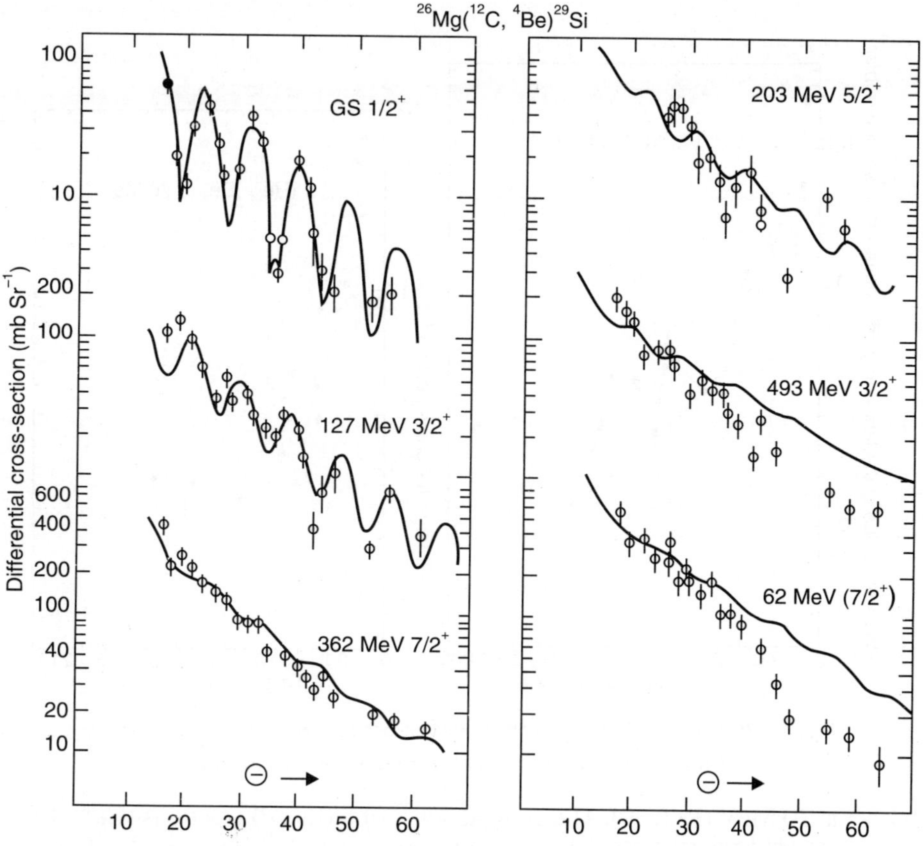

Fig. 17.19 Differential cross-sections for the reaction Mg^{26} (C^{12}, Be^9) Sr^{29}; compared with DWBA calculations, using the cluster approximation (Ref. 38).

It is interesting to observe that a cluster-transfer mechanism, assumes the existence of alpha-clusters in the target. One can determine the probability of the existence of these clusters in the target theoretically (say by SU (3) model) and compare with the experimental data (Ref. 38). Figure 17.19, shows the pick-up of He^3 in Mg^{26}, using (C^{12}, Be^9) reaction. It is complicated to treat the three transferred nucleons individually, but as shown in this figure, the cluster transfer calculations yield results, which reproduce the experiments.

Out of these multi-nucleon transfer reactions; (Li^6, *d*), (Li^7, *t*) or (*d*, Li^6) reactions have been studied very extensively because of their sensitivity to the reaction mechanism or the selection of proper potential. As for example, the reaction of (Li^6, *d*) at 19.5 MeV, on a range of nuclei from B^{10} to Ca^{40} have been studied at 19.5 MeV and in general, except for the lightest target, finite range distorted-wave theory gave good agreement with differential cross-sections. A cluster model wave-function gave the correct absolute cross-sections, compared to the shell-model wave-function. On the other hand, (*d*, Li^6) reaction required the assumption of the presence of multi-step processes, because many unnatural parity states were excited. Similarly (Li^7, *t*) reactions are also explainable on the basis of Coulomb-distorted plane wave model with alpha-cluster wave-functions. Figures 17.20*a* and 17.20*b* show the similarity of alpha transfer dependence on *A*, for (*d*, Li^6) and (He^3, Be^7) selections (Ref. 39). Figure 17.21 shows the comparison of cluster model calculations with the experimental data of Fe^{56} (Li^6, *d*) Ni^{60} and

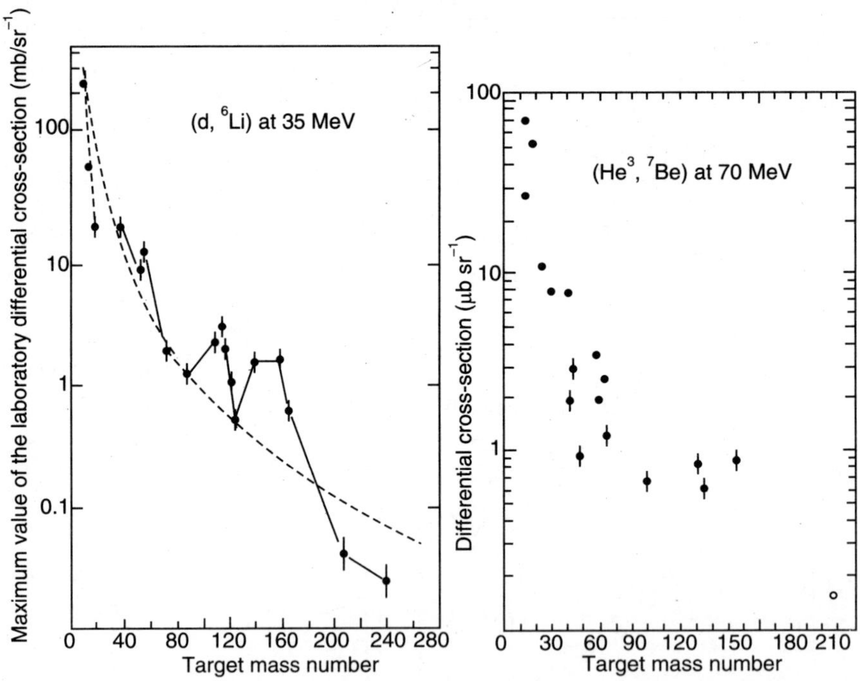

Fig. 17.20 (*a*) (d, Li^6) and (*b*) (He^3, Be^7) cross-sections at 35 MeV and 70 MeV energies as a function of target mass (Ref. 39).

Ni^{58} (Li^6, *d*) Zn^{62} at 30 MeV (Ref. 39). The comparison[37] of a (Li^6, *d*) reactions [Ca^{40} (Li^6, *d*) Ti^{44}] and (O^{16}, C^{12}) reaction [Ca^{40} (O^{16}, C^{12}) Ti^{44}] shows that both the reactions proceed with direct alpha transfer and are explainable on the basis of DWBA, finite range DWBA calculations. A similar detailed study of the usefulness of (Li^6, *d*) at 30 MeV as a tool for spectroscopic studies has been conducted by Fulbright and workers[39]. The angular distributions are sufficiently characteristic of angular momentum transfer. A zero-range, cluster-transfer distorted wave calculation explained the data.

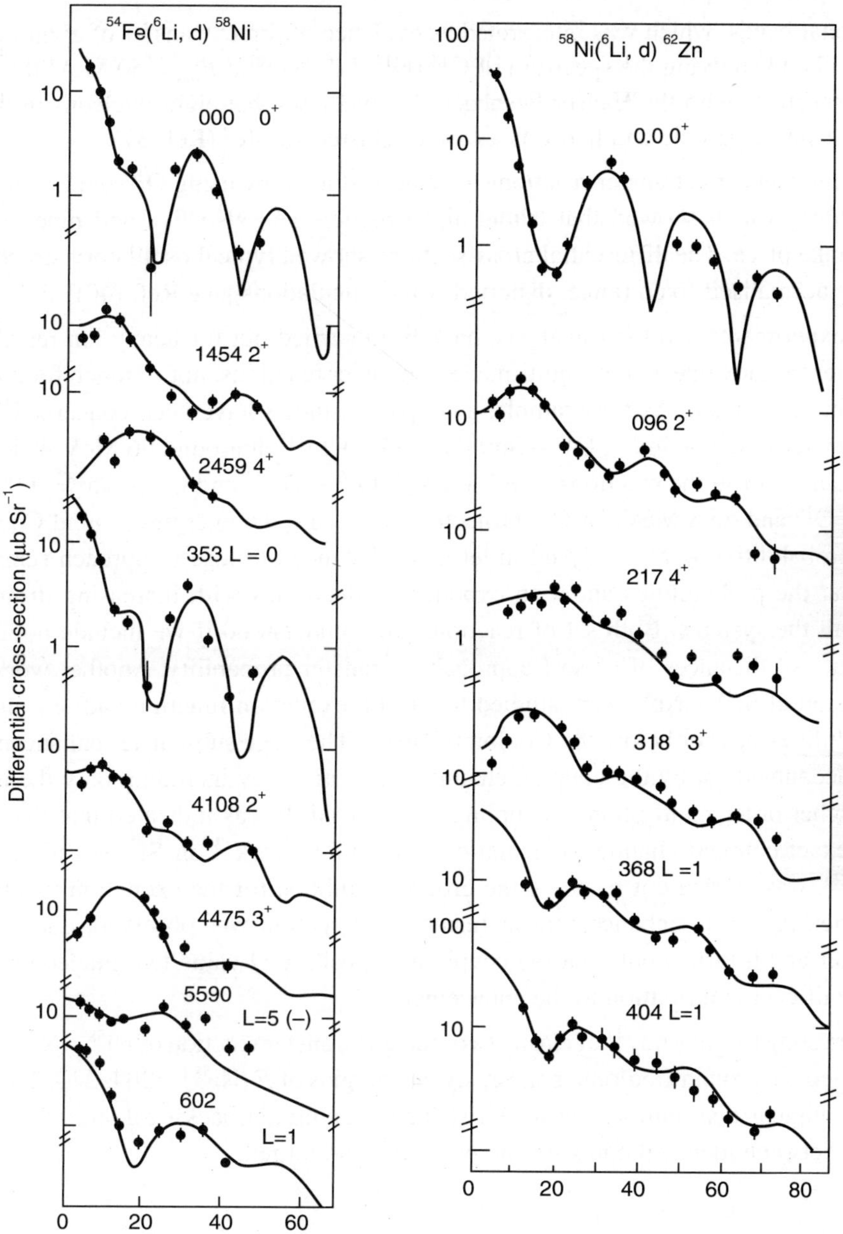

Fig. 17.21 Experimental angular distribution for (Li^6, d) reaction on Fe^{54} and Ni^{58}, at 30 MeV and comparison with zero-range cluster transfer distorted wave calculations (Ref. 39).

Transfer of more than four nucleons and cluster transfer: Typical of such reactions are: C^{12} (C^{12}, α) Ne^{20}, O^{16} (C^{12}, α) Mg^{24}, O^{16} (O^{16}, Be^8) Mg^{24}. All these reactions correspond to multi-transfer. The (C^{12}, α) reaction-studies around 20 MeV show that these can sometimes, take place by a direct process. Both excitation function and angular distribution show the evidence for it. This direct component is found for reactions to 0+, 2+, 4+ and 6+ members of the 8*p*-4*h* band and is thus directly attributable to the additions of eight particles to C^{12} to form 8*p*-4*h* states in Ne^{20}, relative to the O^{16} core. Similarly C^{12} (N^{14}, Li^6) Ne^{20} reaction at 52 MeV showed clear difference in the excitation of the various rotational bands, which was interpreted as evidence of direct transfer of eight nucleons cluster transfer (Ref. 37). In studying the spectrum of C^{12} (N^{14}, Li^6) Ne^{20} at E_{lab} = 50 MeV, the comparison of experimental spectrum with the Hauser-Feshbach theory shows that eight nucleons in Be^8 nucleus are transferred as a single particle and hence is a case of cluster transfer (Ref. 37).

Many multi-nucleon transfer reactions have been studied by using O^{16} and F^{19} on B^{11} and C^{12} at 60 and 68.8 MeV, which showed that triton, alpha-particle and also five and nine nucleon transfer reactions can take place. The differential cross-sections showed typical oscillatory structure and can be reproduced by normalised fixed range, distorted wave calculations [*see* Ref. (40)].

Many experiments and their analysis, have been carried out for heavy ion reaction around the Coulomb barrier to study one, two or multi-nucleon transfer reactions, not so much for nuclear structure but for reaction mechanism. As for example, a couple of studies have been concerned[41] with one and two-proton transfer in C^{12} + Sm^{154}, O^{16} + Sm^{154} and O^{16} + In^{115} at around 96 MeV which is just above the Coulomb barrier or in the reactions of F^{19} + Th^{232}, O^{16} + Th^{232} and C^{12} + Th^{232} at 92 MeV for F^{19}, 86.6 MeV for O^{16} and 56.8 MeV for C^{12} projectile, which are the energies around Coulomb barriers. The transfer probabilities were analysed in terms of distance of closest approach semi-classically. It was found, that the probabilities show an exponential decreases with increasing distance of closest approach for all the systems. Both set of reactions were understood if we include both Coulomb and nuclear branches of distances of closest approach to transfer probability. Another work[42] concerning sub-barrier fusion of Si^{28} + Nb^{93} were studied for fusion excitation function and one and two nucleon transfer probabilities near or below the Coulomb barrier. The large sub-barrier enhancement, observed could not be accounted for by the coupled channel calculations by including only the inelastic states. The role of higher order multi-phonon coupling was proved. It was indicated that there is a need for carrying out exact coupled channels calculations. A similar work[43] on Si^{28} + Zn^{68} system at 65 to 83 MeV of Si^{28}, was carried out by the same group of authors, for the excitation function, below the Coulomb barrier. Again an enhancement in fusion cross-section was observed, due to one and two-nucleon transfer and it turned out, that the coupling of positive Q-value two-nucleon transfer channel results in a significant contribution to the enhacement.

An interesting recent study[44] of elastic two-nucleon transfer reaction of Ni^{58} + Ni^{60} and Ni^{62} + Ni^{64} have been carried out around Coulomb barrier, *e.g.* at energies of E (Ni^{58}) = 204, 220, 236 and 250 MeV. The measured angular distributions show bell-shaped structure, at backward angles, showing neutron pair exchange between identical cores for the projectile and target.

17.6 COMPOUND NUCLEONS FORMATION IN HEAVY ION REACTIONS

Compound nucleus formation and decay have already been discussed in previous chapter for light projectiles. As discussed in the beginning of this chapter, the incident heavy ion particle will form a compound nucleus if $R_L < R_1 + R_2$. Then either many nucleon-transfer or deep inelastic scattering can take place or a compound nucleus is formed resulting into fusion, which may decay through various modes like fission, neutron evaporation and/or gamma decay. Depending on the exact energy, nuclear-transfer or deep inelastic scattering and/or the compound nucleus are formed. The lifetimes of compound nucleus are expected to be 10^{-18} secs – 10^{-16} secs and for nuclear transfer to be 10^{-22} secs. The measured lifetimes range from 10^{-18} secs to 2.5×10^{-21} secs for the excitation energy of the compound state at 75 MeV and 150 MeV respectively for nuclei with $A = 150$. Experimentally, branching technique was used as described in Chapter 2, for the life-time measurements. At the lowest energies only compound nucleus is formed for all non-elastic reactions. As the energy increases, the nuclear transfer process gains in importance.

Many particle spectra at medium and high energies of heavy-ion induced reaction are given in literature[45]. It has three distinct regions. The lower energy part corresponds to compound nucleus formation, the highest energy part belongs to direct reactions including inelastic scattering and transfer reactions and intermediate part belongs to the pre-equilibrium reaction mechanism. We have discussed the various features of these reactions in the previous chapters for light particles. A special difference

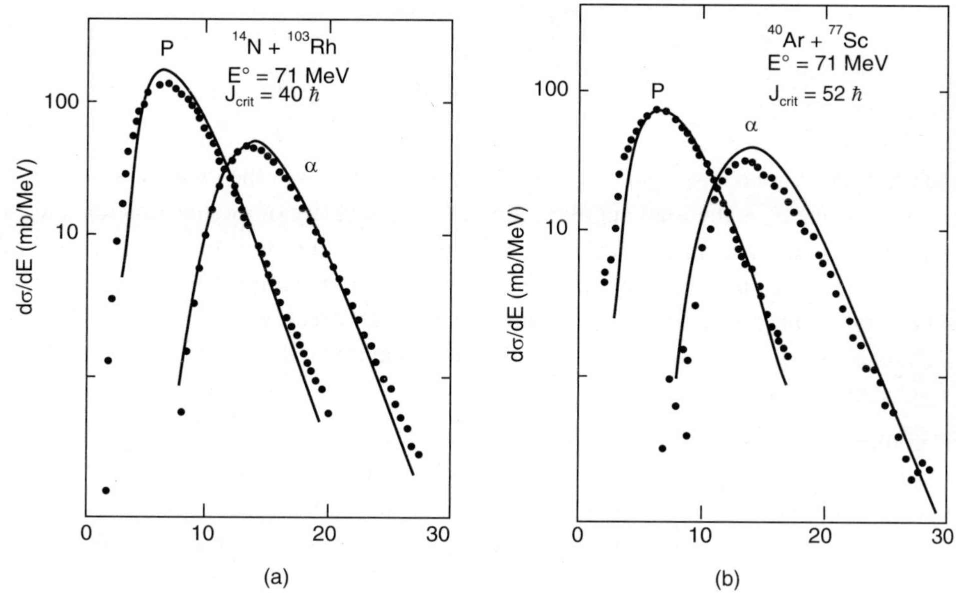

Fig. 17.22 The energy spectrum of emitted particles say protons and alphas, in a heavy ion induced reaction say (*a*) N^{14} + $4Rh^{103}$ at 71 MeV and (*b*) Ar^{40} + Se^{77} at 71 MeV (Ref. 45).

with heavy ion induced reaction will be, that higher angular momenta, can be transferred to the compound nucleus and high spin states and their mode of decay can be suitably studied. Figure 17.22 is a typical compound nucleus formation (Ref. 45), where the calculations include, the shell and pairing effects and a limitation J_{crit} to the spin of, the compound states.

The criteria of compound nucleus formation comes from the angular distribution of the emitted particles. Compound nucleus decay cross-sections are always symmetric around 90°, while direct reactions cross-sections are usually forward peaked, especially at higher energies. One should, however, be careful, because under certain conditions of angular momentum transfer, a direct reaction may also be symmetric around 90°. Also the excitation function has a resonance structure, if it is compound nucleus formation, while the cross-section for direct reactions varies smoothly with incident energy. Again, in heavy ion reactions, the excitation function for direct reaction may show fluctuations, because of the large number of partial waves contributing to the reactions. A detailed and careful analysis is necessary, to distinguish the compound nucleus formation from direct reaction or pre-compound mode.

17.6.1 Theory

The basic principle of the compound nucleus model for heavy nuclear projectile is similar to the one developed for light particles developed in Chapter 13. Lang and Thomas[46] and also, V.F. Weisskopf[47] have developed detailed expression for the probability of emission of a particle with energy E_v and spins, in the direction of θ, using the parameters of the spin of the compound nucleus J, the orbital angular momentum of the emitted particle L and the level density of the residual nucleus at the given excitation energy. For details *see* Reference (46 and 47).

On the other hand, Hauser and Feshbach[48] have developed an alternative formalism based on statistical theory according to which the total cross-sections for a reaction from the incident channel α to the outgoing channel α′ has a general form:

$$\sigma_{\alpha\alpha'} = \pi\lambda_\alpha^2 \frac{2J+1}{(2I+1)(2i+1)} \times \frac{\sum\limits_{LS}\left(T_S^{\alpha}\right)_{J_\pi} \sum\limits_{L'S'}\left(T_{S'}^{\alpha'}\right)_{J_\pi}}{\left[\sum\limits_{\alpha''S''l''} T_{l''}^{c''}\right]_{J\pi}} \qquad \text{...(17.52)}$$

where I and i are the ground state spins of target and projectiles, S is the channel spin, L is the orbital angular momentum and J_π is the total angular momentum and parity of the reaction channel, T_L^α are the transmission coefficients that may be calculated from the appropriate optical potentials and the sum runs over all the open reaction channels. In expression (17.52) the sum in the denominator includes all the channels open to the decay of the compound nucleus. In practice, however, it is not practical to include all channels explicitly. It is, therefore, more convenient to calculate for discrete levels of known levels up to a certain excitation energy.

The compound nucleus theories, assume that the decay of the compound nucleus is independent of the way it is formed. Experimentally this can be proved, as in the famous Ghoshal type experiment, by forming the compound nucleus by two different ways and letting it decay into different channels. In Fig. 17.23 are plotted the fraction of the total reaction cross-sections, which leads to Dy^{156} compound

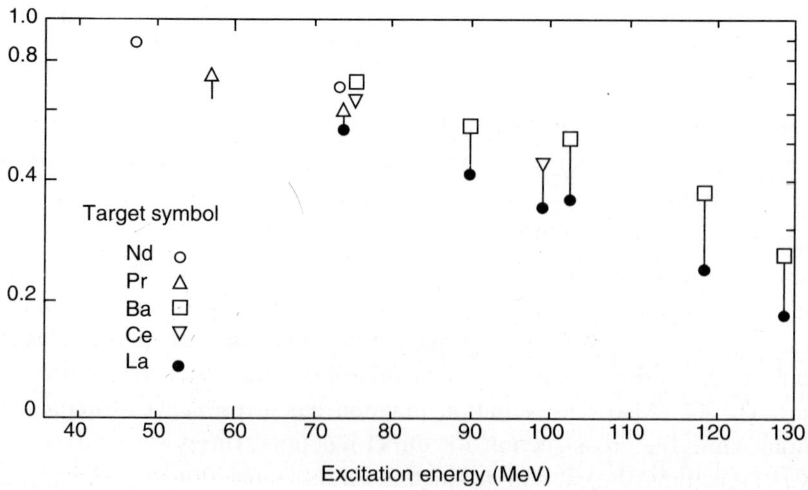

Fig. 17.23 The fraction of the total cross-section due to (HI, xn) reaction, as a function of excitation energy for several targets showing the same fraction for two reactions leading to the same compound nucleus (Ref. 49).

nucleus, from reactions performed by J.H. Alexander and G.M. Simoniff (Ref. 49), *e.g.* $C^{12} + Nd^{144} \rightarrow Dy^{156}$ and $Ne^{20} + Ba^{136} \rightarrow Dy^{156}$ and other reactions for targets of Pr, Ce, La leading to Dy^{156}, as a functions of excitation energy E. Experimentally reactions are detected by the decay (through xn emission) into Dy^{151} ($x = 5$), Dy^{150} ($x = 6$) and Dy^{149} ($x = 7$) using radioactivity methods. It is interesting to see, that all reactions follow very similar curves for f_t (fraction of calculated total cross-section) as a function of excitation energy for various targets and f_t falls for all cases, as the excitation energy rises showing that the decay is independent of the mode of the formation of the compound nucleus.

Both the theories of Weisskopf[47] and Hauser-Feshbach[48] lead to similar features in (*i*) Excitation functions, (*ii*) The shapes of the emission spectra and (*iii*) The angular distribution of the emitted particles.

17.6.2 Experiments

Hauser-Feshbach[48] theory is very extensively used with experimental data and some interesting information may be obtained, about the nuclear structure, apart from the affirmation of the validity of the theory.

Study of the reaction B^{10} (O^{16}, He^4) Na^{22} at $E_{lab} = 46$ MeV, by measuring the angular distribution of alphas, corresponding to different excited states of Na^{22} and comparing it with Hauser-Feshbach theory, could determine the spins of the various states of the excited states of Na^{22} [Fig. 17.24], (Ref. 46). This method has, therefore, become a tool to determine the spins of the excited states of nuclei and many such experiments have been conducted.

Hauser-Feshbach theory[48] has been found to be inadequate, to explain the excitation functions, at higher energies, unless one takes into account the fact, that there is a limit to the angular momentum that can be accepted in actual practice by the compound nucleus set by the density of states as a function of J. So it is quite possible that there are no states of spin J below a certain excitation energy and they may appear only at a higher energy. So while, all the angular momenta up to J_N are ultimately available, but we have to apply a certain limit in our calculation for excitation function to take into account the actual situation. The introduction of a critical value of angular momentum J_{crit} is, therefore required, which is determined semi-empirically to fit the data.

Today, Hauser-Feshbach theory is the standard way of analysing compound-nucleus cross-sections, not only under conditions of weak absorption, but also under strong absorption.

Light particle emission in heavy ion induced nuclear reactions have been studied at either very high energies, *i.e.*, at about 10 GeV/A or at comparatively lower energies say a few MeV/A. The purpose of the two studies is different. The interest in the high energies of the order of GeV arises from the large amount of energy deposited. This creates phenomena, which have been considered to be associated with compression of nuclear matter or heating of nuclear matter resulting in a state of very high baryon density. We will discuss this phenomenon in Section 17.9.8. At lower energies say at 475 MeV and 2 GeV, etc. the intermediate mass fragment (IMF) model holds good, leading to either multi-fragmentation process or binary fission[51]. An interesting case in this energy range is the experiment of neutron yields from 435 MeV/A Nb stopping in Nb and 272 MeV/A Nb isotopes in Nb and Al. The experimental data on neutron yields was obtained by integrating over angular distributions, which where forward peaked. The data[52] was compared to BUU (Boltzmann-Ulhing-Uhlenbach) model calculations with only qualitative agreement (*see* Section 17.9.6).

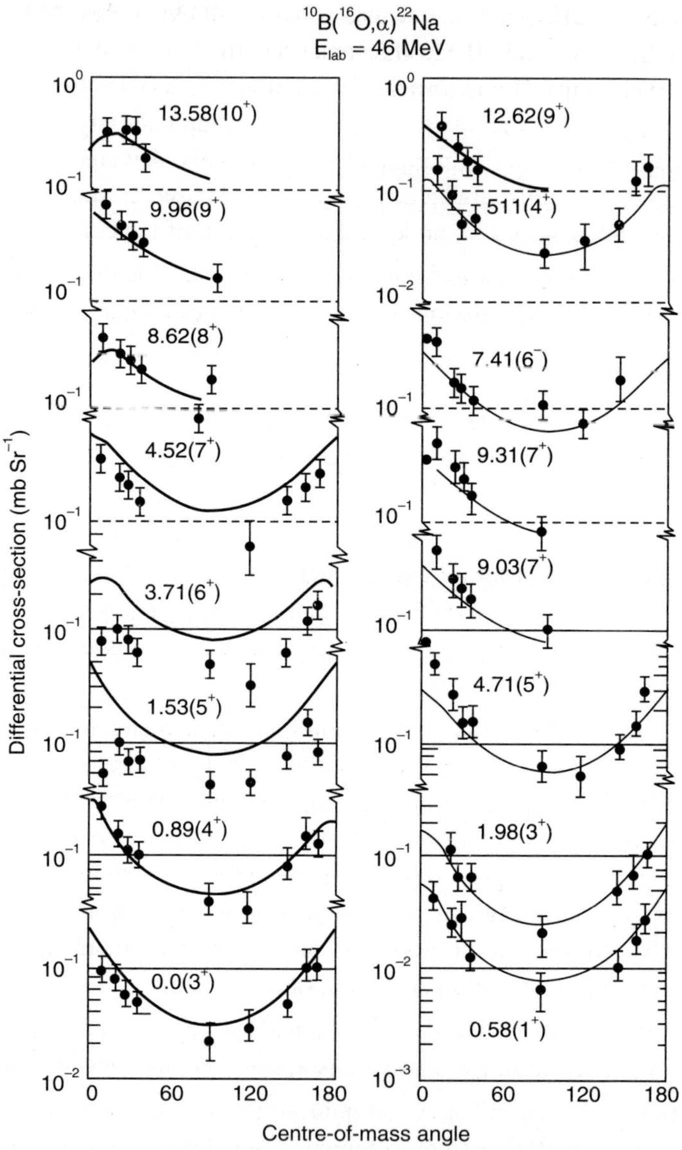

Fig. 17.24 Differential cross-section for reaction B^{10} (O^{16}, α) Na^{22} to various states of Na^{22}, compared with Hauser-Feshbach theory (Ref. 50).

Two interesting studies at somewhat lower energies, say at 35 MeV/A, have shown the possibilities of excitation up to 800–1300 MeV in nuclei with $A \approx 225 - 240$, in the case of $Cu^{63} + Au^{197}$ at 35 MeV/A beam energy[53] and reactions products in Au + Au at 35 MeV/A collision being generated with five different impact parameters[54]. The temperatures of the quasi-projectile systems were found slowly increasing going towards small, impact parameters. These studies bring out the beauty and complex nature of reaction mechanisms at these energies.

This range of energies covers an energy region where direct-reaction, mechanisms like pick-up, break-up, knock-out, inelastic scattering to high lying target states and giant resonance formations and decay take places. Such GR (giant resonance) structure has been seen in $Ca^{40} + Ca^{40}$, $Cu^{63} + Cu^{63}$ at somewhat low energies[55] (7 and 10 MeV/A) and recently[56] at 50 MeV/A. In the later work, evidence was found for the excitation of two known gaint quadrupole resonance characterised by its direct decay scheme.

At lower energies, the measurements of the emission of protons, deuterons, tritons and alpha particles, in coincidence with evaporation residue, are understood only in terms of compound nucleus formation and decay. In typical work[57], collisions induced by 11 MeV/nucleon, $Ni^{58} + Mg^{24}$ reaction was recently studied, which has been understood in terms of the initial formation of a deformed composite system, (di-nuclear), which at a latter stage, in cascade, relaxed into a normal compound nucleus.

At still lower energies, say 5 – 7 MeV/A, it is mostly the statistical model, which governs, the emission of light particles, though some indications are available for the emission of light particles prior to the full relaxation of the compound[58] nucleus. One recent study shows that the α-particle emission in $S^{28} + V^{51}$ at 140 MeV supports this interpretation[59].

17.7 FUSION OF HEAVY IONS

We have discussed till now, the compound nucleus formation and decay in a composite manner, in which the information of reaction mechanism came from the decay products. In Hauser-Feshbach theory, the transmission coefficients of the incident channel—which leads to compound nucleus formations—is calculated from the appropriate optical potential. However, the theories for fusion of two heavy ions, have been specifically developed and compared with the total absorption cross-section of heavy nuclei by taking all the decay channels. So fusion theories yield the total absorption cross-section leading to compound nucleus. It may be realised that interaction of two ions does not always lead to the formation of a compound nucleus, even if their energy is sufficient to overcome the Coulomb barrier. They may stick together and interact to have transfer-reaction, deep inelastic scattering, etc. The process by which the incident kinetic energy is transferred is complicated and requires a model of transport of particle-hole excitations.

Without going through such complicated calculations, one can develop a model of fusion based on classical potential, which was done by Bass[60], who used the liquid drop model to obtain a finite range—two body interaction between heavy ions. First of all we may, however, write a general theory, which expresses the fusion cross-section in terms of this potential.

Following Glass and Mosel[61], we write the fusion cross-section and the reaction cross-section as:

$$\sigma_F = \pi \lambda\!\!\!^{-2} \sum_{l=0}^{\infty} (2l+1)\, T_l P_l \text{ and } \sigma_R = \pi \lambda\!\!\!^{-2} \sum_{l=0}^{\infty} (2l+1)\, T_l \qquad \text{...(17.53)}$$

where T_l is the transmission for barrier (due to Coulomb, centrifugal or any other force) penetration and P_l are probabilities that fusion takes place and not any other reaction, *e.g.* transfer or direct, etc. How do we calculate T_l and P_l? Assuming V_{Bl} as barrier potential for angular momentum l, it can be seen that if V_{Bl} is due to Coulomb and angular-momentum related centrifugal forces, one can write:

$$V_{B1} = V_B + \frac{1\,(I+1)}{2I_B/\hbar^2} \quad ...(17.54)$$

where V_B is Coulomb potential and $I_B = \mu R^2$ is the nuclear moment of inertia. Then following the theory of Hill and Wheeler[63], one can write the transmission coefficient $T_1\,(E)$, as approximated by those of an inverted parabolic barrier, *i.e.*,

$$T_1\,(E) = \frac{1}{1+\exp\{2\pi\,(V_{B1}-E)/\hbar\omega\}} \quad ...(17.55)$$

where $\hbar\omega$ is the frequency of harmonic oscillator potential and E is the energy of the incident particle.

Assuming that fusion takes place only for $l \le l_F$, then for fusion probability P_1 one can write:

$$P_1 = 1 \text{ for } I \le I_F$$
$$= 0 \text{ for } I > I_F \quad ...(17.56)$$

In other words, l_F is the upper limit of angular momentum for fusion, beyond which the nucleus probably disintegrates. Then

$$\sigma_F = \pi\lambda\!\!\!^{-2} \sum_{I=0}^{l_F} \frac{(2I+1)}{1+\exp\,(V_{B1}-E)/\hbar\omega} \quad ...(17.57a)$$

and

$$\sigma_R = \pi\lambda\!\!\!^{-2} \sum_{I=0}^{\infty} \frac{(2l+1)}{1+\exp\,(V_{B1}-E)/\hbar\omega} \quad ...(17.57b)$$

The relationship for $T_1\,(E)$ [Eq. 17.55], shows, that $T_1\,(E)$ falls from unity (when $E \to \infty$), to zero (when $V_B \to \infty$) and 1 passes, in this process through a value l_N, obtained from Eq. 17.54, where we put $E = V_{B1}$, so that,

$$I_N\,(I_N+1) = \frac{2I_B}{\hbar^2}\,(E - V_B) \quad ...(17.58)$$

Upper limit of angular momenta for σ_R can be I_N, while for σ_F, it is I_F. At low energies, all the incident particles are captured, so $I_F = I_N$. As the energy increases, there comes a point where the maximum angular momentum I_N is too large to be accepted by the nucleus, because then nucleus may fly apart. Above that, I_c the critical angular momentum comes into play. We define this angular momentum called the critical angular momentum (l_c) beyond which the compound nucleus will fly apart, as:

$$I_c\,(I_c+1) = \frac{2I_c}{\hbar^2}\,(E - V_c) \quad ...(17.59)$$

The fusion cross-section at low energies is the same as reaction cross-section for which $I_F = I_N$. As the energy goes up to I_c, the fusion cross-section is determined by I_c. So I_F is smaller than I_N at lower energies and smaller than I_c at higher energies.

Figure 17.25 shows[61,62] the plot of experimentally measured values of σ_F and σ_R as function of $1/E_{cm}$ and their comparison with theoretical expressions (Ref. 49) given by Eqs. 17.57*a* and 17.57*b*. It is interesting that while for low energies σ_F agrees with the theory, for high energies fusion is only a small fraction of total reaction cross-sections. The point of discontinuity occurs at $I_N = I_c$.

The difference between these two cross-sections corresponds to higher partial waves, that pass the interaction barrier, but do not reach I_c. It is found that for heavier nuclei—as targets and normal heavy ions as projectiles say for Cl^{35} on Ni^{58} and Ni^{62}, the reaction cross-sections fall below the fusion cross-sections.

Fusion cross-sections, sometimes, show a strong deviation from the above model, the energy dependence of the fusion cross-section for $O^{16} + C^{12}$ between 13 and 27 MeV, shows an oscillation of the cross-sections between these energy points and they have been interpreted as oscillations[64] in the optical potential. They could be due to resonances in the compound states of Si^{28} ($O^{16} + C^{12}$).

One of the most interesting cases of fusion is the production of super heavy elements in the fusion reaction[65] of $Ne^{22} + Cm^{248}$ at beam energies of 121 and 123 MeV, resulting in the production of element 106 (Seaborgium, Sg) yielding isotopes of Sg^{265} and Sg^{266} with emission of 4, 5 neutrons. Refined chemical methods were used for detection, coupled to fast coincidences, from the decay of Sg^{265} and Sg^{266} isotopes.

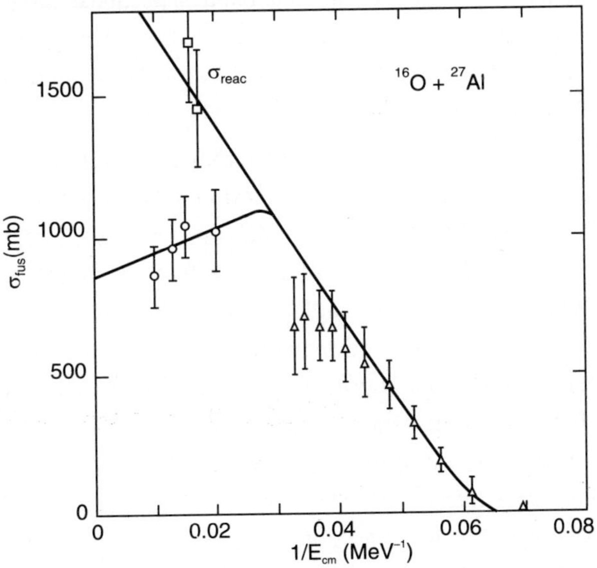

Fig. 17.25 Fusion cross-section of $O^{16} + Al^{27}$ as a function of $1/E_{cm}$ compared with the theory [Eq. 17.57] (Ref. 61 and 62).

There are also many cases of incomplete fusion (ICF) reported in literature[66]. They are basically associated with peripheral collisions, so that part of the reaction goes to quasi elastic-transfer and part goes through incomplete fusion. One such case reported recently[67] is an incomplete fusion in $F^{19} + Nb^{93}$ reaction at 95 MeV of F^{19} beam. From the analysis of the angular distribution of various emitted particles like O^{16}, N^{15}, C^{14}, C^{12}, B and Be, one could separate the contribution of quasi-elastic transfer (QET) from incomplete fusion (ICF).

The incomplete fusion is theoretically understood in terms of reaction with angular moments above the critical angular moment as discussed in Section 17.7. Recently Udgawa and Tamura[68] have explained the shape of particle spectra and angular distribution of α-particles in terms of break-up fusion model based on DWBA. As shown in Ref. 66, the ICF probability depends on entrance channel mass asymmetry. ICF studies have formed an important part of heavy ion reaction for which various models have been proposed[69].

An interesting paper on fusion enhancement, using radioactive beams[70], *e.g.* S^{38} + Ta^{181}, has shown, that the threshold energy with neutron rich beams, has lower value than with the normal beams, using S^{32} + Ta^{181}. It was found to be due to the neck formation between the colliding nuclei.

17.8 INTERMEDIATE STRUCTURE

For high energies, many possibilities exist. One picture of heavy ion interactions is—a quasi-molecule is formed, where each heavy ion interacts with an optical potential with nuclear, Coulomb and centrifugal terms. Such potentials have unbound resonant states, which may represent the gross-structure resonances. Over this gross structure, may ride the band structure, arising out of the formation of a quasi-molecule of the two heavy ions, which may form a molecular band. The quasi molecular band may be quite distinct from the ground state band. The resonant state in this potential of quasi-molecule is about 4 MeV wide and may act as the doorway state responsible for the ground state structure observed in many cases of heavy ion interaction at somewhat higher energies.

In this picture, as the two ions approach each other, they are deformed by their mutual interactions and this increases the radius, at which the nuclear interaction first becomes effective. As they approach more closely, their impenetrability reduces the deformation and hence the nuclear attraction. They further encounter centrifugal barrier. The combinations of these effects produces a second potential barrier. The Pauli Principle also tends to prevent the nuclei from coalescing. At still smaller separation, the potential is strongly attractive. Thus the interaction between the two ions can be represented by a double humped potential. This phenomenon is analogous to the one found in diatomic molecules. A potential of this type has series of broad unbound states (bands) of short time in the region between the two humps and these could account for some of the observed structure in C^{12} + C^{12} type interaction. As the energy increases, deviations occur due to the interaction with the edge of the nuclear field and when energy corresponds to that of one of the unbound states, a quasi molecule can be formed; whose rotational band interacts with ground state band, giving rise to the intermediate structure.

An interesting feature of the cross-sections from C^{12} + C^{12} is the three resonances at 5.68, 6.00 and 6.32 MeV, which occur just below the Coulomb-barrier. Using an interaction potential as given by R.H. Davis[71], it has been possible to show, that this potential gave resonances at 4.9, 5.5 and 5.8 MeV in C^{12} + C^{12}, and agreement with experimental values could be improved by varying parameters. Many calculations have been carried out by using a double hump potential and a quasi-molecular picture, to explain the intermediate or broad resonance structure like intermediate structure with reasonable success.

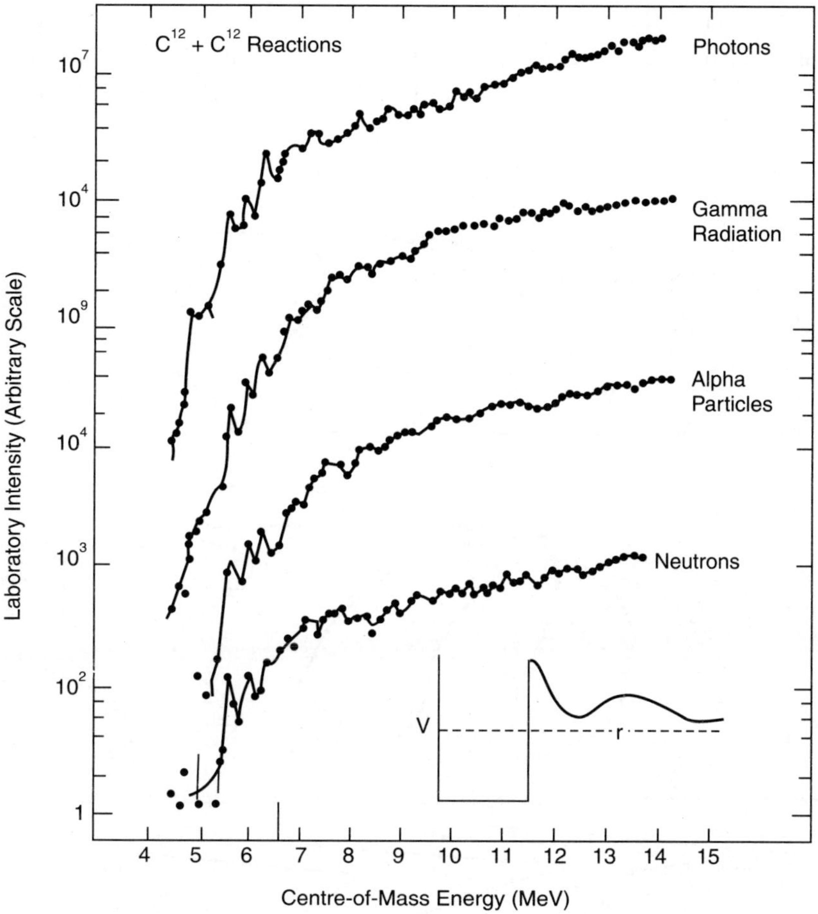

Fig. 17.26 Double humped potential for C^{12} + C^{12} and resonant structure from C^{12} + C^{12} reaction (Ref. 71).

Apart from $C^{12} + C^{12}$, intermediate structure has been found in $C^{12} + O^{16}$ and $O^{16} + O^{16}$ reactions. At higher excited states, generally the level density is high and one should not expect any structure. But in the heavy ion induced reactions high angular momentum are excited, for which the density of high J may be small; hence a structure can be expected. This is especially the case if the density of states of spin J_c at the excitation energy E of the compound nucleus is small, where J_c is the grazing angular momentum. At about $E_\alpha = 23 - 25$ MeV, the excitation energy of Yrast lines for $J = 8$ to 16, overlaps the incident energy and an intermediate structure can be obtained. Figure 17.26 shows, this phenomenon of intermediate structure when the emitted particles are; neutrons, alphas, gammas or lower energy photons like X-rays.

All the cases of intermediate structure can be understood, on the basis of resonance in compound nucleus with two-humped potential, formed in quasi-molecular formation.

The system of $C^{12} + C^{12}$, offers an interesting case of a reaction, where cluster formation plays a very important part. One can have C^{12} (C^{12}, α) Ne^{20}, C^{12} (C^{12}, Be^8) O^{16} or C^{12} (C^{12}, γ) Mg^{24}. Such studies have been carried[72] out by many authors up to an excitation of 25 MeV in Ne^{20}, 20 MeV in O^{16}

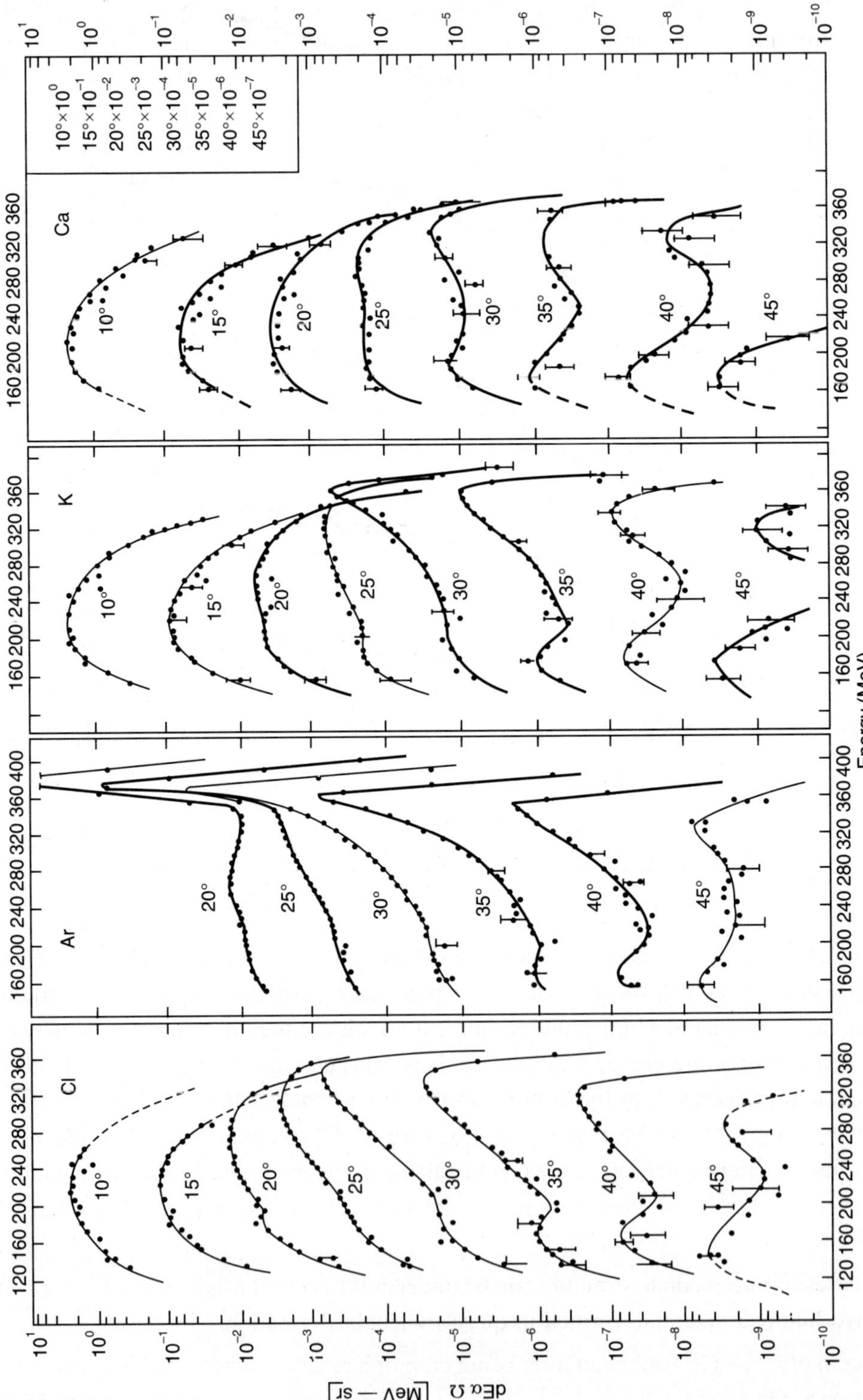

Fig. 17.27 Energy spectra of Cl, Ar, K and Ca fragments for Th^{232} + Ar^{40} at 388 MeV (Ref. 75).

and 38–50 MeV in Mg^{24}. In a detailed work carried out recently[73], it has been found that in all these cases, cluster-resonances play an important role. In the case of Mg^{24} as residual nucleus, it was observed that a cluster of resonances in the excitation region of 39–43 MeV in Mg^{24} decay via α and Be^8 channels. Another cluster of resonances in the region of 44–49 MeV decays predominantly to a possible 4α linear chain band in O^{16} around 18 MeV and to a 20.48 MeV state in Ne^{20}, which is above the 5α break-up threshold. At an excitation of $E = 32.5$ MeV, it was shown in the inelastic scattering that the structure of the spectrum corresponds to 6α linear chain resonance. Another study of $C^{12} + C^{12}$ around 32.5 MeV was carried by S. Szilner et al.[74] and explained in terms of 6α clusters.

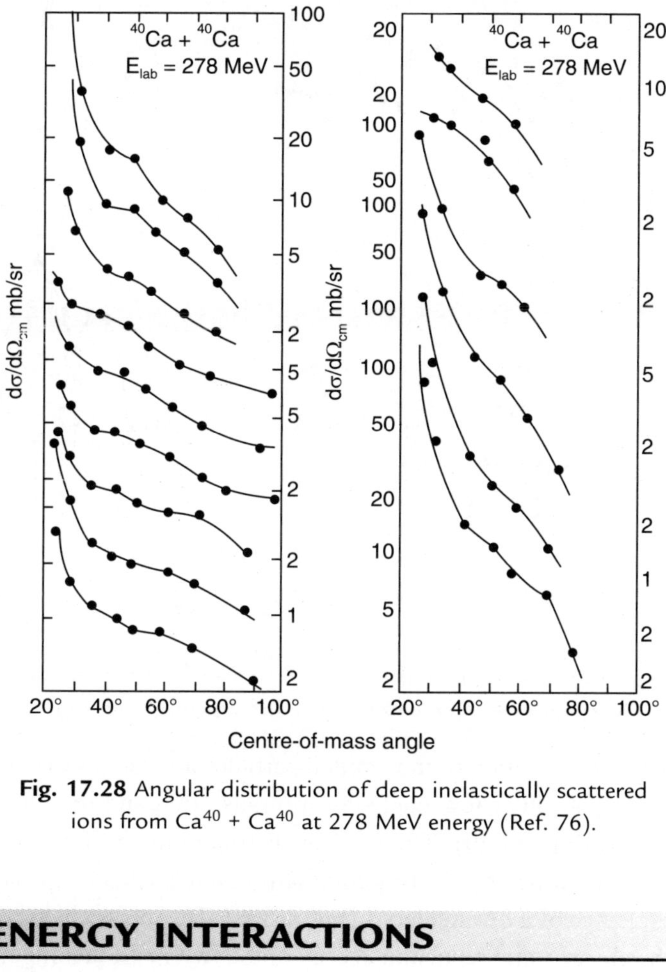

Fig. 17.28 Angular distribution of deep inelastically scattered ions from $Ca^{40} + Ca^{40}$ at 278 MeV energy (Ref. 76).

17.9 HIGH ENERGY INTERACTIONS

17.9.1 Deep Inelastic Scattering

We have seen earlier, that as the energies of the heavy ion projectiles are increased, the reaction passes through different stages; from Coulomb elastic scattering, to Coulomb plus nuclear inelastic scattering; to one nucleon and multi-nucleon transfer and to compound nucleus formation. At still higher energies say above 200 MeV or so, the nuclei may pass through each other losing much energy, but substantially retaining most of their nucleons. This is the process known as deep inelastic scattering.

Figure 17.29 shows typical spectra of nuclei emitted from collisions between heavy ions at energies extending over 200 MeV or more for Ar^{40} + Th^{232}, for $_{19}K$, $_{17}Cl$, $_{18}Ar$ and $_{20}Ca$ as the emitted nuclei. Further Fig. 17.28 shows the angular distribution of deep inelastically scattered ions from Ca^{40} + Ca^{40} at 278 MeV and Fig. 17.29 shows the yield of fragments emitted from Ar^{40} + Th^{232} reactions at 295 MeV as a function of *A* at 18° and 40°. The following characteristics emerge from these experimental facts.

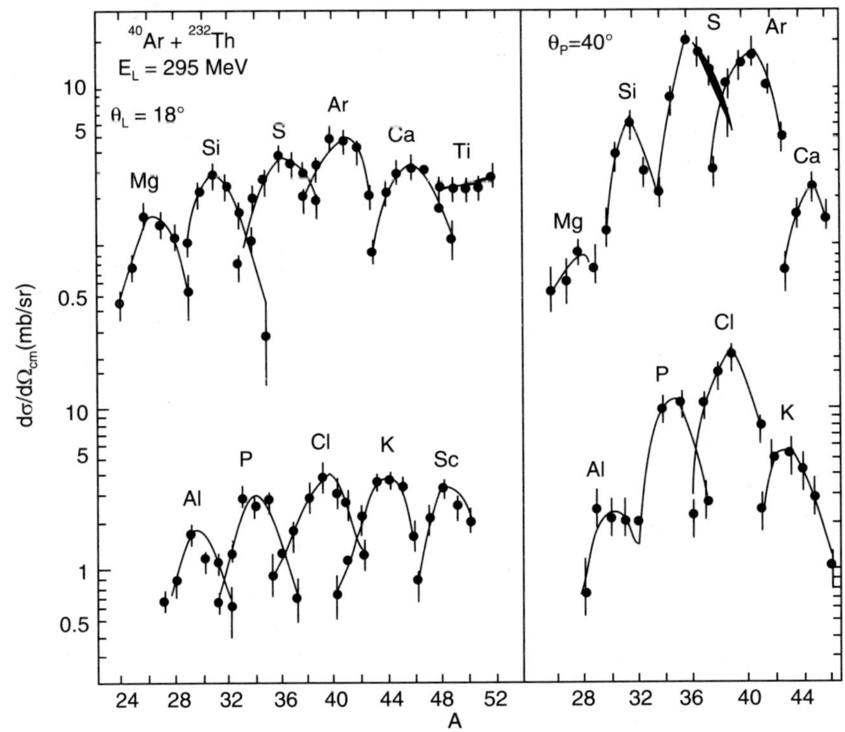

Fig. 17.29 Yield of fragments emitted from Ar^{40} + Th^{232} reaction at 295 MeV as a function of A at 18° and 40° (Ref. 77).

The mass and charge distribution of the emitted particles are very clearly packed in the region of the incident ions, showing that rather few nucleons are transferred and the ions in the initial channel nearly retain their identities [Fig. 17.29]. The angular distributions are forward peaked. This means, that the compound state decays much before its period of rotation, so that the projectile nearly continues its trajectory after the exchange of a few nucleons [Fig. 17.28]. The mean kinetic energy of the outgoing ions is very similar to the electrostatic repulsion energy of the two just-touching nuclei. This means, the interaction is two-body, *i.e.,* the projectile just touches the target and is repelled by it. Sometimes, the kinetic energy is less than expected from electrostatic repulsion, showing the formation of neck [Fig. 17.27]. In some cases, the energies are about twice that expected from Coulomb repulsion, *e.g.* Ne^{20} with Al^{27} at 120 MeV and 66 MeV and this is ascribed to the extra energy due to the rapid rotations of separating system (Ref. 78).

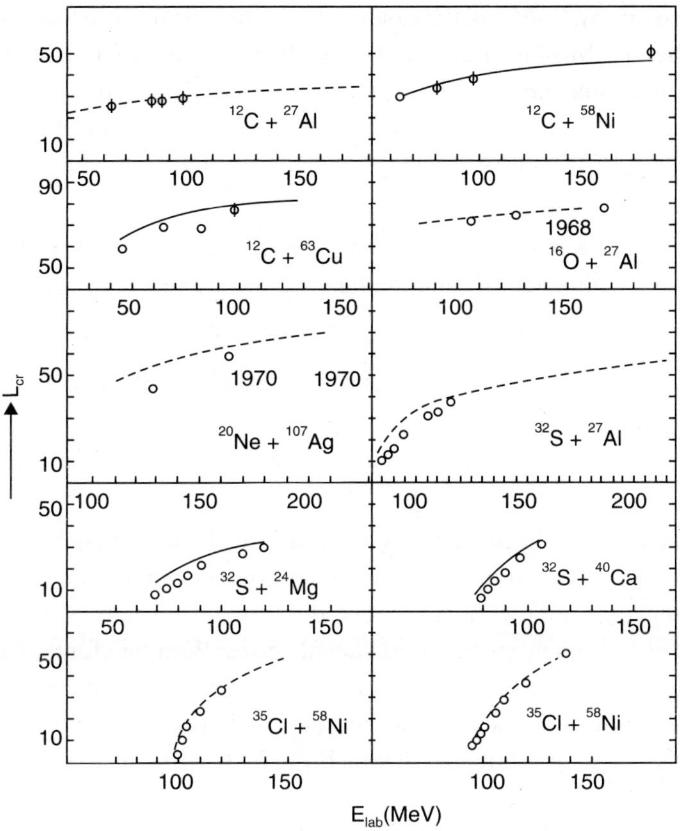

Fig. 17.30 L_{cr} versus E_{lab} (Experimental and theoretical), for heavy ion reactions as a function of energy compared with semi-classical friction model calculations [Eq. 17.59] (Ref. 79).

The case of deep-inelastic scattering has been extensively studied for the last fifteen years[80]. The results are explainable in terms of so-called "Extra push Model", in which the deep inelastic collision are associated with events, where system proceeds behind the interaction barriers, but fails to surpass the conditional saddle ridge[81]. One of the latest studies in this regard have been conducted by deep inelastic scattering in $Xe^{124, 136} + Ni^{58, 64}$ at energies near Coulomb barrier, *i.e.*, 522 – 556 MeV of $Xe^{124, 136}$ by L. Gehring et al. Ref. (80).

17.9.2 Friction Process

These features as mentioned in the previous section, lead one to the assumption that immediately after the initial collision, the two nuclei are brought into rigid contact by frictional forces, while the initial kinetic energy is dissipated into the internal degrees of freedom. Subsequently, a diffusion process leads to an exchange of particles between the two nuclei, giving a time dependent distribution in the symmetry of intermediate system and finally the system decays in a time similar to the period of rotation.

This process can be mathematically described in terms of classical concepts like viscosity and friction. We write the radial motion of the ions as:

$$M\ddot{R} + \nabla_{\nu}\, U\,(R) + \sum_{\mu} C_{\nu\mu}\,(R)\,\dot{R}_{\mu} = 0 \qquad ...(17.60)$$

where M is the reduced mass, $U(R)$ is the potential and the last term represents friction, with $C_{\nu\mu}$ as friction tensor. The friction tensor is related to the imaginary part of the optical potential by the relations for the loss (ΔE) per unit time, *i.e.*,

$$-\sum_{\nu\mu} C_{\nu\mu}\, v_\nu\, v_\mu \approx (\Delta E)\frac{2}{\hbar} W \qquad \text{...(17.61)}$$

In this manner, one can connect $C_{\nu\mu}$ with $g(r)$ of Eq. 17.34 and then the components of the friction tensor, *i.e.* radial component C_r and the tangential component $C_{\phi\phi}$ are connected to $g(r)$ as:

$$C_{rr}(r) = C_r\, g(r); \quad C_{\phi\phi}(r) = C_\phi\, g(r) \qquad \text{...(17.62a)}$$

where $$g(r) = [\nabla U_N(r)]^2 \text{ and } U_N(r) = U(r) - U(r) \text{ coulomb}$$

It has been found that

$$C_r = 4 \times 10^{-23} \text{ secs MeV and } C_\phi = 0.01 \times 10^{-1} \text{ secs MeV}^{-1} \qquad \text{...(17.62b)}$$

Various calculations have been made, using the above classical model to estimate the dependence of the experimental coefficient of friction on the radial distance and on the orbital angular momentum and the values are found consistent with the observed energy loss. The frictional force is closely related to the imaginary part of the optical potential as shown above. With this force, De et al.[79] calculated the critical angular momenta (*see* Eq. 17.59] as a function of energy for many reactions and the results fit the experimental data. Figure 17.30 shows, the comparison between the critical angular momenta (L_{cr}) for several heavy ion reactions, obtained from Eq. 17.59 with the semi-classical friction model calculations, [Eq. 17.60].

It may be realised that the critical angular momenta (L_{cr}) is the angular momentum at which the total potential has a local maximum value equal to the incident energy.

17.9.3 Collective Modes of Mass Transfer

From the experimental and theoretical considerations, it has become quite obvious, that the outcome of nuclear collision depends significantly on the initial projectile-target system. It seems that the mass-asymmetry can be treated as a coordinate, showing collective motion. The possibility that mass and charge-transfer in dampened reactions may be a collective process[82, 83], has been considered by a group led by Greiner. In this model the total energy of the system is approximated by:

$$H(r, \eta, \dot{r}, \dot{\eta}) = \frac{1}{2} m_r\, \dot{r} + \frac{1}{2} m_\eta\, \dot{\eta}^2 + V(R, \eta, r) + \tilde{E}(r) \qquad \text{...(17.63)}$$

where the initial parameters m_r and m_η depend on r and $\eta = (A_1 - A_2)/(A_1 + A_2)$. The initial parameters m_r and m_η are calculated microscopically in the two centre shell model, employing cranking formula, which is applicable for adiabatic, *i.e.* relatively slow motion involving no dissipation of energy into thermal energy.

Mass and charge distribution of fragments from heavy ion interaction has been experimentally measured and compared with theoretical models. Figure 17.31 gives the charge distribution for $Sn^{129} + Xe^{132}$ and $U^{238} + U^{238}$ at high incident energies. On the left are the theoretical curves based on Fokker-Plank approach of Nörenberg and collaborators (Ref. 84).

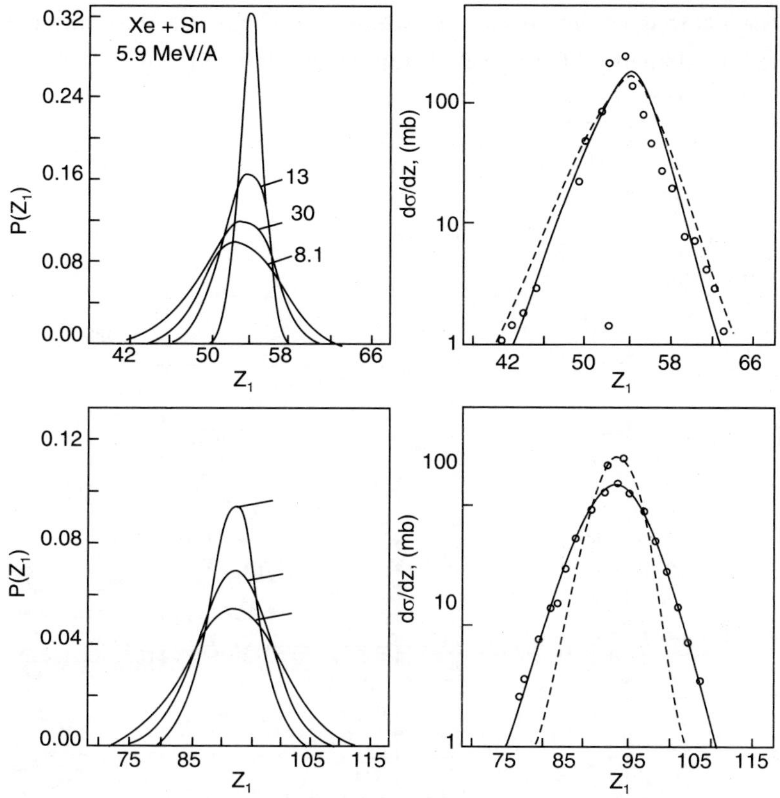

Fig. 17.31 Charge distributions are shown for reactions Sn^{120} + Xe^{132} and U^{238} + U^{238}. The experimental data, in the right side figures, is compared with model predictions based on theoretical transport coefficients. On the left are shown the probability (Ref. 84) distribution P with interaction time.

17.9.4 Transmutation of Angular Momentum

Heavy-ions are capable of imparting very high angular momenta to the system. Understanding this phenomena requires, the knowledge of reaction mechanism and the nuclear structure.

It seems that after formation of a compound system two competing processes come into play, but sequentially. First of all, the nucleus decays through particle emission or fission. We only consider the particle emission which is followed by the γ-emission. Figure 17.32 shows[85] a typical representation of de-excitation cascade starting from an excited Dy^{152} nucleus. The excited states of Dy^{152}, as shown in Fig. 17.32*a*, are the result of heavy ion reaction Pd^{108} (Ca^{40}, 4*n*) Dy^{152}, so that the gamma rays are emitted after the emission of four neutrons.

The heavy ion imparts a collective motion to the nucleus, which is available, even after the emission of particles. In Fig. 17.32*b* is shown the Yrast line, corresponding to low deformation band and super-deformed band. What is a Yrast line? Yrast line is defined as, the lowest energy line for which no level of spin *J*, corresponding to a particular rotational band exists, below the energy $E_J = [\hbar^2 J (J + 1)]/2 I_{\text{rigid}}$. So that, after the formation of compound nucleus, particle emission takes leaving behind a nucleus excited to a highly excited state below the energy threshold of a particle emission. This decays through γ-emission as shown in Fig. 17.32*a*, till it reaches the Yrast line, which corresponds to the lowest energy of the ground state rotational band.

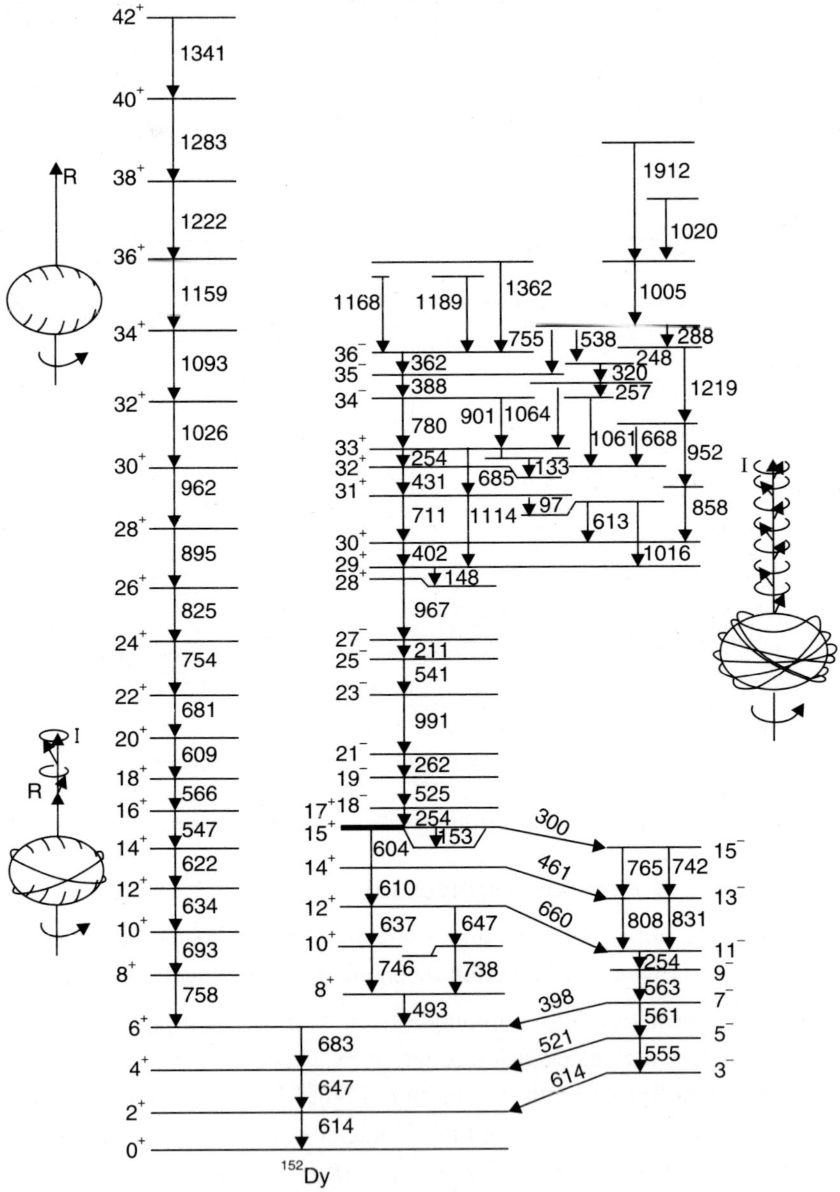

Fig. 17.32a The level scheme of Dy^{152}, produced by reaction Pd^{108} (Ca^{48}, 4n) Dy^{152} at 205 MeV showing the excitation of rotational states up to 42 ħ in Dy^{152}. (85).

In this particular case, which is one of the earliest cases studied, Figures 17.32*a* and 17.32*b* are obtained by analysing the γ-transitions from different excited states, of spins up to 60*ħ* and constructing the decay scheme for each case. The decay scheme for 42*ħ* spin state is shown in Fig. 17.32*a*. The super-deformed and deformed bands follow different relationships of *I* and excitation energy.

(Super-deformed states correspond to $a/c \geq 1.5$ and hyper-deformed states have $a/c \geq 2.0$. These states are high energy states belonging to heavy nucleus). The near Yrast lines states have both positive and negative partition and γ-transitions between them include E_2, E_1, M_1 multipolarities. These properties are signatures of oblate states. Yrast line represents the locus of lowest energy rotational states for a given E and I, so below the Yrast energy, the decay is through vibration or other modes of decay, as there are no states of rotational collective nature below Yrast line.

A typical case of the significance of Yrast line, has been shown in Fig. 17.32*c*, where Ar^{40}, has been used for projectile on target of $A \approx 160$ and four neutrons are emitted. It is somewhat similar to case of Dy^{152}, where Ca^{48} was used as a projectile and four neutrons were emitted. The figure depicts the Yrast line. Under suitable circumstances this Yrast line may be an excited member of the rotational band and from then on, a chain of γ-rays is emitted due to transitions to successively lower members of the rotational band till the nucleus ends up in the ground state. Ideally each of these gamma rays is expected to be of the E_2-type and strongly enhanced due to collective effects.

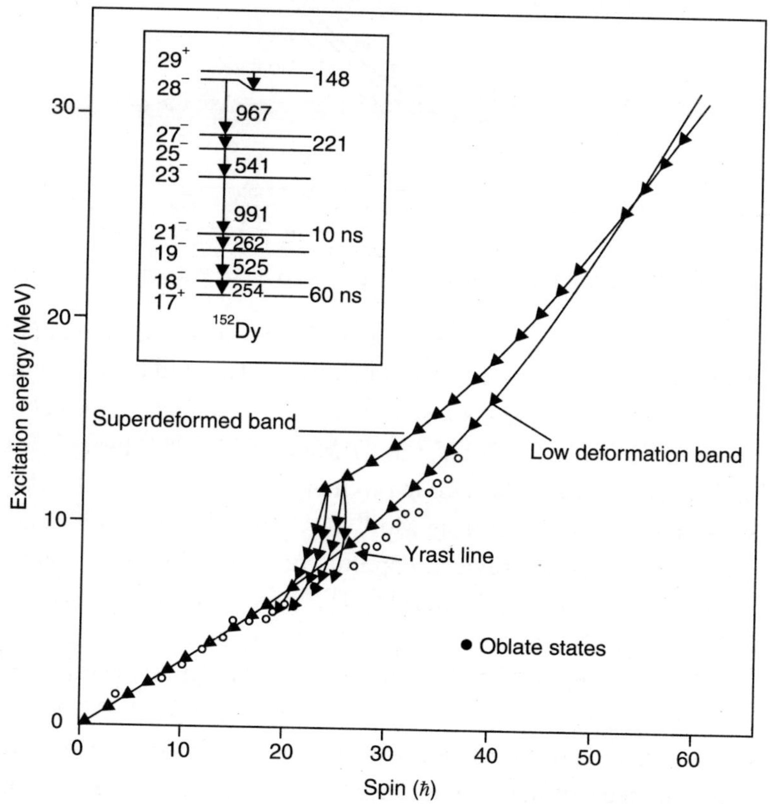

Fig. 17.32b The concept of Yrast line for these states and the super deformed band (Ref. 85).

In general, there are two types of transitions. At relatively high energies above the Yrast line, the nucleus decays mainly by emitting statistically electrical dipolar gamma ray (E_2). The other de-excitations are due to rotational levels. The statistically emitted gamma rays determine the shape of the gamma rays spectrum at low energies.

One of the characteristics of such spectra are the selection of multiplicity with mass number and angular momenta. The average multiplicity $< M_r >$ is defined as the average number of gamma rays per unit decay event, *e.g.* fission or particle emission. It helps in measuring the total angular momentum J of the compound nucleus[86].

Recent studies in this field have extended this field quite a bit. On the lower Z side, ($N - Z$ nuclei in $1f\,7/2$ shell), high spin states have been excited[87], though single particle states were demented. In a recent-most[88] study, on Mn^{50}, created by Si^{28} (Si^{28}, αpn) Mn^{50} with 115 MeV Si^{28} beam, collective mode of excitation has been excited up to 15 $\hbar$. This has established $T = 1$ isobar analogue states of Cr^{50}, up to $I^{\pi} = 4^{+}$ while for higher energies $T = 0$ band has been observed.

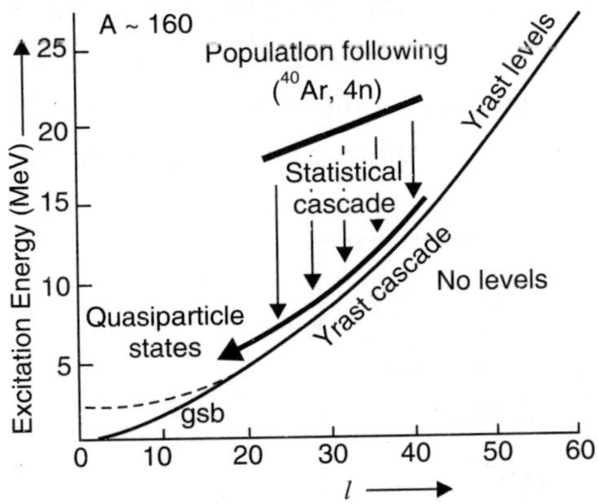

Fig. 17.32c Diagram showing the sequence of γ-emission leading to Yrast line in (Ar^{40}, 4n) reaction.

A somewhat higher mass region of $A \approx 80$ with $N \approx Z$ represents, an ideal opportunity to observe the interplay of collective and single particle degree of freedom. The studies in the nucleon region of 36 and 38 offer such an opportunity of shape coexistence in the ground state deformation[89]. Recent study of Y^{89} through the reaction Fe^{54} (Si^{28}, $1p2n$) Y^{79} at 87 MeV of $Si^{28\,(10+)}$ beam[90] has yielded $K = 5/2$ similar to Sr^{77} and neighbouring Sr isotopes. Going to higher values of A, in the range of $Z \approx 50$, two recent studies[91] of Sn^{113} and In^{107} obtained from Mo^{100} (O^{18}, n) Sn^{113} at 94 MeV and Zn^{66} (Si^{45}, $zp2n$) In^{107} at beam energy of 162 MeV have provided opportunity to study three de-coupled bands in Sn^{113}, up to $(63/2 - 69/2)\,\hbar$ and up to a spin of $33/2\hbar$ in In^{107}. These are good studies of collective and particle interaction. In this region of A, an interesting case of magnetic rotational band has been investigated for Sb^{108}, obtained[92] from the reaction, Fe^{54} (Ni^{58}, 3pn) Sb^{108}, at 243 MeV of beam energy of Ni^{58}. The magnetic dipole bands are characterised by strong M1 transition and weak E_2 cross-over showing that there is low quadrupole deformation B (M1) arising from magnetic rotation, for which a rotating magnetic dipole sector breaks the symmetry of the nucleus. Properties of such bands have been described within Tilted Axis Cranking (TAC) model[93]. In the case of Sb^{108}, the suggested configuration shows good agreement with TAC model.

In the region $A \approx 180 - 210$, which includes $Z = 82$ magic proton number, a large shell energy gap exists for spherical nuclei like Pt isotopes. This is an ideal region to study the onset and evolution of collective motion in nuclei. This region also contains $N = 126$, near which the excitation can be explained in terms of shell model and for large valance neutrons, it can be understood in terms of vibrational motion. This is a region of coexistence of shapes and a phase transition[94] starting from nuclei in the vicinity of Pt^{182} rotational band associated with prolate structure, it comprises the mass majority of known Yrast and near Yrast states[95] above spin $I = 8\hbar$. Investigation[96] for high spin states and band structure in Pt^{182}, obtained from Yb^{170} (O^{16}, $4n$) Pt^{182} and Dy^{162} (Mg^{24}, $4n$) Pt^{182} and Dy^{163} (Mg^{24}, $4n$) Pt^{182} at energies around 95 MeV of O^{16} and 125 MeV of Mg^{24} have yielded eight rotational bands, which can be described within a shape coexistence frame involving normal proton-hole states and intruder particle hole states. Experimental results of Po^{192}, obtained[97] from Er^{164} (S^{32}, $4n$) Po^{192} at a beam energy of 164 MeV was investigated via in-beam γ-rays spectroscopy yielding information about the first three excited states. Theoretically a particle-core mode has been used not only to explain these states but also the states from Po^{192} to Po^{210}, using particle core model[98] (PCM).

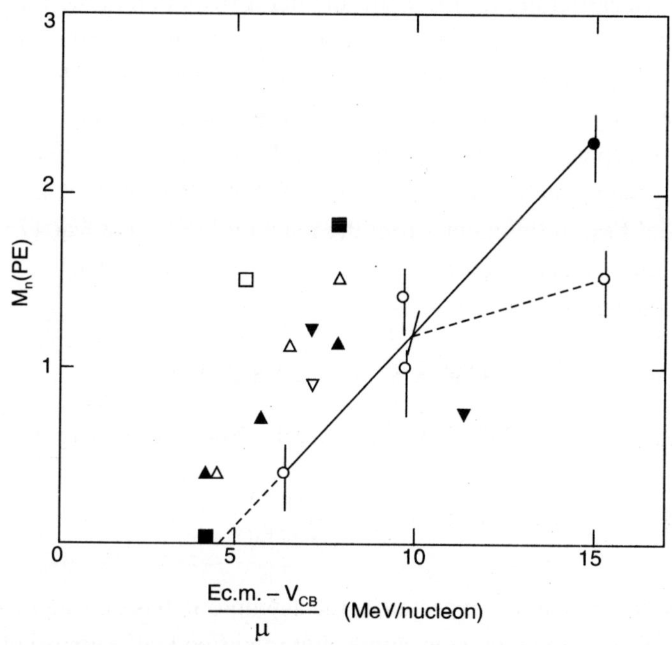

Fig. 17.33 Mean multiplicity, corresponding to pre-equilibrium neutrons, plotted versus the relative incident energies for nucleons above Coulomb barrier (Ref. 99).

17.9.5 Pre-Equilibrium Particle Emission

For light particle emission, we have at higher energies, three phenomenon to look for: (*i*) The evaporation from a compound nucleus. It is called hot spot model. (*ii*) The emission from a source moving with a velocity intermediate between the centre of mass velocity and that of projectile. (*iii*) The pre-compound or pre-equilibrium emission. Figure 16.17 in the last chapter shows the multiplicity of neutrons emitted in the fusion-evaporation reaction Ho^{145} + Ne^{20} at 11, 14.6 and 20 MeV/ nucleon, plotted versus lab-energy for several detection angles. While the low energy part is explainable by evaporation by compound nucleus and moving source, the high energy part is understood by including the pre-compound phenomena. The experimental curves in Fig. 16.17 can be represented by a two component fit using the theoretical model of compound nucleus formation and pre-equilibrium.

In Fig. 17.33, the correlation between the total pre-equilibrium multiplicity M_n (PE) (number of neutrons per fusion event) has been plotted for a variety of reactions, including $Ho^{165} + Ne^{20}$, as a function of incident energy above the Coulomb barrier. There is a general trend of linear increase of M_n (PE) with the kinetic energy per nucleon. Also it is known that mean energy of emitted neutrons increases with incident-energy. This means nuclear matter cannot sustain the high amount of energy, as a temperature.

17.9.6 Fusion-Fission

Fusion is followed by either light particle emission (protons, neutrons, alphas, etc.), or followed by fission. There is a certain difference between the emission of fragments after deep inelastic scattering and emission of fragments after fusion. In the former case, the angular distribution is asymmetrical or anisotopic. We have shown earlier in Fig. 17.28 the angular distribution of the various fragments arising out of deep inelastic scattering. The fusion-fission, however, can only be extracted from the final kinetic energy distributed in the damped reaction induced by heavy projectiles, like Fe^{56}, Kr^{84-85}, Xe^{132}, Rb^{208} and U^{238}. For these systems, the angular dependence of the total kinetic energy spectra are found less informative and emphasis is placed on the dependence of the spectrum on projectile-target combination, on bombarding and final fragmentation. As for example, Fig. 17.35 shows the laboratory kinetic energy spectrum[100], of co-related fission fragments from the reaction $U^{235} + Ne^{20}$ at 175 MeV and 252 MeV. The broad bell shaped energy distribution shown in Fig. 17.35 corresponds to average total kinetic energy of both fission fragments of

$$\langle E_f \rangle = (198 \pm 8) \text{ and } (201 \pm 8) \text{ MeV}$$

for the incident energies of 175 and 252 MeV, respectively. Both these energies (E_f) follows the law:

$$\langle E_f \rangle = \left(22.3 \pm 0.107 \frac{Z^2}{A^{-1/3}} \right) \quad ...(17.64)$$

Qualitatively, the relationship is expected, if the asymptotic fission fragment energies are simply due to acceleration due to Coulomb repulsion, somewhat modified by the contribution to kinetic energy, belonging to the fission degree of freedom which may be present at the saddle.

The phenomenon of fusion-fission is, in general, submerged in the damped reaction energy distribution. It requires a complicated analysis to disentangle it. At higher energies, the statistical scission model explains the angular distribution of the fission fragments in fusion-fission reaction as shown in Fig. 17.34 for $Au^{197} + C^{12}$ at $E_{lab} = 93$ MeV. The symmetric nature of the curve is evident[100]. This model assumes a statistical partition of the initial angular momentum into orbital angular momentum **l** and channel spin **S** of the two fragments, so that **I = l + S**.

After fusion, the possible processes of disintegration of the nucleus are: (*i*) Fission and heavy residues (*ii*) fragmentation (*iii*) spallation as described below:

A. Fission and Heavy Residues: The problems of fission fragments vary with the incident energy. As example, at 5 – 10 MeV/A, there has been found angle dependence of spins of fission fragments in $C^{12} + Th^{232}$, $O^{63} + Th^{232}$ and $F^{19} + Th^{232}$, which corresponds to, indicating the importance of tilting mode of spin excitation[101]. At higher energies say 20 MeV/A, in collisions like Cu^{12} + Au^{197} and A^{127} + Au^{197}, it was found that fission-like events leading to near and trans-gold species are produced. Also the neutron-deficient species possibly the "unknown nuclei" may be formed[102], but at these energies, the reaction U^{208} + Pb^{208} produces neutron-rich fission fragments[103]. This is understood in terms of quasi-elastic or deep inelastic scattering. Such reactions are useful, for generation of neutron rich radioactive-beams by fission of intermediate energy projectile. At still higher energies[104] say 35 MeV/A, using Cu^{63} + Th^{232}, it was found that a hot and heavy nucleus was produced and fission fragments correlation indicated that 70% of the projectile linear momentum can be transferred to the fissioning system. A composite system of mass as high as 275, at an excitation energy of 1 GeV, seem to have been formed.

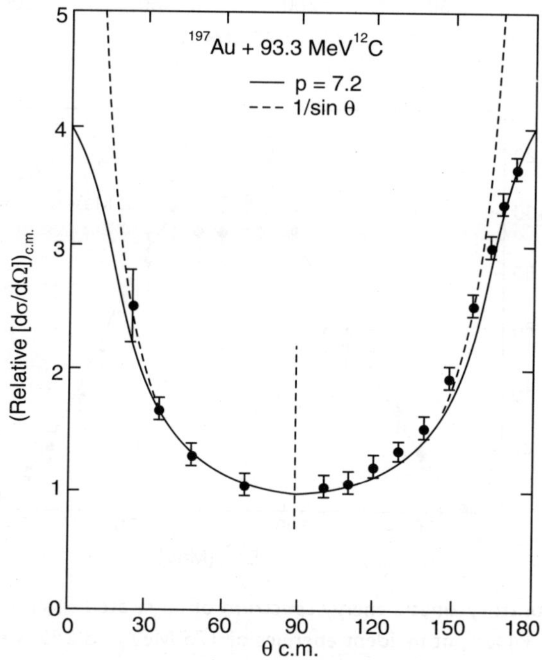

Fig. 17.34 Centre of mass angular distribution of fission fragments from the reaction Au^{197} + C^{12} at E_{lab} = 93.3 MeV. The solid curve is a model fit (Ref. 100).

B. Fragmentation: The phenomenon of multi fragmentation, wherein fragments with charge greater than $Z = 3$ are emitted with large multiplicities (or cross-section) from heavy ion reaction; has been the focus of many measurements[105]. These measurements are compared with the theoretical model, which basically requires that yield of intermediate mass fragments (IMF) beyond $Z = 3$, is predominantly from a mid-rapidity source[106]. Many measurements of fragment emission has been carried out measuring ratios of yield of isobaric pairs of nuclei, as a function N/Z of the fragments as well as the angular distribution of fractional yield. These measurements of mass-asymmetric reaction of light-ions on heavy targets of varying isotopic composition, showed that N/Z ratio of the target is reflected in N/Z ratio of intermediate mass fragments[106]. Such measurements were recently extended to less mass asymmetric heavy ion reactions[107].

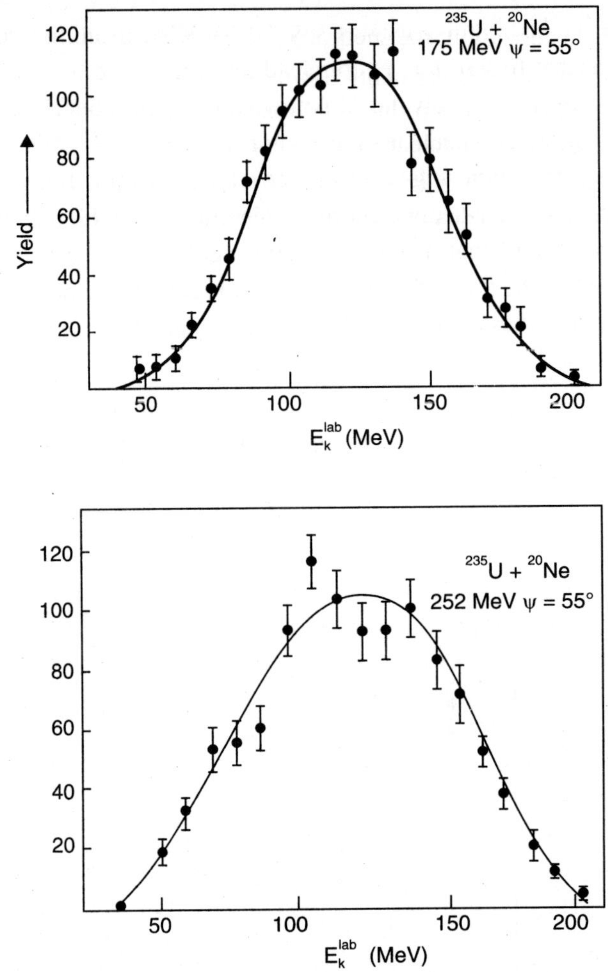

Fig. 17.35 Laboratory kinetic energy spectrum of co-related fission fragments from reaction U^{235} + Ne^{20}, at incident energies of 175 MeV and 252 MeV (Ref. 100).

A recent-most measurement[108] in this field is that of $Xe^{124, 136}$ + $Sn^{112, 124}$. This showed that light fragments from a mid-velocity region were substantially more neutron-rich when compared to fragments from projectile velocity region.

Theoretical models have been developed that treat the formation of nuclear hot matter in heavy-ion collisions in a dynamical way. Two models[109] are especially used:

(*i*) Quantum Molecular Dynamics (QMD) model which takes into account the dynamics in collision phase of the heavy-ion-reaction, followed by the identification of excited clusters and free nucleons from this distribution and the decay of identified hot clusters to nuclear fragements based on the assumption of thermal and statistical equilibrium.

Detailed theoretical calculations by Kumar and Puri have been carried out recently about the role of momentum in fragment formation, in central heavy ion collisions, at the intermediate energies say 150 MeV/nucleon to 600 MeV/nucleon using quantum molecular dynamic method. The results are obtained for fragment multiplicity, rapidity distribution and the stability of fragments. It seems that using three methods of clusterisation, the method of minimum spanning tree is most suitable[109].

(*ii*) A second model called Boltzmann-Ulhing-Uhlenbach (BUU) mean free approach, in which isospin degree of freedom has been introduced by accounting for the difference in the nucleon-nucleon scattering cross-section and in the nuclear potential.

In the measurements[110] of the fragment emission from the mass-asymmetric reactions (Fe^{58}, Ni^{68}) at E_{beam} = 30 MeV/A, ratios of yields of isotopic fragments from hydrogen to nitrogen were measured as a function of lab angle. The calculation based on the QMD model, followed by statistical multi-fragmentation model (SMM)[111] were carried out. It showed that the source emitting the IMF's measured in the mid-rapidity region, is substantially more neutron rich, than that predicted by the models, suggesting a modification of models.

In summary, the various experiments on the measurements of fragments show the importance on the condition in the composite nucleus at the time of fragment formation, which gives subsequent properties of fragmentation emission.

C. Spallations: The mass and charge distribution of reaction products from spallation reaction induced by protons in the energy range from several hundred MeV to several GeV, has been investigated in a number of experiments[112]. One of the interesting outcome of such reaction is the production of stable isotopes. This forms one of the basis of modern accelerators for radioactive nuclei. The process of spallation has also been used for the production of intense beam of fast neutrons. In one experiment, a full mass and charge distribution of the spallation products has been measured at E = 600 MeV for a wide range of target nuclei[113]. In a recent experiment, spallation reactions were studied in Al^{27}, and Fe^{57} induced by 800 MeV protons[114]. This resulted in 13 nucleides ranging from Si^{20} to O^{14} in Al + 800 MeV protons and some 41 nuclides from F^{20} to Co^{56} in Fe^{56} + 800 MeV protons. This resulted in the measurement of production cross-sections of 36 nuclides from proton interaction with Fe^{56} and for 12 nuclides in the case of Al. Theoretically, the calculations based on Quantum Molecular Dynamics (QMD) model[115], along with statistical decay model, (SDM) explained the results.

17.9.7 Still Higher Energies, Shock-Waves

It is expected that at very high energies of the heavy ion projectiles say 10 MeV/N to 100 MeV/N and higher, the nuclear density may rise suddenly in the region, where they first interact, if the relative velocity of the nucleons is greater than the rate of propagation of this density disturbance, through the nucleus. Then a nuclear shock wave may develop. The rate of propagation of the sock wave is the velocity of sound in nuclear matter. In O^{16} – O^{16}, such shock waves are expected even for centre of mass-energies as low as 100 MeV with densities twice the normal values[116]. For energies of 1000 MeV, densities of five times the normal value occur.

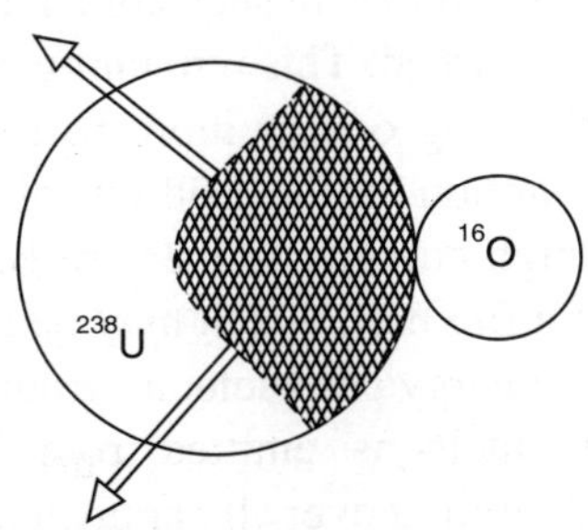

Fig. 17.36a Figurative illustration of shock-wave, when O^{16} of high energy, impinges on a nucleus of U^{238} (Ref. 117).

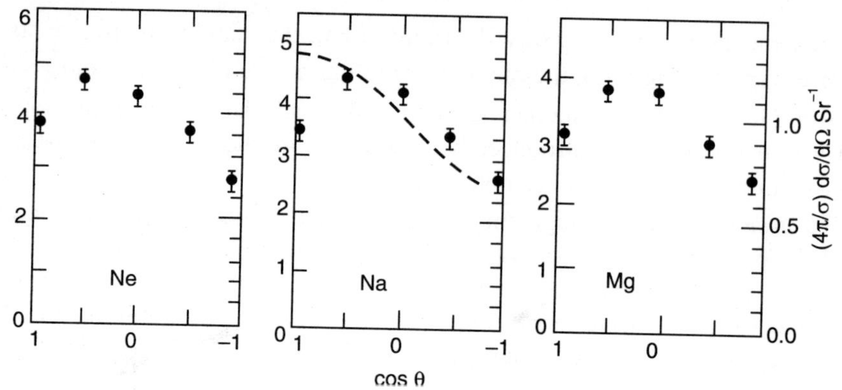

Fig. 17.36b Proof of shock waves. The angular distribution of fragments emitted from uranium irradiated by 28 GeV protons (Ref. 117).

An attempt to detect nuclear shock waves in energetic heavy ion interactions has been made by bombarding nuclei of gold and uranium with 25 GeV protons. One of the effects of the appearance of shock waves is to have a certain broad peak at angles around 70°. Figure 17.36 shows[117] the results for fragments of Ne, Na and Mg which shows the peaks at around this angle. These peaks give an indication that there is some kind of collective motion directly connected with the cascade of particles initiated by the incoming protons. It is also possible that in such cases of shock waves, one may obtain nuclear density isomers of ultra dense nuclei. These are stable, abnormally dense nuclei, in which the energy of compression is compensated by pioncondensation. There are no experimental evidence for such nuclear density isomers; despite a good amount of theoretical conjectures.

17.9.8 Heavy Ion Interactions at Relativistic Energies

At very high energies say above 400 MeV/A, the relativistic kinematics is required to be used and also new types of reaction mechanisms appear. A study of fragments of 29 GeV, N^{14}, in collision with carbon and hydrogen shows, that the fragmentation is insensitive to the target nucleus. This suggests that the role of target nucleons is simply to inject the energy into the projectile causing it into fragments. There is no memory of how the energy is given to projectile. Extensive measurements[118] of the fragmentation of 1.05 and 2.5 GeV per nucleon for C^{12} and 2.1 GeV for O^{16}, on targets from Be to Pb, confirmed that the momentum distributions of the fragments are Gaussian and have no correlation with target mass or beam energy.

A model of Abrasion-Ablation[119] has been developed to explain this phenomenon. The basic idea of this model is that the overlapping volumes in such collision are sheared away, leaving the rest of the projectile to continue relatively undisturbed. This remaining part of the projectile is expected to be highly excited and loses energy by emitting one or more particles—a process called ablation. The relativistic hydrodynamic equations of motion for the head-on collision have been solved. Also a simple model of heavy ion collision at very high energies is the Fire Ball model[119]. It assumes that the two nuclei whose trajectories interact from a fire ball, which may be treated relativistically as are ideal gas whose temperature is determined by the energy per nucleon. Around 300 MeV/A, this gives a Maxwell-Boltzmann energy distribution for the nucleons, emitted from fireball. Assuming isotopic emission from the fireball in its rest frame and integrating over all parameters gives the differential cross-sections. It has been possible to obtain an experimentally determined energy spectra of the emitted nucleons from the interaction of helium and neon nuclei at various relativistic energies with uranium and compared with fire ball model calculations[119]. On the whole, agreement is good, except at the highest energy. At this high energy the assumption that there are separate target and projectile fire balls gives quite a good fit. Hence, in principle, the concept of 'fireball' at these high energies is quite successful in explaining the data.

Recently[120] the emission of protons, deuterons, tritons and pions have been studied for Au + Au reactions at 11.64 GeV/A and 10.20 GeV/A energies using Brookhaven National Laboratory AGS (Alternating Gradient Synchrotron) facility where very high energy ions of gold could be produced. Measurements basically consisted of mean transverse moments $<Px>$ as well as the cross-sections as a function of energy of the emitted particles. Experimental results were compared with two models.

(*i*) A Relativistic Cascade (ARC) which is based[121] on BUU transport model, used for intermediate energy heavy ion collisions.

(*ii*) RQMD (Relativistic Quantum Molecular Dynamic) model[122] combines the classical propagation of particles with the excitation of hadrons into resonances and strings. It is interesting that both these models predict the measured flow when restricted to a pure cascade calculations. The ARC model calculations also indicate that the baryon density is increased by a factor of about 8 and the lifetime of high density region is correspondingly longer.

At externally high (Ultra-relativistic) energies say 200 GeV/A one expects an electromagnetic radiation from the hot system, created in such reactions over and above and undisturbed by the hadronic interactions. Limits of such direct electromagnetic radiation have been measured by R. Albrecht et al.[123] and the background hydraulic interaction have been measured recently by another group[124]. The results confirm such possibilities, when soft-photon production was measured in central 200 GeV/A S^{32} + Au collisions.

17. Heavy Ion Induced Reactions

2000–2008

Heavy ion reactions, forms the largest fraction of research activity in nuclear physics, in these last several years, with the availability of accelerator facilities in many laboratories and universities throughout the world, especially in U.S.A., Europe, America, China and even in Asian countries like India, Pakistan and S. Arabia.

In heavy ion induced reactions, there are many types of interaction *e.g.*,

(*i*) Level structure determination

e.g. ${}_{42}Mo^{92}$ (${}_{32}Ge^{74}$, 2α, $2n$) at 138 MeV of ${}_{32}Ge^{74}$ beam where ${}_{74}W^{166}$ emits 2α, or $2n$ giving rise to ${}_{72}Hf^{158}$ or ${}_{74}W^{164}$ etc.

(*ii*) Fission $Se^{82} + Ba^{134}$ or $Se^{82} + Ba^{138}$ at 225 MeV of Se^{82} beam giving rise to various products after the compound state.

(*iii*) Fragmentation: *e.g.*,

(*a*) Ca + Ca; Ni + Ni, Nb + Nb, Xe + Xe, Er + Er and Au + Au

(*b*) $Ca^{114} + Mo^{92}$ at $E/A = 50$ MeV

(*iv*) Particle production: *e.g.*,

(*i*) $Ni^{58} + C^{12}$ or (*ii*)Ni^{58} + Au at 34.5 MeV of Ni^{58}

(*v*) Scattering and transfer

$Li^7 + Pb^{208}$; $E_{c.m.} = 28.6$ MeV

$Cl^{35} + Pd^{105,\,106,\,110}$ at $E = 107$ MeV – 138 MeV

One of the interesting features, of these efforts is the collaborative nature of these activities, with a large number of authors in a given work; *e.g.* 50 authors in a paper on 'Differential directed flow in Au + Au collision, measured at 90 to 400 MeV. The study presents the first order fourier coefficient for different particles species, from experimental data. [Phy. Rev. C. 64, 041604 (*R*) (2001)]. In about half a doze papers, in 2000; in one paper on set of midevelocity emission in symmetric heavy ion reaction as many as 48 authors were involved in Xe + Sn reaction at 25-50 MeV of Xe ions, using a system of 4π multi-detected system INDRA, at Gail.

In another interesting paper, with 49 authors from Europe, properties of light particles were measured in Ar + Ni collision at 95 MeV; with the process of prompt emission and evaporation [Phy. Rev. C. 64, 034612 (2001)]. In a paper Dynamical multi fragmentation and special correlations, has been reported by J.B. Singh and R.K. Puri from Panjab University, Chandigarh (India) [Phy. Rev. C. 62, 054602 (2000)]. An interesting case of trans-uranium nucleus No^{252} has involved 40 authors from Europe and USA for study, the reaction ($Ca^{48} + Pb^{206}$) $No^{252} + xnyp$ at 215.5 MeV, involving excitation up to 20 *h* spin [Phy. Rev. C 65, 014305 (2002)].

An interesting fusion evaporation study in $Ne^{20} + Tb^{159}$, and $Ni^{28} + Tm^{109}$ has been carried out, between $E/A = 8$ MeV and 16 MeV, by 13 authors. In a time of flight measurements; evaportion; deep-inelastic scattering (DIC), Fission and elastic scattering are quite evident. Theoretical comparison with statistical model enables the authors, to determine the range of fission lifetimes, and contribution of the first fission at different compound nucleus excitation energies [Phy. Rev. C. 68, 034613 (2003)].

In a paper, theoretical model calculations have been carried out for 2500 fusion cross section data, for 165 different systems; for heavy ion elastic scattering and fusion processes. The model provided a good description of the experimental data, without any free parameters [Phy. Rev C. 69, 034603 (2004)].

Earlier S. Kailash et al. have published a paper on 'Statistical Model Calculations' to reproduce α – *spectra in fusion – mechanism* (*Pramanna*, J. of Physics (1990), 439).

In a reaction of $Ge^{76}(C^{13}, 5\eta\gamma)\, Sr^{85}$ gamma rays spectra were measured in a paper by A.K. Jain and other workers. ('Nuclear Physics', A 732 (2004), 13.)

In a paper authored by 20 authors, $Se^{80} + Pb^{208}$ and $Se^{80} + Th^{232}$ reaction have been studied using Se^{80} beam from 470 to 630 MeV, for studying fission and binary fragments. The average number of prompt neutrons are estimated to be $\gamma^{\text{Total}} = 10 \pm 2$, consistent with compound nucleus scenario. For spontaneous fission $\gamma^{\text{Total}} = 15 \pm 2$ [Phy. Rev. C. 75, 024604 (2007)].

Production of nuclei above Sn^{100} in a fusion evaporation reaction between Ni^{58} and Fe^{54} ions was studied at Oak-ridge National Lab. The beam energy was varied to optimize the yields for two-three and four particle evaporation. The results verified the prediction of statistical model code HIVAP. The optimum energy Fe^{54} (Ni^{58}, 4*n*) Xe^{108} reaction channel that allows one to study Xe^{108}, + Tc^{104}, $Sn^{100} \rightarrow \alpha$-decay, is deduced as 240 MeV [Phy. Rev. C. 77, 034301 (2008)].

REFERENCES

1. P.E. Hodgson: Nuclear Heavy Ion-Reaction: Calender Press, Oxford, (1978); H. Bohm, G. Daniel, M.R. Maier, P. Kienler, J.G. Cramner and D. Proctel, Phy. Rev. Letter. 29, 1337, (1972); W.E. Frahn: Fundamentals in Nuclear Theory (eds; A De-Shalit and C. Villi), p. 3. International Atomic Energy Agency, Vienna, (1967); Mechanics, H.S. Hans and S.P. Puri, p. 180, Tata McGraw-Hill Publishing Co., Ltd., New Delhi, (1995); H. Goldstein: Classical Mechanics; Addison-Wesley Publishing Co. June Reacling U.S.A. (1959).
2. W. Nörenberg: J. Phy. Paris, 37, C5 – 141 (1976).
3. P.R. Christensen, V.I. Manko, F.D. Bechetti and R.J. Nickles: Nuclear Physics A 207, 33 (1973).
4. R.J. Ascuitto and N.K. Glendenning: Phy. Letter B48, 6 (1974).
5. R.A. Broglia, L. Landowns and A. Winther: Phy. Letters B. 40, 293 (1972).
6. M.V. Berry: Proc. Phy Soc. 89, 479 (1966).
7. R.da Silveira: Phy. Rev. Letters, B45, 211 (1973).
8. K.W. Ford and J.A. Wheeler: Annual Physics, (NY), 7, 259 (1959).
9. W.E. Frahn: Phy. Rev. Letters, 26, 568 (1971); Annual Physics (NY), 72, 524, (1972); S.D. Baker and J.A. McIntyre: Phy. Rev. 161, 1200 (1967); W.E. Frahn and R.K. Venter: Annual Physics (NY), 24, 243 (1963).
10. W.E. Frahn: Phy. Rev. Letters 26, 568 (1971); Annual Physics (NY), 72, 524, (1972); R.C. Fuller: Phy. Rev. C. 12, 1561 (1975).
11. W.E. Frahn: Heavy Ions, High Spin States and Nuclear Structure, V. 1, International Atomic Energy Agency, Vienna (1975).
12. P.R. Christensen, I. Chernov, D.H.E. Gross, R. Stokstad and F. Videback: Nuclear Physics A 207, 433 (1973).
13. H. Holm, W. Scheid and W. Greiner: Phy. Letters B 29, 473 (1969).
14. F.K. McGowan and P.H. Stelson: Nuclear Spectroscopy and Reactions, Part C; Ed. J. Cerney Academic Press, New York, p. 4 (1974).
15. K. Alder and H. Pauli: Nuclear Physics A. 128, 193 (1969).
16. A. Winther and J.De Boer and K. Alder and A. Winther: Coulomb Excitation: Academic Press, (no 1969); K. Alder and A. Winther: Electromagnetic Excitation, North Holland, Amsterdam (1975); L.C. Biedenhasn and P.J. Brussard, Coulomb Excitations, Clarenden Press, Oxford (1965).
17. A.J. Ferguson, O. Anser, A.B. McDonald and T.K. Alexander: Argonne Conference, p. 187, (1971); W.E. Frahn and K.E. Rehum: Phy. Reports 37C, 1 (1978).
18. J.V. Mehr, M.M. Sachs, R.H. Siemensen, A. Weidinger and D.A. Bromley: Phy. Rev. 188, 1665 (1969).
19. J.C. Heibert and S.T. Garvey: Phy. Rev. B 135, 346 (1964).
20. E. Videback, P.R. Christensen, O. Hanson and K. Ulback: Nuclear Physics, A 256, 301 (1976): J.V. Mehr et al.; Phy. Rev. 188, 1665 (1969).
21. C.S. Dover and J.P. Vary, Heidelberg, Conf. p. (1974); J. Wilezynski: Nuclear Physics, A 216, 386 (1973).
22. J.J. Eck, Phy. Rev. C. 3, 949 (1971); Wilezynski J. and K. Siwesk-Wilezynski, Phy. Letters B. 55, 270 (1975), De Vries and M.H. Clover: Nuclear Physics A. 243, 528 (1975).
23. H. Helm and W. Greiner: Nuclear Physics A. 195, 333 (1972); B. Sinha, Phy. Rev. C. 11, 1546 (1975).
24. R. Siemssen, J.V. Mehr, A. Weidinger and D.A. Bromley: Phy. Rev. Letters, 19, 968 (1967).
25. G.R. Satchler: Nashville Conference, V. 11, p. 175 (1974).

26. Coach J.G., McIntyre J. A. and Heibert J.C.: Phy. Rev. 152, 883 (1960); F.J.A. Butler and Goldfarb L.J.B.: Nuclear Physics, 78, 409, (1966), Ibid, A 115, 461 (1968).
27. F.D. Santos: Nuclear Physics A 212, 341 (1973).
28. P.E. Hodgson: Nuclear Heavy Ion Reaction, Clarenden Press, Oxford (1978); H.J. Korner, G.C. Morrison, L.R. Greenwood and R.H. Siemssen: Phy. Rev. C7, 107 (1973).
29. R.M. De Vries and K.I. Kabu: Phy. Rev. Letters 30, 325 (1973); Ibid, Phy. Rev. C8, 951 (1973).
30. P.E. Hodgson: Nuclear Heavy Ion Reactions: Clarendon Press, p. 348, 351, 377 Oxford (1978); J.L.C. Ford et al.: Phy. Rev. C10, 1429 (1974); C. Chasman et al.: Phy. Rev. Letters, 31, 1074 (1973); P.D. Bond et al.: Phy. Letters, B47, 231 (1973).
31. T. Kamuri: Nuclear Physics A 259, 343 (1976).
32. A.J. Baltz and S. Kahan: Argonne Conference, p. 273 (1973).
33. M.C. Marmaz, J.C. Peng, N. Lisbona and W. Greiner: Phy. Rev. C 15, 307 (1977).
34. J. Bang, C.H. Dasso, F.A. Gareev and B.S. Nilsson: Phy. Letters, B 53, 143 (1974).
35. K. Yagi et al.: Physics Letters, B 29, 647 (1969); Ibid, Phy. Rev. Letters 29, 1334 (1972); Ibid, Phy. Rev. Letters 34, 96 (1975); Ibid, Phy. Rev. C 14, 351 (1976).
36. R.J. Ascuitto and N.K Glendenning: Phy. Rev. B 45 (1973); Argonne Conference, p. 513 (1973).
37. R.M. De Vries: Phy. Rev. Letters, 666 (1973); Nuclear Physics A 212, 207 (1973); N. Marquardt, W. Von Qertzen and R.L. Walter: Phy. Letters, B. 35, 37 (1971).
38. M. Conjeand S. Herar, R.da Silveria and C.Volant: Nuclear Physics, A 250, 182 (1975).
39. F.M. Steele, P.A. Smith, J.E. Fink and G.M. Crawley: Nuclear Physics A. 266, 424 (1976); H.W. Fulbright et al.: Phy. Letters, B 53, 449 (1975); A. Arima et al.: Nuclear Physics A 215, 109 (1973).
40. L.R. Greenwood, K. Katori, R.E. Malmin, T.H. Braid, J.C. Stoltzfus and R.H. Siemsen: Phy. Rev. C6, 2112 (1972); M.L. Halbert, F.E. Durham and A. Vander Wande: Phy. Rev. 162, 899 (1967); Ibid, Phy. Rev. 162, (912) (1967); J.L. Artz, M.B. Greenfield and N.R. Flecher: Phy. Rev. C13, 156 (1976).
41. B.K. Nayak, R.K. Chowdhary, D.C. Biswas, L.M. Pant, A. Saxena, D.M. Kulkarni, and S.S Kapoor: Phy. Rev. C.55, 2951 (1997); D.C. Biswas, R.K. Chowdhary, B.K. Nayak, D.M. Kulkarni and V.S. Ramamurty: Phy. Rev. C 56, 1926 (1997).
42. L.T. Baby, V. Tripathi, D.O. Kataria, J.J. Dass, P. Sugathan, N. Madhavan, A.K. Sinha, M.C. Radhakrishna, N.M. Badiger, N.G. Puttaswamy, A.H. Vinod Kumar and N.C.S.V. Prasad: Phy. Rev. C. 56, p. 1936 (1997).
43. D.O. Kataria et al. (12 authors): Phy. Rev. C. 56, 1902 (1997).
44. Y. Sugiyama, Y. Tormita, Y. Yamanouti, S. Hamada, T. Ikuta, H. Fujita and D.R. Napoli: Phy. Rev. 55, p. R5 (1997).
45. Abul-Magd and El-Abed: Progress Theoretical Physics (Japan) 53, 480 (1975); Y. Eyal, M. Beckerman, R. Chechik, Z. Fraenkel and H. Stecker: Phy. Rev. C13, 1527 (1976); F. Hopkins, J.J. Eck, D.O. Elliott, P. Richard and J.R. White: Phy. Rev. C. 8, 1721 (1973); C.B. Fulmer and D.A Goldberg: Phy. Rev. C11, 50 (1975); J. Galin, B. Gatty, D. Guerreau, V.C. Schlott-Haver-Voss and X. Terrange: Phy. Rev. C. 10, 638 (1974).
46. D.W. Lang: Nuclear Physics 42, 353 (1963); 545 (1966); T.D. Thomas: A. Rev. Nuclear Science, 18, 343 (1968).
47. V.F. Weisskopf: Phy. Rev. 52, 295 (1937); Blatt and Weisskopf: Theoretical Nuclear Physics, Wiley-Interscience, New York, p. 367 (1952).
48. W. Hauser and H. Feshbach: Phy. Rev. 87, 336 (1952).

49. J.H. Alexander and G.N. Simoniff: Phy. Rev. B 133, 93 (1964).
50. J. Gomez del Campo, J.L.C. Ford, R.I. Robinson and P.H. Stelson: Phy. Rev. C.9, 1258 (1974).
51. X. Ledoux et al. (22 authors): Phy. Rev. C. 57, 2375 (1998).
52. L. Heilebsona et al. (18 authors): Phy. Rev. C. V. 58, p. 3451 (1998).
53. R. Wada et al. (17 authors): Phy. Rev. C. V. 55, p. 227 (1997).
54. P.H. Millazzo et al. (25 authors): Phy. Rev. C. V. 58, p. 957 (1998).
55. N. Frascaria, C. Stephen, J.P. Gasson, J.C. Jacmart, M. Riou and L. Tasson-Gt: Phy. Rev. Letter 39, 918 (1977); N. Frascaria et al.: Z. Phy. A. 294, 167 (1980).
56. J.A. Scarpaci et al. (13 authors): Phy. Rev. C. 56, 3187 (1997).
57. D. Shapira, J. Gomez del Comp, M. Kasolija, J. Shea, C.F. Magvire and E. Chavez-Lomeli: Phy. Rev. C. p. 2448 (1997).
58. B. Fornal, G. Prete, G. Nebbio, F. Trotti, G. Viesti, D. Fabris, K. Hagel and J.B. Nowitz: Phy. Rev. C 37, 2624 (1988); J.R. Huizenga, A.N. Behkam, I.M. Govil,, W.U. Schröder and J. Toke: Phy. Rev. C. 40, 668 (1989).
59. I.M. Govil et al. (12 authors): Phy. Rev. C. 57, p. 1269 (1998).
60. R. Bass: Phy. Letters, B47, 139 (1973); Nuclear Physics, A. 231, 45 (1974).
61. D. Glass and U. Mosel: Phy. Rev. C 10, 2620 (1974); Nuclear Physics A. 237, 429 (1975).
62. U. Mosel: Argonne Conference, p. 341 (1976).
63. D.L. Hill and J.A. Wheeler: Phy. Rev. 89, 1102 (1953).
64. P. Sperr, S. Vigdor, V. Essen, W. Henning, D.G. Koner, T.R. Ophel and B. Zeidman: Phy. Rev. Letter. 36, 405 (1976).
65. A. Turler, D. Dressler, B. Eichler, H.W. Gaggeler and D.T. Jost, M. Schadel, W Bruchel, K.E. Grigorich, N. Trantman and S. Taut: Phy. Rev. C. 57, p. 1648 (1998).
66. H. Morgenstern, W. Bohne, G. Galster and K. Gabrisch: Phy. Rev. Letter 52, 1104 (1998); B.S. Tomar, A. Goswami, A. V. R. Reddy, S.K. Das, P.P. Burte, S.B. Manohar and Bency John: Phy. Rev. C 49, 94 (1994); Amit Roy with C. V. K. Baba et al.: N. Physics A 539, 351 (1992).
67. B.S. Tomar, A. Goswami, G.K. Gubbi, A. V. R. Reddy, S.B. Manohar, Bency John and S.K. Kataria: Phy. Rev. C. 58, p. 3478 (1998).
68. T. Udgawa and T. Tamura: Phy. Rev. Letter 45, 1311 (1980).
69. M.I. Sohel, P.J. Siemens, J.P. Bondorf and H.A. Bethe: Nuclear Physics A. 251, 502 (1975); M. Blann: Phy. Rev. C. 31, 1245 (1985).
70. K.E. Zyromski et al. (12 authors): Phy. Rev. C. V. R. 562 (1997).
71. E. Almquist, D.A. Bromley and J.A. Kuehner: Phy. Rev. Letters, 4, 515 (1960); R.H. Davis: Phy. Rev. Letters, 4, 521 (1960).
72. D.A. Bromley, J.A. Kuehner and E. Almqist: Phy. Rev. C 123, 878 (1961); S. Marsh and W.D.M. Rae: Phy. Letters, B. 180, 195 (1986).
73. E.T. Mirgule, M.A. Eswara Suresh Kumar, D.R. Chakarbarty, V.M. Dater, V.K. Pal, H.H. Oza and N.L. Ragoo Wansi: Phy. Rev., C56, p. 1943 (1997).
74. S. Szilner, H. Basra, R.M. Freeman, F. Haas, A. Morsad and C. Beck: Phy. Rev. C. 55, p. 1312 (1997).
75. P. Colombanni, Verennia Conf, p. 429, (1974), A.G. Artukh, G.F. Gridner, V.L. Mikheev, V.V. Volkov and J. Wilozyinski: Nuclear Physics A 215, 91 (1973).

76. P. Colambani, N. Frascaria, J.C. Jacmart, M. Riou, C. Stephens, H. Doabre, N. Poffe and J.C. Royndle: Phy. Letters, B. 55, 45 (1975).

77. J.C. Jacmart, P. Colombani, H. Doubre, N. Frascasia, N. Doffe, M. Ricok, J.C. Royndle, C. Stephan and A. Weidinger: Nuclear Physics A 242, 75 (1975).

78. R. Eggerts, N.N. Namboori, P. Gontheir, K. Geoffroy and J.B. Natowitz: Phy. Rev. Letters, 27, 324 (1976).

79. J.N. DE., D.H.E. Gross and H. Kalmouskie: Crancow Conference p. 121 (1976); B. Giraud J. Le Tourneux and E. Oxness: Phy. Rev. C 11, 82 (1975); D.H.E. Gross and H. Kalinovoski: Phy. Letters, B 48, 302 (1974).

80. A.G. Artukh, G.F. Gridner, V.L. Mikheer, V.V. Volkov and G.W. Icyaski: Nuclear Physics A 215, 91 (1973); K.L. Wolf, J.K. Unik, J.R. Huizenga and J. Birkelund, H. Feisleben and V.E. Viola: Phy. Rev. Letters, 3, 1105 (1974); F. L.H. Wolfs: Phy. Rev. C. 36, 1379 (1987); J. Gehring, B.B. Back, K.C. Chan, M. Freer, D. Hendreson, C.L. Jiang, K.E. Rehum, J.P. Schiffer, M. Wolanski and A.H. Wiosmma: Phy. Rev. C. 56, p. 2959 (1997).

81. K. Toke, K. Bock, G.X. Dai, A. Gobbi, C. Gralla, K.D. Hildenbrand, J. Kuziminski, W.F.J. Muller, A. Olmi, H. Stelzer, B.B. Back and S. Bjorholm: Nuclear Physics A. 440, 327 (1985); A.M. Stefanini, G. Nebbio, S. Lunerdi, G. Montagnoli and A. Vittusi, World Scientific, Singapore (1994).

82. J.A. Maruhn, W. Greiner and W. Sheid: Heavy Ion Collisions (R. Buck ed.). North Holland, Amsterdam (1980); V.Z. p. 399; R.K. Gupta: Physics, A. 281, 159 (1977).

83. R.K. Gupta: Phy. Rev. C. 21, 1278 (1980); W.U. Schröder and J.R. Huizenga: Treatise on Heavy Ion Science (D.A. Bromely, ed.) V. 2, p. 150, Plenum Press, New York (1984).

84. Wolcplin G.: Nukleonika 22, 1165 (1977); G. Wolcplin and W. Nurenberg: Phy. Rev. Letter 41, 691 (1978); W. Nurenberg: Phy. Rev. Letter B 52, 289 (1974).

85. P.J. Twin, B.M. Nyako, A.H. Nelson and J. Simpson and M.A. Bentley, H.W. Canmer Gordon, P.D. Forsyth, D. Howe, A.R. Mokhtar, J.D. Morrison and J.F. Sharpley-Schafer and G. Slatten: Phy. Rev. Letters, V.57, p. 811 (1986); T.L. Khoo et al.: Phy. Rev. Letters 41, 1027 (1978); Stephens F.S. Rev. Mod: Physics, 47, 57 (1975); D. Nisius et al. (25 authors): Phy. Letters 329 B, 18 (1997).

86. M.N. Namboodiri. J.B. Natowitz, P.Kaisraj, R. Eggerts, L. Alder, P. Gonthier, C. Ceruti and S. Simon: Phy. Rev. C. 20, 982 (1979).

87. J.A. Cameron, M.A. Bentley, A.M. Bruce, R.A. Cunningham, W. Geletly, H.S. Price, J. Simpson, D.D. Warner and A.N. James: Phy. Rev. C. 49, 1347 (1994); S. M. Lewzi, et al.: Phy. Rev. C 56, 1313 (1997); J.A. Cameron et al.: Phy. Rev. C. 58, 808 (1998).

88. C.E. Svensson et al. (18 authors): Phy. Rev. C 58, R. 2621 (1998).

89. R.B. Piecy et al.: Phy Rev. Letters 47, 1514 (1981); C.J. Cross et al.: Nuclear Physics A. 501, 367 (1989); C.J. Lister, B.J.Varlay, H.S. Price and J.W. Olness: Phy. Rev. Letters 49, 308 (1982).

90. M.J. Leddy, J.L. Durell, S.J. Freeman and B.J. Varlay, R.A. Bark, C.D. O'Leary and S. Törmänen: Phy. Rev. C. 58, p. 1438 (1998).

91. J.M. Sears et al. (19 authors): Phy. Rev. C. 58, p. 1430 (1998); S.K. Tandel, S.B. Patel, P. Joshi, G. Mukherjee, R.P. Singh, S. Muralithar, P. Das and R.K. Bhowmick: Phy. Rev. C. V. 58, p. 3738 (1998).

92. D.G. Jenkins et al. (15 authors): V C. 58, p. 2703 (1998).

93. S. Frauendorf: Nuclear Physics A. 557, 259 C (1993).

94. J.L. Wood, K. Heyde, W. Nazarewiz. M. Huyse and P. Van Duppen: Phy. Reports, 215, 101 (1992); R.E. Casten, N.V. Zamfir, and D.S. Breuner: Phy. Rev. Letters, 71, 227 (1993).

95. A.J. Larabee et al. (10 authors): Phy. Rev. (Letters 169 B, 21 (1986); M.J.A. de Woigi et al. (10 authors): Nuclear Physics A. 507, 472 (1990); T. Kutsarov et al. (17 authors): Nuclear Physics A 587, 111 (1995).

96. D.G. Popescu et al. (19 authors): Phy. Rev C 55, 1175 (1997).

97. N. Fotiades et al. (21 authors): Phy. Rev. C. 55, p. 1724 (1997).

98. K. Heyde and P.J. Brussard: Nuclear Physics A. 104, 81 (1967).

99. E. Holub, D. Hilscher, G. Ignold, U. Jahanke, H. Orf and H. Rossner: Phy. Rev. C. 28, 252 (1983).

100. Viola Jr. W.E., R.G. Clark, W.G. Meyer, A.M. Zebelman and R.G. Sextro: Nuclear Physics A 261, 174 (1976); G.E. Gordon, A.E. Larsh, T. Sikkeland and G.T. Seaborg: Phy. Rev. 120, 1341 (1960).

101. D.V. Shettly, R.K. Chowdhary, B.K. Nayak, D.M. Nadkarni and S.S. Kapoor: Phy. Rev. C. V. 58, p. R. 616 (1998).

102. G.A. Souliotis, K. Homold, W. Loveland, I. Lhenry, D.J. Morrissey, A. Veck and G.J. Wozniak: Phy. Rev. C. 57, p. 3129 (1998).

103. G.A. Souliotis, W. Loveland, K.E. Zyronski, G.J. Wozniak, J. Morrisey, J.O. Liljenzen and K. Aleklett: Phy. Rev. C 55, R. 2146 (1997).

104. J. Cibor et al. (12 authors): Phy. Rev. C 55, p. 264 (1997).

105. L.G. Moretto and G.J. Wozniak: Annual Review of Nuclear Science 43, 123 (1993).

106. R. Wada et al.: Phy. Rev. Letter 58, 1839 (1987); I. Brzychczyk, D.S. Bracken, K. Kwiatkoeski, K.B. Morlay, E. Renshaw and V.E. Viola: Phy. Rev. C 47, 1553 (1993).

107. H. Johnson: Phy. Rev. Letters B 371, 186 (1996), Phy. Rev. C 56, 1972 (1997).

108. J.F. Dempsey et al.: Phy. Rev. C 54, 1710 (1996).

109. J. Aichelins, G. Pilert, A. Bolnet, A. Rosenthauer, H. Stocker and W. Greiner: Phy. Rev. C 37, 2451 (1988); A. Aichelins: Phy. Rep. 202, 233 (1991); Bav-An Li and S.J. Yennello: Phy. Rev. C 52 R. 1746 (1995): S. Kumar, Rajiv Puri: Phy. Rev. 58, 320 (1998).

110. E. Ramakrishnan, H. Johnson, F. Gimento-Nogues, D.J. Rowland, R. Laforest, Y.W. Lui, S. Ferro, S. Vasal and S.J. Yennello: Phy. Rev. C 57, p. 1803 (1998).

111. J.P. Bondorf, et al.: Phy. Rep. 257, 133 (1995).

112. R. Michel et al.: Nuclear Instruments Methods: Phy. Rev. B 103, 183 (1995).

113. W.R. Weber, J.C. Kish and D.A. Shriev: Phy. Rev. C 41, 547 (1990).

114. H. Vonach, A. Pavlick and A. Wallner, M. Drosej, R.C. Haight and D.M. Drake and S. Chiba: Phy. Rev. C 55, p. 2458 (1997).

115. K. Nuffa et al.: Phy. Rev. C 52, 2620 (1995); S. Chiba et al.: Phy. Rev. C 54, 285 (1996).

116. A.E. Glassgold, W. Heckrotte and K.M. Watson: Annual Physics (NY) 6, 1 (1959); W. Scheid, H. Muller and W. Greiner: Nashville Conference (1974): Phy. Rev. Letter 32, 741 (1974); M.I. Sohel, P.J. Siemens, J.P. Bondorof and H.A. Bethe: Nuclear Physics A 2511, 502 (1975).

117. B.E. Crawford, P.B. Price, J. Stevenson and L.W. Wilson: Phy. Rev. Letters 34, 361 (1975); L.P. Ronsberg and D.G. Perry: Phy. Rev. Letters 35, 361 (1975).

118. D.E. Greiner, P.J. Lindstrom, H.H. Heckman, B. Cork and F.S. Bieser: Phy. Rev. Letters 35, 152 (1975).

119. J. Hafner, K. Schafer and B. Scharman: Phy. Rev. C 12, 1888 (1975); Westfall G.D. et al.: Phy. Rev. Letters 37, 1202 (1975).

120. L. Ahle et al. (67 authors): Phy. Rev. 57, 1461 (1998); L. Ahle et al. (76 authors): Phy. Rev. V. 57, p. R. 466 (1998).

121. Y. Dang, T. Schlage and S.K. Kahan: Nuclear Physics, A 566, 465e (1995); Y. Pang, D.E. Kahan, S.H. Karhana and T.J. Schlagel: Nuclear Physics, A 590, 595 (1995); B.A. Li and C.M. Ko: Phy. Rev. C 52, 2037 (1995).

122. H. Sorge, H. Stocker and W. Greiner: Annual Physics, (NY), 192, 266 (1989); Nuclear Physics A 498, 567C (1989).

123. R. Alberecht et al.: Phy. Rev. Letters 76, 3506 (1996).

124. M.M. Aggarwal, A.L.S. Angelis ...... I.S. Mittra ...... N.K. Rao ...... Y.P. Viyogi, ... G.R. Young (91 authors): Phy. Rev. C 56, 1160 (1997).

PROBLEMS

1. In a 15 UD, tandem accelerator, what are the expected maximum energies for (at the maximum voltage on the terminal) for :

(*a*) (*i*) Protons, (*ii*) deuterons, (*iii*) tritons and (*iv*) He^4 ions, and what are the minimum energies, for

(*b*) Li^6, C^{12}, N^{14}, Al^{27}, Ca^{40}, Ni^{58}, Sn^{120} and Pb^{208}?

2. Write down the values of R_{l_1}, R_{l_2}, R_{l_3} and R_{l_4} for O^{16}, Ca^{40} and Sn^{120} incident heavy ions for energies of 1 MeV/nucleon and 10 MeV/nucleon for target nuclei of Al^{27}, Ni^{60} and Pb^{208}.

3. For O^{16} incident ions on Ni^{60} target, what is the critical angle θ_c for 5.0 MeV incident particle and what are the conditions for Fraunhauffer and Fresnel scattering?

4. Using Fig. 17.7 and Eqs. 17.20 and 17.21, find out the values of σ and δ_N and δ_c.

5. Why do we have such fine oscillations in Si^{28} + Si^{28}, angular distribution for elastic scattering. Compare, the results of Fig. 17.9 with the Mott scattering formula, as derived in Chapter 4, Eq. 4.100*b*.

6. Explain, the physical reason for the energy dependence behaviour of real and imaginary parts of the strength parameter *f*.

7. Angular distribution of two-nucleon transfer reactions Figures 17.16, 17.17 and 17.18 are bell shapes, apart from the oscillatory structure. Explain this phenomenon physically.

8. Using Eqs. 17.53 – 17.59, calculate the fusion cross-section as a function of incident energy say for O^{16} + Al^{27} and compare with the values given in Fig. 17.25. Why for high energies, values of σ_{reac} and σ_{fusion} differ?

9. Using Ref. (53), calculate the energy distribution of say-Argon and Sulpher from Th^{232} + Ar^{40} at 388 MeV as shown in Fig. 17.26.

10. What is the physical significance of $i = 1$ and $i = 2$ in Eq. 16.48 ? Why should there be two types of nuclear temperatures?

APPENDIX A

Theory of Nuclear Matter and Finite Nucleus

1. INTRODUCTION

While discussing the shape of nuclei in Chapter-2, we have observed in Fig. 2.8, that the nucleus has nearly a trapezoid shape. It has a uniform density up to a radius R_o and then the density tapers off. This shape is expressible in terms of nuclear density $\rho(r)$ as a Fermi distribution[1]:

$$\rho(r) = \frac{\rho_o}{1 + \exp\left[(r - R_o)/a\right]} \quad \text{...(A-1)}$$

One obtains this shape from scattering experiments[2], say, of electrons from nuclei. Physically; this shape of the nuclear density implies, that, nucleons in the nucleus, especially in the flat portion, have a uniform density consisting of equal number of neutrons plus protons for per cm^3, when Coulomb force between protons is ignored; and also each particle moves with a constant momentum, except when it interacts with another particle. This is a case of the behaviour of nuclear matter. In nuclear matter, each particle moves in the vicinity of many particles, surrounding it; but these particles are not localized; because of quantum mechanical effect. Hence the energy of a particle cannot depend on its location, but depends on its momentum. Although, in the neighbourhood of a particle, there may be considerable fluctuation in the density of other particles. The distribution of these centres may be uniform and continuous, leading to homogeneous density distribution. This describes the nuclear matter qualitatively.

On the other hand we have seen[3] in 'Shell Model', chapter-10, that one can explain the energy levels of the excited states of nuclei just beyond closed shells, (magic numbers); as if the nucleons are moving in an orbit, which experiences only a common potential:

$$V(r) = V_o(r) + U(r)\, \vec{s}\,.\,\vec{l}\,;$$

where $V_o(r)$ is assumed to have a Woods-Saxon potential shape, (Eq. 10.2*b*):

$$V_o(r) = \frac{U_o}{1 + \exp\left[\dfrac{r - R}{a}\right]} \quad \text{...(A-2)}$$

The r-dependence is similar to Eq. (A-1), and U_o represents the constant density ρ_o of Eq. (A-1).

This behaviour, implies that each nucleon moves in a single particle orbital with constant values of $[\vec{s}]$ and $[\vec{l}]$, [moving in a common potential $V(r)$] and hence its wave-function can be determined, independent of the fact whether another nucleon exists close by. Knowing that nuclear forces are short-ranged; how do we explain it ?

This can be, physically, understood by applying the Pauli Exclusion Principle, according to which, if a quantum state is already occupied; and incoming particle cannot enter that state. So in a closed shell case, all the states are occupied, and hence a nucleon in a quantum state above it will not be able to interact with any nucleon in the closed shell. It gets scattered, as if the closed shell offers a common potential $V_o(r)$.

However, nucleon-nucleon interaction in a free space has been investigated experimentally and theoretically[4]; and the expression for nucleon-nucleon potential have been assigned (*e.g.* from Eq. 6.84-6.89 in Chapter (6) for various nuclear forces. How do these expressions of nuclear forces relate to $V_o(r)$ and U_o of Eq. (A-2) or in other words how do we use, quantum-mechanics to understand 'nuclear matter' ?

2. SOME INITIAL CONCEPTS

(a) Non Local Potential

In the nuclear matter, one can demonstrate the equivalence of the concept of non-local potential and a potential which is momentum-dependent[5]. Let us define vector $\vec{r}$ as the one measured from a fixed centre, then energy at point $\vec{r}'$ can depend, in nuclear matter, only on vector $\vec{r}' - \vec{r}$, connecting the point with other points in its immediate vicinity; so the resultant total density does not depend only on $\vec{r}$ but also on $\vec{r}'$ and $\vec{r}' - \vec{r}$.

The Schrödinger equation, for such a case, in a finite nucleus, becomes

$$-\frac{\hbar^2}{2M}\nabla_r^2\,\Psi(\vec{r}) + \int dr'\,U(\vec{r}, \vec{r}' - \vec{r})\,\Psi(\vec{r}') = e\Psi(\vec{r}) \qquad \text{...(A-3)}$$

where 'e' 'represents' the total energy. If the wave functions are plan waves of constant momentum, and are taken as the first-approximation for an infinite medium, in which U does not depend only on $\vec{r}$; then one can write (A-3) as:

$$e = \frac{p^2}{2M} + \int \vec{\alpha r'}\,U(\vec{r}, \vec{r}' - \vec{r})\exp\left[\frac{i\,\vec{p}.(\vec{r}' - \vec{r})}{\hbar}\right] \qquad \text{...(A-4)}$$

The integral in Eq. A-4, is the Fourier transform of $U(\vec{r'} - \vec{r})$; and we can call it $U(p)$, *i.e.,* a wave function with constant momentum p. Then, one can write:

$$e = \frac{p^2}{2M} + U(p) \qquad \text{...(A-5)}$$

This demonstrates, the equivalence between non-local potential and momentum-dependent formalism.

(*b*) Application of Pauli Exclusion Principle

Classically the probability of the co-existence of any two particles[6] is governed by the rules of classical statistical mechanics. As for example, the probability of finding a particle at $\vec{r_1}$, and another one at $\vec{r_2}$, is given by:

$$P_{Cl}\, d\vec{r_1}\, d\vec{r_2} = \rho(1)\,\rho(2)\, d\vec{r_1}\, d\vec{r_2}$$

$$= \sum_m |\phi_m(1)|^2\, dr_1^2 \times \sum_n |\phi_n(2)|^2\, d\vec{r_2} \qquad \text{...(A-6)}$$

where ρ (1) is the density of particle (1) and ρ (2) is the density of particle (2), ϕ_m (1) represents the wave function of *m*-state of particle (1) and summation Σm represents the summation over all, the occupied levels. Same is true for ϕ_n (2). If these wave functions represent quantum mecanical entities, say indentical Fermi particles, they should obey the Pauli Exclusion Principle. Then for the same charge and spin-state; the quantum statistics requires, that the probability of finding two particles—one at r_1 and the other at r_2, one in state of α and other in state β is given by:

$$\frac{1}{2}\left|\phi_\alpha(1)\,\phi_\beta(2) - \phi_\alpha(2)\,\phi_\beta(1)\right|^2 d\vec{r_1}\, d\vec{r_2} \qquad \text{...(A-7)}$$

Applying this logic to two indentical fermions, the probability of finding one such fermion of the system at $\vec{r_1}$ and another at $\vec{r_2}$, is given by the summing of the above expression over all occupied states *i.e.,*

$$P_F\, d\vec{r_1}\, d\vec{r_2} = \frac{1}{2}\sum_m \sum_n |\phi_m(1)\,\phi_n(2) - \phi_m(2)\,\phi_n(1)|^2\, d\vec{r_1}\, d\vec{r_2} \qquad \text{...(A-8)}$$

Comparing Eqs. (A-6) and (A-8) it is easy to see, that

$$P_F - P_{cl} = -C^2(b_2) = -\,\mathrm{Re}\sum_m \sum_n \phi_m(1)\,\phi_x^*(1)\,\phi_m^*(2)\,\phi_n(2)$$

$$= \left|\sum_m \phi_m(1)\,\phi_m^*(2)\right|^2 \qquad \text{...(A-9)}$$

(c) Infinite Medium-Nuclear Matter

Keeping in mind that:

$$\phi_m \equiv \phi(k_m) = \Omega^{-1/2} \exp(i \vec{k} . \vec{r})$$

represents the wave-function of a particle, with momentum $\vec{k}_m$, normalized to one particle in the volume, which is denoted by Ω. Then for an infinite medium, it is possible to write; [exp. $\vec{k}_m . \vec{r}$ has been converted into $j(\vec{k} . \vec{r})$ function]:

$$\sum_m \phi_m(1)\phi_m^*(2) = \frac{4\pi}{(2\pi)^3}\int_0^{k_F} j_0(kr)\,k^2\,\alpha k$$

$$= \frac{1}{2\pi^2} k_F^3\, j_1(kr)/k_F r \qquad \text{...(A-10)}$$

and

$$P_F - P_{cl} = -C^2(1, 2) = -\left|\frac{1}{2\pi^2} k_F^3 \frac{j(k_F r)}{k_F r}\right|^2 \qquad \text{...(A-11)}$$

or

$$C^2(1, 2) = \frac{k_F^6}{4\pi^4}\left|\frac{j_1(k_F r)}{k_F r}\right|^2$$

We can also write, alternatively:

$$P_{cl}\left(\frac{P_F}{P_{cl}} - 1\right) = -C^2(1, 2) = -P_{cl}(k_F r)$$

so that

$$P_{cl} F_1(k_F r) = \frac{k_F^6}{9 \times 4\pi^4}\left|\frac{3j_1(k_F r)}{k_F r}\right|^2$$

$$= P_{cl}(1 - P_F/P_{cl}) = \rho^2\left|\frac{3j_1(k_F r)}{k_F r}\right|^2 \qquad \text{...(A-12)}$$

where

$$\rho = \frac{k_F^3}{6\pi^2} \qquad \text{...(A-13)}$$

is derived as follows:

If all states up to a Fermion momentum P_F are filled in the neutron-proton Fermi gas (at zero temperature), so that the total number of nucleons is A, then if there in one quantum state for volume h^3

of the phase space, one obtains:

$$A = \frac{4\Omega}{h^3}\int \alpha^3 p = \frac{16\pi}{3}\,\Omega\,(P_F/h)^3 \qquad \text{...(A-14)}$$

where number 4 corresponds to four states of nucleons in nuclear matter-two neutrons (one up and one down) and two proton states (one up and one down).

The writing $k_F = P_F/\hbar = 2\pi\rho_F/h$ we can write nucleon density:

$$\rho = \frac{A}{\Omega} = \frac{16\pi}{3}k_F^3 \times \frac{1}{(2\pi)^3}$$

$$= \frac{2}{3\pi^2}k_F^3$$

for four states.

Hence, for one state, one can write:

$$\rho = \frac{2}{3 \times 4\pi^2}k_F^3 = \frac{k_F^3}{6\pi^2} \qquad \text{...(A-15)}$$

Hence writing

$$F_1 = \left|\frac{3j_1(k_F r)}{k_F r}\right|^2 \qquad \text{...(A-16)}$$

One gets:

$$P_F(1, 2) = \rho^2(1 - F_1(k_F r)) \qquad \text{...(A-17)}$$

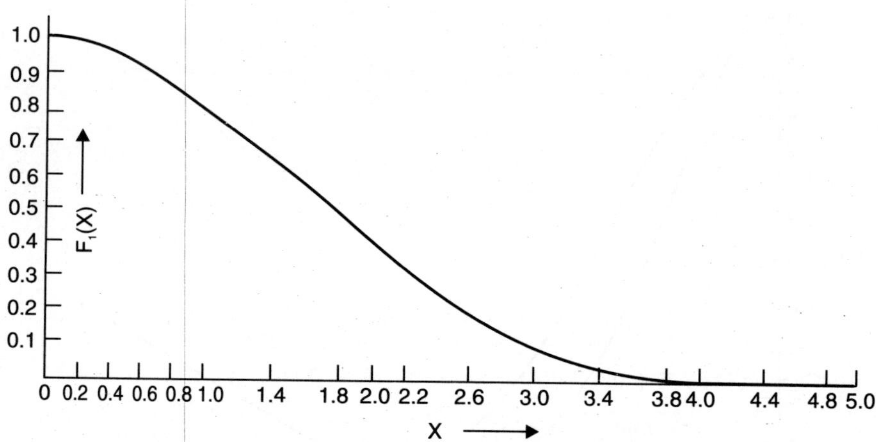

Fig. 1 $F_1(X)$ gives Pauli correlation function for an infinite medium (x = kr).

It is seen from Fig. (1) that for $r = 0$, $F_1(k_F r) = 1$, and hence $P_F(1, 2) = 0$ *i.e.*, quantum mechanically no two particles can co-exist. On the other hand, for $r >> 1$, $F_1(kr) \sim 0$; and $P_F(1, 2) = P_{cl} = \rho^2$ which means that any two particles can exist side by side as in the classical matter. For $0 < kr < 3$ or 4, the effect of Pauli Exclusion correlation can be seen.

As $k_F = 1.36 f_m^{-1}$, $k_F r = 2$ corresponds to $r = 1.5$ fm. And $k_F r = 3$ correspond to $r = 2$ fm. In this range of 1.5 – 20 fm; the Pauli Exclusion Principle is effective and the short range of the two nucleon force which corresponds to repulsive force between identical nucleons will be suppressed, and a small value of P_F (1, 2) will exist. In other words, instead of the full value of repulsive force, there will be an effective force in nuclear matter, much weaker than the free nucleon–nucleon force. The exclusion principle, thus creates a 'wound' or hole in the wave function of other states.

(*d*) Finite Nucleus

For a finite nuclear system, say O^{16}, one can apply the above logic. We can use 1S and 1*P*—the two states filled according to shell mode—so that the states can be described by harmonic oscillator functions and we assume that they are uncorrelated states ϕ_m.

In this case, it has been possible to calculate $C^2(\vec{r}, \vec{R})$ – a function similar to C^2 (1, 2) in Eq. (A-11); and $F_2(\vec{r}, \vec{R})$ similar to $F_1(k_F, r)$ in Eq. (A-17), [(Ref. 5)], then one can write:

$$P_F(1, 2) = \rho(1)\,\rho(2)\left[1 - F_2(\vec{r}, \vec{R})\right] \qquad \text{...(A-18)}$$

For the density function $\rho(r)$ of the nuclear matter in O^{16}, [Eq. (A-1)], we can get $F_2(\vec{r}, \vec{R})$, for the case when centre of mass of the two particles is at the centre of nucleus ($\vec{R} = 0$). This is more probable.

Figure (2) shows, the comparison of the Pauli Correlation formation $F_1(k_1\, r)$ for an infinite nuclear matter and $F_2(\vec{r}, \vec{R})$ for O^{16} as discussed above.[5, 6]

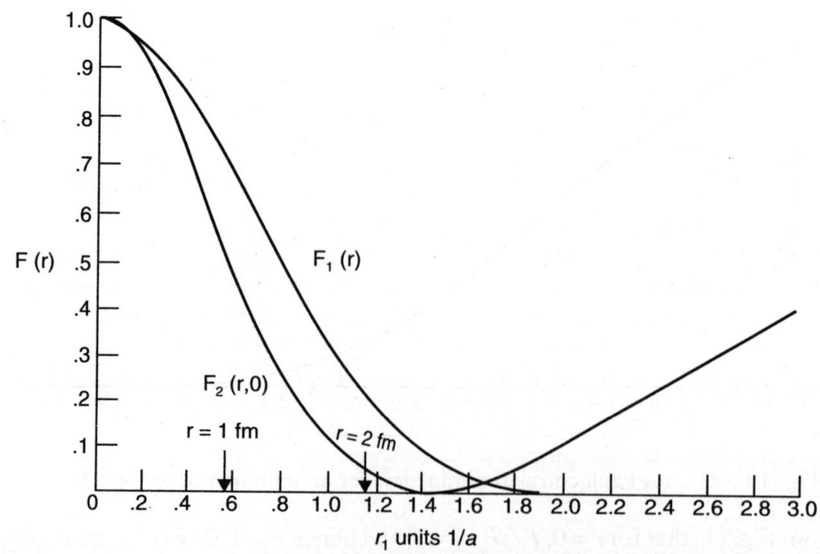

Fig. 2 Comparison of Pauli correlation for an infinite medium and for nuclei like O^{16}.

It is seen from this figure, that at least as far as estimates of Pauli Correlation effects are concerned; calculation with infinite medium would give a very good approximation to the finite case.

We, therefore, have seen that in relating the problem of infinite nuclear matter, to the one of finite nucleus, one requires, the wave function $\phi(r)$ – a model two-body wave function, corresponding to a constant momentum, appropriate as the approximation for the infinite medium and a smooth effective potential G, which relates to the two body actual quantum wave function Ψ and the real two body nucleon-nucleon potential v by the relation:

$$G\phi = \nu\Psi \qquad \text{...(A-19)}$$

The logic of correlation mechanism discussed in the last-para, relates the wave function's Ψ and ϕ, and relates the real two body potential v, with the smooth potential G, as we shall show in the discussion, which follows[7].

Assuming, that only two-particle interaction takes place in nuclear matter *i.e.* three or four body interaction are neglected; (This is called independent pair model), we can write, for nuclear matter, for the whole system:

$$\Psi_{kl} = A\left\{\Psi_{kl}(\vec{r}_1, \vec{r}_2)\,\phi_m(\vec{r}_3)\ldots\phi_n(\vec{r}_A)\right\} \qquad \text{...(A-20)}$$

where A represents anti-symmetrization, $\Psi_{kr}(\vec{r}_1, \vec{r}_2)$ is the two body wave function of the interacting fair, and ϕ's are the wave function of single particles and may be obtained from:

$$\frac{1}{2m}\nabla^2\phi_\alpha(\vec{r}) + E_i\,\phi_\alpha(\vec{r}) = \int d^3\vec{r}\left\langle \vec{r}\,'\,|\,V\,|\,\vec{r}\right\rangle\phi_\alpha(\vec{r})$$

or

$$\left[T(i) + U(i)\right]\phi_\alpha(\vec{r}) = \rho_\alpha\,\phi_\alpha(\vec{r}) = H(i)\,\phi_\alpha(\vec{r}) \qquad \text{...(A-21)}$$

Ψ_{kl}, on the other hand, obeys the relation:

$$[H_0(1) + H_0(2) + V_{12}]\,\psi_{\text{kl}} = e_{kl}\,\Psi_{kl} \qquad \text{...(A.22}a\text{)}$$

where

$$\Psi_{k_v l}(1, 2) = \phi_k(1)\,\phi_l(2) + \sum_{\alpha, b}\alpha_{\alpha b}\,\phi_\alpha(1)\,\phi_b(2) + \sum_{\alpha} a_{k\alpha}\,\phi_k(1)\,\phi_\alpha(2)$$

$$+ \sum_{\alpha}\phi_{l\alpha}\,\phi_\alpha(1)\,\phi_l(2) \qquad \text{...(A-22}b\text{)}$$

where $\quad H_0(1) \equiv [T(1) + U(1)]$ ...(A-23)

and $\quad H_2(2) \equiv [T(2) + U(2)]$

and V_{12} in Eq. (A-22*a*) is the interacting potential between particle 1 and 2; and e_{kl} is the total energy including the interaction energy, so that:

$$e_{kl} \neq e_k + e_l \qquad \text{...(A-24)}$$

Without proof [for which see H.A. Bethe and J. Goldstone, Proceedings Royal Soc. [(London) A 238, 551 (1957) and J. Goldstone Proceeding, Royal Society A-23*a*, 267 (1957)], we write from (A-22), the relation

$$|\psi_{kl}(1,2)\rangle = |\phi_k(1)\,\phi_l(2)\rangle + \sum_{\alpha,\beta}{}' \frac{\langle \alpha,\beta\,|v|\,\Psi_{kl}\rangle}{e_{kl} - e_\alpha - e_\beta} |\phi_\alpha(1)\,\phi_\beta(2)\rangle \qquad \text{...(A-25)}$$

where $\langle \phi k\ \phi 1/\Psi_{kl}\rangle = 1$ and

$$(e_k + e_l)\,\phi_k(1)\,\phi_l(2) + \sum_{\alpha,\beta}{}' a_{\alpha\beta}(e_\alpha + e_\beta)\,\phi_\alpha(1)\,\phi_\beta(2) + v_k\Psi_{kl} = e_{kl}\Psi_{kl} \qquad \text{...(A-26)}$$

Σ' represents summations over single particle states, where

$$\alpha = k,\ a,\ b,\ c,\ \ldots\ldots;\ \beta = l,\ a,\ b,\ c\ \ldots\ldots$$

Then

$$\Psi_{kl}(1,2) = |\phi_k(1)\,\phi_l(2)\rangle + \sum_{\alpha,\beta}{}' \frac{1}{e_{kl} - H_o} |\alpha\beta\rangle\langle\alpha\beta\,|v|\,\Psi_{kl}\rangle \qquad \text{...(A-27)}$$

where $H_o = H_o(1) + H_o(2)$

It is possible to write Eq. (A-27) as:

$$|\Psi_{kl}\rangle = |\phi_k\,\phi_l\rangle + \frac{Q_{kl}}{e_{kl} - H_o} v\,|\Psi_{kl}\rangle \qquad \text{...(A-28)}$$

where Q_{kl} is called the projection operator corresponding to correlation involving Pauli Exclusion Principle. Eq. (A-28) relates, for a multiparticle system, the wave function Ψ_{kl}, which has one pair of interacting particles and the single particle wave function $\phi_k(1)$ and $\phi_l(2)$.

Eq. (A-28) is called Bethe-Goldstone equation.

It should be realized that both ϕ_k or ϕ_l and hence $\phi_k\,\phi_l$ correspond to unperturbed model two body wave function *e.g.* plane wave or oscillator wave function, where we assume a model smooth potential. On the other hand, Ψ_{kl} represents a wave function corresponding to independent-pair interaction. Hence Ψ_{kl} will not be a plane-wave, wave function, but will contain the effect of interaction and may, therefore, be adequately distorted. Then one defines a quantity ξ_{kl}, called the defect wave function given by:

$$\xi_{kl} = \phi_k\,\phi_l - \Psi_{kl} \qquad \text{...(A-29)}$$

Then one can define a 'Wound' based on two-body interaction, in the sea of plane-wave functions. The wound 'K_{kl}' is defined as:

$$K_{kl} = \int |\xi_{kl}|^2\, d\vec{r}_1\, d\vec{r}_2 \qquad \text{...(A-30)}$$

Physically, one can define K_{kl}, the 'wound' as the probability of finding the actual system, excited out of the non-interacting, Fermi sea, relative to the probability of finding it in state *kl*.

The relationship of the wave functions can, now, be extended to the relationship of real two body interaction/potential v and the assumed smooth potential G operating for the unperturbed wave function. Then one can write:

$$G\,\phi_k\,\phi_l = v\,\Psi_{kl} \qquad \text{...(A-31)}$$

Then from Eq. (A-28), one can write:

$$v\left|\Psi_{kl}\right\rangle = G\left|\phi_k\,\phi_l\right\rangle$$

$$= v\left|\phi_k\,\phi_l\right\rangle + v\,\frac{Q_{kl}}{e_{kl} - H_o}\,G\left|\phi_k\,\phi_l\right\rangle \qquad \text{...(A-32)}$$

Symbolically considering that $\left|\phi_k\,\phi_l\right\rangle$ is a common factor in all the terms, we can write:

$$G = v + v\,\frac{Q_{kl}}{e_{kl} - H_o}\,G \qquad \text{...(A-33)}$$

Eq. (A-33) is the operational[8] form of Bethe-Goldstone equation, which connect the smooth potential function G, operating on the model function $\phi_k\,\phi_l$, with v, the actual two body attractive part; and the repulsive part; which is close to core and applies to the wave function Ψ_{kl}, Q_{kl} accounts for Pauli Exclusion Principle operating in the case of, actual two body interaction.

We have, already defined the concept of a 'wound' as defined in Eq. (A-29), and (A-30) for two body interaction. In a general case of nuclear matter, one defines, for A particles a quantity 'K', given by:

$$K = \frac{1}{A}\sum_{k<l} K_{kl} = \frac{1}{A}\sum_{k<l}\left\langle\xi_{kl}\middle|\xi_{kl}\right\rangle \qquad \text{...(A-34)}$$

$$= \frac{A}{2}\left\langle\xi\middle|\xi\right\rangle A v$$

where average is over all pairs k-l. Using the relationship (A-29) for A-particles, it has been found that one can write:

$$K = \frac{1}{8}\rho\sum_{JLS}(2J+1)(2T+1)\left\langle\xi_{JLS}\middle|\xi_{JLS}\right\rangle_{AV(k)} \qquad \text{...(A-35)}$$

where T is defined as in Eq. (6-13) chapter (6). If the average range of ξ is say C; then $(C/r_o)^3$ gives the probability that the motion of particles is correlated with some other particles, where r_o, is the radius of sphere, containing, on the average, one particle so that the density of the correlated matter is given by:

$$\rho_o = \frac{1}{4\pi r_o^3}$$

or

$$\frac{1}{\rho_o} = 4\pi r_o^3$$

is the volume of the sphere containing one particle. In this simple logic, K is proportional to $(C/r_o)^3$.

If K is defined as the probability that the motion of one particle is correlated to that of another, then K^2 is the probability that a particle has one particle within correlation distance. Then if one writes

$\bar{v}$ for an average interaction energy for two correlated particles; we would expect, energy per particle from two particle cluster  as $\frac{1}{2}\,k\bar{v}$; for three particles cluster  as 3/3 ! $k^2\,\bar{v} = \frac{1}{2}\,\bar{v}$, for four particles cluster  as:

$$\frac{6k^3\,\bar{v}}{4\,!} = \frac{k^3\bar{v}}{4}$$

The observed binding energy per particle is 16 MeV; and the average kinetic energy per particle (0.6 e_F) is 23 MeV at $k_F = 1.36\ \text{fm}^{-1}$. Then total energy E is 23 + 16 = 39 MeV. This can be written in terms of K as:

$$\frac{1}{2}\,k\bar{v} + \frac{1}{2}\,k^2\bar{v} + \frac{1}{4}k^3\bar{v}$$

$$= \frac{1}{2}k\bar{v}\,(1 + k + \frac{1}{2}\,k^2) = 39\ \text{MeV} \qquad \text{...(A-36)}$$

If $k = 0.15$, then the above sum becomes

$$0.075\bar{v} + .01\bar{v} + 0.0024\bar{v} = 39\ \text{MeV}$$

or

$$.0852\bar{v} = 39\ \text{MeV}$$

$$\bar{v} = 39/.085 \sim 460\ \text{MeV}$$

For singlet even, for Gammel-Thaler potential; the strength is – 460 MeV.

Then $\quad 1/2k\bar{v} = .075 \times 460 = 33.6$ MeV

$1/2k^2\bar{v} = 5$ MeV

and $\quad 1/2k^3\bar{v} = 0.4$ MeV.

So two-body clusters contribute 33.6 MeV for each particle. Three body clusters contribute 5 MeV for each particle and four body cluster contribute 0.4 MeV for each particle.

k has been calculated by P.J. Siemens[9] (Nuclear Physics *A* 141, 225 (1970)), using Reid potential, given by:

$$v = v\,(LSJ\,,\,r) \qquad \text{...(A-37)}$$

Chosen by Reid[10], where *L, S, J* are the angular momentum of the two interacting nuclei, this form [R.V. Reid, Annals of physics, V. 50, p. 411 (1968)] has the computational advantage that v for *LST* can be directly calculated from the phase shifts for this. In Fig. (3), we show the shapes of Reid hard core and soft core along with Hamada Jonson potential.

Using Reid potential, the value of k was calculated for $k = 0.57\,k_F = 0.86\ \text{fm}^{-1}$ and $k_F = 1.5\ \text{fm}^{-1}$ for various *L, S* and *J*. The sum total was found to be : $k = 0.1569$.

We have seen in Eq. (A-29), that the defect function ξ can be written as the difference between the two-body wave function Ψ_{kl} and the model wave, wave function ϕ_k, ϕ_l.

It has been possible to write ψ in terms of a radial function ${U^M}_{JLS}$ which approaches $kr\,jL\,(Kr)$

asymptotically. Then one can write ξ in Eq. (A-35) as $\chi^M{}_{JLS}$, in such a way that

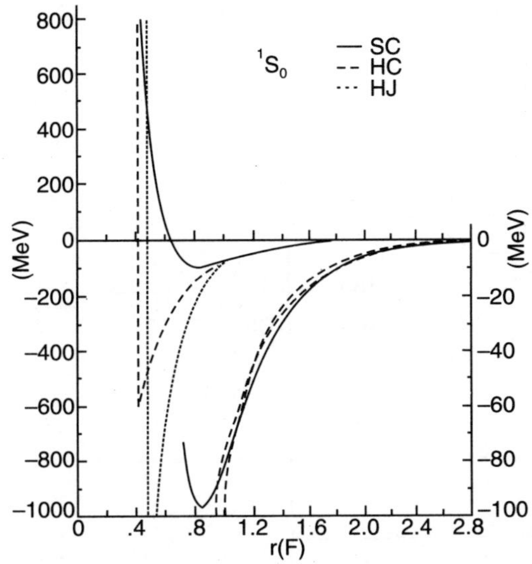

Fig. 3 Potentials in the '$1S_0$' state as functions of the distance between the nucleons in fm --- Hamada-Johnston: ----- Reid hard core; solid curve, Reid soft core. The left-hand energy scale refers to the three curves on the left, and the right-hand scale to the right-hand curves.

$$\chi^M_{JLS} = krj_c(kr) - U^M_{JLS}$$

and $$\chi^{JS}_{L'L} = \delta_{L'L}\, krj_l(kr) - U^{JS}_{L'L} \qquad \text{...(A-38)}$$

χ's are called defect functions, and have physically the same meaning as ξ in Eq. (A-29). Evidently from Eq. (A-3*b*), it is clear that the value of *k* depends on χ or ξ the defect function.

The values of χ (Fig. 4) have been calculated for Reid soft core for different values *L*, *L*′, is for $1S_0$ $(L = 0)$ $3S_1$, $(L' = 0)$, and $3S_1$, $(L' = 2)$; and for $3D_1$ $(L = 0)$ and $3P_1$ $(L' = 1)$. The solid curves in Fig. (4), represents Siemen's calculations, carried out, using Eq. (A-35) and (A-29). The dashed curves are reference approximations, using Eq. (A-38), using unperturbed function $K_0\, rJ_L$, which behaves asymptotically like a sine function of amplitude unity. Thus it can be seen, that the *s*-defect function of maximum magnitude 0.2 – 0.3 is of the same order as the unperturbed function. As – matter of fact, they cancel the unperturbed functions $k_0\, rj_0 = \sin k_0\, r$ at small *r*, which correspond to repulsive core. On the other hand, for 3*D*, $L' = 0$, the maximum is 0.007.

(*e*) Physical Significance of χ-the Defect Function

In nuclear matter, we are assuming, that because of Pauli Exclusion Principle, only wave functions ϕ_k and ϕ_l operate; and the smooth potential is given by the properties of *G*. But when two nucleons interact; a wave function ψ_{kl} describes the wave function of the interacting nucleon, giving rise of a distortion in the smooth wave function $\phi_k\, \phi_l$; so that a 'wound' described by:

$$\xi_{kl} = \Psi_{kl} - \phi_k\, \phi_l$$

is created. This wave-function of the wound is called the defect function as described in Fig. 4.

The shell model of nuclear structure is a good illustration of the relationship of nuclear matter and finite nuclei. For shell model to be valid; it should be possible to consider each nucleon, as moving in a smooth potential, not subject to rapid fluctuation in space.

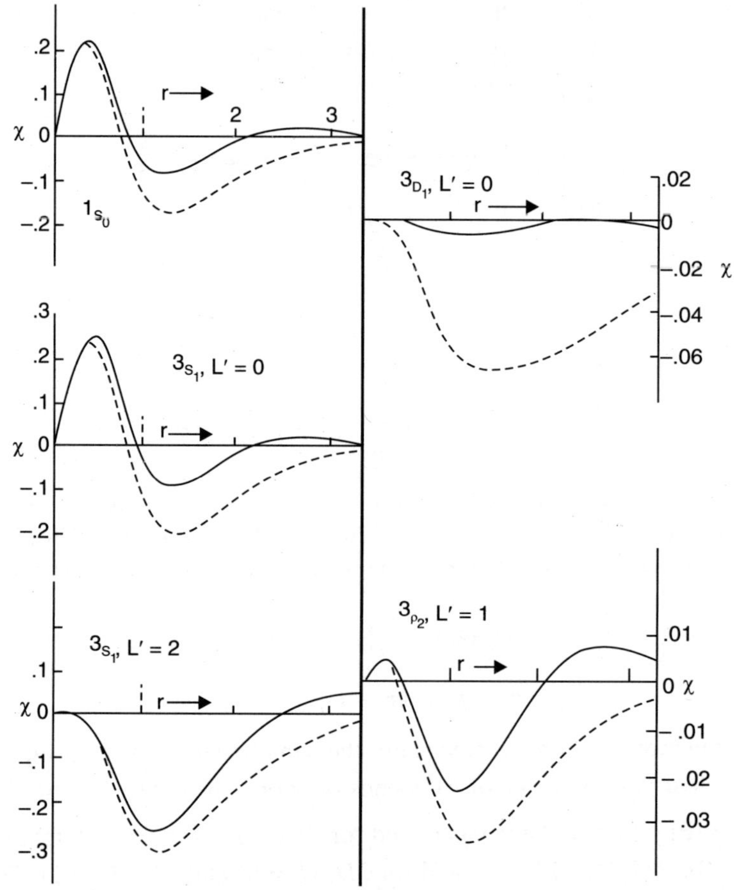

Fig. 4 Defect functions for several two-nucleon states, for k_F=1.36 fm^{-1} and k = 0.75 k_F. Abscissa = distance in fm. Solid curves = exact calculation, dashed = reference spectrum approximation. The entrance channel is noted with each part of the figure; the L′ gives the channel in which the defect function exists.

The above discussion about the properties of defect function, explains this phenomena, reasonably. It is seen in Fig. 4 that defect function causes a 'wound' only over a short distance of two fermis. Since the diameters of most nuclei is of the order of 10 fermis, the wave function for the most part-are just-that shell mode types, in nuclei with one or two nucleus outside the closed shells. Brückner in fifties, when he initiated the theory of nuclear matter, pointed out, that in normal and low excited states of the nucleus, the nucleon appear to move in a smooth potential, while for high energy dependant-phenomenon such as photo effect ≥ (100 MeV, say) and the capture of π-mesons; the short range interaction of the nucleons is important.

3. NUCLEAR MATTER AND FINITE NUCLEI-PHYSICAL QUANTITIES

It is interesting, to find the relationship of the logic, required for the infinite nuclear matter and finite nuclei for the relationship of the physical quantities as are experimentally, measurable in finite nuclei with the logic of infinite nuclear matter.

We, especially, will calculate

(*i*) Nuclear density $\rho\ (r_1, r_2)$

(*ii*) Total binding energy

(*iii*) rms radii

(*iv*) Removal energies

I. Nuclear Density Distribution

The effective interaction: Before we discuss, the above mentioned physical quantities, we will describe the role of G, ν, ϕ and ψ as earlier written in the relation:

$$G\,\phi = \nu\,\psi \qquad \text{...(A-19)}$$

If ν, the interaction between two nucleons depends on an angular momentum; (as the nuclear potential does), then one can write:

$$\phi_l\,(r) = k_0 r\, j_l\,(k_0 r) \qquad \text{...(A-39)}$$

Then, as $G\,\phi_1$ is, a function of r, depending on the parameters k_o, P, W, and l; one can define an effective potential[11] g as:

$$g\,(k_0, P, W, l, r) = \nu_l\, u_l/\phi_l \qquad \text{...(A-40)}$$

or

$$g\,(k_0, P, k_F, W, L, r) = v_L\,(r)\, u_L\,(k_0, P, k_F, W)/\phi_L\,(k_0 T) \qquad \text{...(A-41)}$$

where we have assumed local density approximation, so that G-Matrix (represented by g above); at any place in a finite nucleus is the same as for nuclear matter, at the same density.

The nuclear matter theory, as described in the earlier part of this chapter, when applied to the above formulation Eq. (A-41) has yielded the following results[9].

(*i*) u_L is essentially independent of P.

(*ii*) At given density k_F; U_L/ϕ_L does not depend strongly on k_0.

(*iii*) u_L/ϕ_L depends strongly on k_F, because of Pauli Principle.

(*iv*) u_L/ϕ_L depends on W, the starting energy, $[W = E\,(k_1) + E\,(k_{2+})]$, where k_1 and k_2 are the two initial values of wave propagation numbers $(k = p/\hbar)$.

Nuclear Density Distribution: Using the effective density dependant interaction, g, as given in Eq. (A-38) from which one can write the total energy for two interacting nucleons in occupied states m and n as:

$$W = E_m + E_n \qquad \text{...(A-42)}$$

The eigen-values E_m and E_n in finite nuclei are higher, close to free nucleons, than in nuclear matter of the same density. Hence W is higher and G is more attractive. This extra attraction is important to make the binding energy for infinite nuclei agree with the experiment. It has been possible to write the value of W in terms of ϕ, $k =$ or ρ (r), and v [See Neagle J.W., Phy. Rev. CI, 1200, (1970)[12]]:

$$W = -\frac{\hbar^2}{2M}\sum_i \phi_i^* \nabla^2 \phi \, d\tau + \frac{1}{2}\int d\tau_1 d\tau_2 \, \rho(r_1)\rho(r_2)$$

$$\times\left\{v_s(\vec{r_1}-\vec{r_2}) + v_{Do}(\vec{r_1}-\vec{r_2})\right\} \times \frac{1}{2}\left[k_F(r_1) + k_F(r_2)\right]$$

$$+\int dr_1\, dr_2\, \rho(r_1, r_2)\, v_X(\vec{r_1}-\vec{r_2}) \qquad \text{...(A-43)}$$

Physically the first integral in Eq. (A-43) is interpreted as kinetic energy, the second integral gives the short range (v_S) and direct (v_D) interaction energy, and the last third term gives the exchange interaction (v_X). The density expression ρ (r_1, r_2) in the third term depicts the mixed density given by:

$$\rho(r_1, r_2) = \sum_i \phi_i^*(r_1)\,\phi_i(r_2) \qquad \text{...(A-44)}$$

where the sum goes over all the occupied states, just as in the first term.

Next Negele wrote, the density dependent, Hartree-Fock (DDHF) equation. Taking the variation of Equation (A-43), one writes:

$$-\left(\frac{\hbar^2}{2M}\right)\nabla^2 \phi_i(\vec{r_1}) + \phi_i(\vec{r_1})\int d\tau_2 \rho(\vec{r_2})$$

$$\times\left\{v_S(\vec{r_{12}}) + v_{DO}(\vec{r_{12}}) + \frac{1}{2} vD_1(\vec{r_{12}})\left[\frac{4}{3}k_F(\vec{r_1}) + k_F(\vec{r_2})\right]\right\}$$

$$+\int d\tau_2\, \rho(\vec{r_1}, \vec{r_2})\, v_X(\vec{r_{12}})\,\phi_i(\vec{r_2}) = \varepsilon_i\, \phi_i(\vec{r_1}) \qquad \text{...(A-45)}$$

where ε_i is the Hartree-Fock eigenvalues related to W as:

$$W = \frac{1}{2}\sum_i\left(\varepsilon_i + \langle T_i\rangle\right) - \frac{1}{12}\int \rho(\vec{r_1})\, k_F(\vec{r_1})\, \rho(\vec{r_2})\, d\tau_1 d\tau_2 \qquad \text{...(A-46)}$$

To get the dependence ρ $(\vec{r})$ as a function of r, one must solve Eq. (A-46) self consistently. One should realize that solving Eq. (A-46), which is referred to as Negele's Density Dependent, Hartree Fock (DDHF) equation, gives ϕ_i, and ρ $(\vec{r_1}, \vec{r_2})$ is obtained from (A-44) and (A-39) the sum goes over all occupied states, just as in the first term in Eq. (A-40). Negele, obtained the total density for both neutrons and protons as ρ_N (r) and ρ_P (r). For proton density, one uses the coulomb interaction between protons over and above the average interaction.

Figure (5) gives the charge-density distribution of O^{16}, Ca^{40}, Zr^{90} and Pb^{208} in this manner[13]. Comparison with experimental curves obtained from analysis of electron scattering is quite satisfactory.

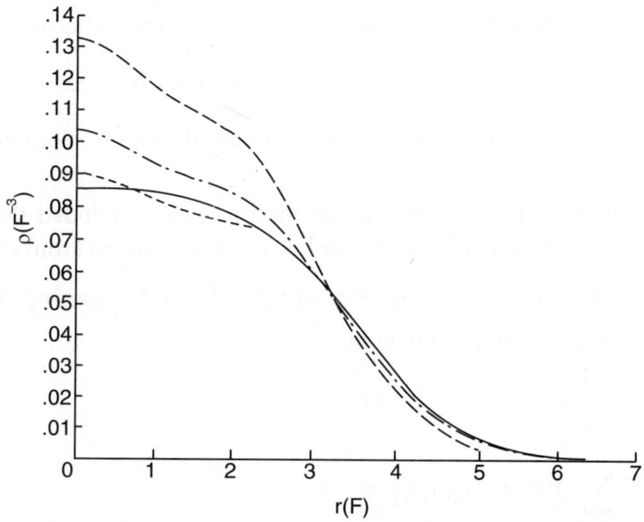

Fig. 5a Distribution of proton charge density in ^{40}Ca, according to Negele. Solid curve: semiempirical fit to electron scattering experiments. Long dashes: calculated with a density-independent force. Dash-dot ---: same with density-dependent force, derived directly from nuclear matter theory. Short dashes: same with force corrected, according to Sec. 12f.

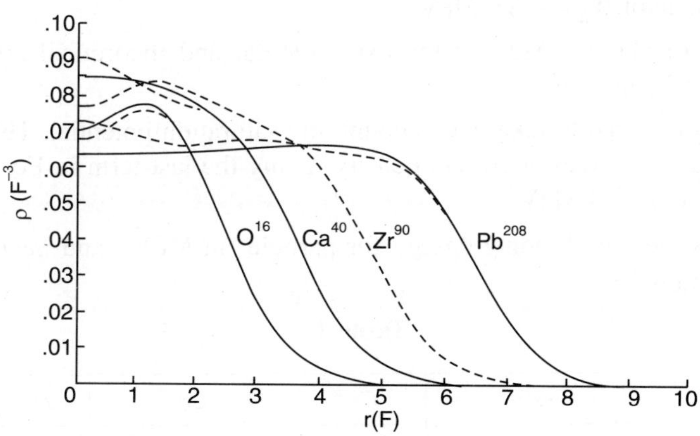

Fig. 5b Distribution of proton charge in some closed-shell nuclei according to the theory of Negele.

II. Total Binding Energy and Nuclear Radii

Eqs. (A-42) and (A-43) give the total energy of a nucleus, where we see the relation of *W*, the total energy, with individual eigenvalues of energies given in (A-46). In this equation; the first two terms *i.e.*, $\frac{1}{2}\sum_i \varepsilon_i$ and $\frac{1}{2}\sum_i \langle T_i \rangle$ are obtained from the solution of Hartree-Fork equation, where the last term:

$$-\frac{1}{12}\int \rho(r_1)\, k_F(r_1)\, \rho(r_2)\, v_{D_1}(r_{12})\, d\tau_1\, d\tau_2$$

corresponds to a correction, which arises from the density dependence of the effective nuclear forces $v_D(r_{12})$. This term is always negative as it corresponds to attractive forces.

For light nuclei, the nucleon wave function are close enough to harmonic oscillator and in this case, center, of mass energy is simply $\frac{3}{4}\hbar\omega$, where ω is the oscillator frequency; and should be eliminated. For heavier nuclei, the correction due to the motion of centre of mass is small and does not require the correction.

Negele[12] has calculated for Ca^{40} the binding energy, which included the correction of the last term in Eq. (A-46). The kinetic energy $1/2\Sigma_i$ (Ti) in Eq. (A-46), is taken from oscillator wave function whose parameter is adjusted to give $\langle r^2 \rangle$ in agreement with electron scattering. Then it is found that the first two terms[14, 15] in Eq. (A-46) are per nucleon.

$$\sum_i \varepsilon_i / 2A = -12.9 \text{ MeV}$$

$$\sum_i T_i / 2A = 8.6 \text{ MeV.}$$

Correction for center of mass = – 0.25 MeV

Theoretical Total W/A = – 12.9 + 8.6 – 0.25

= – 4.5 MeV

But observed value of W_A = – 8.6 MeV

The difference of about 4 MeV in the experimental and theoretical arise from the density dependence of effective interaction.

A term $\partial v/\partial p$ is required to take into account, the saturation potential. This is required to give proper saturation of nuclear forces at correct density. Using the last term in Eq. (A-46) by involving $\partial v/\partial p$, gives this correction of 4 MeV.

Table (1), gives the total binding energy for particle (in MeV), and nuclear radii (in fermi's) using Negele's calculations[12]:

Table 1

Nucleus	O^{16}	Ca^{40}	Ca^{48}	Zn^{95}	Pb^{208}
1. Theoretical binding energy	7.58	7.99	7.96	8.33	7.83
2. Experimental binding energy	7.98	8.55	8.67	8.71	7.87
3. Proton rms radius (theoretical)	2.7	3.41	3.45	4.18	5.44
4. Proton rms radius (experimental)	2.64	3.43	3.42		5.44
5. Neutrons rms radius (theoretical)	2.69	3.37	3.60	4.30	5.67

It can be seen from Table (1) that the experimental and theoretical values of binding energy and radii for (proton) and neutron match with each other.

Summarizing the results about binding energy; one may state that:

(*i*) If the nuclear forces are adjustable to give the correct binding energy for nuclear matter, then they also give correct results for finite nuclei (experimental).

(*ii*) The density dependent Hartree Fock theory reproduces the correct trend of binding energy with mass function.

III. Separation Energies

The separation energies of particles in nucleus, are the opposite of the binding energies in sign. So according to Koopman's theorem, the Hartree Fock equation (A-46) can be used to calculate the separation energies provided the wave functions ϕ_i's of the remaining particles and the interaction between them, remain unchanged.

The separation energy E_{si} is defined[16, 17] as:

$$E_{si} = -\varepsilon_i - \Delta_i \quad \text{...(A-47)}$$

where Δ_i is the differences of the energy of the normal nucleus with state *i* occupied and the nucleus with the particle removed from the state; ε_i is the eigen state energy of the ith state, of the normal nucleus.

Normally separation energy is measured, when i^{th} state corresponds to the outermost particle in the nucleus; Δ_i is very small for the least bound nucleons; but is several MeV for inner nucleons.

Table (2) gives the proton separation energies in Ca^{40} in MeV.

Table 2

State	*OS*	*OP*	*OD*	*IS*	*Binding energies*
ε_i Negele value	47	30	15.2	15.9	7.9
Experimental separation energy	53 ± 11	37 ± 6	16.2	15.7	8.55

18. Theory of Nuclear Matter and Finite Nuclei
2000–2008

In a theoretical paper Tapas Sil et al., have calculated liquid gas phase transitions in nuclei in the relativistic Fermi theory, [Phy. Rev. C. 63, 054604 (2001)].

S. Shlomo from Texas A & M, have published a paper on sound modes in Hot nuclear matter, considering the propagation of iso-scalar and iso-vector sound modes. [Phy. Rev. C. 64, 044304 (2001)].

In an interesting collaboration, theoretical work between India and Brazil, an unfolding method is proposed to extract ground state nuclear scattering data analysis at sub-barrier and intermediate energies. This method can be of value for determining densities of exotic nuclei [Phy. Rev. C. 65, 014602 (2002)].

This is a paper, involving the role of α-particles at low densities in nuclear matter. Starting with a model of α-particles interaction via an effective interaction the suppression of the condensate fraction at zero temperature in considered. The properties of C^{12} and O^{16} are calculated, with this model [Phy. Rev. C. 77, 064312 (2008)].

REFERENCES

1. R. Hofstadter: Reviews of Modern Physics 30, 412 (1958).
2. B. Hahn, D.G. Rovenhall and R. Hofstadter: Phy. Rev. 101, 1131 (1956).
3. M.G. Mayer, J. H. D. Jensen and H.E. Suess: Phy. Rev. 75, 1766 (1949).
4. Chapter 4, chapter 5, and chapter 6.
5. M.A. Preston, R.K. Bhaduri: Physics of Nucleus (1962).
6. H.A. Bethe: Ann. Rev. Nucl. Sci. 21, 93 (1971); and D.W.L. Sprung Advance: Nucl. Physics, 225 (1972).
7. L.C. Gomes, T.D. Waleka and V.F. Weisskopf: Ann. Phy. (N.Y.), 3, 241 (1958).
8. H.A. Bethe and J.Goldstone: Proc. Royal Soc. (London), A 238, 551 (1957); J. Goldstone: Proc. Royal Society, (London), A 239, 267 (1957).
9. P.J. Siemens: Nuclear Phy. A. 14, 225 (1970).
10. R.V. Reid: Ann Phy. 50, 411 (1968).
11. B.H. Brandow, Proc: Int. Sch. Phy. Enrico Fermi Course, 36, 528 (1966).
12. J.W. Negele: Phy. Rev. C. 1. 1260 (1970).
13. J.B. Bellicard, P. Bounin, R.F. Frosch et al.: Phy. Rev. Letters 19, 527 (1967); R.F. Frosch; R. Hofstadter, et al.: Phy. Rev. 174, 1380 (1968).
14. L.R.B. Elton, A Swift: Nuclear Phy. A. 94, 52 (1967).
15. A.N. James, P.T. Andrews, P. Rirkby, B.G. Lewe: Nuclear Phy. A. 138, 1215 (1969).
16. H.A. Bethe, B.H. Brandow, A.G. Pertschek: Phy. Rev. 129, 225 (1963).
17. K.A. Bruckner, D.T. Goldman: Phy. Rev. 117, 207 (1960).

PROBLEMS

1. Write ρ, in Eq. (A-1), in terms of ϕ_m, If ρ is a smooth formation of r, what does it signify, for the properties of ϕ_m?
2. How do we, qualitatively explain the flatness of V_o in (A-2) from the behaviour in ϕ_m in Eq. (A-21)?
3. Prove that in Eq. (A-4), the integral represented by U (p), has a constant value of (p); by calculating the integral in an actual case of U; from chapter 6.
4. What is the physical significance of F_1 (kr) in Eq. (A-17) ? Why should P_F be smaller then P_{cl}? Show how Pauli Exclusion Principle reduces the probability of two nucleons existing side by side?
5. We have defined, for nuclear matter $\frac{P_F}{P_{cl}} = F_1\ (kr)$

Perform, the same calculation for a finite nucleus O^{16}, with 1S and 1P states and prove that if $P_{cl} = \rho(1)\rho(2)$,

Then $$P_F(1, 2) = \rho(1)\rho(2)[1 - F_2(\vec{r}, \vec{R})]$$

where $$F_2(\vec{r}, \vec{R}) P_{cl} = C^2(1, 2)$$

Find the expression for $F_2(\vec{r}, \vec{R})$,

[See Physics of Nucleus, M.A. Preston, and R.K. Bhaduri, page – 250]

6. Consider, the two matrices G_A and G_B of the type given in Bethe – Goldstone equation (A-33), one associated with V_A, Pauli operater Q_A, and energy denominator $e_A = H_{QA} - \omega_A$; and other with similar quantities v_B, Q_B, and E_B. Defining $\Omega = 1 - (Q/R) G$; and $\Psi = \Omega \phi$; and $G_A(\omega_A) = v_A - Q_A/(H - \omega_A) G_A$, show that if G in Hermitian then v and H_0 are also Hermitian.

7. What are the essential properties of the wave-functions of valency electrons which make shell model a good approximation ? How does it match the conditions of nuclear matter; for the nuclear core?

8. Explain, the physical significance of QKk, in Eq. (A-33) by using (A-32) and (A-25).

9. Why is a term $\partial V/\partial \rho$ required to give proper saturation properties of nuclear forces at correct density?

Author Index

Subject Index